Automotive Ergonomics

Heiner Bubb • Klaus Bengler • Rainer E. Grünen
Mark Vollrath
Editors

Automotive Ergonomics

Springer

Editors
Heiner Bubb
Chair of Ergonomics
Technical University Munich
Munich, Bavaria, Germany

Klaus Bengler
Chair of Ergonomics
Technical University Munich
Munich, Bavaria, Germany

Rainer E. Grünen
Consulting Engineer Dipl.-Ing. (FH)
Wiesbaden, Hessen, Germany

Mark Vollrath
Technical University of Braunschweig
Braunschweig, Niedersachsen, Germany

This book is a translation of the original German edition „Automobilergonomie" by Bubb, Heiner, published by Springer Fachmedien Wiesbaden GmbH in 2015. The translation was done with the help of artificial intelligence (machine translation by the service DeepL.com). A subsequent human revision was done primarily in terms of content, so that the book will read stylistically differently from a conventional translation. Springer Nature works continuously to further the development of tools for the production of books and on the related technologies to support the authors.

ISBN 978-3-658-33940-1 ISBN 978-3-658-33941-8 (eBook)
https://doi.org/10.1007/978-3-658-33941-8

This Springer imprint is published by the registered company Springer Fachmedien Wiesbaden GmbH part of Springer Nature.
The registered company address is: Abraham-Lincoln-Str. 46, 65189 Wiesbaden, Germany

Foreword

The automobile is probably the most fascinating industrial product of all. There are various reasons for this. First and foremost, it clearly satisfies the basic need for mobility and even more so for self-determined mobility. It thus expands the natural ability to move, which makes the human being a living being, among other things, and which is therefore essential for survival. In a way, it represents the fulfillment of the dream of seven-league boots, which in mythology are designated boots with magical powers that give the wearer the ability to travel long distances in a short time. Apart from the restrictions that result from the fact that almost everyone can afford this dream in today's highly industrialized regions, the automobile ensures that you can start from (almost) any place at any time and can reach (almost) any place. Yes, even if there are not enough roads, this option is still open – provided you have a suitable vehicle. Of course, these possibilities are practically not available in this ideal form for most people, due to external constraints such as high traffic density, lack of parking space and the like. However, the emotional assessment of the automobile must be based on the total mobility described above as a background feeling that justifies its desirability. In addition, the automobile is also a means for its owner and user to represent his or her own self and personality. One can thus show oneself as serious, sporty, extroverted, pragmatic or modest. It thus fulfils a function similar to that of clothing or the furnishing of one's own home. And like these, it is of course also subject to fashion and fashion trends. Especially the outer form, but increasingly also the interior design, i.e. what is subsumed today under the term Industrial Design, plays an outstanding role for the purchase of a vehicle. However, if you use the vehicle for a longer period of time, you may notice annoying peculiarities. For example, you can't find the right seating position, visibility may be obstructed, you may find one or two things uncomfortable, you may feel that you don't have the vehicle in your hand properly, etc. In short, in everyday-life usability, ergonomic design may play an important role in ensuring that the vehicle is really an extension of your own self. It is often argued that those latter points are, inter alia, the basis for the refusal or repurchase of a vehicle of the previously chosen make. To put it simply, in terms of personal attention to the vehicle, design appeals to "emotion" and ergonomics to "reason".

The automobile traffic also has a negative flip side. Due to the possible high speeds, an uncontrolled crash can result in such a high energy transfer to the passengers that this can lead to considerable, in the worst case, fatal injuries. Even if the injury and death rate is lower in terms of driver performance relative to earlier means of individual transport (riding horse, carriage, cart), the large number of vehicles and high traffic density overall - also seen worldwide - result in a frightening number of injured and killed. Although the so-called passive or secondary safety measures, which in many cases reduce the amount of energy transferred to the passengers to a tolerable level in the event of a crash, have led to a reduction in the number of fatalities in road traffic despite increasing driving performance, the dream of accident-free driving can only be achieved by the so-called active or primary safety, which attempts to avert an imminent accident in its preliminary phase. In addition to the elaborate technical measures of sensors and actuators, ergonomics can also make an essential contribution to active safety. Ergonomic measures ensure fatigue-free driving, prevent visual impairments and distractions and make the effect of assistance systems, which make a significant contribution to active safety, transparent and thus acceptable.

The authors of this book have set themselves the task of comprehensively presenting the application of the broad knowledge in the field of ergonomics to the design of the automobile. In ► Chap. 1, a historical outline shows that an adjustment of the automobile to the needs of humans has already taken place before it was considered a scientific approach. This chapter also contains a brief overview of the various ergonomic topics.

For the scientifically sound ergonomic design of a technical device, the first step must be to describe the *task* that is to be fulfilled with it. This task to be performed is independent of human characteristics and abilities as well as of available technical means. An explanation regarding the tasks to be fulfilled is presented in ► Chap. 2 through the *control loop paradigm* of ergonomics. After that, the information change brought about by the interaction between human and machine is of primary interest.

In order to adapt technology to humans, knowledge of human characteristics and abilities is indispensable. Therefore, ► Chap. 3 presents the *cognitive characteristics* (information reception, processing and conversion) of the driver with special emphasis on the requirements arising from the driving task.

However, not only the purpose for which the tool is used plays a role in the design of human tools - including the car - but also its adaptation to the physical characteristics of the human being. These physical characteristics, summarized under the term *anthropometry* refer on the one hand to the entire physiological apparatus both with a view to the *geometric properties* as well as the possibilities to apply *force.* ► Chapter 4 deals with these two aspects (geometric properties and force) with regard to their use in automobiles.

In order to adapt the "hard" technical tool, i.e. the car to the "soft" characteristics of man, *models of human behavior and properties* have been developed, which make it possible to objectify the degree of adjustment. With today's methods and means of computer technology, a large part of both the cognitive and anthropometric properties of humans can be simulated and modelled in computers. ► Chapter 5 describes the design tools currently used in vehicle development.

With all the knowledge that is now available, it is possible to ergonomically design tools, machines and especially, cars. The first step is to determine the purpose for which the tool is to be used. Are all the necessary resources available for the fulfilment of the task? Is the information understandable? This is described in ► Chap. 6.

This raises the question of the extent to which this tool is adapted to the anatomical characteristics of humans. This essentially refers to *dimensions* – e.g. can the driver even capture the information of the task based on the given geometry? And *strengths* – e.g. is he able to carry out the necessary actions? This and much more will be discussed in ► Chap. 7.

If the above two points have all been correctly completed, the question arises as to whether there may be any external influences during the performance of the task which could affect the driver in any way. An optimal design of these points ensures that the *fitness* of the driver over longer distances is possible. The corresponding influencing factors are described in ► Chap. 8.

The use of *assistance systems* the task of driving a car will partly change considerably compared to what we have been used to so far. From an ergonomic point of view, the question is which partial aspects of the task should be assisted and how the driver should be supported by applying the principles described in ► Chaps. 6 and 7. This will be presented and discussed in ► Chap. 9.

All the findings and demands that have been gathered so far must ultimately be realized. For this reason, ► Chap. 10 deals with *ergonomic vehicle development*, in particular the means used for this purpose (simulators, augmented reality and even customer surveys).

Many aspects, however, cannot be taken into account in advance when developing a new vehicle by applying rules and using computer tools. Therefore, ► Chap. 11 describes the different *measurement techniques* which must be applied in the course of the development process in order to achieve an optimized product.

In particular, studies that use volunteers in any form face the problem of data scattering, which goes far beyond what is usual in the technical field. In order to nevertheless arrive at verifiably solid statements here, certain methods of *statistics* are to be applied that are stored in ► Chap. 12.

Automotive industry leaders recognized the strength of ergonomics a long time ago. For this reason, ergonomic departments have been set up in many companies today – in different organizational forms – whose task it is to apply ergonomic findings in the design of the automotive product.

Today, all technical universities have chairs for ergonomics or work science, and engineers have the opportunity to acquire knowledge in this field during their training. Also, psychologists can obtain a sound technically oriented ergonomic education at many universities. Nevertheless, there is often a lack of suitably trained specialists in industrial practice. For this reason, employees in ergonomic departments often have to acquire the necessary specialist knowledge through "learning by doing".

This book is intended for this group of people. At the same time, it wants to show the manager the possibilities and benefits of ergonomic product design. Last but not least, it also serves as a textbook for students of automotive engineering, ergonomics and engineering psychology. It aims to provide a solid basis for decisions in the development of products in this field by systematically processing ergonomic knowledge with a special focus on vehicle design.

Heiner Bubb
Munich, Bayern, Germany
June 2020

Contents

1 **Inauguration** ... 1

Heiner Bubb

1.1 **Current Main Development Lines of the Automobile** ... 2

1.2 **Brief Historical Overview of Vehicle Development from the User's Point of View** ... 7

1.2.1 Development of the Driving Functions ... 7

1.2.2 Development of the Cockpit ... 11

1.3 **The Importance of Ergonomics for Automotive Development** ... 14

1.3.1 Brief Outline of the Development of Ergonomics ... 14

1.3.2 Micro Ergonomics and Macro Ergonomics ... 14

1.3.3 Subfield of Ergonomics ... 15

1.3.4 Fields of Application of Ergonomics ... 17

1.4 **Hierarchy of the Driving Task** ... 18

1.4.1 Primary Task ... 18

1.4.2 Secondary Tasks ... 20

1.4.3 Tertiary Tasks ... 21

1.5 **Information Flows in Transport** ... 21

References ... 23

2 **The Control Loop Paradigm of Ergonomics** ... 25

Heiner Bubb

2.1 **Driver-Vehicle Interaction** ... 26

2.1.1 The Driver-Vehicle Control Circuit ... 26

2.1.2 Pursuit and Compensatory Task ... 28

2.1.3 Quality and Performance ... 29

2.1.4 Quality of Laterale Dynamics ... 29

2.1.5 Quality of Longitudinal Dynamics ... 30

2.2 **Basic Concepts of System Technology** ... 31

2.3 **System Dynamics** ... 35

2.3.1 Transition Functions and Frequency Response ... 35

2.3.2 Transition Function ... 36

2.3.3 Frequency Response ... 37

2.4 **Quantities Controlled by the Driver** ... 43

2.4.1 Laterale Dynamics ... 43

2.4.2 Longitudinal Dynamics ... 48

2.5 **The Primary Driving Task from a Control Engineering Point of View** ... 49

2.6 **System Reliability** ... 53

2.6.1 Safety, Risk, Border Risk and Protection ... 53

2.6.2 Reliability, Error and Safety ... 54

2.6.3 Human Error and Probability of Accident ... 56

2.6.4 Derivation of Measures from the Fault Tree Analyses ... 57

References ... 60

3 **The Human Being as a Driver** 63
Heiner Bubb, Mark Vollrath, Klaus Reinprecht, Erhard Mayer, and Moritz Körber
3.1 **The Human Being as an Information Processing System** 64
3.1.1 The System Model of the Driver 64
3.1.2 Anatomical-Functional Model 65
3.1.3 Functional-Neurophysiological Model 69
3.2 **Elements of the Information Processing Human Being** 77
3.2.1 Information Reception 77
3.2.2 Information Processing 101
3.2.3 Information Implementation 120
3.3 **Information Processing when Driving a Car** 123
3.3.1 The Feeling for the Time 123
3.3.2 Human Eye Behaviour when Driving a Passenger Car 124
3.3.3 Gaze Behaviour and Inner Models 141
3.3.4 Comfort and Discomfort 143
3.3.5 Stress and Strain 147
3.4 **Driving Error** 149
3.4.1 Human Reliability and Driving Errors 149
3.4.2 Causes of Human Error 151
References 155

4 **Anatomical and Anthropometric Characteristics of the Driver** 161
Rainer E. Grünen, Fabian Günzkofer, and Heiner Bubb
4.1 **Anatomical Basics** 162
4.1.1 The Musculoskeletal System 162
4.1.2 Muscular System 170
4.2 **Anthropometry** 176
4.2.1 Length and Circumference Dimensions 176
4.2.2 Weight 194
4.2.3 Forces 196
4.2.4 Mobility 205
References 213

5 **Human Models** 219
Heiner Bubb
5.1 **Cognitive Human Models** 220
5.1.1 Control Engineering Models 220
5.1.2 The Benefits of Control Engineering Human Models 228
5.1.3 Cognitive Driver Models 229
5.2 **Anthropometric Human Models** 234
5.2.1 Drawing Templates 234
5.2.2 Digital Human Models 237
5.3 **Summary Appreciation of the Benefits of Human Modelling** 250
References 251

6	**System Ergonomics of the Vehicle**	257
	Heiner Bubb, Klaus Bengler, Jurek Breuninger, Christian Gold, and Magnus Helmbrecht	
6.1	**General System Ergonomic Design Guidelines**	258
6.1.1	Function	261
6.1.2	Feedback	266
6.1.3	Compatibility	267
6.2	**Human – Machine – Interaction**	270
6.2.1	Displays	273
6.2.2	Control Elements	289
6.3	**System Ergonomic Recommendations for the Respective Driving Task Levels**	300
6.3.1	Primary Driving Task	300
6.3.2	Secondary Driving Task	308
6.3.3	Tertiary Tasks	308
6.4	**Design of Driving Relevant Characteristics**	311
6.4.1	Lateral Dynamics: The Steering Feel	311
6.4.2	Longitudinal Dynamics	325
6.4.3	X- by-Wire	330
	References	336
7	**Anthropometric Vehicle Design**	343
	Heiner Bubb, Rainer E. Grünen, and Wolfram Remlinger	
7.1	**Vehicle Packaging**	345
7.1.1	Objective of the Anthropometric Package	345
7.1.2	Car Dimension Conception According to SAE	348
7.1.3	Fields of Anthropometric Ergonomics	353
7.2	**Sitting**	358
7.2.1	Consideration of Different Anthropometries	358
7.2.2	Driver	359
7.2.3	Right-Hand Drive Problem	384
7.2.4	Co-Driver	392
7.2.5	Vehicle Rear	392
7.3	**View**	396
7.3.1	Direct View	396
7.3.2	Indirect Vision	407
7.3.3	View of Operating and Display Components	409
7.3.4	Reflections	412
7.4	**Operating and Display Components**	416
7.4.1	Determination of Accessibility Areas According to SAE	416
7.4.2	Functional Gripping Spaces	417
7.4.3	Consideration of Special Operating Requirements	419
7.5	**Space Requirements**	420
7.5.1	Static and Dynamic Space Requirements	420
7.5.2	Shelves	424
7.5.3	Room Feeling	425
7.6	**Entry and Exit**	427
7.6.1	Door Concepts	427
7.6.2	Movement Strategies	431
7.6.3	Evaluation Methods	438

7.6.4	Access to the Second and Third Rows of Seats	449
7.7	**Loading**	449
7.7.1	Geometry	450
7.7.2	Operability	452
7.8	**Consideration of Specific User Groups**	453
7.8.1	Older Vehicle Users	453
7.8.2	Children	460
7.9	**Craftsmanship**	461
	References	465
8	**Design of Condition Safety**	469
	Heiner Bubb	
8.1	**Lighting**	470
8.1.1	Photometric Dimensions	470
8.1.2	Outdoor Lighting	471
8.1.3	Interior Lighting	472
8.2	**Sound**	475
8.2.1	Driving Noises	475
8.2.2	Small Noises	480
8.2.3	Useful Signals	482
8.3	**Vibrations**	483
8.3.1	Vibration Phenomena	483
8.3.2	Perception of Vibrations	483
8.3.3	Vibration Evaluation	485
8.3.4	Comfort and Driving Safety	490
8.3.5	Kinetosis	492
8.4	**Climate**	493
8.4.1	Climate, Performance and Comfort	493
8.4.2	Climatic Comfort	494
8.4.3	Environmental Conditions	499
8.4.4	Technical Requirements	501
8.5	**Odour**	513
	References	515
9	**Driver Assistance**	519
	Heiner Bubb and Klaus Bengler	
9.1	**What Is Assistance?**	520
9.2	**Driver Assistance and Driving Task**	522
9.2.1	Driver Assistance Systems Available Today for the Primary Driving Task	523
9.2.2	Categorisation of Driver Assistance Systems for the Primary Driving Task	548
9.2.3	Fatigue Warning	554
9.3	**Contribution of Driver Assistance Systems to Driving Safety**	558
9.4	**Ergonomic Design**	563
9.4.1	Operation and Display	563
9.4.2	Distinctness of the Modes of a Driver Assistance System	567
	References	570

10 **Methods of Ergonomic Vehicle Development** 575
Heiner Bubb
10.1 **Ergonomics in the Vehicle Development Process** 576
10.2 **Virtual Reality** 580
10.3 **Simulation of Anthropometric Conditions** 581
10.3.1 Seat Box 581
10.3.2 Variable Ergonomics Test Bench 582
10.3.3 Application of Virtual Reality 584
10.4 **Simulation of Dynamic Driving and Traffic Aspects** 589
10.4.1 Motivation for Driving Simulators and their Technical Challenge 589
10.4.2 Simulator Techniques 591
10.4.3 Relevance of Driving Simulators of Different Levels 596
10.4.4 Use of Augmented Reality in Real Vehicles 599
10.5 **Experiments in a Real Vehicle** 601
10.5.1 Trials on Closed off Terrain 601
10.5.2 Experiments on Public Roads 603
10.5.3 Customer Feedback 604
References 605

11 **Measurement Techniques** 607
Klaus Bengler, Heiner Bubb, Christian Lange, Carmen Aringer-Walch, Nicole Trübswetter, Antonia Conti-Kufner, and Markus Zimmermann
11.1 **Necessity of Experiments** 608
11.1.1 Scientific Requirements for Experiments 608
11.1.2 Assessment of Usability 609
11.1.3 Design of Experiments 610
11.2 **Objective Measurements** 611
11.2.1 Recording of Individual Anthropometry 611
11.2.2 Recording of Posture and Body Movement 614
11.2.3 Capturing Contact Forces 617
11.2.4 Eye Movement 619
11.2.5 Performance Measurements 627
11.2.6 Physiological Parameters 632
11.3 **Subjective Measurements** 634
11.3.1 Psychophysics 634
11.3.2 Interview 639
11.3.3 Standardised Questionnaires 647
References 648

12 **Statistical Methods** 653
Mark Vollrath
12.1 **Basic Questions: Distribution Vs. Examination of Differences** 654
12.2 **Expression of Characteristics: Confidence Intervals** 655
12.2.1 Methodology: Sampling 655
12.2.2 Statistics: Determination of Characteristic Value 656

12.3 **Differences Between Conditions: Significance Tests** 659
12.3.1 Methodology: Experimental Designs 659
12.3.2 Statistics: Significance Tests 663
12.3.3 Statistics: Presentation of Results 669
12.4 **External and Internal Validity** 672
References 673

13 **Outlook** 675
Klaus Bengler and Heiner Bubb
13.1 **Electric Mobility** 676
13.2 **Automation** 677
13.3 **Mobility Behaviour** 678
References 678

Supplementary Information
Index 683

Contributors

Carmen Aringer-Walch Technical University Munich, Munich, Bavaria, Germany

Klaus Bengler Chair of Ergonomics, Technical University Munich, Munich, Bavaria, Germany

Jurek Breuninger International University, Munich, Bavaria, Germany

Heiner Bubb Chair of Ergonomics, Technical University Munich, Munich, Bavaria, Germany

Antonia Conti-Kufner UI/UX & Psychology Specialist, ARRK Engineering GmbH, Munich, Bavaria, Germany

Christian Gold BMW Group, Munich, Bavaria, Germany

Rainer E. Grünen Consulting Engineer Dipl.-Ing. (FH), Wiesbaden, Hessen, Germany

Fabian Günzkofer BMW Group, Munich, Bavaria, Germany

Magnus Helmbrecht Area Manager Verification LiDAR, MAGNA, Munich, Bavaria, Germany

Moritz Körber tado GmbH, Munich, Bavaria, Germany

Christian Lange Wacker Nelson Group, Munich, Bavaria, Germany

Erhard Mayer Fraunhofer Institute for Building Physics, Holzkirchen, Bavaria, Germany

Klaus Reinprecht Inspectio Research Institute, Garching near Munich, Bavaria, Germany

Wolfram Remlinger Institute for Engineering Design and Industrial Design, University of Stuttgart, Stuttgart, Baden-Württernberg, Germany

Nicole Trübswetter University of Applied Sciences Landshut, Landshut, Bavaria, Germany

Mark Vollrath Technical University of Braunschweig, Braunschweig, Niedersachsen, Germany

Markus Zimmermann Starship Technologies, Munich, Germany

Inauguration

Heiner Bubb

Contents

1.1 Current Main Development Lines of the Automobile – 2

1.2 Brief Historical Overview of Vehicle Development from the User's Point of View – 7

1.2.1 Development of the Driving Functions – 7

1.2.2 Development of the Cockpit – 11

1.3 The Importance of Ergonomics for Automotive Development – 14

1.3.1 Brief Outline of the Development of Ergonomics – 14

1.3.2 Micro Ergonomics and Macro Ergonomics – 14

1.3.3 Subfield of Ergonomics – 15

1.3.4 Fields of Application of Ergonomics – 17

1.4 Hierarchy of the Driving Task – 18

1.4.1 Primary Task – 18

1.4.2 Secondary Tasks – 20

1.4.3 Tertiary Tasks – 21

1.5 Information Flows in Transport – 21

References – 23

H. Bubb et al. (eds.), *Automotive Ergonomics*, https://doi.org/10.1007/978-3-658-33941-8_1

1

1.1 Current Main Development Lines of the Automobile

Regardless of all politically motivated objections, self-determined mobility, "automobility", will be an essential part of modern life. Even with diminishing energy reserves, which have so far ensured this mobility, the automotive industry is interested in satisfying this basic need. In the past, this desire for mobility was associated with many disadvantages, which have been repeatedly expressed, especially in public criticism. It is the natural interest of all vehicle manufacturers to compensate for these disadvantages while satisfying the needs of consumers. The following fields of research are emerging, which should continue to guarantee the existence of the automotive industry:

- CO_2-reduction

Motors with better efficiency naturally reduce energy consumption. However, according to the basic physical rules of Carnot's cycle, this potential for improvement is limited by the maximum technically controllable combustion temperatures (a practical physical limit is approx. 45%; the efficiency of the engine alone achieved today, not counting the efficiency losses over the entire drive chain, is 20–25%). Engine researchers believe there is still much to be achieved in this field. In particular, the power flow between engine and drive wheels, as well as the design of the wheels themselves, still offers certain possibilities for improvement.

In principle, the energy conversion can be reduced by lightweight construction because less energy is required to accelerate and decelerate the vehicles. Technical measures (among others, see below) can be used to recover some of the energy lost during braking. An aerodynamically optimized body design helps to reduce the energy required to maintain the speed achieved.

- Electromobility

The conversion to electric vehicles represents a great hope for the maintenance of individual mobility, especially in conurbations that are extremely polluted by exhaust fumes. If the electricity required to operate the electric vehicle comes from "clean power plants" that do not produce additional CO_2-emissions into the atmosphere (i.e. no coal and gas-fired power plants), this would help incoming closer to the goal of environmentally friendly mobility. Since electric motors can be operated with a significantly higher efficiency (60–80%) than internal combustion engines and since they can also be used as generators (partial recuperation of the kinetic energy destroyed during braking) in contrast to internal combustion engines, their direct energy consumption in the vehicle is more favourable. However, the entire efficiency chain from primary energy in the power plant to electricity distribution in the power grids and electricity storage in the vehicle must be taken into account in this consideration, so that a comparison of an electric vehicle with electricity generation by fossil fuels with a vehicle with a conventional combustion engine shows no advantage.

The assertiveness of the electric vehicle depends on the development of battery technology. The energy density of a charged battery is only a fraction (about one tenth) of the energy density of fuels from fossil sources in the technologies available today. Despite the improved efficiency of electric motors and the partial recovery of braking energy, this results in significantly shorter ranges for electric vehicles (currently in the range between 100 and 400 km in relation to 450–1000 km for internal combustion engines). In addition, the charging process of the battery requires significantly more time than the fueling process (up to hours in relation to a few minutes). To resolve this dilemma, various options are currently being discussed, tested and researched (rapid charging, battery replacement, battery technologies that enable mechanical charging), with none of these solutions facing any apparent breakthrough. The measures discussed in this context are to use the vehicle's stationary life when parking for loading. An attractive future version is to use the battery capacity of parked vehicles to help store temporary overcapacities in regenerative power generation (especially wind turbines and thermal solar power plants). However, this version can only have a clear effect if enough electric vehicles are in use. The infrastructures necessary for the realization of such visions of the future (for example, battery charging at the car parks or in underground garages) are only in a timid manner today.

Fuel cell technology has also been discussed and tested for a long time. This makes it possible to generate electricity directly for the operation of the electric vehicle from cooled, highly compressed liquid hydrogen in combination with the oxygen in the air. The use of the fuel cell would thus solve the range problem of the electric vehicle due to the high energy density of the hydrogen carried along. However, the infrastructure of a hydrogen supply still has to be built up, which cannot build on the existing infrastructure of the supply by liquid fuels. So it remains to be seen in which direction developments in this sector will move.

Another question that has not yet been satisfactorily resolved is the environmental compatibility of battery building materials in their recycling, the amount of energy required to do so and, in general, for their production. In this field, too, new results from materials research can be expected over a longer period of time.

At present, and probably over a longer period of time, attempts are being made to combine the advantages of the electric motor (recuperation possibility, provision of an almost speed-independent high torque) with the advantages of the combustion engine (long range due to the high energy density of fossil fuels) by means of so-called hybrid concepts. There are various hybrid concepts in use, which will not be discussed in more detail here. The concepts known today are all regulated by the use of information-processing technology in such a way that practically no special new operating behaviour results for the user. One advantage of hybrid technology is that there are hardly any conversion problems for the user because the existing infrastructure can be used. However, given the current continued use of fossil fuels, this concept is by no means climate-neutral and environmentally friendly.

- Safety and security

Traditionally, a distinction is made in automotive engineering between active and passive safety (primary and secondary safety). The latter refers to all measures aimed at reducing the consequences of an accident. Active safety refers to all measures that are intended to make the occurrence of an accident unlikely.

- *Passive safety*

If the vehicle occupants' own energy associated with the speed has to be dissipated within the shortest possible distances during an unscheduled operational sequence (accident), forces and energies arise for the passengers that cannot be endured by their bodies without damage. In this respect, there is theoretically a direct, even quadratic relationship between vehicle speed and accident severity. In individual cases, however, the motion sequences are so different and complex depending on the respective accident situation that an exact prediction of the consequences of the accident seems impossible. Nevertheless, accident research has made it possible to categorise typical accident events, which in turn have led to corresponding technical measures to reduce the consequences of accidents. It is therefore a matter of course today that vehicles provide an inner braking distance by means of the so-called crumple zone, which benefits the passenger through the use of the seat belt. This protective mechanism is supplemented by the airbag, which is capable of absorbing further residual energy and also unfolds a certain protective effect during movement sequences that are otherwise difficult to control (e.g. rollover, crash with lateral traffic). The success of these measures is clearly reflected in the accident statistics (◘ Fig. 1.1). Despite a considerable increase in the number of vehicles on the road, the number of fatal accidents has fallen steadily. It should also be noted, however, that the number of accidents with injuries remains practically the same, although this too, in con-

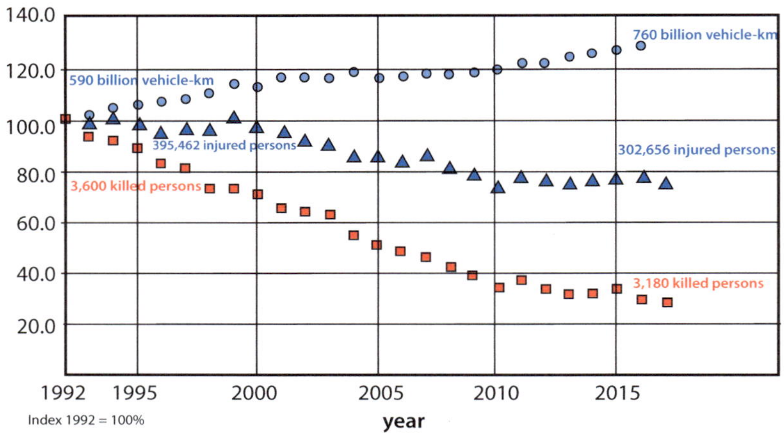

Fig. 1.1 Development of accident statistics in Germany. (Lerner et al., 2013)

junction with the increase in the number of vehicles on the road and the associated increase in traffic density, is to be seen as progress attributable to passive safety measures.

However, passive safety measures are also associated with disadvantages for the user. Due to the required stiff body structure, which should also provide safety during rollovers, thick window columns were necessary. The windows themselves became smaller when the waistline was raised, making the vehicles more difficult to see in normal use situations. The head restraints that reduce the consequences of rear-end crashes contribute significantly to this confusion. The dilemma between the demand for minor restrictions under normal operating conditions and the protective effect of passive safety becomes clear from the fact that the wearing of the seat belt only led to the high rate of use as a result of massive threats of punishment. In countries where this does not happen or where monitoring is less consistent, wearing a seat belt is also much less common.

- *Active safety*

The idea of active safety arose in the 1950s at about the same time as that of passive safety. It was first understood to mean the development of ever better braking systems, which in the course of progress also ensured considerable reductions in braking distances. This is related to the development of tires, which have made significant improvements through systematic research on wheel-road contact, even under unfavourable conditions such as wet and cold.[1] Chassis also contribute to active safety and make evasive manoeuvres possible even in a critical situation, due to sophisticated kinematics.The introduction of electronics into the vehicle added further possibilities. The ABS system should be mentioned, which prevents the wheels from locking during emergency braking and thus keeps the vehicle manoeuvrable even during emergency braking. An important further development is ESP, which uses targeted one-sided braking to bring a vehicle that is in danger of skidding - measured by yaw angle and acceleration sensors - back to the course set by the steering wheel. In all this, it must be made clear that all the active safety measures mentioned so far are of course unable to overcome the limits set by the physical conditions. These have to be taken into account by the

1 In this context, the transition from the more comfortable radial tires to the belted tires, which due to their design provided better grip at higher lateral forces, should be noted. In addition, the design of the treads was increasingly based on scientific methods. As a result of this, deeper and wider so-called drainage channels were created through which the water can be better displaced in wet conditions and "dry" contact can be ensured even at higher speeds. Rubber compounds of different softness levels are matched to the outside temperature to provide better adhesion. This made the use of summer and winter tyres virtually obligatory.

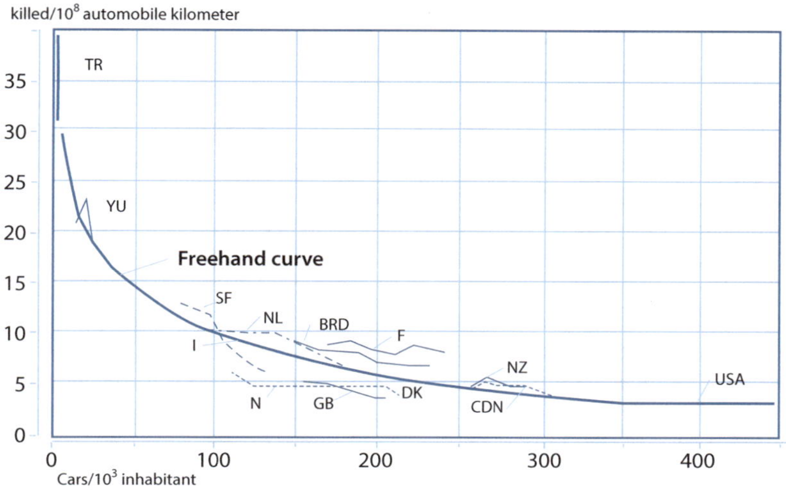

Fig. 1.2 The number of road fatalities per inhabitant per distance travelled. (Source: Die Entwicklung der Verkehrsunfälle in der Bundesrepublik Deutschland und in Berlin (West), BASt, Heft 1, 1974)

driver in the events leading to the critical situation by selecting an appropriate speed and the correct distance between the vehicle in front. More recent developments, such as ACC (Adaptive Cruise Control) and Distronic automatically maintain a safe distance – measured by radar sensors – between the vehicle in front. Today, these systems are so sophisticated that they can act as emergency brake assist systems, automatically protecting the vehicle from a crash or at least from the serious consequences of such a crash. Lane guidance assistants detect when the driver threatens to leave the road and warn him by vibrating the seat or steering wheel. Some systems can also automatically return the vehicle to its original lane. Tiredness alert systems monitor the driver's attention and suggest taking a break if necessary.

The latter system already indicates that the main part of ensuring active safety lies with the driver him or herself. It is, therefore, no coincidence that from the very beginning of motorised traffic, the intention is to influence the driver's behaviour by introducing the driving licence and the associated instruction in "safety-conscious" driving. This is further supported by means of a large number of campaigns, including punitive measures in the event of gross misconduct. In this context, it should not be underestimated that growing up in a world of motorised transport in combination with the measures mentioned above produces an individually rooted, traffic-specific world view that induces appropriate behaviour. This shows a picture taken many years ago by the BASt[2] and has been published (Fig. 1.2). The picture shows: the greater the individual's experience of road transport (expressed by the number of passenger cars/10^3 inhabitants), the lower the individual risk of being killed in road traffic (expressed in fatalities/10^8 vehicle kilometres). Of course, this does not change the fact that the more vehicles there are in a population, the greater the absolute number of fatalities in road traffic assuming a constant accident probability per vehicle (which of course is not the case, see Fig. 1.1). Only related figures make a reasonable statement about what happened in an accident.

The constant number of road traffic injuries over the years (Fig. 1.1), also indicates that the possibilities of passive safety are limited. Included here, is a a high proportion of pedestrians and simultaneously a decrease in

2 BASt: Federal Highway Research Institute: Technical and scientific research institute of the Federal Ministry of Transport, Building and Urban Affairs based in Bergisch Gladbach.

the number of deaths. Today it is undisputed that with the means of electronics and the technical intelligence made possible by them, that active safety measures can be created that largely prevent accidents. Especially in connection with the requirements of CO_2-reduction and electromobility, the weight reduction of vehicles is an outstanding measure. If it were possible to compensate a large part of the weighty technology for passive safety with active safety systems, several goals could be achieved at the same time: any accident that doesn't happen is better than one that goes smoothly. In addition, accidents that have not occurred promote the flow of traffic, which not only prevents further accidents but also helps to save fuel.

However, technical active safety measures necessarily interfere with the driver-vehicle interaction in the sense that the driver is denied certain manoeuvres which he wishes to perform in the free exercise of his will and which he enjoys to some extent (e.g.: Transverse drift is prevented by ESP, close distance which makes it impossible for a slower vehicle to reeve is prevented by ACC and "sporty" cornering is prevented by the lane guidance assistant).

It is a challenge for the future development, operation and effect of such safety-enhancing systems to design them in such a way that the benefits are obvious to the driver at all times and, at the same time, a certain "joy of use" is created, particularly when dealing with the corresponding system.

- *Comfort*

Today, many people spend several hours a day in their cars. Comfort has therefore also become a sales-decisive argument in the lower and middle vehicle classes. In the vehicle, at least the driver is forced into a low-movement position by the driving task alone. Many drivers therefore complain about "uncomfortable seats" that cause them back pain. A comfortable situation can be provided by a consistent design of the so-called anthropometric conditions (i.e. to the individual physical measures and forces adapted), which reaches from the posture to the contact with the seat. However, it is not only the sitting posture that contributes to comfort, but also the climatic conditions in the vehicle and the influence of engine, wind and chassis noise. In addition to the safety aspect when driving at night, good lighting is also a comfort feature, which today leads to the introduction of so-called "ambient lighting"in the interior. This counteracts the unpleasant impression of sitting in a dark cave when driving at night without impairing visibility outside. Good vehicle suspension makes a significant contribution to driving comfort, although today more or less complex active chassis are being developed using the possibilities of microelectronics, which make it possible to achieve an ever better compromise between the demand for less interference from poor roads and good road contact, which is crucial for the safe steering of a vehicle.

In addition to these quasi classical comfort areas, the so-called operating comfort plays an increasingly important role. It is intended to ensure problem-free and goal-oriented operation of the various vehicle functions at all times. The ease of operation is in part closely related to the various entertainment functions that have found their way into the vehicle. Both the operation and the type of entertainment is the subject of great efforts to increase comfort: today, the acoustic design of loudspeaker systems in vehicles has become an important economic sector.

The four areas of future vehicle development mentioned above are all closely linked to user orientation. It is therefore only logical that in recent decades the field that deals professionally with the interaction between man and machine, namely ergonomics, has become increasingly important in vehicle development. The following brief overview is intended to show that this user orientation has been the driving force behind vehicle development from the outset. However, this development, as far as interaction with the user is concerned, was driven more or less intuitively from the technician's point of view.

Fig. 1.3 Mercedes Simplex: On the steering wheel, you can see the lever for the ignition timing adjustment. Next to it on the dashboard are the swabs to supply the lubrication points of the vehicle

1.2 Brief Historical Overview of Vehicle Development from the User's Point of View

1.2.1 Development of the Driving Functions

In the first consolidation phase after the invention of the petrol-driven car, various engine functions were gradually automated in the 20s of the last century.[3] One well-known example is the automatic mechanical ignition timing adjustment by means of centrifugal weights or vacuum, which previously had to be set manually with a lever mounted on the steering wheel, sensitively and by ear while driving (see Fig. 1.3). The annoying pulling of the "choke" when cold starting the engine became superfluous by the introduction of a bimetallic spring, which more or less closed the starter valve in the carburettor according to the temperature level of the cooling medium (air or water). Many other mechanical and electro-mechanical solutions (e.g. the automatic resetting of the direction indicator or the windshield wiper) enabled the driver to pay less attention to the operation of the vehicle and more to the traffic situation.

The increasing traffic density was and is the driving force for further technical developments, which should simplify the use of the car and make it increasingly comfortable. The automatic transmission was invented in the 1930s and largely introduced in cars in the USA during the 1950s. In connection with an increasing traffic density in Europe, especially in Germany, the automatic transmission was increasinlgy preferred in the 80s and 90s. A cruise control system was first used in automobiles in 1958 as "Cruise Control"at Chrysler, in Europe in 1962 at Mercedes-Benz. In the "autobahnfest" vehicles of the pre-war period were still a stand-by cas, which should facilitate driving with constant speed on the (still) empty new motorways.

The economic upswing after the Second World War and the demand of customers for larger and more comfortable touring limousines led to an increase in vehicle weight, which made it necessary to support operating functions (brakes and steering). For this reason, technical servo systems were introduced in the 60s and 70s for various interaction functions. The brake booster uses a vacuum in the intake pipe, whereby the atmospheric pressure on a large diaphragm increases the pedal pressure. The power steering system, which went into series production in 1952, features an engine-driven hydraulic pump to reduce steering torque, especially at low speeds and when parking. These servo systems not only increased driving comfort but also active safety by shortening reaction times. With this development, however, the end of what was feasible with mechanical-hydraulic support systems was also reached. Today, hydraulic power steering is gradually being replaced by electric servo systems. The advantage of this is that energy is only consumed when the system is in use, i.e. when steering, but not when driving straight ahead with virtually no steering movement.

Further innovations in the automotive sector will not be possible without additional

3 Worth mentioning in this context is the valve control of the earliest steam engines, which took place "by hand". It was the groundbreaking invention of James Watt (who is often mistakenly called the inventor of the steam engine instead of Thomas Newcomen), which enabled automatic speed-dependent opening and closing of the steam and water valves, that helped the steam engine to its breakthrough. James Watt et al. used a centrifugal speed limiter to prevent the machine from "going through" without load.

electronic equipment. In 1978, the Bosch Antilock Braking System (ABS) was launched on the market; at the same time, the term "ABS" was legally protected by Bosch.[4] First, the ABS was available for the Mercedes-Benz S-Class and then for the BMW 7 Series. ABS has been standard equipment at Mercedes-Benz passenger cars since 1984. At the end of 2003, around 90 percent of new vehicles registered in Germany were equipped with ABS. ABS is based on the physical knowledge that the maximum braking deceleration is achieved shortly before the wheels lock, i.e. when the tires just barely adhere to the road surface. If the static friction limit is exceeded by too high a braking force, the blocked wheels glide on the road (100 percent slip), with sliding friction typically 15–20 percent below static friction. When the wheels lock, the vehicle is hardly controllable and normal drivers rarely react correctly in this situation (it would be right to reduce the braking force immediately until the wheels turn again; for this reason the so-called stutter brake was then recommended). The maximum braking power is achieved at 50 percent slip. Due to the inertia in the braking system, the ABS control, which is noticeable as a pulsation of the brake pedal, oscillates between 20 and 35 percent slip. The vehicle remains most controllable during braking if the ABS regulates the braking force on each wheel in such a way that slip during braking is as close as possible to these control limits. With the 4-channel ABS standard today there is an inductive sensor on each of the four wheels, which measures the speed of the wheel. The sensor signals are transmitted to an electronic control unit, which closes or opens two fast-acting solenoid valves per wheel when the speed drops too much compared to the other wheels. In this way, the optimum hydraulic pressure per wheel brake cylinder is controlled. Only through fast electronic circuits can approx. 10 control processes per second be achieved in a passenger car[5] and generally does not achieve longer braking distances than with an emergency stop by an experienced driver without ABS.

Another logical step was the introduction of traction control (ASR or ASC for BMW), which prevents the driven wheels from spinning when the vehicle is accelerated. Too much slip when the drive wheels spin means loss of drive torque and lateral guiding force, which can lead to the vehicle slipping away. The simplest technical implementation of the ASR is through brake intervention, where the ABS brakes the spinning wheel and the other wheel gains drive torque. More complex ASR systems intervene in the engine management system, usually in conjunction with an Electronic Stability Program (ESP). This is only possible if there is no mechanical connection between the accelerator pedal and the throttle valve (petrol engine) or injection pump (diesel engine). The task of power reduction is then performed by an "electronic accelerator pedal", which converts the driver's request into an electrical signal by means of a sensor. From this digitized driver's request and other variables, the engine control system adjusts the throttle valve and the injection system via electric motors, whereby ASR commands are given priority over the driver's request, resulting in a reduction of excess drive torque. The ASR control is indicated to the driver by a flashing light in the dashboard so that he can see that he is now on slippery ground. This is therefore also a contribution to active safety.

ASR is the basis for the further developed ESP (Electronic Stabilization Program) systems, which contribute to driving stabilization by means of additional sensors and braking intervention on the non-driven wheels. The pioneers in the development of the system

4 As early as 1936, Bosch granted a patent for a "device for preventing the wheels of a motor vehicle from braking". The devices consisted of about 1000 analog components and were still very unwieldy and slow. Thanks to digital technology, the number of components could be reduced to around 140. This made ABS ready for series production.

5 A system with a higher control frequency could considerably improve the effect. This can be achieved, for example, by a fully electric brake, in which the pedal force is transmitted as an electrical signal to the control unit, which doses the braking force after evaluating all sensor signals.

were the companies Bosch and Mercedes-Benz (ESP is a trademark for Robert Bosch GmbH, which is why other names appear in various companies, such asfor example DSC at BMW, PSM at Porsche). ESP was introduced in 1995. In the case of the Mercedes-Benz A-Class, which was newly launched on the market at the time, the unsuccessful[6] "moose test"led to the introduction of ESP as standard from 1997 onwards in order to improve vehicle behaviour.[7] The ESP system is designed to prevent the vehicle from skidding at its limits by selectively braking individual wheels, thus enabling the driver to control the vehicle in critical situations. In addition to additional sensors, ESP requires the separation of all wheel brake circuits so that each wheel can be braked individually. The system continuously (up to 150 times per second) compares the driver's wishes with the driving status. The steering angle sensor in conjunction with the accelerator pedal position and brake pressure provides the driver with the desired driving direction. The dynamics computer calculates the target curve to be passed through by the vehicle from these variables. The engine management system, the ABS speed sensors, the yaw rate and lateral acceleration sensors provide the signals for interpreting the vehicle behaviour (actual curve). If a significant deviation of the calculated driving condition from the driver's wishes is detected, the system intervenes. Oversteer is corrected by braking the outer front wheel, understeer by braking the inner rear wheel. One-sided brake interventions on the front axle can be felt on the steering wheel, which leads to a reduction in comfort. Therefore, some manufacturers do not allow the front axle to intervene until the rear axle correction proves ineffective. In addition, ESP intervenes in the engine management system and throttles the engine power to reduce the vehicle speed. This intervention is also criticized by some drivers as patronizing.

The ESP can be equipped with additional functions: The Brake Assist (BAS) recognises that the driver wishes to make an emergency stop from a rapid increase in brake pressure and triggers this. If the driver reduces the pedal force again, it switches off. With the Start-up Assistant, the brake holds for a short time from a certain angle of inclination of the vehicle despite the "foot from the brake pedal"in order to enable starting without rolling back. This intervention also represents a considerable simplification for the operator, as the cumbersome combination of handbrake actuation, clutch and accelerator pedal is largely eliminated when "starting up on the hill" with a manually operated vehicle.

In addition to ESP, BMW presented an innovative steering system – Active Steering. It operates as a superimposed steering system with a planetary gear integrated into the split steering column. An electric motor engages in this planetary gear via a self-locking worm gear and generates an additional or reduced steering angle of the front wheels depending on the driving situation. Another component is an adjustable power steering system that controls the steering wheel torque. With these two components, the steering angle of the front wheels and the steering wheel torque can be adapted to the respective driving situation. In critical situations, e.g. during abrupt evasive manoeuvres, the system can selectively change the steering angle of the wheels specified by the driver and stabilise the vehicle faster than the driver by counter-steering, which leads to less noticeable ESP intervention.

The systems mentioned so far intervene when the driver has reached certain physical limits of driving physics. Further develop-

6 In October 1997, the automotive journalist Robert Collin of the Swedish newspaper Aftonbladet had caused the newly presented Mercedes-Benz A-Class to tip over as a result of this driving manoeuvre, causing a huge media response.

7 In this driving dynamics test, a lane change to the left is driven at high speed without braking and after a short straight stretch a lane change to the right. The vehicle should neither break away nor tip over sideways. This test is intended to simulate avoiding a moose that suddenly steps onto the road, a scenario that is quite realistic in Scandinavia. The moose test has now become a standardised driving manoeuvre and is part of the testing of the driving characteristics of new vehicles. The test was described by the VDA under the designation 'VDA lane change test' and was then incorporated into the international standard ISO 3888-2.

ments intervene even before such extreme situations are reached and thus represent a new contribution to accident prevention.

The ACC system (Adaptive Cruise Control) is a functional further development of the cruise control system. In these systems, which have been in use on the European and North American markets since 2000, the position and speed of the vehicle in front is determined by means of a radar or lidar system.[8] The speed and distance are controlled adaptively with motor and brake intervention. The systems available on the market work with three to four radar loops, which measure approximately 150 m ahead at a travel angle of 4° to 8°.[9] As the ACC system maintains a constant distance between the driver and the vehicle in front and accelerates the vehicle in front to the speed required by the driver, the first step towards "automatic driving" has been taken. Adaptive speed control is designed to relieve the strain on the driver and increase comfort, especially on long motorway journeys, by constantly controlling the speed (speed limits!), accelerating and braking by the system.

With additional radar sensors for the close-up range, the advanced ACC systems (ACCplus at Audi and BMW, Distronic Plus at Mercedes-Benz) make it possible to completely brake the vehicle and then accelerate it back to the speed specified by the driver. This "traffic jam assistant" thus further relieves the driver in stop-and-go traffic.

Parallel developments support the lateral control (steering function) of the vehicle. The lane departure warning warns the driver before leaving the lane on the road. Various optical systems are used to communicate the position of the vehicle in the lane to the associated electronics using software. Lane recognition is often achieved by a camera system (e.g. Lane Guard System from MAN) or by infrared sensors on the underbody of the vehicle (e.g. AFIL in the Citroën C4, C5 and C6, system manufacturer Valeo). If the vehicle threatens to drift slowly out of lane, the system warns by "nail band rattling"in the loudspeakers (MAN) or by a seat vibration on the side to which the vehicle drifts (Citroën). The lane keeping support supports the driver in lane keeping through automated permanent steering. The Heading Control (HC) system, developed by BMW as part of the European PROMETHEUS project, records the course of the road with a camera system and calculates the ideal steering wheel angle from image processing data. If the current steering wheel position deviates from this angle, this is indicated to the driver by steering wheel vibration or a correcting steering torque. In the latter case, the driver gets the impression that he is driving in a "barrel-shaped" road.

Lane changing accidents are often due to insufficient observation of the area behind or next to the vehicle ("blind spot"). The lane change assistant warns the driver of impending collisions when changing lanes. The system is activated when the direction indicator is activated whereas the Lane Keeping Assist is deactivated. Vehicles in the neighbouring lane are detected using radar sensors, cameras or laser scanners. There is no automatic intervention to prevent collision with other vehicles; the driver is responsibile for safely operating the vehicle. Warnings are given optically by illuminated displays, usually in the area of the exterior mirrors, acoustically or haptically by vibration of the steering wheel or the indicator lever. A further development is the lane change support, which automatically carries out a lane change at the driver's request.

8 Lidar stands for "Light detection and ranging" and works with laser pulses in which the light scattered back from objects is detected. The distance to the location of the light scattering is calculated from the transit time of the signals and the speed of light. Lidar systems currently still have too much interference when visibility is restricted by weather conditions. Their advantage is however the price difference to the radar systems, since these cost only approx. 1/3 of the radar systems. They are used for vehicles in the lower and middle price range but only allow control up to a speed of 140 km/h.

9 This limited "visibility" causes most manufacturers to limit the control range of the ACC to max 180 km/h (Audi and, more recently, BMW allow speeds up to 210 km/h).

Fig. 1.4 Night-Vision display: here "Night Vision Devices"by BMW

In the future, sensors around the vehicle will continue to increase in order to indicate dangerous situations to the driver that he cannot recognize with his natural senses. Drivers have always feared the dark-dressed pedestrian who walks on the country road at night. Recently, "Night Vision"devices have been offered in luxury class vehicles that use infrared cameras to detect pedestrians, game or other obstacles at a distance up to 300 m in front of the vehicle, long before they become visible to the driver in the headlights (Fig. 1.4). The driver can react in time to the display of the infrared image on the onboard monitor if he has the monitor in view in a dangerous situation. Recently Mercedes and BMW are offering a headlamp system that throws a targeted spotlight at the detected pedestrian and thus making him or her directly visible under all circumstances.

1.2.2 Development of the Cockpit

With the technical development of the automobile and the technical components that support the actual driving process, the human-machine interface for recording and transmitting information also changed. The first vehicles were fitted, for technical reasons, with a thermometer placed directly on the radiator and on some level indicators for the different oil circuits. In many vehicles during this pioneering period, the vehicle had to be lubricated by operating small oil pumps while driving (see Fig. 1.3). Soon the law required a speedometer to be mounted somewhere in the driver's field of vision as an additional instrument. The windscreen wipers, which were originally moved by hand, were soon after motor-driven[10] and direction indicators were added. At the end of the 20's and beginning of the 30's different displays, switches and levers were combined in one dashboard (Fig. 1.5); car radios, however, did not yet belong to the equipment of vehicles.[11] (even in the 50's there was serious discussion about whether listening to the radio while driving was too distracting for the driver). However, the increasing demands of customers for entertainment and information made radio installation unavoidable. Figure 1.6 shows the development stages from the simple tube unit to the high-end "entertainment system" (radio, cassette and CD changer, also TV and video).

The arrangement of an increasing number of individual devices (radio, CD player, climate control unit), displays and control elements on the dashboard (today called instrument panel or I-panel for short) led to the impression of overload in the 1990s. In 2003, BMW launched a revolutionary operating concept in the 7 Series: the i-Drive, whereby – according to BMW – the "i" stands for intelligent and integrated, informative and innovative, intuitive and interactive. The system consists of two components, a large LCD monitor in the center of the I panel and the control knob (controller) in the center console, with which various operating functions can be selected and activated by slide/pull, turn and press (Fig. 1.7). In the starting position, the display showed eight main menus: Communication, on-board computer data, navigation, help, entertainment, settings, air

---

10 Usually this was done with an electric motor. However, other solutions were also in use. In the Opel Olympia Record, for example, the wiper was driven by the camshaft via a flexible shaft until the 1957 model.

11 As early as 1922, however, a radio was installed in a vehicle in the USA. In 1924 Chevrolet already had a factory-installed car radio available. In 1932, Blaupunkt offered a car radio whose volume and reception frequency could be controlled from the steering wheel via Bowden cables (quoted from Spies 2013).

Fig. 1.5 Dashboard of BMW 327 (1939, left) and Mercedes 500 K (1936, right)

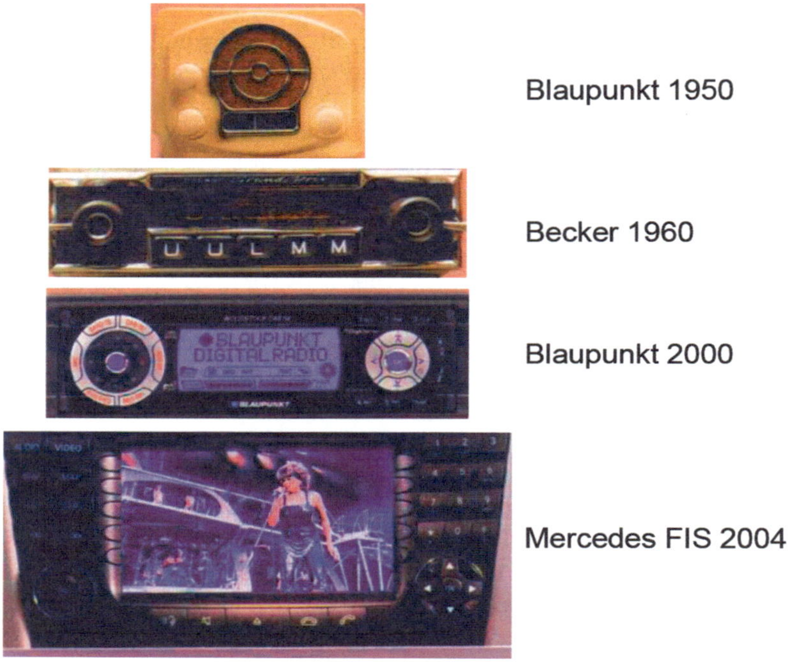

Fig. 1.6 Development of the technology and the design of the radio from the tube radio over the transistor radio to modern entertainment systems

Fig. 1.7 BMW iDrive in the 7 Series from 2003

conditioning and the BMW ASSIST telematics service. They are all controlled by pushing or pulling the controller in the respective direction. The corresponding function is selected by pressing the i-Driver control element (so-called "rotary push button"). At the same time increasing the variety of functions (approx. 700 functions can be operated while the vehicle is in motion), this should make the operation of the vehicle easier. The BMW slogan:

"You won't find anything familiar in the new 7 Series anymore - but you won't miss it" has not been fulfilled for this completely new system for vehicle operation because it takes a long time to get used to and is not easy to operate due to the menu structure. This is especially true for older customers who are less familiar with menu navigation but make up a large part of the target group. The driver who uses i-Drive, for example, to direct only the warm air flow into the footwell is often and longer distracted from the traffic situation.

Due to ongoing criticism i-Drive has been reworked several times. Similar concepts were soon offered by other vehicle manufacturers, such as Audi's MMI system and Mercedes-Benz's *COMAND*-system, some of which attempted to circumvent the obvious disadvantages of the original BMW system in different ways.

In addition to the introduction of new technologies, the information resources on the instrument panel were functionally grouped (see ◘ Fig. 1.8). In the middle of the dashboard, where originally the car radio (standard insert) was located, a (more or less elaborate) screen was installed, which makes new information visible to the driver, especially that of the navigation computer, but also additional information about radio functions, the air conditioning system and in the further development of the telephone and the Internet. In the area of the driver's central field of vision behind the steering wheel, information is offered that is directly related to driving. In addition to the familiar displays (speedometer, rev counter, tank and cooling water display), information about ACC settings and actions as well as ESP interventions are now also displayed.

Some of this information is partially projected into the exterior using Head-Up Display (HUD) technology (Chevrolet Corvette, BMW 3 Series, 5 Series, 6 Series and 7 Series, recently also Audi A6 and A7). The various lighting functions are located to the left of the steering wheel. The driver information system is positioned in the centre of the

◘ **Fig. 1.8** The division of the dashboard area into DAS (Driver Assistance System) and DIS (Driver Information System), which came to a quasi-convention today

I-board and contains the navigation system, entertainment functions (Radio, CD changer, partly also TV), communication functions (telephone, e-mail, Internet), on-board computer information and air-conditioning functions. This system can also be used to adjust driving dynamics, e.g. parameters of the ESP or the Electronic Damping Control (EDC) can be changed.

1.3 The Importance of Ergonomics for Automotive Development

The previous sections show the extent of which the triumphal march of the automobile is also characterised by the fact that the vehicle itself and the possible uses of the vehicle - within the framework of the respective technical possibilities - were oriented towards the needs of the users, indeed that these needs were at the same time the driving force for new technical developments. The technical development today is in a state that - with restrictions - practically "almost everything can be realized". So one is often faced with the question: "What should we develop next- what makes sense?" Under this frame of mind it is worth considering dealing with the needs and abilities of humans from a scientific point of view. Ergonomics deals exactly with this question and through knowledge of human characteristics and abilities, aims to support the technical design of work equipment and working environments. Since about the 1970s, ergonomics has become increasingly established for the development of automobiles and today it goes without saying that every automobile company employs corresponding specialists in their ergonomics departments.

1.3.1 Brief Outline of the Development of Ergonomics

Ergonomics can now look back on a 150-year tradition. The inhumane working conditions of early industrialization and the trust of the nineteenth century mechanistic world view in the producibility of desired "good" conditions through planning intervention made it obvious to establish a scientific discipline of its own also with regard to the world of work. Already in 1857 the Polish scientist Jastrzebowski in the journal "Nature and Industry" made the proposal "... to deal with a scientific approach to the problem of work and even to teach a separate lesson on its (the work's) explanation... so that we may reap the best fruits from this life at the slightest effort with the highest satisfaction for our own and the general well-being and thus do justice to others and our own conscience...". He called this new branch of science "work science"or "ergonomics", although the latter term was subsequently forgotten..

Nevertheless, since the middle of the nineteenth century many scientific examinations of human work have been observed in various countries. According to the prevailing scientific view of the world, it was considered possible in particular to transfer rules from classical physics to all phenomena of nature and thus also to those of human life (see various works by Reuleaux and the psychophysics of Fechner). In the various countries of Europe and in the USA, a field of science emerged falling under what German-speakers call "Arbeitswissenschaft" (industrial science; in Anglo-America the term "human factors" or "human engineering"is used and in Europe, "ergonomics" is used [see below]). In England in 1949 Murrell reinvented the artificial word "ergonomics", composed of the ancient Greek words (εργον = ergon = work; and νομοσ = nomos = law, regularity). From this time on, various international scientific societies were founded and used this name (in Germany 1953: "Gesellschaft für Arbeitswissenschaft, GfA") In 1959 they were brought together under the umbrella of the International Ergonomic Association (IEA).

1.3.2 Micro Ergonomics and Macro Ergonomics

There are different views on the content of the field that has emerged. W. E. Woodson (1981) writes in the introduction to his comprehensive work on this subject: "Human Factors Engineering is the practice of product design in such a way that the user can perform the required use, handling, operation and supporting tasks with minimum stress and maximum efficiency". He also mentions the term "ergonomics", which according to his explanation is generally used interchangeably with the term "human factors engineering". The only difference one could see would be that the term "human factors engineering" is more widely used in the USA than in other countries. In the meantime, the corresponding American scientific society has integrated the name "Ergonomics" and now calls itself "HUMAN FACTORS & ERGONOMICS SOCIETY (HFES)".

M. Helander (1981), who was president of the IEA for many years, formulates in an expanding sense: "Human Factors Engineering attempts to modify work processes and tools in such a way that the physical and psychological abilities and limitations of humans are taken into account". He lists various terms for this discipline, such as "engineering psychology", "technical psychology"and - especially common in Europe - "ergonomics". Engineering and traffic psychology has developed especially in connection with the scientific examination of events in road traffic but also in other areas of traffic. Its focus is more on understanding the psychological processes of human-machine interaction than on the design of work processes and tools. It also deals with questions of traffic design and other topics (e.g. consequences and combating drug use in road traffic). Since design proposals cannot be derived without an understanding of the processes, however, the boundaries are fluid and a sharp separation of the disciplines addressed is practically impossible.

Ergonomics is a multidisciplinary science that draws its basic knowledge from the human sciences, engineering sciences and economic and social sciences. It initially covers the fields of occupational medicine, psychology, education, technology and law as well as sociology of enterprises. Each of these areas deals with human work from its own point of view and thus represents the respective *aspect sciences*. In terms of practical applicability, this basic knowledge is used in so-called *praxeologies* (Luczak et al. 1987). The more socio-scientifically oriented of these is macro-ergonomics, which provides rules for the design of organisations, operating groups and working groups; the more engineering-oriented is micro-ergonomics, which aims to provide rules for the technical design of workplaces and equipment. In both cases, special research focus is placed on the individual person and his or her experience of the situation at work.

1.3.3 Subfield of Ergonomics

One gets an idea of the factors influencing work by looking at the structure of the human-machine system (HMS) (see Fig. 1.9). It is obtained by examining human activity in terms of the information it contains and the associated changes in information. This includes the task(s) and the fulfilment of the task(s) i.e. the corresponding

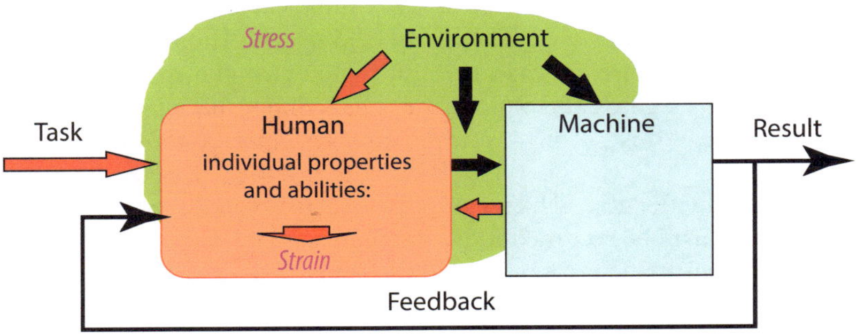

Fig. 1.9 General Structure Scheme of the Human-Machine System (HMS)

result. The arrow indicating the feedback closes the control loop formed by the HMS and shows that the operator is generally able to compare the task and the result. All factors influencing this process are called environmental influences (as far as they do not originate from process or system immanent influences).

A characteristic feature of ergonomics is the so-called stress and strain concept. Then, under stress, all the influences that can influence the working process of a human being are summarized, which are the same for every person who is in this situation (characterized in Figure 19 by the arrows pointing at the person). Strain is the individual reaction to this stress (reaction *in* human in ◘ Fig. 1.9). The task of ergonomic research is, among other things, to investigate the spectrum of these reaction possibilities during varying stress loads. The concept is of great relevance insofar as only the stress can be influenced in the technical design. The stress and strain concept can be applied to the load (stress) caused by the task and the load caused by the environment.

At the *Analysis of the task* one distinguishes between

- Tasks that are predominantly physical in nature (so-called *"physical work"*),
- Tasks with predominantly mental requirement character (so-called *"mental work"),*
- Tasks with mixed requirements character (so-called *"Mixed work"*).

During the analysis of the environmental influences (so-called *"Environmental Ergonomics"*) one distinguishes

- physical environmental influences that can be measured and their effects on humans that can be quantitatively evaluated. These are essentially lighting, noise, mechanical vibration, climate, toxic gases and vapors, radiation exposure, dust, dirt and moisture,
- social environmental influences, which can only in exceptional cases be recorded indirectly via physical-measurement approaches and therefore have to be analysed with other methods (task of the so-called sociology of work, partly also of industrial psychology).

A further subarea of ergonomics is the analysis of the *Human-Machine-Systems (HMS)* in the strict sense of the word. This analysis can be carried out with regard to the geometric design of the workstation and the work equipment (so-called *"anthropometric workplace design."*) and on the other hand with regard to the flow of information in the human-machine system (so-called *"System Ergonomics"* see below).

Anthropometric workplace design refers to the design of the visual, gripping and foot space, of body supports (e.g. seats), as well as the design and arrangement of displays and controls. In addition to knowledge of the relevant sensory physiological limits and conditions (e.g. resolution capacity of the eye, movement accuracy of the extremities), which are necessary for the design of displays and controls, the design of the gripping and foot space and that of body supports, in particular the different sizes of people, plays a primary role. By percentiating the individual body measurements one tries to systematically handle this problem. In order to facilitate the often complex geometric design tasks, computer-generated geometric human models (3D models) were also developed, which allow the design of workstations in CAD.

The basic structure of the integration of the human being into a complex HMS can be defined within the so called *system ergonomics* by means of the methods of system analysis. Their aim is to obtain design requirements for human-machine interaction in the context of the specification of the HMS or indications for possible improvements of existing systems under the aspect of information flow. Since system ergonomics strives to optimize this interaction, it also contributes to reducing the probability of human errors occurring (so-called "active safety") and to increase the reliability of the overall performance of the HMS.

1.3.4 Fields of Application of Ergonomics

Ergonomics is often classified according to its areas of application. A special distinction is made between product ergonomics and production ergonomics.

In *product ergonomics* the primary goal is to offer a user-friendly *product* for daily use of a basically unknown clientele. For the development of such products it is therefore important to know the variability of humans both with regard to their anthropometric and cognitive properties and to take them into account in the design. A current and new field of research in product ergonomics is the scientific recording of what is the *sense of comfort*. Aspects of product ergonomics therefore play a particularly important role in the design of the automobile.

Production ergonomics is about creating humane jobs in production and service companies. The aim here is to reduce the burden on the employee and at the same time optimise the performance output. In most cases, the issue is therefore the question of reasonableness and tolerability. In contrast to the task of product ergonomics, the employees are normally known here and their needs can be met individually. Production ergonomics play an important role in the automotive industry, especially for the design of production facilities. Since both product and production ergonomics use the ergonomic methods described above and since often the "product" of one manufacturer is "work equipment" of the other, an exact separation between these two fields of application is practically impossible.

Priority areas of application in which today ergonomic development is systematically pursued are aviation (especially cockpit design of airplanes, design of radar pilot workstations), vehicle design (cars and trucks: cockpit design, anthropometric design of interiors, so-called packaging, design of new means of information, safety, comfort and individual mobility), maintenance design (chemical plants, power plants, where aspects of human reliability play an important role) and office space (design of screens, office chairs, the whole layout of the workstation, software ergonomics). Since human characteristics are invariant to these areas of application, many mutual suggestions for ergonomic design arise from these areas of application.

This book focuses on the requirements for the technical design of a vehicle that result from knowledge of the physiological and psychological characteristics and abilities of humans when driving motor vehicles. Since the original objective of driving a car was to compensate for the laborious locomotion by physical exertion, aspects of information processing play a primary role in the ergonomic design of the car. The automotive industry has, in part, adopted its own terms for this purpose: for example, *functional ergonomics* is the suitability of the overall system for everyday use, i.e. the utility in connection with concrete usage scenarios, and *cognitive ergonomics* is a system for its adaptation to human cognitive abilities in terms of logic, consistency and conformity with expectations. As explained in Stein (2011), areas such as "cognitive ergonomics", "information ergonomics", "software ergonomics", "communication ergonomics"and "functional ergonomics"can be equated, in whole or in part, with the goal of system ergonomics or can be regarded as one of its sub-disciplines. An essential part of this book therefore deals with the special application of *system ergonomics* of the automobile (▶ Chap. 6).

Since the driver and passengers in a vehicle have to work in a relatively limited space and at the same time often spend a long time (several hours) in this space, so-called *anthropometric ergonomics* plays another important role in vehicle design. Specialists in anthropometric ergonomics adapt the vehicle to the human body's dimensions in terms of space, accessibility and visibility of the controls and displays (▶ Chap. 7).

In addition, comfort plays an important role in the design of the vehicle. Much of the knowledge that is necessary for this is derived from *environmental ergonomics*. The special application of this subfield of ergonomics to

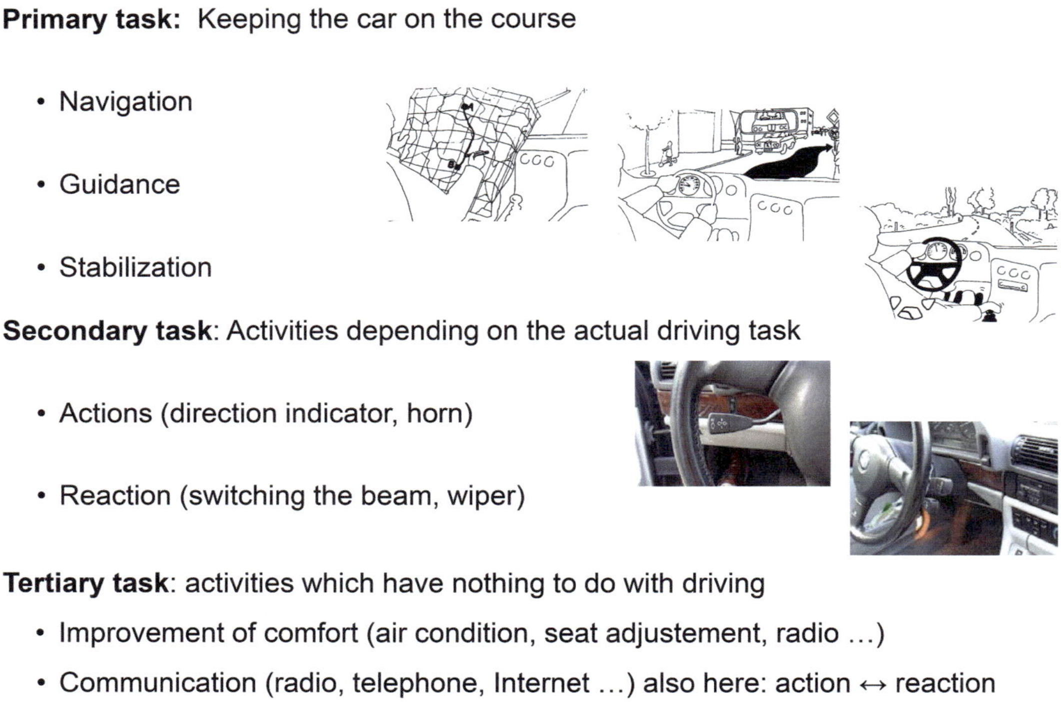

Fig. 1.10 Tasks of the driver

the design of vehicles is another subject of this book (▶ Chap. 8).

1.4 Hierarchy of the Driving Task

The exact description of the task is an absolute prerequisite for any ergonomic analysis. In general, the task consists of several subtasks. Driving a motor vehicle requires the accomplishment of several subtasks, some of which are hierarchically ordered and some of which are independent of each other. A classification suggestion by Geiser (1985) is followed here, which deviates somewhat from the usual division into two task areas.[12] Basically, a distinction can be made between the primary driving task, which requires the driver to keep the vehicle on course at a certain speed, and secondary tasks, which are necessary depending on the respective traffic situation and support the primary task. Tertiary tasks are those operations having nothing to do with actual driving. Figure 1.10 provides an overview of these subtasks.

12 Usually the division into "primary driving tasks" is carried out, which subsumes everything that has to do with driving. In line with Geiser's classification, this summarizes the areas designated as primary and secondary. The term "secondary" is then usually used to describe all tasks that have nothing directly to do with driving. In terms of the Geiser classification used here, these are the tertiary tasks.

1.4.1 Primary Task

The primary driving task is generally defined by the transport of persons and/or goods that are to reach their destination at a certain point in time. This transport task is used to derive the *navigation task* from: the route must be determined, an average speed must be estimated for the distance to be travelled in order to determine the starting time for the journey. The navigation task can be futher subdivided into the *travel planning* and *junction orientation*, where one decides to either deviate from the current road course or not. As is well known, both aspects of this task are now supported by GPS-based navigation systems, with most systems allowing the driver to choose the shortest or fastest connection, secondary routes or highways.

By determining the route to be taken, a global framework for the lower hierarchical level, the *guidance task*, is given.The guiding task requires the driver to determine the target course and the target speed depending on the current traffic situation. These mental reference variables are to be determined according to external conditions such as the course of the road, obstacles on the road, weather conditions, traffic density and the behaviour of other road users. This task can be subdivided in detail: In *maneuver planning* the driver decides on the basis of the given situation and his own inner motivation for a certain action. The manoeuvre "following the course of the road" is trivial and practically permanent. More complex is the manoeuvre "following the vehicle ahead", because in addition to following the course of the road there is also the adjustment of the correct distance. The decision to overtake the vehicle in front is a very complex manoeuvre. Such a maneuver requires one to estimate the capability to accelerate the vehicle, compare the length of the overtaking distance with the visible distance, as well as to estimate the speed of an oncoming vehicle and the distance it will travel. Another manoeuver resulting from the navigation task, for example, can be "turning off the road (left or right)". In this way different manoeuvres can be defined, which are activated depending on the respective situation. If the decision for a particular manoeuvre has been made, then on that basis the *trajectory* to be planned in such a way that contact with standing or moving objects or persons is avoided in any case. The results of scientific investigations show that this task is performed permanently for a spatial distance up to a maximum of 200 m in front of the vehicle and within a time window of approximately 2 s . The driver lays a connecting line from his current position on the road to a destination point in the course of the road that is no further away than 200 m. At obstacles at roads, e.g. parked vehicles, that connecting line is bended correspondingly. The determination of the target course becomes more difficult when the obstacles are moving on the road (vehicle in front, vehicle crossing or overtaking, pedestrian). At this level of the primary driving task, current developments offer support for the driver. The ACC system already introduced maintains a target speed set by the driver and changes this when a slower vehicle drives ahead, whereby a speed-dependent distance is then automatically calculated and maintained. The vehicle is automatically decelerated and accelerated back to the target speed when the vehicle in front brakes or accelerates again. An extension of the ACC system to automatically detect speed limits by networking with the navigation system (unchanging areas with speed limits are defined by GPS coordinates) or by camera systems (e.g. in conjunction with the lane guidance assistant) are now in series production as special equipment. The Lane Guidance Assistant assists in determining the target heading at least to the extent that exceeding tolerance ranges for the target heading is indicated to the driver by corrective steering torques. At a slightly lower level, similar results are achieved by the Lane Departure Warners already described. Special tasks of the guidance task such as lane change are supported by lane change warning and lane change assistant. The further development of assistance systems that serve traffic safety is mainly in the field of guidance tasks.

By operating the control elements, the driver can achieve the following by completing the *stabilisation task* the correspondence of the actual variables (actual course, actual velocity) with the mentally formed target variables within the framework of the reference task. For this two-dimensional stabilisation task (longitudinal and transverse dynamics), five control elements are used in a hand-operated vehicle, whereby four control elements alone have to be operated to control the longitudinal dynamics: Brake, clutch, shift lever, throttle. Vehicles with automatic transmissions offer the advantage that only three control elements are required for stabilisation, which - at least theoretically - ultimately promotes concentration on the traffic situation.

1.4.2 Secondary Tasks

Secondary driving tasks arise depending on the primary driving task, whereby the driver informs other road users of his intentions with actions or reacts to external conditions. Examples of *actions* are the activation of the direction indicator when changing lanes or directions and the activation of the horn to warn other road users and of the headlamp flasher to indicate an intention to overtake or, for example, to indicate priority to a road user waiting at an exit..[13] Examples for *reactions* means switching from the driving beam to the passing beam in the case of oncoming traffic, switching on the wiper in the case of rain or switching on the front fog lamps or rear fog lamps in the case of appropriate weather and visibility conditions. Many of these reactive secondary tasks can be automated, such as in vehicles of the upper middle class or premium class (but mostly only as optional equipment) already realized in the form of intelligent lighting systems.[14] or by rain sensors, via which the wiper can be switched on when it rains and the wiping intervals can be optimally adapted to the windscreen wetting. Shifting gears is also a secondary driving task, either as a reaction to external conditions (e.g. shifting down when driving uphill or downhill) or as an action during certain driving manoeuvres (e.g. shifting down before overtaking). From a purely technical point of view, the clutch and gearbox are necessary in order to enable starting and to adapt the torque of the engine, which depends on the speed, to the requirements of the driving situation. As already explained, starting without using a clutch pedal and shifting gears without a gear lever was automated very early in the USA (1940) by means of an automatic transmission with torque converter, which represents a sensible relief for the driver with regard to the two-dimensional driving task.[15] While American manufacturers produce virtually no manual vehicles for the American market, there are fewer supporters in Europe than opponents of the automatic transmission. The technically adept driver describes the automatic transmission as a "torque destroyer" and the more sporty drivers see a safety gain in the faster manual downshift when overtaking. In fact, the automatic downshifting system only reacts to the kick-down with the accelerator pedal with a certain delay, which requires anticipatory driving. When driving on a steep gradient, the automatic transmission usually does not react to the braking of the vehicle with the speed level that would be necessary for "engine braking". For this reason, the various speeds of automatic

13 One can see from these examples that the coding of these means of communication to the outside world is sometimes not completely clear and can possibly give rise to fatal misunderstandings.

14 The "Intelligent Light System" from Mercedes-Benz consists of variably controllable headlamps with five different lighting functions. At a speed of more than 90 km/h, a motorway light switches on automatically in two stages, whereby the output of the xenon lamps is first increased from 35 to 38 watts and in the second stage from 110 km/h the range of the headlamp inside the road is increased. The result is a uniform cone of light extending up to 120 metres, which illuminates the entire width of the carriageway. As soon as the rear fog lamp is switched on at a visibility of less than 50 metres and a speed below 70 km/h, the left-hand bi-xenon headlamp swivels eight degrees outwards and lowers its beam simultaneously. The extended fog light remains switched on up to a speed of 100 km/h. In the meantime, various manufacturers are offering lighting systems which, controlled by the evaluation of the image of an electronic camera, automatically cut out the area of the oncoming traffic at risk of glare from the light of the main beam. This is done partly via a mechanically inserted diaphragm but meanwhile by controlling LED modules arranged in a matrix in the headlamp.

15 In the 1960s, almost all cars in Germany at that time had the "Saxomat" from Fichtel & Sachs as an alternative to the manual transmission. This was an electro-pneumatically actuated clutch which had a centrifugal clutch for starting. This meant that the clutch pedal could be omitted. The gear changes were controlled by the driver using the accelerator pedal and gear lever. If the driver touched the gear lever, the clutch was disengaged and you could shift gears. The Saxomat was especially popular with older riders but lost its popularity over time as the adjustment was difficult and the susceptibility to interference high. In many cases, vehicles equipped with the Saxomat were converted to normal gear shifting.

transmissions can also be shifted manually. The examples show that automatic transmission is more comfortable for the defensive driver than for certain manoeuvres. Many new developments (e.g. dual-clutch transmissions, automatic transmissions with 7–8 gear steps) have fully compensated the indicated disadvantages, so that today the fuel consumption of some vehicles is even lower with automatic than with the manual version. In addition, the latest developments use information from the navigation system in order to implement certain elements of anticipatory driving (e.g. timely shifting back before a gradient, use of the engine brake on a gradient).

Although an ergonomic requirement to simplify the driving task has been achieved with the automatic transmission, a new technical development brings with it an additional task for the driver, which is not derived from the traffic situation, but from the necessity of the available technical system: With the introduction of hybrid vehicles, the driver has depending on the respective technical design the additional task of choosing between different driving modes (purely electric driving, drive only via the combustion engine, combined drive, different recuperation modes). An ergonomic improvement would already be achieved here if the selection of the appropriate mode were based on information from the navigation system.

1.4.3 Tertiary Tasks

Tertiary driving tasks have nothing to do with the actual driving task. With these tasks the driver regulated comfort settings in the vehicle (e.g. operation of the air conditioning system) or those that provide information, communication and or entertainment (e.g. operation of the radio to listen to traffic reports or weather reports, telephone calls, playing a CD). As with secondary driving tasks, actions and reactions can also be distinguished here. An active task is calling someone on the phone while answering a call is the reaction to the ringing of the telephone. In any case, tertiary tasks distract from the actual driving task. Many studies have shown that the driver generally tolerates avoidance times of up to 2 s. However, in an experimental study Rassl (2004) was able to show that in some cases distraction times of up to 16 s duration occurred when operating the DAS. This leads to the requirement that technical equipment must be designed according to system ergonomic rules in such a way that the resulting distraction potential is kept as low as possible in terms of content and time. If these rules are followed, the probability of the occurrence of extremely long deflection times can be significantly reduced. Various approaches are currently being pursued to manage tertiary driving tasks with a reactive character with the aid of technical equipment. One idea, for example, is based on suppressing a telephone call in a difficult traffic situation (with an automatic announcement to the caller) until safe telephoning appears possible again while driving. The technical problem is how, on the basis of image processing, traffic situations can be categorized with regard to the driver's attention, while the ergonomic problem is how to inform the driver about the system status without distracting him significantly from the traffic situation.

1.5 Information Flows in Transport

The driver alone in his interaction with the vehicle is not sufficient for understanding driving. In order to accomplish the guidance task, special consideration must be given to the interactions with other participants in traffic. ◻ Figure 1.11 compiles the corresponding information flows. The technical information flows in today's conventional traffic are mainly characterized by rigid traffic signs, variable message signs and especially traffic lights. Today, information from traffic is occasionally transferred to the screens of navigation devices (traffic signs, alternative routes in traffic jams). Essentially, however, human information flows play a major role. For example, the future course of the "opposing" vehicle can be deduced from the perceived course of the vehicle. Possibly

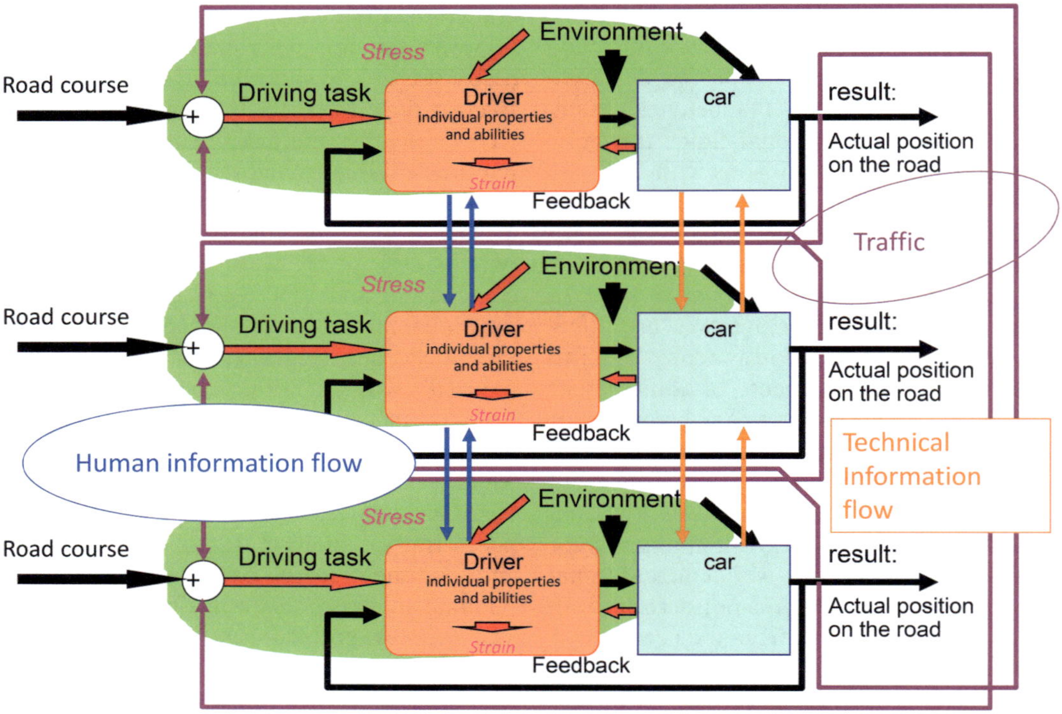

Fig. 1.11 Principle information flows in road traffic

secondary information plays thereby a role such as attitude and appearance of the opposing driver, kind and presentation of the vehicle and similar. Of course, statements to other road users initiated by the driver can also be technically supported and their use is sometimes mandatory (e.g. direction indicator, horn, headlight flasher). In the future, however, technical information flows will gain in importance (car2car), which partly directly intervene in driving behavior, but partly via displays information will be sent to the driver.

As can be seen from Fig. 1.11, in road traffic, in addition to the course of the road, the technical information flows directly influence the driving task of the individual driver, while the human information flows rather have a modifying effect on the individual behaviour (e.g. arousing anger, generous granting of privileges etc.). A distinction can be made between static and dynamic systems in the technical information services used to regulate the flow of traffic. All traffic signs that contain regulations or recommendations for the guidance task are static information (e.g. speed limits, forbidden or required lane uses, overtaking prohibitions, observance of priority regulations, parking prohibitions, information painted on the road and much more). A distinction must be made between these and the dynamic indications, which range from (technically) rather simple traffic light controls and change indicators to dynamic traffic recommendations (e.g. parking guidance systems). Due to their situational adaptability, the dynamic traffic regulations experience a much higher acceptance on the part of the user than the static ones, which are often perceived as inadequate in a given situation. The static signposts are often an indispensable aid for carrying out the navigation task. This indicates a further fundamental distinction between traffic signs: There are traffic signs which point to a situation which is only significant in the short term at the relevant point (e.g. signposts, directional prohibitions or bids, reference to zebra crossings etc.) and those which are valid for a longer driving period (e.g. speed limits, overtaking bans, stopping or parking bans). The correct observance of the latter places corresponding demands on the driver's memory and often leaves him or her in doubt as to whether the order in question is still valid or has already been cancelled. The distinction between traffic signs with a short-term or long-term validity character shows the advantage of "drawing" the relevant information into the DAS or DIS, because there the long-term validity of a regulation can be clearly distinguished from the short-term validity character of another regulation. Since all advertisements there will have a quasi-dynamic character, a higher acceptance of the corresponding regulations is to be expected.

References

Literature Used

Die Entwicklung der Straßenverkehrsunfälle in der Bundesrepublik Deutschland und in Berlin (West), BASt, Heft 1 (1974)

Geiser, G.: Mensch-Maschine-Kommunikation im Kraftfahrzeug. ATZ. **87**, 77–84 (1985)

Helander, M.: Human Factors/Ergonomics for Building and Construction. John Wiley & Sons, New York, Chichester, Brisbane, Toronto (1981)

Lerner, M., Schepers, A., Pöppel-Decker, M., Leipnitz, C., Fitschen, A.: Voraussichtliche Entwicklung von Unfallzahlen und Jahresfahrleistungen in Deutschland – Ergebnisse 2013. Bundesanstalt für Straßenwesen, Dezember (2013)

Luczak, H., Volpert, W., Raeithel, A., Schwier, W.: Arbeitswissenschaft, Kerndefinition Gegenstandskatalog-Forschungsgebiete. Rationalisierungs-Kuratorium der Deutschen Wirtschaft (RKW) e.V., Eschborn (1987)

Rassl, R.: Ablenkwirkung tertiärer Aufgaben im Pkw – Systemergonomische Analyse und Prognose. Dissertation an der Technischen Universität München (2004)

Spies, R.: Entwicklung und Evaluierung eines Touchpadkonzeptes mit adaptiv haptisch veränderlicher Oberfläche zur Menübedienung im Fahrzeug. Dissertation an der Technischen Universität München (2013)

Stein, M.: Informationsergonomie – Ergonomische analyse, Bewertung und Gestaltung von Informationssystemen. Habilitationschrift

Woodson, W.E.: Human Factor Design Handbook. Mc Graw Hill Book Company, New York (1981)

Further Reading

Bubb, H.: Ergonomie. In: Landau, K. (Hrsg.) Lexikon der Arbeitsgestaltung, S. 495–498. Gentner-Verlag, ergonomie Verlag, Stuttgart (2005)

Murrell, F.H.: Ergonomie. Grundlagen und Praxis der Gestaltung optimaler Arbeitsverhältnisse. Chapman and Hall, Econ Verlag, Düsseldorf, Wien, London. (1971)

Rühmann, H., Bubb, H.: Fahrerassistenzsysteme. Ein Gewinn an Sicherheit und Komfort oder elektronischer Schnickschnack? Ergonomie aktuell. **7**, 2–13 (2006)

The Control Loop Paradigm of Ergonomics

Heiner Bubb

Contents

2.1 Driver-Vehicle Interaction – 26
2.1.1 The Driver-Vehicle Control Circuit – 26
2.1.2 Pursuit and Compensatory Task – 28
2.1.3 Quality and Performance – 29
2.1.4 Quality of Laterale Dynamics – 29
2.1.5 Quality of Longitudinal Dynamics – 30

2.2 Basic Concepts of System Technology – 31

2.3 System Dynamics – 35
2.3.1 Transition Functions and Frequency Response – 35
2.3.2 Transition Function – 36
2.3.3 Frequency Response – 37

2.4 Quantities Controlled by the Driver – 43
2.4.1 Laterale Dynamics – 43
2.4.2 Longitudinal Dynamics – 48

2.5 The Primary Driving Task from a Control Engineering Point of View – 49

2.6 System Reliability – 53
2.6.1 Safety, Risk, Border Risk and Protection – 53
2.6.2 Reliability, Error and Safety – 54
2.6.3 Human Error and Probability of Accident – 56
2.6.4 Derivation of Measures from the Fault Tree Analyses – 57

References – 60

H. Bubb et al. (eds.), *Automotive Ergonomics*, https://doi.org/10.1007/978-3-658-33941-8_2

2.1 Driver-Vehicle Interaction

2.1.1 The Driver-Vehicle Control Circuit

In order to understand the driver-vehicle interaction more precisely, the structural image of the human-machine system shown in ◘ Fig. 1.9 must be used. The driver's task - as with any vehicle guidance task, be it for land, air or water vehicles - is hierarchically divided into the areas of navigation, guidance and stabilisation. As already explained in ► Sect. 1.4, the fulfilment of the *navigation task* ("by which route do I want to reach the destination?") is the prerequisite for the *guidance task* to be fulfilled on site ("which exact path do I want to follow to avoid any collision with moving and stationary objects?). Only when the exact nominal course and the appropriate nominal speed have been established in the mind by fulfilling these tasks, can the corresponding actual course and speed be realised by operating the controls via the vehicle in the context of fulfilling the *stabilisation task*. Due to the given vehicle technology, two tasks have to be performed simultaneously on this level, i.e. the steering wheel has to be used to operate the *laterale dynamics* and by the combined operation of accelerator and brake pedal to influence the *longitudinal dynamics*.

In the control engineering sense, disturbance variables are those influences which are not part of the task but which nevertheless affect the process. Although other road users normally determine part of the guidance task with their behaviour, unexpected behaviour may necessitate an evasive manoeuvre which manifests itself in a sudden change in the outcome of the guidance task. Weather influences can also have a similar effect (e.g. fog). However, weather influences can also have a direct effect on the longitudinal or laterale dynamics of the vehicle (e.g. crosswind, strong headwind or tailwind), which makes it necessary for the driver to additionally correct this disturbance.

As already mentioned, in addition to these hierarchically structured tasks there are the secondary tasks to be performed depending on the driving situation and the tertiary tasks, which have nothing directly to do with the driving task. ◘ Figure 2.1 compiles this context once again in a corresponding structural picture. As can be seen from this figure, a large part of the tasks are mental demands to be mastered by the driver, the result of which, however, only becomes visible when the controls are actuated. This mental part is discussed in detail in ► Chap. 3. This chapter will discuss the problems arising from the immediate dynamics of interaction with the vehicle.

Under these conditions, a simplified control cycle screen can be specified, but this now allows a more precise analysis of the dynamic processes. Such a picture is shown in ◘ Fig. 2.2. This picture can be applied analogously to both longitudinal and laterale dynamics.

In general, a control loop is characterized by the *task* and the *result* which in the terminology of control engineering is associated with *command signal and tracking signal* between which the *controller* and the *controlled system* is tense. The controller, in this case the driver, reacts to the deviation between the task and the result and acts on the controlled system, in general this is the machine, in this case the vehicle, in order to minimize the difference between the command and tracking signal. This difference is calculated at the sum point (represented by a circle) at which the command signal with a positive sign and the tracking signal with a negative sign are fed in. The *feedback* of the result characterizes the control cycle. If the feedback is omitted, one speaks of a *open loop control* which naturally reacts more sensitively to external influences than the control system with feedback. In general it can be said that: The advantage of the open loop control is the fast reaction, the advantage of the control with feed back is the precise adaptation of the command sigal and the tracking signal. Driving a car under this aspect is often attributed with a double strategy: fast reaction is often inaccurate in the sense of open loop control (e.g. reaction to a sudden obstacle, fast driving on a winding well-known route), while precise driving is only possible through

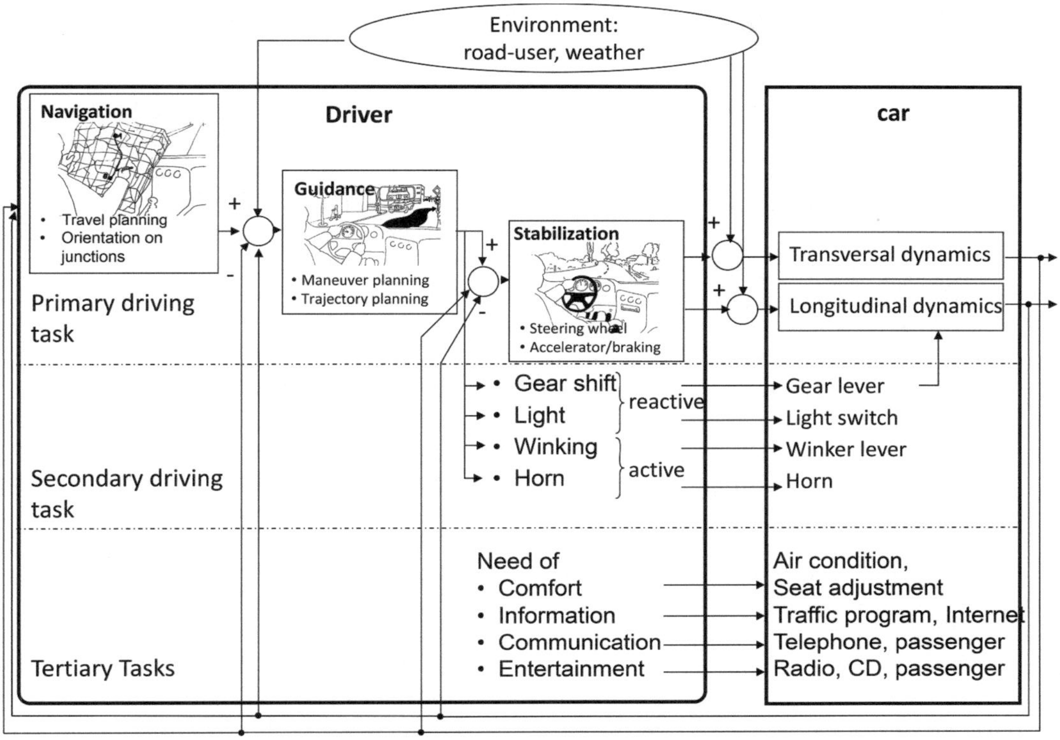

Fig. 2.1 Tasks of the driver in the driver-vehicle control loop

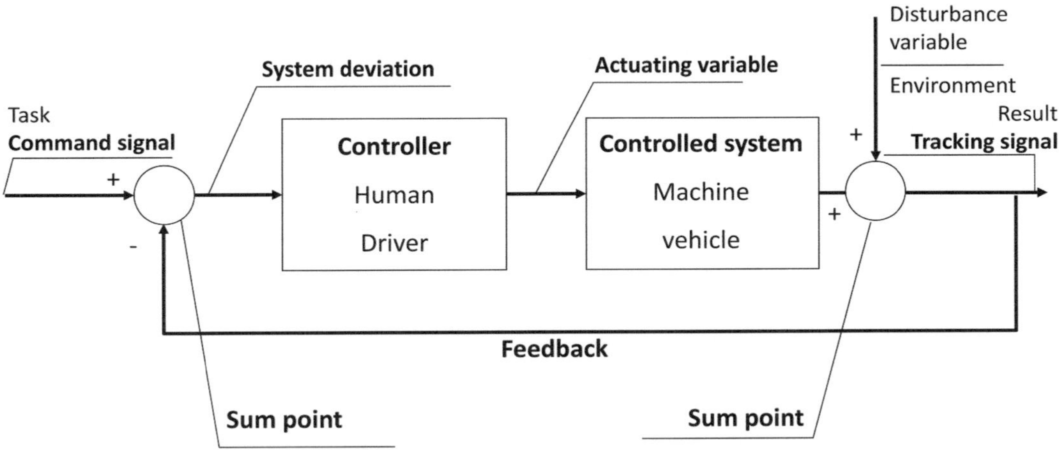

Fig. 2.2 Control loop as paradigm for human-machine interaction here Driver-vehicle interaction

control with feedback (e.g. keeping the lane on the motorway, keeping a distance, manoeuvring). Irrespective of whether a open loop control system or a closed loop control is present, the difference between the command and the tracking signal is calculated as *system deviation* and the intervention of the controller in the controlled system is referred to as the *actuating variable*. The described process can be disturbed by external influences (see above), which are known as *disturbance variables* in the sense of ergonomics generally referred to as *environmental influences*.They are added up over a further sum point, whereby from the mathematical point of view it is irrelevant at which point within the control loop this happens.

2.1.2 Pursuit and Compensatory Task

In the control engineering sense, the difference between the command signal and the tracking signal is formed at the sum point. If the human being acts as a controller, the question already arises how this sum point is realized. Basically there are two possibilities: The sum point is located outside the human being, i.e. the difference between the reference variable and the tracking variable - usually by an instrument - is displayed directly to the user. His task is then to "make zero" the displayed difference. One speaks of a *compensatory task*. The other option is to display the reference variable and the tracking variable separately. The sum point lies then in humans, i.e. he must bring the two sizes mentally to the congruence. This kind of interaction is called *pursuit task*. ◘ Figure 2.3 shows the system-technical representation of the two forms of the interconnection of human and task representation.

What does the situation look like with the "natural" representation of driving a car? The motorist cannot see his position in the world (although to some extent he can see it on the basis of mental performance and imagination), he can only see directly the difference of the position of his vehicle in relation to the two edges of the lane (or an imagined target lane). In this respect, driving a vehicle is a compensatory task. However, when he looks out of the window, he sees not only the distance of his vehicle from the edge of the road, but also the further course of the road, which he of course uses very substantially for the steering process. It is often argued that this anticipation of the road has to a certain extent the character of a pusuit task and serves in particular to react quickly to the course seen there in the sense of steering. In the sense of a closed loop control, the distance to the side of the road (the immaginated nominal course, see ► Sect. 2.5) would then be corrected (for more details, see ► Sect. 2.4.2). In terms of control technology, driving would therefore be a hybrid of open and closed loop control.

The consideration of the form of representation as compensation or pursuit task is important for the technical representation of information related to driving, in particular the navigation display (see ► Chap. 6). In principle, the north-facing representation represents a pusuit task. This is also true, although in all systems the map is readjusted

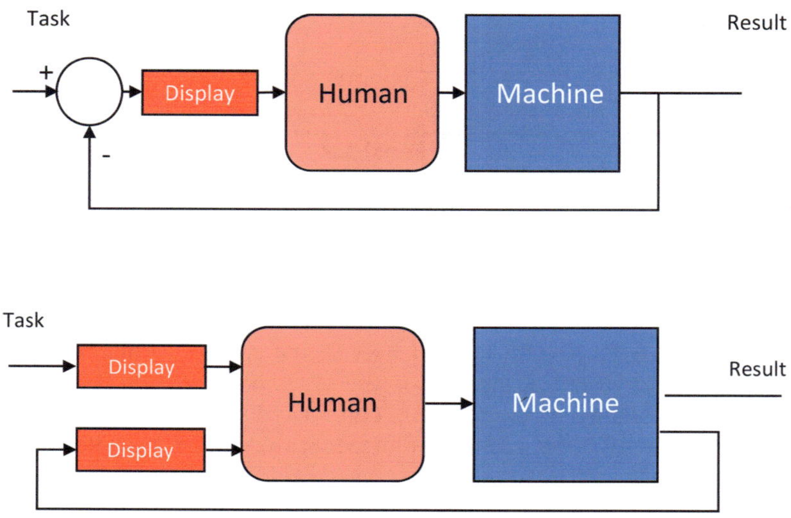

◘ **Fig. 2.3** Structure of compensatory task (*at the top*) and pursuit task (*underneath*)

when the own position reaches the edge of the map. On the other hand, the direction indicating display and especially the so-called "Bird-view display"correspond to the compensation task.[1] This form of presentation is therefore compatible with the "natural" driving task and therefore preferable from an ergonomic point of view. However, it may be of interest for the driver to know "where in the world" he is. For this reason, the ergonomically recommended version would be to show the compensation version of the navigation display in the instrument cluster behind the steering wheel, which gives a recommendation of what action is necessary in the next action area to reach the destination, and to provide a north-facing display on the central screen.

2.1.3 Quality and Performance

The effect achieved by the feedback circuit is that, to a certain extent, which is essentially determined by the characteristics of the controller and the requirement for stability of the entire control loop, the task T and the result R largely coincide at all times and that even the influence of the disturbances can be eliminated.

This property of the control loop can be operationalized by the quotient Q, which can be defined as *quality*[2]:

$$Q = \frac{result}{task} = \frac{|R|}{|T|} \tag{2.1}$$

This definition has general validity. It also applies when the control loop paradigm describes human work in general (see Schmidtke 1993). One then speaks of the *quality of work*.

In tests with the vehicle, often only the difference between task T and result R can be determined (e.g. keeping the vehicle in the middle of the lane). Determine the relative difference D

$$D = \left|\frac{T-R}{T}\right|, \tag{2.2}$$

the quality can also be described as a deviation from the desired maximum quality Q = 1:

$$Q = 1 - |D|. \tag{2.3}$$

If one considers the quality of work achieved in the time t, one obtains the definition of the *Working performance P*:

$$P = \frac{Quality\ of\ work}{time} = \frac{Q}{t}. \tag{2.4}$$

It also makes sense to apply these definitions consistently in connection with the driver's control of the vehicle. Bolte (1991) and Reichart (2001) have developed detailed data and formulas for this purpose, which for reasons of comparability should also be used for the experimental recording of individual driver performance. It should be noted here, that at many driving experiments the dimensions mentioned here are not consistently used for evaluation (see ▶ Sect. 11.2.5). Frequently, a measure of driving quality is satisfied with the deviation from a target lane or from a specified speed. Since these variables vary, the mean values and, if applicable, the standard deviation are used for assessment depending on the research question. In rather rare cases, a frequency analysis is carried out with regard to the mentioned deviation values.

2.1.4 Quality of Laterale Dynamics

2.1.4.1 Straight Ahead Travel

For the *Quality in lateral dynamics* the so-called "lateral guidance quality Q_L" indicates Reichart according to the above formulas:

1 With the Bird-view display, the view is realized from a virtual camera, which is quasi permanently connected to the vehicle behind it in an elevated position.

2 The sizes for task T and result R must be specified in absolute values, since the quality can only assume positive values.

$$Q_L = 1 - \kappa \quad (2.5)$$

where κ is the degree of utilisation of the lateral room for manoeuvre:

$$\kappa = \left| \frac{2 \cdot y_L}{w_{lane} - w_{vehicle}} \right| \quad (2.6)$$

with:
- y_L = Offset of the lane to the middle of the lane,
- w_{lane} = lane width,
- $w_{vehicle}$ = vehicle width.

Figure 2.4 shows the lateral guidance quality according to Eq. 2.5 if one assumes that the driver tends to keep the vehicle approx. 0.5 m offset to the left from the centre of the lane when driving straight ahead (if there is no offset according to Eq. 2.6 in Eq. 2.5, lateral guidance quality = 1).

The driving task is all the more difficult the greater the necessary tracking quality is. Interestingly enough, Reichart (2001) found that the factor of 3.2 by which driving a truck on narrow roads is more difficult than driving on wide roads is almost reflected in the accident figures: on narrow country roads almost 4 times as many truck accidents occur as on wide federal roads.

2.1.4.2 Free Cornering

Free cornering should be characterised by the fact that there is no speed limit and no other road users influencing the journey. Then the task T is characterized by the fact that the lateral acceleration is not greater than the maximum possible lateral acceleration. In this case, the lane keeping quality is therefore determined as

$$Q = 1 - \frac{|a_{lmax} - a_l|}{a_{lmax}} \quad (2.7)$$

with:
- a_{lmax} = maximum possible lateral acceleration
- a_l = lateral acceleration driven

For practical cases, depending on the question, it may make sense to make a safety deduction for the maximum possible lateral acceleration (e.g. only 70% of the physically possible lateral acceleration) or to refer to the known comfort accelerations, which are calculated at $a_{lmax} \leq 0.3\,\text{m/s}^2$ (even lower at higher speeds!).

2.1.5 Quality of Longitudinal Dynamics

2.1.5.1 Driving Without Other Road Users

When driving freely without speed limit, the highest longitudinal guidance quality is naturally achieved at the maximum speed of the vehicle. If there is a speed limit (of whatever kind), the quality results in

	Lane width w_{lane} = 3.0 m		Lane with w_{lane} = 3.75 m	
Situation	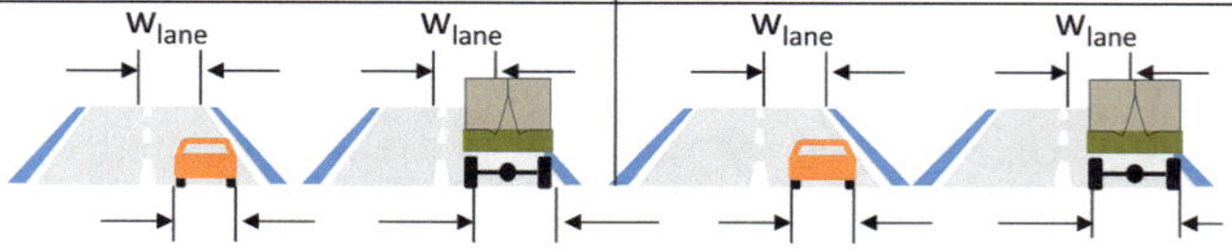			
	$w_{vehicle}$ = 1.62 m	$w_{vehicle}$ = 1.82	$w_{vehicle}$ = 1.62 m	$w_{vehicle}$ = 1.82
Utilization degree κ at 0.5 m offset to the center of the lane	72.5 %	85%	47%	52%
Lateral guidance Quality Q_L	27.5 %	15%	53%	48%

Fig. 2.4 Lateral guidance quality at different constellations (according to Reichart 2001)

$$Q = 1 - \frac{|v_{\max} - v_{actual}|}{v_{\max}} \qquad (2.8)$$

with:
- v_{max} = maximum possible or permitted speed,
- v_{actual} = driven speed.

2.1.5.2 Driving with the Vehicle in Front

For the *Quality in longitudinal direction* can be specified after Bolte (1991):

$$Q = 1 - \frac{x_{shall} - x_{is}}{x_{shall}} \qquad (2.9)$$

with
- x_{shall} = Required safety distance ($x_{shall} = v_{is} \cdot t_s$; v_{is} = driven speed, t_s = safety time, e.g. $t_s = 1{,}5\,s$),
- x_{is} = actual distance to the vehicle in front.

The Eq. 2.9 shall only be valid if $x_{is} \le x_{shall}$. In case $x_{is} \ge x_{shall}$ the case of free travel exists with regard to the quality calculation.

Depending on the respective problem, meaningful quantities are to be defined in terms of formulas 2.7, 2.8, and 2.9. If a test drive is characterized by phases in which several different quality parameters can be formed (e.g. Lane Keeping Quality in Curves and Longitudinal Guidance Quality with Vehicle in Front), the lowest quality always determines the overall quality.

In any case, Eq. 2.4 applies to the performance. By definition, it has the unit 1/sec. In many cases you get a higher performance at higher speed. Since the quality measures are subject to temporary fluctuations in the course of a journey, depending on the respective question, the course can be summarized by calculations such as mean value and standard deviation, modal value, median, maximum values or percentiles or frequency analyses.

2.2 Basic Concepts of System Technology

Systems theory has its origins in communications engineering (Wunsch 1985). In principle, it treats every considered complex of (natural) phenomena according to the same basic scheme. This includes the advantage of being able to treat humans in their interaction with the technical environment using uniform methods. Systems theory is therefore not only a basic discipline in the field of technology (Unbehauen 1993), but also provides the intellectual basis for management methods (Daenzer and Huber 1992). In particular, it is suitable for describing the relationship between the biological being "human" and the technical system "vehicle" with largely uniform model concepts. It deals with the relationship between the (desirable or undesirable) properties of elements, systems and the environment.

Under a *system* is understood to mean the totality of the technical, organisational and/or other means of the highest level of consideration necessary for the independent fulfilment of a complex of tasks. It thus represents a meaningful arrangement of technical units pursuing a certain purpose alone, an arrangement of technical units and humans, or an arrangement only of humans. Each system can be thought of as composed of elements, whereby these elements interact with each other (VDI 4005, Sheet 1). The essential aspect of system theory is that the physical nature of the interaction (e.g. force, electric voltage/current, temperature, airborne or structure-borne sound, etc.) is ignored and only the *information* or the information change caused by the system is taken into account. Information is understood to mean any deviation from the natural (ultimately thermal) distribution state of energy and matter. In practice, the desired change in information can consist both in the change of matter (task of the "working machines") and of energy (task of the "power machines" or

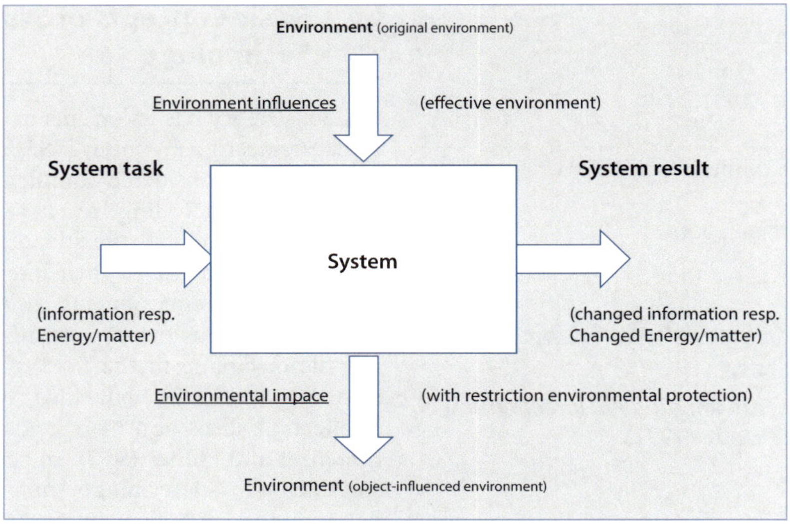

Fig. 2.5 General representation of a system and the system environment

"energy generation"), as well as in the local change of energy/matter (task of the hoists, conveyors, pumps, fans, vehicles) or in the change of matter/energy distributions (task of the information-processing machines, such as computers, data carriers). The operator transfers information to the machine and thus ensures that it achieves the desired effects.

The desired change can only be made if the system is exchanging information with the so-called system environment (so-called "open system", see Fig. 2.5). The system thus has an input side ("Input") through which information from the environment (i.e. energy or matter) is absorbed on one or more channels and an output side ("Output") through which the converted information (i.e. converted energy or matter) is delivered on one or more channels.

All influences that influence the desired process of information transformation through the system are called environment in this sense. This is why we speak of environmental impact. From the multitude of possible *environmental influences* only those are to be considered, however, which influence the function of the system. This is the so-called *"effective environment"* (VDI 4005, Sheet 1). This raises the question of the meaningful drawing of system boundaries (space radiation is currentlye.g. a feature of environmental impact on almost all technical products. However, it is not taken into account in the design of a vehicle, since it is not part of the effective environment of an automobile). All characteristics of the environment that existed before the existence of the system are called *original environment*. Even the mere existence and even more so the operation of the system also changes the environment, so a new one, *object-influenced environment* is created. This influence to the outside will be *environmental impact* which includes environmental protection. Especially in connection with the motor vehicle, this aspect of the object-influenced environment and especially the environmental influence plays an outstanding role in today's discussion and will already have a significant influence on the technology of the vehicle in the near future (see the arguments for the introduction of electromobility).

In order to get a more precise idea of the functions and mechanisms of action of the system, you have to think of the following *elements* and these elements exchange information with each other. Through the *system structure* the interactions between the elements are described. The information channels are represented graphically as arrows, which thus also indicate the direction of action of the information. Within such a structure of effects the parts are called elements which change the incoming information in a characteristic way. A distinction is made between elements that link information with one another or distribute information (sum point, branching point, switch) and elements that change information according to a function, whereby those in which time plays no role ("elements without memory") are separated from those in which time plays a role ("elements with memory"; see ◘ Fig. 2.6). The respective functions of the elements can be

- are described mathematically (e.g. at many technical system elements) or,
- in the form of graphs (e.g. in special cases of so-called nonlinearities) or,
- be presented verbally (e.g. for functions which the human being fulfils).

The presence of elements with memory causes a quite specific *system dynamics* which can be described in simple cases by differential equations or in more complex cases by frequency responses (see ► Sect. 2.3).

From this point of view, both the machine and the human being can be considered as a component of an *"Human-Machine-Systems"*. ◘ Figure 2.2 is an example for such a representation according to rules of system theory. ◘ Figure 2.7 shows the two possible interconnection principles of human and machine. A *serial interconnection* describes the variant in which the human being takes up the information of the task, converts it in an

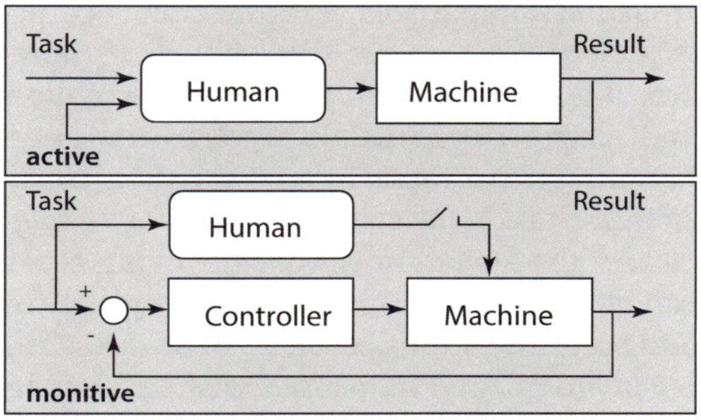

◘ **Fig. 2.7** System structure of a man-machine system *above:* Serial circuit (active system), *below:* parallel circuit (monitive system)

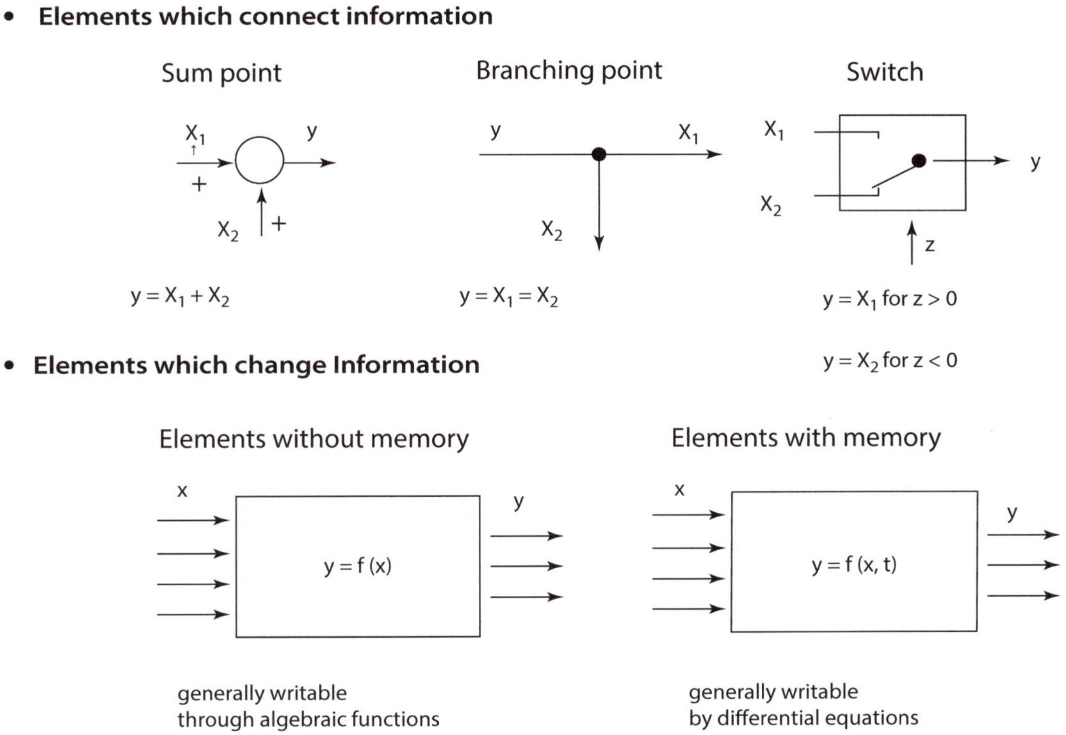

◘ **Fig. 2.6** Compilation of typical system elements and their functions

adequate way and transfers it to the machine via the actuators. This changes - often by converting separately flowing energy - the input information into the intended result (the so-called *active system*). The traditional process of driving a car corresponds to this picture: the task lies in the road guidance and in the situation that presents itself there, the result is the current position of the vehicle on the road. In general, the human can observe the result and derive new interventions via the actuators from the comparison with the task. So this is a *controlled* process (see above). If, however, the time between actuator intervention and receipt of the result becomes too long, he must derive the actuator actuation solely from the task. One speaks now of a *open loop control* (as already mentioned, driving a car can be seen as a mixture of open loop control and regulation). The process can be disturbed by environmental influences, where now is to define more exact, whether this disturbance affects human and his abilities (e.g. noise), transmission between human and machine (e.g. mechanical swinging motions) and machine itself (e.g. cross-wind or disturbance of function by electromagnetic fields).

At the *parallel circuit* of human and machine the human being has an observer activity (so-called *monitive system*). The machine, designed as an automatic machine, independently converts the information describing the task into the desired result. If the human being observes inadmissible deviations or other irregularities, he intervenes in the process (symbolized by the switch) and interrupts the process or takes over the control manually. Other forms of interaction are also conceivable; see also ▶ Sect. 2.6). If necessary, he also adjusts parameters on the system in order to achieve the desired effect (e.g. driving with ACC represents such a parallel connection of driver and vehicle in terms of longitudinal dynamics; if the distances set by the system appear too large to the driver, he can adjust the "safety time" parameter during the control process).[3] The automated process can also be disturbed by the environment, whereby the typical starting points are again human, machine and transmission paths.

There are two fundamentally different ways of treating the system structure. On the one hand, one can use the above-mentioned interconnection and the properties of the individual system elements to calculate the quality of the system, i.e. the degree of agreement between the task and the result. Since the task is generally time-dependent and system elements with memory are also involved, this quality is also time-dependent. So you're looking at the *system dynamics*. The characteristic feature of this approach is the paradigm of the "cause-effect-principle"which is usual in technology. Many of the methods used to calculate this come from the field of the so-called *control engineering*.

Another way of looking at things is to completely disregard the temporal aspect and to be interested in whether the individual system elements function at all. The consideration therefore refers to whether a system element is working incorrectly in the structure of the system structure. One tries to estimate the probability for such an error and in this context the error probability of the entire system. So in this case, it's about the *system reliability* ("How much can I count on everything being all right?"). The methods used in this area are those of the so-called *probabilistics*.

The following applies to both approaches: Regardless of whether one is clear about the implicitly or explicitly chosen system boundaries when dealing with technical processes, this is precisely a problem that should not be underestimated. Although all influences of the system environment that cannot be analysed more precisely can be attributed to the system environment, only those environmental factors whose field of influence and mechanism of action are known allow a prediction of possible effects, consequences or damage.

3 Most operations of the secondary and tertiary driving task are to be understood under this aspect of the parallel connection of human and machine: even if one "only" switches on the light or the windscreen wiper, one trusts that the process initiated by the switching operation runs correctly and independently and virtually only controls the success.

Things that have been "forgotten" are also not taken into account. However, for economic reasons as well as for reasons of feasibility and the current state of knowledge, certain influences are not taken into account in any system analysis. This is certainly a risk that cannot be denied and cannot be assessed.

2.3 System Dynamics

2.3.1 Transition Functions and Frequency Response

The purpose of technical design modelling is a mathematical description of human characteristics, as this makes a calculatory adaptation of the technical elements also described by mathematical methods seem feasible. Control theory offers a formalism for the treatment of continuous informational processes, which can only be dealt with in a suggestive way within the framework of this chapter.

The basis for this description is the simple control loop diagram in ◘ Fig. 2.2. In general, definition 2.1 is also used to characterize the so-called control characteristics of the control loop with reference variable changing over time and disturbance variables fluctuating over time. ◘ Figure 2.2 shows the task (command signal) with w(t), the result (tracking signal) with x(t), and the system deviation with $x_W(t)$ and the actuating variable is indicated by y(t).[4] The generally quite complex characteristics of the controller (driver) are described with $\mathbf{V_R}$ (amplification factor controller) and that of the controlled system (vehicle) with **S** (system parameters of the controlled system), whereby these variables generally contain time-dependent properties of these system elements, which will not be discussed in more detail here. The following descriptions can be carried out with these greatly simplified designations:

1. the actuator actuation y(t) of the driver, caused by the control deviation x_W:

$$y(t) = \mathbf{V_R} \cdot x_w(t)$$

2. the response x(t) (lateral deviation or speed) of the vehicle due to the actuator actuation y(t):

$$S \cdot x(t) = y(t)$$

The properties of the control loop are described by the following formulas:

3. Control deviation:

$$x_w(t) = w(t) - x(t)$$

4. Quality:

$$Q = \frac{x(t)}{w(t)}$$

Elementary mathematics can be used to find by inserting lines (1) and (2) in (3) and (4):

$$Q = \frac{V_R}{S + V_R} \tag{2.10}$$

Equation 2.10 describes a general property of each control loop: for any characteristic of the controlled system, a desired correspondence between the reference variable and the tracking variable can be obtained ($Q \approx 1$) if you make the gain factor of the controller sufficiently large. In practice, however, there are limits to this, which are based on the one hand on the properties of the controller (et al the human controller, for example, has only a limited resolution in the perception of control deviations, which alone prevents an arbitrarily high amplification) and on the other hand on the demand for stability of the control loop, since with increasing amplification of the controller, dynamic properties of the system elements (see below) cause an increase in the control deviation.

The task of the control loop, however, is to follow the command signal that changes with time and/or to compensate for disturbances that change with time. The dynamic proper-

4 Note: the "t" in brackets indicates that the size changes with time.

ties of the individual system elements and the entire control loop thus play a dominant role. In order to describe or test these dynamic properties, two different complementary methods are used, whereby two extremes of the behaviour of the system elements or of the entire control loop are described: by the so-called *transition function* the reaction to a sudden change (e.g. intercepting the vehicle in the event of a sudden gust of wind) and by the *frequency response* the behaviour in the so-called steady state (e.g. consequences of a winding country road). In both methods, the basic idea of quality described above is used for the mathematical description: whether related to the entire control loop or to a single system element, the ratio of output signal to input signal is always formed. In the case of the transition function these quantities are displayed as a function of time and in the case of the frequency response as a function of frequency.

2.3.2 Transition Function

The test function for the transition function is a step change at the input of the element. The temporal response to this is then examined on the output side. The quantities found there are always related to the quantity of the input jump (see above), in order to obtain standardized data independent of its absolute quantity, which only reflect the temporal (= dynamic) behavior. This can be done theoretically with mathematical methods or practically by doing a corresponding experiment. This is illustrated by the example of a pendulum (= jump-shaped change at the entrance; see Fig. 2.8). The pendulum oscillates on the step-like excitation with amplitude decreasing over time (because of the damping). This temporal function related to the size of the input jump represents the transition function. Since the temporal behavior of the pendulum can be described by differential equations, the transition function can also be calculated as a special solution of this differential equation.

A special aspect of the dynamic behaviour of a vehicle is usually also tested on the basis of this consideration: after scribing and subsequently holding the steering wheel, the yaw angle (rotation around the vertical axis), which starts with a certain delay (usually around 200 ms, especially "handy" or "sporty" vehicles press this value on 100 ms), can now be observed, which only increases slowly, then after a certain time with constant speed. Accordingly, the yaw angle velocity must first increase strongly and then move towards a final value, which means constant cornering and is described here by the value "1" according to the standardization mentioned above (see Fig. 2.9). It is the art of the suspension technician to prevent overshooting by damping, as shown in Fig. 2.8, without making the steering too slow.

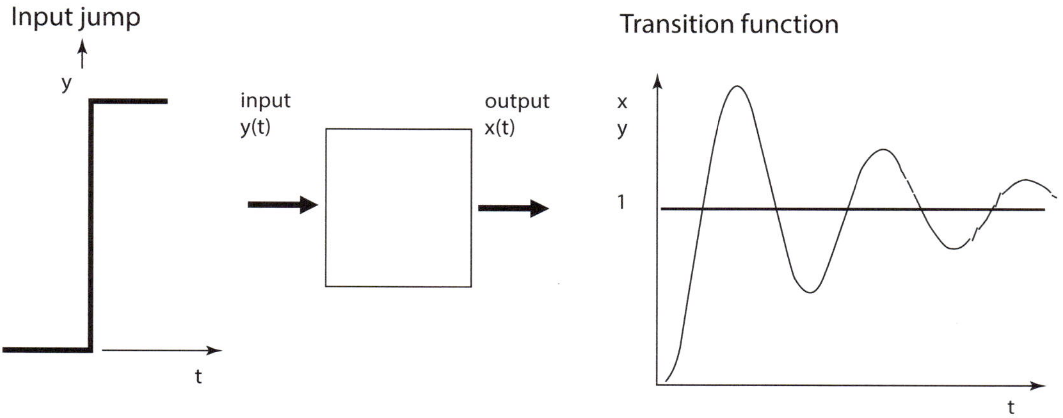

Fig. 2.8 Reaction of a pendulum to a abruptly excitation

Also the behavior of humans on abrupt changes can be tested in this way. Reference is made here to the so-called tracking experiments, which were carried out in many places above all in the 1960s and 1970s of the last century (see Poulton 1974; Sheridan and Ferrel 1974; Wickens 1984). A general finding from these experiments is the typical response reaction of the human regulator, which is shown in ◻ Fig. 2.10: After a reaction time, which is essentially given by the speed of signal transmission in the nerve tracts (typically around 180 ms), there is a "smooth" increase, which can be attributed to the inertia of the hand-arm system. It has "*integrating*" property and is therefore referred to as "I-behavior"for control purposes. In corresponding experiments, in particular when the entry jump occurs absolutely unexpectedly (e.g. reaction to an unexpected obstacle crossing the intended lane), a typical overshoot occurs: the human being only reacts to the change not to the size of the jump. This "*differentiating*" behavior is called D-behavior. In the course of approx. 500 ms the value required by the input variable is then found. Since the human controller now reacts "*proportionally*" to this variable, this is called P-behaviour. All in all, humans show the behaviour of a so-called PID controller for changes, although the respective P, I and D components are subject to strong fluctuations. It must be pointed out again at this point that the response behaviour shown in ◻ Fig. 2.10 represents a "typical" course, i.e. in practice there are considerable individual and situation-dependent deviations from this course.

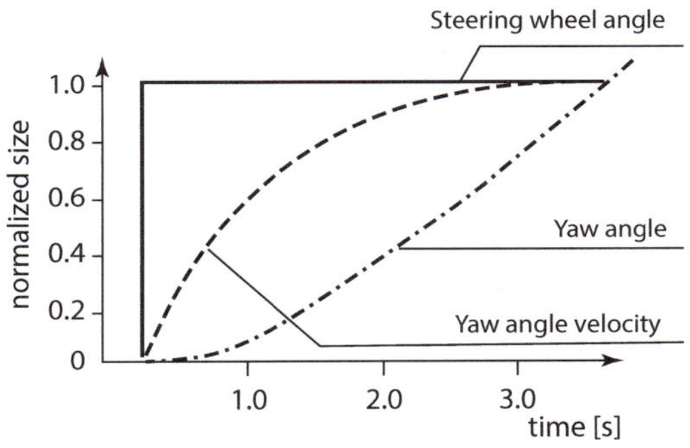

◻ **Fig. 2.9** Response of a motor vehicle to a sudden change in steering wheel angle

2.3.3 Frequency Response

In contrast to the transition function, the frequency response F(p) describes the steady state of a continuously changing input variable that is not characterized by jumps. It is defined by the frequency-dependent output signal y(p) in relation to the frequency-dependent input signal x(p):

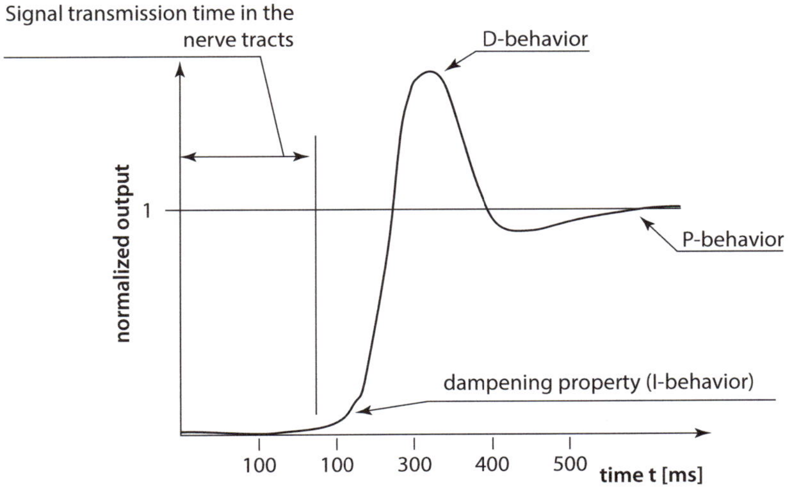

◻ **Fig. 2.10** Typical human response to a jerky input signal

2

$$F(p) = \frac{x(p)}{y(p)} \quad (2.11)$$

The frequency is introduced as the so-called "imaginary frequency", whereby the so-called circular frequency ω (see below) for this purpose with the imaginary number ($i = \sqrt{-1}$) is multiplied:

$$p = i \cdot \omega \quad (2.12)$$

Using the example of the pendulum again, the frequency response would characterize the behavior that occurs when the pendulum is moved continuously, for example in a wobbly ship. In order to describe this behaviour, a sinusoidal excitation is used as a test function at the input, the frequency of which is slowly changed in the range of interest (a technical term in communications engineering for this is "wobbling through"). At the output, the same frequency appears in the steady state but with a different amplitude and with a time offset compared to the input variable.

Here shows the advantage of using the imaginary frequency. If you span a plane from the branches real part *Re* and imaginary part *Im* and lets a point circulate a circular path with the angular velocity $\omega = \phi/t$, then one receives the sine motion visible in reality as a projection of this circular motion on the real axis (see Fig. 2.11a). The frequency ν represents the frequency of a recurring event per time unit. Since the circumference of the unit circle is in the imaginary plane 2π, there is a connection between frequency ν and circular frequency ω too:

$$\omega = 2 \cdot \pi \cdot \nu \quad (2.13)$$

Applied to the example of the pendulum excited with a sinusoidal motion, this means: since the instantaneous value of the sinusoidal input signal can be described by an angular value, the instantaneous value at the output signal can also be described by an angular value. The temporal offset is thus represented by a corresponding angle difference between these two values, the so-called phase angle α.

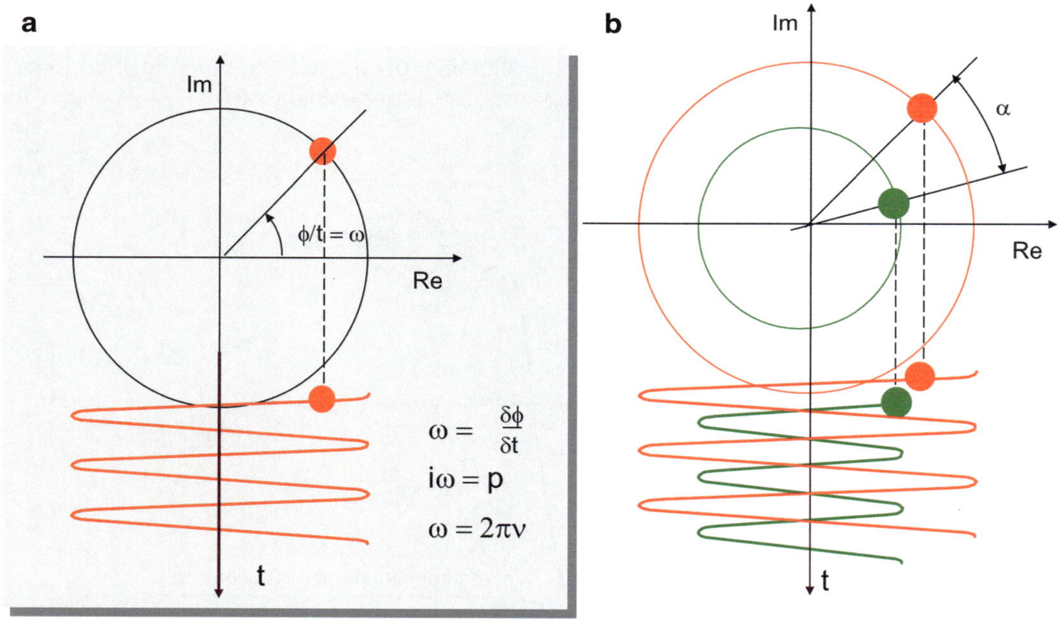

Fig. 2.11 **a** Sinusmotion represented as circular motion in the imaginary plane, **b** Exciting input signal (*red*) of a pendulum (second order dynamic system) and reacting output signal (*green*)

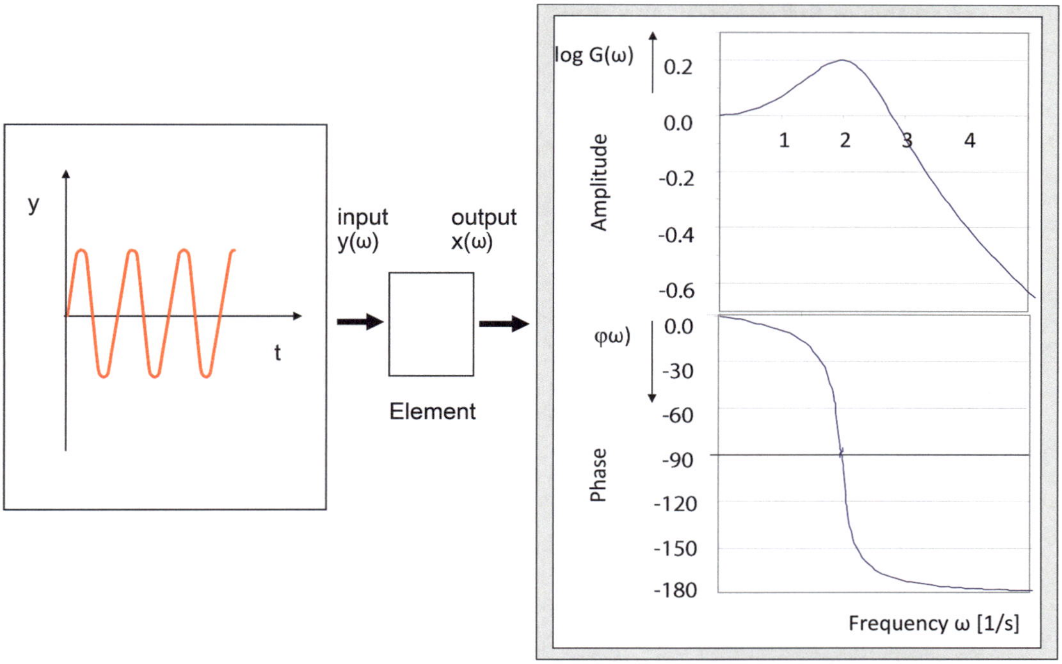

Fig. 2.12 Frequency response of a pendulum, represented by the bodediagram (*right*)

The amplitude of the output frequency also differs from that of the input frequency, which by definition is set to the value "1". The motion of the output signal visible in reality is also characterized by the projection of this motion onto the real axis (Fig. 2.11b).

To illustrate the dependency of input signal and output signal when "sweeping" the frequency, two diagrams are shown one above the other, both containing the frequency on the abscissa (Fig. 2.12). Usually the frequency is represented in the logarithmic scale in the form of the circular frequency ω. The upper diagram shows the amplitude ratio from output to input on the ordinate. It represents the so-called *amplitude characteristic* G(ω). The amplitude response is also usually represented in logarithmic scale. If the respective value is multiplied by the factor 20, the (dimensionless) designation decibel (dB; G(ω) [dB] $= 20 \cdot \log$ (output(ω)/input(ω)) is used. The lower diagram shows the corresponding course of the phase angle α on its ordinate. This combination of the two diagrams is called *Bode diagram*.[5] A Bode diagram can be found in the Fig. 2.12 for the example of the pendulum. Let it illustrate the idea of the frequency response: If one excites the pendulum with a sinusoidal oscillation of very low frequency, the pendulum will follow the excitation with this frequency and also with the same amplitude. In the upper part (amplitude response) of the Bode diagram, this is characterized by the value "0 dB" (log 1 = 0). There is no phase delay. In the lower diagram of the phase response the value $\alpha = 0$ is therefore read. If the frequency is now increased, the value of the amplitude response increases until it has a maximum value in the range of the so-called natural resonance. Here is the phase delay $\alpha = -90°$. Above the natural resonance, the amplitude ratio drops continu-

5 Note: there are other diagram forms which can be used to illustrate the dynamic behavior in the steady state, but which are less vivid for "getting to know each other first".

2

ously. The reason for this is that the pendulum can follow the increasing input frequency less and less due to its own inertia. At the same time the phase delay approaches the value $\alpha \approx -180°$.

The Bode diagram of the Fig. 2.12 can be calculated from the mathematical description of the pendulum using differential equations. This is called the "parametric" procedure, since the corresponding physical parameters (e.g. mass, damping and similar.) are determined directly from the properties of the transmission element.

Abbildung 2.13 shows the corresponding Bode diagram for the reaction of a motor vehicle to the steering movement. It can also be read like the example of the pendulum shown above: the vehicle reacts exactly to very slow steering movements ($G(\omega_{\text{very small}}) \approx 0\,\text{dB}$) and without phase delay ($\varphi_{\text{very small}} \approx 0°$). With increasing frequency of the steering movement, the vehicle cannot follow due to its inertia, it reacts less and less to the movements, whereby the phase delay, as with the pendulum, is equal to the value of $-180°$ approaching. With regard to lateral dynamics, vehicles today are designed in such a way that the (latent) natural frequency is largely damped away; the vehicle has an inherent stability. However, the statement made here is only valid as long as the forces actually occurring can be transmitted to the road due to the given friction conditions. Exceeding the maximum possible friction force, which is known to depend considerably on the respective road and weather conditions, represents a non-linearity in terms of control technology and cannot be dealt with easily using the methods described here (Fig. 2.13).

In many cases a derivation from the individual components of the transmission element is too complex or even impossible, as is the case with humans, where it would be very difficult to attribute the differentiating and integrating properties of information processing to physical elements. In such cases, the frequency response can also be determined experimentally, either by actually "sweeping" the input and measuring the corresponding output signal, or by using a mixed signal with the widest possible frequency range as input

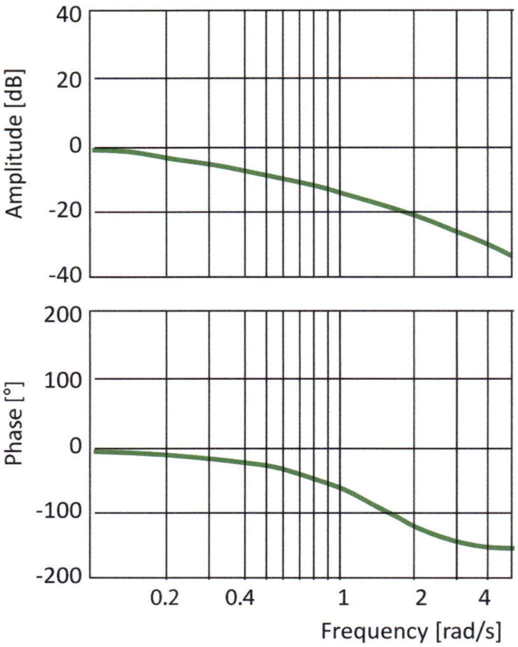

Fig. 2.13 Bode diagramm for the lateral dynamics (reaction to steering motion) of a motor vehicle

variable. By a so-called Fourier decomposition[6] the contained frequencies can be extracted, which describe the seemingly random signal. The same can be done with the experimentally simultaneously measured output signal. Using special correlation methods, it is now possible to derive the Bode diagram from these two signals in graphical form, which describes the dynamic properties in the steady state. This procedure is *"non-parametric method"* called. It delivers only the Bode diagram. It is now the task of the interpreting scientist to use the knowledge of properties of physically known transmission elements to combine them in such a way that they show the same behaviour as that observed in the form of the Bode diagram. Thus, the transmission behavior of the unknown element (i.e. the human) is indirectly modelled. In this way, for example, the control model according to

6 Fourier decomposition makes it possible to represent a frequency mixture by an additive composition of individual frequencies, each with a different amplitude.

Tustin (1944), which is still widely used today, was developed (see ► Sect. 3.3).

Working with frequency responses has a considerable advantage over the transition function. It is also possible to treat complicated links of single frequency responses mathematically. If one would in fact in Eq. 2.10 for the behavior of the controller (human) $\mathbf{V_R}$ and the controlled system **S** the corresponding descriptive differential equations, one would very soon reach the limits of algebraic resolution possibilities. In contrast, quite simple rules apply to the assembly of single-frequency gears into a complex structure. In the context of interest here, only the connection in series and the so-called negative feedback shall be presented. ◘ Figure 2.14 shows the corresponding structure pictures.

With simple mathematics one can calculate the respective resulting frequency responses, if one considers that by definition the respective total frequency response F_{total} is determined by the ratio between the frequency-dependent output signal y(p) and the input signal x(p) (see Eq. 2.11; in order to treat the effect of the phase delay mathematically simply and correctly at the same time, the frequency must be described as an imaginary frequency; see Eq. 2.12).

The result is then for the series connection:

$$F_{total} = F_1(p) \cdot F_2(p) \tag{2.14}$$

and for the negative feedback:

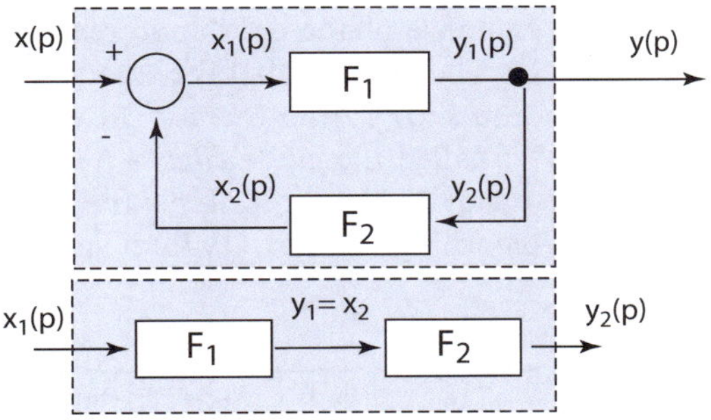

◘ **Fig. 2.14** Serial and parallel interconnection of two frequency responses

$$F_{total} = \frac{F_1(p)}{1 + F_1(p) \cdot F_2(p)} \tag{2.15}$$

As can be seen in ◘ Fig. 2.15, the closed loop consists of connecting the frequency response of the F_R and the controlled system (machine) F_M (so-called "open control loop"), which in Eq. 2.15 corresponds to the frequency response F_1 and frequency independent feedback, which is achieved by $F_2 = 1$ can be described. This results in a frequency response of $F_{closed\ loop}$ of the closed control loop:

$$F_{closed\ loop} = \frac{F_R(p) \cdot F_M(p)}{1 + F_R(p) \cdot F_M(p)} \tag{2.16}$$

With the help of this equation and some parameters found from experimental observation for the modelling of humans (see ► Sect. 3.3) it is possible to determine the closed frequency response of the laterale dynamics consisting of driver and vehicle and to create the corresponding Bode diagram (see Bolte 1991). ◘ Figure 2.16 shows the result, whereby the vehicle behaviour shown in ◘ Fig. 2.13 is the basis. It is noticeable that this frequency response is a clear resonance peak at $\omega \approx 2{,}5\,\text{rad/s}$, which corresponds to a frequency of $\nu \approx 0{,}4\,\text{Hz}$.

The so much feared "skidding of the vehicle" is therefore not, as is generally assumed, a characteristic of the vehicle which, according to ◘ Fig. 2.13, has an inherent stability, but one of the closed control loop, whereby the skidding process itself can certainly be initiated by a non-linear effect (e.g. short-term exceeding of the adhesion limit), but also by unfavourable traffic constellations which give rise to a corresponding hectic reaction. Dreaded building-up however, essentially will

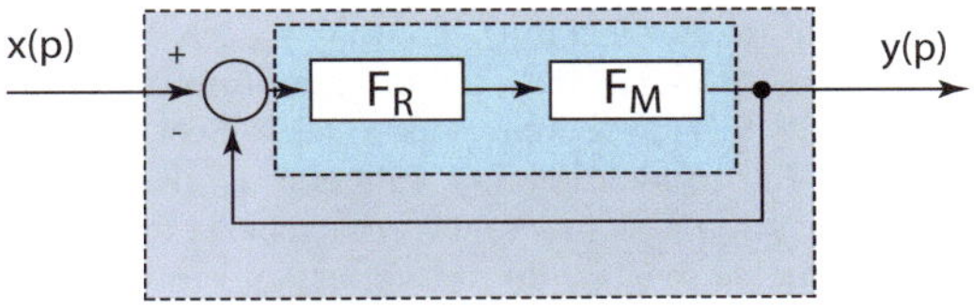

◘ **Fig. 2.15** Circuit diagram for the closed control loop (see also ◘ Fig. 2.2)

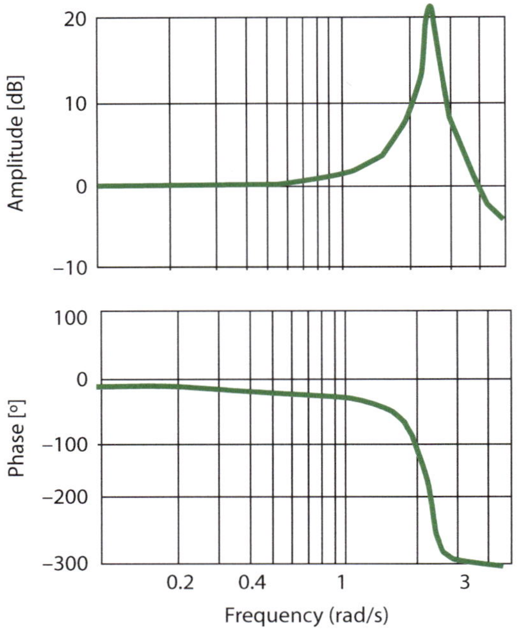

Fig. 2.16 Bode diagramm of the frequency response of the lateral dynamics of the closed driver-vehicle control loop

come up by deceleration of human behavior in connection with vehicle inertia. As Bolte (1991) shows, this effect can be reduced by adequate haptic feedback in the control organ (e.g. corresponding Steer-by-Wire (SbW) design of the steering wheel or so-called "active control element"; see also Huang 2004), since the haptic sensory channel can be controlled via the so-called eigenreflex arc[7] practically by a factor of 4 faster than the optical and kinesthetic feedback (Gillet 1999). The effect of electronic stabilization aids such as ESP[8] must be put into perspective under the aspect of this consideration: it only reduces the effect of skid initiation. However, the clear impact of the introduction of ESP on accident statistics shows just how effective this measure is (Unselt et al. 2004). If it were now possible to "tame" the skid process itself, another significant improvement would undoubtedly be expected.

As the example of the skid process shows, the aim of the control engineering consideration of the human-machine interaction is to avoid resonance peaks in the transmission or at least to shift them into such an area that it is very unlikely to get into it. For example, one can assume for instance that people above frequencies around 2 Hz hardly show any transmission behaviour. In order to assess the proximity to this danger, one uses the consideration of the so-called *cross-over-frequency*. The basis for this is provided by Eq. 2.16 of the frequency response of the closed control loop. For the understanding of the following consideration it is important in this context that - as already mentioned - the frequency ω must be introduced as an imaginary number $p = i \cdot \omega$ for a perfect description for mathematical reasons. This means that in Eq. 2.16, the term $F_R \cdot F_M$ at certain frequencies, at least theoretically, the value "−1" can take. At this point, the frequency response $F_{\text{closed loop}}$ of the closed control loop is indefinitely large. It characterizes the described resonance point. However, this value can only be set if at the same time the absolute value of $|F_R \cdot F_M| = 1$ and the phase angle $\varphi = -180°$. Figure 2.17 shows the possible description of this critical point on the hypothetical frequency response of an open loop (see Eq. 2.14). As *cross-over-frequency* ω_C is the frequency at which the amplitude response of the open loop reaches the value "1". The distance that the associated phase response shows from the critical value $\varphi = -180°$ is called the *phase edge* φ_{edge}. The greater the distance, the more stable the control loop. According to Merz (1973) it should

7 Eigenreflex arc: The muscle length measured by the muscle spindles is directly fed back in the spinal cord with the information coming from the cerebrum (α-Innervation) about the desired muscle length (α-Motoneuron), so that external disturbances without activity of the cerebrum are regulated by this lower control loop (see ► Sect. 3.2.3).

8 ESP. **E**lektronic **S**tability **Program** If the lateral acceleration and yaw rate of the vehicle, measured by suitable sensors, do not correspond to the geometrically fixed lateral acceleration and yaw rate of the vehicle within certain limits due to the measured steering wheel position and speed (see Fig. 2.13), the course of the vehicle is corrected by targeted asymmetric braking. The existing sensors for the wheel speed and the actuators of the ABS system (**A**nti**B**locking**S**ystem) are used.

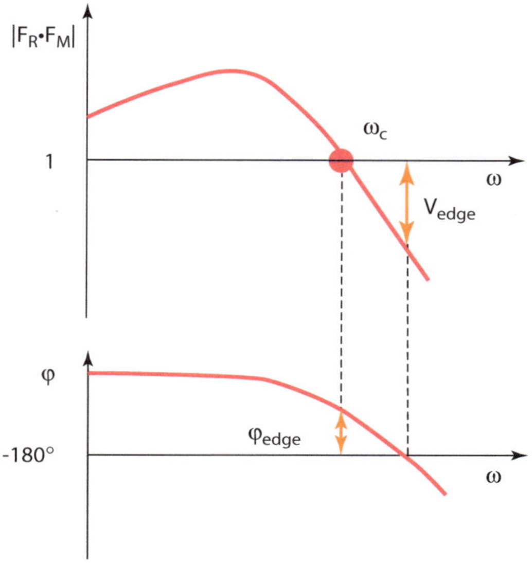

Fig. 2.17 Bodediagram of a hypothetical frequency response and definition of the cross-over frequency ω_C

be >30° and according to Pressler (1967) > 40° – 60°. The gain reserve V_{edge} is the distance of the amplitude value at the critical point φ = – 180° from the critical value "1". It should be > 1,3 according to Pressler and > 2 according to Merz in order to maintain stable conditions.

2.4 Quantities Controlled by the Driver

In order to enable a mathematical modelling as outlined in the ► Sect. 2.3, it is necessary to determine exactly which variables are available to the driver for a regulation. Due to the different technical design and the resulting control elements, this consideration is carried out separately for laterale and longitudinal dynamics, although some of the respective information channels are identical for both dynamics. In the literature, greater attention is paid to laterale dynamics because the reference variable is more clearly defined within narrower limits (see also Fig. 2.4), making control engineering modelling more appropriate.

2.4.1 Laterale Dynamics

Already from Fiala (1966) and Ohno (1966), respectively, in agreement with Crossmann and Szostak (1969), three feedback levels about the vehicle movement for the optical information channel were given:

1. The preliminary information by foresight on the road *("headlamp orientation")*,
2. the direction of the vehicle *("direction orientation")*,
3. the lateral distance *("fog orientation")*.

Only through the information gained from the foresight is the driver able to form an internal target course in the sense of the guidance task for the given situation, which he tries to follow (guidance level of the driving task). Many studies on gaze behaviour (see ► Sect. 3.3.2) also show that this foresight is generally limited to an area about 1 to 2 seconds in front of the vehicle. To generate the necessary steering reaction (stabilization level of the driving task), the driver uses the two other optical feedback levels (direction and fog orientation) and additionally the kinesthetic perception of the lateral acceleration. The derivation of the descriptive differential equations of these three feedback levels is based on Fig. 2.18. It is assumed that the driver deviates slightly from the desired straight course by slightly turning the steering wheel ((in the figure this is exaggerated for reasons of clarity), resulting in a lateral offset from this target course.

The steering wheel angle λ, at a steering ratio u results in a turning angle u•λ on the front wheels. Thus a curve radius r_K is determined with the wheel base l. This applies to small angles:

$$r_K = \frac{1}{u \cdot \lambda} \tag{2.17}$$

If the vehicle passes through an arc of the length $r_K \cdot \varphi_F$ on this radius, the longitudinal axis of the vehicle has changed by the angle φ_F compared to the original direction. At the same time, it shows the lateral misalignment a

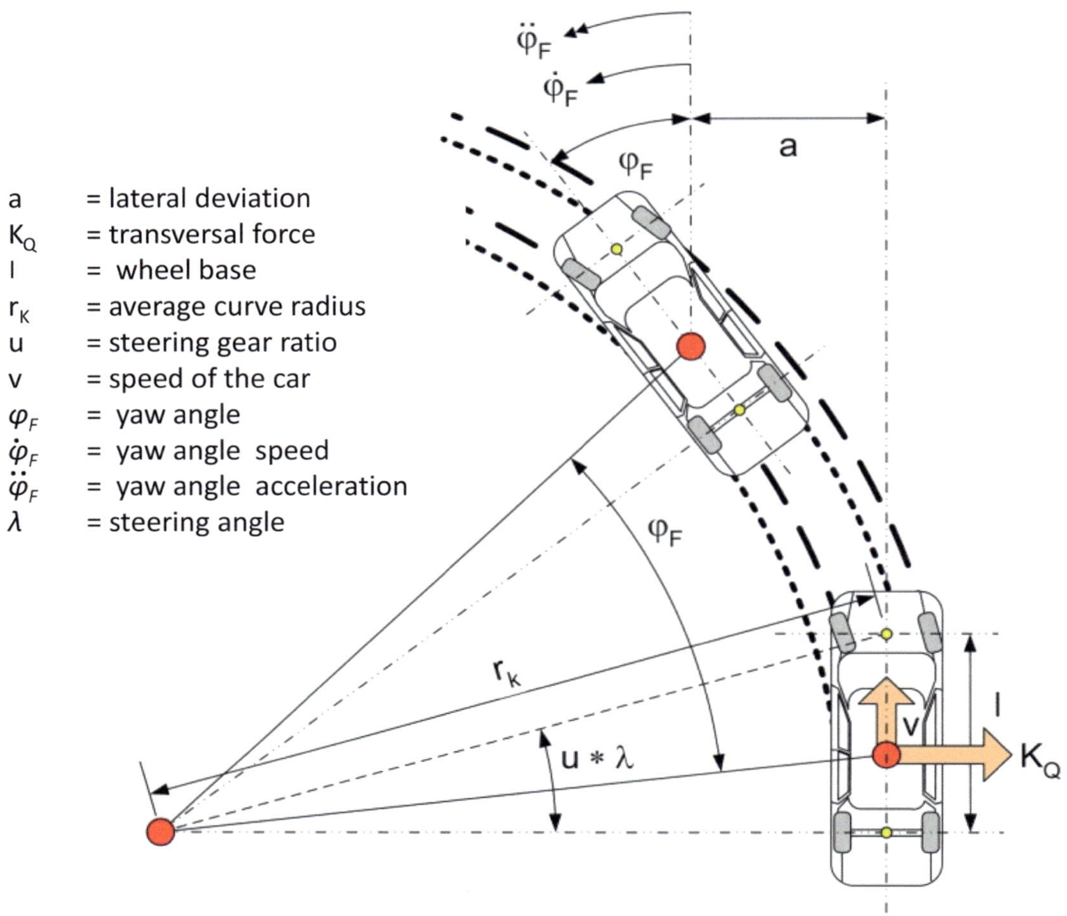

■ **Fig. 2.18** Geometry for cornering a motor vehicle (without consideration of dynamic effects; often also referred to as "bicycle model" or "single-track model")

of the original straight-ahead course. Since this is a consideration that primarily depends on the feedback for the driver in uncritical driving conditions, the description of effects of higher order, such as float angles (in this case, the orientation of the longitudinal axis of the vehicle would show an angle to the tangent of the current curve radius), lateral slip, etc., can be dispensed with. From ■ Fig. 2.18, it is possible to calculate for small angle φ_F the lateral offset of the vehicle from the nominal course:

$$\frac{d^2a}{dt^2} = \frac{v^2 \cdot u}{l} \cdot \lambda \tag{2.18}$$

With the steering wheel angle λ the acceleration $d^2a/dt^2 = \ddot{a}$ of the side offset a is directly proportionally influenced, whereby this influence is still modified quadratically by the speed; i.e., for the driver, the same steering wheel position has a different effect with regard to the side offset depending on the speed. From an ergonomic point of view, a constant driving speed is an acceleration control (see Bubb 1993), which is particularly difficult to handle during compensation tasks (i.e. only the deviation is seen, not the true position of the vehicle on the road, as would be possible from a hypothetical bird's eye view).

For the temporal change of the longitudinal orientation φ_F of the vehicle can be derived from Fig. 2.18:

$$\frac{d\varphi_F}{dt} = \frac{v \cdot u}{l} \cdot \lambda \qquad (2.19)$$

By adjusting the λ of the steering wheel, the speed $d\varphi_F/dt \doteq a$ of the longitudinal orientation φ_F is thus increased. This influence is also dependent on the driving speed v, but only linear. At a constant driving speed, this is a so-called speed control, which is particularly easy to handle for a compensation task (see above).

Since it can be assumed that the driver, based on his experience, has an expectation (see ▶ Chap. 3) about the lateral acceleration as a function of the vehicle speed and the curve radius of the road perceived via the optical channel, the centrifugal force Z acting on the kinesthetic sensation also represents feedback on the state of movement of the vehicle included in the control process. It's on:

$$Z = \frac{m \cdot v^2 \cdot u}{1} \cdot \lambda \qquad (2.20)$$

where m is the mass of the driver. In the case of the required straight course, the driver's task would be to make the perceived lateral acceleration zero.

When controlling according to the kinesthetic perception, the steering wheel angle thus directly determines the magnitude of the lateral acceleration (ergonomically referred to as "position control"). If the driver has a certain idea of the lateral acceleration to be expected, this is in itself a follow-up task (also called "pursuit tracking"; target and actual values are recorded separately) for which position control is particularly easy to handle. This influence is also influenced by the square of the driving speed v as well as the acceleration of the lateral deviation a and by the driver's mass, which is, however, a constant for the respective individual (of course, Eqs. 2.18 and 2.20 are physically equivalent, since the lateral acceleration d^2a/dt^2 determines the centrifugal force according to $Z = m \cdot d^2a/dt^2$; but the driver *sees* one of them via the visual sensory canal and *feels* the other via the kinesthetic sensory channel).

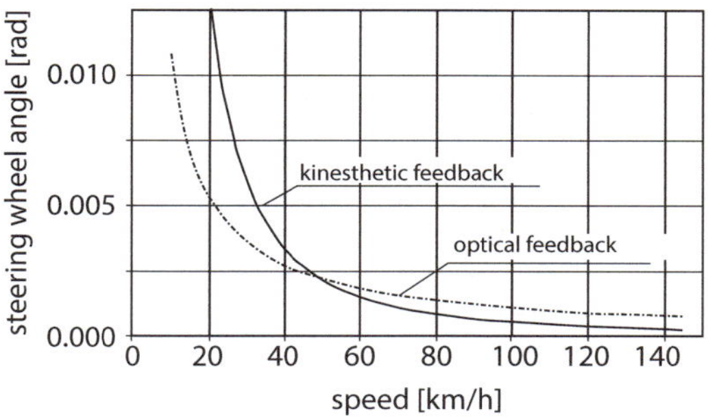

Fig. 2.19 Steering misalignments that lead to an event that can just be felt

Especially the consideration of the problem of straight ahead driving is of special interest. The question is, what does the human being regulate in this case, when virtually all command signals are "zero"? Figure 2.19 shows the steering wheel angle that (with an average design of a mid-range passenger car) just leads to a perceptible event. For the diagram, a dynamic visual resolution for the viewing angle of $\sigma_{min} \approx 2'/s$ (Lindsay and Norman 1972) and a minimally perceptible acceleration of $b_{min} \approx 0{,}008\,m/s^2$ (Steward 1971) was asumed. It can be seen that a deviation from the straight course up to a speed of approx. 50 km/h is first perceived optically, above this kinesthetic perception dominates. This explains the peculiarity found in experiments that drivers also drive a slight zigzag course on completely straight roads (Jürgensohn 2000): The driver sees the straight road, knows that no lateral acceleration is to be expected here and adjusts to lateral acceleration "zero". As the longitudinal direction of the vehicle can deviate slightly from the longitudinal direction of the road, he only perceives the deviation from the course optically after a certain time, corrects it, but then adjusts it again to lateral acceleration "zero", etc. (see also Gothelp 1984). This is also the reason for the excessive steering behaviour in driving simulators without kinesthetic feedback and the observed phenomenon of a tendency to high speeds.

These statements demonstrate how important it is for the driver to have experience that allows a properly coordinated prediction of the different levels of feedback from the *one* moving vehicle. It is also understandable that the inexperienced driver initially orients himself on the lateral offset to be captured with less mental model effort, but the control of which is only possible via the difficult acceleration control. With increasing practice, he is now in a position to include longitudinal orientation and lateral acceleration in the control process, which also makes him safer due to the simpler control modes associated with this.

As explained above, the effect of the steering wheel angle depends significantly on the driving speed. One can assume that the driver has this dependency "in the feeling" due to his experience for normal driving situations. However, gross deviations from the learned context, such as occur during extreme driving manoeuvres (e.g. turning with high lateral acceleration, braking in bends), lead to difficulties in vehicle control. The system parameters are then no longer inversely proportional or inversely quadratically proportional to the speed, but are even more dependent on it. In order to prevent the danger in such borderline situations, today a "neutral steering design" is preferred, i.e. even with extreme lateral accelerations, the speed dependence of the system parameters changes only gradually, not in principle; the driver can use his experience almost unchanged in this situation; he only needs a slightly larger steering wheel angle. This is different for a chassis design with a tendency to understeer or oversteer, characterized by a slip angle greater or smaller than zero (see ► Sect. 6.4.1). In this case, a disproportionately higher or lower steering angle is required at higher speeds to travel through the same curve radius.

As early as 1981 Schulze carried out tests with a speed-dependent steering ratio, whereby the quadratic dependence with regard to lateral deviation and lateral acceleration (Eqs. 2.18 and 2.20) can be reduced to a simple speed dependence and the speed for longitudinal orientation even drops out completely. He was able to show that this makes steering easier. Studies by Honda (2000) and Fleck (2003) as well as the studies on steering with the active control element by Eckstein (2001) and Penka (2000) also show this effect in agreement, although it was found in all of them that below a speed of approx. 40 km/h the steering ratio should be independent of speed. This obviously has to do with the fact that in this area the driver mentally switches from a compensation task to a follow-up task, whereby he imagines the movement from an "objective" bird's eye view, so to speak, as he also needs it to carry out manoeuvres during parking. For safety reasons, in today's production vehicles with such a speed-dependent steering ratio, the following is used.[9] In the upper speed range (above 160–200 km/h) a speed-independent steering ratio is also preferred.

The advantage of a speed-dependent steering ratio is still obvious if the above-mentioned anticipation of the road, the "headlamp orientation", is taken into account. Already Kondo (1953) assumed that the driver steers in such a way that with regard to an imaginary aim point at a certain preview distance the sight point always coincides with the predetermined course. ▫ Figure 2.20 illustrates how easy this anticipatory steering is for the driver when the curve radius is sufficiently large: he has to turn the steering wheel (current position in the picture marked by a bright point) only in the direction of the intended target point in the surroundings. However, it also illustrates the effect of different steering interpretations: A so-called direct steering refers to a closer target point (in the figure by a and k_0), an indirect steering to a more distant target (represented in the figure by a' and k_0' for the same curve radius, a larger steering wheel rotation is necessary here than with direct steering). Since larger distances are measured at higher speeds in the same time, the more indirect steering makes sense there, while at lower speeds the vehicle with the more direct steering appears more manageable. By speed-

9 At present, more and more vehicles are coming onto the market for which such a steering system can be purchased, at least as an optional extra. Not only does it provide a comfort advantage (less steering angle when parking), but it also improves safety by better adapting the dynamic vehicle characteristics to the driver.

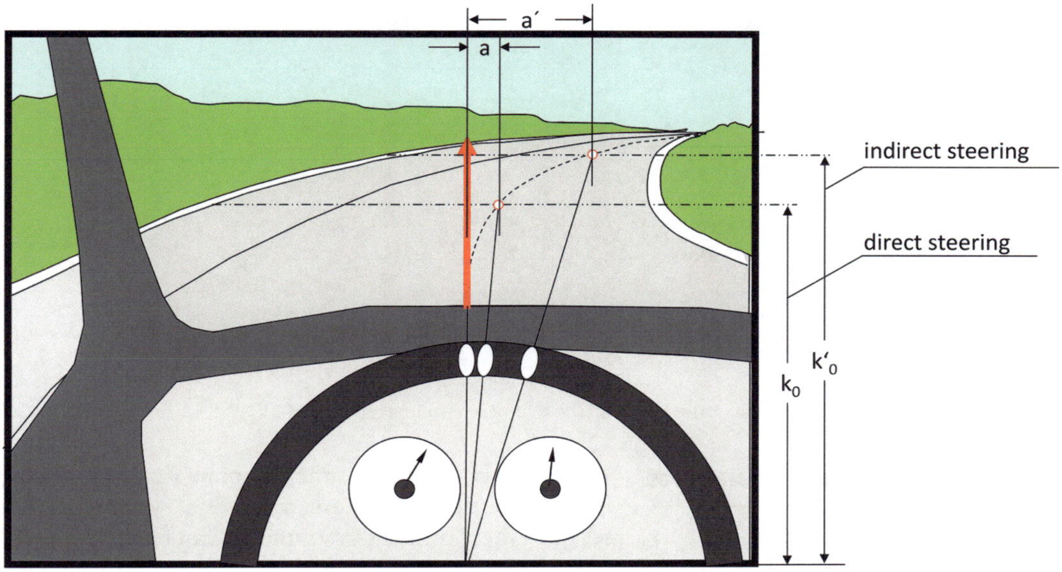

Fig. 2.20 Illustration of anticipatory control

dependent steering, both requirements can be met at the same time (see above).

Various studies have dealt with the necessary foresight. For them, values in the order of 1 s for lateral dynamics and 1.5 s for longitudinal dynamics were successfully used in the simulation of the driver-vehicle control loop (Yuhara et al. 1999; Guan et al. 2000). Donges (1978) came to the conclusion during tests in the driving simulator with motion feedback that the anticipation time or foresight time is independent of the driving speed and lies at approx. 1 s, whereby the driver - as eye-tracking studies show - is generally prepared to avert his gaze from the road up to 2 s (see also the model of Gengenbach, 1997). The distance to the forecast point is increased by the increasing forward speed, as the distance is proportional to the forward speed at a constant forecast time. This was also found objectively in eye examinations by Schweigert (2003) (for further details see ► Sect. 3.3).

The time that elapses until the vehicle reaches or exceeds the edge of the road without changing the steering angle is often used as a measure of the quality of the lateral dynamics control. It is referred to as "Time to line crossing, TLC". It can be calculated from the angle $\Delta\varphi$ between the vehicle orientation and the orientation of the road, the current position d of the vehicle in relation to the lane centre, the vehicle width $B_{vehicle}$, the lane width $B_{trajectory}$ and the speed v of the vehicle according to the following Eq. 2.21 (see Fig. 2.21).[10]

$$TLC = \frac{\frac{1}{2}\cdot\left(W_{lane} - W_{vehicle}\right) - d\cdot sign(\Delta\varphi)}{v\cdot\sin\Delta\varphi} \tag{2.21}$$

In simulator experiments it was found that the average TLC is about 3.7 s. This shows that it is normally not a problem to keep the vehicle on the road. The high frequency of accidents characterised as "leaving the road" is probably largely due to excessive speed (see below) on the one hand and to distraction, inattentiveness or the infamous "microsleep"on the other.

10 All data is always related to one's own lane; in reality, there is often more available if the vehicle is on the right-hand lane of a multi-lane motorway, for example, or on the right-hand side of a lane with oncoming traffic. In the latter case, however, it should be noted that if the traffic situation is appropriate, a collision with another vehicle may occur when leaving one's own lane.

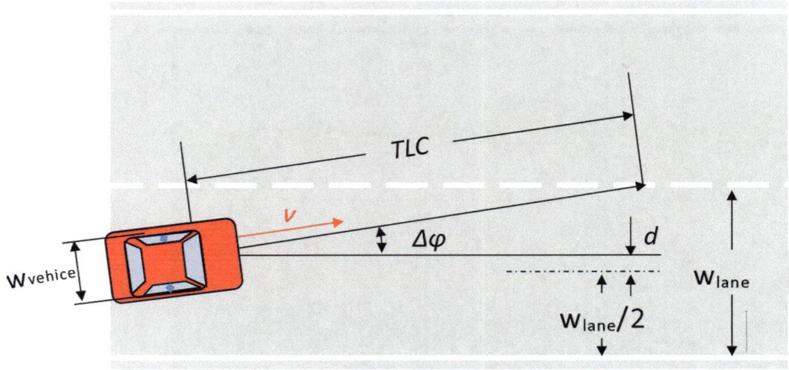

Fig. 2.21 Values required for the calculation of the size "Time to line crossing, TLC"

2.4.2 Longitudinal Dynamics

As already shown by Fig. 2.1, the task in controlling longitudinal dynamics lies in generating the reference variable, i.e. the target velocity, which, in contrast to laterale dynamics, where the roadside and objects on the road are relatively narrow objective boundaries, cannot be derived directly from the view. However, a well-known driving school rule already specifies this target speed: the speed may only be set so high that the vehicle can stop within sight. The stopping distance x_A is calculated from the driving physics:

$$x_A = v \cdot t_R + \frac{v^2}{2\mu g} \tag{2.22}$$

where v is the output speed, t_R is the reaction time of the driver (usually assumed to be 1 s, since it is assumed that the driver needs a minimum time to recognise that a braking manoeuvre is necessary) and μ is the friction coefficient describing the mean maximum deceleration for given tyres on a given road surface. g is the acceleration due to gravity constant. The driver's task would now be to correctly estimate not only the road conditions, but also the square speed dependence of the stopping distance according to Eq. 2.22. Experience has shown that this is practically impossible for him. The driving school therefore teaches simplifying rules of thumb for the stopping distance, but these are also practically not applied in real road traffic, because beside others a numerical estimation of distances from a moving system is even more difficult for humans than from a motionless point of view.[11] If one takes into account the above findings that the driver derives his intentions for action from a time horizon of 1–1.5 s, maximum 2 s, then it is necessary to calculate the estimated time t_v is needed so that the distance to the hypothetical stopping point can be overlooked:

$$t_v = \frac{v}{2\mu g} + t_R \tag{2.23}$$

If one assumes a reaction time t_R from 1 s and assumes a forecast from 2 s, should be on dry road ($\mu \approx 0{,}8$) only a maximum speed of 60 km/h can be driven, on wet road ($\mu \approx 0{,}4$) a speed of 30 km/h and on a winterly slippery road ($\mu \approx 0{,}1$) only from 7 km/h. Such a requirement is, of course, unrealistic and unnecessary, since longer distances corresponding to a longer foresight period are generally available as a visual range (Bubb 1975). However, from Eq. 2.23 it can also be inferred that at a speed of 200 km/h a foresight time of approx. 4 s (which corresponds to a visibility range of 222 m) is necessary. With the same safety reserve, however, you could only drive

11 The rule of thumb for braking distance is often: $x_B = (v[km/h]/10)^2$. This rule of thumb assumes a so-called comfort braking with a maximum of 0.3 g. In the event of emergency braking, the value calculated in this way is therefore halved.

approx. 100 km/h on a wet road and only 25 km/h on a slippery road. This consideration also shows that the real behaviour deviates far from these ideal demands.

However, experience has shown that speed is chosen more for comfort by using the expected lateral acceleration when driving through curves as a yardstick. Thus already Herrin and Neuhardt (1974) found that up to speeds of 130 km/h a maximum lateral acceleration of 0.3 g on average is accepted, at higher speeds this value becomes even substantially smaller.

In column traffic, it can be assumed that only the vehicle at the top of the column needs its own stopping distance, while the following vehicles must maintain such a safe distance that the reaction time of the driver is compensated. The safety distance x_S is simply calculated:

$$x_S = v \cdot t_R \tag{2.24}$$

Usually, this is based on safety times t_R between 1 and 2 s. The road traffic regulations punish a falling below the safety distance of 0,8 s.[12] Experimental investigations by Assmann (1985) show that safety distances actually depend only slightly on speed. It seems more likely that the driver considers an absolute distance to be safe and maintains it almost independently of the speed, with the result that on average safety distances are too large at low speeds and too small at high speeds.

Investigations by Popiv et al. (2009) on motorways with a vehicle equipped with a radar-based distance measuring system show the results found in ◘ Fig. 2.23. It was investigated at what distance the driver of the test vehicle changes lanes when approaching a slower vehicle. Depending on the speed, this distance is characterized by the safety time t_s according to Eq. 2.24. It can be seen from ◘ Fig. 2.22 that a further parameter for this process is the time until the vehicle in front hits the road in the event that the lane is not changed. This time is called time to collision, TTC.

◘ Figure 2.23 shows the values found in this investigation for the safety time t_s and the TTC. According to Fastenmeier et al. (2001) and van der Horst et al. (1994) values of ts <0.6 s and TTC <5 s must be classified as critical.

From all these considerations, Bubb (1975) derives the demand for a contact-analogue display of the braking distance in the Head-Up-Display (HUD). As the investigations by Assmann (1985) with a test vehicle that had been modified according to these ideas show, such a display increases the safety distances even for drivers who refuse such assistance.

2.5 The Primary Driving Task from a Control Engineering Point of View

In ► Chap. 1 the hierarchical structure of the primary driving task into the parts navigation, guidance and stabilization has already been explained. From a control engineering point of view, it must be investigated how and how exactly the respective reference variable is obtained and to what extent the driver is able to meet the resulting requirement.

The navigation task is essentially carried out knowledge based on the basis of rational considerations and considerations (see ► Sect. 6.3.1.3), except for the navigation task resulting from daily routines (e.g. way to work). This applies even if the journey is not planned in detail, but is carried out ad hoc on the basis of rough geographical knowledge and with the help of signposts. As will be shown in the following, the internal reference variable (command signal) must then be formed in the guidance task from estimates of perceived variables and experience as a function of the current situation with regard to location ("target course") and time ("target speed"), which is then realized within the framework of the

12 Since the estimation of distances in time form obviously causes difficulties, the so-called half speedometer distance is demanded in driving schools and also in other recommendations. When the speed in road traffic is given in km/h, this requirement corresponds to a safety distance of $t_s = 3{,}6/2 = 1{,}8$ s.

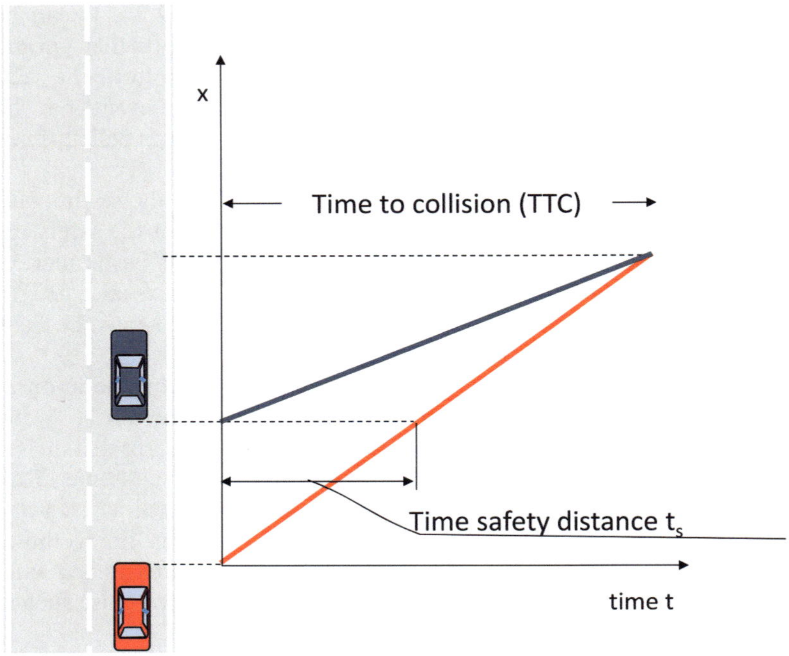

Fig. 2.22 Safety distance measured in safety time t_s and Time to Collision (TTC) if the track is not changed

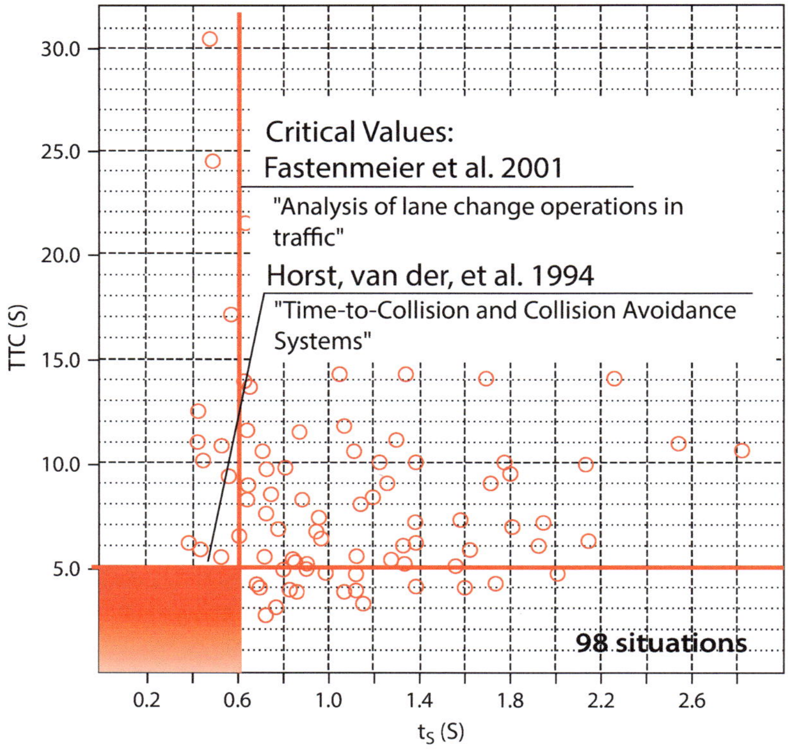

Fig. 2.23 Behaviour when approaching a slower vehicle (Popiv et al. 2009)

stabilization task by actuating the controls of the vehicle. The process of this regulation can largely be described by means of control modelling of humans (see the preceding sections and in particular ▶ Sects. 3.3 and 5.1.1).

In this section, the tasks that the driver has to perform at guidance level are modelled. Since these tasks are extremely varied, this can only be done to a limited extent here. Nevertheless, this approach already opens up important insights with regard to an ergonomic design of the driving task. ◘ Figure 2.24 shows a schematic flowchart of the guidance task.

The determination of the desired guidance variables in the form of the individual desired course x_{des} and the individual desired speed v_{des} to go out. In the latter case, as already mentioned, the desired speed v_{des} of the nominal speed v_{target} the system can be used when no adjustment is required due to front vehicles, speed restrictions, various traffic signs, safety-critical objects, limited visibility or a curvy road surface. In the same sense, the desired course corresponds to x_{des} the target course x_{target} if no adaptation is required due to the course of the road ahead or safety-critical objects. Otherwise, both reference variables are adapted by the driver to the environmental factors of the current situation by means of numerous anticipatory subtasks, which are represented in ◘ Fig. 2.24 by corresponding decision necessities (marked as rhombus) and tasks to be performed (marked as rectangles). As can be seen from this diagram, all these tasks can only be performed by estimating the perceived variables (e.g.: does the perceived distance to a vehicle correspond to the real distance? It is well known, clearly perceptible objects are perceived as nearer than only schematically visible) and dynamic sizes of own vehicle (e.g.: at what distance would my vehicle come to a standstill if I had to brake? At what distance would I be able to change back to the right lane when overtaking?) and other vehicles (e.g. when overtaking: when would the oncoming vehicle pass me?). All these examples refer to the determination of the target speed, which is ultimately based on daring estimates. It should be noted that the limits for the determination of the target course - at least during normal tracking of the

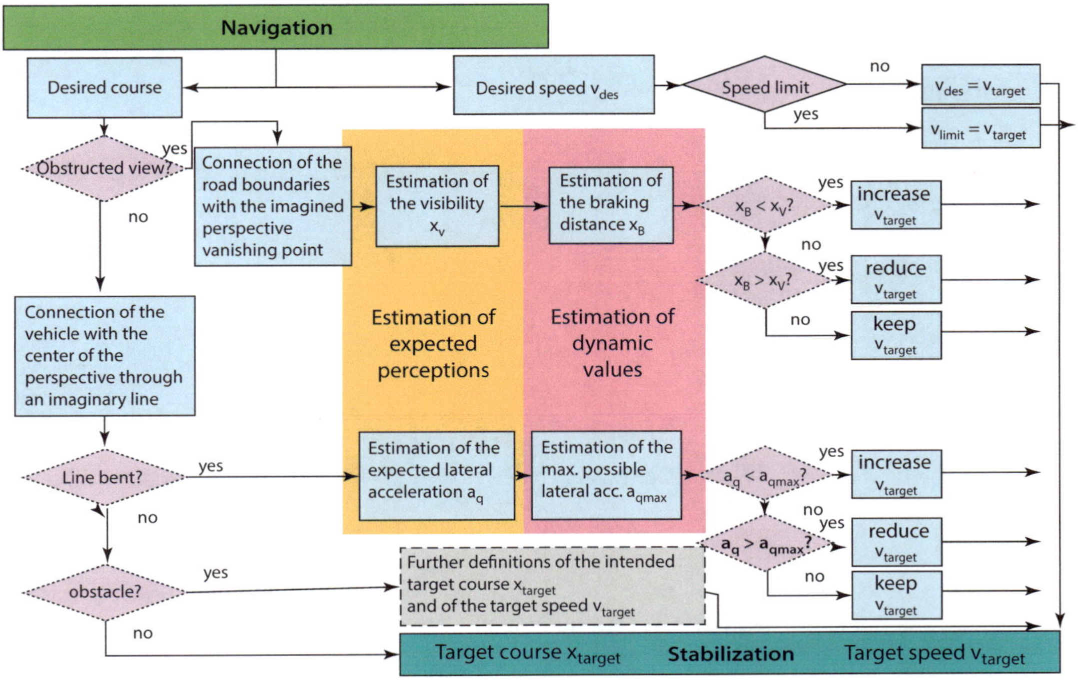

◘ **Fig. 2.24** Decision model of the guidance level (Bubb 1993)

2

course of the road under the condition of ideal visibility conditions - are visually more clearly ascertainable and thus this task is also easier to master. The problem here is that the determination of the target course also imposes restrictions on speed, which are themselves based on estimations.

The difficulties caused by emotional estimates are also reflected in the accident statistics. In the GIDAS research project, carried out in collaboration with the police, the police informs a crew of scientists, consisting of technical, psychological and medical personnel, who conduct their investigations independently of the police, when they are called in for an accident. Thus in the interviews some statements can be accomplished without fear of assigning blame. Gründl (2005) collected and analysed interviews with 528 drivers involved in a total of 312 road accidents. In line with other accident statistics, he found that, at about 30%, accidents in longitudinal traffic occur most frequently when a faster car touches a slower car due to a too small safety distance (see on this and the following ◘ Fig. 2.25).

Approximately 22% of all accidents are so-called single accidents without the involvement of another road user. A carelessness causes the driver to get to the edge of the road; normally he is frightened. It breaks the steering wheel and only then brings the vehicle into a dangerous slipping state. It is this type of accident that can be effectively prevented by the ESP system operating on the stabilisation level (see ▶ Chaps. 1 and 9).

All the following major accidents occur at intersections. Most of them are those where the causative driver turns to the left and has to correctly assess the dynamic behaviour of oncoming or approaching vehicles and correlate it with his own dynamics (approx. 15%). The other accidents are very varied and are summarised here under the term "accidents at junctions" (25%). In most cases they are characterized by the fact that the driver has to accomplish his guidance task on the basis of many simultaneously observable objects. Accidents with standing objects are relatively rare in relation to all this (approx. 2%). The same applies to accidents involving pedestrians (approx. 1%). The rest are various other accidents. From this and similar other analy-

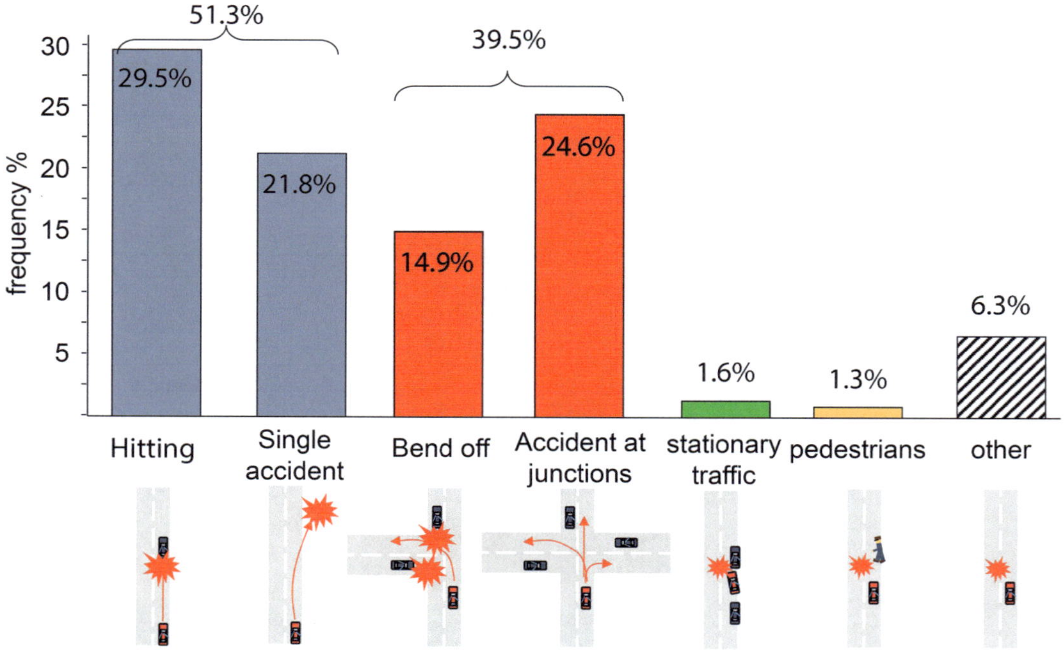

◘ **Fig. 2.25** Distribution of accident types according to Gründl (2005)

ses can be seen: The most important accidents, with approx. 51%, are the so-called longitudinal accidents (rear-end collisions and off-road accidents). In second place or, depending on the country and the accident statistics, even in first place are intersection accidents (here: approx. 40%).

Apart from the fact that some of the off-road accidents are due to distraction and inattention ("momentary sleep"), the above analysis of the guidance task shows how much the accident is dominated by the incorrect estimation of distances and in particular dynamic variables. An ergonomic measure that derives from this can be seen in the objectification and visualisation of these variables, which represents a starting point for the ergonomic design of assistance systems in the vehicle (see ► Chap. 9).

2.6 System Reliability

However, experience now shows, as already indicated in the last section, that human and technical system elements do not function under all circumstances in the sense of the intended purpose. One therefore tries to estimate the probability of the "failure"of system elements (so-called *probabilistic method*). By linking these with the rules of Boolean algebra, taking into account the system structure, and calculating the expected total failure probability, not only can those elements be identified that have a particularly strong influence on the total failure probability, but an improvement in the failure probability can also be achieved by changing the system structure, which of course must be done in such a way that the overall function is retained. Among other things, this procedure - especially if people are included in the consideration - can lead to changes in the organisation and to organisational regulations, which is evident in the area of road traffic and beside others by the large number of traffic regulations.

2.6.1 Safety, Risk, Border Risk and Protection

The 1985 Meyers Großes Universallexikon contains a very general definition of safety: "***Safety*** is the state of non-threatenedness, which is objectively represented by the presence of protective devices or the absence of sources of danger and is subjectively perceived as certainty by individuals or social entities about the reliability of protective facilities and devices". It should be noted in particular that "safety" refers primarily to the behaviour of the system towards humans. Safety is described by the measure of "absence of danger". ***Danger*** of any kind are considered to be impairments of life and health if their effect exceeds a reasonable risk according to the respective state of the art when technical products are used as intended ... (DIN 31000 or VDI 1000).The following definitions shall apply:

- *Safety* is a situation in which the risk is not greater than the limit risk.
- *Limit risk* is the greatest still justifiable risk of a particular technical process or condition. In general, limit risk cannot be quantified, but in road traffic it is generally accepted that any damage to objects and in particular to people cannot be tolerated.
- *Protection* is to reduce the risk through measures that limit either the frequency or extent of the damage, or both. Safety can often only be achieved through the interaction of several such measures.

There is therefore a continuous transition between safety and danger on the axis of increasing risk. Pragmatically, the areas of safety and danger are separated by the limit risk defined above. The risk itself is understood as a probability statement about the expected unfavourable course of an event in accordance with definitions common in economic science, which often multiplicatively link the expected frequency of loss and the expected extent of loss. In road traffic, this

2

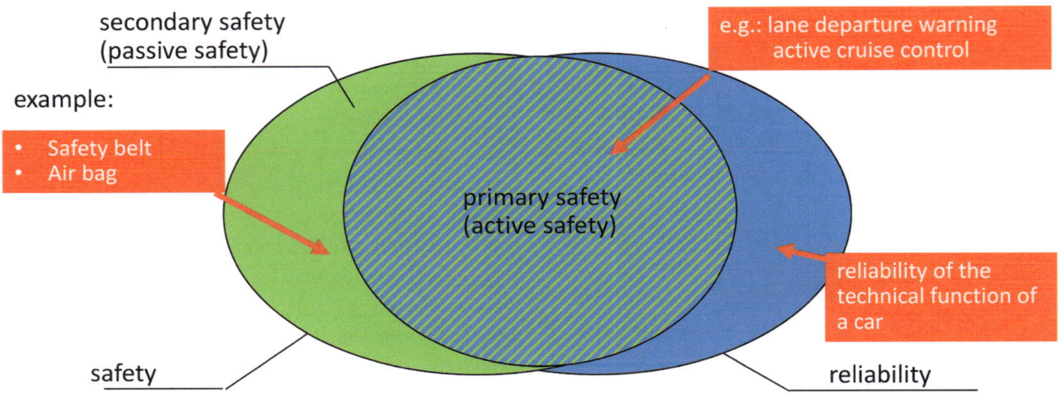

Fig. 2.26 Areas of safety and reliability and their classification under the terms "active"and "passive" safety

definition is often reduced solely to a consideration of the frequency of damage.[13]

2.6.2 Reliability, Error and Safety

In order to understand the relationship between reliability and safety, one can imagine the area of safety and reliability drawn in the form of Venn diagrams (see Fig. 2.26). According to this, there is a central and desirable area where safety is achieved through reliable operation. This is the area known as "primary safety"or "active safety". It is mainly achieved by a reliable interaction of vehicle and driver, but also by assistance systems such as ABS, ESC, ACC and so on. If this common area is left, a certain degree of safety can nevertheless be ensured at least for the passengers of the vehicle outside the (normal) functionality, for example by using safety elements such as the crumple zone, seat belt and airbag. This area is known as "secundary safety" or "passive safety". Here, attention is drawn to a frequently observed confusion: passive safety can be achieved both by passive safety elements, such as the crumple zone, and by active safety elements, such as the airbag. The area outside the safety area is completely undesirable if the (technical) functionality is maintained. For example, it would identify a driverless (not automatically controlled) vehicle that is otherwise technically intact.

According to ISO 8402, all terms related to reliability are derived from quality as defined in Eq. 2.1. After that *dependability* is a summary term describing the availability and its influencing factors functionalty, maintainability and maintenance readiness (ISO 9000, Part 4). *Availability* is the ability of a unit of observation[14] to be in a state in which it can perform a required function under specified conditions at a specified time or during a specified time interval, provided that the required means are provided (IEC 50–191). *Permitted deviations* or *requirements* can be defined for the characteristics of the unit of analysis or the function. If the current condition exceeds these requirements, it is called a *error*, a *malfunction* or *breakdown* (= termination of functionality) or *failure* (= occurrence of a malfunction when the unit is used in an approved manner). These somewhat abstract-looking definitions make it possible to define discrete, countable errors from the continuous

13 This is not generally the case: most accident statistics only record accidents in which (reportable) injured or killed persons have occurred. In the legal sense, however, any damage (including minor material damage) is subject to settlement. This obligation also implies the driver's general duty to carry out his journey in such a way as to avoid any contact between fixed and movable objects and persons.

14 Depending on the purpose of the examination, "units of consideration" can be individual components of the vehicle, the vehicle as a whole, the system consisting of driver and vehicle and the entire traffic system.

temporal course of a condition, which is necessary if one wants to estimate probabilities for failure.

On an experimental basis, the so-called failure or error quote κ is obtained.

$$\kappa = \frac{Number\ of\ observed\ errors}{time\ unit} (e.g.[1/year]) \tag{2.25}$$

It is a so-called a posteriori quantity, which is a good estimate for the theoretical failure or error rate λ with a sufficiently long observation time and a sufficiently large observed number of errors. In order to estimate probabilities in the above sense on the basis of the structure of effects, these quantities must, according to Laplace's a priori concept of probability[15] p

$$p = \frac{number\ of\ observed\ events\, i}{number\ of\ possible\ events} \tag{2.26}$$

or its a posteriori estimated value h

$$h = \frac{observed\ frequency\, N}{Number\ of\ observations} = \hat{p} \tag{2.27}$$

be converted into a dimensionless probability. This is only possible with the assumption of the distribution function of the underlying probability. The simplest form of distribution for this is the exponential distribution.[16] It then applies to the probability R that there are *not* comes to a mistake:

$$R(t) = e^{-\lambda \cdot t} \tag{2.28}$$

15 An a priori probability, for example, is the probability of throwing a certain number when rolling the dice. It is p = 1/6.

16 Other forms of distribution also play an important role in the reliability considerations. In practice, these are essentially the normal distribution, the logarithmic normal distribution (it can be taken into account that no event can occur in negative times) and the largely adaptable Weibull distribution. In the context of this treatise, however, this special field will not be dealt with further.

The quantity R(t) thus indicates the time-dependent reliability of the unit under consideration. Consequently, the error probability function F(t) is the mathematical complement to reliability:

$$F(t) = 1 - R(t) = 1 - e^{-\lambda t} \tag{2.29}$$

The two functions R(t) and F(t) indicate the basic relationship that, over time, the reliability of each system decreases and the probability of an error occurring increases. The size λ can also be referred to as the so-called process constant, which ultimately characterizes how dense events can occur, i.e. how often there is a practical chance of an error. Therefore, their reciprocal value characterizes.

$$1/\lambda = MTTF \tag{2.30}$$

the M̲ean T̲ime T̲o an F̲ailure (MTTF). It is a descriptive value that makes probability values, which are often perceived as abstract, understandable.

The dimensionless probability makes it possible, on the basis of the structural image, to calculate the total probability E_{total} of an event using Boolean algebra. If the events are connected by a ***AND operation***, the overall event only occurs if all n events involved E_i occur at the same time. It's on then:

$$E_{total} = \prod^{n} i = 1 E_i \tag{2.31}$$

With the ***OR operation*** the overall event E_{total} occurs if only one of the individual events becomes reality or several of them or even all.[17] The formula is:

$$E_{total} = 1 - \prod_{i=1}^{n} (1 - E_{i.}) \tag{2.32}$$

17 This mathematical "inclusive or" differs from the colloquial "exclusive or", which is generally understood as "either ... or ...".

2

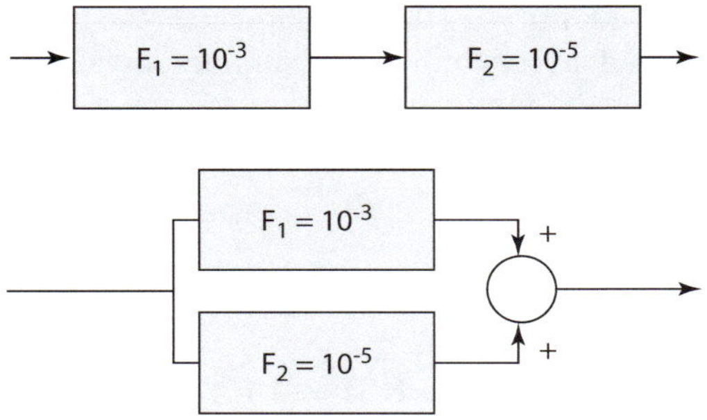

Fig. 2.27 Serial and parallel interconnection of two system elements

If, for example, two system elements are connected in series (see Fig. 2.27 above), each of which has an error probability of $F_1 = 10^{-3}$ and $F_2 = 10^{-5}$ the error of an element always has an effect on the overall error probability, i.e. the OR operation applies:

$$\begin{aligned} F_{total} &= 1-(1-F_1)\cdot(1-F_2) \\ &= 1-\left(1-10^{-3}\right)\cdot\left(1-10^{-5}\right) \\ &= 10^{-3} \end{aligned} \qquad (2.33)$$

If, on the other hand, the system elements are connected in parallel (see Fig. 2.27 below) and both influence the overall system result via a sum point, an error only occurs if both fail at the same time. Thus, the AND operation applies:

$$F_{total} = F_1 \cdot F_2 = 10^{-3} \cdot 10^{-5} = 10^{-8} \qquad (2.34)$$

The comparison of the results in the Eqs. 2.33 and 2.34 shows the great advantage of redundancy, which always contributes to making the overall probability of error smaller or the reliability greater (see Eq. 2.29). The consequences this has for the use of assistance systems are explained in more detail in ▶ Sect. 9.3.

2.6.3 Human Error and Probability of Accident

In terms of Laplace's definition of the a posteriori estimated value, Human Error Probability (HEP) is defined as:

$$HEP = \frac{\textit{Number of erroneously performed tasks of type } i}{\textit{Number of performed tasks of type } i} \qquad (2.35)$$

In general, for highly practiced activities one can average a value of HEP $\approx 10^{-3}$ while for knowledge-based activities the error probability is already between 10^{-1} and 10^{-2} (beside others after Swain and Gutmann, 1983; see also ▶ Sect. 3.4). But what does such information mean from the point of view of the individual? It is logical that those who do not enter the road should not suffer a road accident. It is therefore the so-called process intensity that is at issue if one wants to move from the errors to the rather descriptive mean time between errors (MTBF). For the process intensity one can fall back on studies by Pöppel (2000), who found an internal cycle time of 3 seconds for highly practiced activities and a cycle time of 2 seconds for novel activities (see also ▶ Sect. 3.3.1). This means that for highly practiced activities, such as stress-free normal driving, one must reckon with the fact that on average every 3 seconds one has the "chance" for a mistake. With a probability of 10^{-3} the MTBF will be about 50 min.

But not every small mistake leads to an accident. It will either be corrected in time or it will have no consequences. It remains without consequences if there is no conflict object (e.g.: crossing a red traffic light if there is no cross traffic). Only when a damage event threatens as a result of the error (or several errors) and this can only be avoided by an avoidance action of the participant(s) is a conflict referred to, in the case of driving tasks in particular the traffic conflict (Zimolong 1982). The triggering event for the occurrence of a traffic conflict is either faulty actions by the driver or, in rare cases, faults in the vehicle's technical systems. Only in the presence of stationary or moving conflict objects (e.g. other road users, trees, road boundaries) does a traffic conflict occur. In a single accident, the only possibility of an avoidance reaction lies with the driver causing the conflict. Only the unresolved conflict leads to an accident, the damaging event. Durth and Bald (1987), among others, proposed this funda-

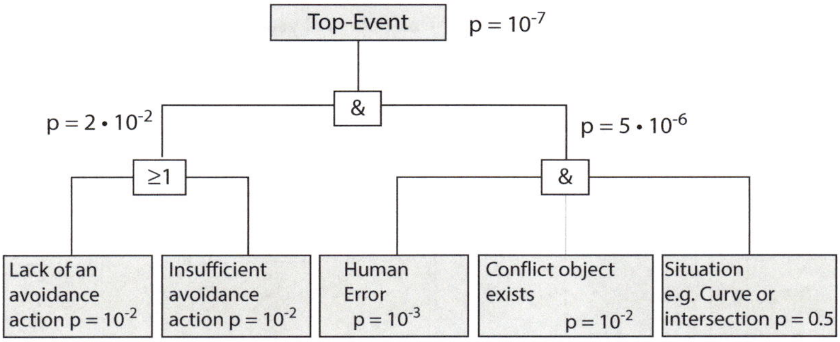

Fig. 2.28 Generic fault tree model of accident occurrence in road traffic

mental consideration of the development of accidents. Reichart (2001) interpreted this model of the origin of an accident in a fault tree representation (see Fig. 2.28). In the sense of the fault tree, the accident or damage event is referred to as a TOP event.[18]

With the value for the human error probability in highly practiced activities as well as some additional assumptions (situations change constantly, which is characterized with a probability of p = 0.5; the occurrence of conflict objects can be realistically estimated with $p = 10^{-2}$ - all these conditions must coincide in the sense of an AND operation), the probability of a conflict can be estimated with $p = 10^{-6}$. Only if now also still another avoidance action fails OR completely fails (since it concerns here for the driver a rather rare event, becomes both to $p = 10^{-2}$ estimated), a TOP event, the accident, takes place. Its probability can therefore be estimated rough of $p = 10^{-7}$.

But what does this mean in practical application? To get a feel for this probability, it is useful to calculate the mean time between errors (MTBF). For this it is necessary to know the process intensity mentioned above. With the value of the human cycle time of 3 seconds and assuming that on average about 1 hour per day is driven, the calculation of the TOP event of Fig. 2.27 at 10^{-7} gives a mean time between accidents (without statement about the gravity of the accident!) of approx. 23 years. For the individual driver the accident is therefore a rather rare event, which means that the subjective experience stands in gross contradiction to the frequency perceived by newspaper reports. This may support, among other things, the subjective feeling that "nothing can happen to me".

18 The TOP event is at the top of a fault tree. Among them are the "branches" that lead to the events that are the prerequisite for the realization of the TOP event. The branches are linked together by AND (&) and OR (≥) joins at the connection nodes, depending on the logic of the overall process.

2.6.4 Derivation of Measures from the Fault Tree Analyses

The fault tree model can be used to evaluate the potential benefits of improvement measures that intervene in the vehicle or infrastructure. In doing so, it is determined which fault events alone or which minimal interrelationships can lead to the TOP event. These are the minimum quantities of fault events that must exist for the TOP event to occur.

The quantitative evaluation of the fault trees is influenced by modeling faults as well as by possible uncertainty of the input data. The incompleteness of models or the uncertainty of input data, however, usually remains of minor importance for the assessment of relative influences of error types on the TOP event or the relative effect of improvement measures. The absolute values determined on the basis of the fault tree calculation do not play a decisive role for the purpose of the procedure, but they should at least roughly approximate the estimates obtained from the accident statistics in the sense of a plausibility check.

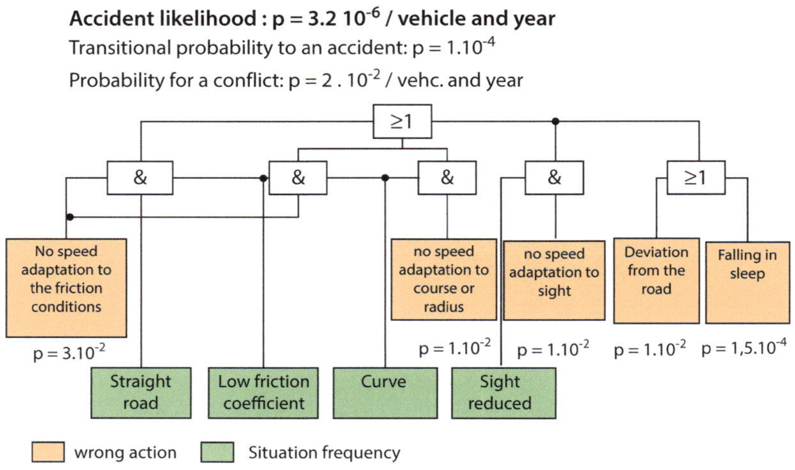

Fig. 2.29 Fault tree for the traffic conflict "Driving accident under the influence of routing or weather"

Using the fault tree analysis, the relative gain of an improvement measure in relation to the original condition can be evaluated and a differentiated selection of measures can be made which promise the greatest relative gain (initially, the simplifying assumption is 100% effectiveness and 100% reliability/availability of the measure). Since, however, technical failures of the improvement measure or their non-use by humans can also be modelled, much more realistic statements can be made about actually achievable improvement potentials than in the usual general assumptions about accident avoidance potentials of systems.

Using the example "Driving accident under the influence of routing or weather conditions", Reichart (2001) uses the fault tree methodology for the selection of improvement measures for active safety. The fault tree model in Fig. 2.29 shows which individual events lead to the TOP event of the fault tree. The corresponding situations are listed in the lower row. The second row indicates the possible wrong actions of the driver. Based on the figures in Fig. 2.28 and estimates made by Reichart, the probability of a conflict is thus $p = 2 \cdot 10^{-2}$/vehicle and year. With the transition probability to an accident of $p = 1 \bullet 10^{-4}$ the unknown probability for the presence of conflict objects and for the failure of driver avoidance actions is captured. As Reichart states, this probability of transition applies to a surprisingly large number of accident types. This finally results in an accident probability of $p = 3.2 \cdot 10^{-4}$ per vehicle and year, which is sufficiently consistent with the official accident statistics from 1994, on the basis of which a probability of $p = 9.2 \cdot 10^{-4}$ (excluding driving accidents when turning) can be calculated.

On the basis of this error tree, you can now perform an import calculation that calculates the contributions of situation probabilities and error combinations that contribute significantly to the TOP event. The results of this calculation are summarized in Table 2.1. Thus, dominant contributions to the probability of occurrence result from the omission of lane control due to distraction and avoidance as well as from the influence of the lack of adaptation of the speed to the routing or to the friction conditions. The "low visibility" situations are very low in terms of their frequency per year, so that the potential for improvement in road safety through technical measures can be assessed as rather low here. In contrast, situations with "low friction" have a higher frequency and play an important role in many other fault tree models of accident types. Information about the frictional connection can therefore be an effective improvement measure (e.g. Bubb 1975; Assmann 1985; Bachmann 1998). Unfortunately, however, there are currently no approaches as to how reliable, forward-looking adhesion information could be realised. One way of car-to-

Table 2.1 Importance of the contribution to the TOP "Driving accidents under the influence of routing or weather conditions" (Reichart 2001)

Situation/error type (minimum cut)	Importance (contribution to TOP)
Failure to check lanes due to distraction	46,6%
No adjustment of the speed to the radius or course and driving on a curve	23,3%
No adjustment of the speed to friction and friction low and driving on a curve	14%
No adjustment of the speed to friction and friction low and driving on a straight stretch of road	14%
Low visibility	1,4%

car communication (Car2Car) would be for a preceding vehicle that detects a reduced coefficient of friction - e.g. via ABS or ESP information - to pass this on to following vehicles by radio technology.

Since almost as many driving accidents occur on straight roads and in bends, measures that provide a preview of the course of the bend or warn of leaving the lane are more effective than information about the appropriate speed of the bend. However, the latter would be very effective against lane departure in curves if the information on adhesion were included. This conclusion can be drawn directly from the quantitative evaluation of the fault tree models with and without this measure.

Tabelle 2.2 provides a compilation of potential improvement measures for the types of defects considered in the fault tree model of Fig. 2.29. For the error type "omission of lane control", an ergonomic cockpit design (e.g.: "The use of HUD technology") reduces the probability of prolonged distractions or deviations from lane control (see Gengenbach 1997). However, more effective protection could only be achieved through TLC warnings or lane guidance assistants (e.g Naab 1998; Penka 2000), which would also be effective against the error type "falling a sleep". An in-depth analysis of the error type "No adaptation of speed to the curve" shows that this is more a problem of a lack of predictive information acquisition than a problem of vehicle stabilization. It follows directly from this that, from the point of view of road safety, a speed recommendation based on knowledge of the curve from a digital road map would be preferable to the use of a stabilisation system. However, if the problem is rather an overestimation of the traction reserve, the stabilisation system is the better measure. This means that ultimately only the combination of both measures would lead to tangible success (Table 2.2).

Table 2.2 Derivation of improvement measures for fault tree of Fig. 2.29

error type	Influence on TOP event	improvement measures
failure to monitor the lane	straight	ergonomic cockpit design (instrumentation in front of the driver or HUD), lane keeping support (heading control), warning in case of imminent departure from the roadway
No adjustment of speed to curve radius/gradient	only for passing through curves	Dynamic stability control, curve from digital road map, display or automatic adjustment of target speed
No adjustment of speed to friction conditions	only with low friction	Force-locking measurement, display or warning
No adjustment of the speed to visibility conditions	only in low visibility	Visibility measurement with display or warning

The quantitative evaluation of the fault tree with all available improvement measures, taking into account their potential failure or their non-use by the drivers, finally provides a probability of occurrence of the TOP event of $p = 7 \bullet 10^{6}$. This means a reduction of the probability of occurrence by a factor of about 2800 compared to the original fault tree calculation without improvement measures. Thus, even if the data uncertainty is comparatively high, a high improvement effect of the measures can be assumed.

The argumentation presented here can be seen as an example of the fact that a reliability analysis, usually in conjunction with the analysis of accident statistics, can lead to interesting conclusions. For further information in this context, please refer to the dissertations of Reichart (2001) and Gründl (2005).

References

Assmann, E.: Untersuchung über den Einfluss einer Bremsweganzeige auf das Fahrverhalten. Dissertation an der Technischen Universität München (1985)

Bachmann, T.: Wechselwirkungen im Prozess der Reibung zwischen Reifen und Fahrbahn. VDI Fortschrittsberichte, Reihe 12, VDI-Verlag, Düsseldorf (1998)

Bolte, U.: Das aktive Stellteil – ein ergonomisches Bedienkonzept Fortschrittsberichte VDI-Reihe 17, vol. Bd. 75. VDI-Verlag, Düsseldorf (1991)

Bubb, H.: Untersuchung über die Anzeige des Bremsweges im Kraftfahrzeug. Dissertation, Technischen Universität München (1975)

Bubb, H.: Systemergonomie. Teil 5. In: Schmidtke, H. (ed.) Ergonomie. Hanser Verlag, München (1993)

Crossmann, E.R.F.W., Szostak, H.: Man-machine-models for car-steering. In: Fourth Annual NASA-University Conference on Manual Control. National Aeronautics and Space Administration, Washington, DC (1969)

Daenzer, W.F., Huber, F. (eds.): Systems engineering: Methodik und Praxis. Industrielle Organisation, Zürich (1992)

Donges, E.: Ein regelungstechnisches Zwei-Ebenen-Modell des menschlichen Lenkverhaltens im Kraftfahrzeug. Zeitschrift für Verkehrssicherheit. **24**, 98–112 (1978)

Durth, W., Bald, S.: Risikoanalysen im Straßenverkehr. Forschungsauftrag FE 02.111 R86A des Bundesministers für Verkehr, Schlussbericht, TH Darmstadt (1987)

Eckstein, L.: Entwicklung und Überprüfung eines Bedienkonzepts und von Algorithmen zum Fahren eines Kraftfahrzeugs mit aktiven Sidesticks Fortschr.-Ber. VDI, Reihe 12, Bd. 471. VDI-Verlag, Düsseldorf (2001)

Fastenmeier, W., Hinderer, J., Lehnig, U., Gstalter, H.: Analyse von Spurwechselvorgängen im Verkehr. Zeitschrift für Arbeitswissenschaft. **1**(55), 15–23 (2001)

Fiala, E.: Lenken von Fahrzeugen als kybernetische Aufgabe. Automob. Z. **68**(5), 156 (1966)

Fleck, R.: "Aktivlenkung" – ein wichtiger erster Schritt zum Steer-by-Wire. Tagung "PKW-Lenksysteme - Vorbereitung auf die Technik von morgen", Haus der Technik e. V., Essen, 02.-03.April. (2003)

Gengenbach, R.: Fahrerverhalten im Pkw mit Head-Up-Display. Gewöhnung und visuelle Aufmerksamkeit. VDI Fortschrittsberichte Reihe 12: Verkehrstechnik/Fahrzeugtechnik. Nr. 330 (1997)

Gillet, G.: Ergonomische Optimierung eines Aktiven Stellteils. Dissertation, an der Technischen Universität München (1999)

Godthelp, H.: Studies on Human Vehicle Control. TNO, Institute for Perception, Soesterberg (1984)

Gründl, M.: Fehler und Fehlverhalten als Ursache von Verkehrsunfällen und Konsequenzen für das Unfallvermeidungspotenzial und die Gestaltung von Fahrerassistenzsystemen. Inauguraldissertation an der Universität Regensburg (2005)

Guan, H., Gao, Z., Guo, K., et al.: An optimal preview acceleration model for velocity control in drivervehicle-environment system. In: Proceeding of AVEC 2000, S. 525–530. Taylor & Francis, Ann Arbor (2000)

Herrin, G.D., Neuhardt, J.B.: An empirical model for automobile driver horicontal curve negotiation. Human Fact. **16**(2) (1974)

Honda Motor Co. 2000: Fact Book, S2000 TypeV. http://www.honda.co.jp/factbook/auto/s2000/ 200007/index.html (auf Japanisch)

Huang, P.: Regelkonzepte zur Fahrzeugführung unter Einbeziehung der Bedienelementeigenschaften. Dissertation an der Technischen Universität München. (2004)

Jürgensohn, T.: Bedienermodellierung. In: Timpe, K.-P., Jürgensohn, T., Kolrep, H. (eds.) Mensch-Maschine-Systemtechnik – Konzepte, Modellierung, Gestaltung. Evaluation. Symposion Publishing, Düsseldorf (2000)

Kondo, M.: Richtungsstabilität (wenn Steuerbewegungen hinzukommen). J. Soc. Automot. Eng. Japan, **7**(5, 6) (auf Japanisch) (1953)

Lindsay, P.H., Norman, D.A.: Human Information Processing. An Introduction to Psychology, New York, London (1972)

Merz, L.: Grundkurs der Regelungstechnik. Einführung in die praktischen und theoretischen Methoden. BI Hochschultaschenbücher, München (1973)

Naab, K.: Heading Control - Ein System zur Fahrerunterstützung bei der Spurhaltung, Vortragsmanuskript, Seminar „Fahrerassistenzsystem" Haus der Technik, Essen (1998)

Ohno, T.: Steering Control on a Curved Course. Japan SAE Review **20**(5) (1966)

Penka, A.: Vergleichende Untersuchung zu Fahrerassistenzsystemen mit unterschiedlichen aktiven

Bedienelementen. Technische Universität München, Dissertation (2000)

Popiv, D., Rakic, M., Bengler, K., Bubb, H.: Bewertung der Antizipationsleistung des Fahrers bei Annäherungs- und Spurwechselvorgängen auf der Autobahn. Proc. 1. Automobiltechnisches Kolloquium, München-Garching, Deutschland, April 2009 (2009)

Pöppel, : Grenzen des Bewusstseins - Wie kommen wir zur Zeit und wie entsteht Wirklichkeit? Insel Taschenbuch, Frankfurt a.M. und Leipzig (2000)

Poulton, E.C.: Tracking Skill and Manual Control. Academic, New York, San Francisco, London (1974)

Pressler, G.: Regelungstechnik, Grundelemente. Bibliographisches Institut Mannheim, Hochschultaschenbücher, Mannheim (1967)

Reichart, G.: Menschliche Zuverlässigkeit beim Führen von Kraftfahrzeugen. Fortschritt-Berichte VDI, Reihe 22 Mensch-Maschine-Systeme. VDI-Verlag, Düsseldorf (2001)

Schmidtke, H.: Der Leistungsbegriff in der Ergonomie. In: Schmidtke, H. (Hrsg.) Ergonomie, 3. Aufl. Carl Hanser-Verlag, München-Wien (1993)

Schulze, B.G.: The application of ergonomic principles in the design of operating controls. In: Oborne, D.J., Jevis, J.A. (eds.) Human Factors in Transport Research, vol. 1, pp. 337–344. Academic Press, New York, London (1981)

Schweigert, M.: Fahrerblickverhalten und Nebenaufgaben. Dissertation an der Technischen Universität München. (2003)

Sheridan, T.B., Ferrel, W.R.: Man-Machine-Systems: Information, Control, and Decision Models of Human Performance. The MIT Press, Cambridge, MA (1974)

Steward, J.D.: Human perception of angular acceleration and implications in motion simulator. J. Aircraft. **8**(4), 248–253 (1971)

Swain, A.D., Guttmann, H.E.: Handbook of Human Reliability Analysis with Emphasis on Nuclear Power Plant Applications. Sandia National Laboratories, Nureg/Cr-1278, Washington, DC (1983)

Tustin, A.: An Investigation of the Operators Response in Manual Control of a Power Driven Gun. C. S. Memorandum, Bd. 169, pp. 172–186. Metropolitan Vickers Electrical Co., Sheffield (1944)

Unbehauen, R.: Systemtheorie. Grundlage für Ingenieure, Oldenbourg, München Wien (1993)

Unselt, T., Breuer, J., Eckstein, L., Frank, P.: Avoidance of “Loss of Control Accidents” Trough the benefit of ESP: FISITA 2004 Congress Proceedings (2004)

van der Horst, R., Hogema, J.: Time-to-Collision and Collision Avoidance Systems. Proceedings of the 6th ICTCT Workshop: Safety Evaluation of Traffic Systems, Salzburg, S. 109–121 (1994)

Wickens, C.D.: In: Charles, E. (ed.) Engineering Psychology and Human Performance. Merrill Publishing Company, Columbus (1984)

Wunsch, G.: Geschichte der Systemtheorie. Dynamische Systeme und Prozesse, Oldenbourg, München Wien (1985)

Yuhara, N., Tajima, L., Sano, S., et al.: Steer-by-wire-oriented steeling system design: concept and examination. Veh. Syst. Dyn.Suppl. **33**, 692–703 (1999)

Zimolong, B.: Verkehrskonflikttechnik, Grundlagen und Anwendungsbeispiele, Unfall- und Sicherheitsforschung Straßenverkehr, Heft 35, Bergisch Gladbach (1982)

Further Literature

Bubb, H.: Systemtheoretische Ansätze. In: Luczak, H., Volpert, W. (eds.) Handbuch der Arbeitswissenschaft, pp. 77–82. Schäffer-Poeschel Verlag, Stuttgart (1997)

The Human Being as a Driver

Heiner Bubb, Mark Vollrath, Klaus Reinprecht, Erhard Mayer, and Moritz Körber

Contents

3.1 The Human Being as an Information Processing System – 64

3.1.1 The System Model of the Driver – 64
3.1.2 Anatomical-Functional Model – 65
3.1.3 Functional-Neurophysiological Model – 69

3.2 Elements of the Information Processing Human Being – 77

3.2.1 Information Reception – 77
3.2.2 Information Processing – 101
3.2.3 Information Implementation – 120

3.3 Information Processing when Driving a Car – 123

3.3.1 The Feeling for the Time – 123
3.3.2 Human Eye Behaviour when Driving a Passenger Car – 124
3.3.3 Gaze Behaviour and Inner Models – 141
3.3.4 Comfort and Discomfort – 143
3.3.5 Stress and Strain – 147

3.4 Driving Error – 149

3.4.1 Human Reliability and Driving Errors – 149
3.4.2 Causes of Human Error – 151

References – 155

H. Bubb et al. (eds.), *Automotive Ergonomics*, https://doi.org/10.1007/978-3-658-33941-8_3

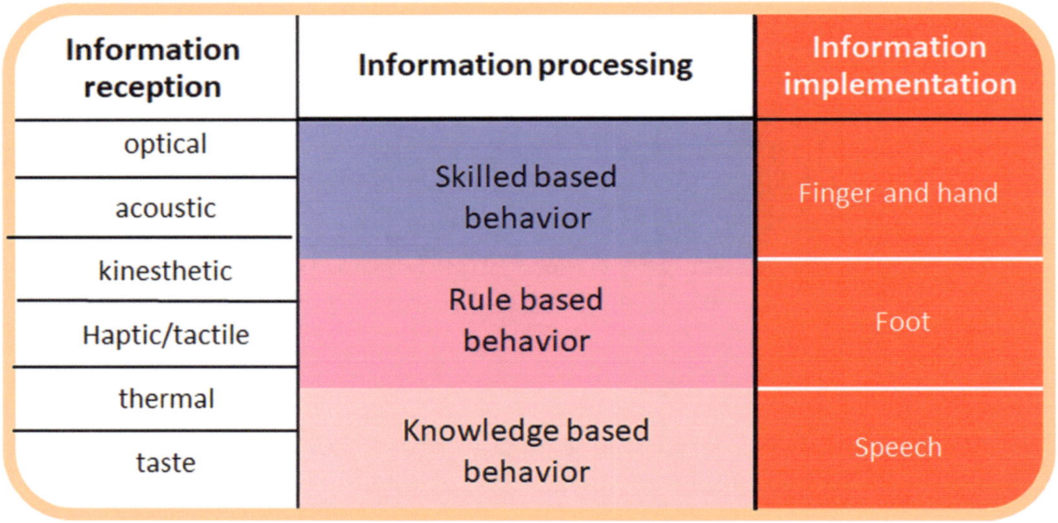

Fig. 3.1 System ergonomically oriented image of the human information processor

3.1 The Human Being as an Information Processing System

3.1.1 The System Model of the Driver

When driving a car, a person has to absorb a great deal of information, evaluate it, make decisions and initiate appropriate actions. From an information technology point of view, the human system element is usually represented graphically as a rectangle with an input and an output side (Fig. 3.1). This then consists of the blocks information reception, information processing and information implementation. The entrance side of the human being, the *information reception* is characterized by the "five senses"[1] whereby the corresponding sensory organs are to be assigned to the individual types of perception in the same way. Optical perception (eye) plays the dominant role in coping with road traffic, but also acoustic perception (ears), kinaesthetic (movement) perception (vestibular organ) and haptic (touch) perception (mechanoreceptors for pressure, touch and vibration as well as the skeletal position receptors). In addition there is the perception of odour, heat or cold.

From the combined processing of these individual sensory channels, our organism forms an "objective" picture of the environment and of our own and other people's actions in this world which is sufficient for normal living conditions. In *information processing*, certain intended actions are derived from this, which are transformed into real events through the *information implementation*. For this information implemantation our human organism has only the mechanical movement available, which is initiated by muscle power. The innervation of the musculature can be used beside others to transfer information to the vehicle by the movement of arms, hands and feet and to modulate information to the air vibrations by the movement of the speech apparatus, in order to establish contact with fellow human beings or, if appropriately equipped, with the vehicle.

In order to understand how people absorb and process information from the environment when driving a car, and how they take appropriate action, two approaches are help-

1 In the classical enumeration of the "five senses", the kinesthetic sense of movement is usually not mentioned, obviously because the sense organ necessary for it is not immediately visible from the outside.

ful which complement each other and must be considered simultaneously for proper understanding.

- On the one hand it has to be considered how the information gets into the brain, where it is processed and how it gets from there to the musculature, so that the imagined actions are converted into externally visible reactions. This view is taken by means of the "*Anatomical-functional model*".
- On the other hand, the content aspect is of interest, as it provides information about what is perceived by the environment and how it is perceived by the sensory organs, to what ideas this leads and how this results in drafts of action. This approach is based on a *"functional neurophysiological model"* of humans.

In the following, these two models are briefly introduced first. The individual sensory organs are then described to the extent necessary for understanding the collection of information relevant to driving. This is followed by a closer look at information processing in the context of driving. From the consideration and knowledge of all these characteristics, sources of error can be found that lead to wrong and ultimately dangerous actions by the driver in road traffic.

3.1.2 Anatomical-Functional Model

Information, which is converted by the sensory organs into electrochemical impulses, is transmitted to the various brain areas via various switching points by means of the neuron cells connected downstream and to each other. A distinction is made between the direct pathways leading to the specific areas, which beside others also serve cognitively conscious perception, and the indirect, non-specific pathways, which show many connections to all possible brain areas and which essentially serve the so-called protopathic perception. From many neurophysiological experiments one knows today the cortices assigned to the sense organs

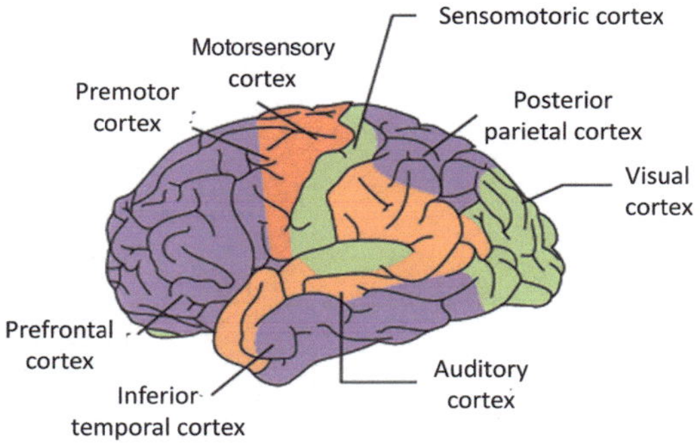

Fig. 3.2 Location and function of different brain areas (eyes and face are left to think)

and the body periphery as well as the specific function of the different brain areas (see in this connection Fig. 3.2). The brain consists – already from an anatomical point of view – of two symmetrical halves, which are connected to each other by the so-called corpus callosum. The two hemispheres of the brain process information largely independently of each other, but transmit it to each other through the callosum. Many experiments show that it takes about 5–40 ms for information to pass from one half of the brain to the other. As the sensory organs and sides of the body, which are also present in duplicate, are cross-connected with the brain hemispheres (i.e. e.g. the left eye is essentially connected to the right visual cortex), there are practical consequences. For example, Wittling (1992) argues that a pedestrian stepping into the lane from the right would have a slightly greater chance of a timely reaction from the driver, since he would first be captured by the right eye, whose information would be played into the left hemisphere of the brain; the reaction realized there would reach the right foot faster than if the pedestrian's image had first been projected into the right hemisphere of the brain.

Experiences of patients with brain injuries, basic examinations with so-called split-brain patients[2] and the examinations possible today

2 At the end of the 1950s, severe cases of epilepsy were treated by surgically cutting the corpus callosum (Sperry 1969).

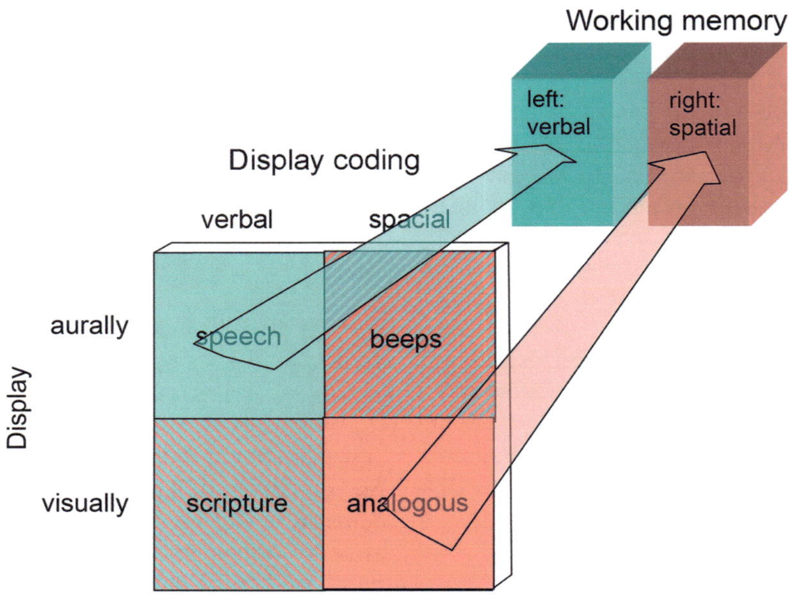

Fig. 3.3 Information simultaneously transferable to both hemispheres of the brain (according to Wickens 1992)

with functional magnetic resonance imaging clearly show that the two halves of the brain also have completely different functions with regard to our abilities and our experience. The ability to think logically, rationally, solve problems and speak is essentially anchored in the left hemisphere of the brain. In it thinking takes place sequentially, analytically with a tendency towards an objective view, but also with a strong focus on details. In contrast, the right hemisphere of the brain is essentially intuitive, holistic and synthesizing, in it spatial-mathematical thinking and musical abilities are anchored, its thinking is strongly subjective and emotional and shaped by a view of the whole. This specialization has consequences for the technical design of information. As will be shown below, the processing capacity of the working memory of our brain is limited. As Wickens (1992) argues, this can be partly compensated if, for example, different information contents can be presented simultaneously on the one hand linguistically auditory and on the other hand visually graphically (analogously, see Fig. 3.3). Therefore, it is possible within limits to carry out the spatial-analogue task of driving a car and the linguistic-digital, auditory task of talking to a passenger virtually independently of each other.

Because of its importance for driving a car, the motosensory (precentral gyrus) or sensomotoric (postcentral gyrus) cortex, on which the body periphery, i.e. both the feeling of the posture and the tactile feeling of contact with other objects (see also ► Sect. 3.1.3) is depicted, and the movement patterns for the activation of the respective muscles are realized, must be briefly discussed.[3] Figure 3.4 shows the spatial distribution of the sensors of the body and the motor control of the muscles. Of particular interest here is the disproportionate representation of the face and hand areas, which make up more than 2/3 of the total surface area of the two cortices, leaving only one third for the rest of the body (see also ► Sect. 3.1.3). From this point of view, it is ergonomically questionable to leave the necessarily sensitive brakes to the foot, an interpretation that has primarily arisen from technical-historical development.

In the following, the principle course of the path of the nerve connection from the body periphery to the brain and from there to

3 The "double name" motorsensory and sensomotoric is used because an exact separation into a receptive and an actuating part is not possible.

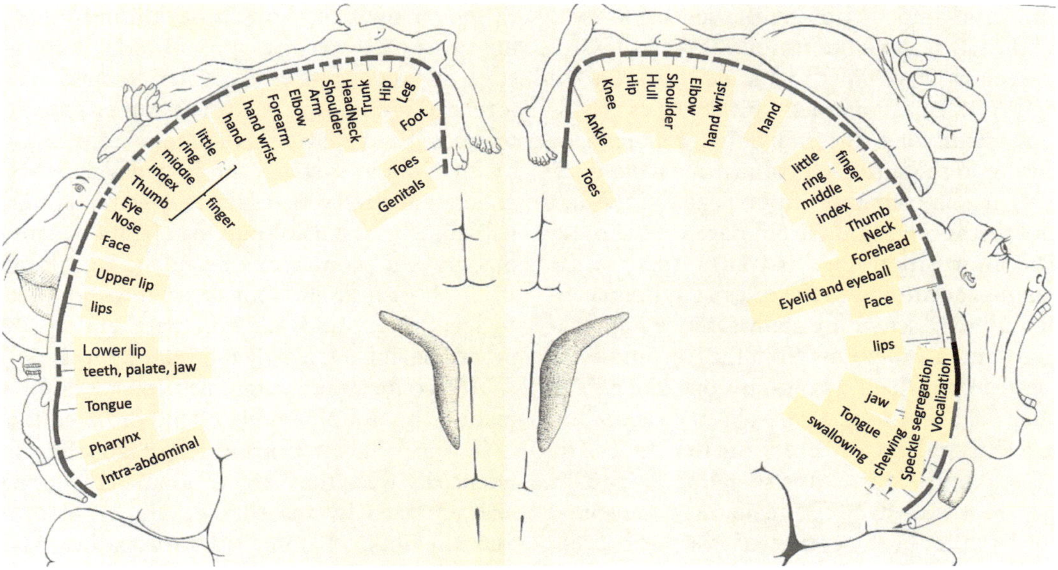

Fig. 3.4 Somatotopic organization of the motorsensory and sensomotoric cortex, illustrated by the "motoric (left) and sensory (right) Homukulus" The Natural History Museum, London (Image Ids: 001915 and 001914)

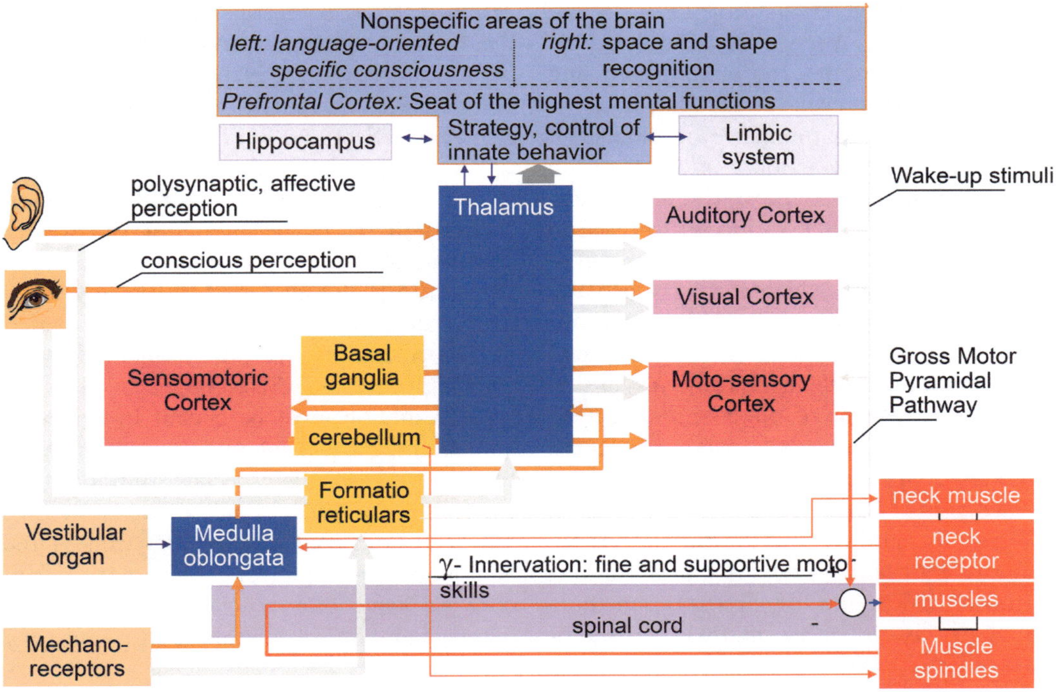

Fig. 3.5 Anatomical-functional model of the interconnection of sensory organs and effectors across the different brain regions

the effectors (muscles) in the body periphery is briefly described. Figure 3.5 shows a very simplified representation of the anatomical interconnection of the nervous system – without consideration of the laterality of the brain described above – starting from the uptake of information by the sensory organs on the left side via the various brain structures (especially thalamus) to the effectors on the right side. In accordance with the rules of system technology, the individual elements of this representation are drawn as rectangles, with the information input page on the left and the information output page on the right. The information itself is transmitted via the ganglia, according to the rules of system technology represented by arrows. Since information can only ever flow from the transmitter to the receiver, these arrows are one-sided. The figure shows the relevant sensory organs on the left, and the effectors on the right. The lower control loop of the so-called "self-reflex arc" (see ► Sect. 3.2.3) realized in the spinal cord (with the exception of the head area) ensures that the target movement (or target force) formed in the brain essentially in the area of the basal ganglia, the cerebellum and the motor sensory and sensorimotor cortex, mediated by the α innervation, is converted into reality. This process is modulated by information from the vestibular organ and from the mechanoreceptors via the γ innervation, so that the maintenance of balance is always guaranteed (for more information on the process of muscle innervation, see ► Sect. 3.2.3). Since the γ innervation is also used to adapt the sensors (especially muscle spindles) in the musculature, which can vary between fine and gross motor skills, the fine motor skills are impaired when driving through heavy terrain or with insufficient suspension not only for reasons of physical inertia reaction, but also for reasons of nervous circuitry. This has a significant impact on the use of touch screens while driving.

The information about the muscle tension, mediated by the different mechanoreceptors, is transmitted beside others via the medulla oblongata to the senso-motoric or motosensory cortex. The medulla oblongata also regulates the heart rate, so that this information can be used to adapt the energy supply to the muscles during physical exertion.[4]

The actual conscious behavior and thinking takes place according to all today knowledge in the prefrontal cortex behind the forehead in interaction with the other specific cortices. The behavior is essentially influenced by the limbic system, through which every action and every decision is always tainted with an emotional coloration. The hippocampus plays a prominent role in the transition from short-term to long-term memory (see ► Sects. 3.1.3 and 3.2.2.3), with emotional mood playing a strongly modulating role.

In connection with the considerations made here, not all details of the presentation in ◘ Fig. 3.5 are important. Especially the difference between the direct rational perception, characterized by the thinner orange arrows and the indirect so-called protopathic perception, represented by the wide grey arrows, should be mentioned. The first provides direct, analog images of the outside world, so to speak, on the respective brain regions assigned to the sensory organs. It serves the rational collection and further processing of information, which can be done both consciously and partly also unconsciously (see also ► Sect. 3.2.2). The latter causes an emotional perception that cannot be avoided in any situation. This emotional perception refers both to the behaviour of other road users and to the situation of the passengers in the vehicle. It also influences the perception of the vehicle as a whole. This is an aspect related to the perception of comfort, which is specifically addressed in ► Sect. 3.3.4.

Information must be coded. In principle, analog and digital coding are conceivable for this purpose. This is especially important on the initial side: first of all we can only give information in analog code: by means of the mentioned lower control circuit of the spinal cord we bring our muscles into position in (almost) any fine gradation or can apply any

4 By measuring the pulse rate, the degree of physical exertion (strain) can be determined very well. Since the organism tries to adjust the heart rate in advance of an expected physical effort, psychological stress also causes a corresponding reaction. Therefore, the pulse rate (and all associated parameters) is a good but somewhat sluggish indicator for mental strain, but does not provide an objective measure of the level of strain.

force in any dosage. This is the way we transfer information to the machine while driving a vehicle. In this way we also control our facial expressions and everyone knows that they are internationally understandable regardless of the respective language. The "analogue" behaviour of a car driver is largely understandable across cultures! It is therefore possible for an experienced driver to "read" the behaviour of other road users and draw conclusions about their reaction in the next few seconds. This is also possible in countries where the driver does not understand the language. However, in the course of historical development, national peculiarities may have developed which can only be understood by "foreigners" after a certain period of experience.

In a digital way we encode information in language. We are able to express and rationally understand very differentiated content, especially abstract content. However, we are bound by a coding agreement, namely the respective language. The transmission of information to the switches of the vehicle, the buttons of the on-board computer or the navigation computer also involves digital coding. Here there must be an agreement between the coding system of the vehicle designer and that of the user. For this reason, the "language" in which information is transmitted to the vehicle or in which the vehicle communicates information about certain operating conditions must be learned in principle. An ergonomic requirement is to make this coding system as easy to learn as possible, which requires the often desired "self-explanatory ability".

3.1.3 Functional-Neurophysiological Model

The functional-neurophysiological model of the human being should provide the understanding for the processing of information. For the system analytical description, it is not necessary to deal with the anatomical and physiological characteristics of the individual receptor channels (see ▶ Sect. 3.2). First of all, it is essential to describe the information transformation in the organism and the basic interconnection types in the individual subsystems of human information processing. The receptors convert external physical stimuli into physiological stimuli in such a way that a corresponding sequence (frequency) of different electrical potentials (so-called impulse rate) is assigned to the respective stimulus intensity. The receptor reacts to the adequate stimulus, i.e. to the physical type of stimulus to which it is sensitive. Even an inadequate stimulation leads to a sensation that corresponds to the respective receptor channel. Conscious or unconscious recognition of the information supplied by the receptors is achieved by measuring the significance. Numerous experiments on animals and humans show that this meaning depends on the location of the arrival of the information in the cortical projection fields. A stimulus in the optical projection field therefore always leads to an optical sensation (Foerster 1936; Iggo 1973).[5] The adequate stimulus for a sensory organ is thus determined not only by the specificity of the receptor, but also by the central processing of the impulse patterns coming from the receptors (Dudel 1976).

The basic mechanisms of information processing are very similar in all sensory channels: the information transmitted by the receptors in the form of impulse rates along the axons (connection between the neuron cells) is linked in the neuron cells with the information of other neighbouring receptor cells via so-called synapses.

However, for information to be processed at all, its intensity must be "supra-threshold":

- "Absolute threshold" refers to the qualitative transition from "not perceived" to "perceptible". In principle, these absolute thresholds were determined for the various senses. It turned out that these are dependent on stimulus characteristics of the context and the person.

5 For example, the mechanical impact on the eye leads to an optical sensation.

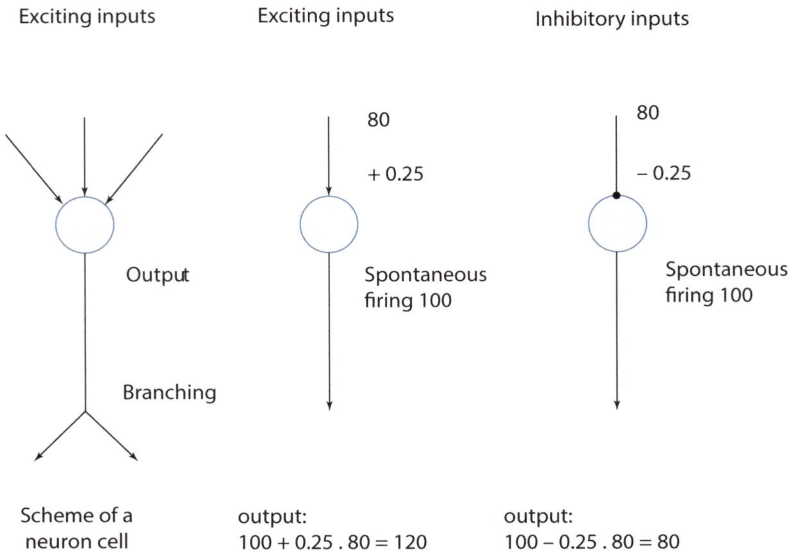

Fig. 3.6 Schematic representation of neuron cells (after Lindsay and Norman 1972)

- The "difference threshold" is probably of greater importance for the practice of driving a car. It is the smallest difference between two stimuli that a person can still perceive straight. In addition to the dependencies that apply to the absolute threshold, it also depends on the stimulus level.[6]

In the following a short consideration of the basic interconnection forms between neurons is made. Since neuron cells fire even when they are not innervated by impulses at the entrance Lindsay and Norman (1972), proposed to simplify the representation by assuming that a neuron cell fires with an arbitrarily assumed frequency of 100 [1/s]. Each neuron cell can have multiple inputs. It has an exit which in turn can branch into several branches (see Fig. 3.6 left). An exciting input increases the frequency at the axon of the excited neuron cell proportionally to its own frequency and a so-called excitation factor, which describes the transmission quality given at the synapse (see Fig. 3.6 middle). Correspondingly, an inhibitory input reduces the frequency at the output of the neuron cell proportionally to the frequency at the input and a factor which in turn describes the quality of the synaptic transmission at the neuron cell (see Fig. 3.6 right). Exciting inputs are marked with an arrow in this representation form, while inhibiting inputs are symbolized by a dot.

The main types of feedback are lateral inhibition, negative feedback and positive feedback. Figure 3.7 shows the calculation of the outputs of neuron cells excited by receptors and inhibited by lateral receptors. For a locally limited stimulus whose stimulus intensity is arbitrarily assumed to be 100, the representation also arbitrarily assumes that the stimulus intensity 0 at the receptors generates the output frequency 0 and that the stimulus intensity 100 at the outputs of the receptors generates the frequency 100. The image of the frequencies measured at the outputs of the neuron cells shows that the lateral interconnection causes an increase in the edge

6 This dependence can be described mathematically and is known as Weber's law: $\frac{\Delta R}{R} = k$. The extent ΔR, by which one must change a stimulus R so that it is experienced as greater or lesser compared to the original stimulus, depends on the size of the initial stimulus. See also ► Sects. 3.2.1.2 and 11.3.1.1.

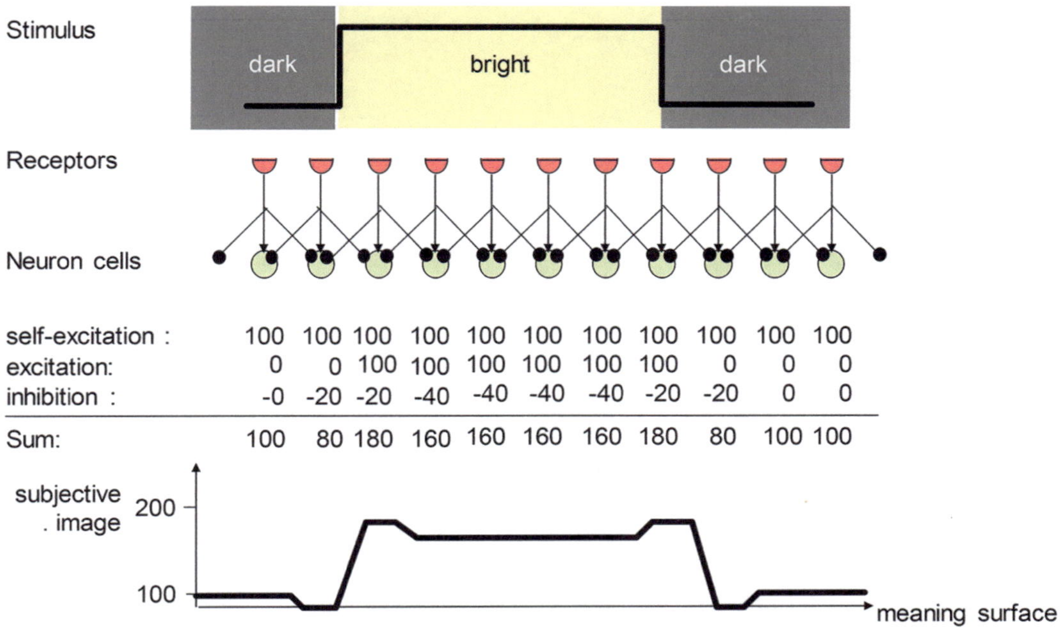

Fig. 3.7 Lateral inhibitory interconnection

intensity. At optical area this occurrence is known as Machian Ring.[7] This system of lateral inhibitory interconnection can be found in all afferent systems. This means that every excitation is surrounded by a zone of inhibition.

In order to understand negative feedback, it should be borne in mind that in each nerve cell a certain time elapses between the excitation at the input and the appearance of the action potential at the output. In negative feedback, the information coming from the receptor passes through one or more nerve cells and is then directly inhibited by one or more nerve cells and returned to a nerve cell further ahead in the chain. In Fig. 3.8, the impulse rate appearing at the output of the system is calculated using the example of two nerve cells (A and B) in the forward branch and one nerve cell (C) in the reverse branch, provided that a suddenly changing stimulus appears at the input of the system. Immediately at the beginning of the process, the incoming pulse rate runs unhindered through the system; the inhibitory effect builds up slowly due to the time delay, so that the pulse rate at the output decreases again over time. The transmission characteristic corresponds to that of a proportional differential element (PD characteristic). This PD characteristic is a general property of human information processing not only at the level of information reception, but also on the further path of processing. A characteristic example is when an unexpected event occurs while driving.

If one compares this negative feedback with the circuit type of lateral inhibition, it can be seen that the former is an *aggravation of the temporal* the latter an *aggravation of the local* contours of a stimulus.

The above-mentioned form of positive feedback interconnection, which can be regarded as the basis of memory, is discussed in the treatment of information processing in ► Sect. 3.2.2.

The circuit types of lateral inhibition and negative feedback are already realized on the lowest level of perception directly in connection with receptor cells. The interconnection

7 This means, for example, that in a correspondingly unfavourable constellation the weak light of a cyclist cannot be detected next to the bright light of a car headlamp if it accidentally falls into the disproportionately brightened area of the car headlamp.

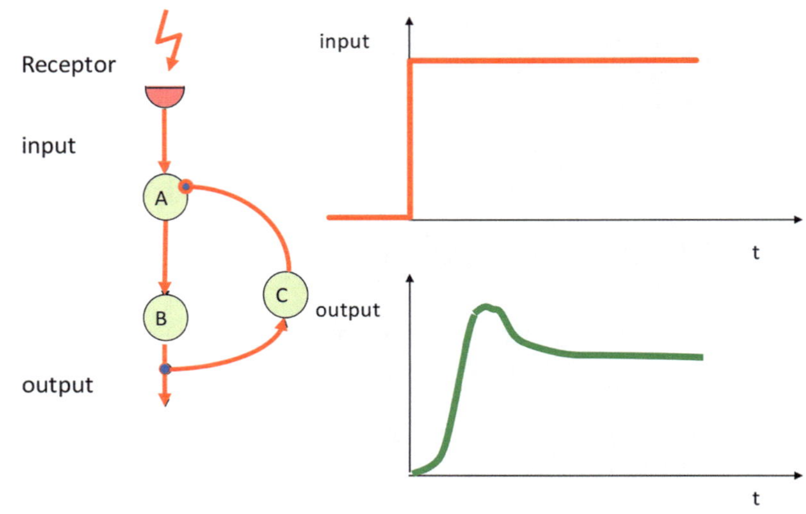

Fig. 3.8 Negative feedback of neuron cells

of several neurons to one neuron cell produces cells that only emit an increased impulse rate if a certain stimulus configuration is present on the sensory surface. We speak of complex cells. These complex cells are further interconnected to form hypercomplex cells, which only show an increased impulse rate when a more detailed stimulus configuration has been defined.

One can imagine that a whole series of specific detectors can be formed in the way shown here. At the level of complex cells, edge detectors and motion detectors have been experimentally discovered, at the level of hypercomplex cells angle detectors and detectors for specific lengths have been discovered (see ► Sect. 3.2.2.1).

An analysis of the discussed neuronal processing on simple, complex and hypercomplex cells shows that from the variety of incoming information more and more specific properties are extracted with increasing complexity of the cells, whereas other properties are ignored. These then lead to the excitation of another specific cell. Fewer and fewer ganglions respond to a given signal in the course of the analysis. The high spontaneous activity rate at the low level of neuronal processing is continuously decreasing towards the higher levels. The cortex is – apart from a permanent "background noise" – at relative rest, which is only sporadically interrupted by short "showers" of activation when relevant external signals appear.[8] It seems to be a basic rule of the nervous system to find changes and differences, to suppress the constant. The latter can be seen as an adaptation mechanism at a higher level.

One can imagine the further course of recognition in this way: The cells, which extract the simple properties in the way discussed above, are switched to even higher cells in such a way that they only react optimally when a

8 Among other things, one tries to measure brain reactions using the method of electroencephalography (EEG). For this purpose, electrodes are attached to the scalp which, like antennas, record the electrical activity associated with nerve activity. This is always unspecific and the signal obtained also depends considerably on the exact position of the electrodes and the electrical resistance between electrode and scalp. One can distinguish very well the state of sleep from the awake state by means of EEG, but hardly different levels of vigilance. The EEG method is suitable for certain experimental projects with some limitations (see ► Sect. 11.2.6.3), but is completely unsuitable for practical use with the aim of monitoring the driver.

There are many attempts to control machines (or also prostheses) "with thoughts" via the EEG. This is possible, but it must be borne in mind that the pacient has to learn very specific patterns of thinking very laboriously, which differ considerably from the patterns of thinking that are necessary for the innervation of the musculature.

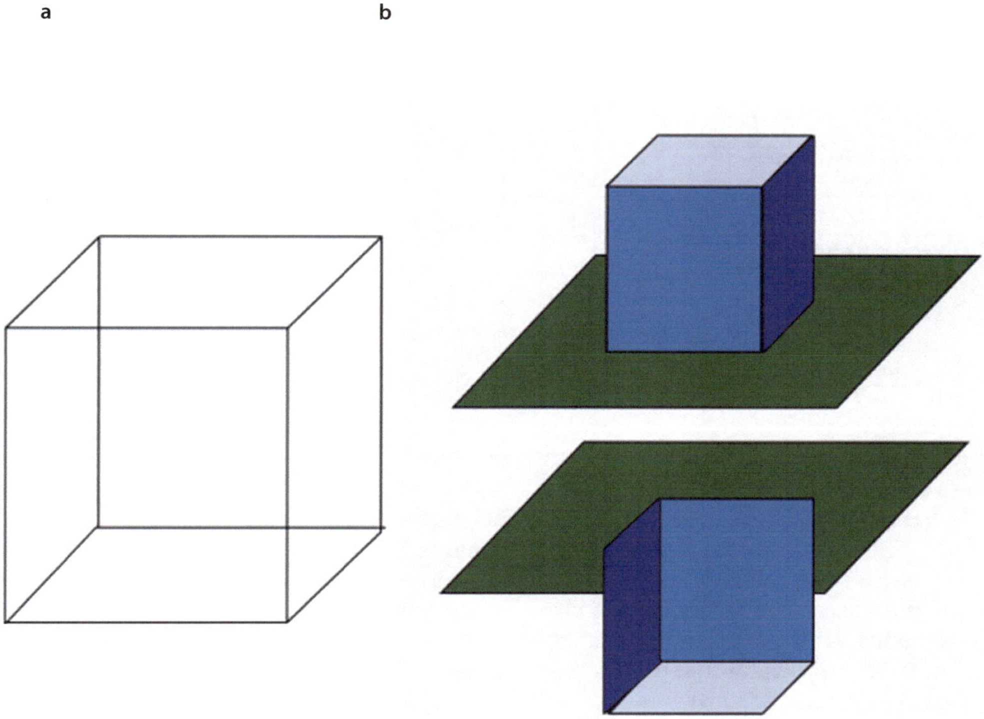

Fig. 3.9 Jumping pattern using the example of the cube **a** and the two possible interpretations **b**

certain pattern given by the switching is mapped on the sensory surface. Several cells can be stimulated at the same time by the same stimulus configuration. For the further recognition process, the one is then selected that allows a self-contained interpretation in accordance with memory contents. One must always decide for only one interpretation. This becomes clear with the example of the well-known jumping pictures. In Fig. 3.9 a line configuration is shown on the left hand side, which we involuntarily spatially recognize as a cube. In fact, the position of the cube can be interpreted in two ways, as it is shown on the right side of Fig. 3.9. The representations there are unambiguous because through redundancy (more information than absolutely necessary; here: Deletion of "invisible" lines, shading of surfaces, representation of a reference surface) the perception is pushed in a clear direction. Redundancy beside others is an important help in the case of a short loss of attention, as it allows additions from the context with the help of stored memory contents. A message that only offers news cannot be followed. In an ambivalent situation, perception always prefers a certain decision to indeterminacy, even if it is wrong to 50% (Sachse 1971). We expect more legality in the course of processes than is often the case in reality. Redundancy is therefore an important design tool for the ergonomic presentation of information (see above all ▸ Chap. 6).

The principles made comprehensible here using the example of optical information recording apply in the same way to all other sensory organs. A model of perception can be derived from this, as shown in Fig. 3.10. Then the stimulus configurations of the outside world (stimulus energy) are subjected to a property analysis by simple, complex and hypercomplex cells (sensory processes). Certain combinations of excited hypercomplex cells lead to the excitation of pattern cells. Using the context, the organism now makes a more or less active decision for certain pattern cells depending on already

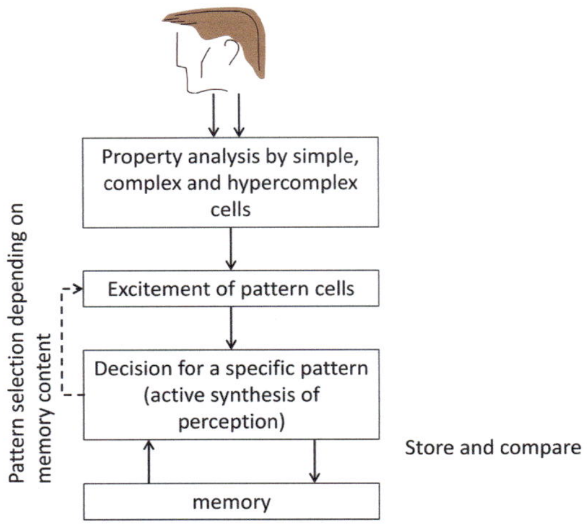

Fig. 3.10 Model of the perception of the environment by the human organism (after Lindsay and Norman 1972)

acquired memory contents. Under certain circumstances, stimulus configurations can be stored in memory, especially if repeated frequently (see also ► Sect. 3.2.2.4).

This model of perception must be supplemented by the integrating processing of information emanating from various sensory channels. The principle of this processing can be explained by the phenomenon of vertical constancy: If you put your head aside, you can still see the world standing upright, even though the image on the retina is twisted.[9] This performance becomes understandable if one considers that the same information of twisting is also provided by the macular organ (see ► Sect. 3.2.1.3). If one now takes the optical information as exciting, and the kinesthetic information as inhibiting interconnected (see **Fig. 3.11**), it is understandable that the resulting information is a fixed environment. By longer rotation around the body axis and sudden stopping, for example, the two information can be separated. Due to the friction-induced rotation of the fluid in the vestibular arches, the kinaesthetic channel provides the information of a rotation, but the visual channel does not provide any rotation

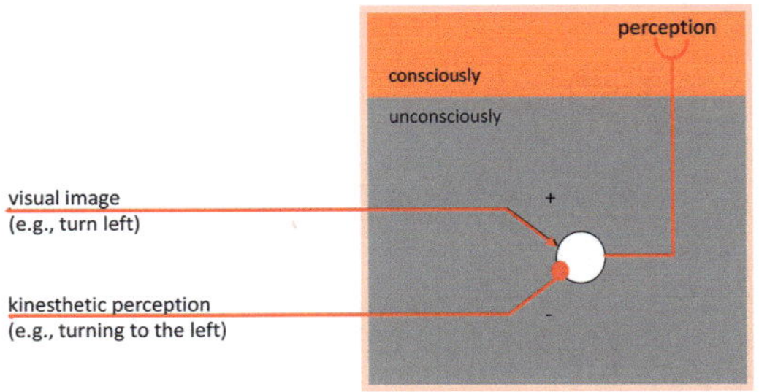

Fig. 3.11 Neuronal interconnection for the power of vertical constancy (Sachse 1971)

of the same direction. The perceived information is therefore a turning of the world in the direction of the original turning.[10]

An unfamiliar relationship between kinesthetic and optical perception causes an individual vegetative effect of varying intensity in all people. This effect is called "seasickness" or correctly "kinetosis". For this reason, many people do not tolerate reading in the car as co-drivers or driving in the back seat where the view of the road is obstructed. All persons who do not tolerate driving as co-drivers do not show kinetosis symptoms when driving themselves, because the optical and kinesthetic stimulus expected from their own activity corresponds to the actual stimulus (otherwise driving would not be possible; see ► Sect. 3.2.2.4). It is interesting to note that even in so-called static driving simulators, which do not perform any movement at all, many people experience massive kinetosis effects.[11]

In addition to the vertical constancy just described, our perception also shows a further

9 This twisting is partly compensated by the fact that the eye is counter-turned accordingly by the innervation of the eye muscles.

10 This also explains the well-known phenomenon of the perception of an apparent movement of one's own when sitting in a resting railway car and observing a train slowly approaching on the side track. The effect only occurs when the train moves very slowly on the sidetrack, so that no stimulus can be expected from the kinesthetic perception.

11 In dynamic driving simulators, even more violent kinetose phenomena often occur. This is probably due to small temporal deviations between the optical information, which can be calculated more quickly, and the slower movement of the simulator's cabin due to the mechanical inertia. Our organism obviously reacts extremely sensitively to such impressions that deviate from our daily experience.

number of constancy performances (here we should mention the color constancy performance, which enables us to recognize the color of objects almost independently of the light color; see ► Sect. 3.2.1.1). In general, it belongs to the special ability of the so-called "constancy performance" of our organism to interpret information from various sensory organs in such a way that object properties are perceived independently of other influencing variables.

Thus, according to the principle described, human builds up the information from all sensory channels into a single perception of his own position and movement in space in relation to stationary and mobile objects. The purpose of this type of processing is to obtain as objective an image as possible of the surrounding outside world inside the organism. From this consideration alone it becomes obvious that frequently encountered recommendations such as e.g. "if the optical channel is overloaded by information, additional information can be transmitted via the acoustic channel" are wrong (see ◘ Fig. 3.3). The information from the different sensory channels must "fit" much more to a self-contained picture of the situation.[12] It is beneficial if the same meaning is transmitted simultaneously via several information channels (e.g. tactile feeling of a switch engaging and audible "clicking"). The statement made here, that by combinatorial processing of the information of the different sensory channels, the human gets a picture of his (possibly supposed) objective position in space, can be generalized beyond the narrower range of environmental perception. In all real actions and mental processes, we always try to get a picture of our situation (also in relation to our fellow human beings, our knowledge, etc.) that reflects the "objective situation". However, this subjective assessment of the "objective situation" does not have to correspond to the objective situation in reality.[13]

Another essential characteristic of human perception, which complements the observation just made, is constructiveness. On the positive side, this describes the amazing ability to understand the environment on the basis of very little sensory information even under adverse circumstances, i.e. to recognise objects or understand words. For vision, this property can be illustrated by ◘ Fig. 3.12. Shown are 8 black circles, in each of which parts of a white grid can be seen. One recognizes however almost immediately the grid of a cube and supplements thereby involuntarily the edges in the white surfaces. This addition is an essential part of the constructiveness (or also of the design law of the continuation of the line; see below). It makes it possible to recognize objects even if only parts of them are visible, which is usually the case in the environment. Another interesting thing about the illustration is that you can see them in several ways and you can arbitrarily switch between the different views. The cube can be interpreted differently in space, so you can see either the left surface in the foreground or the right surface. You can also see the cube against the background of a white area with black spots or vice versa through a white aperture with black holes, i.e. behind the aperture. Both the addition of missing parts and the possibility of interpreting what has been seen in different ways prove that previous knowledge and experience have a decisive influence on perception. The importance of redundancy for the clear interpretation of recorded information has already been pointed out.

Constructiveness is a component of the well-known *Gestalt laws* Psychology, which, through economic information encryption,

12 Through Virtual Reality (VR), which also includes the broad field of simulator technology, technical means are used to give the sensory organs the impression of a reality generated only in the computer. In order to create a particularly good sense of being there, it is necessary to provide as many immersion factors as possible, which is all the more successful the more sensory organs are provided with matching, realistic information (Hofmann 2002).

13 This effect leads besides others to the fact that witness statements e.g. before the traffic court are to be treated with caution, since they do not necessarily represent the actual circumstances, but the circumstances, which the witness arranged himself as the logically correct (unconsciously!).

Fig. 3.12 Example of the constructiveness of visual perception. The cube can be seen in the foreground or behind a perforated surface

"allows us to capture the maximum amount of information by a relative minimum of means and effort" (Kreitler and Kreitler 1980). The Gestalt laws (Knoll 2007; quoted from Goldstein 2002) include

1. that *conciseness law* or law of good form, which states that each stimulus pattern is seen in such a way that the resulting structure is as simple as possible,
2. the law of *resemblance* whereby similar elements are perceived as grouped,
3. the law of *line continuation*,
4. the law of *proximity* ("things that are close together seem to belong together", Palmer 1992),
5. The law of *common destiny* ("Things moving in the same direction appear to belong together") and
6. the law of *intimacy* or law of *significance* ("Things are more likely to form groups when the groups seem familiar or mean something").

Recent studies have added three more, including "common closeness", the "connectedness" factor and the "synchronicity" factor (Palmer 1992).

Although these Gestalt laws can be illustrated particularly for the optical field, they have their outstanding importance as principles according to which we arrange the perceived world, also in the interaction of the information won from different senses. Under this aspect, Knoll (2007) summarizes the laws into the following four principles of our perception, which comprise the above-mentioned Gestalt laws and which he uses as the basis for an aesthetic perception (see also ► Sect. 3.3.4):

1. Maximum effect with minimum effort (principle of simplicity and practicality),
2. Unity in diversity (principle of contrast, harmony of proportions, symmetry and organization),
3. Most Advanced Yet Acceptable (principle of familiarity, curiosity, novelty, complexity),

Table 3.1 Essential principles of human perception and their ergonomic consequences

Principle	Example	Major consequences
Limitation	Viewing: Electromagnetic waves between 400–700 nm Listen: Mechanical oscillations between 20 Hz and approx. 20.000 Hz	Checking the perceptibility of system states Adaptation of signal intensities to threshold values
Psychic dimensionality	Size difference of 1 cm is differently experienced with large and small humans	Verification of the perception of physical differences in signal intensities Selection of signal intensities according to the differences experienced
Constructivity	Object perception despite only partial visibility	Consider previous knowledge, experience and expectations of users
Selectivity	Irrelevant information is ignored Secondary tasks distract when driving	Adapting information output to the driver's resources Support the driver's attention strategies
Constancy performance	Vertical constancy	Ensure road visibility for all passengers. Representations of information on the HUD must be either exactly fixed to the vehicle or (in the case of the so-called contact-analog HUD) exactly fixed to the road.

4. Redundancy and consistency (principle of redundancy and consistency with regard to expectation and reality).

The essential principles of human perception are summarized in Table 3.1. These apply to all human senses, but are differently pronounced there. The relevant characteristics of the individual senses are therefore described in more detail below.

3.2 Elements of the Information Processing Human Being

3.2.1 Information Reception

Tomaske und Fortmüller (2001) has compiled the assignment of sensory channels to the required driver information as shown in Table 3.2. From this alone emerges the dominance of visual perception for driving. Rockwell (1971) and Lachenmayer et al. (1996) also specify that more than 90% of the information important for driving is gained through the optical sense.

The properties of human information reception are quite well understood. For example, the stimulus thresholds and their dependence on the influencing physical parameters are generally known for the various sensory organs. Errors in the recording of information can – generally speaking – occur if the stimuli remain below the stimulus threshold or below the stimulus difference threshold. Errors can also occur, however, if supra-threshold stimuli are not perceived, overlooked or confused in the further processing, which cannot be attributed to the information intake in the narrower sense.

In the following, the individual senses are dealt with and – as far as possible – their properties and limits are presented in connection with the driving task. Due to the connections found in ▶ Sect. 3.1.3, the senses are spoken of decidedly, not primarily the sense organs, because a certain sense often only comes about through the combined interaction of different sense organs. This is especially true for the sense of movement (kinaesthetics), the sense of touch and the sense of warmth. This phenomenon is dealt with in particular there.

Table 3.2 Sense channel assignment of driver information (Tomaske and Fortmüller 2001)

Information	Visual	Vestibular	Tactile	Acoustically
Track deviation	x			
Lateralse velocity	x			
Driving speed	x			x
Longitudinal and lateral acceleration		x	x	
Angle vehicle longitudinal axis nominal course	x			
Yaw rate	x			
Yaw acceleration		x		
Inclination angle	x	x		
Steering angle	x		x	
Forces in actuators			x	
Driving noise				x

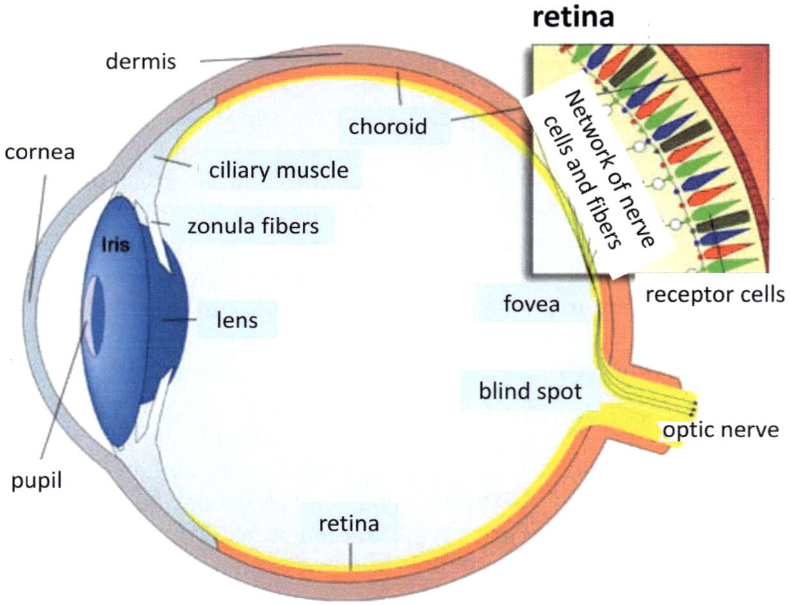

Fig. 3.13 Structure of the human eye (after Funk 2008)

But even in the acoustic sense, for example, which is undoubtedly primarily determined by the impressions of the ear, the entire acoustic sensation is additionally caused by vibration perceptions on the surface of the skin, which are primarily attributable to the haptic sense.

3.2.1.1 Optical Sense

The light waves reflected from the surroundings enter the eye through the cornea, which is about 1 mm thick and acts like a window in the sclera, which is otherwise filled with blood vessels (see Fig. 3.13), are bundled by the refractive properties of the lens and the so-

Fig. 3.14 Quality of the image created on the retina in daylight (Dornhöfer and Pannasch 2000)

called glass body, which forms the shape of the eye, and, according to the laws of geometric optics, create a reduced real image of what is seen on the retina at the back of the eye, standing on its head. Directly in front of the lens is a ring-shaped muscle, the iris, which forms a free opening, the so-called pupil. By innervation of this muscle, which is determined by the intensity of the incidence of light, beside others the diameter of the pupil can change in the range between 1 and 4 mm.[14] Any bad camera would eclipse the image designed there, whereby only in the area of the fovea centralis, which is located on the optical axis directly opposite the lens, a reasonably acceptable image is produced (see Fig. 3.14). The lens itself is connected to the ciliary muscle via ligaments, which pulls it "smoothly" in a relaxed (!) state. With the refractive power of the lens adjusted to the length of the glass body, a sharp image is produced for distant objects on the fundus of the eye within the framework of the conditions.[15] The refractive performance of the lens must be increased in order to sharply image an object in the vicinity (<5 m). Tension of the ciliary muscle causes it to thicken, causing the tension of the ligaments to decrease. Due to the physical properties of the lens, which is composed of a highly liquid gelatinous mass, it now takes on a more spherical shape, shortening its optical focal length. This process is called *accommodation*. As the elasticity of the lens decreases very much with age (effective from about 50 years of age), the effect of the ciliary muscle thickened by the accommodation effort remains ineffective.[16]

The optical system consists of lens and glass body. Like any other optical system, it has the property of refracting light of different wavelengths to different degrees. The short-wave blue light is more strongly refracted than the red long-wave light; i.e. the eye is short-sighted for blue light and far-sighted for red light. This is usually of little

14 Therefore we can see "sharper" in bright sunlight, because the depth of field of the eye is increased by the small pupil like in a camera with a small aperture. If necessary, the otherwise necessary correction of the refractive power by a lens (glasses) can be dispensed with. The iris is also changed by psychological influences ("horror widened eyes"), which can be used for certain experimental projects in a similar way as the already mentioned pulse rate.

15 This applies to so-called normal sight. There are characteristic deviations. If, for example, the refractive power of the relaxed eye lens is too high, the sharp image is produced in front of the retina. This can be corrected by adding a diverging lens (concave lens). Since without this lens only objects close to the eye are sharply imaged, we speak of *nearsightedness*. If the refractive power of the lens is too low, this must be corrected by a focusing lens (convex lens). Since this form of ametropia can be accommodated on distant objects by exertion of the ciliary muscle (see above), one speaks of *farsightedness* (the above-mentioned presbyopia is not to be confused with it). In addition to forms of ametropia, lens shapes that deviate more strongly from the surface of the sphere also have the effect of *astigmatism*. Instead of a focal point, the lens then generates two perpendicular focal lines when light is incident in parallel. This can be corrected by appropriately oriented lenses with a cylindrical surface. All eye corrections should be carried out in such a way that normal vision is achieved. It is defined by a resolution for parallel rays of light from 1′.

16 The advantage of the so-called Head-Up-Display (HUD), in which the picture of a display approx. 3 m-4 m is mirrored in front of the car via the windscreen, is beside others that for older persons the required accommodation performance is significantly lower or is even omitted.

importance, because the eye can adjust to it by slight accommodation effort when viewing objects. However, there are restrictions: When blue objects are on a red background or vice versa, the eye gets accommodation problems. Also monochromatic illumination of instruments, as it is given by the use of LED light sources, can induce reading difficulties – especially with the use of blue illumination.

As shown in ◘ Fig. 3.13, the light-sensitive part of the receptors on the fundus of the eye is directed outwards.[17] This means that the blood supply to the receptor cells and the nerve cords that carry the information from the eye stand in the way of light, so to speak. The nerve cords leave the eye at a common place. No receptor cells can be located at this point. It is called "blind spot." Normally this effect is noticeable by the eye movement (saccades), the fact that the blind spots of both eyes cover different areas of the environment and by the supplementary performance of the brain.

In the retina there are two types of receptors, the cones and rods. The cones are concentrated in the centre of the retina in the central fovea, the site of the sharpest vision (see above). Since there are three types of cones, each with maximum sensitivity for reddish-yellow, greenish and bluish light (three-color theory), colors can only be seen optimally in the fovea. Through the special interconnection of these cone types into antagonistic cells with the bipolar sensations red – green, blue – yellow and black – white, this system can open up the color location of the light source from the sum of the light reflected by different objects in the environment and now change all the captured colors with an internally generated counter color so that the sum of the perceived colors with this counter color result in the color white. With this mechanism, the colours of objects can be correctly recognised to a large extent independently of the colour of the light source. This ability is called color constancy performance.[18]

The rods have a high degree of convergence, i.e. many rods are combined into complex cells. In this way a high light sensitivity is achieved. However, this is accompanied by a loss of resolution and sharpness. Conversely, the cones are less convergent, resulting in very high resolution and sharpness, but lower light sensitivity. The highly concentrated light-sensitive rods in the peripheral area essentially perceive contrast and movement.

As ◘ Fig. 3.14 shows, only in the area of the Fovea Centralis is a satisfactorily sharp image of the environment achieved. This corresponds to a focus range of approx. 2–3°. In order to fix targets that are outside the foveal area, the eye must be turned towards the target objects by jerky movements, the so-called saccades. This can be consciously controlled, but can also happen involuntarily. Another

17 This is a conception conditioned by the developmental history of the eye (see Lamb 2012).

18 This power only works perfectly with a continuous spectrum of wavelengths emitted by the light source in the visible range, as is the case with thermal light sources (e.g. plugs, incandescent lamps and of course the sun). Despite the subjective impression of white light, light from discharge lamps with individual protruding chimneys in the colour spectrum can lead to typical distortions of the colours of the objects seen (example: illumination of road junctions by energy-efficient sodium vapour lamps).

You can easily imagine the effect of the internally generated counter color, if you open the window in a vehicle with green thermal insulation glass (used in almost all vehicles today) only a gap width. The area seen through the gap, which should actually be unadulterated, looks violet (opposite colour to the greenish colour of the thermal insulation pane; see also ◘ Fig. 3.19).

Defects in certain types of cones can lead to hereditary colour blindness, especially in male persons. Red-green blindness is particularly frequent and, despite the impression of an otherwise coloured environment, does not make these two colours distinguishable for the person concerned. As these two colours for "stop" and "drive" were unfortunately chosen for traffic light control by tradition, affected persons can only recognise the display content of a traffic light by the location or even better by a form coding. Since color vision is relatively common, it is important to avoid relying solely on color coding when designing displays.

form of saccade allows "tracking" of the fixed object on the retina. This occurs when the own body or the observed object moves ("optokinetic nystagmus"). A distinction must be made between micro movements of the eye, which can have various causes and characteristics and have an amplitude of a maximum of 10 angular minutes (Joos et al. 2003). An example of this is the so-called tremor, i.e. micro movements of the eye of very high frequency, which presumably serve to stabilize the stimulus on the retina and thus prevent a "fading" of the retinal image (Martinez-Conde and Macknik Hubel 2004). The actual perception takes place during the fixation following a saccade. For the duration of the saccades, however, no information can be gathered, but any characteristics already registered can be processed. The capturing of an image or scene is therefore done by a series of successive saccades and fixations. The finished image is only produced by the processing process in the brain (see ► Sect. 3.2.2.1).

However, the prerequisite for optical information to be processed in the brain at all is the conditions for the recognizability of an object (e.g. of another road user) are given. Provided that possible defective vision of the eye (e.g. short-sightedness, astigmatism etc.) is fully compensated by appropriate visual aids, at least five conditions must be met (Hartmann 1970):

1. The object observed must have a certain minimum contrast with its immediate surroundings: this can be a pure luminance contrast for the same colour or a certain colour contrast for the same luminance. In general, both contrasts will occur at the same time.
2. The object must be imaged in a minimum size on the retina.
3. The object or the surroundings must have a minimum luminance.[19]
4. The eye must be optimally adapted to the currently prevailing facial luminous field density.[20]
5. Objects must be presented for a certain minimum time so that they can be seen.

All these five conditions are interlinked. The coupling of three of the conditions is illustrated by the diagram shown in ◘ Fig. 3.15. On the ordinate, the percentage contrast K is plotted, which, according to Eq. 3.1, is defined as the difference in luminance between object (L_{obj}) and environment L_u related to the luminance L_u of the environment:

$$K = \left(L_{obj} - L_u\right) / L_u \cdot 100\% \tag{3.1}$$

For a dark environment ($L_u = 0$) so a contrast up to $+\infty$ can be given. For a totally dark object ($L_{obj} = 0$) you get a contrast of 100%. On the abscissa, the size of the object is shown in *angular minutes* [']. This takes into account the fact that it is not the actual size or distance of the visual object that matters, but the size of its image on the retina. A parameter is the *environment luminance* where the respective curves represent the border between invisibility (to the left of the curve) and visibility. For a given object size and given contrast, one can create the transition from the invisible to the visible by increasing the environment luminance, that is, practically by increasing the illuminance. On closer inspection, however, it can be seen that the visual performance decreases beyond a certain level when the environment luminance increases. This becomes clear in another illustration of the connection between ◘ Figs. 3.15 and 3.16. The reason for this is the now occurring *glare.* Physiological glare is the term used to describe the fact that scattered light is produced inside the eye (vitreous) at a correspondingly high lighting level, which reduces contrast and thus visual performance. ◘ Figure 3.17 gives a connection according to which at a given glare angle (angle between visual object, eye and glare source) and a given environment luminance (or adaptation luminance), the corneal illuminance caused by the glare source on the eye must not exceed the values given in the diagram in order to avoid physiological glare. Psychological glare, on the other hand, is a subjective disturbance that also depends on the luminance of the glare source, the surroundings, the solid angle at which the glare source is seen and a so-called position factor.

---

19 Luminance L = the light energy reflected into the eye from an object point, measured in cd/m^2.

20 Field of view luminance = total light energy entering the eye, measured in cd/m^2.

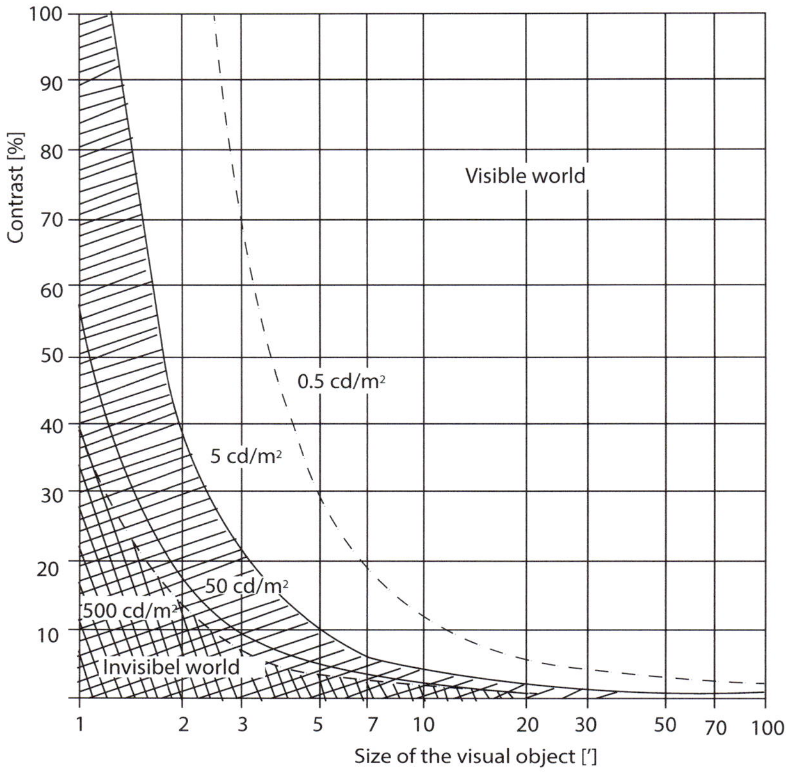

Fig. 3.15 Relationship between contrast, object size and environment luminance (after Hartmann 1970)

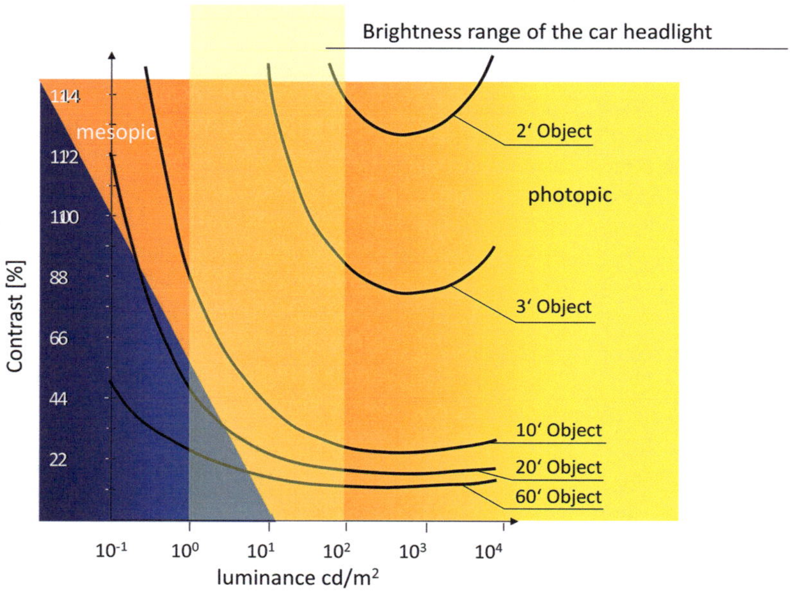

Fig. 3.16 Required object size depending on luminance and contrast. The dark adjustment area (*blue*) and light adjustment (*yellow*) depending on the luminance and brightness range of a car headlamp

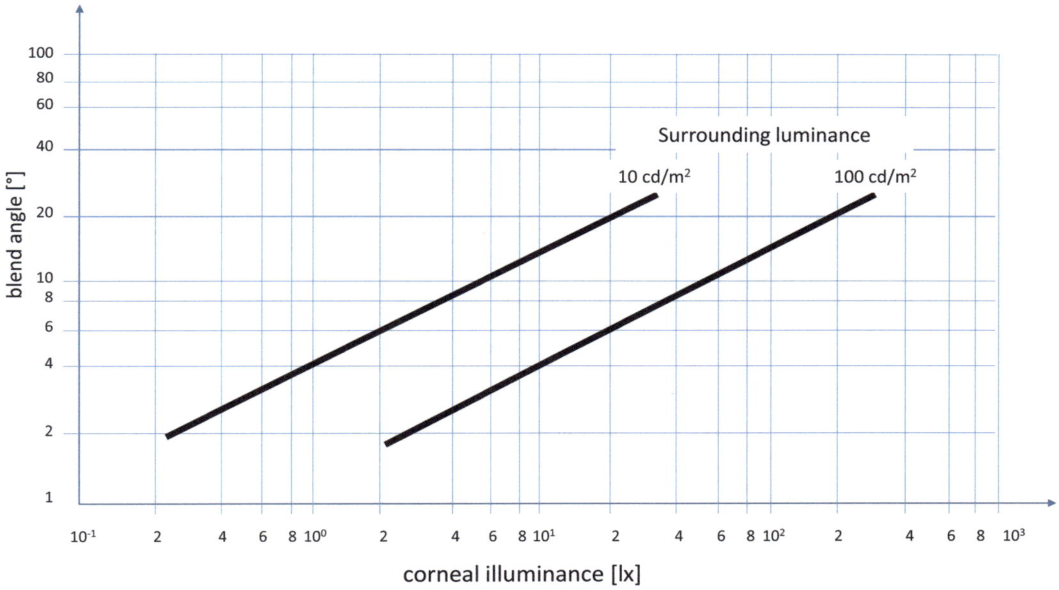

Fig. 3.17 Limit value for the corneal illuminance (lx) caused by a glare source as a function of the glare angle [°] and environment luminance

It usually occurs before the actual physiological glare (for details see Hartmann 1970).[21]

The diagram in Fig. 3.15 shows that even under the best lighting conditions, the resolution of the eye is not better than an angle of 1′ on average. This is due to the distance between the receptors on the retina. In certain individuals, however, a resolution of up to 20′ was observed. The resolution of 1′ defined as "normal vision" corresponds to an object size of approximately 0.1 mm at 33 cm distance, 0.3 mm at 1 m distance and approximately 3 mm at 10 m distance. In practice, however, one should not assume that all people have normal vision. Table 3.3 shows the percentage of a sample of 9468 people who – in terms of distance and proximity – do not reach a resolution before 1′. Important for the interpretation of the table is the indication that the persons who wore glasses at work were also examined with glasses at the respective visual distance. Figure 3.18 demonstrates that wearing glasses does not guarantee optimal correction of visual defects. It follows from all this that ophthalmological

Table 3.3 (according to Hartmann, 1970) Percentage of defective vision in the various age groups; according to Schmidtke and Schober (1967)

Age group (in years)	Percentage of those examined with a visual acuity worse than 1 arc minute	
	Farseeing (8 m)	Close vision (33 cm)
Up to 20 (n = 1329)	28%	15%
21–30 (n = 3193)	V%	32%
31–40 (n = 1857)	39%	35%
41–50 (n = 1497)	55%	53%
51–60 (n = 1203)	63%	68%
Over 60 (n = 389)	65%	56%

21 In most cases of the alleged glare at night driving it concerns the latter. The problem is rather that the mesopic dark adaptation is disturbed and after the strong incidence of light it may take a few seconds until the original adaptation is reached again.

3

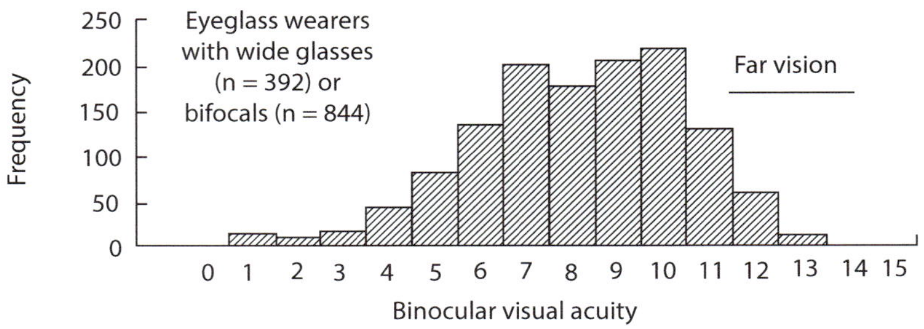

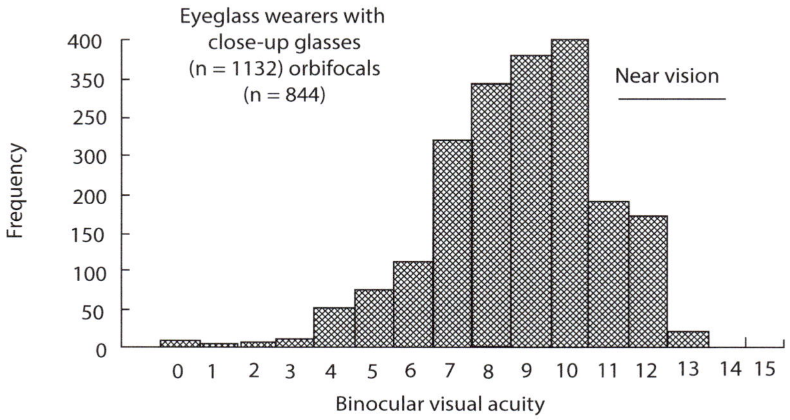

Fig. 3.18 Frequency distribution of binocular visual acuity achieved by spectacle wearers with glasses (according to Schmidtke and Schober 1967), scale value 0 = 10,0′; 5 = 2,0′; 10 = 1,0′; 15 = 0,67′

examination of drivers' visual performance is important.

In addition to the pure brightness contrast, the *colour contrast* to be taken into account. The human eye can distinguish about seven million color valences, which can be divided into two large classes: the colorful and the achromatic colors. The achromatic colours roughly correspond to the already mentioned light-dark contrast. The colorful color valences of the surface colors of objects ("body colors") can be characterized by color tone, saturation and dark or light level. Color tone and saturation are metrically recorded in the known representations of the color triangle (e.g. Standard color chart according to DIN 5033, Fig. 3.19a). From these plates, however, no conclusions can be drawn as to just yet recognizable color differences. A complete analytical recording of this problem is not yet available. However, certain conclusions about such difference thresholds can be drawn on the basis of a deformed colour triangle, as recommended by the CIE in 1976 as a "table of equidistant chromaticity" (see Fig. 3.19b). The color contrast here is characterized by the distance between the respective color locations.

So far only the first three of the five conditions mentioned above have been addressed. The fourth condition contains the optimal *adaptation* of the eye to the current lighting level. In the question of the limits of human performance, the time course for dark or light adaptation is of particular interest here. Figure 3.20 shows the difference in luminance ΔL between the visual sign and its surroundings, which must at least be present so that the visual sign can be recognised at any moment after a sudden transition from a high

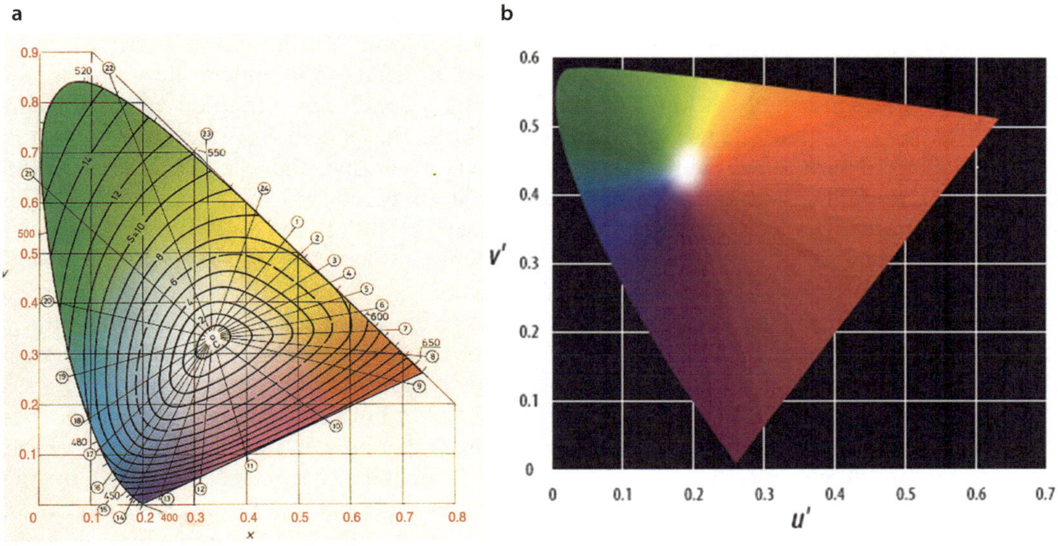

■ **Fig. 3.19** Colour triangle according to DIN 5033 **a** and the UCS colour chart (UCS = Uniform Chromaticy Scale) recommended by the CIE 1976 **b**

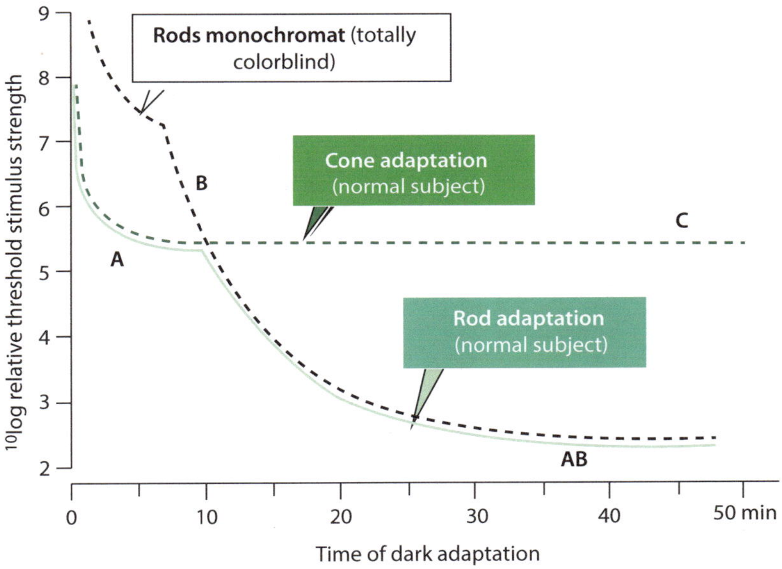

■ **Fig. 3.20** Time course of the dark adaptation of the human eye. The different courses of cones and rods can be recognized

to a low lighting level. The first adaptation is largely due to the cones. After about 10 min, a second adjustment process begins, which is due to the rods. This adaptation takes a little longer and reaches a stable level after about 40 min. In this case, vision is only possible with rods, while weak stimuli can no longer be seen with the cones. In this state, color vision is no longer possible.[22] The course of the curve in ◘ Fig. 3.20 depends not only on the jump height of the adaptation lighting level, but also on the wavelength of the light. In contrast to dark adaptation, the adaptation process runs from dark to light in the range of seconds.[23] The light adaptation takes place mainly via the rods, which are particularly sensitive to blue light.[24]

The perception of brightness also depends on temporal factors, i.e. the question of how long the driver has been in the dark. When driving a car, however, this can change at short notice if the driver is dazzled by other vehicles, is in the urban area with bright lighting, has light in the interior, etc. This adaptation offers advantages for the driver, as it allows for a perception adapted to the respective environmental conditions. From an ergonomic point of view, however, this shifts the optimum depending on the environment, which must be taken into account when designing visual signals and displays. Adaptation processes can be found not only with regard to brightness, but also for the perception of colour and motion.

The fifth condition listed for the recognition of objects refers to the *performance time.* Both by testing the flicker fusion frequency and by observing the drop in electrical activity of the photoreceptors on the retina after exposure, it can be concluded that a minimum exposure time of 150–250 ms is necessary to detect a visual object under otherwise optimal lighting conditions (Lindsay and Norman 1972). If this statement is reconciled with the observation that the eye also changes the viewing direction when viewing moving objects (saccades), the result is that there must be limiting speeds for the recognition of visual objects.

◘ Figure 3.21 shows the experimentally determined relationship between visual angle velocity and necessary size of the visual object. For both vertical and horizontal movement, it can be seen that the necessary object size increases with increasing speed of the viewing angle and thus visual acuity decreases.

In road traffic, the *depth resolution capability* of the visual sense, which is determined in the range up to approx. 20 m (with some persons even up to 40 m) by the binocular seeing. Due to the distance between the two eyes, two different images are produced on the retina of the left and right eye. *(transverse disparation).* Two objects lying one behind the other are seen by the two eyes at different angles. The limit angle up to which depth resolution is still possible is 10″ to 5″ (Trendelenburg and Schütz 1971). Interestingly, such depth resolutions can also be determined if both eyes are optically almost equally incorrectly corrected. In addition to the effect of transverse dispersion, there is also the effect of motion parallax: the movement of the head causes the images of objects at different distances to shift in a characteristic manner, which also promotes depth perception (see also ► Sect. 3.2.2.1). Bergmeier and Bubb (2008), Bergmeier (2008) were able to establish in experiments on the design of so-called contact-analog cHUDs[25] that only from a distance of about 50 m the two effects no longer play a role, but only the size of the represented object in the perspective relation.

22 The observation that in the headlight of one's own vehicle the colour of a vehicle in front or the clothing of a pedestrian can definitely be seen proves that when driving a car, normally no complete dark adaptation is achieved, but only the intermediate form of the so-called mesopic adaptation (see also ◘ Fig. 3.16).

23 The problem of driving into dark tunnels can be seen in this fact. One tries today to defuse this problem by particularly bright lighting at the tunnel entrance. A similar problem occurs when the driver is dazzled by the headlight of an oncoming vehicle at night.

24 Therefore, red or reddish light is often used for instrument lighting (this was also the interior lighting previously used in submarines). In the meantime, however, it has been established that the so-called filtered white light, in which only the blue component is removed by a sharp colour filter, produces considerably better results in terms of readability and prevented light adaptation.

25 Contact-analog Head-Up Displays (cHUDs), which make it possible to display objects at a desired distance, so to speak in contact with the road (see ► Sect. 6.2.1).

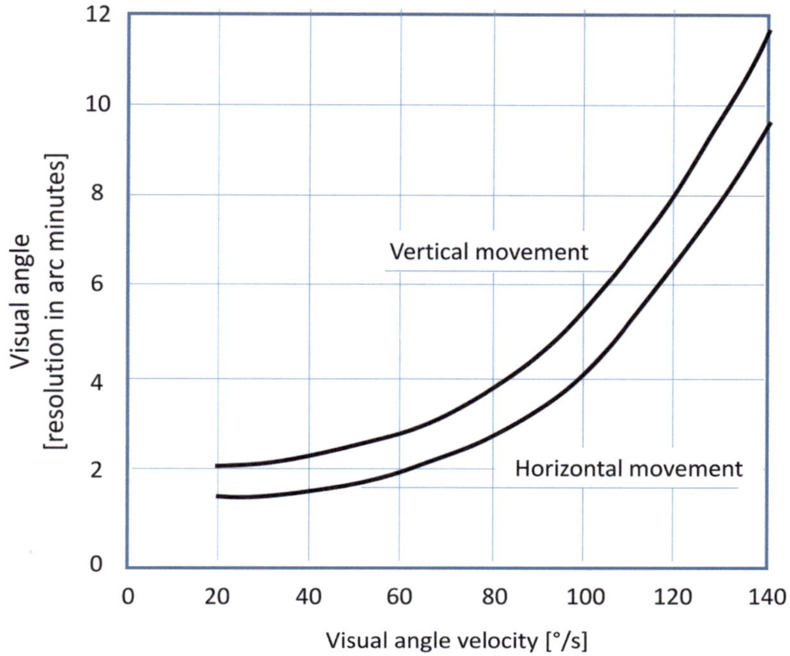

Fig. 3.21 Required object size as a function of the visual angle speed for vertical and horizontal movement

3.2.1.2 Acoustic Sense

With our ear, we are able to sense the frequency and amplitude of sound vibrations (generally airborne sound – through physical contact with vibrating surfaces, structure-borne sound can also be transmitted directly to the inner ear via the bone system) within a certain range. Figure 3.22 shows a cross-section of the human ear in the upper part. In the lower part of the picture a mechanical substitute picture of the ear is shown, which should make it easier to understand its function. The auricle visible from the outside opens into the external auditory canal, which is separated from the middle ear cavity by the eardrum (tympanum). It is connected to the pharyngeal cavity by the so-called Eustachian tube (found by Eustachio 1520–1574). It normally equalises the pressure between the middle ear cavity and the air pressure outside the head.[26] The middle ear cavity contains the ossicle chain, which consists of the elements anvil, hammer and stirrup, so called because of their appearance. As the substitute image illustrates, the bone constellation produces a mechanical transmission ratio. The movement of the eardrum excited by the airborne sound is reduced accordingly in this way and at the same time strengthened by about the ratio 1:22. It acts on the oval window of the cochlea (in the substitute picture the cochlea is drawn rolled up, which is actually not possible because it is embedded in the petrous bone of the bony skull). The inside of the cochlea is divided into three parts by two membranes. The perilymph inside is stimulated to vibrate by the transmission of sound via the ossicular chain. Thus, the middle passage, the so-called membranous labyrinth, which is also filled with liquid, the so-called endolymph, also performs corresponding movements. The actual

26 In some cases it can happen that this compensation cannot take place fast enough, e.g. if one overcomes large differences in altitude quickly by car or cable car or if the Eustachian tube is closed by mucus in the case of a cold. Then the difference in air pressure bulges the in one direction; the effect is that we hear worse, as if through a mountain of cotton wool. By swallowing movements one can restore the balance in most cases.

3

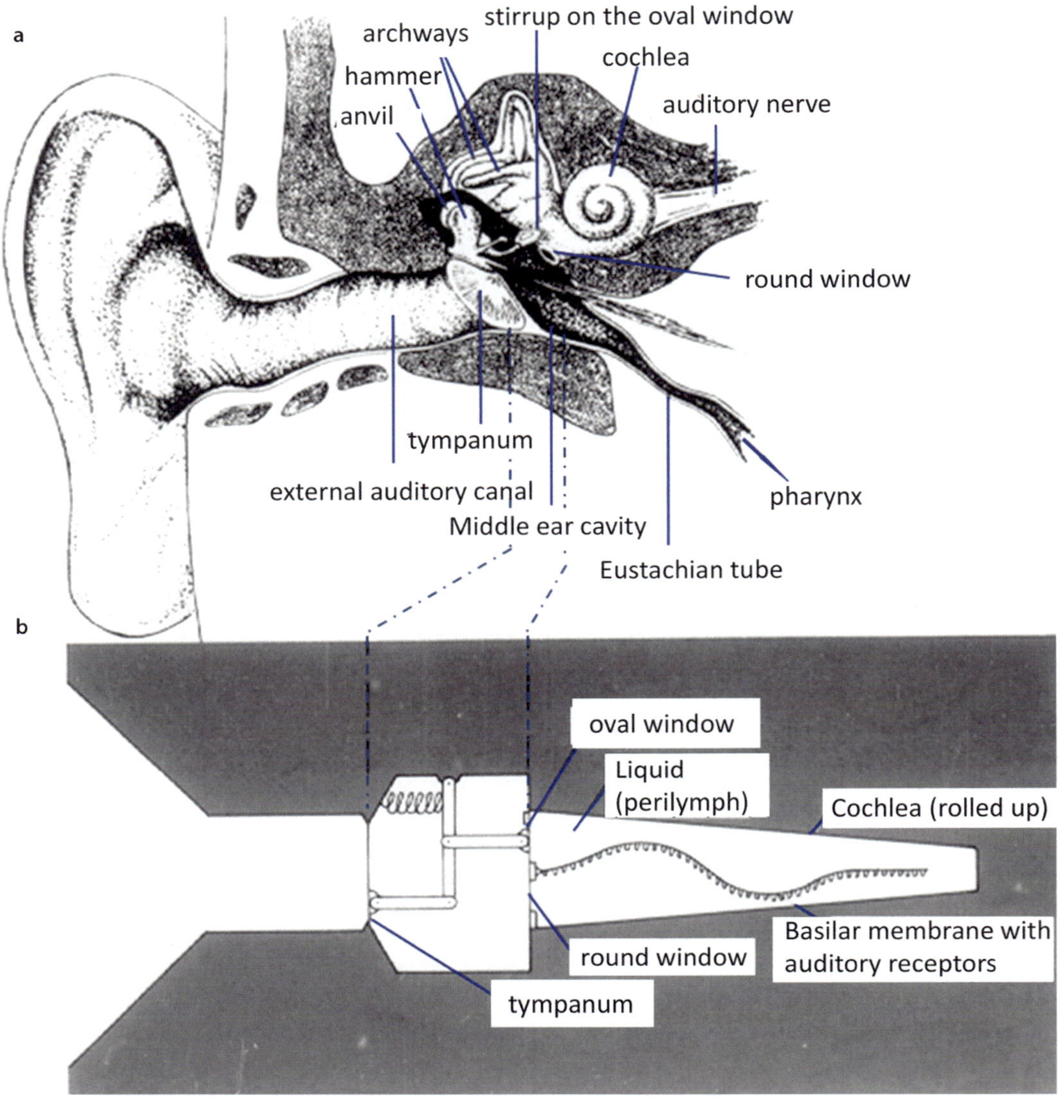

Fig. 3.22 Cross section through the human ear **a** and mechanical model of the ear **b**

hearing organ, the organ of Corti, is located within this central, self-contained corridor, where the receptor cells are surrounded by supporting cells. These are the hair cells that scan the movement of the basilar membrane. Due to a physical process not described in detail here, so-called travelling waves occur on the basilar membrane. The envelope of the incoming and outgoing part of it forms a local maximum depending on the frequency. The maximum high frequencies are located near the oval window and the low frequencies in the cochlea tip. However, this mechanical frequency analysis is very blurred. The information contained in the sound stimulus is recoded several times on its way to the primary auditory cortex, where complex and hypercomplex cells extract different properties.

Figure 3.23 shows the movement on the basilar membrane caused by different sound effects. One can understand that we "see" the local-temporal pattern on the basilar mem-

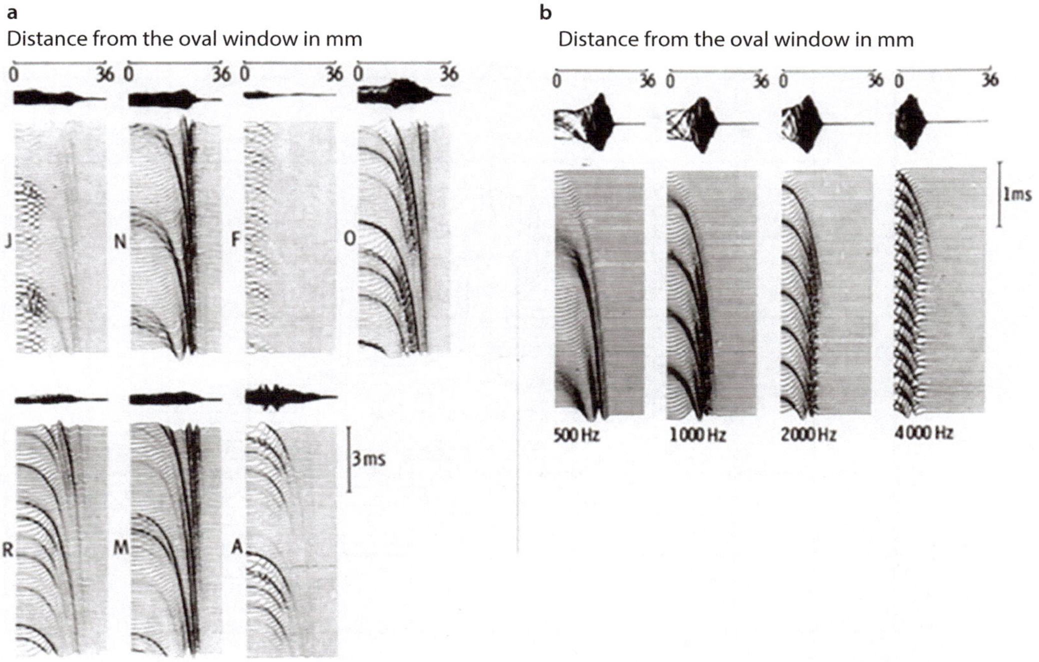

Fig. 3.23 Time course of the basilar membrane deflection during excitation by the sounds "J, N, F, O, R, M, A" of the human language **a** and the frequencies 500 Hz, 100 Hz, 4000 Hz **b** (after David 1972)

brane with our acoustic sensation. The ear thus enables a different differentiation of sound events than is possible by the technically feasible frequency analysis. Among other things, it becomes understandable that the ear recognizes the characteristics of a sound even if the sound is strongly distorted (e.g. differently spoken vowels or even recognizing a musical instrument from a loudspeaker with a moderate frequency response).

The audible range is represented in a diagram, which is spanned by sound pressure amplitude and frequency, as "*Audible area.*" (see Fig. 3.24).

The sound pressure amplitude usually is displayed logarithmically,[27] expressed in dB and called as *Volume level L*:

$$L = 20 \log \Delta p / \Delta p_0 [dB]$$

Δp is the sound pressure amplitude, Δp_0 a reference sound pressure once defined:

$$\Delta p_0 = 2 \cdot 10^{-5} N / m^2;$$

it corresponds approximately to the average hearing threshold at 1000 Hz.

In the low sound pressure range, the audible range is defined by the *hearing threshold* limited. This hearing threshold is frequency-dependent and has a minimum for the healthy

27 This is due to the erroneous assumption of the validity of Weber-Fechner's law. In particular, Stevenson's experiments in the 20s of the last century have shown that, in contrast above 40 dB, an increase in the volume level of 10 dB subjectively corresponds to a doubling of the sensation (referred to as "sone"). This relationship can be described by formula: $E = 2^{0,1(L-40)}$. If you insert the definition of the volume level L into this formula, you will get it after some transformations: $E = \frac{1}{16}\left(\frac{\Delta p}{\Delta p_0}\right)^{0.6}$. In fact, the power law of psychophysics is valid for the description of the connection between perceived loudness E and physical stimulus Δp.

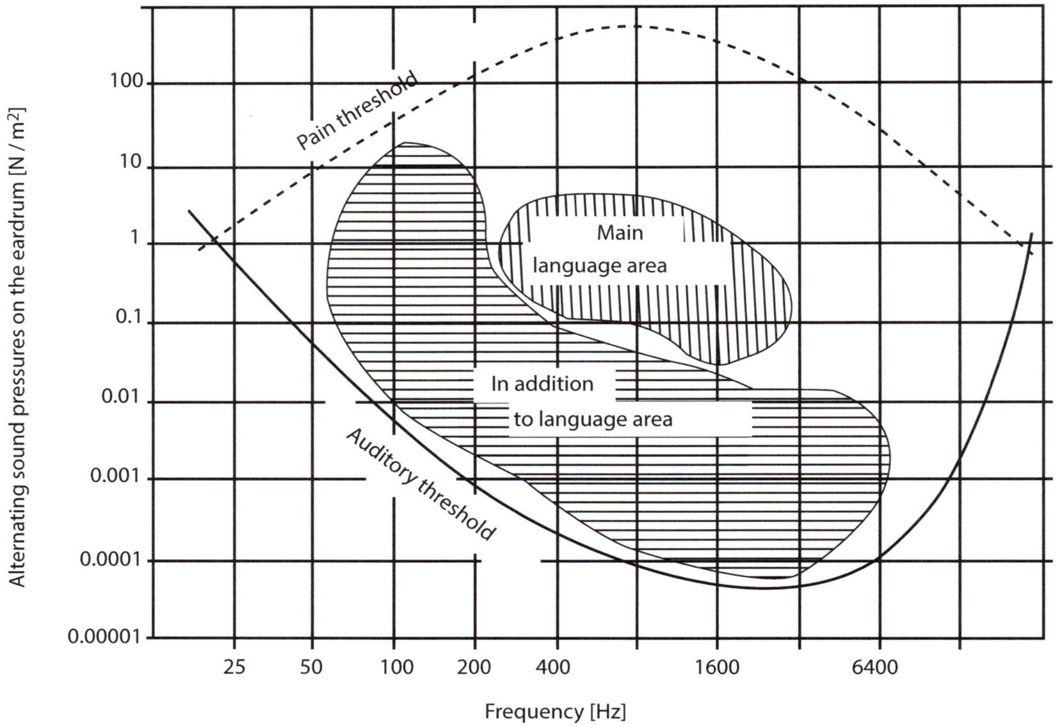

Fig. 3.24 Audible area with the auditory threshold curve and the pain threshold as lower and upper limits respectively

ear of the normal hearing person at approx. 4000 Hz; i.e. the ear is most sensitive there. In the range of high sound pressure, the auditory area is delimited by the pain threshold, where the auditory sensation changes into an unpleasant sensation. The pain threshold is also frequency-dependent. At a tone of approximately 1000 Hz it is approximately 60 N/m^2.

Figure 3.25 shows the volume change in the auditory area (for sinusoidal tones) that was just noticeable. This illustration shows that the volume resolution at very low and high frequencies is relatively worse than at medium frequencies around 1000 Hz. In order to take this frequency dependency at least approximately into account when objectifying volume levels using a sound level meter, weighting filters were defined that attenuate the noises in the various frequency bands in a way that corresponds approximately to the reciprocal of the curves of the same volume. The so-called A-weighting should be used at low volume (20–40 dB), while the C-weighting approximates high volumes (80–90 dB). Irrespective of this dependency, however, it is common today to measure the volume of a vehicle (both interior and exterior noise) using the A-weighting curve (volume specification in dB(A)).

If one offers a test person a second tone at a certain volume of a reference tone (e.g. 1000 Hz) and examines which volume is needed to hear this second tone, the following can be determined: If the second tone has a lower frequency than the reference tone, significantly lower additional volumes are required to perceive this tone than in the opposite case. The second tone is "masked" differently by the first tone depending on this frequency difference. A systematic change of the test tone frequency results in a masking pattern with relatively steep edges to low frequencies and relatively flat edges to high frequencies. This masking effect plays an important role in determining the perceived loudness of noises and is taken into account in the method developed by Zwicker for deter-

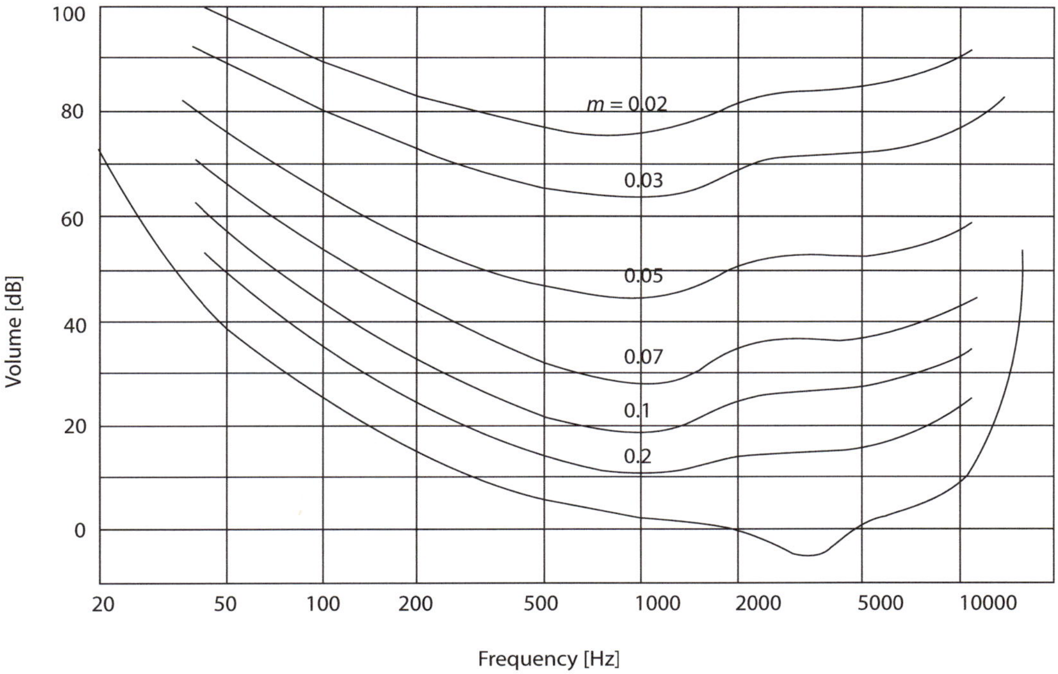

Fig. 3.25 Just now still perceptible volume change m as a function of the frequency at a modulation frequency of 4 Hz; according to Zwicker and Feldkeller (1967)

mining loudness. In the automotive sector, the masking effect plays a significant role, since the relatively low-frequency driving noise can mask numerous undesirable higher-frequency interfering noises (see ▶ Sect. 8.2.2).

The display of further dependencies on the modulation frequency is not required here. Similar correlations to the sensitivity threshold for volume can be found for the *Tone pitch sensitivity threshold* (For details on the whole complex of hearing physiology and psychology see Zwicker and Feldkeller 1967).

Also the *temporal resolution* is a factor limiting the hearing performance. In order to perceive a sound or a noise fully, a time of approx. 180 ms is necessary. Although this time can be explained directly by the physical processes in the inner ear, there is an astonishing agreement with the temporal resolution limit in the visual range, which indicates the coordination of the physical processes in the receptors and the performance limit of nervous processing.

Similarly to the optical channel, the acoustic information channel also has a *direction orientation* possible. Directional hearing is based on the acquisition of time and intensity differences between the two ears. Time differences (approx. 840 μs) from one ear to the other in the range of frequencies <1300 Hz are effective. Volume differences become significant in the frequency range between 1000 Hz and 5000 Hz. By slightly turning the head, it can also be determined whether the sound source is in front of or behind the observer. All in all, a spatial resolution of about 3′ to 5′ is possible in a free sound field.[28]

3.2.1.3 Kinaesthetic and Haptic Sense

The movement of the body is perceived in the interplay between the optical and the haptic information channel by the reaction pressure on the body support surfaces (floor, seat) aris-

28 In the cabin of a vehicle, which is acoustically shielded from the outside world, only very limited acoustic directional orientation is possible, because acoustic signals (e.g. police sirens) cause the surface of the entire vehicle to vibrate, making directional orientation considerably more difficult.

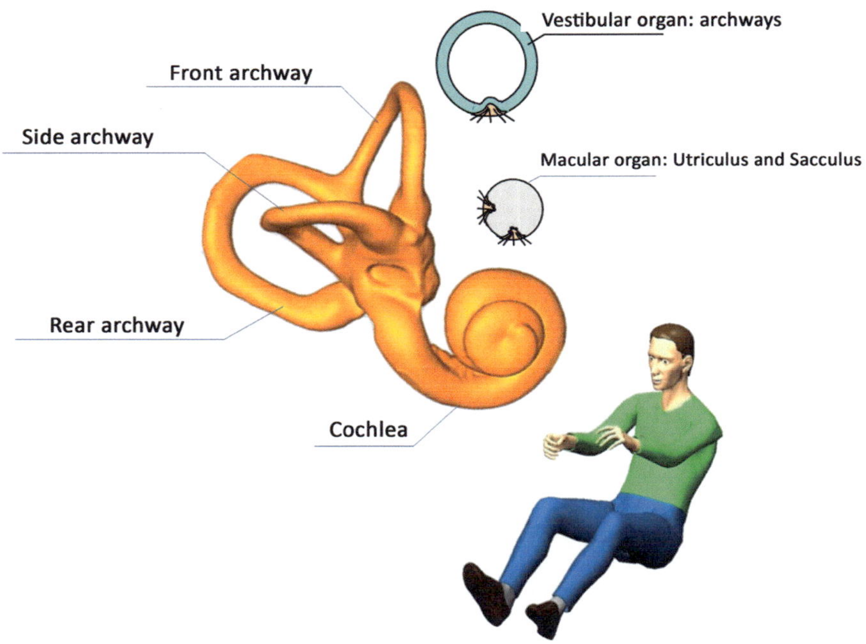

Fig. 3.26 Structure of the sense of balance consisting of vestibular organ and macular organ (computer tomography image of the human inner ear). Source: http: ► www.scientifica.ch

ing during movement and by means of the kinesthetic information channel. The organ of equilibrium (vestibular organ), which provides the adequate information for this purpose, is directly connected to the inner ear. Historically, the ear and the organ of equilibrium have developed from the same source, but then two completely separate sensations can be assigned to the respective partial organs.

Directly in the inner ear, at the foot of the cochlea, are the so-called *macular organs* (statolithic organs; see Figs. 3.22 and 3.26). A statolith organ consists of the otolith membrane, a gelatinous cushion-like mass in which tiny calcite crystals are embedded. On the free surface they carry submicroscopic hairs (cilia) which are connected to sensory cells. Due to the different masses of calcite crystals and otolithic membrane, the sensory epithelium is displaced from its resting position when accelerating forces are applied, thus bending the hairs of the receptors. Two statolithic organs, Utriculus and Sacculus, are located on each side of the head in the inner ear. The macula utriculi being oriented to respond to horizontal acceleration forces and the macula sacculi being sensitive to vertical acceleration forces. In their interplay, they thus enable the perception of translation acceleration in all three spatial directions. In a straight head position, the macula saccules are thus under constant influence of gravitational acceleration. The organism thus experiences information about the position of the head in space via the statolithic apparatus.

The archways are also connected to the inner ear (Fig. 3.26). These are three circularly closed channels filled with endolymphs, which are perpendicular to each other (approximately). At one point, the cupula, which has the same density as the endolymph, protrudes into the interior of the arcade. The cilia of the receptor cells penetrate deep into the cupula. Because of this arrangement, translation accelerations do not influence the organ, but rotational movements, because here the endolymph remains behind due to its

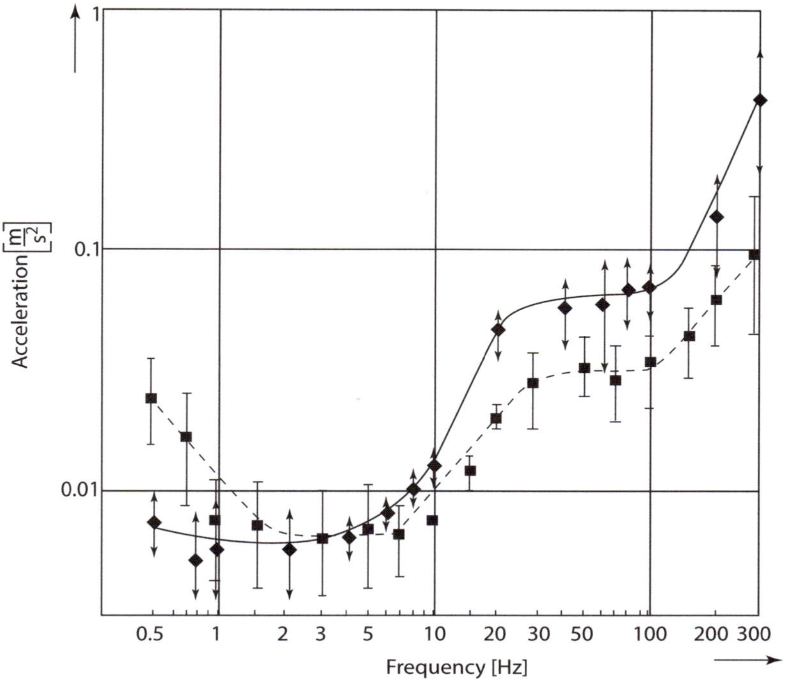

Fig. 3.27 Perception threshold for horizontal (-) and vertical (- -) translation acceleration for the upright seated human being (according to Steward 1971)

inertia in relation to the external movement.[29] However, measurements have shown that the cupula deflection does not correspond to the instantaneous angular acceleration, but to the angular velocity. The endolymph in the archway thus behaves in a first approximation like a torsion pendulum with high damping (Goldberg and Fernandez 1971). Similar to three rotational accelerometers perpendicular to each other, the arc ducts absorb the rotational motion around the three spatial axes.

With this combination of sensory organs, humans are able to detect translational and rotational acceleration forces (including gravity, of course) in size and direction. The threshold values (the lowest acceleration that can just be perceived) for translational acceleration depend on the frequency of movement and the direction of movement. Figure 3.27 shows this dependence for the vertical and horizontal direction of acceleration in an upright person. According to this, humans obviously have a maximum of movement sensation at about 3 Hz.[30]

However, the acceleration thresholds found in the literature are extremely different. Heißing et al. (2000) compiles the threshold values for the three translatory and the three

29 If one is exposed to a rotational movement for a longer period of time (e.g. during a ride in a merry-go-round or by its own rotation, as is the case with a popular children's game), the endolymph remains at rest due to the friction on the walls of the arcade in relation to the latter. If you suddenly stop the movement, the Endolymph will continue to move for a short time because of its inertia. So you perceive a rotation that doesn't actually exist. Because of the neuronal interconnection for the performance of vertical constancy (Fig. 3.11), a rotational movement must now be seen visually which would compensate for the supposed kinesthetically perceived one (because of the nervous interconnection of the information from the vestibular organ, among others, with the eyes, additional nystagmus movements of the eyes occur).

30 When designing the suspension and in particular the upholstery of seats, it is precisely this frequency range that is tried to be damped as well as possible.

Table 3.4 Kinesthetic perception thresholds according to Heißing et al. (2000)

	Movement pattern	Threshold
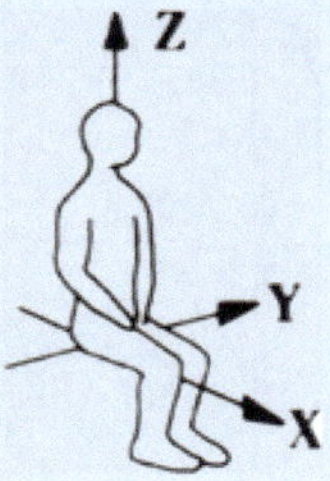	Yaw α_Z	0.05–5 [°/s^2]
	Roll (Wobble) α_X	0.1–0.2 [°/s^2]
	Nod to α_Y	0.1–0.2 [°/s^2]
	Logitudinal X	0.02–0.8 [m/s^2]
	Lateralsal Y	0.05–0.1 [m/s^2]
	Vertical Z	0.02–0.05 [m/s^2]

Table 3.5 Threshold values for yaw acceleration (according to Wolf 2009)

	Minimal	Maximal	Mean
50th percentile (median)	0,2 °/s^2	1,0 °/s^2	0,63 °/s^2
95th percentile	0,990 °/s^2	3,9 °/s^2	1,98 °/s^2
5th percentile	0,035 °/s^2	0,132 °/s^2	0,086 °/s^2

rotational forms of motion in space from a literature review given in Table 3.4.

In the literature, the rotational speed instead of the rotational acceleration is often specified as the threshold value for rotational movements. This is astonishing as the physiological structure of the vestibular organ should only allow for accelerations. Wolf (2009) argues that this is due to the inertia of the perilymph's response to acceleration processes, an effect that is further increased by the friction of this fluid against the vessel walls.

To determine the threshold values of the yaw acceleration (rotation about the vertical axis z), numerous investigations were already carried out at the beginning of the twenties. Wolf (2009) compiles all the values found up to 1965. The result is very different. He summarizes the found values according to Table 3.5. Because of the asymmetric distributions, he calculated the median value and the 5th and 95th percentile for the mean value and the minimum and maximum values respectively.

Regarding the threshold values for the roll acceleration (lateral rotation), Gundry (1978) observed that the greater the roll accelerations, the shorter the detection time. With an exposure time of 100–150 ms, a minimum acceleration of 10°/s^2 is then required necessary to perceive the movement process; with an exposure time of 1 s, this value drops to 1°/s^2. According to observations by Muragushi et al. (2006), the perception thresholds depend considerably on whether the experiments were performed with or without visual feedback (Fig. 3.28). The visual feedback then significantly lowers the threshold of perception for yawing and nodding, while increasing the threshold of perception for lateralse and vertical movement.

In order to provide information about the position of the skeletal system, the muscle spindles, which serve as measuring elements in the muscles of the arbitrary motor system for their elongation, play an important role, as do the joint receptors (see ▶ Sect. 3.2.3). In this way, forces (inertial forces as well as reaction forces) acting on the skeletal system are also recorded. Since these forces can only be transmitted through contact with the environment, the mechanoreceptors of the skin also contribute to the sensation of movement. According to Schmidt (1995) there are four different types of mechanosensors, namely the Merkel cells, the Ruffini corpuscles, the Meissner corpuscles and the Pacini corpuscles. These receptors differ with regard to the

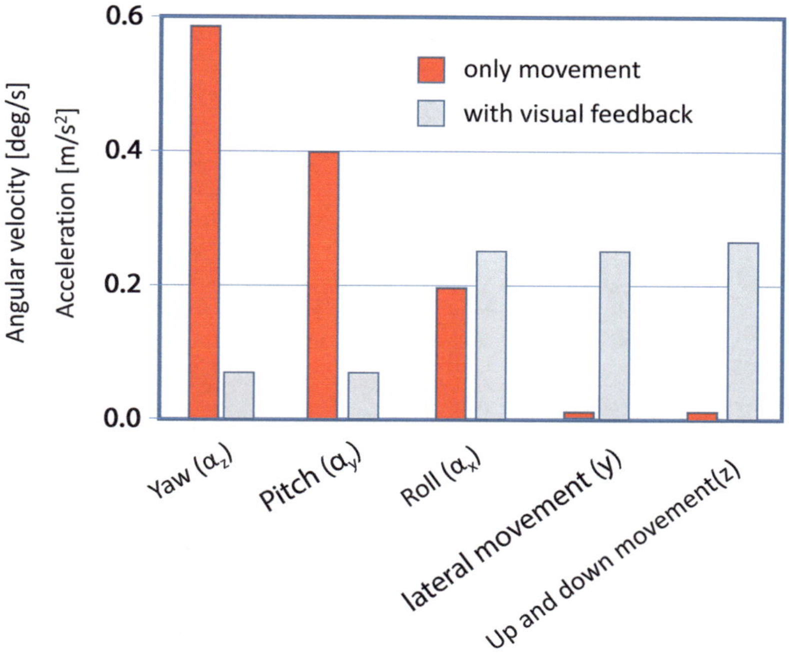

Fig. 3.28 Perception thresholds only based on movement information and with visual feedback (according to Muragushi et al. 2006)

so-called receptive field, i.e. the size of the surface on the skin, which leads to a sensation localized there, and with regard to the adaptation speed. A rough distinction is made between the slowly adapting Merkel and Ruffini receptors, which react constantly to a permanent stimulus (so-called proportional behaviour – P-behaviour) and the rapidly adapting Meissner bodies and Pacini corpuscles, which only respond at the beginning and end of the stimulus. In particular, the Meissner bodies show a simple differentiating behaviour (they only react to pressure changes; differential behaviour, D-behaviour) and the Pacini cells a differential behaviour of second order, i.e. they mainly react to pressure changes (acceleration; D^2-behaviour) and especially on vibration. Beyer and Weiss (2001) assign the haptic qualities and characteristics listed in Table 3.6 to the sensor types. The dynamic behaviour is given according to Schmidt (1995; compiled according to Reisinger 2009).

The threshold for punctiform pressure stimulation is $3 \cdot 19^{-9}$ J (Müller-Limmroth 1981). However, this threshold value is not of great practical interest as it can be exceeded in most cases by arbitrary application of force. In contrast, the tactile spatial resolution is of greater importance, as it can be used to determine up to which resolution surface structures can be captured via the haptic channel. For two punctiform simultaneously mediated stimuli ("simultaneous space threshold"), this limit lies at the tip of the tongue at 1 mm, at the fingertip at 2 mm, at the lips at 4 mm, at the forearm at 40 mm and on the back at 70 mm (Müller-Limmroth 1981; see also Fig. 3.4).[31]

Figure 3.29 shows the course of the threshold for vibration sensation as a function of the vibration amplitude of the skin and the

31 The information is of importance for the haptic differentiability of control elements, e.g. however, when it comes to measuring the pressure distribution on the seat surface (see ► Sect. 11.2.3.1), a relatively coarse grid of measuring points is obviously sufficient.

3

Table 3.6 Properties and assignment of haptic qualities to the various mechanical sensors and their typical transmission behavior

Mechanosensor	Adaptation speed	Haptic qualities	Transfer behavior
Merkel cell	Adapting slowly, small receptive field with sharp boundaries	Sensitive to rising and persistent stimuli, but especially high dynamic sensitivity.	P-behavior
Ruffini corpuscles	Adapting slowly, large receptive field with blurred boundaries	Sensitive to increasing and continuous stimuli, but especially to high dynamic sensitivity. Particularly precise reproduction during the stimulation period.	P-behavior
Meissen corpuscle	Quick adaptation, small receptive field with sharp boundaries	Sensitive to pressure increase on the skin.	D-behavior
Pacini corpuscle	Quick adaptation, large receptive field with blurred boundaries	Sensitive to accelerations or higher discharges such as vibrations. Both deflection and reset of the stimulus.	D^2-behavior

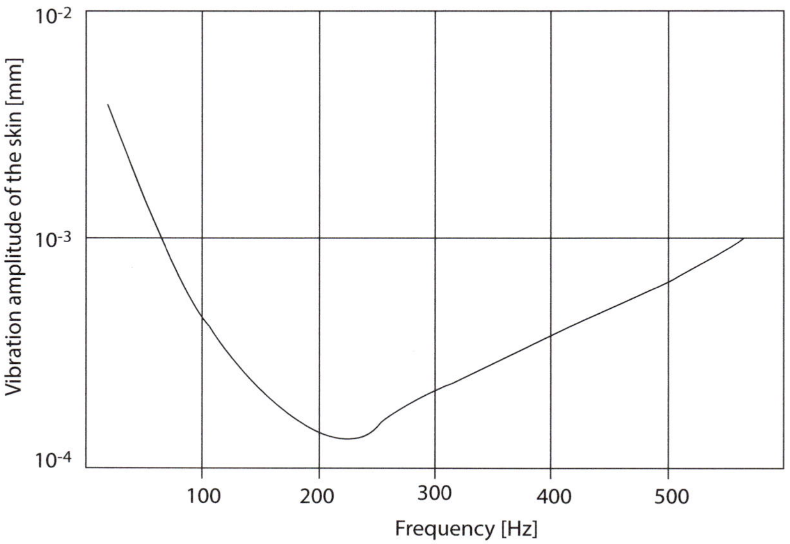

Fig. 3.29 Frequency dependence of the vibration threshold, according to Müller-Limmroth (1981)

frequency. There is a very high vibration sensitivity at 200 Hz.[32]

The sensory organs described here, in their combination, make it possible to sense the mechanical relation of the body to its surroundings. This relation causes different qualities of perception (in analogy to acoustic perception, for example, where we are not able to describe the amplitude and frequency composition of a sound, but qualities of perception such as "grumpy", "bright", "dull", etc.). It therefore makes sense to verbalize the combinatorial effect of these sensory organs with regard to the quality of perception.

32 The transmission of frequencies around 200 Hz to the human body (e.g. by touching the floor or the steering wheel) can therefore be well absorbed, as they lead to a strong discomfort sensation.

In the literature, different conceptual worlds have emerged for the field of sensory perception of mechanical forces. In English psychologically oriented literature, "kinesthetic" is often used to describe the interplay of all positional and movement receptors of the human body (Handwerker 1993). It thus characterizes the perception of the movement of our body parts. In this context, the term "tactile" refers to the mediation of a sensation solely through the skin's touch receptors.

This very widespread terminology has the disadvantage that it only insufficiently takes into account the quality of sensation, i.e. the type of subjective perception of physical force stimuli. From our subjective experience, we can very clearly distinguish between acceleration forces that are based on the *whole body* and which lead to the sensation of a movement of the body in space, from those in which forces only affect *individual body parts* and which lead to the sensation of body posture and movement. When cornering, for example, the acceleration forces are perceived as pressure not only by the vestibular organ (rotational acceleration) and the macular organs (lateralse acceleration), but also by the mechanoreceptors in the buttocks, which in circles of chassis specialists leads to the semi-scientific term "popometer" for the ability to feel the movement of the vehicle through stimuli in the skin surface. If we now actuate the control lever or setting lever of the heating system during this curve travel, then the associated body posture is felt as completely independent of the reaction forces of the movement. On the other hand, we experience the surface quality of objects we touch in combination with the perceived form as a completely separate quality of perception from that described above. The traditional term "sense of touch" describes this type of perception much better: we are able to grasp the shape and type of surface of touched objects independently of movement influences and also of postures. This, too, is a type of constancy performance that Kühner et al. (2011) also demonstrated experimentally. However, both the mechanoreceptors in the skin and the position receptors in the joints and muscles as well as the thermoreceptors interact in a very specific way for this quality of perception. The thermoreceptors, for example, provide information about the thermal conductivity of the objects in contact. This creates the impression that an object feels "cold" (an object with high thermal conductivity, such as metal or an object wetted with liquid) or "warm" (an object with low thermal conductivity, such as wood or polystyrene). This sense of touch comes about in detail in a very complex way: for example, small movements across the surface ("touching") and the frequencies generated play a decisive role in the perception of the surface quality.

It therefore makes sense to define it in terms of ergonomics:

- The *kinesthetic perception* allows the organism to receive the body's own movements in space. Information from the vestibular organ, the macular organs, the position receptors in muscles and joints and the mechanoreceptors in the skin are adequately calculated.

  On the other hand, it should be specified:
- The *depth perception* enables the organism to record body posture and movement independently of external forces. For this, the information from the muscle spindles and the joint receptors with information from the vestibular organ, the macular organ and the touch receptors is processed in a suitable way.

  Finally, it remains to be defined:
- The *haptic perception* allows the organism to detect the shape and surface consistency of touched objects. For this purpose, the information from position receptors in muscles and joints (especially the fingers!), the mechanoreceptors in the skin but also the thermoreceptors in the skin surface (essentially cold receptors) are adequately offset against each other (see Revesz 1950). This results in complex sensations such as soft, hard, smooth, rough, moist, dry and sticky.

3.2.1.4 Thermal Sense

Human being is seen from his tribal history as a tropical being. Accordingly, he is equipped to live in a warm or hot climate. The core temperature must not deviate significantly from approx. 37 °C. The basic physical-physiological measurements of the human heat balance and temperatures at representative body parts by Benzinger (1979) can be summarized as follows:

- For the perception of the thermal state there are no heat flow sensors, but only temperature sensors, so-called thermoreceptors. These emit electrical pulses with different pulse rates. A distinction is made between cold receptors, which respond more strongly below a certain temperature threshold, and warm receptors, which respond more strongly above a certain temperature threshold, each with an increase in the pulse rate. Accordingly, we have a *sense of cold* and a *sense of warmth* (Schmidt 1979). On the skin surface we have far more cold receptors than warm receptors.
- Thermal discomfort caused by cold is perceived via the cold receptors in the body surface (skin) – when the skin temperature falls below a certain threshold value (approx. 34 °C). As the skin cools down, the metabolism increases according to the impulse rate of the skin's cold receptors. At about 17 °C the feeling of cold changes into a feeling of pain. The warm receptors on the skin signal only a pleasant warm sensation when the temperature threshold is exceeded. Pain receptors react to unpleasantly high skin temperatures. Thermal discomfort caused by heat, combined with sweating, is essentially perceived via the warm receptors in the temperature control centre in the brain stem – when its temperature exceeds a certain threshold value (approx. 37 °C). The welding rate follows the pulse rate of the warm receptors in the brain stem. On the other hand, the sweat rate can be reduced if the skin simultaneously falls below the "cold threshold", e.g. by draught.

According to Benzinger (1979), thermal comfort can be defined as "the absence of such impulses from the cold receptors of the skin and the warm receptors of the temperature control centre in the brain stem that give rise to the desire to change the thermal environment". Or to put it simply: "Thermal comfort is given when neither the skin temperature threshold of approx. 34 ° C nor the stem brain temperature of approx. 37 °C is exceeded".

Figure 3.30 shows the skin temperature and eardrum temperature (representative of the stem brain temperature) of an undressed young man at rest, depending on the ambient temperature. The hatched areas indicate the thermally uncomfortable temperatures. The thermal comfort is given both with regard to the too-cold sensation (via the skin temperature) and the too-warm sensation (via the core temperature) at the same ambient temperature of 30 °C: there exists "complete thermal comfort".

Benzinger demonstrated the state of "relative comfort" for a higher degree of activity and otherwise identical experimental conditions, as given in Fig. 3.30. Here the simultaneous and equally strong response of cold receptors of the skin and warm receptors in the brain stem is perceived as more comfortable than the response of only one type of themore receptor (e.g. draught when sweating).

The fact that the ideal ambient temperature measured in Fig. 3.30 is so far removed from the ambient temperatures usual in the vehicle (comfort is achieved in a room with almost no air movement at 50% humidity at approx. 21–22 °C) is essentially due to the fact that the test person was undressed. Our body tries to keep the skin surface temperature at about 34 °C through various climate regulation mechanisms. Essentially, the production of sweat is available for this purpose, with which heat energy can be extracted from the body surface through evaporation and muscle movement, with which heat is generated because of the low efficiency of the muscles of approx. 25% ("cold trembling"!). The heat output is essentially influenced by the given air humidity (at 100% air humidity no evaporation can take place any more) and the so-called wind speed (air movement; with increasing wind speed effects of heat transport increase by convection and evaporation). In addition, there is the effect of heat radiation: If the surface temperature of the environment is significantly lower than the skin surface temperature, heat is emitted, while

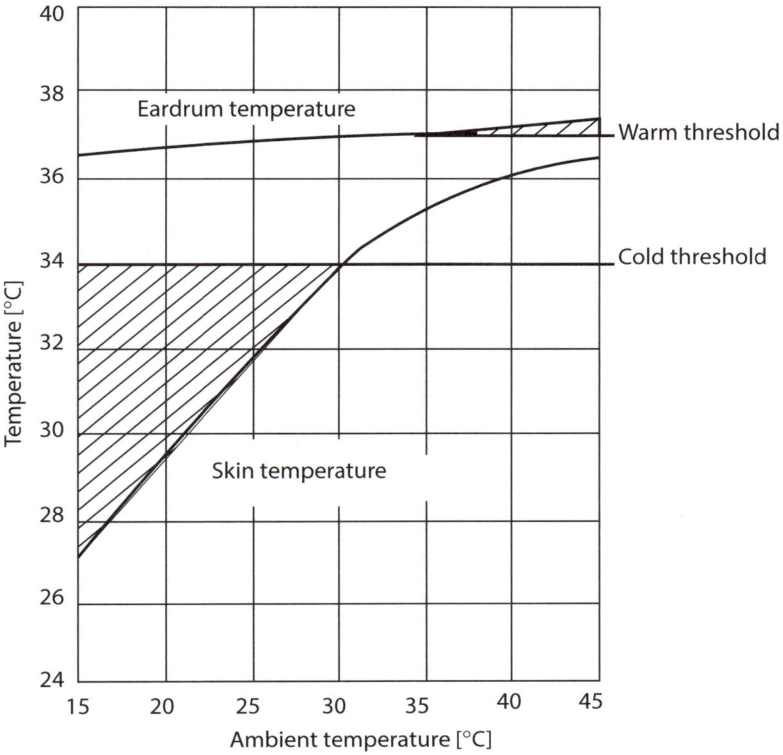

Fig. 3.30 Skin and eardrum temperature as a function of ambient temperature, recorded on a resting, undressed young man (according to Benzinger 1979)

conversely, if the surface temperature of the environment is higher than 34 °C (= 307 K), heat is absorbed. Through clothing we try to compensate the big fluctuations of the outside temperature and to keep the skin surface temperature of 34 °C.

In summary, the following six physical factors influence the human heat balance: physical activity, clothing, air temperature, surrounding surface temperature, air velocity, humidity.

The reason for the often low individual acceptance of given room climate conditions is the fact that the mentioned temperature thresholds are subject to interindividual scattering as well as daytime fluctuations. This was demonstrated by an experiment at the Fraunhofer Institute for Building Physics for four and five persons respectively (Fig. 3.31). The cold threshold values on the skin and the warm threshold values, measured on the eardrum, are shown during the course of the day (more precisely described in Mayer 1986). In addition to the interindividual fluctuations, a pronounced daily variation for the temperature thresholds can be seen.

The significance of skin temperature for the sense of comfort has been tested on about 50 subjects by Mayer and Schwab (1990). For this purpose, these different ambient temperatures were exposed, the neck temperatures were measured and the evaluation of the over-cold or over-warm sensation was queried. The curves shown in Fig. 3.32 show the percentage of dissatisfied (PD) subjects who complained about overcooling or overwarming sensation at the neck, depending on the neck temperature. It can be seen that even very small variations in skin temperature lead to considerable PD values. A temperature fluctuation of only +0.5 °C already corresponds to PD values of 30% (solid curve

3

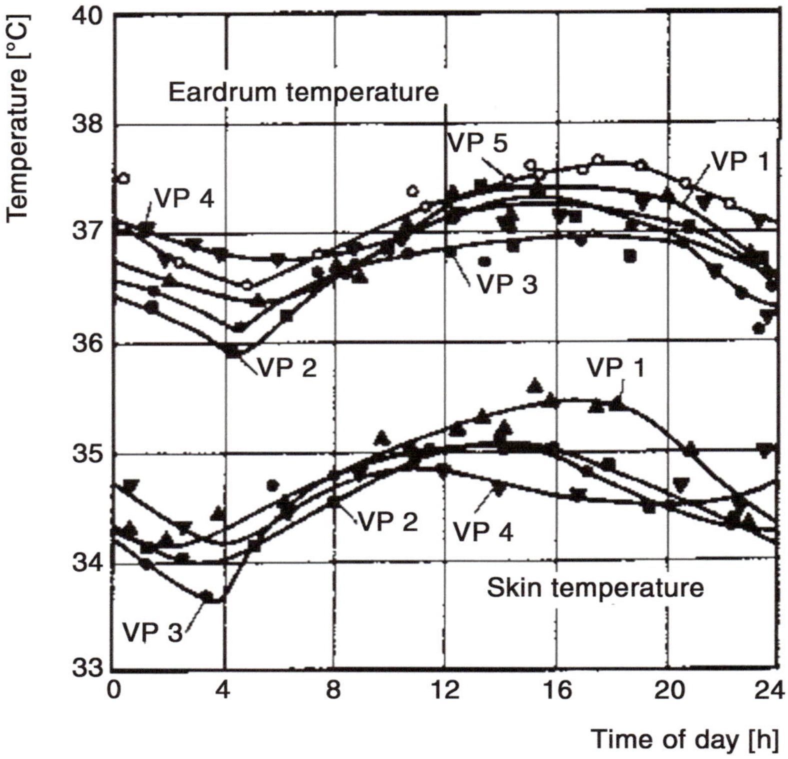

Fig. 3.31 Daily variation of the temperature threshold values for cold sensation, perceived via the skin temperature and for warm sensation, perceived via the core temperature (measured at the eardrum; Mayer 1986)

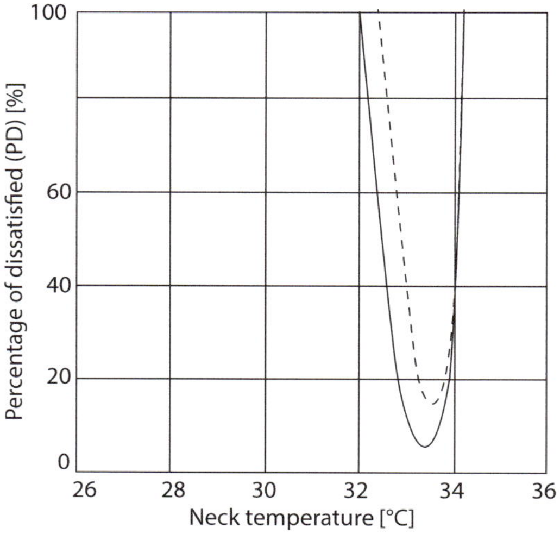

Fig. 3.32 Percentage of test subjects who complained about over-cooling or over-warming sensation at the neck, depending on the neck temperature (continuous curve according to Fanger 1992; dashed curve according to Mayer 1998)

according to Fanger) and 80% (dashed curve according to Mayer et al.).[33]

3.2.1.5 Sense of Taste and Smell

The sensation of smell and taste is one of the oldest sensory systems in the history of development. The receptors for the sense of taste are located in the mouth and throat area. The receptors are taste buds with a clear PD characteristic. There are four taste qualities to distinguish, whose respective receptors are localized as follows: At the tip of the tongue the taste quality is "sweet", at the lateral edges

33 The findings listed here prove the enormous importance of air conditioning systems in vehicles for comfort. On the one hand they show the necessity of individual adjustability, but on the other hand they also show the demand for technical realization, if one considers the mentioned mechanisms of heat transport from and to the human body.

of the tongue "sour", at the tip of the tongue towards the lateral edges of the tongue "salty" and at the base of the tongue (palate) "bitter". The source of the stimulus must be in the immediate vicinity or in contact with the sensory organ. The differentiated taste of food only comes about in connection with the olfactory receptors.

In contrast, the sense of smell is much more differentiated. It reacts to contact with gaseous molecules, whereby its sensitivity is very high. We have approx. ten million receptor cells, which enables us to distinguish and classify several thousand odours, which can be roughly divided into 6 classes (Keidel 1975):[34]

1. spicy (pepper, ginger),
2. flowery (jasmine),
3. fruity (apple ether),
4. resinous (smoking resin)
5. rotten (hydrogen sulphide),
6. flammable (tar).

The olfactory receptors also have distinct PD characteristics. As the air in the olfactory mucosa practically remains stationary during normal breathing, only a few fragrances can reach it. Ventilation ("sniffing") results in a better olfactory performance. When assessing odours, a distinction must be made between the perception threshold at which an unspecific odour sensation is triggered and the recognition threshold at which the odour can be identified. The threshold concentration varies depending on the odorant and other factors such as air temperature and humidity between $4 \cdot 10^{-9}$ and $5 \cdot 10^{-14}$ g/L of air.[35] However, we are not at all sensitive to some gas molecules, such as for example the highly toxic carbon monoxide.

The receptors responsible for the sense of smell, the so-called olfactory cells, are located on the olfactory mucosa in the upper nose area (nasal roof). Nerve fibres emanate from them, which penetrate the base of the skull in the anterior skull fossa. The axons of the receptor cells forward the action potentials to the *olfactory bulb* an immediate protrusion of the brain. Since both the olfactory mucosa and the olfactory bulb have connections to the hypothalamus and the limbic system, the odour is often associated with a genetically determined, strongly emotional, motivating or aversion-inducing reaction.[36] The physiological findings presented here are beside others the basis for the "basic position" of the odour in the comfort pyramid (see ► Sect. 3.3.4).

3.2.2 Information Processing

In information processing, as already mentioned, both a cognitive overall view of the environmental impressions conveyed by the information receiving is achieved and a reaction optimally adapted to these environmental impressions is prepared, which is passed on to the mechanisms of information implementation as a detailed movement draft. The essential prerequisite for information processing is memory, which contains knowledge acquired in the form of so-called "internal models" on the one hand and enables a decision to be made for a concrete action in the form of the "decision mechanism" on the other. The most important question is therefore: how does the human brain link the constantly changing external stimuli with the appropriate actions? In connection with the essential movement in space when driving a vehicle, the question also arises: how do we perceive this movement and how can adequate actions be derived from it? The following paragraphs are intended to provide a basis for understanding.

---

34 Today, there is practically no technical measurement method that can objectively capture odour impressions. So when it comes to capturing odours, specially trained inspectors are called in.

35 Bad air in rooms (also in the vehicle) is almost always due to a high concentration of odorants.

36 Experience from various automobile manufacturers has shown that unfavourable odours, such as those that occur when ink or adhesives dry out, can contribute to a considerable loss of image. Conversely, some manufacturers hope to provide an emotionally positive feeling in their vehicles by providing artificial fragrances.

3.2.2.1 How Does the Information Get Into Our Brain?

When we are in a room, when we move through the room, we always have the impression that we are completely informed about everything, that we have everything in view, so to speak. Without thinking more about it, we have a "right" feeling for the distances and even a more or less precise idea of the space behind us, which we cannot grasp with the field of vision of our eyes. The photographic depiction of a room seems to give us this impression – as far as the richness of detail of the depiction is concerned – sufficiently well, even if the feeling for the spatial distance and the "being in the room" there is inadequate or missing.

In complete contradiction to this subjective feeling stands the objective realization on a physical-physiological basis that our eye is only in the so-called central visual pit (fovea centralis) able to design a sufficiently sharp image of the environment. However, this only corresponds to a viewing angle of approx. 2–3°. This is also where the number of receptors is densest, only there do we have the complete ability to see colour. Towards the so-called periphery, the number of colour receptors (cone) is steadily decreasing (see ▶ Sect. 3.2.1.1). The interconnection of the receptor cells there has been shown to have few complex and hypercomplex neuron cells capable of detecting edges or angles, but several capable of detecting motion. The above-drawn sensation about the "being in the world" is thus only subjectively assembled by successive scanning with the moving eye in our brain. But that also means that we can overlook important things. We consider it admissible to be able to turn our gaze away from an event if we believe, on the basis of experience, that we know "how to proceed" (for static objects this is sufficient in most cases!).

While our environment in which we move has three dimensions, the image on our retina has only two. Despite the two-dimensionality of the retina, we are able to perceive a three-dimensional environment. Thus, most questions of visual perception are directly related to the question of depth perception. The human depth perception is achieved by interpreting the overlapping and slightly different visual fields of both eyes due to the distance between the eyes. ▫ Figure 3.33 shows the geometric constellation of the representation of spatial conditions on a two-dimensional image plane. From the knowledge of the eye distance and the focal length of the eye, the corresponding spatial position can be recalcu-

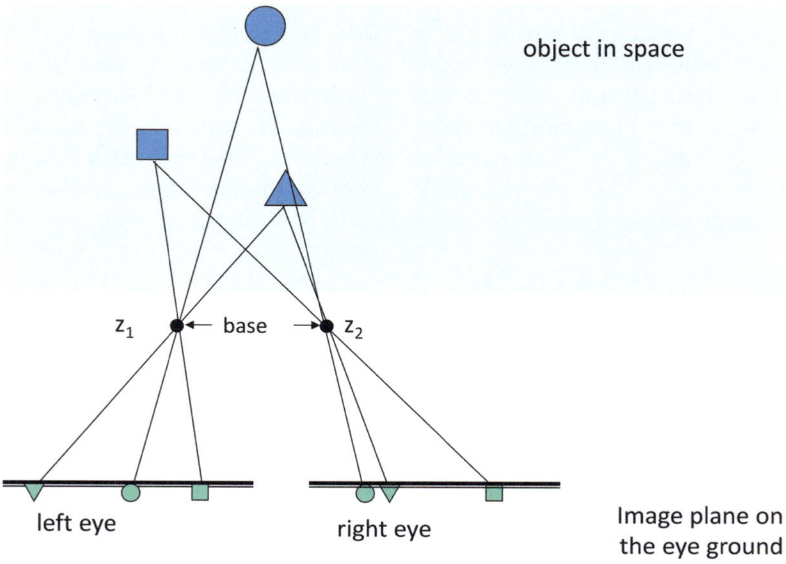

▫ **Fig. 3.33** Principle of stereoscopic imaging of a spatial constellation on two image planes lated by recognizing the respective object point in each of the images. This is a power that our brain is capable, a power that is acquired in the first 2–3 years due to the experience of the time of adolescence. However, spatial vision possible via stereoscopy reaches its limits at a distance of approx. 40 m (Bergmeier 2009).

von Helmholtz (1866) laid the foundation for the description of visual perception with his treatise on physiological optics. These are nowadays sufficiently known as laws of optics in physics and as descriptions of the visual process in biology. However, in order to understand the process of visual information recording when driving a vehicle, the knowledge of a simplified visual process is not sufficient. The aim of the following section is not to describe in detail all processes of visual perception, but rather to describe individual parts of the process in sufficient detail to explain the receiving of visual information when driving a vehicle.

Visual perception begins with the reflection of light rays from a light source (e.g. sun, headlights) on physical objects or of light rays emitted directly by an object (e.g. headlights of oncoming vehicles, rear lights, signal lamps for traffic control). Light bundled by the lens is captured by sensory cells in the retina, where a two-dimensional image of the outside world is drawn as with a camera according to ◘ Fig. 3.33. The energy of the incoming light quanta triggers electrical signals that reach the brain via the optic nerve. The retinal image can be treated like a two-dimensional needle image (Gibson 1950). Lindsay and Norman (1972) gave a good idea of how this happens: The interconnection type of lateral inhibition (see ◘ Fig. 3.7) is already realized at the lowest level of perception directly in connection with the receptor cells. By interconnecting several neurons into one neuron cell – as already discussed in ► Sect. 3.1.3 for the neurophysiological cybernetic process of information treatment – complex cells are formed which only emit an increased impulse rate if a certain stimulus configuration is present on the sensory surface. These complex cells are also interconnected again to form hypercomplex cells, which only show an increased impulse rate when the stimulus configuration is defined in more detail. In the way shown here is formed a whole series of specific detectors, so at the level of complex cells edge detectors and motion detectors and at the level of hypercomplex cells angle detectors and detectors for specific lengths.

Werblin and Roska (2008) find that these hypercomplex cells detect about a dozen very different representations of the visual scene. Each of them contains a different aspect of what happens in front of the eye. The abstractions are continuously updated, adapted to the outside world and contain, among other things, the following information:

1. Object outlines (comparable to a line drawing),
2. speed and direction of movement of the objects in the visual scene,
3. shady or bright areas.

Further contents are difficult to reproduce from the form. Each information of these specialized nerve cells is transmitted in the optic nerve by its own group of nerve fibers, so-called ganglion cells, to higher brain regions. A single ganglion type represents the filtered information of a separate spatio-temporal aspect of various characteristics of the visual scene such as movement, color, depth and form. The individual partial information thus reaches different brain regions separately, where it is further processed partly in conscious processes and partly in unconscious processes. The brain receives only partial visual information for its interpretation. How the brain draws a seamless, convincing picture of reality from the individual information packages is beside others the subject of current research.

How this happens can be illustrated by an analysis by Guzmán (1969), who found typical angular configurations for computer image recognition that suggest a certain interpretation of a two-dimensional image. The human brain already seems to interpret the monocular image on the retina from similar points of view. ◘ Figure 3.34 shows some examples of such configurations and their preferred interpretation. The example in ◘ Fig. 3.35 demonstrates how to apply this

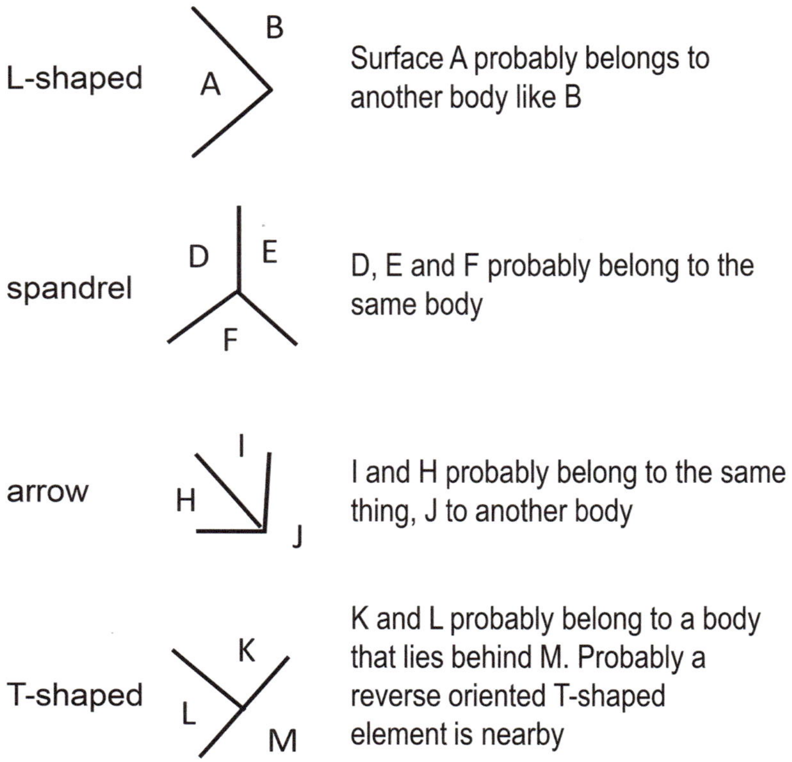

Fig. 3.34 Typical angle configurations and their interpretation

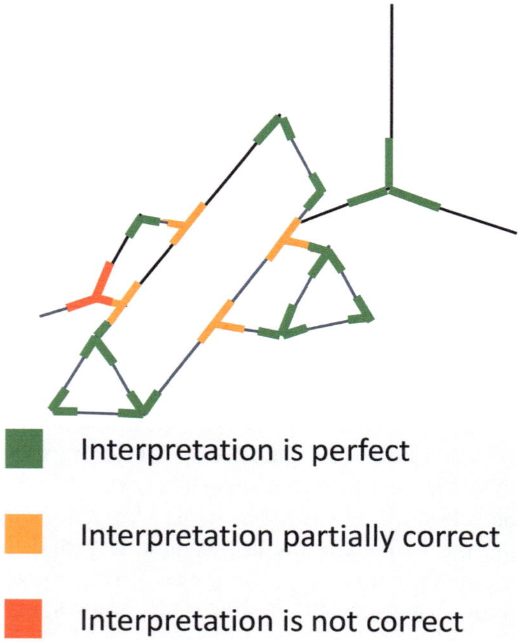

Fig. 3.35 Example for the application of typical angle configurations

interpretation: You can see two logs lying on top of each other in the corner of a room you are looking into. One cannot resist this interpretation even if one intends to see only a tangle of lines.

How does this interpretation come about? The cells, which extract the simple properties in the way discussed above, are switched to even higher cells in such a way that they only react optimally when a certain pattern given by the switching is mapped on the sensory surface. Several cells can be stimulated at the same time by the same stimulus configuration. Then, for the further process of perception, the one is selected that allows a self-contained interpretation that corresponds to memory contents. This becomes clear with the example of Fig. 3.35: The angle configuration marked there as "T-shaped" would actually suggest that the surface of the log below is part of a body that is identical with the room floor (see Fig. 3.34). Due to the weight of the other impressions and in favour of a self-

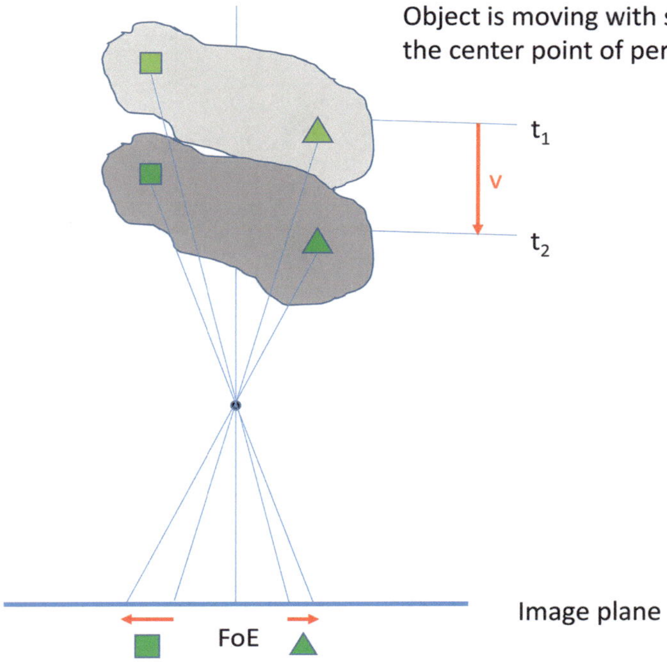

Fig. 3.36 Motion parallax: Object points at different distances from the observer, all moving at the same speed, produce images at different speeds on the back of the eye, from which conclusions can be drawn about the distance. A singular point (focus of expansion, FoE) arises at the point of linear connection between the focus of the central perspective and the point to which the movement occurs

contained perception of the outside world, this interpretation is suppressed.

The process described refers not only to the perception of a static environment, but also to that of movement. Motion perception is of great importance in the context of driving a car; it represents the quality of perception of the modality "vision". The actual perception of movement takes place with the help of "movement-specific" neuron systems. According to Rock (1968), one can understand the optical perception of movement well if one assumes the effect of the following principles:

- The eyes never perceive the individual course of movement of images on the retina, but always mathematical components of it. The principle of motion perception is that the same components form a fixed unit and are thus distinguished from diverging components.
- If, after separating solid units from the image field, motion vectors remain on the retina, new higher-order solid units are formed. These new units are perceived in relative movements to the previously formed units.

von Helmholz (1866) introduced the term motion parallax for the first time. According to its definition, it describes the perspective displacement of distant objects at different depths as a result of a changed observer position. According to von Helmholtz (1910), the motion parallax contains information about the distance between the observer and the object. This applies both to the case that the observer moves towards the objects (see Fig. 3.36) and to the case that he moves past the objects (Fig. 3.37).

3.2.2.2 The Optical Flow

In 1958 Gibson extended Helmholtz's definition of motion parallax and presented his theory of the visual control of locomotion by means of optical flow. Optical flow is the expanding or contracting visual field projected onto the image plane on the retina caused by

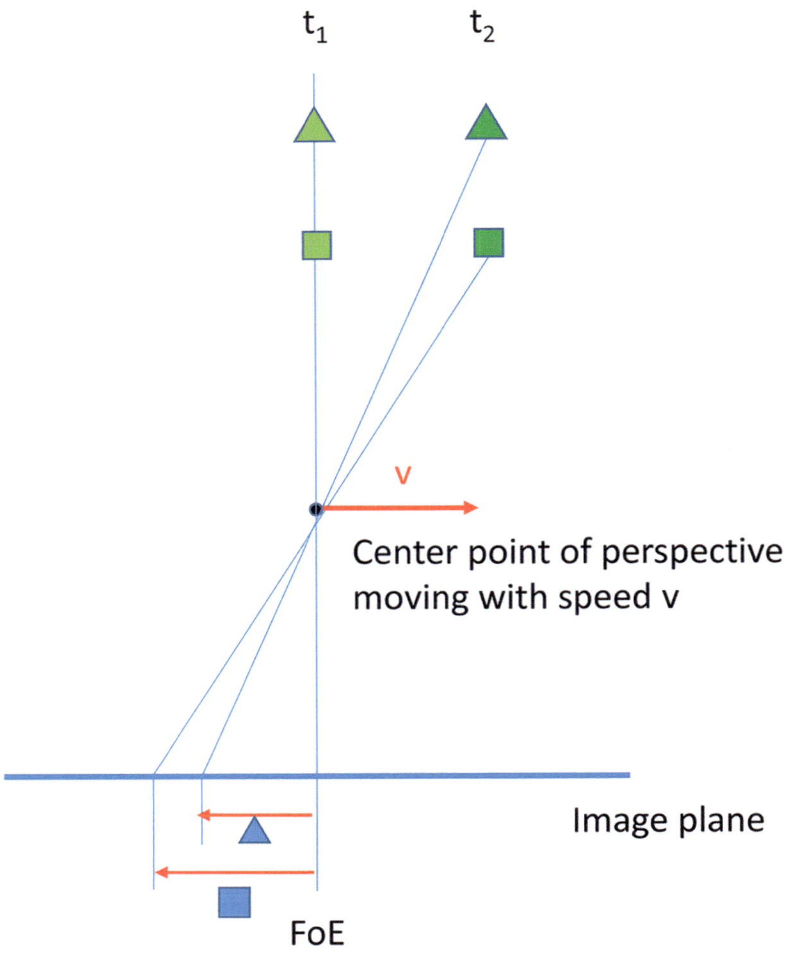

Fig. 3.37 Motion parallax: The observer moves past objects of different distances at a speed v (as e.g. when looking out of the window of a moving train). The further away the observed objects are, the less they move. Looking at a point on the horizon shows no movement at all (Focus of Expansion, FoE)

the observer's own motion (Gibson 1950). The visual field is the representation of the velocity vectors of all visible objects of the visual scene. In other words, optical flow is the transformation of the surfaces of the surrounding optical array during locomotion (Gibson 1966) or the gradient of locomotion (Goldstein 2002). Optical flow is thus the relative movement projected onto the image plane between the observer and the visible points in space (Chatziastros 2003). The literature contains a large number of different definitions of optical flow, which merely reproduce the phenomenon observed by Gibson with other formulations. The present treatise is based on the formulations of Gibson (Gibson 1950; Gibson 1966) and Goldstein (Goldstein 2002).

The optical flow characterizes the totality of the optical changes of the visual scene (Chatziastros 2003) and represents the generalization of the movement parallax between a few objects to all objects of the visual field.

As described above, the optical flow is information produced by the movement of the observer. There is a reciprocal relationship between the locomotion of an observer and the optical flow (Goldstein 2002). The optical flow is generated by the observer's own movement and provides the observer with information that helps him to control and steer his

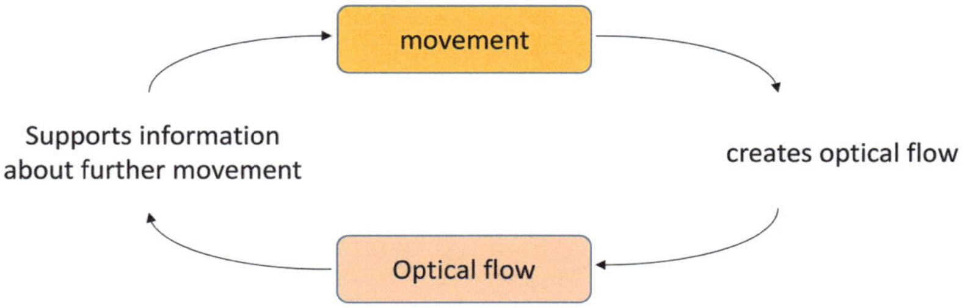

Fig. 3.38 Reciprocal correlation between locomotion and optical flow. The locomotion generates an optical flow, which in turn provides information about the locomotion and thus controls the movement. This is an important principle for our interaction with the environment (see Goldstein 2002)

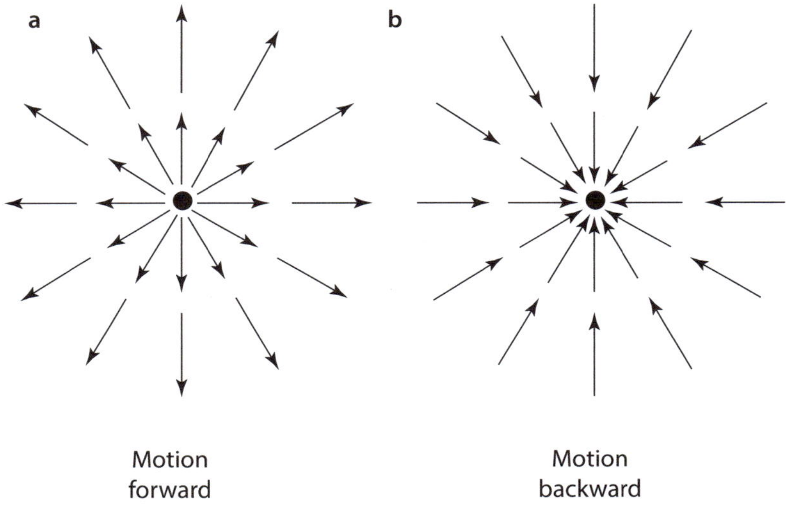

Fig. 3.39 Optical flow **a** during a forward movement and **b** with a backward movement (cf. Goldstein 2002)

further movement. Figure 3.38 illustrates this relationship: Movement is the prerequisite of the perception of the optical flow and its perception is the basis of the experience of a movement. In this context, reference is made to the discussion in connection with Figs. 3.5, 3.47 and 3.48, which are still to be discussed below, which deal with the feedback of the internal flow of information from the intake of information to the implementation of information about the outside world.

Figure 3.39 shows the optical flux generated by a simple motion situation, which can be derived from Helmholtz's considerations on motion parallax in Fig. 3.36. The points represent individual elements of the environment, the lines describe their direction of movement and over the length also the speed of the elements. In the central point of the optical flow, neither a motion nor a motion parallax is visible (Warren et al. 2001). This point is therefore referred to as the Focus of Expansion (FoE) or "singular point" of the flow diagram. Because of the fact that movement can only be perceived optically if the speed of the light stimuli on the back of the eye exceeds a value of approx. 2′/s (Lindsay and Norman 1972), the FoE is actually a field that is smaller the faster you move.[37] Furthermore, the FoE is the point or field from which the velocity vectors of the flow

37 This aspect plays an important role in the context of how over- and understeering can be perceived (see ► Sect. 6.4).

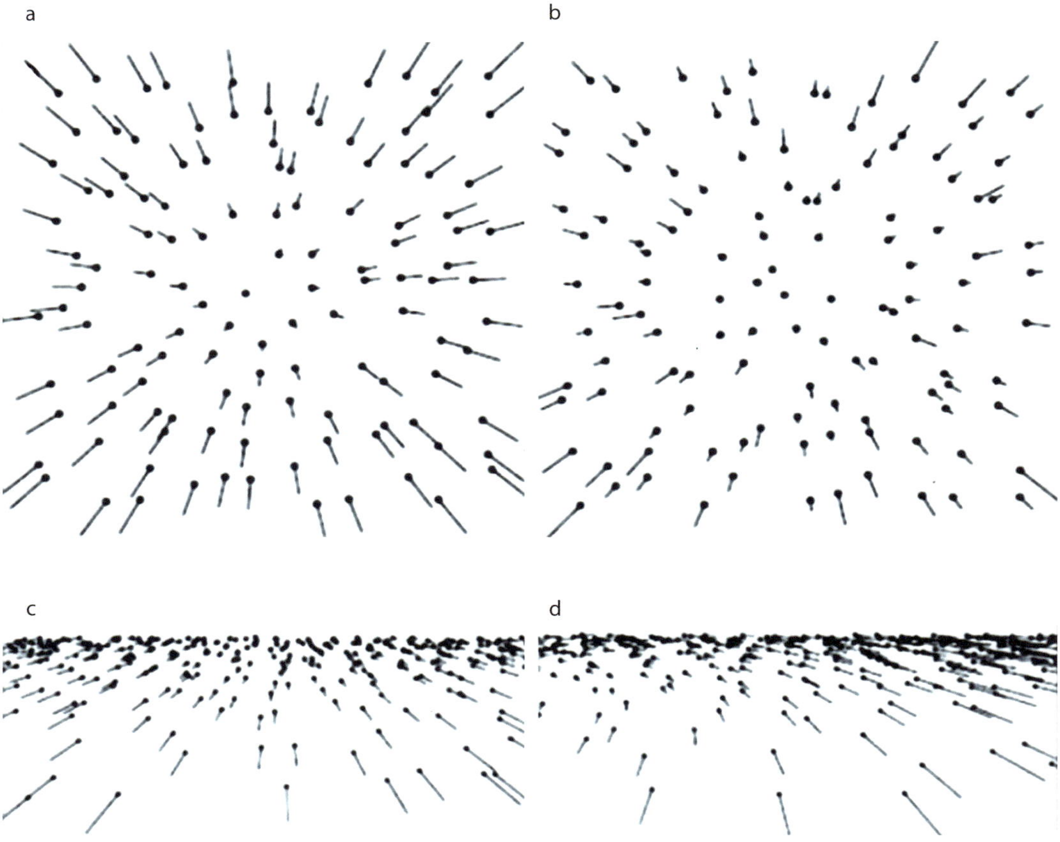

Fig. 3.40 Optical flow in the image plane **a** when driving vertically on a wall, **b** by a point volume, **c** when driving straight ahead, **d** during a turn (according to Chatziastros 2003)

diagram seem to expand radially when moving towards the FoE, or where they converge when moving away from the FoE. The FoE thus indicates the direction of the observer's movement in a linear motion (Gibson 1950; Wann u. Wilkie 2004; see also Fig. 3.36).

The movements of an observer can be divided into a translational and a rotational component (see Fig. 3.40).

The velocity of the individual elements and thus the length of the vectors of the translational flow diagram depend on the distance between the environmental elements and the observer. In a linear translation, the angular velocity of the elements in the optical flow is determined by the following relationship:

$$\frac{d\theta}{dt} = (\sin\theta)^2 \cdot \frac{v}{x}$$

The following is included v the forward speed of the observer, x the distance of the element perpendicular to the direction of movement, and Θ the angle between the direction of movement and the direction to the element of the visual scene (see Fig. 3.41). The smaller the distance of the elements from the observer or the greater the forward speed, the higher the speed in the optical flow.

For the perception of this speed, however, there is a limit, which results from the fact that two or more vectors on the fundus of the eye are stimulated one behind the other in such a short distance that the object appears blurred (Schierz 2001). This can go so far that the object is no longer perceived (e.g. a flying projectile or even an insect at an appropriate distance).

Figure 3.42 shows the course of dynamic visual acuity as a function of the

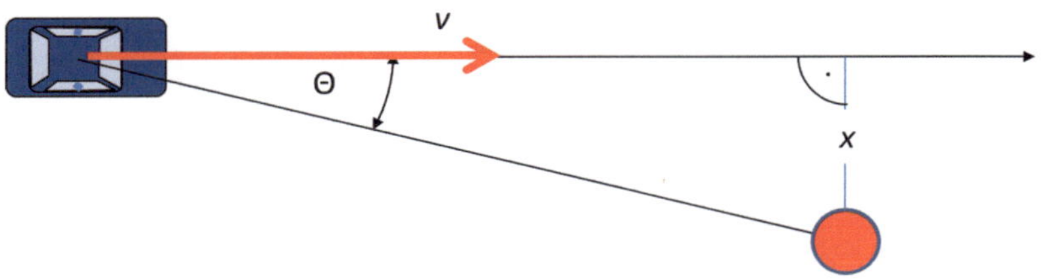

Fig. 3.41 The angular velocity of an object is determined by the forward velocity v of an observer, the passing distance x and the eccentricity Θ of the object (see Chatziastros 2003)

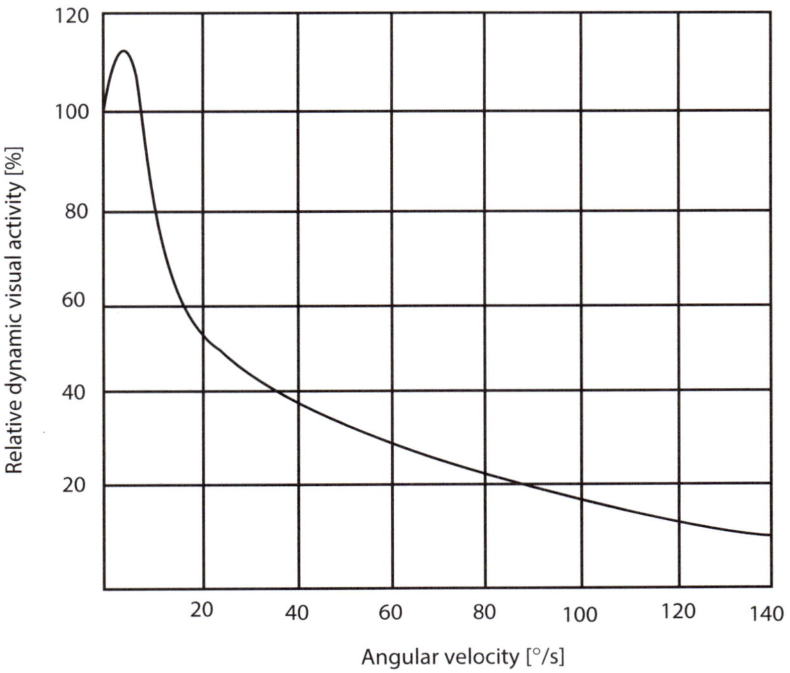

Fig. 3.42 Relative dynamic visual acuity as a function of the speed of movement (angular velocity of an object (Schierz 2001))

angular velocity at which the object is imaged on the back of the eye. When driving, this effect mainly plays a role in the peripheral field of vision.

A further limit is given by the fact that objects that image themselves onto the back of the eye at an angular velocity lower than 2′/s are no longer perceived as moving. At an eye level of 1.2 m above the road, this value is at a speed of 50 km/h at approx. 75 m in front of the driver when driving straight ahead. At 100 km/h it is located at 150 m and at 200 km/h approx. 300 m (the eye level above the roadway plays a marginal role for these values by the way). However, these values apply only to the point immediately in the direction of travel. As Remlinger (2013) shows, these values represent the peak of circular boundary lines that extend into the kilometre range (see Fig. 3.43). Even at inner-city speeds, this boundary to visual tranquillity is so far away that it plays no role in road traffic. However, as shown in ► Sect. 6.4.1.3, it is important for the perception of lateralse dynamic properties of the vehicle, in particular oversteering and understeering.

The flow field generated by a rotary motion – for example a rotary head motion – consists of parallel velocity vectors all pointing in the same direction. The rotatory flow

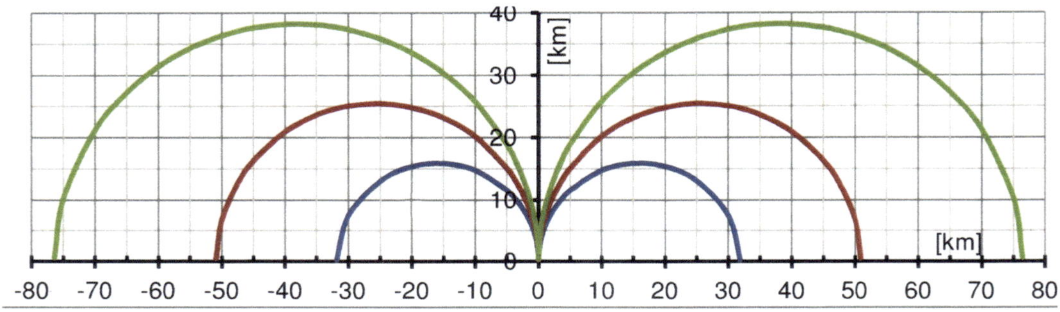

Fig. 3.43 Boundary lines of visual tranquillity in the entire field of vision of the driver for straight ahead travel at the speeds 50 km/h (*blue*), 80 (*red*) and 120 km/h (*green*) (after Remlinger 2013). In the area outside the respective lines no movement would be perceived (this is, by the way, the reason why one does not have the impression of approaching this mountain when approaching a distant mountain range over a long period of time. This effect is also used for the conception of driving simulator images)

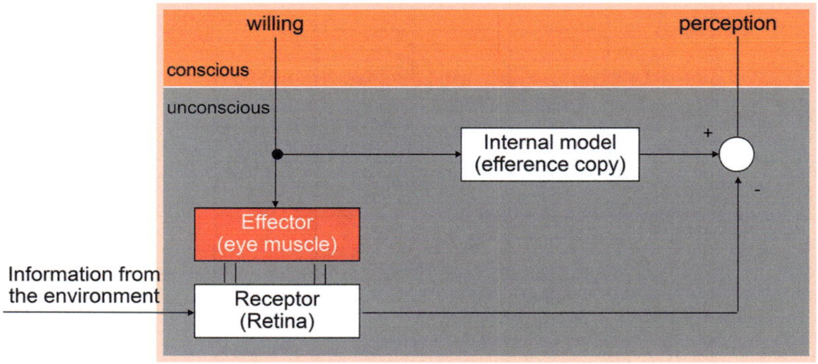

Fig. 3.44 Reafference principle (according to von Holst 1957)

field therefore does not contain an FoE. The length of the flow vectors in the rotational flow field is independent of the distance of the elements to the observer. For this reason, the rotatory flow field, in contrast to the translatory flow field, does not provide any information about the movement of the observer. This must not be confused with the parallax of motion, which is created when looking at an angle to the direction of motion (see Fig. 3.37).

The eye's own movements, e.g. caused by an eye following movement or a head movement, also generate a rotatory flow field (see Fig. 3.45b). However, this is separated from the movement of the entire body according to the efference copy prinziple von Holst (1957) described below, which explains some constancy performances of perception. At the *directional constancy* is to ask what causes the fact that a rotation of the eyeball to the right does not result in a subjective shift of the environment, although the image on the retina is shifted to the left by this movement. According to the efference copy prinziple this effect is clarified by the fact that with each arbitrary movement an internal model exists over the success of this action (efference copy, see Fig. 3.44). When initiating the eye movement to the right, the image is expected to shift to the left. If it really moves to the left, actual movement and expected movement cancel each other out, and the environment is assumed to be dormant. If the image shifts without adequate arbitrary movement, the shift is also perceived as an objective fact, as one can easily convince oneself by slightly twisting the eyeball from the outside with the fingertip. The comparison between the afference of the sensory organ and the image built

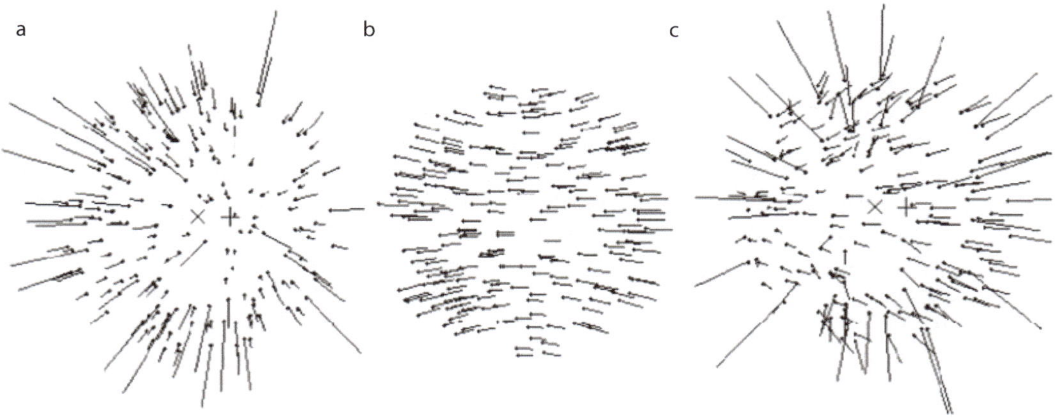

Fig. 3.45 Occurrence of retinal flow (**a**, right) as the sum of the flow field of a translatory forward movement (**c**, left) and the flow field of an eye's own movement (**b**, centre) (Lappe, 2009)

up in the brain is an unconscious process. Only after a deviation does the process become conscious.

The resulting flow field from a translatory movement and an eye movement, from a translatory and a rotational movement or from the combination of all three movements does not differ in form. In all three cases a complex flux field is created (see Fig. 3.45c). Since the optical flow in this case occurs only on the retina, it is called retinal flow.

A human being can derive information relatively easily from a translational flow field, since he can process it without great effort. The addition of a rotational flow field, whether it is the result of the eye's own movement or a rotational movement, significantly changes the structure and appearance of the flow field. This also increases the effort required to process and absorb information from the flow field. In the course of growing up, human learns to separate his own initiative from the movement caused by "external influence" by means of the reafference principle. This means: he is able to filter the effects on the retinal flow field by the movement of the eye in the head or a total movement of the head itself or even his own initiated rotational movement of the whole body and to use the necessary information.

Since the receptor cells on the ocular fundus are interconnected in different ways to complex and hypercomplex cells in the different eye regions (e.g. in the fovea centralis preferably to edge and angle detectors, in the periphery of the eye to motion detectors), it is therefore decisive for the perception on which part of the retina information from the optical flow is incident and processed.

Peripheral or indirect vision describes a sub-area of visual human perception. In contrast to foveal vision, in which the eye must be directed exactly to a certain fixation point in order to exploit maximum visual acuity, peripheral vision provides coarse, blurred and optically distorted visual impressions outside the fixation point (Goldstein 2002). The peripheral system covers more than 999‰ of the visual field. However, only 50% of the optic nerve and 50% of the area of the cortical vision centre are available to the peripheral system for processing the information. The remaining 50% are reserved for the high resolution but very slow foveal system. Due to the unequal ratio of covered visual field and available information processing resources between foveal and peripheral vision, the characteristics of the two types of perception differ. From the central visual pit to the peripheral field of vision, visual acuity, colour perception, sensitivity to light stimuli and perception of contrasts increasingly decrease, while dynamic sensitivity, temporal resolution, perception of light and dark as well as perception of movement and orientation in space significantly increase (Goldstein 2002).

3.2.2.3 Human Memory

Without memory no information processing is possible. A more precise picture of the working principle of memory is therefore necessary for understanding information processing. A rough distinction can be made between sensory, short-term and long-term memory. Sensory memory represents the reverberation of sensory cells and the complex and hypercomplex cells associated with them. Therefore, its capacity corresponds to that of the whole of the sense organs. The retention time is no more than 200 ms. The processes of short-term and long-term memory are very complex and only partially understood. Only a very simplified model of it will be discussed here. However, it is detailed enough to understand the principles of information processing that take place while driving. ◘ Figure 3.46 provides a basis for an easier understanding of this model. Information that is absorbed by the sensory organs and preselected by the complex and hypercomplex cells ("sensory memory") stimulates (in reality, a huge number of) neuron cells that can be distributed across different areas of the brain.

It can happen that some of them are closed to a circle. Such a circle represents the active memory (Palm 1990). As long as one cell stimulates the next, the information represented by this specific circle is active. This state remains between 3 up to 15 s. It represents our active consciousness. In general, it is called short-term, work or primary memory. Through the repeated stimulation of external information – received by the sense organs – or also through internal stimulation, the connections on the synapses are gradually changed. Thus, certain circles representing this repeated content are preferred to other neuron compounds. A structural engram is generated. If these circles are stimulated by a corresponding external or internal (e.g. by active reflection) stimulus configuration, the former experience is "remembered" again. This structural engram is called "long-term memory". Over time, these engrams may fade or be buried. We call this type of long-term

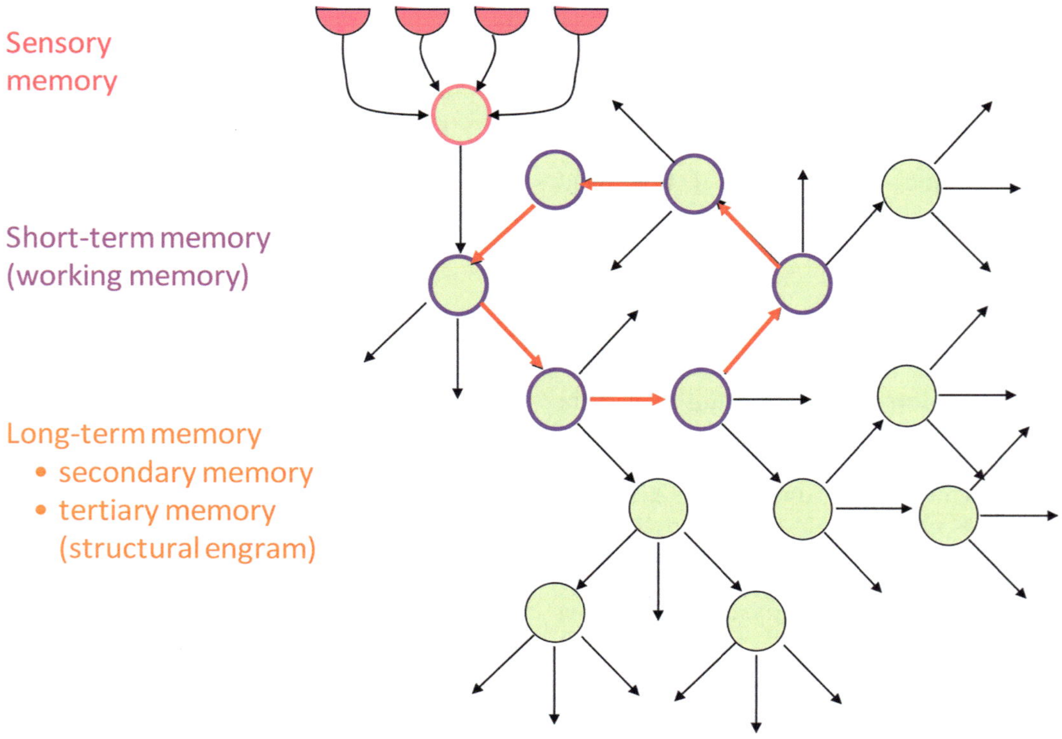

◘ **Fig. 3.46** Principle of human memory

memory "secondary memory". Its retention time can range from half a minute to years. However, there are also contents that will never be forgotten. One's own name, the ability to walk, ride a bicycle or swim, belongs to this kind of memory. This part of long-term memory is called "tertiary memory". For the transition from primary memory to secondary memory, not only repetition but also motivation plays an important role. It is mainly controlled by the brain areas hippocampus and limbic system.

The reaction time of human information processing is particularly important in connection with the dynamic properties of an automobile. Through tracking experiments and the modelling of human properties by means of control engineering methods (in particular the research of McRuer and his colleagues should be mentioned here; see literature index) it is known that in the case of unconscious, so-called skill-based behaviour (Rasmussen et al. 1987) the reaction time is in the range of 200 milliseconds. However, if one wants to understand the interaction between driver and vehicle, not only this reaction time is important, but also the subjective time period within which the driver derives his action activities.

3.2.2.4 Internal Models

As already shown in the previous sections, the categorisation in information reception and processing is somewhat arbitrary. Thus, essential elements of information processing are described by the von Holst principle in the same way as aspects of information reception. The only difference is that a protracted learning process that begins at the beginning of life creates increasingly complex copies of effects that are adequate for certain situations, which are now referred to as "inner models". In ◘ Fig. 3.47, the representation of ◘ Fig. 3.44 has been extended by the model M_B. Otherwise, there is the same situation as with the simple reactor principle. This is illustrated by the following example of driving on a curve: It is by no means a matter of course for an upcoming right turn to know exactly how much steering wheel angle is necessary to

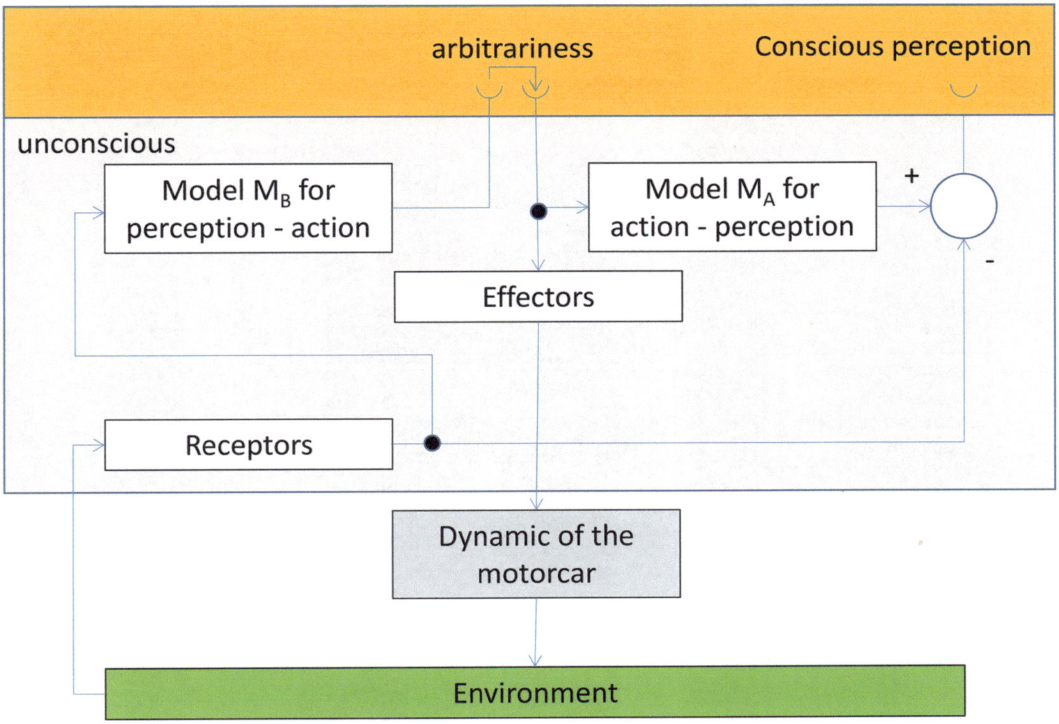

◘ **Fig. 3.47** Behavioural structure of highly practiced activities using the example of motor vehicle driving

adequately solve the driving task. Due to a complicated learning process, which cannot be described here, the driver of a motor vehicle has developed an internal model M_A how its vehicle reacts to steering movements and accelerator pedal movements. He also has a further internal model M_B which tells him what interventions to make in a given traffic situation. Via his receptors, in particular the eyes, but also via the acoustic, kinesthetic and haptic channel, he takes up the given situation and leads it through the M_B given action sequence (The model M_B can only deduce action sequences for simple situations, such as the following of a road course for example; in more complicated cases decisions are necessary; see ▶ Sect. 3.2.2.5). On the basis of this action sequence, he has by use of the model M_A an expectation of changing the information about the outside world. If this change corresponds to the actual change, the whole process normally remains unconscious. Only a deviation between model conception and feedback from the environment (if e.g. the steering behaviour of the vehicle on a slippery road is completely different than usual) penetrates into the consciousness and makes decision processes necessary, which possibly select motor processes, which in the originally used model M_A were not included (see ▶ Sect. 3.2.2.5).

An essential aspect of working according to an inner model is to be seen in the fact that the time required for this process can be estimated. It lies within the range of the physiological reaction time of approx. 200 ms already mentioned. A further point of view is that external stimulus configurations (information reception) can definitely "blur" certain internal models. Among other things, this effect is related to the different degrees of belonging of perceived stimulus configurations to quantities and supersets in the respective mental representation. This "similarity effect" can be the cause of errors by deriving inadequate actions for the situation.

3.2.2.5 Decision Mechanism

A discrepancy between the values obtained by the model M_A and the impressions of the real outside world taken up by the receptors penetrates the consciousness and, if necessary, makes a decision necessary for an action, which also appears possible by the given stimulus configuration (similarity effect). This requires an extension of the information processing model shown in ◘ Fig. 3.47, as shown in ◘ Fig. 3.48. An essential part of this

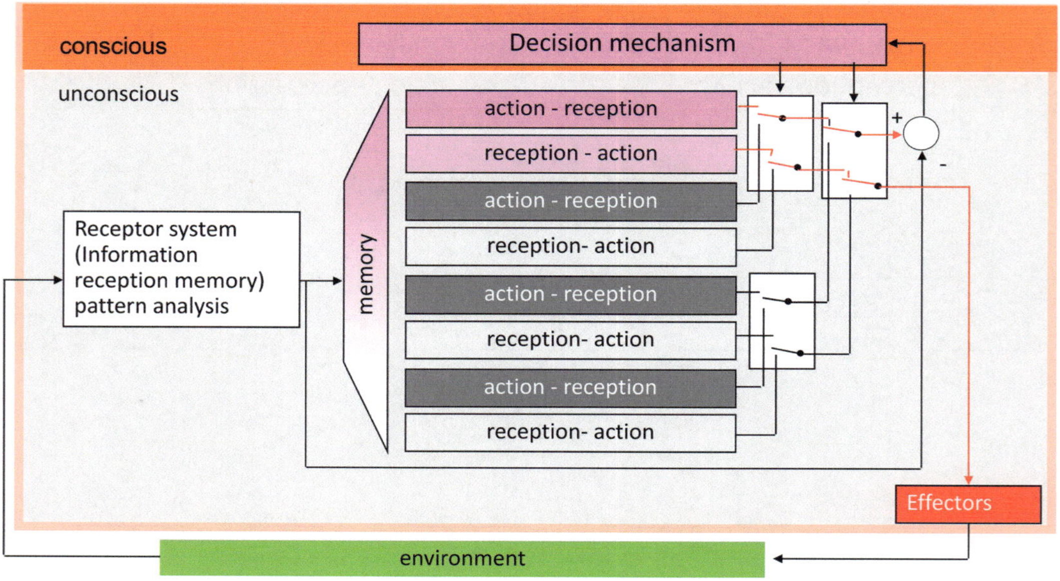

◘ **Fig. 3.48** Model of human information processing

extended model is the decision mechanism, the function of which is to select the most suitable model for the particular moment from the large number of internal models. It corresponds to the working memory mentioned (= short-term memory) and is therefore subject to its limitations (see below).

How can this process be described in more detail? The action remains unconscious as long as the expected information sufficiently matches the observed information. In certain situations we can drive without effort and at the same time perform additional actions such as talking to the passengers, telephoning and listening to the radio. Only when a significant deviation of the observed information from the expected information is perceived does this come to consciousness (in our example, this would happen if the car suddenly comes onto black ice). Now we have to use the active part of our brain, looking for an alternative. We look in our brain for other action-perception models that promise a better result. Once we have found such a model, we direct the corresponding action to the effectors, through which this action is then realized. It is obvious that this procedure takes time (in the case of our example, the car may have already driven into the ditch by the time the better action has been found). This search has yet another limitation, as found in many psychological experiments. Miller et al. (1956) found for the first time that only up to 7 ± 2 so-called chunks can we weigh up, which are identical with the action-perception perception-action pairs. We also call these chunks “psychological units”, “representational units” or “inner models”. These expressions are intentionally a little diffuse, because these chunks themselves are not quite sharp again. Through practice and experience, more and more complex psychological units can be formed, which enable safe and unambiguous action even in difficult situations. For example, this makes the difference between an experienced driver and a beginner (it is worth mentioning here the “compulsion” of the organism, already mentioned in connection with the jumping images, to opt for *one* action). This can also be derived from the course of the accident figures as a function of time. They show that, on average, a low stable level is not reached until about 7 years after obtaining a driving licence (so-called “beginner risk”). This time seems to be necessary until enough experience has been gained so that for (almost) every situation a correct expectation of the probable course is available. As the study by Maycock et al. (1991; see ◻ Fig. 3.49) shows, this effect is independent of the age at which the driving licence was acquired. However, for young (preferably male!) novice drivers, there is also

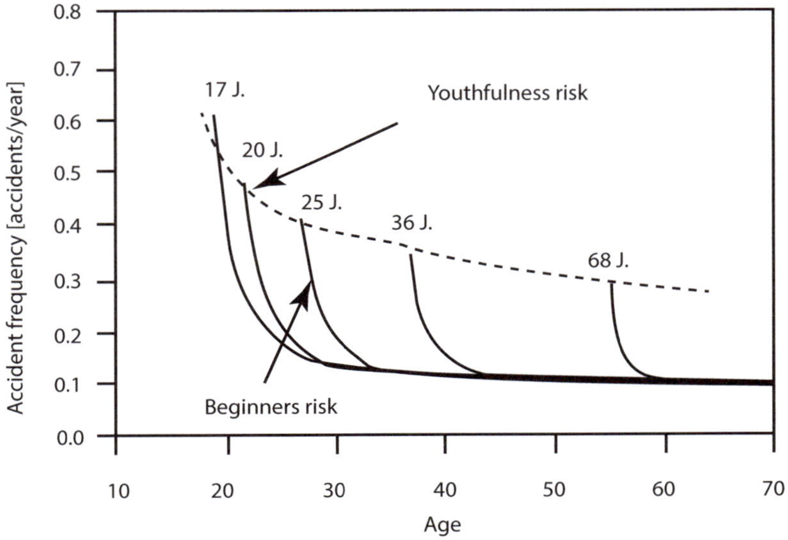

◻ **Fig. 3.49** Youthfulness and beginner risk (after Maycock et al. 1991)

a certain youthful carelessness and possibly also uncontrolled showmanship (so-called "youthfulness risk").

As shown above, the inner models can be divided into two main parts. One part contains the connection of perception configuration and the motor actions derived from it (perception-action-model). The other part contains the idea about the expected perception configuration due to this action (action-perception model). One of these model pairs can be selected via the decision mechanism, which is symbolically indicated in Fig. 3.48 by switches. In the decision-making situation, i.e. when it has not yet been determined which model is to be preferred, the parallel switches for perceptual-action and action-perceptual models are separated in the model-like representation of Fig. 3.48. The decision mechanism then selects ("in thought") sequentially each action-perception model in question and tests the magnitude of the subjective benefit on the basis of the expected perception. The model that promises the greatest subjective benefit under the given circumstances (context) is then selected for the action. In the model-like representation of Fig. 3.48, the switches are now placed on the selected action-perception model as well as on the associated perceptual-action model. Now the way is clear for the innervation of the effectors (musculature), which are supplied with a movement program according to the selected model. The action of the effectors changes the configuration of the environmental stimuli, which is picked up by the receptors and reported to a reference point that compares this feedback with the expectation of the action-perception model. If the action has led to success, i.e. there is no difference at the point of comparison, the selected model pair is stored as particularly suitable for the corresponding stimulus configuration if it is repeated several times, i.e. the respective "switch position" is transferred into the memory as a new, i.e. learned, superordinate action-perception model. Deviations at the point of comparison are consciously perceived and influence the decision mechanism in such a way that when an individual limit is exceeded, the decision structure is changed,

Fig. 3.50 Example of a traffic situation in which a decision is necessary

which can be represented in a new switch combination. If in certain situations the decision for the same action proves to be useful again and again, it is learned together with the corresponding situation as a new quasi superordinate inner model.

In many cases a decision between several possible actions becomes necessary simply because different events are to be expected depending on an assessment of the external circumstances (context, so-called "states of the world"). The human organism is obviously quite capable of estimating objective probabilities of events on the basis of observations of the frequency of events in everyday life (Sheridan and Ferrell, 1974). The expected subjective benefit thus consists of the subjective estimation of the states of the world and the estimation of the benefit.

The connections discussed may be illustrated using the example of Fig. 3.50: The picture shows a traffic situation and the different states of the world, i.e. possible reactions of other road users to be considered. For each of the reactions, the probability of their occurrence is estimated. In Fig. 3.51 the first line contains the hypothetical estimate in case the driver is not in a hurry. In the left column you can see different possible actions. For each of these actions a benefit is indicated. By multiplying the probabilities of the states of the world and the utility of the respective action, the utility of each matrix element can be formally calculated for the following argumentation. An unfavourable out-

Action without haste	Cyclist breaks rank 0.3	Cyclist remains on the right 0.7	Oncoming car takes over 0.3	Oncoming car keeps his lane 0.7	Truck brakes 0.1	Truck keeps speed 0.9	Σ
0.4 straight ahead brake	0.12	0.28	0.12	0.28	0.04	0.36	1.20
0.3 Brake to the left	0.09	0.21	0.09	0.21	0.03	0.27	0.72
0.1 Straight ahead accelerate	- 0.03	0.07	- 0.03	0.07	- 0.01	0.09	0.16
0.2 accelerate to the left	0.06	0.14	0.06	0.14	- 0.0.2	0.18	0.44

Fig. 3.51 Decision matrix for the case of a non-urgent driver

Action with haste	Cyclist breaks rank 0.3	Cyclist remains on the right 0.7	Oncoming car takes over 0.3	Oncoming car keeps his lane 0.7	Truck brakes 0.1	Truck keeps speed 0.9	Σ
0.1 straight ahead brake	0.03	0.07	0.03	0.07	0.01	0.09	0.30
0.2 Brake to the left	0.06	0.14	-0.06	0.14	0.02	0.18	0.48
0.4 Straight ahead accelerate	- 0.12	- 0.28	- 0.12	0.28	- 0.04	0.36	0.64
0.3 accelerate to the left	0.09	0.21	- 0.09	0.21	- 0.03	0.27	0.66

Fig. 3.52 Decision matrix for the case of an urgent driver

come of the action (e.g. a collision with another road user) must be considered as a negative benefit (= damage). The sums in the last column represent the average benefit of a particular action. As Fig. 3.51 shows, the driver's decision is in favour of waiting ("braking straight ahead"). Figure 3.52 shows the same situation in the event that the driver is in a hurry and therefore values more strongly the benefits of actions that guarantee faster progress. The decision is now made in favour of the faster driving ("accelerate to the left"). The comparison of the results shows that the driver, depending on his motivation (= allocation of benefits), comes to completely different actions with otherwise the same assessment of the situation. When analysing a person's decision, it must be borne in mind that the decision is essentially influenced by the internal models (some of which may be wrong), and that the current benefit is the most important and possibly important factors are not taken into account because the current scope of action of the decision mechanism is limited by the capacity of short-term memory (see Fig. 3.53). The information processing of humans is influenced thereby by further factors, like motivation, vigilance, time pressure and so on, whereby the motivation determines finally, what the momentary subjective use is.

According to the model of human information processing presented here, a rough distinction can be made between highly practiced actions that do not require a decision and those that occur during decision-making processes. The former are not constrained in their complexity by the capacity of short-term memory, while the latter are subject to this constraint and take more time. If we derive actions from the inner representations stimulated by the sense organs, the so-called inner models, then the categorization intro-

Action without haste	Cyclist breaks rank 0.3	Cyclist remains on the right 0.7	Oncoming car takes over 0.3	Oncoming car keeps his lane 0.7	Truck brakes 0.1	Truck keeps speed 0.9	Σ
straight ahead brake 0.4							
Brake to the left 0.3			0.09	0.21	0.03	0.27	0.42
Straight ahead accelerate 0.1			- 0.03	0.07	- 0.01	0.09	0.12
accelerate to the left 0.2							

Fig. 3.53 Decision matrix for the case of a non-urgent driver with consideration of limited decision depth, caused by fatigue, time pressure, overstrain and similar. The mentioned limitation of the short-term memory can result in the fact that other influences, as shown in this table, remain accidentally unconsidered

Processing level / Task		Needed time		
		skill based <200ms	rule-based Seconds to minutes	knowledge-based Minutes to hours
available time	***navigation*** *(several seconds to hours)*	daily commute	Choice between familiar ways	Find your way around a foreign city
	guidance *(about 2 seconds)*	Turn off at a familiar intersection	Overtaking other vehicles	Steer on a slippery or icy road
	stabilization *(between 100 ms and max 300 ms)*	to go around a bend	drive an unknown car	Driving student in the first hour of driving

Fig. 3.54 Comparison of the time available for the completion of driving tasks and the time required for this depending on the processing level (according to Hale et al. 1990)

duced by Rasmussen (1986) is helpful because it allows a reference to the processing time to be established. He calls unconscious reacting even to complex stimulus patterns "*skill-based*". The reaction times associated with this are in the range of a few 100 ms (generally approx. 200 ms). Stimulus situations, the coping with of which requires a certain conscious cognitive attention, but which can be dealt with according to "proven patterns", are treated by "*rule-based behaviour*". The processing time for such actions is in the 1–2 s range. For difficult, novel situations, however, solutions can only be found through conscious reflection and consideration of possible consequences. This behavior is called "*knowledge-based*". The time required for this is in the range of at least several seconds up to hours, days and even more. The time required to complete the task due to the respective processing level and the time available for the task may result in error-inducing conflicts, as compiled by Hale et al. (1990; see Fig. 3.54).

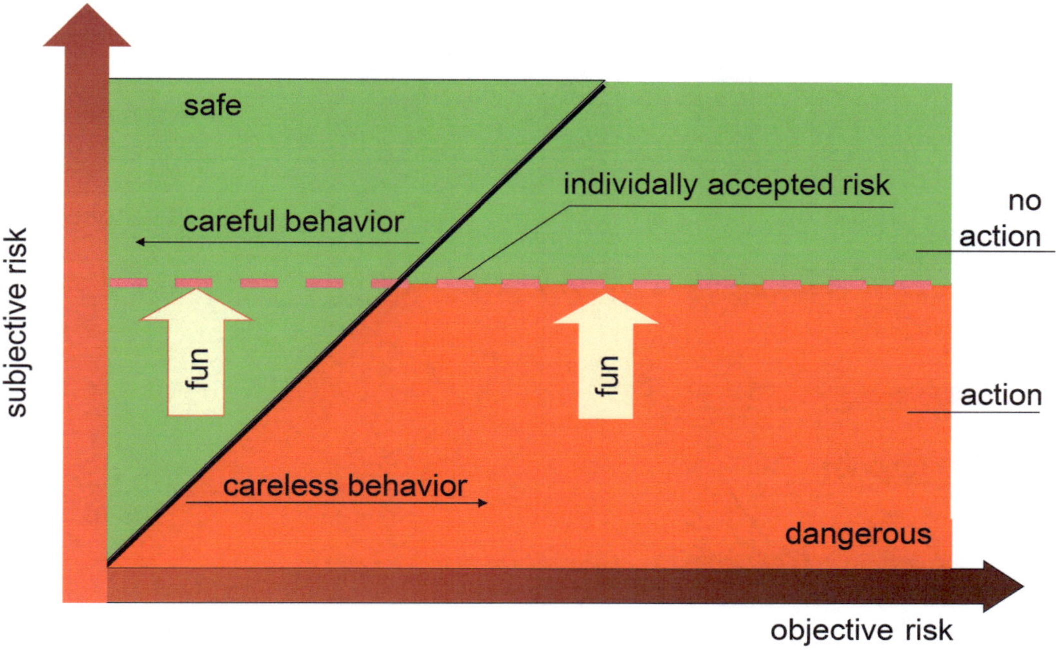

Fig. 3.55 Example of different combinations of subjective estimation of risk, objective risk and resulting safety of action

An essential aspect of a decision-making process is – as shown above – the fact that the action that promises the greatest benefit by varying the external circumstances is selected. This must often be weighed against possible damage (see above = negative benefit). This means that decision-making processes are also influenced by people's attitudes to risk and their behaviour in risk situations. Human subjectively assesses the risk of the outcome of an action that appears desirable to him from the point of view of the expected benefit. He decides to act if this risk is lower than his personally accepted risk. Figure 3.55 shows various possible combinations of subjective estimation of risk and objective risk. In principle, it should be noted that a subjective assessment of the risk that is higher than the objective risk represents a safe state.

The personally accepted risk is subject to short-term and long-term changes. In addition, it is different in the same individual for different behavioural areas (a daredevil motorcyclist does not necessarily risk taking difficult academic exams).

In connection with the risk, two fundamental problem areas are identified with regard to driving behaviour:

- Many forms of misconduct can be traced back to the fact that due to a lack of experience (= stimulus configuration in the information recording) it is not possible to estimate the risk (subjective risk) in a situation-adequate manner (e.g. too small a distance to the preceding driver is subjectively perceived as useful, since it causes the preceding driver to clear the lane; the risk inherent in this would only be experienced if a rear-end collision or at least a near collision were to occur more frequently for the driver).
- According to the theory of "risk homeostasis" (Wilde 1982; O'Neill 1977), if the objective risk is reduced (e.g. through technical measures), the human being changes his behaviour towards "more dangerous" to such an extent that the subjective estimation of the risk again receives the same distance to the personally accepted risk as before the introduction of the measure (e.g.

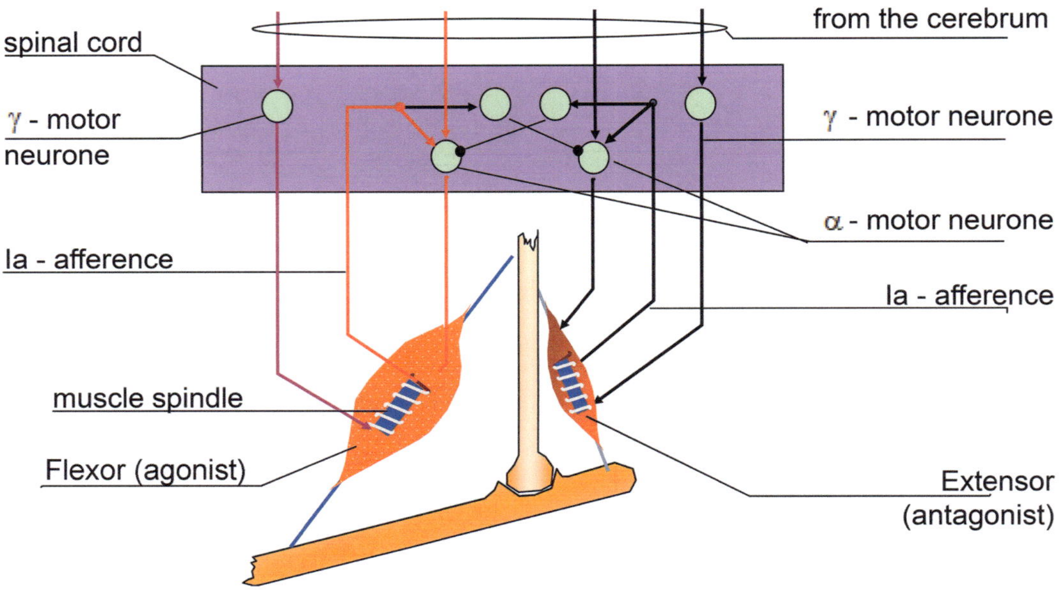

Fig. 3.56 Control loop muscle spinal cord length servo mechanism

through better roads and chassis, higher speeds are generally driven today than in the pioneering age of driving a car[38]).

The question arises in this context: What makes driving fun? If we compare different areas where we have fun, then we see that an essential aspect of fun is to touch the individual limit of one's ability to perform, to show oneself to master the situation. One example is a quote from the motor magazine of "Auto-Motor-und-Sport": "With ESP (the electronic anti-skid system) switched off, you can let the Porsche Turbo drift diagonally full of relish". The intervention of active safety systems can lead to a reduction in driving pleasure. Actually, drivers prefer to be supported in boring or arduous tasks (e.g. backlogs, construction sites or passing bottlenecks), but in situations of free driving they want to act themselves (Totzke et al. 2008). There is a great danger of falling into the area of unlawful behaviour if you want to have fun using the vehicle. All this must be taken into account when developing the Human Machine Interface (HMI) for driver assistance systems. If we want to increase road safety, we must develop solutions to ensure that the primary driving task can also be enjoyed in the area of objective safety.

3.2.3 Information Implementation

The task of information implementation is to convert the action sequences generated in information processing into reality. Only mechanical movement caused by muscle power is available for this purpose. The principles of this implementation are explained by the functioning of the so-called self-reflex bow, which in this form applies to all extremities with the exception of the head area. The same principles also apply to the head area from a regulatory point of view, although the corresponding circuits are not located in the spinal cord.

The movement programs defined by the cerebrum are essentially transmitted via two paths as a reference variable to the subordinate control loop of the reflex arc: Fast movements and gross motor skills are innervated via the α motoneurons, fine motor skills via the γ motoneurons. Figure 3.56 shows schematically a system analytically oriented representation of the implementation of information.

38 This does not necessarily have anything to do with the higher performance of today's vehicles.

The axons of the α motoneurons, which – as far as they innervate the muscles of the body periphery – are located in the spinal cord, form synapses with the striated muscle fibres accessible to arbitrariness. According to the antagonistic principle[39] a basic structure of all highly developed living beings, belongs to every agonistic muscle (e.g. flexor) an antagonistic muscle (e.g. extensor), which is able to cancel the movement of the first. The interconnection pursues on the one hand the purpose of an orderly interaction of agonist and antagonist and on the other hand the orderly conversion of an attitude or movement conceived by the cerebrum into reality. The axon of the motoneuron branches out in the muscle and thus innervates several muscle fibers. All muscle fibers innervated by a α motoneuron form a "motor unit": Since the muscle works according to the "all-or-nothing principle", a dosed muscle contraction is only possible in steps by innervation of a different number of motor units. The muscle spindles are stored in the muscle fibres and serve as length sensors. They react with an increased impulse rate when stretched via the Ia fibres. Since a stretching or compression of the whole muscle also causes a stretching or compression of the muscle spindles, the change in length of the muscle is reported via the Ia afference.

The Ia afferences are excitatory in the spinal cord to the α motoneurons of the same motor unit of the agonistic muscle and inhibitory via an interposed neuron to the α motoneuron of the antagonistic muscle. An excitation of the α motoneuron sent by the cerebrum thus causes a contraction of the flexor, for example. The contraction reduces the stimulating impulse rate of the Ia difference, so that the muscle movement comes to a standstill in a new resting position (this corresponds to a certain posture). This process is supported by the inhibitory connection of the Ia difference to the α motoneuron of a corresponding motor unit of the extensor, where the spontaneously present tension is loosened. If, for example, the extensor is stretched by an external load during a given innervation of the α motoneurons, the excitatory Ia afferences increase the tension of the flexor and their inhibitory circuits reduce the effect of the extensor. The outlined control loop is thus able to keep the position required by the cerebrum constant, largely independent of external loads, or to follow the temporal course of the desired muscle length given by the cerebrum as a reference variable ("length servo mechanism", see Schmidt 1976).

The γ innervation sketched in Fig. 3.56 is able to pre-tension the muscle spindles in such a way that the Ia difference is increased or that larger changes in the Ia difference already result from smaller changes in muscle length (measuring range adjustment). The connection with the α motoneurons causes a tension of the agonist and a corresponding slackening of the antagonist, which leads to a movement as a whole that comes to a standstill at a point determined by the γ innervation. The γ innervation can be used to regulate movement in the fine range, while the α innervation stimulates more rapid and coarse movement.

Since the γ innervation starts from the cerebellum, which itself receives afferences from the vestibular organ again (see also Fig. 3.5), the supporting motor function is also innervated via the cerebellum, whereby a stable upright posture is always aimed at when the body weight is shifted, when the floor is uneven and when acceleration forces are applied in the moving system. Through the way of γ innervation, the supporting motor activity is constantly superimposed on the arbitrary motor activity. Under extreme conditions, the supporting motor system innervated by the cerebellum can interfere with the fine motor system determined via the same path of the γ innervation.[40]

39 So-called "pull characteristic": in contrast to many technical actuators (e.g. hydraulic cylinders) muscles can only contract, i.e. develop forces in one direction (see also ► Sect. 4.1.2.2).

40 In addition to accessibility, this is the main operating problem of touch screens in the vehicle: When fine-motor targeting a button on the touch screen, the position of the hand is not only disturbed by the movement of the vehicle in the form of a physical inertia reaction, but also by the "double task" of the γ innervation.

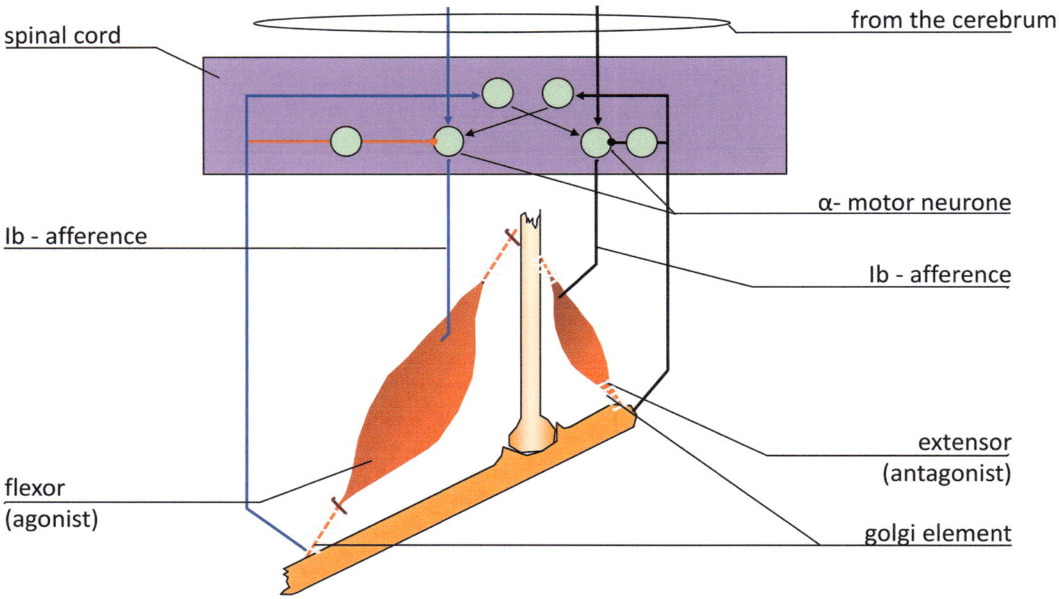

Fig. 3.57 Control loop muscle spinal force servo mechanism

In addition to the control loop of the length servo mechanism described above, a parallel control loop is realized, which can be called a "force servo mechanism". It is shown in Fig. 3.57. In this control loop, the tendon elements (golgi elements) act as sensors which measure the elongation of the tendons connecting the muscles to the skeleton and thus the force transmitted by the muscle. They have an inhibitory effect on the α motoneuron via the Ib fibre and an intermediate neuron.

If the muscle works statically against a non-movable resistance, this control loop via the α innervation transforms a force concept specified by the cerebrum as a reference variable into a defined real force. In addition, this interconnection also has a protective effect in that, in the event of an impermissibly high external force, the Ib-afferences become so large that the α innervation subsides and the muscle thus gives way to the external tension.

In summary, it can be said that our primary goal is to use our musculoskeletal system to realize a course of movement (= action) designed in the cerebrum. We innervate the muscles used for this in such a way that a force is created that makes this movement possible. The achieved position is reported back via the muscle spindles and the position receptors in the joints. The comparison in the spinal cord stops the movement if the "target" given by α motoneurons matches the "actual" measured by the Ia difference. Then only so much force is expended that the "target" can be held. When external forces attempt to change the position of the skeleton as determined by the brain, the "short path" of the spinal cord involuntarily changes the force applied so that the position is maintained. In this application, the haptic reaction is four times faster than the reaction that would occur via the external path of visual feedback (Gillet 1999).[41] Only if the force is applied against a fixed resistance, a desired force can be applied in a metered manner via the force servo mechanism realized by the Ib difference. However, force feedback plays a decisive role in every movement process. If such movement processes have often been repeated, it is precisely through this connection that an inner model emerges which provides information about the correctness of the process and which is drawn into consciousness in the case of an

41 This consideration plays an important role in connection with so-called "steer by wire" steering systems (see further details in ► Sect. 6.4.3).

unusual sequence of the mostly autonomous movement process.

When driving a car, the interconnection mechanism described here plays a decisive role in the steering feel when cornering (see Wolf 2009). The combination of the motion perception via the optical flow, the associated acceleration forces and the restoring forces perceptible at the steering together make up the inner model for cornering. Any over-threshold deviation therefore leads to decision-making processes and thus, in a given traffic situation, to a potentially dangerous delay in the reaction. Force feedback is also very important when operating switches and operating elements, because it provides information about the correctly performed action.

Due to the high degree of exercise, many movement processes can take place practically independently of conscious information processing. Nevertheless, they do not represent an isolated achievement of information translation, since they all refer to movement processes controlled by the eyes. The resolution of the eyes is much better than the pure accuracy of movement, which is supported by Hacker's observation (1967), according to which the movement performance generally deteriorates many times over in the absence of visual control.[42] By suitable haptic feedback, however, a clear improvement can also be achieved in the absence of visual control. Thus, for the accuracy of information implementation, feedback on information reception plays an important performance-enhancing role.

In summary, it can therefore be stated that errors are caused by the implementation of information if the technical design of the vehicle has one or more of the following defects:

- Insufficient adaptation of the controls to the anatomical characteristics of the human being,
- lack of or insufficient haptic distinguishability of the controls,
- missing or incorrect force feedback of a control element (The switching point of a pressure switch should also be haptically noticeable by a sudden change of the reset force, e.g.),
- missing or inadequate feedback of the controlled process (e.g. the "fly by wire" problem: with many mechanically decoupled actuators, which switch servo motors, there is no feedback about the effect caused by the machine, as is natural with mechanical coupling).

3.3 Information Processing when Driving a Car

3.3.1 The Feeling for the Time

Driving a car is movement and movement is change of place in time. The relationship of human being to time is therefore of great interest for understanding action in movement. How do people feel about the current time or the time they have lived through recently? This question is important for all the contexts mentioned, because we live in the now and today, have memories of the past and expectations for the future, aspects that have a lot to do with our memory. Pöppel (2000) has dealt intensively with this question. In accordance with the many neurological brain experiments he found that our consciousness accessible thinking is clocked in 40 ms periods. Of particular interest here, however, are his insights into what we perceive to be the present. In contrast to a strictly logical consideration, according to which the present does not actually exist because it is merely the infinitely short dividing line between the past and the future, we experience the present in reality. Pöppel found that the concept of the present can be divided into three areas: what we immediately perceive as the present corresponds to a period of about 2 seconds. But the feeling of the present also includes the immediate past, which Pöppel calls the present of the past. It corresponds to a time period of a few seconds (for example, we can repeat a short sentence just spoken at any time with

42 A problem that is especially important in connection with the use of touchscreens in the car!

the same wording; this is no longer possible for sentences said more than half a minute ago). But the present also includes the present of the future, that is, the expectation of what will immediately follow. We can only act if we have an inner vision of what we should do in the immediate next moment. The total time from the present of the past to the present of the future is about 10–15 s (Wickens and Hollands 2000). It thus corresponds to the forgetting time that is often attributed to the so-called short-term or working memory, i.e. what was called consciousness in the above-mentioned sense.

The knowledge about the perception of time is of great importance in connection with the understanding of human information processing when driving a car. Since driving a car is a "present experience", most actions, decisions and reactions take place in the time window of the "present of the present" of 2 seconds. A large part of the knowledge about the underlying behaviour can be gained from the gaze behaviour, which will be examined in more detail below.

3.3.2 Human Eye Behaviour when Driving a Passenger Car

3.3.2.1 General Characteristics

Since, as already mentioned, the eye can only perceive a small angular range of 2–3° sharply and only an inner image of the outside world is composed by eye movement in the brain, it is a quite natural everyday experience to grasp the object of a fellow human being's attention by paying attention to what he or she is looking at. By recording the viewing direction, we have, so to speak, a spy tube into the interior of his behavior. Of course, this spy tube does not provide absolutely certain knowledge about what was really seen and which inner models of the observed were actually excited. In addition, a distinction must be made between the external "physiological fovea" and the internal "psychological fovea", which cannot be observed directly but can at best be inquired about (see Zichenko and Virgiles 1972). Both directions of attention do not necessarily have to be identical. The difference can be up to 2° (Kaufmann and Richards 1969), although in general a high correlation with a certain time lag between the two can be assumed.

In addition to this distinction between internal (psychological) attention and external, observable (physiological) attention, the way in which the gaze is directed must also be taken into account (see the Remington 1980 experiments). One speaks of *distraction* if an impulse is triggered by a changing stimulus in the peripheral field of vision, which (on a physiologically relatively low level) causes the gaze to turn there and is triggered by *averting* when, through a deliberate act, the gaze is turned to a new object of immediate attention ("Area of Interest", AOI). Only the latter is actually of interest for the question to be dealt with here, although both are shown in the same way and therefore cannot be separated experimentally from each other exactly.

In spite of the general limitations, gaze analysis is a worthwhile tool to obtain information about the current attention. At least the statement "What is not in the field of vision cannot be captured subjectively and thus plays no role in the generation of action" can be a basis for a practicable experimental procedure.

The time capacity required for an ideal performance of the primary driving task depends on numerous influencing factors such as the individual characteristics of the driver, the traffic density, the road conditions and the weather conditions. The more complex a current driving situation is, the greater the proportion of time required for the primary driving task. The remaining time is available for secondary and tertiary tasks, whereby it is generally desirable to minimise the amount of time required for this in order to achieve maximum security gains at the same time.

Schweigert (2003) distinguishes between three levels of basic visual tasks, which must be performed within the framework of the guidance and stabilization task:

1. Continuous control of the own movement on the roadway

The movement of the own vehicle must be visually monitored. For experienced drivers, this happens to a large extent via peripheral vision. Obviously, information in the near area (approx. 6 m in front of the vehicle) is important (Chatziastros et al. 1999), whereby the lower half of the face is clearly more efficient than the upper half.[43] In addition, however, orienting foveal glances are always necessary, because the foveal share increases with decreasing width of the road or when passing obstacles as well as with inexperienced drivers.

2. Continuous anticipation of the behaviour of other road users

The area in front must be continuously monitored in order to be able to recognize events in good time that require one's own reactions. This corresponds to the explorative scanning behaviour postulated by Cohen (1976). This is only possible with the help of foveal vision. Also according to Rensink et al. (1997, cited after McCarley et al. 2002) the driver must actively scan the driving environment, mentally process objects ("encoding") and notice changes in them. If the behaviour of another road user, who could enter the sphere of influence of his own vehicle, cannot be extrapolated and predicted with sufficient accuracy, he must be observed (processing behaviour, according to Cohen 1976). Foveal vision also predominates here. Likewise the traffic behind the own vehicle must be observed continuously and not only situatively (see point 3). This is necessary in order to be able to make a sufficiently precise internal representation of the rear events in good time.

3. Situatively required gaze behaviour

In addition, there are driving situations which, in contrast to the tasks discussed above, are not necessary continuously while driving, but require a special gaze behaviour adapted to the respective situation. This is the anticipation of the behaviour of other road users at junctions and obstacles.

When observing gaze behavior during driving using an eye-tracking system, Schweigert (2003) was able to clearly distinguish between "scanning" and "processing" gaze patterns (see ◘ Fig. 3.58). Scanning views are characterized by a fairly short duration of an average of 400 ms. They are used to record road edges, other road users and traffic signs on a quasi-continuous basis. Obviously, they serve to record the data necessary for generating the internally generated target course within the framework of the guidance task (see also ▸ Sect. 3.3.3). Part of this information necessary for guidance can also be obtained by capturing the optical flow in the peripheral area of the image capture. Through the targeted processing views, so-called areas of interest (AOI's) are specifically fixed. They are particularly often used to gather information for anticipating the future behaviour of other road users. But it can also be the display instruments, the mirrors, the central screen, or other interesting but not traffic-relevant objects in the surroundings. The processing views require on average twice the time of the scanning views (see ◘ Fig. 3.59).

In this context, it should be pointed out that Underwood et al. (2003) found that the gaze behaviour of beginners differs significantly from that of experienced drivers. While beginners generally fix for longer and immediately direct the view from each AOI back to the fixation point on the road, experienced drivers scan much more the road, especially in distant areas, with a processing view dedicated to specific AOIs (see ◘ Fig. 3.60).

For the evaluation of the gaze behavior it is to be distinguished basically between the *traffic situation* which represents the objectively given spatial and temporal constellation of the traffic-related influencing variables, the *driving situation* which represents the resulting principally perceptible driver's view and the *driver situation* which describes the subjective driving situation actually perceived by the driver (Reichart 2000). If it is at all possible to define a target viewing behaviour, this can

43 Summala et al. (1996) carried out field trials in which the test persons had to look at a display in the vehicle so that the road could only be perceived peripherally. The display was presented with an eccentricity of 7°, 23°, and 38° in relation to the horizontal line. With increasing eccentricity, the tracking quality deteriorated.

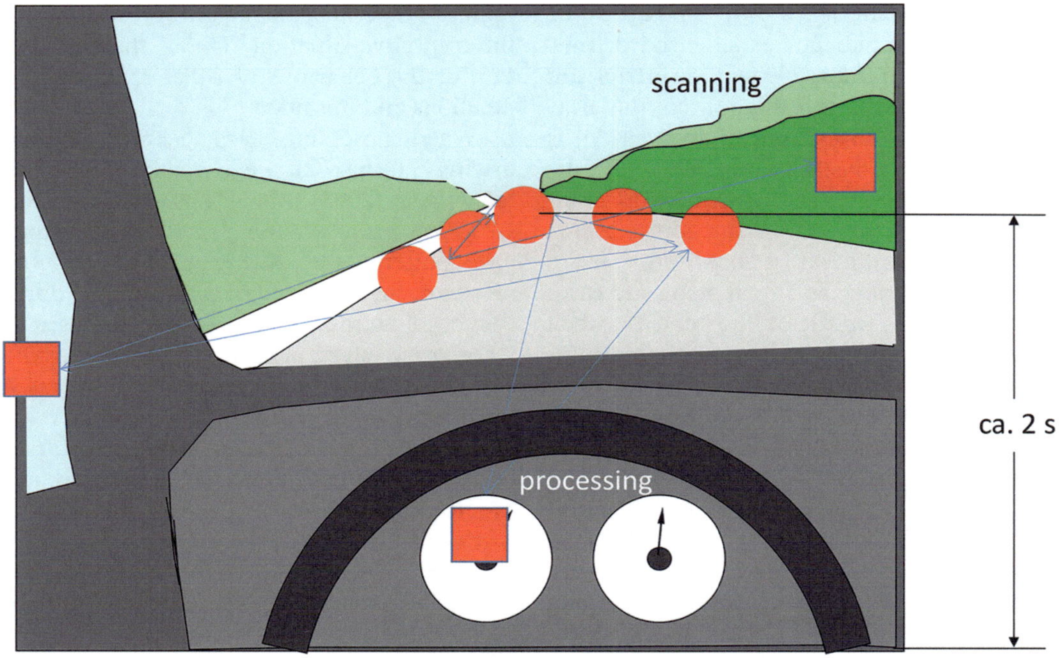

Fig. 3.58 Gaze behavior: "Scanning" views capture the course of the road. "Processing" looks at certain areas of interest (AoI's)

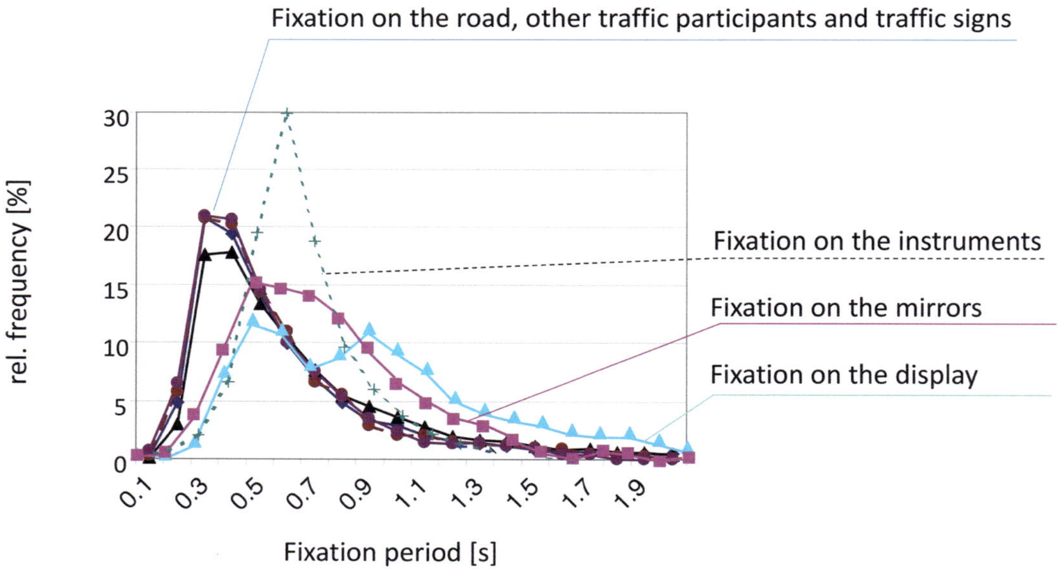

Fig. 3.59 Experimental results of gaze detection studies (Schweigert 2003)

only be done at the level of the driving situation. The taxonomy required for this is based on that of Fastenmeier (1995), which is supplemented by a number of subgroups that primarily take the driver's view into account. Table 3.7 gives examples of different complexity classes and lists situations used in Schweigert's (2003) study.

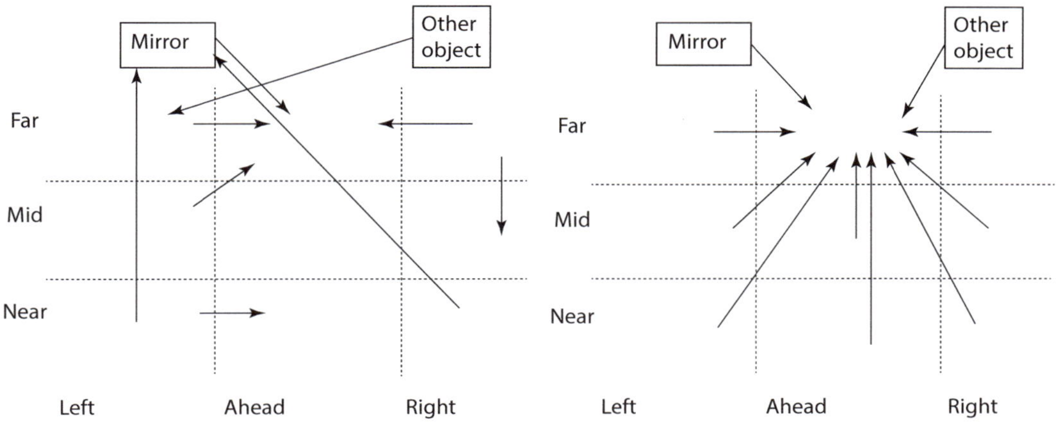

Fig. 3.60 Typical sequence of views for experienced drivers **a** and beginners **b** (Underwood et al. 2003)

Table 3.7 Examples of complexity classes and situations used in the studies by Schweigert (2003)

Com-plexity	Example	Selected situations
Low	Highway, little traffic Country road, little traffic City with little traffic, knot-free road	Single-lane country road, low traffic, speed 80 km/h or 60 km/h
Medium	City with signposted intersections, traffic lights, bottlenecks and curves Motorway junctions Highway with curves and slopes	1st motorway, 3 lanes, speed 80 km/h, two entrances and exits, medium traffic 2nd city, two lanes, traffic light controlled intersection, medium traffic
High	City, sign-posted junction with Waiting Motorway junctions	City, Tempo 30 km/h, right-left before, bottlenecks by parked cars

Since the present case deals with visual complexity, Schweigert tries to determine the degree of complexity of the task more precisely. Obviously, the presence of oncoming and preceding vehicles, as well as junctions and bends, plays a major role. A more far-reaching approach presents the complexity factors of a route section that are relevant for the gaze behaviour individually. Figure 3.61 shows the composition of this complexity index. The number of these objects is used for the influencing factors "curves" and "nodes". For the "presence of preceding vehicles", the mean proportion of time at which preceding vehicles were looked at in the specific section of the line is determined as a benchmark. The "presence of oncoming vehicles" shall be measured by the average number of such vehicles. This is a pragmatic and easy to understand approach that does not weigh the factors.

On the basis of these considerations, Schweigert carried out tests on gaze behaviour with a total of 30 test persons on public roads and subjected the results to a detailed evaluation.[44] Of interest were the "normal" gaze behaviour and its change through additional tasks. The relative frequency of the fixation duration here shows a uniform left-hand distribution with a modal value between 0.3 and 0.4 s (see also Fig. 3.59). More complex sections of track result in narrower distributions and a more pronounced and at the

44 The evaluation is based on extensive investigations: 10 h Video material was evaluated, 40,000 fixations, approx. 12,000 glances/fixations analysed and 3000 visual tasks quantified.

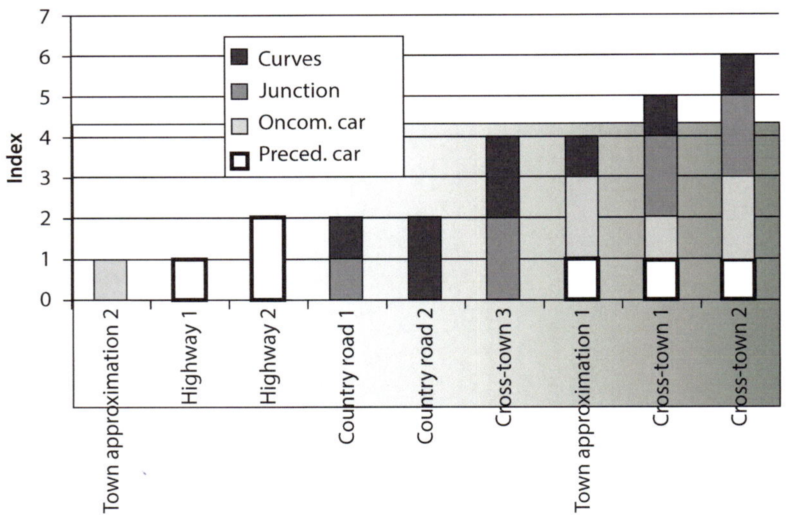

Fig. 3.61 Composition of the complexity index for individual route sections (according to Schweigert 2003)

same time smaller modal value. Depending on the objects viewed, there are significant differences. On average, the shortest time spent looking at traffic signs (0.5 s) and the longest time spent looking at the vehicle in front (0.73 s). The viewing time during processing (rear vehicle, speedometer) is increased by a factor of about 2. Predominantly preceding vehicles (62.6%) and the roadway are looked at longer than 2 s. The objects that have been looked at for a long time are always those with essential relevance for receiving traffic events. Fixations on the vanishing point and on the vehicle in front take up approx. 50% of the fixation time, whereby the respective relative ratio depends strongly on the route section. On the busy motorway section, the fixations on the vanishing point dominate with 31%, followed by 24% on vehicles in the other lanes and 19% on the vehicle in front. On the winding but little frequented through road 42% of the viewing times are directed to the vanishing point, 14% are used for lane control and only 9% are dedicated to the vehicle in front. On the motorway section with the highest average speed, 54% of the fixation time is directed to the vanishing point. From these data, however, no "target" for the correct gaze behavior can be derived. Nevertheless, these values are used in the following as average behaviour for the further evaluation of gaze behaviour in additional tasks.

On this basis, a measure of quality can be given for some tasks in the form of "number of correctly performed tasks in relation to the number of tasks to be performed". It is noticeable that only the tasks

- Control of a vehicle in front,
- Control of bottlenecks,
- Observation of pedestrians entering the traffic area,
- View into the rear-view mirror during certain manoeuvres

will be completely fulfilled. The so-called shoulder look is only performed to 76%, traffic signs to the speed regulation and speedometer only to 64%. Traffic lights are considered to 90% and danger signs only to 45%. Many other traffic signs are even less noticed.

The viewing behaviour depends to a large extent on the section of road that is driven through and the current local traffic situation. On average, more complex route sections have shorter fixation times and more frequent changes, as more information per time unit obviously has to be received here. Particularly short (approx. 0.4 s) are the eye sequences on objects irrelevant to the driving task and searching fixations (scanning), while targeted fixations (processing), which are particularly often used to gather information for anticipating the future behaviour of other road users, are longer (approx. 0.8 s and more). However, all fixation durations remain below the limit set by 2 s. This also applies to the eye-off-road-times. It can also be seen that there is no need to look into the rear-view mirror at all on sections of road with only one lane, where overtaking is virtually impossible; in convoy traffic, regular monitoring of the speedometer plays a very subordinate role. In continuous visual tasks, drivers seem to be able to derive sufficient information from peripheral vision to avoid irrelevant glances (e.g. observing no parking signs when not looking for a parking space). Even at junctions where you have the right of way, you are less likely to look for other road users.

By means of an optical or acoustic additional task[45] the influence of the operation or the observance of additional devices on the gaze behaviour should be examined. The fulfilment of the driving task may also impair the fulfilment of the additional task. However, the review shows that the quality of the additional task did not differ when the vehicle was stationary or in motion and that it was always between 92% and 98%. With regard to the main task of driving, the additional auditory task shows a sharp drop in quality only in the securing behaviour (e.g. shoulder view perform). In the additional visual task, on the other hand, the ride quality deteriorates in all categories. For compensation, the longitudinal distance is increased for 30% of journeys with an additional auditory task and for 39% of journeys with an additional visual task. Changes in fixation and gaze duration are only apparent in the auditory tasks with an extremely high proportion of fixations above 2.5 s, which was not observed in the additional visual task. Otherwise, there are no changes in the distribution forms of gaze frequencies. However, when processing the visual task, the mean fixation times in all categories decrease by 26%, since the test subjects obviously try to gain time for fulfilling the additional task. With the additional auditory task, on the other hand, the viewing time in a straight line direction increases by approx. 50%. A remarkably high proportion of fixations can be observed on all route sections, which last longer than 2 s. Fixations of more than 2 s in the case of the additional auditory task, as well as in the case of journeys without an additional task, are directed primarily at the vehicle in front, while in the case of the additional visual task they are dedicated to the display. The averting to the display with the visual task shows for very complex traffic situations a pronounced modal value of 0,5 s. For low-complex sections (definition see ◘ Fig. 3.61), two peaks are found at 0.5 and 1 s, for very low complexity index of the traffic situation even several maxima at 0.5, 0.9, 1.1, 1.3 and 1.6 s. Obviously, at very low complexity, the driver is inclined to perform not only one but even several tertiary tasks one after the other.

In addition to the continuous driving task just discussed, which includes the regular observation of one's own lane keeping and the behaviour of other road users, there are also a number of situation-specific tasks, such as observing the vehicle ahead, looking at the shoulder before overtaking and driving through a narrow lane, which are characterised by the fact that there is only a narrow time window within which the fixations, glances or sequences of glances must take place. In connection with an additional task, the degree of fulfillment of the undisturbed task is only reached in a few cases. Particularly heavy losses are recorded in tasks such as observing pedestrians and searching for a possible vehicle at a junction.

In general, it can be stated that the additional task reduces the amount of time spent on irrelevant viewing objects, whereby this reduction amounts to 65% for the acoustic task and even 96% for the visual task. This proves that through the additional task, so to speak, "nothing is left for other looks". The proportion of journeys without a single mir-

45 A visual task was conceived, in which the driver had to recognize a certain geometric figure on a display mounted in the middle of the dashboard and had to confirm it by pressing a button on the steering wheel, while the next task was left to the test subject by pressing another button. The acoustic task consisted of recognizing by keystroke, from randomly presented monosyllabic words, the one that represents a living being. Also in this case, the frequency of the task presentation could be controlled by the test person himself and interrupted at any time in difficult traffic situations.

ror image increases in both additional tasks. The difference between an additional task and "normal driving" does not refer to how often you look in the mirror, but to whether you look in the mirror at all. The same applies to the speedometer view.

A relatively high correlation ($r^2 = 0{,}84$) can be found between the driving error "inaccurate lane guidance" and the visual measurements of the maximum or average turnaround time, whereby even average turnaround times of 0.3–0.6 s lead to inaccurate lane guidance for 60% of the journeys, while lane guidance errors can be observed for all journeys (100%) for gaze averting times of 1.2–1.5 s. However, there is no connection between the duration of the turnaround and jerky steering corrections.

Additional tasks, whether presented visually or acoustically, have a negative effect in different ways on the viewing behaviour with regard to the driving task. In the case of the additional optical task, the main advantage is that the viewing time required for it must now be shared with that required for the driving task; in the case of the additional acoustic task, on the other hand, there is a narrowing of the viewing behaviour to only those parts relevant to driving.

Drivers don't just look at traffic-relevant objects. The investigation of Schweigert shows beside others that with undisturbed, less complex driving almost 90% of the glances belong to things not relevant to driving. This proportion is reduced to only 68% for the acoustic secondary task and even to 17% for the optical secondary task. The non-driving-relevant glances can almost be seen as an indicator of the complexity of a driving situation, an indicator which, however, only has an informative value on average over longer test periods, not in the acute situation. With increasing complexity, the non-driving-relevant looks decrease more and more. This can be illustrated in the shell-shaped compensation model shown in ◘ Fig. 3.62. From the outside to the inside these shells with increasing complexity index of the situation show the following compensatory behaviour:

1. Decrease in the attention paid to objects not relevant to traffic,
2. Reduce the viewing time on certain AOI's (areas of interest),

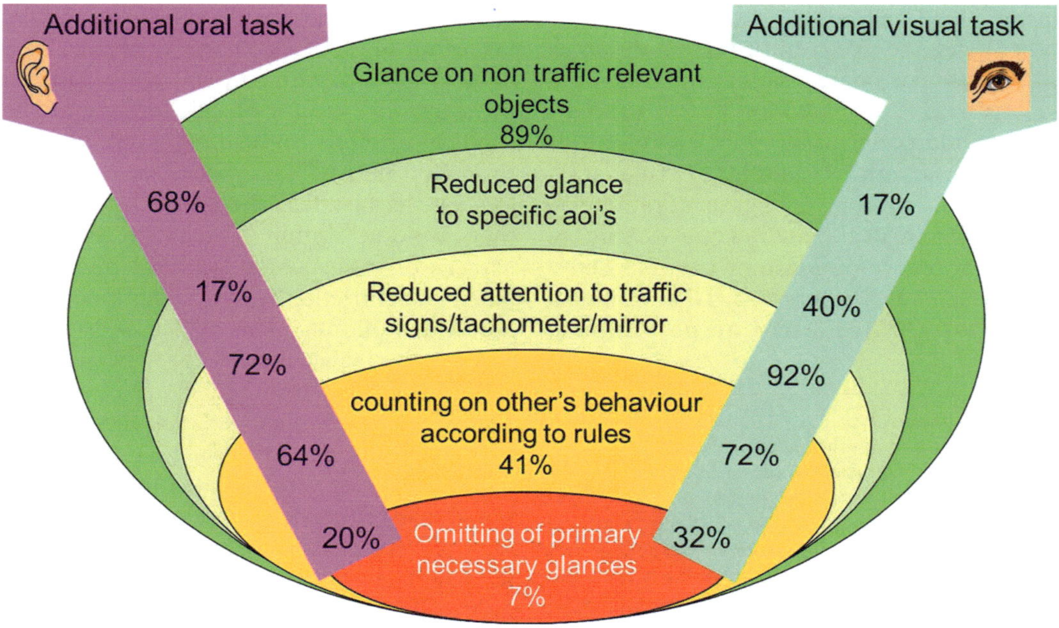

◘ **Fig. 3.62** Percentage of faulty gaze behaviour (Schweigert 2003)

3. Less control of traffic signs, speedometer and mirrors,
4. Relying on rule-compliant behaviour of other road users,
5. Refrain from primarily traffic-relevant glances.

Depending on the additional optical or acoustic task, these compensatory actions increase even further, whereby the percentage for the additional acoustic task is slightly lower overall.

From all this it can be seen that the presence of the present, which lasts about 2 seconds, is of great importance for our gaze behaviour. All experiments on viewing behaviour when driving a car show that normally only at a distance of 1–1.5 s, maximum 2 s information is sampled in advance (distance = speed × preview time; Donges 1978; Yuhara et al. 1999; Guan et al. 2000; Schweigert 2003). And there are many experimental results that show that we normally accept to take the view up to 2 s off the road (e.g. Zwahlen et al. 1988; Gengenbach 1997; Schweigert 2003). Gengenbach (1997) was able to show that after turning away the perceived loss of information is compensated by an increased sampling rate (see ◘ Fig. 3.63).

In a further study, Rassl (2004) examined in detail the influence of the tertiary task and various operating layouts of such tasks on gaze behaviour. Among others, his subjects had to select one of 3, 5, 8, or 14 alternatives while driving by means of a rotary pusher (similar to the BMW-I-Drive). It is of interest that the selection took on average about 1.2 s and there is no significant difference between these deviation times depending on the number of alternatives presented. However, if one considers the maximum duration of averted views, the case of 14 alternatives differs significantly from the others. The average maximum deflection time for 14 alternatives is only 2.2 s on average. However, as ◘ Fig. 3.64 shows, in this case even sporadic deflection times of 12 s were observed.

This is not a singular event. In Rassl's investigations, in another part of the experiment even the deflection of 16 s was observed and in further unpublished experiments on the operation of the ACC system, deflection times of up to 12 s were also found. The psychological explanation for these long periods of darkness lies in the contemporary experience already mentioned. Normally, after 2 s of distraction, we are worried and look back at the road. However, if at the beginning of the distraction from the scene we had the impression that no major changes were to be expected, and if in addition the distracting task – for whatever reason – becomes attrac-

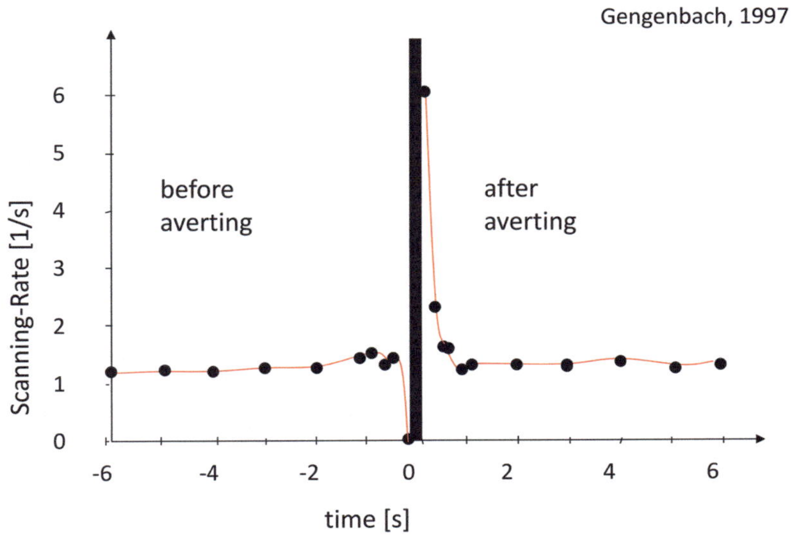

◘ **Fig. 3.63** Scanning behaviour before and after turning away from the road (after Gengenbach 1997)

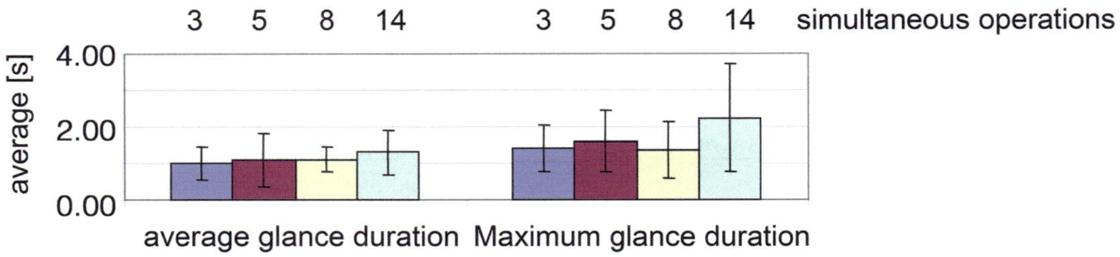

Fig. 3.64 Gaze behaviour in connection with tertiary driving tasks (according to Rassl 2004)

Fig. 3.65 Typical sequence in a left- and right-hand bend

tive, we rarely use the total period between the present of the past and the present of the future. In none of these cases do we subjectively perceive the long distraction time from the street, because through the inner model the initially assumed course of the scene on the street is subjectively present to us.

3.3.2.2 Visual Behaviour Without Lateral Traffic

Bubb and Wohlfahrter (2012) took a closer look at the scanning and processing looks when cornering. The experiments were carried out in the simulator of the Institute of Ergonomics at the Techical Univerity Munich (TUM) which uses the Silab software and a 180° projection to provide the test driver with a very realistic picture. The Silab software allows the test subject to be brought into a specific situation, regardless of their previously uncovered behavior. The sequences of glances, which at first appear random, both individually and due to the current situation, allow the regularity, which is nevertheless evident in all test subjects, to become visible by means of Markov analysis with the significance test according to Liu (1998). The most important anchor points were the area of the tangent point, the near, middle and distant road area, the speed display, traffic signs and, if available, a preceding vehicle, oncoming traffic and a breakdown vehicle present in the simulation.

When cornering, a rough distinction can be made between the sections "approaching a curve and orientation", "gaze behaviour in the curve" and "gaze behaviour when leaving the curve". The typical behaviour is illustrated in Fig. 3.65. The driver then looks at the speedometer during the running-in phase to make sure that the speed is adequate for the estimated curve radius. The next view is directed to the tangential point of the curve (in a left turn this is – normally – the central reservation, in a right turn this is the right side of the road). Among other things, it serves to anticipate the curve and is fixed on average 2 s before entering the curve. The anticipation time increases (i.e. the tangent point is fixed earlier) especially with increasing restriction of visibility by any objects, e.g. a truck in front. The next look checks the distance to the right side of the road (in the case of the right-hand bend, this can also be the distance to the left central reservation for some test subjects). The next look then assures itself of the distance to the respective opposite side. Now a control look at the speedometer follows again.

In a further simulator study on visual behaviour, Remlinger (2013) found that there are significant deviations from this "standard visual behaviour". On the basis of experiments with 40 test persons, he was able to observe four types of gaze behaviour, namely

- View type "Inside", which reflects the behaviour described above and in other literature, according to which the gaze is directed to the inner edge of the curve when cornering (60% of test subjects),
- View type "outside", which characterizes persons who orient themselves principally to the respective outer edge of the curve (12.5% of the test persons),
- View type "Left", represented by persons who – independent of the direction of the curve – orient themselves in principle at the left edge of the lane (10% of the test persons),
- View type "right", represented by persons who orient themselves principally at the right edge of the lane (17.5% of the test persons).

According to Bubb and Wohlfahrter (2012), follow-on driving situations are characterised by a pronounced visual link to the vehicle in front. This is particularly evident in the case of a truck in front (see Fig. 3.66).

Fig. 3.66 Typical sequence of vision when driving behind a vehicle in front

In a further study in the driving simulator of the Institute of Ergonomics, 35 test persons were guided through an overland and city circuit (Bubb and Wohlfahrter 2012). Among other things, eye examinations regarding peripheral vision were carried out. After special scenes, the test persons were asked about the task. In one scene, the test persons had to read a text on an advertising board with the additional task of maintaining their own driving speed in accordance with the current speed limit. The scene is in a slight right turn. Figure 3.67 shows this scene. On the right hand side there is a traffic sign ("60 km/h in wet conditions").

The limit "60 km/h in wet conditions" is only correctly recognised by 14.3% (5 test persons) of the drivers, although the traffic sign is easily visible, free-standing and never concealed (see Fig. 3.67). The workload is obviously very high for the majority of the test persons in this scene. Within the gaze analysis it is noticeable that during the fixation of the billboards hardly any control gaze on the road takes place, as can be expected according to the "normal" gaze behaviour shown in Fig. 3.58. Although this confirms that lane keeping via peripheral perception is relatively problem-free despite cornering, it also increases the angular distance to the traffic sign and thus apparently reduces the probability of perception.

13 test persons have fixed the 60 km/h limitation by one or more glances. However, only 40% of these participants correctly recognized the additional sign "if wet". From this it can be deduced that the additional sign "if wet" is not identified as an important key element. Surprisingly, the drivers who only noticed the speed limit did not even notice the existence of an additional sign (questionnaire after the scene). This information was thus apparently sorted out as irrelevant by the perception system. This example shows that for the perception of details it is absolutely necessary to turn one's gaze towards them.[46]

46 The results of this experiment show what effective support a traffic sign recognition and the representation of the pictogram in the I-Combi or Head-Up-Display would provide.

Fig. 3.67 On the left side is the advertising board. On the right you can see the speed limit sign. It is completely free-standing and can be viewed at any time

In a further study, 23 subjects were observed in a field trial with a gaze detection system (dikablis). The test persons went through selected sections of the route and had to follow a given route. The route sections were very varied and ranged from small right-before-left situations to crossing large intersections in the city area. Especially the evaluation of narrow spots (single-lane side streets with parking vehicles) shows how individual information units at the roadside have to be refreshed at shorter intervals in order not to overlook anything important. Figure 3.68 shows the saccade frequencies for the different track types.

The same number of saccade jumps is required on the "motorway" and "multi-lane with markings" sections (significance $\alpha = 0{,}356$). They differ significantly from each other on all other types of carriageway. The saccade frequency therefore depends on the visual requirements of the track type. The higher the visual demand and thus the higher the need to receive information, the higher the saccade frequency (this is a result that has already become visible when looking away and turning the gaze back to the street; see Fig. 3.63). For motorways – as well as for multi-lane urban traffic with lane markings – no additional information other than the lane itself and the surrounding traffic is required to define the driving strategy. This means that the situation is clearly regulated by road markings and the surrounding traffic flow. On the one hand, the available driving space is narrower when driving through towns, i.e. the potential threat from other road users increases. Therefore, information from the environment needs to be refreshed with higher frequency. On the other hand it is necessary to collect a lot of secondary information like pedestrians, oncoming traffic, traffic lights etc.. The driver must therefore consider several sources of information to assess the situation. Another aspect that determines mental stress is the everyday nature of a situation. If, for example, the driver finds himself in unfamiliar situations in which he lacks important sources of information (e.g. missing central strip), he looks in the environment for visual aids to help him assess the situation. The search views result in an increased saccade frequency, as can be found with the track types " multi-track without marking" and, " single-track without marking". The comparatively high saccade frequency of 2 saccades per second at the bottleneck is due to the fact that the driver constantly alternates between the left and right edges of the bottleneck in order to avoid collisions. A slight yaw movement of the vehicle can also be observed objectively.[47]

47 This behaviour causes the drivers to change their control behaviour. The lateral distance and no longer the longitudinal orientation of the own vehicle is controlled. In line with the considerations in ▶ Sect. 2.4.1, the driving task for the driver changes from more easily manageable speed control to more difficult acceleration control. It could be shown that this effect is omitted if a vehicle drives ahead in the bottleneck, whereby the concentration on the longitudi-

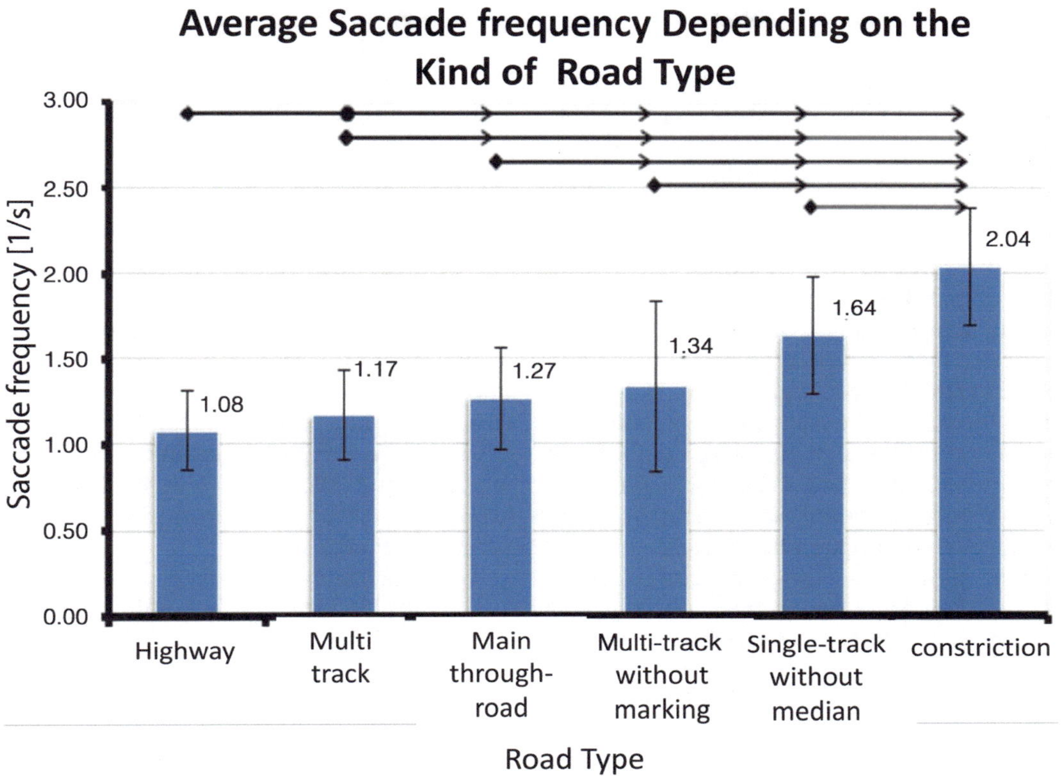

Fig. 3.68 Mean saccade frequency over all test persons depending on the type of track. The arrows in the upper part of the figure indicate between which categories a significant ($\alpha < 5\%$) difference has been observed

3.3.2.3 Viewing Behaviour at Crossings

In addition to accidents in purely longitudinal traffic, which essentially consist of departures from the road with possible consequences and rear-end collisions, accidents at junctions also play an equally important role. In Europe, depending on the country, intersections are responsible for 30–60% of all accidents resulting in injury (Intersafe 2009). Of these accidents happen outside of town to astonishing 42% at easily visible intersections.

nal orientation is achieved again. Experiments by Israel (2012) also show that this effect can be achieved by displaying the safety distance in the form of a safety bar at the corresponding distance ($x_s = v \cdot t_s$ with $t_s = 1.5$ s) is made available on a permanent basis in the so-called contact-analog cHUD (Bubb 1981) as a virtual transverse beam lying on the carriageway (see ► Sect. 6.3.1.2).

One of the few experiments dealing with eye movements at crossroads is by Langham (2006). Among other things, he observed the differences between experienced drivers and beginners, as Underwood et al. (2003) found for driving on country roads, namely that experienced drivers fix less concrete points than beginners, but rather scan wider areas. Also in the fixation sequence there are differences in the form, that beginners rather fix objects in a clear order, while experienced fixators let the glances jump between different fixation directions. However, the speed of the scanning plays a prominent role. Plavšic et al. (2010) showed in simulator experiments that an average scanning speed is most favourable. In the experiments, two groups of volunteers who caused a crossing accident could be separated, those who scanned the scene relatively slowly and those who scanned the scene ineffectively fast. Obviously too many and too fast saccades lead to suppression of visual perception and too few saccades to an inefficient capture of the scene.

Plavšic (2010) has dealt in detail with the behaviour and especially the gaze behaviour at intersections, whereby the above mentioned driving simulator of the institute of Ergonomics of the TUM was used again. The object of their study was to examine the influence of right-of-way regulations, the type of driving manoeuvre required (straight ahead, left/right turn), the influence of a vehicle in front and the influence of time pressure. These factors were derived from a theoretical analysis as the main influencing factor on accidents at crossings (Plavšic et al. 2009). On the basis of these investigations, a test track consisting of 10 crossing situations was constructed. The manoeuvres (straight ahead, right and left turns), the right-of-way regulations (have right of way, grant right of way, stop sign and right-to-left) and the presence of a vehicle in front were varied for this purpose.

On this basis and the further classification of the necessary views according to Schweigert (2003) into "essential", "important", "foresighted" and "irrelevant", trigger points are set for the individual scenarios which determine a normative view behaviour. The observed behaviour is then relayed to it.

In order to define the task, the traffic situations are bundled in "chunks" according to the considerations of Miller et al. (1956). Chunks are road users who move almost uniformly and can therefore be treated as a single unit from the driver's point of view. The more such independent chunks a traffic situation has to be described, the more complex it is for the driver.[48] In addition, a distinction has to be made between chunks to which the driver has to react immediately and chunks which he only has to observe. On the basis of this idea, Plavšic is developing a method to describe the complexity of a traffic situation. This makes it possible to rank the difficulty of the scenarios used in the experiments. The intersection situations investigated and their evaluation are outlined in ◘ Fig. 3.69.

In general, ideal driving behaviour can be described as behaviour that complies with legal regulations. Nevertheless, there is a range of correct behaviour for every traffic situation. To define all this, the intersection area was divided into five segments: Approach, delay, drive through, turn and leave the intersection. A rule-based decision is necessary in every area. Both the decisions and the tasks and their importance were determined for each intersection and each area. A fault analysis is carried out by comparing defined and actual driving behaviour. The detailed tasks to be performed in each segment are described in detail in Plavšic (2010).

With the simulator it was possible to control the traffic situation depending on the behaviour of the test person. This guaranteed that every test person was in the same traffic situation.

The experiments were carried out with 24 subjects (average age 27 years, 3 women, 21 men). After the test subjects had gone through the first test rounds, the individual scenes were presented to them for assessment. Then the experiment was repeated under time pressure. This was followed by a subjective assessment of the individual scenarios.

The results of the tests are divided into subjective data – obtained from the evaluation of the questionnaires – and objective data (driving simulator, eye tracking system). With the *subjective data* the NASA TLX questionnaire (see ► Sect. 11.3.3.1) shows that crossing an intersection is a very challenging task. There is also a significant difference in the load between the baseline and the journey under time pressure. Otherwise, practically no differences could be found in the subjective assessment of the task difficulty, the orientation, the misjudgement and the risk of a collision in the various scenarios. The test subjects would like an assistance system for coping with intersections, but have no idea what such a thing could look like.

The *objective data* refer mainly to the recording and analysis of eye movements, especially the deviation between the observed

48 The difference to Schweigert's complexity definition is that in a simulator experiment the complexity can be defined as described. In a real live experiment, on the other hand, it is extremely difficult to categorize the complexity of situations according to Plavšic's definition.

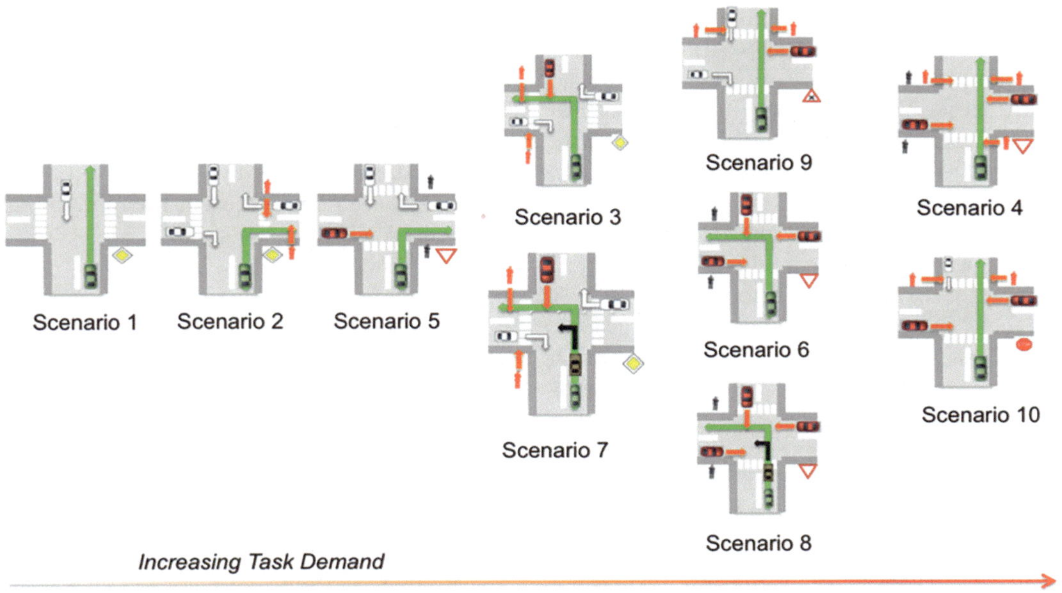

Fig. 3.69 Crossing scenarios used in the study of Plavšic are sorted according to the degree of difficulty

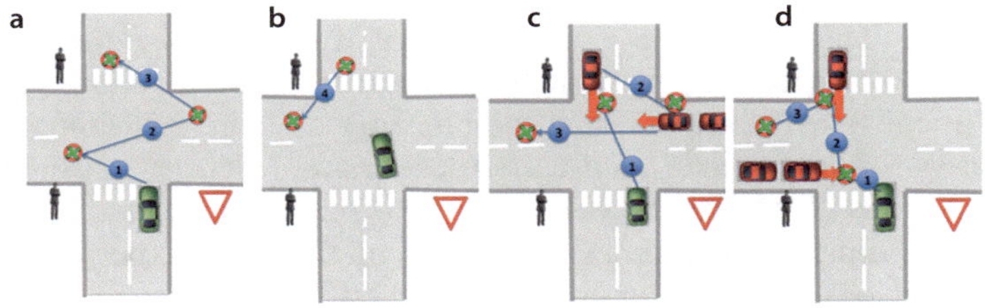

Fig. 3.70 Typical view sequence in Scenario 6 for **a** Drive through if there are no other road users, **b** Turn when there are no other road users, **c** Drive through with oncoming traffic crossing from the right, **d** Drive through with oncoming traffic crossing from the left

behaviour and the theoretical ideal behaviour. For each of the 10 crossing scenarios, the most critical behaviour is presented in detail and differences are recorded according to statistical rules. The general result is that gaze behaviour is essentially determined by the presence of other road users, but not by any crossing parameters. If there are more than three or four independent road users or groups of road users in the scene, the behaviour is even determined exclusively by them. Nevertheless, the active scanning of the scene is determined by the inner models of the driver and thus by the intersection parameters (see ► Sect. 3.3.3). The intended manoeuvre, which serves to stabilise the vehicle, has a greater influence than objective traffic control. The gaze behaviour is independent of the manoeuvre carried out or the right of way regulation. The consequence is that in complex scenarios, glances into directions with right of way were often omitted. Figure 3.70 shows the difference between the eye sequences without and with foreign traffic for the case of scenario 6 from Fig. 3.69. It is also worth mentioning that very often the same test subjects committed the same type of error, be it the lack of focus of the traffic signs, the omission of important subtasks or the typical number of glances in certain phases. This supports the hypothesis that a high proportion of errors is of a systematic nature and could therefore be prevented.

A description of the most important results per segment is presented below. Detailed results can be found at Plavsic (2010).

Approach phase In the approach phase, the drivers usually focus on a middle (1–2 s) or further (>2 s) driving hose. In about 60% of cases, the focus was only on the right side of the lane. The most important tasks in the approach phase are anticipatory tasks. The perception of traffic signs and the corresponding adaptation of driving behaviour has the highest priority. However, only about 60% of the test subjects focused the regulating traffic sign foveally, with the exception of the stop sign. However, the most serious error was the unadjusted speed (see ► Sect. 3.4.2.2: "Structural error").

Delay phase The accidents that occurred in the third and fourth phases were mostly characterised by errors committed during the delay phase. The discrepancy between ideal and actual behaviour was very high in this phase. The most serious errors were the omission of views towards weaker road users (pedestrians, regardless of whether they are right or wrong, and non-privileged vehicles). Only 15% of the test persons were satisfied that the foreign traffic with right of way obligation complied with the right of way regulation. The influence of time pressure was strongest in this phase, which often led to the omission of the most important tasks.

Turning phase The gaze behavior in this phase is characterized by the focusing of the point located in the middle of the roadway (anchor point). The focusing on this point serves to better stabilize the vehicle during the turning maneuver (see also ◘ Fig. 3.65). The typical sequence of views when turning right was a quick look to the left and then a look to the right. In 40% of the cases there was also a look in the opposite direction. The typical sequence of views when turning left was a view to the left, to the right, then in the oncoming direction and then again to the left. When the test persons had right of way, the view to the left was only present in about 25% of cases. The influence of other road users was very strong in this segment. Drivers often carried out the wrong eye sequences, which often resulted in overlooking the right-of-way traffic. Crossing pedestrians are mainly seen in this phase, but only one, even if there are several.

Leaving the intersection The errors committed in this segment were not as critical as those committed in the other phases. However, in this phase the drivers no longer reassured themselves about the rule-compliant behaviour of other road users.

Remlinger (2013) has dealt in detail with the influence of the A-pillar view cover in intersection situations. The available technology of the simulator at the Institute of Ergonomics made it possible to construct various intersection situations (always with a road with right of way crossing at an angle of 90°) in which crossing traffic was held in a defined manner in the visual shadow of the left or right A-pillar. The test control stopped the colliding vehicle abruptly shortly before the collision, so that no "fright effect" arose for the test subject. A skilful scenario was put together to conceal the subject of the study. The experiments were carried out with 46 volunteers of different ages and body sizes. The results show that the test subjects can be clearly divided into two groups, namely the so-called "crossing brakeman", which brake the vehicle strongly before an intersection, almost to a standstill (<30 km/h), and the "crossing scurry man", which observe the traffic far in advance, but then drive over the intersection almost without speed reduction. The "crossing brakeman" did not cause a single collision, while the "crossing scurry man" only caused collisions with those (2 test persons!) who were characterized by strong lateralse head movement in front of the crossroads. It is interesting to note that even if the crossing vehicle is visible, but no relative movement between the vehicle and the A-pillar can be registered from the driver's point of view, the "standing bearing"to the right A-pillar is not perceived by the majority of the test persons.[49] (see also ▶ Sect. 7.3.1).

The results of the intersection experiments show that drivers can benefit most from an assistance system that provides appropriate support during the delay phase. A simple system would show the driver a suitable speed already in the deceleration phase. Another support option is to reduce the burden of support in the most demanding subtasks. This is, for example, the knowledge about the current traffic regulation, especially in critical moments and in situations with a large number of third-party vehicles. Therefore, an assistance system that shows the driver the right of way regulation at critical moments would significantly relieve the driver. This visualization could be done in the Head-Up Display (see ▶ Sect. 6.2.1.1). In this way, the prioritisation of the most relevant information for the driver has already taken place. Thus, the sensors and map information already available in the vehicle could provide the necessary information to reduce the most dangerous errors during driving maneuvers in the intersection area. Such systems would promote driver competence. Nevertheless, the execution of the action was left to the driver. Compared to existing warning systems, such an approach can increase user acceptance and would also be cost-effective.

3.3.2.4 Eye Behaviour During Simulated and Real Night Driving

A part of the described investigations took place in various driving simulators. It is still unclear to what extent the results found in driving simulators can be transferred to eye behaviour in a real driving environment. Reinprecht (2011) investigated how gaze behaviour presents itself in darkness/dawn and whether there is a difference between a night ride in a driving simulator and a real night ride. For the realization a real existing track of approx. 20 km length was constructed in the driving simulator. The route consisted of country roads and urban areas. A total of 14 subjects (4 female, 10 male) with an average age of 28.6 years (SD = 8.06) took part in the study, which was designed as a repetitive measurement design. The task of the participants was to drive through the given route once in a driving simulator and once in normal road traffic according to a permuted experimental plan. The gaze behaviour was measured with a head-mounted eye tracking system.

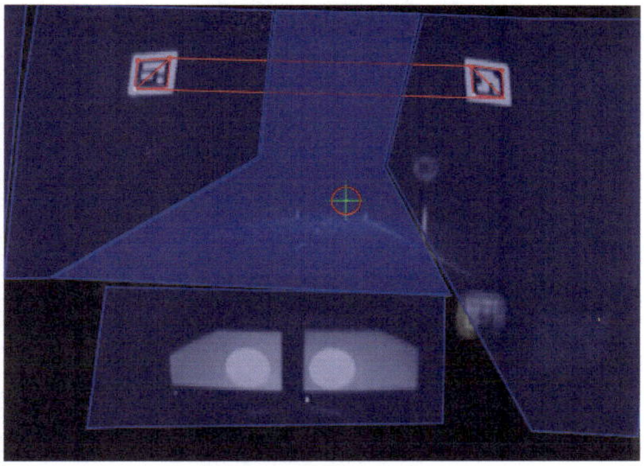

Fig. 3.71 Area of interest

The evaluation of the gaze behaviour was carried out on the basis of predefined Area of Interests (AOI) for the entire journey (**Fig. 3.71**). A total of four AOIs were defined:

- Street,
- Combi instrument (display),
- Left side of the street,
- Right side of the street.

If you look at the *medial* (**Fig. 3.72**), the AOIs road (simulator = 6,7 s; real trip = 8,3 s), combi instrument (simulator = 1,5 s; real trip = 1,6 s) and right-hand side of the road (simulator = 1,8 s; real trip = 1,9 s) there is no significant difference between the two routes travelled. Significant differences, on the other hand, can be found in the average viewing time on the left side of the street. Here the

49 The experiments have a learning effect, so that fewer collisions occur at the end of the test drive. From the experiments Remlinger (2013) deduces the necessity of an appropriate training in the driving schools (if possible with a simulator!), by which the strategy of the "crossing brakeman" and the head transverse movement – similar to the "shoulder look" – is to be acquired.

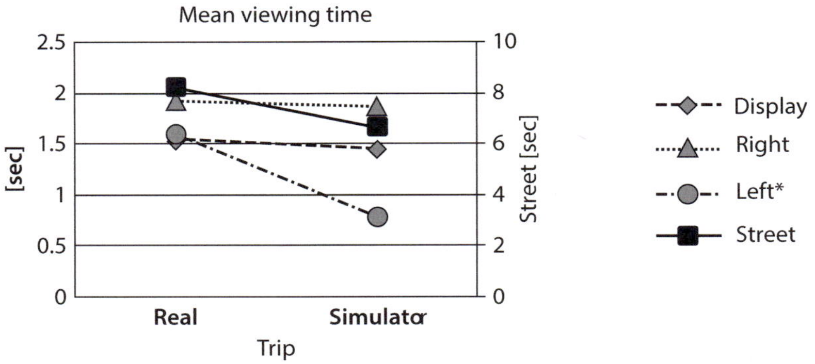

Fig. 3.72 Average viewing time on the four AOI depending on the completed trip. Note: * significant difference

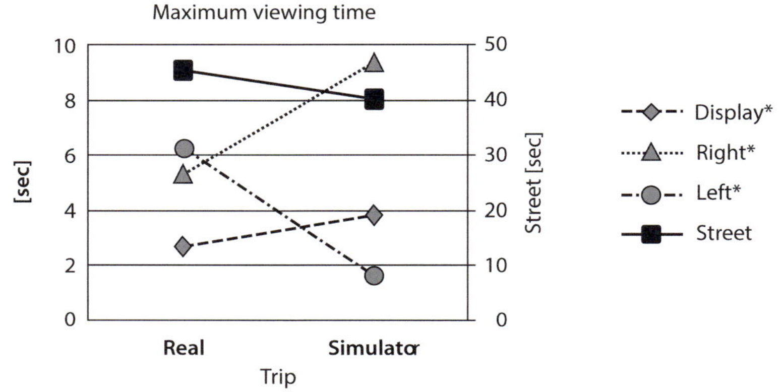

Fig. 3.73 Maximum viewing time on the four AOI depending on the completed trip. Note: *significant difference

average observation time in the real run (1,6 s) is significantly longer compared to the simulator run (0,7 s).

The drivers therefore look at the AOIs road, combi instrument and right-hand side of the road for the same average length of time under both test conditions. In real traffic, however, compared to the simulator, a longer view of the left side of the road is possible. This can cause objects that cross the road from the left to be perceived later in the simulator. Due to the delayed perception, the driver in certain circumstances have less time to react appropriately to the crossing object.

A similar picture can also be seen for the *maximum* looking at the individual AOIs (Fig. 3.73). The maximum duration of the observation of the left side of the road is shorter in the simulator than in the real driving environment. Similarly, significant differences are observed in the maximum viewing time for the AOIs on the right side of the street and for the instrument cluster. Here, both AOIs are considered significantly longer in the simulated drive than in the real driving environment.

This changed eye behaviour in the driving simulator can lead to the fact that the accident probability in the driving simulator is higher than it actually is.

A direct transfer of the observed gaze behaviour in a simulated driving environment to a real driving situation can therefore only be carried out with caution. Particularly when it comes to testing driver reactions to lateral objects, differences can arise between the simulated and real environment, which could be attributed, among other things, to the different viewing behaviour in the two driving environments. Despite these apparently negative

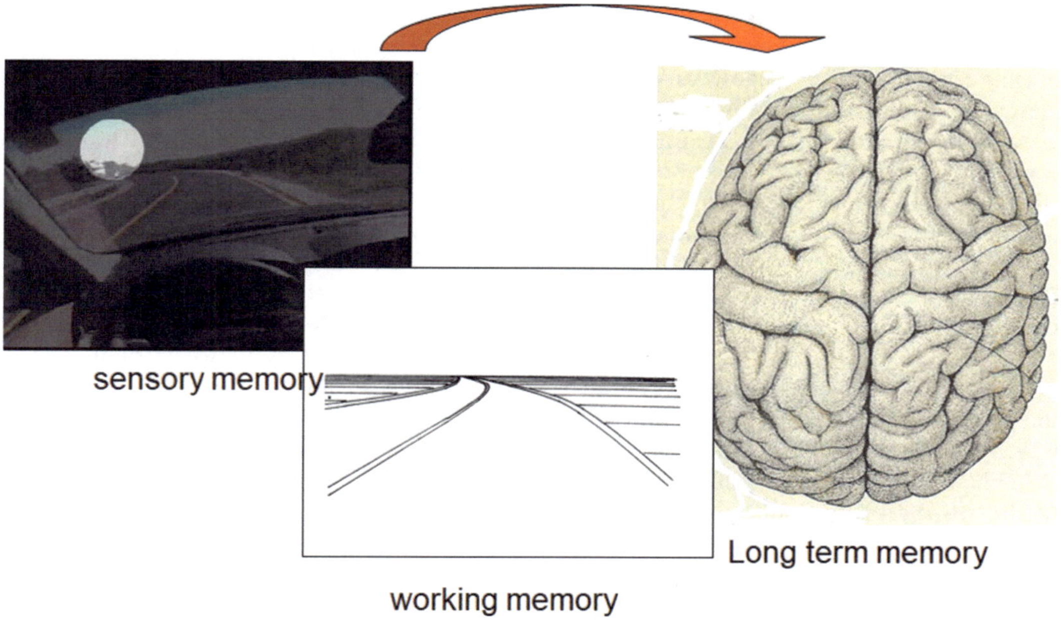

Fig. 3.74 Stimulation of internal models by an external situation

findings, it is almost impossible to do without tests in a driving simulator, because for ethical reasons special, especially dangerous traffic situations cannot be created in the real situation and because only in a simulator can comparable situations be created for different test persons.

3.3.3 Gaze Behaviour and Inner Models

As already explained in ▶ Sect. 3.1, the primary goal of human information processing is to get a spatial-temporal picture of the surrounding situation in order to draw the right consequences for acting in space and time, promising a safe and pleasurable life. Most aspects of abstract thinking are also dealt with in these categories because of this spatio-temporal way of thinking, which is explained by our history of development. Perhaps even the attractiveness of driving a car lies in the fact that it realizes this original aspiration to change over time in space and even far exceeds all limits given by natural conditions. So if we want to understand driving a car, we have to see it in the context of this prerequisite of human thought and action.

How can we complete our picture of information reception and processing with these results? Figure 3.74 is intended to help understand this process. Through scanning, information is recorded at the level of sensory memory to stimulate internal models that are stored in our long-term memory as an overall concept in the form of a structural engram.

This suggestion "awakens" the corresponding inner model and thus – as already mentioned in ▶ Sects. 3.1.3 or 3.2.2.3 – becomes part of the working memory. This process is to be understood more precisely by the example of a drive on a road bent to the left. The information acquired by the scanning stimulates a general inner model of a concept of a road bent to the left. By adapting this concept to the current stimuli in the working memory, we recognize the real width and the real curvature of this street. Further sampling stimuli – the processing glances – give us information about further details of this street. The virtually empty image of the street is thus filled with objects from the surroundings (trees, bushes, hills, mountains in the background). We have the feeling that we are moving in exactly this place. In this way, moving objects (other vehicles, pedestrians, possibly animals) are also captured by the processing views and "inserted" into the captured scenery. The information of these objects also stimulates internal models about their behaviour. In this way, only a brief glance is sufficient to capture the speed and heading of an oncoming car or the expected behaviour of a pedestrian without constantly focusing on the object in question. The combination of all this information gives us a sense of the presence of the situation. This inner image created in this way is reality for us.

This orientation and behaviour in the world, which is explained by our history of development and which is adapted to the speed ranges of the original natural environment of homo sapiens, often reaches its limits in the technical world created by himself..[50] Directly related to this is the fact that the immediately experienced present comprises about 2 seconds. Thus we are ready when driving to avert the view from the road for these 2 seconds, because the construction of our inner models suggests that "everything remains the way it was" or "everything develops the way it has developed" as we have grasped it before. So it can happen that we have looked into the rear-view mirror for example, have not discovered a vehicle with a corresponding manoeuvre there, but change lanes a little later (within the 2-s period or even a little longer) in the sure conviction that there can be no vehicle on the secondary lane. Since our gaze can only scan the scene in one sequence, it is of course also possible that relevant objects are not observed and therefore subjectively not present. Changing objects attract the foveal gaze when they are imaged in the peripheral field of vision. If – as already mentioned – an objectively visible vehicle is located in a standing bearing behind the A-pillar (i.e. does not move in relation to the pillar), there is no averting stimulus into the periphery and the vehicle remains undiscovered when the gaze is directed towards the approaching intersection. As already mentioned, inner models refer not only to static objects, but also to the expectation of movement. For example, it can happen for example that a driver who wants to drive straight ahead at a right turn notices the acceleration of a turning vehicle, directs his attention to his own lane and no longer notices that the turning vehicle suddenly brakes again in the meantime due to a crossing pedestrian.

All of the above examples show how the subjective presence of the present can play a trick on us with the mentioned time span of 2 seconds, if something different from the usual happens within this time period. It is therefore extremely important for road safety that sudden changes are avoided where possible because they may not be covered by other road users. In particular, the rule-compliant behaviour expected of other road users – apart from exceptions[51] – represents a contribution to road safety that should not be underestimated (see also the observations on gaze behaviour by Schweigert, 2003; ◘ Fig. 6.62). Inner models decide not only what we perceive, but also how we act. The operation of a vehicle is a skill-based, in some cases rule-based action for the experienced driver with regard to most interactions. Skills, however, are achieved only after a long practice period. This is partly the dilemma for the introduction of new techniques that change the operation, even if they actually improve simplify the

50 Evolution has not found perfect solutions in any area; rather, such solutions, which moreover all had to develop from existing preconditions, have prevailed which provided a sufficiently high probability of survival for individual beings so that reproduction and thus the preservation of the species was guaranteed (species for which these preconditions no longer existed, e.g. under changed living conditions, disappeared again).

51 Conflicts can also arise here: e.g. on many speed-limited routes it is common practice to exceed the maximum permitted speed by a certain amount (e.g. on routes with a limit of 80 km/h, trucks are almost invariably driven at the technical speed limiter, which is usually set to 90 km/h. The "law-abiding" driver is then often harassed by such a lorry driver). It may therefore be better to speak of "expected habitual behaviour of other road users" rather than just of "conforming to the rules".

operation.[52] Weinberger (2001) was able to show in an extensive habitat study that it takes more than 4 weeks to get used to the handling of the ACC system (Active Cruice Control). Gengenbach (1997) also found these values for gaze behaviour in connection with Head-Up-Displays (HUD). Only after this time was information read from the HUD just as often as from the conventional combi instrument, and only then did the advantage of less gaze aversion become effective. In this context there is also the problem of so-called migration, which has to be considered when introducing new operating procedures: a radically changed type of operation is not accepted (e.g. steering the vehicle with joystick instead of steering wheel). If a new type of operation has undeniable theoretical and experimental advantages, an introduction can only promise the desired success by slowly changing the previous operation to the new one.

3.3.4 Comfort and Discomfort

As already indicated in ▶ Sect. 3.1.2, the experience of driving is not only a rational matter, but also an emotional one. In this context, the question of experiencing comfort is repeatedly raised. It is therefore necessary to answer the question of what comfort might mean. The definition of comfort in Wikipedia gives the answer: "Comfort (or being comfortable) is the perception of physical or psychological lightness, often characterized as a lack of hardness. A certain level of psychological comfort can be achieved through experiences associated with pleasant memories, by handling familiar objects and enjoying tasty foods. …"Comfort is therefore a rather diffuse term: in today's language it means as much as *comfort, convenience and satisfaction* but also refers to the evaluation of the *luxury* of an outfit. In the Third International Dictionary of the English Language (1981), for example, comfort is defined as "a state of relief, encouragement and enjoyment".

However, in order to make comfort technically operable, it is necessary to define more precisely which areas are covered by the term comfort. It is worthwhile to look at the areas with which human beiing deals cognitively and mentally, and which have always been the subject of consideration, not least in high schools. It's these:

- *Science.* It deals with all conceivable aspects of function and conditionality, in short with the structure of the world, including human existence. A distinction is made roughly between the humanities, which – to put it simply – deal with the question of "what for" or "why" under the aforementioned aspect, and the natural sciences, the aim of which is to uncover the "how".
- *Art.* It offers direct impressions for the sensory organs eye and ear, which should fall or shake you awake on a psychological/emotional level and make you emotionally aware of certain problems.
- *Sports and games:* They provide rules and considerations on how to achieve a positive fulfilment of life, at least temporarily.

Of these partial aspects of human life, comfort refers to a border area between the scientific "how", the artistic "pleasure" and the positive fulfilment of life that is conveyed through sport and play. From this, the not completely scientific ascertainability of comfort already becomes obvious. However, comfort – like any other sensation – must come about through different impressions on the human sensory organs, whereby the result is more than the sum of the parts.

Zhang et al. (1996) have shown on the basis of survey experiments, the results of which they subjected to a cluster analysis (see ◻ Fig. 3.75), that comfort is composed of two independent influencing variables: namely the aspect of the *pleasing* and the aspect of *suffering*. They called these two sizes – a little unhappy – *comfort* and *discomfort*. The difference can be seen in the example of a sports car with hard suspension: although it can strongly cause the "suffering" of hard shocks, little

52 Just think in this context of the rejection that many drivers have of the automatic transmission, which is actually better from an ergonomic point of view.

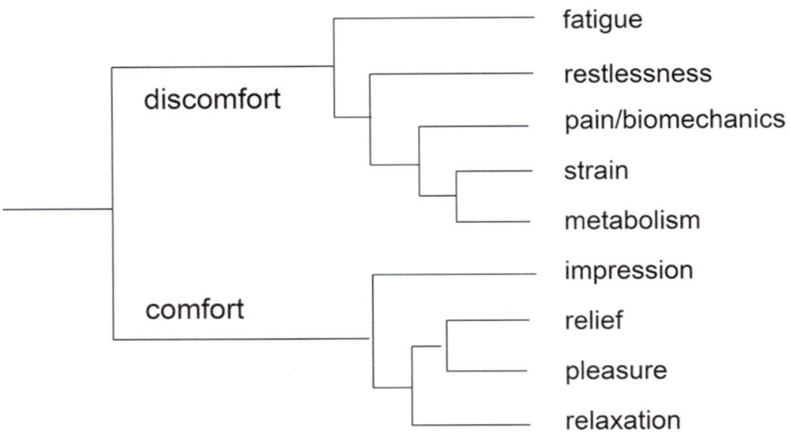

Fig. 3.75 Result of the cluster analysis in the experiments of Zhang et al. (1996)

headroom and possibly an unfavourable posture, it is still very "pleasing" because of its form and the image it promises.

The *pleasing* has a lot to do with *fashion and zeitgeist* and thus initially eludes strict scientific research methods. Nevertheless, it is presumed that it also follows objective rules that the philosophy of aesthetics has endeavored to establish since ancient times. Knoll (2007) explains why the description of gestalt aesthetics ends with Kant: no one before him has characterized "aesthetics" as an interaction between subject and object, a view that has survived to this day. The vast majority of philosophers since Pythagoras have attributed "beauty" to the question of order and proportion (cf. "Golden Section") and saw it objectively founded in the perfection of the composition of the parts of a whole. Kant's opinion, on the other hand, that beauty arises from the interaction of object and observing subject, is still held unanimously today. Immanuel Kant was the first in the history of aesthetics to attach importance to judgement in his critique of beauty, in which he stated that beauty does not originate from any particular property of the object, but from an individual judgement that relates to the taste and pleasure of the judging subject. The taste judgement is therefore aesthetic and not logical, it is subject-related – in contrast to the knowledge judgement – because it is based on the subjective feeling of pleasure and aversion. Kant clearly rejects an ideal of beauty. In summary, it should be noted that in the formation of taste judgments, the form and content of the object must be strictly separated. The form is judged ("without interest") by the taste, the content by the mind ("of interest"). Since Kant, the role of the subject in aesthetics has gained in importance and the aesthetic value judgement has developed from generally valid rules of beauty in the object to criteria of subjective experience.

Nevertheless, especially since the twentieth century there have been various attempts to fathom beauty and aesthetics with scientific methods. Evolutionary aesthetics, for example, assumes that certain preferences have manifested themselves in our genome. Neuroscientific studies try to find out what happens in the brain when we find something beautiful. However, a "beauty centre" has not yet been found. Cognitive aesthetics argues that objects that stimulate the brain with a certain – but not too great – complexity and thus stimulate pattern formation (inner models) are perceived as beautiful. Obviously, what is most appealing is what has a high recognition value (Kersten 2006). This would initially suggest that averageness has the highest attractiveness (see Langois and Roggman 1990 and Rhodes and Tremaran 1996). Studies by Braun et al. (2001) on the attractiveness of faces show, however, that system-

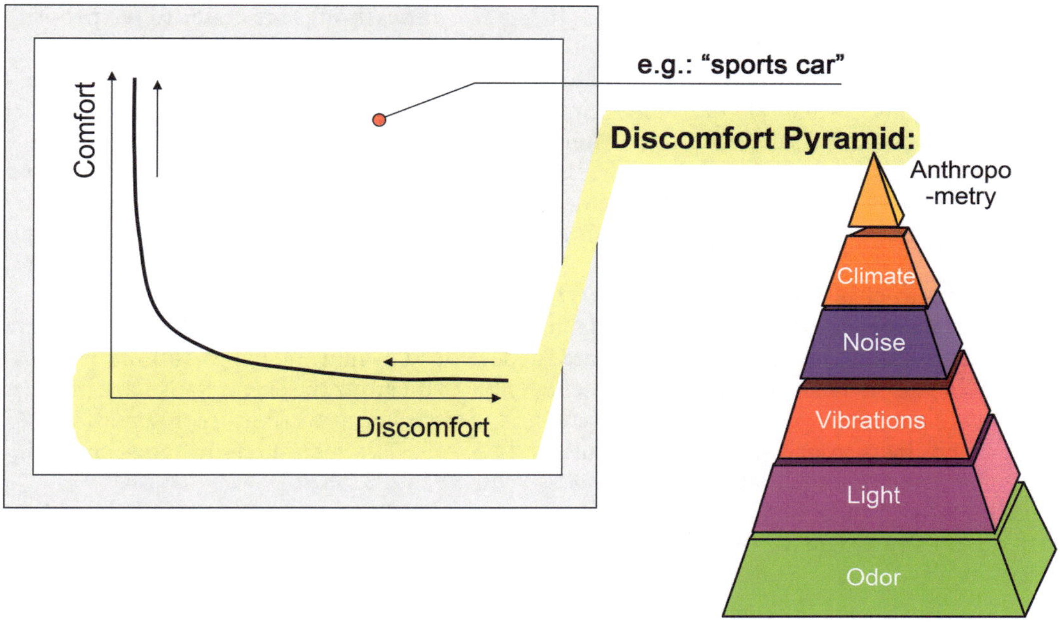

Fig. 3.76 Comfort as a paradigm for "pleasure" and discomfort for "suffering"in connection with the comfort pyramid after Krist (1993)

atic deviations from the average increase the attractiveness of faces..[53]

The *suffering* can be examined with the *classical methods of psychophysics*. Psychophysics investigates the lawful relationship between a physical stimulus and individual sensation (most of the findings on sensory organ sensitivity have been researched using psychophysical methods, see ▶ Sect. 3.2.1). It represents a fundamental paradigm for scientific ergonomics. With regard to these stimuli, we must now further ask which have a significant influence on suffering and which do not. Krist (1993) asked test persons what they thought was important for comfort. Interestingly, they very often mentioned things like spatial comfort and confinement, which is scientifically called anthropometric design, and then secondly climate, less often noise, even less often oscillations, very little lighting and no smell at all. This makes you wonder: because the olfactory nerves are directly connected to brain centres that generate strong emotional moods. They should therefore be particularly important for comfort. The solution to this apparent contradiction lies in the following: the things that are in order from the (dis-)comfort point of view are not mentioned either. In accordance with the results of Zhang et al. (1996) and with other peculiarities of human perception, which generally point to the effectiveness of a masking effect, this allows the comfort pyramid to be defined: at the bottom is the fulfilment of the needs which are very important. If these are satisfied, the next highest importance is given to them (see also Fig. 3.76).

The increasing perfection that the electronification of products brings with it has led to the demand for ergonomic design *manageability* and let them step out of the way. Especially from the commercial point of view of the attractiveness of a product, the question of the so-called ease of use is connected with this. It should also be seen under the

53 This effect can also be observed in the likelihood of automobile design. It is noticeable that – of course also due to the technical possibilities of the design – there are time-dependent fashion directions (e.g. pontoon shape, trapezoidal line, one-box design, SUVs etc.), which on the one hand cause a high similarity of the vehicles, whereby the attraction lies in slight deviations from the mainstream.

aspect of the comfort/discomfort model. It can also be assumed here that scientific methods can only reduce the discomfort of inadequate operation, while at the same time demanding aesthetic satisfaction. In summary, the illustration in Fig. 3.76 shows the connection between scientifically-oriented *ergonomics* and artistically oriented *industrial design*.

The construct of Hassenzahl et al. (2007) of the User Experience (UX) corresponds in principle to the model of comfort presented here: It spans a field of two independent dimensions for the subjective quality assessment of products, consisting of the so-called hedonistic quality, which largely corresponds to the "*pleasure"* of Zhang et al. (called "comfort" there) and pragmatic quality, which corresponds to Zhang's "suffering". Especially the *hedonistic quality* it subdivides again into the subgroups

- *Stimulation:* "creative", "original", "challenging",
- *Identity:* "brings me closer to the people", "professional", "binding",
- *Attractiveness:* "good", "attractive", "pleasant".

The *pragmatic quality* refers to "the perceived ability of a product to achieve its objective by providing useful and usable functions". Typical product attributes are: "Practical," "predictable," "straightforward."

Using the AttraktDiff2 questionnaire developed by him and his co-authors, it is possible to illustrate the extent to which a product is assessed in a portfolio representation (see Fig. 3.77), which shows the great similarity with Fig. 3.76.

In addition to the two dimensions "pragmatic quality" (= "non-suffering" or "discomfort") and "hedonistic quality" (= "pleasure" or "comfort"), a third largely independent dimension has to be added, namely that of "wanting to have". It can be that a product or the execution of a partial aspect of a product is

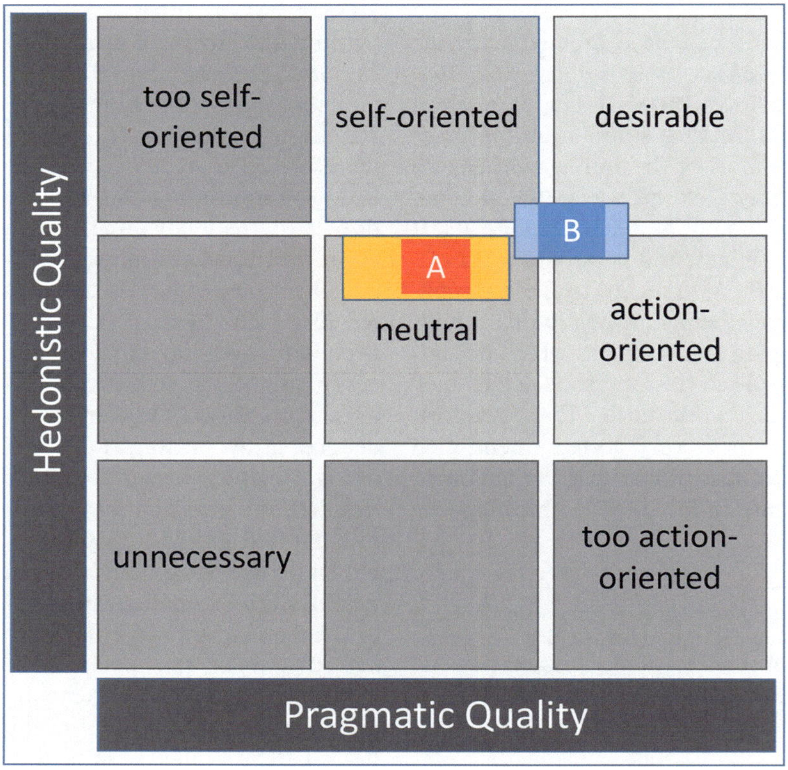

Fig. 3.77 Portfolio presentation of the "Comparison Product A – Product B" (Hassenzahl et al. 2007)

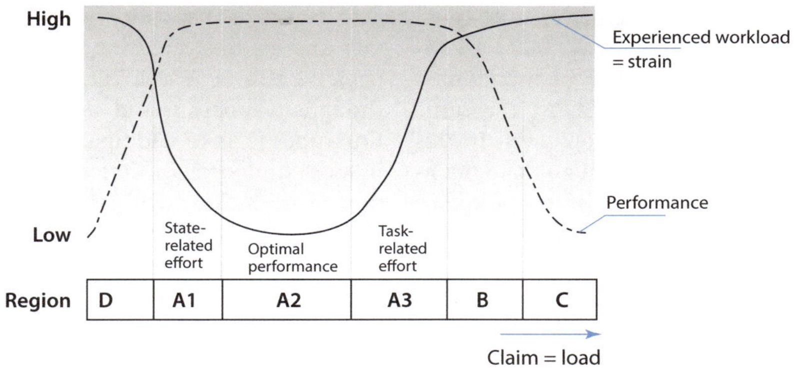

Fig. 3.78 Relationship between workload (load) and performance (possible performance due to load) according to De Waard (1996)

"liked" and at the same time also considered as "practical", that one does not "want" the product however, because it does not fit – independently of the economic acquisitionability – to the view of the own personality, i.e. the own image. Conversely, it is quite common to adorn oneself with an object that is actually impractical and may not be particularly pleasing even in the longer term, but which is currently "in". From a personal point of view, a product can therefore occupy any position in the three-dimensional space defined in this way.

3.3.5 Stress and Strain

As already mentioned in ▶ Sect. 1.3, the concept of stress and strain represents a fundamental paradigm of ergonomics. Stress – often referred to in the same sense as ***"workload"*** – characterizes the objective situation (work task, environmental conditions, available machine, anthropometric conditions) and strain the individual reaction to it. The word strain must not be misunderstood in the sense that it is a negative reaction in every case, much more the subjective experience as well as many scientific studies (see below) point out that there is an optimal stress/workload that leads to minimum strain and at the same time highest ***performance***. While it seems logical due to the limited processing capacity of the working memory that an excessively high requirement leads to a reduction in performance, early studies (e.g. Schmidtke 1965) have already shown that even a very low requirement leads to low performance. Schmidtke (1965) has assigned the following keyword to this area ***overstrain due to understrain*** characterized. The management of this situation is charcterised by the term ***vigilance***. This is understood to be "the state or degree of readiness to recognize and respond to small changes that occur in the environment at random intervals" (Wirtz, 2013) or "the ability to maintain attentional focus and be alert to stimuli over time" (Warm et al., 2008)..

De Waard (1996) derives the relationship between workload and performance from an extensive literature study (Fig. 3.78). Workload and performance then pass through six regions. The claim (= load/stress) increases from left to right. In region D, the performance is low and the *experienced* workload is high because the driver has trouble maintaining attention even though little performance is required. This constellation can be found, for example, in simple and monotonous car journeys and corresponds to an underchallenge of the driver. If the claim is increased, the experienced workload decreases and the performance increases (region A1), since the driver no longer has to concentrate under effort. By increasing the excitation level, the performance can be increased (Helton et al. 2009).[54]

54 This effect is matched by the observation that many drivers increase their speed when they feel an emerging fatigue.

In the middle area (A2) the driver is adapted to the requirements of the task. Here a constant level of performance can be maintained even with increasing demands. At the same time, the stress remains at a low level. In this optimum range, the driver is thus able to maintain a consistently high level of performance over a longer period of time and to compensate for slight fluctuations in requirements. In the third region (A3), the workload experienced increases again without any loss of performance. This is made possible by mobilizing the effort (tasks related effort). However, in this area a constant performance can only be maintained for a limited period of time before exhaustion occurs. If the entitlement continues to increase, the workload experienced now increases again (region B), but the performance level is still maintained. In region C the person is overwhelmed and the performance drops to a minimum. Overstretching effects lead to an increase in reaction times (Conti et al. 2013) and to tunnel vision (Rantanen and Goldberg 1999; see also the results of Schweigert 2003 and ◘ Fig. 6.62). But monotony and understrain also have demonstrable negative effects. A well known route leads to a thought wandering and thus to longer reaction times (Yanko and Spalek 2013). This also induces tunnel vision (He et al. 2011). In connection with the use of assistance systems (see ▸ Chap. 9), these aspects of overstrain and understrain play an important role. On the one hand, overstraining situations can be absorbed by automatic functions. On the other hand, there is a danger that the effect of monotony will be further increased in low-irritant environments. In addition, the workload is an extremely subjective matter. It is therefore hardly possible to predict the effect that the use of a certain automatic function will have on the individual driver, especially since this is also subject to strong intra-individual temporal fluctuations. However, in summary of the findings described in the literature, a rough timeline can be drawn up describing the emergence of a lack of attention (Körber 2014, ◘ Fig. 3.79). Accordingly, endogenous fac-

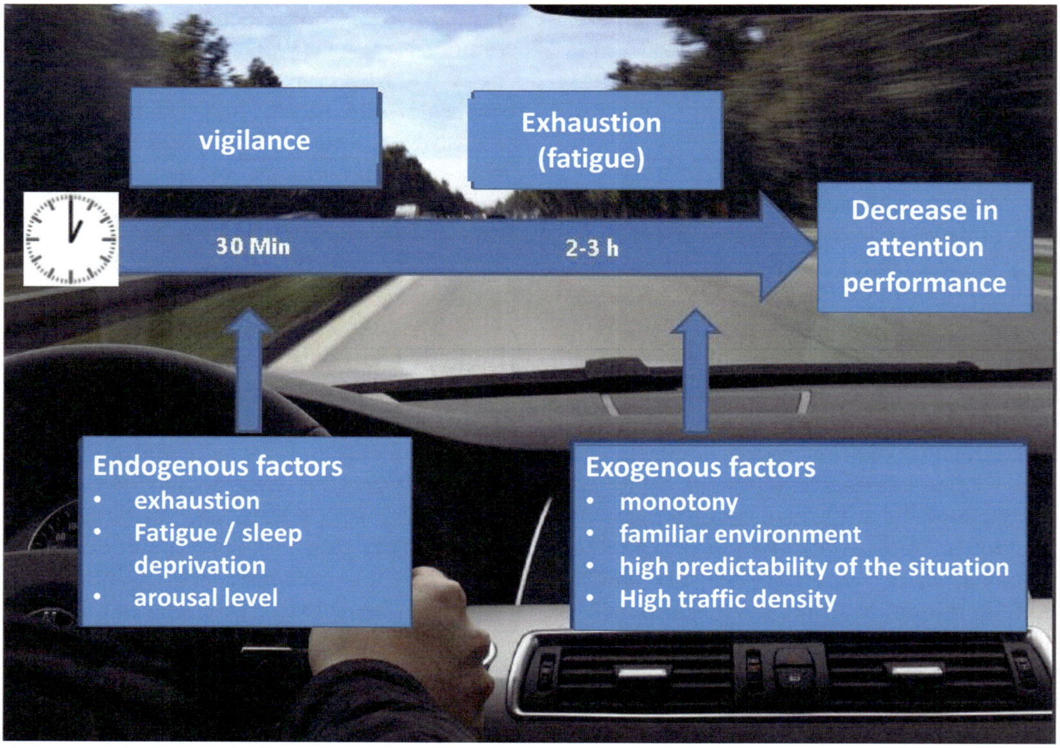

◘ **Fig. 3.79** Endogenous and exogenous factors leading to a decrease in attention performance (Körber 2014)

tors such as exhaustion, fatigue – possibly caused by sleep deprivation and a low level of arousal – tend to cause a lack of attention in the short term. Exogenous factors that cause low stimulus such as monotony, driving in a familiar environment and a high predictability of the situation lead to understrain situations in the long term (area D in Fig. 3.78), while long-lasting stressful car journeys at high traffic density create overstrain situations (areas B and C in Fig. 3.78). These long-term attention deficits can only be compensated by a short sleep phase of approx. 20 min.

3.4 Driving Error

3.4.1 Human Reliability and Driving Errors

In the previous chapters, many aspects have already been addressed which could be the cause of incorrect driver behaviour. As already explained in ► Sect. 2.6, the notion of reliability is reciprocally related to the error. Reliability is also defined as freedom from error ("impeccability"). According to Reason (1994), a rough distinction has to be made between intended *infringements* and unintended *errors,* a distinction that of course also applies to the driving error. The same distinction can therefore be found in the concept of human reliability. It's understood to mean

1. the individual's positive basic attitude towards taking responsibility for a course of action that does not harm the other person but rather results in his or her well-being,
2. the avoidance of accidental malfunctions or malfunctions due to unfavourable design which impair the functionality of a technical system – in this case the vehicle or the transport system.

While point 1 refers to the individual's attitude – which must also be assessed ethically/moral – point 2 addresses unintentional action which, however, has unacceptable consequences.

Ever since vehicle accidents have occurred and been investigated, attempts have been made to improve human reliability in order to avoid them. Essentially, this is done through training measures designed to learn and practice behaviour that avoids accidents and to promote insight into the futility of selfish behaviour that violates prohibitions. The latter aspect will be emphasised by appropriate sanctions.[55] From the technical side, one tries to make wrong actions less likely by the humane design of the vehicle (see ► Chap. 9).

As described in ► Chap. 2, the relationship between the current actual condition of the vehicle (position on the road in relation to the roadside and stationary and moving objects, direction of travel and speed) and the driving task (target condition) is described as quality. A fault can therefore be defined as one in which a required level of quality is not achieved. If this error is caused by the driver, it is a driving error. With regard to these driving errors, the following hierarchically graded tolerance limits are distinguishable (free according to Rigby 1970):

- *Physically existing acceptance limits* represent real physical barriers, such as obstacles, barriers, crash barriers, nail strips, the edge of the road or the kerbstone, the non-compliance with which is clearly perceptible (impact noise, scratching, grinding noises, nail strip rattling, jerking or vibrations of the vehicle) and can lead to material damage or personal injury if the course is unfavourable.
- *warning limits* are acceptance limits which have been issued in the sense of preventive protection for physical and technical reasons. Exceeding them is possible, but is currently actively signalled by warning signals or can be clearly identified by markings or other signs (e.g. red light of a traffic light system, no overtaking, no parking, etc.). On the vehicle side, this includes, for example, tyre pressure warning or distance

55 For pragmatic reasons, the intentional violation is assumed for each observed error, which may impair the acceptance of sanctions.

warnings (e.g. Park Distance Control, PDC).
- *Empirical acceptance limits* are based on the experience of individual drivers or on the formation of standards based on the experience of many as social conventions. The problem is that crossing these borders is not immediately perceived by the senses. Their perception presupposes the interpretation of the situation on the basis of experience, memory and willingness to see the technical-operational or behavioural-psychological connections, in other words to "think along" and safety-oriented behaviour. These limits include, for example, a empirical "if-then-regulations" (e.g.: the adequate behaviour when passing small children or old people or the rule to leave the way clear for a crossing side street).
- *Forensic boundaries* are acceptance limits that primarily refer to legal provisions or regulations that arise from standards, regulations (e.g. speed limits). The problem of their observance is partly due to the lack of understanding of their validity (ban on mobile phone use without hands-free equipment) and/or the blurriness of their definition (choice of speed depending on road conditions) but also partly due to the fact that it often means more effort to comply with them than to circumvent the regulation. A classic example is the lack of acceptance of the use of seat belts (in some countries), which, as is well known, could only be enforced by an increased threat of punishment.

In many practical cases, however, these quality requirements are not explicitly stipulated, but only implicitly in the form of an appeal to the driver "to behave in such a way that no one else is harmed, endangered or harassed more than is unavoidable, obstructed or annoyed under the circumstances" (§ 1 (2) of the StVO; German road traffic act). Frequently, human failure (synonymous with human error) is defined only after the accident has occurred, i.e. when the area of safety has been objectively left. In fact, driving errors can only be observed at two points within the driver-vehicle interaction structure:

1. Directly by observing the human action itself; this is possible for example by observing the gaze behavior and the operation of the control elements. However, in order to detect errors in the above sense, it would be necessary to know the "right" way of looking or the "right" way of operating elements for every conceivable situation. With the exception of artificial experimental situations (e.g. also simulator experiments), this observation site is excluded in practical cases.
2. Indirect to the result; in this case, the deviation of the ride result from the required quality tolerance is referred to as a defect. This type of observation also brings with it difficulties in the practical assessment of errors, particularly in determining the correct target values for each situation. Violation of a legal provision and, in extreme cases, the accident is a clear violation of the acceptance limits. Accident research is thus an essential resource for human error research and thus also for the humane design of the technical system. In addition, the observation of "near-accidents" is becoming increasingly important in scientific research today.

One of the main difficulties in the reliability assessment of driving tasks is the determination of situation frequencies and the estimation of error probabilities. Situation frequencies can be obtained from statistical data from road construction offices (e.g. Lippold and Mattheß 1994). Most of the data available today on human reliability have been collected in the field of nuclear installations or are based on the experience and assessment of experts (for an overview see Swain and Guttmann 1983). As Reichart (2000) showed, however, these data can be successfully applied to road traffic behaviour. On the basis of the information available today, the data summarised in ◘ Table 3.8 can be given.

Table 3.8 Probabilities of errors during driving estimated from the data of Rasmussen 1982; Reason 1994; Swain and Guttmann 1983

Type of task	Example	Probability of an error
Highly practiced activities	Insert the ignition key into the ignition switch (without drugs/alcoholisation), Follow the road by car.	$p = 10^{-4} \dots 10^{-5}$
Practiced simple activities	Switching, operating the wrong control element (e.g. indicator instead of wiper).	$p = 10^{-2} \dots 10^{-4}$
Rule-based activities	Speed adjustment in wet conditions.	$p = 10^{-1} \dots 10^{-3}$
Knowledge-based activities	Normal evasive reaction, assumption of time gaps.	$p = 10^{-1} \dots 10^{-2}$
Activity under high stress	Avoidance reaction in the event of imminent accident.	$p = 0{,}1 \dots 1$

3.4.2 Causes of Human Error

If you want to avoid accidents, you need to understand the cause of human error. In this context, the above distinction between intentional infringements and unintentional errors is necessary, bearing in mind that smooth transitions between them are also possible.

3.4.2.1 Infringements

Intentional infringements (violation) can also be understood from the information processing model (▶ Sect. 3.2.2.5, Fig. 3.48) and in particular from the idea of how decisions are made (Figs. 3.50 and 3.48). It must also be assumed that the acting driver more or less correctly estimates the situation (conditions of the world) according to his possibilities. The problem lies in the allocation of benefits to the individual actions that appear available to him. If the benefit of his action seems particularly high to him (e.g. faster progress, fun with the action, avoiding an expensive taxi ride) and at the same time the risk of being captured by the police seems low, he will carry out the action. According to Reason (1994), an additional distinction must be made between exceptional offences and routine offences. He writes:

> "In this vast hinterland of intentional but not malicious rule violations, one can make a further rough division into routine violations and exceptional violations. Routine violations are largely habitual and form a fixed part of an individual's behavioural repertoire; exceptional violations are individual infractions that occur under certain circumstances. The field of road traffic offers many examples of routine violations". (p. 242, quoted after Gründl, 2005)

The entries in the central traffic register of the Federal Motor Vehicle Office of Germany can be regarded as an indicator of which erroneous actions appear particularly attractive, even if in individual cases these actions may correspond to unintentional errors (e.g. overlooking a traffic sign for speed limitation). In 2011, out of a total of almost 4.8 million entries, 61% were speeding violations, 8% were right-of-way violations, 4% were driving under the influence of alcohol and/or drugs, 2% were driving without a driver's license and 7‰ was unauthorized removal from the scene of the accident. Especially in the case of the most frequent infringements mentioned, an extremely high number of unreported cases (estimated at least at 1:800) is to be assumed, so that the actual infringements are much more frequent than these figures suggest. The fact that speeding is in most cases a deliberate act is indirectly shown by the fact that for example proposals to technically limit the speed of a vehicle (it would be possible for example that the speed limiter automatically takes over the local limit recorded by the traffic sign recognizer or stored in the navigation system) are rejected both by the responsible persons in the vehicle companies (because of the consideration of losing customers as a result) and by test persons in appropriate experiments. Here, as for many other violations (e.g. telephoning with a mobile phone while driving, no waiting in front of a red traffic light at an obviously low-traffic time, or in front of a hazard flasher at a level crossing when no train is visibly arriving; driving close to to encourage the vehicle in front to clear the roadway, traffic-impaired stopping or oarking; and much more), the current benefit (possibly also fun in the action) is considered high and the possible damage (police detection, accident) is considered low.

Measures against intentional infringements are only possible through behavioural influence, i.e. educational measures. Although this is achieved in driving school lessons, the learning effect of everyday life may favour behaviour that is completely different from that which is recognised as "correct" (see also ▶ Sect. 3.2.2.5). After a repeated training is omitted, a learning effect can only be achieved by sanctions of observed and provable offences. In most cases, however, the sanctions are imposed with such a time lag that the intended learning success is questionable. In addition, infringements and errors that do not lead to an accident are only observed with low frequency in relation to their occurrence, as already mentioned. This approach, which does not take sufficient account of the psychology of the learning effect of behaviour to avoid misconduct, largely robs sanctions of the desired effect. From this reasoning alone, the value of situation-related regular television commercials or short newspaper articles on the correct objective of driving becomes obvious.

The field of infringements also includes the abuse of assistance systems, which is sometimes cited as an argument against their introduction. A prominent example of this is the introduction of the ABS brake, the use of which was originally "rewarded" by the insurance companies with a lower premium because of the presumed safety benefit. Unfortunately, it turned out that many drivers abused the supposed "super brake" to a more aggressive and therefore more accident prone driving style. Similar consequences are feared for assistance systems such as Adaptive Cruise Control (ACC), Lane Guidance Assist, Lane Change Assist, etc. By designing the operation of such systems, such abuse possibilities can be prevented, which poses a very special challenge to operator ergonomics.

3.4.2.2 Errors

In order to understand the unintended error, a number of models have been developed which have a certain similarity to each other and which partly complement each other. Gründl (2005) provides a very clear overview. The essential basis for this is the Wickens model (1992), which in principle corresponds to the model ideas developed in ▶ Sect. 3.2. While this model can explain any successful action, i.e., the model by Rasmussen (1982) specifically addresses internal errors. Norman (1981, 1986) already pointed out that a distinction must be made between unintentional errors in the implementation of information (so-called "slips") and the actual "mistakes" that occur in information processing. This error model complex is supplemented by the considerations of Hacker (1987), who separates between the main branches "objective lack of information", as it can come about through concealment, poor visibility and the like, and "lack of use of objectively available information", whereby this can then be described according to Rasmussen's categorization. ◘ Figure 3.80 gives an overview of these different error cause models.

Gründl (2005) uses the Rasmussen model, which he originally developed for aircraft accidents and which Zimmer (2001) adapted for driving vehicles, in his in-depth accident analysis[56] to specify the frequency of error causes (see ◘ Fig. 3.81). In detail, the following categories are to be named on the basis of the above-mentioned error models:

56 In this form of analysis, the police, who were called to an accident, informed a team of scientists, consisting of technicians, psychologists and doctors, who made their investigations independent of the police. The accident victims who agreed to be interviewed had previously been assured that the statements would be confidential and would not be passed on to the public accident recording bodies or the subsequent negotiating partners in court. In this way, statements could be recorded and observations made which are closer to the real events than the official police versions.

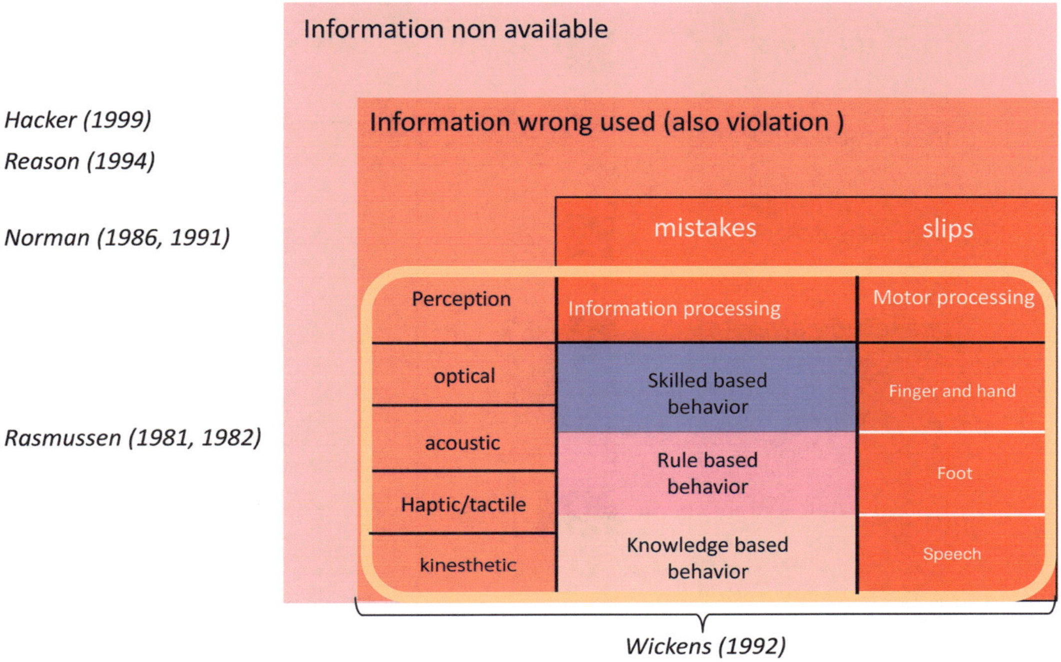

Fig. 3.80 Cause-oriented classification of human errors

Structural error:	At the time of the occurrence of the error of action, there is no longer any possibility of immediate action. However, the error here usually lies in the run-up to the action, e.g.: Non-adaptation of the speed or too small safety distance.
Information error:	Traffic-relevant information is not noticed or is noticed too late.
Diagnostic error:	Incorrect assessment of the situation, e.g. of time gaps, incorrect interpretation of the intention of others to act, etc.
Goal-setting errors:	Intention not appropriate to the situation, e.g. avoidance of a small animal.
Method error:	Wrong choice of action for given action alternatives (this case normally hardly ever occurs in road traffic because of the small number of action alternatives, which are ultimately reduced to braking, accelerating and steering).
Errors of action:	Insufficient transformation of the chosen approach into an action; e.g. tearing the steering wheel.
Operating errors:	Operation of the controls not appropriate to the selected action; e.g. slipping off the brake pedal.
The last mentioned handling and operating errors are summarized by Norman to the term "slips" (see above)	

As can be seen from Fig. 3.81 *information errors* is by far the most common cause of accidents. The information error can occur in different ways. The main reason for this is the fact that the human eye only maps a small angular range of 2–3° on the so-called central visual pit with real sharpness (► Sects. 3.2.1.1 and 3.2.2.1). Only objects that are mapped in this area can be cognitively captured and assigned to memory content. The perception in the periphery serves, as shown under ► Sect. 3.2.2.2, rather to capture movement, speed and also the direction of movement. In order to gain a quasi inner picture of the out-

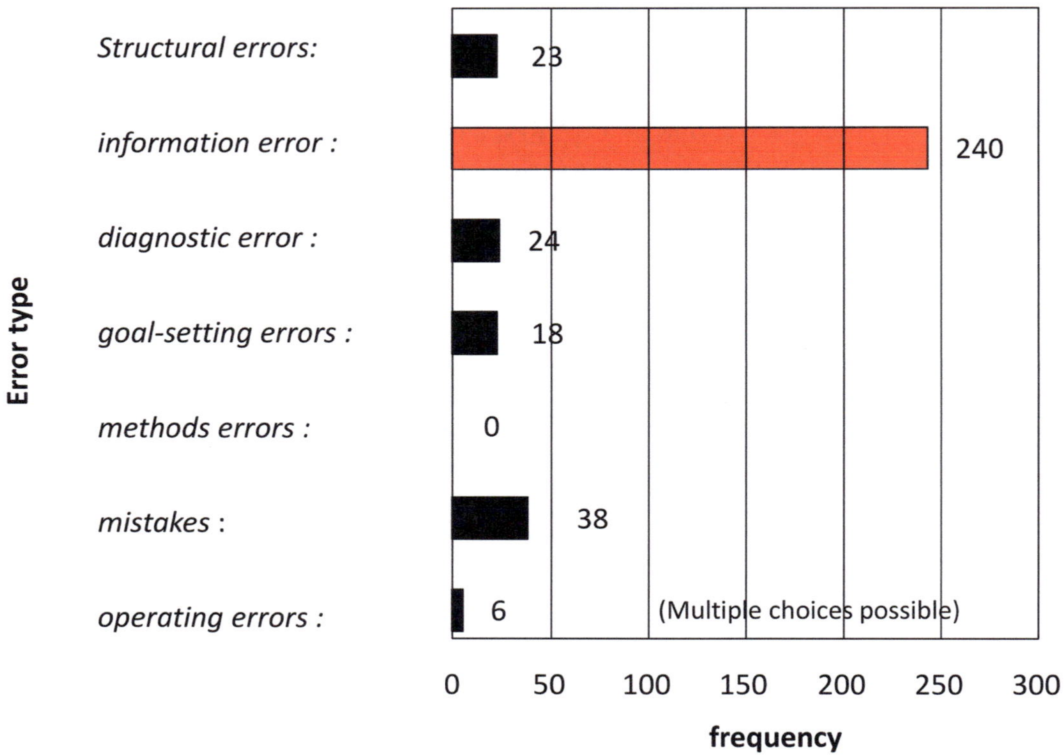

Fig. 3.81 Frequency of internal causes of accidents in a total of 312 accidents analysed (Gründl 2005)

side world, the eye is forced to scan the surroundings through constant movement – the saccades. What was not captured by chance is also not present in the inner image of the outside world. Therefore, the main source of information errors are the distractions caused by operating the vehicle's internal controls (e.g. radio, CD player, air conditioning, searching for places on the map on the passenger seat, entering destinations into the navigation system, operating the mobile phone, etc.). But also masking by objects in the outside world (e.g.: wrongly parked vehicles at junctions, bushes, trees, billboards) can be the cause for the error "Information (temporarily) not available".

The next most common errors are *errors of action*. Most of it come off by a fright reaction to a suddenly occurring danger situation (according to Gründl nearly 90% of all action errors). More than half of these errors of action are preceded by an information error. By far the most common error is the tearing of the steering wheel (after deviation to the banquet or touching the kerbstone, after deviation to the oncoming lane, after awakening from a moment's sleep, after deer on the road or too strong counter-steering in a lateral gust of wind). However, too weak steering, such strong braking and too early or too strong acceleration are also observed.

According to Gründl's investigations, the third place in the ranking of errors is occupied by the *diagnostic errors*. They are mainly caused by incorrect estimation of distances and speeds as well as wrong interpretation of the intentions of other road users, errors which occur particularly frequently at junctions, but also when changing lanes and when overtaking oncoming traffic. In this context, reference is made to ▶ Fig. 2.22, which shows that many driver requirements can only be met by estimates but not by comparison with concrete information.

Structural defects are present when the driver is no longer able to avoid the accident by acting on his own in the respective situation. This is the case if there are either no objective signs of an external hazard for him or if these signs occur so shortly before the critical situation that he can no longer react even with the highest attention and fastest reaction. Such errors are caused among other things by technical defects (e.g. blowouts). In most cases, however, the driver has made a mistake in the run-up to the test by selecting a speed too high for the road conditions (e.g. with black ice, aquaplaning) or a too small distance to the vehicle in front (diagnostic error).

Goal-setting errors are characterized by the fact that an objective of the action that is not appropriate to the situation is set (e.g. avoidance to protect the life of an animal or fear of a collision with a wide lorry, although there is sufficient roadway width and similar). They often arise from a sudden situation that does not correspond to the repertoire of actions that the driver has built up over the course of his experience.

Operating errors are mainly observed in the form of the foot slipping off the brake pedal (due to wet shoe sole). In this context, however, the spectacular cases of "unexpected acceleration" should also be mentioned, where the driver (only automatic vehicles!) suddenly accelerated in connection with a manoeuvre, although he was convinced that he was keeping his foot firmly on the brake. It could be shown that this was a pedal mix-up (Wierwille 1991), which was obviously caused by a combination of the illogical operation of the transmission selector lever (reverse travel by selector lever forward – forward travel by selector lever backward) and braking or accelerating each time by actuating a pedal by moving forward (in this context reference is made to the compatibility considerations in ► Sect. 6.1.3).

References

Literature Used

Benzinger, T.H.: The physiological basis for thermal comfort. Indoor Climate, pp. 441–474. Danish Buiding Research Institute, Copenhagen (1979)

Bergmeier, U.: Methode zur kontaktanalogen Visualisierung von Fahrerassistenzinformationen unter automotive-tauglichen Gesichtspunkten. Jahresdokumentation 2008. Bericht zum 54. Kongress der Gesellschaft für Arbeitswissenschaft, München, 9–11 April 2008. GFA-Press, Dortmund (2008)

Bergmeier, U.: Kontaktanalog markiertes Nachtsichtsystem – Entwicklung und experimentelle Absicherung. Dissertation an der Technischen Universität München (2009)

Bergmeier, U., Bubb, H.: Augmented reality. In: Vehicle – Technical realisation of a contact analogue head-up-display under automotive capable aspects; usefulness exemplified through night vision systems. FISITA 2008 World Automotive Congress (2008)

Beyer, L., Weiss, T.: Elementareinheiten des somatosensorischen Systems als physiologische Basis der taktil-haptischen Wahrnehmung. In: Grunwald, M., Beyer, L. (Hrsg.) Der bewegte Sinn. Birkhäuser, Basel (2001)

Braun, C., Gründl, M., Marberger, C., Scherber, C.: Beautycheck – Ursachen und Folgen von Attraktivität. Projektabschlussbericht. [pdf-Dokument] (2001). http://www.beautycheck.de/bericht/bericht.htm

Bubb, H.: The influence of braking distance indication on driver's behaviour. In: Oborne, D.J., Levis, J.A. (eds.) Human Factors in Transport Research, vol. II, pp. 388–346. Academic, London and New York (1981)

Bubb, H., Wohlfahrter, M.: Eye-tracking data analysis and neuroergonomics. In: Fafrowicz, M., Marek, T., Karwowski, W., Schmorrow, D. (Hrsg.) Neuroadaptive Systems, S. 255–311. CRC Press, Taylor & Francis Group, LLC, Boca Raton, London, New York (2012)

Chatziastros, A.: Visuelle Kontrolle der Lokomotion. Dissertation. Justus-Liebig-Universität Gießen (2003)

Chatziastros, A., Wallis, G.M., Bülthoff, H.H., et al.: The effect of view and surface texture on driver steering performance. In: Gale, A.G. (Hrsg.) Vision in Vehicles VII, S. 83–91. Elsevier Science Publisher B.V. (North Holland), Amsterdam (1999)

Cohen, A.: Augenbewegungen des Autofahrers beim Vorbeifahren an unvorhersehbaren Hindernissen und auf freier Strecke. Zeitschrift für Verkehrssicherheit. **22**, 68–76 (1976)

Conti, A., Dlugosch, C., Vilimek, R., Keinatz, A., Bengler, K.: An assessment of cognitive workload using detection response tasks. In: Stanton, N.A. (Hrsg.) Advances in human factors and ergonomics series. Advances in human aspects of road and rail transportation, S. 735–743. CRC Press, Boca Raton (2013)

David, E.: Elektronisches Analogmodell der Verarbeitung akustischer Information in Organismen. Habilitationschrift, Erlangen (1972)

De Waard, D.: The Measurement of Driver's Mental Workload. Proefschrift, Dissertationsschrift an der Rijksuniversiteit Groningen (1996)

Donges, E.: Ein regelungstechnisches Zwei-Ebenen-Modell des menschlichen Lenkverhaltens im

3

Kraftfahrzeug. Zeitschrift für Verkehrssicherheit. **24**, 98–112 (1978)
Dornhöfer, S., Pannasch, S.: Risky Business: Der Gefahr ins Auge geblickt! In: Preisträgerbroschüre des zweiten Deutschen Studienpreises der Körber-Stiftung (2000)
Dudel, J.: Allgemeine sinnesphysiologie. In: Schmidt, R.F., Thews, T. (Hrsg.) Einführung in die Physiologie des Menschen. Springer, Berlin, Heidelberg, New York (1976)
Fanger, P.O.: Thermal Comfort. Robert E. Krieger publishing company, Malabar (1992)
Fastenmeier, W.: Autofahrer und Verkehrssituation. (Band-Hrsg. In: Häcker, H. (Hrsg.) Reihe Mensch-Fahrzeug-Umwelt, Bd. 33, Verlag TÜV Rheinland, Deutscher Psychologen-Verlag, Bonn (1995)
Foerster, O.: Sensible corticale Felder. In: Bunke, O., Foerster, O. (Hrsg.) Handbuch der Neurologie, Bd. 6, Soringer, Berlin (1936)
Funk, G.: Digital Media for Artists. [Online] Kunst Universität Linz, 12. Mai 2008. [Zitat vom: 23. Juni 2008.] (2008). http://www.dma.ufg.ac.at
Gengenbach, R.: Fahrerverhalten im Pkw mit Head-Up-Display. Gewöhnung und visuelle Aufmerksamkeit VDI Fortschrittsberichte Reihe 12: Verkehrstechnik/Fahrzeugtechnik, Bd. 330 (1997)
Gibson, J.J.: The Perception of the Visual World. Houghton-Mifflin, Boston (1950)
Gibson, J.J.: The Senses Considered as Perceptual Systems. Houghton-Mifflin, Boston (1966)
Gillet, G.: Ergonomische Optimierung eines Aktiven Stellteils. Dissertation, Lehrstuhl für Ergonomie, TU München (1999)
Goldberg, J.M., Fernandez, C.: Physiology of peripheral neurons innervating semicircular canals of the squirrel monkey. I–II. J. Neurophysiol. **34**, 635 (1971)
Goldstein, E.B.: Wahrnehmung und aktive Motorik. In: Wahrnehmungspsychologie, S. 327–343. Spektrum Akademischer Verlag, (2002)
Gründl, M.: Fehler und Fehlverhalten als Ursache von Verkehrsunfällen und Konsequenzen für das Unfallvermeidungspotenzial und die Gestaltung von Fahrerassistenzsystemen. Inauguraldissertation Universität Regensburg (2005)
Guan, H., Gao, Z., Guo, K., et al.: An Optimal Preview Acceleration Model for Velocity Control in DriverVehicle-Environment System. In: Proceeding of AVEC 2000, S. 525–530. Taylor & Francis, Ann Arbor (2000)
Gundry, A.J.: Experiments on the detection of roll motion. Aviat. Space Environ. Med. **57**(5) (1978)
Guzmán, A.: Decomposition of a Visual Scene into Three-Dimension Bodies. In: Gresselli, A. (Hrsg.) Automatic Interpretation and Classification of Images. New York (1969)
Hacker, W.: Grundlagen der Regulation von Arbeitsbewegungen. Probl. Erg. Psychol (Beiheft 1), (1967)
Hacker, W.: Fehlhandlungen und Arbeitsfehler. In: Hacker, W. (Hrsg.) Arbeitspsychologie, Schriften zur Arbeitspsychologie, Nr, Bd. 41, H. Huber, Stuttgart (1987)
Hale, A.R., Stoop, J., Hommels, J.: Human error models as predictor of accident scenarios for designer in road transport systems. Ergonomics. **33**, 1377–1387 (1990)
Handwerker, H.O.: Somatosensorik. In: Schmidt, R.F. (Hrsg.) Neuro- und Sinnesphysiologie (1993)
Hartmann, E.: Beleuchtung und Sehen am Arbeitsplatz. Goldmann, München (1970)
Hassenzahl, M., Burmeister, M., Koller, F. (2007): Der Unser experience (UX) auf der Spur. Zum Einsatz von www.attrakdiff.de
He, J., Becic, E., Lee, Y.-C., McCarley, J.S.: Mind wandering behind the wheel: performance and oculomotor correlates. Hum. Factors. **53**(1), 13–21 (2011). https://doi.org/10.1177/0018720810391530
Heißing, B., Kudritzki, D., Schindlmaister, R., Mauter, G.: Menschgerechte Auslegung des dynamischen Verhaltens von Pkw. In: Bubb, H. (Hrsg.) Ergonomie und Verkehrssicherheit. Beiträge der Herbstkonferenz der Gesellschaft für Arbeitswissenschaft, S. 1–31. Herbert Utz Verlag, München (2000)
Helton, W.S., Matthews, G., Warm, J.S.: Stress state mediation between environmental variables and performance: the case of noise and vigilance. Acta Psychol. **130**(3), 204–213 (2009). https://doi.org/10.1016/j.actpsy.2008.12.006
Hofmann, J.: Raumwahrnehmung in virtuellen Umgebungen – Der Einfluss des Präsenzempfindens in Virtual Reality – Anwendungen für den industriellen Einsatz. Deutscher Universitätsverlag Kognitionswissenschaft, Wiesbaden (2002)
Iggo, A.: Somatosensory system. Handbook of Sensory Physiology Bd. 2. Springer, Berlin, Heidelberg, New York (1973)
Intersafe, (2009): INTERSAFE. Subproject of PReVENT (Preventive and Active Safety Applications Integrated Project)
Israel, B.: Potenziale eines kontaktanalogen Head-up Displays für den Serieneinsatz. Dissertation an der Technischen Universität München (2012)
Joos, M., Rötting, M., Velichkovsky, B.: Psycholinguistik/Psycholinguistics. Ein internationales Handbuch/An International Handbook. In: Die Bewegungen des menschlichen Auges – Fakten, Methoden, innovative Anwendungen, S. 142–168. de Gruyter, Berlin & New York (2003)
Kaufman, L., Richards, W.: Spontaneous fixation tendencies for visual forms. Percept. Psychophys. **5**, 85–88 (1969)
Keidel, W.D.: Physiologie, 4. Aufl. (1975)
Kersten, B.: Was uns gefällt. Gehirn und Geist, 3. Aufl. Verlag Spektrum der Wissenschaften, S. 50–57 (2006)
Knoll, C.: Einfluss des visuellen Urteils auf den physisch erlebten Komfort am Beispiel von Sitzen. Dissertation an der Technischen Universität München. (2007)
Körber, M.: Hypovigilanz durch automatisiertes Fahren (2014). Seminarvortrag am Lehrstuhl für Ergonomie der Technischen Universität München am 24.01.2014

Kreitler, H., Kreitler, S.: Psychologie der Kunst. Kohlhammer, Stuttgart, S. 88 (1980)
Krist R.: Modellierung des Sitzkomforts, eine experimentelle Studie. Katholische Universität Eichstätt. Dissertation (1993)
Kühner, M., Wild, J., Bubb, H.: Haptische Konstanzleistung bei Drucktastern. Ergonomie Aktuell. **012**, 28–32 (2011)
Lachenmayer, B., et al.: Sehstörungen als Unfallursache Bericht der Bundesanstalt für Straßenwesen, Mensch und Sicherheit, Bd. M65. (1996)
Lamb, T.D.: Das Auge – Organ mit Vergangenheit. Spektrum der Wissenschaft. **12**(2), 23–28 (2012)
Langham, M.: What Do Drivers Do at Junctions? 71st RoSPA Road Safety Congress, 2006 (2006)
Langois, J.H., Roggman, L.A.: Attractive faces are only average. Psychol. Sci. **1**, 115–121 (1990)
Lappe, M.: Visuelle Wahrnehmung von Bewegung. Universität Münster (2009). http://www.Psy.uni-muenster.de/imperia/md/content/psycologie_institut_2/ae_lappe/freie_dokumente/einführung1.pdf
Lindsay, P.H., Norman, D.A.: Human Information Processing. An Introduction to Psychology. Academic, New York, London (1972)
Lippold, C., Mattheß, V.: Anwendung von Risikoanalysen, Anwendungsbeispiel "Hindernis auf der Fahrbahn" Forschung Straßenbau und Verkehrstechnik, Bd. 689. Bundesminister Für Verkehr, Bonn (1994)
Liu, A.: What the Driver's Eye Tells the Car's Brain. In: Eye Guidance in Reading and Scene Perception, S. 431–452. Elsevier Science Ltd, Kidlington, Oxford (1998)
Martinez-Conde, S., Macknik Hubel, S.D.: The role of fixational eye movements in visual perception. Nat. Rev. Neurosci. **5**, 229–240 (2004)
Maycock, G., Lockwood, C., Lester, J.: The Accident Liability of Car Drivers Department of Transport TRL Report RR, vol. Bd. 315. Transport Research Laboratory, Crowthorne (1991)
Mayer, E.: Tagesgang für thermisches Behaglichkeitsempfinden. Gesundheits-Ingenieur – gi. **107**(3), 173–176 (1986)
Mayer, E.: Ist die bisherige Zuordnung von PMV und PPD noch richtig? KI Luft- und Kältetechnik. **34**(12), 575–577 (1998)
Mayer, E., Schwab, R.: Untersuchung der physikalischen Ursachen von Zugluft. Gesundheits-Ingenieur – gi. **111**(1), 17–30 (1990)
McCarley, J.S., Vais, M., Pringle, H., Kramer, A.F., Irwin, D.E., Strayer, D.L.: Conversation disrupts visual scanning of traffic scenes. In: Gale, A.G. (Hrsg.) Vision in Vehicles IX. Elsevier North Holland Press, Amsterdam (2002)
Miller, G.A., et al.: The magical number seven plus or minus two: some limits on our capacity for processing information. Psychol. Rev. **63**, 81–97 (1956)
Müller-Limmroth, W.: Sinnesorgane. In: Schmidtke, (Hrsg.) Lehrbuch der Ergonomie, 2. Aufl. Carl Hanser Verlag, München, Wien (1981)
Muragushi, Y., Fukui, K., Asaga, Y., Ono, E.: Development of Human Sensitivity Evaluation System for Vehicle Dynamics 31. International Congress FISITA, Yokohama, 22–27. Oktober, 2006 (2006)
Norman, D.A.: Categorization of Action Slips. Psychological Review. **88**, 1–15 (1981)
Norman, D.A.: New views in information processing: implications for intelligent decision support systems. In: Hollnagel, E., Manchini, G., Woods, D.D. (Hrsg.) Intelligent Decision Support in Process Environments. Springer-Verlag, Berlin (1986)
O'Neill, B.: A decision-theory model of danger compensation. Acc. Anal. Prev. **9**, 157–165 (1977)
Palm, G.: Assoziatives Gedächtnis und Gehirntheorie. Gehirn und Kognition. Spektrum der Wissenschaft Verlagsgesellschaft, Heidelberg (1990)
Palmer, S.E.: Common region: a new principle of perceptual grouping. Cogn. Psychol. **24**, 436–447 (1992)
Plavšic, M.: Analysis and modelling of driver's behavior for assitance systems at road intersections. Dissertation an der Technischen Universität München. (2010)
Plavšic, M., Duschl, M., Tönnis, M., Bubb, H., Klinker, G.: Ergonomic Design and Evaluation of Augmented Reality Based Cautionary Warnings for Driving Assistance in Urban Environments The 17th World Congress on Ergonomics (International Ergonomics Association, IEA), Beijing, China, Aug. 9–14, 2009 (2009)
Plavsic, M., Bengler, K., Bubb, H.: Analysis of Glance Movements in Critical Intersection Scenarios 3rd International Conference on Applied Human Factors and Ergonomics, ami, Flori. Juli. 2010 (2010)
Pöppel: Grenzen des Bewußtseins – Wie kommen wir zur Zeit und wie entsteht Wirklichkeit? Insel Taschenbuch, Frankfurt a.M. und Leipzig (2000)
Rantanen, E.M., Goldberg, J.H.: The effect of mental workload on the visual field size and shape. Ergonomics. **42**(6), 816–834 (1999)
Rasmussen, J.: Human errors: a taxonomy for describing human malfunction in industrial installations. J. Occup. Accid. **4**, 311–333 (1982)
Rasmussen, J.: Information Processing and Human-Machine Interaction. An Approach to Cognitive Engineering, North-Holland, New York (1986)
Rasmussen, J., et al.: The definition of human error and a taxonomy for technical system design. In: Rasmussen, J., Duncan, K., Lepat, J. (Ed.) New Technology and Human Error, Chichester, Wiley (1987)
Rassl, R.: Ablenkungswirkung tertiärer Aufgaben im Pkw – Systemergonomische Analyse und Prognose. Dissertation an der Technischen Universität München (2004)
Reason, J.: Menschliches Versagen: Psychologische Risikofaktoren und moderne Technologien. Spektrum, Heidelberg (1994)
Reichart, G.: Menschliche Zuverlässigkeit beim Führen von Kraftfahrzeugen – Möglichkeiten der Analyse und Bewertung. Dissertation am Lehrstuhl für Ergonomie der TU München (2000)

3

Reinprecht, K.: Blicke in der Nacht – Ist das real? Tagung experimentell arbeitender Psychologen, Halle (Saale) (2011)
Reisinger, J.: Parametrisierung der Haptik von handbetätigten Stellteilen. Dissertation an der Technischen Universität München (2009)
Remington, R.W.: Attention and saccadic eye movements. J. Exp. Psychol. Hum. Percept. Perform. **6**, 726–744 (1980)
Remlinger, W.: Analyse von Sichtverdeckungseffekte im Fahrzeug. Dissertation an der Technischen Universität München (2013)
Rensink, R.A., O'Regan, J.K., Clark, J.J.: To see or not to see: the need for attention to perceive changes in scenes. Psychol. Sci. **8**(5), 368–373 (1997)
Revesz, G.: Psychology and Art of the Blind. Longmans Green, New York (1950)
Rhodes, G., Tremaran, T.: Averageness, exaggeration and facial attractiviness. Psychol. Sci. **7**, 105–110 (1996)
Rigby, L. V.: The Nature of Human Error, Sandia National Laboratories Report Sc-R70-4318, Albuquerque (1970)
Rock, I.: The basis of position-constancy during passive movement. Am. J. Psychol. **81**, 262–265 (1968)
Rockwell, T.H.: Eye Movement Analysis of Visual Information Aquisition in Driving: An Overview. North Carolina State University, Raleigh (1971)
Sachse, H.: Einführung in die Kybernetik. Uni-Text, Vieweg, Braunschweig (1971)
Schierz, C.: Sehen und Bildschirm. Vorlesungsmanuskript, Institut für Hygiene und Arbeitsphysiologie. ETH Zürich, Zürich, S. 23 (2001)
Schmidt, R.F.: Integrative Funktion des Nervensystems. In: Schmidt, R.F., Thews, G. (Hrsg.) Physiologie des Menschen. Springer-Verlag, Berlin, Heidelberg, New York (1976)
Schmidt, R.F. (ed.): Grundriß der Neurophysiologie. Springer, Berlin (1979)
Schmidt, R.F.: Physiologie des Menschen, 26. Aufl. Springer-Verlag, Berlin (1995)
Schmidtke, H.: Die Ermüdung. Symptome – Theorien – Meßversuche. Huber Verl., Bern und Stuttgart (1965)
Schmidtke, H., Schober, H.: Sehanforderungen bei der Arbeit. Geutner-Verlag, Stuttgart (1967)
Schweigert, M.: Fahrerblickverhalten und Nebenaufgaben Dissertation an der Technischen Universität München (2003)
Sheridan, T.B., Ferrell, W.R.: Man-Machine-Systems, Information, Control and Decision. Models of Human Performance. The MIT Press, Cambridge, London (1974)
Sperry, R.: A modified concept of consciousness. Physiol. Rev. **76**, 532 (1969)
Steward, J.D.: Human perception of angular acceleration and implications in motion simulator. J. Aircraft. **8**(4), 248 (1971)
Summala, H., Nieminen, T., Punto, M.: Maintaining lane position with peripheral vision during invehicle tasks. Hum. Factors. **38**(3), 442–451 (1996)
Swain, A.D., Guttmann, H.E.: Handbook of Human Reliability Analysis with Emphasis on Nuclear Power Plant Applications Sandia National Laboratories, Bd. Nureg/Cr-1278. Washington, DC (1983)
Tomaske, W., Fortmüller, T.: Der Einfluss von Wahrnehmungsschwellwerten auf die Auslegung von Bewegungssystemen in der Fahrsimulation. In: Deutsche Gesellschaft für Luft- und Raumfahrt e.V. (DGLR), Bonn (2001)
Totzke, I., Huth, V., Krüger, H.-P., Bengler, K.: Overriding the ACC by keys at the steering wheel: Positive effects on driving and drivers' acceptance. In: Human Factors for Assistance and Automation, S. 153–164. Shaker, Maastricht (2008)
Trendelenburg, W., Schütz, E.: Der Gesichtssinn. Springer, Berlin Göttingen, Heidelberg (1971)
Underwood, G., Chapman, P., Brocklehurst, N., Underwood, J., Crundall, D.: Visual attention while driving: sequences of eye fixations made by experienced and novice drivers. Ergonomics. **46**(6), 629–646 (2003)
von Helmholtz, H.: Anatomische Beschreibung des Auges Handbuch der physiologischen Optik, Bd. 1. Voss, Hamburg (1866)
von Helmholtz, H.: Die Lehre von den Gesichtswahrnehmungen Handbuch der physiologischen Optik, Bd. 3. Voss, Hamburg (1910)
von Holst, E.: Aktive Leistungen der menschlichen Gesichtswahrnehmung. Stud. Gen. **10**(4), 232 (1957)
Wann, J.P., Wilkie, M.: How do we control high speed steering. In: Vaina, L.M., Beardsley, S.A., Rushton, S.K. (Hrsg.) Optic Flow and Beyond, S. 401–417. Kluwer, Dordrecht (2004)
Warm, J.S., Parasuraman, R., Matthews, G.: Vigilance requires hard mental work and is stressful. Hum. Factors. **50**(3), 433–441 (2008). https://doi.org/10.1518/001872008X312152
Warren, W., Kay, B., Zosh, W., Duchon Sahuc, A.S.: Optic flow is used to control human walking. Nat. Rev. Neurosci. **4**, 213–216 (2001)
Weinberger, M.: Einfluss von ACC-Systemen auf das Fahrverhalten, Dissertation. Technische Universität München (2001)
Werblin, F., Roska, B.: Wie das Auge die Welt verfilmt. Spektrum der Wissenschaft. **5**, 41–47 (2008)
Wicken, C.D., Hollands, J.G.: Engineering Psychology and Human Performance, 3. Aufl. Prentice Hall, Upper Saddle River (2000)
Wickens, C.D.: Engineering Psychology and Human Performance. Harper Collins Publishers, New York (1992)
Wierwille, W.W.: Driver error in using automotive pedals. In: Peters, G.A., Peters, B.J. (Hrsg.) Automotive Engineering and Litigation. John Wiley & Sons, New York (1991)
Wilde, G.J.S.: The theory of risk homeostasis: implications for safety and health. Risk Anal. **2**(2), 209 (1982)
Wirtz, M.A.: Lexikon der Psychologie: Dorsch – 2014/2015. Dorsch Lexikon der Psychologie. Hans Huber-Hogrefe AG, Bern (2013)

Wittling, W.: Hemisphäreninteraktion, Leistung und Belastung. Was „nutzt" die Hirnforschung der Arbeitswissenschaft? In: Bubb, H., von Eiff, W. (Hrsg.) Innovative Arbeitsgestaltung – Mensch, Organisation, Information und Technik in der Wertschöpfungskette. Wirtschaftsverlag Bachem, Köln (1992)
Wolf, H.: Ergonomische Untersuchung des Lenkgefühls von Personenkraftwagen. Dissertation an der Technischen Universität München (2009)
Yanko, M.R., Spalek, T.M.: Route familiarity breeds inattention: a driving simulator study. Accid. Anal. Prev. **57**, 80–86 (2013). https://doi.org/10.1016/j.aap.2013.04.003
Yuhara, N., Tajima, L., Sano, S., et al.: Steer-by-wire-oriented steeling system design: concept and examination. Vehicle System Dynamics Supplement. **33**, 692–703 (1999)
Zhang, L., Helander, M.G., Drury, C.G.: Identifying factors of comfort and discomfort in sitting. Hum. Factors. **38**(3), 377–389 (1996)
Zichenko, V.P., Virgiles, N.Y.: Formation of visual images: studies of stabilized retinal images. In: Gale, A.G. (Hrsg.) Vision in Vehicles II, S. 335–344. Elsevier North Holland Press, Amsterdam (1972). Translated by Consultants Bureau, New York, Plenum Press
Zimmer, A.: Wie intelligent darf/muss ein Auto sein? Anmerkungen aus ingenieurspsychologischer Sicht. In: Jürgensohn, T., Timpe, K.-P. (Hrsg.) Kraftfahrzeugführung. Springer, Berlin (2001)
Zwahlen, H.T., Adams Jr., C.C., DeBald, D.P.: Safety aspects of CRT touch panel controls in automobiles. In: Gale, A.G. (Hrsg.) Vision in Vehicles II, S. 335–344. Elsevier North Holland Press, Amsterdam (1988)
Zwicker, E., Feldkeller, R.: Das Ohr als Nachrichtenempfänger. 2. Neubearbeitete Auflage. Hirzel-Verlag, Stuttgart (1967)

Further Literature

Bubb, H., Schmidtke, H.: Physiologische und Psychologische Grenzen menschlichen Leistungsvermögens. Kap. 38. In: Masing, M. (Hrsg.) Handbusch der Qualitätssicherung. Carl Hanser Verlag, München, Wien (1988)
Fikus, M.: Visuelle Wahrnehmung und Bewegungskoordination. Deutsch, Frankfurt am Main (1989)
Goldstein, E.B.: In: Ritter, M. (Hrsg.) Wahrnehmungspsychologie, 2. Aufl., S. 190. Spektrum Akademischer Verlag, Heidelberg, Berlin, Oxford (2002)
Ilgmann, W.: Ergonomische Untersuchungen über die Einwirkung rotatorischer Schwingungen Forschungsberichte aus der Wehrtechnik, Bd. 79-33. Dokumentationszentrum der Bundeswehr, Bonn (1979)
Liao, J., Moray, N.: A simulation study of human performance deterioration and mental workload. Le Travail Humain. **56**, 321–344 (1993)
Mühlendyck, H.: Augenbewegung und visuelle Wahrnehmung. Enke, Stuttgart (1990)
Sanders, M.S., McCormick, E.J.: Human factors in engineering and design, 7. Aufl. McGraw-Hill, New York (1993)
Schmidtke, H.: Mentale Beanspruchung. In: Schmidtke, H. (Hrsg.) Lehrbuch der Ergonomie, 2. Aufl. Carl Hanser Verlag, München, Wien (1981)
Third International Dictionary of the English Language (1981): by Gove, Philip Babcock, 1902–1972; G. & C. G. & C. Merriam Co.; Springfield, Mass.: G. & C. Merriam Co

Anatomical and Anthropometric Characteristics of the Driver

Rainer E. Grünen, Fabian Günzkofer, and Heiner Bubb

Contents

4.1 Anatomical Basics – 162

4.1.1 The Musculoskeletal System – 162

4.1.2 Muscular System – 170

4.2 Anthropometry – 176

4.2.1 Length and Circumference Dimensions – 176

4.2.2 Weight – 194

4.2.3 Forces – 196

4.2.4 Mobility – 205

References – 213

H. Bubb et al. (eds.), *Automotive Ergonomics*, https://doi.org/10.1007/978-3-658-33941-8_4

4

4.1 Anatomical Basics

Rainer E. Grünen and Fabian Günzkofer

For the ergonomic design of motor vehicles, knowledge of the human body is necessary because this provides many approaches to solving design problems. The *Anatomy as the study of the structure of the human body* provides the necessary terms and contexts. It is also of particular importance for the correct representation of mechanical conditions in all models and simulations.

The human body with its various functions and tasks represents a complex unit that is coherent and indivisible, and whose mechanical and biological processes are closely intertwined. Morphologically the following components are distinguished: The **locomotor apparatus** with bones, joints and the musculature, the **nervous system**[1] and the metabolic organs, such as the **respiration**- and **digestion** system. The **sensory organs** have a special significance for the reception of information. In addition **skin**, **tissue**, **heart** and **vascular system**, **blood,** the **lymphatic system** and the **endocrine system, urinary tract,** and the **genital apparatus** are anatomically described.

This chapter confines itself to the organs and functions relevant to the anthropometric design of the motor vehicle.

4.1.1 The Musculoskeletal System

In anatomy, the musculoskeletal system is understood as the components and effective structure of bones and tissues that enable the human body to move on its own. However, the musculoskeletal system is also responsible for all resting postures, especially sitting, due to its supporting function. The skeleton is divided by the joints and supports the entire body. The musculature generates the body forces and transmits them to the supporting structure of the skeleton. The tendons transmit the tensile forces of the musculature to the skeleton, while the ligaments stabilize the loosely connected surfaces of the joints. The soft tissue between the joint surfaces provides shock absorption and pressure distribution in the joints. The supply of nutrients to the joints is ensured by the synovial fluid.

[1] The nervous system is divided into the following sub-areas: Central nervous system (brain, spinal cord), peripheral nervous system (network of nerves) and vegetative nervous system.

4.1.1.1 Terminology

Axes and Planes

To designate the orientations on the limbs, three main axes are usually defined which are perpendicular to each other (Fig. 4.1): The **longitudinal axis** of the body is perpendicular to the standing plane when in an upright position. The **transverse axis** runs from the left to the right half of the body and is perpendicular to the longitudinal axis. The **sagittal axis** runs from the rear to the front half of the body and is perpendicular to the other two axes. The three main planes can be assigned to these three main axes: The **sagittal plane** is aligned along the sagittal axis and divides the body into a right and a left half. The **frontal plane** (Coronal plane) runs in the direction of the longitudinal axis and is aligned parallel to the forehead. The **transverse plane** is transverse to the body and thus perpendicular to the longitudinal axis.

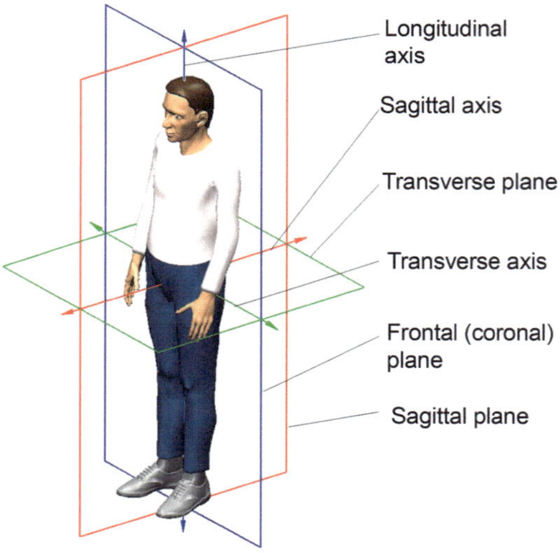

Fig. 4.1 Body axes

Location and Direction

The following definitions are usual for the exact description of the position and the directions: body parts which are aligned to the head are marked with **cranial** (*superior*), while body parts that are aligned towards the the breech are termed as **caudal** (*inferior*). An alignment to the abdominal side is called **ventral** (*anterior*), towards the back surface as **dorsal** (*posterior*). At the extremities **distal** parts of the body are those that are at the end of the limbs, while **proximal** are oriented towards the core of the body.

General Anatomy of the Musculoskeletal System

The various and complex movements that are possible for locomotion, but also for the performance of mechanical work and thus also for driving and operating motor vehicles, are represented by the interaction of skeleton and muscles. The skeleton with its bony supporting elements the joints connected by connective tissue, is called the passive musculoskeletal system, while the muscles represent the active part. Together with the joints and muscles, the skeletal elements form the locomotor organs, which are also responsible for the upright posture when sitting.

4.1.1.2 The Skeletal System

Bones

At the beginning of its life, the human being consists of about 300 single bones, which grow together in the course of the progressive growth partially to the 206–214 bones of an adult. Apart from the bones with a protective function, such as the skull bone and the ribs of the thorax, the bones connected to the joints provide the passive musculoskeletal system due to their supporting and levering properties. According to the type of external shape one distinguishes *long*, *short*, *flat*, and *irregular bones*. The size of the bones mainly depends on the mechanical strain and function in the musculoskeletal system. The bones of the free extremities (arms and legs), with the exception of the hand and tarsal bones, are long bones (*Ossa longa*) (see Fig. 4.2). The long shaft (*diaphysis*) is limited on both sides by the bone ends (*epiphyses*), which are

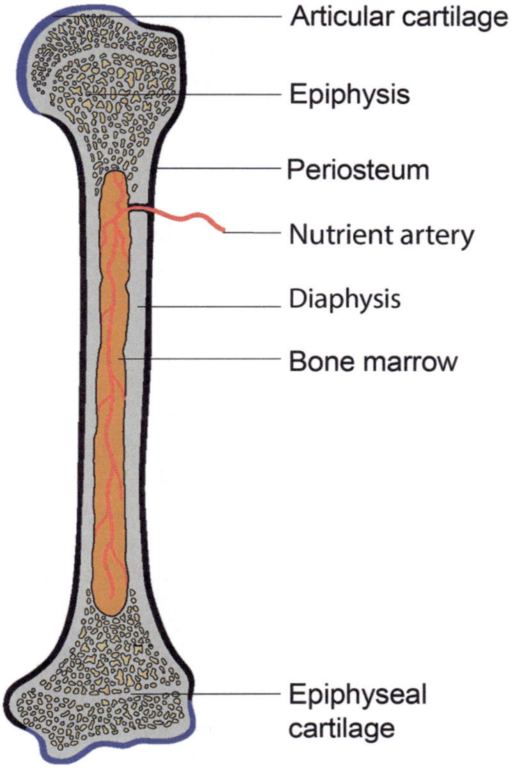

Fig. 4.2 Hollow bone

characterized by the functional surfaces of the joints. Between the epiphysis and the diaphysis there is the epiphyseal cartilage until the end of the growth phase, in which the longitudinal growth of the long bone and thus of the body takes place by continuously dividing cells. The shaft is filled with bone marrow, the primary function of which is to form the blood cells of the human body. For this reason, blood vessels with good blood circulation are located on and in the bones. Flat bones (*Ossa plana*) like the skull bones, the shoulder blade or the pelvis are characterized by a flat and arched shape and have above all protective functions. The short bones, such as the hand and tarsal bones, are small and irregular in shape and have several joint surfaces. The sesame bones, such as the kneecap or the pisiform carpal on the outside of the edge of the hand, are embedded in a tendon and thus ensure a greater distance between the muscle and the joint, resulting in greater strength or increased mobility. In addition, a distinction is made between irregular bones, such as the vertebrae of the spinal column or the lower jaw, which have a special structure due to their special function.

Joints

The individual bones of the passive musculoskeletal system are connected by joints. **Synarthrosis joints** are bone pairings that are linked with different types of cartilage or connective tissue.

In **Synovial joints**, the adjacent bones are separated by a synovial cavity filled with synovial fluid inside the joint capsule. The joint capsule is formed from the continuation of the periosteum surrounding the bone. The respective head ends of the bones are formed into concave or convex joint surfaces and covered with a hyaline joint cartilage. This joint cartilage, which is about 2–8 mm thick, has a smooth surface and has supported by the joint fluid a shock-absorbing and gliding effect. Since the joint cartilage has no blood vessels, the joint fluid must be pressed into the cartilage by regular movement. Low mobility and immobilization of the joints thus inhibits the supply of nutrients to the joint cartilage and at the same time the surrounding joint capsule loses flexibility, which narrows the joint cavity. This leads to reduced mobility and painful joint movements, especially for elderly people. These phenomena tend to already be observed when driving motor vehicles in a sitting posture for a longer period of time, which is why a break with a change of posture and above all movement is advisable after about two hours of driving in order to stimulate the metabolism and blood circulation and to set the joints in motion again in order to guarantee the supply.

The basic structure of the joints (◘ Fig. 4.3) shows that a convex joint head is usually opposed to a convex head by a concave socket. The shape of the joint surfaces is determined by the task of the body part and determines the degrees of freedom of movement. The joint types described in more detail below are assigned to the individual joints of the human body in ◘ Fig. 4.4 and ◘ Table 4.1. **Ball-and-socket joints** have, characteristically, an almost spherically shaped joint head and

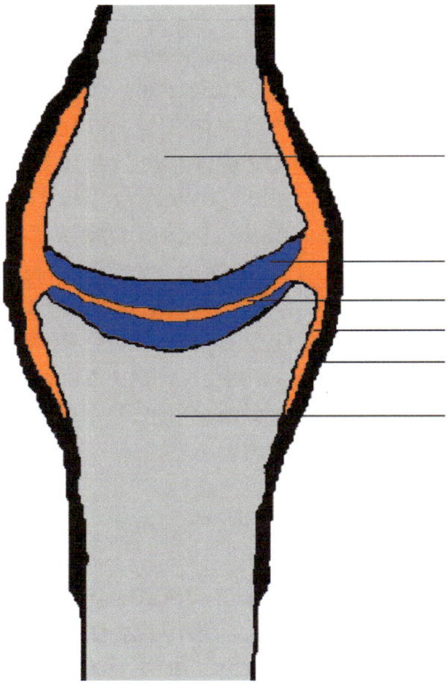

◘ **Fig. 4.3** Basic structure of a joint

correspondingly a joint socket. Thus they have three main axes perpendicular to each other and consequently six main movements. The various movement possibilities of the hip and shoulder joint result from the characteristics of the ball-and-socket joint. Similar to the ball joint is the **condyloid joint** (or ellipsoid joint) which is elliptically deformed and thus has one major axis less. Rotation around the normal axis is not possible. A typical representative of this type is the first upper cervical joint between the skull and the atlas vertebra that allows the head to nod ("yes" saying). A **hinge joint** engages in a hollow cylindrical depression formed by a channel-shaped skeletal element. An example of this is the elbow joint, which has only two main directions of movement. The rotation of the hand and forearm does not take place in the elbow joint, but is caused by the twisting of the *ulna* in relation to the radius. With **pivot joints** one joint partner is enclosed in a ring by the other, so that mobility is only possible in two opposing main directions. An example of this is the proximal radioulnar joint, which allows only rotational movements due to the ring ligament. Thus the elbow joint is a composite joint consisting of three partial joints, each with different types. Another example of a

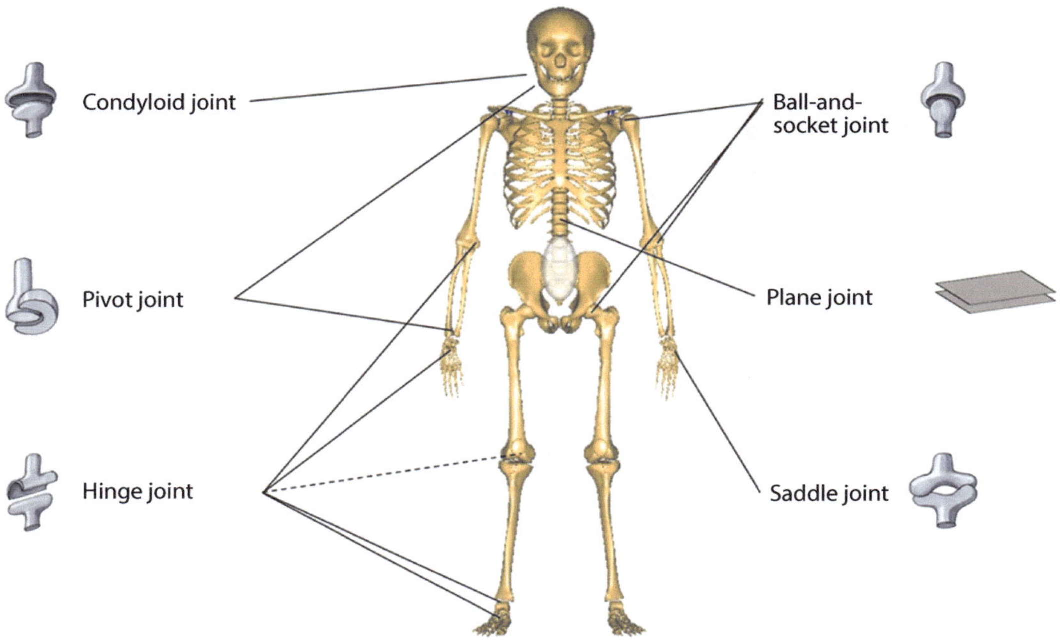

Fig. 4.4 Assignment of joint types to selected joints of the human body. Display of joint types by Produnis [GFDL (▶ http://www.gnu.org/copyleft/fdl.html) or CC-BY-SA-3.0 (▶ http://creativecommons.org/licenses/by-sa/3.0/)], via Wikimedia Commons

pivotjoint is the second upper cervical joint between the atlas and axis vertebrae. The cone formed in the axis vertebra reaches into a pit in the atlas vertebra and thus mainly allows rotational movements of the head ("no" sayings). Approx. 70% of the lateral head turning movements are made possible by this joint alone. **Saddle joints** are characterized by the fact that one joint side has a concave sliding surface, while the other has a convex sliding surface that can move towards each other. Saddle joints thus have two main axes of motion which are perpendicular to each other and thus allow four degrees of freedom of movement. Such a saddle joint is, for example, the thumb saddle joint, which allows the thumb to spread out from the hand and from a flat position next to the palm to face the other four fingers (*opposition* and *reduction*). **Plane joints** have no pronounced joint recesses and thus permit translative displacements parallel to the joint gap. Thus such a joint is mainly a pressure-absorbing supporting element that allows small lateral movements of the joint partners. The small vertebral joints of the spine fall under this category.

Joint Mechanics

The degrees of freedom of a joint and the shape of the joint surfaces only determine the basic mobility of the limbs. The joint movements are influenced by the existing muscles and ligaments, which determine the direction and relation of mobility. Although the phalanges can be passively twisted along the longitudinal axis of the toe, this mobility is not possible solely through activation by the existing musculature. Thus, no real hinge movements or complete rotations are possible, but only limited pendulum movements between the inhibiting limits. The mobility of the joints ranges from a few (1° to 2°) degrees in the small vertebrae to large areas (>100°) in the shoulder joint. With simple hinge joints with basically two degrees of freedom, two opposite directions are possible around the axis of motion: flexion and extension, abduction and adduction as well as internal and external rotation. Thus the twisting of the forearm towards the body is called *pronation* while the external rotation away from the body is referred to as *supination*. With the saddle and condyloid joints comprising four degrees of

Table 4.1 List of the most important joints of the musculoskeletal system and their degrees of freedom

Joint	Partner		Shape	Degrees of freedom
Upper cervical joint	Skull (Occiput)	First cervical vertebra (Atlas)	Condyloid joint	4
Lower head joint	First cervical vertebra (Atlas)	Second cervical vertebra (Axis)	Pivot joint	2
Vertebral column	Small vertebrae	Small vertebrae	Plane joint	4
Shoulder joint	Shoulder blade (Scapula)	Humerus	Ball-and-socket joint	6
Elbow joint 1	Humerus	Ulna	Hinge joint	2
Elbow joint 2	Humerus	Radius	Ball-and-socket joint	2
Verrückt Elbow joint 3	Ulna proximal	Radius proximal	Pivot joint	2
Proximal wrist joint	Radius) distal	3 Carpal bones (Os scaphoideum, Os lunatum, Os triquetrum)	Condyloid joint	4
Distal wrist joint	2 Carpal bones (Os scaphoideum, Os lunatum)	3 Carpal bones (Os trapezium, Os trapezoideum, Os capitatum)	Hinge joint	2
Finger base joint(s) "ankle	Carpal bone	Finger Phalanges	Ball-and-socket joint	4
Finger middle joint(s)	Basal phalanges (phalanx proximalis)	Finger middle phalanx (Phalanx media)	Hinge joint	2
Finger end joint(s)	Finger middle phalanx (Phalanx media)	Finger end phalanx (distal phalanx)	Hinge joint	2
Thumb saddle joint	First metacarpal bone	Trapezium	Saddle joint	4
Hip joint	Pelvis	Thigh bone (femur)	Ball-and-socket joint	6
Knee joint	Thigh bone (femur)	Shinbone (Tibia), Kneecap (Patella)	Combined joint	4
Upper ankle joint	Shin bone (tibia) Calf bone (fibula)	Ankle bone talus	Hinge joint	2
Lower ankle joint	Ankle bone (talus)	Heel bone (Calcaneus) Navicular bone (Os naviculare)	Hinge joint	2
Metatarsophalangeal joint(s)	Metatarsal bone (Ossi metatarsi)	Phalanx bones (Phalanges proximalis)	Ball-and-socket joint(s)	6
Interphalangeal jointsjoint(s)	Phalanx bones (Phalanges proximalis)	Phalanx bones (Phalanges media/distales)	Ball-and-socket joint	6

freedom, the movements combine and can overlap. At some joints the degrees of freedom are dependent from the position of the joint. Thus the index finger can be stretched out and spread to the side, while this movement is no longer possible when rolled up. Even though the ball-and-socket joints in the shoulder and hip appear to be the most flexible with six degrees of freedom, the many combined joints in the foot and especially in the hand allow for very complex movements. Alone in one hand 27 bones (8 carpal bones, 5 metacarpal bones, 14 long finger bones) unite, which are interconnected by ligaments and thus allow the multiple movement possibilities and manipulations of the hand.

Vertebral Column

The spine (Fig. 4.5) forms the connecting and supporting central structure of the musculoskeletal system. The vertebral bodies of the individual sections of the spine are similar in principle, with the exception of the atlas vertebra with its special function as part of the upper cervical joint. The spine is divided into 5 parts: The cervical spine with 7 *cervical vertebrae* (C1–C7), the thoracic spine with 12 *thoracic vertebrae* (Th1–Th12) and the lumbar spine with 5 *lumbar vertebrae* (L1–L5). These 24 vertebrae form the so-called free (*presacral*) spine, the vertebrae of which are connected with intervertebral discs . The free spine corresponds to about 35% of the body length and is therefore 53–70 cm long in adults. Below the free spinal column is the sacrum (*Os sacrum*), which consists of 5 fused sacral vertebrae, and the coccyx (*Os coccygeum*), which consists of 4–5 rudimentary *coccygeal vertebrae*. When standing upright, the spinal column in the sagittal plane has a double S-shaped curve. The cervical and lumbar spine are curved convexly forward (*lordosis*) while the thoracic and sacral spine are concavely curved backwards (*kyphosis*). In principle, this tendency to curvature is maintained even in a seated position, but there is a tendency to kyphosis, especially in the lumbar spine area - especially after long journeys - which greatly increases the stress in the vertebrae L3/L4 and L4/L5.

A comparison of the spine standing (Fig. 4.6 left) and sitting (Fig. 4.6 middle) clearly shows the double S-shaped form, with a stronger curvature in the neck area (cervical spine) sitting and a flatter lordosis in the area of the lumbar spine. In particular, the lower vertebrae of the lumbar spine and the sacrum are stressed, so that pain or even a herniated disc (*nucleus pulposus prolapse*) can be the consequence. The support of the lumbar spine in sitting position with a lumbar support, as it is available in many vehicle seats, leads to a restoration of the lordosis and relieves the one-sided pressure on the intervertebral discs of the lumbar spine (Fig. 4.6 right). The steeper position of the seat backrest and the use of the lumbar support thus have a multitude of positive and health-promoting effects, which alleviate the physical disadvantages of the basically unhealthy sitting posture (see below and the essay in ▶ Sect. 7.2.2.1).

The structure of the free vertebrae (Fig. 4.7) is characterized by the central vertebral body (*corpus vertebrae*) which surrounds the intervertebral disc and at its edge

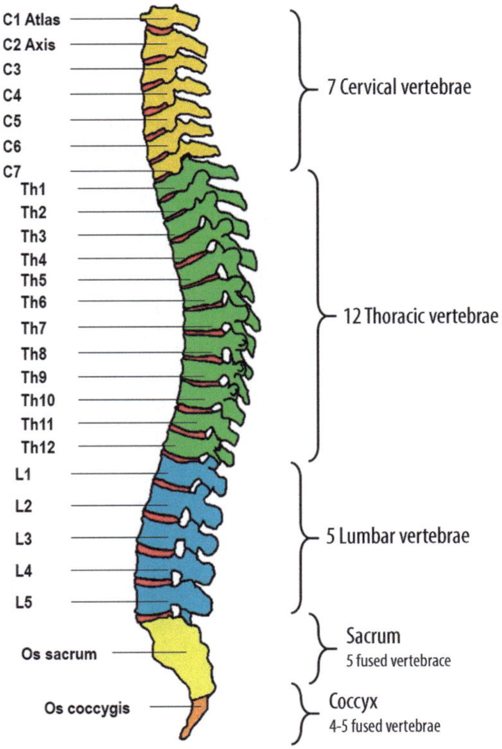

Fig. 4.5 Vertebral column

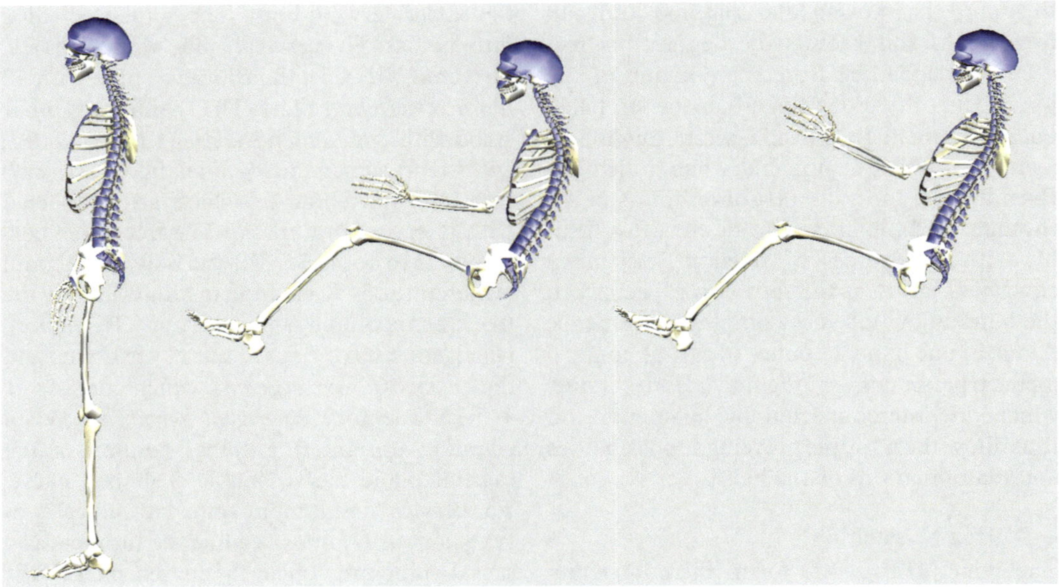

Fig. 4.6 Comparison of standing and sitting posture

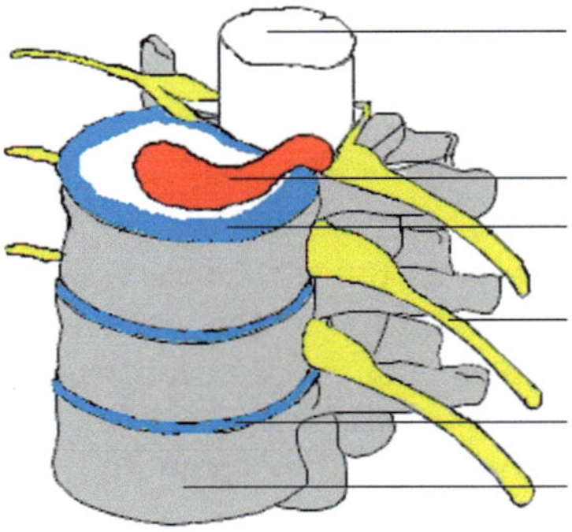

Fig. 4.7 Structure of a vertebra

the fibrous ring (*anulus fibrosus*) . Dorsally the vertebral body is extended by a *transversal process*, which represents the articulated connection to the rib arches, and centrally the *spinosus process*, which is clearly visible from the outside. These processes are separated by the vertebral arch (*arcus vertebrae*), which connects the vertebral body to the vertebral hole (*foramen vertebrae*), the entirety of which is called the spinal cord canal and which absorbs the spinal cord. At the vertebral arch there are still the four (two upper and two lower) joint processes (*articular processus*). Two adjacent vertebrae form through the upper incision (*Incisura vertebralis superior*) the intervertebral hole from which spinal nerve cords emerge laterally. The extraordinarily important task of the spine can be seen in the complex mobility of its elements. Thus the joint surfaces of the processes are real joints (*diarthroses*) and, as flat joints, allow lateral displacements and very high load absorption in the normal direction of action. The intervertebral disc connecting the vertebral bodies with the fibrous ring, which has grown together with the adjacent vertebral bodies, represents a synarthrosis, which allows slight angle changes and fixes the position of the vertebral bodies to each other. The intervertebral disc with its gelatinous inner soft core (nucleus pulposus) has the task of shock absorption in the vertical direction. The great flexibility of the spinal column with all the possibilities of bending and stretching the upper body and also the lateral rotation of the torso relative to the pelvis produces very different and one-sided loads on the vertebral joints, the balancing effect of which can only be achieved by a viscous medium due to the homogeneous distribution of the supporting forces. A special load situation for the intervertebral discs is sudden changes in movement (sudden bending or turning or lifting a too large load) or permanent angular positions at the range of motion (sitting or crouching with simultaneous loading of the shoulder girdle), in which the fibre ring is extraordinarily loaded, so that cracks can occur, through which the nucleus can be squeezed out due to the additional internal pressure in the soft core. This is referred to as a herniated disc (*prolapsus nuclei pulposi*), which can be very painful when nerve strands are pushed through the emerging nucleus. This can even lead to local paralysis of the lower extremities.

An unfavourable driving posture can be a cause for the excessive strain on the intervertebral discs in the lumbar vertebra area. Even if driving a car in a seated position cannot be the sole reason for a herniated disc, a permanent incorrect posture of the spinal column due to the driving posture is a frequent situation in which prolapse can occur if the intervertebral disc is damaged due to an undersupply of nutrients (dehydration). Since the intervertebral discs, as well as the joint cartilage surfaces, are not supplied with nutrients from the blood circulation, but by diffusion, an adequate water and nutrient-rich supply is necessary through constant movement of the vertebrae. Conversely, insufficient water absorption and lack of exercise are responsible for the pre-damage of the joint and cartilage surfaces and thus also the intervertebral discs. Many vehicle users tend to have a too flat backrest inclination, accompanied by the strong curvature of the spine in the lumbar region. However, this kyphotic, "casual" posture stretches the intervertebral discs of the lumbar spine strongly and can lead to damage. It can cause the ligaments to creep in the long term, which reduces the stability of the musculoskeletal system and means that the back muscles have to take over the task of the ligaments. In addition, prolonged hyperflexion due to the strong tensile stresses in the collagen matrix of the fibrous ring can lead to a herniated disc. On the other hand, Wilke (2004) was able to show that moderate flexion of the lumbar spine is accompanied by a reduction in load transfer via the facet joints and a significant reduction in pressure in the intervertebral discs (see ◘ Fig. 4.8). Thus it could be shown that the so-called "casual lazy posture" enjoyed such a bad reputation wrongly. Previously, it was considered healthy to take an upright posture instead, which, however, according to ◘ Fig. 4.8, is associated with a significantly higher intravertebral disc pressure. This is due to the fact that the trunk muscles alone are responsible for maintaining the upper body without external support. Thus a constant tension of the musculature exists in front of (abdominal musculature) and behind (back musculature) the intervertebral disc. Consequently, the intervertebral disc is compressed even more than already caused by gravity (Wilke 2004; Dolan 2006). On the other hand, an upright sitting posture with a steep backrest and good support in the lumbar vertebra area (lumbar support) is said to have a relieving effect on the intervertebral discs. At least the upper part of the body is raised, which also raises the eyepoint for a better overview in traffic. At the same time, the pressure on the internal organs is reduced, which facilitates breathing and promotes blood circulation (Faller 1999). The distance to the steering wheel is reduced, which promotes the controllability of the vehicle. This posture is usually recommended and practiced in driving safety training by professional driving instructors.

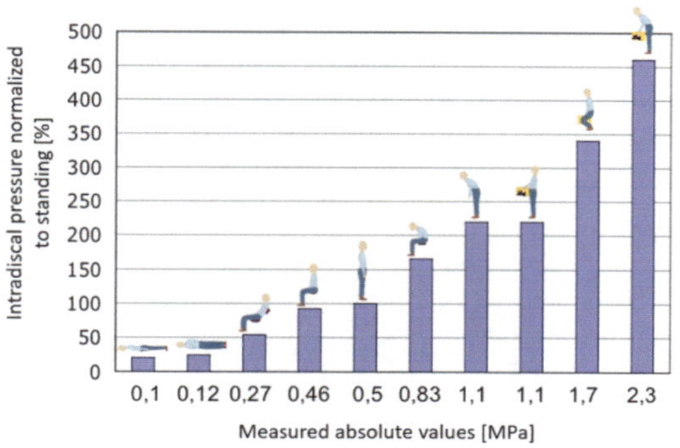

◘ **Fig. 4.8** Representation of the intradiscal disc pressure depending on the posture. (Taken from Wilke et al. (1999))

The intervertebral discs themselves have no pain receptors. Pain in the area of the back can therefore essentially only be caused by the pinching pressure of an emerging nucleus on a nerve cord or the pain receptors in the back muscles. Obviously even slight muscle ten-

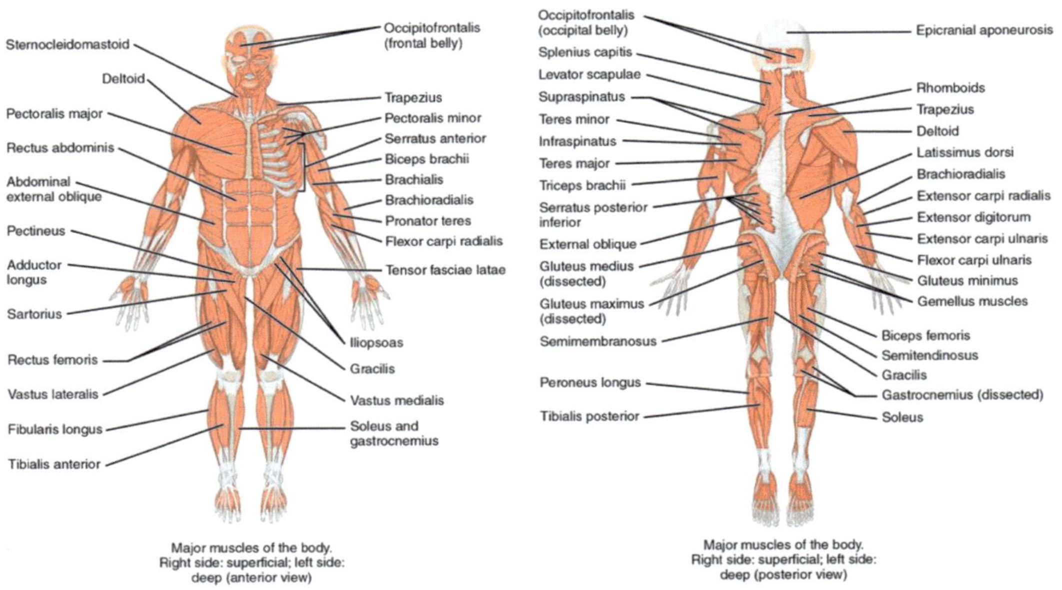

Fig. 4.9 The Skeletal Musculature (This file is licensed under the ► Creative Commons ► Attribution 4.0 International license. (OPenStaxx: ► https://cnx.org/contents/FPtK1zmh@8.25:fEI3C8Ot@10/Preface))

sions caused by an unfavourable posture can lead to such pain if they last longer. The individually correct posture, which avoids unfavourable minimal muscle tensions, should therefore be given special attention in addition to a sufficient amount of sports in order to prevent back pain (see also ► Sect. 7.2.2.1).

Especially in the area of the spine there are often rigid views (poor lumbar posture, superiority of an upright posture, correct adjustment of a lumbar support). Wilke (2004) states: "Dogmas should not exist in back school. What is good for one patient with disc problems may be bad for another with arthrosis in the small vertebral joints. In any case, it is not advisable to sit in a static position that is too long without changing posture. This forces the entire fluid out of the intervertebral disc and makes efficient load absorption impossible (Pope 2006; Adams 2006)".

4.1.2 Muscular System

Fabian Günzkofer

Beside the passive musculoskeletal system with its bones and joints, the transversely skeletal muscles[2] together with their auxiliary organs such as tendons, ligaments and sesamoid bones[3] as an active musculoskeletal system are decisive for mobility (see Fig. 4.9). In general, the muscle belly, which represents the accumulation of individual muscle cells (muscle fibres), is connected to the skeleton via one or more tendons. In contrast to the rhythmically working heart muscle and the smooth musculature of the intestine controlled by the autonomic nervous system, the skeletal musculature is conscious and arbitrarily controllable by humans. However, this control is always task and purpose-bound and can bring about targeted movements, even if the isolated purpose-free contraction of singular muscle fibres is not possible. The task of the muscles consists on the one hand of dynamic contractions in order to perform mechanical work with the passive musculoskeletal system and on the other hand of static

2 The transversely striated musculature of the active locomotor system, which is accessible to the voluntary movement, is to be distinguished from the smooth musculature of the internal hollow organs (except the heart) as contractile tissue.

3 Sesamoid bones are bones embedded in tendons.

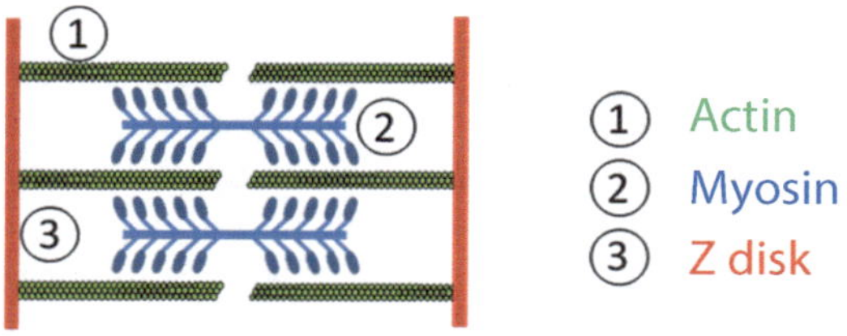

■ **Fig. 4.10** Illustration of the components of a sarcomere

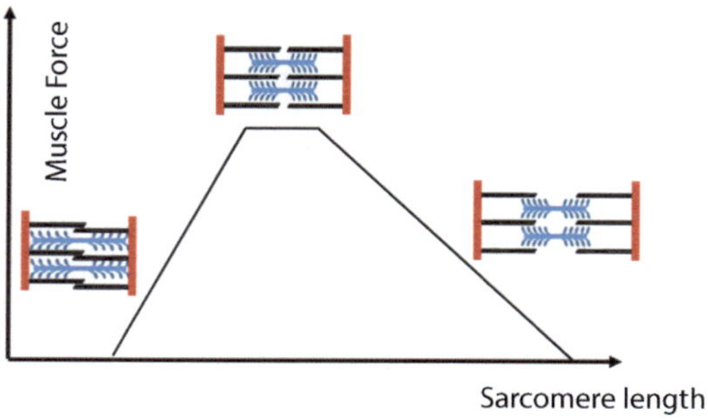

■ **Fig. 4.11** Qualitative representation of the muscle strength-muscle length relation

stiffening processes in order to support preferred postures.

4.1.2.1 Muscular Forces

A muscle has origin and attachment to different bones via tendons and crosses at least one joint. Thus, muscle forces exerting torques on the joints result in relative rotations of the bones. In order to understand the function of a muscle, a basic understanding of the basic structure is necessary. A muscle fiber visible to the naked eye is composed microscopically of individual myofibrils, which form a sequence of individual sarcomeres separated by Z -discs (Campbell and Reece 2005; Tittel 2003). Sarcomeres consist of the two protein filaments actin and myosin and represent the actual contractile unit (see ■ Fig. 4.10). Six thin actin filaments each, which are attached to the Z disc, surround a thick myosin filament. The myosin filaments are long protein filaments with myosin heads aligned in the direction of the actin filaments.

4.1.2.2 Muscle Contraction

A muscle contraction is initiated by stimuli of the nervous system, whereby Ca^{2+} is released inside of the muscle cell (cytosol) (Sherwood 2012). This exposes the binding sites of the actin for the myosin heads. The actual contraction requires energy in the form of adenosine triphosphate (ATP), which binds to the myosin heads. Through hydrolysis of the ATP to adenosine diphosphate (ADP) and to an inorganic phosphate, the myosin head is tilted and forms a cross bridge to the actin filament. When the reaction products are released, the head tilts back, resulting in a relative movement between actin and myosin (Campbell and Reece 2005). In sum, several sequential microscopic "rowing movements" result in a macroscopically perceptible muscle shortening. The contraction described above results in muscle strength in the longitudinal direction of the muscle fibre. The amount of the resulting force corresponds to the number of cross bridges involved.

■ Figure 4.11 illustrates that muscle strength initially increases with increasing muscle length, reaches its maximum on a plateau and finally decreases again. This is related to the geometric interaction of the proteins myosin and actin. If the muscle length is very short, the individual actin filaments superimpose themselves and thus partly hinder the tying of the myosin heads in question. In extreme cases, the myosin filaments collide with the Z discs, preventing further shortening. In the area of the plateau one speaks of an ideal muscle length, since the highest number of cross bridges can be developed and thus also the highest muscle strength can be produced. As the sarcomere length continues to increase, the area of overlapping protein filaments begins to decrease. As a result, muscle strength decreases again in an idealized linear manner. If one does not consider only a single sarcomere, however, it can be determined in the overall muscle system that the total strength of the muscle increases with further stretching. This is due to the passive restoring forces of the overstretched parallel elastic components in the form of connective tissue and muscle fiber membrane (Winter 2005; Knudson 2007).

In addition to muscle length, muscle strength also depends on contraction speed

(Hill 1938). In dynamic muscle work, a distinction must be made between concentric and eccentric types of contraction (Chaffin et al. 1999). In the case of muscle shortening under force application, e.g. when lifting a load, one speaks of the concentric load case. The eccentric fall occurs when the muscle extends under load, e.g. when a load is put down in a slowly controlled way.

In the concentric case it is assumed that the force decreases parabolically with the angular velocity of the joint rotation compared to the isometric case. This can be explained by the fact that fewer cross bridges can be active at the same time. Furthermore, friction effects due to an increase of viscosity are mentioned in the technical literature (Winter 2005). In the case of the eccentric load case there is still disagreement in science. The majority of existing studies describe a maximum force increase of up to twice that of the isometric situation. This could be explained by the fact that breaking open the cross bridges requires more energy than maintaining them in the isometric case (Winter 2005). However, there are other studies that do not assume any force potentiation in the eccentric case. Neuronal inhibition effects could limit the force to an isometric level to protect the musculoskeletal system (Hahn 2008). The focus of the previous consideration was on the structure and function of individual sarcomeres. At this point we want to show the interrelations, how a global joint torque is created by this (◘ Fig. 4.12). As already discussed, the producible muscle strength depends significantly on the muscle length. Transferred to the entire musculoskeletal system, this is influenced by the respective joint angle. Furthermore, the force depends on the total number of parallel sarcomeres of all muscle fibers involved. This is indicated by the physiological muscle cross-sectional area (PCSA), which does not necessarily have to correspond to the anatomical muscle cross section as it depends on the muscle pennation angle. If the muscle fibres run at an angle to the longitudinal axis of the muscle and are attached obliquely to the tendons, the number of parallel sarcomeres is increased with the same anatomical muscle cross-section. This increases strength compared to non-pennate muscles, but reduces the speed of contraction. The PCSA thus depends on the pennation angle, which in turn changes slightly with posture (dotted dependence in ◘ Fig. 4.12).

The muscle strength produced is finally transferred to the adjacent bones via tendons. A corresponding joint torque is generated via the virtual lever arm between the muscle-tendon unit and the joint rotation axis. The amount of the lever arm will therefore change with the joint angle.

The global joint torque is a product of the muscle forces and lever arms of all synergists involved in the movement. In summary, the joint torque depends on the joint angle in two ways. On the one hand by the influence on the muscle length and on the other hand by the influence on the lever arm.

A special feature of muscles is that they can only contract and cannot actively lengthen again. Thus a fine motor adjustment of postures and movements can only be done by antagonists, who compensate this missing possibility of the synergists by their contraction ability. Only the interaction of agonist and antagonist allows a fine adjustment of postures and movements (see ► Sect. 3.2.3). The modelling of joint torque-joint angle

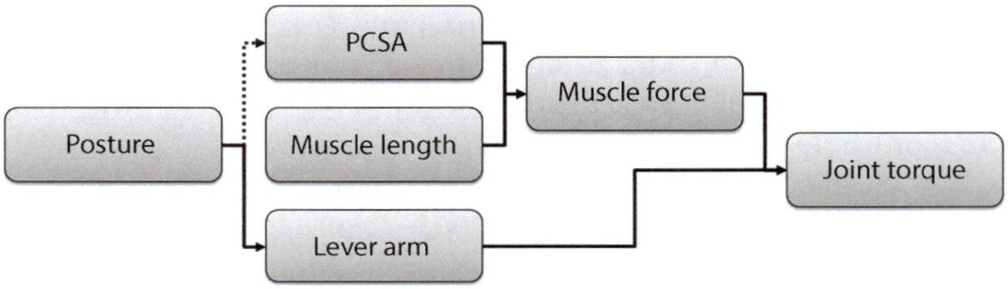

◘ **Fig. 4.12** Display of the relationship between posture, PCSA and joint torque

Flexion	Extension	Internal Rotation	External Rotation
m. semimembranosus	m. rectus femoris	m. semimembranosus	m. biceps femoris
m. semitendinosus	m. vastus lateralis	m. semitendinosus	
m. biceps femoris	m. vastus medialis	m. gracilis	
m. gracilis	m. vastus intermedius	m. sartorius	
m. sartorius		m. popliteus	
m. gastrocnemius			
m. popliteus			
m. plantaris			

Fig. 4.13 Muscles involved in different knee joint movement directions (red = multifunctional muscles, blue = biarticular muscles)

functions is usually based on empirical tests, in which test persons apply maximum forces in different measuring positions. For example, the knee flexion angle is varied in discrete steps taking into account the force-length relation for the measurement of knee extension torques.

In order to understand the holistic relationship between joint angles and joint torques, however, the multifunctionality and multi-joint nature of muscles must be considered. In order to be able to understand this problem clearly, the muscles of the knee joint are shown in Fig. 4.13 as examples.

Figure 4.13 shows that certain muscles (e.g. m. semimembranosus) have not only a flexing but also an internal rotating effect. Consequently, their muscle length changes not only by flexion, but also by internal rotation. This means that at the same flexion angle, a different internal rotation angle can lead to a different muscle length of certain flexors. Thus it can generally be concluded that the maximum torque of a degree of freedom in the case of multifunctional muscles also depends on further degrees of freedom of the same joint. In addition, the involvement of two-joint (biarticular) muscles must also be taken into account. For example, the musculus biceps femoris originates from the ischium of the pelvis (tuber ischiadicum) and attaches to the fibula (caput fibulae). Thus this muscle crosses two joints and causes a flexion in the knee as well as an extension in the hip. Consequently, the muscle length depends not only on the knee flexion, but also on the hip flexion angle. In order to predict maximum knee flexion torques for any posture, it is necessary to carry out theoretical experiments under different knee flexion, knee rotation and hip flexion angles. The results of Günzkofer et al. (2011a) show that in practice the observation of different flexion angles of knee and hip is sufficient. Figure 4.14 shows a compilation of the complex formation of body strengthes.

Similar effects occur with the extremely complex mobility of the hand. Figure 4.15 shows the possible maximum forces for different types of actuation. This compilation is to be understood as an illustration for the actuation-dependent variability of the forces; because in fact there are considerable individual variations compared to the data given there.

4.1.2.3 Individual Variability

Physical forces are subject to a high degree of interindividual variability. This includes above all motivation, handiness and training condition. For vehicle design, the influences of age and gender must be taken into account.

- **Sex**

On average, women have less physical strength than men. Diverse sources prove, these differences can differ depending on the joint. According to Churchill et al. (1978), women own 70% of the strength of men in the lower extremities, 63% in the trunk and 55% in the upper extremities. These results are well in line with the data provided by Hosler and Morrow

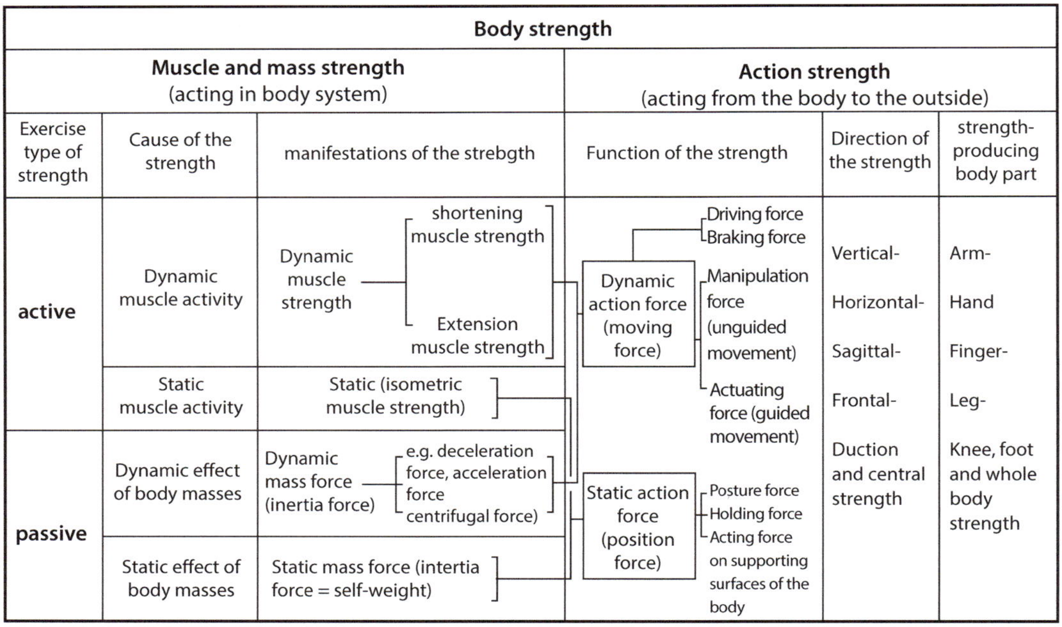

Fig. 4.14 Overview of muscle strengthes (according to DIN 33 411, Part 1)

(1982). These describe a ratio of 60% for arm forces and 74% for leg forces. Lee and Bruckner (1991) report very similarly that women have 49–55% of the arm power of men. D'Souza et al. (2011) also recorded a ratio of 60% for knee extension torques and 49% for elbow flexion torques in their experiments. Consequently, it can be stated that different studies come to the conclusion that the differences in the upper extremities are (considerably) greater than in the lower extremities (Hafez et al. 1982). Due to these clear differences, the question arises whether these differences are purely determined by gender or whether this correlation is not more likely to be influenced causally by other variables. Indeed, Hosler and Morrow (1982) were able to show that the significant force differences became marginal once anthropometric effects were considered. Finally, only 2% of the variance in arm forces and 1% of the variance in leg forces could be explained by the gender factor. Similarly, Schantz et al. (1983) and Bishop et al. (1987) prove that the differences in strength are rather due to different muscle volumes.

Regardless of the reasons for these differences, Wakula (2009) states as a simple rule of thumb that women can muster on average half the strength of men. Daams (1994) broadens the range somewhat and increases the interval to 0.5–0.75 due to the strong heterogeneity of available studies.

Age

Another important influence factor is the loss of strength due to ageing (Samuel and Rowe 2009; Hurley 1995; Goodpaster et al. 2001; Herzog 2004). It is generally accepted that physical forces increase up to an age between 25 and 30 years and then decrease again (Hettinger 1968; Rohmert and Hettinger 1963; Hurley 1995; Morgan et al. 1963; VanCott and Kinkade 1972). Morgan et al. (1963) and VanCott and Kinkade (1972) describe that the forces after the peak remain almost constant up to an age of about 40 years. With 50 years still approximately 85% and with 60 years still approximately 80% of the original forces are preserved (Morgan et al. 1963). Hurley (1995) quantifies the power loss simplified with 12–15% per decade, starting at the age of 50. Various studies show that the lower extremities are more affected by the ageing process than the

Stress		Maximum Force [N]
Fist connection around a cylinder of 40 mm diameter		410
Pressure of the thumb against four fingers		190
Actuation of a pressure strip through the ball of the thumb		180
Pressure of thumb against index finger		120
Actuation of a knob by the thumb, counter pressure by index finger		100
Actuation of a knob by the thumb		100
Actuation of a knob by the index finger		60
Closing of pliers		316 (Women) 613 (Men)

Fig. 4.15 Types of hand actuation and average possible forces

upper extremities (Thompson 1994; Frontera et al. 1991; VanCott and Kinkade 1972).

The reason for the decrease in maximum forces is mainly due to atrophy of the musculature (Herzog 2004; Goodpaster et al. 2001). In addition, an increased proportion of cocontractions of the antagonists could also be responsible for the reduction of the resulting forces (Duchateau et al. 2006). Furthermore, neuromuscular functions are affected by aging (Bassey and Short 1990).

An enormous amount of studies is available in the literature which deal with age effects on maximum forces for each joint. In order to obtain a deeper insight the reader should refer to Daams (1994) and Kulig et al. (1984). Only two works should be mentioned here. Stoll et al. (2000) measured the forces of all joints of the body in 290 women (20–82 years) and 253 men (21–79 years). Based on the results, regression equations were published that can be used to predict the strength of a joint at any age. Furthermore, reference is made to one of the most recent series measurements of elbow flexion and knee extension torques by D'Souza et al. (2011). In this study, 141 male and 142 female subjects between 50 and 79 years of age were measured with the IsoMed 2000 isokinetic test device. With the help of the resulting joint torques, linear regression equations were formed, which indicate the maximum joint torques for knee extension and elbow flexion with the help of the predictors sex, age and forearm mass or thigh mass (see also ▶ Sect. 4.2.3.3).

4.2 Anthropometry

In addition to anatomy, anthropometry[4] provides the dimensions of the body and the methods for determining them. The ergonomic objectives of anthropometry are a user-centred design of workplaces and products and the definition of safety measures. For this purpose, length and circumference measurements, body part weights, movement spaces, reaching spaces, visual data (visual axes, visual fields) and forces are considered as dimensions. A special focus of anthropometry is the determination of the dimensions of the extremities and their position in relation to each other. The mobility and range of the extremities as well as their strength are also considered. Of particular interest, of course, are the position and movements of the joints and the kinematics of living people. Here, however, only the effect on the externally visible body can be observed and unfortunately not the exact position of the joint partners in the movement, since these locations are covered by tissue, muscle and skin. Two methods have been established for measuring geometric anthropometric measures, which are applied depending on technical complexity and availability.

4 Anthropometry is an artificial word composed of the ancient Greek words anthropos = man and metrein = to measure. As a science of human measurements, it is considered to be a subfield of anthropology. The scientific anthropology deals with the origin of man, his development, differentiations and expansion. Philosophical anthropology seeks to shed light on the nature and destiny of man.

4.2.1 Length and Circumference Dimensions

Rainer E. Grünen and Fabian Günzkofer

4.2.1.1 Recording of Body Measurements

A decisive requirement for the measurement of body dimensions is a reliable reproducibility and comparability of different measurement series. Martin (1914) has established a standardization in which length measurements are basically recorded as distances from bone point to bone point. Bone points are prominent bony sites near the body surface that are easy to find and, unlike soft tissues, do not depend on the contact pressure of the measuring instrument. Further measuring rules are compiled in DIN 33402 Part 1. These say that measurements in prescribed positions on naked people are only carried out on the right side of the body. However, it should be noted that the requirement of nudity is a rather theoretical requirement, which is not often considered in practice. For the practical recording of body measurements, the manual method has established itself worldwide with the help of Martin's measuring instruments. On the other hand, optical methods have recently been developed, so-called body scanners, which allow non-contact measurement of the test persons.

■ Measuring Method According to Martin

Martin's measuring instruments consist of a collection of pluggable scales and callipers with which external measurements can be taken. Included is a four-part pluggable tube scale from 0 to 1950 mm with which length measurements such as body height and sitting height can be determined. Also included is a pair of compasses with 450 mm measuring range as well as straight and curved scales. A slide compass for thickness measurement up to 200 mm as well as a 2000 mm long tape measure complete the set (◘ Fig. 4.16).

The body height shall be measured in accordance with the measurement procedure in a maximum stretched posture with the head held in the "Frankfurt horizontal" position

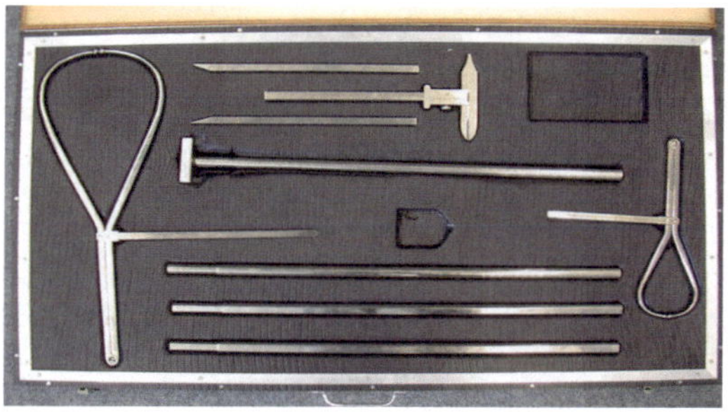

■ **Fig. 4.16** Martin's measuring instruments

(horizontal alignment of the connecting line from the highest point of the auditory canal and the lowest point of the eye socket). It should be noted that the practical relevance of all measures must not be lost sight of. Although the specification of the maximum aspect ratio allows reproducible measurement, it does not usually correspond to the more physiologically determined working postures. Consequently, for example, the eye height when standing in a relaxed position would be significantly overestimated by a stringent anthropometric measurement. Martin's method has been applied to many classical body measurements and is heavily dependent on the experience and care of the measuring personnel. Often, for reasons of protection of personality, the data were only collected from persons dressed in clothing, which makes it even more difficult to identify the corresponding measuring points. Measured values from different institutes are often not or only to a limited extent comparable with each other. A traceability or a second check is not possible here, since the data alone represent a momentary situation. Thus, in the case of a prompt second measurement of the same person, considerable deviations can often be observed. Today, this manual measurement is only used in most cases if only a small number of body measurements are to be recorded per test person. Otherwise, an enormous amount of personnel and time is required. In addition, the manual measurement represents a strong intervention in the personality of the test persons due to the requirement for little clothing.

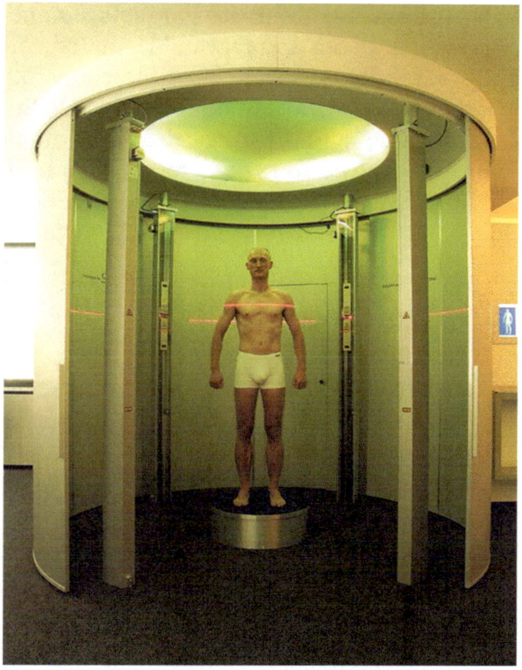

■ **Fig. 4.17** Body scanning (example Vitus from Human Solutions GmbH)

■ Body Scanning

The current state of the art is the use of so-called body scanners (■ Fig. 4.17). The body surface is recorded three-dimensionally with the aid of laser beams (see also ► Sect. 11.2.1). The use of appropriate software allows the extraction of relevant body measurements. In Germany, a large-scale series measurement called SizeGERMANY was carried out with this technology in the years 2007–2009 in which over 13,000 men, women and children aged 6–87 were measured at approximately 30 locations throughout Germany (Seidl et al. 2009). Following the same pattern, extensive measurements have been carried out 2012–2013 in Italy (SizeITALY), 2016–2019 in America and Canada (Size NorthAmerica). The data obtained in this way have led to new findings which are of great importance for the future development of clothing, but also for the automotive industry. Compared with the data from earlier surveys, statements can be made on population trends and future trends in body shape. The accuracy of the data acquisition with 1–2 mm is many times higher than with manual measurement methods and allows an in-depth and repeatable analysis of body data. In this way, circumference and length measurements can be determined without touching and in great detail, without influencing the person's posture or position. It is even possible to automatically remove data points, which ensures a uniform approach.

4.2.1.2 Characterization of Body Measurements

■ **Percentiles**

In order to be able to make valid statements on the distribution of body measurements within a population, a representative sample of suitable size is drawn. The localization of individual body measurements within the distribution occurs via the so-called percentile. Here a body dimension value of the xth percentile indicates that x% of the population falls below this level. For example, a length of the 5th percentile means that only 5% of the population has a smaller measure, but 95% has a larger measure. The 50th percentile corresponds to the median and, in the particular case of normal distribution, to the arithmetic mean of the sample. An advantage for the subsequent calculation is that body measurements are usually normally distributed (Geuß 1994). Thus the indication of mean value $\overline{x}$ and standard deviation are sufficient to describe and calculate the total distribution of a measure within the population. ◘ Figure 4.18 shows a graphic representation of the 5th and 95th percentile.

A simple assignment between body dimension and assigned percentile is made by the z-transformation, which converts the distribution into a standardized normal distribution (Eq. 4.1).

$$x_{perz} = \overline{x} + z_{perz} \cdot s\left(z - \text{Tranformation}\right) \quad (4.1)$$

with:

$z_{95\%} = 1645$ and $z_{5\%} = -1645$
$z_{97,5\%} = 1960$ and $z_{2,5\%} = -1960$

Finally, the 5th, 50th and 95th percentiles are typically stored in tabular form in the evaluation of each body measurement, differentiated according to gender and age groups (see DIN 33402 Part 2). Special attention must be paid to the fact that all body measurements are individually percentiated and thus no connection to the individuals of the measurement acquisition exists any more. People usually do not consist of body measurements of exactly the same percentile. This is evident from the frequently noticeable differences in proportions between the trunk of the body (head tip to buttocks in sitting position) and the legs. Thus it is not expedient to add individual measures of the same percentile to a synthetic human of a certain total percentile.

■ **Application of Body Neasurements**

After the statistical evaluation, the error is often committed that only the mean value is used to describe the data set. A pure use of the 50th percentile, however, leads to clear problems, as the following example from building technology shows. If the average body height were used for the design of a door frame, half of the target group would no longer be able to pass through the door in an upright gait. Consequently, it is crucial to mention the

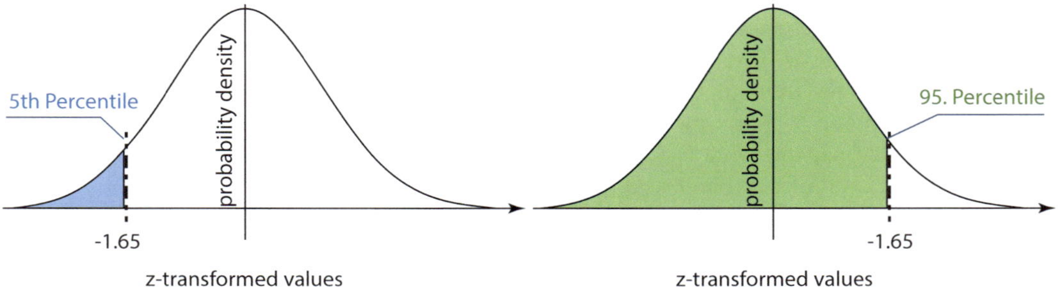

◘ **Fig. 4.18** Representation of 5th and 95th percentile in standardized normal distributions

mean value only in connection with a dispersion measure in order to take into account the shape of the distribution. In ergonomics, the 5th percentile woman and the 95th percentile man are often used as border percentiles. A further expansion would lead to unjustifiably high additional economic costs. The remaining 5% of the population are referred to special solutions. The 1% and 99% percentile or the usual percentiles with safety margins are only used for safety-relevant designs (among others also in the abovementioned example of the door frame).

In addition to the mostly faulty design according to mean values, there is often another problem in product design. The designer adapts the product to his own anthropometry by trying it himself, thus making himself the centre of interpretation and neglecting the effect on other body dimensions.

In order to circumvent these problems, anthropometry classified the design into so-called inner and outer dimensions. Inner measures here mean that the human being with his dimensions must find space within the product. These can be the lower leg length under a table and the trunk length under the roof in the vehicle. This differs from outer measures, which usually represent accessibility measures, such as the maximum height of a bookshelf or the maximum distance from a control element. In order to ensure anthropometric product adaptation to all percentiles to be considered, inner measurements are based on the largest person (e.g. 95th percentile man) and outer measurements on the smallest person (e.g. 5th percentile woman).

The use of table dimensions represents the simplest form for product design. However, this can only be used for one-dimensional questions, e.g. door height, door width, chair height. In the case of a more complex design such as a vehicle interior, tables can only be used to a limited extent. It is therefore not sufficient to look up the shortest arm length in a table to determine the steering wheel position. For this one would have to know where the shoulder is and thus the arm begins. The shoulder position, however, depends on the H-point when the torso angle is fixed, which in turn depends on the leg length, knee angle and pedal position (see ▶ Sect. 7.2.2). Therefore not only arm length, but also shoulder height in sitting, thigh length etc. are needed. As mentioned above, however, it is not permitted to add up individual body measurement percentiles. This need for a meaningful, realistic aggregation of body measurements has led to the development of human templates that are easy to use and cheap to manufacture (see ▶ Sect. 5.2.1). Various disadvantages as well as the widespread use of computers have finally led to the use of digital human models, which represent the state of the art in product design.

4.2.1.3 Influence Factors on Body Dimensions

It is widely known that humans differ strongly in the absolute measures as well as in the proportions of their body measurements. Many of these differences are due to factors such as gender, age and region. The knowledge of the most important differentiation features allows a differentiation of user groups in order to define market-specific test collectives.

▪ Body Types

In constitutional biology as a further branch of anthropology (Grimm 1966), different *somatic constitution types* are distinguised. Sheldon (1954) has created three categories of somatotypes that are distinguished purely by external body shape. At *Pycnian Type* the physique is rather small with short arms and legs, round face and soft musculature. People of this type which is also called *endomorphic* represent about 7% of the German population. The *Leptosome Type* is characterized by a short upper body with long legs and arms, slender hands and feet. People of this *ectomorphic* type are mostly tall and make up about 9% of the German population. The *Athletic Type* with its powerful chest, broad shoulders and strong musculature of corresponding muscle strength is also called a *mesomorphic type* and represents 12% of the Germans. The remaining approximately 70% are mixed forms resulting from combinations of the pure forms. A distinction is made between skeletal structure, muscle mass and fat storage in the tissue. Conclusions on behaviours are not possible due to the somatotype and an application of this differentiation in ergonomics is unusual. Even the outdated classification of humans into three (or four) main groups according to Boyd (1950) in Europids, Negrides and Mongolids (as well as Australids), which is based only on skin colour, hairiness and skull shape, is not suitable for ergonomics. A division into ethnic groups should be replaced in favour of a detailed target group definition and market-specific customer requirements. The individual regions of the automotive world market are characterised by an ethnic mix, which is, however, more strongly influenced by typical regional behaviour patterns and customs than by ethnic origin. According to Geuß (1994), the typology realized in the human model RAMSIS has proved to be very practicable, especially for the design of interior and exterior dimensions of automobiles. It is based on the three determining variables body length, proportions (sitting height to body length) and obesity (see section on differences in proportions).

Regional Differences

The physique and thus also its dimensions vary very strongly in the individual regions. This already applies to regions in the same country (e.g. North to South Germany, or Tyrol and Sicily). In particular, however, a significant difference in absolute measures and proportions can be seen in a global comparison. Figure 4.19 shows an exemplary comparison of the 5th, 50th and 95th percentiles of body height of selected nations for both sexes. The differences in the average value between Germany and Mexico alone are about 14 cm . A very small woman in Germany is with approximately 1,55 m bigger than half of all Mexican women. Figure 4.20 reflects the average height of men in different nations. Also within Europe there is a big difference in body height between the small Southern

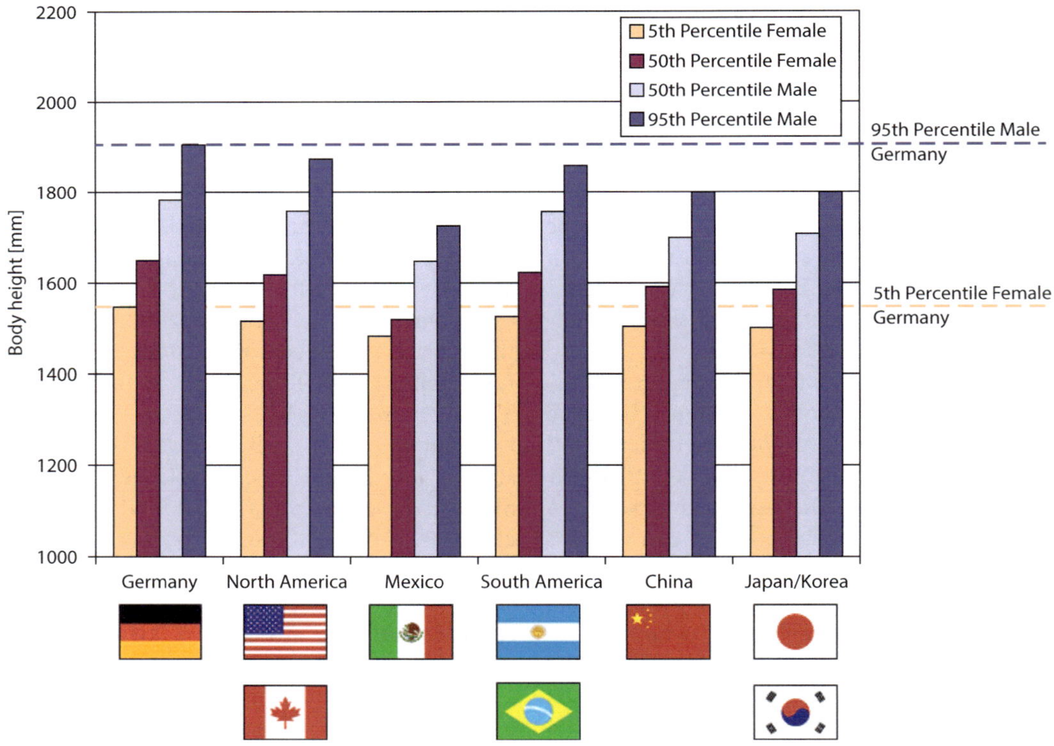

Fig. 4.19 Exemplary comparison of discrete body height percentiles of selected nations (Kaiser 2011)

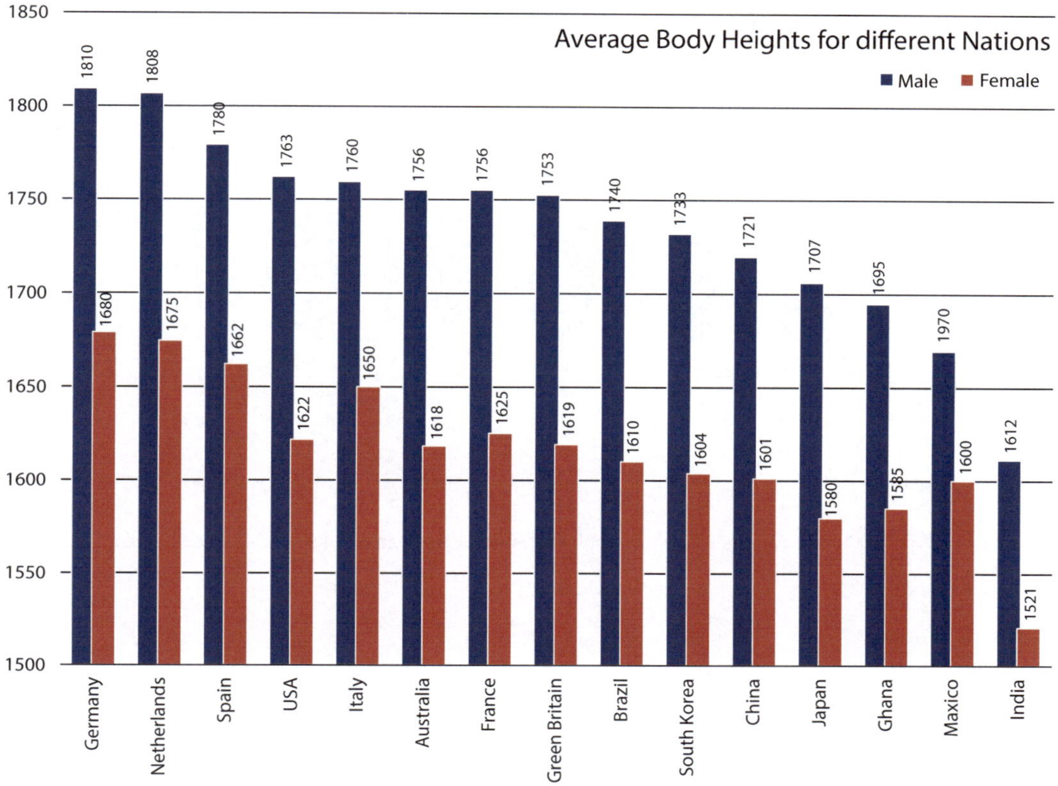

Fig. 4.20 Average body size (50th percentile) of men and women in different nations

Europeans and the big Northern Europeans (the biggest people are found in Holland).

The main reasons for reaching full - genetically possible - body size depend primarily on socio-cultural factors. The availability of energy-rich food (adequate calorie intake) and clean water is just as important as basic medical care and a prepuberant growth phase that is above all free of severe physical stress and disease. But even in a region with relatively homogeneous living conditions there are regional differences, as the results from Siedl et al. (2009) show (Figs. 4.21 and 4.22).

Differences in Proportions

In everyday life, body height is often regarded as the primary anthropometric differentiator between people. Greil (1972) has carried out correlation analyses to various length and circumference measurements. It turns out that above all height and length measurements have a large correlation. What is striking, however, is how strongly the composition of a certain body height can vary sitting height and leg length (see also the investigations by Jürgens et al. 1971). This phenomenon is commonly referred to as the *seated giant* (long trunk, short legs) and *seated dwarf* (short trunk, long legs) (Fig. 4.23). "From a practical point of view, the body height is therefore suitable as an indicator feature when it comes to the definitions concerning the leg area or [...] the arm area. On the other hand, it is useful to use the sitting height as an indicator measure for determining measurements for seated people" (Bullinger 2013).

These differences in proportions are remarkable already within a largely homogeneous population, but especially in a global comparison. An exemplary comparison of the proportions of average German and Chinese men impressively shows the enormous difference in proportions. A German

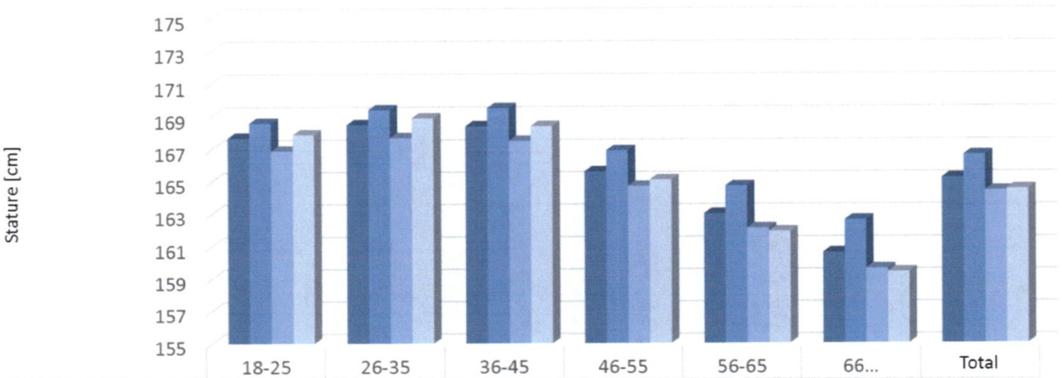

	18-25	26-35	36-45	46-55	56-65	66...	Total
Germany	167,7	168,5	168,4	165,6	163	160,6	165,2
Northern	168,6	169,4	169,5	166,9	164,7	162,6	166,6
Southern	166,9	167,7	167,5	164,7	162,1	159,6	164,4
Newly-formed states	167,9	168,9	168,4	165,1	161,9	159,4	164,5

Fig. 4.21 Average body height of all female volunteers by age group in Germany and regions. (From Siedl et al. 2009)

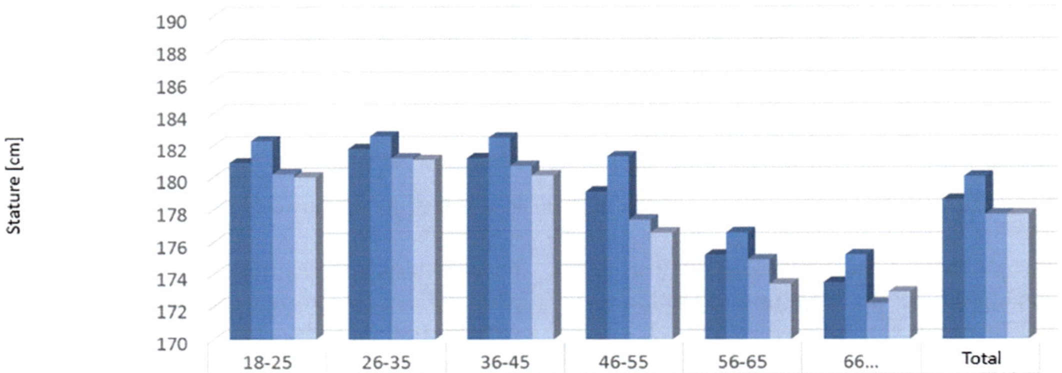

	18-25	26-35	36-45	46-55	56-65	66...	Total
Germany	180,9	181,8	181,2	179,1	175,2	173,5	178,6
Northern	182,3	182,6	182,5	181,3	176,6	175,2	180
Southern	180,2	181,2	180,7	177,4	174,9	172,2	177,7
Newly-formed states	180	181,1	180,1	176,6	173,4	172,9	177,7

Fig. 4.22 Average body height of all male subjects by age group in Germany and regions. (From Siedl et al. 2009)

man of the 50th percentile is 1,78 m tall and has a sitting height of about 92 cm in the 50th percentile. The upper body thus accounts for less than 52% of the total size. The average Chinese man is smaller in absolute terms, namely 1,68 m, but his sitting height in the same percentile is 91 cm (GB10.000-88). This means that he is 10 cm smaller when standing, but only one centimetre smaller when sitting. His upper body accounts for 54% of his total body height. These basic proportions are typical for demarcated regions and also area-dependent in population-rich areas such as India, China and Russia. If, when considering trunk length and body size, the proportion of an individual exceeds the average value by more than 1%, one speaks of a seated giant, if the measure falls below the average value by 1%, one calls the person a seated dwarf (see above.). For German women, the normal percentile range is between 52% and 54%, while for men it is between 51% and 53%. This

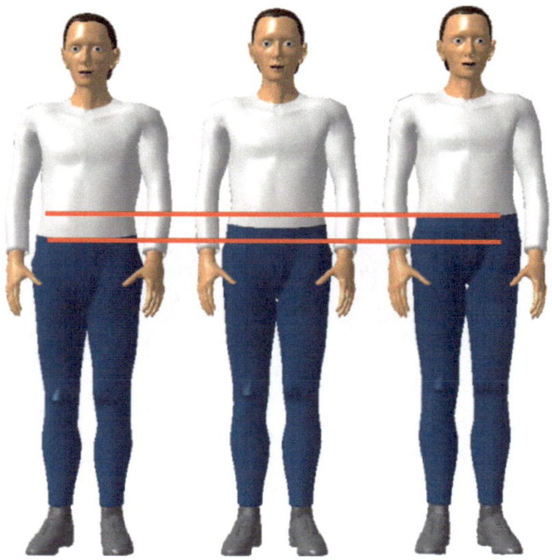

Fig. 4.23 Representation of the variation of the leg length or sitting height with same body height for seated giant (*on the left*), normally proportioned person and seated dwarf (*right*)

makes it clear that even with men and women of the same size, men have shorter upper bodies and consequently longer legs. This difference is partly due to the strong differences between the male and female pelvis and the different position of the hip joint (CCD angle[5]), which is smaller for women than for men. Compared to Europeans, Asians also appear to be seated giants due to the longer proportions of their upper bodies. For example, the normally-proportioned range for Chinese men is 53–55% and for Chinese women 53.5–55.5%. As a result, due to the inevitably shorter lower extremities, a sitting position close to the steering wheel must be chosen in order to reach the pedals, while the long upper body rises high in the restricted headroom of the roof frame. The design of the interior is geared to the target market and takes into account the anthropological needs of the customers.

5 Centrum-Collum-Diaphyseal Angle is the projected angle between the neck and shaft of the femur. It is about 130°–140° for men and about 120°–130° for women. With increasing age, the CCD angle becomes more and more acute (up to 115°) due to decreasing bone strength (osteoporosis).

Sex

In addition to the described differences in proportions between men and women with regard to stature, other attributes between the sexes must generally be distinguished. As a rule, the absolute difference in size between men and women is about 10 cm in the same percentile, which is mainly due to the different growth process during puberty (see also Fig. 4.22). Whereas in girls the growth in length almost comes to a standstill with the onset of menstruation at the age of 12–14 years[6] young men grow up to an age of over 20 years. The physique differs in many ways. The gender-typical shape of the pelvis, for example, is responsible not only for the different proportions, but above all for seriously deviating sitting postures and differently perceived support needs. While in men the lumbar support of the seat back is sometimes perceived as uncomfortably hard and disturbing, for women an even more extensive support may be desirable. The position of the pelvis in the sitting position and the required support of the lumbar spine until a noticeable relief of the back span a large area[7] which is technically feasible. Since the woman's wider pelvis is related to the hip width when sitting, the shape of the seat shell, cushion and backrest is decisive for the acceptance and comfort assessment of the entire vehicle. While a narrow[8] seat can be judged by men as sporty and comfortable, even average percentile women can reject the same seat as unacceptably tight, since the proportions of men and women in the thigh area sometimes differ drastically. The contour of the seat is not only decisive in the hip area, but must also be considered up to the thigh area. The main reason for this is the different distribution of obesity. While the abdomen of corpulent men increases locally,

6 The growth phase after puberty can be considered complete if the annual growth is less than 1 cm. This is the case for girls in Germany at 14–17 years of age and for boys at 16–20 years of age.
7 The adjustment range of the lumbar support can be up to 80 mm depending on vehicle type and seat.
8 A vehicle seat is perceived as "narrow" when the thighs are noticeably guided laterally by the seat cheeks.

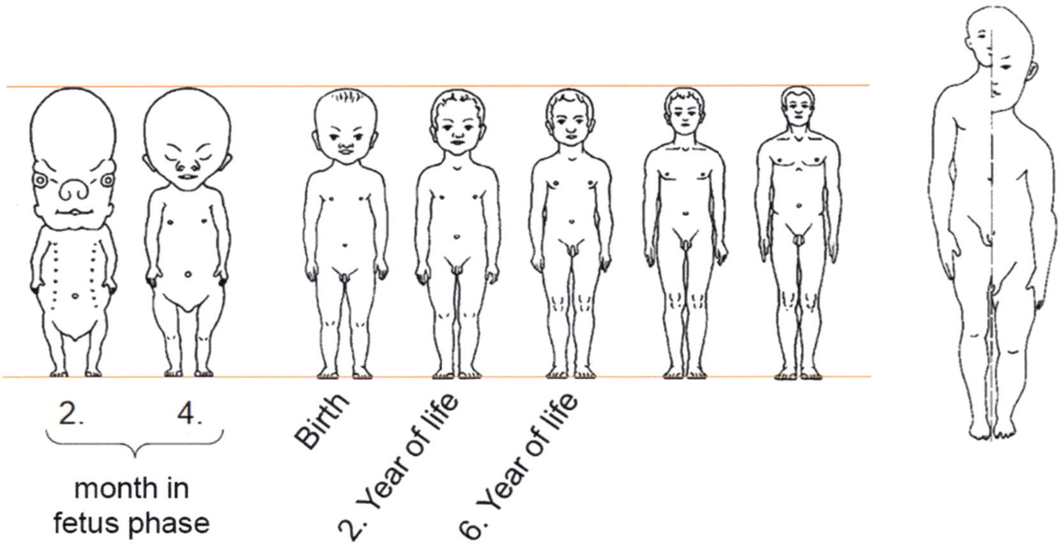

Fig. 4.24 Comparison of body proportions in newborns, adolescents and adults. (From Schmidtke 1993 and lecture manuscript Rühmann, Produktionsergonomie)

the female body prefers to store fat on the buttocks, hips and thighs. Furthermore, women usually have narrower shoulders and shorter extremities (arms, legs and fingers). The resulting need for the largely individual adaptability of a seat is dealt with specifically in ► Sect. 7.2.2.1. With the smaller physique of the woman also a reduced muscle mass goes along, which leads to a reduced strength ability of approx. 40% (see ► Sect. 4.2.3).

Age Changes

During the growth phase, which essentially ends after the age of 20 for men and 17 for women, there are significant changes in body proportions (Bullinger 2013, Prader 1981). Figure 4.24 shows the change in body proportions from embryo/infant to adult. It is clearly visible how the amount of the head decreases and the amount of the legs increases. It should be noted, for example, that body proportions of children and adolescents cannot be calculated back on the basis of data from adults using simple scaling (Bullinger 2013).

The decrease in body height with increasing age is known from everyday life. This is mainly due to a flatter Centrum-Collum diaphysis angle (CCD angle between femur and femoral neck), a lowering of the arch of the foot and a decrease in the thickness of cartilage including the intervertebral discs (Faller 1999). In contrast, there is often a significant increase in body weight and circumference (e.g. chest circumference, waist circumference). Especially with regard to internal dimensions (see ► Sect. 7.8.1) regarding the space requirements of persons, age plays a decisive role for a correct anthropometric interpretation.

Secular Acceleration

In addition to the examination of individual development during the pubertal growth phase, a socially typical change in the population average can be observed. The further development of linear growth can be seen above all in prospering development zones, where the improvement in the overall economic conditions is having an impact on the majority of the population. Secular acceleration refers to the increase in body length of adult persons as a function of the time of collection. In addition to the availability of adequate food and clean water, medical care and a stress-free growth phase in childhood are signs of an improvement in general living conditions, which affect the anthropometric indicators of a population. An increase in

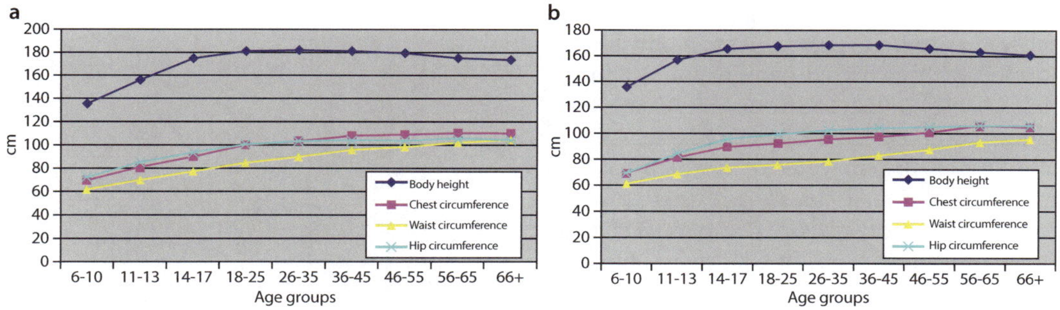

Fig. 4.25 Body height, chest, waist and hip circumference of men **a** and women **b** according to the surveys of Siedl et al. (2009)

length growth (acceleration) is usually assumed by 1 cm for a period of 10 years. However, due to the great logistical effort of a national series measurement, often only local sample measurements are available, which reflect only to a limited extent the conditions in the total population (see also Figs. 4.21 and 4.22). Many national surveys for the static determination of body measurements were taken at intervals of 30 years, so that only a rough estimate of acceleration is possible. In Central Europe, the trend in linear measures has stagnated, while circumferential measures have increased considerably in some cases (Fig. 4.25). In so-called emerging markets such as Brazil, and to some extent also in China and India, an acceleration of the measures of length is likely due to the economic upturn. While longitudinal growth depends on bone structure and thus on longer-term factors and is slower, circumferential growth depends primarily on muscle and fat tissue and can increase more rapidly both individually and socioculturally.

Handedness

A phenomenon that can be observed beyond humans in all primates is the consistently preferred use of one side of the body for fine motor activities. This dominance is called *handedness* and refers to the distinction between right-handed and left-handed people. Surprisingly, this dominance is not evenly distributed, but the majority is right-handed (Cashmore et al. 2008). So only about 10–15% of people are left-handed (according to SizeGERMANY about 6,%).[9] This number is cross-cultural and also constant in terms of development history. This dominance is mainly related to dexterity, as the preferred hand is used for motorically complicated and demanding tasks (e.g. writing or sewing). Due to the preferred use of the dominant side, this also leads to a slight advantage in strength, which is why the term "strong side" is also used. However, the asymmetry of the power capability is relatively low at 5–10% and can also be read externally from the approximately evenly distributed muscle mass.

In certain constellations, the distribution of the operating devices in the automobile has advantages or disadvantages, depending on the handiness with which the user encounters a particular vehicle. For example, right-handed people in left-hand drive vehicles have an advantage when operating the switches in the centre console, such as radio or climate control, because the arrangement is particularly conducive to fine motor skills here. The operation of the electric window regulator switches or the main light switch to the right of the steering wheel is easy for right-handed people in right hand drive vehicles, because here too the arrangement of the dominant hand is con-

9 Whether someone is called left-handed is not entirely clear. There are different test or questioning procedures, through which a classification should be possible. To a certain extent, there is a continuous transition between right- and left-handedness. The SizeGERMANY data mentioned above are self-assessments of the test subjects.

venient. If a left-handed person encounters the same situation, these tasks can only be performed with increased concentration and are more likely to distract from the driving task. Under certain circumstances, cross-over operating procedures may occur, in which the fine-motor-conditioned hand is used on the opposite side of the body. The manual engagement of gears in left-hand drive vehicles is convenient for right-handed drivers and vice versa in right-hand drive vehicles for left-handed drivers. A critical special case is always the manual parking brake. Especially in right hand-drive markets, many small women are confronted with an abundance of disadvantages. Due to their smaller stature, they are required to position the seat very close to the steering wheel in order to be able to operate the pedals. This causes the handbrake lever to move further back relative to the seat position, creating an unfavourable arm position for operating the lever. Often the backrest with the side seat bolsters is still in the way or the armrest between the front seats blocks free access to the lever. Due to the relative strength disadvantage of women compared to men of about 30–40%, the absolute strength of small persons is very limited. If a right-handed little woman has to operate the handbrake lever with her clumsy, weak left hand, the necessary technical force cannot be applied in extreme situations (parking the vehicle on a slope or even with a trailer) and the vehicle cannot be adequately secured. If this is required in an emergency situation, a critical situation may arise in the event of panic, resulting in the vehicle rolling away. Often the decisive amount of force can only be produced by using the second hand or by shifting the sitting position to the lever. The increasingly frequent use of electric parking brakes has the great advantage of being a self-reinforcing system that compensates for disadvantages suffered by small, elderly or weak persons and can therefore be classified as an assistance system (for a definition of assistance, see ▶ Chap. 9).

Strictly speaking, handiness refers not only to the hands, but also in a broader sense to the feet. However, the practiced dexterity is of much greater importance here. While one foot has learnt to "accelerate" and brake sensitively, the other is only expected to operate the clutch pedal as quickly and powerfully as possible. Occasionally it can be observed that a driver thinks he is in a clutch vehicle, but sits in an automatic vehicle that misses the clutch pedal and hits the brake pedal in the usual manner. The extremely violent vehicle reaction shows how different right and left feet can be conditioned.

Fig. 4.26 Test method for eye dominance (after Remlinger 2013)

It should be mentioned here that there is also a "handiness" for the dominance of the eyes. Remlinger (2013) shows how this preference can easily be discovered: One fixes with both eyes a further distant object through a nearby "keyhole" (this can be formed, for example, by the two crossing hands; Fig. 4.26). By alternately closing the two eyes, you now check with which eye you actually recognize the object.

Figure 4.27 shows the distribution of handiness and eyeiness according to a non-representative preliminary study by Remlinger (2013) with 91 persons. Concrete effects on the driving of vehicles, particularly with regard to the asymmetrical design of the driver's position and traffic flows, are not known in the literature. However, it can be assumed that eye dominance has an influence on the sequence of side-directed movements, such as getting in and out of the vehicle, the perception of intersection traffic or the processing of secondary tasks in the vehicle's centre console (Remlinger 2013).

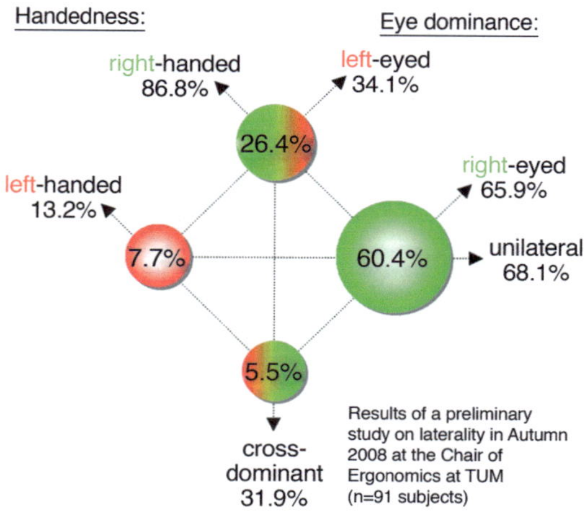

Fig. 4.27 Distribution of Laterality of Eye and Hand (after Remlinger 2013)

4.2.1.4 Important Body Distance Measures

For reasons of comparability and traceability, the measurement of human bodies suggests the use of identical measurement methods. Irrespective of the technology used with manual callipers or electronic body scanners, the postures and planes at which the measurement is to be taken must be defined. As a rule, the measurements are made on the uncovered or lightly[10] clothed body. For the determination of special situations, clothing with normal[11] or heavier[12] clothes may be necessary. In the following sections, individual measures are presented which are important for the development of motor vehicles. Not only the dimensions of the vehicle seat in length and width of the seat cushion, but also the entire interior design with available space, entry and exit openings, ranges and visibility zones must be determined on the basis of the current body dimension data of the target markets.

Body Height (Stature)

The person to be measured stands upright with her/his legs closed (Fig. 4.28 left). The head is held in such a way that the upper edge of the ears forms a horizontal line with the eyes (so-called Frankfurt horizontal). The vertical distance from the standing plane to the tip of the head is measured. This measure is an easy-to-quantify parameter that most people know about themselves quite well, although respondents often tend to indicate height including shoes, or too high values. This parameter should therefore be determined with priority and not only asked for. As a measure, the body height is only used as a reference or for special situations, such as standing upright under the open tailgate when loading the trunk.

Sitting Height

The person to be measured sits with fully supported thighs on a flat, non-deformable surface, while the lower legs hang freely and the feet are not supported (Fig. 4.28, centre). The upper body is stretched straight, similar to the standing posture. Here, too, the head is to be held in the sense of the "Frankfurt horizontal". The vertical distance from the seat plane to the tip of the head is measured. The sitting height is of decisive importance for the design of the driver's position. However, the sitting height serves as an anthropometric initial measure only for referencing and percentile calculation, since a completely upright and stretched posture of the upper body hardly occurs in practice. The driving posture in trucks and buses comes relatively close to this basic posture, even though the upper part of the body sinks somewhat due to movement and vibration. In passenger cars, a slightly reclined posture of the upper part of the body[13] is to observe. Along with the sinking-in behavior of the upper body, the spinal column inclination and curvature of the stretched

---

10 Lightly dressed is primarily understood to mean a person dressed only in underwear, but who is unclothed in most parts of the body, especially the main joints.

11 Normal clothing is understood to be street clothes covering all parts of the body.

12 Heavy clothing usually refers to military combat equipment or heavy protective clothing worn by firefighters or other emergency services.

13 Most vehicles are designed for a torso angle of 22°–25°. In practice, however, much flatter and steeper upper body postures can be observed, ranging from 5° to 35°. The average value in a passenger car is approx. 20° and is strongly dependent on the height of the seat above the heel (dimension H30) (see ► Sect. 7.2.2.1 and ► Fig. 7.28).

4

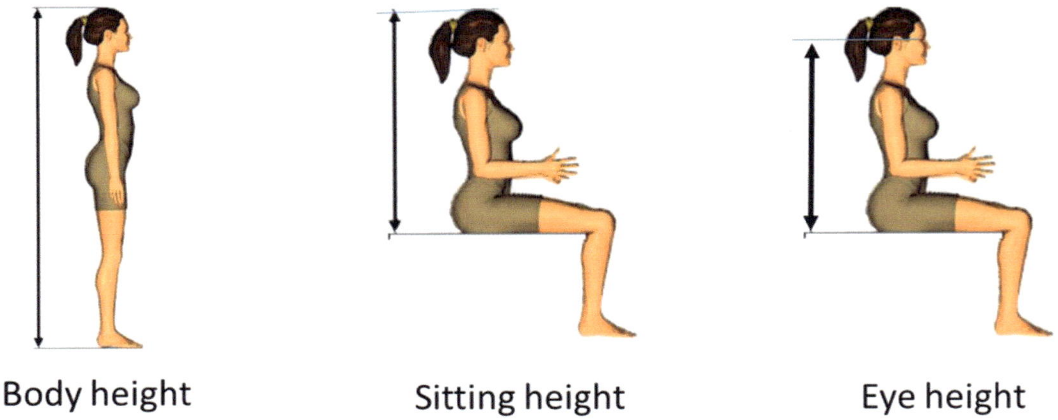

Fig. 4.28 Definition of the measured variables body height, trunk length and eye height

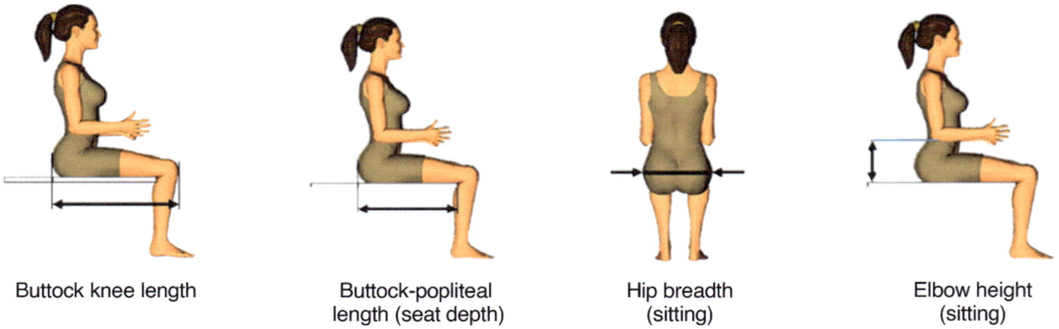

Fig. 4.29 Definition of the measurands seat-knee depth, seat depth, seat width and height of the elbow over the seat surface

posture causes approximately 40 mm to be lost. An upper body that has an anthropometric sitting height of 900 mm is thus only 860 mm high when it has collapsed under a certain seat back inclination. If the interior now also offers a reference height of 900 mm, the driver has 40 mm headroom, which allows him to sit upright in the seat. For determining the necessary headroom in cars it has to be taken into account that the sitting height is measured in relation to the skull contour, which means that hair and hairstyle must also be considered, which normally rise above the tip of the head at least at 30 mm, as well as any headgear (see ► Sect. 4.2.1.6).

Eye Height in Sitting Position

The measuring posture is similar to that used to determine the sitting height. The distance between the seat and the pupil of the right eye is measured (Fig. 4.28 right). The eye level is actually of essential importance for the conception of the visibility conditions from the vehicle cabin and on the instruments. However, the lower eye height caused by the normal sitting posture due to a discount between 5% and 10% must be taken into account (see also the comments on sitting height).

Buttock Knee Length

The measuring posture is similar to that used to determine the sitting height. However, the projected horizontal distance between the foremost point of the kneecap and the rearmost point of contact of the buttocks with a measuring block is determined (Fig. 4.29, left). This measurement describes the space required between the vehicle seat and instru-

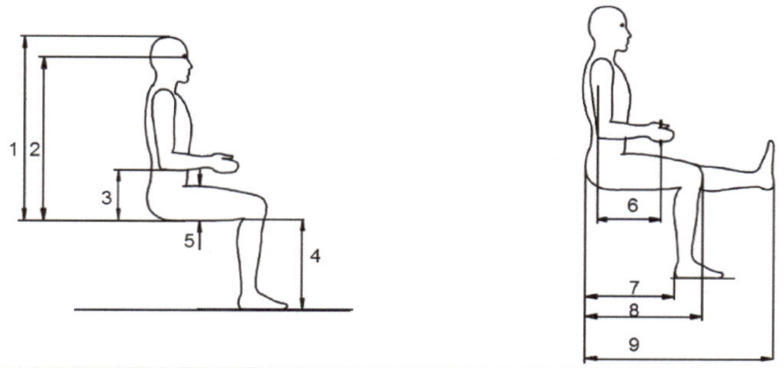

Dimensions in cm	Percentile male 5%	male 50%	male 95%	female 5%	female 50%	female 95%
1 Sitting Height)	84.9	90.7	96.2	80.5	85.7	91.4
2 Eye Height (sitting)	73.9	79.0	84.4	68.8	73.5	78.5
3 Elbow Height (sitting)	19.3	23.0	28.0	19.1	23.3	27.8
4 Lower Leg Length	39.9	44.2	48.0	35.1	39.5	43.4
Thigh Clearing						
5 Forward Grip Reach	11.7	13.6	15.7	11.8	14.4	17.3
6 Buttocks-popliteal length	32.7	36.2	38.9	29.2	32.2	36.4
7 Buttocks Knee length	45.2	50.0	55.2	42.6	48.4	53.2
8 Buttock Length	55.4	59.9	64.5	53.0	58.7	63.1
9 Leg Length	96.4	103.5	112.5	95.5	104.4	112.6

Fig. 4.30 Human body measurements according to DIN 33 402

ment panel as a combination of thigh length and established buttock contour. Due to the freedom of movement required to operate the pedals, an extra charge must be taken into account. It must also be taken into account that, due to the position of the lower leg, local points of the tibia can also determine the anterior boundary.

Seat Depth

The seat depth is determined in the same supported and extended sitting position. However, here the projected distance between the hollow of the knee and the rearmost point of contact of the buttocks is measured with a measuring block (Fig. 4.29, middle left). As the designation already indicates, the dimension is an orientation for the necessary seat support length. If the length of the seat cushion exceeds the individual seat depth, the front edge of the seat pushes the hollow of the knee forward, whereby the lower back and especially the lumbar vertebrae lose contact with the backrest contour and, due to the lack of support, run the risk of falling into a kyphotic posture, which is detrimental to comfort and ultimately also to health. On the other hand, sufficient support of the thigh is necessary for large persons. The variation width of the seat depth (see Fig. 4.30) thus requires an adjustability of the seat surface.

Seat Width (Seated Seat Width)

Here, too, the person to be measured sits upright on a flat surface with fully supported thighs. The knees of the freely hanging lower legs touch each other slightly. The largest horizontal distance of the undeformed hip volume is measured (Fig. 4.29, centre right). Even if the contact points are often located in the area of the pelvis or hip joints, this measurement gives an indication of the necessary seat width, i.e. the inner dimension of the seat surface between the laterally bounding seat sidebolsters. Due to the anatomical differences between men and women in the pelvic area, this size is the outstanding measure at which female data exceed male data even for smaller body sizes. Although this measurement is based on the skeletal hip bone width, it depends to a considerable extent on the fat-related soft tissue mass. In this area in particular, there has been a considerable increase in Central Europe and, above all, North America for some time now. The mean value of the mixed population was still in 1999[14] at 380 mm, this value has increased in ten years by 30 mm. In the same period, the women's lead over men melted from 30 to 10 mm. Men and women have increased considerably in size despite stagnation in length growth (see also ◘ Fig. 4.25), with fewer and fewer people enjoying the sporty appearance of the vehicle seat, but rather perceiving the narrowness as uncomfortable and unpleasant. This requires, if one wants to do without individual adaptability, a flatter design of the seat cushion contour or seats with wider or softer upholstery.

▪ Elbow Height (Sitting)

This measurement is also taken in the stretched sitting position. The upper arms hang down as freely as possible, the forearms are bent forward at right angles. The vertical distance from the seat plane to the lowest point of the elbow is measured (◘ Fig. 4.29, right). This dimension gives indications of the arrangement of possible support geometries for the arms, such as the armrest in the door panel and tunnel console. However, it should be noted that this elbow height also increases with increasing angle of spread of the upper arm. This measurement also changes with the backward inclination of the upper body and with the position of the hands. A compromise between the needs of large and small positions and the different postures of the driver and passenger seat is difficult and requires targeted, user-oriented solutions (see also ► Sect. 7.2.2.5).

▪ Length of Extremities

Sub-sections of the body such as upper or lower arm length, lower leg length with foot, as well as head dimensions such as head height, width and depth are frequently listed in anthropometric tables and are of interest in the application of automotive ergonomics as reference dimensions for derived measures such as reach of the hand or field of activity of the feet. ◘ Figure 4.30 gives an example from DIN 33 402. However, specific investigations are necessary to determine the accessibility based on body size, sitting posture and surrounding geometry in order to give a final assessment of comfort. The range also depends on the type of gripping task, e.g.whether a lever should be gripped with a fully encompassing hand or whether a switch with an extended index finger can be pressed.

▪ Dimensions of the Hand

The size of the hands is proportional to the height of the human body due to the tubular bones of the fingers. However, width, thickness and shape can vary greatly. An application in ergonomics finds above all the boundary percentiles[15] which determine which customer groups are still able to operate certain operating devices safely. The dimensions of the fingers and the hand are therefore of particular importance in the context of the hand posture and type of grip, as they determine the necessary clearance around an operating element (◘ Table 4.2).

4.2.1.5 Body Measurement Tables

It is not possible to reprint complete body dimension tables here, as all databases and tables are contained in standards, publications or online portals with restricted access. The fact that it is important to orientate oneself on current data is illustrated by the information on the body height of the male population in ◘ Table 4.3.

The following list shows an excerpt of possible sources of information:

- Flügel, B., Greil, h. und Sommer, K.: Anthropologischer Atlas. Grundlagen und

14 Mixed population of 50% males and 50% females, each the 50th percentile.

15 By default, the fifth female and 95th male percentile is considered the limiting percentile, occasionally the 2.5 to 97.5th percentile is interpreted. More rarely, a design from the first to the 99th percentile is found, which ensures a large coverage of potential customers in the population, but can also lead to a considerable expansion of the technically necessary free space and adjustment ranges.

Daten, Alter- und Geschlechtsvariabilität des Menschen. Edition Wörtzel, Frankfurt/M (1986)
- DIN 33402 Part 2: Specification of 69 body measurements "based on statistically confirmed measurements of persons living in the territory of the Federal Republic of Germany" (DIN 33402 Part 2)
- EN ISO 7250 Essential measures of the human body for technical design - Part 2: Anthropometric databases of individual national populations
- Database iSize (▶ https://portal.i-size.net/SizeWeb/pages/home.seam): International body dimension portal from Human Solutions to Germany (SizeGERMANY), France, Netherlands, Switzerland, Japan, Korea, USA, China
- WEAR database (▶ https://wear.istdayton.com/WearHome/Login/Login.aspx): Compilation of many measurement campaigns mainly from the North American military. Among others, the series measurements CAESAR (first series measurement with body scanners of North American, Dutch and Italian persons), NHANES III (measurement of the North American civilian population 1970) and ANSUR (measurement from the North American military system 1988) are included.

Table 4.2 Dimensions of the human hand [acc. to DIN 33 402, part 2]

Section	5. percentile female	95th percentile male
Fingertip index finger	13 mm	20 mm
Hand thickness	21 mm	32 mm
Palm length	91 mm	117 mm
Hand length	159 mm	201 mm
Hand width with thumb	82 mm	116 mm

4.2.1.6 Special Dimensional Derivatives

Foot and Shoe

The anthropometric dimensions of the foot are statistically recorded and approximately normally distributed. As expected, women usually have shorter and narrower feet than men. Although any length or width can be statistically proven, the shoe industry has agreed on uniform measurement systems and, above all, on a graduation of shoe sizes, which allows a more rational production of ready-made sizes. Regardless of this, the foot length of the left foot can deviate considerably (up to 5%) from that of the right foot and the length-width ratio of a foot as well as the height and

Table 4.3 Male body height for different survey periods (Germany)

	DIN (1968–1977)			DDR (1971–1986)			HdE (88–89)			Siedl et al. (2007–09)		
	Percentile			Percentile			Percentile			Percentile		
Age	5.	50.	95.	5.	50.	95.	5.	50.	95.	5.	50.	95.
18–19	167.7	177.1	189.3	162.0	175.8	187.6	168.8	180.9	193.3		180.8	
20–25	165.2	175.8	186.5	164.0	176.4	188.2	168.0	179.0	193.0		181.8	
26–40	164.5	174.5	185.2	162.5	173.8	186.1	167.5	177.5	188.0	165.3	178.2	192.9

curvature of the back of the foot can also deviate greatly from the ideal shape, so that it is not possible to wear a ready-made (shoe) size and custom-made shoes (made-to-measure shoes) are required for anthropometric reasons. Common shoe size systems are the *"Pariser Stich"* (Continental Europe), *barley corn* (GB, USA) and the *monopoint system* (Japan). Men's and women's footwear do not differ in the classification of shoe size and foot length. A separate marking of the shoe width is generally rare, although some shoe manufacturers have introduced different width gradations, in particular for children's shoes. Thus, in Germany the WMS system is quite popular. In Asia, a width-classification with letters is usual. The Brannock system is used in the USA. ▫ Table 4.4 provides an overview of the various systems in use and ▫ Table 4.5 related dimensions.

In general, the shoe size is derived from the foot length and an allowance is added so that the foot can glide in the shoe during the rolling movement of walking/running. The inside length of the shoe results from the foot length plus the allowance. The outer length results from the design of the sole and the shoe type. The shoe size, however, mainly refers to the shoe length, but basically the last measurement is decisive as the measurement of the underlying archetype of the foot representation.

The "Pariser Stich" specifies a gradation of 2/3 cm (approx. 6,67 mm). A 28 cm long foot needs with 1 cm thrust, the material thickness of the upper material and the sole overhang of 2 cm a shoe, which is approximately 320 mm long (size 45). This shoe is then usually about 110 mm wide. In order to ensure that the pedals are operated, movement allowances must be made in addition to the pure length and width dimensions. In addition, for vehicles mainly used with workwear, it must be taken into account that work safety shoes with larger dimensions and free spaces must be taken into account. The choice of the footrest surface should also be such that the entire required foot length can be accommodated collision-free and the foot width is supported at least 2/3, so that a tilt-free support is guaranteed.

▫ Table 4.4 Shoe size systems in various markets

Region	Gradation
Germany	W (wide), M (medium), S (narrow)
Germany	F ("Normal width" - slim), G ("comfort width" - normal), H ("comfort width" - strong), J ("comfort width" - oversize), K, L, M ("special widths")
GB	N (narrow - narrow), M (medium) or R (regular), W (wide - wide)
Asia	A (narrow), B, C, D, E, EE, EEE, EEEE, F, G (wide)
USA	4 A (narrow), 3 A, 2 A, A, B, C, D, E, 2E, 3E, 4E, 5E, 6E (wide)

▪ Headgear

The wearing of headgear in public and thus also while driving depends on the region, the culture, fashionable trends and possibly on the social position or professional function. To interpret the head space of a vehicle solely according to the anthropometric sitting height of the human being without headgear would be too short-sighted neglecting the social context and thus considering the corresponding headgear. The economic success of certain vehicles often depends on such additional special requirements. A prominent example of such a requirement is the continued success of the Hindustan Ambassador in India, a vehicle that dates back to a 1956 development of the Morris Oxford Series III and is still being built in India. It is particularly popular as a taxi, because the strongly arched roof provides enough interior height for the driver and passenger to use the vehicle comfortably with a turban. Occupationally, as with shoes, special headgear may be necessary or prescribed, such as the cap of a chauffeur or a special work safety helmet, which must also be worn while driving. ▫ Table 4.6 shows surcharges that are necessary for the various application areas.

Table 4.5 Description of Shoe sizes depending on the dimensions of the foot

Shoe length	Shoe width	Foot length	Foot width	Percentile	Percentile				GB, USA	USA	
mm	mm	mm	mm	masculine	feminine	D/EU	F	JP	gentlemen	ladies	E
225	80	210	75		5.	34	34	21 ½	2	3 ½	–
230						34/35	35	22	2	4	–
235	85	220	80			35	36	22 ½	2 ½	4 ½	34
240						35/36	37	23	3	5	–
245	95	230	90			36	37/38	23 ½	3 ½	5 ½	35
250						36/37	38	24	4	6	–
255	100	245	95		50.	37	38/39	24 ½	4 ½	6 ½	36
260						37/38	39	25	5	7	37
265	105	250	100	5.		38/39	40	–	5 ½	7 ½	–
270						39	41	–	6	8	38
275						39/40	41/42	–	6 ½	8 ½	–
280	105	255	100			40	42	–	7	9	39
285						40/41	–	–	7 ½	9 ½	–
290						41	–	–	8	10	–
295						41/42	–	–	8 ½	10 ½	–
300	110	260	105			42/43	–	–	9	11	–
305	110	265	105		95.	43	–	–	9 ½	11 ½	–
310	110	270	105	50.		43/44	–	–	10	12	–
315						44	–	–	10 ½	12 ½	–
320	110	280	110			45			11	13	–
325	115	290	110	95.		46	–	–	11 ½	13 ½	–
330	115	300	115			46/47	–	–	12	14	–

Table 4.6 Typical add-ons for selected headgear

Headgear	Add-ons for Sitting Height, Head Clearance
Peaked cap/	20–80 mm
Peaked cap/chauffeur's cap	20–60 mm
Safety helmet Construction helmet (EN 397)	30–50 mm
Turban	40–150 mm
Hat (Bowler)	10–30 mm

Table 4.7 Body part weights (Source Siedl et al. 2009)

Body part	5. percentile female	95th percentile male
Hull (60%)	31 kg	66.2 kg
Head (6%)	3.1 kg	6.6 kg
Leg (12.5%)	6.4 kg	13.8 kg
Arm (4.5%)	2.3 kg	5.0 kg
Total (100%)	51.5 kg	110.4 kg

Additional Garments

Depending on the location or use of the vehicle, special items of clothing or additional equipment must be taken into account that restrict freedom of movement or require additional space. For emergency services such as police or personal security, carrying small arms in the belt holster requires more hip space than usual, which must not be restricted by the contour of the seat cushion or backrest. If necessary, the vehicles may have to be converted for the intended purpose at great expense and equipped with special seats that provide the necessary space.

The hands occupy a special position, since many vehicles, especially during operation in winter, have to be operated with seasonal clothing. Doors, flaps, hoods and service openings should also be operable with gloves. The clothing allowance for the diameter and thickness of fingers and hands up to 1–2 cm must be taken into account here, which may make it necessary to increase the size of the openings for the hands by 50–100%.

4.2.2 Weight

Rainer E. Grünen

Another important body measure is the body weight and thus also the body part weights of the individual extremities. The body weight of adult humans lies between 46 kg and 130 kg, only 1% of the population exhibits a lower or higher body weight. However, since the body proportions are typical for humans as a species, the body part weights behave according to a continuous similarity, with slight deviations due to obesity. Table 4.7 shows the body part weights for the 5th percentile female and the 95th percentile male according to the results of Siedl et al. (2009).

The weight distribution has changed regionally - and in particular in countries of the western cultural area by energy-rich food[16] - to corpulent types. The correlation between body size and body weight is limited (see Fig. 4.31). Thus the body weight is distributed from 80 kg over body sizes from 1,55 m to 2,00 m. People with a stature of 1.78 m can weigh from 60 kg to 150 kg.

In order to be able to classify the strongly varying body weights with the body size into comparison classes, the Body Mass Index (BMI[17]) was internationally established. However, there are different calculation approaches, of which the ratio of body mass

16 According to ICD-10 of the World Health Organisation (WHO), obesity is divided into obesity due to excessive calorie intake, drug-induced obesity, excessive obesity with alveolar hypoventilation, other and unspecified obesity.

17 BMI: Body Mass Index also known as Quelet-Caup-Index

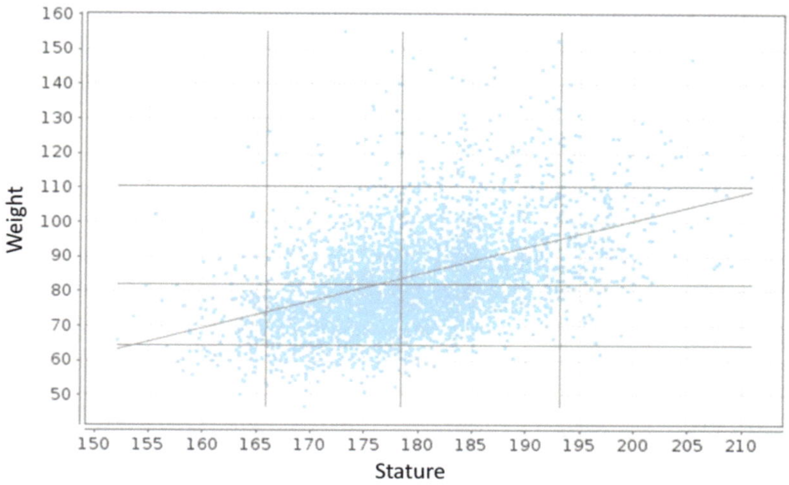

Fig. 4.31 Dependence of weight on body height [Size Germany 2009] Correlation $r^2 = 0{,}45$

to body height in square is the most common (4.2):

$$\mathrm{BMI} = \frac{m}{l^2} \tag{4.2}$$

with: m = Body mass in kilograms (kg) and l = Stature in meters (m)[18].

The corresponding dimensionless key indicator *BMI*[19] is divided into different obesity classes. The "normal weight" of a person is defined with a BMI of 18.5–24.9. Persons with a BMI between 25.0 and 29.9 are referred to as overweight.[20] In addition, three obesity classes are described: Grade I obesity (Obesity Class I) BMI 30.0–34.9, grade II obesity (Obesity Class II) BMI 35.0–39.9 and grade III obesity, super obesity (Obesity Class III) with a BMI of 40 and above. Below the normal weight, three classes of underweight are defined: Light (17.0–18.5), moderate (16.0–17.0) and severe underweight (less than 16.0). Figure 4.32 shows these weight classes as a function of body height.

The 2009 microcensus showed the following values for Germany: 44.4% of men and 29.1% of women are overweight, 15.7% of men and 13.8% of women have a BMI of 30 or more and are therefore obese. Almost half of Germans are of normal weight (men: 39.2%, women: 53.7%) and only a few are underweight (men: 0.7%, women: 3.4%). A completely different picture emerges in North America, where, according to a study by the CDC[21] 35.7% of adults and even 16.9% of children are obese (BMI > 30). A special challenge for automobile manufacturers derives from this changed constitution of people and thus of vehicle users. Until a few years ago, design values with a body weight of 90 kg for the durability of seats (seat structures and foams) were up to date, but these values have to be greatly increased due to the changed body mass of vehicle users in order to prevent mechanical failures, which can only be attributed to higher mechanical stress. Not only the direct body support on components (seats, armrests, centre console, boot) requires tech-

18 BMI based on imperial units of measurement: BMI = (body mass in pounds (lb) * 703)/(body height in inches (in))^2

19 The correct unit of the BMI is kg/m^2. However, it is usually omitted. The BMI is usually given as a unitless number

20 Interestingly enough, according to the latest studies, it is precisely this weight class that is considered to have the highest life expectancy, so that the previously common designation is considered worth considering.

21 CDC: Center of Disease Control and Prevention, Division of Nutrition, Physical Activity and Obesity (DNPAO): NCHS Data Brief, Prevalence of Obesity in the United States, 2009–2010.

Stature [m] \ Weight [kg]	50	55	60	65	70	75	80	85	90	95	100	105	110	115	120
2.00	12.5	13.8	15.6	16.3	17.5	18.8	20.0	21.3	22.5	23.8	25.0	26.3	27.5	28.8	30.0
1.95	13.1	14.5	15.8	17.1	18.4	19.7	21.0	22.4	23.7	25.0	26.3	27.6	28.9	30.2	31.6
1.90	13.9	15.2	16.6	18.0	19.4	20.8	22.2	23.5	24.9	26.3	27.7	29.1	30.5	31.9	33.2
1.85	14.6	16.1	17.5	19.0	20.5	21.9	23.4	24.8	26.3	27.8	29.2	30.7	32.1	33.6	35.1
1.80	15.4	17.0	18.5	20.1	21.6	23.1	24.7	26.2	27.8	29.3	30.9	32.4	34.0	35.5	37.0
1.75	16.3	18.0	19.6	21.2	22.9	24.5	26.1	27.8	29.4	31.0	32.7	34.3	35.9	37.6	39.2
1.70	17.3	19.0	20.8	22.5	24.2	26.0	27.7	29.4	31.1	32.9	34.6	36.3	38.1	39.8	41.5
1.65	18.4	20.2	22.0	23.9	25.7	27.5	29.4	31.2	33.1	34.9	36.7	38.6	40.4	42.2	4.1
1.60	19.5	21.5	23.4	25.4	27.3	29.3	31.3	33.2	35.2	37.1	39.1	41.0	43.0	44.9	46.9
1.55	20.8	22.9	25.0	27.1	29.1	31.2	33.3	35.4	37.5	39.5	41.6	43.7	45.8	47.9	49.9
1.50	22.2	24.4	26.7	28.9	31.1	33.3	35.6	37.8	40.0	42.2	44.4	46.7	48.9	51.1	53.3

<= 16.0	Very severely underweight	25.0 - 29.99	Overweight
16.0 - 16.99	Severely underweight	30.0 - 34.99	Obesity Class I
17.0 - 18.49	Underweight	35.0 - 39.99	Obesity Class II
18.5 - 24.99	Normal	<40.0	Obesity Class III

Fig. 4.32 Weight classes as a function of body height

nical adaptations, but also cover fabrics and trim parts, which are subject to increased abrasion and wear due to the higher load. This not only leads to a structural increase in weight and thus to higher fuel consumption, but also to increased manufacturing costs and thus to higher product costs. The weight spiral that emanates from the increased weight of the user thus continues with the vehicle.

4.2.3 Forces

Fabian Günzkofer

Knowledge of available and maximum forces plays an important role in the anthropometric design of a motor vehicle. On the one hand, there is a large number of activities in the context of automobile use that require different levels of effort (entry/exit, clutch, shifting, braking, handbrake pulling, trunk loading). On the other hand, the knowledge of posture-specific maximum forces also makes it possible to evaluate submaximal force tasks. Zacher and Bubb (2004) were able to show that the discomfort (see ► Sect. 3.3.4) increases linearly with the degree of utilisation of the maximum force. This allows posture evaluations to be derived from force curves in the movement space. Other design parameters, such as acceptable force levels, are often expressed as a percentage of the maximum force (MacKinnon 1998). The level of maximum force in a posture therefore serves as an indicator of discomfort for the application of submaximal forces.

Daams (1994) cites another argument on the need to know forces. For an adequate product design, the designer/engineer needs the forces of the weakest so that they can use the product safely and comfortably, and the forces of the strongest so that they do not accidentally destroy the product. She also argues for a sensible consideration of human forces, as this could save unnecessary servo systems.

4.2.3.1 Approaches to Measuring and Categorising Forces

Force measurement and categorization is a very difficult task due to the will-dependent individual disability, which is by far not solved as doubtlessly as that of length and circumfer-

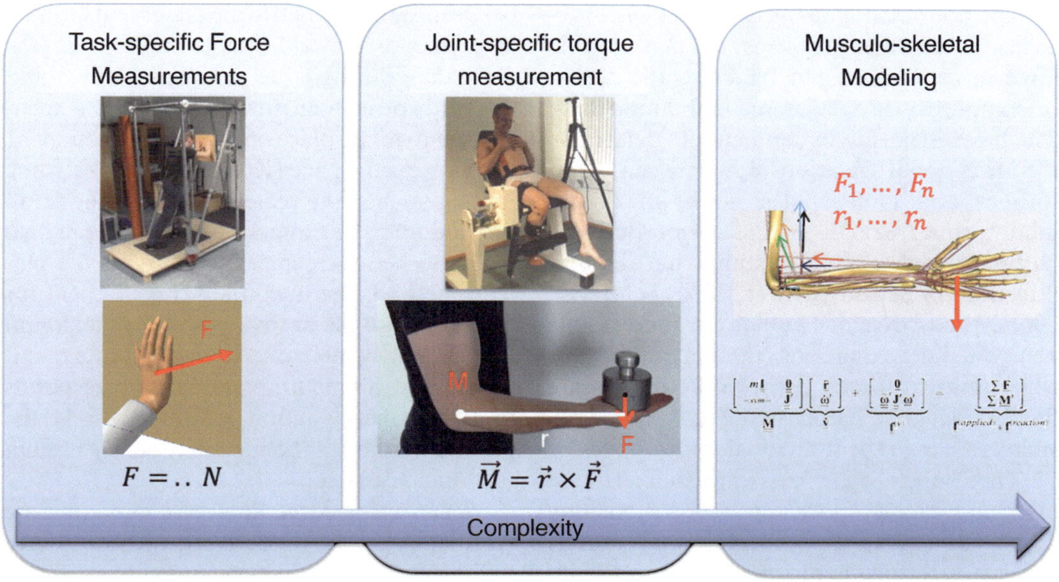

Fig. 4.33 Different measurement and modelling approaches for force modelling

ence measurements. In principle, three different approaches for the measurement and modelling of forces can be distinguished (Fig. 4.33).

In the simplest case, force cases of interest can be measured directly. This corresponds to the procedure *task-specific force measurements* in which the posture is either freely selectable or clearly defined by the task. An example of this would be the force measurement when pulling the handbrake in a mock-up. Rühmann (1992), for example, carried out a long-term series measurement of isometric maximum forces with more than 3000 test persons as part of the funded project "Humanization of Working Life" under the research project "Human Body Forces". The corresponding data are contained in the Handbook of Ergonomics; Volume 3 (Schmidtke 1993, 2013). The task-specific force data collection has been updated and expanded by the "Montagespezifischer Kraftatlas" (assembly-specific force atlas) (Wakula et al. 2009). This includes action forces of the whole body as well as the finger, hand and arm system for realistic postures in industry. The great advantage of task-specific force measurements lies in the very fast acquisition of maximum forces of different test persons. A tabulation and possibly percentile calculation of measured values allows easy application for practical cases that correspond exactly to the test conditions. However, this approach neglects two crucial parameters. It was already mentioned in ► Sect. 4.1.2.1 that the force to be applied by a joint depends on the posture. However, this is not specifically specified for task-specific force measurements. Thus, the application of the results is only valid if the posture to be evaluated corresponds as closely as possible to the measurement posture. Furthermore, the anthropometry of the test subjects also plays an important role, which is often not specifically considered in this approach (Engstler 2012). Note that with the same experimental setup, e.g. pressing against a plate at a certain height, different postures result for people of different body dimensions. Thus humans of the same power capacity, but different anthropometry, will reach different power values. In this respect, in addition to gender, the values would also have to be divided according to relevant body measurements. In summary, it can be stated that task-specific measurements can be carried out very quickly and are easy to apply, but cannot be applied to any situation.

The approach *joint-specific force measurements* solves the problems of task-specific force measurement, but increases the degree of complexity of measurement and modeling. The basic idea is to experimentally determine the functional relationship between joint torques and joint angles. Thus, maximum joint torques can be predicted for different anthropometries and postures in different directions of action (Schwarz 1997; Seitz et al. 2005). Using dynamic equations, for example Newton-Euler equations, the maximum possible outward forces at the end effectors are finally calculated based on the maximum possible torques in the individual joints.

The most complex approach is the *musculo-skeletal modeling*. The modelling depth is increased and individual muscle forces, which together form a joint torque, are considered. In the inverse approach, the necessary muscle forces for a dynamic equilibrium are calculated for a certain movement (e.g. from marker trajectories of a motion tracking system; Damsgaard et al. 2006). For correct results the lever arms from the muscles to the joints, origin and approach of the muscles, pennation angles of the muscles, maximum force with ideal muscle length, muscle cross sections and selected muscle models (e.g. Hill-muscle) must be exactly known. In addition, according to Bernstein's theory, there is the problem of the redundancy of the human musculoskeletal system (Bernstein 1967). The equation system for muscle recruitment is clearly overdetermined, as many muscles are available for a single degree of freedom. An attempt is made to simulate muscle recruitment by the central nervous system using various recruitment algorithms (Crowninshield 1978; Dul et al. 1984; Jongen et al. 1989; Rasmussen et al. 2002; Abdel-Malek et al. 2006). However, the search for the only true algorithm has not yet been completed (Rasmussen et al. 2001). In spite of all the aforementioned optimization needs, musculoskeletal models offer an ideal opportunity to reveal reactions inside the body to external stress.

In general, different measurement methods are differentiated for all force measurements. On the one hand, there is a difference between static isometric measurements, which keep the posture and thus the muscle lengths constant during the measurement, and dynamic measurements, in which the measurement takes place in a certain section of the movement space (Kumar 2004). For static measurements, the plateau method, the ramp method and the momentum method are also distinguished (Kroemer 1977). With the plateau method, the maximum force is held for 4 s, while with the ramp method the maximum force is slowly increased to the absolute maximum. The momentum method corresponds to a short, jerky action. Of all methods, the plateau method is the most frequently used in scientific research.

When using force data, the user must be familiar with the underlying method, since different types of measurement can sometimes lead to different results (Engstler 2012).

4.2.3.2 Force Measurement

Force measurements are time-consuming and the presentation of results is also quite complex. In the following, we will briefly explain how a static maximum torque measurement works conventionally and what the presentation of results can look like. Isometric, in comparison to isokinetic (dynamic), measurements have the special feature that no movement takes place during the measurement. Therefore, force measuring devices must be used which can be adapted to different anthropometries of the test subjects and fixed in different positions. The measuring principle of every device is that a torque sensor is aligned exactly to the anatomical axis of rotation of a joint. Thus the joint torque can be sensed directly without the need of an indirect calculation by force and lever arm. In order to allow maximum strength to be developed and to maintain the desired posture exactly, the test person must be sufficiently stabilised using pads and/or straps (Günzkofer et al. 2012a). ◘ Figure 4.34 shows two examples of joint torque measurements of the elbow and the knee joint.

The most important elements of a measurement consist of an adequate warm-up phase, at least two force measurements per posture, a sufficient regeneration time of two

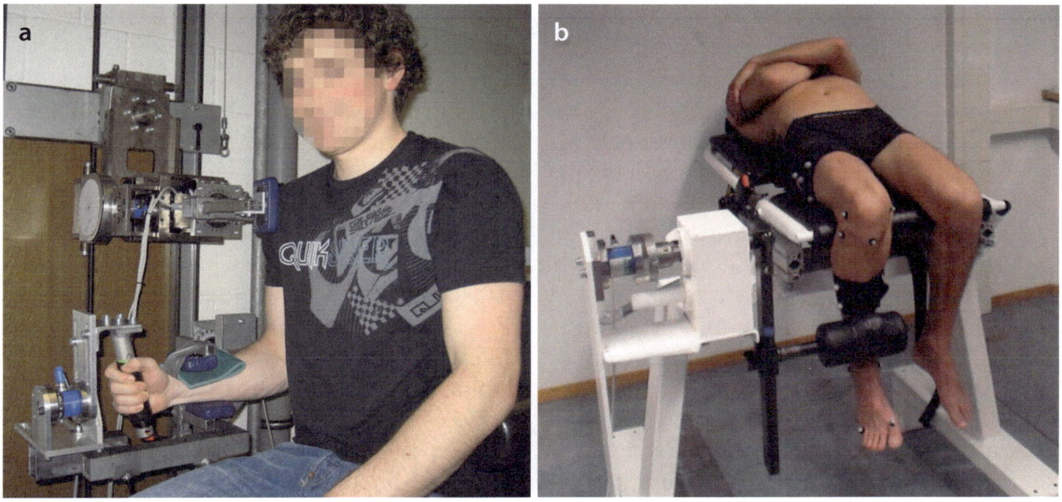

Fig. 4.34 Examples of torque measuring devices for elbow flexion/extension **a** and knee flexion/extension **b**

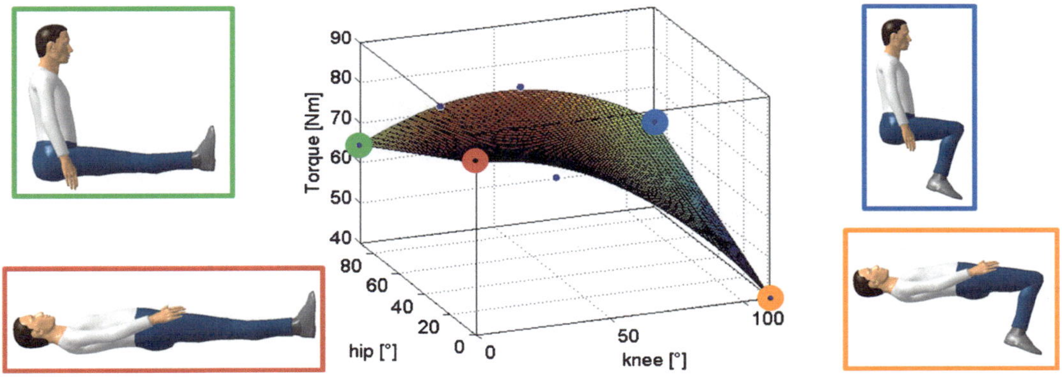

Fig. 4.35 Functional relationship for maximum knee flexion moments as a function of hip and knee flexion from Günzkofer et al. (2012c)

minutes between the experiments and a reasonable limitation of the number of maximum torque measurements of a joint per measuring day (Smidt and Rogers 1982; Kumar 2004; Mital and Kumar 1998; Brown and Weir 2001).

The results of maximum torque measurements with a sufficiently large sample of test persons are then statistically evaluated. For each attitude the descriptive indication of mean value and standard deviation takes place. Usually, linear, multiple regressions are used to model the relationships between joint torque and joint angles. In addition, inferential statistical methods such as e.g. variance analyses (ANOVA) can be used to check significant differences in force due to gender, age or posture (see ▶ Chap. 12).

The relationship between knee flexion torque and knee flexion and hip flexion angle serves as an example for a three-dimensional result plot (Günzkofer et al. 2011a). The graph shows that the highest knee flexion torque is achieved with the leg and hip stretched (Fig. 4.35). The torque decreases with increasing hip and knee angle as shown in the graph.

An example of a two-dimensional diagram is the representation of maximum elbow flexion torques for different shoulder and elbow flexion angles (Günzkofer et al. 2012b). It shows that the highest elbow flexion force can be applied with the arm positioned at right angles and the upper arm hanging down (Fig. 4.36). If the shoulder angle is increased, the maximum elbow flexion torque decreases with a constant elbow flexion angle. Furthermore, the increasing behavior up to the maximum at 90° elbow flexion is clearly shown, followed by a decrease towards more bent postures.

Up to this point, only maximum torques in the respective main directions were considered. Figure 4.37 illustrates how maximum flexion torques are measured in a plane of different flexion angles as well as maximum supination/pronation torques for different forearm rotations.

From typical force measurements only the respective maximum torque values along the main axes are obtained. It is obvious, however, that the required joint torques resulting from external loads can be directed in any direction. Thus, for force modelling, the description of maximum joint torques must also be taken into account in intermediate directions. Due to the multifunctionality of different muscles, e.g. biceps as flexor and supinator, a pure vector addition as shown in Fig. 4.38 cannot be assumed.

Günzkofer et al. (2012c) have investigated this phenomenon within the framework of test subjects and have come to the conclusion that the relationship in the two-dimensional

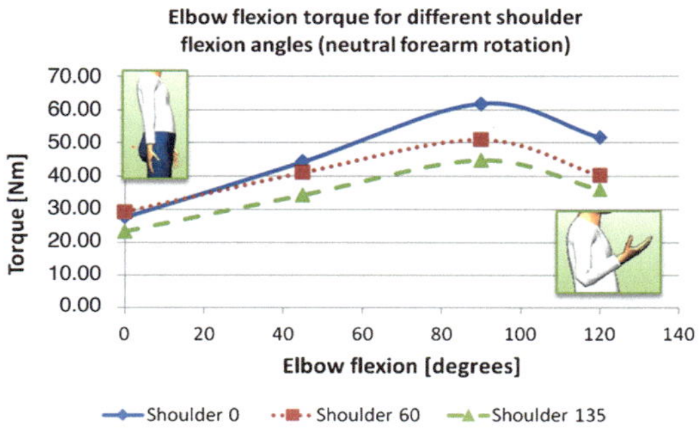

Fig. 4.36 Maximum elbow flexion moment as a function of shoulder and elbow flexion according to Günzkofer et al. (2012b)

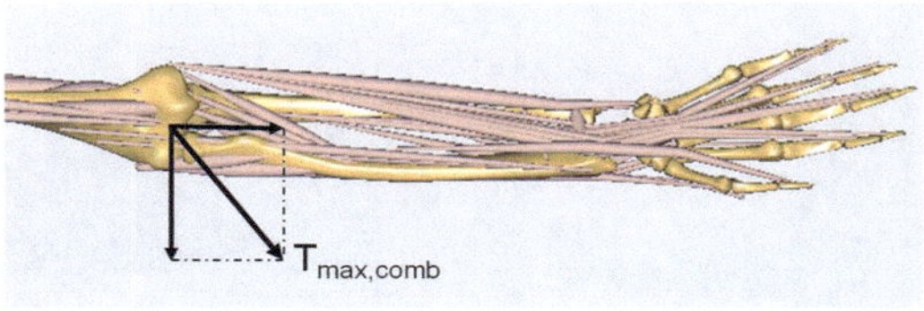

Fig. 4.38 Sketch of a required joint torque in an intermediate direction with the question, how this can be derived from the joint torques in main directions (The arm model corresponds to the digital human model AnyBody, AnyBody Technology, Aalborg, Denmark)

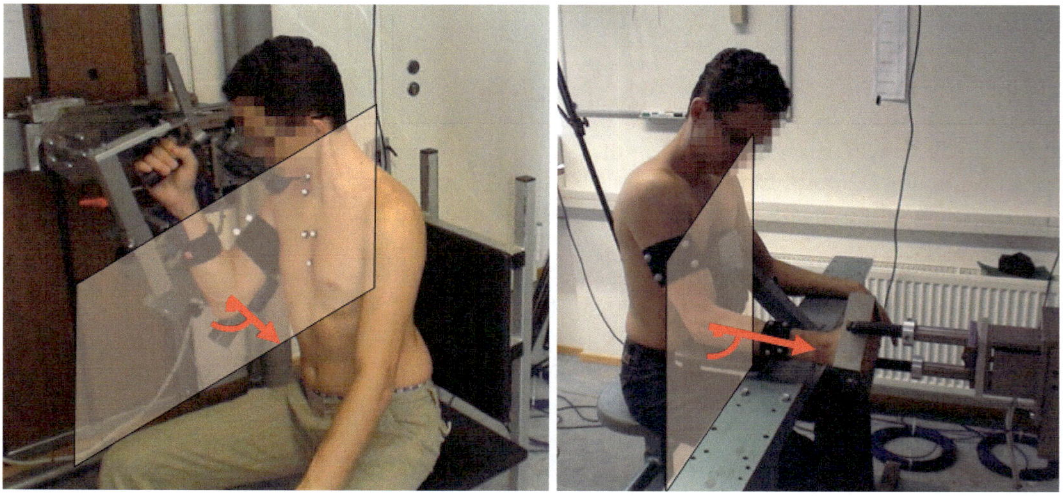

Fig. 4.37 Representation of conventional force measurements in which measurements are only taken in main directions

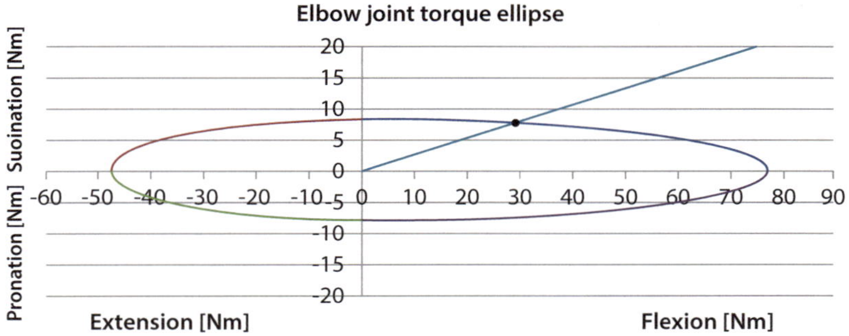

Fig. 4.39 Exemplary illustration of maximum elbow joint moments in any direction based on the elliptical approach

Table 4.8 Maximum forces on control valves (according to Schmidtke and Rühmann 1989d, e, B-4.3.1 and B-4.3.3)

			Maximum forces [N]	
Control lever (central)	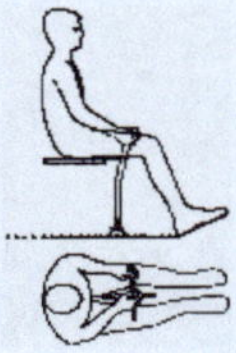		Women	Men
		Forward (pressure)	242.0 ± 76.2	535.8 ± 153.5
		Reverse (train)	284.9 ± 58.9	470.4 ± 76.2
		Right (abduction)	206.5 ± 61.2	363.3 ± 98.5
		Left adduction	164.8 ± 51.8	335.7 ± 93.2
Steering horn	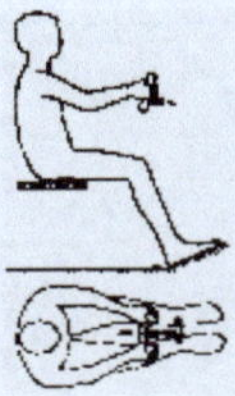	Forward (pressure)	878.3 ± 245.1	1623.4 ± 270.5
		Reverse (train)	485.2 ± 118.3	929.9 ± 217.3
		Clockwise rotation	121.7 ± 30.2	239.4 ± 44.6
		Left turn	134.3 ± 39.4	254.6 ± 48.6

case (e.g.Elbow joint) can best be modelled by a sectional ellipse (Fig. 4.39). The exact way of modelling and application can be found in Günzkofer (2013).

4.2.3.3 Force Prediction Methods

The different approaches to force measurements have led to corresponding prediction methods, with the help of which it is possible to estimate the maximum possible action forces. A distinction is therefore made between force atlases for task-specific data, joint-specific data and musculo-skeletal force modelling. In this context, however, it should be noted that many processes are about estimating forces for production processes. The transfer to the application in the vehicle is therefore only possible to a limited extent and with appropriate expert knowledge. In order to remedy this disadvantage, work has been going on for years to prepare or create corresponding force models for digital human models.

- **Power Atlases**

As already mentioned, the values measured by Rühmann (1992) are published in the Handbuch der Ergonomie (Schmidtke 1993, 2013). Tables 4.8, 4.9, and 4.10 show excerpts from these tables for applications

Table 4.9 Maximum torque on star knobs and rotary knobs as well as forces on push buttons (according to Schmidtke and Rühmann 1989a, c, B-4.36 and B-4.4.2)

		Maximum torque [Nm]	
		Women	Men
Star knobs (vertical)	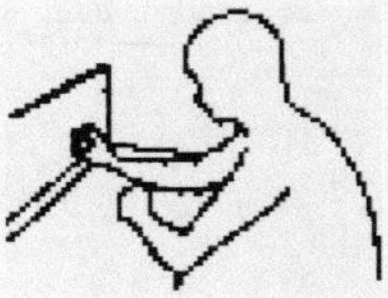	1.7 ± 0.53	2.9 ± 0.98
Rotary knobs (vertical)		1.5 ± 0.53	2.8 ± 1.06
Pushbutton (horizontal) index finger		58.8 ± 14.0	98.4 ± 28.6
Thumbs		66.4 ± 18.1	145.2 ± 33.1
		Maximum forces [N]	
Push button (vertical)		78.8 ± 20.4	156.5 ± 28.9

Table 4.10 Maximum forces on foot pedals at a knee angle of 120° (according to Schmidtke and Rühmann 1989b, B-4.5.1)

		Maximum forces [N]	
		Women	**Men**
Pedal (30°) pivot point at rear end of pedal	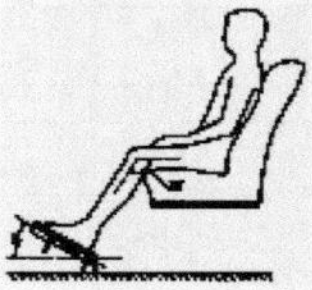	642.2 ± 196.7	972.4 ± 228.6
Pressure pedal (30°) without rotational axis	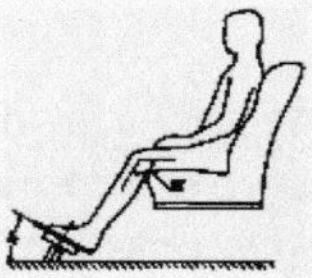	578.2 ± 156.2	855.8 ± 213.1
Pedal (30°) pivot point at the front end of the pedal	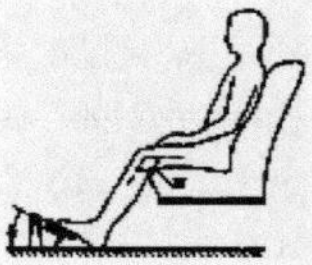	328,7 ± 104,7	457,8 ± 128,1
Horizontal pedal pivot point at front end of pedal	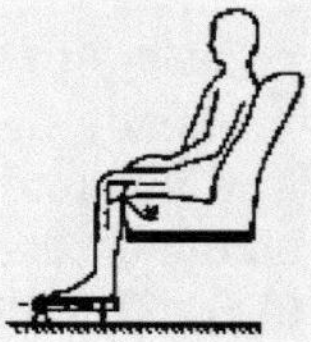	316,8 ± 89,7	513,9 ± 119,2

that have a certain relation to motor vehicles. The tables also show the standard deviation. According to Eq. 4.1, any percentile can be calculated from this.

Further information on maximum possible forces for different force directions can be found in DIN EN 1005; Rohmert et al. (1994) and Wakula et al. (2009). In particular, DIN 33411, Part 5 contains information on maximum static action forces in percentile form. When dimensioning action forces, the designer should therefore use the lower percentile values (5th to 15th percentile) as a basis in order to to enable also weaker people to operate it. When it comes to the dimensioning of the component, orientation to the upper percentile values is necessary, whereby even safety margins (usually 10–20%) have to be added here (Schlick et al. 2010).

Joint Specific Procedures

Especially with the aim of a modelling approach that can be applied in connection with digital human models, a large number of studies on joint specific prediction methods have been carried out and partly published. The method developed by Burandt and Schultetus (1978) for the estimation of permissible forces and torques of the hand-arm, hand-finger system and the legs can be counted among the joint-specific methods, although it was not originally planned for application in digital human models.[22] The method is based on maximum force measurements. As explained in Schlick et al. (2010), at that time it was not yet customary to present forces in percentile form, so that the published values must be based on mean values. In principle, the procedure consists in extracting the corresponding reference force f from the values contained in the tables, which is modified for sitting or standing activity depending on the position to the body height, the distance from the body and the direction of the force. This force is modified as a function of various influencing variables (age and gender, training, type of stress). In order to obtain rough reference values (e.g. for maximum actuating forces of different levers and for the actuation of buttons in different spatial directions), this procedure can also be used with some restrictions for the product design of a motor vehicle, although its scientific justification is partly doubted (Schlick et al. 2010). In the opinion of its authors Burandt and Schultetus, however, the procedure has proven itself in practice to a large extent. It was implemented in the Siemens PLM software system for factory planning.

Due to the complex handling, joint-specific procedures are actually only realized in digital human models. For the human model RAMSIS, spatially oriented maximum torque prediction spheres were developed on the basis of 29 young female test subjects by Schaefer et al. (2000), which are called "torque potato" because of their shape (Fig. 4.40). Synthetic distributions allow a force prediction for predefined compositions of gender and age groups (Schaefer et al. 1997).

Michigan University began developing the 3D Static Strength Prediction Program in the 1970s, the latest version of which was released as 3DSSPP 7.0.6 in July 2011 (The University of Michigan, 2011; Fig. 4.41). As with the system implemented in RAMSIS, the torques to be applied to the joints are calculated using inverse kinematics on the basis of the force to

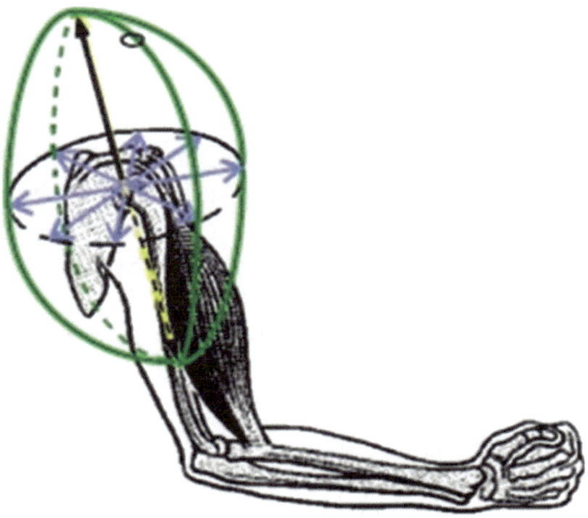

Fig. 4.40 Possible joint torques visualized as potatoes (Schwarz 1997)

22 The process was developed in the Siemens ergonomics laboratories. Later it was adopted with modifications by REFA and the VDI.

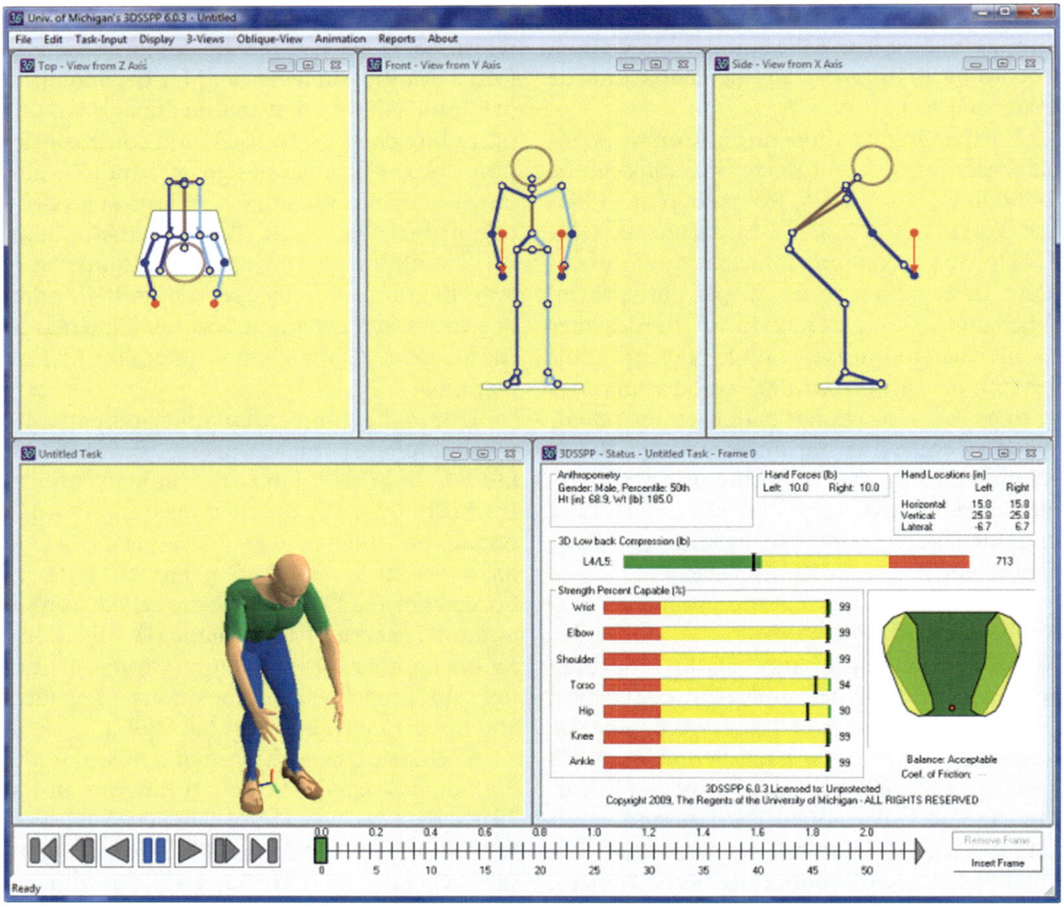

Fig. 4.41 Main program window of 3DSSPP. (© 1990 The Regents of the University of Michigan, ▶ http://umich.edu/~ioe/3DSSPP/index.html)

be applied at a certain posture and compared with the maximum torques. These maximum torques are based on values obtained in over 2000 subjects in various studies (Schanne 1972; Clarke 1966; Burgraaff 1972; Chaffin 2001). Compared to RAMSIS, however, 3DSSPP does not consider spatial torque potatoes, but uses the system only for torque vectors perpendicular to the sagittal plane. 3DSSPP is the basis for the Jack Static Strengh Prediction tool (JSSP) and for the Ford Ergonomic Static Prediction Solver (FSSPS; (Chiang et al. 2006; Kajaks et al. 2011).

Scaling Techniques

The methods discussed so far are based on the assumption that detailed force measurements are carried out on a large number of test persons in order to obtain knowledge about the distribution of forces. The effort involved is immense, which means, among other things, that the experiments are distributed among different institutions and are sometimes difficult to bring into congruence with one another. A fundamentally different approach is based on the fact that there is a high correlation between muscle cross section and muscle strength. From the knowledge of the body part volumes, which are available via the human models, it is therefore necessary to scale to the respective muscle strength. This procedure is used in the biomechanically oriented human model AnyBody, which also has a detailed modelling of individual muscles (see also ▶ Sect. 5.2.2.3).

At the *University of Maastricht* in cooperation with the *Research Department of Ford*

(Aachen) an approach has been developed that allows the scaling of muscle forces in the arm and thigh depending on age, sex and fat content. Different scaling approaches were tested, mass alone, mass and percentage of fat, multiple regression and the so-called cumulative approximation (a variant of the application of an artificial neural network). The last two mentioned methods lead to the best results.

D'Souza (2014) notes that the original scaling algorithms in Anybody lack the influence of age and gender as well as the addressing of different functional muscle groups. Therefore she deals specifically with this aspect and limits her research also for a better measurability of the muscle volumes - to the modelling of the elbow and knee muscles. The trials were conducted with age groups 50–59, 60–69 and 70–79 with 100 subjects each (equally male and female) in each group. For length and mass, there was a significant difference between men and women in both upper arm and lower leg, with the highest age group showing lower values in both cases. The maximum torques at the elbow joint as well as at the knee joint decrease with increasing age, whereby the decrease is stronger in men than in women. However, there are considerable individual variations with regard to these values. A factor analysis shows gender and body element length (50%) as the main influencing factors for both elbow and knee joint forces, followed by body mass (16%) and age (12%). Using multiple regression, equations were developed that allow a prediction of elbow torque or knee joint torque. The application of the found correlations in the modelling of AnyBody brings an improvement of the prediction of the torques around approximately 20% in relation to the procedure implemented so far. In accordance with the results otherwise described in the literature, women show only about 50% of the corresponding forces of men. However, the decrease in strength over age is stronger in men than in women. The knee forces are more affected than the elbow forces. Also the prediction of the forces with the main influencing factors gender, body mass and age is in very good agreement with literature data. The influence of body mass is discussed critically, because for various reasons the present study assumes a fixed density value, regardless of the fact that fat has a lower density than muscles. The assumption of linearity in multiple regression is also critically discussed, but assumed to be acceptable for the target age range of 50–80 years.[23]

4.2.4 Mobility

Heiner Bubb and Fabian Günzkofer

4.2.4.1 Measurement of Movement Spaces

Similar to the categorization of forces, one can also differentiate between task-specific movement spaces and joint-specific movement spaces when it comes to mobility. Various standards include *task-specific ranges of motion* for the most diverse tasks (e.g.work consoles, movement spaces on work tables, foot free spaces, etc.). In ◘ Fig. 4.42, the accessibility area according to SAE J 287 and the test facility on which it is based are shown as an example for the automotive sector.

With the joint specific motion ranges one must distinguish between active and passive. In the case of active motion range, the subject uses his own muscles to bring the respective joint into the possible extreme position, while in the case of passive motion range the respective extremity of the test subject is usually clamped in an apparatus and the joint of interest is deflected to the pain threshold by an external person. "Passive mobility is generally greater than active mobility" (Dietrich and Lehnertz 1993, p. 215). There are a large number of scientific studies on joint mobility, but the most varied questions are the reason for the investigations, mostly only selected joints are investigated and therefore applicability to the automotive field is sometimes very difficult. A detailed overview of the various literature references on joint mobility can be found in Amereller (2014).

23 D'Souza provides a detailed overview of force measurements described in the literature and the various modelling approaches.

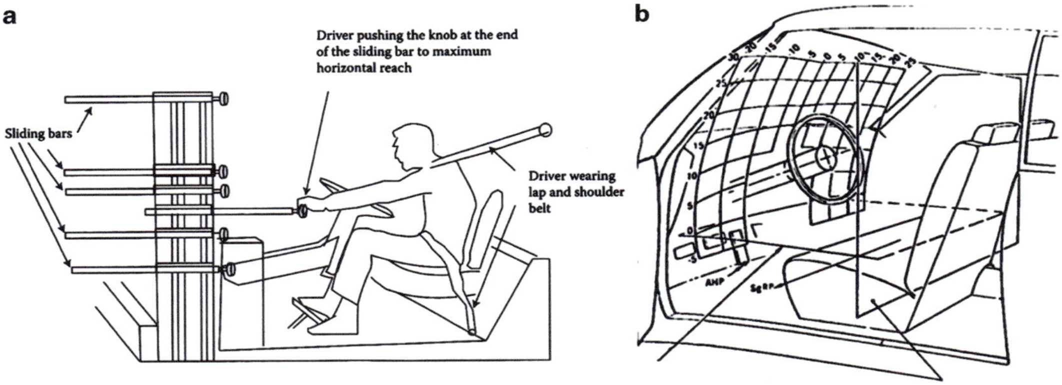

Fig. 4.42 Test setup for measurements of the maximum hand range **a** and accessibility areas **b** according to SAE J 287. (Quoted from Bhise 2012)

4.2.4.2 Passive Joint-Specific Movement Spaces

Relatively complete studies on passive mobility were presented by Damon et al. (1966), Ahlberg et al. (1988), Beissner et al. (2000), and Soucie et al. (2011), whereby Damon's data are also available in percentile form (see Figs. 4.43 and 4.44).

4.2.4.3 Active Joint-Specific Movement Spaces

Of the studies on active mobility, the publication by Kapandji (1985) should be mentioned here in particular. In order to provide reasonably realistic information on mobility, the following sections show ranges of maximum joint angles, which have been taken from relevant studies (Boone and Azen 1979; Youm et al. 1979; Ahlberg et al. 1988; Roach and Miles 1991; Stubbs et al. 1993; Department of Defense 1997; Schwegler 1998; Zatsiorsky 1998; Escalante et al. 1999; Tittel 2003; Doriot and Wang 2006; Hu et al. 2006; Kunsch and Kunsch 2006; Chung and Wang 2009; Kapandji 2009; Tillmann 2010).

Elbow

Figure 4.45 shows the typical movement spaces for the defined degrees of freedom of the elbow.

Shoulder

The typical movement spaces for the shoulder joint are listed in Fig. 4.46. The visualization of the movement directions flexion and extension is already listed in Fig. 4.45.

Hip

Figure 4.47 shows the typical movement spaces for the hip joint.

In contrast to the degrees of freedom listed so far, it is evident that hip flexion is indicated with additional information on knee flexion. This is related to the existence of biarticular musculature, which has been explained in ► Sect. 4.1.2.1. The biarticular musculature on the back of the thigh, which is antagonistic to the hip flexion, changes its length and thus its state of tension depending on hip and knee flexion. For this reason, a much higher counterforce of the antagonists must be overcome for hip flexion when the leg is extended. The table in Fig. 4.47 thus contains two different specifications for maximum hip flexion. Engstler et al. (2011) have developed a formula-based relationship between maximum hip flexion and predominant knee flexion within the framework of an experimental study. It has been shown that on average only 60% of the maximum possible hip flexion is possible with perfect leg extension (see Fig. 4.48).

4.2.4.4 Gender and Age Dependence of Mobility

Buying and using a car today is done equally by men and women. With regard to the development of the age pyramid, it is also obvious

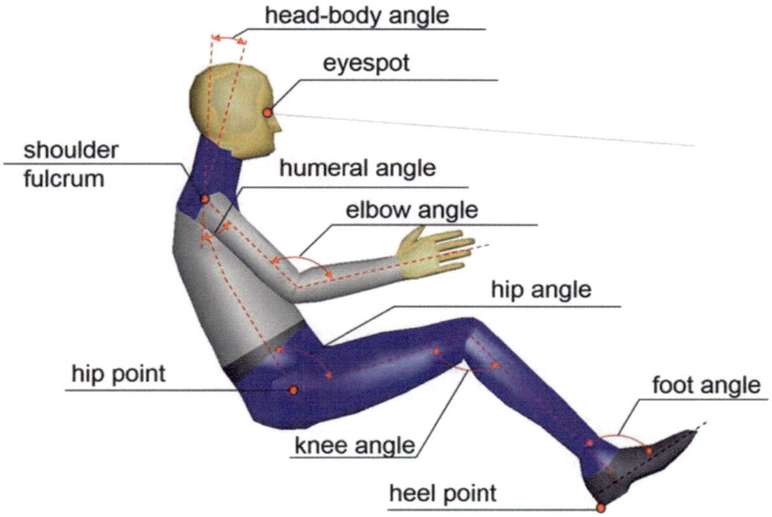

angle	percentile			physiological better area
	5.	50.	95.	
Head-body angle	52° to -60°	67° to -77°	82° to -93°	standing 15° sit 25°
Wrist angle	35° to -12°	47° to -27°	59° to -42°	0°
Elbow angle	54° to 180°	38° to 180°	22° to 180°	115°
Upper arm angle	38° to -168°	61° to -188°	84° to -208°	28°
Upper torso angle	comfortable position 9° ⇒ corresponds bent position			
Lower hull angle	comfortable position 17° ⇒ corresponds bent position			
hip angle	88° to 180°	67° to 180°	46° to 180°	100°
knee angle	88° to 180°	67 to 180°	46° to 180°	110° to 140°
foot angle	67° to 108°	55° to 128°	43° to 148°	90° ± 10°

Fig. 4.43 Passive joint angle ranges according to Damon et al. (1966; compiled by Rühmann 2000)

that the average car buyer is getting older. The average age of new car buyers today is 50.6 years, with 29% older than 60 years (Duttenhöffer 2008). This must therefore be taken into account in the development of vehicles. A special aspect is the consideration of age-related changes in mobility. Amereller (2014) has dealt with this problem in extensive experiments. 346 test subjects aged between 18 and 85 years were measured in cooperation with BMW and the Department of Biomechanics in Sport at the Technical University of Munich with the aid of specially developed equipment, whereby 30 persons were planned for each age group. The rotation of the cervical spine, the mobility of the shoulder in flexion direction, the mobility of the hip abduction and the mobility of the foot pronation are shown as examples (Figs.4.49, 4.50, 4.51, and 4.52). In all cases there is a

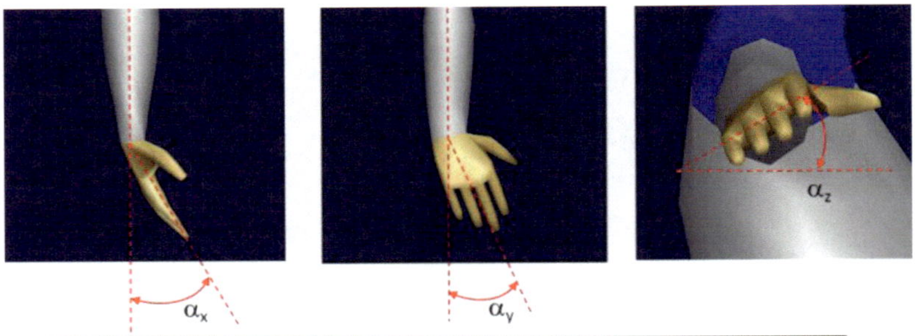

angle	percentile			physiologica better area
	5.	50.	95.	
αx	78° to -70°	99° to -90°	120° to -110°	0°
αy	35° to -12°	47° to -27°	59° to -42°	0°
αz	77to -38°	113° to -77°	149° to -116°	0° 30° with loosened forearm

Fig. 4.44 Passive joint angle ranges of the wrist according to Damon (1966; compiled by Rühmann 2000)

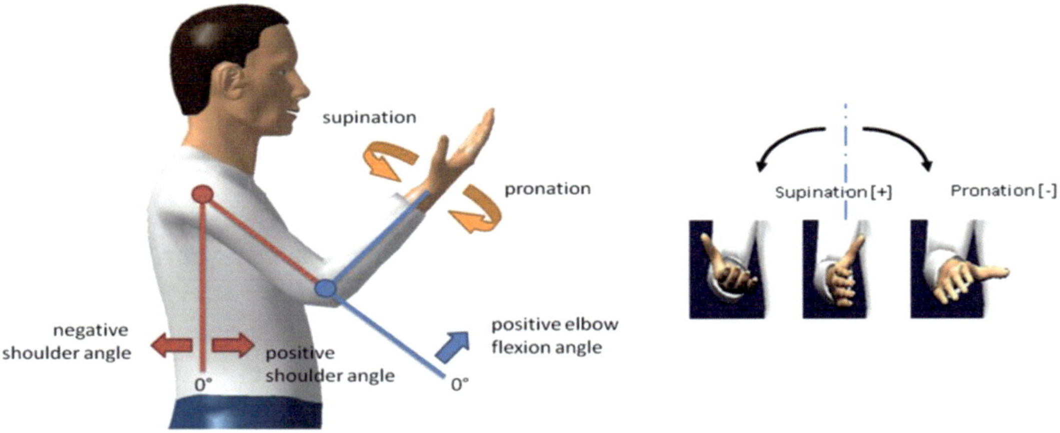

elbow diffraction (flexion)	140° to 159°
elbow extension	Maximum to -10°
subination	80° to 107°
pronation	70° to 105°

Fig. 4.45 Movement directions and movement ranges of the elbow (Günzkofer 2013)

decrease in mobility with age and a difference between men and women (women are usually more mobile). However, this age dependency is very strongly dependent on the joints. The foot pronation shows e.g. almost no dependence on age. The standard deviation shown in the pictures also indicates that persons with high mobility are still comparable with young persons (18–25 years), even at the highest age group, who, however, have low mobility.

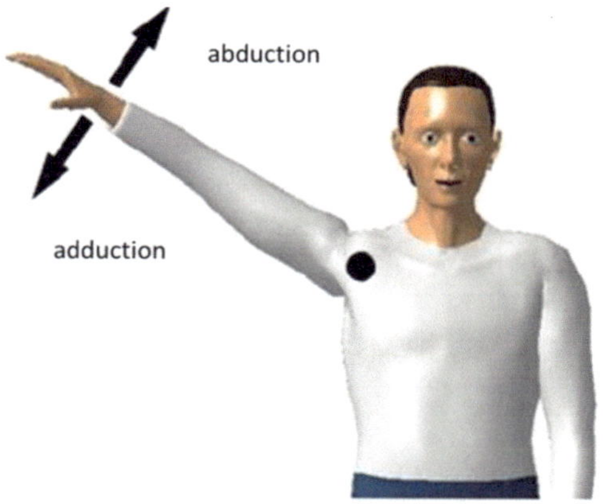

Adduction	20° to 50°
Abduction	125° to 180°
Extension	40° to 70°
Flexion	150° to 190°

Fig. 4.46 Representation of the movement directions of the shoulder (Hartmann 2012)

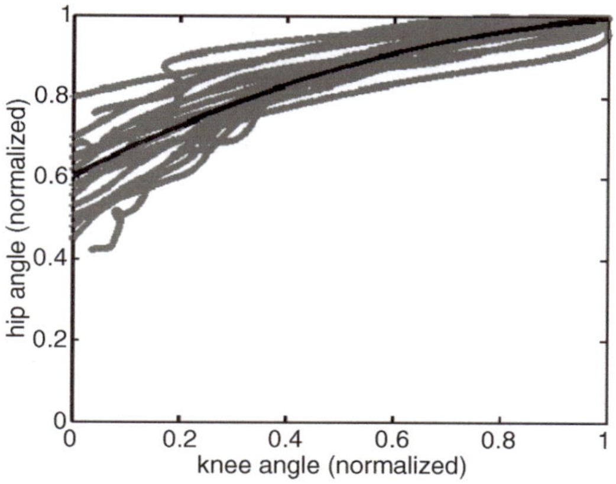

Fig. 4.48 Quantitative dependence between maximum hip flexion and knee flexion. Investigation of 18 male and female volunteers of two age groups (20–35 years; >65 years)

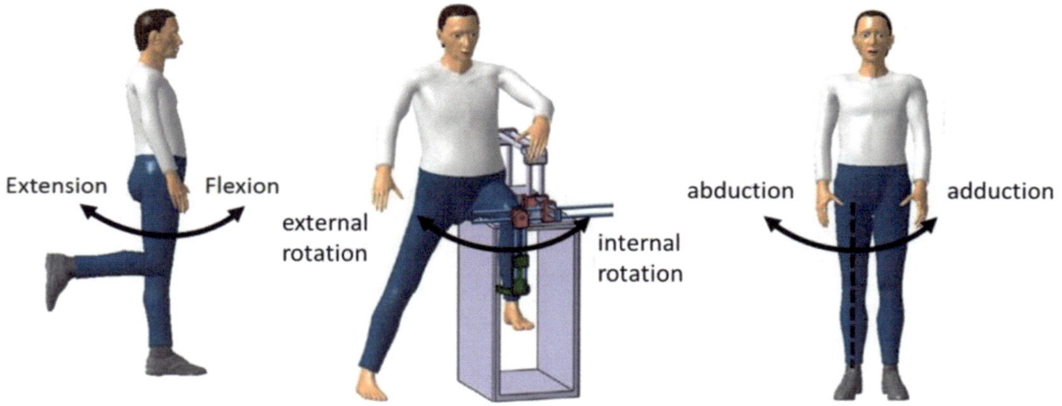

Adduction	20° to 30°
Abduction	45° to 53°
Extension	18° to 43°
Flexion (stretched leg)	66° to 90°
Flexion (bent leg)	110° to 130°
Internal rotation	40° to 50°
External rotation	30° to 45°

Fig. 4.47 Representation of the movement directions of the hip; middle picture from Amereller (2014)

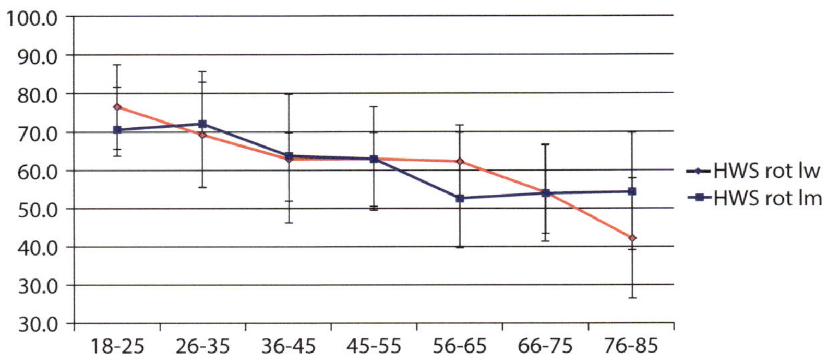

Fig. 4.49 Cervical rotation to the left depending on age and sex (Amereller 2014)

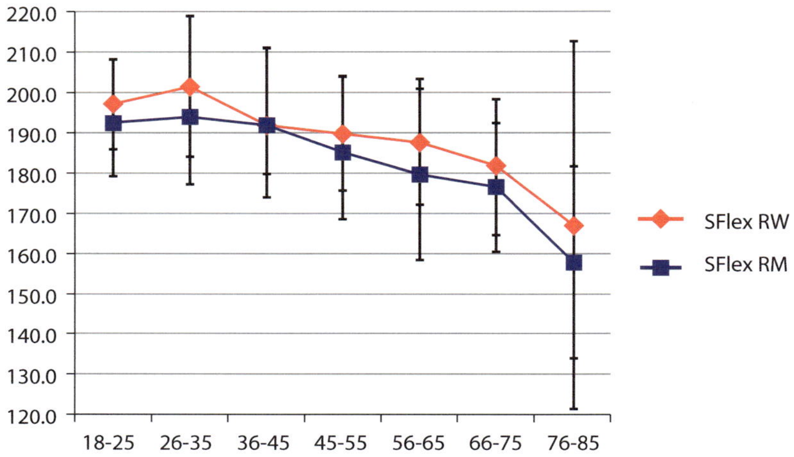

Fig. 4.50 Shoulder flexion to the right depending on age and sex (Amereller 2014)

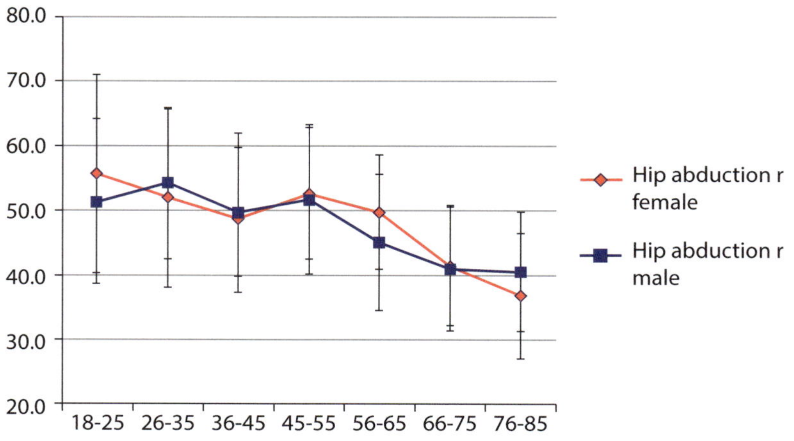

Fig. 4.51 Hip abduction to the right depending on age and sex (Amereller 2014)

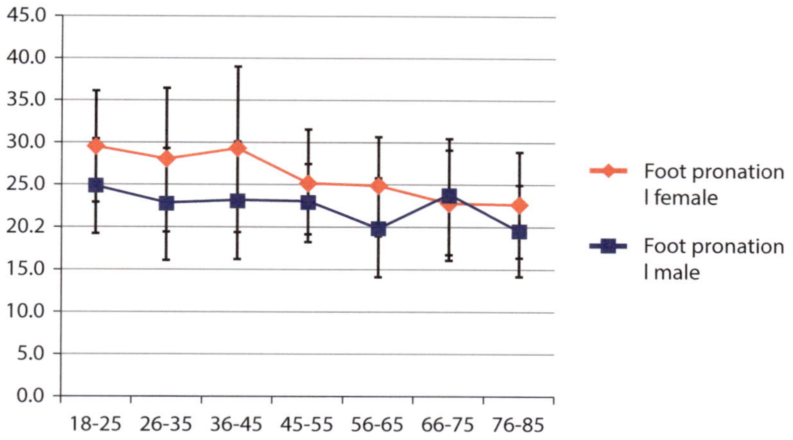

Fig. 4.52 Foot pronation to the left depending on age and sex (Amereller 2014)

Among other things, this is also an indication of the well-known demand for "design for all", in which an age- or gender-dependent design of vehicles is rejected, although the "weakest" must be taken as a guide in all questions (see also ► Sect. 7.8.1).

4.2.4.5 Biomechanics of Visual Movements

For the design of direct and indirect vision from a vehicle (see ► Sect. 7.3), the mobility of the eye in conjunction with cervical spine rotation and shoulder flexion is of crucial importance. The validity of the biomechanical findings is not only limited to large muscles of the primary joints of humans, but also applies to smaller muscles such as the six outer eye muscles (Fig. 4.53).

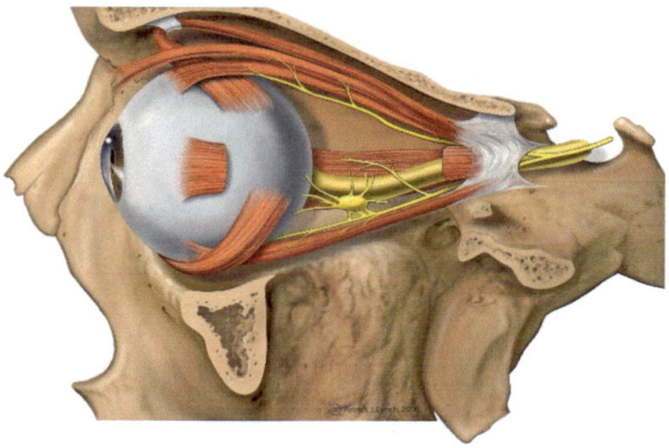

Fig. 4.53 Representation of the eye and the surrounding outer eye muscles (By Lateral_orbit_nerves.jpg: Patrick J. Lynch, medical illustrator derivative work: Anka Friedrich (Lateral_orbit_nerves.jpg) [CC-BY-2.5 (► http://creativecommons.org/licenses/by/2.5)], via Wikimedia Commons)

With increasing distance from the ideal muscle length - in this case the neutral position of the eyes - discomfort increases. According to Schmidtke (1993), neutral position means that the head is inclined by 10°–15° and the eyes by 15°–20° downwards. In the case of an upright standing or seated person, the line of sight is thus inclined downwards by 25°–35°. With a fixed head and fixed eyes (visual field), objects can only be perceived sharply a few degrees around the fixation point. Consequently, for objects outside this range, the eyes (field of vision), then the entire head (field of vision) and finally the entire torso (fixation field) must be moved with increasing distance. An overview of the visual ranges offers Table 4.11.

The angle at which the transition from one visual area to the next occurs with increased physical effort can also depend on the duration of the fixation. If only a brief change of gaze is required in a motor vehicle, only eye movements will take place within the maximum movement range. In order to avoid a static eye posture in the border area, however, the further kinematic chain is also moved after a certain period of a gaze turn. According to Schmidt (1988), head rotations follow from a deviation of approx. 10° from the neutral

Table 4.11 Definition of visual ranges according to Schmidtke (1993) and Hudelmaier (2003)

				Maximum range		Optimal range for bright light stimuli	
	Eye	Head	Torso	Horizontal	Vertical	Horizontal	Vertical
Facial field	fix	fix	fix	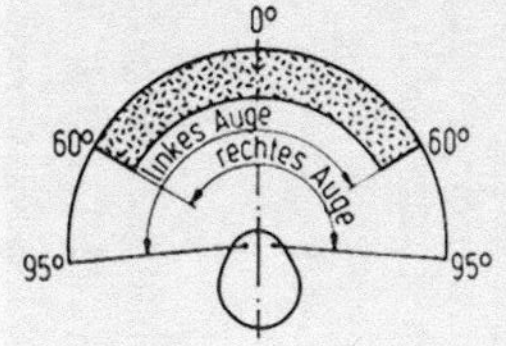	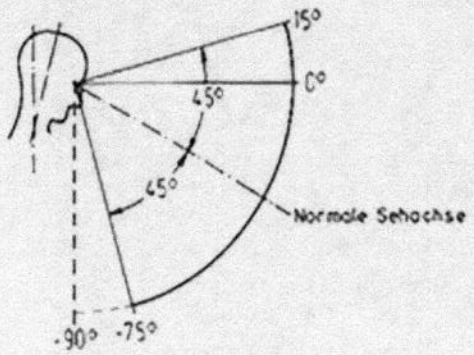	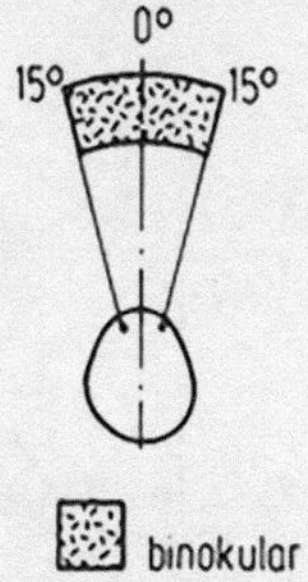	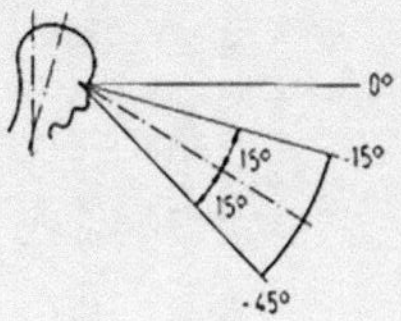
Field of vision	free	fix	fix	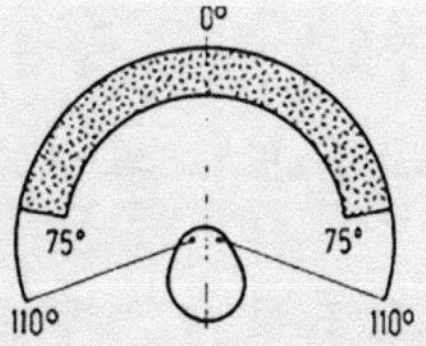	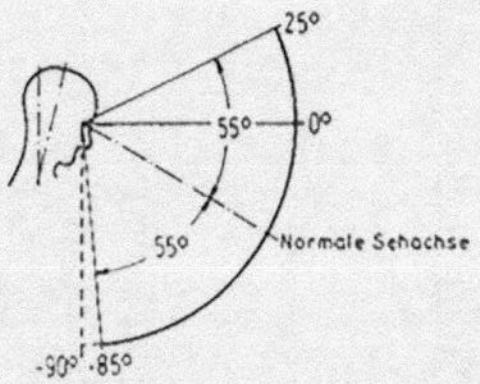	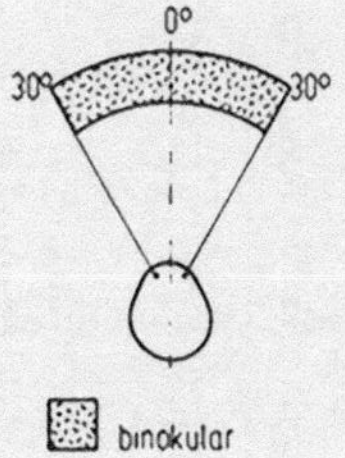	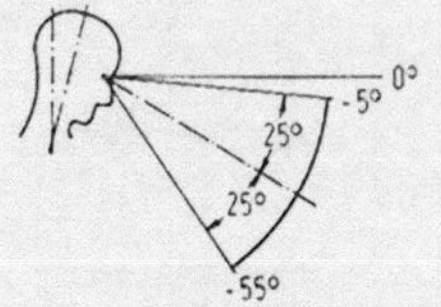
View field of vision	free	free	fix	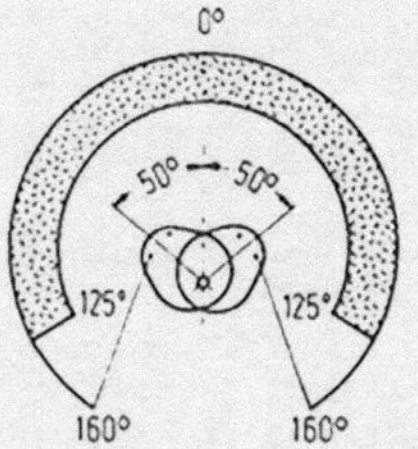	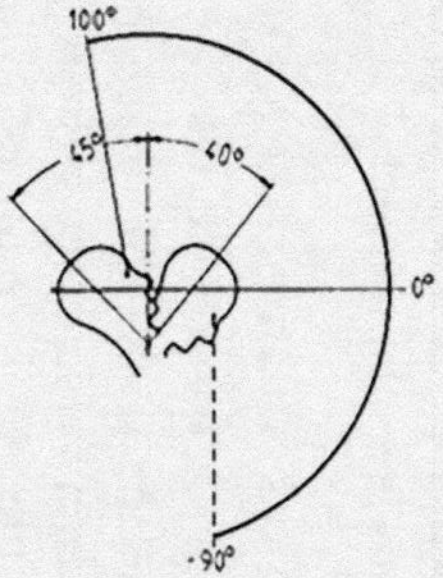	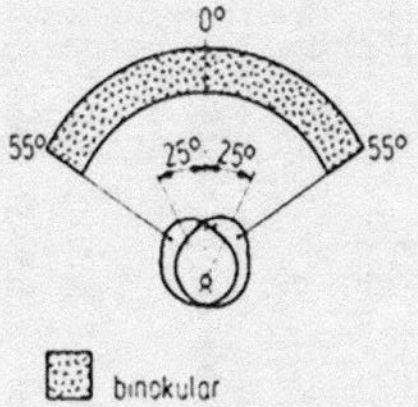	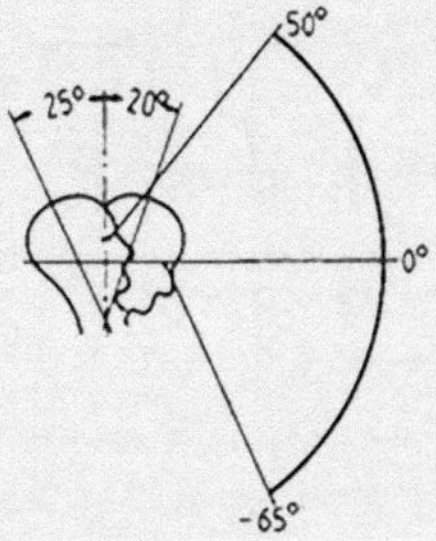

line of sight. In practical terms, this means, for example, that the brief reading of the speed from the speedometer will only be accompanied by eye movements, the setting of a destination in the navigation device with at least superimposed head movements.

These correlations must be taken into account when designing the vehicle. Due to the different speed of eye and head movements, different time requirements arise, depending on the distance of the object to be viewed from the neutral visual beam. These findings are taken into account in the digital human model RAMSIS in the form of isochrones (lines of equal time requirements) in the module "RAMSIS cognitive".

4.2.4.6 Application

An application of maximum movement spaces for vehicle design based on table values no longer corresponds to the state of the art. Similar to body measurements and body forces, a meaningful application is the integration in digital human models. For movement analyses such as entry/exit and posture analyses (gripping shells, shoulder gaze, etc.), the knowledge presented here is implicitly taken into account. A special form of evaluability is the correlation of the percentage utilization of the maximum range of motion and the discomfort sensation. A future potential for digital human models lies in the consideration of biarticular and age-related effects (Engstler et al. 2011; Amereller 2014).

References

Literature Used

Abdel-Malek, K., Yang, J., Marler, T., Beck, S., Mathai, A., Zhou, X.: Towards a new generation of virtual humans. Int. J. Human Factors Model. Simul. **1**(1), 2–39 (2006). http://inderscience.metapress.com/content/ETB04X9U9C863V45

Adams, M.A.: Wie beeinflusst die Sitzposition die Lastverteilung in der Bandscheibe und ihre Ernährung. In: Ergomechanics Interdisziplinärer Kongress Wirbelsäulenforschung, p. 2. Shaker Verlag, Aachen (2006)

Ahlberg, A., Moussa, M., Al-Nahdi, M.: On geographical variations in the normal range of joint motion. Clin. Orthop. Relat. Res. **234**, 229–231 (1988)

Amereller, M.: Die Gelenkbeweglichkeit des Menschen im Altersgang als Fokus wissenschaftlicher Forschung im automobilen Kontext. Entwicklung, Absicherung und Anwendung einer neuen Messmethode. Dissertation, Technische Universität München (2014)

Bassey, E.J., Short, A.H.: A new method for measuring power output in a single leg extension – feasibility, reliability and validity. Eur. J. Appl. Physiol. Occup. Physiol. **60**(5), 385–390 (1990)

Beissner, K.L., Collins, J.E., Holmes, H.: Muscle force and range of motion predictors of function in older adults. Phys. Ther. **80**(6), 556–563 (2000)

Bernstein, N.A.: The Co-ordination and Regulation of Movements. Pergamon Press, Washington D.C (1967). http://books.google.de/books?id=F9dqAAAAMAAJ

Bhise, V.D.: Ergonomics in the Automotive Design Process. CRC Press Taylor & Francis Group, Boca Raton, London, New York (2012)

Bishop, P., Cureton, K., Collins, M.: Sex difference in muscular strength in equally-trained men and women. Ergonomics. **30**(4), 675–687 (1987)

Boone, D.C., Azen, S.P.: Normal range of motion of joints in male subjects. J. Bone Joint Surg. **61**, 756–759 (1979)

Boyd, W.C.: Genetics and the races of man: an introduction to modern physical anthropology (1950)

Brown, L.E., Weir, J.P.: ASEP procedures recommendation 1: accurate assessment of muscular strength and power. J. Exerc. Physiol.. (online). **4**(3), 1–21 (2001)

Burandt, U., Schultetus, W.: Ermittlung zulässiger Grenzwerte für Kräfte und Drehmomente. Siemens firmeninterne Schulungsunterlagen zur Arbeitsgestaltung (1978)

Burgraaff, J. D.: An Isometric Biomechanical Model for Sagittal Plane Leg Extension: Unpublished MS thesis (1972)

Campbell, N.A., Reece, J.B.: Biology, 7th edn. Pearson, Benjamin Cummings, San Francisco (2005)

Chaffin, D.B.: Digital Human Modeling for Vehicle and Workplace Design. Society of Automotive Engineers, Warrendale (2001)

Chaffin, D.B., Andersson, G.B.J., Martin, B.J.: Occupational Biomechanics, 3rd edn. Wiley-Blackwell, New York (1999)

Chiang, J., Stephens, A., Potvin, J.: Retooling Jack's Static Strength Prediction Tool. SAE International, Warrendale (2006)

Chung, M.-J., Wang, M.-J.J.: The effect of age and gender on joint range of motion of worker population in Taiwan. Int. J. Ind. Ergon. **39**, 596–600 (2009)

Churchill, E., Tebbetts, I.E., McConville, J.T., Laubach, L.L.: Anthropometry for designers Anthropometric source book (1978)

Clarke, H.H.: Muscular Strength and Endurance in Men. Prentice Hall, Englewood Cliffs (1966)

Crowninshield, R.D.: Use of optimization techniques to predict muscle forces. J. Biomech. Eng. **100**, 88–92.

4

Retrieved from http://ci.nii.ac.jp/naid/80014980426/en/ (1978)
D'Souza, S. F.: Physical efficiency as a function of age: development of an age scaling model for biomechanical simulation. Dissertation Technischen Universität München (2014)
D'Souza, S.F., Brueckner, B., Rasmussen, J., Schwirtz, A.: Development of Age and Gender Based Strength Scaled Equations for Use in Simulation Models First International Symposium on Digital Human Modeling, Lyon, France (2011)
Daams, B.J.: Human force exertion in user-product interaction. backgrounds for design. Series in physical ergonomics Bd. 2. Fac. of Industrial Design Engineering, Delft Univ. of Technology, Delft (1994)
Damon, A., Stoudt, H., McFarland, R.: The Human Body in Equipment Design. Harvard University Press, Cambridge (1966)
Damsgaard, M., Rasmussen, J., Christensen, S.T., Surma, E., de Zee, M.: Analysis of musculoskeletal systems in the AnyBody Modeling System. Simul. Model. Pract. Theory. **14**(8), 1100–1111 (2006). https://doi.org/10.1016/j.simpat.2006.09.001
Department of Defense: Military Handbook 759 C Notice 1 (1997)
Dietrich, M.C., Lehnertz, K.: Handbuch Trainingslehre, 2nd edn. Hoffmann, Schondorf (1993)
Dolan, P.: Welchen Einfluss hat die Sitzhaltung auf die Muskelaktivität? In: Ergomechanics 2–2. Interdisziplinärer Kongress Wirbelsäulenforschung. Shaker Verlag, Aachen (2006)
Doriot, N., Wang, X.: Effects of age and gender on maximum voluntary range of motion of the upper body joints. Ergonomics. **49**(3), 269–281 (2006)
Duchateau, J., Klass, M., Baudry, S.: Évolution et adaptations à l'entraînement du système neuromusculaire au cours du vieillissement. Santé, sport et innovations – Bruxelles 27–28 mai 2005 – Séminaire organisé par le CNAPS avec le concours de l'Université libre de Bruxelles. In: Science & Sports, Jg. 21, H. 4, S. 199–203 (2006)
Dul, J., Townsend, M A., Shiavi, R., Johnson, G.E.: Muscular synergism – I. On criteria for load sharing between synergistic muscles. J. Biomech. **17**(9), 663–673 (1984)
Duttenhöffer, F.: Demographische Entwicklung und schrumpfender Automarkt. Jahrbuch der Absatz und Verbrauchsforschung. **54**(1), 81–90 (2008)
Engstler, F.: Perzentilierung maximaler Gelenkmomente des Menschen. Dissertation, Technische Universität München (2012)
Engstler, F., Günzkofer, F., Bubb, H., Bengler, K.: Lower Limb Joint Range of Motion Considering Inter-joint Dependencies. Lyon, Lyon, S. ID2163 (2011). DHM 2011
Escalante, A., Lichtenstein, M.J., Hazuda, H.P.: Determinants of shoulder and elbow flexion range: results from the San Antonio Longitudinal Study of Aging. Arthrtitis Care Res. **12**(4), 277–286 (1999)
Faller, A.: Der Körper des Menschen: Einführung in Bau und Funktion, 13th edn. Georg Thieme Verlag, Stuttgart (1999)
Frontera, W.R., Hughes, V.A., Lutz, K.J., Evans, W.J.: A cross-sectional study of muscle strength and mass in 45-year-old to 78-yr-old men and women. J. Appl. Physiol. **71**(2), 644–650 (1991)
Geuß, H.: Entwicklung eines anthropometrischen Meßverfahrens für das CAD-Menschmodell RAMSIS. Dissertation, Technische Universität München (1994)
Goodpaster, B.H., Carlson, C.L., Visser, M., de Kelley, D.E., Scherzinger, A., Harris, T.B., et al.: Attenuation of skeletal muscle and strength in the elderly: the Health ABC Study. J. Appl. Physiol. **90**(6), 2157–2165 (2001)
Greil, H.: Der Körperbau von Männern und Frauen aus der Deutschen Demokratischen Republik: Konstitutionsbiologische Ergebnisse einer 1967 für die Bekleidungsindustrie durchgeführten Querschnituntersuchung, Dissertation, Humboldt-Universität Berlin (1972)
Grimm, H.: Grundriss der Konstitutionsbiologie und Anthropologie, 3rd edn. VEB Verlag Volk und Gesundheit, Berlin (1966)
Günzkofer, F., Bubb, H., Bengler, K.: Maximum elbow joint torques for digital human models. Int. J. Human Factors Model. Simul. **3**(2), 109–132 (2012a)
Günzkofer, F., Bubb, H., Bengler, K.: Elbow torque ellipses: investigation of the mutual influences of rotation, flexion, and extension torques. Work J. Prevent. Assessment Rehabil. **41**(Supplement 1), 2260–2267 (2012b)
Günzkofer, F., Bubb, H., Senner, V., Bengler, K.: Dependency of knee extension torque on different types of stabilisation. Int. J. Human Factors Model. Simul. **3**(1), 1–15 (2012c)
Günzkofer, F., Engstler, F., Bubb, H., Bengler, K.: Joint Torque Modeling of Knee Extension and Flexion. In: Proceedings of the 14th International Conference on Human-Computer Interaction Orlando, Florida, USA (2011a)
Günzkofer, F.: Elbow Strength Modelling for Digital Human Models. Dissertation, Technische Universität München (2013)
Hafez, H.A., Gidcumb, C.F., Reeder, M.J., Beshir, M.Y., Ayoub, M.M.: Development of a human atlas of strengths. In: Proceedings of the Human Factors Society 26th Annual Meeting Santa Monica, CA. S. 575–579. The Human Factors Society, Kongress-Proceedings (1982)
Hahn, D.: Biomechanische Analyse der Kraft-Längen- und Kraft-Geschwindigkeits-Relation bei der mehrgelenkigen Beinstreckung in vivo zur Beurteilung der muskulären Funktion, 1st edn. Sierke, Göttingen (2008)
Hartmann, F.: Versuchsplanung und Auswertung isometrischer Maximalkraftmessungen des Schultergelenks: Studienarbeit am Lehrstuhl für Ergonomie der Technischen Universität München (2012)

Herzog, W.: Determinants of muscle strength. In: MUSCLE STRENGTH, S. 45–82 (2004)
Hettinger, T.: Isometrisches Muskeltraining. Georg Thieme Verlag, Stuttgart (1968)
Hill, A.V.: The heat of shortening and the dynamic constants of muscle. Proc. R. Soc. B Biol. Sci. **126**(843), 136–195 (1938). https://doi.org/10.1098/rspb.1938.0050
Hosler, W.W., Morrow, J.R.: Arm and leg strength compared between young women and men after following for differences in body size and composition. Ergonomics. **25**(4), 309–313 (1982)
Hu, H., Li, Z., Yan, J., Wang, X., Xiao, H., Duan, J., et al.: Measurement of voluntary joint range of motion of Chinese elderly living in Beijing area by a photographic method. Int. J. Ind. Ergon. **36**, 861–867 (2006)
Hudelmaier, J.: Sichtanalyse im Pkw unter Berücksichtigung von Bewegungen und individuellen Körpercharakteristika. Dissertationsschrift an der Technischen Universität München (2003)
Hurley, B.F.: Age, gender, and muscular strength. J. Gerontol. **50**, 41–44 (1995)
Jongen, H.A., van der Denier Gon, J.J., Gielen, C.C.: Activation of human arm muscles during flexion/extension and supination/pronation tasks: a theory on muscle coordination. Biol. Cybern. **61**(1), 1–9 (1989)
Jürgens, H.W., Habicht-Benthin, B., Lengsfeld, W.: Körpermaße 20jähriger Männer als Grundlage für die Gestaltung von Arbeitsgerät, Ausrüstung und Arbeitsplatz. Bundesministerium für Verteidigung, Bonn (1971)
Kaiser, R.: Anthropometrie – Neue Erkenntnisse und Anwendungsbeispiele RAMSIS User Conference, Kaiserslautern (2011)
Kajaks, T., Stephens, A., Potvin, J.R.: The effect of manikin anthropometrics and posturing guidelines on proactive ergonomic assessments using digital human models. Int. J. Human Factors Model. Simul.2(3), 236 (2011). https://doi.org/10.1504/IJHFMS.2011.044512
Kapandji, I.A.: Funktionelle Anatomie der Gelenke. In: Otte, P., Schlegel, K.F. (eds.) Bücherei der Orthopäden, Bd. 47. Untere Extremität. Enke, Stuttgart (1985)
Kapandji, I.A.: Funktionelle Anatomie der Gelenke. Obere Extremität – Untere Extremität – Rumpf und Wirbelsäule: Schematisierte und kommentierte Zeichnungen zur menschlichen Biomechanik, 5th edn. Thieme, Stuttgart (2009)
Knudson, D.: Fundamentals of Biomechanics, 2nd edn. Springer, New York (2007). http://www.gbv.de/dms/hbz/toc/ht014955463.pdf
Kroemer, K.H.E.: Die Messung der Muskelstärke des Menschen: Methoden und Techniken Forschungsbericht/Bundesanstalt für Arbeitsschutz und Unfallforschung, Bd. 161. Wirtschaftsverl, NW, Bremerhaven (1977)
Kulig, K., Andrews, J.G., Hay, J.G.: Human strength curves. In: Terjung, R.L. (ed.) Exercise and Sport Science Reviews, pp. 417–466. The Collamore Press, Lexington, Massachusetts (1984). 12
Kumar, S.: Muscle Strength. CRC Press, Boca Raton (2004)
Kunsch, K., Kunsch, S.: Der Mensch in Zahlen. Spektrum Akademischer Verlag (2006)
Lee, K., Bruckner, A.: Human lifting strength in different postures. J. Safety Res. **22**(1), 11–19 (1991)
MacKinnon, S.N.: Isometric pull forces in the sagittal plane. Appl. Ergon. **29**(5), 319–324 (1998)
Martin, R.: Lehrbuch der Anthropologie, 3rd edn, pp. 1956–1958. Karl Saller, Fischer, Stuttgart (1914)
Mital, A., Kumar, S.: Human muscle strength definitions, measurement, and usage: part I - guidelines for the practitioner. Int. J. Ind. Ergon. **22**(1–2), 101–121 (1998)
Morgan, C.T., Cook, J.S., Chapanis, A., Lund, M.W. (eds.): Human Engineering Guide to Equipment Design. McGraw-Hill, New York (1963)
Pope, M.H.: Wasser in der Bandscheibe: Neuer Messansatz mittels Kernspintomographie. In: Ergomechanics 2–2. Interdisziplinärer Kongress Wirbelsäulenforschung. Shaker Verlag, Aachen (2006)
Rasmussen, J., Damsgaard, M., Christensen, S., Surma, E.: Design optimization with respect to ergonomic properties. Struct. Multidiscip. Optim. **24**(2), 89–97 (2002). https://doi.org/10.1007/s00158-002-0219-x
Rasmussen, J., Damsgaard, M., Voigt, M.: Muscle recruitment by the min/max criterion – a comparative numerical study. J. Biomech. **34**(3), 409–415 (2001). https://doi.org/10.1016/S0021-9290(00)00191-3
Remlinger W. Analyse von Sichteinschränkungen im Fahrzeug. Dissertation Technischen Universität München (2013)
Roach, K.E., Miles, T.P.: Normal hip and knee active range of motion: the relationship to age. Phys. Ther. **71**(9), 656–665 (1991)
Rohmert, W., Berg, K., Bruder, R., Schaub, K.: Kräfteatlas Teil 1: Datenauswertung statischer Aktionskräfte, Teil 2: Literaturzusammenstellung Schriftenreihe der Bundesanstalt für Arbeitsmedizin. Forschung Fb 09.004. Verlag für neue Wissenschaft GmbH, Wirtschaftsverlag, Bremen (1994)
Rohmert, W., Hettinger, T.: Körperkräfte im Bewegungsraum. Beuth Vertrieb GmbH, Berlin (1963)
Rühmann, H.: Körperkräfte des Menschen. Dokumentation Arbeitswissenschaft Bd. 31. O. Schmidt, Köln (1992)
Rühmann, H.: Vorlesungsskriptum zu Produktionsergonomie. Lehrstuhl für Ergonomie, Technische Universität München, Vorlesungsskriptum im Internet verfügbar! (2000)
Samuel, D., Rowe, P.J.: Effect of ageing on isometric strength through joint range at knee and hip joints in three age groups of older adults. Gerontology. **55**(6), 621–629 (2009)
Schanne, F.: A Three Dimensional Hand Force Capability Force Model for the Seated Operator: Unpublished doctoral dissertation (1972)
Schaefer, P., Rudolph, H., Schwarz, W.: Digital Man Models and Physical Strength – A New Approach

in Strength Simulation Proceedings of the SAE DHMC (Digital Human Modeling Conference). SAE 2000-01-2168 (2000)
Schaefer, P., Rudolph, H., Schwarz, W.: Variable Force Limits for Optional Target Populations - a New Approach realized in CEN - Standardization. In: Proceedings of the 13th IEA Conference Tampere. S. 533–535 (1997)
Schantz, P., Randall-Fox, E., Hutchison, W., Tyden, A., Astrand, P.O.: Muscle fibre type distribution, muscle cross-sectional area and maximal voluntary strength in humans. Acta Physiol. Scand. **117**(2), 219–226 (1983)
Schlick, C., Bruder, H., Luczak, H.: Arbeitswissenschaft. Kap. 3.2.9, Heidelberg, Dordrecht, London, New York (2010)
Schmidt, R. A.: Motor Control and Leaming. A behavioral emphasis, Champaign (1988)
Schmidtke, H.: Ergonomie, 3rd edn. Carl Hanser Verlag, München (1993)
Schmidtke, H., Rühmann, H.: Kräfte an Drucktastern B-4.3.2. In: Handbuch der Ergonomie (HdE) mit ergonomischen Konstruktionsrichtlinien. Bundesamt für Wehrtechnik und Beschaffung (Hrsg.), Koblenz. Karl Hanser Verlag, München, Wien (1989a)
Schmidtke, H., Rühmann, H.: Kräfte an Fußstellteilen B-4.5.1. In: Handbuch der Ergonomie (HdE) mit ergonomischen Konstruktionsrichtlinien. Bundesamt für Wehrtechnik und Beschaffung (Hrsg.), Koblenz. Karl Hanser Verlag, München, Wien (1989b)
Schmidtke, H., Rühmann, H.: Kräfte an Stern- und Drehknöpfen B-4.3.6. In: Handbuch der Ergonomie (HdE) mit ergonomischen Konstruktionsrichtlinien. Bundesamt für Wehrtechnik und Beschaffung (Hrsg.), Koblenz. Karl Hanser Verlag, München, Wien (1989c)
Schmidtke, H., Rühmann, H.: Kräfte an Steuerhebeln B-4.3.1. In: Handbuch der Ergonomie (HdE) mit ergonomischen Konstruktionsrichtlinien. Bundesamt für Wehrtechnik und Beschaffung (Hrsg.), Koblenz. Karl Hanser Verlag, München, Wien (1989d)
Schmidtke, H., Rühmann, H.: Kräfte an Steuerhörnern B-4.3.3. In: Handbuch der Ergonomie (HdE) mit ergonomischen Konstruktionsrichtlinien. Herausgegeben vom Bundesamt für Wehrtechnik und Beschaffung, Koblenz. Karl Hanser Verlag, München, Wien (1989e)
Schmidtke, H.: Wissenschaftliche Bearbeitung: Handbuch der Ergonomie (HdE) mit ergonomischen Konstruktionsrichtlinien. Bundesamt für Wehrtechnik und Beschaffung (Hrsg.), Koblenz. Karl Hanser Verlag, München, Wien (2013)
Schwarz, W.: 3D-Video-Belastungsanalyse: Ein neuer Ansatz zur Kraft- und Haltungsanalyse Fortschritt-Berichte VDI-Reihe 17, Biotechnik/Medizintechnik, Bd. 166. VDI-Verl., Düsseldorf (1997). Als Ms. gedr.
Schwegler, J.S.: Der Mensch – Anatomie und Physiologie, 2nd edn. Thieme, Stuttgart (1998)
Seidl, A., Trieb, R., Wirsching, H.-J.: SizeGERMANY - the new German Anthropometric Survey: Conceptual Design, Implementation and Results. In: IEA 17th World Congress on Ergonomics (2009)
Seitz, T., Recluta, D., Zimmermann, D., Wirsching, H.-J.: FOCOPP - An approach for a human posture prediction model using internal/external forces and discomfort Proceedings of the SAE DHMC (Digital Human Modeling Conference), Bd. SAE 2005-01-2694 (2005)
Sheldon, W.H.: Atlas of Men, A Guide and Handbook of Somatotyping. Harper, New York (1954)
Sherwood, L.: Human Physiology: from Cells to Systems, 8th edn. Brooks/Cole, Cengage Learning, Belmont (2012)
Smidt, G.L., Rogers, M.W.: Factors contributing to the regulation and clinical assessment of muscular strength. Phys. Ther. **62**(9), 1283–1290 (1982)
Soucie, J.M., Wang, C., Forsyth, A., Funk, S., Denny, M., Roach, K.E., et al.: Range of motion measurements: reference values an a database for comparision studies. Haemophilia. **17**, 500–507 (2011)
Stoll, T., Huber, E., Seifert, B., Michel, B., Stucki, G.: Maximal isometric muscle strength: normative values and gender specific relation to age. Clin. Rheumatol. **19**(2), 105–113 (2000)
Stubbs, N.B., Fernandez, J.E., Glenn, W.M.: Normative data on joint ranges of motion of 25- to 54-year-old males. Int. J. Ind. Ergon. **12**, 265–272 (1993)
The University of Michigan. 3D Static Strength Prediction Program Version 6.0.5: User's Manual. Ann Arbor, MI (2011). Retrieved from http://www.engin.umich.edu/dept/ioe/3DSSPP/download.html
Thompson, L.V.: Effects of age and training on skeletal muscle physiology and performance. Phys. Ther. **71**(1), 71–81 (1994)
Tillmann, B.: Atlas der Anatomie des Menschen: [mit Muskeltrainer], 2nd edn. Springer, Heidelberg (2010)
Tittel, K.: Beschreibende und funktionelle Anatomie des Menschen, 14th edn. Urban & Fischer, München (2003)
VanCott, H.P., Kinkade, R.G. (eds.): Human Engineering Guide to Equipment Design. USGPO, Washington (1972)
Wakula, J., Berg, K., Schaub, K., Bruder, R.: Der montagespezifische Kraftatlas BGIA-Report, Bd. 3/2009. BGIA; Technische Informationsbibliothek u. Universitätsbibliothek, Sankt Augustin, Hannover (2009)
Wilke, H.J., Neef P., Caimi M., Hoogland T., Claes LE.: New in vivo measurements of pressures in the intervertebral disc in daily life. Spine. **24**(8), 755–762 (1999)
Wilke, H.-J.: Möglichkeiten zur Bestimmung der Wirbelsäulenbelastungen und Konsequenzen für die Empfehlungen für das Sitzen. In: Ergomechanics – Interdisziplinärer Kongress Wirbelsäulenforschung. Shaker Verlag, Aachen (2004)

Winter, D.A.: Biomechanics and Motor Control of Human Movement, 3rd edn. Wiley, Hoboken (2005)

Youm, Y., Dryer, R.F., Thambyrajah, K., Flatt, A.E., Sprague, B.L.: Biomechanical analyses of forearm pronation-supination and elbow flexion-extension. J. Biomech. **12**(4), 245–255 (1979)

Zacher, I., Bubb, H.: Strength Based Discomfort Model of Posture and Movement Proceedings of the SAE DHMC (Digital Human Modeling Conference), Bd. SAE 04DHM-66 (2004)

Zatsiorsky, V.M.: Kinematics of Human Motion. Human Kinetics, Champaign (1998). http://www.gbv.de/dms/bowker/toc/9780880116763.pdf

Further Literature

Bullinger H.-J. (Ed.): Ergonomie -Produkt und Arbeitsplatzgestaltung, Vieweg-Teuber Verlag, Wiesbaden (2013)

Cashmore, L., Uomini, N., Chapelain, A.: The evolution of handedness in humans and great apes: a review and current issues. J. Anthropol. Sci. **86**, 7–35 (2008)

DIN 33402-1. Ergonomie – Körpermaße des Menschen – Teil 1: Begriffe, Messverfahren: Beuth

DIN 33402-2. Ergonomie – Körpermaße des Menschen – Teil 2: Werte: Beuth

Flügel, B., Greil, H., Sommer, K.: Anthropologischer Atlas. Minerva – Edition Wötzel, Frankfurt/Main (1986)

Günzkofer, F., Engstler, F., Bubb, H., Bengler, K.: Isometric elbow flexion and extension joint torque measurements considering biomechanical aspects. In: Proceedings of the First International Symposium on Digital Human Modeling Lyon, France (2011b)

Prader, A.: Normales Wachstum und Wachstumsstörungen bei Kindern und Jugendlichen. Wochenschr. **59**, 977–984., Springer Verlag (1981)

Human Models

Heiner Bubb

Contents

5.1 Cognitive Human Models – 220
5.1.1 Control Engineering Models – 220
5.1.2 The Benefits of Control Engineering Human Models – 228
5.1.3 Cognitive Driver Models – 229

5.2 Anthropometric Human Models – 234
5.2.1 Drawing Templates – 234
5.2.2 Digital Human Models – 237

5.3 Summary Appreciation of the Benefits of Human Modelling – 250

References – 251

H. Bubb et al. (eds.), *Automotive Ergonomics*, https://doi.org/10.1007/978-3-658-33941-8_5

5

5.1 Cognitive Human Models

Very early attempts were made to describe the interaction between human interaction and machine using control engineering methods. These approaches were mainly used in the aircraft industry to compare the dynamic behaviour of an aircraft with the capabilities of the pilot. Later, these methods were also used to adapt the dynamic properties of a vehicle to the capabilities of the driver. Since driving a car does not only consist of keeping the vehicle on the road within the framework of the stabilisation task (see ▶ Sect. 1.4), but also of finding decisions between different alternatives for action, these control engineering models were later joined by models that are intended to reflect the decision-making behaviour of people. The latest approaches relate to combining the control engineering models with the decision model.

5.1.1 Control Engineering Models

It causes considerable difficulties to depict the human being as a technical controller, since an analytical description of the subsystems of the system element "human being" is usually not possible. In order to nevertheless achieve the desired goal of modelling, the results of neurophysiological research are to be converted into technical analogies and a control structure of the human information processing channel is to be established from this. Based on results from various tasks, it is now possible to determine the individual parameters of the respective model using the methods of nonparametric methods (see ▶ Sect. 2.3.3). Such linearizing descriptions of human beings are only adequate, however, if they are highly practiced activities and the system is in a steady state, as is the case, for example, when following a route with moderate curves.

5.1.1.1 Quasi-Linear Human Model

The first and still quoted design of such a quasi-linear human model was by Tustin (1944). Originally it did not refer to the modelling of the driver, but to the behaviour when following a target with a rifle. In the period that followed, this model was mainly used to describe the behaviour of aircraft pilots (so-called "paper pilot"). The design presents a very simple description of the human control characteristics in the form of the frequency response F_{RM} where the observations, which were already prerequisites for the representation in ▶ Fig. 2.10, were the basis for the modelling. Information acquisition and information processing are treated as one unit, which is modelled under purely technical aspects by a "universal controller":

$$F_{RM}(p) = \frac{K \cdot (1 + pT_A)}{(1 + p \cdot T_I)(1 + pT_N)} \cdot e^{-\tau p} \qquad (5.1)$$

The information recording information processing block is here amplified by the amplification K and a derivative element T_A which reflects the ability of humans to react as PD regulators. The so-called delay of the fitting term T_I stands for the ability of humans to show a certain smoothing (regulatory "integrating") effect in the transmission of signals. A further smoothing effect is achieved by the inertia of the hand-arm system, which can also be simplified by an integrating system element. This is reflected by the neuromuscular delay time T_N. The term $e^{-\tau p}$ describes a delay in the sense of a runtime and generally corresponds to the physiological reaction time caused by the nerve runtime.

The human transfer function described here is based on many investigations (see also McRuer and Krendel 1957; McRuer and Jex 1967; Elkind 1956, 1964). After nonparametric methods (see ▶ Sect. 2.3.3) have been used to determine the human frequency response in so-called tracking experiments[1] was determined in the form of a graph of a soil diagram, the parameters of the model of eq. 5.1 are adapted in such a way that they correspond optimally with the found courses.

1 The test subject's task is to bring a target mark moving stochastically on a screen (one- or two-dimensional) into line with the actual mark.

According to Senders (1964), the parameters move in the following range:

Dead time (reaction time) τ:	0.2	to	0.3 s
Time constant of the neuromuscular system T_N:	0.1	to	0.16 s
Gain factor K:	1	to	100 s
Preservation of the fitting term T_A:	0	to	25 s
Delay of the fitting term T_I:	0	to	20 s

The large variation range of the parameters is explained on the one hand by the interindividual differences but also by the different conditions. For example, the difference between capturing a curve with a radius of curvature from 30 m, a lateral acceleration of 0.3 g and a forecast from 2 s (the deviation appears at an angle of about 15°) and perceiving a slight deviation from the straight course (sensitivity threshold <1′) causes a difference of ≈ 1: 900.

McRuer and Jex (1967) carried out a much more detailed modelling, more oriented towards physiological conditions, in the form of the "precision model"(see McRuer and Krendel 1974; Young 1973; Stein 1974; Stein and Pioch 1975), which is hardly suitable for prognostic modelling due to its many freely definable parameters. However, as McRuer et al. (1967) show, there are clear deviations between the simplified model according to Tustin and the precision model, especially in the area $\omega = 0.6$ rad/s and $\nu = 0.11$ Hz, which are to the disadvantage of the Tustin model in relation to the performance of test subjects. Schweizer (1975) argues that the neuromuscular delay time and the delay time of the fitting term only significantly influence the transmission behavior for frequencies larger than 1 Hz. Therefore, the human transmission behavior for the frequency range of $0.2 < \nu < 1.0$ Hz, which is essential for manual control, can be specified by the following simplified description function:

$$F_{RM}(p) = K \cdot (1 + T_A p) \cdot e^{-\tau p} \qquad (5.2)$$

This corresponds to the transfer function of a PD controller with the dead time τ.

None of the transfer functions described so far makes it possible to reproduce the feedback of a control unit on the haptic channel. In the model analogy, it is always assumed that the control element is simply connected downstream of the human and only provides feedback proportional to the position, i.e. no feedback from the dynamics of the machine (this assumption would be correct for accelerator or brake pedals, for example). However, this does not cover all cases of reality where the mechanical coupling also allows reaction forces in the machine's structure of action to be sensed as feedback (e.g. steering wheel in motor vehicle). Even the precision model does not offer the possibility to include the feedback signal at the actuator. The information structure model (Bolte and Bubb 1990) offers consistent consideration of this feedback. However, the number of freely selectable parameters increases to 17, so that the same note applies as for the precision model.

5.1.1.2 Cross-Over Model

Regardless of which of the models described above is used, the respective parameters must be determined together with the control description of the vehicle dynamics (depending on the application: lateralse or longitudinal dynamics). The basis for this is McRuer's cross-over theory (1967). He has found in numerous experiments that humans can largely adapt to arbitrary dynamics of controlled systems. From a control point of view, this adaptation can be described by his observation that in the range of the cross-over frequency (see also ▶ Sect. 2.3.3) the open control loop consisting of human and controlled system of different dynamics shows a decrease of the amplitude response from 20 dB over a decimal power of the frequency. This is described by Eq. 5.3 for the open loop:

$$F_{RM} \cdot F_S = \frac{\omega_c}{p} e^{-\tau_e p} \qquad (5.3)$$

The cutting frequency ω_c and the dead time τ_e, which simulate all time constants, i.e. the reaction behaviour of the human being, his delay characteristics and those of the machine depend on the dynamics of the controlled system and the bandwidth of the reference variable (difficulty of the task). McRuer and Jex (1967) provide detailed information about this, which will not be discussed here. The model given in eq. 5.3 is valid in the frequency range between 0.08 and 0.32 Hz.

In principle, the cross-over model can only be used in connection with human regulatory activities for so-called compensation tasks. These are tasks in which the operator only sees the difference between the reference variable and the tracking variable. This is actually basically the case with the vehicle, since here all the variables seen are observed from the position of the vehicle (= tracking variable). However, anticipation turns the control system into a kind of control system for experienced drivers (see "two-level model"). Nevertheless, the cross-over model can be used to carry out successful stability studies of the driver-vehicle system. This can be seen by looking at the position of the cutting frequency and the effective dead time: control of the controlled system is simplified for the driver if the cutting frequency is higher and the effective dead time is lower. This is achieved by the kinesthetic feedback, as experiments by McRuer et al. (1977) and Hosman et al. (1990) have shown. Conversely, a more difficult task (= bandwidth of the reference variable; e.g. fast travel on winding roads) increases the cutting frequency and reduces the effective dead time. Yuhara et al. (1992, 1994) were thus able to show that the haptic feedback of the reaction of the controlled system (restoring forces at the steering wheel, which are dependent on the driving condition, or active control element, in which parameters of the driving condition such as lateral acceleration or yaw angle speed are displayed as restoring forces) at the control element simplifies the control activity. Overall, the cutting frequency for manual control tasks is between 2 rad/s and 10 rad/s or 0.3 Hz and 1.6 Hz (Johannsen et al. 1977). When driving a vehicle, the cutting frequency varies depending on the driving task and driving situation (Appel and Mitschke 1997):

- When driving straight ahead, it lies between 0.1 and 0.2 Hz,
- for curves at approx. 0.35 Hz and
- in critical situations (e.g. evasive manoeuvres) at approx. 0.5 Hz.

The cutting frequency of the longitudinal dynamics, at 0.046 Hz, is a power of ten lower than that of the lateralse control (Donges 1982). Therefore, the control of the longitudinal dynamics is a comparatively simple task. Experiments by Eckstein (2001) with an active control element also show that the feedback of driving states of the longitudinal dynamics, in contrast to that of the lateralse dynamics, does not have a positive effect on their controllability.

As described above, the stability of the control loop can be evaluated by the phase reserve in the cutting frequency range. It is directly related to the effective dead time. In contrast to the large value range of the cutting frequency at different driving situations of the lateral dynamics, the phase reserve range is relatively constant between 30° and 60° (Appel and Mitschke 1997).

5.1.1.3 Two-Level Model

It should be pointed out in particular that all the so-called quasilinear human control models described so far assume that only a single control deviation affects humans. However, as shown in ► Sect. 2.4, it must be assumed that the driver processes different forms of feedback at the same time. In particular, it can be assumed that a target course will be set in the area of the route overlooked in advance as part of the fulfilment of the guidance task. In the so-called "two-level model" it is assumed that anticipation of the course of the road takes place through foresight ("headlamp orientation"), which determines a target course at the guidance level in the form of a control system (= open circle, without feedback). This is then regulated at the stabilization level in the form of a compensatory control (see Donges 1978; Willumeit and Jürgensohn 1997). With regard to a forward look-ahead point (see Kondo 1953, ► Sect. 3.3.2), the curvature of the curve, which is required at a later time for the control of the vehicle, is estimated (Appel and Mitschke 1997; Donges 1978; Neculau 1992; Reichelt and Strackerjan 1992; Prokop 2001). This anticipatory control reduces track deviation and makes it possible to stabilize the entire driver-vehicle control loop (McRuer et al. 1977). For example, the driver can turn

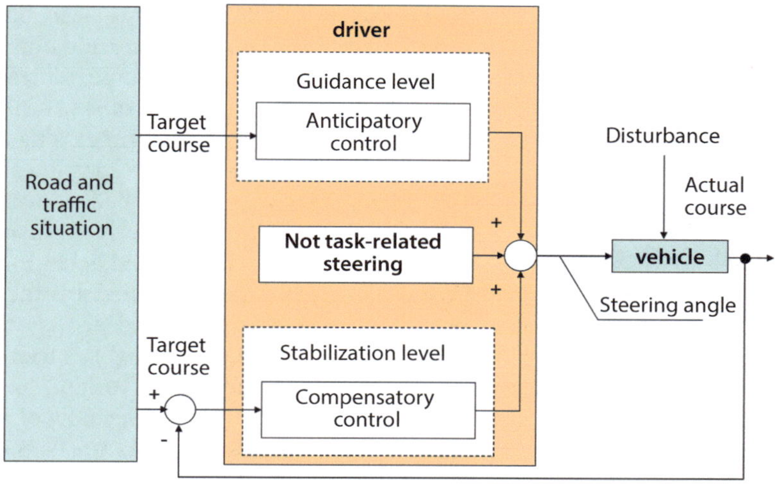

Fig. 5.1 Two-level model after Donges (1978)

in shortly before a bend to compensate for the effective dead time of the driver and vehicle (Donges 1978; Reichelt 1990). This increases the phase reserve of the driver-vehicle control loop. The estimated road curvature is also made available for the longitudinal dynamics and the driver can decide from his own experience in the current driving situation whether the driving speed must be adjusted (Neculau 1992). Figure 5.1 gives this model conception of the two-plane model concretized by Donges (1978): on the level of the path guidance, a (fast) anticipatory control reacts to the mentally conceived nominal course, and on the level of stabilization, the still observed deviation of the vehicle direction from the curve tangent ("direction orientation") and the lateralse deviation ("fog orientation") are made "to zero" in the form of a compensatory control, which can be modelled with the methods described above. Deviations of this mechanistic model from actually observed reactions are taken into account by a non-task related steering (so-called additive noise). As Jürgensohn (1997) points out, Donges' model was exemplary for a number of derived developments (e.g. Horn 1986; Plöchl and Lugner 1994; Bösch 1991 and Reid et al. 1981) and was also used for practical questions (Braun 1986).

In addition to these classical linear model types, there are also derived models whose structure can change during simulation. As an example, Jürgensohn cites McRuer's dual-mode model (1967, 1980), in which, depending on the control situation, a switchover takes place between control and programmed control. These structurally variable models also include adaptive models in which the model changes with a sudden change of environment (e.g. black ice) (Reichelt 1990; Nagai 1983). In Jürgensohn (1997) further models are listed which are characterized by a number of variations, such as different dead times, addition of nonlinear elements (e.g. hysteresis), perception thresholds (e.g. Carson et al. 1978) and time-discrete elements, such as sample-and-hold links (Crossman et al. 1966; Carson et al. 1978; Hayhoe 1979; Kroll and Roland 1970; Reid et al. 1981). In most models the driver's starting point is the steering wheel angle λ, but also the steering speed (Sheirko 1972; Hayhoe 1979) or the steering torque (Fiala 1966; Wohl 1961). It should be noted, however, that the "natural" initial value is the steering wheel angle, since the aim of the self-reflex arc already mentioned is to apply such a force (= relative to the steering column axis: moment) that the desired path (= angle) is achieved (see Schmidt and Thews 1990). The target value is almost always the lateral offset a often used in conjunction with the yaw angle φ, rarely the road curvature (e.g. Radonjic 1990). Sometimes other state variables are also used as input variables, such as the roll angle (Niemann 1972) e.g. Reichelt (1990) lists in a summary a total of 21 different information variables for the driver, from the swimming angle velocity (Braess 1970) to the third derivation of the nominal curvature (Fiala 1966).

5.1.1.4 Optimal Theoretical Models

The previous models, in particular the crossover model, the application of which is practically indispensable for determining at least some of the many parameters of all quasilinear controller-human models dealt with so far, have triggered a number of criticisms both from the point of view of psychological research and from the point of view of engineering application. From a psychological point of view, the complete detachment of the input-output model from physiological-psychological findings was criticized. The engineer commissioned with the task of system optimization measures the easy transferability to multi-variable control loops, the modelability of time-variable problems and the lack of extensibility (Jürgensohn 1997). Together with the system theoretical tools established in the 1960s, these criticisms led to the development of the optimal theoretical model (Baron and Kleinmann 1969; Elkind et al. 1968; Baron and Levison 1977, 1980; Kleinmann 1969, Kleinmann et al. 1971. "The human controller has the ability to self-adjust its structure and parameters with the aim of optimising the transmission characteristics". This statement by Oppelt and Vossius (1970) can be regarded as the core idea of the optimal theoretical model. In the present context, only its structure (see ◘ Fig. 5.2) is to be explained, since the mathematical treatment would go beyond the scope of this presentation. A detailed and easy to understand description can be found in Donges (1977).

In the optimal theoretical model, the technical display and the information recording (observation process) of the human being are combined into a subsystem, whereby the ability of the human being not only to record the explicitly displayed quantities, but also their temporal derivation, can be taken into account. Shifted by one dead time, these are fed to the central processing unit as perceived variables. The dead time represents a summary representation of all sensory, central and motor response, processing and running times, similar to the quasilinear controller-human models described above. In "central processing", the perceived variable is converted into the intended actuating movement, which in turn, modified by the mechanical properties of the hand arm system ("neuromuscular delay element"), is transformed into the actual actuating variable intervening in the machine. Within this flow of information, statistical disturbances are superimposed at two points, which are referred to as "observation noise" and "motor noise" according to their location. This additive addition of statistical signals is not to be understood in the sense of a part of the human positioning movements not reproducible by the model as in the quasilinear models, but as an effect of a randomness inherent in the human control behaviour, which superimposes itself on the useful part of the positioning movement as a random disturbance variable. It is to be interpreted as a direct consequence of the limited accuracy of human sensory perception and motor output of information. Their size is generally influenced by the properties of the displays and controls used.

The essential part of the optimal theoretical model is the part referred to as "central processing" in ◘ Fig. 5.2, which simulates the human ability to develop an optimal control strategy for the task at hand, taking into account his or her own characteristics and given machine dynamics. According to the optimal estimation theory, this is made possible by the fact that the fitting element derives an intended positioning movement from the perceived variables in such a way that a quantitative cost criterion is minimized. In the case of a linear system disturbed by white noise, the solution is to connect a predictor and a Kalman filter in series. The Kalman filter achieves an optimal estimate for the perceived size shifted by the dead time τ that eliminates the disturbance. The predictor provides an estimate of the current size that optimally compensates for the dead time. As the mathe-

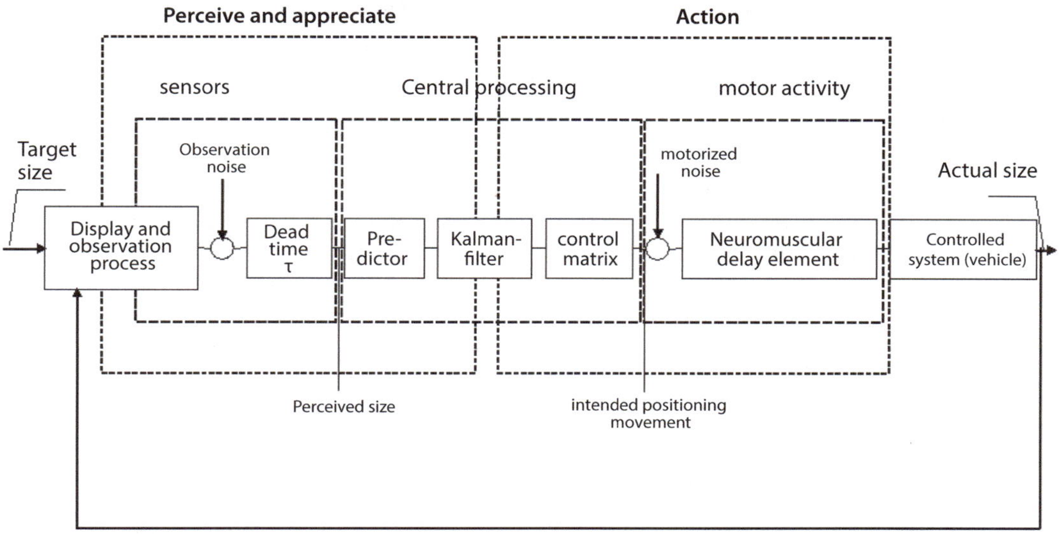

Fig. 5.2 Structure of the optimal theoretical model

matical treatment shows, both system elements contain exact models of the dynamics of the controlled system. The Kalman filter and the predictor require an *complete* knowledge of the controlled system and provide the dynamic behavior of the state variables. As Jürgensohn (1997) notes, "the admissibility of this prerequisite is based on the high predictive power of the human being, which can only be explained if the human being has an 'internal model' or a 'representation' of the route dynamics to be regulated. These concepts were hotly debated in cognitive psychology and action theory at the height of the optimal theory model. While McRuer (1980) also interprets the crossover model as a structure with an 'implicit' internal model of distance and operator near the crossover frequency, the explicit modeling of the internal model had a much higher suggestive power and was welcomed by the psychological side."

In the period that followed, a number of criticisms arose of the optimal theoretical model which, on the one hand, called into question the assumptions necessary for mathematical derivation and, on the other, concerned the difficulties of practical application. Even the optimal theoretical model can only be used to a limited extent for predicting behaviour, whereby according to Jürgensohn (1997) the discrepancy between mathematical severity and depth and the apparent "incomprehensibility" of human action appears to be the most problematic. Dey and Kirchhoff (1975) even go so far as to deny the optimal theoretical model (and also the other linear models) the ability to reproduce human behaviour. Nevertheless, as Jürgensohn explains, the ideas of "internal model" and "superordinate principle of action" work their way into today's model approaches and can also be found in many non-linear algorithmic models.

5.1.1.5 Shift Models, Bang-Bang Models, Hybrid Models

The so far treated so-called linear human models always cause a smooth continuous course of the output variable independent of the course of the input variable, although the human being acts non-linearly and discontinuously in most cases after all observation. Tustin (1947) already described the "actually not smooth" course of operator movements in his early model proposals. This behavior is particularly evident in the control of lines described by differential equations with higher derivatives after time and in tasks that give room for free decision. This discontinuity becomes visible not only with respect to temporal behavior, but also with respect to amplitude levels (see also Poulton 1974; Rühmann 1993). Pew's (1963) experiments show that these are well-trained switching strategies. He proved that the same control power was obtained for second order lines (the dynamic behaviour is characterized, for example, by the pendulum described in ► Fig. 2.8) if, instead of a control element with continuous input possibility, a control element with only two switching points was used. The "switching models" (discrete action in fixed or stochastically variable time periods) and the "Bang-Bang model"(optimum time behaviour, whereby inputs are always made at the edge of the steering range; for a car this would be a change between full throttle and full braking; for more details see Jürgensohn 1997) can, however, only be applied to very limited extent to driving a car. "Hybrid approaches", on the other hand, which combine discrete decision-making processes with quasilinear models, represent the state of the art today, also with regard to modern computer technology. In this context, reference is made to the early "three-level-model" approach of Crossman and Szostak (1969), which then only modeled the quasilinear part, as well as the approach of Kroll and Roland (1970) and the model of Reddy and Ellis (1982), which combines a production system (similar to a flowchart) with geometric considerations and thus attempts a first approach for the modelling of gaze strategies. The first complex model that can describe the driver in different driving situations was DRIVEM (Wolf and Barret 1978; Lieberman and Goldblatt 1981). It was designed by the NHSA (National Highway Safety Administration of the USA) for the preliminary analysis of accidents and as an aid to combating accidents (Jürgensohn 1997). Psychologists criticised the excessive simplification of driver behaviour (Perel 1982).

5.1.1.6 Fuzzy Control Models

A big disadvantage of all linear models is the fact that the information acquisition is reduced to a single real quantity, e.g. the lateral offset or the deviation of the yaw angle, and is not modelled as it comes to the acquisition of these quantities. Already Wierwille et al. (1967) pointed out that the driver uses the entire visual field of vision as a source of information. Neculau et al. (1990) and Neculau (1992) attempted to take this into account by introducing several lateralse deviations at different foresight distances as driver input variables. Kramer and Rohr (1982a), Kramer (1985) and Willumeit et al. (1983) include the whole field of view of the driver (driver display) as information source and use in their fuzzy driver model the average of the fuzzy cross correlation to derive a steering angle λ from the "driver display" (see ◘ Fig. 5.3) It is interesting that this and similar approaches beside others are also used to develop on a technical basis driver assistance systems for lateral control (Feraric et al. 1992; Feraric and Onken 1995; Feraric 1996). The ALVINN model from Carnegie Mellon University (Pommerleau 1991), which uses artificial neural networks to simulate the pattern recognition process of the driver, follows a similar approach to Kramer and Rohr. Jürgensohn (1997) also uses fuzzy control theory in his own modelling approach, assuming that the driver sees the road "as a whole". He assumes that what is seen is assigned to rough internal concepts such as "straight line", "curve", "exchange rate curve (S curve)" and that the behavior (here the velocity behavior) is derived from this. For example, a curve with a very large radius can fall into the "straight" category from the driver's point of view with regard to the control of longitudinal dynamics if it does not influence the speed. At the same time, however, the same sensory impression can be assigned to a different concept with regard to lateral control (here, reference should be made to the results of experimental observations presented in ► Sect. 3.3.2 and to the relationship between gaze behaviour and internal model derived from this in ► Sect. 3.3.3).

The fuzzy assignment of sensory stimuli or extracted stimulus patterns to categories or concepts, which is characteristic of fuzzy control approaches, is the very basis of current modelling approaches. For example, the purely qualitative model of Rasmussen et al. (1987); see ► Sect. 3.2.2.5) in Newell's unified

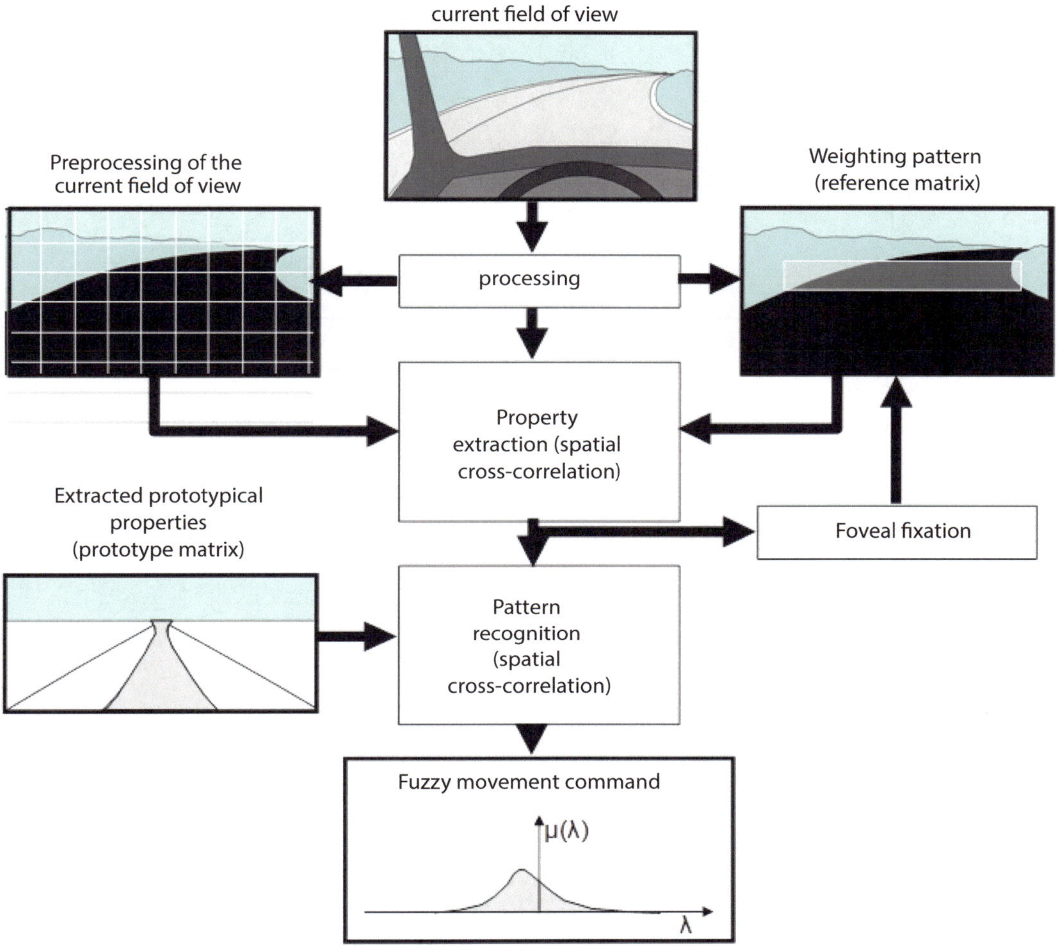

Fig. 5.3 Driver model that derives the action from the driver's field of vision using methods of fuzzy mathematics. (Modified after Kramer and Rohr 1982a)

cognitive theory (1990) is presented in the form of four so-called "cognitive bands" so that a computer-aided simulation of cognitive processes becomes possible. The so-called biological band mainly comprises neuronal functions with a time constant in the millisecond range, the cognitive band works symbol-oriented and comprises automated actions in the range of 100 ms, the band of rational behaviour, on the other hand, is characterized by real knowledge action and comprises times between 10 and 10,000 seconds. In contrast to Rasmussen's proposal, Newell adds a fourth so-called social band in which human cooperation is modeled as a distributed set of rationally acting actors, an approach that has not yet received sufficient attention in the modelling of the driver's traffic behaviour. In these approaches, Wickens and Hollands (2000) speak rather of a regulatory continuum, which is probably best described by Newell, in that the Rasmussen planes are described there, following corresponding models in solid-state physics, with "cognitive bands", which are characterized by certain potential walls, the transgression of which requires endogenously or exogenously generated "activation energy". Also more recent modelling approaches, whose aim is to develop controllers with "human-like" behaviour for automatic vehicle guidance, refer to the three-level model of Rasmussen, whereby the mod-

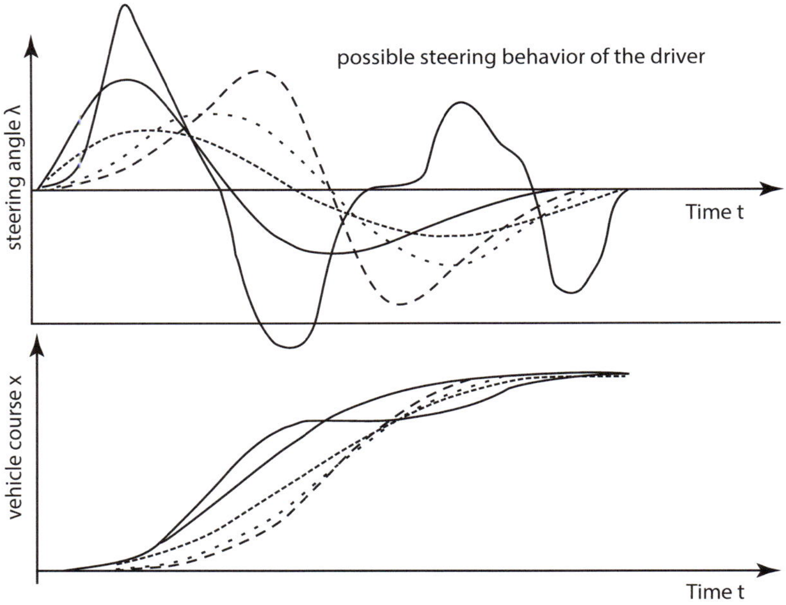

Fig. 5.4 Steering angle course λ(t) and resulting vehicle course x(t) at a lane change. (According to Jürgensohn 2002)

elling is limited to the low skill-based and rule-based level (von Garrel et al. 2001).

5.1.2 The Benefits of Control Engineering Human Models

If one has the requirement to model the driver in order to obtain a complete, precise and unambiguous prognosis of the behaviour depending on the respective situation, then all the proposals made so far are disappointing. This is not only due to the inadequacy of the models, which are always able to cover only partial aspects of human behaviour, but is also the result of the great interindividual and intraindividual variability of the driver, which a model in this form is unable to reproduce at all. Figure 5.4 shows as a simple example different steering behaviour of drivers when changing lanes and the resulting behaviour of the vehicle (it should be pointed out here that the vehicle shows a strongly integrating behaviour with regard to the "input" steering movement). With regard to the experimental survey of driver behaviour, this indicates, by the way, how little can be said about the driver's intentions from the interpretation of driving condition parameters such as speed and longitudinal or lateral acceleration or yaw angle change. In general, however, it can be said that the technical simulation of the human being in the respective situations to which it refers is just as capable of following curves and, if necessary, adjusting speeds as the real driver is. The aim of control engineering or other mathematical simulation of human behaviour cannot be to make exact predictions about individual behaviour. Rather, the value of the analysis lies in making the respective influences and their fundamental effects transparent and in deriving possible technical aids for the driver from them. The background of such consideration is simple: the information needed by the model to accomplish the task is also needed by the person who in reality takes his place. In this sense, certain tendencies can be derived as target specifications:

The detention times generated by the driver should be as short as possible, it follows:

- the dead time of the vehicle should be as short as possible (< 100 ms),

- Different human characteristics must be taken into account for different driving tasks,[2]
- Driving task and vehicle must be considered in combination,
- Input and feedback at the controlled system vehicle must be considered together and
- higher-frequency control activities to stabilize the vehicle are to be carried out by electronic control systems (e.g. ABS, ESP).

5.1.3 Cognitive Driver Models

The description of the speed selection already shows the limits of control engineering driver models. Jürgensohn (1997) states, for example, that control engineering human models can only ever reflect a limited range of driver activity, and that in particular the two-level model, which provides for a separation between guidance and stabilisation tasks, appears "artificial" and should be supplemented by "hybrid models". In particular, models from the psychological field should be combined with more technically oriented models, as is done to some extent by fuzzy control models but also by other approaches.

In the following, a brief overview is given of relevant cognitive driver models that are based on so-called "cognitive driver models". According to Anderson (2007), cognitive architecture is defined as "a specification of the structure of the brain at a level of abstraction that explains how it achieves the function of the mind". It thus represents a structure that is thought to reflect the structure of human thought. Models based on cognitive architectures are considered psychologically valid, but this is bought at the price of a high degree of complexity. The basis for such driver models are therefore often models that are supposed to reproduce human behaviour in general in situations that demand cognition. Specific driver models are then derived from these models (for more details see Plavšic 2010).

5.1.3.1 ACT-R

One of the most famous models of this kind is the one developed by John Anderson at Carnegie Melon University. *ACT-R* (Adaptive Control of thought-Rational), whose original intention was to generate a user model that enables human interaction with different interfaces (Anderson et al. 2004). It is composed of different modules which represent the corresponding cortical regions. Thus there is beside others a visual module, a module, which is responsible for actions, as well as modules for memory and language. ACR-R also takes into account the limited processing capacity of 7 ± 2 chunks mentioned in ▶ Sect. 3.2.2.5.

From this model a special ACT-R Driver Model was derived, which allows to model driving on three-lane highways including lane changes, as well as simultaneous telephoning. Other road users may be involved if they travel in the same direction. The model is based on the two-level model by Donges (1978). With regard to lateral dynamics it is coupled with a simple single-track vehicle model ("bicycle model") and with regard to longitudinal dynamics with an energy balance model. Steering is implemented on a control engineering basis by a PID controller (see ▶ Sect. 2.3.3). Using the attention module EMMA (Eye Movements and Movement of Attention; Salvucci 2000), the next eye movement can be predicted, which alternates between a so-called near point and a far point. The nearest point is approx. 0.5 s prospects on the road. Depending on the situation, there are three alternatives for the distance point: the vehicle in front, a tangent point on the curve or, on an empty road, a distance point that is a maximum of 4 seconds ahead of the driver. ◻ Figure 5.5 shows the structure of the ACT-R driver model. It largely takes into account the experimental results described in ▶ Sect. 3.3.2, but the jumping between these two extreme points is also criticized (Plavšic et al. 2010). However, ACT-R reaches its

2 Of particular interest here is the fuzzy control model by Jürgensohn (1997), which, for example, predicts different behaviour for the control of longitudinal and lateralse dynamics, a result which also came to light in other investigations, e.g. Eckstein (2001) and Huang (2004). This knowledge may need to be extended to specific traffic situations.

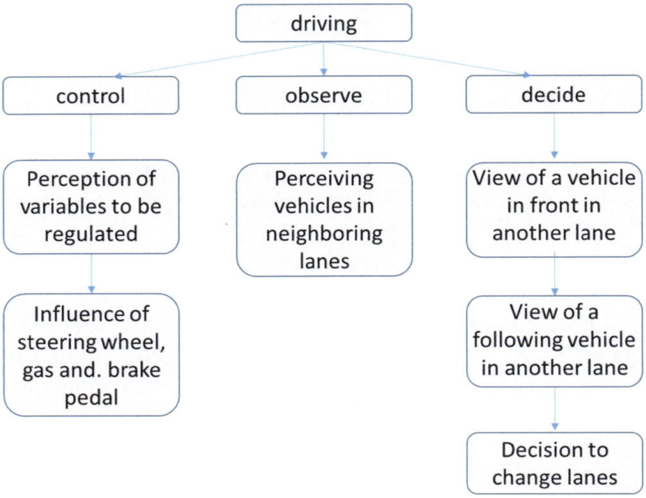

Fig. 5.5 The driver model of ACT-R

limits due to the enormous computational effort when complicated situations such as intersections have to be simulated, which excludes an online connection to a driving simulator with current computer technology (Plavšic 2010).

5.1.3.2 Soar

Another model that has to be assigned to cognitive architecture is the model *Soar* (State, Operator And Result), which is also based on Newell's (1990) fundamental work at Carnie Mellon University and is being further developed at the University of Michigan. It contains mechanisms for problem solving, learning, motor behaviour, visual orientation and a multi-task mechanism. In particular, it distinguishes between long-term memory and working memory, which is the active part and which is limited in its working capacity to 7–8 chunks per time unit.

For Soar, the model ***DRIVER*** was developed, which consists of various modules such as visual orientation, navigation, speed control, etc. (Aasman 1995). At the lowest level, DRIVER contains operators that represent the virtual body of the driver and that allow to simulate body movements and estimate the times necessary for them. These operators pick up commands from working memory. There are modules for carrying out the navigation task, the guidance task and at the lowest level of the stabilisation task, where it is even possible to initiate a gear change, for example, depending on the route and engine situation. The model has an inner representation of the ideal course used to calculate path deviations. There is then a separate module for steering, which is more precise than that of ACT-R. The steering angle is calculated based on several parameters such as angle and lateral deviation from the internal representation of the ideal course and TLC (Time to Line Crossing; see ► Sect. 2.4.1; this parameter is the most important parameter within the "steering" module). Speed control is also more complex than with ACT-R. It is based on stimuli that are important for the correct speed selection, such as speedometer value, time to intersection (TTI) and engine noise. In addition, the mental model of the situation has importance for deciding on the appropriate speed.

The most important module of the cognitive driver model is the basic perception. It is responsible for object recognition, attention and basic eye and head movement control. A distinction is made between the Functional Visual Field (FVF), which has an angular range of approx. 20°, and the Peripheral Visual Field (PVF), which has a detection range of 210° horizontally and 90° vertically (the sizes of the fields were determined according to Miura 1986). Objects within the functional visual field have 100% chance, objects within the peripheral visual field have less chance to be perceived. The information in the peripheral field is used to initiate eye and head movements. Items in the functional field of view are assigned the following attributes: Existence, movement, direction, size, colour, shape and type of object. In contrast, items in the peripheral visual field are only provided with the first four attributes. The Elemental Perception module also contains several limitations regarding visual processing such as that no new visual information can enter the working memory during eye movement, that necessary eye movements cause head movements beyond a certain level, or that the time required for eye and head movements depends on the speed and distance of the observed moving objects.

The Visual Orientation Module is based on the Elemental Perception Module. It is the part of DRIVER that was originally not present in Soar. It regulates the orientation at intersections. The visual control is realized by means of a so-called attention operator, which applied to a viewed object marks it as soon as it has been noticed. Both intentional (top-down control) and involuntary (data-controlled) eye movements are possible. The normal visual activity is data controlled.

The Visual Orientation module includes several top-down rules for orientation at intersections. They were all obtained from experimental investigations by Harsenhorst and Lourens (1988). A distinction is made between default orientation rules and maneuver orientation rules as well as between local and global scan paths. Local scan paths represent the bottom-up (involuntary) controlled eye movements, and global scan paths indicate that a search is planned and therefore top-down controlled. In non-critical situations, the default rules prevail and the eyes are mostly directed to the objects in the functional visual field. In critical situations, such as approaching an intersection, implemented intersection rules cause the active search in the environment through global scanning. If there is a moving object in the periphery, the operator is applied to it. Differences between experienced and beginners are reflected by rules such as:

- experienced drivers use fixed strategies when approaching intersections and change the main field of view less frequently, or
- experienced drivers look at the relevant things in the traffic environment and pay attention to the relevant objects earlier than beginners, and
- experienced drivers can rely on peripheral vision to a much greater extent and are able to intentionally update the position of moving objects.

DRIVER is the only cognitive architecture driver model that can simulate driving through intersections. Unfortunately there are no publications about the further development of DRIVER since 1995 despite the very agile further development of the Soar architecture itself. Some significant problems of this model are also described (Plavšic 2010). The most important criticism relates to the key element of DRIVER, visual orientation strategies, according to which orientation based on the sharp separation of FVF and PVF seems all too simple. The accuracy of eye movements and the lack of learning in visual orientation are also criticised. In addition, DRIVER has a problem with memory size and forgetting information. New Soar developments have overcome these problems, but nothing has been changed in the DRIVER model itself.

5.1.3.3 QN-MHP

QN-MHP (Queuing Network - Model Human Processor) is a computer architecture that combines mathematical theories and simulation methods of queuing (QN) with a processor of human behavior (MHP). The latter is again based on the GOMS model (Goals, Operator, Methods, Selection Rules). QN-MHP is based on the commercially available software Promodell. On the basis of a network of 20 process units, various cortical areas and corresponding functional modules for human information acquisition, processing and implementation are simulated. Because of this “brain-like” structure, QN-MHP enables the visualization of internal information flows during the simulation of corresponding activities.

A driver model was also derived from QN-MHP which is real-time capable and largely oriented to the known information processing of the driver. It was tested in conjunction with a driving simulator environment (Tsimhoni and Liu 2003). Similar to ACT-R, a far point on the road is assumed to be 2–4 seconds ahead of the driver for steering. The lateral position is controlled by a near point one second in front of the driver. For curves, the tangent point of the curve is used. QN-MHP enables the simulation of a second (i.e. third) task. Future developments should relate to speed control, taking into account influencing traffic, vestibular organs and auditory inputs. Currently, however, the model operates at a fixed speed of 72 km/h. It can only be used by the University of Michigan, where it has also been developed.

5.1.3.4 COSMODRIVE

In addition to these models based on general cognitive structures, there are also some so-called stand-alone cognitive driver models, which are briefly discussed below.

COSMODRIVE (COgnitive Simulation MOdel of the DRIVEr) was developed in France at INRETS on the basis of the programming language *SmallTalk* (Bellet et al. 2011; Bornard et al. 2011). It was developed less with the aim of assessing assistance systems, but more than ACT-R and Soar to explain the information processing of the driver. As with the latter, mental models play a decisive role at COSMODRIVE. They are all represented within the "Tactical Module". This contains a so-called anticipation agent, which contains various anticipation representations (AR) that contain possible developments on the basis of the current situation. The decision agent selects one of these different anticipation representations, based specifically on the risk associated with the action and the time saved. The success of the action is compared by the operation module with the expectation and in the case of a difference new *Current Tactical Representations* (CTRs) were formed. The described process is then repeated, whereby a learning process is also simulated by forming new higher-level anticipation representations. Ultimately, COSMODRIVE thus represents a programmatically implemented simulation of what is described in more detail in ► Sect. 3.2. Unfortunately, COSMODRIVE is currently not publicly available, apart from the fact that most of the literature is only available in French.

5.1.3.5 PADRIC

PADRIC (PATH DRIVer Cognitive) is based on the structure of COSMODRIVE. It was developed in cooperation between the Institute of Transportation Studies (ITS), the University of California and Caltrans (see bibliography for reference). In particular, it can simulate critical situations caused by visual distraction. Otherwise, the same restrictions apply to this system as to the program on which it is based.

5.1.3.6 ACME

ACME was developed at DLR with the aim of providing a model of the driver in real-time simulation that allows critical driver states to be modelled in various traffic situations and that can be used for the development of assistance systems. The biggest disadvantage of the system is that it is not open to external users. In addition, it is still in a state of development.

5.1.3.7 PELOPS

The program *PELOPS* was developed by BMW in cooperation with RWTH Aachen University. Its purpose is to simulate the interaction of driver, vehicle and environment. In particular, it can simulate traffic flow with high accuracy even in conjunction with complex stop & go situations. However, it is entirely specialized in longitudinal traffic.

5.1.3.8 SSDRIVE

Also the program *SSDRIVE* is to model the driver-vehicle interaction in real time with a special focus on driver errors. It is also not open to the public.

5.1.3.9 Plavšic's Model Proposal

Based on experience with existing driver models, Plavšic et al. (2010) develops recommendations for the conception of a new driver model. The object of this model is to support the development of driver assistance systems

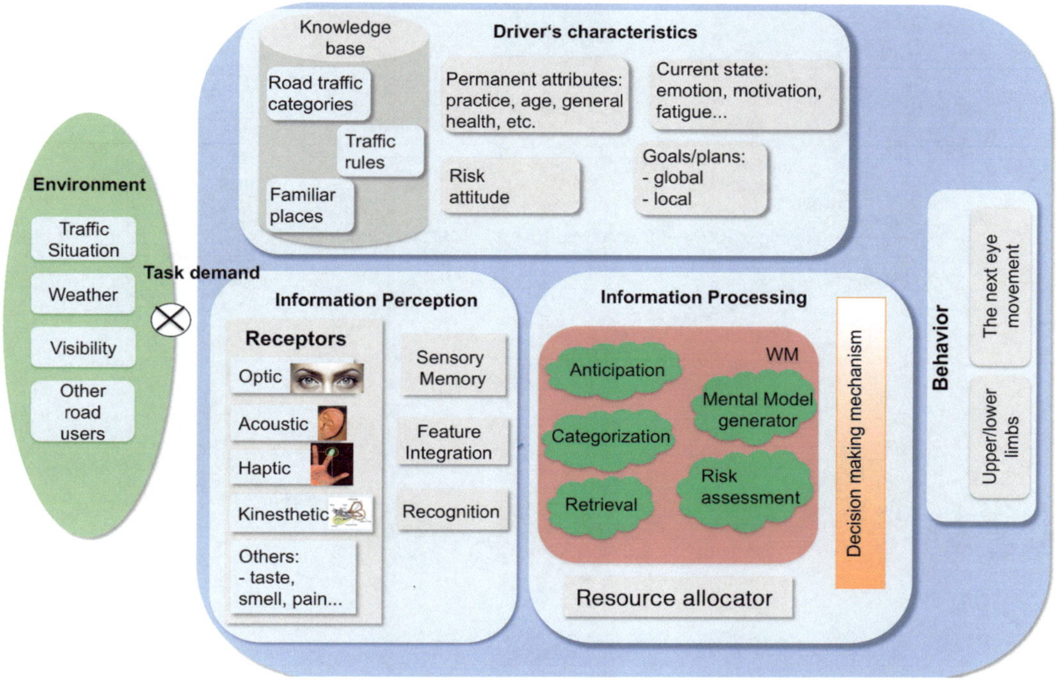

Fig. 5.6 Multi-agent system. (Plavšic 2010)

at guidance level. It should be able to simulate and predict driving errors and at the same time explain why they happen. Recommendations are made on the level of detail and complexity of the programming effort. Here, the ergonomic requirement for easy to learn programming is raised, especially with regard to further development capability. In Plavšic's opinion, the so-called multi-agent system fulfils the requirements very well (Fig. 5.6). The following description also illustrates the procedure of the models described above.

The agents themselves can be divided into three groups:

- Perception agent: optical, acoustic and vestibular information
- Information processing agent: categorizing, capturing, generating mental models, anticipation, risk assessment and decision making
- Information conversion agent: eye movement, movement of the upper and lower extremities.

Each of the individual agents in Fig. 5.6 must be implemented individually. As an

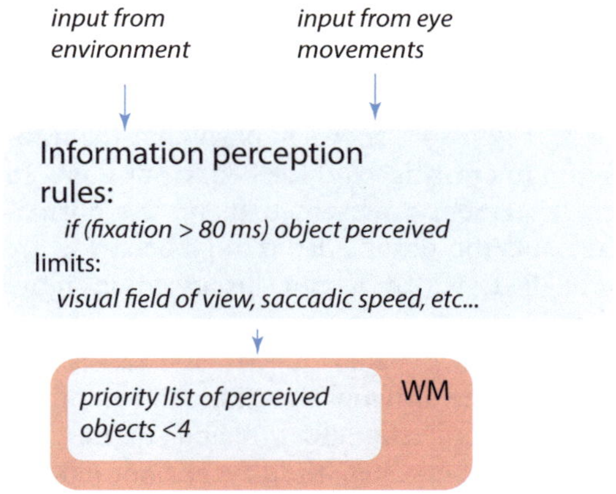

Fig. 5.7 Modelling of information acquisition agents. (Plavšic 2010)

example, the modeling of the perception agent is shown in Fig. 5.7. There is also an agent for driver characteristics, which contains aspects such as knowledge, permanent attributes, current state etc.

A driving process can then be represented in the form of a cycle with different transition options (Fig. 5.8). For each of the states a separate agent has to be assumed again. This

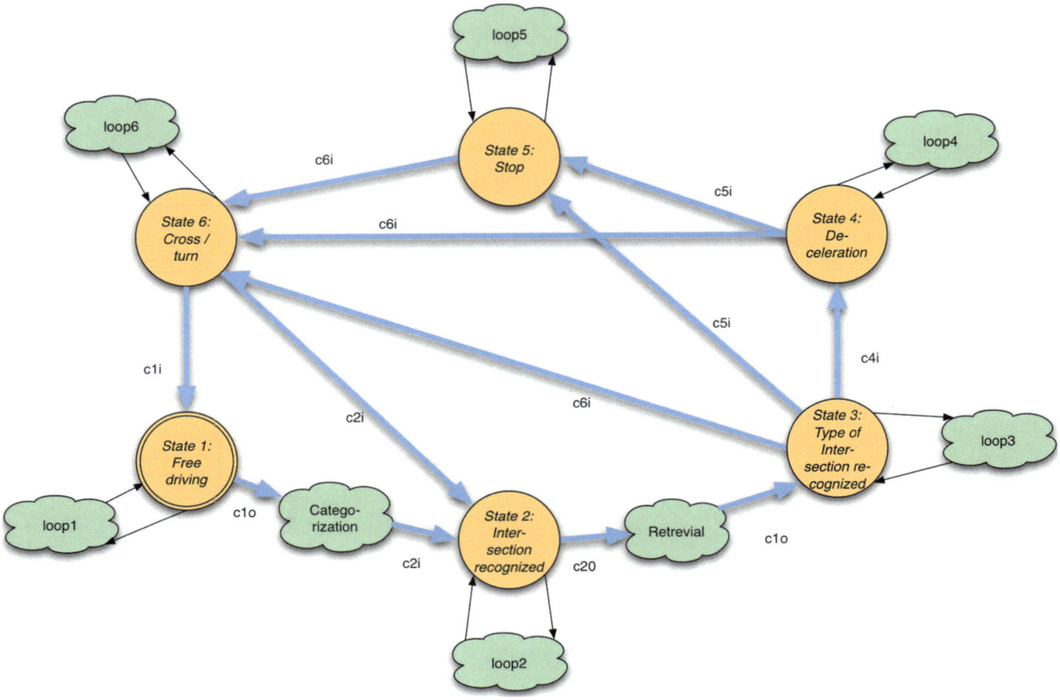

5

■ **Fig. 5.8** Temporal and random distribution of states using the example of crossing a crossroads. (According to Plavšic 2010)

is shown as an example for the case of free travel (State 1) in ■ Fig. 5.9.

■ Figure 5.8 shows in particular the transition to crossings and the behaviour there. In each segment represented there, the normative and the observed ordinary behavior are modelled, which is not discussed in more detail here.

The driver behaviour is modeled at the level of the driving situation and contains both deterministic and probabilistic descriptions. In the first step, the driver's knowledge is modelled as a fuzzy value that influences the other functions of the model. In the following, the behavior of the eye movement agent and the output rules for each of the possible inputs are presented in detail for each sequence.

Since the Plavšic model is particularly concerned with driving errors, special attention is paid to the modelling of eye movement in all the segments previously examined in detail. In particular, a so-called error watcher is implemented in this context, which is based on the difference between the normative behavior and the behavior observed in experiments.

The model described here is currently under construction at the Institute of Ergonomics at TUM. It contains many features that also appear in the aforementioned models. Detailed information can be found in Plavšic et al. (2010).

5.2 Anthropometric Human Models

5.2.1 Drawing Templates

The use of anthropometric tables quickly reaches its limits in connection with the geometric design of vehicle interiors (so-called packaging). These limits can be partially overcome with suitable body outline templates. Müller (2010) quotes Seeger (2007), according to which as early as 1928 body outline templates were used in Zeppelin airship construction whose anthropometric body measurements were based on company-specific specialist knowledge and whose genesis can

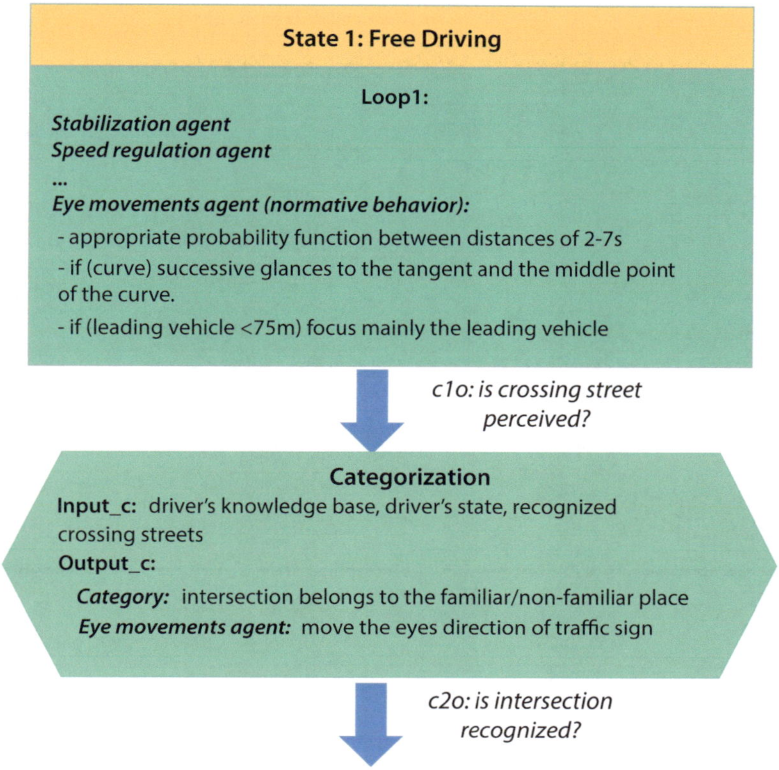

Fig. 5.9 Cognitive state 1: Free driving. The loop encompasses all agents within a state and is so named because of its repeatability. The categorization agent belongs to state 1. (According to Plavšic 2010)

no longer be traced today. He further reports that, on the basis of fundamental work in the field of anthropometry, body measurements of men, women and children of different ages (6, 8, 11 and 14 years) were presented for the first time in the form of line drawings under the names Joe, Josephie and Joe Junior and published in the reference book "The Measure of Man" (Dreyfuss 1959).

In conjunction with the H-point measuring machine, which is of essential importance for SAE-compliant packaging (see ▸ Sect. 7.1.2), a drawing template (SAE J826) was developed as early as 1962, consisting of the lower leg segment, thigh segment and torso. Remarkably, this outline template has no head (see Fig. 5.10), which must be seen in connection with the eye ellipse defined in SAE J941 (see ▸ Sect. 7.3.1). In its original form, this body outline template represented the dimensions of a 50-percentile man based on the measurements of Geoffrey (1961). In addition, in 1969 the U.S. Department, Education and Welfare added data for the lower leg and thigh of the tenth and 95th percentile man.

In the 1960s, the drawing stencil set according to Jenik (1973, DIN 33416) was realised for the ergonomic workplace design, which even permitted the three-dimensional construction of workplaces by applying basic geometric knowledge. However, this system has not become established for the design of vehicle interiors. A much greater significance in Germany has the so-called "Kiel Doll" ("Kieler Puppe") developed by Helbig and Jürgens in 1977. *Kiel doll* has been standardized as a two-dimensional template in three views in DIN 33408 since 1981 (see Fig. 5.11). However, the three-dimensional construction with these templates is extremely complex, so that in practice usually only the two-dimensional side view is used. The *Kiel doll* is available in the sizes of the first and fifth

5

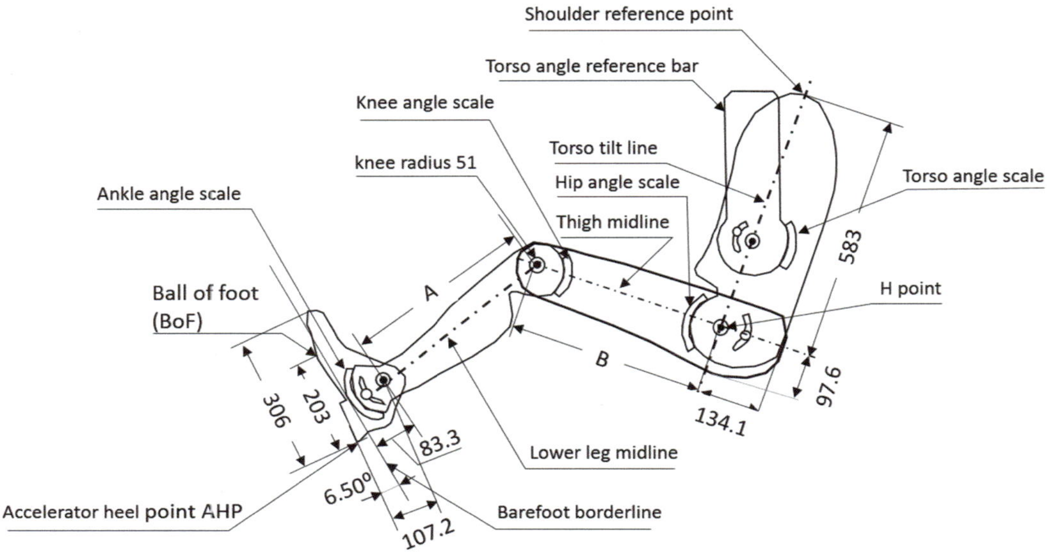

	10% M	50% M	95% M Body outline template
Lower leg A	392.7 mm	417.1 mm	459.1 mm
Thigh B	407.7 mm	431.5 mm	456.0 mm

Fig. 5.10 SAE body outline template

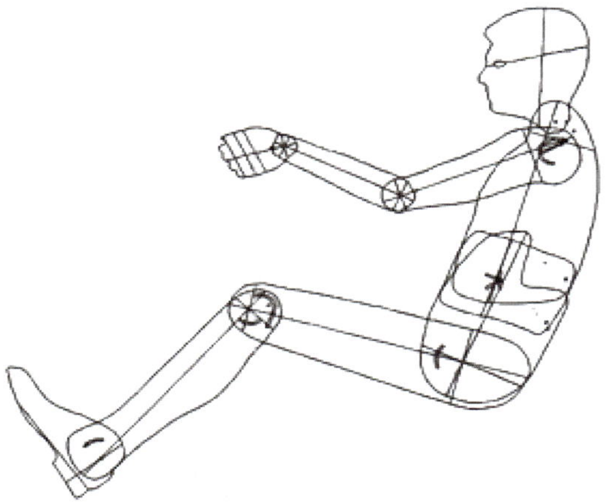

Fig. 5.11 Drawing template of the human figure according to DIN 33408 ("Kiel doll")

percentile woman, as well as the 50th, 95th and 99th percentile man. Different postures can be displayed by adjusting the joints on the drawing templates. In order not to produce too arbitrary postures here, DIN 33408-1 specifies so-called comfort angles (see Fig. 5.12), the experimental basis of which is questionable.

At least Bubb (1992a) gave a large variation range of the angle values for the adjustment of the Kieler Puppe on the basis of an experimental analysis on more than 40 test subjects (see Table. 5.1). It was obtained by placing the Kiel doll as a measuring instrument over the photographs of the test subjects in a variable vehicle mock-up.

Since the above mentioned SAE drawing templates and the Kiel doll are compatible in their basic structure, they represent today indispensable means in the automobile industry, with whose help a fast and standard-compliant conception of the driver's workplace is possible.

The relatively simple handling of the drawing templates, however, is opposed by the aforementioned limitation to a two-

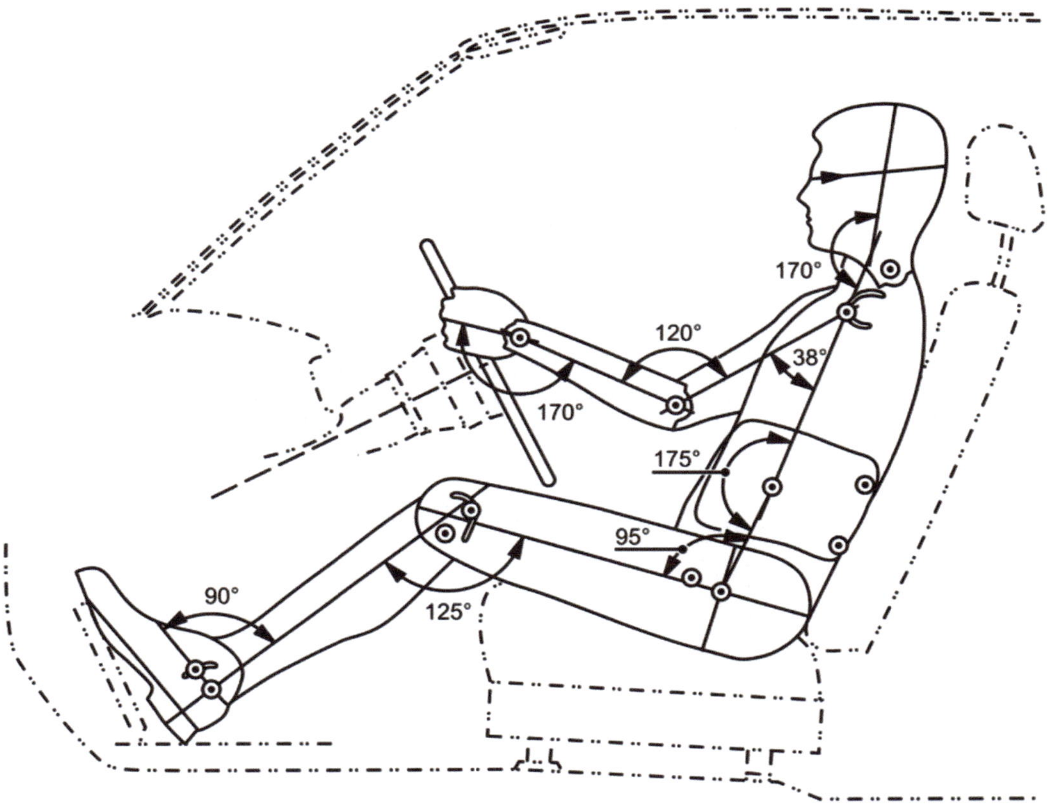

Fig. 5.12 Recommendations for the settings of drawing templates

Table 5.1 Posture taken by test subjects represented by the Kiel doll (Bubb 1992a)

Torso	Shoulder	Elbow	Hip	Knee	Ankle
25° ± 3°	39° ± 12°	146° ± 17°	107° ± 7°	122° ± 8°	84° ± 16°

dimensional representation and their inflexibility with regard to the representation of different postures, the consideration of movements, the sinking into the seat and much more. In addition, there is the problem that correlative relationships between body measurements cannot be represented in principle. For this reason, the body dimensions shown always represent a compromise, since the percentile data always refer to certain body elements, but not to the individual combination within a person.

5.2.2 Digital Human Models

5.2.2.1 Development and Areas of Application

With the advent of computer technology, which in the form of CAD techniques has brought about a revolution in automotive design, there has also been an effort to create a 3-dimensional image of the human figure in the computer. The first attempts in this direction were to transfer the existing drawing templates into the computer area. A seri-

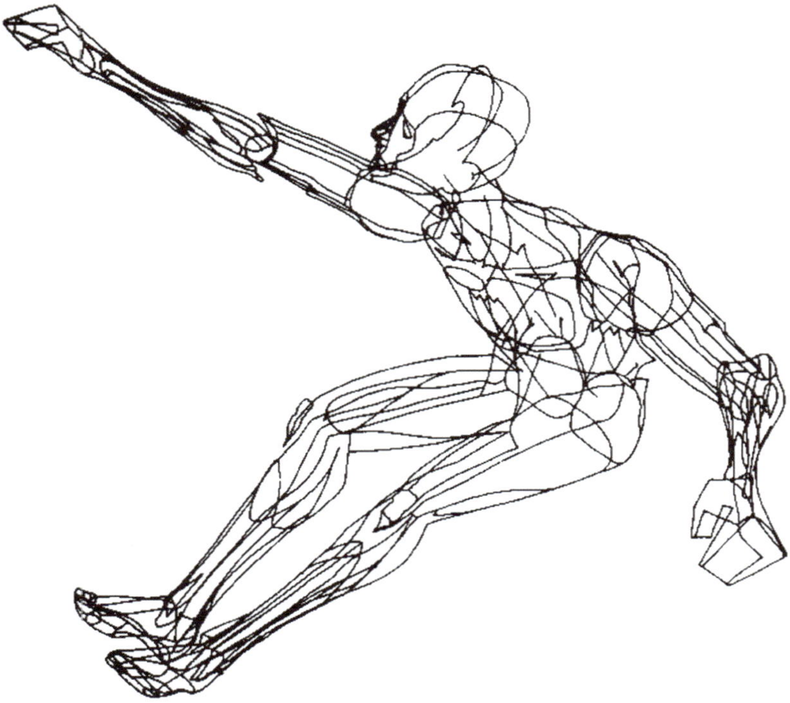

Fig. 5.13 The manikin First Man

5

ous development of human computer models had its starting point in American aerospace research. Initially, the aim was to map the mechanical behaviour of humans in the form of physical models in the computer. A more general biomechanical model for predicting population-dependent forces was developed by Chaffin et al. in the late 60s. The main goal of this work was the simulation of static forces to be applied by astronauts who have to lift, pull or push objects during their stay in space or on the moon. For this purpose, force measurements were made on over 2000 people. The result was the program 3DSSPP (3D Static Strength Prediction Program), which is still used today by authorities and organizations, especially in the USA. Almost at the same time an independent computer-based human modeling began with the development of the model BOEMAN for the simulation of accessibility for an average human being in fighter planes. Fetter realized the computer model *First Man* for this in 1967, which, based on the anthropometric dimensions of a 50-percentile man, was scalable to different body sizes (see Fig. 5.13). It contained a program for the optimization of postures on a non-linear basis, which, however, was so time-consuming for the possibilities of computer technology at that time that it could practically not be used for everyday questions. In the 70s BOEMAN was taken over by the American Air Force Medical Research Laboratory for Aerospace (ARML). The posture simulation model was simplified and the possibility was implemented to simulate a larger number of male and female anthropometries that could be integrated into different types of military aircraft. The human model resulting from ARML's research became known under the name COMBIMAN.

The first anthropometric human model of importance created especially for the computer, which was also generally available and usable, and which is used, among other things, for the conception of automobiles, is probably SAMMIE (System for Aiding Man-Machine Interaction Evaluation; Bonney et al. 1969). It was developed at the University of Nottigham and is primarily used to design vehicle interi-

Fig. 5.14 SAMMIE

ors as well as other activities around the vehicle. The model is still available today in a refined and further developed form (see Fig. 5.14). With its "Fit" module, people of different nationalities, different ages and different body heights can be modelled. The "Vision" and "Mirror" modules make it possible to answer questions of direct and indirect vision. With the module "Reach" gripping spaces and with the module "Posture" postures can be generated and evaluated.

At about the same time (1974) the biomechanical software system CALSPAN-3D-CVS (Crash-Victim-Simulator) was developed. The integrated dynamic analysis functions are suitable for studies of safety systems (airbags, belts) in vehicles and for collision analyses between cars, pedestrians and motorcycles (Hickey et al. 1985a). These two systems represent, so to speak, the starting point for two lines of development, which now appear more or less parallel and independent of each other.

Figure 5.15 shows the temporal development of the human models used in the automotive industry. The lower line characterizes models which serve the construction and evaluation of the anthropometric design of workplaces. Named here here models like ERGOMAN from the Laboratorie d'Anthropologie Appliguée et d'Ecole Hamaine in Paris, WERNER from the Institut für Arbeitsphysiologie at the University of Dortmund, Tommy from the TU Dresden, HEINER from the TU Darmstadt, as well as ANYBODY and ANTHROPOS from the German company IST and Man3D developed at INRETS (today IFSTAR) in France. Since the beginning of the 1990s, these models have been joined by human models that simulate work processes and thus allow a simplified calculation of the working time required depending on the geometry of the workplace according to the MTM method (e.g. Tecnomatix: after an original connection with its own human model eHuman first realized via RAMSIS, later via Jack under the name "Siemens PLM"). In addition to the human models developed for automotive development, there are around 130 human models worldwide, most of which were developed for general use in workplace design, product design, safety checks or the documentation of planning results (Aune and Jürgens 1989; Waldhier 1989; Hickey et al. 1985a). In the anthropometric data sets they differ only slightly from those used in the automotive industry. However, the analysis functions of the individual systems are strongly adapted to the application focus.

A relatively new development is the model *Santos* which, however, is not yet available as

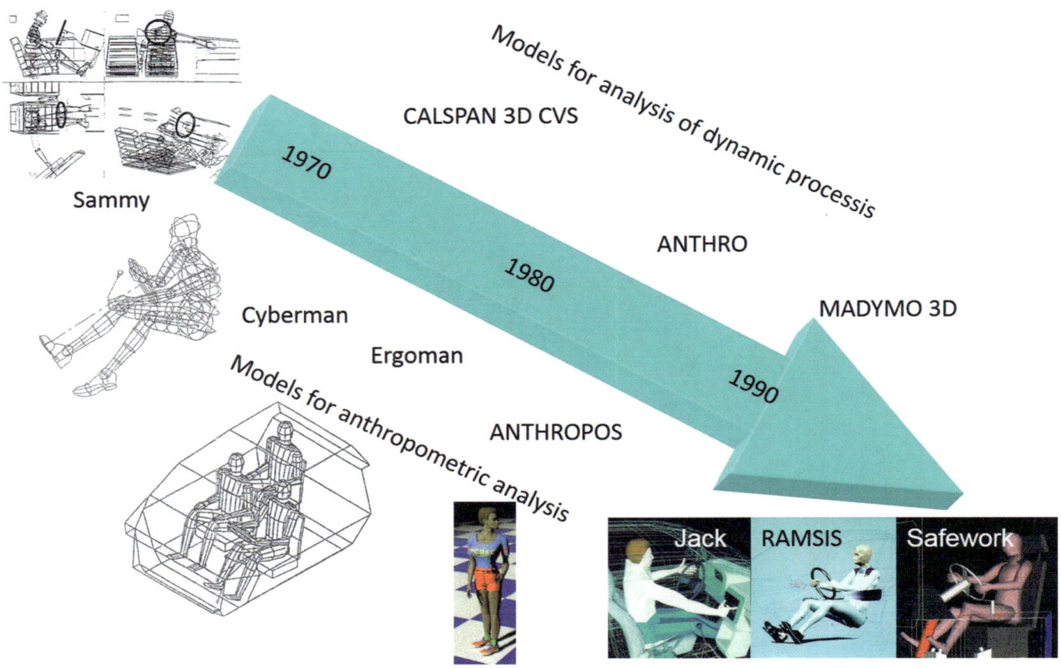

Fig. 5.15 Development of human models used in the automotive industry

commercial software. It comes from the Virtual Soldier Research Program of the University of Iowa. The main objective of this program is the simulation of a soldier in order to evaluate systems, components and products in the military field. The partial aspects published so far, however, also show some interesting properties for commercial applications. The model has accurate biomechanics, precise models of muscles including contraction, deformable skin and simulation of vital values (heartbeat, blood pressure, metabolism). Furthermore, it is possible to check visibility, forces and moments at load handling as well as the effect of clothing including internal thermodynamics. A module for artificial intelligence will provide willful perception as well as interaction and simulation of groups (quoted from Mühlstedt et al. 2008).

The upper line in Fig. 5.14 represents the branch that deals with the simulation of passive dynamic processes. The above-mentioned CALSPAN-3D-CVS model, which is used, among other things, for studies of safety systems in vehicles (airbags, belts) and for collision analyses between automobiles, pedestrians and motorcycles, deserves special mention here. It was further developed and used by various companies (Hickey et al. 1985a). In 1987, a processor for the dynamic modelling of humans was developed for the Adams calculation system, a multi-body simulation system widely used in the automotive industry, with a similar aim. In 1988 the company TNO in the Netherlands developed the dynamic crash analysis system MADYMO-3D (Mathematic Dynamic Models), whose main task is to represent hardware dummies (e.g. Hybrid III) in the computer. In each of the examples mentioned, the behavior of the human body, which would otherwise be performed using crash tests with hardware dummies, is already to be simulated in the virtual computer experiment. This has enormous cost advantages if one takes into account the expenditure for such crash tests (Seidl 1997). However, with regard to the legal requirements to be fulfilled, more emphasis is placed on the realistic simulation of the crash dummies than on those of the living human organism.

5.2.2.2 The Most Important Geometrically Oriented Human Models for Vehicle Design

All computer models are based on the assumption that by using the three percentiles (fifth, 50th and 95th percentile), separated according to gender, the consideration of human dimensions can be sufficiently ensured. Although it is often referred to the possibility provided by computers that the models can be adapted to individual anthropometric data, a coherent human model concept, consistent from data collection to simulation and computer analysis, did not exist until the early 1990s (Gärtner and Schweingruber 1992). Only the more recent developments by RAMSIS, Safework/ Human Builder and Jack are endeavouring to remedy these shortcomings (see below) and furthermore to open the way to a new future for anthropometric tools. Mühlstedt et al. (2008), Seidl (1997), Bubb (2007), and Bubb et al. 2019) provide an overview of human models and their areas of application.

The structure of an anthropometric digital human model is characterized by the outer skin model, which gives it a realistic appearance, and by the inner skeleton model. This inner model has the task of reproducing all postural and movement functions of the human being with as few joints as possible (see Fig. 5.16): It can be observed that with the continuously improved computer capacity this restriction falls more and more in favour of a realistic representation of the functionality of the skeleton. The inner model serves as a framework for the skin, which is firmly or elastically connected to it via a mathematical algorithm. The mathematical algorithm must take care that with an animation of the dummy harmonic, realistic transitions and deformations of the skin are achieved. At present, work is being done in various places to reproduce the deformity of the skin and the underlying muscle and fat tissue by using finite element methods (FEM) to correctly simulate contact with seats. (e.g.: Casimir, DYNAMICUS).

A particular challenge for anthropometric digital human models is the correct reproduction of the dimensions of body parts, mobility and movement spaces. A simple transfer of values from anthropometric tables, whose values have been collected with conventional measuring methods, is actually forbidden, since no individual is characterized by identical percentile values in all body dimensions and therefore, on the other hand, the stringing together of identical percentile measurements for the construction of a human model would lead to contradictory results. Nevertheless, the JACK model developed in the USA in the mid-1980s under the auspices of NASA (see below) uses table values adapted for anthropometry. The human model SAFEWORK (developed in the 80/90s at the University of Montreal) uses correlation coefficients, which are derived from the percentile values of the individual body measurements given in tables, a pragmatic solution, which is statistically incorrect, but leads to usable results. For the human model RAMSIS (1986–93 developed for the German automotive industry as part of a FAT project), the actual correlations that were obtained in anthropometric surveys in the 1980s in the former GDR (East German Republic) were used for modelling. In the new conception of

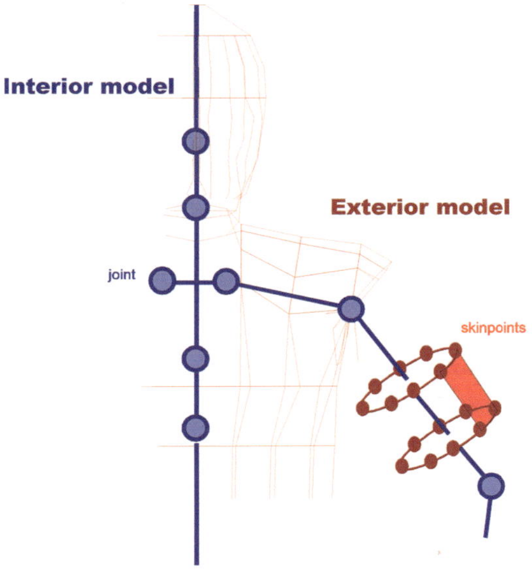

Fig. 5.16 Inner ("skeleton") and outer (skin) model of a digital human model

this model (RAMSIS Nextgeneration), the values obtained in the "Size-GERMANY" measurement survey (see below) and the correlations known from it are be used for the modelling, which lead to more precise predictions of the anthropometric dimensions of this model.

Today, anthropometric models are also used to model muscle forces. There are competing methods of reproducing the anatomical-functional properties of the muscle and anchoring it to the skeleton (e.g. the modelling of ANYBODY) with purely functional models in the form of direction-dependent torque values for each joint, but which are based on measurements on test subjects (currently implemented in this form at RAMSIS and at Santos – there somewhat modified) and those which, without modelling the occurrence of the forces on the basis of regression analyses, predict the force values that are released to the outside (force analysis at Jack).

From the application requirements for such computer models of humans, the necessity of modelling movement arises, among other things. A wide variety of approaches are known for this, ranging from a pure *animation* (= tracing of a movement recorded by computer-aided measurement) to the various *simulation approaches* who partially work with animation tools (e.g. visualization of running, see also the movement simulation), but also try an independent movement simulation. For the latter, the most common approaches are to minimize the sum of the energetic expenditure in the individual joints with regard to a movement goal. Another promising approach is to simulate the motion design formed in the brain with regard to the motion goal by calculating the trajectories for "leading body parts" of the motion design and calculating the remaining motion of the body on the one hand on the basis of minimizing the force/energy expenditure and on the other hand on the avoidance of collisions (proposed for RAMSIS).

Anthropometric human models are mainly used to design the dimensions of workplaces (particularly frequent field of application: design of vehicle cabins, but also of other workplaces, e.g. assembly workplaces) as well as the arrangement of displays and control elements in dependence on the expected distribution of the physical dimensions of the user population already in the design and construction phase.

In the following, the anthropometric human models most frequently used today for vehicle design will be briefly presented and characterized.

■ Safework/Human Builder

During the 1980s, the SAFEWORK model was developed in Canada at the Ecole Polytechnique Montreal (Carrier et al. 1987). The first version was realized on PC and used modules for anthropometry, motion and the representation of the environment (Seidl 1997). SAFEWORK was acquired in 2000 by the French company Dessault-Systemes. Due to the implementation in the CAD system CATIA, this human model is widely used today. It is also available in the Enovia and Delmiay product families. In all cases, today it is known under the name *Human Builder*. Anthropometric data from various countries (France, America, Canada, Japan) are available to the model. As mentioned above, by simply using the standard deviation in these tables, a correlative relationship between the individual body measurements is estimated that allows a reasonably correct reproduction of the proportions. Spine, shoulder and neck are modelled in detail. The natural limitation of the joint angles is taken into account. By using prepositioned postures and inverse kinematics[3] a natural posture of the model can be created. Predefined postures are used. With the help of further modules the functionality

3 In contrast to "forward kinematics", in which a movement is calculated with knowledge of the geometric conditions and the moments of inertia of the individual body elements by assuming forces, or in the static case an attitude is calculated, with the "inverse kinematics" from knowledge of the geometric and mechanical boundary conditions with knowledge of the observed movement or attitude is calculated, which forces are necessary with their respective effective directions, in order to explain the observations.

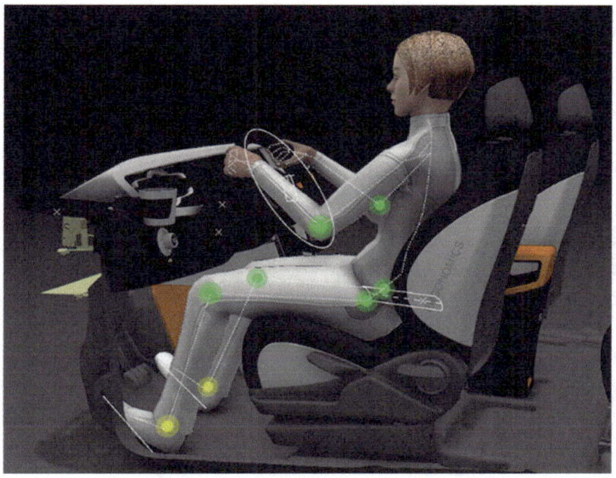

■ **Fig. 5.17** Human Builder

of the application can be extended. The module "Vehikel Occupant Accomodation" is of particular importance for car design (Müller 2010; example ■ Fig. 5.17). However, there are also tools with the help of which static posture analyses, lifting and carrying according to Niosh, pushing and pulling according to Snook & Ciriello, as well as hand-arm movements (RULA) can be examined. The "Human Posture Analysis" allows the quantitative and qualitative posture analysis, whereby for each joint comfort and discomfort areas (this model approach does not agree with the ideas about comfort and discomfort given in the ► Sect. 3.3.4!) are defined by the user, which in turn allow an optical control by a colour highlighting. Automatic posture optimisation is possible. "Human Task Simulation" allows action analyses for work processes to check ergonomic, health and safety aspects (e.g. climbing stairs, using a ladder, creating macro-like movement paths, etc.). An interface with the "Delmia Process Engineer" program also allows time analyses (quoted from Mühlstedt et al. 2008).

■ Jack

The JACK model was developed under the auspices of NASA together with the University of Pennsylvania at the Center for Human Modeling and Simulation in the mid-1980s. Originally, this model, which initially bore the name TEMPUS, was designed for work planning during the assembly of today's ISS space station. Questions should be clarified how an

■ **Fig. 5.18** JACK

astronaut can best reach and fasten objects, but also how the astronaut's view is with a space suit. This system is an animation tool based on methods of robotics, which realizes dynamic sequences almost in real time with a very high-quality visualization. Over the years, JACK has also been used for analyses on military aircraft and other vehicles. JACK uses the anthropometric database Ansur. In the meantime JACK is complemented by its female counterpart called JILL. JACK has a movable spine and joints with natural conditions simulating limitations. In "Classik JACK", the "Occupant Packaging Toolkit" module uses inverse kinematics to enable positioning in a car interior (■ Fig. 5.18). This tool can also be used to model the driver's visibility. A number of further analysis functions are already integrated into the human model, so that e.g. power-guided posture and movement simulation enables the calculation of even more complex tasks. As Seidl (1997) states, Jack is not a self-contained ergonomics tool. His strength is that the user can integrate his own methods and procedures through open interfaces and present them vividly with Jack. This is also its great significance today; because of its value for money, it is particularly attractive as a visualization tool for scientific institutions to illustrate developments made there. Often these developments are

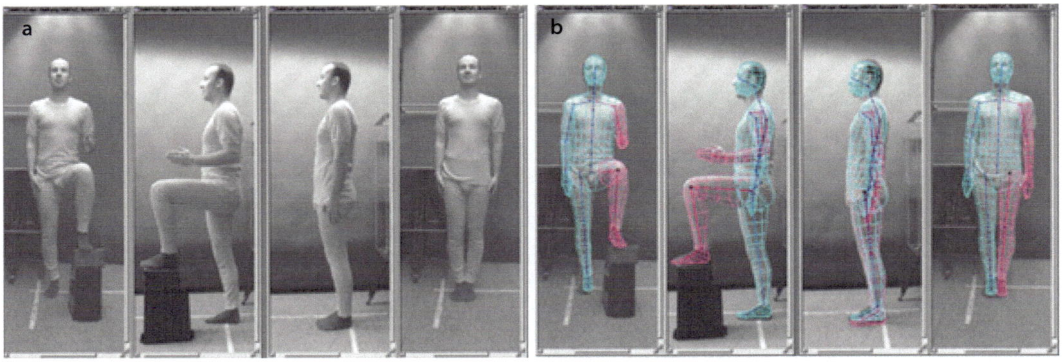

■ **Fig. 5.19** Photogrammetric recording of the test subject **a** and overlay with the computer dummy **b**

taken over by the customers directly into the product or production process.

■ RAMSIS

Between 1987 and 1994, in cooperation with the German automotive industry and some important suppliers, Tecmath, the Catholic University of Eichstätt, and the Institute of Ergonomics at the Technical University of Munich, the human model RAMSIS (**R**ealistic **A**nthropological **M**athematical **S**ystem for **I**nterior-comfort **S**imulation) was developed. Tecmath subsequently took over the marketing and further development of RAMSIS, while many more scientifically oriented investigations continued to be carried out by the Institute of Ergonomics at the Technical University of Munich. In 2002 the company Human Solutions GmbH, still based in Kaiserslautern, separated from Tecmath and now develops and distributes RAMSIS. Due to its entire history, RAMSIS is initially tailored specifically to the needs of the automotive industry. Because of its importance for the automotive industry, this model is described in more detail below.

A special feature of the project right from the start was that the model was used both for the measurement and for the design. This should ensure that the model always remains very close to the real person. Using a non-contact (optical) measuring method, individual anthropometric data were collected in such a way that a good anthropometric calculation model of the respective test subject can be built up on the basis of the dummy described. The procedure consists of recording the subject by two cameras whose optical axes are perpendicular to each other. The simple values that can be captured by these cameras (e.g. body height, fingertip height etc.) are used to adapt the dummy to the proportions of the test subjects by means of superposition (see ■ Fig. 5.19). The dummy required for this superposition was originally programmed on a PC, which is why it was given the name PCMAN. PCMAN has the same geometrical properties as RAMSIS. The data acquired by PCMAN can be transferred directly to RAMSIS via a corresponding interface. The further anthropometric measuring program provides for various fixed postures, through which the exact position of the joints and, to a certain extent, their variation depending on the posture are recorded. By modifying the outer model of the dummy so that its contour is as close as possible to the contour of the persons photographed, a good image of the dummy is obtained in the computer.

RAMSIS enables, not least on the basis of this data acquisition, the percentile and correct representation of humans based on suitable populations with the guiding dimensions body length, corpulence and proportion (ratio of body height to stem length).

Later, the described measuring procedure was refined with the help of a body scanner developed specifically by Tecmath/Human Solutions. Between 2000 and 2010 the garment industry became aware of the possibilities of the body scanner. In cooperation between the Hohenstein Institutes, the

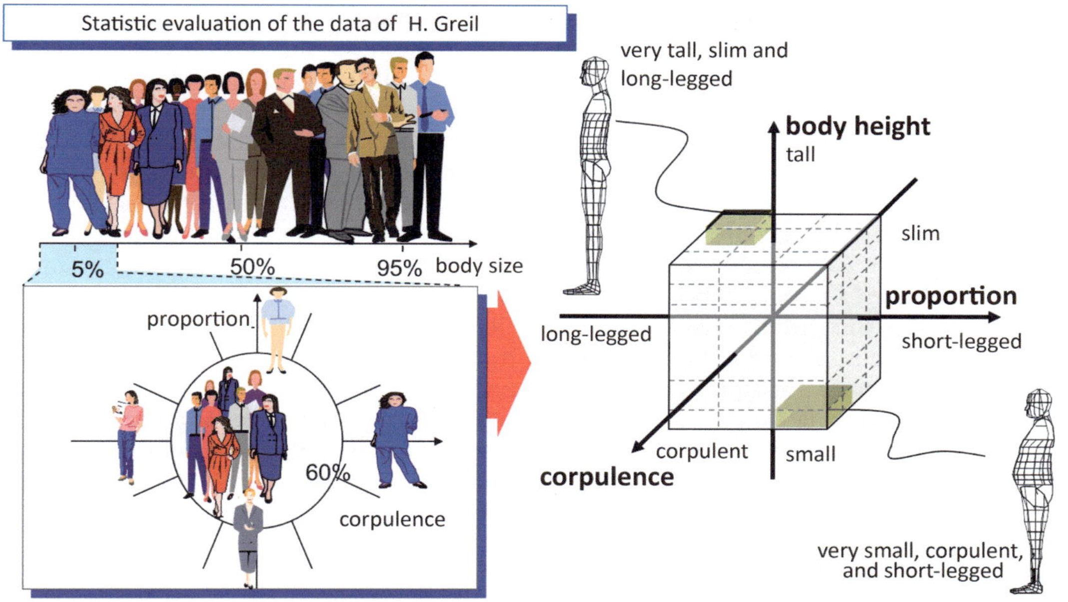

Fig. 5.20 Statistical method for generating the 45 RAMSIS types

German automotive industry and Human Solutions, the SizeGERMANY project was launched on this basis, in the course of which representative series measurements were carried out on 12,000 women, men and children in Germany. The result was a unique set of anthropometric data that was previously unavailable worldwide with such precision.[4] On the basis of this data set, a new edition of the RAMSIS is now developed, the so-called Ramsis Next Generation, which provides an even better adaptation to different proportions of the human appearance. Currently, a "SizeITALY" and "SizeAMERICA" measurement series based on SizeGERMANY is being developed. Similar data are available for France and Spain. In this way, a world-encompassing anthropometry is successively created, which goes far beyond all existing measurements.

For the anthropometric modelling of the collected data, reference was made to factor-analytical results, according to which body height and corpulence represent two largely independent factors. Because of the problems that arise in narrow vehicle cabins, the third factor added was proportion, which is defined by the ratio of trunk length to body height. Individual body measurements can thus be classified into the three-dimensional space spanned by these axes (Geuß 1995). Starting from a size type determined by body length (e.g. fifth percentile with a range from 2.5 to 7.5th percentile; see also Fig. 5.20, left), a mean expression and eight extreme expressions are now defined for the remaining dimensions "corpulence" and "proportion". In this way, with the five body size types selected, 45 types are created for each sex, whose proportions correspond to those found in an elaborate study by Greil (1993) (see Fig. 5.20). Once all the relevant correlations have been calculated for these investigations, RAMSIS can then develop a further

4 If one takes random samples (or otherwise stratified samples according to certain criteria) in order to obtain anthropometric data, as normally happens, the small and the large are in principle underrepresented due to the distribution of body sizes, which can be described with good approximation by the Gaussian distribution. However, since the distribution of body size types is known, SizeGermany has made sure that the same number of body size groups are measured. This results in much more precise values, especially for the marginal groups, than is possible with the usual data collection.

stage for special investigations in which extreme types can be put together by the so-called "body builder"- starting from any guide dimensions - and the probability of this type occurring can be specified in each case.

Various experiments and studies were carried out to create a model of the postural comfort prognosis (Seidl 1994). The most important tool was a variable vehicle mock-up, which allowed a flexible adjustment of the vehicle interior dimensions. Pedal, seat and steering wheel positions were freely configurable so that vehicle concepts ranging from sports cars to vans could be presented. Various settings were made in the experiments. The test subjects had a driving task in the driver's platform, which was presented to them via a simple simulator. During the experiment, the subjects were recorded with video cameras and their posture was recorded using the PCMAN method described above. After the experiments, the subjects had to complete questionnaires on postural comfort, fatigue and tension of individual body parts. Further experiments complemented this fundamental investigation:

- *Accessibility studies:* Here the test subjects had to manipulate different control elements typical for motor vehicles. The postures during operation were recorded according to the method described.
- *Screening tests:* The test subjects were asked to look at illuminated lamps that were mounted around the driver's platform and controlled in random order. The postures adopted in the process were also registered. This test is an essential basis for the visual simulation in the RAMSIS CAD ergonomics tool. From this it was possible to deduce which visual tasks are to be solved from the corners of the eye and when and how strongly the head must be turned in order to still recognize visual targets.
- Furthermore, the sitting position of test persons in trucks was investigated. Mercedes Benz AG provided an experimental vehicle for this purpose, which, similar to the driver's platform described above, was freely adjustable in terms of its dimensional concept.

The posture of the body was recorded in all the studies mentioned for the creation of the posture model. This is available in the form of space angles for each body element. The analysis of the angular distributions shows that for some body elements very "sharp" distribution curves result. This means that with these body elements the human being always wants to set a certain angle largely independent of the experimental constellation, which he also finds tolerable (e.g. in the hip). With other body elements, on the other hand, the human being is uncritical with regard to the body element angle, the angle distribution is, apart from the edge areas, rather flat: no special angle was preferred or classified as particularly uncomfortable by the test subjects.

By transforming these measurement results into mathematical formulas (which must be constantly differentiable), so-called probability pots are created for each joint (▫ Fig. 5.21). Using a special optimization procedure, the software then always searches for the lowest point within the multidimensional mountain range of angular probabilities thus created in the application. In this way, the system always calculates the most probable posture that the persons would adopt under the given boundary conditions.

A series of validation tests were carried out which clearly demonstrated the relevance of the probability model for the predictions of postures especially in motor vehicles (e.g. Kolling 1997).

The experiments described and the modelling of their results form the basis for the vehicle-specific capabilities of RAMSIS, namely:

- the simulation of realistic and statistically verified postures depending on the given geometric limitations,
- the evaluation of the sense of comfort for a given posture,
- ergonomic analysis functions e.g. for the view or the belt run.

These basic RAMSIS functions are supplemented by additional modules. The "RAMSIS-dynamic" module enables the simulation of hand and foot movements of the

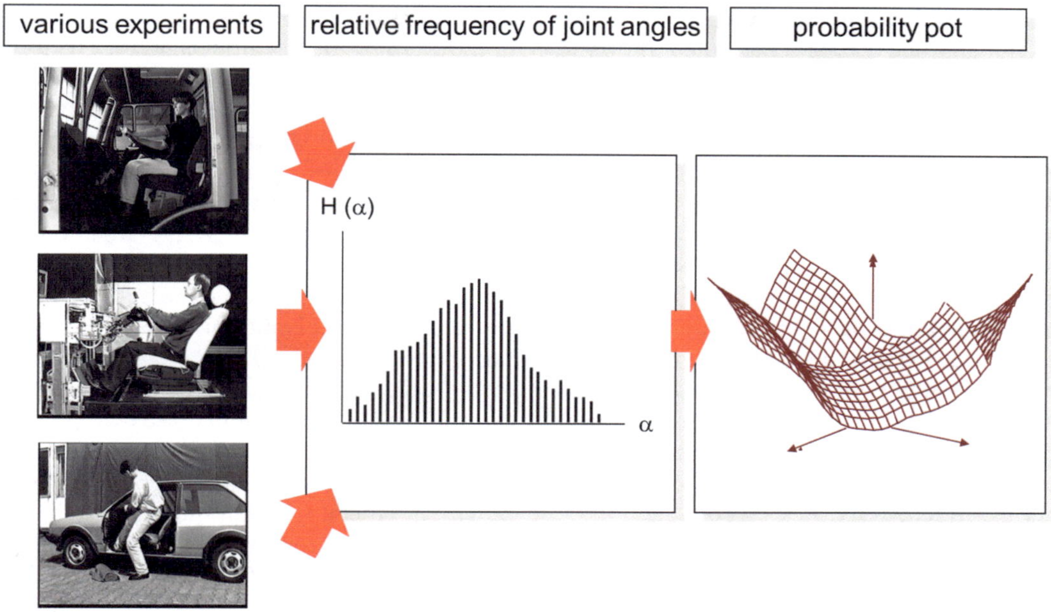

Fig. 5.21 Test procedure for determining the probability plug in the joints (example shoulder joint)

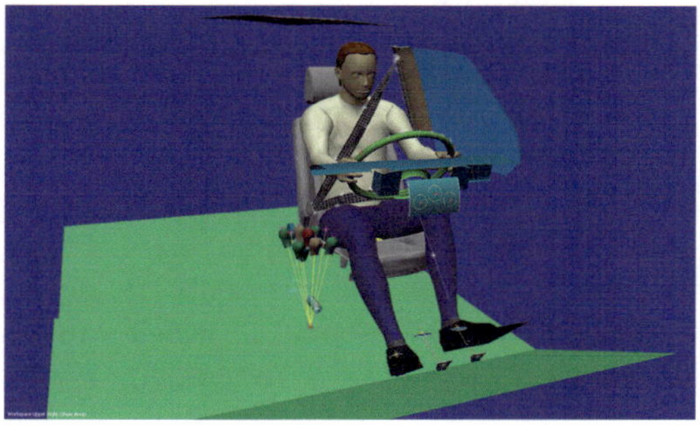

Fig. 5.22 RAMSIS

human model sitting in the vehicle on an experimental basis. The "Package Designer" module is available especially for car dimensional concepts (see Fig. 5.22). In addition to various other options, the "RAMSIS sits" module provides a largely correct positioning of the human model in a seat described by physical properties. A significant improvement in the functionalities associated with visibility has been achieved by the recently developed module "RAMSIS cognitive" by Remlinger (2013). RAMSIS was supplemented by important analysis and design functions for taking visual perception into account: The aim was beside others to consider the factually given interlocking of the system ergonomic design of display and control elements and the geometric localisation, which is determined by anthropometric conditions. The analysis functions refer to the field of vision, eyeglass vision, physiological and psychological glare, aspects of accommodation, visual acuity and reduced information intake due to aging. The advantage of calculating the situation from the point of view of drivers of different anthropometries also affects the analysis functions with regard to the geometric-optical limits of visual perception. Special functions were developed for the calculation of geometric masking (A-pillar problems), the limited viewing angles of LCDs and the positioning of the optics of HUDs. A further analysis function relates to the physiological limits that influence the direct view of traffic events. The new RAMSIS functions are supplemented by the so-called Daimler-Scholly method (see ▶ Sect. 6.3), which provides for an evaluation of visual ranges based on experiments.

Not least because of these very specific vehicle-related possibilities, RAMSIS is now used almost worldwide (according to Human Solutions 70%) in the automotive industry for package evaluation and design, since almost all populations can be represented anthropometrically (e.g. Germans, Japanese, Koreans or Americans). In addition to adults, children can also be simulated. RAMSIS is also used in the design of motorcycles, aircraft, construction machinery, industrial trucks and other workplaces. A further field of application arises in connection with Virtual Reality, since with this technique using the so-called "helmet mounted display" the operating operator cannot see his own body (e.g. the hands). The corresponding body elements are then visualised by RAMSIS, whereby they are controlled by the position of the real body parts of the test person to which the corresponding markers are attached (see ► Sect. 10.3.3.2).

5.2.2.3 Biomechanical Human Models

In contrast to anthropometric models, biomechanical models place less emphasis on the precise reproduction or modelling of physical dimensions than on the mechanical-dynamic properties. One is often even satisfied with the three important percentiles for the respective sex. Biomechanical models use computer programs to realize multi-body mechanics, which themselves are based either on the d'Alembert principle (realization of the equilibrium of forces under consideration of static and dynamic forces) or on the Lagrange function (a system is completely described by the energy of motion and the potential energy). The program system ADAMS, which was developed in general for the calculation of mechanical problems and which uses the d'Alembert principle, provides a biomechanical human model. With the help of the SIMPACK system, which uses the Lagrange function, a biomechanical human model was also developed, with the help of which various detailed problems (e.g. mechanics of movement of the knee) were investigated. Of great importance today is the human model MADYMO developed at TNO (Netherlands) (see ◘ Fig. 5.23 left), which was initially designed to simulate the properties of crash dummies in the computer and therefore reproduces them in its anthropometric properties. Recently, however, adaptations to the anthropometry provided by RAMSIS have been made in connection with various questions. A model that enables the continuity from the individual test subject, measured with PCMAN via RAMSIS, to a biomechanical representation, is the model DYNAMICUS, which is based on the multibody system program *alaska* developed at the University of Chemnitz (◘ Fig. 5.23 on the right). In this way, at least the individual dimensions of the joint to joint distances as well as the masses and moments of inertia of the individual body parts, which can be calculated on the basis of the largely correct

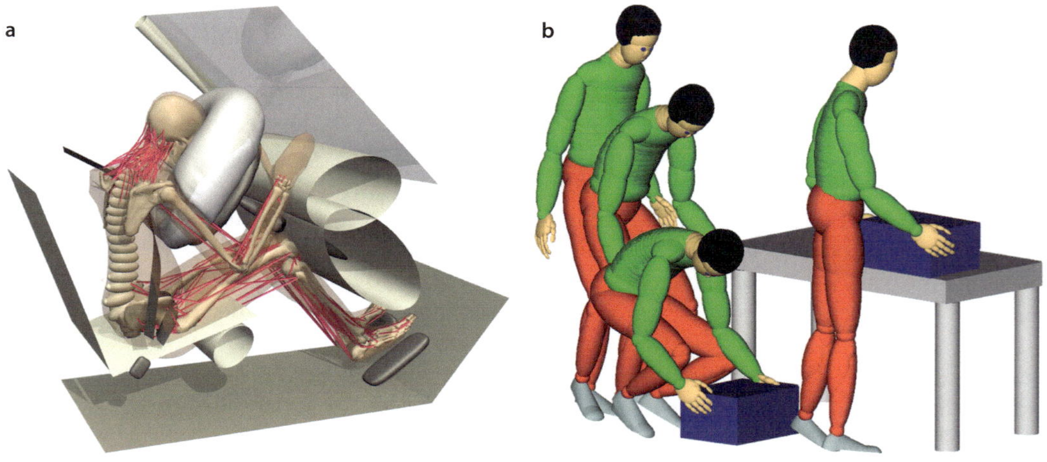

◘ **Fig. 5.23** Biomechanical model MADYMO (**a**; ► www.tassinternational.com, TASS International and TNO; Meijer et al. 2012, 2013). (Courtesy and DYNAMICUS **b**)

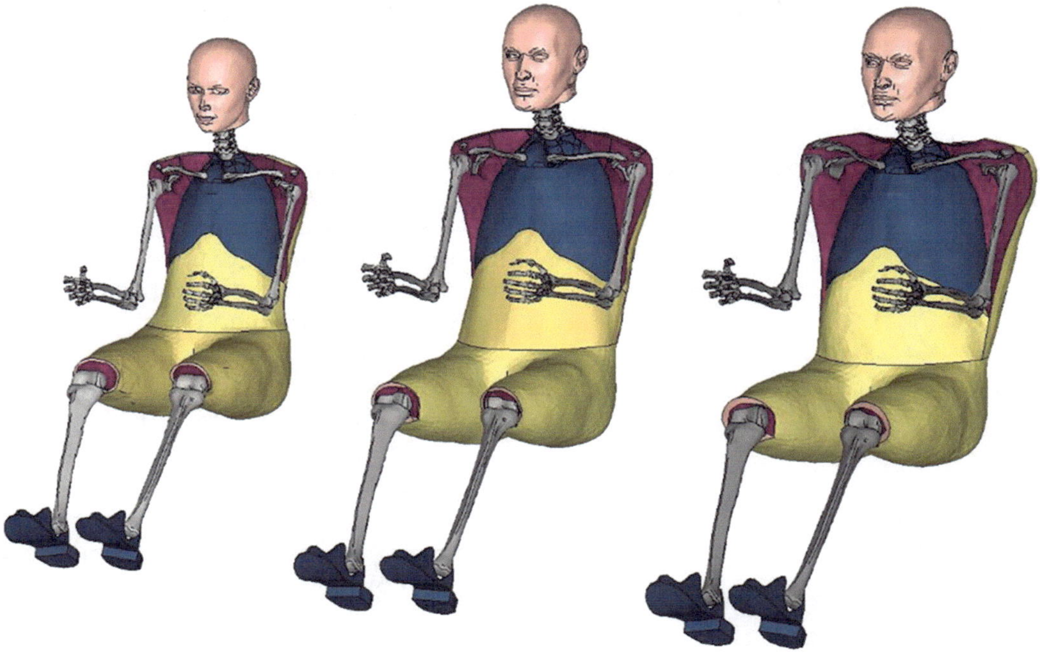

Fig. 5.24 The biomechanical human model CASIMIR in three different percentiles

RAMSIS geometry and assumption of an average specific weight of the person, can be correctly transferred to the model. The determination of joint stiffness, description of the musculature by spring damper elements and the like always remains a detail task left to the skill of the user.

Biomechanical models can also be extended to include FEM methods to calculate the surface pressure distribution in contact with the environment (e.g. a seat), whereby elastic properties are modelled for both the seat and the corresponding body part. On this basis, the CASIMIR model was developed, which is to be used specifically for calculating the pressure distribution between seat and body at an early stage of development (Fig. 5.24). According to Knauer (2010), CASIMIR currently represents the most comprehensive biomechanical model of the human being, which specifically depicts the complete anatomy of the seated person, in particular all soft tissues of the thighs, buttocks and back, in an FE model. Abdominal and back muscles are reproduced via non-linear and frequency-dependent spring/damper elements. Muscle activity is determined via an optimising mechanism with the aim of optimising the energy the body provides.

Biomechanical models are used to investigate the passive reaction of humans to external forces. The calculation of reaction forces in crashes but also in vibration excitations on seats are frequent areas of application. In the field of sports science, it has become established to introduce active forces into these models and thus to examine the sequence of complex sports movements from a theoretical point of view. One promising method for the future is to use computer-based measurement to record the movements of real test subjects and to measure the active forces in biomechanical models using a mathematical optimization procedure so that the movements of the model correspond to the movements measured. In this way one could indirectly measure the forces applied. The prerequisite for such a procedure is the combination of biomechanical and anthropometric models, which are adapted to the individual measurements of the respective test subject. A first approach to such an approach has already been carried out within the framework of the European research project REALMAN. The AnyBody model developed at the Danish University of Aalborg, which provides a very precise modelling of the muscular connection to the skeleton (see ◘ Fig. 5.25), is also used today in the automotive industry to objectify questions concerning comfort. There is a cooperation process between this model and RAMSIS so that AnyBody can benefit from its good anthropometric data. Similarly, there was a cooperation project between CASIMIR (Wölfel und Partner), RAMSIS (HumanSolutions GmbH) and the Institute of Ergonomics (TUM) with the aim of measuring seat pressure distributions already in the concept phase and evaluating them under discomfort aspects.

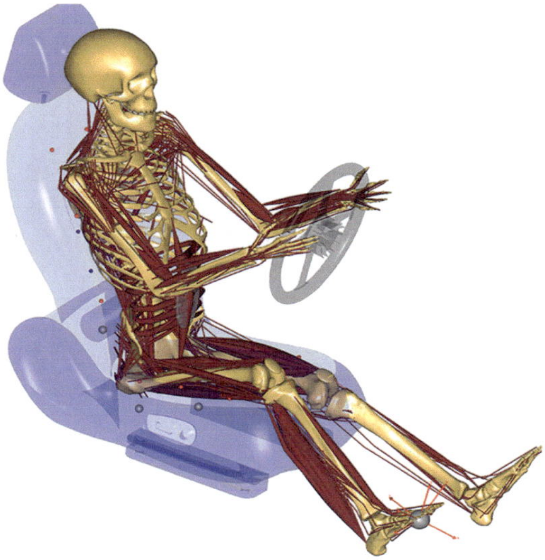

◘ **Fig. 5.25** The AnyBody model, which enables the most precise simulation of muscle use to date

5.3 Summary Appreciation of the Benefits of Human Modelling

As already indicated at the beginning, it can be a promising future development of modelling to combine not only control engineering and cognitive models, but also anthropometric and biomechanical models (see Sträter 2001). Especially through the methods of multi-body simulation and inverse kinematics it is possible to predict for example the influence of posture on steering precision, which in turn depends on different body types (see also Jürgensohn 2002). ◘ Figure 5.26 shows the

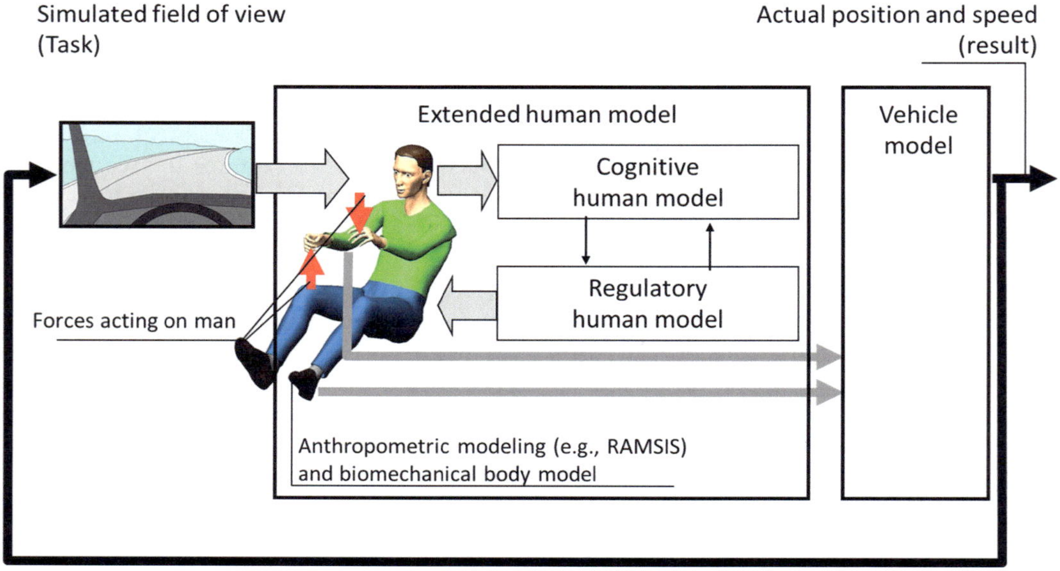

◘ **Fig. 5.26** Overall structure of a biomechanical-anthropometric-cognitive human model for the purpose of adapting vehicle characteristics to general driver characteristics

concept of such an integrated model, as suggested by Bubb (2002). Such modelling also has the aim of deriving technical measures through which ever better adaptation to general human characteristics can be achieved. Individual adaptations to anthropometric conditions and possibly even individual adaptations to human cognitive characteristics are another future field of research.

References

Literature Used

Aasman, J.: Modelling Driver Behaviour in Soar. Dissertation,. Rijksuniversiteit te Groningen (1995)

Anderson, J.: How Can the Human Mind Occur in the Physical Universe? Oxford University Press, New York (2007)

Anderson, J.R., Byrne, M.D., Douglass, S., Lebiere, C., Qin, Y.: An integrated theory of the mind. Psychol. Rev. **111**(4), 1036–1050 (2004)

Appel, A., Mitschke, M.: Adjusting vehicle characteristics by means of driver models. Int. J. Veh. Des. **18**(6), 583–596 (1997)

Aune, I.A., Jürgens, H.W.: Computermodelle des menschlichen Körpers. Bericht Nr. 27 – Auftragsstudie des Bundesamtes für Wehrtechnik und Beschaffung. Dokumentationszentrum der Bundeswehr, Koblenz (1989)

Baron, S., Kleinmann, D.L.: The human as an optimal controller and information processor. IEEE Trans. Man Mach. Syst. **10**(1), 9–17 (1969)

Baron, S., Levison, W.H.: Display analysis with an optimal control model of the human operator. Hum. Factors. **19**, 437–457 (1977)

Baron, S., Levison, W.H.: The optimal control model: status and future directions. IEEE Trans. Man. Mach. Syst. **21**, 90–100 (1980)

Bellet, T., Bornard, J.C., Mayenobe, P., Gruyer, D.: A computational model for car drivers Situation Awareness simulation: COSMODRIVE. In: Proceedings of the First International Symposium on Digital Human Modeling Lyon, France, June 14–16. (2011)

Bolte, U., Bubb, H.: Regelungstechnische Simulation der Schnittstelle Mensch-Maschine. Zeitschrift für Arbeitswissenschaft. **44**, 6–13 (1990)

Bonney, M.C., Evershed, D.G., Roberts, E.A.: Sammie – A computer model of man and his environment. In: Proceedings of the 1969 Annual Scientific Meeting of the Ergonomics Research Society. Bristol (1969)

Bornard, J.-C., Bellet, T., Mayenobe, P., Gruyer, D., Claverie, B.: A perception module for car drivers' visual strategies modeling and visual distraction effect simulation. In: Proceedings of the First International Symposium on Digital Human Modeling Lyon, France, June 14–16, 2011. (2011)

Bösch, P.: Der Fahrer als Regler. Dissertation, TU Wien (1991)

Braess, H.-H.: Untersuchung des Seitenwindverhaltens des Systems Fahrer-Fahrzeug Deutsche Kraftfahrtforschung, Bd. 206. (1970)

Braun, H.: Rechnerische Simulation der Fahrdynamik von Nutzfahrzeugen unter Einschluss von Fahrer und Umwelt VDI-Broschüre, Bd. 613. VDI-Verlag, Düsseldorf (1986)

Bubb, H.: Menschliche Zuverlässigkeit. – Definitionen, Zusammenhänge, Bewertung. Ecomed, Landsberg (1992a)

Bubb, H.: Computer aided tools of ergonomics and system design. Hum. Factor. Ergon. Man. **12**(3), 249–265 (2002)

Bubb, H.: Digitale Mensch-Modelle. In: Landau, K. (Hrsg.) Lexikon der Arbeitsgestaltung, S. 454–458. Gentner-Verlag, ergonomie Verlag, Stuttgart (2007)

Bubb, H.: Why do we need digital human models? Chapter 2. In: Scataglini, S., Paul, G. (eds.) DHM and posturography, pp. 6–32. Elsevier Inc, ISBN: 978-0-12-816713-7. (2019)

Carrier, R., Gilbert, R., Schiettekatte, J., Belanger, A., De Chamlain, B., Fortin, C.: A micro computer-aided workstation design and analysis. In: Proceedings of the 20th Annual Conference of Human Factors Association of Canada October 14 – 17, 1987. Human Factors Association of Canada, Mississanga, ON (1987)

Carson, J.M., Wierwille, W.W., Eastman, C.: Development of a strategy model of the driver in lane keeping. Veh. Syst. Dyn. **7**(4), 233–253 (1978)

Crossman, E. R. F. W., Szostak, H., Cesa, T. L.: Steering Performance of Automobile Drivers in Real in Contact-Analog Simulated Tasks. Human Factors Society 10th Annual Meeting, Anaheim, Kalifornien, (zitiert nach McRuer und Klein, 1975) 1968 (1966)

Crossman, E.R.F.W., Szostak, H.: Man-Machine-Models for Car-Steering. Forth Annual NASA – University Conference on Manual Control. National Aeronautics and Space Administration, Washington (1969)

Dey, D., Kirchhoff, U.: Beschreibungsmöglichkeiten für Mehrgrößenregeleigenschaften des Menschen bei der Fahrzeugführung ILR-Bericht, Bd. 6. Institut für Luft und Raumfahrt, Technische Universität Berlin, Nr (1975)

Donges, E.: Ein Zwei-Ebenen-Modell des menschlichen Lenkverhaltens im Kraftfahrzeug Bericht, Bd. 27. Gesellschaft für angewandte Naturwissenschaften, Forschungsinstitut für Anthropotechnik, Meckenheim (1977)

Donges, E.: Ein regelungstechnisches Zwei-Ebenen-Modell des menschlichen Lenkverhaltens im Kraftfahrzeug. Zeitschrift für Verkehrssicherheit. **24**, 98–112 (1978)

Donges, E.: Aspekte der Aktiven Sicherheit bei der Führung von Personenkraftwagen. Automobil-Industrie. **27**(2), 183–190 (1982)

Dreyfuss, H.: The Measure of Man: Human Factors in Design. Whitney Library of Design, New York (1959)

Eckstein, L.: Entwicklung und Überprüfung eines Bedienkonzepts und von Algorithmen zum Fahren eines Kraftfahrzeugs mit aktiven Sidesticks Fortschr.-Ber. VDI, Reihe 12, Bd. 471. VDI-Verlag, Düsseldorf (2001)

Elkind, J.I.: Characteristics of Simple Manual Control Systems TR, Bd. 111. MIT (1956)

Elkind, J.I.: A survey of the development of models for the human controller. In: Longford, R. (Hrsg.) C. Guidance and Control II. Munde, New York. (1964)

Elkind, J.I., Fall, P.L., Baron, S., Kleinmann, D.L., Levison, W.H.: An Optimal Control Method for Predicting Control Characteristics and Displays. Requirements of Manned Vehicle Systems (1968). AFFDL-TR-67-187

Feraric, J.P.: Echtzeitfähige Modellierung des individuellen Fahrerverhaltens zur Realisierung adaptiver Unterstützungsfunktionen in einem Monitor- und Warnsystem. Dissertation, Universität der Bundeswehr München, 1996 (1996)

Feraric, J.P., Onken, R.: DAISY- A driver assisting system which adapts to the driver. In: Proceedings of the Man-Machine-Systems (MMS '95) Cambridge, MA, 27–29. Juni 1995 (1995)

Feraric, J.P., Onken, R., Kopf, M.: Modellierung eines Fahrernormalverhaltens mit Hilfe Neuronaler Netze und Interpretation der Abweichung in einem Monitor- und Warnsystem. In: Gärtner, K.-P. (Hrsg.) Menschliche Zuverlässigkeit. Maßnahmen zur Eingrenzung menschlichen "Versagens" DGLR-Bericht, Bd. 92-04, Deutsche Gesellschaft für Luft- und Raumfahrt e. V., Bonn (1992)

Fiala, E.: Lenken von Fahrzeugen als kybernetische Aufgabe. Automob. Z. **68**(5), 156 (1966)

von Garrel, U., Otto, H.J., Onken, R.: Adaptive Modellierung des fertigkeits- und regelbasierten Fahrerführungsverhaltens VDI-Berichte, Bd. 1613., S. 239–262 (2001)

Gärtner, K.P., Schweingruber, J.: Innenraumgestaltung mit rechnergestützten Mensch-Modellen VDI-Bericht, Bd. 948. VDI-Verlag GmbH, Düsseldorf (1992)

Geoffrey, S.P.: A 2-D Manikin – The Inside Story Paper, Bd. 267 A. Society of Automotive Engineers, Inc., Warrendale (1961)

Geuß, H.: Entwicklung eines anthropometrischen Meßverfahrens für das CAD-Menschmodell RAMSIS; Dissertation am Lehrstuhl für Ergonomie der TU-München (1995)

Greil, H.: "Körperbaugerechte Mittelwerte und Variabilitätsbereiche von Körpermaßen im Erwachsenenalter zur Regenerierung mit dem Programmsystem RAMSIS", Abschlußbericht einer Auftragsstudie vom 5.3.93, unveröffentlicht (1993)

Harsenhorst, J.V., Lourens, P.: Classification of Driver Errors and Analysis of Driver Performance Parameters Technical Report, Bd. 87-25. University of Groningen, Traffic Research Center, The Netherlands (1988)

Hayhoe, G.E.: A driver model based on the cerebellar modell articulation controller. Veh. Syst. Dyn. **8**(8), 49–72 (1979)

Helbig, K., Jürgens, H.W.: Forschungsbericht/ Bundesanstalt für Arbeitsschutz und Unfallforschung Entwicklung einer praxisgerechten Körperumrißschablone des sitzenden Menschen, Bd. 187. Wirtschaftsverl. NW in Komm., Dortmund. Bremerhaven (1977)

Hickey, D., Pierrynowski, M.R., Rothwell, P.: Man-modeling CAD- programs for workspace evaluations. University of Toronto, Kanada (1985a). Vortragsmanuskript.

Horn, A.: Fahrer-Fahrzeug-Kurvenfahrt auf trockener Straße. Dissertation, Institut für Fahrzeugtechnik, TU Braunschweig 1986

Hosman, R.J., Benard, B., Frourquet, H.: Active and passive sidestick controllers in manual aircraft control. In: Proceeding of the IEEE, International Conference on Systems, Man and Cybernetics L.A. (CA) USA, Nov. 4 – 7, 1990. S. 527–529. (1990)

Huang, P.: Regelkonzepte zur Fahrzeugführung unter Einbeziehung der Bedienelementeigenschaften, Dissertation an der Technischen Universität München (2004)

Jenik, P.: Biomechanische Analyse ausgewählter Arbeitsbewegungen des Armes. Beuth-Verlag, Berlin, Köln, Frankfurt (1973)

Johannsen, G., et al.: Parametrische quasi-lineare Modelle für den Menschen als Regler. In: Johannsen, G., Boller, H.E., Donges, E. (Hrsg.) Der Mensch im Regelkreis – Lineare Modelle. R. Oldenbourg Verlag, München Wien (1977)

Jürgensohn, G.: Hybride Fahrermodelle. Dissertation an der Technischen Universität Berlin ZMMS Spektrum, Bd. 4. Pro Universitate Verlag, Würzburg (1997)

Jürgensohn, T.: Bedienermodellierung. In: Timpe, K.-P., Jürgensohn, T., Kolrep, H. (Hrsg.) Mensch-Maschine-Systemtechnik - Konzepte, Modellierung, Gestaltung, Evaluation. Symposion Publishing, Düsseldorf (2002)

Kleinmann, D.L.: Optimal control of linear systems with time-delay and observation noise. IEEE Trans. Autom. Control. **AC-14**, 524–527 (1969)

Kleinmann, D.L., Baron, S., Levison, W.H.: A control theoretic approach to manned-vehicle analysis. IEEE Trans. Autom. Control. **AC-16**(6), 824–832 (1971)

Knauer, P.: Objektivierung des Schwingungskomforts bei instationärer Fahrbahnanregung. Dissertationsschrift an der Technischen Universität München (2010)

Kolling, J.: Validierung und Weiterentwicklung eines CAD-Menschmodells für die Fahrzeuggestaltung, Dissertation an der Technischen Universität München, 1997 (1997)

Kondo, M.: Richtungsstabilität (wenn Steuerbewegungen hinzukommen). J. Soc. Automot. Eng. Japan. **7**(5, 6), 104–106 (1953). auf Japanisch

Kramer, U.: On the application of fuzzy sets to the analysis of the system driver-vehicle-environment. Automatica. **21**(1), 101–107 (1985)

Kramer, U., Rohr, G.: A model of driver behaviour. Ergonomics. **25**(10), 891–907 (1982a)

Kroll, C.V., Roland, R.D.: A Preview-Predictor Model of Driver Behaviour in Emergency Situations Report, Bd. VJ-2251-V-6. Cornell Aeronautical Laboratory (1970)

Lieberman, E.B., Goldblatt, R.A.: A Review of Driver-Vehicle Effectiveness Model Final Report under NHTSA Contract, Bd. DTNH22-80-C-07082 (DOT-HS 806110) (1981)

McRuer, D.T.: Human dynamics in man-machine-systems. Automatica. **16**, 237–253 (1980)

McRuer, D.T., Jex, H.R.: A review of quasi-linear pilot models. IEEE Trans. Hum. Factor. Electron. **8**(3), 231–249 (1967). NASA CR-1212

McRuer, D.T., Krendel, E.S.: Dynamic Response of Human Operator (1957). WADC-TR-56-523

McRuer, D.T., Krendel, E.S.: Mathematic Models of Human Pilot Behaviour AGAR Dograph, Bd. 188. Advisory Group for Aerospace Research and Development, Neuilly sur Seine, France (1974)

McRuer, D.T., Allen, R.W., Weir, D.H., et al.: New results in driver steering control models. Hum. Factors. **19**(4), 381–397 (1977)

McRuer, D.T., Jex, H.R., Clement, W.F., Graham, D.: Development of a Systems Analysis Theory of Manual Control Displays Report, Bd. TR-163-1. System Technology Inc., (1967)

Meijer, R., Elrofai, H., Broos, J., van Hassel, E.: Evaluation of an Active Multi-Body Human Model for Braking and Frontal Crash Events, 23rd Enhanced Safety of Vehicles (ESV) Conference, Seoul, Korea (2013)

Meijer, R., van Hassel, E., Broos, J., Elrofai, H., van Rooij, L., van Hooijdonk, P.: Development of a Multi-Body Human Model that Predicts Active and Passive Human Behaviour, International Research Council on Biomechanics of Injury (IRCOBI) Conference, Dublin, Ireland (2012)

Miura, T.: Coping with Situational Demands: A Study of Eye-Movements and Peripheral Vision Performance. North-Holland, Amsterdam, S. 126–137 (1986)

Mühlstedt, J., Kaußler, H., Spanner-Ulmer, B.: Programme in Menschengestalt: Digitale Menschmodelle für CAx und PLM-Systeme. ZfA. **2–2008**, 79–86 (2008)

Müller, A.: Systematische und nutzerzentrierte Generierung des Pkw-Maßkonzepts als Grundlage des Interior- und Exteriordesign. Dissertation an der Universität Stuttgart (2010)

Nagai, M.: Adaptive behaviour of driver-car systems in critical situations. In: Proceedings of the International AMSE Summer Conference "Modelling and Simulation" Nice, France (1983)

Neculau, M.: Modellierung des Fahrerverhaltens: Regel- und Steuerstrategien. In: VDI-Gesellschaft Fahrzeugtechnik (Hrsg.) Das Mensch-Maschine-System im Verkehr VDI Berichte, Bd. 948, VDI-Verlag GmbH, Düsseldorf (1992)

Neculau, M., Jürgensohn, T., Willumeit, H.-P.: Model of the Driver Lateral Control behaviour and scanning pattern. In: Proc. of the 9th European Annual Conference on Human Decision Making and Manual Control Lyngby, Denmark, 1989. (1990)

Newell, A.: Unified Theories of Cognition. Harvard University Press, Cambridge, MA (1990)

Niemann, K.: Messungen und Berechnungen über das Regelverhalten von Autofahrern. Dissertation, TU Braunschweig, Institut für Fahrzeugtechnik, 1972 (1972)

Oppelt, W., Vossius, G.: Der Mensch als Regler. VEB Verlag Technik, Berlin (1970)

Perel, M.: The development of a computer simulation model of driver performance to predict accident probability. In: Proceedings of the Human Factors Society, 26th Annual Meeting, S. 239–244 (1982)

Pew, R.W.: Temporal Organization in Skilled Performance. Ph.D. Thesis, Univ. of Michigan (1963)

Plavšic, M.: Analysis and modeling of driver behavior for assistance systems at road intersections. Dissertation an der Technischen Universität München (2010)

Plavšic, M., Guzman, E., Bengler, K.: Modelling control behavior for the cognitive driver model. In: The Second European Conference on Human Centered Design for Intelligent Transport System, HUMANIST Berlin, Germany, April, 2010 (2010)

Plöchl, M., Lugner, P.: Theoretical Investigation of the Interaction Driver – Feedback-Controlled Automobile SAE Paper, Bd. 9438006 (1994)

Pommerleau, D.A.: Efficient training of artificial neural networks for autonomous navigation. Neural Computation (3), (1991)

Poulton, E.C.: Tracking Skill and Manual Control. Academic, New York (1974)

Prokop, G.: Fahrermodellierung durch Modelprädiktive Optimierung. In: Der Fahrer im 21. Jahrhundert VDI-Berichte, Bd. 1613, S. 219–238 (2001)

Radonjic, R.: Definition and measurement of vehicle handling characteristics. Int. J. Veh. Des. **11**(2), 166–170 (1990)

Rasmussen, J., et al.: The definition of human error and a taxonomy for technical system design. In: Rasmussen, J. (Hrsg.) New Technology and Human Error. Wiley, New York (1987)

Reddy, R.N., Ellis, J.R.: Contribution to the Simulation of Driver Vehicle Road System SAE Paper, Bd. 810513 (1982)

Reichelt, W.: Ein adaptives Fahrermodell zur Bewertung der Fahrdynamik von Pkw in kritischen Situationen. Dissertation an der Technischen Universität Braunschweig (1990)

Reichelt, W., Strackerjan, B.: Bewertung der Fahrdynamik von Pkw im geschlossenen Regelkreis mit Hilfe von Fahrsimulatoren und Fahrermodellen. In: VDI-Gesellschaft Fahrzeugtechnik (Hrsg.) Das Mensch-Maschine-System im Verkehr VDI Berichte, Bd. 948, VDI-Verlag GmbH, Düsseldorf (1992)

Reid, L.D., Solowka, E.N., Billing, A.M.: A systematic study of driver steering behavior. Ergonomics. **24**(6), 447–462 (1981)
Remlinger, W.: Analyse von Sichteinschränkungen im Fahrzeug. Dissertation an der TUM (2013)
Rühmann, H.: Schnittstellen in Mensch-Maschine-Systemen. In: Schmidtke, H. (Hrsg.) Ergonomie. Hanser Verlag, München (1993)
SAE J826. 2008–11. Devices for Use in Defining and Measuring Vehicle Seating Accommodation. Warrendale: Society of Automotive Engineers, Inc
SAE J941: 2002–09. Motor Vehicle Drivers' Eye Locations. Society of Automotive Engineers, Inc, Warrendale (2002)
Salvucci, D.: A model of eye movements and visual attention. In: Proceedings of the International Conference on Cognitive Modeling, pp. S. 252–S. 259 (2000)
Schmidt, R.F., Thews, G.: Physiologie des Menschen, 24. Aufl. Springer, Berlin, Heidelberg, New York (1990)
Schweizer, G.: Das Systemglied "Mensch" in der Simulation. In: Schöne, A. (Hrsg.) Simulation stetiger Systeme. München, Wien (1975)
Seeger, H.: Vom Königsschiff zum Basic Car: Entwicklungslinien und Fallstudien des Fahrzeugdesign. Wasmuth, Tübingen (2007)
Seidl, A.: Das Menschmodell RAMSIS: Analyse, Synthese und Simulation dreidimensionaler Körperhaltungen des Menschen. Dissertation an der Technischen Universität München, 1994, (1994)
Seidl, A.: Computer man models in ergonomic design. In: Schmidtke, H. (Hrsg.) Handbuch der Ergonomie (Part A – 3.3.3). Hanser Verlag, Koblenz (1997). Bundesamt für Wehrtechnik und Beschaffung
Senders, J.W.: The human operator. In: NASA (Hrsg.) Bioastronautics Data Book, Section 18, NASA SP-3006. National Aeronautics and Space Administration, Washington DC (1964)
Sheirko, F.: Experimentelle Untersuchung des Geradeauslaufs von Straßenfahrzeugen unter dem Einfluss von Seitenwind und Straßenunebenheiten. Automobil-Industrie 1, (1972)
Stein, W.: Vergleichende Untersuchung einachsiger manueller Regelvorgänge und Beurteilung ihrer linearen Modellierung Bericht, Bd. 21. Forschungsinstitut für Anthropotechnik, Meckenheim (1974)
Stein, W., Pioch, E.: Zur Anwendung von Regelaufgaben bei ergonomischen Untersuchungen Bericht, Bd. 23. Forschungsinstitut für Anthropotechnik, Meckenheim (1975)
Sträter, O.: Modellierung und Bewertung kognitiver Aspekte der Aufgaben des Fahrers VDI-Berichte, Bd. 1613., S. 263–274 (2001)
Tsimhoni, O., Liu, Y.: Modeling steering using the Queueing Network–Model Human Processor (Qn-Mhp). In: Proceedings of the Human Factors and Ergonomics Society 47th Annual Meeting, S. 1875–1879 (2003)
Tustin, A.: An Investigation of the Operators Response in Manual Control of a Power Driven Gun C. S. Memorandum, Bd. 169. Metropolitan Vickers Electrical Co., Sheffield, England, S. 172–186 (1944)
Tustin, A.: The nature of the operator's response in manual control and its implications for controller design. J. IEE. **94**(Part II A2), 190–202 (1947)
Waldhier, T.: Menschmodellierende Verfahren zur rechnergestützten Arbeitsplatzgestaltung – Eine Übersicht. Marktanalyse am Lehrstuhl für Arbeitswissenschaft und Betriebsorganisation der TH Karlsruhe (1989)
Wickens, C.D., Hollands, J.G.: Engineering Psychology and Human Performance, 3. Aufl. Prentice Hall, Upper Saddle River (2000)
Wierwille, W.W., Gagné, G.A., Knight, J.R.: An experimental study of human operator models and closed-loop analysis methods for high-speed automobile. IEEE Trans. Hum. Factor. Electron. **8**(3), 187–201 (1967)
Willumeit, H.-P., Kramer, U., Rohr, G.: Closed-loop simulation of the driver-vehicle-environment system. Veh. Dyn. **12**(1), 1 (1983)
Wohl, J.G.: Man-Machine Steering Dynamics. Human Factors (3), (1961)
Wolf, J.D., Barret, M.F.: Driver-Vehicle Effectiveness model NHTSA Report, Bd. 78113. (1978)
Young, L.R.: Human control capabilities. In: NASA (Hrsg.) Bioastronautics, Data Book, 2. Aufl. NASASP, Bd. 3606, Scientific and Technical Information Division, National Aeronautics and Space Adminstration, Washington, DC (1973)
Yuhara, N., Horiuchi, S., Asanuma, N., et al.: Improvement of vehicle handling quality through active control of steering reaction torque. In: Proceeding of AVEC Jokohama, 1992. S. 407–412. (1992)
Yuhara, N., Horiuchi, S., Asanuma, N., et al.: Improvements of vehicle handling qualities by means of kinesthetic information. In: Proceeding of the Trienial Congress of the international Ergonomics Association Toronto, 1994. Bd. 5. (1994)

Further Literature

ACT-R Theory and Architecture of Cognition. http://act-r.psy.cmu.edu/
Assmann, E.: Untersuchung über den Einfluss einer Bremsweganzeige auf das Fahrerverhalten. Dissertation an der Technischen Universität München (1985)
Bachmann, T.: Wechselwirkungen im Prozess der Reibung zwischen Reifen und Fahrbahn VDI Fortschrittsberichte, Reihe 12. VDI Verlag, Düsseldorf (1998)
Bernotat, R.: Anthropotechnik in der Fahrzeugführung. Ergonomics. **13**(3), 353–377 (1970)
Bollere, H.E., Krüger, W.: Untersuchung eines Bedienelements mit Krafteingabe und Wegrückmeldung bei der manuellen Lenkung von Unterwasserfahrzeugen Bericht, Bd. 38. Forschungsinstitut für Anthropotechnik, Meckenheim (1978)
Bolte, U.: Das aktive Stellteil – ein ergonomisches Bedienkonzept Fortschrittsberichte VDI-Reihe 17, Bd. 75. VDI-Verlag, Düsseldorf (1991)

Breitling, T., Breuer, J., Petersen, U.: Enhancing traffic safety by active safety innovations. In: CONVERGENCE 2004 Proceedings SAE Paper, Bd. 2004-21-0056. (2004)

Bubb, H. Untersuchung über die Anzeige des Bremsweges im Kraftfahrzeug. Dissertation, Technische Universität München (1975)

Bubb, H.: Analyse der Geschwindigkeitsschätzung im Kraftfahrzeug. Zeitschrift für Arbeitswissenschaft. **31**, 103 (1977)

Bubb, H.: Informationswandel durch das System. Kap. 5.2. In: Schmidtke, H. (Hrsg.) Ergonomie, 3. Aufl. Carl Hanser-Verlag, München-Wien (1993a)

Bubb, H.: Systemergonomische Gestaltung. Kap. 5.3. In: Schmidtke, H. (Hrsg.) Ergonomie, 3. Aufl. Carl Hanser-Verlag, München-Wien (1993b)

Bubb, H., Schmidtke, H.: Systemstruktur. Kap. 5.1. In: Schmidtke, H. (Hrsg.) Ergonomie, 3. Aufl. Carl Hanser-Verlag, München-Wien (1993)

Bubb, P.: Ergonomie in der Fahrzeugentwicklung. In: Verein Deutscher Ingenieure (Hrsg.) Entwicklung im Karosseriebau. VDI Berichte, Bd. 968, S. 481–501. Gesellschaft Fahrzeugtechnik. Entwicklungen im Karosseriebau, Hamburg (1992b). 25. und 26. Mai. Düsseldorf

Caltrans http://www.dot.ca.gov/aboutcaltrans.htm

Chaffin, D.: Digital Human Modeling for Vehicle and Workplace Design. Society of Automotive Engineers, Inc. (2001)

Dutke, S.: Mentale Modelle: Konstrukte des Wissens und Verstehens. Verlag für Angewandte Psychologie, Göttingen (1994)

Engels, A.: Geradeauslaufkriterien für PKW und deren Bewertung, Dissertation, TU Braunschweig (1994)

Autofahrer und Verkehrssituation. In: Fastenmeier, W., Häcker, H. (Hrsg.) Reihe Mensch-Fahrzeug-Umwelt, Bd. 33, Verlag TÜV Rheinland, Köln (1995)

Fleck, R.: "Aktivlenkung" – ein wichtiger erster Schritt zum Steer-by-Wire Tagung "PKW-Lenksysteme – Vorbereitung auf die Technik von morgen", Haus der Technik e. V., Essen, 02–03.April (2003)

Geiser, G.: Mensch-Maschine-Kommunikation im Kraftfahrzeug. ATZ. **87**, 77–84 (1985)

Gengenbach, R.: Fahrerverhalten im Pkw mit Head-Up-Display. Gewöhnung und visuelle Aufmerksamkeit VDI Fortschrittsberichte Reihe 12: Verkehrstechnik/Fahrzeugtechnik, Bd. 330 (1997)

Gillet, G.: Ergonomische Optimierung eines Aktiven Stellteils. Dissertation, an der Technischen Universität München (1999)

Godthelp, H.: Studies on Human Vehicle Control. TNO, Institute for Perception, Soesterberg (1984)

Guan, H., Gao, Z., Guo, K., et al.: An optimal preview acceleration model for velocity control in driver-vehicle-environment system. In: Proceeding of AVEC Ann Arbor, 2000. S. 525–530 (2000)

Hale, A.R., Stoop, J., Hommels, J.: Human error models as predictor of accident scenarios for designer in road transport systems. Ergonomics. **33**, 1377–1387 (1990)

Helander, M.: Handbook of Human-Computer Interaction. North-Holland, Amsterdam (1988)

Herrin, G.D., Neuhardt, J.B.: An empirical model for automobile driver horizontal curve negotiation. Hum. Factors. **16**(2), 129–133 (1974)

Hickey, D., Pierrynowski, M.R., Rothwell, P.: Man-Modeling CAD- Programs for Workspace Evaluations. University of Toronto, Kanada (1985b). Vortragsmanuskript

Honda Motor Co: Fact Book, S2000 TypeV. (2000). auf Japanisch. http://www.honda.co.jp/factbook/auto/s2000/200007/index.html

Kramer, U., Rohr, G.: A fuzzy model of driver behaviour: Computer simulation and experimental results. In: Proceedings of the IFAC Congress on Analysis, Design and Evaluation of Man-Machine-System Baden-Baden, Germany, 1982 (1982b)

Lindsay, P.H., Norman, D.H.: Human Information Processing. An Introduction to Psychology. Academic Press, New York, London (1972)

McRuer, D.T., Magdaleno, R.E., Moore, G.P.: A Neuromuscular Actuation System Model Third Annual NASA-University Conference on Manual Control., S. 281 (1967). NASA SP-144

Merz, L.: Grundkurs der Regelungstechnik. Einführung in die praktischen und theoretischen Methoden. Hochschultaschenbücherei, München, Wien (1973)

Michon, J.A.: A critical view of driver behaviour models. What do we know, what should we do? In: Evans, L., Schwing, R. (Hrsg.) Human Behavior and Traffic Safety. Plenum Press, New York (1985)

Mitschke, M.: Beurteilung des Geradeauslaufverhaltens und des Center Point Feelings IfF-Bericht, Bd. 631. Institut für Fahrzeugtechnik, TU Braunschweig, Braunschweig (1984)

Ohno, T.: Steering control on a curved course. Japan SAE Rev. **20**(5), 413–419 (1966)

Penka, A.: Vergleichende Untersuchung zu Fahrerassistenzsystemen mit unterschiedlichen aktiven Bedienelementen. Dissertation an der Technische Universität München (2001)

Pressler, G.: Regelungstechnik, Grundelemente Hochschultaschenbücher. Bibliographisches Institut Mannheim, Mannheim (1967)

Reichart, G.: Menschliche Zuverlässigkeit beim Führen von Kraftfahrzeugen Fortschritt-Berichte, Reihe 22, Mensch-Maschine-Systeme, Bd. 7. VDI-Verlag, Düsseldorf (2001)

Salvatore, S.: The Influence of Sensory Pattern and Alcohol on Vehicular Velocity Sensing. Public Health Service and Mental Health Administration, Sudbury, Ontario, London (1972)

Schmidtke, H.: Der Leistungsbegriff in der Ergonomie. In: Schmidtke, H. (Hrsg.) Ergonomie, 3. Aufl. Carl Hanser-Verlag, München-Wien (1993)

Schulze, B.G.: The application of ergonomic principles in the design of operating controls. In: Human Factors in Transport Research, Bd. 1, S. 337–344. (1981)

Schweigert, M., Bubb, H.: Eye-movements, performance and interference when driving a car and performing secondary tasks. In: Gale, A.G. (Hrsg.) Vision in Vehicles, Bd. IX, Elsevier North Holland Press, Amsterdam (2002)

Shannon, C.E., Weaver, W.: The Mathematical Theory of Communication. University of Illinois Press, Urbana (1949)

Sheridan, T.B., Ferrel, W.R.: Man-Machine-Systems: Information, Control, and Decision Models of Human Performance. The MIT Press, Cambridge, MA (1974)

Stachowiak, H.: Allgemeine Modelltheorie (1973)

Steward, J.D.: Human perception of angular acceleration and implications on motion simulator. J. Aircraft. **8**(4), 248–253 (1971)

Unselt, T., Breuer, J., Eckstein, L., Frank, P.: Avoidance of "loss of control accidents" through the benefit of ESP. In: FISITA 2004 Congress Proceedings (2004)

Weiser, M.: The computer for the 21st century. Sci. Am. **265**(3), 94–104 (1991)

Wickens, C.D.: Engineering Psychology and Human Performance. Charles E. Merrill Publishing Company, Ohio (1984)

Wiener, N.: Cybernetics or Control and Communication in the Animal and the Machine. The Massachusetts Institute of Technology, New York (1948)

Wierwille, W.W.: Tijerina: Eine Analyse von Unfallberichten als Mittel zur Bestimmung von Problemen, die durch Verteilung der visuellen Aufmerksamkeit und der visuellen Belastung innerhalb des Fahrzeuges verursacht werden. Zeitschrift Für Verkehrssicherheit. **41**, 164–168 (1995)

Willumeit, H.-P., Jürgensohn, T.: Fahrermodelle – ein kritischer Überblick – Teil I und Teil II. Automobiltechnische Zeitschrift (ATZ). **99**(9), 424–428 (1997)

Yuhara, N., Tajima, L., Sano, S., et al.: Steer-by-wire-oriented steeling system design: concept and examination. Veh. Syst. Dyn. Suppl. **33**, 692–703 (1999)

Zwahlen, H.T., Adams Jr., C.C., DeBald, D.P.: Safety aspects of CRT touch panel controls in automobiles. In: Gale, A.G. (Hrsg.) Vision in Vehicles II, S. 335–344. Elsevier North Holland Press, Amsterdam (1988)

System Ergonomics of the Vehicle

Heiner Bubb, Klaus Bengler, Jurek Breuninger, Christian Gold, and Magnus Helmbrecht

Contents

6.1 General System Ergonomic Design Guidelines – 258
6.1.1 Function – 261
6.1.2 Feedback – 266
6.1.3 Compatibility – 267

6.2 Human – Machine – Interaction – 270
6.2.1 Displays – 273
6.2.2 Control Elements – 289

6.3 System Ergonomic Recommendations for the Respective Driving Task Levels – 300
6.3.1 Primary Driving Task – 300
6.3.2 Secondary Driving Task – 308
6.3.3 Tertiary Tasks – 308

6.4 Design of Driving Relevant Characteristics – 311
6.4.1 Lateral Dynamics: The Steering Feel – 311
6.4.2 Longitudinal Dynamics – 325
6.4.3 X- by-Wire – 330

References – 336

H. Bubb et al. (eds.), *Automotive Ergonomics*, https://doi.org/10.1007/978-3-658-33941-8_6

6.1 General System Ergonomic Design Guidelines

As already explained in ▶ Chap. 1, the flow of information between driver and vehicle plays a dominant role in dealing with the automobile and thus in its design. The investigation and description of this flow of information is the subject of so-called system ergonomics (Bubb 1993). The starting point of any system ergonomic analysis is the information on which the task is based. In the sense of the control loop diagram in ▶ Fig. 1.9, the interaction between driver and vehicle transfers the information of the task into that of the result. This thought applies to every form of task. However, there are many different forms of task when driving a car. For the ergonomic design of a vehicle, it is therefore first necessary to examine for what purpose it is intended. ◘ Table 6.1 provides an exemplary overview with some examples in the sense of usage scenarios and can also be extended at will.

6

For each of the driving emissions listed in ◘ Table 6.1, the driving history shown in ◘ Fig. 6.1, which is subdivided into individual sub-sections, is included. The system ergonomic rules listed under ◘ Fig. 6.3 are in principle to be applied to each of these sub-sections individually. In each subsection, therefore, there is a special task whose degree of fulfillment can be described with the respective result. One or more specific technical systems are used to carry out the task (e.g. the car for carrying out the driving task, the wiper for carrying out the task "clear view in rain"). The respective technical system is described by its function. Taking into account the basic idea of the control loop (▶ Fig. 1.9), it is possible to determine:

▪▪ Definition

The ***function*** of a technical system or system element represents the defined conversion of the input variables into the converted output variables, taking parameters into account. In this capacity, it is part of a human-machine system from which the ergonomic requirements for the technical system are derived, taking into account the context of use.

Some striking examples are given below in order to illustrate the diversity of the design necessities resulting from these considerations. For example, it can be assumed that a business trip – especially if it is carried out by a group of people – is planned outside the vehicle, for example in the office. It then makes sense to carry out the navigation planning on the PC

◘ **Table 6.1** General driving missions and examples

Parties mission	Single person	Group of persons	Object
Official	Trip to the work Trip too an conversation/ treatment appointment Customer visit (e.g. sale representative)	Voyage too an meeting appointment Various aims for the voyagers Carpool with joint target	Transport one object Transport of several objects
Private	Visit with relatives, friends Provision Pick-up/bringing by objects Round trip Vacation with an destination Vacation with several destinations	Related group Group from adults, children or mixed Unknown carpool Acquaintances Round trip Vacation with an destination Vacation with several destinations Various destinations for the voyagers	Transport one object Relocation Pick-up/bringing objects Transport of foreigm objects Transport of sports equipment (Golf bag, Bicycles, Ski, boat etc.)

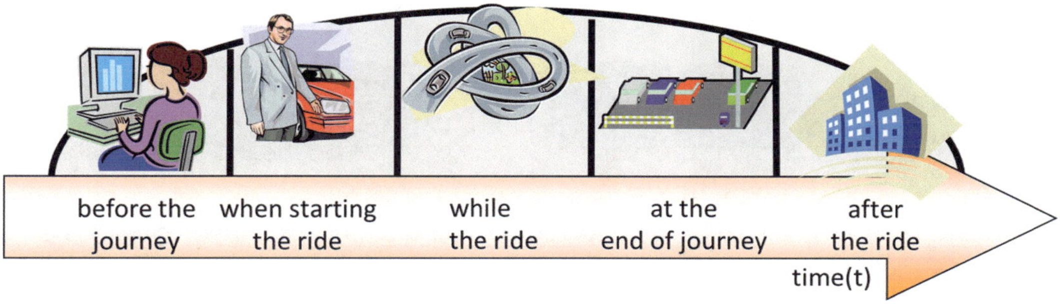

Fig. 6.1 trip progression

and to transfer the result to the vehicle's navigation system in a suitable manner (e.g. via radio; USB stick, storage medium). It can be quite different, for example, with a tour that is carried out by a single person for fun. In the meantime, round trips are available for download and exchange between users via social networks. The planning task therefore becomes more of a search and decision task. Then navigation wishes can arise spontaneously during the journey, which were not even thought of before the start of the journey, and which can then be fulfilled via the "point of interest" function. In general, the mission "driving for fun" or the desire to actively use the vehicle represents a very special challenge for ergonomic design. While for example – as mentioned several times – from the point of view of information change the necessity of shifting should be eliminated, this can be a cause for joy for a committed driver who experiences mastery of the machine as competence. This factor should also not be underestimated for vehicles and missions used for business purposes. Finding the right compromise between the requirements of comfort (= "pleasure") and the avoidance of discomfort (avoidance of "suffering"; see also ▶ Sect. 3.3.4) is a recurring challenge that arises in the system ergonomic design of the interaction between driver and vehicle, but also in anthropometric design.

If a mission of the journey is the transport of objects, the question of comfortable loading (and later at the end of the journey of unloading) plays an important role. This is a special part of the anthropometric design of the vehicle, which in this case of course also depends on the nature of the objects notified. So there is a difference whether handy suitcases or crates are to be transported in the luggage compartment of the vehicle or possibilities for transporting bulky sports equipment have to be created. For physical reasons, the loading condition changes the driving characteristics of the vehicle. An ergonomic requirement is that these changes do not require any changes in the driver's behaviour. Depending on the driving mission, very specific tasks and thus demands on the vehicle also arise during the journey. For example, when the remaining range of some vehicles and systems is nearing diminishing, information about available filling stations (charging stations or battery change stations for electric vehicles) is now automatically displayed in the navigation device. For international journeys, the country-specific traffic regulations – in particular speed regulations – should be displayed directly at the border, which is often not clearly recognisable as such today, and should not only be accessible to the user. The tasks of searching for a parking space and parking are assigned to the end of the journey. This may be quite different for a family trip to relatives than for a business trip. While in the first case the vehicle should provide assistance in finding a free parking space in a residential area, in the last case a company parking space is available or a nearby multi-storey car park and its capacity must be found in an inner city. Corresponding information is now available in a modern vehicle-integrated navigation system. Completely new scenarios arise in connection with Car-Sharing, which is currently the subject of intense debate. In all the con-

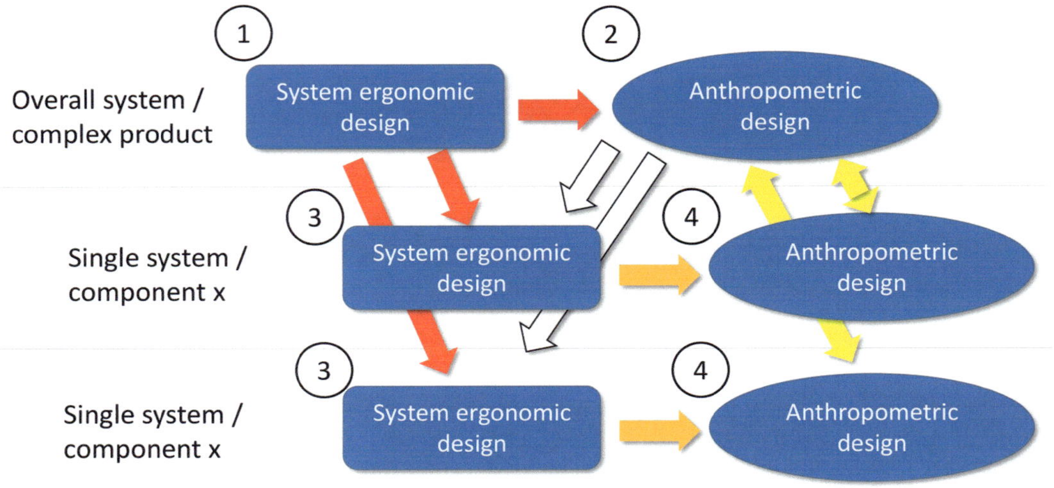

Fig. 6.2 Sequence of ergonomic design of complex products according to Remlinger and Bubb (2007)

texts outlined here, the basic ergonomic approach is always to consider the corresponding scenarios specifically from the user's point of view and needs. However, many of the information functions mentioned now lead to increased requirements for their use while driving in order to keep driver distraction to a minimum.

From system ergonomic research results design rules can be derived, through the observance of which it is possible to give concrete hints for the interpretation of this flow of information, i.e. in particular for the design of controls and displays. In the ideal case, the anthropometric design of the vehicle, i.e. the geometric assignment of displays, controls and other objects necessary for driving to the driver and also to the other passengers, takes place on the basis of the system ergonomically defined design. After the anthropometric determination, ideally the environmental ergonomic design should take place, i.e. concretely the determination of lighting, sound and climate conditions as well as the mechanical vibrations (vibration damping). In practice, however, the design does not take place in this ideal order, but represents an iterative, interactive process of the departments concerned with the respective subareas. In the rarest of cases, the development of a new vehicle also happens on the "tabula rasa", but rather takes its starting point from an already developed predecessor model. Remlinger and Bubb (2007) put together a model of the processing sequence for the design of the overall system and the individual systems or components in Fig. 6.2, where the numbering from 1 to 4 shows the sequence of the processing of the individual work steps and the arrows show which steps influence subsequent ones.

In this chapter, the system ergonomic design guidelines in general are presented first and then their effects on the various driving tasks. In particular, the design of the displays and operating elements serves to ensure operating safety. The aspects of anthropometric design are dealt with in ► Chap. 7. A separate chapter is devoted to the important aspect of the view. The above-mentioned environmental design is carried out under the heading "condition safety", since environmental conditions that cause as little discomfort as possible basically prevent driver fatigue.

The following basic questions must be answered satisfactorily for the ergonomic design of the task resulting from the system order, the specially selected design of the system and the system components (e.g. design of the operating elements, the displays, design of the software):

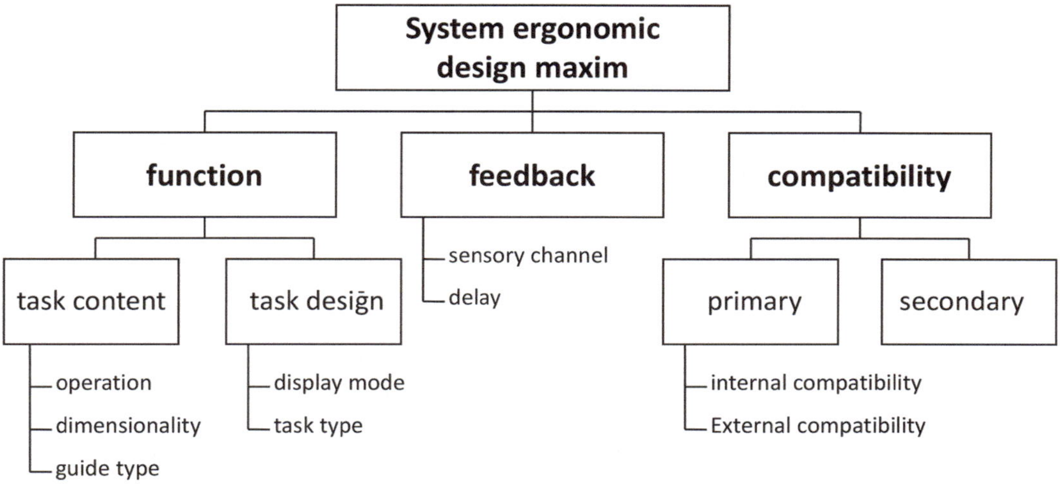

Fig. 6.3 Overview of the system ergonomic design maxims Function, feedback and compatibility

(a) *Function:* What does the driver/user intend to do and to what extent does the technical system support him?
(b) *Feedback:* Can the driver/user see if he has done something and what was or will be the result of his action?"[1]
(c) *Compatibility:* What is the effort for the driver/user to recode between different information channels?

Figure 6.3 gives an overview of how the system ergonomic design maxims function, feedback and compatibility are further subdivided. These individual aspects will be dealt with in detail below.

6.1.1 Function

The *function* refers on the one hand to the actual *task content* which is fixed and essentially describes the information processing required by humans, and on the other hand to the *task design* which can be influenced by the system designer within the scope of the respective technical possibilities. The operator's mental strain is thus influenced both by his previous knowledge and the complexity of the task and by those aspects of the task which are determined by the technical system provided.[2] It is therefore the task of the vehicle designer to reduce the operating difficulty caused by the design to such an extent that, ideally, only the inherent difficulty of the task remains.

6.1.1.1 Task Content

Operation Each task can be described by its temporal and spatial order of the activities necessary for its solution. For example, when overtaking on a country road, the driver must first assess whether the available free distance is sufficient for overtaking, then check in the rearview mirror that no faster vehicle has just started the overtaking procedure, then set the left turn signal, estimate the available free distance again and finally accelerate the vehicle to

1 In doing so, undesirable side effects on the vehicle itself or on the environment may also have to be reported.

2 For example, for the control of the longitudinal dynamics of the vehicle (speed), the driving and traffic situation (the course of the road and the movements of other vehicles and road users) plays a major role, from which the driver must derive the appropriate speed. Using the accelerator pedal and if necessary the brake pedal, he tries to maintain this mentally acquired speed. In a manually-shifted vehicle, the correct gear selection thus additionally complicates the accomplishment of tasks. The difficulty of accomplishing the tasks thus results not only from the task content, but also from the technical design of the vehicle.

maximum speed and change lanes to the left lane at the same time.

Temporal aspects can be categorized by distinguishing between simultaneous and sequential operation. If the sequence of the necessary work steps is predetermined by time, as in the above example, this is called sequential operation. It describes the fact that, for factual reasons not caused by the technology used, certain operations can only be carried out in this order. If, on the other hand, there is no factually specified chronological sequence of the work steps, there is a simultaneous process. Simultaneously, an equal arrangement of different choices expresses the fact that different tasks are to be performed at the same time and that, for factual reasons, the order in which the tasks are to be performed is at the pleasure of the operator. For the programming of a navigation computer, for example, it is irrelevant whether the driver wants to enter the destination first or the general route selection first (fast, economical or shortest route).

As soon as the user is confronted with several options, a simultaneous operation is required. On the basis of these decision options, three different types can be distinguished, namely simultaneous operation of a mandatory, varying and diverging nature (Rassl 2004). In the case of the mandatory type, several work steps have to be carried out at the same time to make a decision. All must be performed to fulfill the task. A typical pattern for a simultaneous compulsory operation is the complete completion of a personnel sheet. The task is to specify all the data. However, whether the first or last name is entered first in the sheet is irrelevant for the result. Even with the varying type, different work steps have to be decided at the same time. However, not all need to be edited because every possible step leads to the goal. To fulfil the task, it is sufficient to carry out only one work step. An example is the selection of different route suggestions (motorway yes/no, toll roads avoid yes/no, fastest route/shortest route etc.), which all lead to the same destination. Simultaneous operation of a divergent type occurs when the selection leads to different results. An example of this is the selection of destinations from the address memory of a navigation system. ◘ Table 6.2 shows the three types of simultaneous operation, each with a schematic flowchart.

The temporal order of tasks is appropriately represented by a flowchart (see also ▶ Fig. 2.3). In this flow chart, the information needs of the machine are characterized by rhombuses (the machine cannot know what the operator wants, so it asks him a "question") and by rectangles, actions of the operator through which he transfers information to the machine. In the first step of a system ergonomic solution cycle, a so-called target flow diagram is to be created, which represents the necessary temporal order of the information transfer to the machine from the operator's point of view without taking into account any technical implementation possibilities. From this it can then be deduced which tasks are simultaneous and which are sequential. In the case of simultaneous tasks, the operator must be free to choose which operating step he chooses first. In contrast, the sequence of an objectively necessary sequence should be fixed in the software program or by the arrangement of the control elements, whereby the operator must have appropriate feedback on the current step of the system.

The analysis of the operation is fundamental for every system ergonomic design. In the next analysis steps, the respective ergonomic design must now be carried out for each of the individual actions characterized by rectangles in the flowchart.

Dimensionality The demand for the spatial order of the task increases with the number of dimensions on which the user must exert influence. A task is simple if only *one dimension* (e.g. the setting of a pointer on an analog instrument) or *two dimensions* (e.g. targeted positioning of the mouse pointer or driving a motor vehicle) must be checked. Even a three-dimensional task is still simple from a content point of view, but can become difficult due to restrictions such as displaying it on a two-dimensional screen. When controlling an aircraft, for example, the difficulty lies above all in the control of flight dynamics, not in the three-

Table 6.2 Summary of the three subtypes of simultaneous operation. Depending on the predominant simultaneous selection options, this is presented in different ways

Manda-tory type	Several steps are required for simultaneous decision	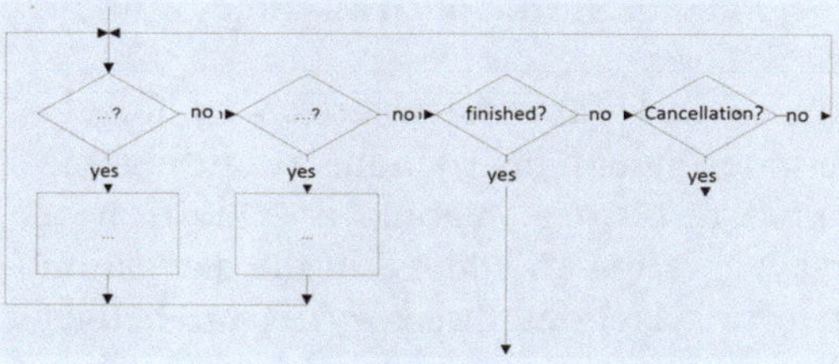
	Every work step must be carried out	
	Order of execution is irrelevant for the result	
Varying type	Several steps are available for simultaneous decision as alternatives	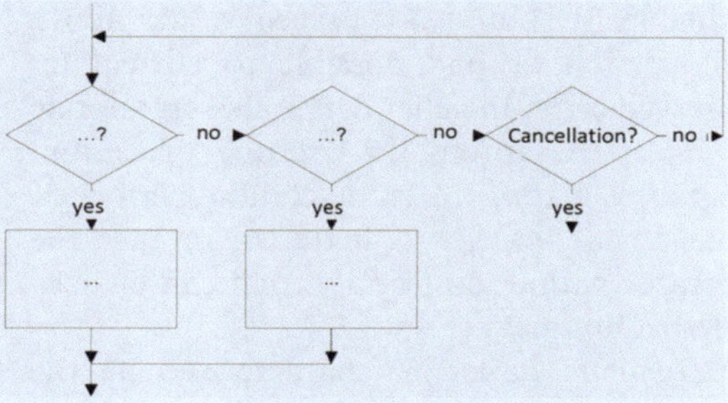
	Only one work step must / can be carried out to fulfill the task	
	Every step leads to the result / goal	
Diver-gent type	Several steps are required for simultaneous decision	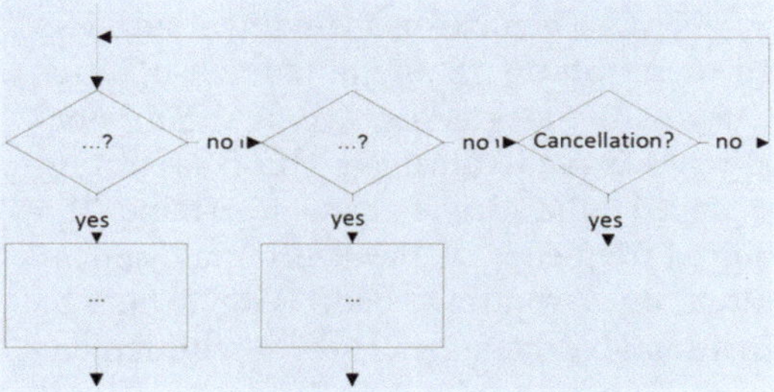
	Not every work step has to be carried out	
	The individual work steps lead to different results	

dimensionality of the task. In any case, it is more difficult when it becomes necessary, *four dimensions* (e.g. the handling of a gantry crane in which both the longitudinal, lateral and vertical movements and the orientation of the load around the vertical axis can be influenced) and particularly difficult when *six dimensions* (e.g. positioning of a component for the purpose of assembly, among other things also inserting an object into the equipment carrier of a vehicle).

Ergonomic improvements are achieved if the dimensionality of the actuator corresponds as closely as possible to the number of influenced dimensions. An example of this is the movement of a pointer on the two-dimensional screen surface by the two-dimensionally movable mouse. If, on the other hand, this two-dimensional task is performed by two separate one-dimensional control elements or sequentially by the rotary pusher, which is often the case in vehicles, it is artificially made more difficult by the type of technical design. By using guide rails, a multi-dimensional task, such as the assembly of an object (e.g. assembly of the luggage cover roller blind in a station wagon, threading a drinking bottle into a cupholder), can be reduced in size and thus simplified. Similarly, meeting positions on a touch screen is a three-dimensional task compared to selecting list items with a rotary knob (one-dimensional).

Guide type In some cases, the task must be completed within a specified time window, which may create time pressure, or within a specified location window (location window: the limited surface of the screen often makes a shift of the content necessary). This leads to very specific compatibility problems (see ► Sect. 6.1.3). A large time window characterizes *static tasks* and a small time window *dynamic tasks.*

Static tasks are characterized by largely time-independent instructions regarding the required result (e.g. reading a value from a display; entering an address in the navigation system or telephone directory). Time budget is the ratio of the time required to the time available. It should not exceed values above 0.5. Then it is just possible to correct an observed error. In general, it is also ergonomic to reject the automatic resetting of certain operating settings after a certain period of time. In any case, it is better to provide the operator with a reset option that can be activated voluntarily.

Dynamic tasks are characterized by the continuous operation of a machine. (e.g. steering a vehicle on a winding road). For their description, the required cut-off frequency of the task is decisive. It should normally not exceed values >1 Hz. By selecting the speed, the driver can determine the required frequency of the steering movement himself on a winding road. With regard to longitudinal dynamics, the requirement changes are normally also below this limit value. Here, too, the requirements profile can be largely determined by the choice of speed. In addition to the dynamics required by the task, the dynamics of the machine also play an important role. In principle, the dynamic version of the vehicle is optimally suited: Both the lateral dynamics and the longitudinal dynamics can be characterised from an ergonomic point of view as so-called speed control, i.e. that the speed of the respective output variable is determined by the actuator actuation – at least in the initial phase. In addition, driving a car is a compensation task, which in combination with speed control is an ideal combination (for more details see Bubb 1993). The two-dimensional driving task itself is perceived as separate in many situations due to the extremely different time behaviour in the lateral and longitudinal dynamics (lateral dynamics in the 100 ms area, longitudinal dynamics in the seconds area!) from a subjective point of view, which is also supported by the completely different control elements. In the event of jerky, counter-rotating steering movements in short succession, as may occur in the case of shock reactions, the control circuit consisting of the driver and the vehicle may, in extreme cases, start to oscillate, whereby the adhesion limit of the wheels need not necessarily be exceeded (see also ► Fig. 2.15). The cause of this build-up is mainly due to the delayed reactivity of the driver, which is above the above mentioned limit frequency of 1–1.5 Hz.

6

6.1.1.2 Task Design

Display mode The difficulty of the task can also be influenced by the way the task and result are displayed. In the case of a technical display, the task and the result can either be displayed separately, or only the difference between the two.

In the case of a separate display, this is referred to as a *pursuit task*. It is recommended in observation situations which are fixed to the ground and installed outside vehicles (e.g. airplanes on the radar screen of air traffic control or position of a marker in a CAD drawing; this form of display may also be recommended in the vehicle if, for example, the current position of the vehicle "in the world" is displayed there on a north-facing electronic map). With this form of *pusuit display* the operator can obtain information about the movement or the change of the task and the result independently of each other. Therefore he can make short-term predictions about the future movement of both and consequently react in time. In addition, it gives the user a correct image of reality on the display – provided that compatibility is observed (see ► Sect. 6.1.3) within the scope of the design – which makes orientation easier.

If only the difference between task and result is displayed, you speak of a *compensation task* or *compensation display*. In technical systems, this is often preferred because the display gain (size of the scale) can be freely selected. The driving direction indicator in a navigation display represents such a compensation display: the entire scenery of the map moves around the fixed vehicle shown in the lower edge of the display. The so-called bird-view display is also a form of such a compensation display. In this case, a perspective view of the map from a fixed position behind the vehicle is calculated, which may facilitate orientation for the driver.

In some navigation systems, a mixture of both principles is implemented so that the vehicle signal moves moderately in a limited area within the map, which also moves within the screen.

In connection with the user interfaces of software, one generally has to deal with the representation of the pursuit task (by means of the mouse, the rotary pushbutton or the finger on a touch screen one can move to the individual positions on the screen, whereby their position does not depend on the mouse position). Since the image section is limited by the size of the monitor, it is often necessary to shift the display. If you use the scroll bar, a compatibility problem occurs: Does the scroll bar move the position of the section or the position of the object behind the section? The problem can be solved by switching to the display form of a pusuit task: by moving the mouse pointer to a free space (background) of the display, the screen content can be moved in a direction compatible way by holding down the mouse button. This is also the case when using a touch screen. The automatic change to a compensation display when the mouse pointer is at the edge of the image is also intuitively correctly understood in many cases.

Controlling a vehicle via the natural view in the traffic environment is always a compensation task, since only the difference between one's own position and the desired position in the outside world can be perceived. However, the highly dynamic driving task would not be manageable without the foresight of the road. This foresight enables the driver to give the character a pusuit task to driving a vehicle through mental effort. This is particularly the case in the low speed range for manoeuvring manoeuvres. From the point of view of information ergonomics, therefore, in the higher speed range (experience has shown that >40 km/h) in a vehicle, sizes that have to do with the outside world (e.g. representation of the correct course in a navigation device at an intersection) must be represented in the form of a compensation display (direction-indicating or bird-view; see above). In the low speed range, on the other hand, the display in the form of a pusuit task may be preferable (e.g. display of a parking aid).

Task type For the person operating the vehicle, it is a fundamental difference whether he/she himself/herself is actively involved in the work process or whether he only has to define the basic settings of an automated process and then observe how it works. One speaks of *active* or *monitive* task of human. Since it is almost the domain of the computer to control or regulate even a complex process intelligently, such automated processes are gaining more and more importance in the vehicle (a well-known example is the distance control cruise control; ACC).

Well-known characteristics of human and machine can be used to decide whether to select automatic or manual control: Automation must generally be recommended when problems arise with human limitations in terms of accuracy, speed and reliability. However, this recommendation is only valid if only situations occur for which the designer has taken precautions. However, as this cannot be guaranteed in road traffic, every automated partial task is accompanied by the human being as an observer (monitor).

Monitive systems have disadvantages that arise from the risk of monotony and thus lead to a loss of the operator's vigilance.[3] In addition, the driver may lose practice in handling the system elements and their modes of operation due to the constantly operating automatic system. In particular, there is the disadvantage that the driver has difficulties with the system control due to his lack of familiarity with the system in the event of a failure of the automatic system. Due to the complexity of situations and the difficulty of recording and processing adequate technical

3 However, even a boring, low-event, actively performed driving task can lead to monotony effects. Then the automated driving process may even be a safety benefit.

signals, complete automation has not been possible at all in road traffic in particular. Damböck (2013) was able to show that the driver needs at least 6 to 8 seconds to correctly detect the now unknown situation of the traffic outside world from a complete distraction and to react appropriately to the situation (see also Gold et al. 2013; Petermann-Stock et al. 2013).

For areas in which partial automation or automation is feasible (today e.g. for automatic distance keeping or for lane guidance on well-marked roads), it is therefore essential with regard to the overall reliability of the system to ensure that the human being is somehow integrated into the human-machine system in such a way that his attention and training condition are maintained.[4] The advantages of a monitive task design for reliable operation in pre-planned situations can be maintained in connection with this requirement if only the local and temporal limits within which the driver has to keep the system variables to be influenced are reliably determined by the machine. The driver must then always operate the vehicle within these limits. If he touches the limits, he will be adequately informed by the vehicle (see ▶ Sect. 6.1.2 "Feedback") and he can decide whether to follow the recommendation of the automatic or take over the regulation himself. However, since there are many technical/physical conditions under which such a system cannot operate reliably, the driver must be given feedback on the current system status (so-called mode awareness).

6.1.2 Feedback

The feedback of what has been achieved to the driver is one of the most important factors through which a coherent understanding of the system status can be conveyed. If, in addition, the information about the state of the system is conveyed by different sensory organs, this redundancy is generally positive. The human "situation awareness" is improved if the same information is perceived by at least two, preferably even more sensory organs at the same time (e.g. reference to a danger by a control lamp lighting up *and* of an acoustic signal). Another particularly important aspect is the time interval between the input information on the actuator and the reaction of the system on the output side. If this time exceeds 200 ms (duration of human information reception), this leads to confusion and disorientation of the driver, because the reference to his own action is lost. If such a time delay cannot be avoided for technical reasons, this must be indicated (e.g. in the most primitive way by the well-known "hourglass sign" or similar). If the time delay is more than 2 sec, the process to be regulated appears to the operator as a open control system. It then requires at least immediate feedback via the activated system input (e.g. the button lights up). In this case, however, it is better to give exact feedback on the progress of the process (e.g. progress display via a growing bar and indication of the remaining time to be expected when calling a computer program).

The change of position of a control unit is an important feedback about the current switching status of a device. Götz (2007) argues that actuators that change their form during operation are not yet known. However, if novel materials are able to change shape during or immediately (< 200 ms!) after actuation (e.g. compound materials), it would be conceivable to use this property for feedback. Variable key shapes (e.g. angular, round) would be able to inform about functional changes in this way.

For all types of feedback, the corresponding sensory modality must be designed to be supra-threshold and the necessary signal-to-noise ratio must be maintained. Whatever technical means are used, a well-designed feedback must always allow the driver to answer the questions:

- "What have I done?"
- "What's the state of the system?"

4 For reasons of "saleability", the argumentation today attributes greater attractiveness to fully automatic driving because it allows passengers – like rail passengers – to deal with other things on boring routes. Unfortunately, this is also linked to the argument of increasing road safety, but this is highly questionable (see ▶ Sects. 2.6 and 9.3).

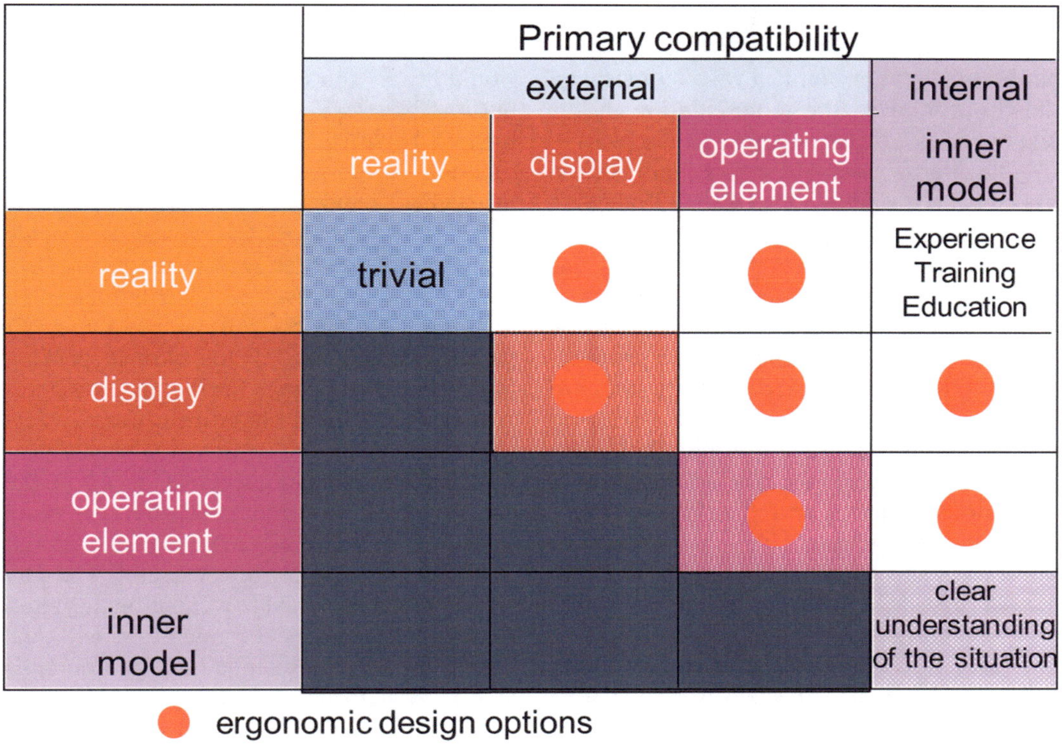

Fig. 6.4 Ergonomic design areas of compatibility

6.1.3 Compatibility

Compatibility describes the ease with which an operator can transcode information between different information channels. A distinction must be made between primary and secondary compatibility.

Primary compatibility refers to the possible combinations of different information areas such as reality, displays, actuators and internal models of the operator (see Fig. 6.4). Within primary compatibility, a distinction can also be made between external and internal compatibility: *External compatibility* refers to the interaction between human and vehicle with respect to the outside world ("reality"; an example of this is the compatibility of steering wheel rotation and vehicle movement), whereas *internal compatibility* the interaction between the outside world and the corresponding inner models of the human being. As shown in Fig. 6.4, ergonomic design is only possible in certain areas. For example, the compatibility of different information is trivial in reality. On the other hand, compatibility between reality and inner models can only be achieved through experience, training and education. Compatibility between different internal models corresponds to a consistent understanding of the situation. In this context, it should be noted that each person has certain inconsistencies with regard to the ideas in different areas of his memory, which, beside others also leads to the fact that a certain action carried out under given circumstances is not understood by an outsider. The remaining areas in Fig. 6.4 can be ergonomically designed in such a way that, for example, a movement forwards or to the right in reality corresponds to a movement forwards or to the right at the pointer or at the control device, etc. This so-called spatial compatibility plays a prominent role with view on the connection of the representation of the real outside world on a display (e.g. screen).[5]

The internal compatibility is characterized by so-called stereotypes (see ◘ Fig. 6.6), which for certain contexts may only have this meaning in the western culture (probably due to the writing habit from left to right and from top to bottom – however, it must be assumed that due to the dominance of the technical devices primarily developed in the western culture, this assignment has also been adopted in other cultures, at least for technical objects). SAE J1139 describes the modes of operation of various controls and their **stereotypical operation**. In general, sliding forward or upwards means switching on or enlarging (see also ◘ Figs. 6.5 and 6.6).

The continuous repetition of stereotypical operation also outside the automotive application leads to an expectation of the user, which presupposes a certain mode of operation for the control elements. If the technical implementation meets this expectation, one speaks of "*expectation conformity*". In the course of the advancing technical development and the internationalisation of the approval regulations, there have also been variations on stereotypical operating rules, as well as partly also brand-typical operating devices, which have entered into the wealth of experience of the brand-loyal clientele. A prominent example here is the electric parking brake button. The stereotypical operation of the switch to turn on the locking function would be, as with other switches, to press the button. To release (open the brake) it would have to be lifted ("released"). However, the function is derived historically from the mechanical handbrake lever, which must be pulled upwards to lock the vehicle. This mode of operation has been transferred to the electrical switch and can lead to irritations when learning the function. Part of the somewhat sluggish acceptance of the electric parking brake is certainly due to the fact that it is operated in a way that deviates from stereotypical operation. The more consistently such deviations or special arrangements occur in the products of a brand manufacturer, the more this form and mode of action is perceived and expected to be typical of the brand. The customer is conditioned to functionality and associates this mode of operation with the brand. However, a change in conditioning is associated with faulty learning processes and can lead to irritation or even annoyance on the part of the user due to incorrect operation. A change of operation and above all the deviation of conditioning and stereotypical operation should be carried out extremely sensitively during model maintenance or change.

Secondary compatibility means that an internal contradiction between partial aspects of compatibility is avoided. For example, a hanging pointer is secondary incompatible, since the movement from "left = little" to "right = much" is associated with an incompatible left turn. Also the variant of fixed pointer moving scale is secondary incompatible, because here the direction of movement of the scale is always incompatible with the arrangement of the ranking of the digits. Since the control elements also have an indicator character in principle, what has been said also applies to them. For example, attaching inscriptions to a rotary switch that moves relative to a fixed marker is equivalent to the "fixed pointer – moving scale" variant mentioned above. Secondary incompatibility, however, does not only refer to contradictory rotational

5 The design of the selector lever of an automatic transmission on the centre console represents a historically created incompatibility of this kind: in order to *engage the reverse gear*, the selector lever must be moved *forward*, while for the *forward direction*, the lever is to move *backwards*. Under certain circumstances, one reason for the "unexpected acceleration accidents" that became virulent in the USA in the 1980s can be found in this form of interpretation. During a manoeuvring manoeuvre, completely incompatible movements must be carried out between the movement of the selector lever and the movement of the foot on the pedals.

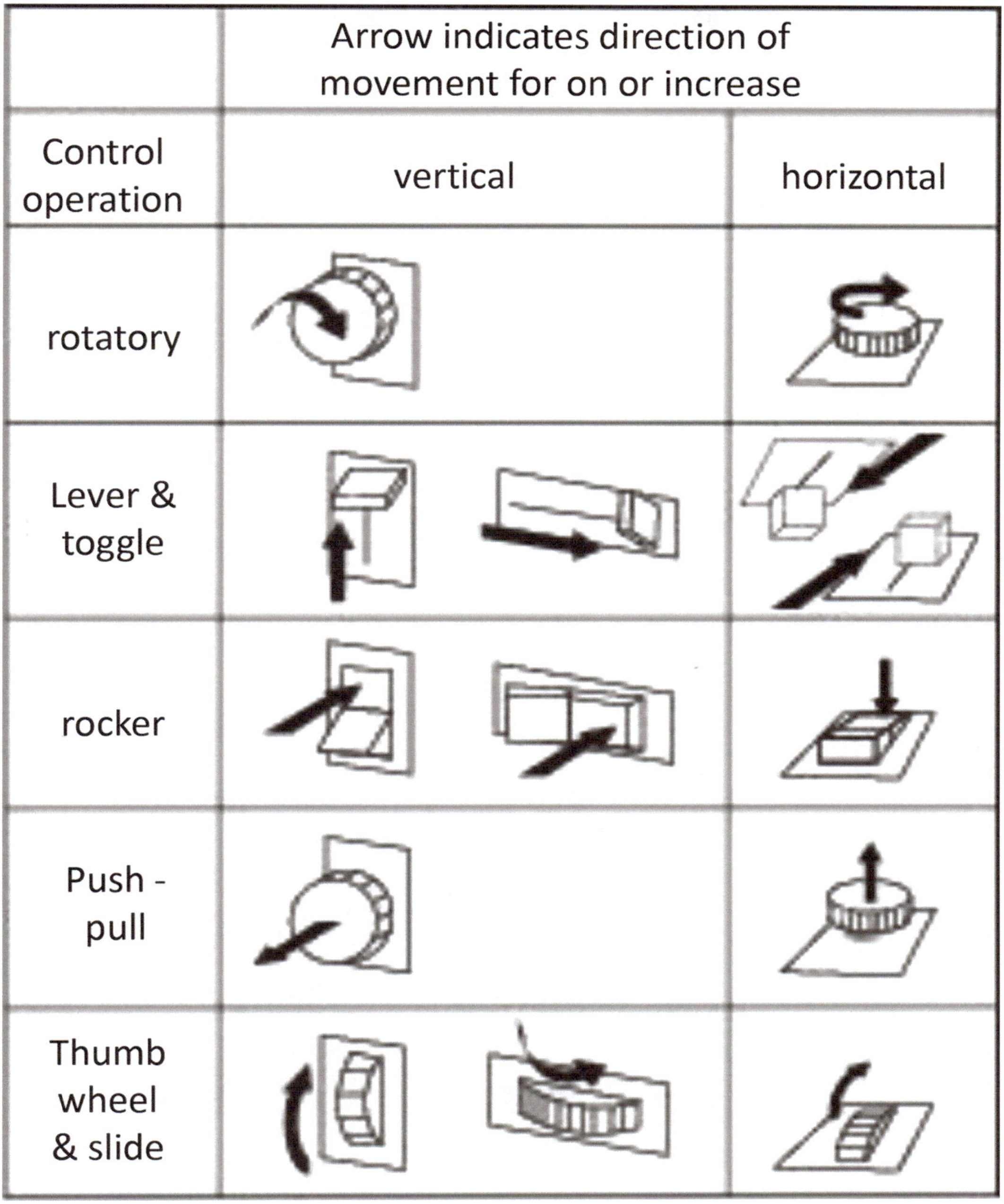

Fig. 6.5 Stereotypical operation according to SAE J1139

movements, but generally to the fact that different compatibility requirements are contradictory. An example of this is the arrangement of the indicator in the inner angle of the headlight. In this case, the position of the flashing light in the headlamp housing is incompatible with the position of the headlamp in the vehicle. Examples are shown in Fig. 6.7. These contradictions are usually not noticeable when looking at them in the design office, because "it is clear how it is meant" when rational considerations are considered. In real traffic, however, direct, doubt-free and rapid interpretation may be important, and here secondary incom-

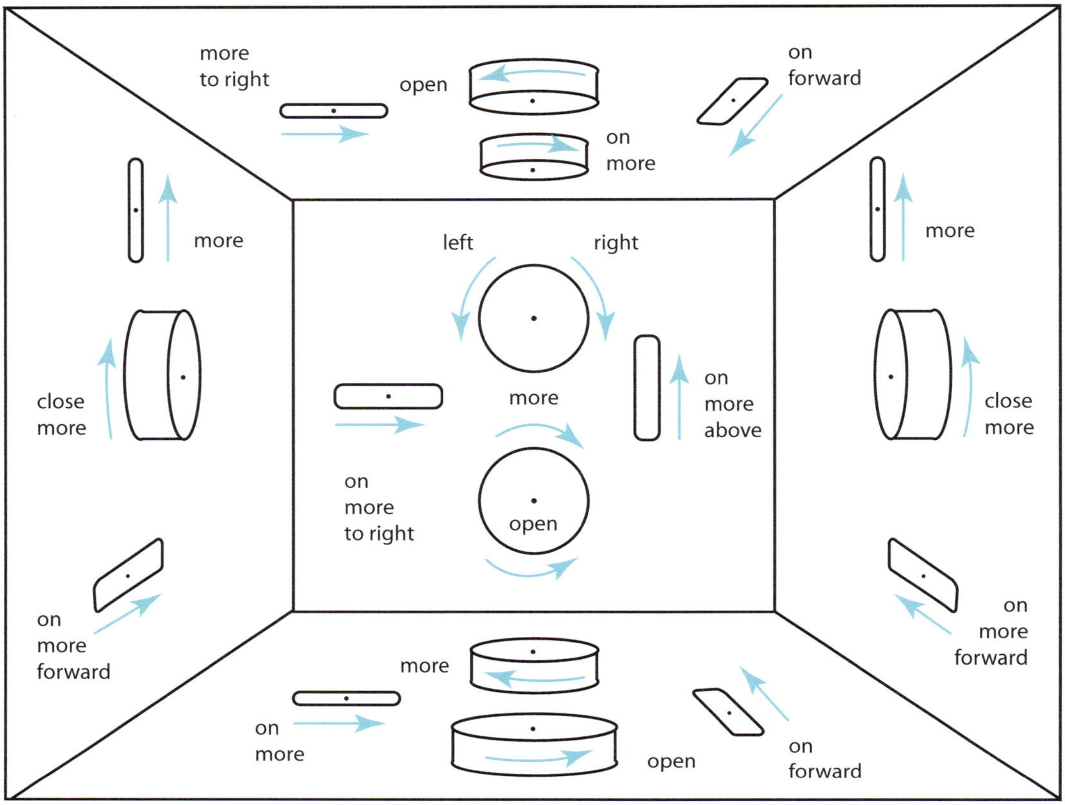

Fig. 6.6 Stereotypes according to Götz (2007) and Woodson (1992). Care must be taken with the rotary knob in the right wall of **Fig. 6.6**, as this is a secondary incompatibility!

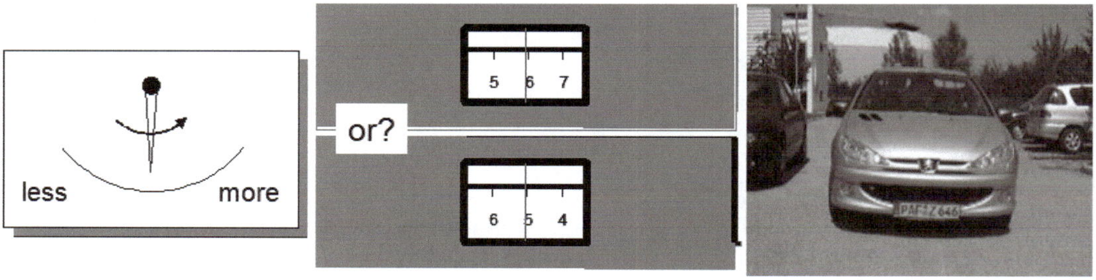

Fig. 6.7 Examples of secondary incompatibility

patibility plays an important role because it can be the cause of misinterpretation. However, this is precisely why their significance is experimentally difficult to prove. However, inconsistent compatibility rules and assignments (different directions of rotation in different menus) within a device must be avoided at all costs.

6.2 Human – Machine – Interaction

The places where information is transferred from the human being to the machine, i.e. to the vehicle and where, conversely, information is transferred from the vehicle to the human being, are referred to as the interface between the human being and the machine. The term "interface" is correct insofar as the human being can move away from the vehicle and thus the transmission of information is separated. On the other hand, this term is also misleading because it suggests that the vehicle and the driver are completely separated in terms of information technology when driving. But this is not correct: a successful operation of the vehicle by the driver is only possible if he has a certain idea of the technical function of the vehicle, if he at least knows what effect a certain action will have. Conversely, "the vehicle also has an idea of the driver", which has been implemented by the vehicle designer, because he had to get an idea of the ideas of the future user during the conception of the operation.[6] When designing the interaction between driver and vehicle, this aspect must always be taken into account. Meyer-Eppler (1959) describes this phenomenon in such a way that the designer of a product sends a message of which – depending on the quality of the product language – the user understands only a subset as receiver. However, he interprets further messages into the product which the designer had not even thought of in this form. "The task of the designer in his function as an expert in the field of design is to translate the various functions of a product into signs in such a way that they are understood by the potential user. For this it is necessary that the designer is particularly aware of the character repertoires of the respective users" (Götz 2007, see also the term "affordance"coined by Norman in 1988). The fundamental telecommunications principle must be taken into account: Meaning does not arise at the transmitter (this is the specific design here), but only at the receiver (this is the user of the vehicle here). Misleading and ambiguous design can cause meanings in the user that lead to incorrect operation and rejection of a product (Götz 2007; Bengler et al. 2012).

In general, all technical elements are described as ***displays***[7] through which information is transferred to humans in a manner that is structurally appropriate. All sensory organs are involved, so that a distinction is made between *visual, acoustic and tactile displays*. Thermal displays are unusual. The kinaesthetic also provides the driver with relevant information. In general, however, this is not understood as a display, although the design of the driving-relevant characteristics provides the driver with more or less targeted information about the dynamic driving status in different ways (for more details, see ► Sect. 6.4).

As ***control elements*** is the name given to the technical equipment through which information is transferred from a human being to the machine. Traditionally, all the elements that are mechanically operated with the extremities, i.e. the fingers, hands and legs, play a role here. Essentially, a distinction is therefore made between *foot control units* (pedals) and *manual control parts* (levers, knobs, buttons, etc.). With the increasing possibilities of information processing, sensors (microphone, camera, IR sensor) are also increasingly being used as control elements, via which information is transmitted to a detection unit and interpretation unit (speech recognition, mimic recognition, gesture recognition) technically implemented in the computer, which then triggers corresponding functions (Bengler 2001, 2005; Bengler et al. 2012).

The information coming from the vehicle is to be coded by displays in such a way that it can be assigned meaning by the user in the sense of semantics. The control element is used to code the user's intentions in such a way that the vehicle executes them. In any case, code systems must be used that enable the information to be understood by as large a

6 He often derives this idea of the future driver's world of ideas from his own ideas, which, however, depend essentially on his own training and the development process that has taken place up to that point. A consistent application of system ergonomic rules should prevent such "short circuits".

7 As Bernotat (1993) points out, the proposal of VDI/VDE Directive No. 2172 E to designate the technical elements transmitting information as "indicators" has not been able to prevail.

proportion of the total population as possible (Rühmann 1993). In addition to Götz's proposal (2007), the following code systems can be considered in principle:

- *Position coding:* Orientation in space is a fundamental human ability and thus the main prerequisite for largely error-free orientation and operation.
- *Motion coding:* The kind of movement required to operate an control element makes it possible to clearly distinguish between operating operations. In many cases the external compatibility has to be considered (e.g. "movement forward = effect forward" – Example: Switching between high and low beam). If interference occurs between different requirements, this can be countered by a different position coding or form coding (e.g. the requirement to implement the speed selection of the cruise control by moving "forward = more" can compete with the above example of the driving light function).
- *Form coding:* The control element but also the displays should differ by their form in order to ensure a – learned – but secure assignment of the presented or transmitted function to task-relevant contents or intended intentions. In particular, haptic feedback plays an important role in the design of control elements.
- *Size coding:* In a way, size coding is a special type of form coding. The importance can be coded by it (e.g. small round instrument for the oil pressure – big round instrument for speedometer or rev counter; big rotary knob for the switching of different possibilities in the central control panel – small rotary knob for the radio volume).
- *Colour coding:* The selected colour should take into account the meaning of everyday experience (e.g.: Red = Danger, Yellow = Caution, Green = Normal/Harmless) The colour coding principles for display and control element are described in detail in DIN EN 60073 and DIN 04844.
- *Character encoding:* A well-done character encoding significantly facilitates the learning of functions in a vehicle unknown to the user or of previously unknown functions. Even for rarely used functions or rarely presented information, an appropriate coding that captures the user's knowledge can be very helpful. Alphanumeric characters allow a very clear assignment of meaning, especially when using modern screen-based display systems in which whole sentences can be displayed. The word coding, however, is language-bound[8] and should therefore preferably be used in screen-based display systems, where different languages can be stored in computer memory. Pictograms, on the other hand, are language-independent, but should only be limited to a few operating and display functions that are also frequently used (e.g. lighting functions, windscreen wipers, horn). They help the occasional user to quickly find his way around an unknown vehicle. Because they are easier to grasp, it makes sense to always choose a combination of pictograms and lettering if circumstances permit (Bengler et al. 2012).

For the mentioned codings a ranking of importance is valid, which is represented by the above enumeration. That is position coding is always more important than motion coding and this more important than shape coding etc. It is important to realize that none of the above codings is self-explanatory. Each must be learned. However, learning processes are considerably facilitated if existing knowledge can be tapped and if, by observing the principles of compatibility and feedback, what has been experienced can be quickly consolidated.[9]

The exact localisation of displays and operating elements is a complex task that must be solved within the framework of the anthropometric design of the vehicle (packaging). In any case, however, the labelling of dis-

8 It is a misconception to assume that most technical terms in English are generally understood!

9 Even the famous wiping movements when using smartphones are not self-explanatory. However, since they take into account principles of compatibility and feedback well and follow on from everyday experiences, they are learned so quickly that this is no longer perceived as "tedious learning" at all.

plays and controls shall be such that the relevant information can be read even by the tall man seated far back, taking into account normal vision. Control elements must also be installed in the gripping space available for both the small woman seated far in front and the large man seated far behind.

As already explained in ► Sect. 1.2, an arrangement of displays and control elements in the vehicle has been established in the historical development in the sense of a scheme, which provides for the lighting function in the left-hand dashboard area, all primary and secondary driving functions in the area around and behind the steering wheel and the functions of the so-called vehicle information system (DIS, radio, navigation system) and other tertiary comfort functions (e.g. heating/air conditioning setting) in the centre of the vehicle (see also ◘ Fig. 1.8). Schmid and Maier (2012) divide the cockpit space into the following zones:

- primary area: central field of vision/both hand zone
- secondary area: peripheral field of vision/one-hand zone
- Tertiary area: outside the field of vision/extended one-handed zone.

The control elements and displays of the primary, secondary and tertiary driving tasks are to be accommodated in these zones. Based on these considerations, they develop a basic cockpit that largely meets ergonomic requirements, contains a high degree of "self-explanatory ability" (see the comments in the previous section) and thus makes a significant contribution to active safety (◘ Fig. 6.8).

6.2.1 Displays

Displays are intended to make an image of the outside world or technical details (measured values) available to the user. As already mentioned, this can be done visually, acoustically and tactilely and in appropriate combinations (multimodal). In the sense of the control loop diagram in ◘ Fig. 1.9, all displays that convey information about the technical condition of the vehicle and its movement also have feedback character. Operating elements that are mechanically operated by the user can also provide feedback about their operation via the haptic sensory quality. In particular, many control elements that are operated using the hands or fingers have a display character due to the new position they have taken on. In principle, therefore, the rules for visual displays set out below apply to them, insofar as they are applicable.

6.2.1.1 Visual Indicators

Depending on the degree of abstraction of the representation, a distinction is made between digital, analogue, pictorial situation-analogue, photorealistic and contact-analogue displays for visual displays. Using the navigation display as an example, Israel (2013) illustrated these stages and at the same time estimated the mental performance necessary to bring the display content into congruence with reality (◘ Fig. 6.9).

Digital displays A continuum of values related to the real world is divided into segments and only the value of the current segment is displayed. For example, displaying the status off/on in the form of two or even just one light is already a very simple form of digital display. Normally this form is used for the feedback of an operating status. No more than three states should be represented in the vehicle – normally additionally colored or coded by an adequate symbol. In most cases, however, the corresponding value is represented in the form of alphanumeric digits. The number of digits can be used to determine the accuracy with which the corresponding value can be read (the technical accuracy of the measuring instrument should be at least a factor of two, or better a power of ten, greater than the display accuracy used). No more than three, max. four digits should be used in the vehicle. In principle, the error rate is very low compared to all other forms of display, even when reading a digital display quickly. Digital displays are always recommended when the correct reading of the numerical value is of decisive importance.

For a long time, the digital display was realized in the form of the so-called seven-segment display, because this technology was

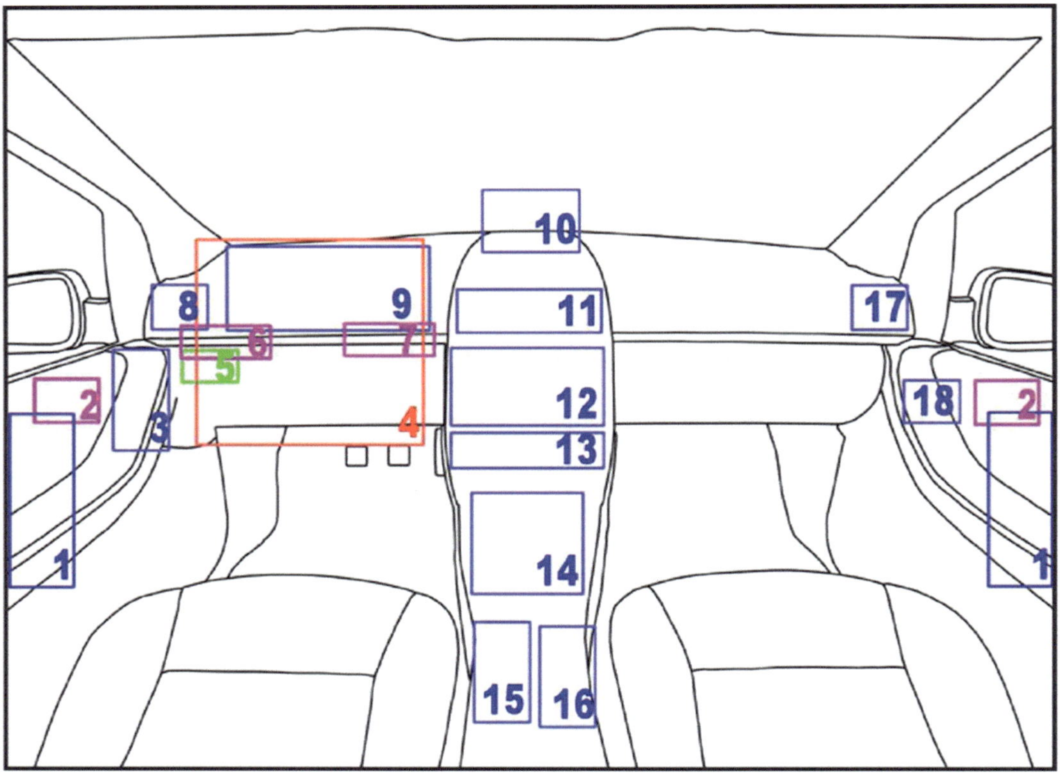

1: Door handle left & right; 2: Door opener left & right; 3: Window regulator, mirror adjustment, central locking 4: Multifunction steering wheel; 5: Light switch; 6: Left-hand steering column lever; 7: Right-hand steering column lever; 8: Left-hand air nozzle & associated controls; 9: Instrument cluster; 10: Central display; 11: central air nozzles & associated control; 12: climate control module; 13: entertain module; 14: gearshift lever; 15: control unit for central display; 16: handbrake; 17: right air nozzle & associated controls; 18: window regulator for passenger side

Fig. 6.8 Arrangement areas of the control elements in the basic cockpit according to Schmid and Maier (2012)

available at a relatively low price. The fact that such displays are also used in cheap watches probably led, among other things, to the fact that the aspect of liking has turned out to be to the disadvantage of the digital display (Fig. 6.10). Today, TFT or OLED displays are used, which enable a very fine resolution of the display (at least a 7 × 11 matrix for each digit/letter) and thus also ensure an attractive appearance from the design point of view.

Analogue displays By analog displays the continuum of reality is represented by a corresponding continuum of the angular position or position of a pointer or a variable element in its length (e.g. bar indicator of speed of the 60s, bar display on a screen). The pointer position or a displayed length triggers perceptual processes at a very low level, as this addresses the specific angle or length receptors that are already realized on the retina (see ► Sect. 3.1.3).[10] Therefore, this display is par-

10 Since it has become common practice today to place an additional display in the middle of an analog circular display that can display other information, the pointer is therefore very short. However, this counteracts to a certain extent the mentioned advantage that angle detectors on the retina are already addressed at a very low level (see also Fig. 6.10 left).

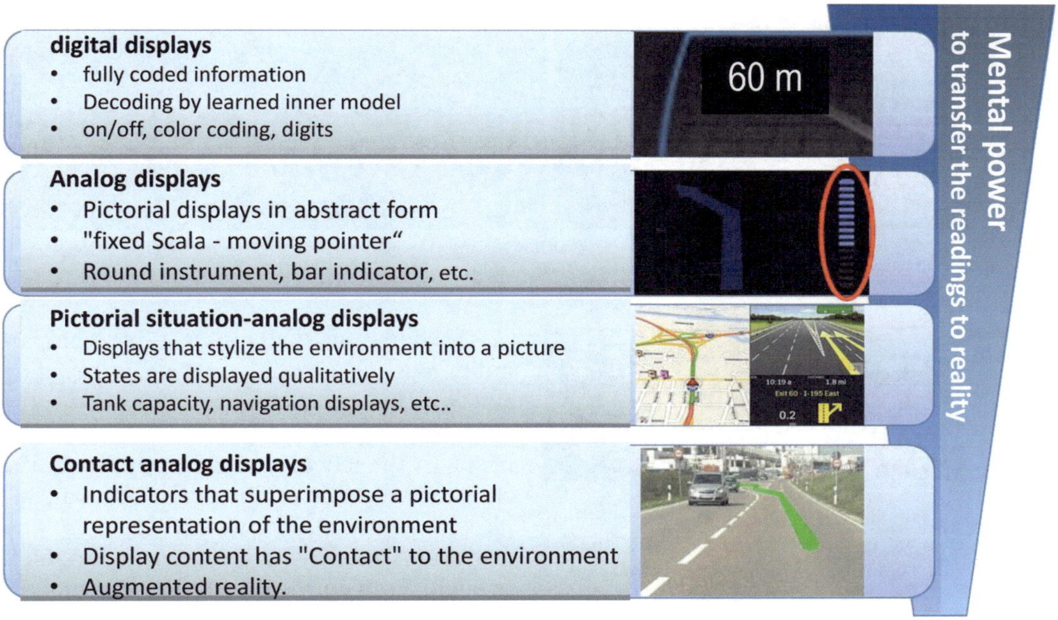

Fig. 6.9 Visual displays and mental performance using navigation as an example (Israel 2013)

Fig. 6.10 Example of a digital speed display and mileage of the vehicle in 7-segment design

ticularly suitable if two values are to be compared, because then – assuming the same design of the analog display – different angular positions or lengths can be perceived very well (Treisman and Gelade 1980; Goldstein 2009). In order, for example, to check whether “everything is in order” in monitoring tasks, analog indicators are aligned in such a way that this state is represented on all instruments by the “12 o’clock position” or equal lengths in the sense of the Gestalt laws. A deviation is immediately noticeable (the bar graphs on the right side of the cockpit shown in Fig. 6.10 provide an inadequate example!). In the application for the vehicle, of course, the question must be asked whether this case is really relevant at all.[11] Another advantageous use of the analogue display is that the pointer strokes over a colour-coded scale that not only characterises different operating states, but also, through the movement of the pointer, which is perceived very simply, the approach to a desired or undesired operating state. The same applies, of course, to the design of a variable bar length. This display version is useful for the rev counter, where the range and the approximation to unfavourable too low and too high revs can be indicated or also for a display of the current energy conversion (fuel consumption/battery discharge). If, from the point of view of the task, it is necessary to also record the numerical value of the measurement, the user is required

11 For example, it is possible to display in this form indications of vehicle operating conditions such as battery charge, engine operating temperature (water and/or oil). However, the question must be asked as to whether, in view of modern computer technology, it would not be better to draw attention to deviations from the normal operating state digitally in verbal plain text, possibly with an indication of how the imminent damage can be averted.

Fig. 6.11 Design of analogue displays: Pointer conceals digit **a**. Perfect reading possible **b**

Fig. 6.12 Poor interpolatable nonlinear scale division. The example also shows that the center of the circular scale must coincide with the pointer pivot point

to make a mental assignment between the pointer position and the numerical scale. The numerical value is of course read most accurately when the pointer points exactly to the corresponding digit. For this it is necessary that the pointer does not cover the digit (see Fig. 6.11, it is always about the quick glance that provokes such a misreading – if there is enough time, it is of course no problem to carry out the necessary interpolation).[12] Murrel (1969) also recommends that the pointer be designed in such a way that it just reaches the scale. If the pointer is positioned between the digits, interpolation is required from the user. This is most accurate in the middle position. Depending on the task, this results in the necessary number of inscriptions and additional scale divisions between the inscribed scale lines. In order to make interpolation as safe as possible, only all multiples of either 1, 2 or 5 or their powers of ten should be used as scale values. A perfect interpolation is only possible if the scale division is absolutely linear (see a historical, bad example in Fig. 6.12).

12 Unfortunately today mostly inside numbers are preferred, which at the one hand is justified by better aesthetics, at the other hand however allows most long pointer, which obviously seems to show an emotional increase of speed resp. revolutions while accelerating.

Of course, the rules for secondary compatibility must also be observed when designing analog displays. Figure 6.13 shows the optimum display range of a round instrument. In the left half of the instrument, the clockwise rotation of the pointer is compatible with a movement from bottom to top and in the upper half with a movement from left to right. Both correspond to the inner model of an increase. The upper right segment and, if necessary, also the lower right segment can also be used, because the history of the pointer's course is included in the observation. For instruments that only use one segment, however, it is advisable to avoid hanging pointers and pointers whose pivot point is to the left of the scale (see also Fig. 6.14).

There are extensive ergonomic studies on the usability of the various instrument variants depending on the respective application. Table 6.3 provides a compilation for the question "Analog or digital display" in particular. This shows that the variant "fixed pointer – moving scale" is basically unfavourable. As already mentioned, this is due to the secondary incompatibility. The famous Citroën-magnifier speedometer, which was used in the DS, ID and CX models of the 50s to 70s, is an example of this design (Fig. 6.15). Recently, this variant has reappeared in the form of scrolling through menus in the computer-generated screen displays in DIS.

The suitability of the digital tachometer can also be read from Table 6.3. The question to be asked is what is the actual task of reading it: is it a question of determining the speed level and the speed changes – this can be determined much more directly from the immediate view to the outside and from the reaction of the vehicle (see ▶ Sect. 3.2.1) – or

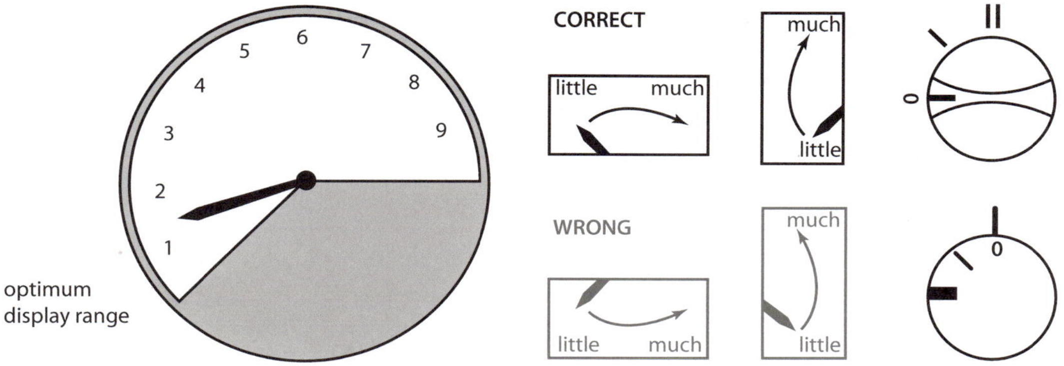

Fig. 6.13 Observing secondary compatibility when designing analog displays

Fig. 6.14 Insufficient (**a** fuel gauge) and correct **b** Observance of secondary compatibility. The picture on the left shows the designer's wish to orient the point of rotation of the hands as far as possible to the centre of the instrument

is it a question of reading the current value quickly and reliably at given speed regulations? From an ergonomic point of view, the digital speedometer is therefore recommended according to Table 6.3. This also corresponds to the findings of Bouis et al. (1983). The situation is completely different with the rev counter, where the aim is to capture the area in which the current engine speed is located and the proximity to the technically determined maximum speed by a quick glance. The creative flexibility that results from freely programmable and fully graphic-capable combination elements should therefore be used to facilitate the reception of information, especially in view of these design principles.

Pictorial displays: Pictorial displays provide a stylized picture of the environment (Rühmann 1993) and thus facilitate the mental process of linking the display content with reality. Pictorial displays are usually *two-dimensionally*. In Fig. 6.14b, between the two analogue displays, an example of a simple pictorial display is shown, which shows whether and which door is open on the vehicle. With the introduction of computer technology into the motor vehicle, pictorial displays have gained in importance. For example, a visual display of the distance to stationary obstacles makes parking easier (Fig. 6.16a). Similar displays can also be used on an ACC system to show which target distance (Fig. 6.16b, example BMW) is set or whether the distance to the vehicle ahead is in the safe (green) or unsafe (red) range (Fig. 6.16c, used by Audi). As already shown in Fig. 6.9, pictorial displays are used in connection with navigation devices. A simple form of the pictorial display are the direction arrow and the distance bar as a turn indicator. Indicated cross roads at the direction arrow facilitate the identification of the target road. The Birdview display on the stored digital map gives the driver a much better orientation adapted to the situation. Orientation can be further improved by displaying a stylized view of the corresponding

Table 6.3 Suitability of analog and digital displays for various applications

Application	Digital indicator	Analog indicator	
		Moving scale	Moving pointer
1. Quantitative reading	**Good** Minimum reading time and error for recording numerical values.	**Moderate**	**Moderate**
2. Qualitative reading	**Unfavourable** Numbers must be read. Position changes are badly noticed.	**Unfavourable** The direction and magnitude of the deviation are difficult to judge without reading the scale values.	**Good** Pointer position easily recognizable. Scale values do not have to be read. Position changes are quickly noticed.
3. Setting values	**Good** Accurate monitoring of the numerical setting.	**Moderate** Misleading relation to the movement of the control element. No change of the pointer position as a monitoring aid. Difficult to read with fast settings	**Good** Clear relationship between the movement of the pointer and the control element. Changing the pointer position facilitates monitoring. Fast adjustment possible.
4. Adjust	**Unfavourable** Position changes are missing for monitoring tasks. Difficult to understand relation to the movement of the control element. Difficult to read with fast changes.	**Moderate** There are no noticeable position changes for monitoring tasks. Conditionally understandable relationship to the movement of the control element. Changes difficult to read.	**Good** The pointer position is easy to monitor and control. Easy to understand relation to the movement of the control element.

After Baker and Grether, quoted from Bernotat (1993)

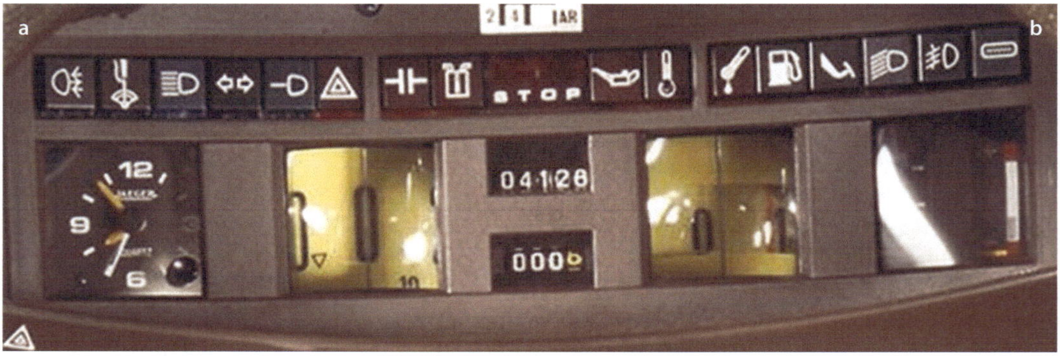

Fig. 6.15 Magnifying glass tachometer **a** and rev counter **b** in the Citroën CX

road section, as is the case today with many navigation devices for motorway exits, but also for certain stretches of road in cities. Since the screen on which such displays become visible is itself two-dimensional, the display is also two-dimensional, even if a three-dimensional impression is created by the calculation of a central-perspective representation. One therefore often speaks of a pseudo-three-dimensionality.

The use of pictorial displays in vehicles is extremely diverse and will continue to increase

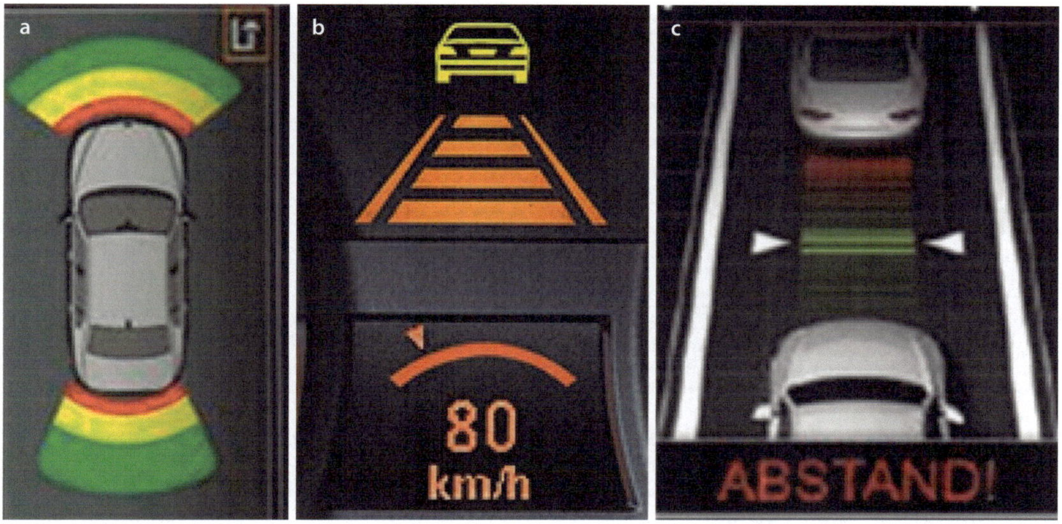

Fig. 6.16 Pictorial display of the distance to stationary obstacles during parking **a** and distance cruise control (**b** BMW, **c** Audi)

in the future in conjunction with the corresponding sensors and displays. Various displays have been implemented to show the energy flow in hybrid vehicles. Lorenz (2011) has given an example of how a tutorial for correct seat adjustment can be supported by pictorial displays.

Contact analog displays: Contact analog displays are characterized by the fact that artificial information in the sense of augmented reality is introduced into a real image of the environment – today generally into an image captured by an electronic camera (so-called augmented reality). By means of the HUD technology this is also possible with regard to the real environment. To speak of a "contact analog" is only permissible, however, if this information is directly (locally and temporally) assigned to objects of reality.[13] It is conceivable, for example, that instead of the aforementioned stylized image of a motorway exit or a street situation in a city, the current image of a video camera is displayed, in which the corresponding navigation information is directly superimposed at a fixed location. However, such a technology requires a very precise positioning. Using the so-called contact-analog head-up display (Bubb 1975, 1981; Schneid 2009; Bergmeier 2009; Israel 2013) it would even be possible to display such information directly in the real environment. The lower illustration in Fig. 6.9 gives an impression of this. Also for night vision devices, which can make objects visible by means of infrared technology even without direct lighting, which stand out from the environment by their heat radiation, it is possible by image recognition to detect critical objects (for example pedestrians or animals) and make them recognizable by means of a colored frame in the image designed by the infrared camera. As Bergmeier (2009) proved, however, this form of technical implementation is not very effective in terms of road safety, as the driver has to look at the display in the dashboard area, especially in hazardous situations. According to his investigations, the use of the contact-analog head-up display would also be advantageous here (Fig. 6.17).

A particularly simple form of a contact-analogue image is provided by the mirror, which is used almost exclusively in the form of a rear-view mirror in the vehicle. However,

13 The HUD, which is often offered today as an optional extra, is not a contact analog display, since vehicle-related data (speed, action status of the assistance systems, conventional navigation instructions) are generally superimposed on reality here without direct reference to the outside world.

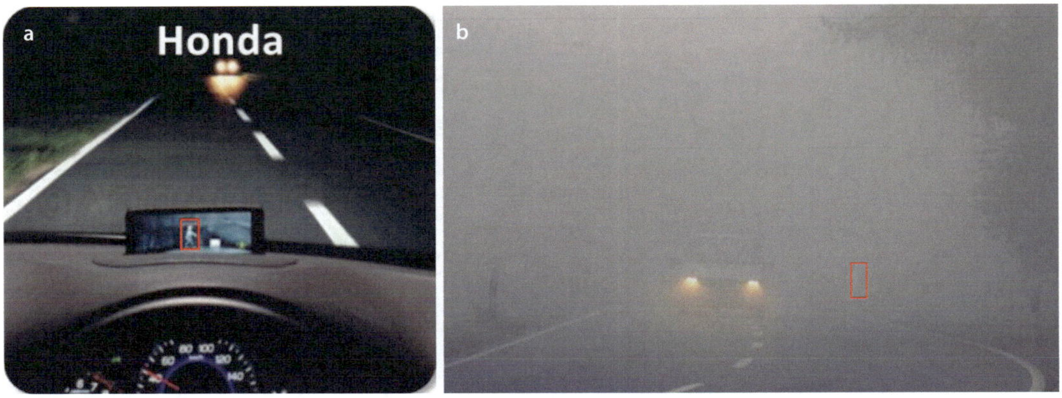

6

Fig. 6.17 Contact analog display to identify a pedestrian by means of an infrared night vision device, **a** on the display of a screen (example Honda), **b** directly in the real view according to a proposal by Bergmeier (2009)

according to the above definition, it is not entirely correct to speak of "contact analog" because no additional information is artificially introduced. Nevertheless, the mirror is a technical element that manipulates the view of the real environment. The flat mirror causes the least change. As Egger (1990) has shown, only with the plane mirror is it possible to estimate distances (to some extent) correctly and thus also changes in speed. The combination of left and right exterior mirrors and interior mirrors results in a configuration in which dangerous so-called dead angles are created when flat mirrors are used (and also in the convex mirrors commonly used today). Especially on the driver's side, such a dead angle occurs, which when changing lanes may make it possible to overlook a vehicle located there (therefore, as is generally known, it is recommended to look at the shoulder before changing lanes). The exact position of the blind spot also depends on the position of the eye points, which in turn are significantly influenced by body size. Today, convexly curved mirrors are often used because the viewing angle can be considerably extended by them. However, this is bought by the fact that the distances are not only displayed in a smaller size, but also in a scale dependent on the distance in accordance with the laws of optical imaging. In principle, mirrors have the advantage over the combination of camera and screen, which is often recommended as a future option for aerodynamic reasons, that the viewing area can be changed by changing the head position. In addition, the shifts of images of objects at different distances from each other are used for distance estimation. However, due to the different curvature of the right and left wing mirrors and the lack of experience of users with curved mirrors, these effects hardly play a role in distance estimation.[14] Nevertheless, the use of camera and screen as a mirror replacement is an alternative to be considered in the future, since useful, technically available information can be superimposed on the camera image. This would make it possible not only to compensate for the disadvantage of reduced distance estimation, but also to bypass the blind spot effect if the camera systems were correctly designed. In the form of the rear view camera, into which the lanes for a parking process are faded in, a model has already been created for this (Fig. 6.18). General information is to be observed:

- Augmenting representations in vehicles must be oriented to the requirements of the primary driving task.
- Moving images should be avoided as far as possible or are forbidden in certain contexts; especially if they are not displayed as contact analog.

14 An exact conception of the mirror view, especially for persons of different body sizes, is possible using the RAMSIS human model (see Sect. ► 7.3.2).

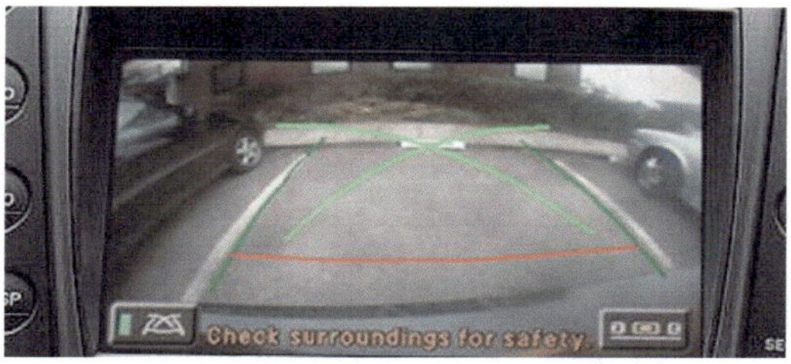

Fig. 6.18 Reversing assistance with faded-in auxiliary lanes for parking, which takes into account the above recommendations

- The overlaying of additional information must not lead to masking of real objects.
- Above all, the improvement of driving performance must be the indicator of the efficiency of additional information.
- Side effects such as cognitive capture and tunnel vision must be avoided.

Design of visual displays: Visual displays must be designed in such a way that the information they display from the driver's position can be perceived safely at all times. In particular, this places demands on the size of the used visual signs and the contrast with which they stand out against the background. The following applies not only to displays in the narrower sense, but also to the labeling of control elements, which also have a display character.

According to general ergonomic rules, a visual sign should be presented at least from an angle of 15′ (ISO 15008 of 2009). In fact, however, practically twice the size is recommended: Although the visual performance of young people (average age 25.8 years) was significantly better than that of older people in one study, the same effect was observed in the latter: The visual performance of young people also increased up to a character size of 33′ and then even decreased again.[15] Incidentally, the inability of older people to accommodate on close objects due to the ageing process (see ► Sect. 3.2.1.1) can only be compensated to a certain extent by larger visual signs. Above all, an appropriate reading distance plays an important role with increasing age. This is one of the reasons why HUD is so widely accepted, especially among older people, because the HUDs realized today display information in sufficient size and on a virtual level, which is located at a distance of 2.5 to 3 m in front of the driver's eye, thus practically eliminating the need for accommodation.

Dark writing on a light background can be read better by a factor of 1.2 to 1.3 than light writing on a dark background (Reinig 1997). However, a bright display can be very annoying, especially in a dark environment, where the actual driving task is presented. Therefore, if it is not possible to change the background colour of the display, it is preferable to use the variant "light writing on dark background" in the vehicle. However, to the same extent that LCD displays find their way into the instrument cluster, this recommendation should be reconsidered and, if possible, an adjustment should be preferred depending on the outside brightness, as is already the case with many navigation devices today. According to ISO 15008, a contrast between symbol and background of 3: 1 should be maintained for daylight conditions.

Although the graphic design of the display is the most important aspect of information transmission, colours also play an important role, especially in displays. Here you have to distinguish between the background color and the color of a font or an eye sign. According to DIN-IEC 73/VDE 0199, a white background characterizes neutral, general information. Also black and grey background have no special meaning. The colour red indicates immediate danger, yellow caution, attention and green safety and dangerousness. In particular, "red" can also be used for labelling to indicate a dangerous condition (e.g. too low air pressure in a tyre).[16] Otherwise it is rather confusing to use different colors in a text. For

15 The "RAMSIS cognitive" module illustrates the visibility of visual signs by projecting a Landold ring onto the object viewed by the virtual human model, thus providing a clear aid to the design of displays.

16 In this context, it is important that no monochromatic light is used for graphic visual objects; because the eye is farsighted for red light and short-sighted for blue light, it cannot see at least two visual signs in these colours simultaneously.

advertisements directly related to the traffic situation, the original layout of the traffic signs in colour and shape shall be used. For displays that are not directly related to the traffic situation, graphic representations reminiscent of traffic signs should rather be avoided.

6.2.1.2 Acoustic Indications

Acoustic displays can in principle be divided into two categories: on the one hand, those which only give notes or warnings and, on the other hand, those which are able to convey detailed information about language.

With the ***notes or warnings*** the information is coded by the frequency of the tone or the frequency composition of the sound and the temporal sequence (repetition frequency). Since the human ear has the highest sensitivity at 4000 Hz, warning tones with a frequency spectrum between 1000 and 5.000 Hz are particularly effective. In principle, they should drown out the existing volume level in the vehicle interior by more than 5 dB, better up to 15 dB (DIN EN 457; with an interior volume of approx. 70 dB(A), the warning tone should therefore be around 85 dB(A)). Only in critical traffic situations (e.g. when there is a danger of driving onto a much slower person in front or a standing obstacle, when a pedestrian suddenly enters the road, when there is a danger of leaving the road, etc.) should attention be drawn by a warning tone. Although it is common to use pure tones for this purpose, a study by Fricke (2009) showed that the introduction of a sound that mimics brake squealing leads to significantly reduced reaction times. The urgency of the reaction can be coded by the repetition frequency of the warning tone. A repetition frequency of one 1/2 Hertz (tone length 0.7 to 1 s) is much less urgent than a repetition frequency of 1 to 2 Hz. Fricke (2009) also examined whether a spatial coding (implemented by the loudspeakers of the audio system distributed throughout the interior) could be used, for example, to warn against the approach of a crossing vehicle. This purely acoustic spatially oriented warning has proved ineffective, whereas a combination with an optical warning (LED array under the windscreen, where the locally illuminated LED is reflected) offers advantages. However, there are other findings where spatial coding has advantages.

It is common today to transmit the measurement results of ultrasonic-based parking distance devices to the driver by means of coded acoustic signals (an additional visual or contact-analogue optical representation improves the interpretation in principle, see above). Different pitches are used for the front and rear of the car (the assignment is irrelevant, because it must be learned by the driver in any case). The distance is indicated by the repetition frequency: low frequency means object detected, safe distance; with decreasing distance the repetition frequency increases; shortly before touching the obstacle the acoustic signal changes to a continuous tone.

Events that result in less urgency than any damage to the outside world are better indicated by a discreet acoustic signal. An artificially generated gong has been established for this purpose. Different, but no more than three different signals would be conceivable. In this context, it should be noted that the warning and warning signals mentioned above occur so often in daily driving practice that the assignment to the corresponding events and thus their significance is quickly learned. The latter less urgent events usually occur rarely. It is therefore indispensable that the audible signal that attracts attention is combined with appropriate visual information that explains the event in more detail.

The latter is superfluous if one uses synthetic or human-voice stored ***speech signals***. However, long sentences, in particular subordinate clauses, should be avoided. It has become common today to optionally convey navigation information in this way (in older radio-based navigation devices, the information is even only given acoustically). This method can also be used to provide indications of any occurring conditions of the vehicle. In both cases, however, it must be taken into account that the acoustic signal and thus also the spoken language is always transient, i.e. if the driver was mentally distracted at the moment of delivery, the message may not be recognized. It is therefore recommended that a redundant optical "fallback level" is available in parallel to the acoustic voice message (by the way it would be an advantage if traffic announcements transmitted via radio were also displayed on the map of the navigation system. It makes perfect sense if, as with TMC, such announcements are electronically coded so that they are stored on an internal memory and can therefore be retrieved at any time). A word rate of 160 words/min should not be exceeded when voice messages are output (Byblow 1990). Further recommendations can be found in Bengler et al. (2012).

Although acoustic signals of any kind have a high, attention-grabbing character and are therefore very effective in principle, they should still be used with caution. The signal is heard not only by the driver, but also by the passengers. The driver can then feel "embarrassed" in front of them because of his driving style or other measures and thus reject such signals from the ground up (Thoma 2010). Increasingly, information management approaches are also being discussed to prevent low priority messages from being presented in complex traffic situations.

6.2.1.3 Haptic and Tactile Information/Indications

In haptics, a distinction must be made between *surface haptics* (feeling the material quality of the object touched = tactile) and *operating or actuating haptics* (reaction of a mechanism on human perception = haptic). The former can be influenced by the material used and the shape, the latter additionally by the force-displacement process during actuation. The haptics not only provide feedback in the sense of ▶ Fig. 1.9, but also have a considerable influence on the perceived quality of the vehicle. It thus significantly enhances the pleasure aspect and thus the perceived comfort.

For the ***surface haptics*** the used *material* plays an essential role. Heat-conducting metal, perceived as "cool", combined with visually detectable metallic shine, is perceived by many as valuable. Less heat conducting material with preferably polished wood surface also gives a "valuable" impression. The

Fig. 6.19 Haptically and optically coded seat adjustment unit at Mercedes

special luxurious "feeling" of leather (with visible seams[17]) and also its optics is known to be artificially difficult to imitate. On the other hand, poorly heat-conducting plastic with a rough surface, on which there may still be marks of scratching, is perceived as "cheap". As can be seen from this description, the optical impression must not be forgotten for the assessment. But it must match the touch sensitivity. For example, plastic parts that are coated with a thin metal surface and feel "warm" disappoint the optical expectations.

The *shape* of the touched objects, in addition to the above-mentioned value aspect, also has a feedback character, since this – after a learning process through practice – provides information about the touched control element. However, this often contradicts aesthetic requirements and economic aspects (use of identical parts). Nevertheless, steering column levers, especially when several levers are mounted on one side behind the steering wheel, should not only differ significantly in shape and position (e.g. by levers of different lengths), but also in the tactile feel of the surface (e.g. one smooth and the other corrugated). In this way, mix-ups can possibly be caught at the last moment before actuation. Similar requirements apply to buttons mounted in the centre console, which often give the impression of a "keyboard" contrary to this requirement. The operation of controls that are not used very often is always accom-

17 This statement should be made with caution: The valuable "rustic" character of seams obviously only has this meaning in the European cultural sphere!

panied by a turning of the gaze (Arlt, 1999, McKenzie, 1989, McKenzie et al., 1990). For this reason, such control devices must be mounted in such a way that they can also be detected optically. A classic example is the complaint of many users about the positions of the control elements for the electric seat adjustment in the invisible area to the left or right of the seat below the upholstery and on the other hand the positive assessment of the position of this adjustment unit in the interior door trim of Mercedes (see ◻ Fig. 6.19).

The surface haptics also plays an important role for the ***actuating haptics*** – at most for pedals with certain restrictions – of course, since the control element has to be touched with every movement. The actuating haptic itself is essentially characterised by the force-displacement curve. The ideal course depends on the intended use and the type of control element. At this point, we will not go into the tactile handling of the primary controls (steering wheel, clutch, accelerator pedal, brake, gear lever), although we can sometimes speak of haptic display in the narrower sense here, because haptic information intentionally generated by servomotors may be conveyed here (e.g. On the steering wheel, see ► Sect. 6.4.3). Since these controls are always associated with kinesthetic feedback because of the vehicle movement they influence, this is explained in more detail in ► Sects. 6.4.1 and 6.4.2. The control elements used for secondary and tertiary driving tasks must be distinguished between translational and rotational control elements with regard to haptic feedback. In addition, a distinction must be made between the sensing and locking versions. With touching control elements, the control element returns to its initial position after actuation, while locking control elements remain there. For rotary control elements, the term "rotary encoder"has become established for the touching variant, while the term "rotary switch"stands for the locking variant. In the case of translatory switches, the designation is for momentary pushbutton and locking pushbutton (Reisinger and Wild 2007). While tactile operating elements are characterized in principle by a restoring force or moment increasing with the deflection, the locking ones, on the other hand, show a more complex characteristic. Only these will be discussed below.

Reisinger (2009) and Kühner (2014) have systematically dealt with the haptic feedback of control elements. The generation of the haptic feedback takes place with *rotatory control elements* generally by spring elements acting on a gate. In principle, there are two different designs for this (see ◻ Fig. 6.20).

a b

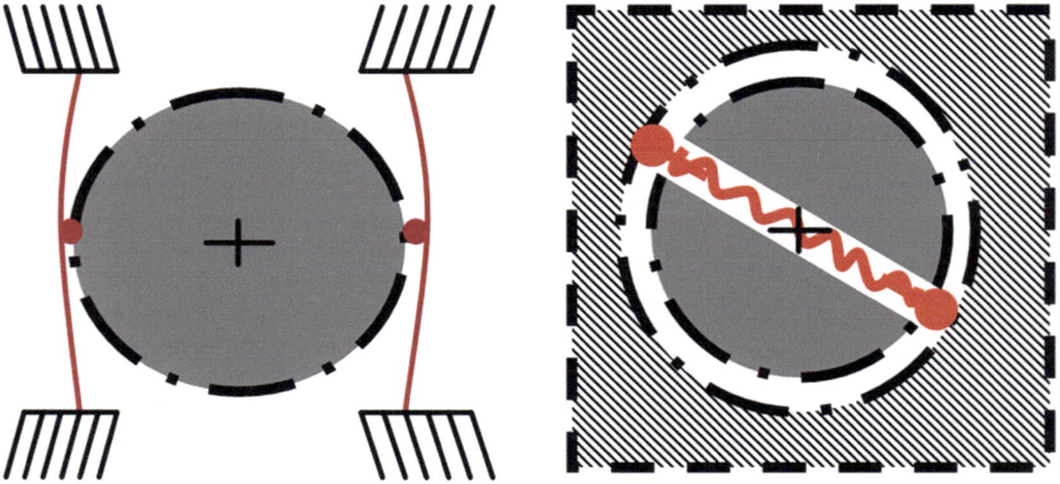

◻ **Fig. 6.20** Principle of the generation of haptic feedback with rotary control elements: Radial direction of action of spring elements, **a** outside with leaf spring, **b** Inside lying with spiral spring. (After Reisinger 2009 and Kühner et al. 2011)

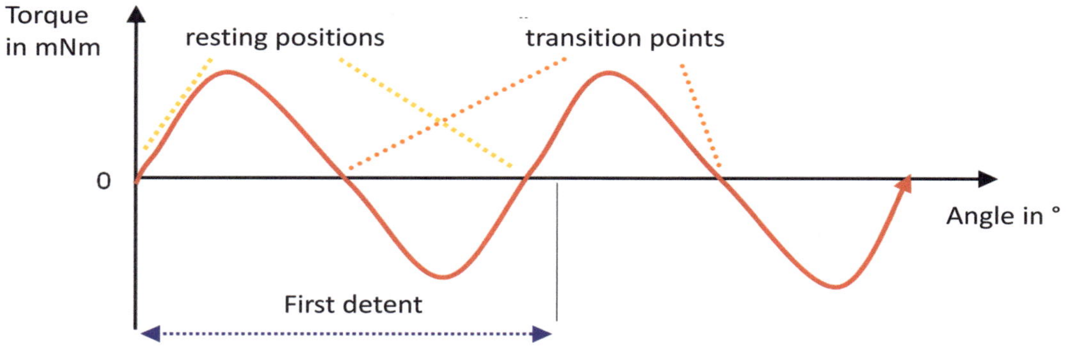

Fig. 6.21 Systematic representation of a torque-angle of rotation characteristic curve. Two detents with rest positions and transition points are shown. The position of the rest position or transition points can only be described qualitatively in this graph due to Coulomb friction, which is generally directed against the direction of action of the user (Reisinger 2009)

For a more precise description satisfying ergonomic requirements, however, the usual technical parameters of the torque-torsional angle curve, which only contain the maximum torque and the number of so-called detents, i.e. the number of detent positions, must be supplemented with the position of the electrical switching point, the rest position, the transition point (= change from positive to negative torque) and the friction offset (= friction hysteresis) (see Fig. 6.21).

Reisinger has developed a special haptic simulator for investigations of haptic sensation, with which any torque/angle of rotation characteristics can be tested against each other within wide limits. He examined not only sinusoidal progressions, but also triangular and sawtooth progressions. An astonishing result is that not only was hardly any difference subjectively perceived between these sequences, but also that the felt position of the rest position is not felt at the moment zero points at which the control element actually comes to a standstill without external force, but at the positions of greatest negative torque. It turns out that by integrating the torque over the angle of rotation all observations can be explained sufficiently well. By means of an idealized mechanics of a locking rotatory control element it can be shown that the user obviously opens up the underlying idealized locking disc, which is explained by the integral (= work expended) (see Fig. 6.22). It is the basic task of haptic sensory perception to make the form of palpated objects accessible through the combination of movement and tactile stimulus. Turning a switch is like stroking your fingers over a (rough) surface. According to Reisinger's investigations, the decisive factor for haptic sensation is not the torque, but the force felt at the contact surface between fingers and control element. With a larger knob diameter, a larger restoring torque must be provided in order to create the same force impression as a smaller control knob.

In a further series of tests, Reisinger identified and quantified relevant technical parameters and adjective pairs based on the torque/angle of rotation characteristic curve. He was able to extract the two factors "mobility" and "value". Their quite complex dependence on the technical descriptive variables is described in detail in his work. Especially in connection with the integral representation, it can be seen that the area of the torque representation is directly related to the sensation of hardness or stiffness. In a further, even more detailed study, Kühner (2014) investigated how the parameters of mass inertia, damping and friction influence each other in rotary actuators. In the interpretation of the very complex results, he states very briefly summarized here:

- The differentiation capability for the parameter mass moment of inertia decreases with pronounced friction and/or damping, whereby the influence of damping is greater.

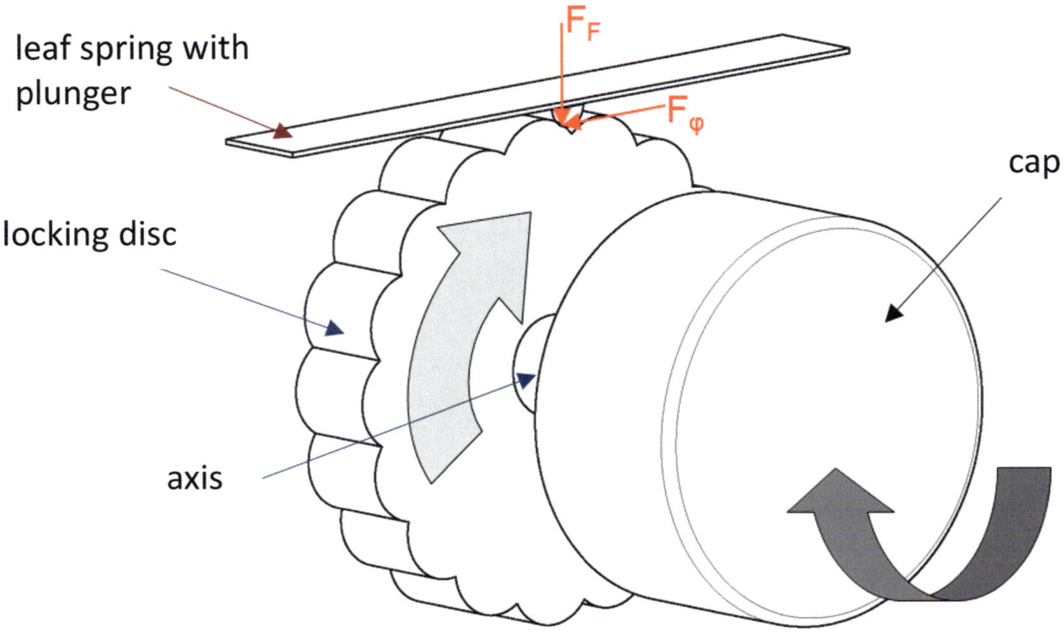

Fig. 6.22 Simplified representation of a locking rotary actuator, as it probably corresponds to the inner model of most users

- The parameter mass inertia has – at least up to an average value – no masking[18] effect on the damping and friction parameters.
- The differentiation capability for the parameter friction is particularly reduced by the parameter damping, also because at a constant angular velocity both parameters cannot be distinguished. The parameter inertia, on the other hand, only has a masking effect if it is very pronounced.
- A detent basically has a masking effect on all mechanical parameters.[19]

For *translational controls* (pushbutton) Reisinger (2009) first shows common technical principles of effects in connection with typical force-displacement processes (Fig. 6.23). In order to generalize the characteristics given in Fig. 6.23, he developed a systematic description of the technical designations, which also makes reference to DIN EN 196000 (2001). The properties of the pushbutton are essentially described by the following characteristic values (see Fig. 6.24).

- The *spring preload* represents a force value that must first be overcome before the control element moves (Weir et al. 2004). The first force increase can be described as the "intensity of the touch" and the slope as the "initial resistance" ("initial response" according to Osumi et al. 1990). The higher the spring preload is, the deeper the impression (deep, Kosaka and Watanabe 1996), i.e. the greater the stroke of the control element is perceived.
- The *idle stroke* is before the actual increase to spring preload. It comes about through tolerance chains. It can be recognized by the almost axially parallel course of the characteristic curve. Experience has shown that it is rather rated negatively (it is not shown in Fig. 6.24).

18 Masking characterizes an effect of perception that can be observed in different (within and between) sensory organs: A sufficiently large stimulus of one quality makes the perception of another disappear.

19 For detailed questions, in particular for information on technical parameters of the very complex interrelationships, please refer to the work of Reisinger (2009) and Kühner (2014).

- The *force jump* refers to the often abrupt drop in force at a certain force level. The maximum of the curve represents the starting point of the force jump and is usually used to specify pushbuttons. According to Kosaka and Watanabe (1996), its triggering force (peak) is responsible for the majority of haptic properties. At the same time, the electrical switching point and thus the triggering of the function should be linked to it (DIN EN 196000, 2001). The ratio of spring preload to force jump level can also be referred to as sensitivity.
- The *flanks before and after the force jump* can vary in their form at will. They can be convex, concave or linear, independently of each other.
- The *mechanical end stop* limits the travel of the pushbutton and is therefore decisive for its stroke. It usually represents the steepest and stiffest section of the characteristic curve. Overpressing would destroy the control element.

Reisinger also developed a simulator for translatory control elements, which was able to reproduce force-displacement curves in any combination within the framework of the computer-controlled characteristics shown in ◘ Fig. 6.23. Although real force-displacement curves deviate from the ideal of ◘ Fig. 6.24,

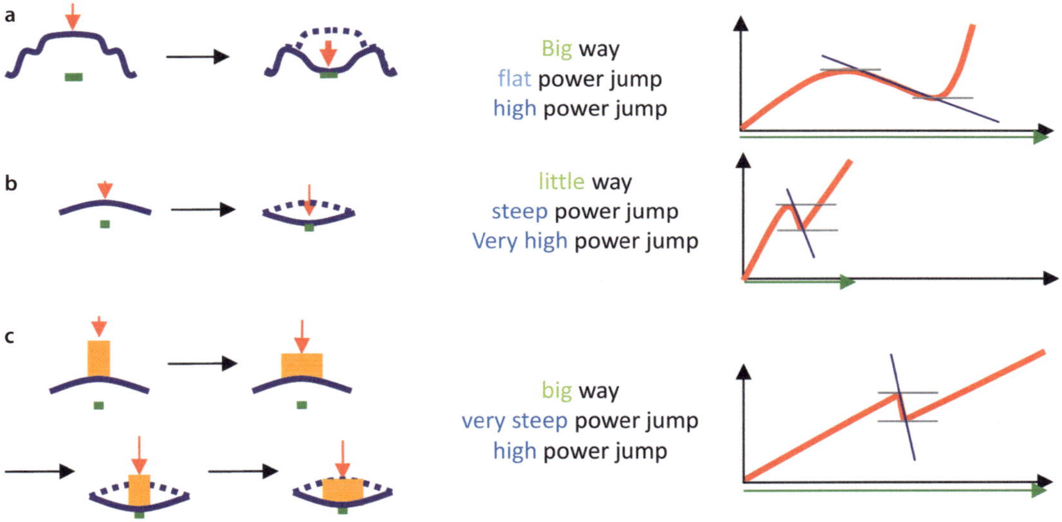

◘ **Fig. 6.23** Common principles of haptic feedback in pushbuttons with typical features of **a** Silicone keypad, **b** Spring washer, **c** Microswitch as combination of spring washer with silicone plunger (Reisinger 2009)

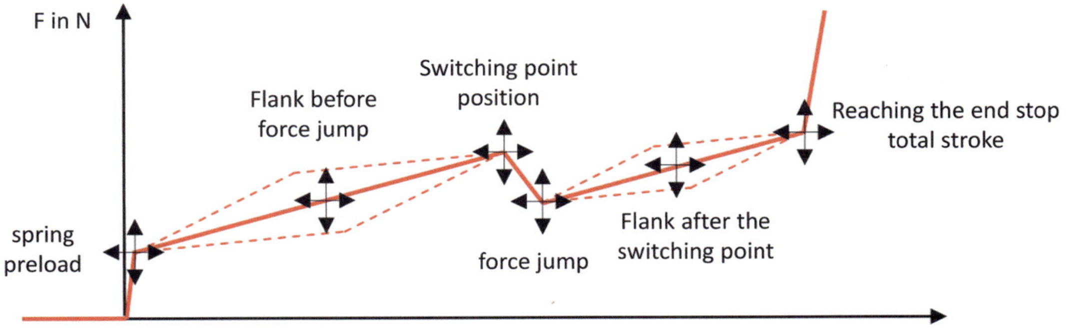

◘ **Fig. 6.24** Systematic representation of a force-displacement characteristic as well as the basic variation possibilities of the courses (Reisinger 2009)

they are not the same. However, the gradient of the force jump has a significant influence on the haptic sensation. All in all, it can be seen that in contrast to the rotary control element, there is no clear description possibility for the haptic feeling of translatory control elements. This is probably also due to the fact that there is no uniform inner model of how such control elements work. Reisinger describes the observed phenomena as "event-oriented perception", according to which forces and force progressions are perceived in a way that can be easily estimated until an event – e.g. in the form of a blow (in this case a shift jump) – comes to the fore and thus makes the force progression itself appear unimportant. The equivalent of the translational control element from everyday experience is, for example, the displacement of an object, in which the correct course of the resistance force is felt, but an impact against an obstacle makes the course in front of it appear unimportant.

In the test persons' subjective perception tests, the two factors "value and liking" and "mobility and hardness" could be filtered out as determining factors. It turns out that a switching point must be perceptible for the assessment "high quality". With the dependencies and values found by Reisinger, the designer of a translational control element has values at his fingertips that enable him to achieve a specific combination of judgements.[20]

Under *constancy performance* one generally understands the fact that humans are able to recognize object properties independently of external circumstances. In the present context, it should therefore be asked whether the haptic sensation depends on the mounting position, since the force required to actuate controls varies with it (e.g. a different force is objectively required to actuate a pressure switch in the headliner than in the mounting position on the centre console). Reisinger (2007) has already been able to demonstrate a consistent performance of human perception in the sense that the properties of the control element are perceived independently of the installation position. Kühner (2014) conducted the same experiment for translational control elements. From the point of view of the objective of the investigation, this is even the more general case. Two different force/displacement curves occurring in automotive practice were realized. He was also able to show that the mounting position has no influence on the sensation. This also applies if the test persons have been informed about the aim of the test. It can therefore be assumed that there is a constancy performance in the sense that human perception of haptic stimuli is largely independent of the forces necessary for maintaining equilibrium or reaching the control device.

A special feature are ***active controls*** because through them a *artificial haptic feedback* can be generated with respect to the force-displacement curve. ◘ Figure 6.25 shows the principle of such a control element. As with any other control element, the operator acts on it with a certain force, the force is measured and transmitted to a computer which, in many cases, can be suitably connected to the machine actually being operated. The computer determines an appropriate path from the incoming force – possibly taking into account the current state of the machine – which is implemented via a servo motor on the control element. Thus the person feels the characteristic curve generated by the computer on this control element in a haptic way.[21] The simulator developed by Reisinger for both rotary and translatory control elements works according to this principle. The controller (central rotary pusher) used by BMW

20 As far as the exact dates are concerned, reference is made to the work of Reisinger (2009).

21 Instead of the force, the displacement can also be measured as an input variable. In this case, a torque motor is used as the servomotor, which converts the result of the computer into a corresponding torque or force felt by the user. According to the considerations of Gillet (1998), both possibilities are of equal value if the engines used produce sufficient power and the system can be characterised by "infinite" mechanical rigidity. See ▶ Sect. 6.4.3 for further application possibilities of the active operating element.

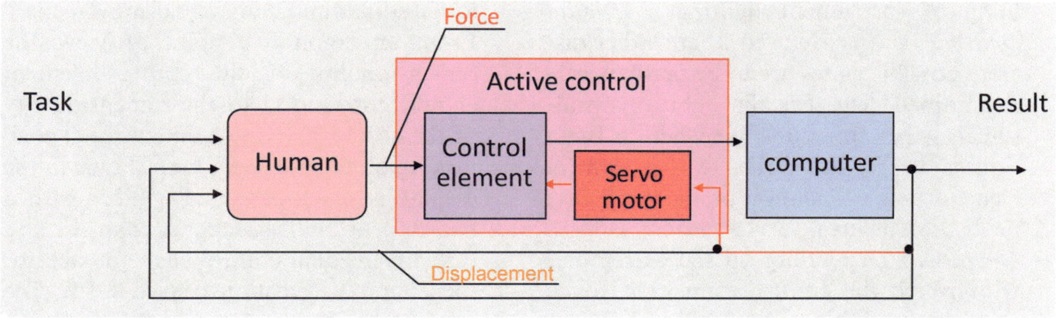

Fig. 6.25 Principle of the active control element for generating an artificial haptic feedback

in the first generation of the i-Drive is also based on this. The advantage of the active control in this application is that the number of detents can be currently adjusted to the number of choices given in the respective menu. A disadvantage is that for the realization of the end stops – theoretically – infinitely high energy would be necessary. The operating element thus feels "soft" compared to a mechanically constructed operating element. This negative feeling of many customers, which can actually be attributed to the liking aspect, was the reason to use a mechanical controller in the successor models and to dispense with the advantage of correct haptic feedback.

6.2.2 Control Elements

For controlling, regulating, switching, driving and triggering functions, muscle forces are transferred to machine parts via the conventional control elements, which serve to implement the intentions developed from human information processing in reality. This is done using controls that are operated with the fingers, the hand or the foot. As already mentioned, it is also possible to pick up speech information from microphones and convert it into corresponding functions coded via a computer. Recently it has also been discussed whether the recording of gestures (hand movements without touching objects, head and upper body movements and eye movements) can be used to trigger certain functions on the vehicle. Especially in the latter approach, but also already in the use of speech information, a fundamental problem – apart from the reliability of recognition technology – is to distinguish between random actions, which then possibly trigger unwanted functions, and intentional ones. Of course, this possibility of error also occurs with conventional control elements, but it is not so common there (after Norman 1986, 1981 so-called "slips"). With the progressive development of technical possibilities, so-called multimodal interactions are discussed and partly realized today. In principle, this is in keeping with the human ability to transmit information not only via one channel, but via several channels (for example, speaking in conjunction with gestures and possibly even touching the other person).

6.2.2.1 Categorisation of the Control Elements

On the basis of the categorization by Rühmann (1993), the following criteria are presented for conventionally mechanically operated control elements, which can in principle be combined with each other as desired:

- *Mode of action:* As with the displays, the information can also be transmitted digitally or in analog coded form in the control elements, depending on the design. *Analog control elements* are used where continuously changing information has to be transmitted for a control process (e.g. steering wheel, accelerator and brake pedal, volume control of the radio, dimming of instrument lighting).[22] *Digital control elements* are used where either discrete possible states are to be permanently maintained (e.g. certain light or windscreen wiper functions) or where a function is to be called up by the control element (e.g. calculation of the route in a navigation device).

- *Operation:* Depending on the extremities with which the control element is to be operated, a distinction is made between finger (e.g. light switch), hand (e.g. gear lever), foot (e.g. accelerator pedal) and leg (e.g. clutch pedal) controls.
- *Movement type (movement coding):* A distinction must be made between *translational and rotational movements* (an example of a rotational movement with the legs is the pedal crank of the bicycle, which, however, is hardly used in the vehicle[23]). Although both types of movement can be used for different purposes, it must be taken into account that a translatory operating element – in contrast to a rotary operating element, where this is not necessarily the case – always has two end stops, which should correspond to a correspondingly restricted range in reality (e.g.: time interval to which an ACC system regulates).
- *Dimensionality:* Dimensionality describes the degrees of freedom in which the control element can be moved. A knob or pushbutton is one-dimensional because it can only be moved in one degree of freedom (one axis of rotation, one direction of translation). The gear selector switch of the transmission is two-dimensional, since the well-known H-scheme – even if guided by a backplate (digital!) – permits both longitudinal and transversal movement.[24] From an ergonomic point of view, the dimensionality of the control element should correspond to the dimensionality of the task (e.g.: the two-dimensional positioning of a cursor on a map shown in the display is much easier to achieve with a two-dimensional touchpad than with a one-dimensional control element each for the X and Y directions; Spies 2013). The virulent problem of mastering the two-dimensional task of driving a car with up to five actuators is dealt with separately in ► Sect. 6.4.3.
- *Lockability*: Operating elements can be locked in the position in which they have been moved by different mechanisms. If such a lockability makes sense, because it should be ensured that the actuator does not change its position (e.g. gearbox selector lever), the question arises as to how the lockability can be released again. Possible options here are: increased effort (given with each control element with a detent), changed direction of movement or actuation of an unlocking actuator (e.g. handbrake lever). Unintentional unlocking is increasingly prevented in the order shown, but is bought by the fact that the operation appears cumbersome and, in combination with a certain clumsiness or ignorance about the type of operation, may not even succeed. A special form of locking are on/off switches, which change between these two states each time they are actuated. If such a control element is used, an optical feedback of the switching status is indispensable.

In the case of non-locking controls, a distinction must be made between those which do not provide position-dependent feedback (feedback only during movement by friction or viscous damping) and those which return

22 Analog control elements can also be described as those which objectively have a very narrow grid. The decisive factor here is whether the individual stages are perceived by the operator as discrete or quasi-continuous.

23 In research, such vehicles whose propulsion is supported by an electric motor similar to that of a pedelec are being investigated as an alternative to purely urban vehicles.

24 Foot controls are actually always one-dimensional. In an experimental vehicle at Daimler-Benz, a two-dimensional pedal for accelerator and brake actuation was investigated. The implementation time was reduced by up to 50%, but at the same time acceptance was considerably limited (Braun 1993).

to their neutral position by spring force or a more complex mechanism acting in a similar manner. In the case of the former, the said feedback serves to perceive the movement and to ensure that the actuator remains in its position after actuation (e.g. volume control). The latter play a role primarily in connection with the driving process (e.g. steering wheel, accelerator, brake and clutch pedal). The restoring force provides direct haptic information about the degree of operating influence. The corresponding special features are described in more detail in ▶ Sect. 6.4.

- *Integration:* One speaks of integrated control elements, if in *one* control *several* operating functions can be combined for alternative, sequential or simultaneous actuation. Integration can be achieved by mounting one or more additional control elements on one control element (e.g. an additional rotary switch for preselecting the interval time on the steering column lever responsible for wiper actuation) or by assigning different functions to the dimensions of a two-dimensional or multi-dimensional control element. From an ergonomic point of view, it is advisable to combine thematically related elements in such an integrated control element (e.g. lighting functions: Moving the left steering column lever forwards and backwards = dimming up and down of the driving light, moving the steering column lever still up or down = turn signals right or left).[25] Integrated control elements are useful if certain functions can be called up relatively frequently from a more or less fixed driver posture depending on the driving task. The rotary pusher, which is frequently used today, can be regarded as a two-dimensional control element, whereby the two dimensions "rotation" and "translation" trigger different actions: Select a function by turning – trigger the selected function by pressing. Today's rotary push actuators installed in the vehicle can often also be moved translatorically in the plane (X-Y plane), which then calls up additional functions. In this case, however, it is helpful to use a suitable display on the screen to make it clear to which movement certain functions are assigned. Similar assignments of different functions are observed in the different versions of steering column levers (some are even four-dimensional: translation up-down, forwards-back, pushing and turning around the steering column axis). None of these function assignments are self-explanatory. The learning process necessary for correct operation is, however, facilitated if thematically related functions are accommodated in such a control element (e.g. only lighting functions, only wiper functions). ◘ Figure 6.26 provides an illustrative overview of the categorisation presented.

In the following, the principles mentioned for the pedals and hand control parts most frequently found in the vehicle are explained in more detail.

6.2.2.2 Pedals

Pedals in vehicles today are actually only used as so-called accelerator pedals for controlling engine torque or engine power and as brake pedals for reducing the kinetic energy of the entire vehicle. Vehicles with a combustion engine and manual transmission also have the clutch pedal. With regard to the arrangement of these pedals, after an initial development process, which took about 3 decades, the position coding clutch left, accelerator pedal right and brake pedal in the middle of the footwell in front of the driver has prevailed (interestingly also for vehicles with right-hand drive). It is not possible to visually check the position of the pedals. The feedback is only haptic. The fact that the functions "regulation of engine power", "braking" and "clutch" are performed via pedals is primarily due to historical and technical reasons. The clutch's high actuating forces, which were not supported by a servo until the 1950s and which

25 The integration of the on-board computer control into the steering column lever for the lighting function – triggered, for example, by pressing the lever in the axial direction – therefore makes less sense from an ergonomic point of view.

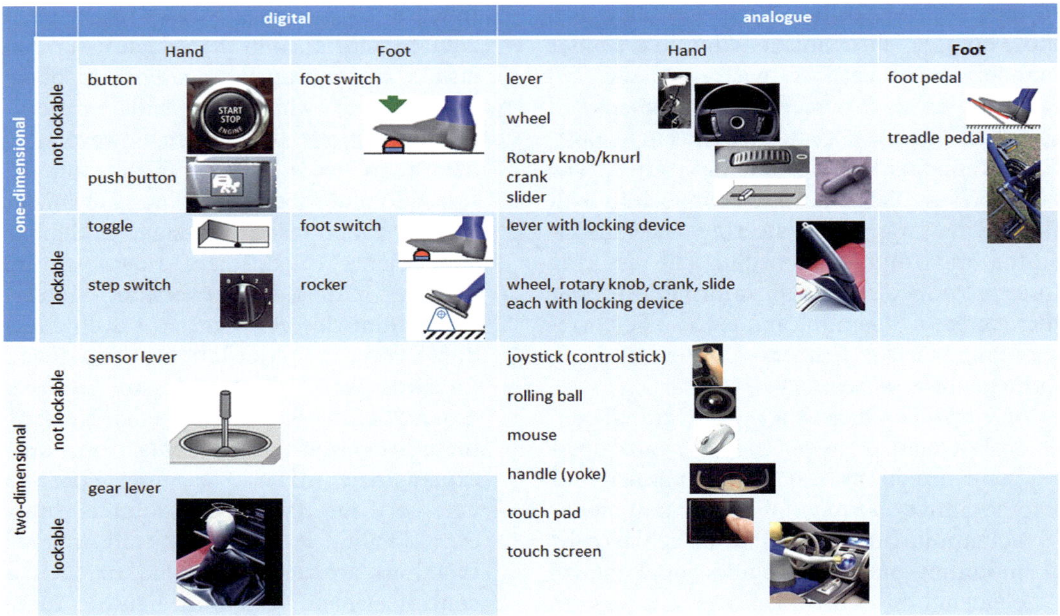

Fig. 6.26 Categorization of control elements

6

are all the greater the greater the engine power and the brake's force, which increases with the vehicle weight and the possible driving speed, could only be mastered by the higher foot forces compared with manual actuation. Especially as far as the sensitive actuation of the brake, which is necessary under certain circumstances (e.g. black ice) is concerned, the hand control would actually be preferable.

Pedals can be designed standing (pivot point near the vehicle floor) or hanging (pivot point well above the tip of the foot). This is irrelevant for the transmission of information. Hanging pedals offer a certain protection against slipping of the foot upwards (not to the side!).

Some vehicle manufacturers have made it common for automatic vehicles to provide an additional pedal for the parking brake to the left of the (empty) space for the clutch pedal. For the manual versions offered in parallel, however, many drivers have conflicts when "starting off on the hill", since the hand-operated release button of the parking brake must be pulled at the right time and there is no possibility for corrective readjustment in the event of operating errors. The operation of the headlights on and off or the windscreen washer by footswitches, as provided for in historical vehicles, has rightly not been able to prevail, since the need to operate three pedals with two feet even in manually operated vehicles leads to overstraining effects in special driving situations.[26]

6.2.2.3 Manually Operated Control Elements

Movement characteristics: In addition to the position coding already mentioned, motion coding plays a primary role in the manually operated control elements. Essentially, a distinction is made here, as already mentioned, *rotatory* (e.g. knob) and *translational* (e.g. slider control) actuators. From an ergonomic point of view, however, the term does not refer first and foremost to the technical design, but to the action to be performed manually. In this respect one assigns levers (e.g. gear selector lever) and toggle switches, which actually move around a

26 Only recently has a hill start aid been offered for some vehicles, which, after releasing the handbrake, does not actually release the brake until the drive wheels experience a moment in the forward direction. This is actually an indispensable ergonomic measure!

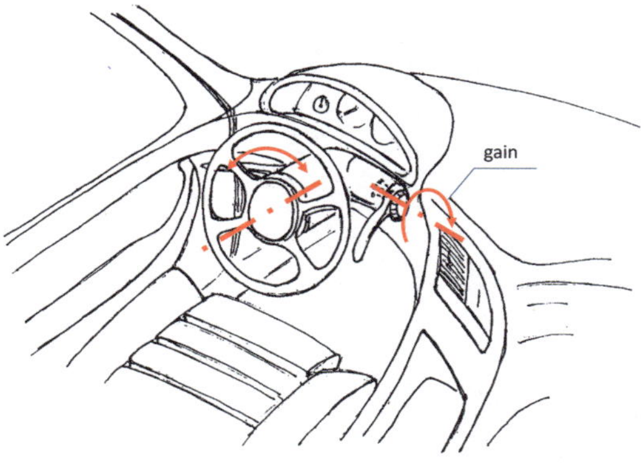

Fig. 6.27 Right-hand steering column lever arrangement to avoid secondary incompatibility

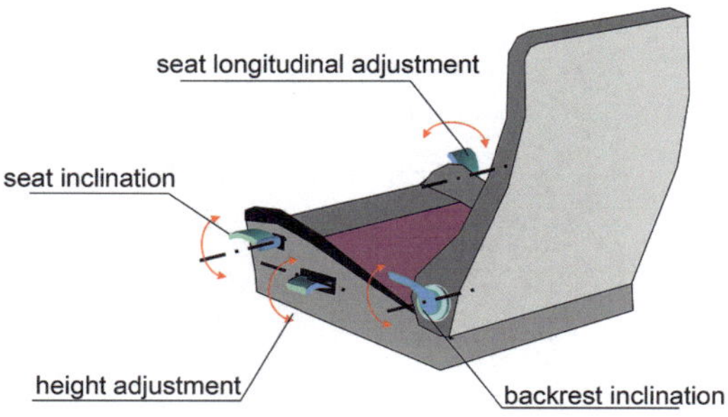

Fig. 6.28 Positioning in line with expectations and movement of control elements compatible with reality for mechanical seat adjustment (using the driver's seat on a left-hand drive as an example)

rotary axis, to a translational movement. Problematic with this kind of assignment are the steering column levers, which suggest a rotary movement because of the clear reference to the steering axle (e.g. would not be understood correctly without such an assignment of the indicator levers). This leads to secondary incompatibility with the right-hand steering column lever, which is often used for wiper operation: is the movement of the lever upwards as a "more/on" or as a left turn (this is the direction of movement of the wiper)?[27]with the meaning "less/off"? Different manufacturers come here to quite different interpretation! The dilemma can only be solved by a differently oriented rotation axis, as illustrated in Fig. 6.27.

From a study work, which had the positioning of the control elements for the mechanical seat adjustment in relation to the object assumed by the test persons (Bieniek 1990), a proposal for the positioning and movement coding of these control elements was developed, which is compatible with the movement of the seat initiated by it. In this proposal, a double-acting ratchet actuation is consistently provided for each adjustment process, which makes the adjustment possible even during travel without the risk of a seat suddenly slipping away or a backrest failing to provide support (Fig. 6.28).

Avoiding secondary incompatibility is particularly challenging when a control on a screen is intended to change something. Fig. 6.29a shows an example: The clockwise rotation of the controller means "more"/"next". The cursor moves up ("back"). This is justified by the graphic, which suggests a central pivot point. This problem can be solved by a reference centre of rotation shown on the left of the screen and a radius in the graphical representation (Fig. 6.29b).

Touchscreens are a special feature with regard to motion coding. Most of today's vehicle versions offer virtual buttons, which can be touched (contact handle, see below) to trigger a function directly or to open a new menu that offers additional buttons. The advantage of this version of the touch screen is that a large number of functions can be called up in a limited local area. This is bought by the disadvantage that there is no haptic feedback about the touch of such a button, so that a gaze control is indispensable that is on average longer than whenOperation:simultaneous operating a physical switch.[28] Spies (2013) also found significantly longer viewing times for touchscreen

27 In the case of wipers where the pivot point is located to the left of the centre of the vehicle, or in the case of so-called butterfly wipers, it is often argued that the movement of the wiper lever upwards or the rotation to the left is compatible with the corresponding movement of the left wiper.

28 A study by the Institute of Ergonomics at TUM (2006) also shows that the selection of functions from a menu is faster and safer with a rotary push-button switch than with the touch screen. The touch screen operation has clear advantages only for the two-dimensional displacement of objects (e.g. adjusting the map).

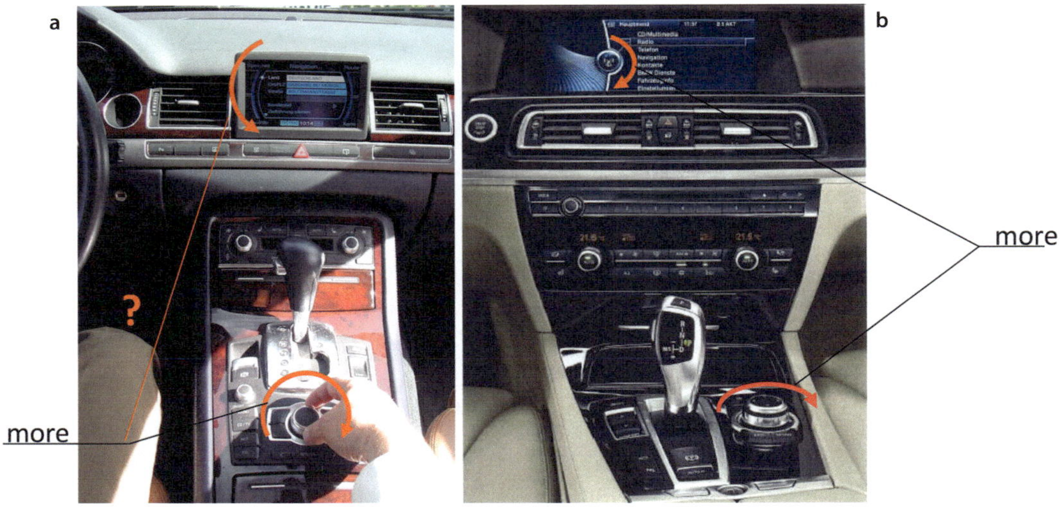

Fig. 6.29 Incompatibility **a** between the rotation of the controller and the reaction on the screen and the solution of the conflict **b**

6

operation than for all other alternatives. The rules for simultaneous operation (see ► Sect. 6.1.1.1) must be observed for the exact design of the buttons offered and the menu sequences. A gesture-based operating form that has become popular through the use of smartphones is "wiping" with one finger and "zooming" with the movement of two fingers, which is now also increasingly being used in the touchscreens installed in the vehicle. A special feature is the writing of alphanumeric characters on a specially designed surface (so-called touchpad), which are converted into characters that can be read by the on-board computer using appropriate detection software. The latter form of information transmission has proven to be particularly cost-effective and less interfering with the actual driving task (Hamberger and Gößmann 2009), especially for the input of data in navigation and telephone applications.

In the meantime, handwriting recognition technology has reached series maturity and is developing its potential especially in tertiary activities, as the input of alphanumeric characters for telephone numbers, destinations or addresses can generally be implemented very well. Current technologies allow fingertip input without a pen or stylus and accept simple standard letters (Hamberger 2010). Several experiments have shown that this manual-visual interaction has significant advantages compared to voice input, touch screen input or input by means of turn/press (Bechstedt et al. 2005). The decisive factor, however, is the positioning of the touchpad in a position that offers good accessibility and storage possibilities for the ball of the hand or wrist and a correspondingly fast processing and feedback of the recognized input to the user by means of an equally optimally positioned screen, which can now be separated from the touchpad.

Gripping methods: For the contact between the hand and the control element, the most different variations are conceivable (■ Fig. 6.30). Practically, there's a distinction:

- *Contact grip:* A vertical force is transmitted from the hand to the control element via a finger – normally the index finger. Mechanical operating elements are generally buttons and switches of different types. For their design, the various aspects of haptic feedback must be considered in addition to size (► Sect. 6.2.1.3). A touch screen can also be operated via the contact handle. Functions are normally triggered here by a simple touch. For technical reasons, the object cannot be selected beforehand. Many studies (Spies 2013) show that the real problem is not the feedback on

Fig. 6.30 Types of grip (according to Götz 2007); contact (index finger), grasping (two or three fingers) and grasping grip

touching the button and not the feedback on its operation. The latter can be done sufficiently effectively also acoustically. Measures to report the execution by vibration have proven to be uncomfortable and sometimes confusing because frightening. This solution is also technically difficult to implement, since a touch is necessary to feel the feedback, but a release of the finger may trigger the function.

- *Grasp:* The handle with two or three fingers creates a frictional connection or, with the appropriate design, a positive connection between the control element and the hand. The rotary actuator and slide switch are operated via the handle. The feedback of the rotary actuators in particular is described in ► Sect. 6.2.1.3.
- *Grasping grip:* The wrap-around handle also allows higher forces to be transmitted. A distinction is made between orthogonal and tangential force application. Orthogonal force application practically produces a positive connection. The transmittable forces are greater (e.g. handbrake levers, door handles etc.). The tangential application of force essentially leads to a frictional connection, which can be supported by form-fitting elements, if necessary, by means of appropriate shapes (e.g. steering wheel with "finger-friendly" contouring on the back).

Display function of the control element: Götz (2007) explains that most control elements are characterized by the elements gripping element, motion transmission element, reference surface and display (see Fig. 6.31). The display not only provides feedback on the current state of the control element, but also, in conjunction with the reference area, information on how it can be operated (affordance). It is already clear from this statement that the lack of display and reference surface, as is the case for example with the steering column levers requires some compensatory measures (e.g. unambiguous labelling, feedback on the operating status via indicator lights in the instrument cluster) in order to eliminate this ergonomic shortcoming.

Götz (2007) investigated the essential influences that cause a certain operating intuition. He developed a total of 88 models from a system of control elements (see Fig. 6.32), which he virtually presented to the test persons using a so-called power wall. The subjects were asked what their first intuition was about the movement and meaning of these controls. Figures 6.33 and 6.34 give some important results of this study (for further results see Götz 2007).

Touchscreen As already mentioned, the operation of the conventional touch-sensitive touch screen in a motor vehicle poses a particular challenge, because neither before nor during the actual operation can it be haptically confirmed whether a virtually displayed key has been touched. This is particularly important under the stricter conditions of the main task "driving", as processes that take longer than 200 ms processing time must be avoided at all costs. If the touch screen is positioned in the optimum gripping space, its visual detection requires too much averting of the view; however, if it is in a favourable field of vision, it can only be achieved by leaning forward and stretching the arm. During the journey, reliable operation can no longer be guaranteed due to the influence of vibrations. Spies et al. (2010) has proposed a way out of this dilemma in a fundamental study. In his experiments, he used a matrix arrangement of pin elements originally developed for the reproduction of Braille,

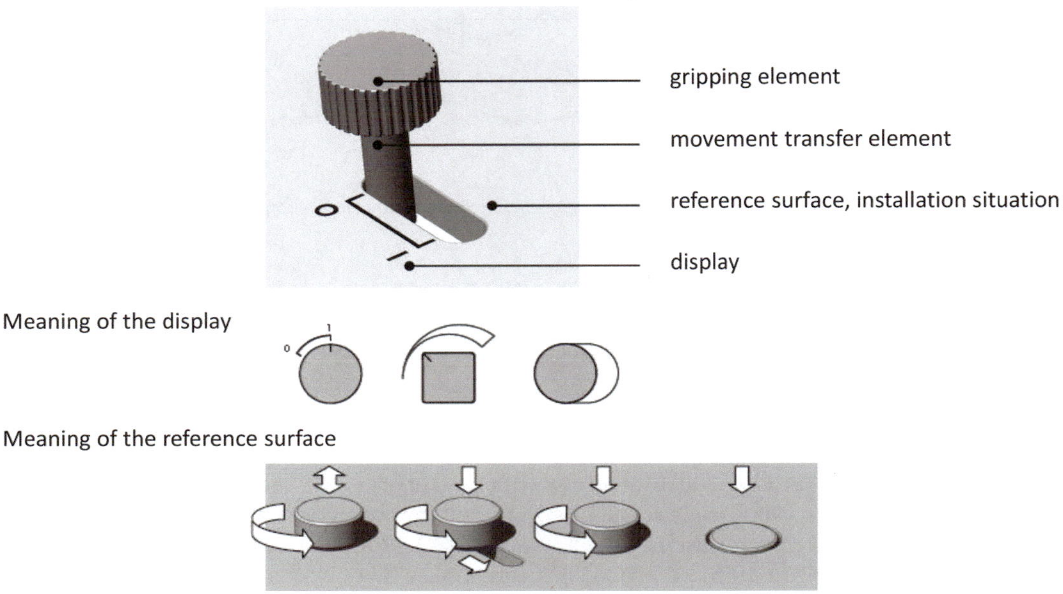

Fig. 6.31 Main components of a control element, the control element as display and the meaning of the reference area

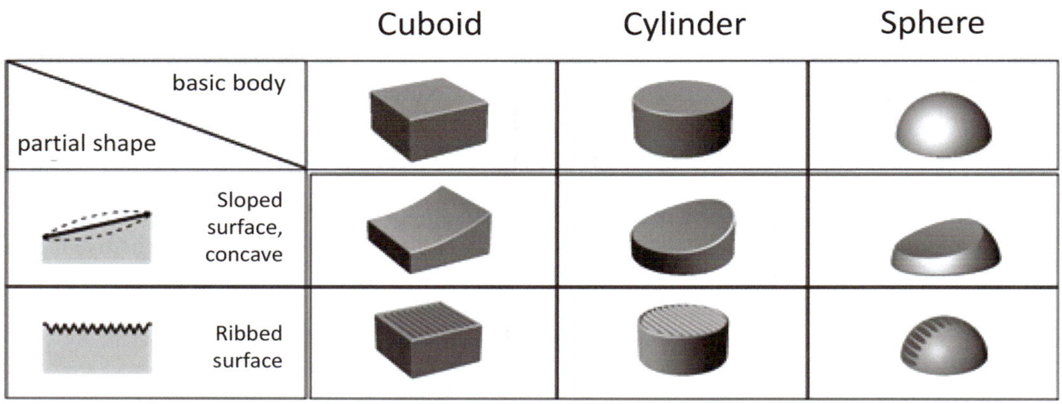

Fig. 6.32 Systematics of the operating elements from which a total of 88 individual models can be derived (Götz 2007)

moved by piezo elements, which are mounted in the favourable gripping space in the centre console. The central screen is located at the position required in ◘ Fig. 6.8 (see ◘ Fig. 6.35). The position of the buttons displayed on the screen can thus be felt on the haptic touchpad. This means, for example, that it is no longer necessary to jump sequentially to a certain function, as is the case with the rotary pushbutton, but the corresponding button can be pressed directly. Of course, two-dimensional inputs, such as moving the map, can be carried out with this touchpad using direct manipulation. An improved version also allows letters to be written. The test in the driving simulator clearly shows the superiority of this new human-machine interaction element with regard to all performance criteria (operating time, track deviation, average viewing time on the display). It is interesting to note that the display of a cursor on the screen is obviously counterproductive and increases the operating

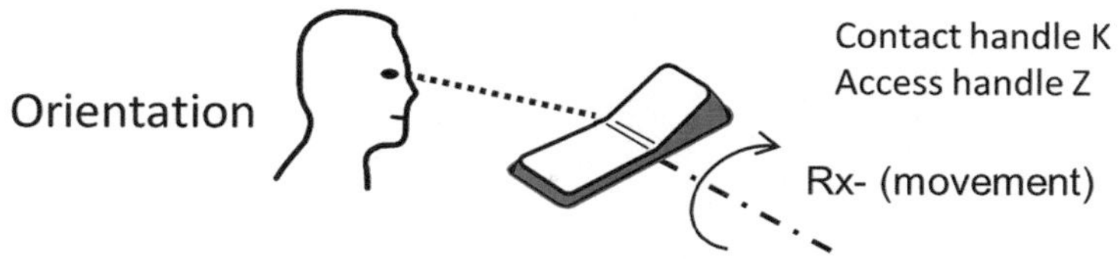

Type: On/Off	motion	gripping type	orientation
Surface segment, curved (concave, convex) asymmetrical to rectangle	Rx+	K	cambered surface element to the user
inclined surface (concave, flat, convex)	Rx-	K	low side to the user
Chamfer circumferential	Tz-	K	arbitrarily

Fig. 6.33 Control elements that convey the intuition ON/OFF. x = axis planar to body, y = axis away from body/towards body, z = up/down; R = rotatory; T = translatory; K = contact handle; Z = grasp handle

Type: more/less	motion	gripping type	orientation
Cylinder segment corrugated, axis horizontal	Rx/Tz	K	Corrugation at right angles to the user
…			
Side surface inclined on both sides, inwards (concave, convex, flat)	Ty	Z	arbitrarily
…			
surface vertical corrugated	Rz	Z	arbitrarily

Fig. 6.34 Control elements that convey more or less intuition

Fig. 6.35 Concept of the haptic touchpad (Spies et al. 2010; Spies 2013)

time as well as the necessary viewing time. Spies' experiments, however, can only be seen as a pilot study to illuminate the potential of such interaction. The operating feeling with the piezoelectrically activated Braille writing elements is technically quite spongy. The further development of this approach envisages a tactile variable haptic feedback on the touchpad in order to further increase the degree of blind operability (Blattner et al. 2013).

6.2.2.4 Voice Control

Since in many situations the input of values, names or commands requires many individual manual inputs, voice commands – due to the different main capabilities of the right and left hemispheres of the brain (see Chap. 3.1.2, ► Fig. 3.3) within certain limits[29] – offer in principle a high potential to convey action requests to the vehicle parallel to the driving task. Speech is a digital information transmission and is therefore only suitable for the transmission of specific digital commands. It is rather unsuitable for the transmission of general analog commands (e.g. "louder", "further left"). The technical challenge of filtering out speech from the background of the driving noise, taking into account the different speech colouring also caused by the dialect, has now been very well solved. There is also the problem of clearly separating the voice command addressed to the vehicle from elements of normal conversation between passengers. An appropriate microphone characteristic and microphone placement can contribute a great deal to error avoidance. An activation button ("Push to Talk") is also used, whereby the driver announces his intention to communicate with the vehicle.[30] However, if after pressing this key only *one* command (e.g. "Light on") is possible, pressing a corresponding key could be the better option. This restriction does not apply if, for objective reasons, several commands specifying the desired function follow after the first command (e.g.: "Navigation – Munich – Maximiliansplatz – five"). The biggest problem of technical speech interac-

29 However, the left brain is also involved in problem solving processes and language binds the same resources. It depends on when and which actions are triggered by voice commands. If the left side of the brain is occupied with problems such as navigation or music selection, voice control slows down the processing of these tasks.

30 Recently there is also the possibility that the start of the conversation is initiated by a fixed spoken keyword (for example "Hi Mercedes").

tion is that – at least at the current stage of development – misrecognition may also occur. These can have different causes: The user speaks in the wrong way, he speaks the wrong commands or at the wrong point of time. Due to such individual modifications, the detection system often finds itself in dialogue dead ends (with the consequence of feedback such as e.g. "please repeat" up to "abort"), which are inscrutable for the user because he does not know or remember the way to the position of the communication abort. The above-mentioned advantage of linguistic interaction is then reversed, because the now necessary orientation, reorientation or repetition of the whole process requires mental resources, which can considerably impair the driving task. Under such conditions up to 80% operating errors occur, which are also caused by technical recognition error rates of up to 20%. Overall, however, acceptance tests showed a clear preference of participants for speech in all driving scenarios. For adjustment procedures (e.g. seat adjustment) voice input is clearly rejected.

Particularly when considering possible technical developments, great importance must be attached to voice interaction. However, it should also be borne in mind that the use of commands requires the user to know which terms (vocabulary) are to be used. Multimodal interaction also plays an important role here, as it shows that users prefer to use the terms for voice input, which are also displayed on the screen in written form. In contrast to this, the classic operation by means of control elements – with ergonomic design by means of lettering and arrangement – provides a kind of user guidance, which can also facilitate the spontaneous use of a foreign vehicle.

6.2.2.5 Gesture Control

As the technical possibilities of gesture recognition (especially in the field of video games) are constantly being advanced, the question arises as to whether this can also be used with benefit for in-vehicle interaction. In this context, it should first be borne in mind that gestures are used spontaneously and mostly unconsciously in everyday life to support the deliberate delivery of information (usually when speaking), especially connotatively. Of course, gesture is also used as a means of communication in interpersonal relationships. If this is to be used for the secure transmission of information (usually visually over long distances, e.g. instructions for parking), however, the corresponding gestures must be learned. There is also the technical problem of separating involuntary gestures from arbitrary ones. Mostly gesture recognition is discussed in connection with screen operation. For example, when the fingers approach the touch screen, the targeted button is displayed larger. However, this measure does not compensate for the above-mentioned need for a longer period of observation in this type of operation. You can also use the familiar smartphone zoom in and zoom out gesture when your hand is near the screen. At present, however, there are no valid scientific findings on the benefits of such applications in vehicles. Even though already published research results report very positive results, the use of gesture recognition should be viewed with skepticism until it has been fully integrated into an interaction concept.

The examples of character recognition, speech recognition and gesture control also show that general design recommendations must be observed for every type of recognition-based interaction: For the user, it is often not clear which inputs at which location in the vehicle are recognised as permissible by the system. Here, both the preliminary information and the screen design play a special role.

Based on the everyday experience with these forms of communication in interpersonal communication, users expect immediate feedback from the system. Already delays of the feedback around more than 100 ms lead to misleading repetitions of the input, in each case however to confusion and lacking acceptance. The feedback does not necessarily have to be the recognition result, but at least a system reaction (see also ► Sect. 6.1.2).

To avoid detection errors, most detection systems are not permanently active. In many cases, the system must then be activated by pressing a key, which is often not comprehensible or traceable for many users. Here too, the

design of visible information (key labels, displays) represents an important ergonomic design task. This is illustrated by the example of speech interaction: The following questions of the user are to be answered here by the targeted design (Bengler 2000):
- When should I speak?
- How should I speak?
- What do you want me to say?

In general, a multimodal interaction approach (Oviatt 1999, 2000) that supports user guidance and interaction with the recognizer through graphical displays, audio signals, and speech output is helpful here (Bengler 2001 and Spies et al. 2009).

6.3 System Ergonomic Recommendations for the Respective Driving Task Levels

6.3.1 Primary Driving Task

6.3.1.1 Stabilisation Task

The *operating elements* for the stabilization task are steering wheel, accelerator pedal and brake pedal. The clutch pedal and shift lever are necessary for technical reasons, since the speed and torque range of the combustion engine cannot completely cover the speed and drive force range of the vehicle. The development of the various forms of automatic transmission and in particular the possibilities of a future electric drive make these control elements superfluous. Because of the great importance for the driving experience and the ability of the driver to "keep the vehicle under control", the feedback of the driving or stabilisation condition, in particular from the steering wheel, is dealt with separately in ► Sect. 6.4.

Only the speedometer, which shows the actual speed, is actually available as a display reporting the state of the stabilization. In sports cars and off-road vehicles, the longitudinal and lateral acceleration of the vehicle and its lean angle are occasionally displayed. The latter displays should be interpreted as analog displays, whereby physical limits (e.g. the maximum permitted inclination) can also be displayed in the sense of a range value display. Within the PROMETHEUS project, proposals were also developed to measure the maximum force flow of the wheels to the ground. According to a proposal by Porsche, the result could be displayed in a two-dimensional display of the longitudinal and lateral acceleration as a dynamic limit value (see Fig. 6.36).

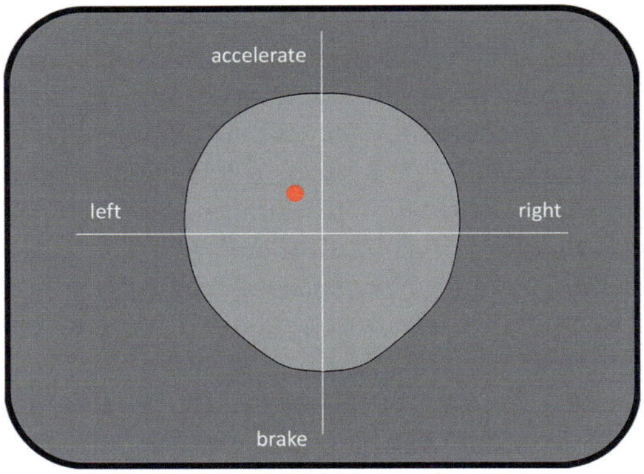

Fig. 6.36 Representation of the actual acceleration in longitudinal and lateral direction as well as the currently measured limit values (according to a proposal by Porsche)

As already mentioned, from an ergonomic point of view the digital display is preferable for the speedometer, as the regulations given by speed limits are also in digital form.

Most of the other instruments installed in the vehicle today are located in the hierarchical order below the stabilization task. The same applies to the already mentioned clutch pedal and gear lever. All of them refer to actual conditions of the vehicle's technology. On the instrument side, there are the tank capacity, engine speed, cooling water or oil temperature, battery charging current, etc.[31] While the display of the tank content is

31 It should be pointed out in this context that such instruments only became common property when the corresponding information for the driver had not become as urgently necessary from a technical point of view as in earlier historical vehicles. For cost reasons, up to the luxury class (apart from sports cars) only speedometer, cooling water display and display of tank contents were found there. For a long time, even the latter has been dispensed with by switching the fuel supply to "reserve".

directly related to the navigation task (see ► Sect. 6.3.1.3), the engine speed of an automatic transmission can practically not be influenced independently of the speed, so the indication of it is largely superfluous in this context. Cooling water and oil temperatures cannot normally be influenced either. However, the information about this is useful in order to recommend a correspondingly gentle driving style when the vehicle is still cold. The charging current of the battery cannot be influenced either. Its display provides indirect information about the condition of the battery and whether any electrical loads should be switched off. Without additional information, most drivers are unlikely to know what to do with this display. The same applies to most of the indicator lights found in vehicles today, unless they are combined with additional information for the driver (see below). If, from a rational ergonomic point of view, the value of the controls and displays serving the technology is doubted here, however, the "aspect of favourability" (see Chap. 3.3.5) must not be disregarded. It is not by chance that the subjective value of a vehicle is shown by the number (and aesthetic beauty) of the instruments presented – both yesterday and today. They give the user the feeling of being master of a complex technology. As already mentioned, the joy of clutching and shifting is also part of this, as it gives the impression of being able to intervene directly in the vehicle's technology. The occasional glance at these instruments requires an average fixation time of about 600 ms (Schweigert 2003; see also ◘ Fig. 3.55). Especially in view of the ever more complex traffic and the associated demands on the driver, it must be considered whether the distraction effect caused by these purely technology-oriented applications may be problematic.

Recently, displays have been added to support safe driving operation, such as indicators of falling below a certain outside temperature, which may mean a risk of black ice, underinflated tyres, worn brake pads, a defect in the engine control system, etc. Such advertisements should only appear if the relevant condition is met. In many cases, the driver should also be given direct advice (e.g. as a typeface in the central display instrument or as an acoustic word announcement), which must be done in the case in question. In the case of manually-operated vehicles, switching recommendations are sometimes given today in order to reduce fuel consumption. As a study by Lange (2010) shows, these displays are hardly noticed in practical driving situations, especially in connection with complex traffic situations, as they only become visible when the gaze is accidentally directed at the instrument panel. A haptic display by the active accelerator pedal, which gives the driver a "knock on the door", similar to a double-click when operating the PC, the prompt to shift gears, has proven to be much more effective.[32]

Particularly with regard to resource consumption, the associated increase in CO_2-emissions or the limited range of electrically powered vehicles, the display of the current energy consumption is of increased importance. As early as the 1980s, displays of current fuel consumption based on different measuring principles (measurement of the vacuum in the intake manifold or direct measurement of fuel flow) appeared. In practical operation, however, these displays hardly cause a change in driving style because they would have to be constantly observed simultaneously with the observation of the traffic situation that determines vehicle driving. Also in connection with hybrid vehicles and those with purely electric drive, the consideration of the current consumption of the energy reserve is of decisive importance for the driving style. Various pictorial displays have been presented which show the energy flow to the drive units and the energy recovery during braking manoeuvres (see below). Here, too, it should be noted that it is unlikely that these displays will be continuously taken into account in accordance with the current traffic situation and the guidance task resulting from it. A cer-

32 It should also be pointed out in this connection that, with today's technology of automatic transmissions, in particular the use of the double-clutch transmission, the fuel consumption of the driving style of an average driver in practical driving mode is largely similar to that of the manual transmission.

Fig. 6.37 Display of speed, current energy consumption and range (example: BMW i3)

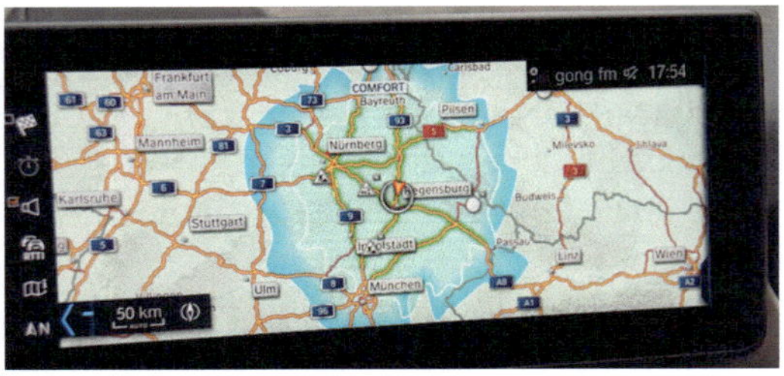

Fig. 6.38 Representation of the current range on the navigation map, taking into account different route-dependent driving resistances. The light blue area indicates the range for normal driving, the dark blue area indicates the range for particularly economical driving (example: BMW i3)

tain "educational" effect cannot be denied, however, if the driver recognises through such displays that rare and moderate braking manoeuvres lead to better energy management and thus to longer ranges. In the future it would be necessary to investigate whether a corresponding haptic display via the active accelerator pedal leads to a more sustainable effect.[33]

Due to the significantly lower range of vehicles with purely electric drive compared to vehicles with combustion engines, the currently still significantly lower availability of charging stations and the longer time spent there (from half an hour in the best case to a maximum of several hours), special displays are needed to enable drivers to set up a charging management system that matches their driving habits. Especially standstill and parking times e.g. during working hours can be used for charging and benefit from charging management and connectivity functions.

As an example, Fig. 6.37 shows the display as it is realized in the BMW i3. The position of the pointer below the digital speed display indicates whether energy is currently being extracted from the battery (ePOWER) or whether the battery is being charged by the braking effect of the electric motor (CHARGE). In addition, the bar graph below the digital speed indicator shows the currently available range in both analog and digital form. Depending on the actual energy consumption, this display changes simultaneously. When the electric vehicle is switched on, the word READY indicates that the vehicle is ready to drive because, unlike a vehicle with a combustion engine, there is no feedback about the running engine.

For many vehicles, the range is also shown in the navigation display (Fig. 6.38). If a destination has been entered via the navigation computer, possible charging stations and the route to them are displayed in case the destination cannot be reached with the current range.

Since the energy supply in an electric vehicle requires far more attention than in a vehicle with a combustion engine, various displays are also required to support energy management. For example, a charging schedule can be set in the tariff window in order to charge at an acceptably price. If the vehicle is charged in the domestic garage, it is useful to define the departure time as well as the air conditioning for the departure, since the energy required for cooling and heating is taken from the public electrical system and does not reduce the range (see Fig. 6.39).

Also while driving, displays show how the range can be increased if certain consumers would be switched off (Fig. 6.40).

Apart from the uniform charging options that are offered exclusively for Tesla vehicles, different settings can be made during charging depending on the available options of a cur-

33 As early as the 1950s, DKW attempted to curb the high fuel consumption of two-stroke engines in the 3 = 6 series by means of a second return spring, which became effective from a certain accelerator pedal position. The author is not aware of any studies that have verified this effect.

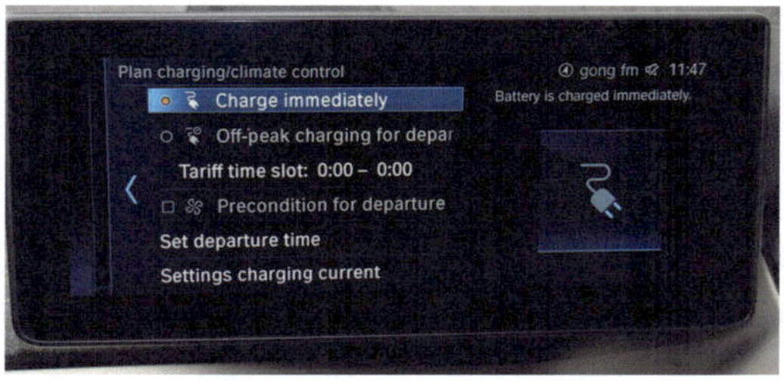

Fig. 6.39 Display of the load management (example BMW i3)

Fig. 6.40 Display of additional range potential by switching off consumers (example BMW i3)

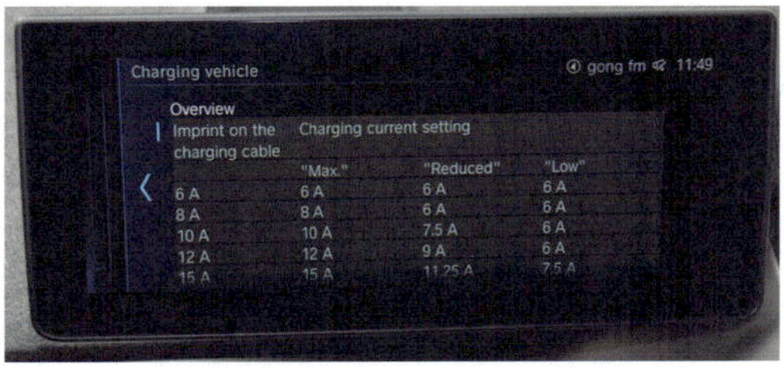

Fig. 6.41 Reference to possible settings on the charging station (example BMW i3)

rent charging station (Fig. 6.41). From an ergonomic point of view, it is unreasonable for the user to calculate the necessary charging time for a desired range himself from this and from the charging current values (given in amperes). Expected charging times and ranges resulting from a given charging situation and his/her driving habits should be presented as a preview to the user following the general idea of usability.

In general novice users should have the possibility to review their driving style and resulting energy consumption as well as their recuperation behavior. These informations are important to optimize the driving style and driving strageties using an electric vehicle.

From the described expenditure, which is necessary for the enterprise of an electric vehicle, it becomes clear that still substantial ergonomic organization expenditure is necessary, in order to keep the operation of such a vehicle as simple as this became possible in the course of the long evolution of the vehicles with a combustion engine.

6.3.1.2 Guidance Task

The information for the guidance task is obtained from the view of the natural environment, which contains the course of the road and the current constantly changing traffic situation. As can be seen from the analysis shown in Fig. 2.23, the generation of the target values for lateral and longitudinal direction is based on the perception of this situation, the estimation of the dynamic behaviour of the road users that may influence the own journey and the estimation of the current dynamic characteristics of the own vehicle by the driver, all estimations that may be inaccurate or even erroneous by nature. Due to the human characteristics of information processing, this estimate reaches a maximum of two seconds into the future, although a longer period of time would often be necessary, especially at higher speeds or adverse conditions (on a dry road, for example, a vehicle from a speed of 200 km/h comes to a standstill after approx. 4 s!). Night and fog accidents can be caused by the fact that in a given situation the stopping distance is too large in relation to the headlight and/or restricted fog visibility in connection with the road conditions. It is precisely under such conditions that improved headlamp systems, which, for example, only blank out the area for oncoming traffic when the main beam is switched on and which, supported by the navigation system, follow the course of the bend with foresight, offer great help.

Until recently, there were practically no displays in the actual sense of the word that would facilitate the guidance task and help to keep its execution within safe bounds. As a first step for displays at the guidance level, the display of the current speed limit ("traffic sign assistant") detected by cameras and processed by software – also with the aid of information stored in the navigation computer – can be viewed. At the very least, this display helps the driver, similar to an overtaking prohibition display based on the same principle, to be sure whether the regulation is still valid or has already been repealed. The prerequisite for the driver's confidence in such a display, however, is a sufficiently high detection rate and a low error rate (better than 95% or lower than 5%). The haptic display of the so-called "Lane Departure Warning" (LDW), which, depending on the design, uses the prepared information of a front camera or simply the reflection of laser beams at the road boundary markings and thus indicates the approach or already the crossing of the road boundary in the lateral direction by vibration or counter torque at the steering wheel or seat, can also be regarded as a limit value display of the lateral deviation. In the Peugeot 5008, a display developed by Continental was presented which measures the distance x to the vehicle in front on the basis of radar measurement and displays the time value t_S (t_S = x/v) in a simple HUD[34] is output. The task of the driver is then to adjust the speed v and the distance x in such a way that values < 0,8 s, which would result in a police warning, are avoided (if possible, time distance values between 1.2 and 2 s should be observed!). Even the night vision devices offered today in the luxury class belong to the category of displays at the guiding level, as they are able to detect a possibly hardly visible creature in front of the vehicle by using thermal radiation. Today, the information obtained in this way is shown on the central display. In a similar way to what was already shown for the gear shift recommendation, the restriction that this information can only be used if the display is looked at randomly in the given situation applies. However, this is unlikely in difficult viewing conditions. A study by Bergmeier (2009) showed that the information displayed on the central display is the most insecure, even a warning in the conventional HUD is relatively ineffective, because it is not clear to the driver where the endangered creature is and how it behaves. Only a local marking in the contact-analog HUD results in a significantly safer perception (◘ Fig. 6.17).

Since the task of driving a vehicle in the real field of vision is presented in front of the vehicle, displays that offer help in this respect would also have to be displayed there. Only the contact analog HUD can be used for this. Bubb had already made proposals for this in 1985. Then the so-called "driving hose"is displayed in the contact-analogue HUD, i.e. the course the vehicle would take if no changes were made to the steering wheel. Depending on its design, it is sufficient to display a crossbar which is located virtually in front of the vehicle on the road surface at a distance of between 1.2 and a maximum of 2 seconds and whose lateral position depends on the current steering wheel position (◘ Fig. 6.42). Simulator tests have shown that it does not matter whether the driving hose as a whole is visible, i.e., that the connecting lines between the vehicle and the distance bar are visualised or not (Lange 2007). It is important that the position of the bar is always determined by the steering wheel position (and possibly other dynamic characteristics of the vehicle). Israel (2013) was able to show in simulator experiments that it is now much easier and safer to navigate through bottlenecks such as

◘ **Fig. 6.42** Display of the driving hose in the contact-analog HUD; the lateral position of the reflected bar indicates where the vehicle is in 1.5 seconds, if no corrosive control element interventions are made

34 In this case, the HUD is displayed via a small folding combiner (semipermeable mirror), which avoids the otherwise elaborately optical compensation of the windscreen curvature. By the way, the display of the time distance would not necessarily have to be done in a HUD, but offers the highest probability to be considered in the normal traffic flow.

construction sites. Assmann (1985) demonstrated with a real vehicle equipped with a contact-analog HUD that the times in which a too small distance to the vehicle in front is driven are significantly reduced. In addition to the display of the driving hose, Bubb (1985) also proposed to display the necessary overtaking distance. However, with the large distances that occur (500 m and more), one reaches the technical limits of representability in a contact-analog HUD.

Apart from normal driving, there are also special forms of the guidance task. This includes all manoeuvres, especially parking. During the general treatment of the different display variants (▶ Sect. 6.2.1.), displays were presented which facilitate this process. Acoustic indicators are now widely used to indicate approaching objects by different pitches (differentiating between obstacles in front of or behind the car) and by increasing intermittent frequency of the respective tones. The manoeuvring task is much better supported by visual displays on a display that visually show the distance to obstacles (◘ Fig. 6.16). A further increase for the vividness is given by computer-technically prepared video pictures, which show the vehicle from the bird's eye view including the real environment. The task is further facilitated by a quality level if the "driving hose" given by the current steering wheel position is displayed in the image of a rear view camera (◘ Fig. 6.18).

The advertisements for the leadership task discussed in the last two sections do not presuppose the existence of assistance systems. With the exception of parking aids, all variables can be calculated directly from values available on the vehicle (essentially speed, lateral acceleration or steering wheel position). The dynamic sphere of influence of the vehicle is visualized, so to speak. The task of the driver is then to compare this dynamic influence with the limits given by reality and to adjust it by selecting the appropriate steering wheel position and speed. In contrast, assistance systems are characterized by the fact that they take measurements of external variables and on this basis intervene in the driving process or, on the basis of a virtual control system, issue limits in the form of warnings.[35] Nevertheless, the displays discussed here can be combined with assistance systems, which would provide the driver with a better understanding of how the assistance system works (see Sect. 9.4).

6.3.1.3 Navigation Task

Without technical aids, the navigation task must be performed from memory and/or with the aid of maps. At the latest when changing the known route and especially when it has to be carried out ad hoc during the journey, its fulfilment quickly reaches its limits. Today it is technically supported to a large extent perfectly by navigation computers. Today, all navigation systems, including those from the secondary market, have the characteristic that if they deviate from the originally calculated route, they report the deviation to the driver within a very short time and calculate a new route. Ergonomic requirements must, however, be met in particular by the programming of the route and the presentation of the navigation recommendations.

Programming the route: for programming with *stationary vehicle* the system ergonomic rules for sequential and simultaneous operation described in ▶ Sect. 6.1.1.1 shall be applied. Irrespective of whether the information is entered via touch screen, rotary pushbutton or touch pad (character recognition), it is crucial that no unnecessary operating steps are provided (for example, is a "start guidance" button necessary if all address entries have been made, or is a backspace key better that allows incorrect entries to be corrected?). All rules regarding feedback and compatibility must also be observed. For example, the destination address should always be displayed after the address has been entered – even during the journey. If a destination that is not available in the standard destination memory of the navigation com-

35 The above-mentioned system implemented in the Peugeot 5008 as well as the parking distance indicators represent a hybrid position in this respect, since the distance to the vehicle in front or other obstacles is measured, but is not used for a technical regulation.

puter was finally found via conventional navigation (map material or interviews with locals), it should also be possible to add this to the individual address memory later. As already mentioned, it should be possible to carry out complex route planning on a PC or smartphone outside the permanently installed navigation system and finally transfer it to it.

As Sacher (2009) observed, in general the navigation computer is often programmed while driving (observed order of priority: first drive off, then think: where to?). This means that the *Programming while driving* must be as reduced as possible or as simple and less distracting as possible. In addition to remote programming of the vehicle via smartphone or the Internet, one measure is to store freely assignable "station keys" or lists of favourites on the known, frequently visited destinations individually. However, no more than eight such keys should be provided. Under certain circumstances, addresses stored in the telephone directory can be transferred to the navigation system via voice input or touchpad. This would, for example, be an extension of the station keys mentioned above. However, destinations that vary greatly or appear ad hoc during the journey (sightseeing, searching for a restaurant, resting place, etc.) make it necessary to enter more complex destinations. It is also possible to enter speech or letters using the touchpad. In this context, however, research still needs to be invested in both input variants, in particular to solve the problem of finding a way out of a technical "dead end".

There are also goals that result directly from the condition of the vehicle. This includes in particular the search for filling stations or power distribution points for electric vehicles. If the remaining range calculated from the tank capacity or the remaining capacity of the battery falls below a certain value, the navigation system should automatically display the corresponding energy delivery stations together with the specifications (price, distance, fast charging station, availability, etc.). A simple operating step (e.g. touching the touchscreen) should then navigate to this waypoint without deleting the navigation to the original main destination.

▪ ▪ Navigation display

The display can be at different levels. The first navigation systems only used the *acoustic indication* (e.g. "next crossroad right ... now please right"). Only with clear local conditions can an error-free transmission of information take place. The acoustic display parallel to the visual display is still useful, as it attracts attention. Since this display is often perceived as annoying, it should be possible to switch it off in a simple way (one action!). Optical displays can also be displayed at different levels. As already mentioned, it makes sense to display an instruction to deviate from the previous route in the next few seconds in the instrument cluster behind the steering wheel or in the HUD. The central display, on the other hand, should ideally be reserved for a general north-facing map display.[36] The simplest optical display is achieved by a *arrow icon*. Unfortunately, it has become common practice to display a second bar next to the turn symbol, which symbolizes the distance to the action. The driver must look back and forth between the two displays for correct interpretation. It would be easier to integrate the distance indicator directly into the turn arrow (see ▫ Fig. 6.43).

Far more descriptive than the abstract arrow symbol, however, is the *Bird-view display* of the intersection situation, where the vehicle symbol indicates the current own position, possibly supplemented by surrounding streets and signage (Bengler et al. 1994). The display of turning lanes offers the driver a further improvement in safe orientation. The simplicity of orientation is increased once again by the *contact-analogue display* of the turn recommendation. This can be done by fading into the image of a front camera or in the very best case in the contact-analog HUD (see also ▫ Fig. 6.44). Initial tests with such a display indicate that the contact-analog insertion is only useful in the range of 2 to 3 seconds before the actual turn. In the time period before this, a static symbol for preparing the

36 However, since individual preferences differ here, it should be possible to choose between pointing north and pointing in the direction of travel.

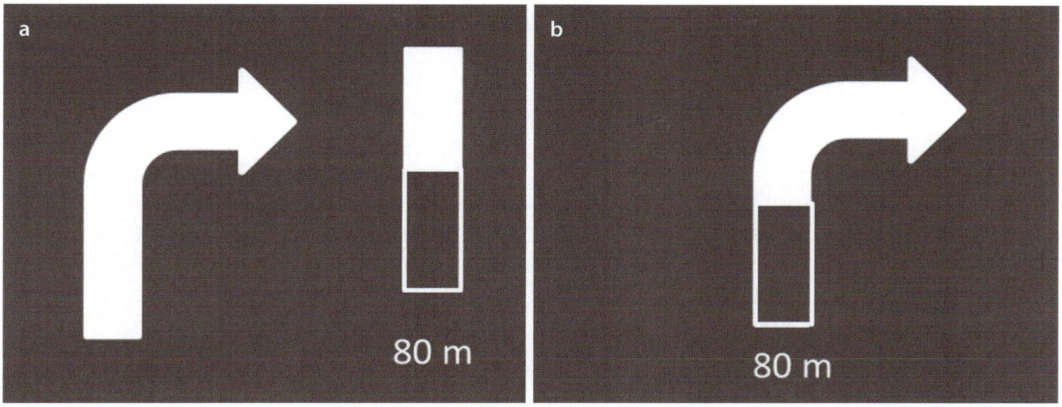

Fig. 6.43 Navigation information through pictorial turn symbol, **a** with separate distance display, **b** with integrated distance indicator

Fig. 6.44 Navigation display of a complex intersection situation in contact-analog HUD (the intersection situation is shown below right)

turning process (e.g. recommendation for lane change) seems to be more advantageous as a display. Israel (2013) was able to show in the simulator experiment that a contact-analogue navigation display in the HUD significantly reduces the misinterpretation even in very complex crossing situations compared to a conventional HUD (Fig. 6.45). However, his experiments also indicate that further research is required for this display version. The output of the navigation information should also be graphical and linguistic. However, it regularly appears that voice output from users is switched off. Design measures that help to avoid the phonetic quality of expenditure and its monotony could rem-

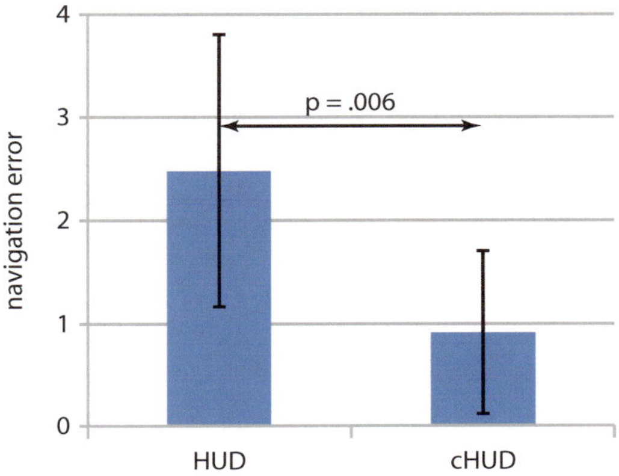

Fig. 6.45 Average number of navigation errors in the contact-analog HUD in relation to a conventional HUD in complex intersection situations (Israel 2013)

edy this situation. Even minor changes in the way of speaking and smaller variations of the texts would be helpful in this sense without affecting the comprehensibility or meaning of the messages.

6.3.2 Secondary Driving Task

As already described in ► Chap. 1, tasks that are directly related to the driving process but whose non-fulfilment does not directly influence the driving process are referred to as secondary. These tasks must be further differentiated according to whether they are active, i.e. whether they contain a driver's expression of will, or whether they are reactive, i.e. whether they are to be carried out depending on a given situation. Active tasks are the setting of all warning signals (horn and headlight flasher) as well as the direction indicators ("indicators") and recently the initiation of assistance functions (lane departure warning, cruise control, ACC, etc.). The reactive tasks can be automated in principle, which is also recommended from an ergonomic point of view in view of the increasingly complex traffic situations. It is currently a question of the manufacturer's respective "interaction philosophy" whether the driver initiates the automatic process or whether the automatic process is always active. However, this fact leads to intransparency and high operating errors for users of different brands. Classic reactive tasks are the switching of the various lighting functions, in particular the switching on and off of the headlights and the up and down lights, as well as the wiper functions (in the case of manually switched vehicles, this also includes the clutch and switching). Since no automatic system can cover every situation and also does not always meet all expectations of the driver, it is necessary to make corrective actions as simple as possible and, if possible, conform to a familiar operation "by hand". Irrespective of whether the tasks are active or reactive, the control elements are arranged around the central control element for the primary driving task, namely the steering wheel. Today, this is almost exclusively done using steering column levers[37] or by control buttons mounted directly on the steering wheel. In any case, it is necessary to provide feedback on the switching status of a function at all times. This can be done by the (visible!) position of the control element, a light on the control element or by switching a symbol in the instrument cluster. The anthropometric design (see Sect. 7.3.3) must ensure that this feedback is visible from every driver posture and for every driver's stature (anthropometry).

6.3.3 Tertiary Tasks

Tertiary tasks have nothing to do with driving, neither directly nor indirectly, but are nevertheless performed while driving. In principle, they have a distracting character. This also applies to the traditional operating procedures, such as adjusting the heating/air conditioning, operating the windows, the sunroof and the radio. These operations have become partly more complex, as higher comfort requirements have to be satisfied (e.g. radio with switching to different sound sources and their operation, selecting and possibly compiling the sequence of special pieces of music). In some cases, however, they have also been simplified by technical measures (e.g. station keys and radio station search; automatic climate control, electric windows). With the extension of the radio with a navigation computer[38] and the possibility to make phone calls in the car, the scope of operation in connection with all the previous tertiary tasks was once again significantly increased. In addition, the vehicle has recently been integrated into the internet. The pressure to have all

37 By using today's possibilities of computer-aided information processing, the operating satellites used by Citroen in their CX models of the 70s are definitely worth a new consideration, considering fundamental ergonomic requirements.

38 It should be pointed out once again that the operation of the navigation computer is an aspect of the primary driving task. Only by the mentioned history of the technical development this is today assigned to the central screen unit and thus integrated into the operation of the remaining tertiary tasks.

these interaction possibilities in the vehicle, and the fact that the diversity associated with them can hardly be operated with the switches and buttons previously used, has led to special operating concepts being developed for this purpose. The touch screen, rotary pushbutton, touch pad, joystick and mixed forms from these input devices are essentially used today. All these input devices navigate in a menu structure to trigger the individual functions. Already Michon 1993 and Parkes and Franzen 1993 point out the potentials of these systems but also the necessity of their ergonomic design.

Rassl (2004) examined in detail the system ergonomic relevance for the design of such operator interventions during journeys (in low-traffic environments) in the real vehicle. The operation was carried out consistently by a rotary pushbutton in connection with a uniformly mounted TFT display. Nevertheless, the results can also be transferred to the other control elements mentioned above, since the cognitive binding through finding one's way around the menu tree was the subject of the investigation. A total of 36 individual tasks were realized for the investigation, with one good and one bad variant with the entire associated menu structure being realized for each task from the point of view of system ergonomics. A wrong navigation in the menu tree could be corrected by a "back" menu item, which caused a jump to the higher menu level. With a separate control button it was possible to get to the beginning of the menu tree.

The following rules can be derived from the results of the tests:

1. The user may not be offered more than nine (simultaneous) choices at the same time. If necessary, these are to be distributed among sequential steps.
2. If possible, simultaneously displayed choices shall be arranged according to their importance.
3. In the sense of the system ergonomic requirements (6.1), a maximum of only *one* unnecessary operating step. This sequential operating step, which in principle is unnecessary, must at least fit into the logic of the operating sequence.
4. Simultaneous operation must not be displayed sequentially.[39] Only when more than nine choices are presented is an exception allowed under Rule 1.
5. Sequential operation must not be displayed simultaneously.
6. Tertiary tasks are to be designed statically, i.e. a quasidynamic effect, such as that caused by the automatic return to the initial state when not being operated, should also be avoided wherever possible. If this return is unavoidable, the once selected setting should be held for more than 3 seconds.
7. The feedback must be sent within 200 ms.

Tertiary tasks obviously require the driver's attention to a much greater extent than is generally assumed. On average, Rassl's tests took longer to look at the display (1,35 s ± 0,54 s) than at the road (0,67 s ± 0,32 s). The average minimum viewing time on traffic events was 0,24 s ± 0,14 s, the average maximum viewing time on the display was 2,72 s ± 1,47 s, with extremes of up to 16 s occurring. Even with seemingly simple tasks, such as setting up the radio, maximum gaze averting durations of up to 10 seconds were measured. Unpublished experiments by the Institute of Ergonomics of the TUM on the operation of the air-conditioning system showed similar results. These tests were also carried out on public roads. The mean gaze averting times were in the same order of magnitude as at Rassl. This shows an interesting effect: Systems well designed according to system ergonomic rules showed that *on average* hardly better values than worse designed systems. However, the standard deviation of the bad systems is significantly higher. The maximum values differ significantly once again. Even with the well-designed systems, there were gaze aversions of 4 seconds. With the poor systems, even gaze aversion times of 12 seconds were observed.

39 This result prohibits the frequently encountered "switch-through" of the various on-board computer functions, which is so popular with manufacturers because it requires only one button for many functions.

Both in these and in Rassl's experiments, it was noticeable that the test persons were not aware of these long turnaround times. Obviously, under certain circumstances, the tertiary task becomes the primary task for the subject. However, by consistently observing system ergonomic rules, at least the probability of such a rebalancing can be significantly reduced.

Further data on gaze aversion in tertiary activities and its effect on blinks, steering and distance behaviour were determined in the ADAM project in various driving simulators (Bengler et al. 2002, 2003, 2004; Breuer et al. 2003). The activities investigated include interaction with menu systems as well as with mobile devices and socially accepted tasks such as searching for objects or sweets.

One of the biggest problems of operating systems whose possible functions are organized in a tree structure is finding one's way around this menu tree. In principle, the tree structure does not correspond to the associative organizational structure of human memory. This would, for example, make the possibility of jumping from one branch of the menu pole to another desirable. Which jumps should be possible depends on the context of the use.

Ablassmeier (2009) argues that such a system should behave like an "intelligent passenger" who supports the driver largely autonomously in his tasks. The use of "intelligent" information agents on the basis of conditional probabilities is intended to relieve the driver, particularly in the case of frequently recurring action sequences, and to pre-filter relevant information (Hofmann et al. 2001). His proposal is based on the assumption that the functions are not arranged hierarchically at all. Rather, an efficient dialogue should guide the driver according to the task at hand. For his experiments he used the examples of tank, contact and appointment agents as well as restaurant receivers (as further conceivable agents he mentions: congestion agent, arrival agent, climate agent and train/flight information agent), for which he developed solutions in each case. As mathematical possibilities for the selection he proposes the Bayesian nets and the neuronal nets and here especially the probabilistic neuronal nets, whereby the Bayesian net is used for the tank agent and the contact agent and the probabilistic neuronal net for the restaurant-receiver. The weightings necessary for Bayesian net were determined by interviewing the test subjects. As a result of driving simulator tests it can be stated that information agents in the vehicle represent suitable methods to support the driver. However, it should also be noted that further research is needed to investigate the acceptance and, above all, the hit rate of such agents by the user in more detail.

In order to design such systems, the scenario technique mentioned at the beginning of this chapter must be used in any case. Irrespective of the feasibility of the proposals mentioned above, an indispensable condition for the so-called *robustness against errors* is the existence of a button that cancels the last step if one has "got lost" in the menu tree or the system, and a separate button that brings the user back to the root of the menu tree.

Due to the fact that finding one's way in the menu tree of the available functions or on the internet requires mental abilities similar to those of the primary navigation task (it is no coincidence that this activity is described by the term "navigation"), this can lead to interference with the primary driving task. Only by providing assistance systems for the primary driving task can the danger caused by averting the tertiary task be effectively countered. However, this measure also involves the danger that the driver will now turn away completely from the main task, possibly relying on the automatic machine and mentally devoting himself only to the secondary task. A solution to this dilemma is possible through the calculation of a safety corridor by the assistance system, within which the driver has to drive the vehicle, as already mentioned in ► Sect. 2.6.

6.4 Design of Driving Relevant Characteristics

The often advertised "joy of driving" is ultimately achieved in the primary driving task on the level of stabilization through the perfect interaction between driver and vehicle. It will therefore be dealt with separately in the following separate chapter. The driver sees his vehicle as an extension of his own body and thus of his own abilities. This effect is all the more pronounced the better, the more directly the driver's commands are implemented by the vehicle and the reactions of the vehicle are reported back to the driver. As a general rule, changes to the controls (steering, driving and brake pedals) can be felt somehow via the sensory organs within a time window of 100 ms, at most 200 ms. In addition to influencing engine performance, particular attention must also be paid to steering. It has therefore been the subject of constant research throughout the long history of the vehicle. Sometimes the focus is on technical aspects, sometimes on the psychological aspects of experience. With the possibilities provided by electrical/electronic control, this topic takes on new significance. With electronically controlled steering gears, the transmission between the steering wheel and the position of the steered front wheels is already influenced in addition to the driver's interventions. By completely separating the steering wheel control element from the actuator on the steering gear, as it is known under the name Steer-by-Wire, the possibilities of a tailor-made steering system are further increased. With all these possibilities, the driver's feeling of being one with the vehicle and feeling how it is connected to the road must not be lost to ensure the joy of stabilisation.

6.4.1 Lateral Dynamics: The Steering Feel

The steering feel is defined by Pfeffer (2013) as follows:

▪▪ Definition

Steering feeling is the sum of the optical, kinesthetic and haptic sensations of the driver when driving a vehicle and corresponds to a subjectively perceived, complex experience.

6.4.1.1 Road Contact

In many test reports, good steering is expected to provide feedback on the quality of road contact in order to react to adverse road conditions with correspondingly low longitudinal and lateral accelerations. In fact, from a driving mechanics and dynamics perspective, the steering wheel provides information about the reaction forces between the steered front wheels and the road. However, the significance of this feedback can only be evaluated by considering the force ratios of a wheel when cornering. If a vehicle passes through an arc of a circle, a lateral force (centrifugal force) increases with the square of the speed and must be transmitted by the wheels by means of static friction. For the restoring force noticeable at the steering wheel, the transmission of the lateral force in the contact area of the front wheels (also known as tyre slip due to the elastic deformation of the tyre due to the wheel load) is of particular importance. When forces are transmitted from the tyre transversely to the direction of travel, the tyre is braced from the beginning of the contact area until the transversely directional tangential stress is limited by the maximum value given by the coefficient of friction (so-called " crabwalk ", ◘ Fig. 6.46; Schallamach 1961). In principle, there must therefore be an angle between the driving direction of the wheel and its roller plane (rolling direction), the so-called slip angle, so that forces can be transmitted at all. With increasing transmitted lateral force S, this force increase is steeper within the contact area. Finally, the local force may reach a value limited by the effective coefficient of friction (see ◘ Fig. 6.47). The transmitted force is determined in its magnitude by the area enclosed by the actual tangential stress pattern, the point of contact being given by the centre of gravity of this area. This creates a lever arm between this starting point of the resulting lateral force S and the starting point of the resulting upward reaction of the surface pressure in the contact surface (◘ Fig. 6.48). It is called tyre follow-up n_R. Due to the reactions described, the starting

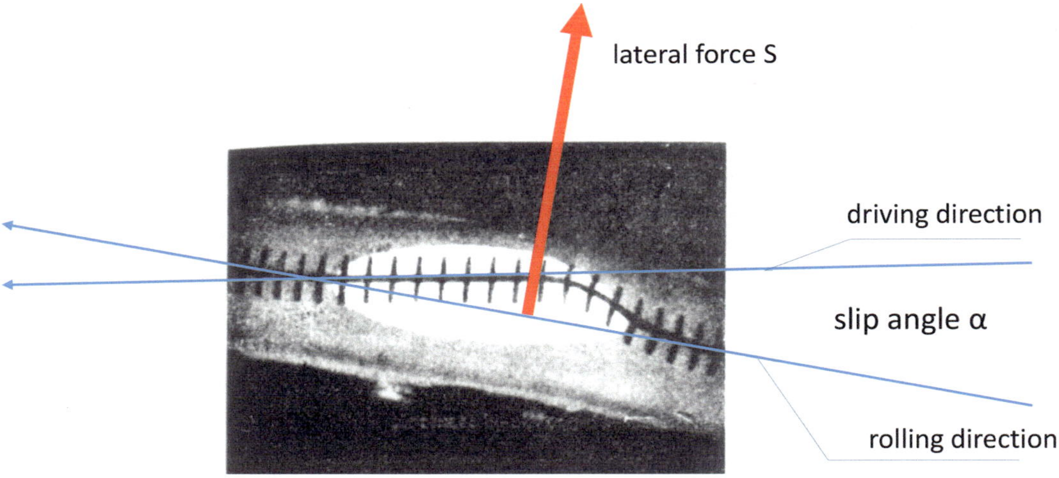

■ **Fig. 6.46** "Crabwalk" of an inclined wheel. (From Schallamach 1961)

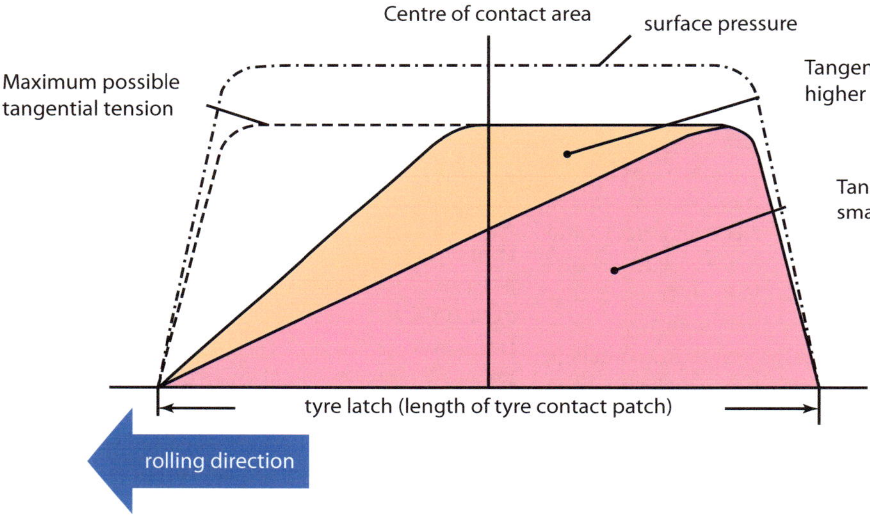

■ **Fig. 6.47** Tension curve in the tyre contact area due to lateral forces. (After Gough 1962)

point of the lateral force S – with the exception of the wheel sliding as a whole – is always such that the lateral force is a moment M_R which tries to turn the tyre from the skew position in the direction of travel (see ■ Fig. 6.48).

At the steering wheel, however, not only the restoring torque, which is influenced by this tire trailing of the front wheels, is noticeable, but also a restoring torque, which is caused by the constructive trailing (so-called "tea trolley effect") and other parameters. The constructive castor n_K is the offset between the

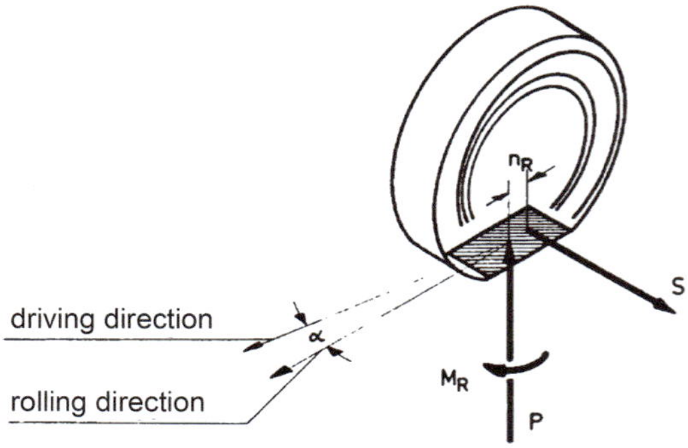

■ **Fig. 6.48** Reaction of the lateral force S and the wheel load P in the contact surface

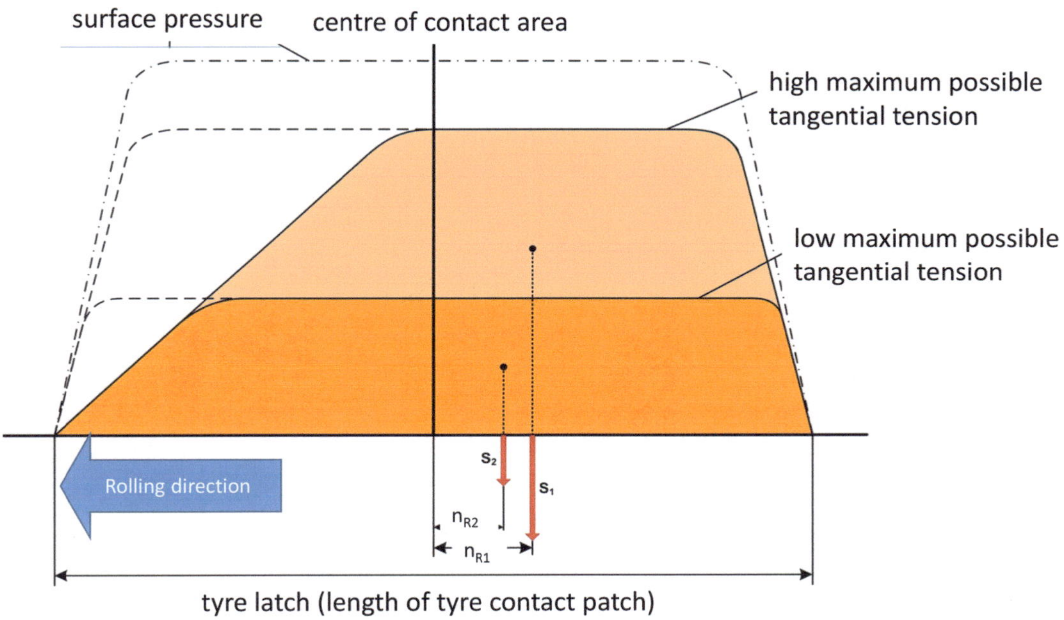

Fig. 6.49 Effect of the reduction of the maximum possible friction on the tyre follow-up n_R (after Gough 1962 and Gengenbach 1968), S = lateral force, n = lever arm of the restoring torque

wheel axle and the rotary axle around which the front wheels are swivelled during steering. In contrast to the tyre follow-up discussed here, it is a variable independent of road contact. Figure 6.44 shows that the tyre follow-up n_R changes with the friction coefficient μ (Gengenbach, 1968). The diagrams shown in Figs. 6.47 and 6.49 give a clear explanation for the division of the ground contact area into contact zone (adhesion zone) and slip zone by many authors.

- Maximum exploitation of a high coefficient of friction: Restoring torque $M_1 = S_1 \cdot n_{R1}$.
- Maximum exploitation of a low coefficient of friction: Restoring torque $M_2 = S_2 \cdot n_{R2}$. $M_1 > M_{2:}$ Low coefficient of friction is felt due to lower tyre reset torque.

If there is water on the road surface, the model described must be modified further. The tyre contact area can then be divided into 3 zones (3-zone theory). In the first zone, which is located in the front part of the tyre contact patch, the so-called approach zone, the tyre has no contact with the road surface. It's carried here by a water film. Some of the water can be absorbed through drainage channels in the tread pattern. In the second adjoining zone, the so-called transition zone, the water film is broken through by the studs of the tyre tread or by the roughness of the road surface. In the third zone, in the rear part of the tyre contact patch, the water is completely displaced and there is dry road contact. The conditions here are then similar to those in the two-zone theory for friction of pneumatic tyres on dry roads described above (see Fig. 6.50).

Due to the influence of the approach zone, in whose area almost no tangential stresses occur due to the low possible friction forces, the tensioning of the tyre begins practically only in the third zone of dry road contact. At higher speeds, this start of bracing is shifted even further into the rear part of the contact zone. This results in a very complicated dependence on the maximum coefficient of friction, water height and speed for the size of the tyre follow-up n_R (Gengenbach 1968). Weber and Persch (1975) point out that the tyre behaviour discussed so far only applies to stationary tyres. In transient operation, i.e. when, for example, the steering angle of the front wheels

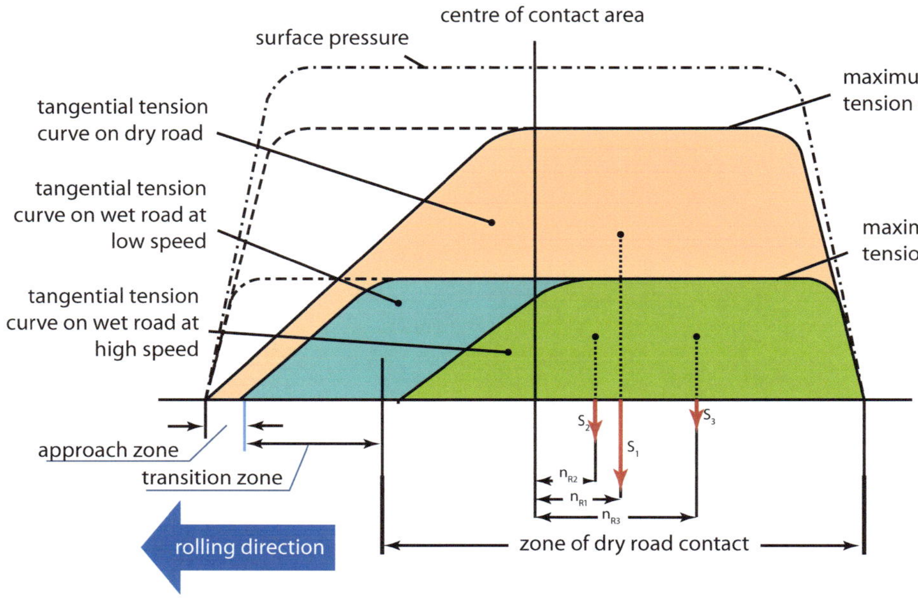

Fig. 6.50 Transversal tension curve on wet and dry road (Bubb 1975). Utilization of the coefficient of friction in dry road conditions: Restoring torque $M_1 = S_1 \cdot n_{R1}$. Utilization of friction coefficient on wet road and low speed: Restoring torque $M_2 = S_2 \cdot n_{R2}$. Utilization of the coefficient of friction on wet roads and at high speeds: Restoring torque $M_3 = S_3 \cdot n_{R3}$. $M_1 > M_3 > M_2$. Wet road and high speed may suggest better road contact than wet road and low speed before floating

is varied within a certain period of time and thus the forced skew angle of the wheels is changed over time, characteristic hysteresis-like shifts of the aforementioned correlations occur.

A restoring torque is generated at the steering wheel, which depends on the steering angle, the speed and the coefficient of friction, which is non-linearly connected to the influencing variables in several respects. Only after a great deal of experience, which also presupposes frequent driving in the so-called border area under different road conditions and road conditions, whereby the boundary to instability would also have to be crossed frequently, should it be possible to develop a feeling for this complex connection. This requirement may apply to professional rally drivers or trained test drivers in automobile factories. For most drivers, however, who can be regarded as experienced due to many years of accident-free driving experience, a much simpler experience should be stored, precisely because they normally avoid the transition to an unstable driving condition:

» If, when turning the steering wheel after initial increase, the restoring torque suddenly decreases, the road is smooth or slippery.

According to the above explanation of the occurrence of the restoring torque, this occurs when the maximum possible lateral stress is exceeded in practically all tyre contact patch. However, experience does not include any warning prior to the occurrence of this condition. The steering wheel torque does not provide sufficient information about the road condition to ensure safety.

Donges (1982) points out that although today's vehicle designs leave plenty of room for extreme manoeuvres, a maximum of 40–50% of the objectively available reserves are actually used in "normal" road traffic. These reserves are described by Kamms' circle, whereby it is assumed that the maximum force transmittable by friction is the same in the longitudinal and transverse directions. However, as Braess and Donges (2006) show, these limits are still exceeded under unfavourable road conditions (Fig. 6.51). Due to the mechanical correlations, the feedback for the

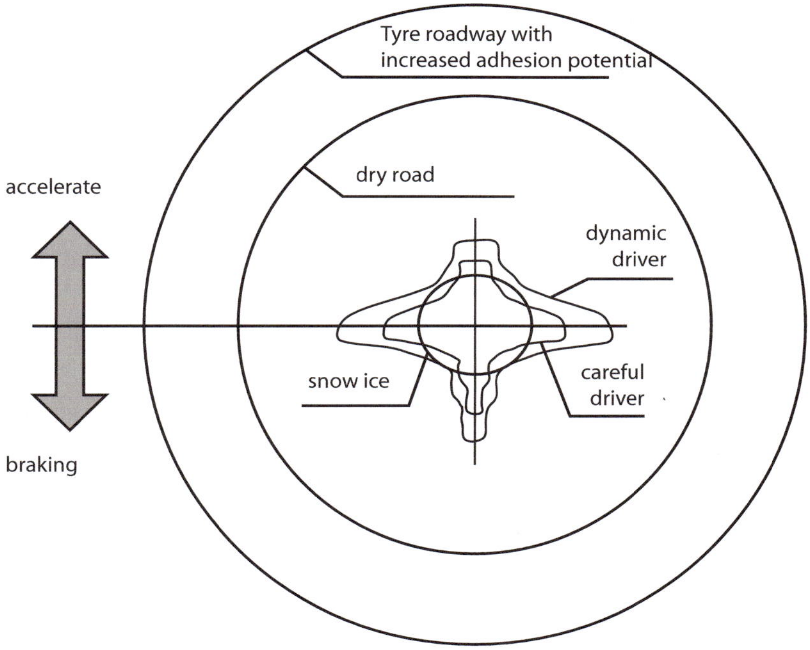

Fig. 6.51 Utilisation of the adhesion potential of normal drivers on differently adhesive surfaces on the basis of the Kamms Circle (not to scale) (Braess and Donges 2006)

driver in these extreme driving dynamic ranges is non-linear. For example, during circular travel with a constant curve radius (so-called stationary circular travel), the necessary steering wheel angle initially increases linearly for almost all vehicles with increasing lateral acceleration (here achieved by increasing speed), but then increases disproportionately at the limit. The steering wheel return torque also rises quasilinear at first, but at the limit the behaviour is then reversed in relation to the steering wheel angle (see above). The driver is thus overwhelmed by the attempt to extrapolate from the linear range of experience to the limit range.

Tyres – including those on the non-steered rear axle – must in principle run at an angle to the direction of travel when transmitting lateral forces due to the relationships described. This angle is called the skew angle. If this skew angle is the same on the front and rear axles, the so-called *neutral handling* is given. If this is larger at the front axle than at the rear axle, the vehicle must be forced practically more strongly into the curve. This is called *understeer*. Conversely, it is when the skew angles at the rear axle are larger. Then the vehicle reacts quasi "overzealously" on the steering angle. One speaks of *oversteer*.

6.4.1.2 The Steering Feel in the Narrower Sense

Today, steering systems in all companies are optimized on the basis of expert judgements by test drivers. Assessment scales have been established which provide for a categorisation of the subjective test driver assessments in a scale of 10 (1 = safety risk, 10 = outstanding). The development objective is usually described as achieved if the rating "8" is assigned. In addition to this evaluation, the test drivers make a statement on the justification for a possible deviation from an optimal evaluation. In general, assessments are carried out on the following criteria (quoted after Pfeffer 2013):

- *Steering torque level during parking* (Goal: low power level for the driver),
- *Steering torque curve during parking* (Goal: Avoid oscillating progressions over the entire steering wheel angle range),

- *Steering torque curve around the centre* (Goal: restoring forces dependent on driving speed),
- *Centricity – Centering* (Goal: complete resetting of the steering wheel for small steering wheel angle inputs when driving straight ahead),
- *Steering wheel torque curve – Linking* (Goal: moderate but steady increase in steering wheel torque as a function of lateral acceleration up to the forewarning and limit range),
- *Steering wheel torque during cornering* (Goal: constant steering wheel torque depending on driving speed and lateral acceleration during stationary cornering with a noticeable stopping point).

6

In all of this, it must be ensured that the expert judgement is sufficiently accurate to correspond to the client's judgement. An interesting contribution to this was made in an experimental observation by Wolf (2009): According to this, even the "normal driver" can describe the specific steering characteristics of vehicles in a relatively precise and differentiated manner, whereby the use of certain words does not differ significantly from the language used in the press. Of particular interest here is the fact that no difference in language usage between readers and non-readers of motor press could be observed.

Wolf has generally dealt in great detail with the steering feel from an ergonomic point of view in order to be able to make design statements that already take into account relevant human judgements in the run-up to a prototype assessment.[40] He distinguishes between the steering feel in the narrower sense (hand-arm-system – steering wheel) and in the broader sense (entire control loop driver – vehicle in connection with the stabilisation task to be performed in accordance with the requirements of the guidance task). The steering feel in the narrower sense refers to investigations which, from a purely technical point of view, are prerequisites for "good" steering.

40 The detailed compilations by Wolf (2009) can only be reproduced here in very abbreviated form.

Wolf has compiled recommendations from the literature on the course of the steering torque over the steering angle and lateral acceleration (◘ Figs. 6.52 and 6.53). As he rightly points out, one should actually quote the steering force, since the same restoring torque, depending on the steering wheel diameter, requires different forces to be applied by the driver and thus noticeable forces. From a technical point of view, however, the representation of the moment has prevailed. The information given below therefore refers to steering wheel diameters between 320 and 410 mm which are common today. In conjunction with the steering ratio and wheelbase (see also ► Eq. 2.17), this results in considerable differences. Wolf observes a turning radius between 117.80 m (Opel Speedster) and 226.87 m (Rolls-Royce Pantom) on the steering wheel rim of 50 mm during a steering travel. The average value for current passenger cars is 163 m with a standard deviation of ± 16,8 m. The small values essentially represent the sporty vehicles, the large limousines and SUVs.

Of particular importance for driving safety is the course of the ***Steering wheel torque in the limit range.*** This has already been mentioned in the comments on road contact in ► Sect. 6.4.1.1. Two diametrically opposed views are at odds here: Like Heißing and Brandl (2002), Bielaczek (1998) rates the acceptance of early soft steering as very high. In contrast, Webhofer (1991) and Bubb (1985) report that a steering torque that becomes stiffer at the limit is best assessed as an approximation of a limit. Bubb and Bolte (1987) interpret these different results as meaning that many drivers have not yet experienced the non-linearities at the limits and therefore cannot make a reliable statement about the behaviour of the steering torque there. Huang (2004) also argues that soft steering is best for most drivers because this steering torque curve corresponds to today's vehicles. A stiffening steering, on the other hand, means a barrier for the driver, which he may not cross, but against which he "fights" particularly strongly. This would mean that in a critical situation it would remain hard at this limit and possibly exceed the physical limits instead

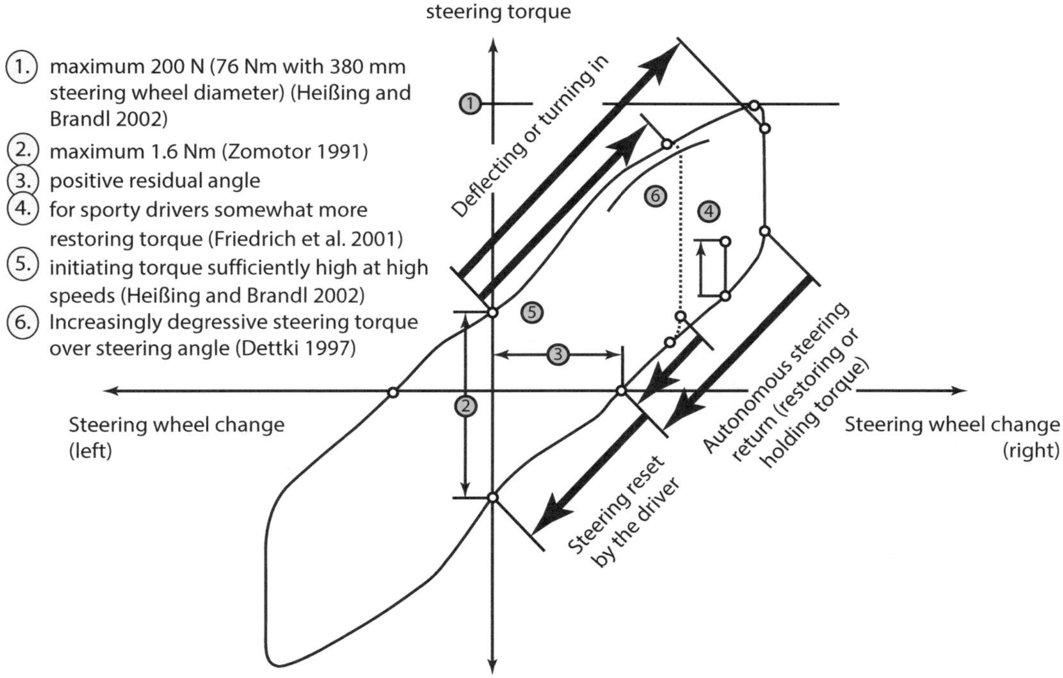

Fig. 6.52 Recommended optimum values for the course of the steering torque over the steering angle. (From Wolf 2009)

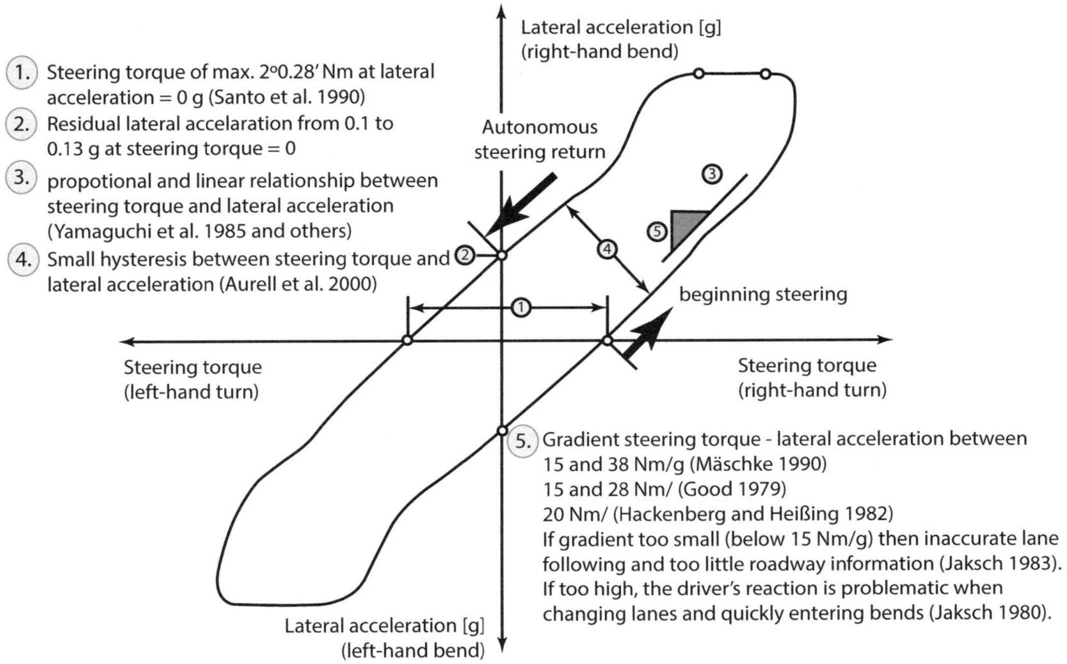

Fig. 6.53 Recommended optimum values for the lateral acceleration curve over the steering torque (quasi-stationary)

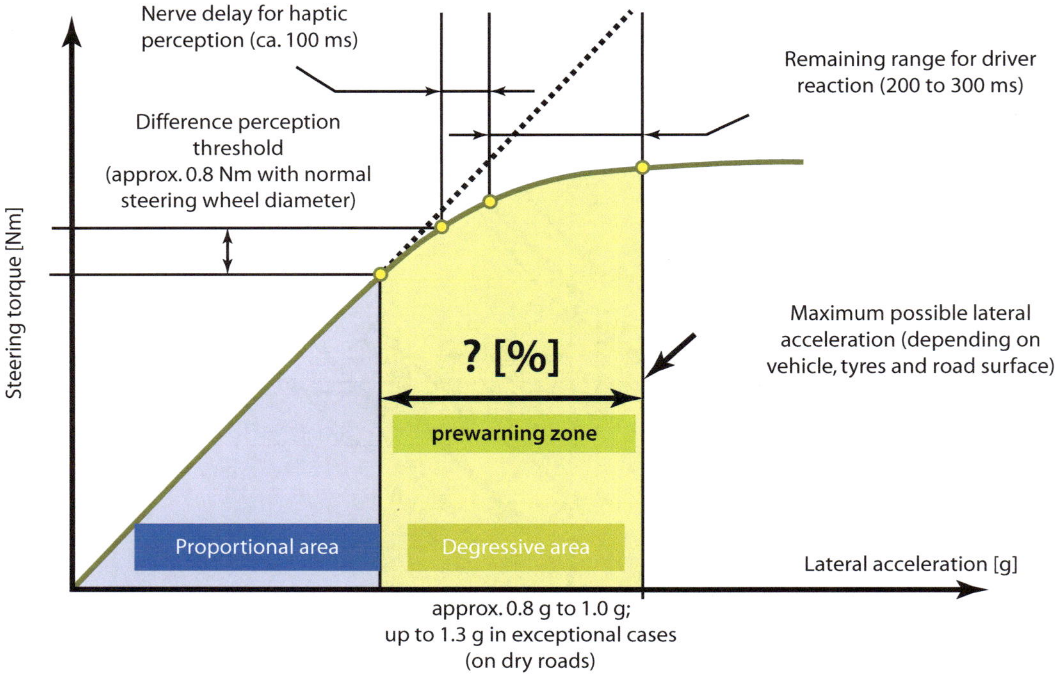

6

Fig. 6.54 Fundamental and qualitative course of the steering torque over the lateral acceleration with degressive early warning range (Wolf 2009)

of moving away from this limit. From all this, Wolf (2009) draws the conclusion that with today's design there should be a pre-warning zone in which a degressive increase in steering torque with lateral acceleration points to the impending system limit (see Fig. 6.54). If, however, in the future the approach to the limit could be measured by appropriate sensors in the tires, then according to Huang (2004) the driver should be adequately informed of this condition by suitable acoustic and optical means (especially HUD).

Viscose Damping limits the maximum possible steering angle speed and angular acceleration. However, only very few research results are available on this subject. According to Poulton (1974), viscous damping in the event of vibration effects against unintentional actuator movements by humans works better than static friction or an increase in the mass of the operating element. Rühmann (1993) shows that the combination of viscous damping and spring return (which is provided by the steering design itself) produces the best results in terms of control power.

Mechanical slackness is characterised by the fact that a change of angle within the slack at the steering wheel does not cause a change of direction of the vehicle. Mechanical slackness also deteriorates the control power (Rühmann 1993). Today's car steering systems are generally regarded as slackness-free.

Dead zone: The "steering torque dead zone" corresponds to the mechanical slackness around the zero position (Rühmann 1993). As a result, the zero position can no longer be felt exactly and thus worsens the control power (Rühmann 1993). In principle, the dead zone should be avoided in vehicle steering designs (see the comments on the center-point feeling, Fig. 6.55).

Coulomb friction has both negative and positive effects depending on type and strength (Harrer 2013). According to the findings of the ergonomics of the human-machine system, Coulomb's friction causes a deterioration of the control power. In particular, it impedes the exact dosage of the steering movement, particularly in the area around the centre position. In addition, breakaway or

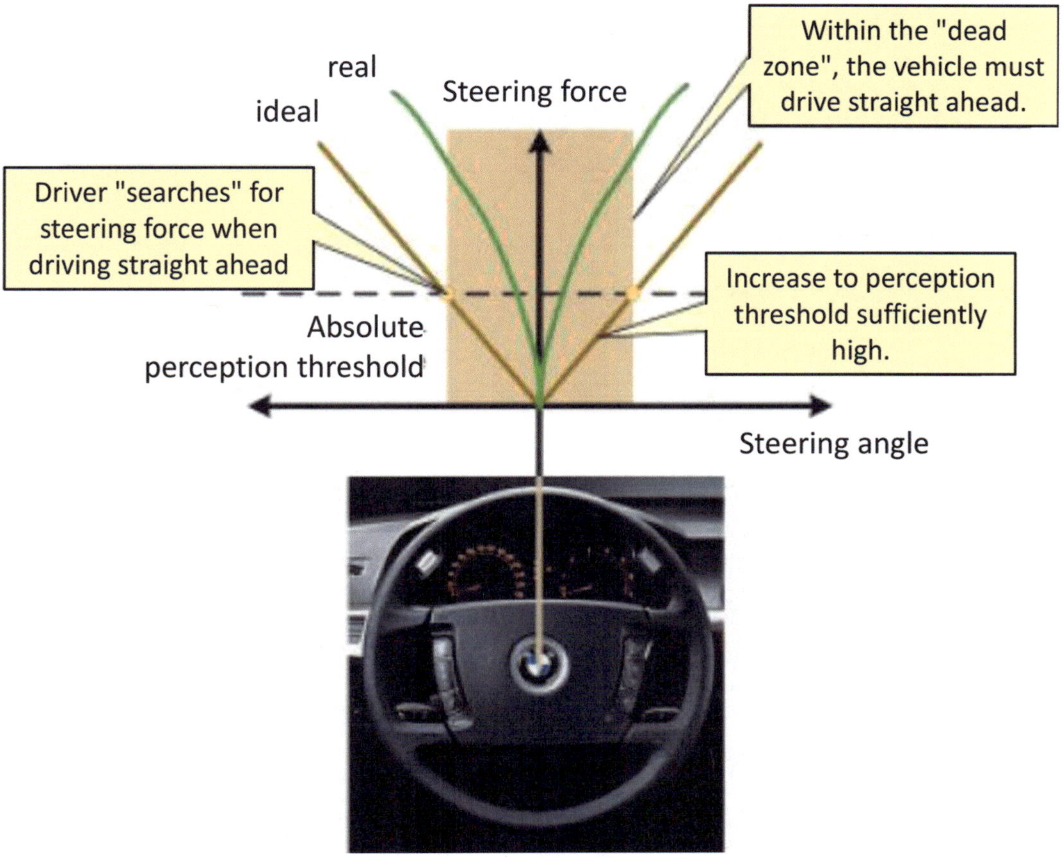

Fig. 6.55 Design of the Center-point-feeling according to Wolf's suggestions (2009)

sticking of the steering should be avoided due to excessive static friction when the steering is attached from the centre. However, friction also leads to hysteresis of the steering wheel torque over the steering angle (see also Fig. 6.52). According to Harrer (2013), this hysteresis behaviour is essential for precise cornering, as it leads to a working point with a noticeable increase in steering torque when the steering wheel angle is increased and a significant reduction in steering torque when the steering angle is reduced. In addition, a high coefficient of friction can also contribute to the suppression of disturbance information such as shocks and periodic excitations due to wheel imbalance, brake force fluctuations, etc.

According to most authors (e.g. Wolf, 2009; Harrer 2013), the so-called ***center-point-feeling***, i.e. the tactile steering center position, plays an essential role for a good steering feel. Steering around the center layer is referred to as on-center handling, whereby slight angle changes around the center layer are also to be included. The center-point-feeling is so important because the normal driver is in the on-center area most of the time (Harrer 2006) and because the driver can better fulfill his control task (Buschardt 2003). However, the fundamental problem here is that the driver lacks haptic feedback about the steering when driving straight ahead in an ideal direction, i.e., straight ahead without interference from the environment, such as uneven road surfaces or side winds. Under these circumstances, the driver is virtually forced to "search" for such feedback. The test results of the SANTOS project also indicated that a lane keeping support design was preferred, which generated a slight counter-torque already during normal driving in the subjec-

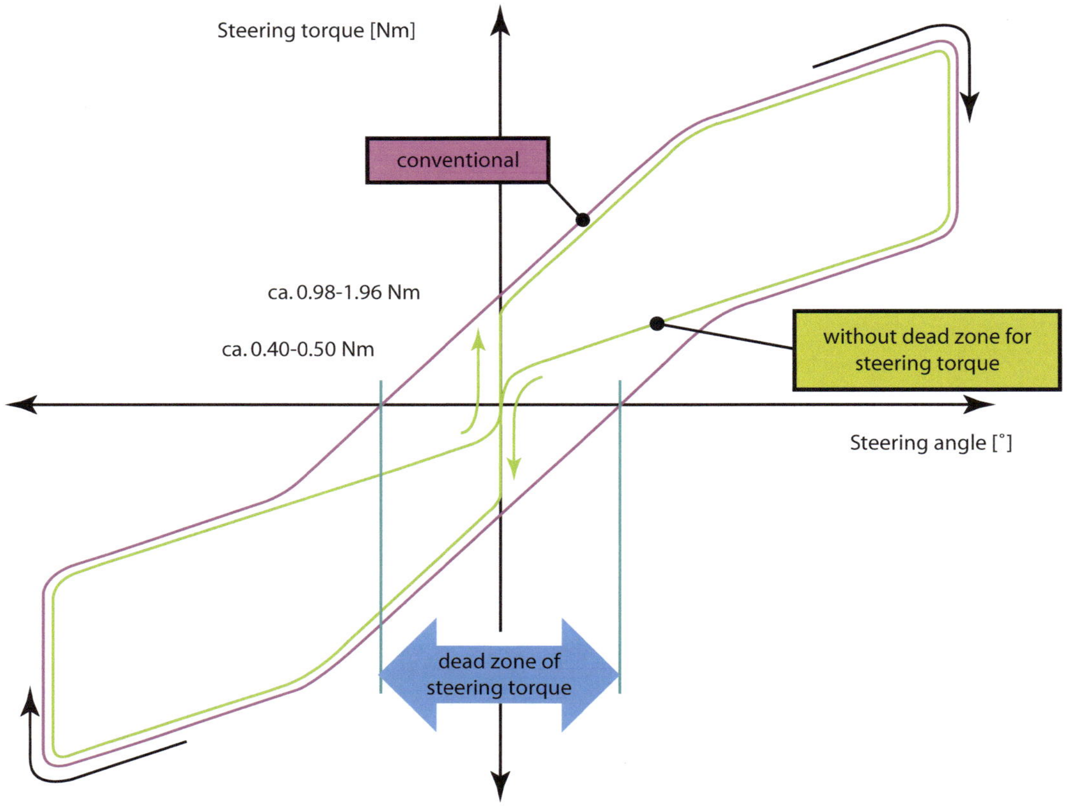

Fig. 6.56 Principle of a tactile steering center position without steering wheel torque dead zone. (From Wolf 2009)

6

tive vehicle lane centre (König et al. 2000). Wolf (2009) makes the following recommendation from these results, which he illustrates by the brown line ("ideal") in Fig. 6.57:

"The driver would look for a steering force on one side when driving straight ahead. To do this, the absolute perception threshold must first be exceeded. The 'search' of the steering force is done by changing the steering angle. The vehicle should still drive straight ahead despite a change in steering angle due to the driver's expectation that the vehicle will run straight ahead. However, this is opposed by the goal of developing a steering system that is as stiff and as free of play as possible. Consequently, the rise to a perceptible steering force must be steep enough so that the driver does not apply too much steering angle (see also Kushiro et al. 2008)".

A further proposal by Wolf (2009) provides for a situation-dependent centre-point feeling, which could compensate for the disadvantage of the dead zone of the first proposal, which now exists again. The proposal is illustrated by Fig. 6.56 (green line). The test results of Sato et al. (1990) are adopted, according to which the steering torque should be between 0.98 and 1.96 Nm for the linkage. According to Buschardt (2003), safe steering torque perception is achieved at a value of 0.8 Newton metres. Wolf doubts, however, that such an interpretation is appropriate under all circumstances. He explains that such a design – controlled by appropriate sensor technology – may make sense especially for construction site driving, severe visibility restrictions (e.g. heavy rain, fog) and night driving with low beam, while for city driving, winding country roads or even off-road driving, a conventional design is more likely to be preferred.

According to Wolf's research, the description of the interaction of the driver's hand-

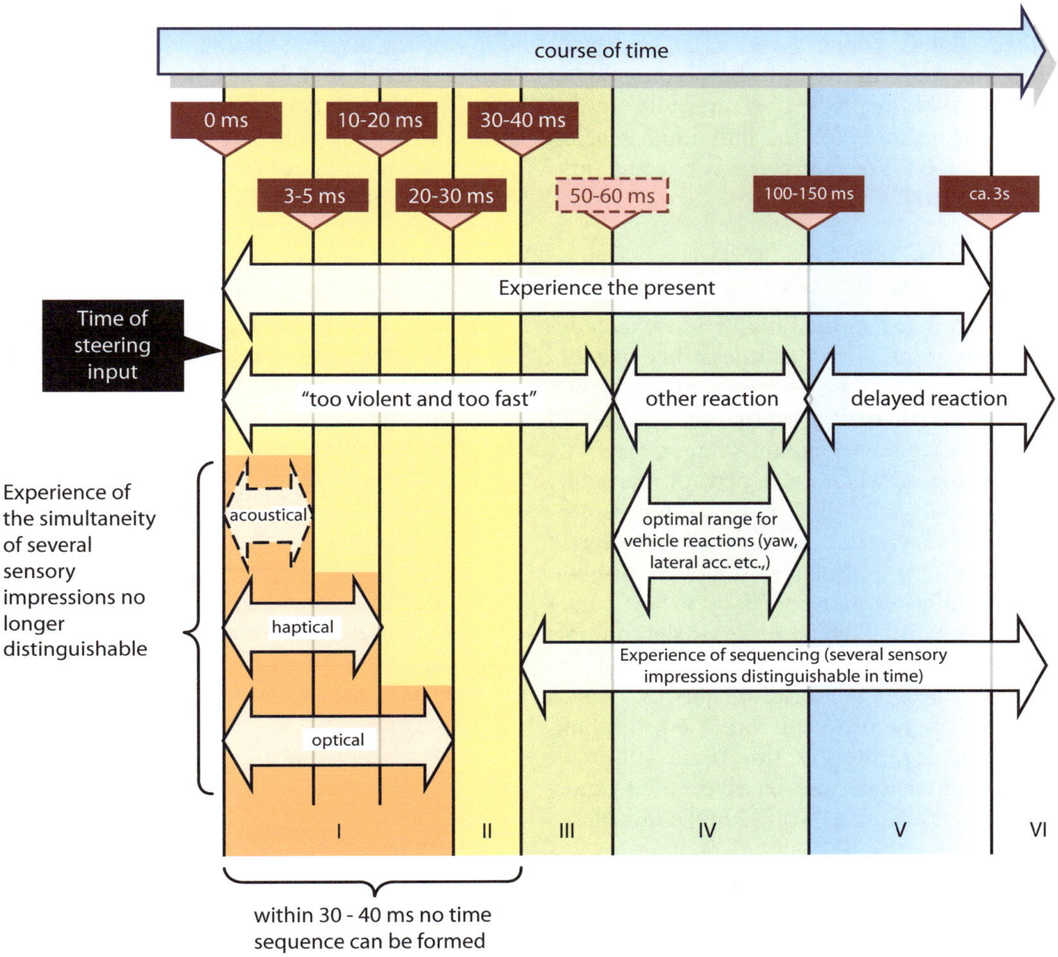

Fig. 6.57 Temporal experience of sensory impressions compiled by Wolf (2009) on the basis of publications by Vaas (2005) and Pöppel (2000)

arm system with the steering wheel using driver models is of little use for an adequate interpretation of the steering feel. In general, he states in summary that scientific work that deals with a correlation between objectively measurable parameters and subjective assessments of steering behaviour often shows only weak or no correlations at all. Stronger correlations were only found for the ***temporal aspects*** of perception. In this context, the experience of time described in ▶ Sect. 3.3.1 plays an essential role. On the basis of the works of Pöppel (2000) and Vaas (2005), Wolf has made a compilation of the impressions on the various sensory organs during steering, which is reproduced in Fig. 6.57. After that, it is practically impossible to distinguish between time-shifted signals or to perceive their sequence (zone I and II) in a time window of 30 to 40 ms. Some studies on vehicle handling also suggest that there is a lower limit of 40 to 60 ms for the vehicle reaction (Fujinami et al. 1995) below which the reaction would be perceived as "too violent and fast" (time zone III). Reactions in the range between 60 and 150 ms can be assigned chronologically correctly, but are so close to the input that they are experienced as "immediate". They are the ideal area for experiencing the reaction of yaw angle and lateral acceleration to a steering wheel movement (zone IV). Everything that is clearly above 150 ms would be experienced as "strongly delayed" to "no longer related" (Zone V and VI). The prerequisite for such reactions to be perceived is in any case that the perception thresholds or difference thresholds for the individual sensory perceptions are exceeded (see also ► Sect. 3.2.1).

6.4.1.3 The Steering Feel in the Broader Sense

The steering feel in the broader sense includes the driving task. All aspects described in detail in ► Chap. 3 can be considered. In particular, the importance of the internal models should be emphasized once again. A "good" steering feel is achieved when the expectation given by inner models is fulfilled by the actual perception. However, the orthogonality between "comfort = pleasure" and "discomfort = suffering"mentioned in ► Sect. 3.3.4 also comes into play. By observing ergonomic rules, largely unpleasant sensations on the steering wheel (e.g. vibration, shocks, uneven build-up of the restoring force) can be eliminated, but regardless of this, users will prefer different interpretations based on their expectations. For example, Wolf (2009) conducted a road test to find out what "normal drivers" think of the steering characteristics of different vehicles, how they assess steering behaviour and whether there are significant differences to so-called test drivers. The following results can be summarised briefly:

- Due to the external appearance, certain driving and steering characteristics are assigned to the vehicles (e.g. a roadster triggers the expectation of particularly precise, direct steering).
- It was possible to create a "map" showing the influence of technical properties on the subjective assessments of the test persons (◘ Fig. 6.58).
- Criteria such as directional stability, correction effort, responsiveness and targeting accuracy show a high degree of agreement between customer and expert judgements; they are acceptable for the criteria impact, resetting behaviour, steering force progression and steering force level; very poor is the agreement on centre feel and feedback.

The internal models in connection with the human time constants (see above) as well as

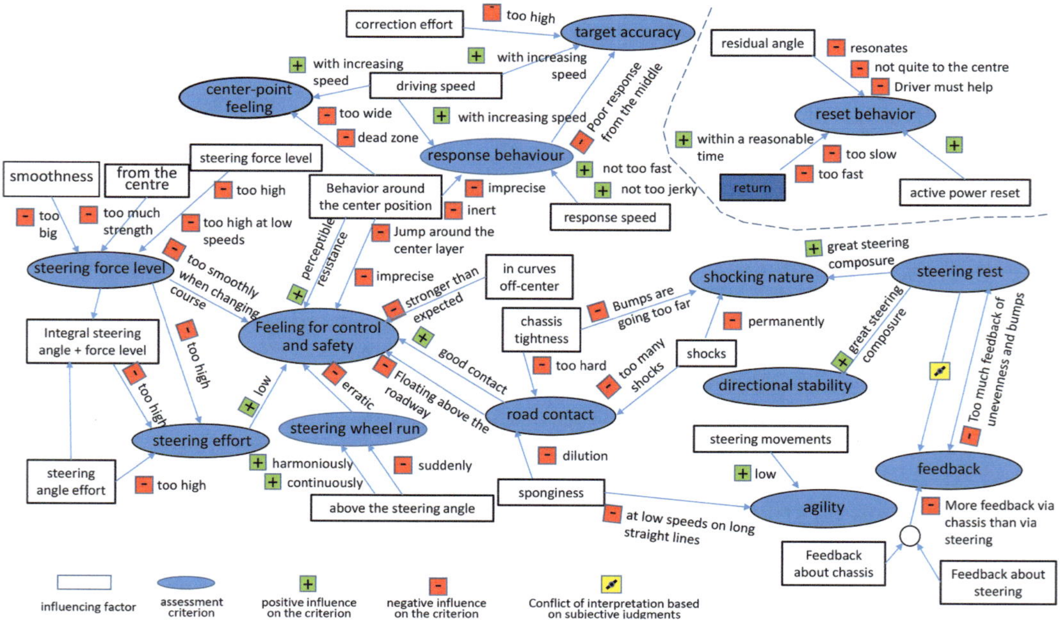

◘ **Fig. 6.58** Influences on steering feel based on defined criteria and customer judgements (Wolf 2009; only valid for vehicles with rear-wheel drive)

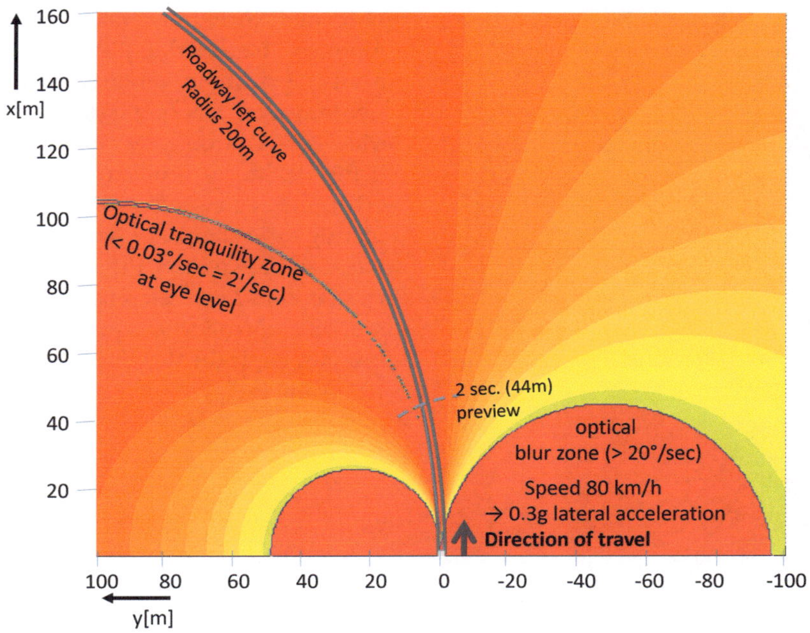

Fig. 6.59 Optical flow in the driver's eye at a speed of 80 km/h and a left turn with a radius of 200 m (lateral acceleration: 0,3 g)

various control engineering approaches provide a deeper insight into the development of steering feeling in the broader sense. The image shift vectors from the driver's point of view are of particular importance here when cornering, as already mentioned in ► Sect. 3.2.2.2.

Remlinger (2013) has calculated the dynamic perception limits for cornering. He took into account that the human eye is no longer able to detect displacement vectors below 2′/s as "moved" and that displacement vectors above 20°/s lead to a blurred, blurred image impression. Figure 6.59 reproduces the optical flow perceived by the driver in plan view. The position of the vehicle is located at x = y = 0 The calculation shows a curved line of optical rest to the left of the left-hand bend serving as an example. The increasing optical flow velocity is represented by the false color representation starting from the zone of rest. It increases (represented by the transition from red to yellow to green) to the range in which the flow velocity is too high for sharp vision. This area to the left and right of the vehicle is shown in dark red. In comparison to a straight line, where the center of the optical flow and the intersection of the road boundaries coincide, there is a clear offset to the left between the perceived orientation of the orbital curve and the center of the optical flow.

Figure 6.60 shows a corresponding representation for speeds of 120 and 150 km/h. This makes it clear that the offset becomes larger with increasing lateral acceleration. In addition, the figure shows how a "tunnel view" is created with increasing speed, as the zone of blur to the left and right of the driver becomes larger and larger. With regard to the steering feel, the illustrations can explain which stimuli the driver can use to differentiate between ***understeering, neutral and oversteering vehicle behaviour***. For at the same speed and curve radius, i.e. the same lateral acceleration, these differences cannot be perceived via the kinesthetic sensation (despite the repeatedly asserted existence of the "popometer"). Rather, this is only possible via optical perception, which is of course coupled with kinesthetic and haptic sensation: with an oversteering vehicle, the perceived vehicle's longitudinal axis is more aligned with the area of optical rest, whereas with an understeering

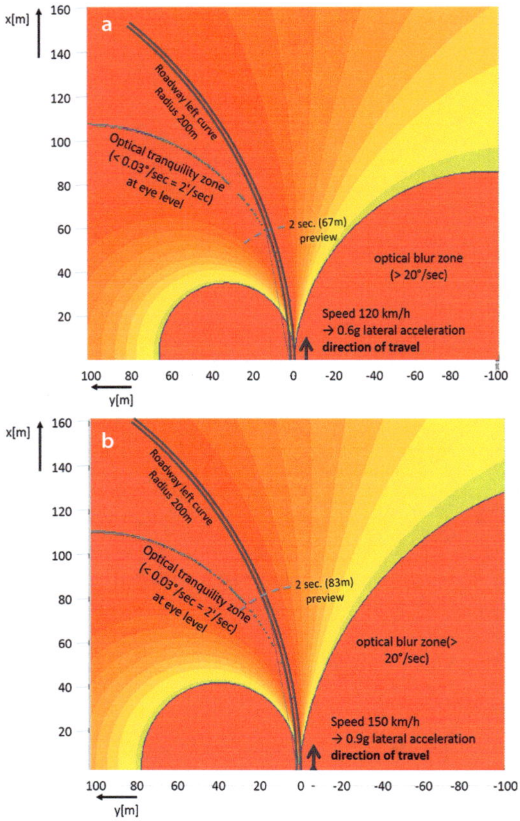

Fig. 6.60 Optical flow in the driver's eye at a speed of 120 km/h **a** and 150 km/h **b** each in a left-hand bend with a radius of 200 m (lateral acceleration: 0.9 and 0.6 g)

vehicle, this is oriented more in the direction of the perceived curve.[41]

A further parameter that has a significant influence on the steering feel is the ***steering ratio***. As already mentioned under ► Sect. 6.4.1.2, the distance required on the steering wheel rim in relation to the radius of the curve set by this plays a perceptible role for the driver. For sporty vehicles, a much smaller value (so-called direct steering) is envisaged than for luxurious comfort limousines (indirect steering). As already explained in ► Sect. 2.4.1, a speed-dependent steering ratio, which provides a direct ratio at low speed and an indirect ratio at higher speed, improves the steering feel considerably. By means of various variants of the so-called superimposed steering system, this can be achieved electronically controlled without further technical problems (Reuter and Saal 2013). From an ergonomic point of view, the steering ratio should be selected in such a way that – as shown in ► Fig. 2.20 – the steering wheel is aimed at a reference point at 2-second intervals. Remlinger (2013) calculates the foresight angle for this purpose δ according to the formula:

$$\partial = 90^\circ \cdot \frac{a_y \cdot t_v}{\pi \cdot v}$$

with:

a_y = lateral acceleration, t_v = estimated time, v = velocity [m/s].

The corresponding steering wheel position and thus the exact steering ratio depends on the anthropometric position of the steering wheel and driver in the given vehicle. Since all of this is a guideline (the time of expectation from 2 s is only a guideline!), the calculation can be based on the eye position of a medium-sized driver. Provided that the normal driver does not select lateral accelerations greater than 0.3 g, a similar effect can be achieved – with restrictions – if a non-linear steering ratio is selected which, according to these considerations, becomes indirect in the middle range and increasingly direct to the edge ranges (Hackenberg and Heißing 1982; see also Friedrich et al. 2001). In conjunction with the display of the driving hose in the HUD (see ► Sect. 6.3.1.2), this would result in a completely new and ultimately simpler driving task, which transfers aspects that are already present today in the steering wheel-dependent movement tracks faded into the rear view camera to the entire driving process.

An electronically controllable superimposed steering system, as mentioned first, has the advantage that various assistance functions can be implemented. Since the driver for low speeds (< 40 km/h) obviously has the inner model of a curve radius determined by the steering wheel position and for higher speeds the simple inner model of targeting a target

41 A similar optical effect also occurs on a straight track with the so-called "dachshund gait", which is caused by a warped chassis.

point, one can correspond to the speed-dependent steering ratio influenced in a corresponding way. Due to the so-called length servo mechanism (▶ Sect. 3.2.3), the driver exerts a force on the steering wheel in such a way that the desired steering wheel position is achieved by the corresponding inner model. The superimposed steering is therefore to be realized in the sense of an active control element. This ensures that any external disturbance (e.g. side wind, inclined road) is felt, but the steering wheel angle once selected is maintained and the correct compensating torque is transferred to the steering system. In addition, a desired vehicle characteristic (neutral, over- or understeering at the limit) can be programmed. In principle, also the approach to the border area is representable (probably according to previous experience by a decreasing steering momentum). However, there is still work to be done to find out which parameters can reliably predict this approximation. A superimposed steering system designed in this way offers the advantage of providing the driver with information on other assistance functions such as Lane Departure Warning, Lane Keeping Assist and Lane Change Assistant in an ergonomically compatible manner.

6.4.2 Longitudinal Dynamics

To date, there have been no explicit studies on the experience of longitudinal vehicle control comparable to that of steering feel. Eberl (2014) identified various experience dimensions of longitudinal vehicle guidance based on a literature search, which he concretized and operationalized on the basis of an expert workshop and two explorative test subjects. According to this, in addition to the familiar experience dimensions of safety, discomfort and comfort, a newly defined dimension, namely the feeling of energy, can also be described.

- Regarding the local, i.e. subjectively experienced ***Safety*** (Jürgensohn 1997), the feeling of controlling a situation plays an essential role. According to Haider (1977), this is described by the hierarchically interdependent levels of transparency, predictability and influenceability (Grote 1997). In particular, the influenceability is described by the dosability, i.e., the precision with which a change in the vehicle reaction can be set, and the directness – that is the degree of temporal influence with which a change in the vehicle speed is brought about.
- The ***discomfort*** characterizes the physical and mental strain associated with the control of longitudinal dynamics. The physical strain is essentially determined by the necessary change between driving and brake pedal and by the anthropometric foot posture. The psychological strain arises from the cognitive stress of selecting the current need for action from the traffic environment.
- The ***comfort*** is, on the other hand, connected with the positive emotional experience, which is characterized by an overall evaluation of the experience triggered by the driving activity as well as a state subjectively characterized by the absence of excitement, tension, nervousness or by the presence of relaxation, calmness and inner peace (Petermann 2009). In addition, there is the hedonistic aspect of distinguishing oneself from others through the vehicle or driving behaviour, or of outdoing others (Dick 2002). These hedonistic attributes are described in more detail by the aspects of competence as the state of the human being capable of feeling capable of an activity (Hassenzahl et al. 2009), stimulation, which represents the stimulation acting on the driver by the characteristics of the vehicle and autonomy as the need to be able to decide and execute things freely and self-determined (Diefenbach and Hassenzahl 2010).
- Especially in connection with rising fuel prices but also especially in connection with electromobility, which will become more important in the future, the aspect of ***feeling of energy*** plays an essential role. By

means of corresponding displays[42] this feeling is supported by the experienced extent of energy recovery through recuperation and the estimation of influencing energy consumption through the driving action.

The two control elements which essentially influence the longitudinal dynamics of the vehicle, namely the accelerator pedal and the brake pedal (apart from the control elements "clutch" and "gear lever", which only operate the "technology" of the vehicle), differ considerably from the steering wheel which influences the lateral dynamics in that, due to the technology used, they do not give the driver any feedback about the success of the initiated action with regard to the vehicle movement. They only provide more or less precise information (see below) about the position of the pedal and the associated restoring force.

6

6.4.2.1 Acceleration Behaviour

The design of the restoring forces for the ***accelerator pedal*** is comparatively simple, as it only depends on the strength of the return spring. In their investigations, Wang et al. (1996) dealt with the subjective perception of forces on the basis of the evaluation according to the Borg scale (Borg 1982). For short-term actuation for "very strong" forces, these are at 48 N, which, according to the Borg scale, lies between the designation "weak" and "moderate". However, only 15% of this force, i.e. 7.2 N, can be expected for permanent operation. If one considers, however, that according to the above-mentioned survey the average resting force of the foot is 20 N, it seems acceptable to set a value of approx. 27 N for continuous operation at 2/3 of the accelerator pedal position. This also corresponds to the value realised in most motor vehicles. This also corresponds well with the guideline values of the HdE (Handbuch der Ergonomie, Schmidtke und Rühmann 1989), which consider it unacceptable for women to exceed 26 N in the case of safety-critical pedal operation.

In conjunction with the so-called *electronic accelerator pedal*[43] there is also the possibility to design the connection between accelerator pedal position and throttle valve non-linearly and also relatively freely depending on other parameters (e.g. engine speed, load condition). This opens up the opportunity to virtually linearise the non-linear relationship between throttle valve position or injection quantity, engine speed and torque for the driver's perception. This gives him the feeling that the engine seems to be more "on the throttle". Different tactics can be used: On the one hand, one can try to achieve as linear a relationship as possible between the position of the accelerator pedal and the torque, whereby one refers to the speed with the respective torque peak. From an ergonomic point of view, this design is preferable because it meets the basic human assumption of a linear relationship between actuator movement and the effect achieved. However, a non-linear characteristic can also be preferred in order to conceal a possible weakness in the engine's acceleration. The engine of the vehicle then appears more "lively" than it is at the expense of consumption. By shifting to an "eco-characteristic curve", only a low torque is called up even with a large accelerator pedal position, which is intended to push the driver towards an economical driving style.

More recent developments go one step further: they lead to the so-called *"active accelerator pedal"*. In extreme cases, the restoring force is produced completely synthetically. This makes it possible to report additional driving dynamics information via the accelerator pedal. An application of the active accel-

42 It should be pointed out once again that such advertisements can only be considered useful in a situation that is not very demanding from the traffic point of view.

43 With the so-called electronic accelerator pedal, the position of the accelerator pedal is converted into an electrical quantity via a potentiometer or a technology with a corresponding effect. The motor is influenced (for example via the throttle valve) by means of a servomotor. Therefore, it is in principle possible to design the connection between the accelerator pedal position and the servomotor position by means of a freely definable characteristic curve.

erator pedal can be, for example, to convey strategies for consumption-optimized driving to the vehicle driver according to the situation (also taking into account differences in altitude, influenced by the information of the navigation computer, Samper and Kuhn 2001). As corresponding tests show, such displays can lead to a reduction in fuel consumption of approx. 10%, but only if the driver is willing to follow the recommendations, which prescribe a rather quiet driving style.

In any case, the driver expects a corresponding feedback of the vehicle movements (acceleration or deceleration) with a change of the accelerator pedal position. In principle, the same time considerations apply that are also important for the steering characteristics and are shown in ◘ Fig. 6.57: i.e. a noticeable reaction between 50 and 150 ms would be perceived as "immediate", anything that lasts longer as "delayed". For kinesthetic perception, it is decisive that at least the stimulus threshold for translation acceleration must be exceeded. Heißing et al. (2000) specify a range from 0.02 to 0.8 m/s^2 as the perception threshold. Rockwell and Snider (1965) were able to determine a difference threshold of 0.15 m/s^2 in experiments within the framework of various real vehicle field studies. Müller et al. (2013, 2014) found a 50% threshold of 0.1 m/s^2 in a double Staircase experiment, which remains constant even under different test conditions (speed, engine noise, driver stress). From the point of view of sensory physiology, it is quite possible to conceal a delayed response of the drive unit by means of a correspondingly early acoustic feedback. However, the delay between the acoustic swelling of the engine noise and the perceptible kinesthetic sensation must not exceed 200 ms. The driver also tolerates little if the change in acoustically perceived frequencies (synonymous with engine speed) is not sufficiently synchronous with the increase in speed (= kinaesthetically perceived acceleration), as was the case with automatic transmissions with high torque converter slip and the earlier designs of continuously variable transmissions (so-called wrap-around transmissions). Here, too, the tolerance threshold should be <200 ms for a time delay between the perceptible changes of these two stimuli.

In addition to the response behaviour, the absolute level of the longitudinal acceleration possible due to the drive also plays an important role. If, in the sense of the control loop (► Figs. 2.1 and 2.2), the vehicle is understood as an amplifier of the driver's will, every movement of the control unit (= accelerator pedal) should be converted immediately and without time delay into a corresponding output signal (= speed). For control reasons, however, this is only possible with an infinitely high power of the transmission element (= vehicle).[44] However, a natural limit for this performance requirement is that under normal circumstances the coefficient of friction at the wheels hardly exceeds the value "1". This limits the maximum accelerations and decelerations to 1 g. The time often used as a performance indicator, which elapses before the vehicle reaches a speed of 100 km/h, would be approximately 3 s (these are values that are already reached or even undercut by extremely powerful sports cars today!). It is obvious that this value still lies within the experience of the present. Even the values between 8 and 13 s possible today with most vehicles are still in the range of the "present of the past", i.e. they are experienced very directly. In contrast, historical values (e.g. 36 s for the VW 1200 of 1956) are clearly outside this range. The behaviour of these vehicles was therefore experienced as "agonizingly slow". The constant increase in the engine power of the vehicles observed in the historical development has a reason here. For the perception of the performance behaviour of the vehicle while driving, however, the times that are needed to accelerate e.g. during overtaking (from 80 to 120 km/h) play a preferred role. These times should also be at least in the range of the "present of the future" or the "present of the

44 It should be noted here that for an acoustic hi-fi system, the high power is not necessary to produce the highest possible volume, but – for the same reason – to ensure that the acoustic output signal follows the specifications of the sound carrier as closely as possible.

past" (i.e. in the range of 10–15 s) in order to enable a safe assessment of an overtaking process at least under this aspect.

Since humans assume a linear transmission behaviour of the controlled system, the motor characteristic should be described by a torque curve that is independent of the speed. Today's supercharged combustion engines meet this demand for a wide speed range (often however bought by a delayed response), not to mention the qualities of the electric drive, which largely fulfils this demand ideally.

In general, the ***electric drive*** completely new challenges in designing the characteristics of the accelerator pedal. If in previous hybrid vehicles only a slight recuperation (recovery) of the kinetic energy takes place in the deceleration phase in the form of charging of the starter batteries, a complete electric drive offers the possibility not only of designing the propulsion but also the braking electrically and thus, reduced only by the efficiency of the drive motor and the charge losses of the battery, of recovering a large percentage of kinetic energy during deceleration. There are basically two conceptual approaches to integrate the recuperation function into the longitudinal dynamics interaction. On the one hand, the control of the recuperation can be combined with the mechanical brake (also present in the electric vehicle) in the conventional brake pedal. On the other hand, the realization of a one-pedal interaction – the recuperation function is integrated into the accelerator pedal – offers the possibility to control the acceleration and deceleration of the vehicle without moving the foot solely by the position of the accelerator pedal (so-called *one-pedal drive*). Since no deceleration greater than 0.3 g occurs during normal driving, the power of the drive motor(s) is (are) sufficient for a braking process within this range. Now the question is: Should the deceleration up to this size already take place by completely depressing the accelerator pedal? This would result in a completely new driving experience, as the position of the accelerator pedal would now, after a certain delay time, determine the speed of the vehicle almost directly (by means of the active accelerator pedal and a corresponding electronic control, this could even be done completely correctly in the sense of an active control element). Or, in analogy to the previous driver-vehicle interaction, by taking back the accelerator pedal one should leave the vehicle completely without any drive (so-called sailing), whereby – if the driver incorporates this into his driving style in a sensible way – a significant reduction in energy consumption would be achieved. In this case, recuperative braking would only take place via the brake pedal. If a particular pressure point is exceeded, one could then signal that a braking process has been initiated, which converts kinetic energy into thermal energy.

Field tests for handling a one-pedal operating concept (one-pedal drive), such as used in the MINI-E, have shown that this form of longitudinal dynamic interaction becomes accustomed within minutes and that experienced drivers are not at a disadvantage compared to conventional combustion engines with regard to efficient driving. The single-pedal operating concept thus enables both a dynamic-sport driving style and a uniformly efficient driving style and is therefore recommended as an ideal form of longitudinal dynamic interaction for electric vehicles. This interaction concept is also technically less complex due to the separation of the mechanical brake pedal and the electronic control unit for controlling the electric machine. A first detailed analysis was presented by Eberl (2014). It was carried out in conjunction with a Mini converted to electric drive. However, this study could only be carried out with authorised employees of the company – i.e. rather with test persons with an affinity for technology – as the test vehicle at that time did not yet have general road approval. Nevertheless, one can assume that the results described in the following have a certain general validity. In accordance with reports from journalists who had initial contact with electric vehicles, Eberl also generally rates the spontaneous behaviour of an electric vehicle in the acceleration phase as positive. The delay behaviour is of particular interest. He has defined four different levels for this purpose, implemented them in terms of experiments and had them evaluated by his test persons. The results can be summarised as follows:

1. Sailing is characterized by an extremely low deceleration, which is caused solely by the driving resistances (tyres, air resistance, internal friction losses, $\approx -0{,}1$ m/s^2). It is primarily experienced as a loss of control of the vehicle's longitudinal control, which leads to a high level of discomfort due to the need for frequent pedal changes. Even with a pronounced foresight, it is hardly possible to adjust correctly to traffic situations only by releasing the accelerator pedal due to the low deceleration.
2. The drag torque of the combustion engine (so-called engine brake) corresponds to a deceleration of approx. -0.8 m/s^2. It is largely neutrally evaluated in the various experience dimensions, since it corresponds to the usual control of the vehicle's longitudinal control, even with a neutrally experienced feeling of energy. Because of this habitual behaviour, however, it is also experienced as boring with little stimulation and little potential to expand driver competence.
3. A so-called experiential drag torque was generated with a delay of $-1{,}5$ m/s^2. In many traffic situations, it enables a sufficient delay to be triggered simply by releasing the accelerator pedal (one-pedal operation). The test persons evaluate this with a significantly increased control and a positive experience of energy recovery. The necessary foresight in highly dynamic driving situations and the resulting comparison of the delay when releasing the accelerator pedal with the conditions given by the traffic situation is described as demanding. On the other hand, one-pedal operation is experienced as stimulating.
4. The high drag torque caused a delay of $-2{,}3$ m/s^2 and thus corresponds to a value that would still be perceived as comfortable braking during normal driving in conventional vehicles. This design is subjectively experienced as particularly safe, as it allows symmetrical interaction with the vehicle via the accelerator pedal by triggering similarly pronounced accelerations and decelerations. However, the necessary precise pedal interaction is also perceived as psychologically demanding. The basic concept of high drag torque offers the highest potential for the development of a longitudinal vehicle guidance experience and thus a pronounced stimulation and extension of driver competence through single-pedal operation.

In an overall view, Eberl assumes that one has to provide a choice between the experiencable and the high towing moment in order to do justice to individually different needs but also to different traffic conditions. From a subjective point of view, the negative performance of sailing, which today is also seen in connection with conventional propulsion systems as a way of reducing energy consumption, obviously does not meet the technically justified expectations.

6.4.2.2 Delay Behaviour

In principle, the following applies for the delay behavior and thus for the properties of the ***brake pedals.*** The same requirements as for the accelerator pedal, especially with regard to the response behaviour. Due to the technical realization, the degree of deceleration via the brake pedal is actually only determined by the pressure on the pedal. This effect was realised in the brake actuation of the Citroën DS and the later Citroën CX by actuating a ball-shaped rubber element directly connected to the hydraulic brake system with the foot – without any pedal travel. This brake actuation, however, was so different from the systems implemented in the other vehicles that this led to a large rejection on the part of the customers. Obviously, due to the path servo mechanism (see ▶ Sect. 3.2.3), man needs the feeling of "applying a force in such a way that a desired path is achieved". This effect is achieved more or less specifically by elasticity in the brake system.

▫ Figure 6.61 shows the measured force-displacement curve of a conventional servo brake for different brake actuations. Due to the mechanical and hydraulic properties of the system, this is always characterised by a strong hysteresis, regardless of the type of actuation. In addition, this characteristic

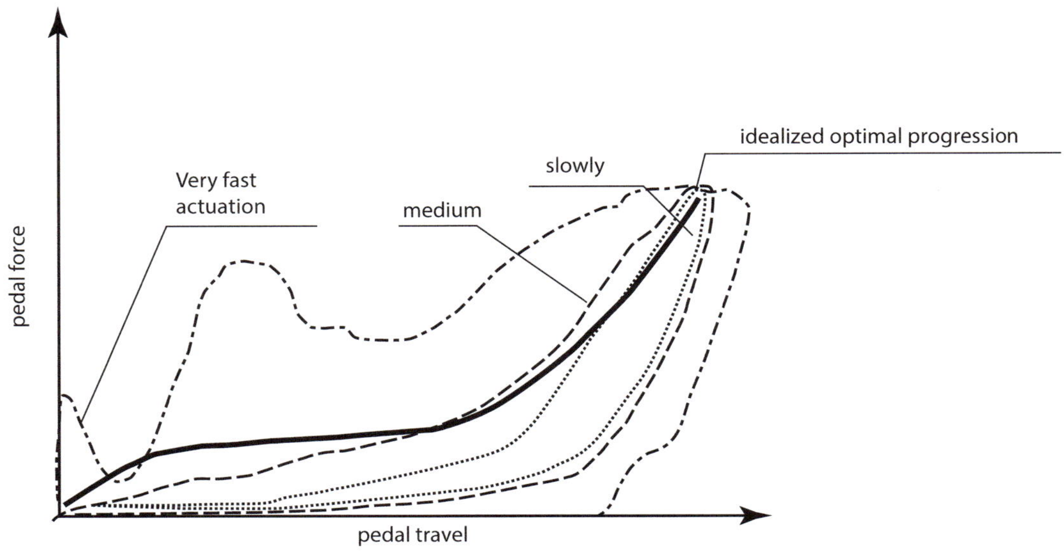

Fig. 6.61 Travel-force curve on the brake pedal of a modern passenger car with servo brake

changes considerably depending on the actuating speed. In view of the fundamental human expectation of linear behaviour already mentioned, this is of course a disadvantage. In recent times, there has therefore also been discussion about how to create the braking sensation by introducing an electric brake. The hysteresis characteristic can be compensated with the aid of electric "bywire brakes"with electronic control. Figure 6.61 also shows an electronically implemented optimum curve, which shows neither a hysteresis nor a dependence on the actuating speed. Its progressive course can be traced back to works by Göktan (1987), in which this characteristic was found to be the best compromise in simulated follow-up tests and downhill drives. The result is that the driver expects a response path that provides only low restoring forces for light to medium deceleration (plateau in the middle pedal travel range). This is followed by a quasi linear force-displacement curve which is effective for high deceleration values (> 0,4 g). However, as Bill et al. (1999) emphasize, the feel of the pedal also depends considerably on the type of vehicle (e.g. car or van). There is also a need for further research into this and in particular in connection with the future electric drive.

6.4.3 X- by-Wire

X-by-Wire generally means that the mechanical connection between control element and executive organ is completely replaced by an electrical/electronic connection. This always requires a sensor on the control element that detects the driver's operating request and an actuator on the executing organ. The case of the electronic accelerator pedal represents such a variant of the X-by-Wire (here with "*drive-by-wire*"). As mentioned several times, a haptic feedback is necessary for the driver at least during the operating procedure. In the case of the electronic accelerator pedal, this is done mechanically by a spring. In more elaborated cases, however, this feedback should also provide information about the desired effect. For this reason, an actuator is also required on the control element. With the active accelerator pedal, this feedback is provided by a torque motor, which can generate restoring forces controlled by electronics within certain limits.

Such x-by-wire technologies have also been discussed for some time for steering, as they would offer advantages especially in connection with driver assistance systems (e.g.: controllable functionality, better prerequisites for passive safety, variant reduction, simpler axle geometry, specifically designed feedback, design; quoted after Pruckner 2013). This kind of steering is called *"Steer-by-Wire"*. Steer-by-wire, for example, requires a torque sensor on the steering wheel to record the driver's wishes and a servomotor to set the steering wheel angle calculated by the electronics. The reaction of the vehicle is then measured either directly on the steering track rod or globally by acceleration sensors on the vehicle and reported back to the driver via a torque motor on the steering wheel. Siegel and Bundorf (1966) already presented a test vehicle in which the steering column was completely separated mechanically from a servo device for adjusting the track rod. A torque motor artificially provided a restoring force for the driver. This experimental arrangement had already anticipated modern ideas of "steer-by-wire". The steering wheel position had a direct influence on the servo device for the track rod. On the torque motor, which simulates the restoring force, however, information on roll angle and roll angle change, yaw angle, yaw angle change, lateral acceleration, vehicle speed and restoring forces in the tie rod are given with adjustable weight. The aim was to synthesize an optimal steering feel experimentally. With today's expanded possibilities of "steer-by-wire"and the prospect of a legal change that will make it possible to move away from the mechanically rigid connection of steering wheel and tie rod, tests of the type described are being carried out in all vehicle companies. Friedrich et al. (2001) reported about a VW bus equipped with such a steering system. Experimental drivers could not see the difference to a conventional steering system if the system was adjusted accordingly.

As already indicated, the "by-wire control" offers the driver new possibilities to allow an intuitive and thus more reliable operation of the vehicle than is possible with the conventional concept, which ultimately came about due to the historical-technical development of the vehicle and not due to fundamental considerations of operation. The argument is: With the possibilities of modern microelectronics in combination with the sophisticated potential of mechanics (keyword: *mechatronics*), almost any form of operation can be technically realized. Therefore, it is now possible to design this operation from the point of view of the physiological and psychological characteristics of the human being. In particular, this opens up the opportunity to fulfil an ergonomic requirement, namely to accomplish the two-dimensional task of driving a car using a two-dimensional control element. The joystick is the most suitable control element for this. As early as 1958, General Motors presented a test vehicle (Chevrilet Impala) that had such a control element for steering the vehicle ("Unicontrol", ◘ Fig. 6.62).[45] Only with the possibilities of mechatronics and the realization of this operating element as an "active control element" (see ◘ Fig. 6.30) could this type of motion control become interesting for the vehicle. Experiments in this direction were first carried out by Bolte (1991) on the simulator. Eckstein (2001) installed the "Active Control Element" for the first time in various real vehicles (Mercedes-Benz 200, 500 SE and 500 SL) and Friedrich et al. (2001) report on the construction of similar test vehicles at VW and Audi. In all cases, it is preferable to equip a vehicle in pairs with such a *'sidestick'* (◘ Fig. 6.63),

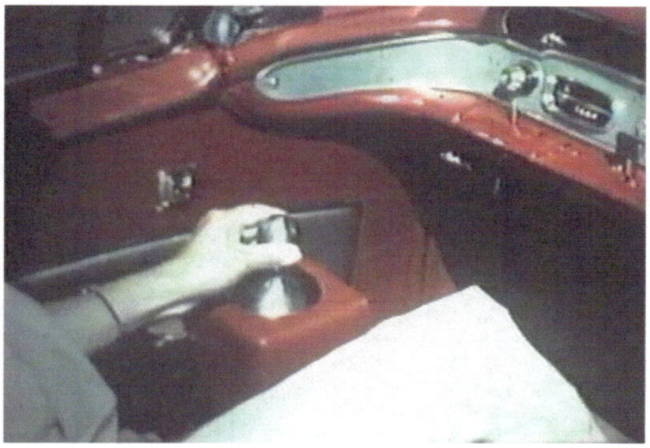

◘ **Fig. 6.62** Sidestick operation of the vehicle in a GM 1958 test vehicle

45 The feedback was given here in a simple mechanical way, with the joystick in the handle being equipped with a heavy weight, which made the mass inertia forces during acceleration, deceleration and cornering perceptible to the driver at the control elements.

Fig. 6.63 Example for test vehicle with sidestick control (Mercedes 500 SL; Eckstein 2001)

whereby, as with steering wheel operation, it is possible to switch at will between two-handed and one-handed operation.

A particular advantage of the active control element is that the feedback on the driving status can be freely designed and does not depend on conditions that are determined by mechanical conditions. Huang (2004) has dealt with these possibilities of feedback in detail. Regarding the ***longitudinal dynamics*** he recommends a concept with force specifications for longitudinal acceleration. In driving tests it should be checked whether a feedback about the reached driving speed is given by a control element with displacement or whether a purely isometric control element (i.e. without displacement) is used. This also depends on the design of the operating element: In the case of only minor possible angular deflections, the isometric variant shows advantages. Eckstein (2001) has opted for the isometric variant in his design of a real vehicle (Mercedes-Benz 500 SL, see Fig. 6.63), whereby the speed achieved in each case is maintained by a cruise control. Recent attempts to combine such a concept with assistance systems (see below), however, show advantages for the control element with speed feedback.

In connection with the control of the longitudinal dynamics by means of a hand-held control element, the question of directional compatibility is frequently raised. It is feared that with the ergonomically compatible assignment "Movement of the control element forward = acceleration – Movement of the control element backward = braking", the upper body falling forward during braking could prevent the braking process. Bolte (1991) clearly answers this question by saying that the feet are no longer occupied with pedal movements, but can be used to apply supporting forces and thus keep the upper body in the seat. Finally, the human organism is designed for reactions in the field of 1 g acceleration and is therefore able to support the upper body accordingly by stiffening the leg, hip and back muscles. In addition, the driver initiates the braking process on his own initiative, i.e. he knows what to expect from his inner model. After initial tests with the alternative concept, the various real vehicles built by Eckstein were also realized according to the system ergonomically correct concept.

Another problem is reverse driving. One *directionally compatible* assignment would mean that after "engaging reverse gear" a backward push of the control element accelerates the vehicle to the rear and a forward push decelerates the vehicle. In contrast to this, the *functional* assignment conceivable, in which – even after "engaging reverse gear" – the force pushing forward fundamentally influences the engine power and the force exerted to the rear always responds to braking. This question is practically unresolvable at the moment. Müller (1999) found in a static simulator that for driving 38 test persons backwards 57% preferred the direction-compatible assignment and 38% the functional assignment. Penka (2000) also received similar results: before a test drive in a static simulator, 81% of the 37 test subjects preferred the direction-compatible assignment; after the test, this preference was reduced to 70% of the test subjects. Eckstein found in 12 seventeen-year-olds without driving experience in the field test with the stick equipped car that initially 6 test persons preferred the direction-compatible assignment, but after the second drive 8 out of 11 test persons (1 failure) were in favour of the functional assignment.

Especially for the ***lateral dynamics***, which are practically 10 times faster than longitudinal dynamics, the question of the concept of the active control element is of outstanding importance. Provided there is no delay in transmission, the concept of displacement-force-feedback and force-displaycement-feedback is equivalent (Gillet, 1998, Huang, 2004). Since the delay between the specification and the vehicle reaction is not as great in lateral dynamics as in longitudinal dynamics, both control concepts can be used for vehicle steering. According to Huang (2004), however, a low-pass filter is required in the default setting at low driving speeds. For this reason, the control concept of force-displaycement-feedback is more advantageous for reasons of flexibility and stability. An additional argument in favour of the force-displaycement-feedback concept is that in reality no system is free of delay, which entails the risk of additional instability for the displacement-force-feedback concept.

In the case of lateral dynamics, many individual findings for low speeds of up to 25 km/h mean that the steering default for the curve radius κ_{target} can be set for and at higher speeds (< 45 km/h) that of the yaw speed ψ_{target} with a continuous transition zone between these two extremes are recommended. The driving task at low speeds can assume the character of a subsequent task, since the driver can imagine the position and movement of his vehicle "in the world" on the basis of his inner models. This is no longer possible at medium and higher speeds. Instead, the driver wants to reduce or control the yaw angle error by controlling the yaw rate. The driver aims for a "target" in the real environment. The motion task is therefore clearly a compensation task.

Eckstein (2001) was able to show in the dynamic driving simulator from Daimler-Benz that the driving quality with the active sidestick control concept is clearly superior to the vehicle steering with steering wheel and pedal in terms of longitudinal dynamics, which is essentially due to the elimination of the time required to move the foot from the accelerator pedal to the brake pedal and the continuous use of cruise control. From a lateral dynamic point of view, too, the active control component concept achieves almost the same driving quality as the steering wheel and pedal. Under the influence of cross winds, the test persons achieved significantly better driving qualities with the active control device. The fundamental difference between the two operating concepts was demonstrated in the analysis of reaction-critical situations on motorway sections: During driving manoeuvres that required both a reaction in longitudinal and lateral dynamics, the sidestick test persons were clearly superior to the test persons with the conventional operating concept, as they were accustomed to simultaneous influencing even under these conditions. The test persons driving with the conventional operating concept reacted almost without exception sequentially, i.e. they operated either the brake pedal or the steering wheel, i.e. they actually do not use the advantage of simultaneous steerability when braking an ABS vehicle, an effect that is also otherwise known from accident analyses (Petit et al. 1993).

Bolte (1991) describes the handling advantages of such a new type of vehicle control by means of an active control element. It shows, for example, that the lateral dynamics of the closed driver-vehicle control loop in conventional steering have a resonance point at approx. 0.4 Hz (see Donges 1982), which disappears completely in the active control component (◘ Fig. 6.64; see also ► Sect. 2.3.3). This is explained by the considerably shorter reaction time of the haptic sensory canal, which, in contrast to the other sensory organs (eyes, but also macula/vestibular organs), has its own lower control loop via the spinal cord (so-called "self-reflex bow"), which allows an approximately four times faster reaction. Eckstein (2001) found – also in simulator experiments – a shorter adaptation time for novice drivers, which can be traced back to the fact that it is easier to fall back on an inner model already formed in the growing up phase for learning such control. In addition, he also describes clear familiarization problems for older people who have become accustomed to the conventional system. This again points to the effect of the inner models formed.

However, the active control element has fewer advantages in real driving tests than in static driving simulators. The reason for this is the effect of the kinesthetic feedback, which from a dynamic point of view has a similar

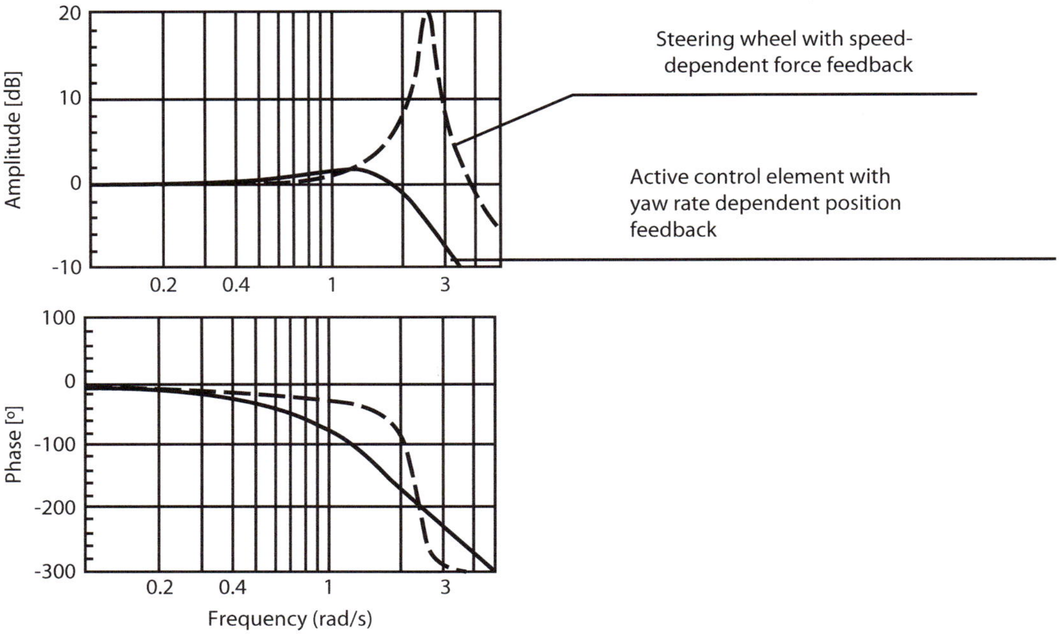

Fig. 6.64 Bodediagramm of the effect of the Active Control Element on the lateral dynamics in relation to the operation by a conventional steering wheel (Bolte 1991)

effect as the haptic feedback of the active actuator. Even Eckstein (2001), with his real vehicle with active stick control under extreme driving conditions (black ice), could not find any advantage of this concept for the control in the dynamic limit range. He therefore recommends that in any case the autonomous control at stabilisation level should be underpinned by split brake intervention, as is the case with ESP, for example.

The combination of the two dimensions of driving in a two-dimensional control element in the form of a sidestick also has disadvantages. Penka (2000), for example, describes that the stick attached to the right of the driver in his experiments regularly observed the vehicle pulling to the left while braking. He attributes this to the fact that in an emergency situation the driver tends to pull the stick on himself (see Fig. 6.65).

Alternative suggestions to the conventional steering wheel operation were made again and again, which take into account the two-dimensionality without the disadvantages of the embroidery operation. A proposal by Lammel was quoted as early as 1985

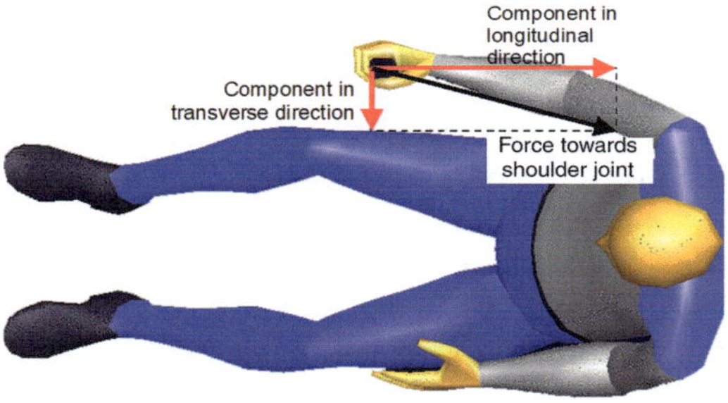

Fig. 6.65 Reaction direction of the driver in critical situations (Penka 2000)

(Bubb 1985), for example, in which the two mechanically connected rotating plate-like handles ensure lateral guidance; longitudinal guidance is achieved by shifting the lever, which is movably guided in the centre console (Figs. 6.65 and 6.66). Another example is the Filo drive-by-wire concept vehicle developed by Bertone, SKF and Brembo. With its operating element it is possible to steer with two handles mechanically coupled in the same direction (Fig. 6.67). Turning the handle accelerates like on a motorcycle. The vehicle is braked by pressing the handles together. The

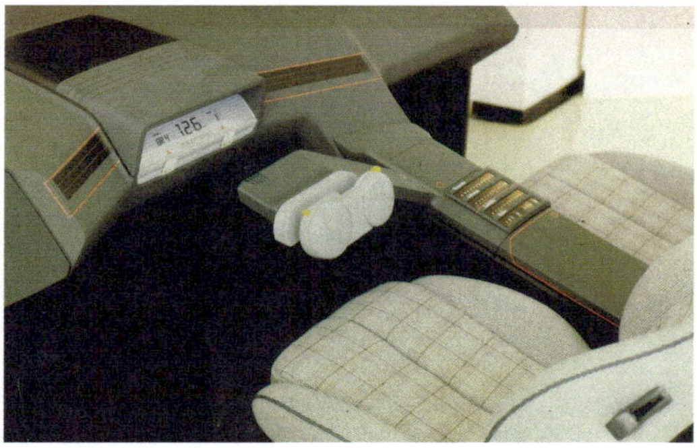

Fig. 6.66 Driver's workstation with active control element according to a design suggestion by Lammel (Bubb 1985)

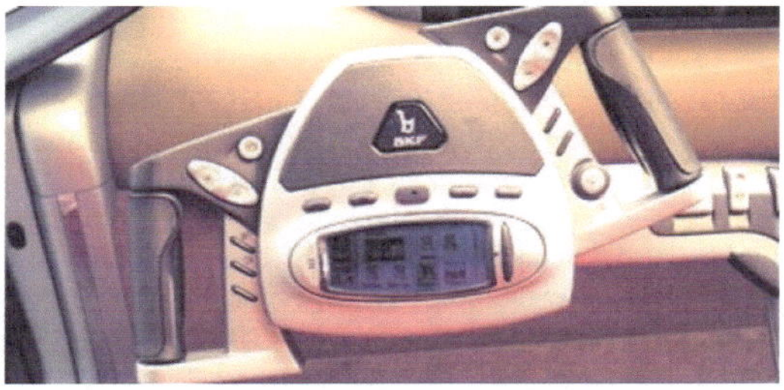

Fig. 6.67 Filo control element for longitudinal and lateral guidance of Bertone and SKF (2001)

control element is active in the lateral guide and passive in the longitudinal guide (quoted from Huang 2004). At present, a system – called "Yoke" – is being tested at the Institute of Ergonomics at the TUM, which can be turned like the control horn known from the aircraft and pushed forwards and backwards for longitudinal guidance. Both operating dimensions are actively designed. This control achieves subjectively similar acceptance with the test subjects as the steering wheel, shows comparable performance with it and is clearly preferred to the joystick (Kienle 2014).

In view of the increasing development of assistance systems, theoretical advantages can be attributed to such systems integrating longitudinal and lateral guidance (Naab and Reichart 1994). For example, radar-supported automatic spacer systems (ACC, Distronic) have the system ergonomic disadvantage of being unable to provide haptic feedback on the drive or braking status of the vehicle. The already mentioned "active accelerator pedal" only provides information about the necessity to "take off the throttle", an initiated braking process cannot be displayed (just as today, when using the ACC, the driver is left in the dark as to whether only the engine brake is effective or is already being actively braked.[46]). In contrast, the Active Control Unit would convey this information haptically and directly and moreover, if from a human point of view the recommendation of the assistance system has to be acted against, it would simply be to suppress it in a directionally compatible way. This consideration becomes even more important in conjunction with a lane keeping assistance system. Penka (2000) has therefore used a simulator to investigate various situations that make it necessary to override assistance systems. They show that under such conditions the active control device generally receives greater acceptance than conventional operation. However, there is a general distrust of the assistance systems, so that conventional operation – because of its greater familiarity – is still preferred.

In spite of these limiting findings, it is worthwhile, with a view to the future development of assistance systems, through which an artificial dynamic protective wall is built up around the vehicle in motion (see Labahn and Boehlau 2001), to deal with the active control element in this or that design form, as this gives the opportunity to always report the penetration of one's own vehicle into this protective wall via the haptic sensory channel to the driver in the same way and adequate to the situation, irrespective of the respective driving situation. As explained above, this not only enables faster reactions, it also makes them intuitively correct because they conform to the inner models of everyday life. In addition, the technical recommendations can be overridden at any time without any rethinking. The driver therefore remains in control of the situation despite assistance.

Pfeffer and Harrer (2013) also point out, however, that with the exception of the last-mentioned restrictions with regard to

46 By the way, it would be an advantage if the lighting of the brake lights in the cockpit would be reported back.

longitudinal guidance, the same effects can be achieved for the operating feeling by means of the superimposed steering system as with the active control element. With regard to the questionable legal release of pure steer-by-wire technology and customer acceptance, they see no application for the x-by-wire design described here in the near future.

From an ergonomic point of view, however, optical feedback (if possible in contact-analog HUD) on the type of intervention is also necessary, because, as Lange (2007) states as a summary of his experiments, the driver should *tactile* mediated, *what* he should do and *optical, why* this is necessary.

6

References

Ablassmeier, M.: Multimodales, kontextadaptives Informationsmanagement im Automobil. Dissertation an der Technischen Universität München (2009)

Arlt, F.: Untersuchung zielgerichteter Bewegungen zur Simulation mit einem CAD-Menschmodell. Dissertation an der Technischen Universität München (1999)

Assmann, E.: Untersuchung über den Einfluss einer Bremsweganzeige auf das Fahrerverhalten. Dissertation an der Technischen Universität München (1985)

Bechstedt, U., Bengler, K., Thüring, M.: Randbedingungen für die Entwicklung eines idealen Nutzermodells mit Hilfe von GOMS für die Eingabe von alphanumerischen Zeichen im Fahrzeug. In: Urbas, L., Steffens, C. (Hrsg.) Zustandserkennung und Systemgestaltung VDI-Fortschritt-Berichte, Bd. Reihe 22, Bd. 22, S. 125–130. VDI Verlag GmbH, Düsseldorf (2005)

Bengler, K.: Automotive speech-recognition – success conditions beyond recognition rate. Proceedings of LREC 2000. European Language Resources Association, Athens, S. 1357–1360 (2000)

Bengler, K.: Aspekte der multimodalen Bedienung und Anzeige im Automobil. In: Jürgensohn, T., Timpe, K.-P. (Hrsg.) Kraftfahrzeugführung, S. 195–206. Springer, Berlin, Heidelberg (2001)

Bengler, K.: Potenzial von MMK-Technologien für zukünftige Fahrzeug-Bedienkonzepte. In: Fastl, H., Fruhmann, M. (Hrsg.) Fortschritte der Akustik: Tagungsband und CD-ROM der 31. Deutschen Jahrestagung für Akustik, DAGA 2005, München, Bd. I und II, DEGA, Berlin (2005)

Bengler, K., Haller, R., Zimmer, A.: Experimental optimization of route-guidance information using context information. In: ERTICO (Hrsg.) Proceedings of the First World Congress on Applications of Transport Telematics and Intelligent Vehicle-Highway Systems, S. 1758–1765. Artech House, Boston (1994)

Bengler, K., Noszko, T., Neuss, R.: In: Usability of multimodal human-machine-interaction while driving proceedings of the 9-th world congress on ITS, CDRom, 9th World Congress on Intelligent Transport Systems, ITS America, Bd. 10/2002 (2002)

Bengler, K., Huesmann, A., Praxenthaler, M.: Investigation of Visual Demand in a Static Driving Simulator within the ADAM Project Proceedings of the Annual Spring Conference of the Gesellschaft für Arbeitswissenschaft e.V. (GfA), pp. 49–52. ergonomia Verlag, München (2003)

Bengler, K., Praxenthaler, M., Theofanou, D., Eckstein, L.: Investigation of visual demand in different driving simulators within the ADAM project driving simulator conference 2004 Europe, Paris. Proceedings., S. 91–104 (2004)

Bengler, K., Bubb, H., Totzke, I., Schumann, J., Flemisch, F.: Automotive. In: Sandl, P., Stein, M. (Hrsg.) Information Ergonomics – A Theoretical Approach and Practical Experience in Transportation. Springer, Berlin, Heidelberg (2012)

Bergmeier, U.: Kontaktanalog markierendes Nachtsichtsystem Entwicklung und experimentelle Absicherung. Technische Universität München, Dissertation (2009)

Bernotat, R.: Anzeigengestaltung. In: Schmidtke, H. (Hrsg.) Ergonomie. Hanser Verlag, München (1993)

Bertone; SKF: Filo-Präsentations-CD (2001)

Bielaczek, C.: Untersuchung zur Auswirkung einer aktiven Fahrerbeeinflussung auf die Fahrsicherheit beim Pkw-Fahren im realen Straßenverkehr Fortschritts-Bericht VDI, Reihe 12, Bd. 35. VDI-Verlag, Düsseldorf (1998)

Bieniek, E.-M.: Untersuchung zur Akzeptanz verschiedener mechanischer Fahrersitzverstellmöglichkeiten in Personenkraftwagen. Diplomarbeit an der Philosophisch-Pädagogischen Fakultät der Katholischen Universität Eichstätt (1990)

Bill, K.H., Leber, M., Becker, H., Breuer, B.: Forschungswerkzeug zur Untersuchung der Schnittstelle Fahrer/Bremspedal. Automob. Z. **101**(2), 86–93 (1999)

Blattner, A., Bengler, K., Hamberger, W.: Optimized combination of operating modalities and menu tasks for the interaction between driver and infotainment-system using a touchpad with haptic feedback. In: Stanton, N. (Hrsg.) Advances in Human Aspects of Road and Rail Transportation, S. 13–19. CRC Press, Boca Raton (2013)

Bolte, U.: Das aktive Stellteil – ein ergonomisches Bedienkonzept Fortschritts-Berichte VDI, Reihe 17 "Biotechnik". VDI-Verlag, Düsseldorf (1991)

Borg, G.: A category scale with ratio properties for intermodal and interindividual comparison. In: Geisler, H.G., Petzold, P. (Hrsg.) Psychophysical Judgement and the Process of Perception, S. 25–34. VEB Deutscher Verlag der Wissenschaft, Berlin (1982)

Bouis, D., Haller, R., Geiser, G., Heintz, F., Knoll, P.M.: Ergonomische Optimierung von LCD-Anzeigen im Kraftfahrzeug. Proceedings of ISATA (1983)

Braess, H.-H., Donges, E.: Technologien zur aktiven Sicherheit von Personenkraftwagen – „Konsumierbare“ oder echte Verbesserungen? Tagung „Aktive Sicherheit durch Fahrerassistenz“, Garching b. München, 4.-5. April, 200 (2006)

Braun, M.: Ergonomische Untersuchung zur Akzeptanz eines kombinierten Fahr-/Bremspedals. Institut für Nachrichtensysteme der Universität Karlsruhe, durchgeführt in der Forschungsabteilung F1M/Fa der Daimler-Benz AG. Stuttgart (1993)

Breuer, J., Bengler, K., Heinrich, C., Reichelt, W.: Development of advanced driver attention metrics (ADAM). In: Strasser, H., Kluth, K., Rausch, H., Bubb, H. (Hrsg.) Quality of Work and Products in Enterprises of the Future Proceedings of the 50th – Anniversary Conference of the GfA and the XVII Annual ISOES Conference in Munich, 2003. S. 37–40. Ergonomia, Stuttgart (2003)

Bubb, H.: Untersuchung über die Anzeige des Bremsweges im Kraftfahrzeug. Dissertation, Technische Universität Müncehen (1975)

Bubb, H.: The influence of braking distance indication on driver's behaviour. In: Oborne, D.J., Levis, J.A. (Hrsg.) Human Factors in Transport Research, Bd. II, S. 388–346. Academic, London and New York (1981)

Bubb, H.: Arbeitsplatz Fahrer – Eine ergonomische Studie. Automob. Ind. **30**, 265–275 (1985)

Bubb, H., Bolte, U.: Einfluss der fahrdynamischen Kräfte auf die haptische Rückmeldung bei konventionellen Bedienelementen und bei Aktiven Bedienelementen. Mitteilung Lehrstuhl für Ergonomie. TU München, (Interner Bericht) (1987)

Buschardt, B.: Synthetische Lenkmomente Fortschritts-Berichte VDI, Reihe 22, Bd. 12. VDI-Verlag, Düsseldorf (2003)

Byblow, W.D.: Effects of redundancy in the comparison of speech and pictorial display in the cockpit environment. Appl. Ergon. **21**(2), 121–128 (1990)

Damböck, D.: Automationseffekte im Fahrzeug – von der Reaktion zur Übernahme, Dissertation, Technische Universität München (2013)

Dick, M.: Auf den Spuren der Motive. Auto zu fahren – Die Perspektive der Fahrenden. Verkehrszeichen. **18**(4), 9–16 (2002)

Diefenbach, S., Hassenzahl, M. (Hrsg.): Handbuch zur Fun-ni Toolbox – User Experience Evaluation auf drei Ebenen. Abschlussbericht, Förderkennzeichen 01ISO9007. Universität der Künste, Folkwang (2010)

DIN EN 19 60000: DIN EN 196000:1992/A1: 2001: Deutsche Industrie Norm, Beuth-Verlag, 2001 (2001)

Donges, E.: Aspekte der Aktiven Sicherheit bei der Führung von Personenkraftwagen. Automobil-Industrie. **27**(2), 183–190 (1982)

Eberl, T.: Charakterisierung und Gestaltung des Fahr-Erlebens der Längsführung von Elektrofahrzeugen. Dissertationsschrift an der Technischen Universität München (2014)

Eckstein, L.: Entwicklung und Überprüfung eines Bedienkonzepts und von Algorithmen zum Fahren eines Kraftfahrzeugs mit aktiven Sidesticks Fortschr.-Ber. VDI, Reihe 12, Bd. 471. VDI-Verlag, Düsseldorf (2001)

Egger, T.: Berechnungen und experimentelle Untersuchungen zur indirekten Sicht aus Kraftfahrzeugen. Dissertation an der Technischen Universität München (1990)

Fricke, N.: Gestaltung zeit- und sicherheitskritischer Warnungen im Fahrzeug. Dissertationsschrift Technische Universität Berlin (2009)

Friedrich, H., Hoffmann, J., Kreft, J., Semmler, C., Witte, B.: Auf dem Weg zum intelligenten Auto – Steer-by-Wire als Basis zukünftiger Assistenzfunktionen. In: VDI (Hrsg.) Der Fahrer im 21. Jahrhundert VDI-Berichte, Bd. 1613, S. 277–296. VDI-Verlag GmbH, Düsseldorf (2001)

Fujinami, H., Yamamoto, M., Hisaoka, Y.: Vehicle Response and Drivers' Feeling. Toyota Technical Review **45**(1) (1995)

Gengenbach, W.: Experimentelle Untersuchung von Reifen auf nasser Fahrbahn. ATZ. **70**(8 und 9), 83–89 (1968)

Gillet, G.: Ergonomische Optimierung eines Aktiven Stellteils. Dissertation an der Technischen Universität München (1998)

Göktan, A.G.: Optimale Bremspedal-Kraft-Weg-Charakteristiken bei Pkw-Bremsanlagen. Automobil-Industrie. **32**(2), 161–169 (1987)

Gold, C., Damböck, D., Lorenz, L., Bengler, K.: "Take over!" how long does it take to get the driver back into the loop? Proceedings of the Human Factors and Ergonomics Society Annual Meeting September. **57**, 1938–1942 (2013). https://doi.org/10.1177/1541931213571433

Goldstein, E.B.: Sensation and Perception. Wadsworth Cengage Learning, Belmont, CA (2009)

Götz, M.: Die Gestaltung von Bedienelementen unter dem Aspekt ihrer kommunikativen Funktion. Dissertation an der Technischen Universität München (2007)

Gough, V.E.: Reibung von Gummi. Straßenbau und Straßenverkehrstechnik (19), 22 (1962)

Grote, G.: Autonomie und Kontrolle: Zur Gestaltung automatisierter und risikoreicher Systeme. Vdf, Hochschulverlag an der ETH Zürich, Zürich (1997)

Hackenberg, U., Heißing, B.: Die fahrdynamischen Leistungen des Fahrer-Fahrzeug-Systems im Straßenverkeh. Automobiltechnische Zeitschrift (ATZ). **84**(7/8), 341–345 (1982)

Haider, E.: Beurteilung von Belastung und zeitvarianter Beanspruchung des Menschen bei kompensatorischen Regeltätigkeiten: Simulationen- Feldstudien-Modelle Fortschritt-Berichte der VDI-Zeitschriften. Biotechnik, Bd. 17, pp. 3–18. VDI-Verlag, Düsseldorf (1977)

Hamberger, W.: MMI touch – new technologies for new control concepts. In: IQPC (Hrsg.) Automotive

Cockpit HMI. Steigenberger Graf Zeppelin, Stuttgart (2010)

Hamberger, W., Gößmann, E.: Bedienkonzept Audi: die nächste Generation. In: VDI Wissenforum GmbH (Hrsg.) Elektronik im Kraftfahrzeug VDI-Berichte, Bd. 2075, S. 677–686. VDI-Verlag, Düsseldorf (2009)

Harrer, M.: Steering System Development in Premium Car Segment (2006). SAE 2006-01-0935

Harrer, M.: Auslegung von Lenkungen. In: Pfeffer, P., Harrer, M. (Hrsg.) Lenkungshandbuch – Lenksysteme, Lenkgefühl, Fahrdynamik von Kraftfahrzeugen, 2. Aufl ATZ/MTZ-Fachbuch, Springer, Wiesbaden (2013)

Hassenzahl, M., Eckoldt, K., Thielsch, M.T., et al.: User Experience und Experience Design– Konzepte und Herausforderungen. In: Brau, H. (Hrsg.) Usability Professionals 2009. Fraunhofer Verlag, Stuttgart (2009)

Heißing, B., Brandl, J.: Subjektive Beurteilung des Fahrverhaltens. Vogel, Würzburg (2002)

Heißing, B., Kudritzki, D., Schindlmaister, R., Mauter, G.: Menschengerechte Auslegung des dynamischen verhaltens von Pkw. Erschienen. In: Bubb, H. (Hrsg.) Ergonomie und Verkehrssicherheit: Beiträge der Herbstkonferenz der Gesellschaft für Arbeitswissenschaft. Herbert Utz Verlag GmbH, München (2000)

Hofmann, M., Bengler, K., Lang, M.: Ein Assistenzsystem zur fahrer- und situationsadaptiven Prädiktion potenzieller Zielorte für eine robuste Interaktion mit sprachgesteuerten Navigationssystemen Elektronik im Kraftfahrzeug. VDI-Berichte, Bd. 1646. VDI-Verlag, Düsseldorf (2001)

Huang, P.: Regelkonzepte zur Fahrzeugführung unter Einbeziehung der Bedienelementeigenschaften. Dissertation, Lehrstuhl für Ergonomie, TU München (2004)

ISO 15008: „Ergonomic aspects of transport information and control systems—visual presentation of information," (2009)

Israel, B.: Potenziale eines kontaktanalogen Head-up Displays für den Serieneinsatz. Dissertation an der Technischen Universität München. (2013)

Jürgensohn, T.: Hybride Fahrermodelle, 1. Aufl. Pro-Universitate-Verlag, Sinzheim (1997)

Kienle, M.: Anzeige-Bedienkonzept hochautomatisierter Fahrzeuge – vom manuellen zum hochautomatisierten Autofahren. Dissertationsschrift an der Technischen Universität München (2014)

König, W., Weiss, K.E., Gehrke, H., Haller, R.: S.A.N.T.O.S Situationsangepasste und nutzertypzentrierte Optimierung von Systemen zur Fahrerunterstützung GfA Herbstkonferenz über Ergonomie und Verkehrssicherheit, Technische Universität München, 12.-13.10.2000 (2000)

Kosaka, H., Watanabe, K.: Reaction forces of switches and push feeling. automotive design avancements in human factors: improving drivers´ comfort and performance. SAE, Warrendale, S. 169–174. Publication No.SP-1155 (1996)

Kühner, M.: Haptische Unterscheidbarkeit mechanischer Parameter bei rotatorischen Bedienelementen. Dissertation an der Technischen Universität München (2014)

Kühner, M., Wild, J., Bubb, H., Bengler, K., Schneider, J.: Haptic perception of viscous friction of rotary switches. In: Jones, L.A., Harders, M., Yokokohji, Y. (Hrsg.) IEEE World Haptics Conference WHC 2011, Istanbul, Turkey, 21–24 June 2011. S. 587–591. IEEE Computer Society, Piscataway (2011)

Kushiro, I. et al. A new approach in the study on-center-handling. Paper 20080432, AVEC (zitiert aus Wolf, 2009) (2008)

Labahn, N., Boehlau, C.: Unterstützung des Fahrers durch Umweltsensorik und intelligente Lichtsteuerung. In: VDI (Hrsg.) Der Fahrer im 21. Jahrhundert VDI-Berichte, Bd. 1613, S. 315–332. VDI Verlag GmbH, Düsseldorf (2001)

Lange, C.: Wirkung von Fahrerassistenz auf der Führungsebene in Abhängigkeit der Modalität und des Automatisierungsgrades. Dissertation an der Technischen Universität München. (2007)

Lange, C., Schmitt, G., Arcati, A., Bengler, K., Bubb, H.: Haptic gear shifting indication: Naturalistic driving study for parametrization, selection of variants and to determine the potential for fuel consumption reduction, In: Proceedings of the 3rd International Conference on Applied Human Factors and Ergonomics, Miami (17.–20. Juli 2010)

Lorenz, S.: Assistenzsystem zur Optimierung des Sitzkomforts im Fahrzeug. Dissertation an der Technischen Universität München (2011)

McKenzie, C.L.: A note on information-theoretic basic for Fitt's law. J. Mot. Behav. **21**, 323–330 (1989)

McKenzie, C.L., Paulignan, Y., Marteniuk, R., Jeannerod, M.: The coupling of arm and finger movements during prehension. Exp. Brain Res. **79**, 432–435 (1990)

Meyer-Eppler: Grundlagen und Anwendungen der Informationstheorie. Springer, Berlin (1959)

Michon, J.A.: Generic intelligent driver support. Taylor & Francis, London (1993)

Müller, F.: Realisierung eines Fahrzeugführungskonzeptes mit Steuerknüppel im BMW Fahrsimulator. Diplomarbeit, Lehrstuhl für Ergonomie, TU München (1999)

Müller, T., Hajek, H., Radić-Weißenfeld, L., Bengler, K.: Can you feel the difference? the just noticeable difference of longitudinal acceleration September 2013. Proceedings of the Human Factors and Ergonomics Society Annual Meeting., S. 1219–1223 (2013)

Müller, T., Hajek, H., Frank, T., Bengler, K.: Das menschliche Auflösungsvermögen von Dynamikveränderungen im Fahrzeug Conference on Future Automotive Technology, Garching b. München, 17–18. März 2014 (2014)

Murrel, K.: Ergonomics. Chapman and Hall, London (1969)
Naab, K., Reichart, G.: Driver assistance system for lateral and longitudinal vehicle guidance—heading control and active cruise support. In: Proceedings of International Symposium on Advanced Vehicle Control (AVEC '94) Tsukuba, Japan. S. S449–S454 (1994)
Norman, D.: Categorization of action slips. Psychol. Rev. **88**, 1–15 (1981)
Norman, D.: New views in information processing: implications for intelligent decision support systems. In: Hollnagel, E., Manchini, G., Woods, D.D. (Hrsg.) Intelligent Decision Support in Process Environments. Springer, Berlin (1986)
Norman, D.: The Psychology of Everyday Things, pp. 8–10. Basic Books, New York (1988)
Osumi, Y., Inuzuka, Y., Hideichi, I.: Operability of car audio controls. In: Countdown to the 21st Century Proceedings of the Human Factors Society 34th Annual Meeting, Orlando (USA), Bd. 1, pp. 618–622. The Human Factors Society, Santa Monica (1990)
Oviatt, S.: Ten myths of multimodal interactio. Commun. ACM. **42**(11), 74–81 (1999)
Oviatt, S.: Multimodal system processing in mobile environments. In: Proceedings of 6th International Conference on Spoken Language Processing (ICSLP 2000) Beijing, China, October 16–20, 2000 (2000)
Parkes, A., Franzen, S.: Driving Future Vehicles. Taylor & Francis, London (1993)
Penka, A.: Vergleichende Untersuchung zu Fahrerassistenzsystemen mit unterschiedlichen aktiven Bedienelementen. Dissertation an der Technischen Universität, München (2000)
Petermann, F.: Entspannungsverfahren, 4. Aufl. Beltz, PVU, Weinheim (2009)
Petermann-Stock, I., Hackenberg, L., Muhr, T., Mergl, C.: Wie lange braucht der Fahrer?: Eine Analyse zu Übernahmezeiten aus verschiedenen Nebentätigkeiten während einer hochautomatisierten Staufahrt 6. Tagung Fahrerassistenzsysteme, München (2013)
Petit, C., Priez, A., Brigout, C., et al.: Driver's use of ABS in emergency. A Physiological an Behavioral Trail Proceedings of the ISATA Conference (1993)
Pfeffer, P.: Lenkgefühl, Interaktion Fahrer- Fahrzeug. In: Pfeffer, P., Harrer, M. (Hrsg.) Lenkungshandbuch – Lenksysteme, Lenkgefühl, Fahrdynamik von Kraftfahrzeugen, 2. Aufl. ATZ/MTZ-Fachbuch, Springer, Wiesbaden (2013)
Pfeffer, P., Harrer, M.: Ausblick – Zukunft der Lenkung im Automobil. In: Pfeffer, P., Harrer, M. (Hrsg.) Lenkungshandbuch – Lenksysteme, Lenkgefühl, Fahrdynamik von Kraftfahrzeugen, 2. Aufl. ATZ/MTZ-Fachbuch, Springer, Wiesbaden (2013)
Pöppel, E.: Grenzen des Bewußtseins – Wie kommen wir zur Zeit und wie entsteht Wirklichkeit? Insel Taschenbuch, Frankfurt a.M. und Leipzig (2000)
Poulton, E.C.: Tracking Skill and Manual Control. Academic, New York (1974)
Pruckner, A.: Steer-by-wire. In: Pfeffer, P., Harrer, M. (Hrsg.) Lenkungshandbuch – Lenksysteme, Lenkgefühl, Fahrdynamik von Kraftfahrzeugen, 2. Aufl, ATZ/MTZ-Fachbuch, Springer, Wiesbaden (2013)
Rassl, R.: Ablenkungswirkung tertiärer Aufgaben im Pkw – Systemergonomische Analyse und Prognose. Dissertation an der Technischen Universität München. (2004)
Reinig, H.J.: Anthropometrische Untersuchung von Schriftzeichen im öffentlichen Personenverkehr. ETR. **12**(26), 843–847 (1997)
Reisinger, J.: Haptik von Bedienelementen. Ergonomie Aktuell. **07**, 29–34 (2007)
Reisinger, J. : Parametrisierung der Haptik von handbetätigten Stellteilen. Dissertation an der Technischen Universität München (2009)
Reisinger, J., Wild, J.: Haptische Bedienschnittstelle. In: Meroth, A., Tolg, B. (Hrsg.) Informationssysteme im Kraftfahrzeug. Grundlagen, Komponenten, Systeme und Anwendungen, 1. Aufl. Vieweg, Wiesbaden (2007). Kap. 6
Remlinger, W.M.: Analyse von Sichteinschränkungen im Fahrzeug. Dissertation an der Technischen Universität, München (2013)
Remlinger, W., Bubb, H., et al.: Erweiterte Möglichkeiten der Sichtanalyse mit dem Menschmodell RAMSIS. In: Grandt, M., Bauch, A. (Hrsg.) Simulationsgestützte Systemgestaltung DGLR-Bericht, Bd. 2007–04, S. 189–198. Deutsche Gesellschaft für Luft- und Raumfahrt e.V., Bonn (2007)
Reuter, M., Saal, A.: Überlagerungslenkung. In: Pfeffer, P., Harrer, M. (Hrsg.) Lenkungshandbuch – Lenksysteme, Lenkgefühl, Fahrdynamik von Kraftfahrzeugen, 2. Aufl, ATZ/MTZ-Fachbuch, Springer, Wiesbaden (2013)
Rockwell, T., Snider, J.: An Investigation of Variability in Driving Performance on the Highway: Final Report. Ohio State University, Columbus (1965)
Rühmann, H.: Schnittstellen in Mensch-Maschine-Systemen. In: Schmidtke, H. (Hrsg.) Ergonomie. Hanser Verlag, München (1993)
Sacher, H.: Gesamtheitliche Analyse des Bedienverhaltens von Fahrzeugfunktionen in der täglichen Nutzung. Dissertation an der Technischen Universität München (2009)
Samper, K., Kuhn, K.-P.: Reduktion des Kraftstoffverbrauchs durch ein vorausschauendes Assistenzsystem. In: VDI (Hrsg.) Der Fahrer im 21. Jahrhundert VDI-Berichte, Bd. 1613, S. 79–94. VDI Verlag GmbH, Düsseldorf (2001)
Sato, H., Haraguchi, J., Osawa, I.: Experimental analysis of steering response characteristics and steering feeling. Japan SAE Review. **44**(3), 52–58 (1990)
Schallamach, A.: Mechanik der Reifenabnutzung. ATZ. **63**(2), 56 (1961)
Schmid, M., Maier, T.: Ergonomie versus Ästhetik? Die funktionale und formale Integration im Fahrzeugcockpit. Zeitschrift für Arbeitswissenschaft. **2**(3), 129–139 (2012). Ergonomia Verlag Stuttgar

Schmidtke, H., Rühmann, H.: Zentralverschlußhebel und Zentralverschlußräder. In: BWB (Hrsg.) Handbuch der Ergonomie. Bundesamt für Wehrtechnik und Beschaffung, München. Abschnitt C-11.1.2 und C-11.1.3 (1989)

Schneid, M.: Entwicklung und Erprobung eines kontaktanalogen Head-up-Displays im Fahrzeug. Dissertation an der Technischen Universität München (2009)

Schweigert, M.: Fahrerblickverhalten und Nebenaufgaben. Dissertation an der Technischen Universität München (2003)

Siegel, L., Bundorf, R.T.: Variable stability auto tests behavior of vehicles before they are built. SAE J. **06–01**, 88–91 (1966)

Spies, R.: Entwicklung und Evaluierung eines Touchpadbedienkonzeptes mit adaptiv haptisch veränderlicher Oberfläche zur Menübedienung im Fahrzeug. Dissertation an der Technischen Universität München (2013)

Spies, R., Blattner, A., Horna, G., Bubb, H., Hamberger, W.: Entwicklung einer aufgabenorientierten Menüstruktur zur Infotainmentbedienung während der Fahrt. In: Lichtenstein, A., Stößel, C., Clemens, C. (Hrsg.) Der Mensch im Mittelpunkt technischer Systeme VDI-Fortschritt-Berichte Reihe 22, Bd. 29, S. 411–416. VDI-Verl., Düsseldorf (2009)

Spies, R., Horna, G., Bubb, H., Bengler, K., Hamberger, W.: Haptisches Touchpad – Zentrales Infotainmentbedienteil mit adaptiv haptisch veränderlicher Oberfläche. Neuer Arbeits- und Lebenswelten 56. Kongress der Gesellschaft für Arbeitswissenschaft. GfA Press, Dortmund (2010)

Thoma, S.: Mensch-Maschine-Interaktionskonzepte für Fahrerassistenzsysteme im Kreuzungsbereich. Dissertation an der Technischen Universität München (2010)

Treisman, A., Gelade, G.: A feature integration theory of attention. Cogn. Psychol. **12**, 97–136 (1980)

Vaas, R.: Zeit und Gehirn. Online-Lexikon der Neurowissenschaft (2005). http://www.wissenschaftonline.de/abo/lexikon/neuro/14651. http://www.wissenschaft-online.de/abo/lexikon/neuro/14651, Zugegriffen: 27.06.2005

Wang, X.G., Mick, F., Vernet, M., Fraigneau, F.: Experimental investigation of the relationship between the sensation and the force applied to a pedal. In: Ozok, A.F., Salvendy, G. (Hrsg.) Advances in Applied Ergonomics, S. 572–575. USA Publishing, West Lafayette, India, USA (1996)

Weber, R., Persch, H.G.: Seitenkraftfrequenzgänge von Luftreifen – Ein Beitrag zum Verhalten bei instationärem Schräglauf. ATZ. **77**(2), 44 (1975)

Webhofer, G.: Implementation und Inbetriebnahme einer Regelung für ein Aktives Lenksystem. Diplomarbeit, TU München, Lehrstuhl für Verbrennungskraftmaschinen und Kraftfahrzeuge (1991)

Weir, D., Colgate, J., Peshkin, M.: The haptic profile: capturing the feel of switches. In: Proceedings of the 12th International Symposium on Haptic Interfaces for Virtual Environment and Teleoperator Systems (HAPTICS'04) (2004)

Wolf, H.: Ergonomische Untersuchung des Lenkgefühls an Personenkraftwagen. Dissertation an der Technischen Universität München (2009)

Woodson, W.E.: Human Factors Design Handbook. Verlag McGraw-Hill, New York (1992)

6

Further Literature

Akamatsu, M., Green, P., Bengler, K.: Automotive technology and human factors research: past, present, and future. International Journal of Vehicular Technology. **2013**(2013), 526180 (2013), 27 pages

Aurell, J., Fröjd, N., Nordmark, S.: Correlation between objective handling characteristics and subjective perception of handling quality of heavy vehicle Proceedings of the International Symposium on Advanced Vehicle Control AVEC '00. Ann Arbor, S. 105–113 (2000)

Baker, C.A., Grether, W.F.: Visual Presentation of Information. WADC.Report No. TR 54-160

Bernstein, A., Bader, B., Bengler, K., Kuenzner, H.: Visual-haptic interfaces in car design at BMW. In: Grunwald, M. (Hrsg.) Human Haptic Perception – Basics and Applications. Birkhäuser, Berlin, Basel, Boston (2008)

Blattner, A.: Bedienkonzeptentwicklung für Fahrerinformationssysteme basierend auf einem Touchpad mit haptischer Rückmeldung. Dissertation, Technische Universität München (2014)

Dettki, F.: Methoden zur Bewertung des Geradeauslaufs von Pkw VDI-Berichte, Bd. 1133. VDI-Verlag, Düsseldorf (1997)

Haslbeck, A., Popova, S., Krause, M., Pecot, K., Mayer, J., Bengler, K.: Experimental evaluations of touch interaction considering automotive requirements. In: Jacko, J.A. (Hrsg.) Human-Computer Interaction. Interaction Techniques and Environments 14th International Conference, HCI International 2011, Orlando, FL, USA, July 9–14, 2011. Lecture Notes in Computer Science, Bd. 6762, S. 23–32. Springer, New York (2011). Proceedings, Part II

Ho, C., Spence, C.: The Multisensory Driver: Implications for Ergonomic Car Interface Design. Ashgate Pub Co., Aldershot (2008)

Jaksch, F.O.: The Steering Characteristics of the Volvo Concept Car VIII. ESV-Conference, Wolfsburg, 1980 (1980)

Jaksch, F.O.: Vehicle parameter influence on steering control characteristics. Int. J. Veh. Des. **4**(2), 171–194 (1983)

Lange, C.: Fähigkeiten und Defizite der "Silver-Generation". Unveröffentlichter Bericht für die Audi-AG (2008)

Lange, C., Arcati, A., Bubb, H., Bengler, K.: Haptic gear shifting indication: naturalistic driving study for parametrization, selection of variants and to determine the potential for fuel consumption reduction. In: Proceedings of the 3rd International Conference

on Applied Human Factors and Ergonomics Miami, 17.-20. Juli 2010. (2010)
Totzke, I., Huth, V., Krüger, H.-P., Bengler, K.: Overriding the ACC by keys at the steering wheel: positive effects on driving and drivers' acceptance. In: Human Factors for Assistance and Automation, pp. 153–164. Shaker, Maastricht (2008)
Trübswetter, N., Bengler, K.: Why Should I Use ADAS? Advanced Driver Assistance Systems and the Elderly: Knowledge, Experience and Usage Barriers. In: Proceedings of the Seventh International Driving Symposium on Human Factors in Driver Assessment, Training, and Vehicle Design (2013)

Anthropometric Vehicle Design

Heiner Bubb, Rainer E. Grünen, and Wolfram Remlinger

Contents

7.1 Vehicle Packaging – 345
7.1.1 Objective of the Anthropometric Package – 345
7.1.2 Car Dimension Conception According to SAE – 348
7.1.3 Fields of Anthropometric Ergonomics – 353

7.2 Sitting – 358
7.2.1 Consideration of Different Anthropometries – 358
7.2.2 Driver – 359
7.2.3 Right-Hand Drive Problem – 384
7.2.4 Co-Driver – 392
7.2.5 Vehicle Rear – 392

7.3 View – 396
7.3.1 Direct View – 396
7.3.2 Indirect Vision – 407
7.3.3 View of Operating and Display Components – 409
7.3.4 Reflections – 412

7.4 Operating and Display Components – 416
7.4.1 Determination of Accessibility Areas According to SAE – 416
7.4.2 Functional Gripping Spaces – 417
7.4.3 Consideration of Special Operating Requirements – 419

7.5 Space Requirements – 420
7.5.1 Static and Dynamic Space Requirements – 420
7.5.2 Shelves – 424
7.5.3 Room Feeling – 425

H. Bubb et al. (eds.), *Automotive Ergonomics*, https://doi.org/10.1007/978-3-658-33941-8_7

7.6 Entry and Exit – 427
7.6.1 Door Concepts – 427
7.6.2 Movement Strategies – 431
7.6.3 Evaluation Methods – 438
7.6.4 Access to the Second and Third Rows of Seats – 449

7.7 Loading – 449
7.7.1 Geometry – 450
7.7.2 Operability – 452

7.8 Consideration of Specific User Groups – 453
7.8.1 Older Vehicle Users – 453
7.8.2 Children – 460

7.9 Craftsmanship – 461

References – 465

7.1 Vehicle Packaging

7.1.1 Objective of the Anthropometric Package

The compilation of all vehicle components and assemblies, taking into account their relative movements and assembly freedom, is the classic discipline of vehicle packaging.[1] Based on the vehicle concept of the superstructure, the installation spaces of the components such as engine, tank and load compartment are arranged in principle (▪ Fig. 7.1). With the same vehicle architecture (substructure/platform), different vehicle superstructures ("*bodystyles*") can be used (see also ▪ Table 7.1). From a four-door notchback saloon, a five-door compact saloon, a station wagon and a cabriolet are derived.

The human being and the dimensions of his or her extremities, the necessary freedom of movement and the limits to which they can be reached, as well as his or her visual conditions, represent the ergonomic side of the package and are defined in the vehicle architecture at the beginning of the dimensional concept. The vehicle interior, which is in principle reserved for vehicle occupants, is restricted by necessary components such as seats, steering wheel and pedals, as well as instrument panel, centre console and storage compartments. In the remaining rooms without components, drivers and passengers must not only find sufficient space for accommodation, but also all free space to ensure comfort. The target group of the market, in particular the gender distribution of the buyers, is a decisive factor. In the age of globalization, however, consideration of all relevant global customers - from small Asians to large Central Europeans - is a basic prerequisite for economic vehicle platforms. To this end, a range of anthropometries must be defined that will allow the complete accommodation of all targeted customers in all relevant markets. Based on a vehicle specification, the number of seats or the vehicle type and segment, occupant accommodation is determined for each individual seat. A distinction is made between full and compromised seats (▪ Fig. 7.2).

Fully-fledged seats are fully usable for all target customers, while compromised seats as emergency or demand seats have limitations in terms of comfort and usability. Here, the width of the seat, the legroom provided or the headroom above the seat surface may be restricted and thus only usable for a limited group of customers. Coupés or convertibles are often called "*2 + 2-Seater*" if, in addition to the two fully-fledged front seats in the rear row, two demand seats are available. These on-demand seats are often foldable or retractable and can increase the number of seats available in the vehicle at the expense of the boot volume. For example, large-capacity sedans are equipped with two additional seats in the boot floor, which can easily be brought into the position for use if required. Occasionally[2] the demand seats are also arranged against the direction of travel in order to keep the overall vehicle length to a minimum by using the headroom of the adjacent rows of seats in a back-to-back arrangement. It should be noted, however, that driving with one's back to the direction of travel in such places can cause nausea to passengers, especially children (see ▸ Sect. 3.1.3 and ▸ Fig. 3.11; Simulator Sickness). Seven-seater vehicles ("MPV"[3]or "Microvans") are often based on five-seater station wagons in which two folding demand seats are installed in the boot. In addition to the consideration of the occupant accommodation with its static space requirements, the dynamic movement requirements

1 The term package in automotive engineering must be separated from that of packaging & labelling in the packaging industry, where packaging refers to the outer shell, whereas in automotive engineering it refers to the arrangement of the inner components.

2 Examples of vehicles with seats arranged in the opposite direction of travel: Zündapp Janus second row of seats (1957), Mercedes-Benz E-Class T Model third row of seats (BR210 1995), Loremo second row of seats (concept 2006).

3 MPV: Multi Purpose Vehicle.

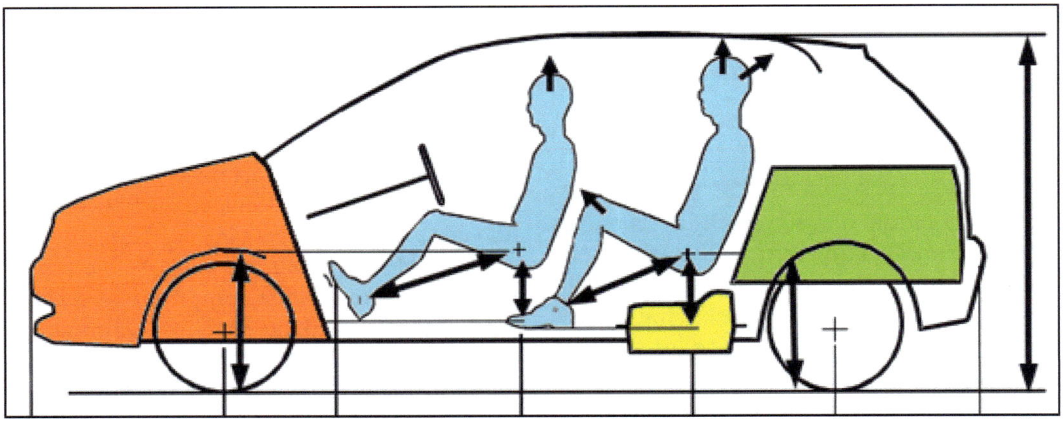

Fig. 7.1 Basic vehicle package

Table 7.1 Vehicle types

Concept	Doors	Seats	Rear	Roof
Coupé (coupe)	2	2 or 2 + 2	Notchback, hatchback	Fixed, over occupants
Cabriolet (convertible)	2	2 or 2 + 2	Notchback	Mobile, over occupants
Limousine (4 door notchback)	4	5	Notchback	Fixed, over occupants
Limousine (5 door hatchback)	5	5	Hatchback, hatchback	Fixed, over occupants and luggage
Station wagon (estate limousine)	5	5 or 7	Hatchback	Fixed, over occupants and luggage
Off-road vehicle (sport utility vehicle SUV)	5	5	Hatchback	Fixed, over occupants and luggage
Minibus/minivan (multi purpose vehicle MPV)	5	7	Hatchback	Fixed, over occupants and luggage
Car platform (light truck, pickup)	2 or 4	2 or 5	Open loading area	Fixed, over occupants sep. loading area cover

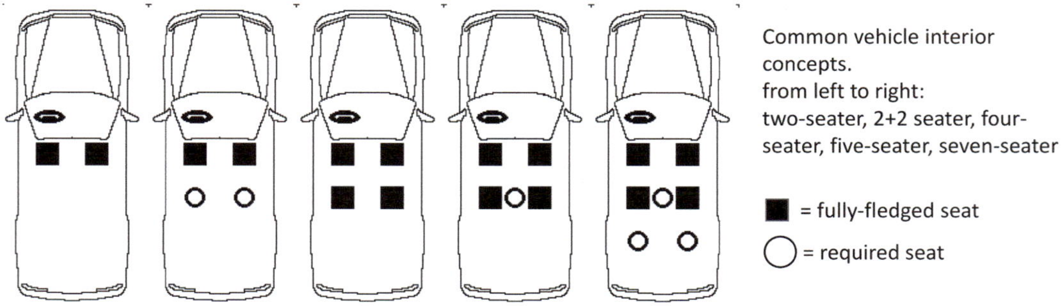

Fig. 7.2 Common vehicle interior concepts

during the journey, as well as the quasi-static[4] to take into account entry and exit procedures. A static occupant posture, as it is suggested in schematic representations and vehicle package plans or seating plans, hardly occurs in reality. The need for dynamic vehicle control requires operation of the steering wheel, pedals and gear lever. For this it is necessary that the driver moves towards the control unit and actuates it over a technically conditioned path. For example, when steering, rotary motion at the steering wheel is necessary, where depending on the steering angle, the hands on the rim of the steering wheel may reach over or over. All installations and attachments on the seat or door trim and dashboard must not restrict or impede this movement. As a result of this movement, not only the hands, but above all the elbows are in motion. Depending on demanded power input, real space-grasping motions can come up. When there was no servo assistance in trucks, the cabs around the large-diameter steering wheels were very spacious to allow manoeuvring. The operation of the pedals requires besides the freedom of movement of the feet with all shoe sizes and types of course also a freedom of movement of the shinbone and the knees to the steering wheel, steering column fairing, instrument panel as well as the door fairing. The gearshift lever, with its mechanically determined shifting paths, requires on the one hand the freedom of movement of the leading hand, which must not be trapped between the front gearshift lever position and the instrument panel. On the other hand, neither the seat nor the armrest of the centre console should prevent the lever from pulling back into the rear gear positions. Body part movements are indispensable not only for operation, but also for orientation. Thus, the head is always in motion when the necessary averting of the gaze goes beyond the movement of the eyes, when looking into the side mirror or the shoulder gaze to the surrounding traffic. The oscillating head movement during dynamic cornering or when manoeuvring the vehicle in confusing traffic situations also requires corresponding freedom of movement, especially in the area of the interior roof lining, the roof handles and the sun visors. Complex movement processes, such as getting in and out of the vehicle (see ► Sect. 7.6) are dependent on many independent parameters[5] and require different free spaces and arrangements depending on the situation. For example, a wide-opening door is advantageous when getting into a vehicle, as it is less obstructive to access. If the driver is seated in the seat, the same door may be inaccessible from a seated position due to the large opening angle, with the result that it cannot be closed. As in this example, there are many such conflicts of objectives, which can only be successfully weighed up if the ergonomic consequences are known before the technical context. This is the task and challenge for ergonomics in vehicle packaging.

In all this, the driver's workplace is the first priority for design. Even though visions of the future propagate autonomous driving and automatic parking is already an available option for some standard vehicles, the driver's seat remains the first priority in the vehicle. Here, all operating and control elements for driving the vehicle are centrally arranged and many auxiliary functions are also within the direct or indirect reach of the driver. A high prioritisation of the driver's seat is also necessary for marketing reasons, as the purchase decision is primarily determined by the appearance of the exterior, but is also influ-

4 Static processes are mainly postures which, in contrast to dynamic movements, do not change significantly in terms of time or location. The term quasi-static processes is borrowed from thermodynamics and actually refers to dynamic processes that are simulated or observed on the basis of discrete static situations. The decomposition of a fast or complex motion process is often a practicable solution for considering and describing individual aspects of an overall context.

5 Parameters of the entry/exit are the height of the person, the vehicle geometry (in particular the size of the entry opening), the entry situation (e.g. constriction by neighbouring vehicles), but also behaviour-dependent habits and unconscious movement strategies.

Table 7.2 Importance of seats for the purchase decision

seat	Coupé or sports car	sedan	station wagon	Van/Minibus
Driving position	70%	60%	50%	50%
Passenger seat	20%	15%	15%	15%
Luggage compartment	5%	10%	20%	10%
Second row of seats outside	5%	10%	10%	10%
Second row of seats in the middle	0%	5%	5%	5%
Third row of seats outside	0%	0%	0%	10%

enced by the appearance and functionality of the interior in the driver's seat. Depending on the intended use, the further priorities shift in favour of the second or third row of seats or to the load compartment (Table 7.2).

7.1.2 Car Dimension Conception According to SAE

Until 1973, vehicle dimensions were not standardized. In September 1973, the SAE[6] standard, drawn up by the Human Factors Engineering Committee, was adopted J1100 for the first time. This standard, or the DIN 70020-1 standard based on it, still represents an essential basis for passenger car dimension concepts today. In SAE J 1100 the following coordinate system is defined with reference to SAE J182: the x-axis runs in the longitudinal direction of the vehicle, the y-axis perpendicular to it in the lateral direction of the vehicle to the right and the z-axis in the vertical direction to the top. The coordinate origin is located in the middle of the vehicle near the front axle (Fig. 7.3). Planes orthogonal to these axes are used to describe the dimensions defined in SAE J1100.

Deviations from this standard result from the fact that the coordinate origin of the US American manufacturers is preferred at the front of the vehicle. The geometric designs of the vehicle components are mainly oriented towards the definition of coordinates according to SAE J1100. On the other hand, coordinate systems are selected for the functional calculations of vehicle movements as part of computer simulations and also for electronic control units in the vehicle, in which the x-axis points in the direction of travel and the y-axis points to the left. The origin of the coordinates is arranged as close as possible to the centre of gravity of the vehicle. The use of these different coordinate systems leads to necessary conversions and transformation requirements for each vehicle development.

A large number of legal texts, directives and other standards refer to vehicle dimensions according to SAE J1100. Among other things, the vehicle dimensions defined in this way are used by vehicle manufacturers to create the so-called GCIE exchange lists (Global Car Manufacturers Information Exchange Group - formerly ECIE, European Car Manufacturers Information Exchange Group) and car package plans, enabling the dimensions of different vehicle manufacturers to be compared directly with one another. This is intended to reduce the enormous effort that every vehicle manufacturer used to put into measuring the special design aspects of a competing product. The vehicle dimensions

6 The SAE (Society of Automotiv Engineers) issues guidelines similar to the VDI (Association of German Engineers). As the "state of the art", they play an important role for the approval of vehicles, as do the VDI guidelines and ISO/DIN standards.

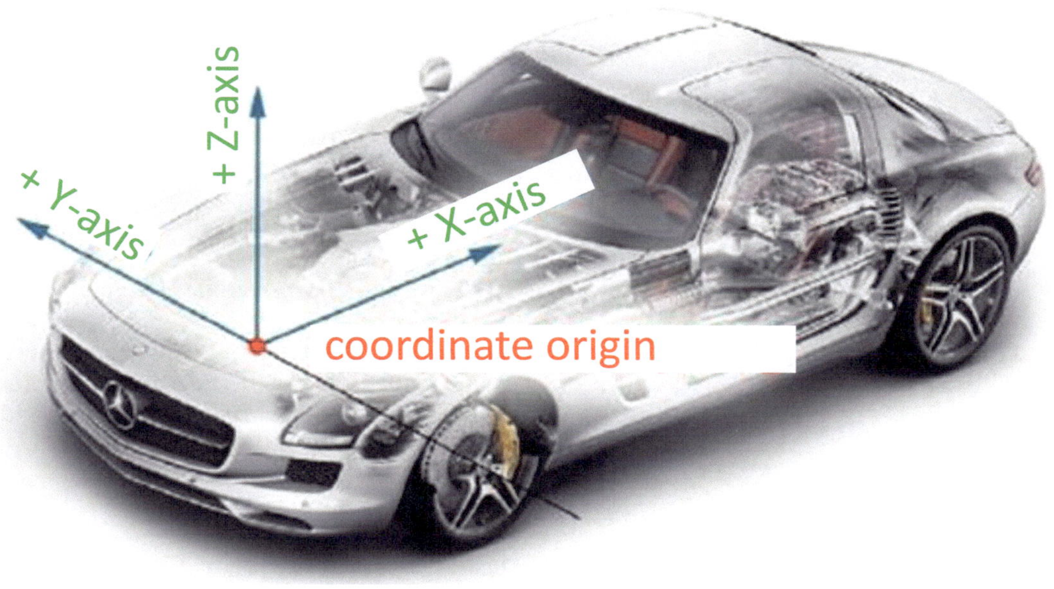

Fig. 7.3 Vehicle coordinate system according to SAE J182. (From Bothe 2010)

are coded by a combination of letters and numbers. Distances parallel to the x-axis are designated with "L" (length), to the y-axis with W (width), and to the z-axis with H (height). Numerical values between 1 and 99 define interior dimensions and numerical values between 100–199 exterior dimensions (Müller 2010). All angles are characterized with the prefix "A" (angle) and volumes with "V" (volume) (SAE J110 and SAE Handbook).

One of the aims of the Human Factors Engineering Committee was to create a body of standards that would ensure that vehicles were designed in a humane manner. The starting point for this was the body outline template first published in 1962 in SAE J826 (see ► Sect. 5.2.1, ► Fig. 5.10). A large number of other SAE standards referring to it were presented in subsequent years. Today, these are the standard for automotive development, especially since a vehicle can only be registered on the US market if it complies with them. They are revised and modified at regular intervals as part of the annual revision of the SAE Standards Catalogue. A variation of this definition of vehicle dimensions is also regularly defined by the GCIE. In this working group the dimensional concept engineers of the largest automobile manufacturers are represented. For this reason, this standard variant is particularly practical and well defined.

The position of the driver's workplace in the vehicle seat is of essential importance for its humane design. From a design engineer's point of view, the seat poses a problem because of its padding and the associated inaccurate position of the driver in it. Therefore in SAE J1100 the seat reference point (**SgRP**) was defined. In relation to the "hard" components of the seat (e.g. seat rail), it indicates the point at which the hip point (H point) of a defined person[7] comes to rest. It is found by the H-point measuring machine according to SAE J826.[8] It is a test specimen that mimics

7 The SAE is based on a procedure that assumes the 95th percentile of a male population with 95% leg length and 95% shoe length (see SAE J826, SAE J1100, SAE J1516 and SAE J1517).

8 The hip point is defined as the pivot point of the thigh in the torso of the H-point measuring machine. It should be pointed out that the hip point measured with the H-point measuring machine does not correspond to that of a real person, as this person adopts significantly different postures depending on individual anthropometric conditions and preferences (differences were measured in the area of 60 mm!).

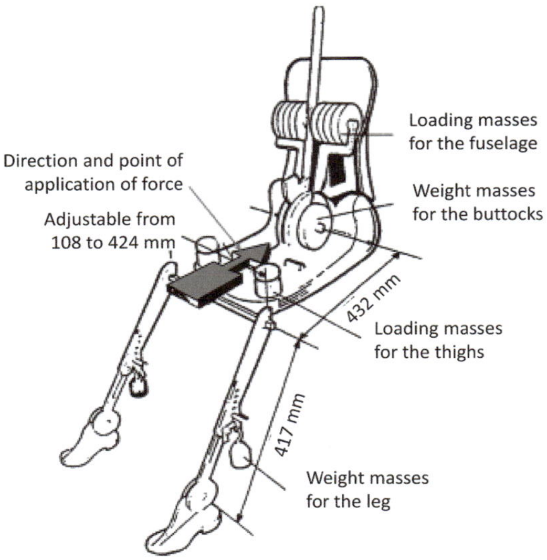

Fig. 7.4 H-point measuring machine according to SAE J826

the buttocks of a tall man (95th percentile), is equipped with mechanical models of the thigh, lower leg and foot (95th percentile U.S. man each) and is loaded with the weight of a 50 percentile U.S. man (76 kg) (Fig. 7.4). This impactor is placed in the seat intended for the subsequent construction of the vehicle in accordance with a very specific requirement. In many cases, the SgRP or the H-point adjustment field is determined in advance at an early stage of the design. It is then the seat manufacturer's task to design the seat in such a way that its H-point adjustment field corresponds to the design specification. The fundamental elaboration of these standards took place in the late 1950s, so that some of the body measurements used in them still refer to American surveys from this period.

The SgRP is the central reference point for the anthropometric design of a vehicle according to SAE. This vehicle-related SgRP, often referred to as the seat reference point or R-point, defines its (theoretical) position for a 95-percentile man. Its vertical distance from the ground plane of the vehicle (more precisely the height of the heel point, Accelerator Heel Point, AHP) is the H30 dimension. Through its influence on the driver's sitting posture, it essentially determines the character of a vehicle. It can range from 140 mm for a low sports car to 400 mm for a VAN/minibus (see Fig. 7.5; for competitive racing cars even negative H30 dimensions are common). Based on the given dimensions of the thigh and lower leg of the H-point measuring machine, the trajectory of the 95-percentile H-point results (Fig. 7.6). This results in further geometric dependencies: The distance X_{95} between the 95-percentile H point and the pedal reference point PRP (more precisely defined as ball of foot, BOF; the distance from the ball of foot BOF to the heel point AHP is given in SAE J1517 with 203 mm and in SAE J4004 with 200 mm) is calculated:

$$X_{95} = 913{,}7 + 0{,}672316Z + 0{,}00195530Z^2 \quad (7.1)$$

The angle A47 of the pedal plane to the plane of the vehicle floor is:

$$A47 = 789{,}6 - 0{,}015Z - 0{,}00173Z^2 \quad (7.2)$$

Figure 7.7 shows some dimensions and reference points defined according to SAE J1100 based on these basic dimensions.

The horizontal distance L53 between the 95-percentile H-point and the heel point AHP results in too:

$$L53 = X_{95} - 203 \cdot \cos(A47) \quad (7.3)$$

The vehicle-related SgRP is the basis for the application of the other SAE J1100 and SAE J826b related standards. Figure 7.8 compiles the main SAE regulations used in vehicle design.

The view from the vehicle is essentially determined by the seating position. Therefore, SAE J941 describes how to determine eye-point positions and eye ellipses based on the SgRP (see ▶ Sect. 7.3.1). SAE J1052 is used to determine the head positions and head contours of drivers and passengers. SAE J287 enables the determination of gripping spaces using the G-factor. The G-factor describes the proportion of men to women taking the type of belt into account. Arm ranges can be read out from tables on the basis of the G-factor. In SAE J1516, the heel point (AHP) and foot

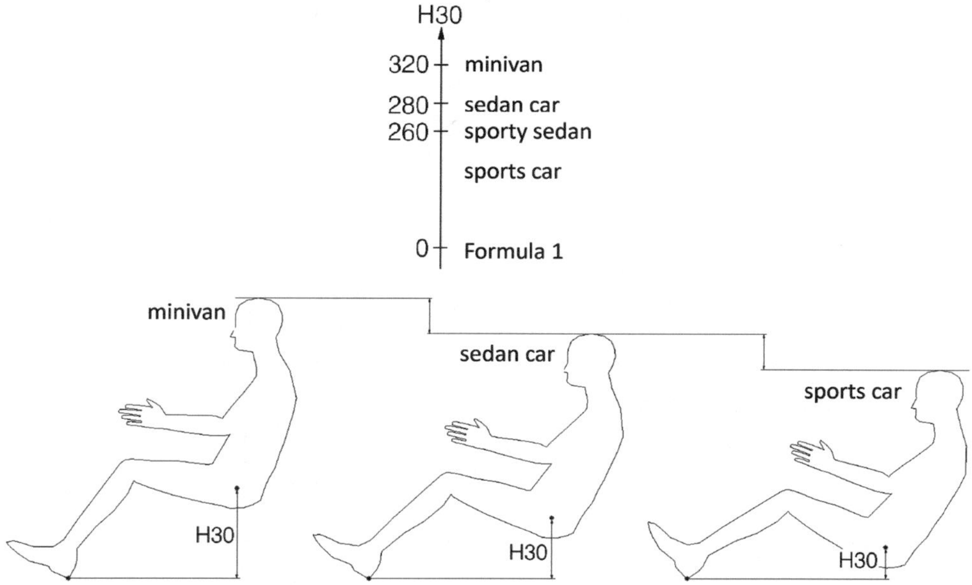

Fig. 7.5

Fig. 7.5 Variation range of the H30 dimension, which essentially influences the character of a vehicle. (From Vogt 2003)

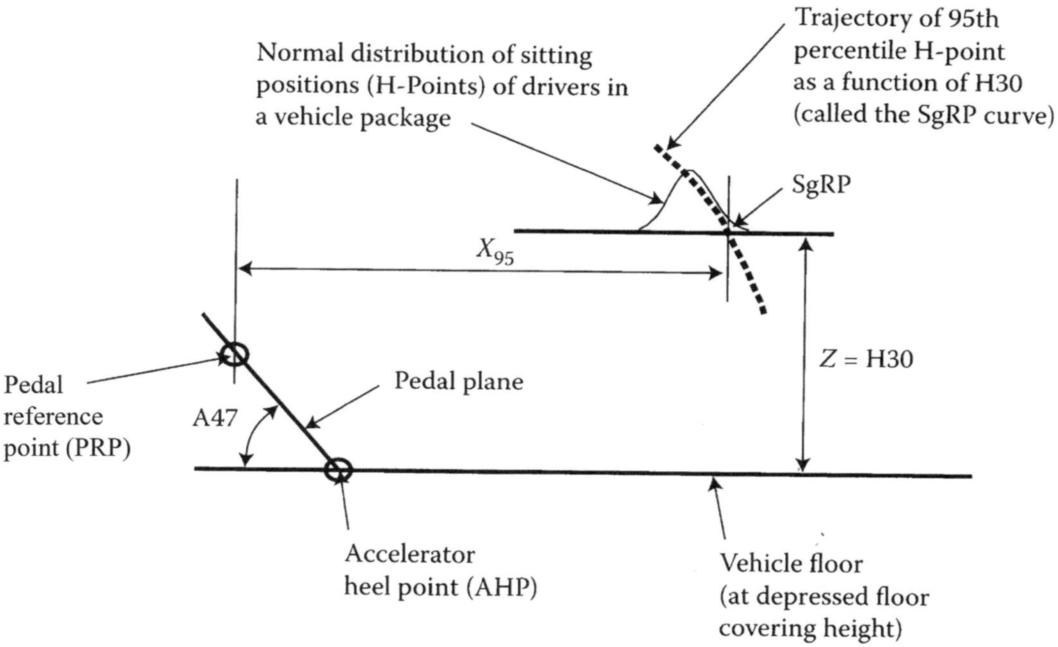

Fig. 7.6 95-percentile H-point position according to SAE. (Representation from Bhise 2012)

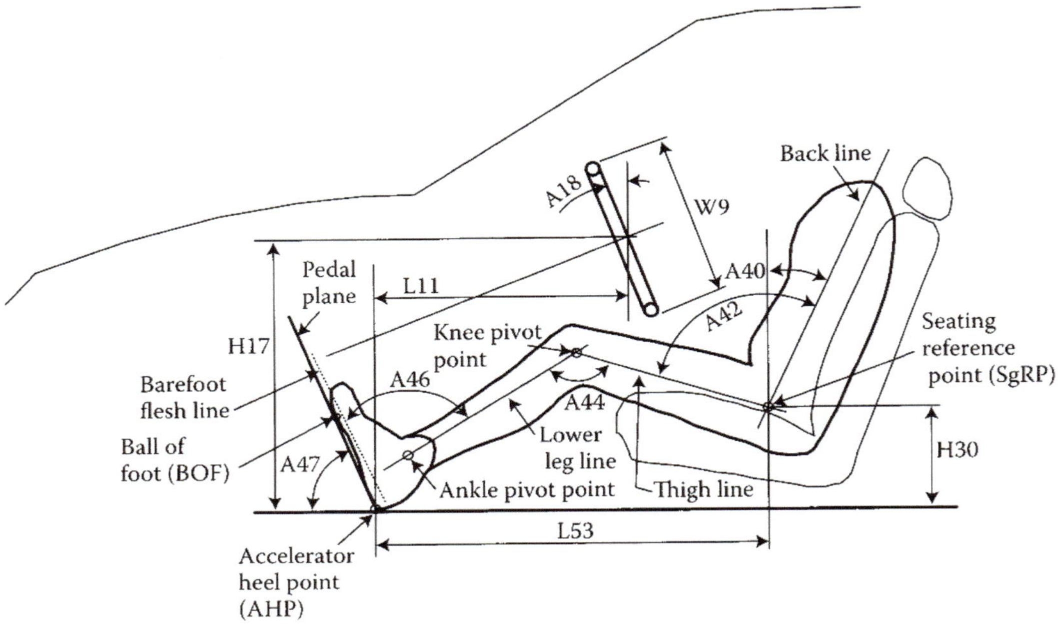

Fig. 7.7 Vehicle dimensions and reference points according to SAE J1100. (From Bhise 2012)

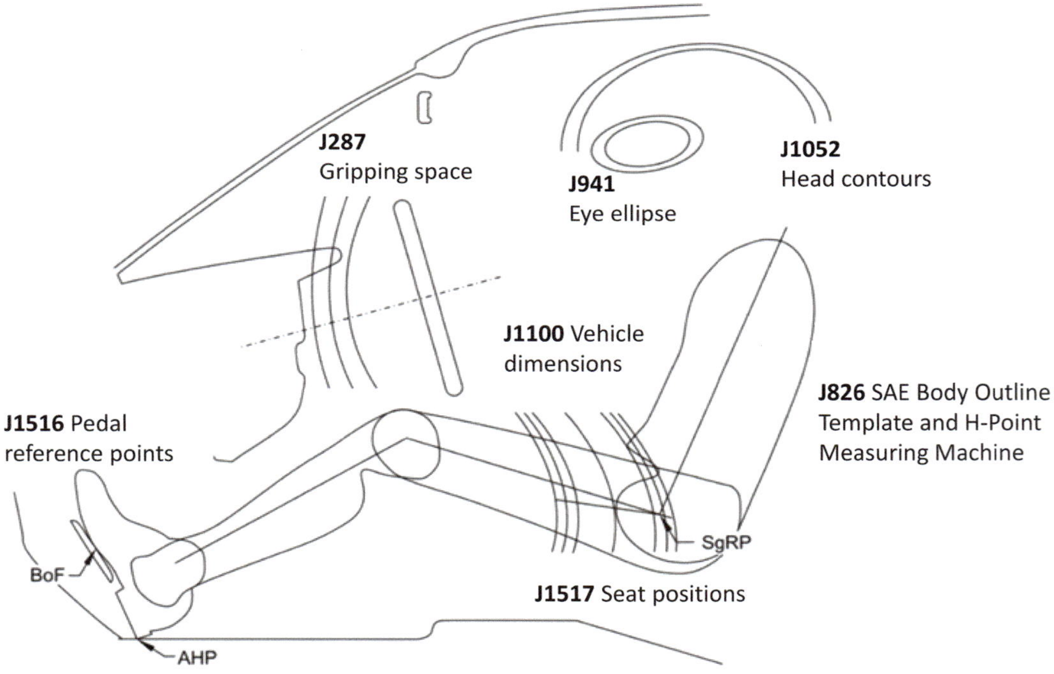

Fig. 7.8 SAE standards relevant to the driver's cab layout. (From Müller 2010)

ball point (BoF) are defined as reference points. In SAE J1517, depending on the seat height (H30–1),[9] the following are calculated the adjustment field of the SgRP-1 is defined numerically. This distance determination of the SgRP from the pedals is not binding according to the European approval regulations. Some vehicle manufacturers choose different stretches of the driver's leg of the template. Depending on the reference points and the seat adjustment position, the angles of the pedals must then be defined.

7.1.3 Fields of Anthropometric Ergonomics

When developing vehicle packaging, several topics have to be considered which influence each other with regard to the comfort or discomfort felt by the user, safety and user friendliness. Their treatment is partly covered by the SAE regulations, all of which have been developed by ergonomics experts on the basis of scientific research - largely on the basis of vehicle models of the US market existing at the time. However, the feedback from customers and the experience of manufacturers sometimes results in deviations from these regulations, which is explained by the technical progress that has now been made. According to Bothe (2010) the following topics are to be treated:

1. **Sitting:** Driving can in principle only take place while seated and since long distances are often covered, the highest demands must be placed on the seats and seating position in view of the generally limited freedom of movement available. In particular, a correct and fatigue-free seating position for the driver is essential for reasons of comfort and safety. The position of the driver is essentially determined by the position of the steering wheel and the pedals. The seat adjustment is intended to enable adaptation to individual conditions to a large extent. In order to cope with long journeys without fatigue, appropriate supports must be designed. The above points also apply, with certain restrictions, to the other passengers in the vehicle. Since the use and wearing of a safety belt is mandatory to ensure passive safety, special attention must be paid to its correct handling and course, taking into account different anthropometric conditions.
2. **Sight:** Via 90% of the information to be received for driving is via visibility. Therefore, the design of the technical elements that can impair or support visibility, as well as the aforementioned sitting, is of elementary importance. This design primarily refers to the so-called greenhouse[10] (also called glass house), which comprises the windscreen, rear and side windows as well as the columns separating them from each other (starting at the front of the vehicle referred to as A, B, etc. columns) and the roof. The view cannot be seen independently of sitting. The driver often adjusts his seating position in such a way that he/she - subjectively - has an optimum view of the driving environment. He/she may then compromise his/her position on the steering wheel, pedals and other controls, which can lead to discomfort during long journeys. A distinction has to be made between the direct forward view of the roadway, traffic signs and signal systems and the rearward view through the rear window e.g. for reversing and the indirect view through interior and exterior mirrors. When viewing operating and display components, the readability of text and symbols must be observed. The instrument cluster, warning displays and the central display in the centre of the vehicle are important visual targets. In particular, the concealment by the steering wheel and steering column lever must be taken into

9 "-1" means that this is the 1st (front) row of seats.

10 The term "Greenhouse"comes from the field of horticulture: glass roofs over greenhouses are called this way. The term has extended to the field of architecture, where artificial landscapes created inside a building are separated from the outside world by a large glass roof.

account. Unfavourable constellations of external light sources, shiny surfaces in the dashboard area and inadequate instrument covers can lead to unwanted reflections which can irritate the driver or prevent information from being picked up.

3. **Operating and display components:** The primary focus here is on the ergonomic design of the instrument panel, centre console, door, seat and headlining, as well as all controls around the steering wheel, pedals, gearshift lever and handbrake. The main focus is on the driver. In the development of a chauffeur vehicle with a high-quality equipped rear area, the design must also be extended to this area. The criteria are accessibility and usability. This means that different anthropometric conditions have to be considered as well as the anatomical/biomechanical conditions of movement possibilities and limitations.
4. **Room feeling:** As already mentioned, freedom of movement in the narrow vehicle cabin is restricted in principle. Therefore, special attention must be paid to the shoulder and head area as well as the leg and foot areas for the driver. Due to the symmetrical design of the greenhouse, the conditions found for the driver also apply to the passenger, with the exception of the glove box opening into the knee area. Depending on the exterior design, the space required by the upper part of the body, especially the shoulder and head space, may be impaired by the roof retraction and the side retraction of the greenhouse, especially in the case of saloons and coupés. In the second row, as in the third, the room feeling is significantly influenced by the distance between the knee and the seat back and the possibility of placing the feet under the front seat. If the seat is installed against the direction of travel, components such as the tailgate, the rear body end and the like must be taken into account. For vehicles with a standard drive system, the access level is an important criterion, especially in vehicles with a generously dimensioned rear end.
5. **Entry and exit:** This area is particularly important in view of the aging clientele. Decisive criteria here are the available range of motion, especially with regard to foot, knee, buttock, shoulder and headroom. The height and lateral position of the door sill play an important role, as does the position and inclination of the A and B pillars for the front seats. In a four-door vehicle, access to the rear is significantly influenced by the position of the B-pillar in the foot area and also by the height and lateral position of the door sill as well as the position of the access opening restricted by the roof. In addition, there are the size and opening angle of the doors, whereby in the case of unimpeded entry and exit and the same procedure in narrow parking spaces, different requirements may have to be met. A special topic here in a two-door vehicle is access to the rear seats. The controls used to operate the doors must also be ergonomically optimised.
6. **Loading:** Access to the luggage compartment of a combined vehicle is essentially determined by geometric characteristics such as the height of the loading edge, the loading floor, the opening angle of the boot lid or the rear wall door. In addition, ease of use plays an important role, both in terms of the type of handles used and the movement the user makes when opening and closing the door.
7. **Service:** Certain service tasks must be performed by the user of the vehicle. In this context, the position and ease of handling of the fuel filler cap (or, in the case of electric vehicles, the corresponding device for attaching the charging cable) is of primary importance. However, some activities must also be carried out in the engine compartment area. Therefore, opening the hood should be treated in a similar way to accessing the luggage compartment. In the interior of the engine, the service objects to be used by the driver must be installed optimally from an operating point of view, in particular the filler neck for the wind-

screen washer water and for the engine oil. Checking the oil level in the engine using the oil rod can also be a problem for people of different sizes and mobility.

Some of the issues addressed are highly interdependent. In order to take this into account, various methodologies have been developed for car measurement concepts. Müller (2010) provides an overview that also takes historical aspects into account and presents an approach to car dimensional design that places the ergonomic and anthropometric requirements of vehicle users at the centre of vehicle development. With its application, vehicles are designed centrifugally around the vehicle occupants in nine work steps, taking into account the main requirements. Vehicle dimensions of reference vehicles to be defined as well as ergonomic specifications form a non-binding design framework. The car dimension concepts generated in this way are to form the basis for interior and exterior design.

From today's perspective, the car measurement concept of the ASPECT program (Automotive Seat and Package Evaluation and Comparision Tool) is particularly worth mentioning, the research results of which have been documented in numerous publications. Reed et al. (1999a) provides a detailed overview of this. On the basis of the ASPECT programme and own experiments, Reed et al. (1999b, 2002) has developed the methodology of the "ASPECT" programme "***cascade prediction model***"for driver seat design, which is summarised in Fig. 7.9.

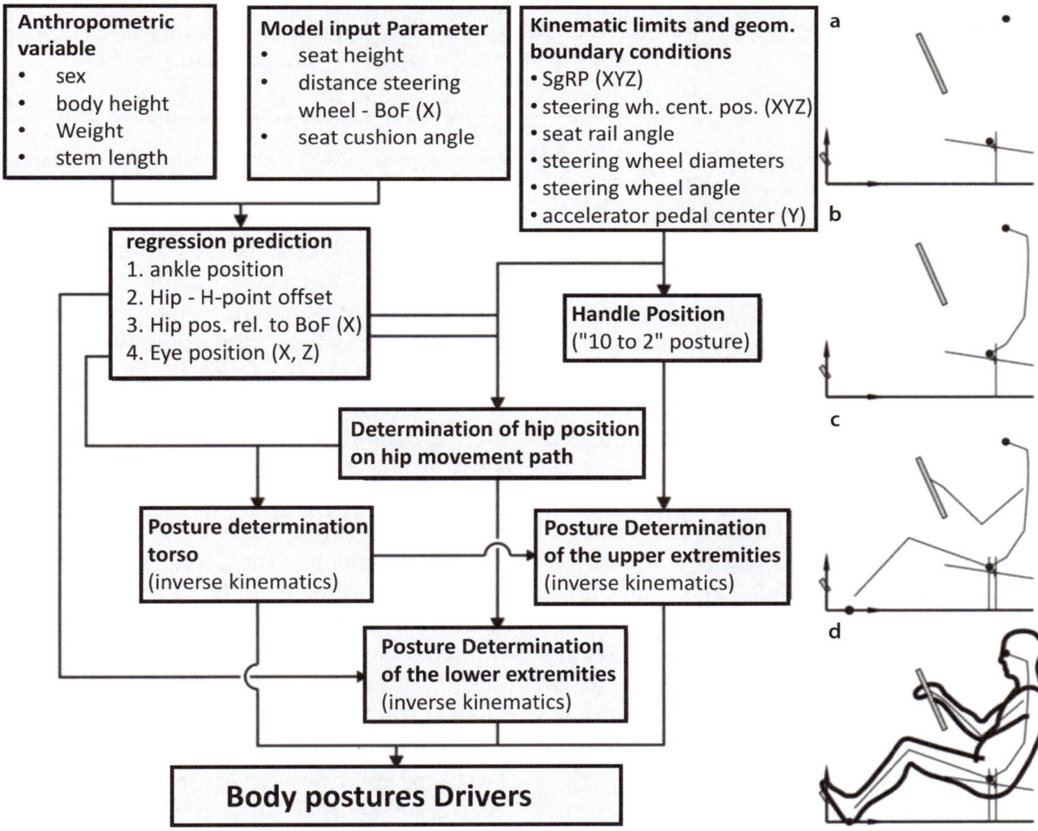

Fig. 7.9 Flowchart and workflow of the "cascade prediction model"(after Reed et al. 2002; quoted from Müller 2010). **a** Prediction of hip and eye position, **b**: by inverse kinematics determined suitable torso posture, **c** by inverse kinematics, the appropriate posture of the extremities is determined, **d**: complete human modelling in a suitable posture

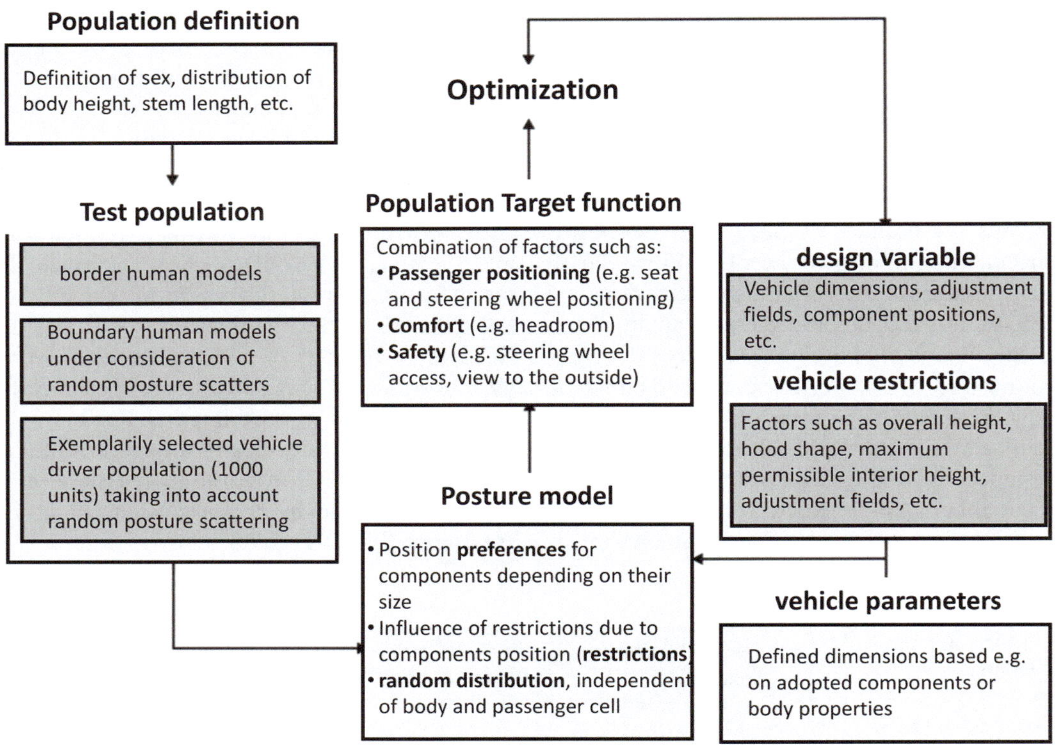

Fig. 7.10 Schematic representation of the car dimension concept with optimization methods according to Parkinson et al. (2006; quoted from Müller 2010)

On the basis of the cascade prediction model, Parkinson et al. (2006) developed the so called "*optimization method*". It links different boundary conditions and is particularly suitable for integration into software applications for digital human modeling. Figure 7.10 gives an overview of the modules of this method and their mutual connection.

Various software applications have been developed which, on the basis of human models and controlling parameters, enable an automatic derivation of results and thus allow different concept variants to be compared with each other particularly quickly. These software systems are described in the following using the compilation by Müller (2010). The most important systems in this context are those that are directly linked to the human models already used in automobile development.

The digital human model **RAMSIS** represents the module "**Package Designer**"for the anthropometric design of the driver's workplace (Human Solutions 2000, 2002, 2008). The parameters used are a parameterizable seat, a parameterizable steering wheel, heel and pedal point. Seat adjustment and steering wheel adjustment are freely selectable for the user. Depending on the steering wheel angle and seat reference point, the vehicle headliner, floor and footrest are then determined. All resulting dimensions are output according to SAE J1100. RAMSIS makes it possible, on the basis of extensive experimental surveys, to calculate the posture that is most likely to be adopted under the given conditions. The posture calculated in this way is evaluated for specific parts of the body. A module "Seat Belt Design"allows a very detailed analysis and design of the seat belt course depending on the posture. The newly developed module "RAMSIS cognitive"also enables an analysis of all visual conditions, including glare, HUD design, influence of varifocals and similar, which goes well beyond the usual visual calculations. With the module "Standards & Regulations"the most important regulations can be evaluated according to the SAE regulations.

Also for the human model **Jack** there is a program for the car dimension conception: The software application "Classic Jack" can be extended by the module "***Occupational Packaging Toolkit***". The version "*Standard*"immediately implements the SAE standards. By means of the "*Enhanced*" version, the postures of vehicle occupants can be calculated using the cascade prediction model (Siemens 2010). The model for calculating body posture takes into account body heights and vehicle type-specific dimensions. In addition, comfort can be assessed by comparing posture with qualified posture models. The software also provides a tool for visual analysis. A further application in this context is the "NX General Packaging"of Siemens PLM (Siemens 2009), which supports the user in checking a large number of SAE standards and in complying with directives relating to vehicle registration. In addition to human modelling on the basis of SAE J826, the Jack human model also offers the possibility of modelling the postures of vehicle occupants and evaluating them on the basis of stored databases (Xiaoxiang 2007).

The CAD system CATIA, which is widely used in the automotive industry, provides the optimally implementable software module "***Vehicle Occupant Accommodation***" at your disposal. It uses the human model implemented in CATIA "***Human Builder***"(Dassault Systems 2009).[11] The measurement concept is based on the application of the cascade prediction model described above and the optimization method. The company Transcat provides the software CAVA (CATIA Automotive-Extensions Vehicle Architecture) as a supplement. This is mainly used to check compliance with the law during vehicle development. It is composed of the following modules: "CAVA OVA" (Overall Vehicle Architecture) allows you to determine the dimensions of the floor assembly according to specific requirements. The "CAVA Manikin" module is used for vehicle interior design using the SAE body outline template defined in SAE J826. The module "CAVA Vision" allows to check requirements related to the direct and indirect vision of the driver. The "CAVA Safety" module enables aspects of passive safety, such as pedestrian protection (design of the hood and windscreen) and vehicle occupants, to be checked virtually. The "CAVA Wiper"module can be used to model the windscreen wiper movement and, in combination with the CAVA Vision module, to examine the relationship to the field of vision.

In addition to these systems which are directly connected to the established human models, a number of software developments which are independent of them were also presented.

The automotive supplier company **Visteon,** which was spun off from Ford Motor Company in 2000, provides the software "GENPAD" (GENeric PArametric Design), which takes ergonomic aspects into account and provides over 50 prefabricated car interior measurement concepts (McGuire et al. 2002). The software is integrated into the CAD system *pro engineer* so that the prefabricated car interior dimension concepts can be easily modified. The 3-D human model of Pro-Engineer is used. The software allows to evaluate visual obstructions, reflections, gripping spaces and the position of the gear selector lever. By specifying safety zones in the area of the pelvis, head and knees, passive accident protection can also be evaluated.

At the University of Michigan-Dearborn, a software application called "A Parametric Model for Automotive Packaging and Ergonomic Design"was developed, where a selection of vehicle exterior-specific properties and dimensions is the basis for positioning passengers in the vehicle interior. Using a simple, self-developed human model, ergonomic analyses can then be carried out, especially with regard to the gripping space

11 Negotiations are currently underway between Dassault and Human Solutions with a view to further cooperation and development with a view to integrating RAMSIS into the CATIA system. RAMSIS has far more extensive tools for vehicle development than Human Builder and is widely used in the automotive industry in particular.

and visibility (Bhise et al. 2004; Bhise and Pillai 2006).

A comprehensive geometry generation for vehicle development was presented by the Institute of Automotive Engineering of the TU-Graz and the company Magna (Hirz et al. 2008a, 2008b; Rosbacher et al. 2009). Essentially, it describes how vehicle concepts can be parameterized using modern CAD applications. Geometric dimensions, ergonomic data, safety-relevant data and data guaranteeing vehicle registration are retrieved from external databases. The SAE body outline template defined in SAE J826 and the standards associated with it form the basis for the car dimensional concept.

7.2 Sitting

7.2.1 Consideration of Different Anthropometries

Already in the thirties one began to consider the individual anthropometric conditions by the mechanism of a seat adjustment. Using scientific anthropometry, which categorizes the body elements in the form of percentiles, it seemed reasonable to define a range of percentiles in order to consider a certain percentage of the population as potential users of the vehicle. In the field of ergonomics it is often recommended to consider the range from the 5-percentile-woman to the 95th percentile-man. In this way one theoretically covers 95% of the population. 97.5% is taken into account when extending the range from 2.5% woman to 97.5% man.

According to SAE 1517, the distance between the frontmost SgRP of the little woman and the ball of the foot point BOP is then[12]:

$$X_{2,5} = 687{,}1 + 0{,}895336Z - 0{,}00210494Z^2 \tag{7.4}$$

and analogously for the distance between the rearmost SgRP of the tall man and the ball of the foot point BOP

$$X_{97,5} = 936{,}6 + 0{,}613879Z - 0{,}00186247Z^2 \tag{7.5}$$

the necessary adjustment range to be taken into account to satisfy 97.5% of the population:

$$TL23 = X_{97,5} - X_{2,5} \tag{7.6}$$

The percentiles mentioned refer to the body height. However, in narrow vehicle cabins in particular, special problems arise because there are individual differences in proportions which have a significant effect on the dimensional concept. For example, a long-legged little woman has to move the seat a little further up to still have a sufficiently good view of the road because of her short upper body. It may collide with the lower edge of the steering wheel. A short-legged little woman, on the other hand, has to push the seat very far forward in order to operate the pedals correctly. Her upper body collides with the steering wheel if it cannot be moved sufficiently in the longitudinal position. Similar problems arise for the large short-legged or long-legged man. This consideration alone shows that a mere longitudinal shift of the seat is not sufficient to take account of different anthropometric conditions.

Because of the many influencing variables that determine posture, the use of human models is very helpful and is becoming more and more important with the use of CAD techniques in design (see the overview of software in ▶ Sect. 7.1.3). However, many procedures in vehicle development are still based on company-specific empirical values and procedures that have evolved over time. Human models are used in industry less for initial design than for the verification of concepts. It is therefore desirable to develop an objectively comprehensible procedure that takes ergonomic criteria into account from the outset without neglecting practical limitations. Taking into account the importance within the automotive industry, the RAMSIS

12 The formulas published in SAE are based on percentile values from American tables from the 1970s.

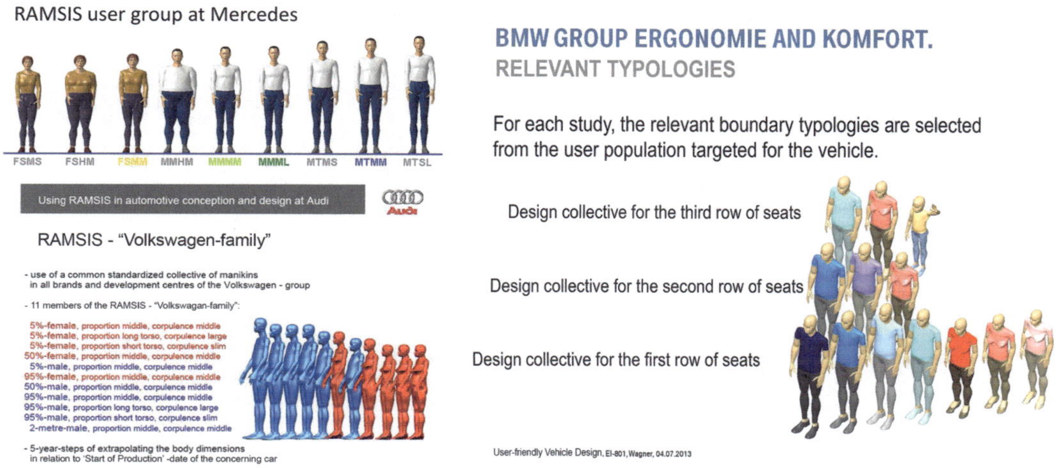

Fig. 7.11 Examples for "RAMSIS families" as used by Volkswagen/Audi, BMW and Mercedes

human model will be used in the following, although many of the applications presented here can also be implemented with other human models.

The human model RAMSIS allows a selection according to the anthropometric criteria body height, corpulence and proportion. Without considering possible intermediate types (which can be created by the body-builder), 45 different male and 45 different female manikins are immediately available for design tasks (see ▶ Sect. 5.2.2). From this collection, each car company puts together a so-called RAMSIS family, which is used for checking and designing. If necessary, this family also receives modifications within a manufacturer depending on the knowledge of marketing about the buyer characteristics of specific model series. Figure 7.11 shows examples of the composition of this RAMSIS family at different vehicle companies.

Table 7.3 gives an example of a compilation of such RAMSIS manikins, including the important anthropometric data, which is also partly used in the following illustrations. To characterize the manikins, however, different short names are used in the different companies (see also Fig. 7.11). In most cases, the extreme types are of particular interest (short designation according to Table. 7.3).

- Very large long-legged man of medium obesity (MTMM),
- Very large short-legged fat man (MTHS),
- medium tall man of medium obesity with medium proportions (MMMM) or very tall woman of medium obesity with medium proportions (FTMM),
- very small, fat, short-legged woman (FSHS)
- very small, thin, long-legged woman (FSSL)

In their body size, the manikin "very big man" corresponds to the 95th percentile, the manikin "medium-sized man" or "very big woman" to the 50-percentile man and the manikin "very small woman" to the 5-percentile woman.

7.2.2 Driver

7.2.2.1 Seat

Sitting Posture

Due to the severely restricted freedom of movement and the resulting posture, which is often almost unchanged over a long period of time, the highest attention must be paid to the sitting position, especially for the driver's workplace. The problem with this is that people are very tolerant with regard to their posture - apart from individual movement restrictions - and barely notice incorrect postures, at least in their first perception. In research in this context, special attention is paid to the so-called comfort angles.

Table 7.3 Compilation of frequently used RAMSIS manikins in connection with driver position analyses (short names according to Rekittke and Brückner 2010)

	Size	Corpulence	Proportion	Body height [mm]	Stem length [mm]
Man (M)	Very large (T)	Medium (M)	Medium (M)	1878	974
		Thick (H)	Short-legged (S)	1880	1010
		Thin (S)	Long-legged (L)	1896	953
	Medium (M)	Medium (M)	Medium (M)	1766	929
	Very small (S)	Medium (M)	Medium (M)	1651	879
Woman (F)	Very large (T)	Medium (M)	Medium (M)	1750	919
	Medium (M)	Medium (M)	Medium (M)	1647	876
	Very small (S)	Medium (M)	Medium (M)	1550	833
		Thick (H)	Short-legged (S)	1539	856
		Thin (S)	Long-legged (L)	1549	808

7

Objectively these find their reason in the fact that there is a neutral position for each muscle, in which it is neither contracted nor extracted. In principle, these angles can only be found experimentally in a relaxed situation in a weightless state. The angle values published by NASA are therefore often used as a reference (Fig. 7.12). Dirlich (2010) carried out special underwater experiments with untrained test persons, whereby a relaxed state of the test persons was brought about by a special experimental technique. There are already clear differences to the values published by NASA. Table 7.4 provides a compilation of comfort angle values published in the literature on an experimental basis. Although there are common intersections between these angle values, the respective ranges of variation show considerable differences. The values specified for RAMSIS are the most probable values found in the observation. An optimization algorithm, taking into account the entered restrictions (e.g. hands on the steering wheel, feet on pedals and foot support, buttocks on the H-point, see below for more details) and the interdependence of the angles, ensures that the most probable position is found. This position is then evaluated with regard to the comfort to be expected, whereby multiple regressions between the angular positions of the joints observed in the experiment and the comfort statements in standardized questionnaires were used here as a prediction model (Krist 1993). The investigation by Lorenz (2011) shows the corresponding range of values that can be found in practical experiments. The posture adopted by RAMSIS, which is not restricted by any restrictions (Fig. 7.13), is therefore not the most physiologically favourable under all circumstances. In any case, this is the attitude probably assumed by test persons under the given restrictions, which has also been confirmed by evaluation experiments (Seidl 1993; Kolling 1998; Nilsson 1999). It may be advisable to calculate the body posture using the RAMSIS H-30 module, which supplements the original seat posture model by interpolating between the car and truck seat posture modules so that correct body postures are also possible at H30 dimensions and significantly greater than 260 mm (e.g. for the design of the driver posture in VANs and SUVs).

Despite the limitations mentioned above, the RAMSIS body posture is used for the first of the following considerations, which results if no external restrictions are set (see Fig. 7.13).

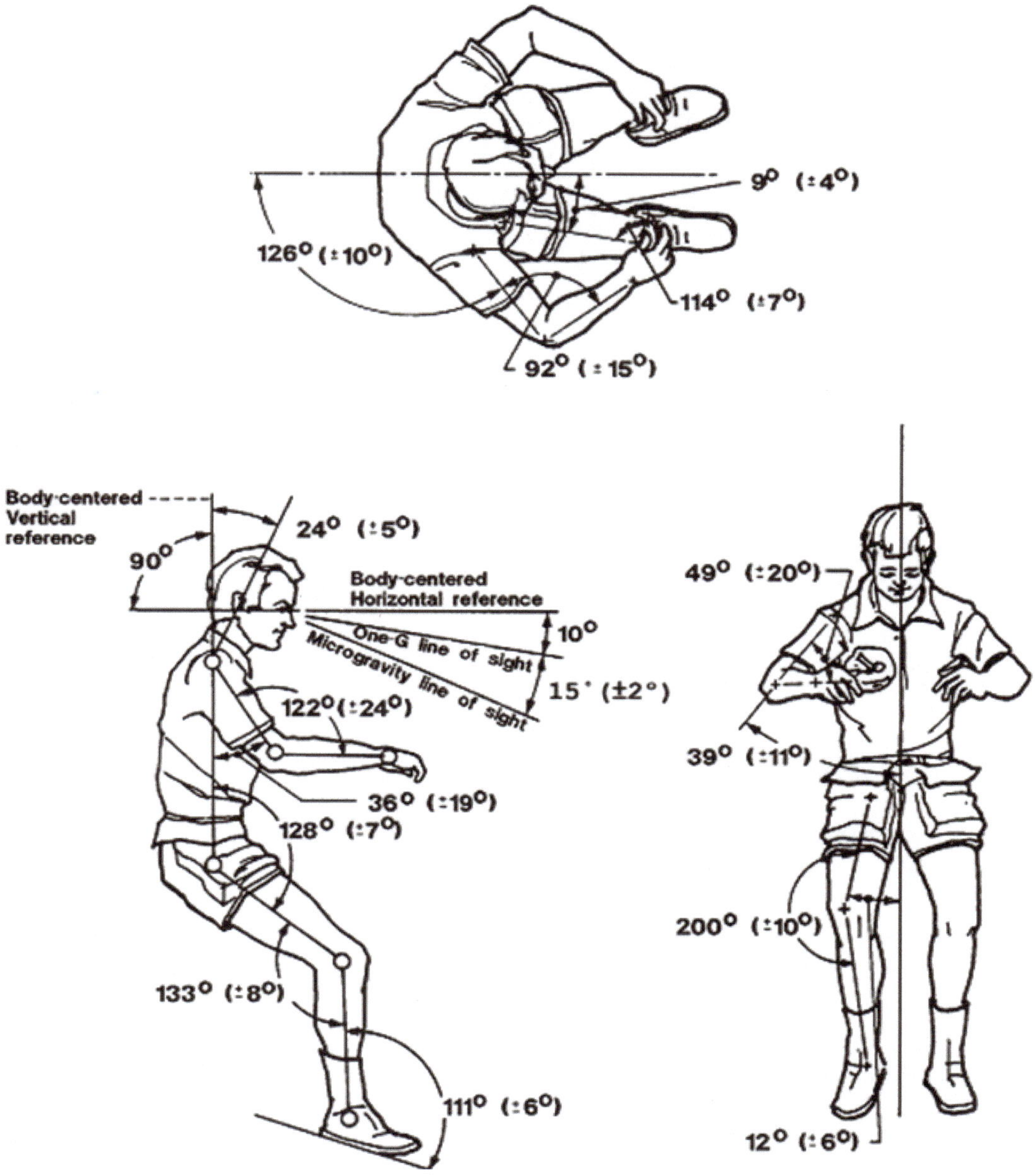

Fig. 7.12 From NASA reported angles observed on three astronauts in a relaxed posture

Adjustment Range

By adjusting the seat, i.e. in the first step by varying the H-point, it is possible to meet the requirements of different anthropometric conditions. In order to carry out a correct design in this respect using a human model, it is necessary to determine body points on this model, the variation of which reflects the necessary adjustment ranges for adaptation to the various anthropometries. Figure 7.14 uses RAMSIS as an example to show the body points that are absolutely necessary for an initial application.

In order to arrive at a solution that largely takes ergonomic requirements into account and is independent of a given vehicle concept, it is necessary to fix one of these body points and then investigate which variations result for

Table 7.4 Compilation of the comfort values found experimentally by various researchers angle values in side view

investigator	torso	shoulder	elbow	hip	knee	Footg.
NASA (1995)	–	36° ± 19°	122° ± 24°	128° ± 7°	133° ± 8°	–
Dirlich (2010)	–	27° ± 8°	105° ± 5°	145° ± 4°	144° ± 6°	–
Dreßel and Kain (1985)	25° ± 3°	39° ± 12°	146° ± 17°	107° ± 7°	122° ± 8°	84° ± 16°
Wallentowitz (1995)	20°–30°	28°	105°–115°	100°–105°	110°–130°	90°
RAMSIS	27°	22°	127°	99°	119°	103°
Kahlmeier and Marek (2000)	15°–25°	15°–35°	85°–110°	85°–110°	95°–120°	85°–95°
Hirao et al. (2006)	31° ± 4°	–	–	111° ± 5°	125° ± 9°	158° ± 12°
Lorenz (2011)	27° ± 4°	33° ± 10°	124° ± 18°	100° ± 6°	111° ± 7°	92° ± 8°

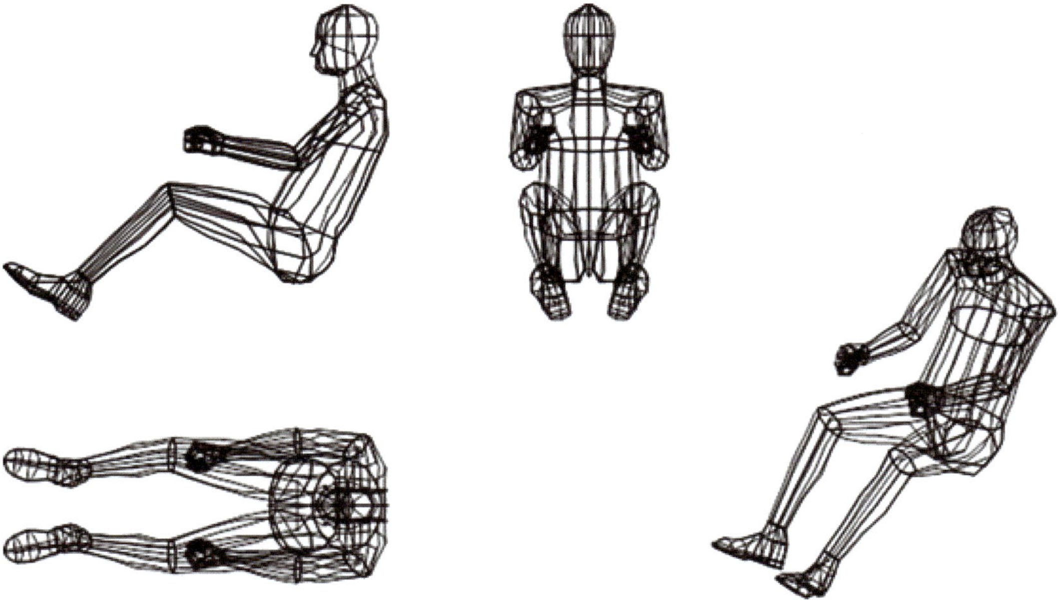

Fig. 7.13 Four views of RAMSIS in restriction-free position

the other body points with otherwise free posture optimisation. Vogt et al. (2005) have done this with the body points defined in Fig. 7.14. Figure 7.15 shows the results for the two most practice-relevant variants. The eye-point-fixed variant, which by the way is used for the design of aircraft cockpits, would have the advantage that drivers of any height and proportions would always have the same visibility conditions, which could then be optimized for this purpose. As can be seen from Fig. 7.15, this would, however, be associated with a large adjustment range, especially of the pedals, which is ruled out for many practical reasons. With the fixed heel point variant, which is common in today's vehicles, the result is what

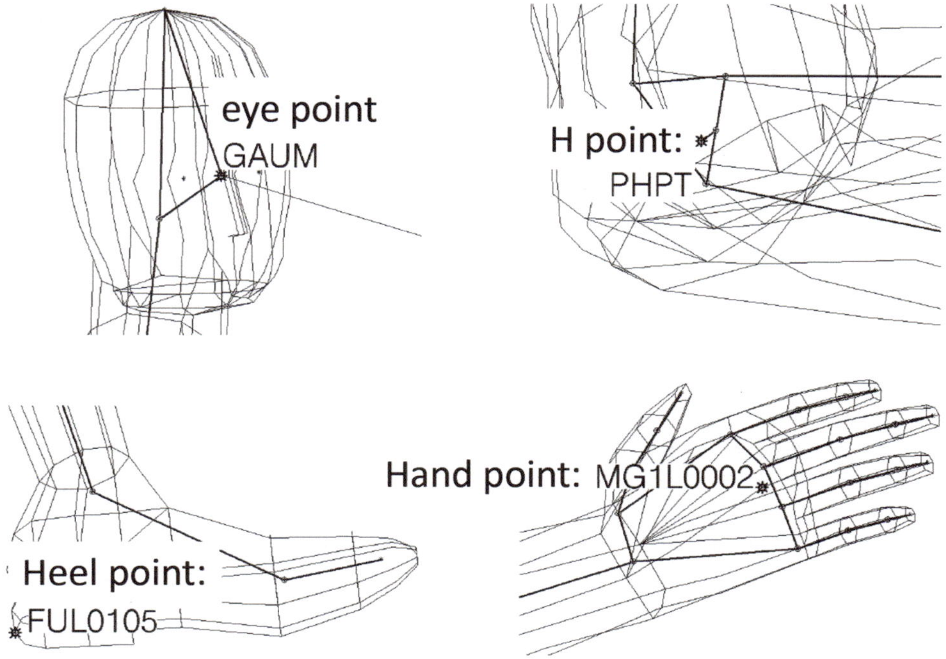

Fig. 7.14 RAMSIS body points. (From Vogt 2003) absolutely necessary for the design of the driving area

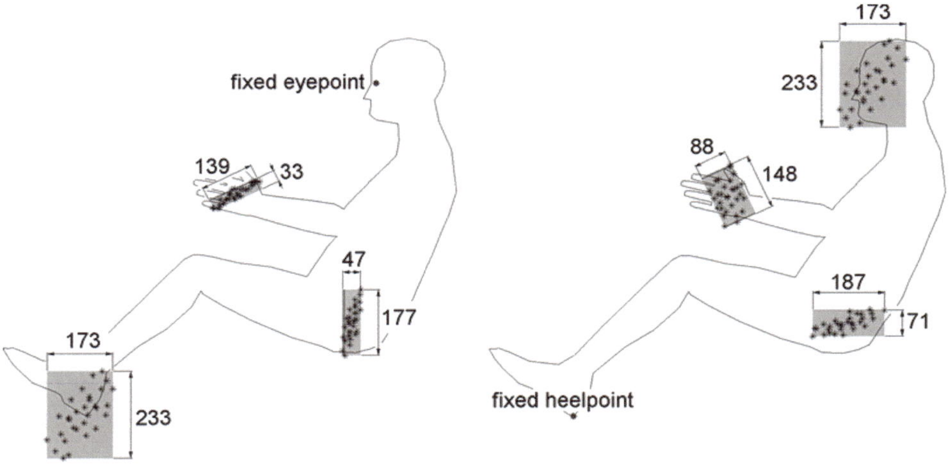

Fig. 7.15 Eye point and heel point fixed arrangement of 30 different RAMSIS manikins without further restrictions

at first glance seems to be a curious and apparently contradictory insight: small people sit lower (and of course further forward) and large people sit higher (and of course further back, see Fig. 7.15 right) in an optimal position. In practice, the opposite strategy is adopted, since the eye of small people has to be raised in relation to larger people so that their line of vision is above the steering wheel and bonnet. Ultimately, the last result presented means nothing other than that in order to maintain equal conditions - at least theoretically - small people needed smaller cars and larger people needed larger cars. The completely free setting from external boundary conditions in the experiments of Lorenz (2013) also shows that people set the dimensions of the driver's seat quasi according to their own body size, i.e. the dimensions of the driver's seat grow linearly with the respective anthropometric dimensions.

In practice, such a solution, which only fulfils ergonomic requirements, is not feasible. This is due to the fact that, as mentioned above, the buying behaviour of customers is essentially influenced by the attractiveness of the exterior. In extreme cases, it may even be necessary to ensure that the customer group fits - somehow - into the attractive exterior design. In fact, however, there are generally acceptable compromises that are made in mutual agreement between design and ergonomics - and of course the representatives of various other technical requirements (e.g. from the field of passive safety). In order to realise such compromises, RAMSIS software sets so-called restrictions which influence the calculation of the posture. Various studies have dealt with the restrictions that have to be imposed. Kolling (1998) found, for example, that a minimum distance of 50 mm is maintained from the roof, particularly by tall people, with this subjective requirement being met by adjusting the backrest if necessary. The following restrictions are set for all calculations:

- Heel points right and left on the heel line; the right foot keeps the accelerator pedal depressed 1/3.
- H-point in seat centre plane
- Fix the pelvis laterally against tilting and twisting
- Head point below the roof level with 50 mm headroom
- Fix the viewing direction in the neutral position.

The SgRP is found by the SAE man implemented in the RAMSIS software, whose anthropometry is modelled on the H-point measuring machine. For this SAE man, in addition to the above restrictions, the H point is placed on a plane at H30 height. In the next step, the posture is calculated for the manikin "very tall man, medium obesity, short-legged" (MTMS) with the specified restrictions, resulting in the highest eye point position. From this point of view, a tangent is now placed over the hood (roof, hood as well as the already mentioned H30 line are sizes determined by the exterior design), defining the optimum view of the road under the given conditions. For all other manikins with their different anthropometries this line is defined as further restrictions for the eye point. The result is shown in ◻ Fig. 7.16.

There are two striking features compared to solutions that have actually been implemented:

- The necessary adjustment field for the steering wheel in the longitudinal direction is much larger than normally realized. In addition, it should be noted that the fre-

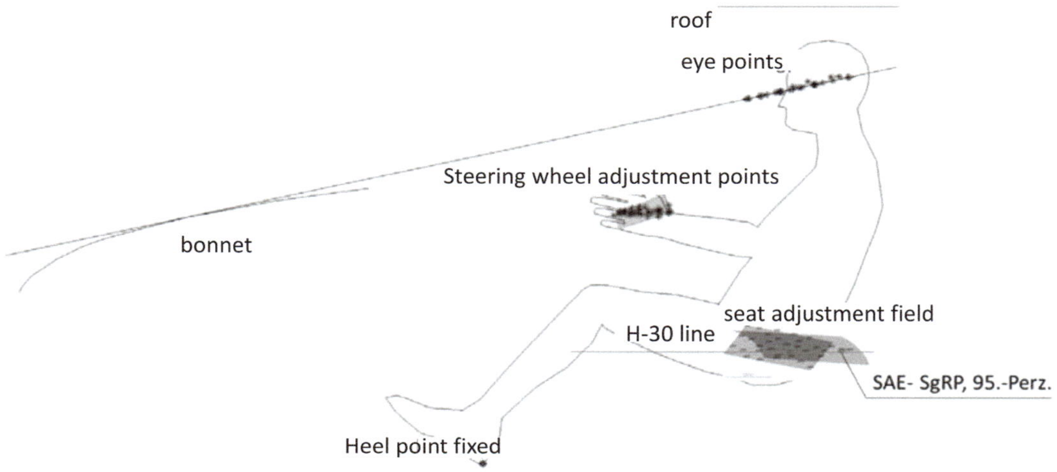

◻ **Fig. 7.16** Eye point optimized posture and necessary adjustment fields

Fig. 7.17 visibility conditions affecting driver posture. (Lorenz 2013)

quently realised steering wheel adjustment, which consists of swivelling the steering wheel around a line aligned with the y-axis, is far from sufficient to satisfy people of different body sizes and proportions.

- The alignment of the necessary seat adjustment field is exactly the opposite to the realised solutions. This may be due to the fact that the technical realisation of a seat height adjustment with a parallelogram mechanism prefers the reverse solution.

In the previous paper only a few specifications are made, namely heel point, height of the H-30 line, roof height and position of the hood. In practice, the seat adjustment is made on the basis of further restrictions, some of which are also due to the existing technical implementation. Lorenz (2013) found out both by interviewing the test persons and in his experimental observation that the accessibility of the pedals across all body heights is the leading criterion for the orientation of the sitting posture. In addition, the quality of the visual conditions plays an important role for test persons who have grown small. With increasing body size, however, the relevance of the view to the outside decreases again, whereas the importance of the space requirement increases. In detail, the following results can be seen: The position of the wind running point (see also Fig. 7.17) has a recognizable weight for small and medium persons for the seat adjustment, but not for large persons. For small persons, the road view has a high weight, which decreases with increasing body height. On the other hand, the roof frame position only takes on a certain weight with large persons for seat adjustment. None of the passenger groups perceives the height of the parapet as an influence on the seat adjustment, although subjectively a lower one is desired than that offered by a mid-range car such as the Audi A6. Even different road viewing angles have no influence on the sitting posture. However, if possible, all persons select the seating position so that they can look at the hood in order to better estimate the dimensions of the vehicle. Lorenz (2013) defined a special hood vision factor κ as the measure for this (Fig. 7.18).

The following recommendations result from these investigations for the design of the vehicle dimensions taking into account the visibility conditions, which, if ignored, have a clear influence on the adoption of an unfavourable sitting posture:

- Vertical distance wind running point H_W - H point: $H_W < 500$ mm
- Vertical distance parapet height H_B - H point: $H_B < 500$ mm

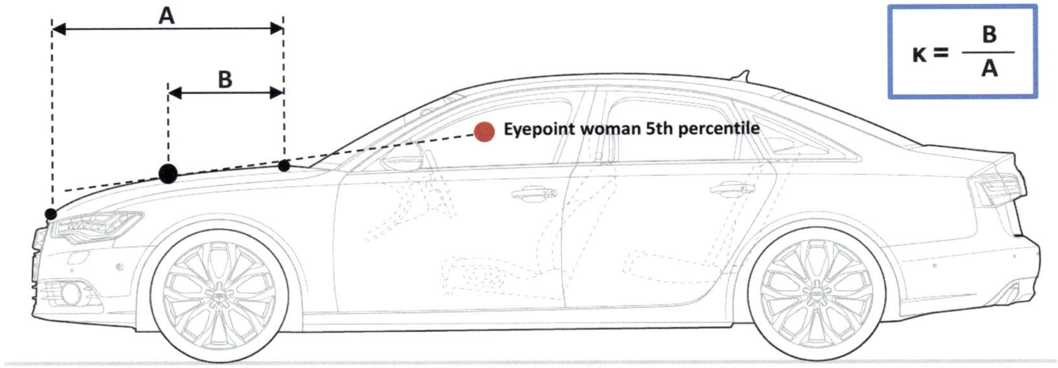

Fig. 7.18 Definition of the hood vision factor κ (The given dimensions are measured from the wind running point; according to Lorenz 2013)

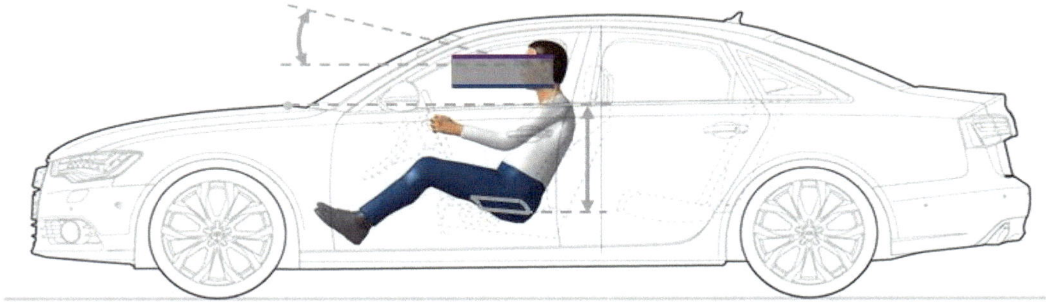

Fig. 7.19 Area in which postural changes do not occur as a result of visual inspection

- Roadway angle of vision in relation to the eye position of the 5-percentile woman: >8°.
- Traffic light angle with respect to the eye position of the 95th percentile man >13°
- Hood vision factor $\kappa \geq 0{,}5$

Since the wind running point and the roof frame position are the dominant influencing parameters for the choice of seat position according to the results of this work, an upper and lower limit plane is derived which is used for positioning the RAMSIS manikin in order to calculate the most probable seat position taking into account the visibility conditions in CAD (Fig. 7.19). These limit levels represent additional restrictions to those mentioned above which have already been used. Their calculation can be integrated into the RAMSIS software (Fig. 7.20).

There is another problem for the adaptation of the RAMSIS manikin to a given vehicle geometry: when the RAMSIS manikin is individually adapted to a test person (for example, by applying the PCMAN method or the bodybuilder offered by the RAMSIS software), a much more realistic position of the hip pivot point in the seat is obtained than with the SAE H point measuring machine. This results in an offset between the individual hip pivot point and the seat related SgRP. Since the upholstery properties of each seat differ in a complex way, this offset must be measured anew for each seat type with a corresponding number of test persons. The result of many such measurements can be summarized as follows: The z-component of the offset has the greatest influence on the RAMSIS seating position. It depends essentially on the sex, the hip circumference of the test person and, of course, the characteristics of the seat. People with a larger volume, who are generally also heavier, obviously sink less into the seat than narrow, slim people. They float on the

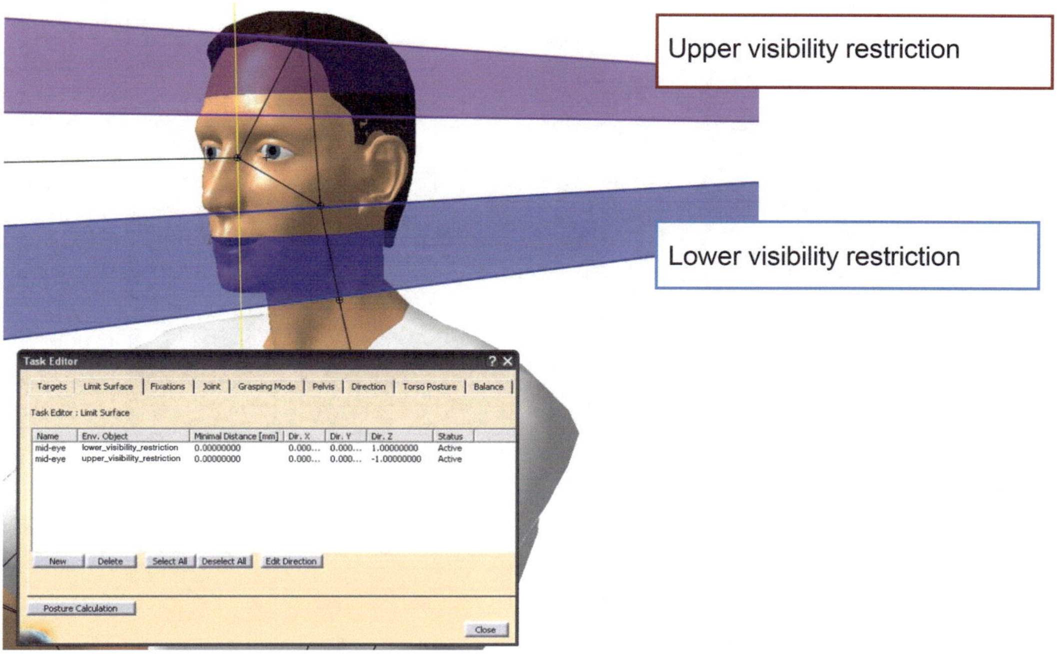

Fig. 7.20 Visual interfaces according to Lorenz implemented in RAMSIS (2013)

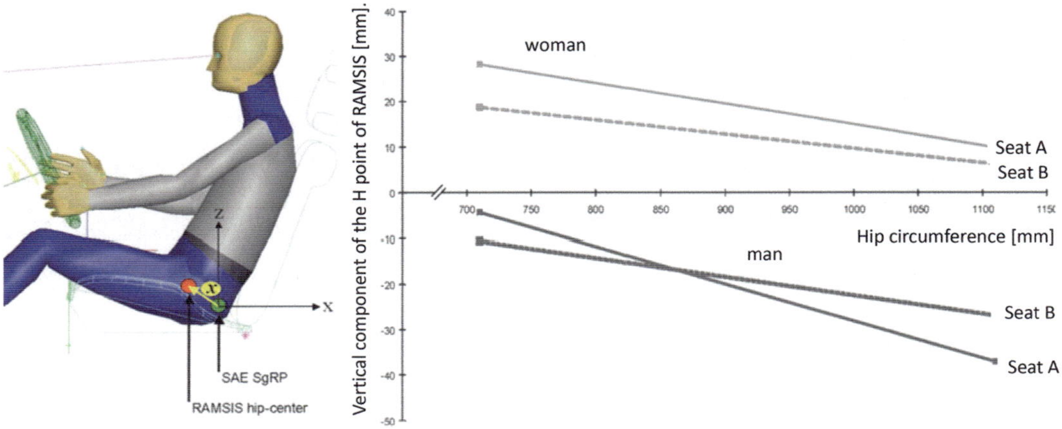

Fig. 7.21 Offset between RAMSIS hip pivot and the SAE SGRP

contour of the seat (Fig. 7.21). So if you want to determine the seat adjustment range from the positioning of different RAMSIS types, this offset must also be taken into account. Bothe (2010) therefore designates the point "Positioning point PHPT"pointing from the SgRP to the RAMSIS hip centre.

In addition to the aforementioned restrictions, it defines further restrictions that control the posture of the RAMSIS manikin depending on a given vehicle concept in order to arrive at a realistic forecast of the posture of different driver types. This means that surfaces such as the inner floor or the accelerator pedal surface must be touched by the heel point or sole point. Interior boundaries such as roof linings, instrument panels or tunnel linings must not be penetrated. The position-

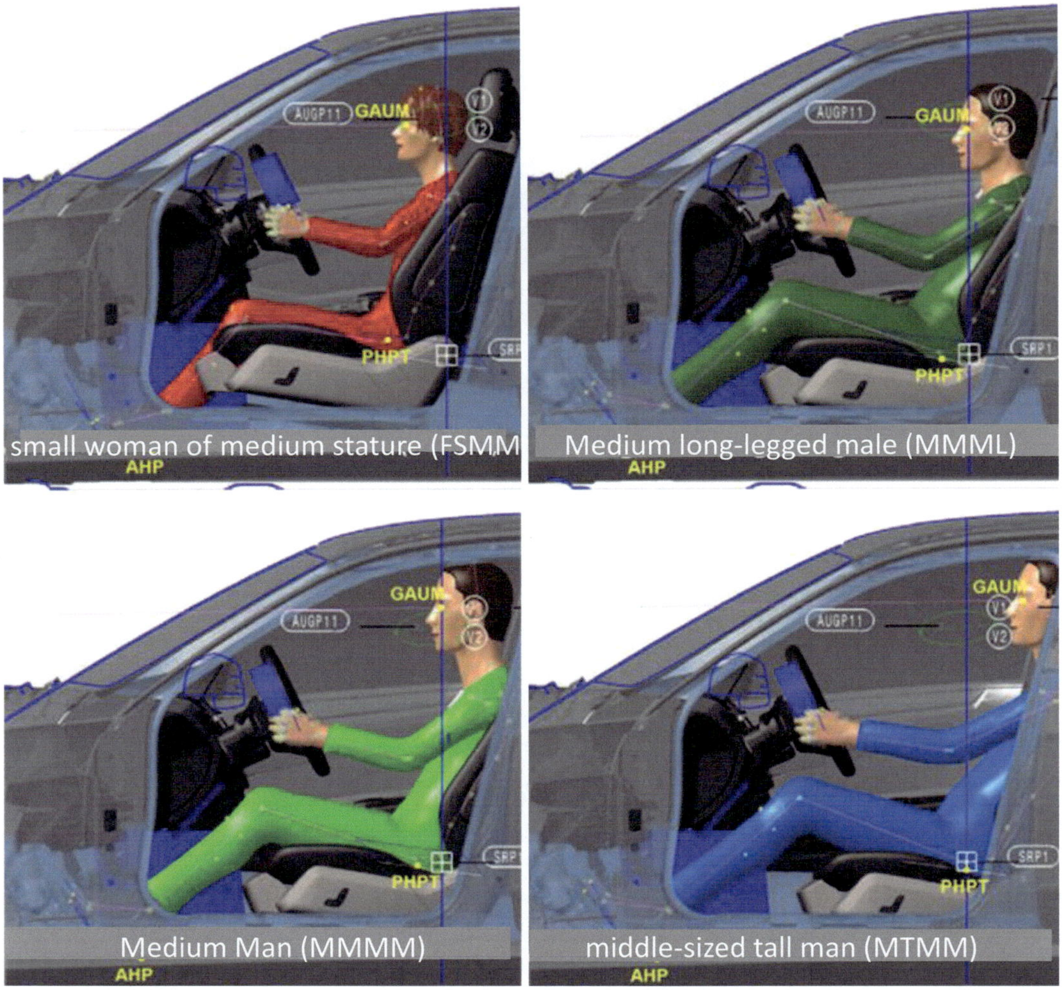

Fig. 7.22 Driver attitude of different design manikins in a given middle class car. (From Bothe 2010)

ing point described above must be within the area of the given seat adjustment field. Defined skin points of the palm of the hand have a contact relationship to the resulting areas of the possible steering wheel positions (see also ► Sect. 7.2.2.2).

Figure 7.22 shows the driver attitude of various RAMSIS-Mannikin in a given mid-range concept (Daimler W212), which were calculated according to this method. Although no extreme types have been used here as examples, some anomalies also observed in practice are apparent (Bothe 2010):

- The little woman (FSMM) sits very upright and far in front in the vehicle, because she has to reach the fixed pedal system with her feet urgently. In order to obtain a sufficient view over the steering wheel and instrument panel at the same time, she must prefer an upright sitting position.
- The medium sized man with short upper body and long legs (MMML) shows a typical behaviour by preferring a relaxed sitting position to a good view from the vehicle. He uses the lower part of the seat adjustment field.
- The medium man of medium stature (MMMM) can use the full range of adjustment possibilities. He finds an upright and relaxed sitting posture. His eye point is clearly higher compared to the long-legged middle man.

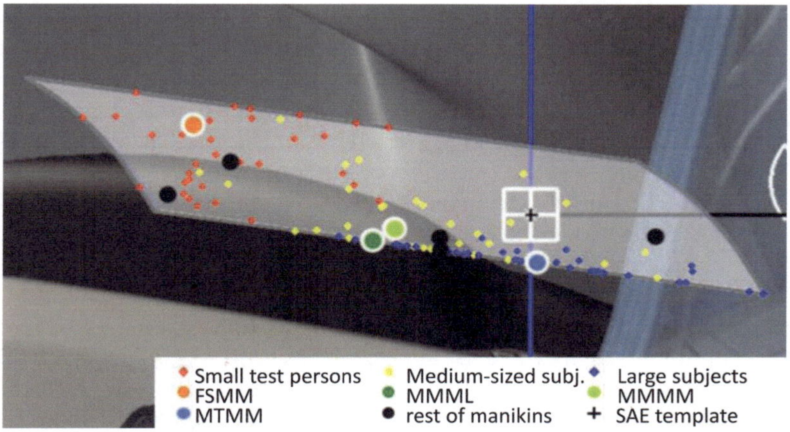

Fig. 7.23 Location of the hip points of test persons and the positioning point PHPT for the manikins from the RAMSIS family used. (From Bothe 2010)

- The tall man of medium stature (MTMM) cannot sit upright because of the limited headroom, but must place the backrest to the rear. This results in a flat sitting posture with outstretched arms. His eye points are positioned very far back.

The eye points found when using all manikins of the RAMSIS family are the basis for the conception of the vision (for further information see ► Sect. 7.3). From the positioning of the various manikins, the range of use of the seat adjustment field can also be analysed. Figure 7.23 shows the corresponding result for the RAMSIS family used at Bothe (2010) as well as for subject positions.

The adjustment field is therefore fully used by the small test persons in the anterior region. The middle and large test persons prefer to use only the lower drainage level of the adjustment field in the rear area (see also Lorenz 2013). As observed in other investigations, the RAMSIS manikins do not exploit the existing seat adjustment to the same extent as real test persons.[13]

13 The difference to the results in Fig. 7.16 is essentially due to the fact that a given vehicle concept was assumed here, while a driver attitude concept was developed there with only a few restrictions, quasi on the tabula rasa.

Support

The calculated seating postures must be supported by an appropriate seat construction. The design of this seat is of great importance both in terms of minimizing discomfort and in terms of a healthy and fatigue-free posture. At the same time, it becomes clear from what has previously been discussed that the seat can always only be designed or assessed in conjunction with the respective sitting posture, which is also determined by the respective boundary conditions. In particular, the seat design is intended to prevent the development of back pain, which occurs during prolonged, quasi motionless sitting. In short, the cause of back pain can be summarised by the following points:

- Insufficient supply of the intervertebral discs due to insufficient movement of the spine: This cause can actually only be counteracted by movement, although foreign movement (including massage) has only a minor effect due to the lack of activation of the muscles involved. However, effective movement is only possible by interrupting the journey and doing appropriate gymnastic exercises.
- Creep of the ligaments due to hyperflexion: This effect occurs primarily when an extremely unfavorable posture is adopted, for example, extremely small body angles or a particularly strong kyphotic (C-shaped) posture of the spine.

- Static holding work of the musculature: A certain amount of muscle effort is always necessary to maintain a body posture. Normally when sitting, this is clearly below the fatigue limit of 15% of the maximum strength of the respective muscle package. But even long-lasting submaximal low forces combined with a lack of movement can lead to tension, which is probably the main cause of muscle pain felt as back pain.
- Atrophied back muscles: if there is an excessive deterioration of the back muscles in the area of life outside the vehicle due to illness or low muscle stress, the aforementioned aspect plays an increased role.
- Whole-body vibrations: long-term predominantly vertical exposure to whole-body vibrations in sitting can lead to disc-related lumbar spine disease, whereby the duration of exposure appears to have a more significant influence than the level of exposure, which indicates a cumulative effect (Nordin 2003). Based on investigations by Bitter (2005) it can be concluded, however, that this effect is less avoided by particularly good damping of the seat cushion properties than by the suspension system of the entire vehicle.

A frequently used criterion for assessing seat quality is the distribution of pressure in the seat and backrest surface, because pressure on skin areas with little fat and muscles causes higher pressures there (e.g., especially among the so-called ischial tuberosities). The result is a reduction in blood circulation and thus in the supply of nutrients to the tissue as well as a slowing down of the nerve conduction velocity. Already pressures of 20 to 33 mmHg (2,6 to 4,3 kPa) can lead to this effect. If the body posture is unfavourable, shear forces can also cause tangential shifts in the skin layers, which also cause compression of the blood vessels and thus undersupply of the corresponding tissue. Because they are virtually impossible to measure, shear forces are now regarded as a misunderstood cause of discomfort when sitting.

The task of the upholstery is to reduce high pressures in the contact surface between body and seat and at the same time to absorb different body shapes. A simple physical consideration shows that a precise adaptation of the seat contour to that of the seated person leads to a complete compensation of the pressures, but only if there is no movement of the seated person.[14] Franz (2010) and Franz et al. (2011) have taken up this idea for a lightweight seat by skilfully averaging different seat and back contours. As will be shown below, however, a completely soft seat that produces practically no locally distributed seat pressure profile would lead to considerable discomfort during prolonged sitting. The contouring of the aforementioned lightweight seat was also carried out in such a way that a largely optimum seat pressure profile was created. So the question is: what is the optimum seat pressure distribution? Hartung (2005) and Mergl (2005) carried out extensive investigations with a specially designed variable research chair, which led to a seat pressure profile that produces minimal discomfort. The aim of the experiments was to find a connection between physically measurable quantities and subjective perception. For this purpose, the contact surface of the body with the seat was divided into segments, with respect to which the test persons expressed their subjective discomfort sensation during the experiments on a modified CP-50 scale (◘ Fig. 7.24).

According to the studies mentioned above, there are three parameters in the pressure distribution that have a direct influence on discomfort. It's these:

- The percentage load distribution (= force on the relevant part of the body/body weight)
- The maximum pressure
- The gradient of the pressure rise

In the experiments, numerical values could be found for most of the body areas mentioned

14 For racing cars, almost cushionless seat shells are used which are exactly adapted to the individual body profile of the racing driver.

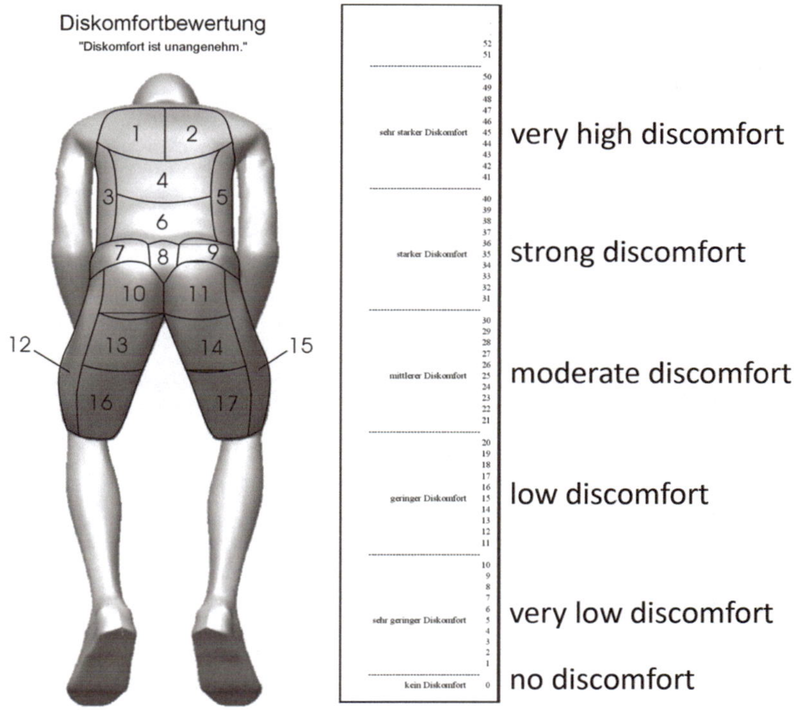

Fig. 7.24 Body regions and CP 50 scale according to Hartung (2005) and Mergl (2005)

in Fig. 7.24, which ensure minimal discomfort. These values only apply individually. For a given seat, however, people of different anthropometry experience different seat pressures. In addition, even a slight variation of the posture changes the pressure distribution considerably. There is therefore no ideal seat, but only an ideal combination between a seat and an individual.[15] This means that the seat must be adapted to the individual conditions of the person.

The values shown in Fig. 7.25 were evaluated in numerous experiments (Mergl 2005; Zenk 2008; Lorenz 2011). With a 42-year-old test subject (weight 83 kg, ~50th percentile man), who voluntarily had flexible pressure sensors inserted into the intervertebral disc between the vertebral elements L4 and L5 and L5 and S1 for a test day, it could be shown in a real driving test that an extremely low intervertebral disc pressure of 0.5 bar was generated with an adjustment of the seat that produced the values of Fig. 7.25. If this person's load under the front of the thigh was increased from 6 to 10.6%, the tape disc pressure increased to 0.95 bar. With an extreme relief under the thighs to 0.4% load, the band disc pressure even increased to 1.5 bar (Zenk et al. 2007). The astonishing result of all these experiments is that back pain is more likely to be caused by unfavourable pressure distribution in the buttock area than in the back area. In the experiments by Hartung (2005) and Mergl (2005), only weak correlations could be found between pressure distribution in the back area and subjective discomfort. This can be explained by the fact that the backrest absorbs only max. 15% of the body weight in a sitting posture, which is usually adopted by drivers. When sitting, the driver also has much more freedom of movement with regard to the back than in the buttocks. Mergl explained this dominant influence of the pressure distribu-

15 By analogy, no one would expect there to be a shoe that fits everyone!

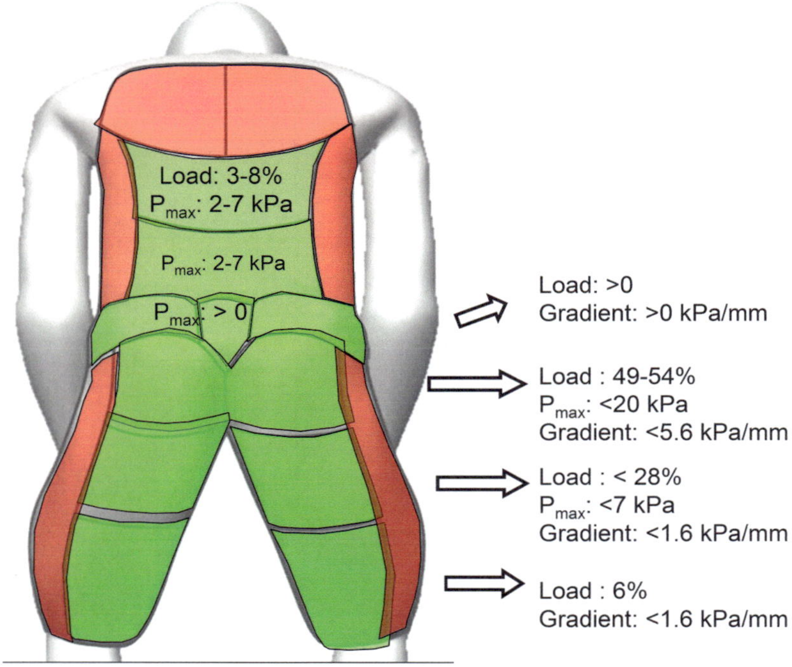

Fig. 7.25 Load, maximum pressure and gradient of the increase in pressure which, according to the investigations by Hartung (2005) and Mergl (2005), leads to a minimal discomfort

tion in the buttocks on the perceived discomfort in the back with the muscle group of the iliopsous muscle (Psoas minor, Psoas major, Illiacus), which moves from the femur to the spinal region (Fig. 7.26). One tries to compensate an unfavourable pressure distribution obviously by light far submaximale muscle tension to compensate, which leads however with longer not by movement compensable tension to back pain. A lumbar support in the backrest area can obviously produce a similar effect. By preventing the pelvis from tilting, a kyphotic posture can be largely prevented. According to Wilke (2004), this itself is not unfavourable. If the appropriate support is lacking, the above-mentioned submaximal muscular effort, which causes pain in the long run, also occurs.

The adaptation to different anthropometric conditions, through which the above-mentioned optimum values are achieved individually, can only be achieved by a variable seat surface length and an adjustable seat surface inclination. The length of the seat in

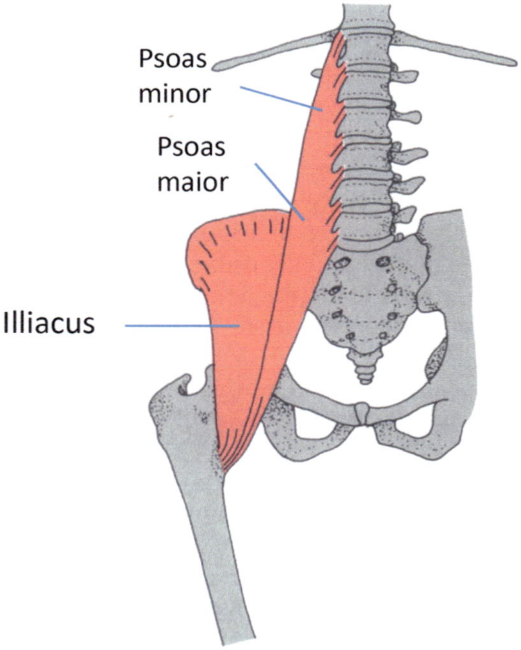

Fig. 7.26 Musculus iliopsoas. (Psoas minor, Psoas major, Illiacus, after Weineck 2001)

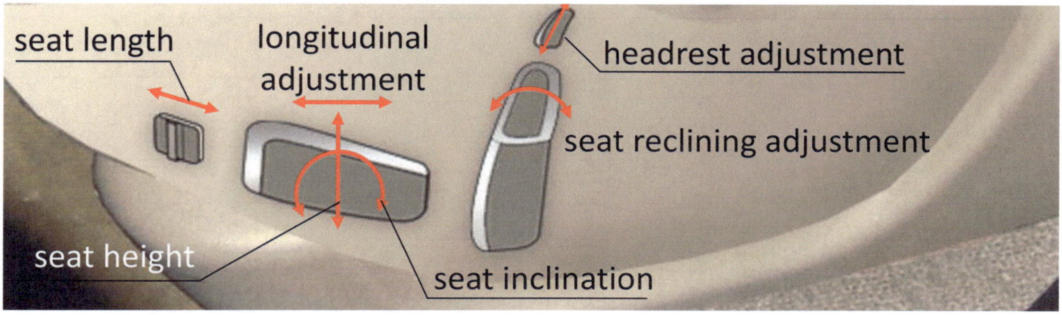

Fig. 7.27 Adjustment unit attached to the side of the seat, which can be reached without changing the driver's posture, but is difficult for some drivers to understand due to the haptic feedback only

particular is a problem because it must cover the area from the short-legged small woman to the long-legged tall man (approx. 430–520 mm). Overall, however, these results also show the enormous influence that external conditions have on the effort the driver has to put into maintaining his posture. It would therefore be desirable to have model-based knowledge of these interrelationships already in the concept and development phase of a new vehicle.

Both the experiments with the voluntary test subject and the evaluation experiments by Zenk (2008) and Lorenz (2011) have shown that the drivers are not able to make the optimal seat adjustment due to the immediate subjective discomfort sensation. Lorenz even showed that test persons later rejected their own preferred seating position if it was offered to them without this knowledge as one recommended on the basis of supposedly "scientific" results. Zenk (2008) therefore developed an automatic seat adjustment system in which the seat pressure distribution in the seat area was measured and used to optimally position the seat. Through long-distance tests (travel time > 3 hours), he was able to show that even in test subjects who initially rejected this automatic setting, the discomfort notation decreased to a very low level with increasing test time. Lorenz (2011) developed a tutorial in the driver information display that shows the driver in a structured form what to look out for in each of the setting steps, in addition to the automatic setting. The settings achieved in this way came close to the required optimum settings. It is essential that the adjustment is made from the same position as when driving. This places demands on the location of the control unit (example Fig. 7.27). Due to the various adjustment possibilities of a seat, particular importance is to be attached to the ergonomic design of the operation of the seat adjustment. In this context, reference is made in particular to ► Sect. 6.2.2.3 and ► Fig. 6.27.

On the one hand, drivers select the backrest adjustment angle depending on the vehicle parameters, but on the other hand, this choice is also strongly influenced by the personal preferences and habits of the vehicle users. In addition, the informative value of the backrest inclination with regard to the actually adopted torso posture is strongly limited due to the large variation of the individual back contours. Even when the human back makes extensive contact with the surface of the seat back, the torso angle often deviates considerably from that of the backrest tilt. The higher the seat is arranged in the vehicle (dimension H30), the steeper the seat back is adjusted. The arrangement of the headrest also influences the selected backrest angle (Kolich 2010). If the occupant's height forces him or her to move the seat very close to the instrument panel in order to reach and operate the pedals, the distance between the steering wheel rim and the seat backrest cheek is inevitably reduced as well (see below). In this constellation, the driver's freedom of move-

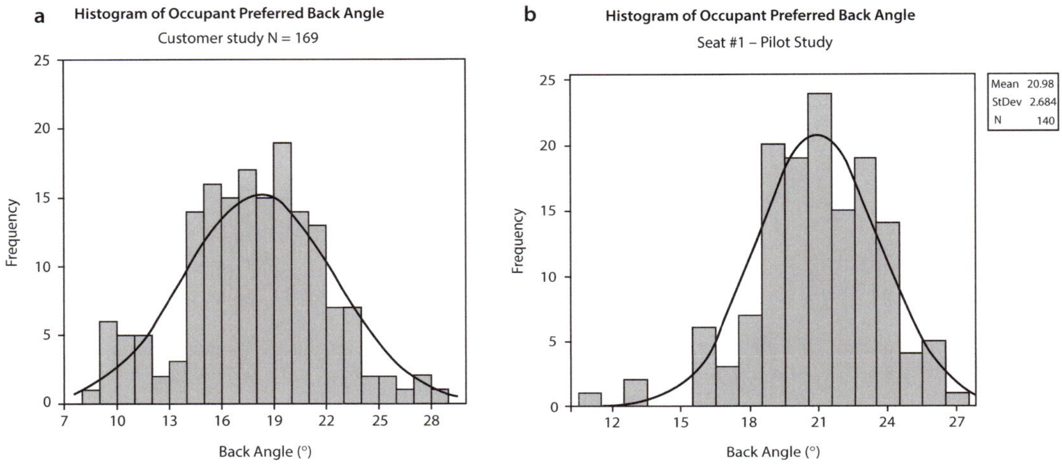

Fig. 7.28 Observe the fief angle of two different vehicles: **a** Adam Opel AG, **b** Ford Motor Comp. (According to Kolich 2010)

ment may be restricted, making the steering wheel and gear selector difficult to operate. This can be remedied by shifting the seat backwards, which, however, leads to a reduction in the comfort of the resting seat or the more pronounced inclination of the backrest, as discussed above. A certain number of vehicle users prefer a very flat seat back position, which only provides contact and support in the seat area. As already mentioned, the subjective feeling of well-being does not necessarily have to coincide with objective demands adapted to the individual. The statistical distribution of the preferred backrest adjustment shows an approximately normally distributed accumulation around the mean value, which deviates more or less from the design torso angle of 25° depending on the vehicle concept (Fig. 7.28). Far less than half of the persons examined use this angle, with a bandwidth of almost 10° varying upwards and downwards.

The previous remarks do not deal with the lateral support that a seat should provide when driving a car. Unpublished studies by the Institute of Ergonomics (TU-Munich) indicate on the one hand that an extensive adaptation to the individual anthropometric conditions is necessary and on the other hand that this adaptation also depends on the driving conditions. For example, when driving a dedicated car on a twisty road, a relatively strict lateral support is expected, while on a relaxed motorway drive this lateral support should be reduced to allow minor compensatory movements. From an ergonomic point of view, this raises the question of whether it would make sense to carry out an adjustment controlled by the navigation system. At the very least, the side support should be modified in such a way that it can be easily operated by the driver while the vehicle is in motion.

7.2.2.2 Pedals

Distance Between Pedals

Just as the driver's posture is influenced by the design of the pedal arrangements as well as by the need to be able to observe the traffic situation through the windscreen sufficiently, the driver's posture is also influenced to a large extent by the design of the pedal arrangements. If RAMSIS calculates the posture on the basis of the heel point, the position of the accelerator pedal in relation to this heel point must be clarified. Kolling (1998) found in practical driving tests that the average contact point in women is x = 10 mm before and y = 50 mm to the left of the right edge of the accelerator pedal. For men, the corresponding coordinates are x = 40 mm and y = 90 mm. If necessary, the raised or chamfered heel of the shoe must also be taken into account.

Although the longitudinal seat adjustment enables the driver to assume the correct position to the heel point of the accelerator pedal or the parking area for the left foot, the distance between the pedals is fixed by the vehicle design. Various authors have dealt with the question of pedal positioning. Particularly worth mentioning here is the investigation by Bäumler (1992), who used a force plate instead of real pedals in a vehicle simulator experiment and caused the test persons by the simulation program to switch between the self-selected accelerator pedal position and a spontaneously selected brake pedal position to the left of it. In agreement with the authors Brackett et al. (1989) and Dreßel und Kain (1985), he found that the optimal position of the brake pedal should be approx. 38 mm and that of the accelerator pedal approx. 161 mm to the right of the centre of the seat. The free distance between the pedals should be approx. 60–61 mm (◘ Fig. 7.29 left). In general, the entire pedal arrangement preferred by the test persons is shifted to the right by the wheel arch mounted on the left or to the left by the gear tunnel mounted on the right, depending on the foot space restriction. The aforementioned effect is further enhanced if the steering wheel is also shifted out of the centre of the seat and also rotates in relation to the seat orientation (see ◘ Fig. 7.29 on the right; see also the treatment in ► Sect. 7.2.3).

In general, both Brackett et al. (1989) and Bäumler (1992) state that an off-centre arrangement of the steering wheel results in torsion of the upper body, which must be compensated for by slight muscle tension, which can lead to back pain in the long run in a similar way to an unfavourable driver posture in longitudinal section. So if possible, an absolutely symmetrical sitting position should be aimed for.[16]

This ideal orientation is very often deviated from in order to take sufficient account of technical boundary conditions. The arrangement of the pedals is often shifted towards the centre of the vehicle to ensure freedom of movement to the wheel arch. Especially in right-hand drive vehicles, this procedure is frequently encountered and sometimes unavoidable, since it is not the stationary foot rest, as in left-hand drive vehicles, but the accelerator pedal with its actuating travel and the necessary freedom of movement of the foot that is positioned here. A moderate (<40 mm) shift of the pedals towards the centre of the vehicle is usually inconspicuous, but in long-distance traffic it is associated with comfort restrictions (see above). A shift of the pedal centre in the direction of the wheel arch is very strongly perceived by the driver, since he is turned out of the so-called arrow (towards the centre of the lane). The steering column, too, is often moved slightly (approx. 10 to 40 mm) from the ideal position to the centre of the vehicle in order to take account of the unevenness of the cardanic steering column and the entry point in the steering gear. Here, too, larger values are conspicuous and comfort-limiting due to the parallax that occurs, and an outwardly offset arrangement is perceived by the driver as irritating. An eccentric position of the steering wheel is visually perceptible at the position of the instrument. Therefore, if the steering wheel is in an offset position, the instrument cluster must follow this offset in the same direction and tend to be larger. The voting is quite complex, since it has to be secured from dif-

16 Unfortunately, this rule is often violated today due to modular vehicle concepts. Because of the medium-term high flexibility and insensitivity of the human body, deviations in design assessments and driving tests are not initially noticed.

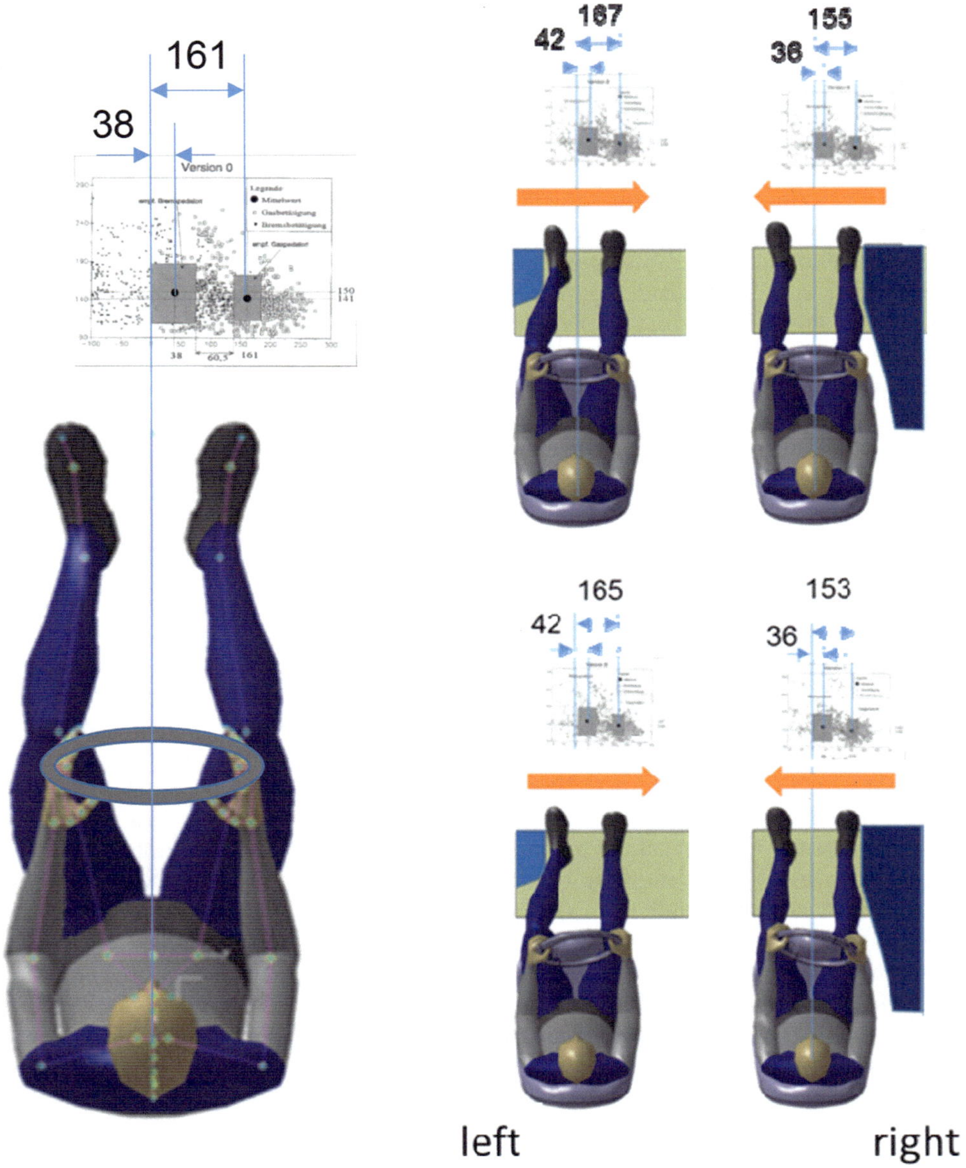

Fig. 7.29 Illustration of the results from Bäumler's study (1992)

ferent eye points and their distances (approx. 600 mm to1000 mm) to the display level (see ► Sect. 7.3.3).

Both the investigations by Brackett et al. (1989) and by Bäumler (1992) were carried out with a flat surface representing the pedal level. This makes no statement about a possibly necessary differentiation in pedal height with regard to brake and accelerator pedals. This so-called pedal jump is the subject of various discussions. For reasons of rapid movement between the driving and brake pedals, it would be advisable not to provide for a level difference between these pedals. However, a noticeable jump in the pedal would have the advantage that the driver would become more aware of whether he is using the driving or brake pedal. This aspect plays an important role especially in connection with the phenomenon of so-called "unexpected acceleration"

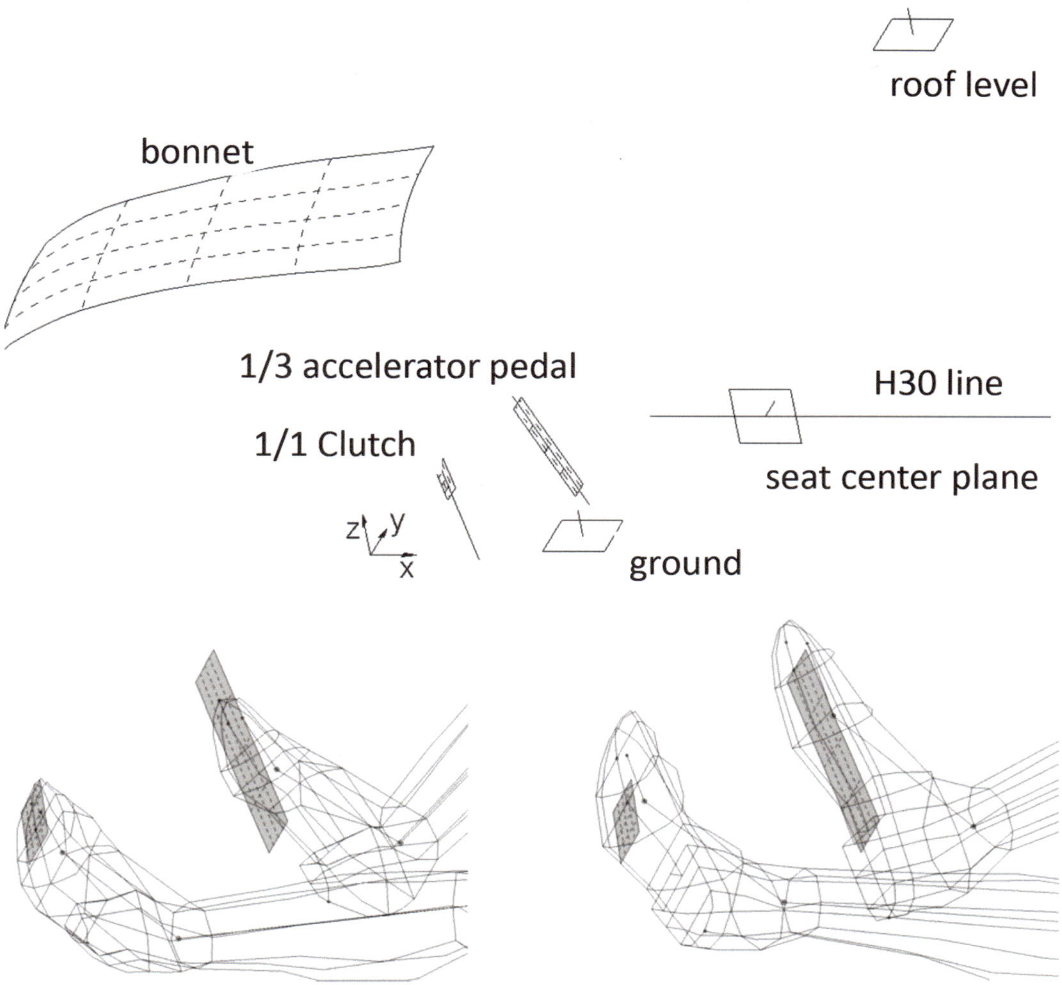

Fig. 7.30 Geometric boundary conditions for the calculation of the RAMSIS posture model including the clutch pedal (*at the top*) as well as foot positions on the pedals. (From Vogt 2003)

(actually "unintendent acceleration"), in which the driver, supposedly standing on the brake pedal, fully presses the accelerator pedal. A pedal jump between 20 and 30 mm is therefore recommended. Irrespective of this geometric condition, Kolling (1998) observed two different implementation strategies from the accelerator pedal to the brake pedal. During light braking manoeuvres, the foot is only turned around the heel contact point without lifting off the ground (this is especially true for tall people). When braking more strongly, the hoe is always moved in front of the brake pedal. When moving back to the accelerator pedal, large persons stay in the brake position with their heels for approx. 3 to10 seconds, although they are already accelerating again. For more detailed layout analyses, these aspects must also be taken into account.

Accessibility of the Clutch Pedal

The clutch pedal imposes further restrictions on the posture calculation and thus on the layout of the seat longitudinal adjustment. First of all, it is necessary to obtain information from the technical side about the position of the clutch pedal in the depressed state (Fig. 7.30 above). In practice, it has been found that realistic postures can only be achieved by assuming different points of con-

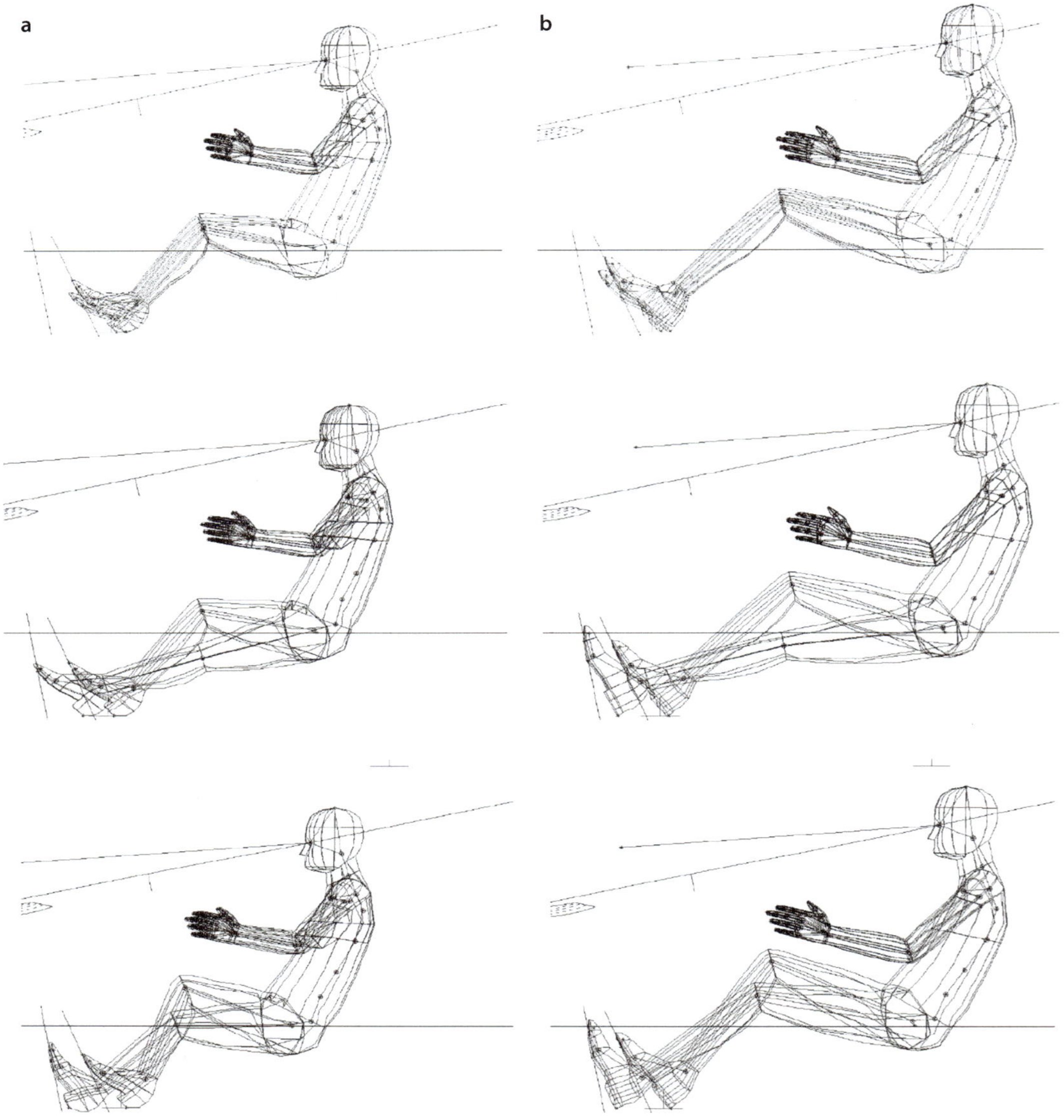

Fig. 7.31 Calculations of the probable posture for the short-legged little woman (FSMS, **a**) and the long-legged big man (MTML, **b**)

tact at the soles of the feet for men and women (Vogt 2003; Fig. 7.30 below).

In the following, only the extreme types "small short-legged woman" and "large long-legged man" will be considered. In Fig. 7.31 the posture is calculated for the case that the feet are each standing on the accelerator pedal (1/3 depressed) or on the left footrest (this is the basis for the posture calculations that led to the result of Fig. 7.16). The second row shows how far the left foot has to be pushed through so that the clutch pedal can be depressed under this condition. This results in unrealistically large angles. The lower row shows the posture that results when RAMSIS assumes the most probable posture

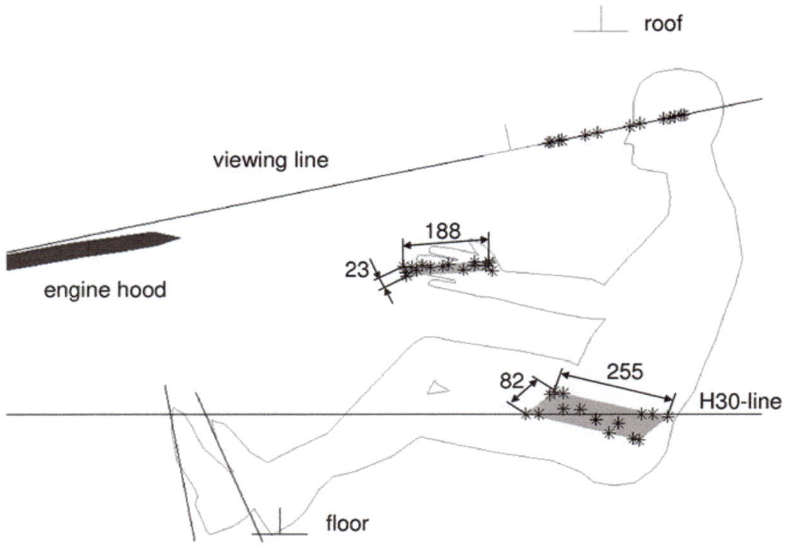

Fig. 7.32 Eye-point-optimised position of the Fig. 7.16 with additional consideration of clutch actuation

for the accelerator pedal to be depressed on the one hand and the clutch to be fully depressed on the other. In this case, the right foot will have an uncomfortable posture with strongly angulated knees. This case is also unrealistic because the posture model assumes that the respective posture will be adopted over a longer period of time. However, the coupling is only carried out for a short time.

In order to arrive at realistic assumptions, Mergl et al. (2006) carried out a mock-up of the preferred position for clutch pedal actuation and measured the respective knee angle using a goniometer. Knee angles between 125° and 155° were measured. The investigations suggest that the preferred knee angle for clutch actuation is subject to strong individual variation. Therefore, for the ergonomic design of a vehicle with clutch, it is proposed to calculate each of the considered design manikins in the position of the lower row of Fig. 7.31 and additionally the restriction "no knee angle >155°". By this measure one can assume that one develops an attitude variation range, which considers each possible individual preference. Figure 7.32 shows the result. The illustration shows the necessary adjustment ranges for seat and steering wheel. The comparison with the results in Fig. 7.16 also points to the problem of the hand-operated car from the anthropometric point of view. In order to ensure that the clutch pedal is actuated, compromises must be made which are at the expense of a relaxed driver posture for the longest period of the driving process. The consideration of unrestricted clutch actuation also has a safety effect for the actuation of the brake pedal. If one of the two brake circuits of the vehicle fails, the brake pedal has an empty travel so that for emergency braking the brake pedal travel is comparable to that of the clutch pedal. In this way, the unrestricted actuation possibility in such an emergency is implicitly ensured by the seated position in relation to the clutch pedal. The driver's seat position, which is frequently shifted to the rear in automatic vehicles, restricts sufficient brake actuation in the event of failure of a brake circuit.

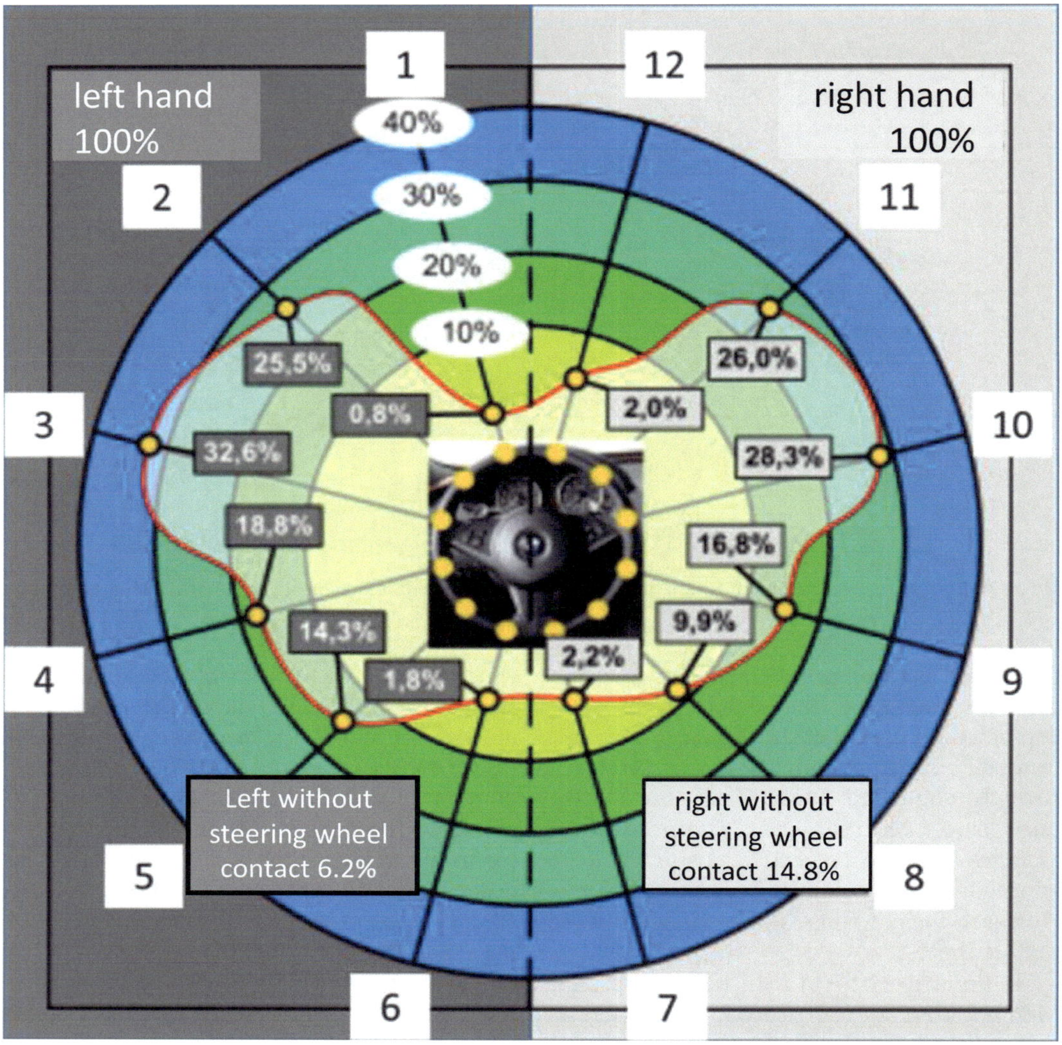

7

Fig. 7.33 Distribution of hands on the steering wheel. (After Kolling 1998, modified by Wolf 2009)

7.2.2.3 Steering Wheel

Steering Wheel Position

The steering wheel is the most important interaction medium for the driver and thus significantly determines the character of the vehicle (see also ► Sect. 6.4.1). Kolling (1998) has made observations in connection with the modelling of driver posture by the human model RAMSIS on the *position of hands on the steering wheel* in real driving operation. He divides the steering wheel into twelve positions. Figure 7.33 shows the summary of the results. According to this, the most frequently used grip locations for the right and left hand are positions 3 (32.6%) and 10 (28.3%). One reason for this is that the hands are hooked into the spoke ring with the thumb in order to support their own weight. It has also been observed that small women often touch the steering wheel further up, which is probably due to the relatively small distance from seat to steering wheel. If you want to perform a positioning in a given vehicle, these additional RAMSIS type-dependent restrictions must be observed. Since the multifunction steering wheel is also operated while driving, the manual position in positions 4 and 9 is necessary. Overall, it was observed that up to 21% of the total driving time was driven with one hand (15% of women and 26% of men). The left hand is not at the steering wheel about 6.2% of the time, the right hand even 15%. Kolling observed a "regulator hand" and a "power hand": The "regulator hand" is usually supported at an environmental point (knee, arm rest, parapet) and touches only loosely the steering wheel rim. Actual steering movements are introduced into the steering wheel via the "power hand" (in most subjects via the right hand). Many test persons grab the steering wheel asymmetrically, others change frequently or drive constantly the whole distance in one posture. Kolling recommends constructing a point at the height of the upper spoke on the outer rim of the steering wheel for the posture calculation, which is brought into contact with the RAMSIS hand point (see also ◘ Fig. 7.14).

The ***steering wheel diameter*** has, due to the kinematics of the hand-arm system, a considerable influence on the steering forces, steering angles and possible temporal behaviour that can be transmitted by humans, whereby these effects depend on the anthropometric conditions and the position of the driver behind the steering wheel. From the hand positions which RAMSIS assumes without entering any restrictions, the optimum steering wheel diameter can be determined by assuming that the hands are held in positions 3 and 10 and calculating the circle to be entered under these conditions. The diameter of this circle varies from 300 to 427 mm depending on the RAMSIS type used. As a compromise, a diameter of 363 mm can be derived from this. Wolf (2009) found an average of 376.3 mm in an overview of common passenger cars in Europe for 2002, 2003 and 2004, with the smallest diameter being 320 (Opel Speedster) and the largest being 410 (Rolls-Royce Phantom).

The ***Thickness of the steering wheel rim*** has no direct influence on the manageability. Wolf (2009) also recommends a cross-section of 20 mm under the aspect of "liking".

The ***Inclination of the steering wheel relative to the horizontal plane*** should not be changed even when using the adjustment range (see below) due to the anatomically relatively fixed optimal angle of the wrist around the z-axis (ulnar and radial abduction). For driver positions in a limousine it should be 85.5°.

Adjustment Range

◘ Figure 7.32 shows that in the driver's seat concept for the steering wheel carried out there, an adjustment range of 188 mm in the longitudinal direction and of 23 mm in the vertical direction must be provided. This adjustment range also results from driver seat concepts which are carried out according to other criteria. If one does not, as here, ensure the same visibility conditions for tall and short people, but assumes that the eye point remains at approximately the same height, in the vertical direction even a larger necessary adjustment range of approx. 130 mm results. The recommended adjustment ranges of the steering wheel result in an elbow angle of about 124° (Lorenz 2011). As studies by Schmidt et al. (2014) show, an angle range between 95° and 120° is recommended for fast and precise steering wheel movement from a biomechanical point of view. For ergonomic reasons, a coupling between seat adjustment and steering wheel would therefore be desirable, as it gives the driver a favourable and healthy position from the outset. The deviations from this function, which are caused by proportions and corpulence deviating from the average, could then be taken into account by a small, individually adjustable steering wheel adjustment in the range of ±30 mm in x-direction and z-direction.

The adjustment of the steering wheel in current cars is too small compared to the layout suggestion with RAMSIS shown here. In particular, the adjustment range in the horizontal direction is not sufficient to ensure optimal posture for people of all anthropometries. Reasons for this are the regulations for crash behaviour. For small persons there must be enough space between the steering wheel (airbag) and the body. With a large adjustment range from 188 mm this cannot be achieved. Other restrictions arise for cost reasons and other technical restrictions. Since the

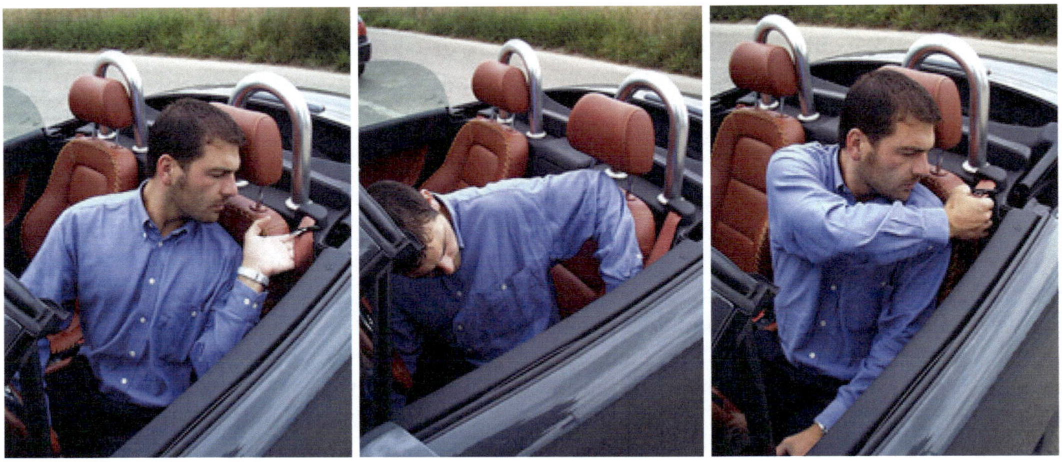

Fig. 7.34 Various belt strategies. (Remlinger 2013)

tolerance range for good comfort at elbow (80°-158°) and shoulder joints (9°-69°) is very large, many people do not perceive a suboptimal posture (see also the comments in ▶ Sect. 7.2.2.1). In any case, it should be noted that the solution frequently found today - for cost reasons - of swivelling the steering wheel in a universal joint only about a fixed axis parallel to the y-axis does not in any way meet ergonomic requirements.

7.2.2.4 Belt

The compulsory use of seat belts, which today applies in practically all countries where vehicles are registered, means that the handling of the belt with regard to the application process and the discomfort caused by wearing the belt is an important aspect of interior design.

Reachability

In particular, Monnier (2004) carried out extensive basic investigations into the buckling process. In addition to the distance of the upper belt attachment to the driver's H-point in the x and z direction, he found the method of attachment of the lower attachment to be relevant as the main influencing factors for the seat belt application process. For the fastening process itself, various movement strategies for gripping the belt plate were observed during his investigations. The so-called "right-hand strategy" is of particular importance because of its frequency. The driver grabs the belt tongue on the left with his right hand, then pulls it to the right side in front of his body and finally puts it into the belt buckle. In particular, when the belt tongue is easily accessible, this movement strategy is preferred by 78% of drivers. If the base plate is incorrectly attached and, in particular, the user is very corpulent, a much more complex movement strategy takes place: in this case, the driver tries to grasp the belt plate with his left hand in the upper left-hand corner, then pulls it forward until he can grasp it with his right hand and finally inserts it into the belt buckle (see also Fig. 7.34). In both strategies, in many cases it is observed that the driver is trying to visually check the process. The initial posture of this process can be simulated with the RAMSIS human model by setting a contact restriction for the right hand to the belt plate from the driver posture. The various design manikins must then be used to check whether the process can be carried out at all (Fig. 7.35).

Belt Routing

RAMSIS offers a sophisticated belt routing model which can be used to model both the position of the belt tongue in the initial position and that of the belt buckle. By means of the different anthropometries, it must then be checked to what extent an adjustment of these two points is necessary, taking into account the resulting belt run (Fig. 7.36). In most

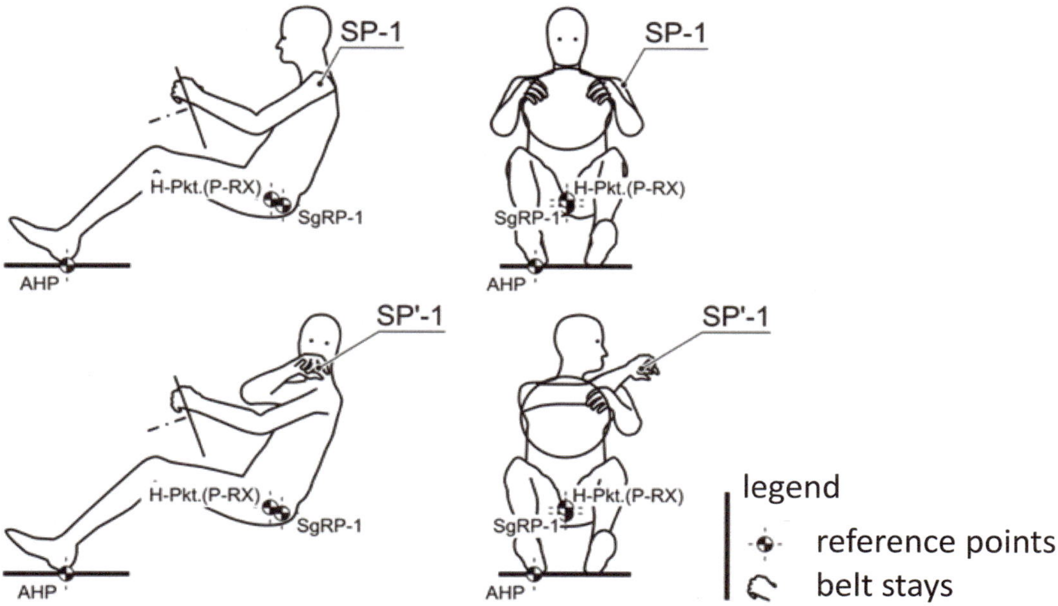

Fig. 7.35 Modelling of the support geometry for the seatbelt. (From Müller 2010)

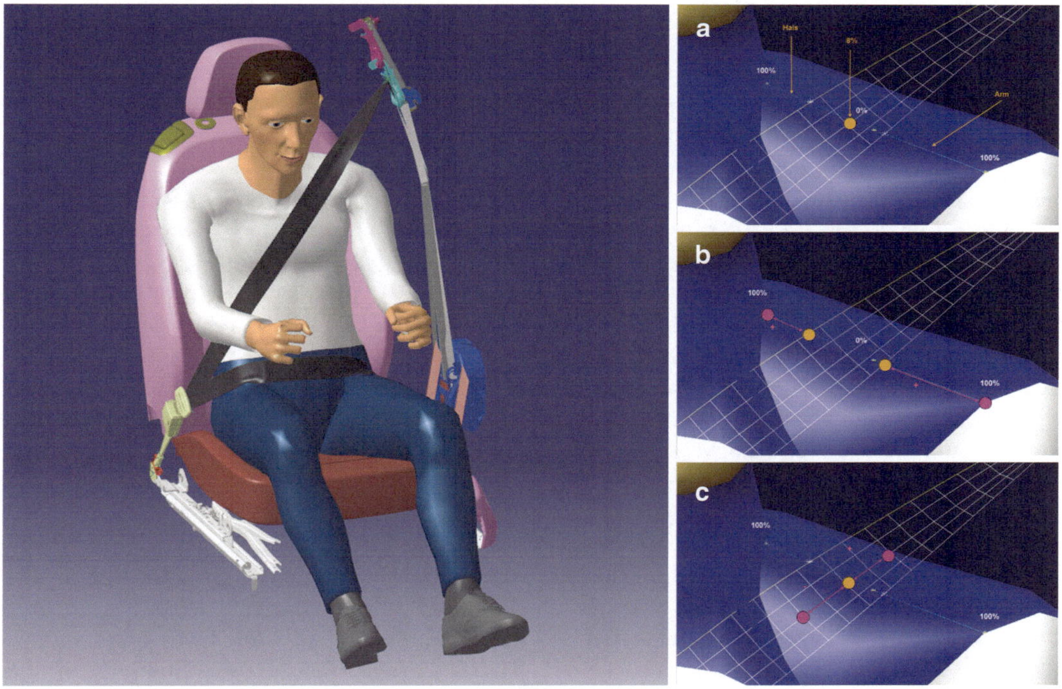

Fig. 7.36 Calculations of belt routing including evaluation of the basis on the shoulder **a**, the edge of the belt **b** and the belt release point **c** From RAMSIS 3.8 to CATIA V5 Manual

cases an adjustment through the upper belt anchorage point is therefore necessary. From an ergonomic point of view, it is also recommended here to automatically change this point depending on the seat and backrest position, whereby additional individual adjustment appears unavoidable.

7.2.2.5 Armrests

As already explained, long-lasting, even clearly submaximal muscle efforts lead not only to discomfort, but also to muscle pain in combination with only minor possible movement. This can be prevented by appropriate body support. Arm rests in the middle of the vehicle and in the door area are particularly suitable for this purpose. Even if such armrests are not available, the driver will still seek support by using the window sill for the left arm, for example, or by placing his hand on the gear shift lever for the right arm. A correct positioning of armrests, which can be used by people of different anthropometries, is therefore useful for fatigue-free driving.

Bothe (2010) describes an interesting method of how a compromise can be reached using the RAMSIS-implemented comfort assessment and how its limits can be recognised. The procedure is briefly described below using the armrest in the left door panel as an example. First you have to differentiate whether the armrest should be used with or without steering wheel contact. Using the example of the large man (MTMM) and the small woman (FSMM), ◘ Fig. 7.37 shows, in addition to the neutral static driver posture, three armrest heights (here 100 mm, 300 mm and 500 mm respectively) above the vehicle-fixed seat reference point SgRP1, as well as the discomfort evaluation calculated by the RAMSIS software to the left of each image. The discomfort values also depend on the lateral distance of the armrest. In this example, a distance of $\Delta y = 350$ mm from the center of the seat is selected.

◘ Figure 7.38 shows the difference between the discomfort and the neutral sitting posture as a function of different armrest heights for the three human models large man (MTMM), middle man (MMMM) and small woman (FSMM). If one accepts a deviation of only one digit of the discomfort (see also Krist 1993), it turns out that for the tall man there are hardly any problems regarding an armrest height, while for the middle man in the chosen example a minimum armrest height of 330 to 400 mm above the SgRP would be necessary. For the little woman there are actually never favourable values. If one carries out the described procedure for the case "no steering wheel contact", then one obtains the result of the ◘ Fig. 7.39. Here one shows a common acceptable range from 220 to 260 mm over the SgRP for person types between the large and the medium-sized man. As expected, the optimal range for the small woman, 300 to 460 mm, is significantly higher. These results indicate that the armrest height should be individually adjustable. The same results are obtained by performing the described procedure for the right arm. The results depend considerably on the distance Δy of the armrest from the middle of the seat and of course also on the selected H30 dimension. From this it can be concluded that the procedure described here must be carried out anew for each fixed vehicle concept.

7.2.3 Right-Hand Drive Problem

In all countries with automobile individual traffic, the road side to be used regularly is regulated by law. In 152 countries, for example, the right-hand side of the road is mandatory, while 63 countries drive on the left-hand side (◘ Fig. 7.40). What these countries have in common is that the driver's side is on the inside of the road, i.e. the side facing oncoming traffic. Accordingly, in vehicles designed for the preferred use of the right-hand side of the road, the steering wheel is located on the left-hand side of the vehicle and is accordingly referred to as a left-hand drive vehicle (*LHD = Left Hand Drive*). Similarly, if the vehicle is intended for left-hand traffic, the steering wheel and the driver's position for right-hand drive vehicles (*RHD*) are on the right-hand side of the vehicle. As a result, vehicles for right-hand drive and left-hand

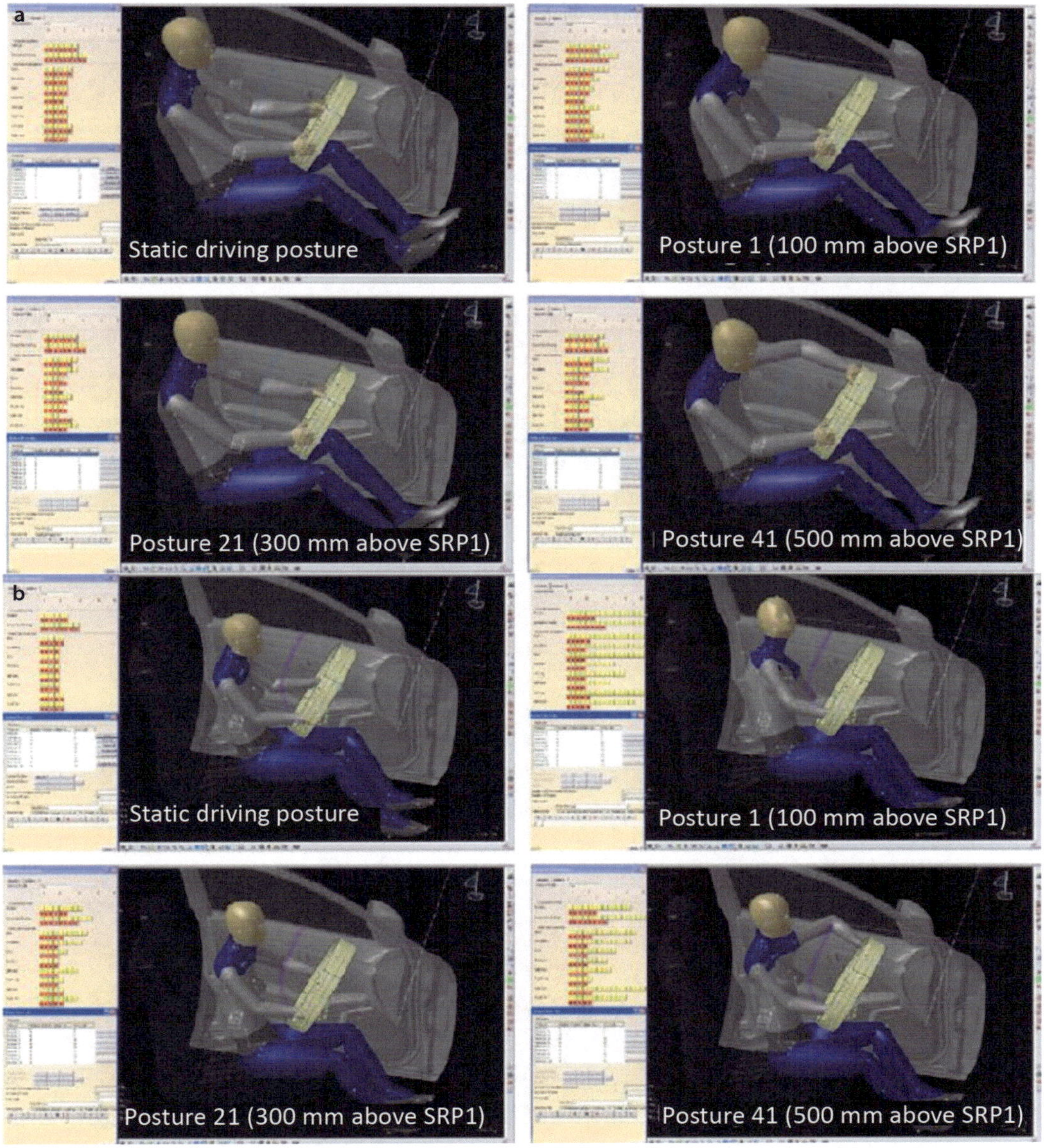

Fig. 7.37 Simulation of the use of the armrest in the door trim with steering wheel contact. Example: double row **a** big man (MTMM), double row **b**: little woman (FSMM). The left-hand side of each sample image shows the corresponding RAMSIS comfort rating

drive markets must have a different package of driver position controls. Most vehicle manufacturers supply both markets with their products and are confronted with the unavoidable fact of finding an economic compromise in order to keep as many vehicle components as possible the same and still demonstrate the compatibility of right-hand and left-hand drive vehicles. This is how we try to keep as many parts of the body shell, the chassis and the drive train as possible unchanged for all LHD and RHD versions, such as engines and transmissions.

In addition to the regulation of right-hand and left-hand traffic, some countries of the Anglo-American cultural area also use the English mile (*mile*) instead of the metric kilometer specification (*km, km/h, litres/100 km*)

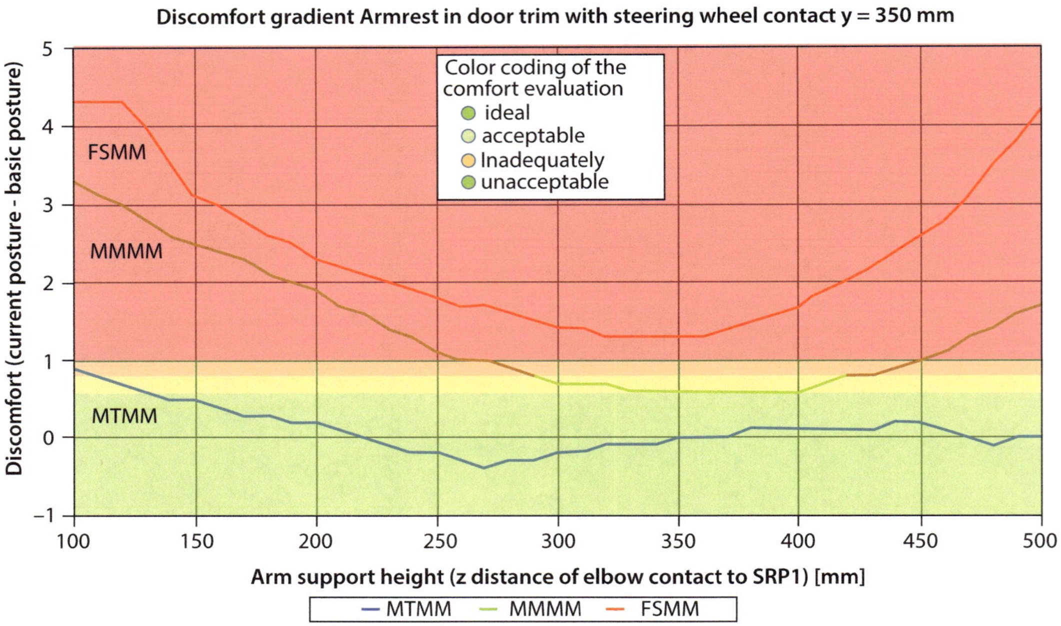

Fig. 7.38 Discomfort curve for the armrest in the door trim for steering wheel contact and a distance Δy = 350 mm of the armrest from the centre of the seat

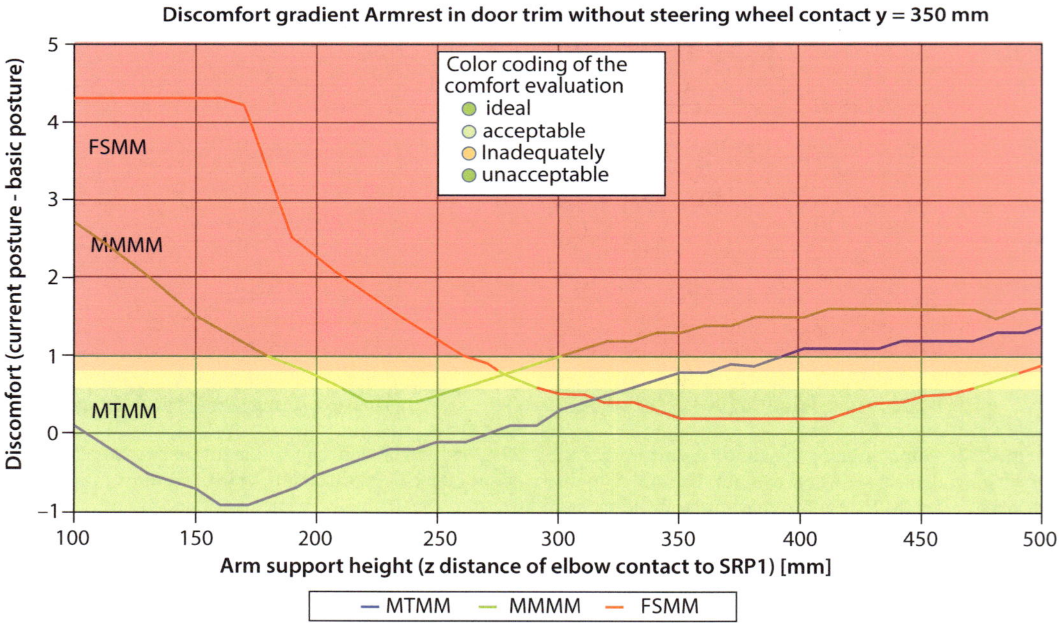

Fig. 7.39 Discomfort gradient for the armrest in the door trim without steering wheel contact and a distance Δy = 350 mm of the armrest from the centre of the seat

and units derived from them (*mph* miles per hour for the speed, *mpg* mile per gallon for the consumption of the route). Some countries with right-hand traffic, especially in Africa, also use km indications in right-hand drive vehicles (see Table. 7.5).

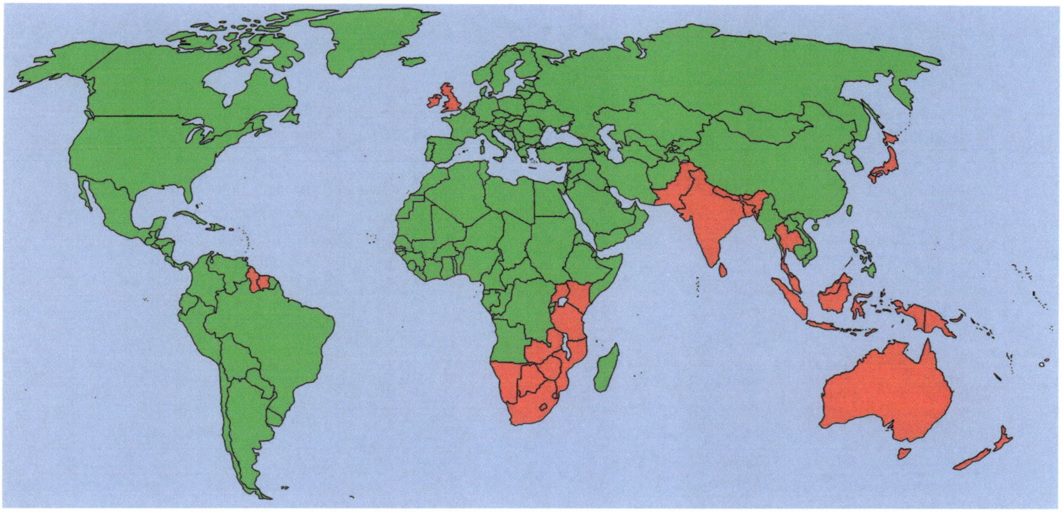

Fig. 7.40 Left-hand drive (green) and right-hand drive (red) markets

However, the arrangement of the components is not mirrored along the longitudinal axis of the vehicle, but mainly shifted parallel from one side of the vehicle to the other. In both markets the accelerator pedal is always operated with the right foot, which means that in left-hand drive markets it is necessary to position the accelerator pedal at the centre tunnel in the area of the heating system, while in right-hand drive the movement of the accelerator pedal is mainly influenced by the wheel arch (Fig. 7.41). Since the **pedals** are the interface of many vehicle functions (acceleration, braking, clutch for gear changes), they connect components (pedals) in the driver's footwell with components (master brake cylinder and brake booster) in the engine compartment. As these components are voluminous and the space available in the engine compartment is very limited, these units cannot easily be moved to the other side. In the case of the booster, several solutions are possible to avoid a collision with the drivetrain (engine, gearbox and auxiliaries). Since the mirrored arrangement of the motor is not necessary for economic reasons, an adaptation design for the pedal system and the connected aggregates is generally preferred, especially as a symmetrical arrangement is not possible due to the displacement and also the aspects of the production assembly as well as the removal for maintenance purposes must be taken into account in the package. If there is sufficient package space in the engine compartment, the pedal system is shifted as far to the right as the wheel arch and wheel envelope in the interior and the bodyshell side member and damper dome in the engine compartment allow.[17] Ideally, this arrangement will not cause any collisions with powertrain units. Afterwards, all electrical lines, as well as any cooling water hoses and lines of the air-conditioning system must be newly laid, as must the electrical and hydraulic supply lines to the pedals. Often, however, the installation space for the basic variant is so exhausted that it cannot be adopted unchanged and the pedals have to avoid the components of the drive train. In this case, the entire pedal system can be arranged further towards the occupant, which is primarily associated with a restriction of the free space between the vehicle seat and pedals. If the target market is a country with smaller people than in the original market, it can be assumed that the leg lengths are

17 This approach is based on the assumption that a vehicle for the left-hand drive market should be converted to an arrangement for a right-hand drive market. In the reverse case, other limitations and restrictions may occur and other assemblies and parts may be affected.

Table 7.5 List of the main markets in terms of traffic regulation and route units used

LHD Left-hand-drive markets Right-hand traffic	Unit	RHD Right-hand-drive markets Left-hand traffic	Unit
Afghanistan, Albania, Algeria, Andorra, Angola, Egypt, Equartional Guinea, Argentina, Armenia, Aruba, Azerbaijan, Ethiopia, Azores, Bahrain, Belgium, Belize, Benin, Bolivia, Bosnia, Bosnia, Brazil, Bulgaria, Burkina Faso, Burma, Burundi, Chile, Rep. China, PR China, Costa Rica, Curacao, Denmark, Germany, Dominican Republic, Djibouti, Ivory Coast, Ecuador, El Salvador, Eritrea, Estonia, Finland, France, France, French Riviera, United Kingdom, Guyana, Gabon, Gambia, Georgia, Ghana, Greece, Guadeloupe, Guatemala, Guinea, Guinea-Bissau, Haiti, Honduras, Iraq, Iran, Ireland, Iceland, Israel, Italy, Yemen, Jordan, Canada, Cambodia, Cameroon, Cape Verdean Islands, Kazakhstan, Qatar, Kyrgyzstan, Colombia, Comoros, Congo, Rep. Congo, Croatia, Cuba, Kuwait, Laos, Latvia, Lebanon, Liberia, Libya, Liechtenstein, Lithuania, Luxembourg, Morocco, Macedonia, Madagascar, Mali, Martinique, Mauritania, Mexico, Moldova, Monaco, Mongolia, Montenegro, Morocco, Myanmar, Nauru, Namibia, Nicaragua, Netherlands, Netherlands Antilles, Niger, Nigeria, North Korea, Norway, Oman, Austria, Panama, Paraguay, Peru, Philippines, Poland, Portugal, Puerto Rico, Réunion, Rwanda, Netherlands Antilles, Romania, Russia, Sao Tome & Príncipe, San Marino, Saudi Arabia, Sweden, Switzerland, Senegal, Serbia, Sierra Leone, Somalia, Slovakia, Slovenia, Spain, South Korea, Syria, Taiwan, Tajikistan, Togo, Czech Republic, Chad, Tunisia, Turkey, Turkmenistan, Tuvalu, Ukraine, Hungary, Uruguay, Uzbekistan, Vanuatu, Venezuela, United Arab Emirates, Vietnam, Wallis & Futuna, Belarus, Central African Republic [150]	Km, km/h	Botswana, Guyana, Japan, Kenya, Lesotho, Malawi, Namibia, Zambia, Suriname, Swaziland, Tanzania, Uganda, Zimbabwe, Cyprus, Zambia, Malawi [14]	Km, km/h
USA [2]	Miles, mph	Anguilla, Antigua & Barbuda, Ascension, Australia, Bahamas, Bangladesh, Barbados, Bermuda, Bhutan, Brunei, Cayman Islands, Cook Islands, Dominica, Falkland Islands, Fiji, Grenada, Hong Kong, India, Indonesia, Jamaica, Virgin Islands, United Kingdom, Kiribati, Maldives, Malta, Malaysia, Mauritius, Montserrat, Mozambique, Nepal, New Zealand, East Timor, Pakistan, Papua New Guinea, Solomon Islands, Samoa, Seychelles, Singapore, South Africa, Sri Lanka, St. Helena, St. Kitts & Nevis, St. Lucia, St. Vincent & Grenadine, Thailand, Tokelau, Tonga, Trinidad & Tobago, Turks & Caicos Islands [49]	Miles, mph

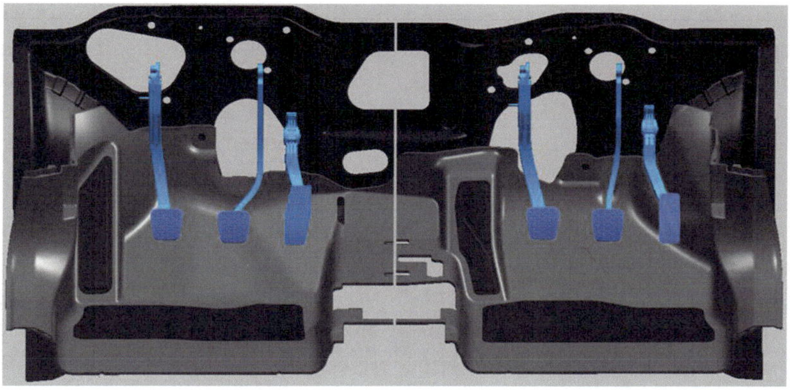

Fig. 7.41 Comparison of an LHD and RHD pedal gallery [Opel Astra]

also smaller and less leg space is needed. In this case, a moderate restriction is an acceptable compromise. If the target market includes larger persons with longer extremities, such a restriction would not be recommended and would lead to insufficient accommodation.

In vehicles that have been cost-effectively converted to the complementary market, there is a solution in which the brake booster remains on the original side but the pedals are moved to the other side and the transmission of the pedal forces is ensured via a coupling rod that runs from one side to the other.

In addition to the undesired torsion capability of the coupling rod and the very massive body-in-white bearing arrangements, the question of passage from one side to the other must also be clarified, which is always a challenge in a tight package. A transfer of the entire brake booster from the engine compartment to the vehicle interior is also a technical possibility, but this requires massive pedal mounts and deflection rods. Frequently, solutions are also found where smaller diameters of brake boosters are used or where an electrical support is used instead of a pneumatic one. The central starting point for such an adjustment is always the pedals with the safety-relevant brake system. The adaptation of the clutch actuation and the connection of the accelerator pedal by means of bowden cable transmission represents a lower effort compared to the aspects of the brake system.

The distances between the pedal plates are based on the width of the driver's foot (see ▶ Sect. 4.2.1.6). The width of the footwell shall be such that, irrespective of the use (or absence) of a clutch pedal or footrest, the minimum freedom of movement of both feet is adequately ensured. For example, the free distance between the A-pillar and the central tunnel lining should comply with the 'four-leg rule', i.e. a minimum width from 440 mm (4 × 110 mm) to 500 mm (4 × 125 mm).

Steering adjustment is a relatively elegant solution nowadays with the widespread use of front wall mounted rack and pinion steering gears and short safety steering columns, while the long steering rods running along the engine block with elaborate reversing levers are a thing of the past. The wiper system is often a more or less mirrored arrangement, which represents a complete assembly variant and is accommodated in the so-called water box on the lower windscreen cross beam. As a rule, the basic version of the water tank is designed in such a way that it is able to accommodate both the LHD and RHD variants. The mechanical gear selector lever, when mounted on the centre tunnel, is also arranged so that it can be used for both the LHD and RHD variants. In the case of driver-oriented alignment, an adjustment must also be made, but this can produce extensive component variants such as tunnel console, armrest and shelves. Particularly in automatic transmissions, a shift diagram with the display of the selected shift step is arranged on the driver's side. The visibility requires that, even with the shift lever in the same position, the gear selection indicator for the complementary market must be moved to the driver's side of the lever.

If the vehicle is fitted with a hand-operated parking brake, the lever is always best assigned to the driver to minimise loss of power due to range. Especially in right-hand drive vehicles, this necessity arises from the fact that the handbrake lever is operated with the left arm, which is clumsier and possibly weaker for about 90% of the population (see comments on handiness at the end of this section and in ► Sect. 4.2.1.3). An economic compromise is often to place the handbrake lever centered on the center tunnel to avoid right and left-hand drive variants of the lever, but this is at the expense of the storage space in the tunnel console. ◘ Table 7.6 shows the necessary arrangement of control elements depending on the target market.

The adaptation of components that are not directly related to the vehicle operation is often not necessary for economic reasons, since the conversion of the pre-assembly or manufacturing processes, a second tool set or even a whole series of subsequent changes are associated with it. As a rule, for example, the tank opening flap is arranged so that it is on the driver's side. If a vehicle designed in this way is then converted for the complementary market, the opening remains on the original side so that the entire underbody package can be left unchanged. Here approx. 40 assemblies including the fuel system with innumerable parts, as well as the entire lateral outer skin of the body and their long-running tools would be affected. For vehicles with a laterally hinged rear door (*safari door*), a cost-intensive adaptation is waived for the same reasons, even if this solution has a use-relevant disadvantage. If the vehicle is parked parallel to the road, it is not possible to unload the trunk on the sidewalk side because the open rear door blocks the way (◘ Fig. 7.42). In asymmetrically divided and foldable rear seats, the larger section of the backrest is often arranged on the passenger‘s side, as this gives the driver a larger throughloading volume, while one person can take a seat on the remaining single seat. This configuration will then be retained unchanged when the company switches to the complementary market,. which may lead to limited usability in the transport sector. Also, the controls originally facing the driver are not reflected in the instrument panel, such as the buttons and switches of the radio and heating controls, although this creates disadvantages for the secondary market customer in terms of range. At this point it is necessary to return to the handbrake lever for the mechanical parking brake. Adaptations for the complementary market usually include moving the lever to the other side, unless the vehicle is developed for a market where the usability and accessibility of the *cup holder* is much higher prioritised by customers than that of the hand brake lever. This is particularly the case in the North American market, where, due to the predominance of automatic transmissions with their integrated parking lock and the widespread habit of regularly carrying large beverage cups, the handbrake lever is only used in exceptionally inclined positions. An elegant solution to this plight seems to be the electrically operated parking brake, as the switch required for this requires far less installation space and can therefore be more easily moved to another arrangement. Unfortunately, this variant enjoys only limited acceptance among many customers in numerous markets, so that the additional technical and financial costs are difficult to market as functional added value.

Some assemblies necessarily need to be adapted for the complementary market as they concern registration requirements or vehicle safety. For example, the headlamps must always be adjusted due to the technically usual asymmetrical dipped beam, the windscreen wiper system due to the wiper fields relevant for approval and the exterior mirrors due to different national regulations. Fortunately, the required scope of modification is limited to a very limited module. If there are window regulator switches for all vehicle side windows in the driver's side door panel, this arrangement is symmetrically mirrored, which causes its own component and additional tool variants, but is of little effort in development due to the computer-aided design. This also applies to the main light switch if it is located in the instrument panel. Occasionally one finds vehicles, with which

Table 7.6 Overview of control elements depending on target market (LHD and RHD)

	LHD **Left-hand-drive car**	**RHD** **Right-hand-drive car**
Used street side	Right-hand traffic	Left-hand traffic
Steering wheel	Left	Right
Pedals (drive, brake, clutch pedal)	Left	Right
Driver's foot rest	Left footwell, sillside	Right footwell, tunnel side
Handbrake lever	Centrally on the tunnel console or facing the driver on the left side of the tunnel console*	Centrally on the tunnel console or facing the driver on the right-hand side of the tunnel console*
Foot operated parking brake	Left footwell, sillside	Right footwell, tunnel side
Switch for electrically operated parking brake	Centrally on the tunnel console or facing the driver on the left side of the tunnel console*	Centrally on the tunnel console or facing the driver on the right-hand side of the tunnel console*
Shift lever (manual and automatic transmission)	Centrally on the tunnel console or facing the driver on the left-hand side of the tunnel console	Centrally on the tunnel console or facing the driver on the right-hand side of the tunnel console
Shift diagram/gear display	Facing the driver on the left side of the gear lever	Facing the driver on the right side of the gear lever
Steering column lever for direction indicator "turn signal lever	Left of steering wheel	Left of steering wheel (UK/JAPAN) right of steering wheel (India/ Australia)
Steering column lever for wiper/washing system "wiper lever	To the right of the steering wheel	Right of steering wheel (UK/ JAPAN) left of steering wheel (India/ Australia)
Hood release lever	Left footwell, sillside	Right footwell, sillside
Button for hazard warning lights	Centre of vehicle on instrument panel or driver's side (left) near Centre of vehicle on instrument panel	Centre of vehicle on instrument panel or driver's side (right) near Centre of vehicle on instrument panel
Volume knob or radio on/off button	On the driver's side (left) near the Centre of the vehicle on the instrument panel	Driver side (right) near the Centre of the vehicle on the instrument panel
Arm rest on the tunnel console	Centrally on the tunnel console or the driver facing on the left side of the tunnel console	Centrally on the tunnel console or the driver facing on the right side of the tunnel console*
Cylinder lock for emergency unlocking	Left front door (driver's door)	Right front door (driver's door)
Fuel filler flap	Left vehicle side (driver's side)	Right vehicle side (driver's side)
Tank flap release	Left footwell, sillside or left door panel	Right footwell, sillside or right door trim
Rear door release (handle, lock)	Right side of the tailgate	Left side of rear door

aThe handbrake lever is positioned on the side facing away from the driver when cupholders are given higher priority by the market

Fig. 7.42 Right hinged rear door

the hood release lever was left on the original vehicle side and is now in the passenger footwell. However, many customers are irritated by this practice and search in vain in the wrong place in emergencies. Even if it means an additional design and economic effort, it is advisable to adapt to the expectations of the customer and the maintenance personnel.

Not to be confused with the right-hand-drive problem is the question of the *handedness*. This means the preferred use of one side of the body, in particular the hands, for the purpose of operation or writing. Regardless of the direction of travel regulated by law, all markets have in common that the distribution of right-handed to left-handed is the same regardless of culture and anthropology. It is assumed that 10–15% of the population are left-handed and that the power of the clumsier hand is slightly less than on the preferred site. As a result, in left-hand drive markets, the gearshift and handbrake levers and most controls in the centre console can be operated with the skilful and stronger hand, whereas in right-hand drive markets the drivers have to perform these operations with the untrained hand. Undoubtedly, the effect of practice and habituation is the decisive factor here, and it can be seen that changing to a complementary market vehicle poses greater problems for users than operating the vehicle with the "wrong" hand.

7.2.4 Co-Driver

The considerations set out in detail for the driver also apply to the passenger with some restrictions. For the calculation of the posture, of course, a steering wheel contact is not to be assumed here, but rather, for example, a support on armrests which have been determined according to the procedure described above. Usually the same seat is used for the passenger as for the driver. From an ergonomic point of view this is not entirely correct. In contrast to the driver, the co-driver can adopt different postures, which in principle can prevent postural discomfort. By using the driver's seat, however, he is forced into a relatively rigid posture. This can be an advantage when cornering in a committed manner. However, a seat that allows more freedom of movement is desirable for long motorway journeys with low lateral accelerations. Even more so than for the driver's seat, the side bolsters of the front passenger seat may be changed automatically depending on navigation information and acceleration values measured on the vehicle, or at least be easily adjusted by the user himself.

7.2.5 Vehicle Rear

7.2.5.1 Second Row of Seats

An H30 dimension can also be defined for the second row of seats (H30–2). The difference between the corresponding SgRP 2 point and the SgRP 1 point of the driver's seat (position of the 95-percentile SAE manikin) is essential to describe the space available on the rear seat. Another important descriptive variable is the distance between this SgRP-2 and the outer contour of the backrest of the front seat (knee space). An essential, often neglected size is the foot space under the front seat, which, together with the knee space for the rear passenger, determines the possibility of adopting a relaxed posture (Fig. 7.43).

It is often assumed that the second row of seats is not important because each vehicle has an average of only 1.5 drivers. Apart from two-seater sports cars, the advantage of a vehicle with a second row of seats is that it can occasionally be driven by several people. This is mainly the case over long distances. It is noticeable that the second row of seats is obviously used more frequently in compact cars than in limousines that are better suited for this purpose (see Fig. 7.44). Also interest-

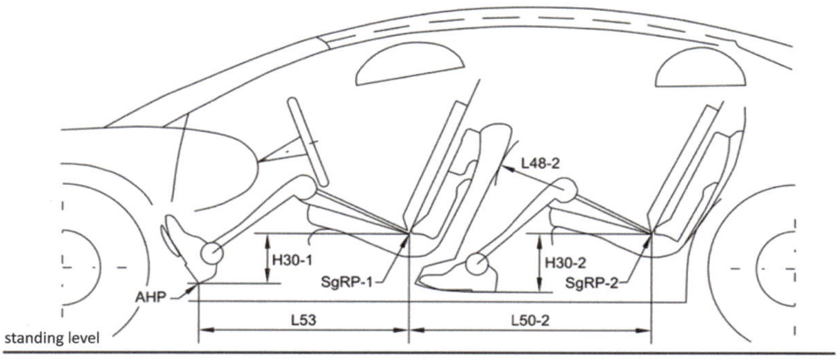

Fig. 7.43 Important package dimensions for the second row of seats

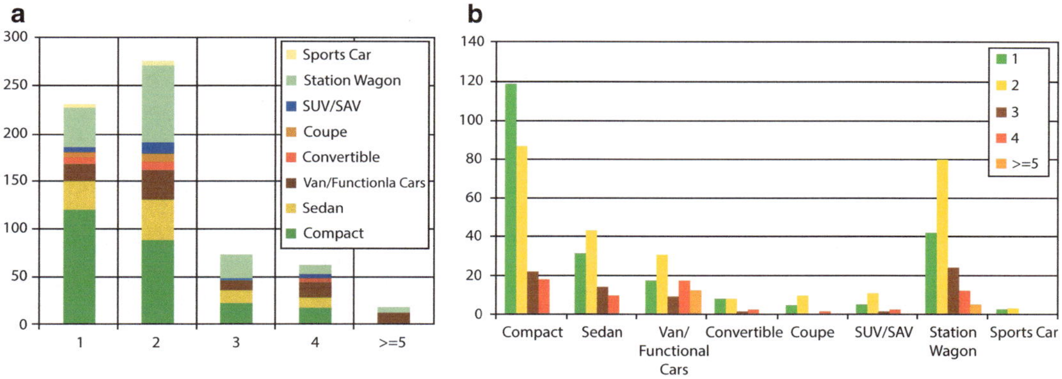

Fig. 7.44 Results of an exemplary traffic analysis on a Sunday on a route between Munich and Delft (850 km). The diagram **a** shows the frequency distribution of the number of persons in different vehicle types, the diagram **b** the respective vehicle types with the corresponding number of persons. (Kilincsoy et al. 2014)

ing is the relatively high number of uses of the second row of seats in convertibles. These are all reasons to also consider the second row of seats with high priority in vehicle development and assessment.

In principle, the layout of the second row of seats can be designed in a similar way to that of the driver's seat using human models (e.g. RAMSIS). It is based on the pessimistic situation for the rear-seat passenger, i.e. the front seat is to be brought into a position for the tall long-legged man. Different automobile companies have developed different philosophies with regard to the layout of the vehicle fund. For example, depending on the model series (compact car, lower middle class, middle class, luxury class), it can be determined up to which percentile space in the rear seat is to be provided under these conditions. It can also be advantageous to make the reference point SgRP-2 moveable similar to that of the driver's seat, as is possible with some vehicle concepts (SUV's, VAN's, combination vehicles).

In the back seat, different postures are adopted, as the posture prescribed by the driving task is no longer required. Kilincsoy et al. (2014) have extracted three seating postures from observations of seating postures in railway trains (Kamp et al. 2011) which are primarily suitable for rear-seat passengers and which are illustrated in Fig. 7.45. The position on the left gives a *upright position* for short journeys, which is characterized by the passenger, for example, observing the surroundings, using the mobile phone, talking to other rear-seat passengers or eating. The slightly relaxed, slightly *standard posture* needs a little more space. The passenger is awake and is dedicated, for example, to listen-

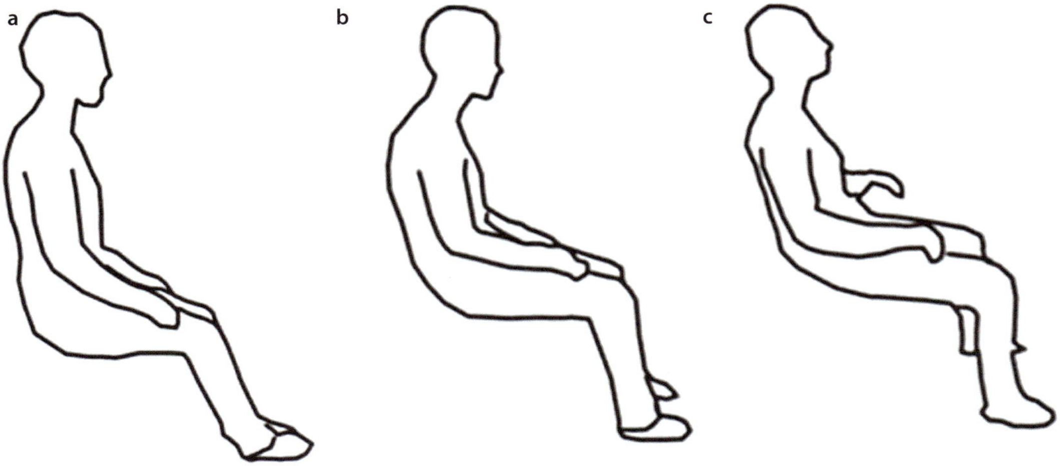

Fig. 7.45 Typical seating postures for rear-seat passengers, **a** upright posture for various activities, **b** Slightly relaxed posture, **c** resting posture

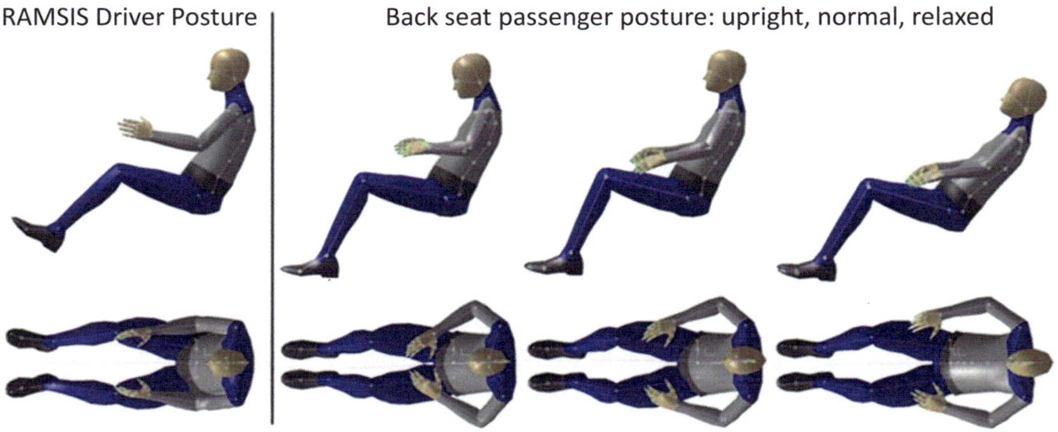

Fig. 7.46 Representation in RAMSIS of the postures found by Kilincsoy et al. (2014) in comparison with the driver's posture (left)

ing to music. The picture on the right *relaxed posture* is relevant for larger cars. The passenger undertakes a long journey, perhaps sleeps for a longer time or is otherwise largely relaxed.

Kilincsoy et al. (2014) had test persons adopt the three postures mentioned in a suitable vehicle mock-up (see ► Sect. 10.3.1), recorded the respective three-dimensional body angles using the PCMAN method (see ► Sect. 11.2.1.3) and reproduced the corresponding body postures in RAMSIS (Fig. 7.46). In order to allow a comparison with the data in Table 7.4, the projections of the angles found on the sagittal plane including the standard deviation are shown in Table 7.7. The shoulder angle is conspicuous in the upright posture, which deviates from the other two postures. It is attributable to the activity carried out there (e.g. mobile phone use). The hip angle in the relaxed posture is clearly different from the other two postures, especially also from the driver posture. The ankle angles differ only

Table 7.7 Mean value and standard deviation of the three typical rear seat passenger postures in 2-D projection (Kilincsoy et al. 2014)

posture	angles						
	Head Nape of the neck	Nuchal torso	Shoulder	Elbow	Hip	Knee	Ankle
Upright	177,5 ± 4,6	130,0 ± 3,5	32,4 ± 13,3	113,1 ± 11,7	105,5 ± 5,5	103,4 ± 12,5	104,9 ± 5,8
Standard	187,2 ± 3,9	139,5 ± 0,7	0,6 ± 12,6	128,5 ± 14,1	104,2 ± 7,6	99,5 ± 9,9	104,7 ± 4,6
Relaxed	185,3 ± 4,3	142,7 ± 3,5	1,0 ± 11,8	139,9 ± 11,8	118,9 ± 10,5	104,9 ± 11,9	107,9 ± 8,2

slightly, which indicates that when designing the second row of seats, particular attention should be paid to the foot space under the front seat.

The postures shown above can be used to optimize the rear seat space. With regard to seat design, support and belt routing, criteria similar to those developed for drivers and passengers apply. However, it should be borne in mind that a much greater freedom of movement must be provided for the passenger in the rear seat. In particular, this requires compromises to be made with regard to belt guidance. The additional facilities in luxury vehicles (telephone, entertainment, individual air-conditioning, massage seats etc.), in which the rear-seat passenger plays a special role, are only referred to in this context.

As a result of the ever-increasing progress in passive safety regulations and techniques, an important issue for the well-being of rear-seat passengers is increasingly being overlooked. As already explained in ► Sect. 3.1.3, a major reason for the occurrence of kinetosis ("seasickness") is the unusual divergence of optical and kinesthetic information. This is particularly the case when the passenger is unable to follow the course of the road (keyword: "looking to the horizon"). But it is precisely this effect that is created by the headrests, which increasingly adjust the view. Under this aspect, it should be examined to what extent the development of kinetosis can be prevented by a different design of the headrests or by a different positioning of the rear-seat passengers (for example more towards the centre of the vehicle).

7.2.5.2 Third Row of Seats

In some vehicle types a third row of seats is also provided. In principle, the same conditions apply for this row of seats as for the second row of seats. However, these requirements can only be met for a few vehicles (larger SUV's and VAN's). In most vehicles, this row of seats is designed as an emergency seat or as a seat for younger children. With some combination vehicles, a third row of seats can also be mounted in the luggage compartment against the direction of travel. In any case, it is a matter of determining in the specifications for which target group these seats are to be designed. In all cases, it is possible to design such seats in more detail using human models as described in ► Sect. 7.2.5.1. In particular, the aspects of the support provided by the seat and the belt run must also be taken into account here.

A problem specific to the third row and in many cases also to the second row is the design of the folding mechanisms with which the seats can be lowered into the vehicle floor. Sometimes it is also planned to remove the seat as a whole from the vehicle. Then appropriate release mechanisms and handholds must be designed to meet ergonomic requirements (comprehensibility, low actuating forces). In particular, human models can help with the size of the controls required and the ability to carry the seat as a whole. In corresponding ergonomic standard works (e.g.: Schmidtke: Handbuch der Ergonomie n.d.) or software tools (e.g.: EKIDES) you will find general, but thoroughly detailed information about this, which, however, must first be adapted to the corresponding application.

7.3 View

7.3.1 Direct View

7.3.1.1 Forward View

Guidelines and Regulations

As explained in ► Sect. 7.2.2, driver posture is significantly influenced by forward vision. There are a number of legal regulations of different nations or economic areas to ensure a required minimum quality of visibility conditions in every vehicle registered for road traffic. The essential regulations for direct and indirect vision in the regions Europe, North America and Australia were compiled by Hudelmaier (2003) and are reproduced with updates and additions in ◘ Table. 7.8. There are also separate rules for Japan, India and the Commonwealth of Independent States (CIS). The content and form of most of the regulations in these countries are based on those in Europe and North America. In addition, there are standards of engineering associations such as DIN, VDI, SAE or JAMA which describe the "state of the art" and are therefore relevant, but not necessarily legally binding (Remlinger 2013).

Only the most important of these provisions are briefly explained below, as a comprehensive presentation would go far beyond the framework provided here. Reference is made in this context to the individual regulations.

In practical terms, the concept of the ellipse of the eye according to SAE J941 plays an important role in vehicle design.The position of this ellipse is determined in size and inclination of its main axes by the 95% seat reference point SgRP. From the ◘ Fig. 7.47 it can be deduced that this is actually an ellipsoid body composed of two ellipsoids (one for the left eye and one for the right eye) which penetrate each other. The distance between the seat reference point SgRP and the rear focal point of this ellipse in side view is 635 mm. The ellipse of the eye is defined by its history as follows: American convertibles built in 1963 were used to measure the two-dimensional distribution of eye point positions in side view and top view with subjects of different body sizes (Meldrum 1965).[18] At these points, straight lines in any direction were created in such a way that 95% of the eye points are on one side and the remaining 5% on the other side of these straight lines. The envelope of all these straight lines is an ellipse (in the example shown the 95% eye ellipse; Hudelmaier 2003). The tangents to the ocular ellipse drawn in ◘ Fig. 7.47 thus include an area that can be seen by 95% of the users. In detail, there are many additional rules for the use of the eye ellipse, which according to their definition contain only eye points with the head in a fixed position.

Many recent studies show that the actual eye positions often clearly exceed the range of the ocular ellipse (◘ Fig. 7.48), which may also be due to the vehicles available when they were created and to the population of test subjects used in the experiments. Remlinger (2013) shows that the real eyepoint positions can be predicted very well by the RAMSIS extreme types already in the conception phase (◘ Fig. 7.49).

For the registration of vehicles in the European Union, the verification of binocular vision screens on the basis of the registration regulations 77/649/EEC (2008) plays an important role. This rule is based on the SAE SgRP and uses a simplified model of the head represented by a triangle consisting of the two eyes and a nuchal pivot (see Remlinger 2013 for details). This regulation defines two eye points, V1 and V2, which are important for the fulfilment of a number of EEC directives and the certification of the vehicle dependent on them.

The German regulation StVZO specifies in § 35b additionally the visibility areas to the front and to the side necessary for a vehicle. It is based on a fixed eye point defined in a verti-

18 The vehicles used had seats with a relatively small adjustment range between 114 and 137 mm from today's point of view. They were also equipped with a bench seat with rigid backrest angles between 22.5° and 26.5°. The seat adjustment was inclined to the horizontal with an angle between 7° and 15°. Moreover, these vehicles had no height adjustment of the seats and no adjustment possibility for the steering wheel.

Table 7.8 Important regulations and guidelines regarding visibility from a passenger car (updated and supplemented according to Hudelmaier 2003)

visual field	Windscreen wiper and washer	safety glass	rear-view mirrors
StVZO §35	**FMVSS** 104	BSAU 178a Specifications	**STVO** §56
ISO 7397 Measurement and evaluation	SAE j903 wiper field	**ECER**43	**ADR** 12 Leveled off
SAE J1050**a** description and measurement of the driver's field of vision	**ISO** 3469 Testing method	**ISO** 3637 Material	ADR 14
77/649 **EWG** → 90/630 **EWG** minimum forward field of vision covers	EWG 78/318 Wiping field, function	**ISO** 3538 Optical properties	2003/97**EG** → 2005/27 **EG** Description and measurement of the driver's field of vision
FMVSS 128 Minimum field of vision masking sun visors	**ADR** 16	**ISO** 3917 Testing procedures	**ECER**46
ECER 125 Ont field of view	Wind screen defrosting	**FMVSS** 205	**FMVSS** 111 Arrangement Field of view Operation
JP/061 Field of view at close range	**EWG** 78/317 Areas, time	**ADR** 8	**FMVSS** 107 Withdrawn
	SAE J902, **SAE J**381 Defrosting fog removal	Windscreens air jets equipment	**ISO** 5740 Test method reflection factor
	ISO 3468, **ISO/NP** 3470 Testing method	**ISO** 3470 Testing procedures	
Shadow bands	**FMVSS** 103 Defrosting fog removal	Rear window wiper and washer	
SAE J100	**ISO** 5898 Rear window defrosting	**ISO** 6255 Demands	

cal plane located in the centre of the seat 130 mm in front of the front edge of the backrest and 700 mm above the unloaded driver's seat in the centre position. From this point lines of sight are drawn onto the road surface in front of the vehicle (experimentally this can be done by using a point-shaped lamp and measuring the shadow crack of the vehicle on the road surface). In particular, forward visibility is deemed to be assured if a radius of at least 9.5 m is maintained in a semicircle of 12 m (Fig. 7.50).

The procedure given by the StVZO can also be extended to the whole vehicle. In order to determine the floor area covered by the vehicle body, the window openings are projected from the defined eye point onto the standing plane and the surface area is determined limited by the resulting area at defined boundary distances to the vehicle (Fig. 7.51). The resulting floor area serves as a benchmark between two competing concepts or competing vehicles. This method is suitable for determining the projection surface

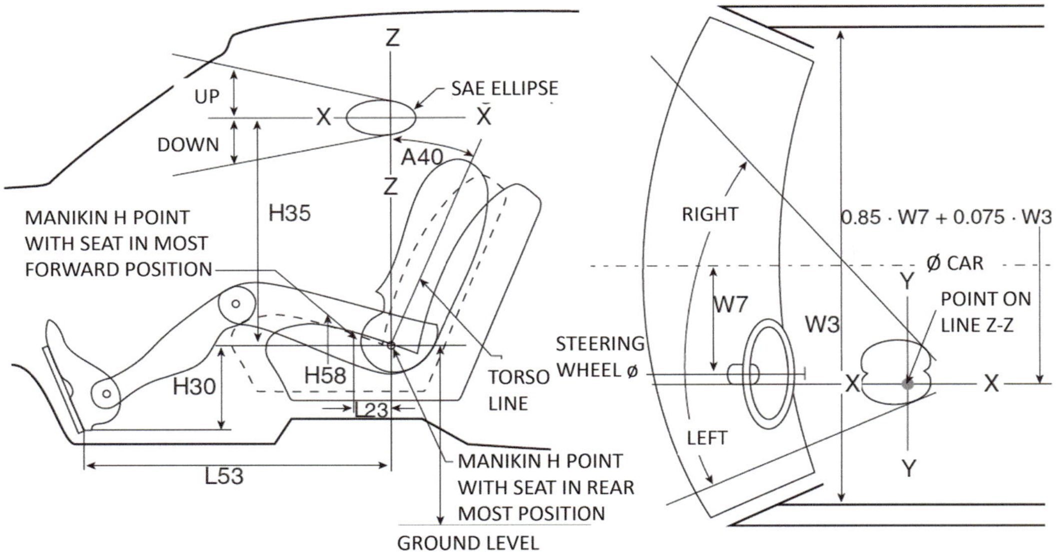

Fig. 7.47 Position of the eye ellipse with respect to the SgRP in the side view **a** and supervision **b** (SAE J941)

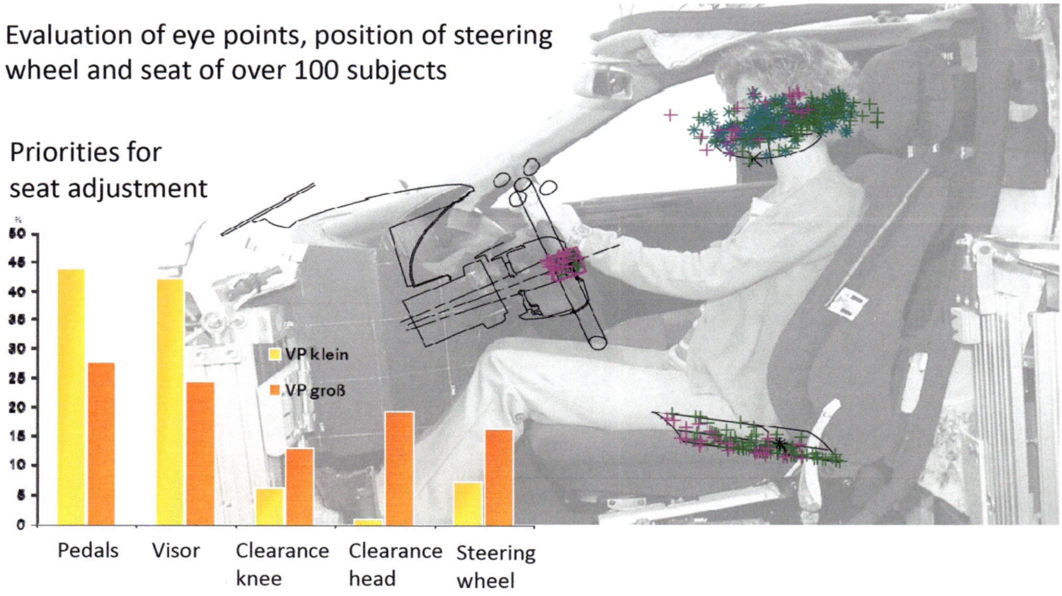

Fig. 7.48 Variation of the eyepoint positions, the H-point positions and the steering wheel position with more than 100 test subjects. (Brückner 2011)

on the basis of a virtual CAD model as well as for physical evaluation with vehicles for which no mathematically described data is available. From the eye point, the "cut-off line" can also be projected onto the stand plane with a laser beam, or a bright light source can be attached to this point and the shadow cast on the floor surface documented photographically or with marking lines.

With this measuring method, the connection between visible traffic space and some vehicle parameters becomes clear. If the eyepoint is raised slightly, the shadow area is reduced because the beam angle over the door

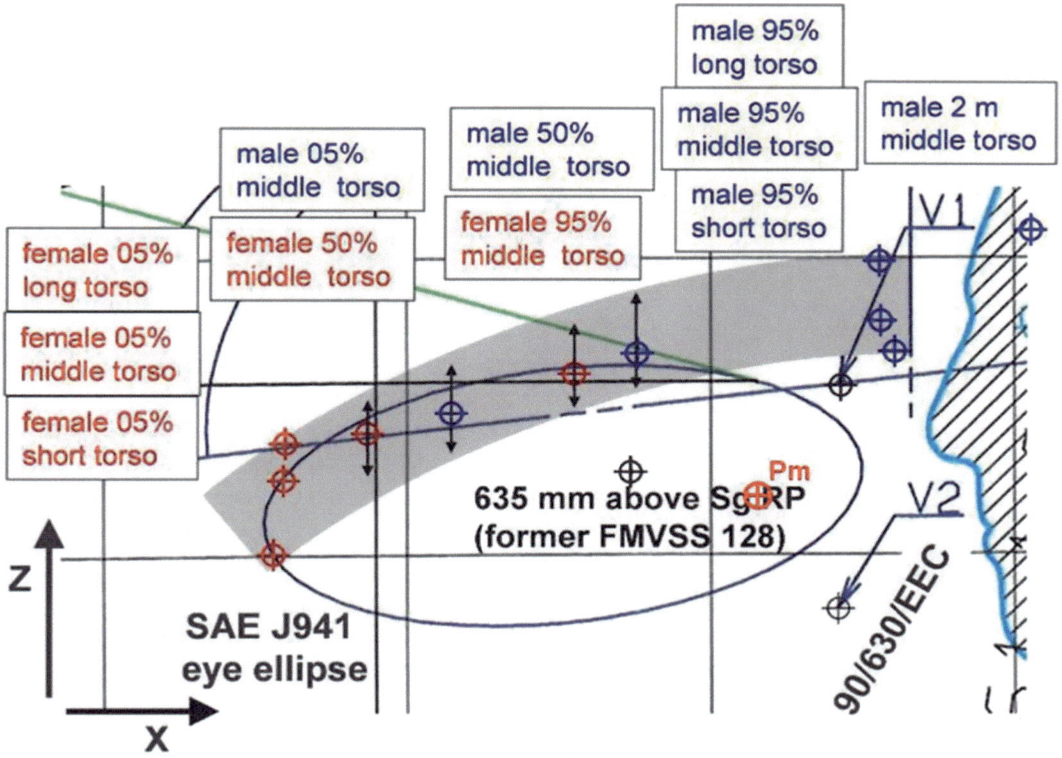

Fig. 7.49 Eyepoint positions according to SAE standard compared to RAMSIS simulation. (From Remlinger 2013)

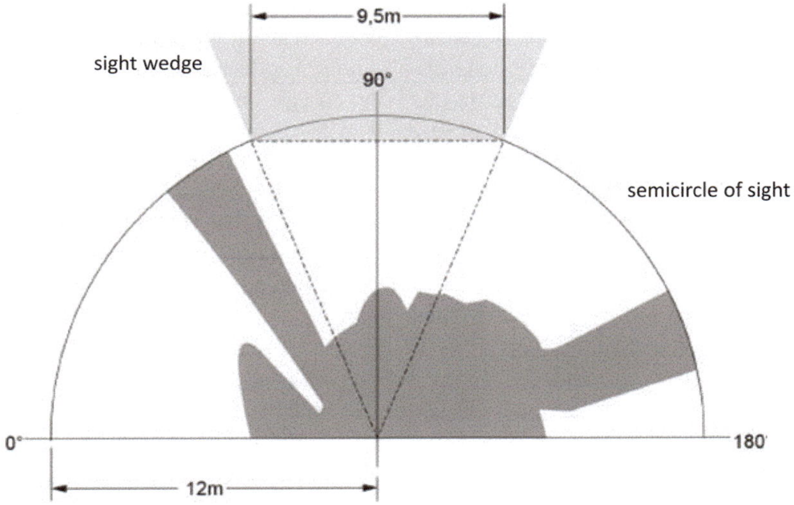

Fig. 7.50 Semicircle and wedge according to StVZO § 35b

parapets becomes steeper, bringing the point of view closer to the vehicle. The same applies to a lowering of the waistline. The deeper the parapet line lies under the eye point, the better the surroundings can be seen. The same applies to the view over the hood and the view

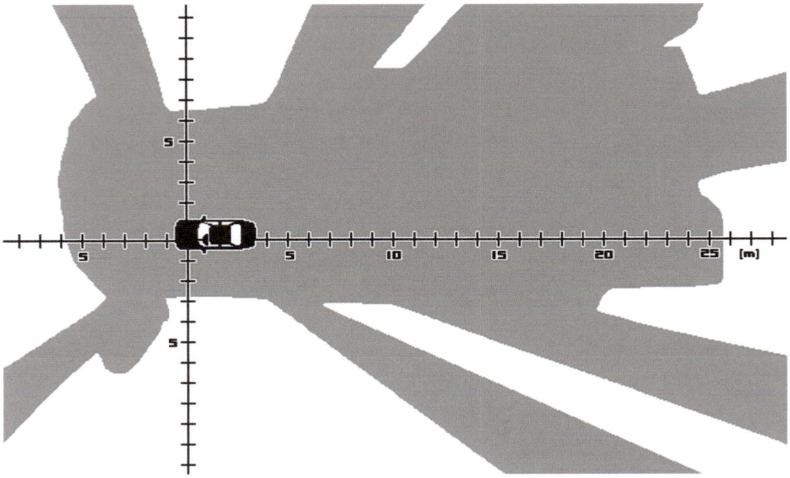

Fig. 7.51 Concealed area projected onto the stand level

to the rear of the vehicle. Visibility of the leading edge of the hood promotes the estimation of vehicle dimensions, but should not be so voluminous as to impair visibility of the road ahead (see also Fig. 7.18).

The measurement method presented here makes it easy to determine the point closest to the vehicle at which the driver is able to see the road beyond the edge of the hood from a defined position of the eye point. However, in order to determine the range of the ground view from all eyepoint positions, it is necessary to make a variation over all possible driver positions. This is best done with the help of the design manikins, which have been defined using a human model. From this point of view it becomes clear that tall drivers who sit far behind and at the same time very flat in the seat adjustment field may only be able to see the road only very far in front of the vehicle. The same applies to drivers with very short upper bodies and correspondingly low seating positions. On the other hand, a high eyepoint position, which is equally accessible for both tall and short drivers, makes it possible to see the road much closer. For example, with only a slight raising of the eyepoint by a few millimetres, an improvement in road visibility by a few metres is possible. Ground visibility is often only a result of the selected seat position and not the sole determining criterion for the choice of seating position; the accessibility of the pedals and the adjustment range of the steering wheel often force small riders into a lower position, which can impair visibility.

Ergonomic Analyses

The RAMSIS manikin offers the following special features with the program module *RAMSIS cognitive*: a set of design and analysis capabilities that enable a realistic visual inspection at the CAD level that goes far beyond what is required by regulations and guidelines. In addition, the respective fields of vision can be calculated from the driver position of the various RAMSIS manikins, giving the CAD user a realistic impression of the visual impression of different types of persons (see Fig. 7.52). In many cases it is sufficient for such analyses to start from the so-called centre eye in order to get an impression of what the driver's visual impression is. In some cases - especially at close range - it makes sense to calculate separately for the left and right eye and to determine by superimposing the two images which area is binocular and which area only monocular. Monocular visibility is often sufficient for the recognition of objects. Therefore the areas that are hidden binocularly are critical for masking effects.

For many design tasks, it is important to know how long the view application times are for capturing certain objects. This becomes obvious when comparing the middle and right

Fig. 7.52 Maximum driver field of view: **a** View of a small woman (FSMM), **b** View of a medium-sized man (MMMM), **c** Vision of a medium-sized man with varifocals. (From Remlinger 2013)

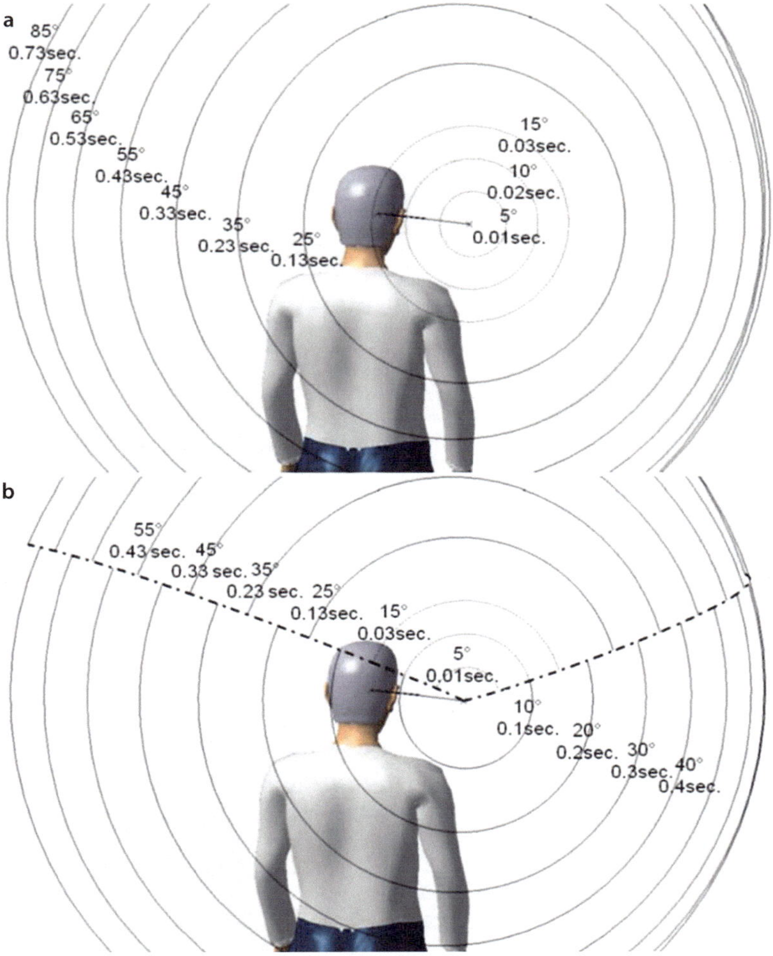

Fig. 7.53 View averting times as isochrones of a normal-sighted **a** and a wearer of varifocals **b**. (From Remlinger 2013)

picture in Fig. 7.52: while the normal medium sized man only has to move his eyes when looking from the street to the instruments, the presbyopic medium sized man with progressive glasses has to move his head. Remlinger has compiled areas of equal expenditure of time (isochrones) from literature data (Fig. 7.53). In RAMSIS, the corresponding data can be displayed in conjunction with the vehicle geometry.

A visualisation of the various fields of vision (see also ► Table 4.11) in the form of a polar diagram (▣ Fig. 7.54) can be superimposed on the vehicle geometry or an additional traffic situation if necessary. Various aspects of presbyopia can be visualized and converted into vehicle-relevant representations (▣ Fig. 7.55). In particular, it is possible to construct traffic situations and thus assess aspects such as masking by the A-pillars, rear-view mirrors and similar in a practice-oriented manner from the perspective of different manikins (▣ Fig. 7.56).

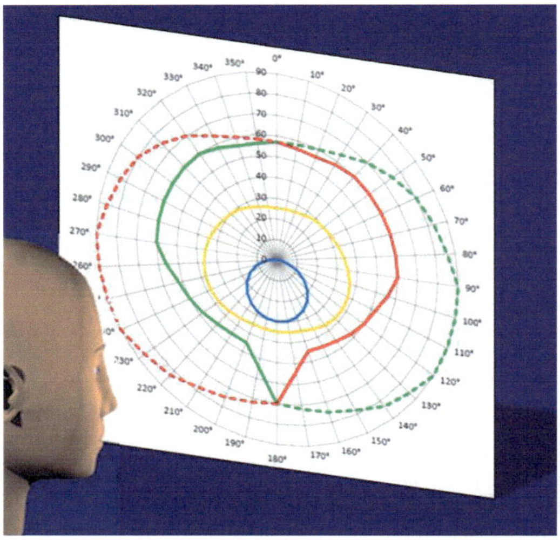

▣ **Fig. 7.54** Polar diagram of the spatial range of the human visual field monocular: left eye (red)/right eye (green), binocular visual field: solid red-green lines, ambinocular visual field: dotted lines (according to Schober 1970 or Flügel et al. 1986), optimal (blue) and maximum (yellow) visual field (according to Lange 2008). (From Remlinger 2013)

In particular, the influence of vehicle geometry on visual obstructions is an important method of analysis and assessment that goes beyond the purely geometric conditions set out in the regulations. At Daimler AG, Scholly (2006) carried out systematic eye movement analyses in real traffic situations. In order to achieve neutral observations of the density of forward visibility, the tests were carried out with a test vehicle without a blind cover, the so-called glass dome car. One result of the tests is that at junctions and on winding roads, the visual cover provided by the

▣ **Fig. 7.56** Parametrically adjustable traffic situation with road users and objects for visual analysis in connection with a given 3-D CAD geometry. (From Remlinger 2013)

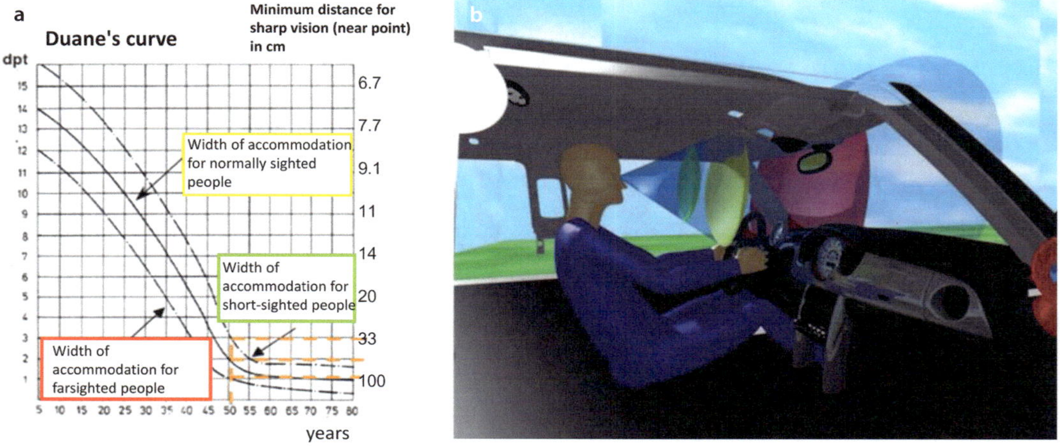

▣ **Fig. 7.55** Minimum viewing distances in the vehicle: **a** age-related loss of accommodation, **b** Distances in the vehicle for 50-year-old mid-size driver. (From Remlinger 2013)

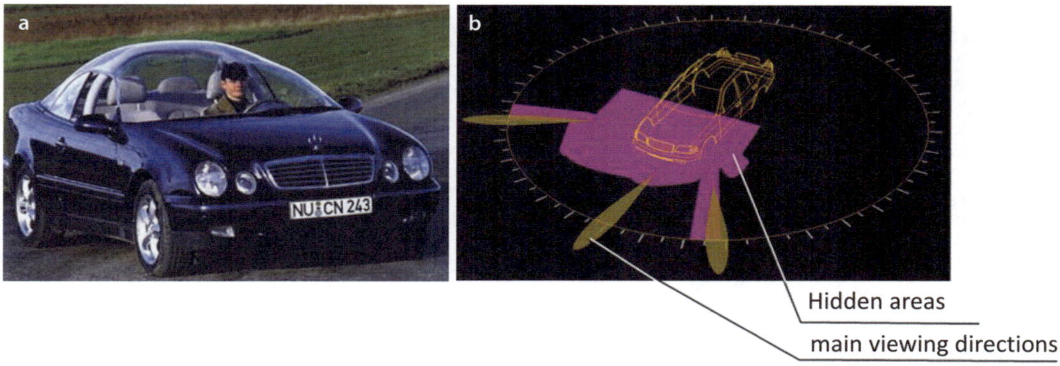

Fig. 7.57 Scholly's "Glass Dome Car" **a** and **b** areas hidden by the A-pillar and main viewing directions

Fig. 7.58 View from the driver's position in a modern car into a junction

A-pillars often coincides with the main directions of vision (see Fig. 7.57). The visibility discomfort is apparently caused less by the width of the hood than by the position of the A-pillar, as the drivers tend to keep the view within the windscreen area. Figure 7.58 shows the masking effect of the A-pillar in a modern car.

Based on the ocular density distribution found in the glass dome car during the eye-tracking tests, Scholly developed an evaluation procedure which can be used to derive weighted evaluation variables on the basis of the CAD design data of the vehicle. These evaluation variables make statements about the visual occlusion of the future vehicle and indicate in which areas of the view these occlusions have a particularly disturbing and impairing effect. Figure 7.59 shows the development of a conical projection of the greenhouse and the evaluation zones of the Daimler-Scholly method for the forward view, calculated from the eye point ("middle eye") of a RAMSIS manikin. Depending on the manikins used, there are different projections. If the calculation for the left and right eye is carried out separately, the masking areas can be determined which cannot be seen even when both eyes are used, because only objects in the binocular masking field remain invisible. The different color zones shown in Fig. 7.59 represent areas of visual occlusion weighted differently. This allows the transparent and non-transparent portions of the sight zones to be accounted for numerically and combined to form a weighted overall result. Remlinger (2013) comments: "This visual assessment procedure has already been agreed within the European Automobile Manufacturers Association (ACEA) and also with representatives of the Japanese Automobile Manufacturers Association (JAMA) as a uniform procedure. The procedure harmonizes well with the methods of French and Japanese manufacturers. It was introduced as a joint manufacturer proposal for the determination of visibility conditions within the framework of a new test procedure of the independent association Euro-NCAP for the active safety of vehicles. This initiative to test Active Safety has not yet been implemented by Euro-NCAP."

However, visual obstructions must not only be assessed statically, but also have a very dangerous dynamic component. Zaindl (2009) conducted simulator experiments in which, by

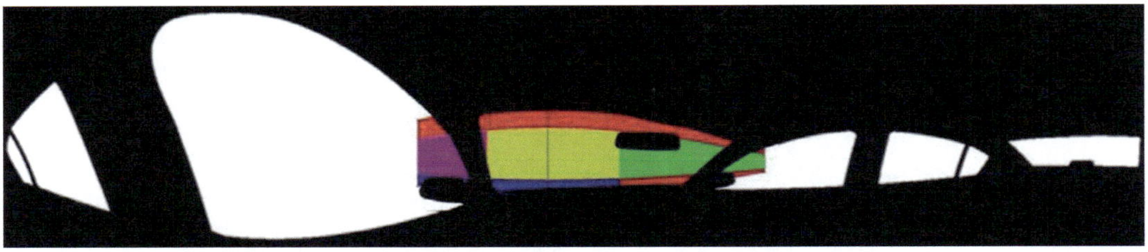

Fig. 7.59 Projection of the driver's field of vision superimposed with the evaluation zones of the Daimler-Scholly method. (From Remlinger 2013)

continuously tracking the driver's eye position, the crossing vehicles were controlled in such a way that they were in the visual shadow of the left or right A-pillar. These vehicles remained undiscovered, even if they were slightly offset from the A-pillar's visual shadow, but did not move relative to it. This phenomenon, known in ship and aircraft technology as "standing bearing", can be the cause of many intersection accidents, if one considers that the inner image of the driver of the outside world is only created by a brief glance, which may not have captured the vehicle crossing at that particular moment.

Remlinger (2013) has calculated the visual impairment caused by the A-pillar on the basis of the prediction angle, which can be determined with the generally occurring prediction time of 2 s, as a function of lateral acceleration and speed, and has developed a diagram for this purpose that is of great value for the safety-conscious design of cabs (Fig. 7.60).[19] There are some serious fundamental problems:

- The probability of masking increases with increasing speed,
- It is more problematic for more tall persons seated further back, although the percentage hood area is larger for smaller persons; i.e. smaller drivers have to struggle with larger hood areas, while more tall ones are more frequently confronted with the dynamic hood effects.
- A sporty driving style with higher lateral accelerations - interestingly enough especially in the low speed range - increases the risk of the hood being covered.

As a consequence of these considerations, Remlinger asked the question whether the A-pillars, which are becoming increasingly flatter today for fashionable and aerodynamic reasons and which have their roots further forward in the vehicle, could be carried for safety reasons and whether it would not be worth considering a return to the panoramic window with steeper A-pillars and those with their roots further back among the current possibilities of glass technology.

7.3.1.2 Rear View

There are practically no regulations that would affect the direct view to the rear. All available standards and legislation refer only to the areas that need to be covered by the rear-view mirrors. This means that there are no direct requirements for the vehicle manufacturer that relate, for example, to difficult driving manoeuvres such as threading into a road with traffic approaching at an angle from behind or reversing (Hudelmaier 2003). This lack of regulation is partly extremely exploited by the current vehicle design, as a comparison of the direct rear view of vehicles of different design generations shows (Fig. 7.61). Also the already mentioned shadow casting method or the calculated viewing areas on the basis of a conical projection (Fig. 7.59) and similar methods do not provide sufficient realistic information for the view to the rear. The main shortcoming in all procedures is that the overall panoramic view from the vehicle is viewed from a static fixed point. In reality, however,

19 The boundary lines drawn in Fig. 7.60 for tall men and small women may have to be calculated specifically for the vehicle under consideration according to Remlinger's data, but should not deviate significantly from those shown here in current vehicle designs.

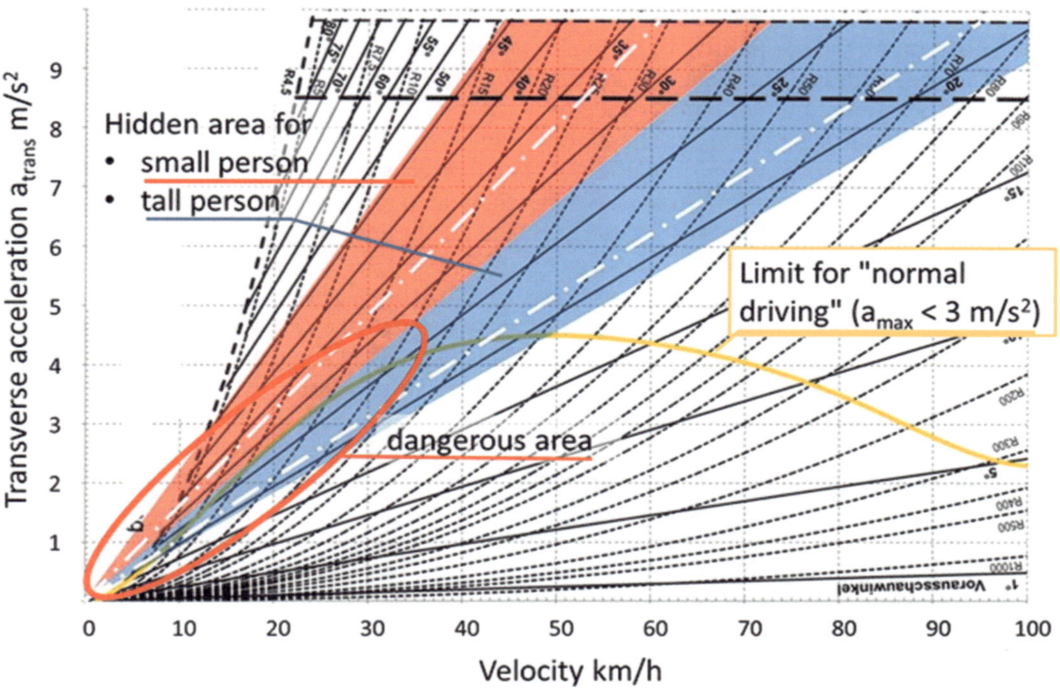

Fig. 7.60 Forecast angle of 2 s in a left-hand bend as a function of bend radius, velocity v and lateral acceleration a_{trans}. Below the yellow line is the area for normal driving with $a_{trans,\ max}$ < 3 m/s². (The drawn border lines for small woman and big man are strictly speaking only valid for a 6 series BMW [E64]; from Remlinger 2013)

Fig. 7.61 View diagonally backwards (centre) and backwards (right) for three vehicles of different design generations. **a** BMW 326 year of construction 1938, **b** BMW 2000 year of construction 1968, **c** BMW 520i Year of construction 2002

head and torso movements cause large shifts in the position of the eye points, for example when looking through the rear window in reverse. The function contained in the RAMSIS simulation software is as follows *"move eye and head"* contains a simulation of human behaviour based on test persons (Seidl 1993; Göhler 1999), which makes it possible to set an eye target based on the normal driver posture and thus automatically generate a corresponding head and body movement. In his experiments, Hudelmaier (2003) found out that from a viewing angle of 110° compared to the view in the direction of travel, both in reality and in the RAMSIS simulation, the backrest contour deviates from the seat back contour to such an extent that this serves as a switchover point for the application of a special reverse driving posture model. He can show that very realistic posture predictions can be obtained for this particular application (▫ Fig. 7.62).

▫ Figure 7.63 shows the direct rearward view through the rear window of an E-Class Mercedes (W212) calculated on the basis of the posture data from Göhler (1999).

▫ **Fig. 7.62** Calculated posture when looking backwards using the Move-eye-and-Tool (left, from Bothe 2010) and the RAMSIS reverse driving posture model by Hudelmaier (2003; right)

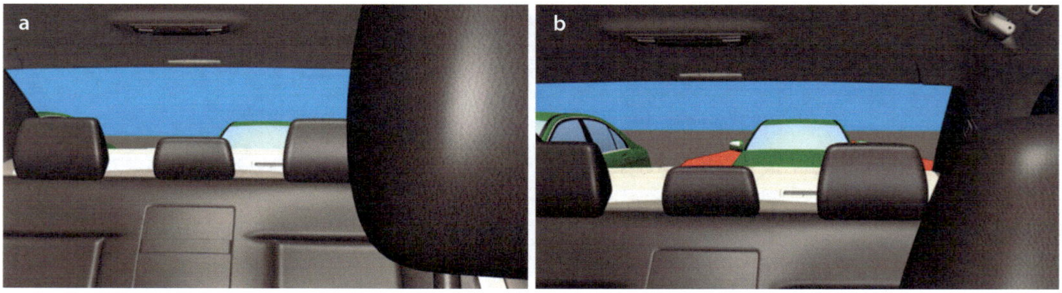

▫ **Fig. 7.63** Direct view to the rear from the centre eye **a** of the little woman (FSMM), **b** of the great man (MTMM). (From Bothe 2010)

Reference vehicles are set up in the vicinity of the virtual vehicle for assessment purposes (see also ◘ Fig. 7.56). It can be clearly seen that due to the right C-pillar and the rear headrests, the view is clearly worse for the small woman than for the large man. Similarly, for the various design manikins, views from the side windows on the left and right and diagonally to the rear can also be simulated, so that inadequacies can be discovered and, if necessary, corrected during the CAD development phase of a vehicle.

7.3.2 Indirect Vision

The indirect view refers to all information from the outside world recorded via mirrors. The EEC Directive 71/127 and ECE-R46 combine design rules that define the minimum driving areas that can be seen via the rear-view mirror. ◘ Figures 7.64 and 7.65 provide a summary of these rules. However, the field of vision above the interior mirror shown in ◘ Fig. 7.64 shall only be ensured if the design of the vehicle makes an interior mirror appear reasonable. According to Hudelmaier's (2003) eye studies, however, rear-end traffic is observed while driving through the interior mirror rather than through the exterior mirror. This can be explained by the lower movement of the head between looking straight ahead and looking into the interior mirror in relation to the corresponding views into the exterior mirror. However, it must be borne in mind that a view of the rear-view mirror realised on the basis of the regulations is by no means sufficient to safely carry out lane change manoeuvres, for example. The "blind" area would be far too big for that. But even the blind spots left free by the exterior mirrors can be a reason for dangerous driving manoeuvres; therefore the shoulder view is strongly recommended for a lane change. Technical remedies are provided by assistance systems (see ► Chap. 9).

It is therefore useful to define a (static) traffic scenario for the design of rear-view mirrors which occupies the remaining blind spots under EEC regulations with vehicles. This scenario is then observed via the mirror function of the RAMSIS software from the perspective of the RAMSIS design manikins (◘ Fig. 7.66). The Move-eye-and-head function is also used for this calculation, since a glance in the rear-view mirror is always associated with a combined eye-head movement.

The calculation is based on the current eye point position. This initially creates a pyramid of vision which, on the one hand, allows the mirror to be adjusted and, on the other hand, allows immediate recognition when objects obstructing the view of the mirror are in the beam path (◘ Fig. 7.67). For better orientation, the exterior mirrors are usually adjusted in such a way that the contour of the vehicle is

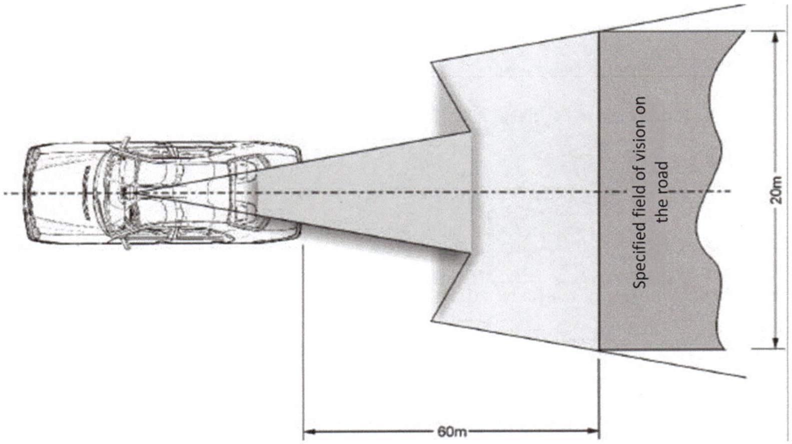

◘ **Fig. 7.64** Specified field of vision for interior rear-view mirrors according to EWG 71/127 or ECE-R46. (not to scale; from Hudelmaier 2003)

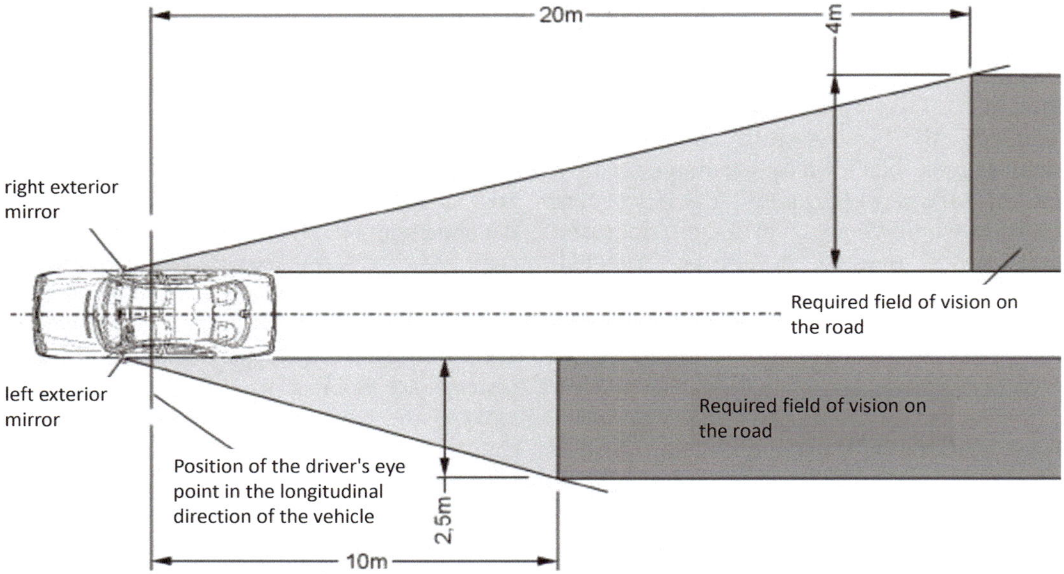

Fig. 7.65 Required fields of vision for left and right exterior mirrors according to EWG 71/127 or ECE-R46 for vehicles of category M1 and N1 with a mass up to 2 t (not to scale; from Hudelmaier 2003); for vehicles of category and with a vehicle mass via 2 t slightly different regulations apply

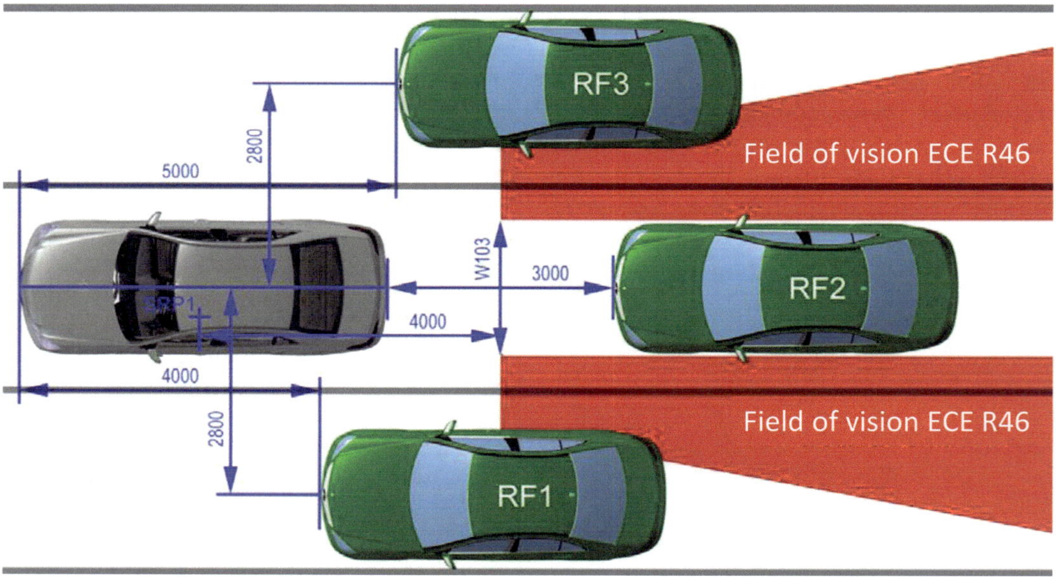

Fig. 7.66 Parametrically adjusted traffic situation for the assessment of the rear-view mirror view. (From Bothe 2010)

still visible at the edge. Since the software also allows reflections on free-form surfaces, this function can also be used to investigate the extent to which aspherically curved mirrors can be used to reduce blind spots. Using the function described in Fig. 7.53, it is also possible to estimate from the position of the mirrors the time required to look at the cor-

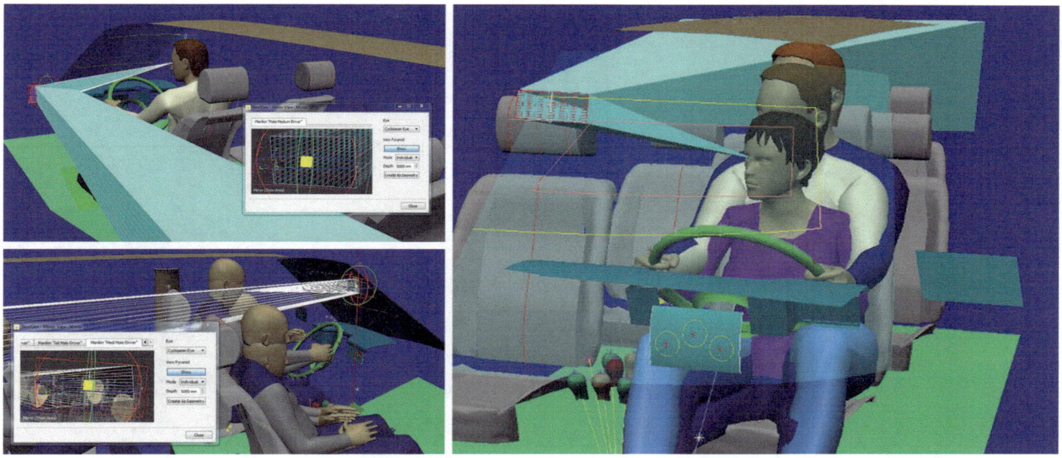

■ **Fig. 7.67** RAMSIS user interface for setting the dynamic mirror pyramid. (Human Solutions 2011)

■ **Fig. 7.68** Calculated mirror view of the traffic scenario of ■ Fig. 7.66 for the little woman (FSMM; **a**) and the big man (MTMM; **b** Mercedes E-Class, W 212; from Bothe 2010)

responding mirror. The time required to record the information seen is then added to this allocation time (on average in the range of 0.8 seconds; see ► Fig. 3.55). ■ Figure 7.68 shows that in the example chosen, the mirror view is roughly equivalent for tall and small persons, although it can also be seen that the small woman overlooks a larger angular range than the tall man because of her closer position on the mirrors.

7.3.3 View of Operating and Display Components

The need to ensure an unrestricted view of displays is trivial. But also the view on control elements is necessary, because every operation - apart from a few exceptions (e.g. operation of the turn signal lever, the steering column lever for fade-in and fade-out, the gear shift lever) - is connected with a gaze turn (Arlt and Bubb 1999). The calculation of the view of operating and display components can in principle be carried out with the same tools as those used for the design and evaluation of the direct and indirect view. An essential question to be answered is the obscuring of the view especially by the steering wheel, but also steering column levers, gearshift levers, filled cup holders etc. A special function has been developed for RAMSIS cognitive. Starting from the driver's eye, visual beams are drawn onto the geometry in ques-

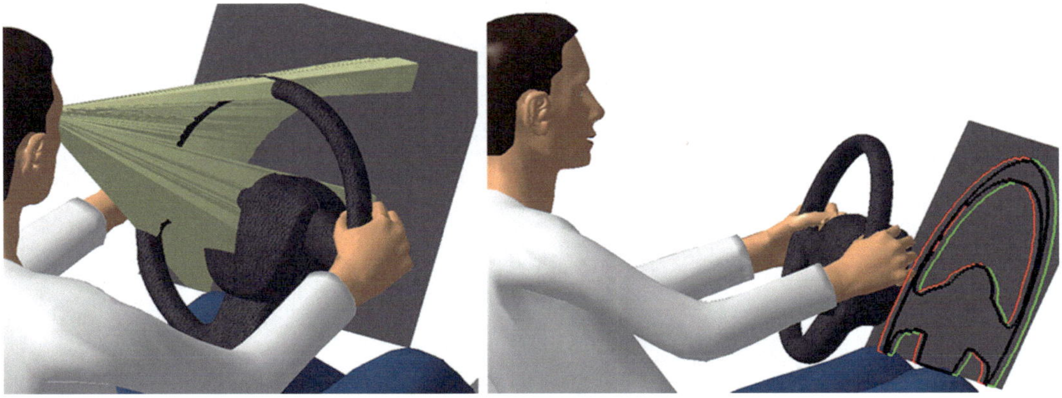

Fig. 7.69 Analysis of the visibility shadow of the steering wheel for the view of the instrument cluster. Representation of the monocular and binocular visual limits at the dial level (so-called "steering wheel banana", Remlinger 2013)

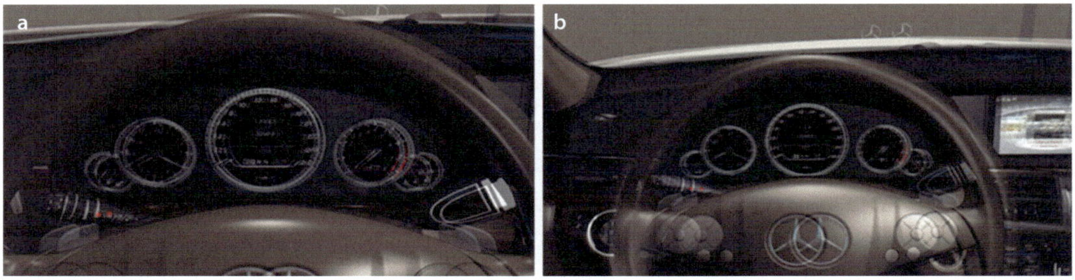

Fig. 7.70 Superimposition of the calculated images of the view from the left and right eye on the instrument cluster for **a** little woman (FSMM), **b** tall man (MTMM; Mercedes E-Class, W 212). (From Bothe 2010)

tion. The rays that hit a shadow-giving object are limited by it, all rays of sight that hit the background geometry directly are only limited there. The borders of the shadow boundaries can then be explicitly highlighted (Fig. 7.69, Remlinger 2013). It is recommended to perform such examinations binocularly, i.e. separately for each eye. This is particularly important in connection with small steering wheel diameters. Figure 7.70 shows an example. During the evaluation, it must be ensured that displays, control devices and inscriptions on control devices remain visible from at least one eye.

RAMSIS cognitively offers the possibility of projecting a Landoltring onto the viewed surface via a defined visual beam in order to be able to assess the required size of visual signs on displays as well as on the labelling of control elements. The size of the Landoltring can be parameterized to take different visual conditions into account (see Fig. 7.71). The visibility can be estimated by comparing the Landoltring with the displayed labels.

It is often assumed that one can compensate the decreasing visual acuity in old age (correct: decreasing ability of the eye to accommodate for different visual distances) by larger visual signs. This is fundamentally wrong, because the inability to accommodate to the proximity always leads to an optically blurred picture, which is however under circumstances with sufficient habituation to this condition subjectively not perceived at all. This deficiency can only be partially compensated by using reading glasses or varifocals (see Fig. 7.55). An advantage of the Head-Up-Display (HUD), especially for the age-sighted driver, is that the virtual distance of the display lies between 2 and 3 m, an area for

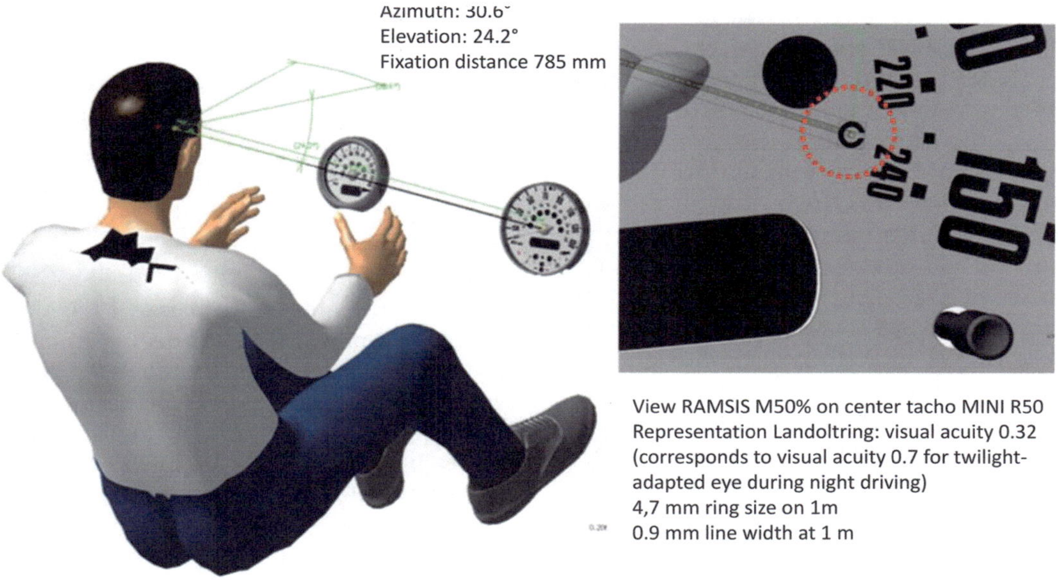

Fig. 7.71 Visual acuity analysis for vehicle instrumentation and labelling of control elements. (Remlinger 2013)

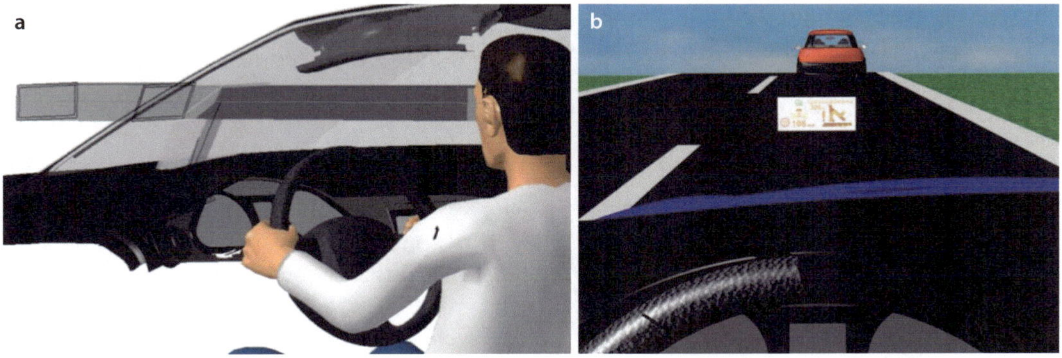

Fig. 7.72 Design of a Head-Up-Display (HUD), **a** Beam path analysis, **b** Image position from the driver's point of view

which accommodation is no longer necessary under adequate lighting conditions. The superimposition of the instrument view on the close range of the road scene reduces the accommodation and view application times for drivers of all age groups.

Information is displayed in the HUD by mirroring it on the windshield. The system driver - eye - windscreen and HUD unit thus form an optical system. The corresponding elements must be coordinated according to the laws of optical radiation physics. RAMSIS-cognitiv offers a tool for this (Fig. 7.72). By tilting the HUD unit or the last deflecting mirror in the unit, the HUD display can be adjusted to different heights of the driver's eye or the height of the image can be adjusted in the driving environment. Since, on the one hand, too much HUD information must be avoided in order to avoid masking important traffic events and, on the other hand, different eye positions must be taken into account, the HUD tool in RAMSIS-cognitiv represents an important tool for taking these complex dependencies into account at the planning stage.

The use of Liquid Crystal Displays (LCDs) has become widespread in recent years, especially for use in navigation and infotainment systems. LCDs are now also used for flexible design in instrument clusters. For the display in the middle of the vehicle (DIZ), so-called dual-view displays are already in use, which enable different image representations for the driver's and passenger's perspective. In order to fulfil the high contrast properties even under the conditions of the generally high brightness level in the vehicle interior, special technical measures were taken for the LCD in vehicle use, which have their effect only from limited angles due to the LCD physics. Another tool available as part of RAMSIS-cognitiv serves to coordinate the optical properties of the displays and their arrangement relative to the viewing direction of the occupants as well as possible. For this purpose, the goniometrically experimentally determined optical limits of the display are entered as parameters into the corresponding function. On this basis, a funnel-shaped geometric body (in technical jargon referred to as a "tulip", ◻ Fig. 7.73) is then represented, representing the space segment from which a vehicle occupant can look at the display without having to experience any restrictions. It is now easy to check whether the eyepoint positions of the various RAMSIS manikins correspond to the "tulip".

Of course, the functions described above can also be used to calculate an image in the opposite direction, which can be seen from the driver's or co-driver's perspective.

◻ **Fig. 7.73** Viewing area ("tulip") of a point on the LCD in the vehicle. (half-side view; Remlinger 2013)

7.3.4 Reflections

7.3.4.1 The Origin of Reflections

HUD, LCD displays or even very simple indicator lights are so-called "self-chandeliers". The luminance of these self- chandeliers must always be higher than the ambient luminance so that the information can be reliably detected (recommended 10 to 1; see also ▶ Sect. 3.2.1.1). If the ambient luminance is too high, the contrast provided by the display technology may no longer be sufficient for the information displayed to stand out sufficiently from the background (see ▶ Fig. 3.15). The situation is different for most objects in the environment, including conventional analogue indicators, which are referred to as "extraneous chandeliers". Here the striking light is reflected from the surface. The rays reaching the eye give the impression of the object being looked at. The reflection of the incident light on the surface can be diffuse, directed or mixed diffuse-directed. Also the directed part of a mixed diffuse-directed reflection always obeys the law "angle of incidence = angle of reflection", whereby the angles to the normal of the respective surface point are measured. The proportion of directionally reflected light in the total reflected light determines whether an object appears "shiny" or "matt".

Due to the design of the glass surfaces of the greenhouse, a large part of the interior of the vehicle is exposed to direct sunlight in appropriate weather conditions, which can enter the vehicle from any angle above the horizon, depending on the vehicle position. Looking into the low-lying sun causes direct glare. If the sun hits high-gloss surfaces in the vehicle interior, the mostly curved surfaces can also reflect rays directly into the driver's eye. This is referred to as reflected glare due to daylight reflection. Surfaces that can contribute significantly to this risk of glare are high-gloss interior surfaces such as trim strips or application surfaces and the cover glasses of the displays. The direct or reflection-induced strong incidence of light into the eye leads to a brightness adaptation adapted to it, whereby no more information can be obtained from the incidence of light of the environment. One speaks of *physiological glare* (disability glare). Even if the luminance of the light striking the eye does not yet lead to adaptation, the driver may feel irritated by the high difference in the light stimuli on the retina. This is called *psychological glare* (discomfort glare). The use of high-gloss applications and metallic-bright surfaces as applications for the instrument panel and tunnel console is desired as a modern and high-quality accessory and is widely used as a fashionable upgrading element in vehicle interiors. If the use goes beyond small accent parts, larger components with flat high-gloss surfaces produce a reflective surface that can also direct light incident from the side, be it from a bright summer sky or from direct sunlight, to the driver's eye point. The rapid change from very bright light reflexes to dark shaded positions leads to a "lively" perception. These rapidly changing lighting conditions suggest movement to the eye and are suitable for diverting attention from the surrounding traffic at least briefly to the vehicle interior. Even below the physiological glare, the intensity of the light reflected at the cover glasses of the displays can be at least partially higher than the intensity of the light reflected at the display surface. Reading the information is then impossible. However, bright reflections of this kind do not necessarily have to be produced by sunlight, but can also be caused by light-coloured clothing if the cover lenses are positioned unfavourably ("white shirt reflection"). Psychological glare effects also occur when looking through the windscreen, when very bright objects (e.g. white paper on the instrument panel, light dashboard surface covering) or surface sections with very contrasting colours are reflected. Even at night, annoying reflections can occur on the windscreen. They are often caused by backlit displays such as displays, instruments or switch lighting, primarily in the center console area. This phenomenon is referred to as the *night reflection.*

It is a time-consuming business to discover the many possibilities described above for the creation of reflections at an early stage on the basis of CAD data and to avoid them accordingly, but it is worth it, since glare effects that are only discovered late in the development process can usually only be averted inadequately. RAMSIS cognitive therefore contains a function that can be used to geometrically track the course of reflected light rays in the vehicle interior and determine the possible origin of the reflected light. In principle, a beam of radiation is generated from the respective driver's eye (or centre eye), which is directed at the surface of interest and reflected there according to the reflection properties stored in CAD. The constructed calculation shows the beam path from the respective eye point over the reflecting surface to the area where a dazzling light source could be located. This geometric representation of the light path reveals possible options for eliminating this glare effect (Fig. 7.74).

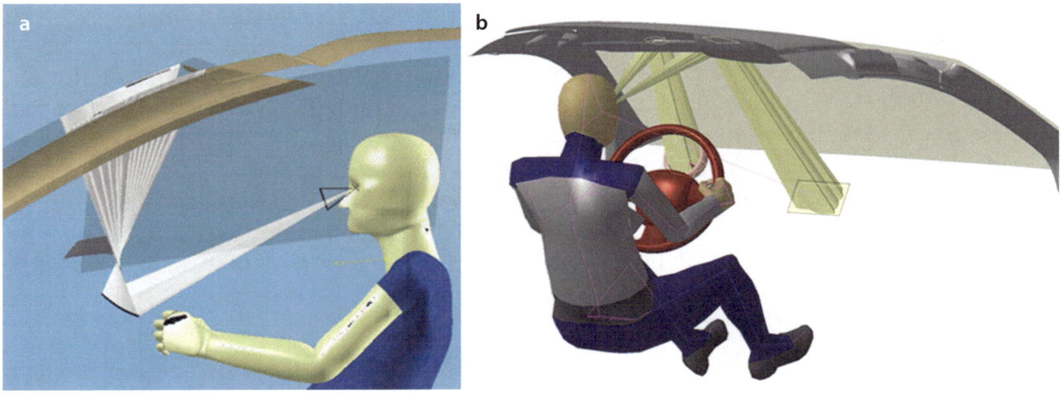

Fig. 7.74 Simulated reflection beam path of glare effects **a** Daylight reflection of sunlight, **b** Night mirroring of illuminated displays in the windscreen. (From Remlinger 2013)

Fig. 7.75 Calculations of the reflection of the dashboard surface in the windshield for the little woman (**a** FSMM), the medium-sized, long-legged man (**b** MMML) and the tall man (**c** MTMM). The viewing cone is directed towards the so-called A-field (dark green rectangle), which must be free of any reflection according to the various legal requirements. However, smaller reflections may occur in the surrounding B-field (blue rectangle). These two fields are shown in the pictures. (From Bothe 2010)

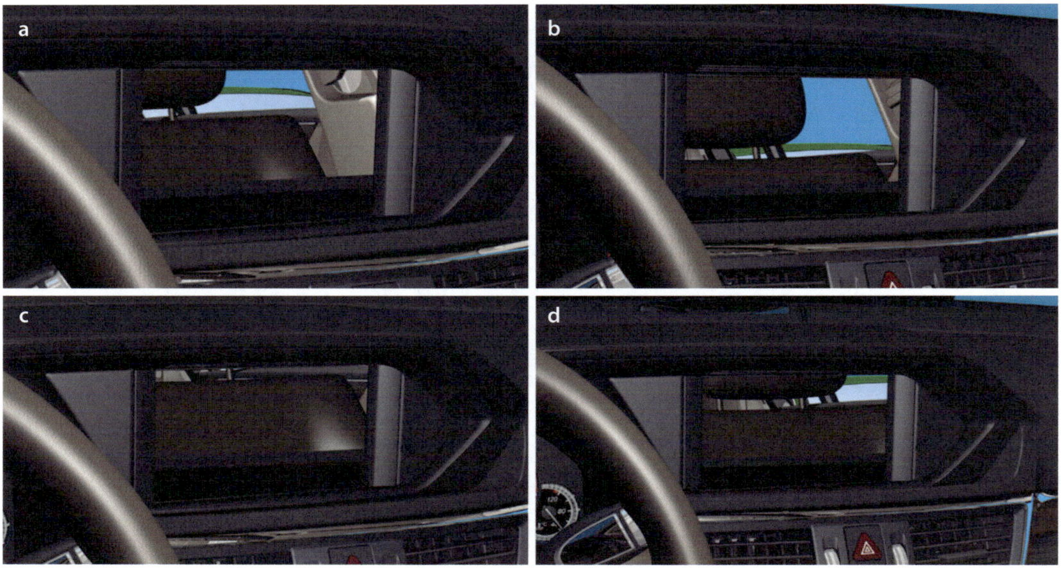

Fig. 7.76 Calculations of the reflection in the central display **a** little woman (FSMM), **b** medium long legged man (MMML), **c** middle man (MMMM), **d** tall man. (MTMM; from Bothe 2010)

These calculations must also be carried out for different eye point positions, which are determined by the design manikin. With the viewing angle of a selected eye point, in conjunction with the CAD system used, it is then possible to calculate the reflection image that is available to the driver and estimate whether this can really be classified as disturbing (Fig. 7.75).

Figure 7.76 shows the calculation of the reflections in the central display for four different design manikins. Here you can see how important such a calculation is for different eye positions. The seemingly perfect situation for the middle man turns out to be considerably less favourable for other types of persons.

7.3.4.2 Shielding

Direct glare from sunlight is one of the most dangerous situations that can occur due to environmental influences. While the reflection glare of the sunlight on a wet road surface can only be controlled by polarizing sunglasses[20] direct glare can be combated by sun visors. In principle, it is the task of the sun visor to avoid direct sunlight hitting the eyes under all circumstances, i.e. to keep them in the shadow

20 In principle, the reflected light is polarized parallel to the reflecting surface, which is why the largest proportion of reflected light can be filtered out by polarization glasses with blocking in this direction.

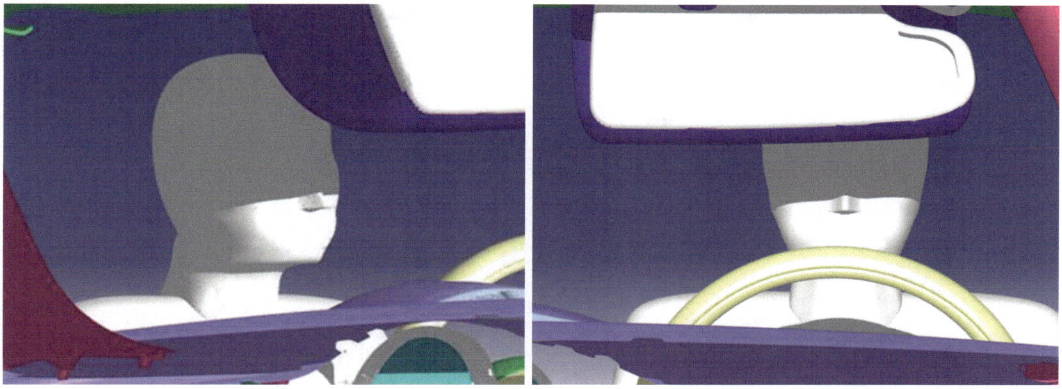

■ **Fig. 7.77** Principle effect of the sun visor

area (■ Fig. 7.77). This applies to both lateral and frontal incidence of light. This requirement can only be achieved by a high mobility of the sun visor. With the help of the design family of the human model used, it must be checked for the different eye point positions whether this requirement is guaranteed for all possible directions of the sun's rays.

The highest requirement in this respect arises for the situation where the rising or setting sun on the horizon is at the direct vanishing point of the road being travelled on. The lower edge of the maximum sun visor must then be exactly at eye point height. The smallest distance to the constructive H-30 position therefore results for the small long-legged woman (FSMS). If the upper edge of the sun visor does not leave any gap between the light and the windscreen, people with a higher eye position can always find a favourable position by tilting the sun visor accordingly. With today's highly inclined windscreens, where the position of the roof frame may cause a short distance to the driver's head, problems may arise when folding the sun visor. These problems occur more often when the sun visor has to be folded to the side to protect it from light from the side window. This problem results in complex geometric requirements that can only be solved satisfactorily with the help of correctly positioned human models.

Using the function shown in ■ Fig. 7.74, shadows above the instruments (instrument cluster and central display) can also be designed in such a way that bright light falling through the windows or emitted by the occupants' bright clothing is not reflected into the driver's eye. Here, too, the eye-point positions of the family of interpretation must be taken into account (see also ■ Fig. 7.76). One method of effectively avoiding such reflections is to adjust the inclination of the cover glass for the instruments to the length of the scoop. Even cone-shaped or curved covers that are matched to this can prevent undesired reflections if the design is correct. Recently it has unfortunately become fashionable again to dispense with such simple measures using geometric optics and instead use plane or - even worse in terms of capturing reflections - slightly convex instrument covers. In order to realize this fashion effect, however, the question arises again and again whether reflections can be avoided by using the $\lambda/4$ effect. ■ Figure 7.78 a shows the effect of the $\lambda/4$ coating in the case of vertical incidence of light.[21] An extinction of reflected light waves is therefore only possible for *one* wavelength of light. One normally chooses the wavelength at which the sunlight shows maximum intensity (λ = 520 nm). The illustration in ■ Fig. 7.78 also shows that this effect can only be valid for one direction of view. With the given direction of view, the observation of a $\lambda/4$ coated surface under the influence of incident white light results in rainbow-colored

21 Of course, the $\lambda/4$ coating can also be calculated correctly for a different angle of incidence of the light.

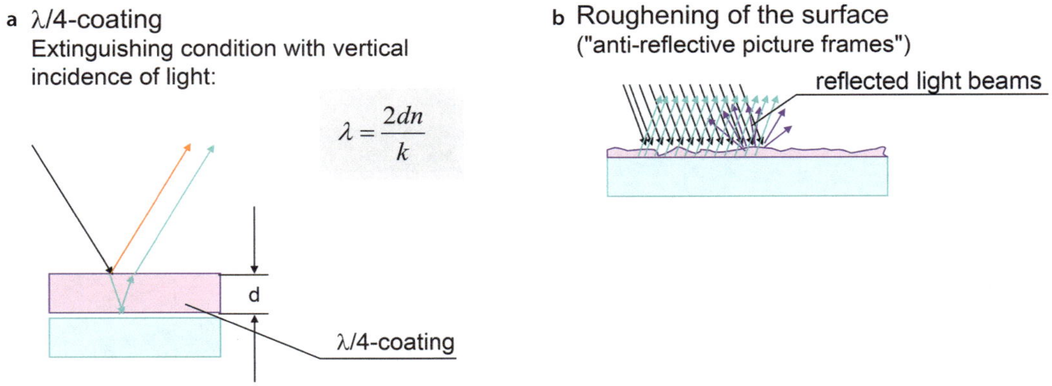

Fig. 7.78 Possibilities of reflection reduction by surface treatment (k = 1,2,3 etc.)

impressions, which also vary depending on the eye position. Since the methods of linear optics described above for avoiding reflections in connection with touch screen technology cannot be realized, it is precisely here that the desire arises to solve this problem by applying the λ/4 coating. Due to the inherent necessity of this technology to touch the screen surface with the finger, a layer of grease is always transferred which destroys the whole λ/4 effect.

As an alternative to reducing unwanted reflections, a roughened surface (frosted glass-like), which diffusely reflects the incident light (Fig. 7.78b), is therefore a good choice. Apart from the fact that this reduces the sharpness of the visual signs displayed, a high level of directional reflection cannot be avoided because of the necessarily flat side supporting the luminescent layer. The use of touchscreens in motor vehicles, for example, places high demands on correct positioning in terms of reflection avoidance.

7.4 Operating and Display Components

7.4.1 Determination of Accessibility Areas According to SAE

In addition to visibility, accessibility also plays a dominant role, especially for the operating components. There are also recommendations under SAE J287 on accessibility issues. A set of rules has been published there based on experimental studies by Hammond and Roe (1972) and Hammond et al. (1975) on maximum and preferred accessibility. According to this set of rules, the horizontal distance is determined by a vertical reference plane, which is itself determined by the distance HR to the heel reference point AHP:

$$HR = 786 - 99 \cdot G \tag{7.6}$$

G is a general package factor, which is calculated according to:

$$G = 0{,}00327 \cdot (H30) + 0{,}00285 \cdot (H17) - 3{,}21 \tag{7.7}$$

H17 is the height of the centre of the steering wheel above the heel point (see Fig. 7.79). The package factor varies from −1.3 for the package of a sports car to +1.3 for that of a heavy truck. If the calculated distance HR is greater than the distance L53, the reference plane is placed in the 95 percentile seat reference point SgRP. Depending on the variables "restraint system used by the driver" (only abdominal belt, abdominal belt and shoulder belt), "G-factor"and "male-to-female portion", the coordinates of the corresponding gripping area can be read off in tables.

Different gripping spaces are issued for the left and right hand. Examples of this can be found in Fig. 7.79.

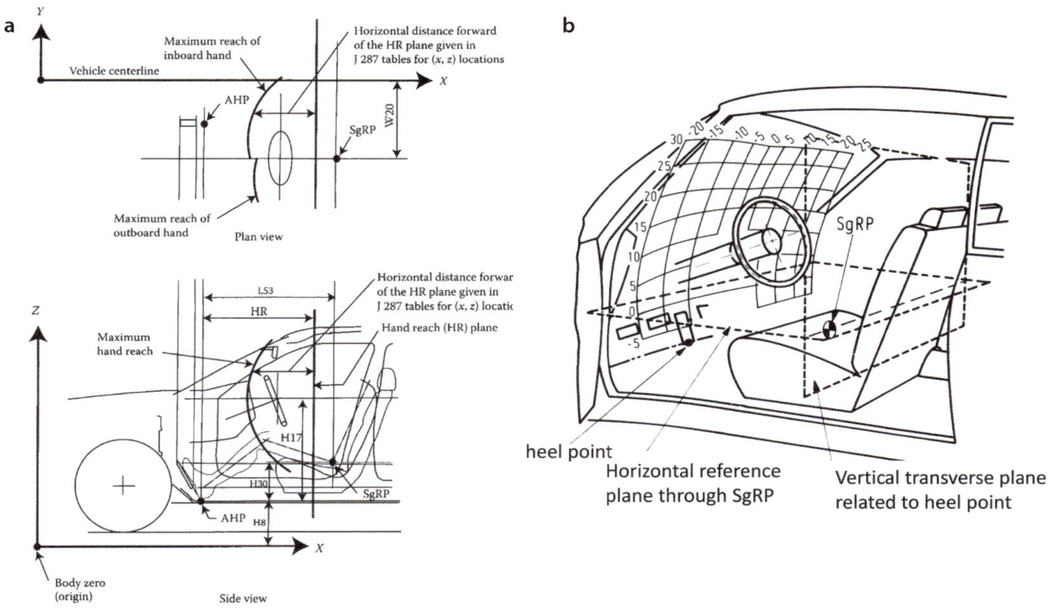

■ **Fig. 7.79** **a** Ground plan and outline of the vehicle constellation for determining the reference plane at distance HR according to SAE J287, **b** Illustration of the accessibility areas based on it

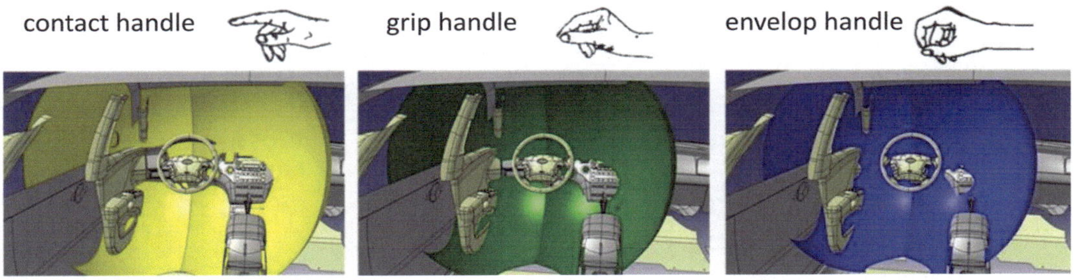

■ **Fig. 7.80** Functional gripping spaces in the vehicle cockpit

7.4.2 Functional Gripping Spaces

Especially in the combination of a CAD system with the application of human models, it is possible today to design the driving environment much more precisely than recommended by SAE. The so-called functional gripping space, which depends on the necessary function to be performed, is defined for this purpose. In connection with the application in the motor vehicle, the contact handle (pressing a button), the grip handle (often defined as a three-finger handle with which a corresponding control element is pulled or turned) and the envelop handle (for the transmission of larger forces) are essentially possible (■ Fig. 7.80, see also ► Fig. 6.29).

The gripping spaces thus defined must now be superimposed for the various design manikins (it is usually sufficient to consider the gripping spaces of the short-legged small woman and the long-legged large man; see ■ Fig. 7.81). The area closest to the driver's seat is then important for further design, but these areas must be modified in some cases, since the range of movement to the left and right of the driver's seat shortly before or even behind the seat backrest is additionally restricted by further anatomical conditions that can be taken into account by the human model.

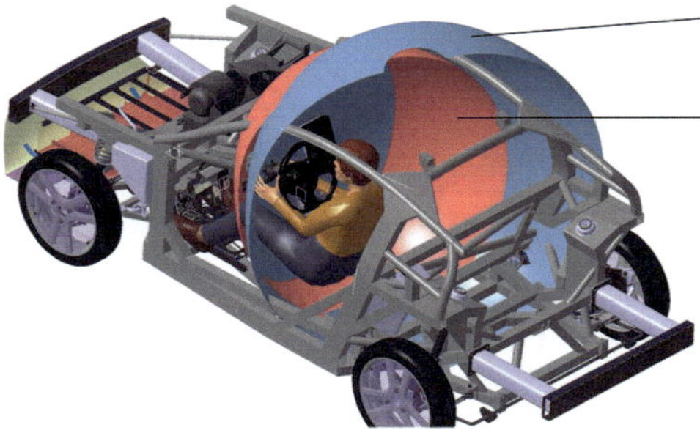

Fig. 7.81 Superposition of the gripping space of the short-legged small woman (FSMS; red) and the long-legged tall man (MTML; blue)

7

Table 7.9 Priority classification of control elements according to the "accessibility" criterion

Priority I	Priority II	Priority III
Direction indicators Low beam / high beam / flasher light Horn Windscreen wiper actuation Sun visor Switch for assistance systems (cruise control, speed limiter, ACC, lane departure warning / holder) Gearshift Exterior mirror adjustment Seat adjustment	Hazard lights Parking brake Main light switch Fog lamps front/rear Heatable rear window Instrument lighting Air conditioning Radio: Volume, station selection Side window Ashtray/cigar lighter Sunroof Spotlight (driver/co-driver)	Ignition, start button Glove compartment Cup holders Storage compartments in side doors / Centre console Hood release Interior lighting

In many cases, the areas defined in this way are not sufficient to accommodate the control elements and furnishings to be installed in a modern vehicle. It therefore makes sense to define additional gripping spaces of different priority, e.g.:

- *Priority I* can be reached with a three-finger grip without moving the shoulder joint points forward.
- *Priority II* with forward displacement of the respective shoulder joint point with three-finger grip (upper body remains in rest position).
- *Priority III* upper body displacement permitted without limitation.

The individual control elements are assigned to priority classes according to their significance for vehicle control (examples are shown in Table 7.9).

The consideration of the differently defined gripping spaces mentioned can become very confusing. It therefore makes sense to divide the areas themselves into zones again after the overlapping of the areas mentioned (in many cases a "traffic light rating" is sufficient: green = easily accessible; yellow = accessible; red = poorly accessible; everything outside the fields defined in this way is "not accessible"). In the next step, these zones can be intersected with the interior sur-

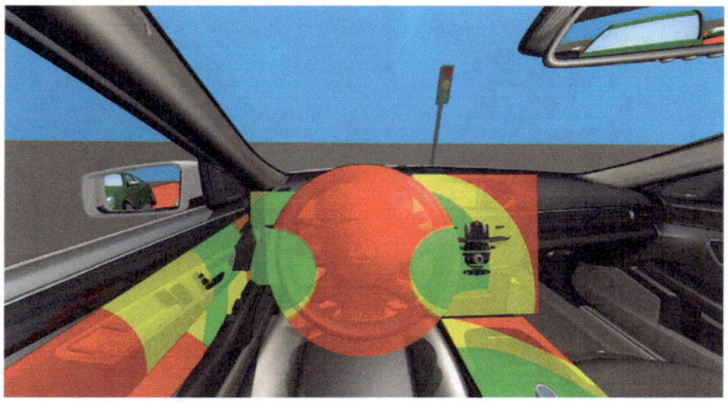

Fig. 7.82 Evaluated accessibility areas in a given vehicle cockpit. (Example Mercedes E-Class, W 212; from Bothe 2010)

faces of the respective vehicle. In this way, evaluated zones are created on the interior surfaces, see Fig. 7.82.

7.4.3 Consideration of Special Operating Requirements

When considering the displayed areas, not only the ease of accessibility but also the type of operation must be taken into account. This may have to be checked by animating the manikin from the layout family. A well-known example of this is the operation of the parking brake (handbrake). The fixed localisation of this control element due to its design means that not only the absolute force required for operation must be taken into account, but also the individual anthropometric situation. Human models that have a force simulation can help to identify the space required for this and arrive at a good compromise that guarantees safe actuation for persons with different proportions (Fig. 7.83). From an ergonomic point of view, the solution that is increasingly favoured for automatic vehicles today, i.e. to apply the parking brake at the same time as the parking lock "P", is to be rated positively.[22] However, in a similar way to foot operation of the parking brake, this excludes the possibility that in an emergency the co-driver can also apply a brake (dosed!).

In the case of manually-shifted vehicles, special attention must be paid to the gearshift lever from the point of view of accessibility. In particular, the accessibility of all positions of the so-called "control bouquet" (summarizing representation of all possible positions of the control lever; Bothe 2010) must be checked. The driver's position with both hands on the steering wheel also serves as the starting position here. By placing the right hand in the position of first to sixth gear, including reverse gear, discomfort can be determined for each of these positions. This procedure is repeated for all design manikins. It is now possible to determine a weighted reachability area for the gearshift lever in a similar way as for the previously described weighted reachability areas and thus find an optimised position for the gearshift lever (Fig. 7.84).

The range of views on the instruments can also be evaluated according to the scheme discussed here under the aspect of necessary accommodation (Fig. 7.85).

In addition to the cases described above, quite complex situations may arise when taking the operating effort into account: e.g. operating a control unit mounted behind the gear shift lever, concealing a control unit with a beverage bottle in the bottle holder. Complex operating procedures may also have to be taken into account, such as the side flap of the sun visor already mentioned, the removal of a beverage bottle etc.

As mentioned several times, control elements also have a display character, that means it must also be checked whether the corresponding control elements are visible from the respective eye positions. This results in a specific problem in connection with touchscreens, especially for drivers with age vision. It can be seen from Fig. 7.86 that the distance at which a 65-year-old driver can still accommodate (near vision) is outside the reach of a tall man, let alone a small woman. This means: without the appropriate visual aid (in this context only progressive glasses can be used) he is not able to clearly see the content displayed on the touch screen when

22 According to a SAE study (Becker 2013), drivers of an automatic vehicle only use 40% of the parking brake anyway. If a foot parking brake is present, this percentage even drops to 20%.

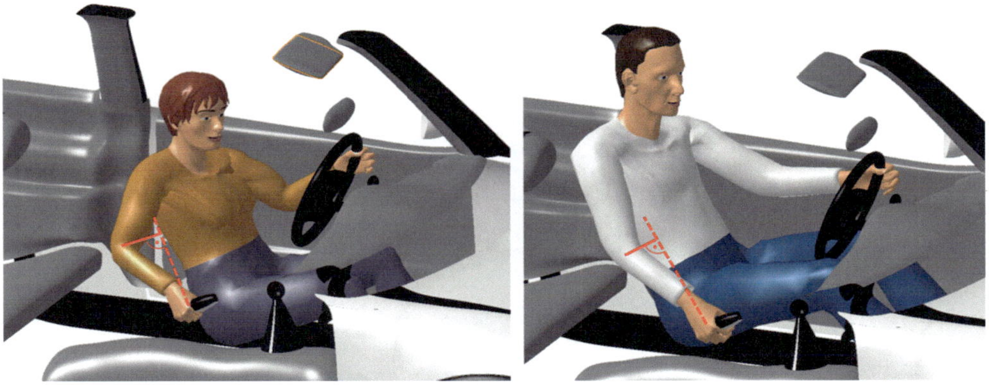

Fig. 7.83 Predicted posture of handbrake pulling for a fifth female and a 95th male body height percentile using the RAMSIS force posture model. Lines drawn in red show lever arms between elbow joint and direction of force when pulling the handbrake

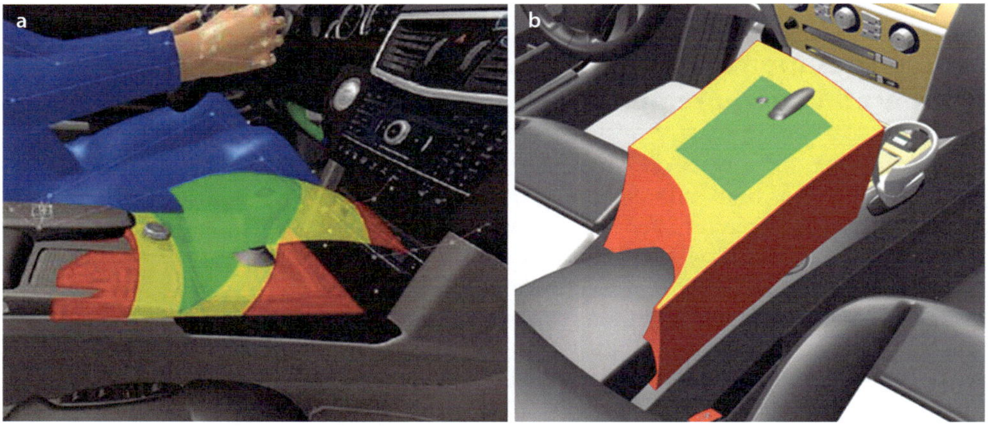

Fig. 7.84 Evaluated accessibility area for the hand lever (**a** Example Mercedes E Class, W212; from Bothe 2010, **b** BMW, Wagner 2013)

Fig. 7.85 Evaluation of the viewing distance on the instruments. (Wagner 2013)

operating the device. For the 20–30-year-old (who may have designed the appropriate layout during development) this is not a problem.

7.5 Space Requirements

7.5.1 Static and Dynamic Space Requirements

A distinction is made between static and dynamic space requirements. The overlapping of these two results in the resulting space requirement, which is decisive for the design of the vehicle.

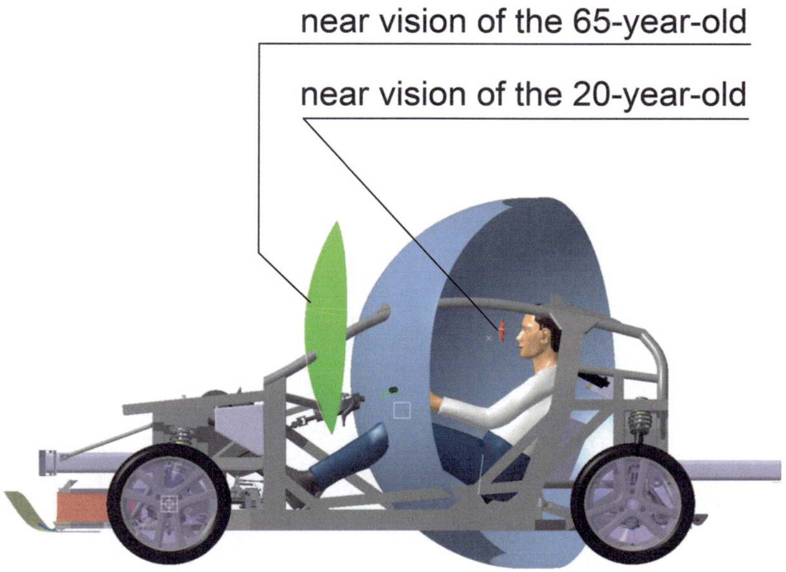

Fig. 7.86 Gripping space of the long-legged tall man and maximum accommodation (near vision) of a 65-year-old and a 20-year-old driver

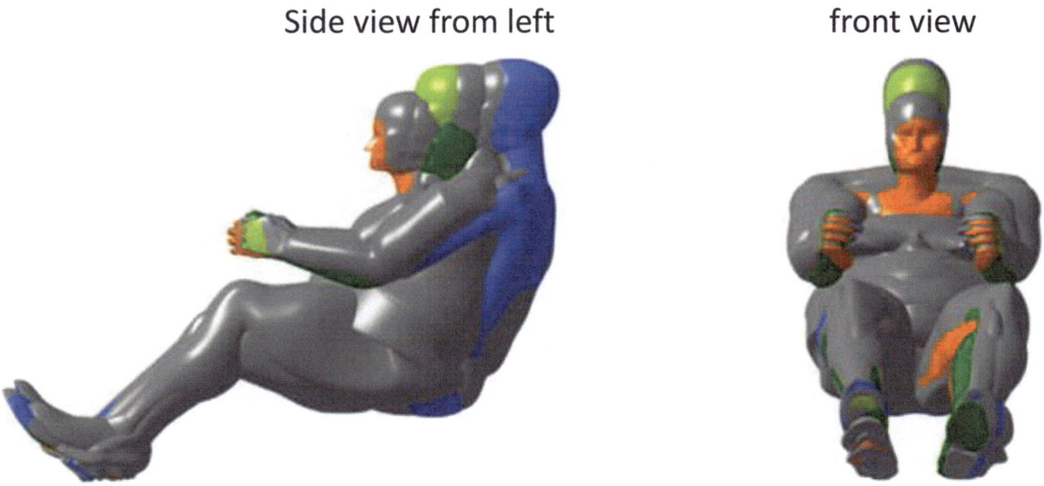

Fig. 7.87 Static space requirement of the test collective. (From Bothe 2010)

The SAE J941c guideline describes the static space requirement of the head in the form of an envelope curve. These guidelines are based on Meldrum's investigations (1965), which led to the SAE eye ellipse. From this, the head contour was developed by adding head contours to the eye positions of Meldrum, which were taken from relevant specialist literature (Mies 1987). Thus the head envelope curve published in SAE J1052 is bound to the ellipse of the eye, which again does not reproduce the actual eye positions sufficiently well. For the static space requirement, a more realistic result is obtained if the design manikins used are stacked in the driver position and only the envelope is considered (Fig. 7.87).

For the dynamic space requirement, experimental investigations must first be used. Mies (1987) investigated the space required by the

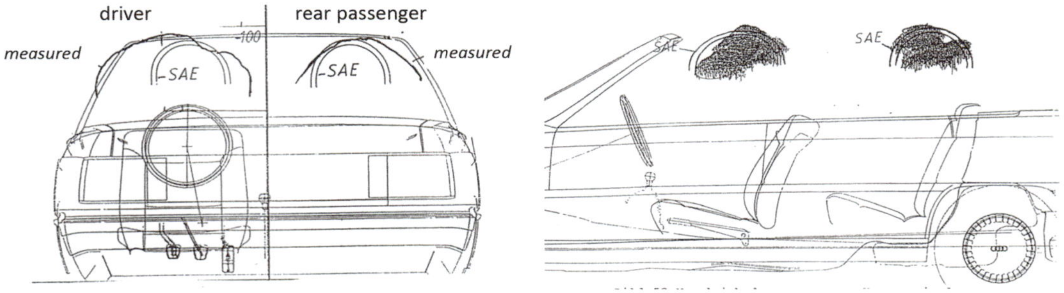

■ **Fig. 7.88** Comparison of the experimentally determined head envelope curves (Mies 1987) with those according to SAE J 1052

7

driver during dynamic cornering by observing the head position of the test persons in a test vehicle with a cut-off roof and open side windows using a stereo camera. The driver's head movement then consists of the steering movement of the upper body and the change of line of sight to control the vehicle. Contrary to expectations, the driver's upper body does not follow the direction of the centrifugal force, but is directed in the opposite direction, since the driver knows what forces are coming to him. It is noticeable that smaller persons have to use their upper body much more to perform the required steering work than larger persons, who are generally able to handle the required movement of the steering wheel with their arms alone.

The behaviour of the rear passengers differs completely from that of the drivers (Mies 1987). When cornering, the upper body slips outwards. The inclination of the head also follows the direction of the centrifugal force. The passenger tries to keep the direction of view and the head in the original position and aligns the posture accordingly. ■ Figure 7.88 shows a comparison of the mentioned experimentally collected head envelopes and the corresponding SAE envelopes. In the longitudinal direction of the vehicle, the experimental curves are in very good agreement with those according to SAE. In the lateral direction of the vehicle, the dynamic curves, in particular those of the driver, clearly exceed the SAE specifications and at the same time show just how much a narrow roof intake restricts the dynamic space requirements.

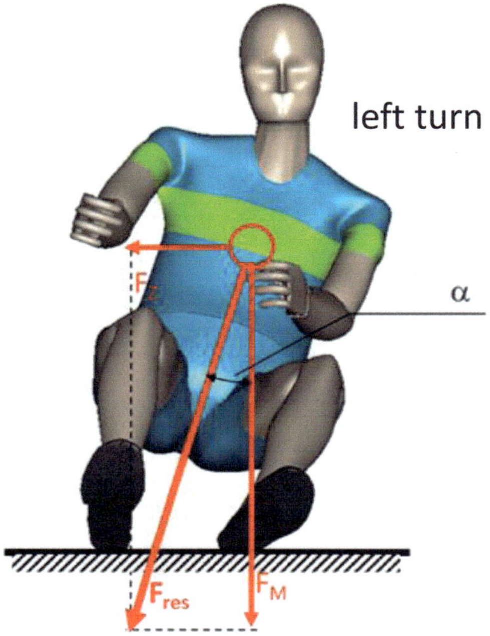

■ **Fig. 7.89** Driver behaviour when cornering. (Bubb and Hudelmaier 2001)

The observations of Mies (1987) can be modelled (Bubb and Hudelmaier 2001): the driver tries to counteract the centrifugal force with his upper body, but keeps his head constant in the vertical direction, thereby directing his gaze towards the visual target (■ Fig. 7.89). If one assumes a comfort acceleration of 0.2 g, then after this consideration an angle of inclination of the upper body of α = 20° results. Bothe (2010) assumes in his considerations an inclination of the upper body of up to 10° to both sides. The head

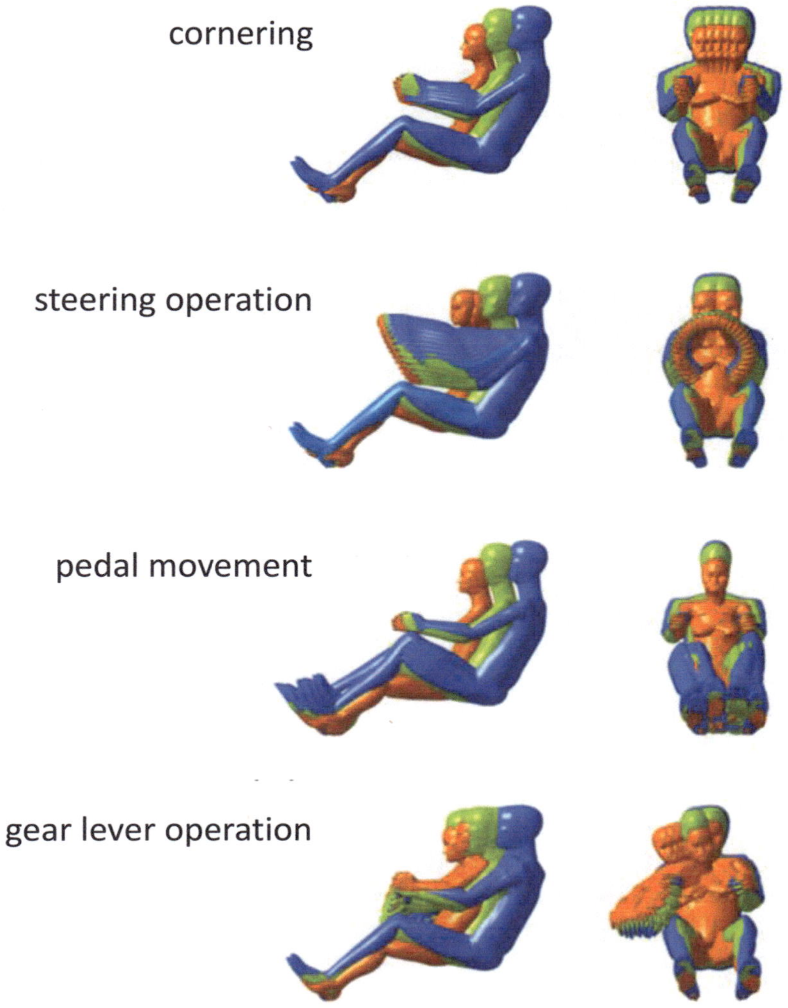

Fig. 7.90 Space requirements of different scenarios due to superimposition of extreme design manikins. (According to Bothe 2010)

compensates the torsional inclination analogous to Fig. 7.89.

According to Bothe (2010), for the simulation of dynamic space requirements, the movements for cornering, the steering process, the pedal movements and the shifting process must first be calculated (Fig. 7.90). Based on the driver posture and the given vehicle geometry, these motion sequences are in principle calculated in the following way. First, trajectories are defined which describe the trajectories of the body points. On these trajectories, grid points are placed as target points with which it is possible to calculate the corresponding posture using the posture probability model. The RAMSIS Motion Recorder is used to record and archive these supporting postures. This results in a database of all motion sequences of the selected human models. By superimposing all the images of the calculating space requirements including the static space requirement, the resulting space requirement is obtained (Fig. 7.91). The grey areas, which reflect the result of the static space requirement, show that this is also decisive for certain areas. From the analysis of the available data compiled by Bothe (2010) for the Mercedes E-Class can be recorded:

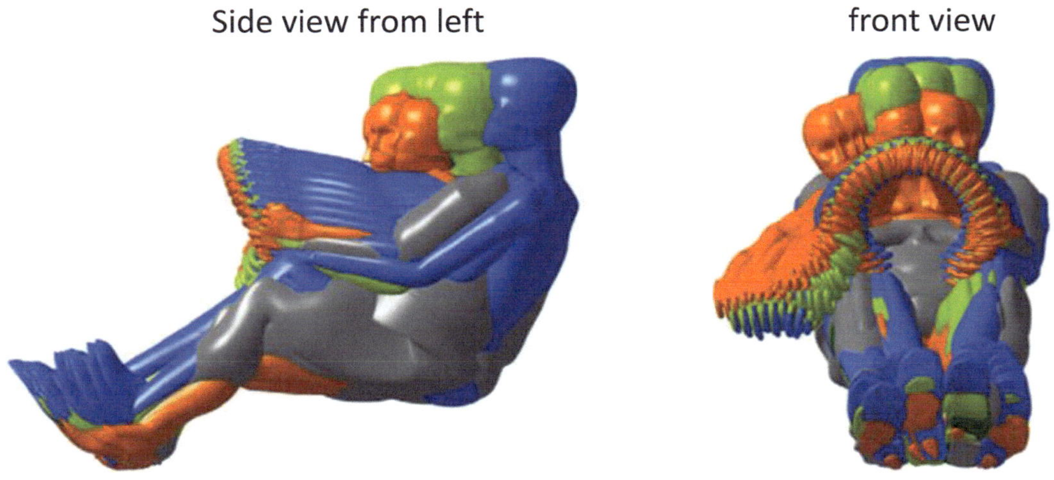

7

Fig. 7.91 Resulting space requirement of the test collective. (Bothe 2010)

The foot and leg space is determined in the upper area especially by the tall man (MTMM) because of his maximum foot and lower leg length. The small woman (FSMM) determines the lower area by her sitting position and the short lower leg length. The lateral area is defined by the small fat woman (FSHM) and the middle fat man (MMHM). The big man (MTMM) and the middle fat man (MMHM) together determine the shoulder space. The middle man (MMMM) requires the necessary headroom in the front and the tall man (MTMM) in the back.

The spheres of influence of the various representatives of the test collective compiled here do not have to have general validity. In the case of a different vehicle concept, deviating results can certainly be observed. As Bothe (2010) remarks, such investigations have revealed, for example, the areas of residence of the lower extremities during the driving task. This is of interest when the contact surface for door and tunnel cladding in the calf and knee areas of design models is to be designed, because in these areas neither sharp edges nor disturbing design elements may influence the comfortable sitting posture. The early visualization and communication of such comfort relevant areas gives designers the opportunity to adapt their designs to the customer's requirements at an early stage, thus avoiding possible costly correction loops during the model and prototype phase.

The considerations presented here about the driver's workplace are to be carried out in a correspondingly modified form and in principle also considerably less complex for the second and possibly third row of seats.

7.5.2 Shelves

The interior of a vehicle also represents living space. It is therefore necessary to provide for options for depositing items needed before, after and, in particular, during the journey. In order to gain ideas for this, it makes sense to start from the scenario technique explained in ► Sect. 6.1. Depending on the intended use of the vehicle, very different requirements may arise here. The positions given in ◘ Table 7.10 can only serve as a suggestion for further consideration. In particular, these positions should also be considered for the second and third rows of seats. Depending on the intended use of the vehicle (e.g. family vehicle, company vehicle, chauffeur limousine), very different requirements arise. Suggestions for this can also be gained from the work of Michel (2014), who, however, dealt with corresponding questions for the design of truck workplaces.

Table 7.10 Examples of to provide the shelves

Before the trip	During the journey	After the trip	At a standstill
Storing Clothing (e.g. jacket) Briefcase (wet!) umbrella Shoes (e.g. exchange of high heels for flat shoes) Vehicle paper/driving licence parking ticket	Indispensable for the journey: Sunglasses Mobile phone including holder Power supply for accompanying electronic devices Desirable: Music CDs (tapes?) Drinks Dishes Toys for children For special cases: Atlas in book form Operating manual	Taking back the deposited objects Key for garage Access authorization to company car park Fuel card	Storage for laptop to work with Storage possibility for small snack (picnic)

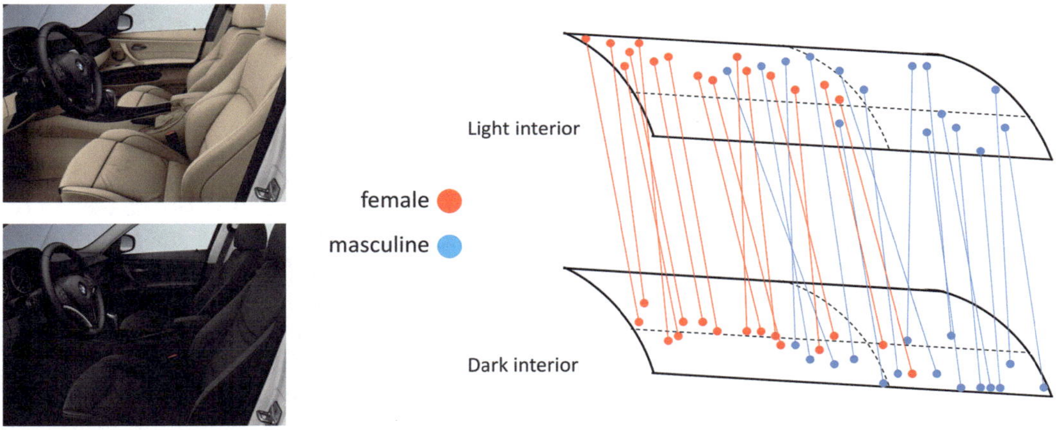

Fig. 7.92 Seat setting selected depending on the colour scheme of the interior. (e.g. BMW 3 Series, E90)

7.5.3 Room Feeling

In addition to objective parameters such as available space and the freedom of movement given by the special constellation, many subjective influences also play an important role in the so-called room feeling. These include aspects such as colouring, brightness or haptics, influences which, according to the comfort-discomfort model described in ► Sect. 3.3.5, are to be assigned to the "pleasure" aspect. Wagner (2013) states: "Room feeling is the emotional reaction to the subjective impression through perception and perception of a space. It is influenced positively or negatively by the interplay of form and colour within the space. It is also slightly dependent on the current mental condition". In an experiment with 15 male and 15 female test persons of different body sizes, he was able to observe on a vehicle (BMW E90) with identical interior fittings, but different colours, that in the dark interior all persons within the given seat adjustment range tended to choose a lower and more rear-facing seating position (Fig. 7.92). The subjective feeling is illustrated by the statements made in ◘ Table 7.11. Among other things, one can infer from this that the concept and thus the expectation of a vehicle plays an important role. For example, a sports car can or should even convey a rather limited sense of space (Wagner 2013).

◘ **Table 7.11** Subjective evaluation of a vehicle interior as a function of colour (Wagner 2013)

Bright interior	Dark interior
"I feel..."	
"Like in my living room"	"Sporty"
"Welcome"	"Dynamic
"Well"	"Enclosed"
"Kindly received"	"Right in the car inside"
"Relaxed"	"Focused on driving"
"Special"	"Salvaged"
"Warm"	
But also:	But also:
"Old"	"Constricted"
"Bored"	"Depressed"
"Worried that the vehicle will get dirty."	"Sad"
	"Cold"
"The interior can be described as..."	
"Better"	"Just right"
"Open"	"Optimal would be dark sky"
"Generous"	"As desired, the room hardly seems so that driving comes to the fore"
"Airy"	"Compact
"Spacious"	"Optimally enclosing"
"Exceeding my expectations"	
But also:	But also:
(tall people) "constricting"	(big and small people) "constricting"
(and even) "too much room for a 3-series BMW"	"Small", "dark", "oppressive", "closed"

Not only the colour design has an effect on the seating position, but also the design of the interior under design aspects. Wagner (2013) further reports on a comparative experiment on driver seat adjustment between two vehicles with almost identical dimensional concepts. The predecessor vehicle (BMW 5 Series, E39) was characterised by a curved dashboard that gave a rather airy impression, while the successor (BMW 5 Series, E60) made a more voluminous impression from the optical point of view. As can be seen from ◘ Fig. 7.93, the two versions do not differ from the anthropometric conditions, especially the gripping space. Nevertheless, all test persons (12 female and 11 male aged between 24 and 60 years) in the new vehicle tend to move the driver's seat a little more backwards with the same seat height adjustment, obviously in order to get more distance from the dashboard, which feels more massive.

The colour design of the upper vehicle interior ("interior ceiling") also has a major influence on the perception of the feeling of space. For example, dark (brown, dark grey, black) colours are perceived as deeper, narrower, more oppressive or more "sporty", for example as light (white, beige, light grey) trim parts (see also ◘ Table 7.11) if the geometric position remains unchanged. An optical opening of the roof by means of large glass surfaces expands the view to the outside, allows more light into the interior and thus creates a wide and open feeling of space, even if the distance to the occupants is somewhat less than with dark panelling (◘ Fig. 7.94). A special effect is provided by so-called panorama windscreens, in which the front windscreen merges into the glazed roof surface without any attachment (without upper roof frame) (◘ Fig. 7.95). This results in a very impressive visual effect, which conveys a great feeling of space and at the same time a good view to the outside. Situations in which the sun is in the opposite direction of travel or very high in the sky pose a particular challenge for the driver. A shading system that goes beyond the concept

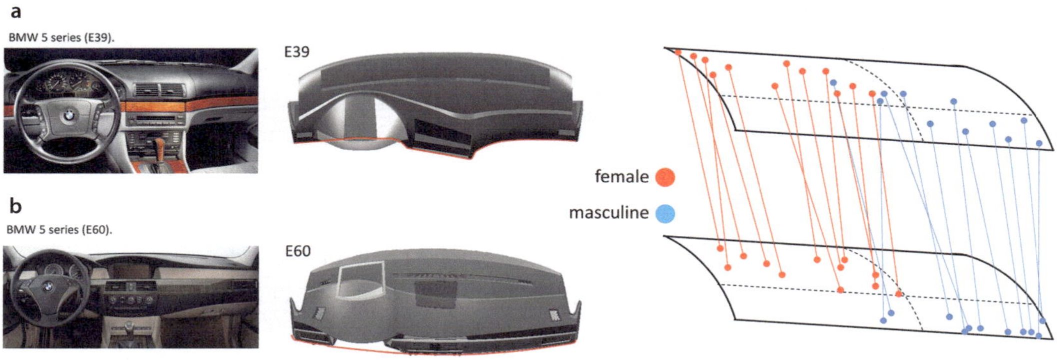

■ **Fig. 7.93** Seat setting selected depending on the design of the dashboard (**a** BMW 5 Series E39, **b** BMW 5 Series E60; Wagner 2013)

■ **Fig. 7.94** Opel Meriva with glass roof

■ **Fig. 7.95** Opel Zafira with panorama screen

of the usual sun visors is indispensable here. A further problem of the generous glazing arises with regard to the air-conditioning of the interior: thermal radiation heats up the surfaces of the interior through the glass surfaces and thus creates greater demands on the air-conditioning system on the one hand and on the other hand temperature differences on the skin of the occupants that cannot be technically compensated.

7.6 Entry and Exit

7.6.1 Door Concepts

Getting into a vehicle is particularly important from an ergonomic point of view, as this extremely complex and dynamic process is influenced by many anthropometric and vehicle geometric boundary conditions. As in the chapter heading, the entry with exit is often called in one breath, but these two movements are different and subject to their own principles, as will be explained in the following section.

In general, entry and exit are influenced by three parameter groups: anthropometry, vehicle geometry and motion strategy. The **anthropometry** describes the size, proportion and obesity of the person entering the vehicle and it is obvious that the dimensions of the body can promote or impede entry into the vehicle. Above all weight and obesity, but also mobility (mobility of the joints) and muscular fitness influence the agility, balance and speed of the movement sequence. Thus it is decisive for the movement and the smooth flow of the boarding process whether, where and to what extent the boarding person seeks, finds and needs support for his or her body on the vehicle. The fact that a person is small, light and

■ **Fig. 7.96** Mercedes-Benz F100 research vehicle with centrally positioned driver's seat and sill that enters the car floor to facilitate access to the vehicle

slim does not make it easier to get into a vehicle per se, but a pronounced obesity is a hindrance in most situations.

7

The **vehicle geometry** has an influence on the sequence of movements in that as a rule all persons bend down to get on the car[23] and both feet over a threshold[24] to lift it. Analogue to stair steps[25] as they are necessary for overcoming height differences in buildings, a combination of height and step length is also necessary for motor vehicles in order to reach the seat comfortably[26] to make this possible (■ Fig. 7.96).

23 Most passenger cars have a height of about 150 cm. Thus the lower edge of the roof frame, under which all persons must move their heads into the vehicle interior, is still below the fifth percentile of body height. Therefore a bend is unavoidable at the beginning. For SUVs and minibuses, the vehicle height can be found at 200 cm. If the roof frame of such vehicles lies further inside, the boarding passenger is less required to bend down than in lower vehicle concepts.

24 The entry sill, which defines the lower step for getting in, is often at the level of the wheel centre and thus in a range of 200–400 mm. Smaller vehicles tend to have lower entrance heights, whereas in SUVs and minibuses this step can be so high that an intermediate step is necessary and a step must be provided in the entrance.

25 A step in the building has a height (slope) of about 160 to 200 mm.

26 In individual cases, vehicle concepts are presented which, with a conventional vehicle width around 1800 mm, provide for a driver's seat arranged in the middle of the vehicle. This means that the seat is located at a distance from the outside of the vehicle via 600 mm and can no longer be reached in a single step. This calls for a special conceptual solution to the accessibility problem (e.g. Mercedes-Benz F100, 1991; BMW Z13, 1992; Mia electric, 2010; see ■ Fig. 7.96).

A significant influence on entry and exit has the constructively used **door concept** of the vehicle. The vast majority of all production vehicles use conventionally front hinged doors, i.e. the front doors for driver and front passenger are hinged on the A-pillar and open on the B-pillar, the rear or rear doors are hinged on the B-pillars and open on the C-pillar. The unlocking device (external door handle with unlocking button and, if necessary, locking cylinder) is located close to the opening flange edge and is usually operated in the opening direction. The lifting plate for opening the door, which was preferred for some time for aerodynamic reasons, has not been able to assert itself for safety reasons because the force required to open a door even after an accident cannot be transmitted sufficiently (■ Fig. 7.97). In general, the strength, accessibility and operating direction of the rear doors in particular must also be adapted to suit smaller persons in order to enable children to board the vehicle independently (see also ► Sect. 7.8.2). The localisation of the inner door handles results from the interaction with the analysis of the accessibility areas (■ Fig. 7.82). The analysis of the gripping spaces (taking into account the forward inclined upper body) can also be used to determine the grips for closing the door in conjunction with its maximum opening angle.

A collision of the upper window frame with the head (neck, chin, face or glasses) of the boarding passenger should be avoided by the shape of the frame and the arrangement of the door handle. The deeper the handle is arranged, the more a person will lean towards the handle with his upper body and is in danger of getting too close to the window frame if the handle is opened quickly and possibly carelessly. Human models can also be used to help clarify all these questions.

Rear hinged doors open against the direction of travel. Historically, front and rear doors were attached to the B-pillar with common hinge bolts[27], so that the driver's doors had the opening flange on the A-pillar. If the door was unlocked accidentally or intentionally during the journey, the wind caused the door to open abruptly, which could even tear

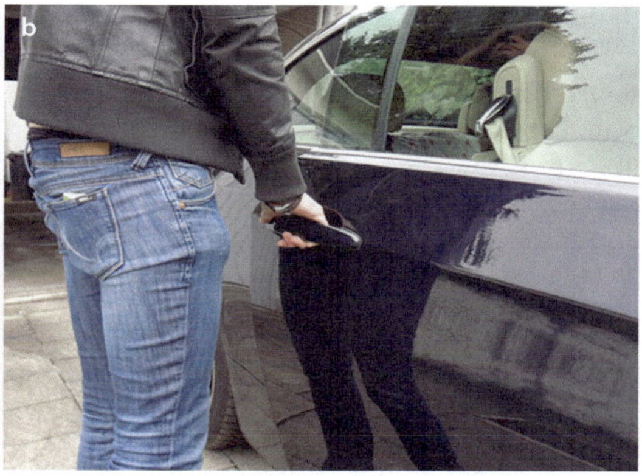

Fig. 7.97 Door handle as lifting plate (**a**) or as a handle (**b**)

Fig. 7.98 **a** Opel Meriva B (2010), **b** Mazda RX-8 (2003–12)

the door off and cause the driver or passenger without a seat belt to fall out. Due to the risk of accidents, this type of construction was banned in many countries and may only be used under the strictest conditions since then.[27] A great advantage of the rear hinged door, however, is undoubtedly the comfortable and unhindered access, as the door does not block the large access area in front of the seat (it leads to the sagging strategy when getting in and to the lifting strategy when getting out; see ▶ Sect. 7.6.2). It is also easier to install a child seat and check the child seated in the child seat before closing the rear door if the door gap is facing forward. These last two aspects played a major role in the development of portal doors[28] of the Opel Meriva B[29] (Fig. 7.98). A further advantage can be gained when getting out if a vertical handle is attached to the B-pillar. In this way, the person getting out can hold on to the handle and lift easily out of the seat by pulling himself forward. This also prevents him from holding on to the B-pillar and possibly being injured if the person sitting in front slams his door shut. The Mazda RX8 sports coupé also dispenses with a B-pillar and locks the front doors in the rear doors. When all doors are fully open, the rear seats can be accessed more comfortably

27 For passenger cars, a safety device shall be provided to prevent the doors from being opened while the vehicle is in motion. There are other exceptions for slow-moving tractors, construction machinery and material handling vehicles.

28 The first series of the VW Transporter (VW type 21,950–1967) had for the entrance to the passenger compartment of the second and third row of seats two central pillarless portal doors, whereby the front door had to be opened first. These were replaced by a sliding door from the T2 series onwards.

29 The Opel Meriva B (2010) has a front hinged door and a rear hinged door. Both doors lock at the B-pillar and can be opened and closed independently of each other.

than would be possible with pure two-door coupés over the folded backrests of the front seats. However, with this concept, the rear door cannot be opened without the front door already being opened. Especially in cramped parking situations, it is hardly possible for rear passengers to leave the vehicle.[30] Since the safety belts are usually attached to the B-pillar and deflected in the direction of the front occupants, a special constructive solution must also be found for passive safety.[31]

A further door concept used in particular for minibuses is the sliding door, the great advantage of which is that it can be opened and closed without the need for a separate door.[32] Usually allows a complete opening of the door cutout even in cramped parking situations. As soon as the door has jumped open from the locking position, it can be pushed back parallel to the vehicle and is therefore hardly applied laterally. Here, too, operation of the actuating device (inside and outside) in the opening direction makes ergonomic sense. A combination with doors hinged at the front and a missing B-pillar is also suitable for the sliding door to facilitate access.[33]

Wing doors[34] allow the vehicle to be approached from the front or rear with almost unhindered access and even offer an umbrella effect that keeps the entrance free from precipitation in adverse weather conditions. However, the maximum required lateral opening width of the door must be completely available, otherwise it is almost impossible to get in or out (◘ Fig. 7.99).

◘ **Fig. 7.99** Wing doors on the Opel Monza concept car (2013)

The situation is similar with the butterfly door,[35] which swings a little more forward and allows access to the vehicle only from behind. With a scissor door[36] it is possible to open the vehicle entrance without taking up any lateral space. However, here (analogous to wing and butterfly doors) the accessibility of the fully opened door from the driver's seat is almost impossible and must be supported by auxiliary equipment. The aspects of passive safety are certainly also essential for the choice of a particular door concept, especially the question of leaving the vehicle if the vehicle has been involved in an accident and is possibly lying on the roof. For these reasons and due to the limited distribution of these alternative door concepts, the conventional door concept of the front hinged doors is used in the following considerations was taken as a basis.[37]

30 The Mini Clubman (2008) also uses the concept of the B-pillar-free portal door, but only on the passenger side.

31 If the belt is not to run in the door opening (Mini Clubman, Mazda RX-8), it must be integrated into the seat, which places increased rigidity requirements on the seat back.

32 In addition to the advantages, the sliding door has the design challenge of providing a longitudinal guide for the sliding length. The total weight of the sliding door is higher than that of a conventional door and the centre of gravity is also higher.

33 The Peugeot 1007 (2005–2009) used only two sliding doors for access to the first and second rows of seats. The Ford B-Max (2012) combines a front-hinged front door with a sliding door for the rear passengers and does without the B-pillar

34 Wing door: Door hinged to the side roof frame and opening upwards (DeLorean DMC-12; Opel Monza Concept 2013). The wing door allows access when the sills are very wide and high (Mercedes-Benz 300 SL W194/198 (1952–57)) or are a very eye-catching and spectacular design feature with an effective show effect.

35 Butterfly door: door hinged to the front window frame (A-pillar) and opening upwards and forwards (McLaren F1, Ferrari Enzo Ferrari)

36 Scissor door: door hinged to the A-pillar and opening upwards (e.g.: Lamborghini Countach, Diabolo, Murcielago, Reventon, Aventador; Bugatti EB110)

37 Front door hinged to the A-pillar and locked to the B-pillar, rear door hinged to the B-pillar and locked to the C-pillar

7.6.2 Movement Strategies

With the movement strategies you have to differentiate between entry and exit. The *entry* usually takes place from a standing position in a lateral forward movement across the sill step and under the roof frame. The body moves mainly forward through the door opening, which is bounded by the columns and the door panel. As a result, the limiting geometry is usually in the direct range of vision. Once the outer vehicle geometry has been overcome[38] the driver still has to master the edge of the seat cushion, the steering wheel and the pedals. Depending on seat height[39] the body can be lowered onto the seat during the insertion process or must be lifted onto the seat by lifting the hips and buttocks. Lowering the pelvis to the seat is perceived as more comfortable than lifting, as gravity "helps" the body fall. Lifting the body requires additional effort, which is perceived as strenuous climbing. However, a lowering of the buttocks to a very low level (for example in a sports car) is also perceived as very uncomfortable, as here a large curvature of the spine and great holding forces are necessary to keep the body in balance and lower it in a controlled manner to the seat surface.

At *egress* the conditions are reversed and can thus induce other comfort evaluations in the perception. The exit takes place from sitting or bent over crouching posture sideways or even sideways backwards into the stand. For this the body must find a stable support of at least one foot outside the vehicle before the head and upper body are lifted again under the roof frame through the door opening into the open air. For vehicles which allow the body to be lowered onto the seat when boarding, the body shall be lifted in the opposite direction to gravity when disembarking. On the other hand, in vehicles that require the pelvis to be lifted when getting in, the body can glide down from the seat to the standing level following gravity when getting out, which is perceived as more comfortable. For example, higher seating concepts are generally more uncomfortable when getting in, but more comfortable when getting out. Although lower vehicle concepts make it easier to get on, they require more effort to get off. A good compromise is offered by vehicles with a slightly raised seating position approximately at the height of the buttocks in relation to the road surface, as is the case in SUVs and off-road vehicles due to their design. The exit in cramped conditions[40] poses a particular challenge, as a minimum width is required to allow passengers to get in and out of the vehicle. The minimum opening width is determined on the vehicle side by the thickness of the door leaf (approx. 100–200 mm) and on the human side by the basin depth (approx. 250–450 mm). Under these assumptions, a door must open at least 350 to 650 mm beyond the outer skin to provide a sufficient gap to allow the pelvis to put it out of the vehicle.[41]

Although the aspects described above are basically valid, different entry and exit strategies are observed depending on the boundary conditions. The reason for the choice of one or the other strategy are alternative strategies in addition to a self-taught way of movement, if the usual procedure does not lead to success. Here the vehicle geometry plays the decisive role if the vehicle differs from the previously used one at certain points. On the one hand this can make the situation (narrow parking situation) necessary, on the other hand it can result from the door concept of the vehicle or from the found position of the seat and its relative position to the steering

38 External vehicle geometry of the access opening: All components beyond the door flange seal (A-pillar, B-pillar, if necessary C-pillar, access sill, roof frame above, window frame, door panel with window guide, outer skin and inner door trim).

39 ISO dimension H5.

40 Cramped conditions prevail in situations where vehicles are so close together that the door can only be opened a small distance, e.g. in car parks or multi-storey car parks or single-piece production halls. Unfortunately, most building regulations date back to times when vehicles were considerably narrower, and so parking spaces are now perceived as narrow due to the increased width of vehicles.

41 As a rule, the pelvis, the hip area or the abdomen is the limiting part of the body when getting out.

wheel and pedals. An increase in obesity as well as a change in mobility in old age can lead to a change in strategy. The greater the free space in and around the vehicle, the easier it is for the vehicle user to apply his preferred behaviour pattern. "The perfect ingress and therefore also the perfect egress as an absolute size does not exist" (Cherednichenko 2007, p. 142). Nevertheless, certain configurations are suitable for more or less fulfilling the expectations of the target customers with regard to entry/exit comfort.

7.6.2.1 Entry Strategies

In Rigel (2005) one finds references to the parameters of the motion sequence, which is characterized by so-called *"leading body parts"* and compliance with the requirements described therein *"protective clearances"*. Afterwards the movement is controlled by the movement paths of special body parts. In a targeted movement, such as the grasping of a hand for an object, the hand is moved by the muscles of the arm over bones and joints, but the brain "controls" the hand as the leading part of the body. Direction and posture of arm and lastly also of body are controlled by downside control loops (for maintenance of balance), so intended movements are realized. The model of protective distances describes that when certain body parts move past objects, a certain minimum distance is maintained depending on the respective body part, which is intended to prevent the body from colliding with the surrounding object. According to Arlt (1998), for example, this protective distance is smaller for the feet and hands than for the head. The condition of the surrounding object (soft or pointed) also has an influence on the distances maintained. These protective distances are, of course, subject to intra-[42] and interindividual[43] variance and cannot prevent the boarding person from hitting a part of the body under unfavourable boundary conditions or rapid movements (see ◘ Fig. 7.100).

When getting into a car, clear deviations can be observed in the motion sequence. In addition to anthropometric measurements such as body height and trunk length, other individual factors such as weight, age and mobility can also be considered. The chosen strategy is also influenced by the perception and evaluation of the situation (Rigel 2005). In addition, in an experimental environment there is the effect of learning.[44]This is due to the fact that the principle optimizations and learned bypass strategies and findings are applied when repeated several times. Rigel (2005) has 225 people[45] at the age of 22 to 68 years and differentiated between four main strategies based on the work of Höllrich (1992) and Layer (1992). The following descriptions refer to the entry to the driver's seat of a left-hand drive vehicle with the door already fully open.

Entry: "Slipping"Strategy

The person entering the vehicle transfers the body weight to the left leg as a supporting leg in order to lift the right leg facing the vehicle over the side skirts into the interior of the vehicle (◘ Fig. 7.101). However, the right knee remains on the left side of the steering wheel and the steering column fairing. Now bend the knee of the standing leg and lower the buttocks to the seat. This process takes place dynamically with or without support of the body on the steering wheel rim. The left leg is then lifted over the sill into the interior

42 Intraindividual (within) variance: variance that occurs in one and the same subject when that subject repeats the same experiment several times. Rigel observed an average variance of about 20 mm for individuals with individual deviations of about 50 mm.

43 Interindividual (between) variance: variance that occurs in different subjects with the same or very similar body dimensions when these subjects perform the same test. Rigel determined an average variance of about 40 mm with maximum deviations of up to 100 mm.

44 At the beginning of a new task, many mistakes are made that may prolong the completion of the task. During later learning phases, the identified and resolved errors decrease and are taken into account from the start of the new cycle, which means that each further process can be carried out more quickly and safely.

45 225 persons: 30 women, 195 men, limousine drivers (H5 between 470 and 515 mm, H30 between 245 and 285 mm; Rigel 2005).

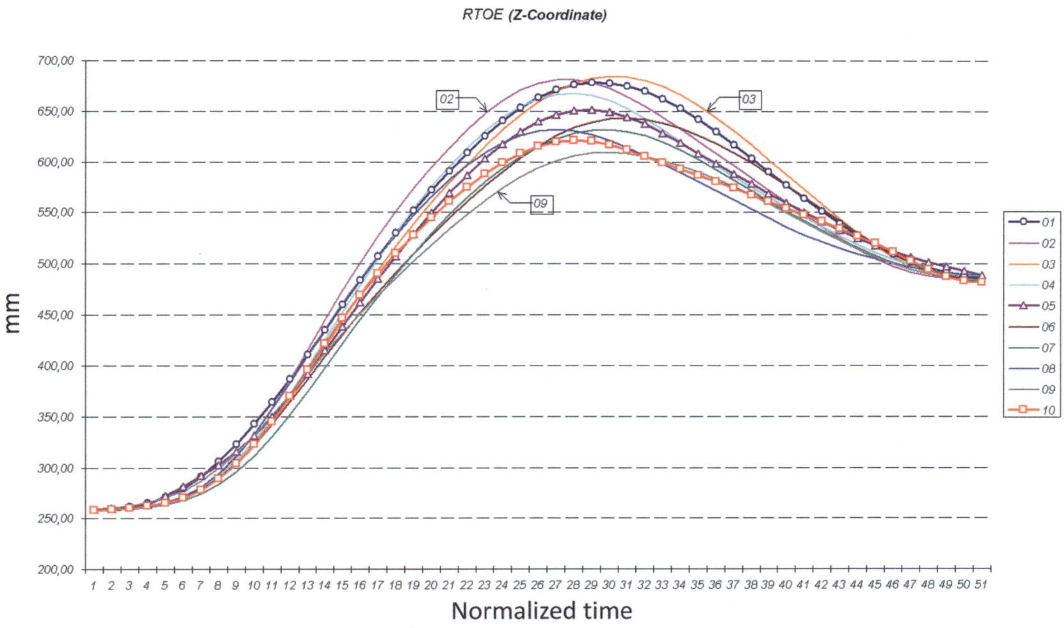

Fig. 7.100 Interindividual (between) variance of a single joint point. (Cherednichenko 2007)

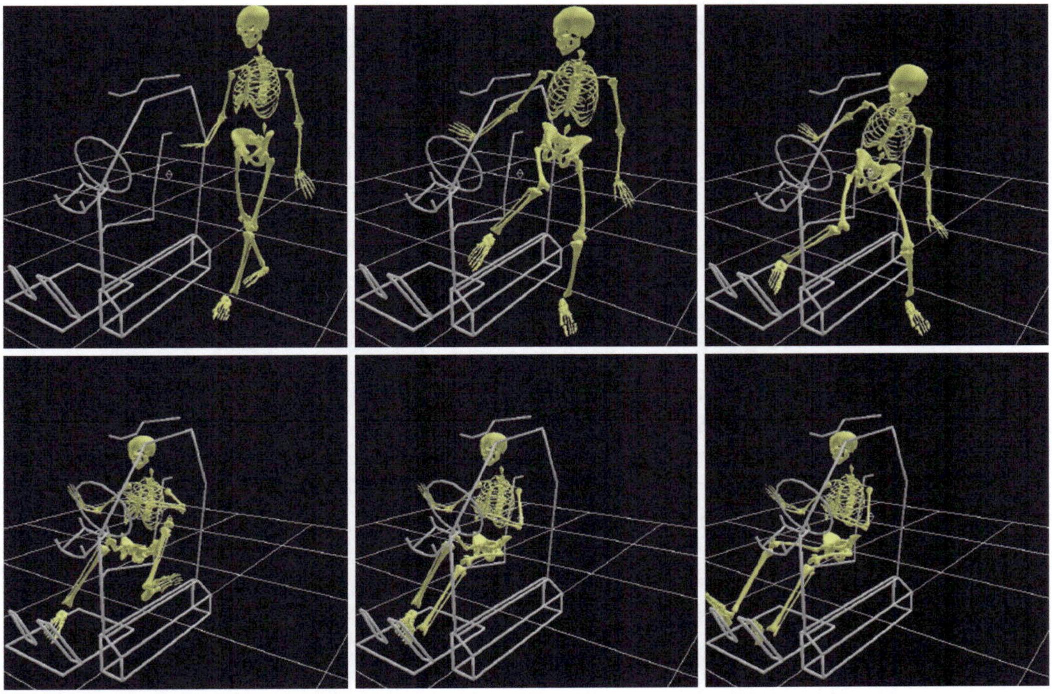

Fig. 7.101 "Slipping" strategy

of the vehicle, at the same time or subsequently the right knee is moved (slipped) under the steering column fairing and the right foot is placed on the accelerator pedal. Sub variants of the slipping strategy result from the moment when the knee is moved

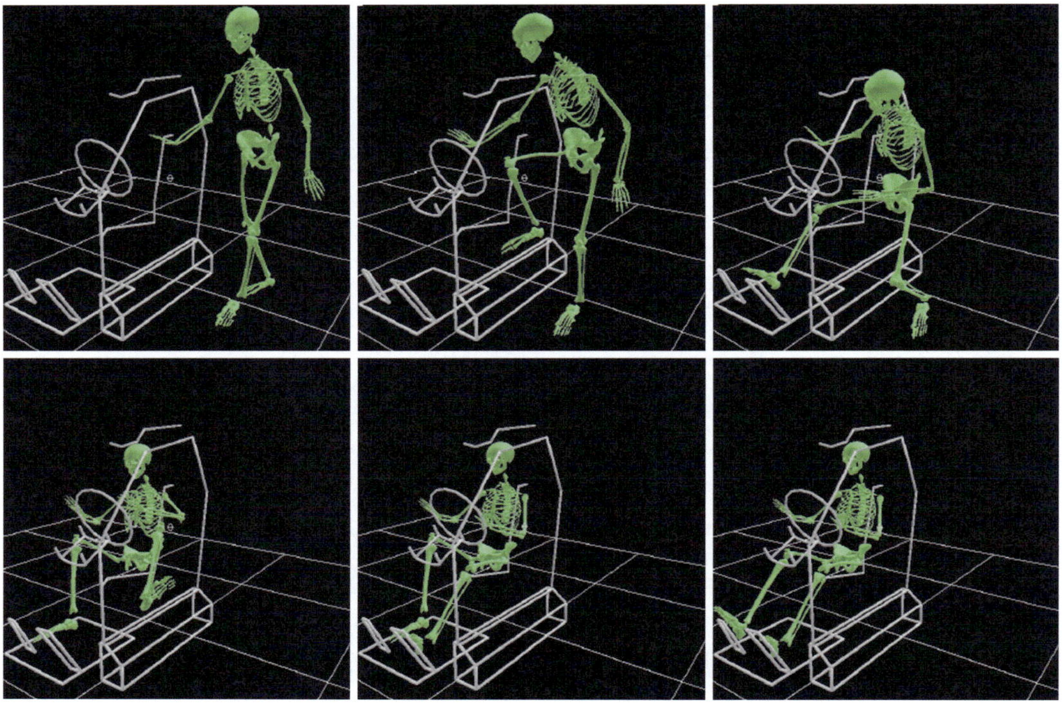

Fig. 7.102 "Fiddling" strategy

under the steering wheel (before or after lifting the left leg into the vehicle). There are also the variants that the right foot must be lifted when threading through and that the left foot is placed on or in front of the storage surface after being lifted into the vehicle.

Entry: "Fiddling" Strategy

In the fiddling strategy, the person entering the vehicle also shifts the body weight to the left leg and lifts the free leg into the vehicle interior (Fig. 7.102). However, the right thigh is immediately spread outwards so that the knee can be threaded under the steering wheel. The person is already supported by the steering wheel rim when the head is still outside the vehicle. The right foot is placed in front of or on the accelerator pedal, then the buttocks are lowered onto the seat and the head is swivelled under the roof frame into the interior of the vehicle. Finally, the left leg is lifted over the sill into the vehicle interior. This concludes the movement process. Also here there is the variant that the left foot is set down after lifting in front of or on the tray.

Entry: "Sagging" Strategy

In contrast to the two strategies mentioned above, the boarding passenger turns his back towards the vehicle opening and remains in a fixed position with both feet (Fig. 7.103). Now, support is often sought with both hands on the door or window frame or on the B-pillar and both knees are bent at the same time. With a fuselage movement away from the vehicle, only the buttocks are lowered to the side cheek of the seat. In this already seated position, the body is brought into the forward driving position with a rotary movement, the head is swivelled into the vehicle and the right leg is moved knee forward under the steering wheel. The left leg is then actively or passively lifted into the vehicle. Especially elderly and corpulent people prefer this strategy.

Entry: "Hurdle Jumping" Strategy

In the fourth entry variant, the person entering the vehicle stands on both legs facing the door opening (Fig. 7.104). Similar to a hurdler, the head is moved ahead and one leg is moved into the vehicle interior at the same time. The

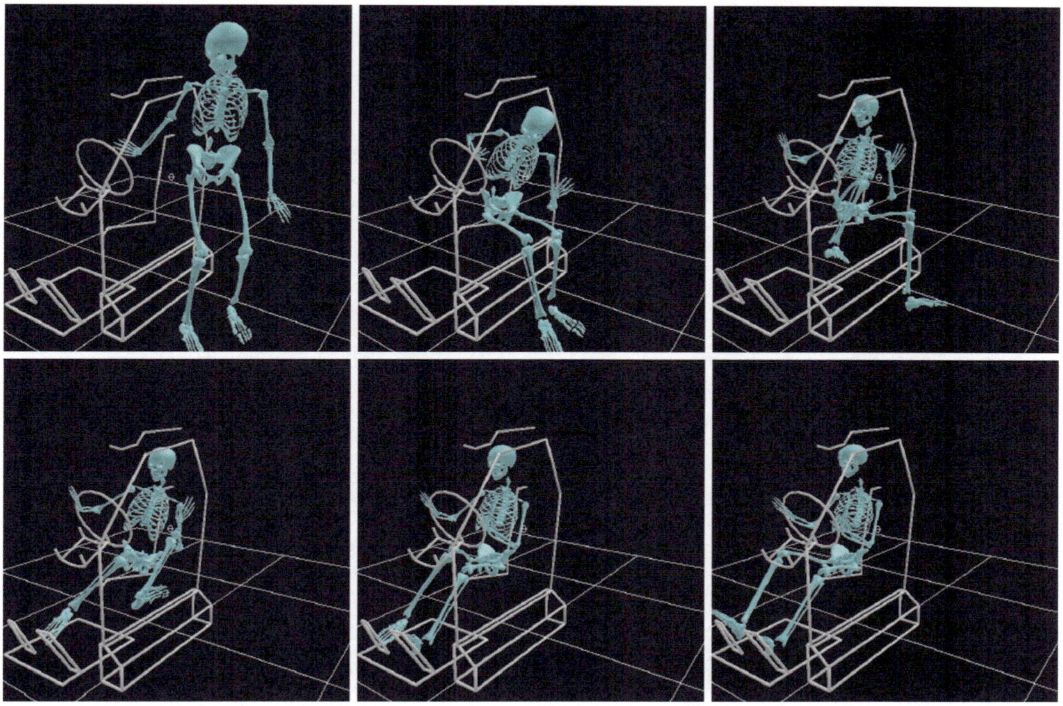

Fig. 7.103 "Sagging" strategy

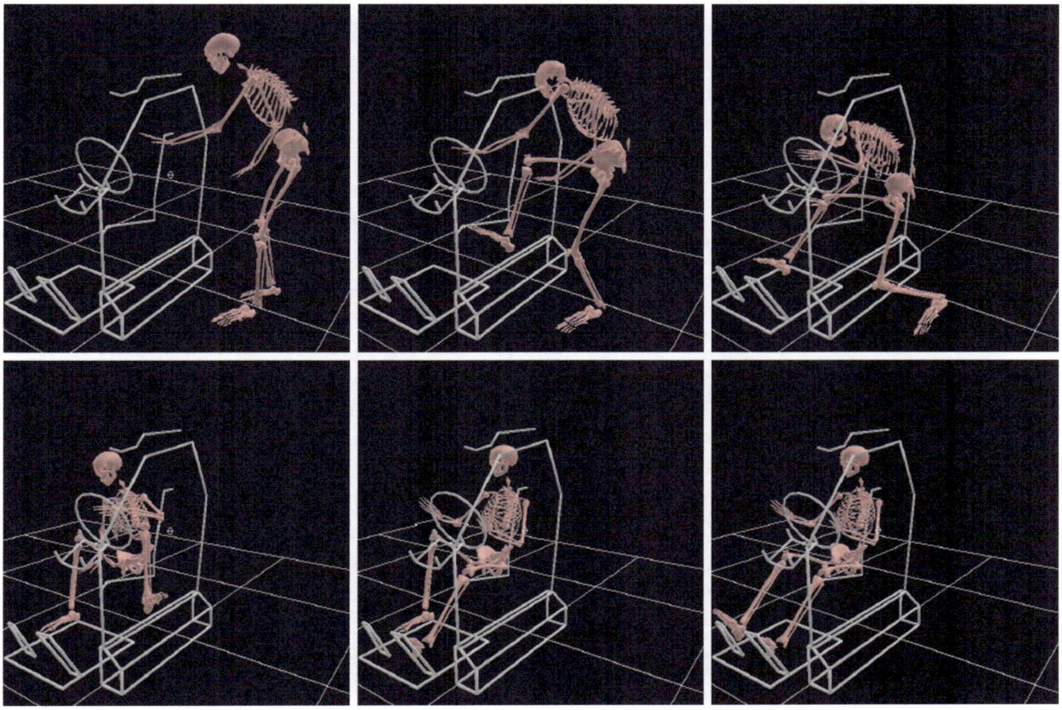

Fig. 7.104 "Hurdle-jumping" strategy

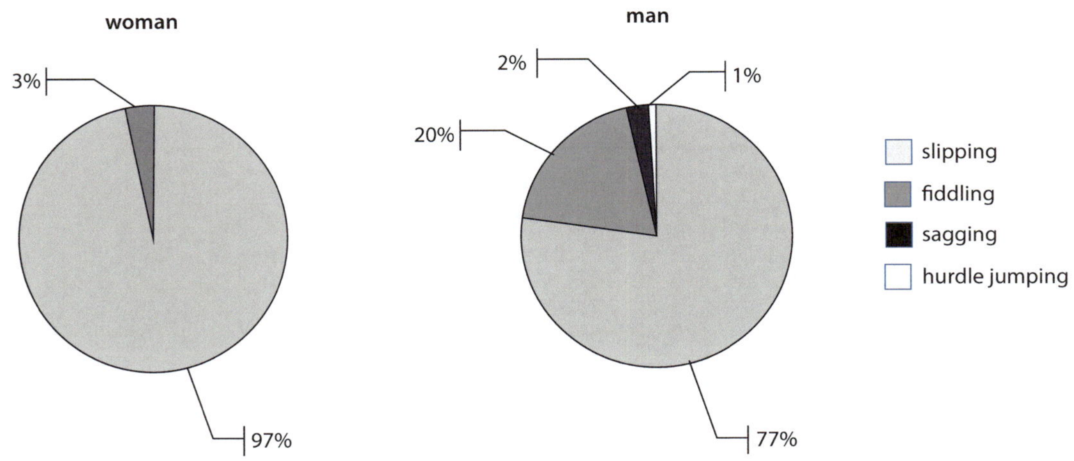

Fig. 7.105 Observed frequencies of the different entry strategies. (Rigel 2005)

right leg is moved between the seat and steering wheel and placed on the ground, the body is supported with the left hand on the steering wheel rim or window frame and with the right hand on the seat back or centre console. The seat is swivelled over the seat and lowered with a twisting movement. Lifting the left foot into the vehicle completes the movement process.

Although Rigel's investigations were also carried out on limousines, under changed boundary conditions mixed forms of the described strategies result. In minibuses with an intermediate step, for example, fiddling and hurdle strategies merge. For sports cars[46] with high side skirts and/or short door cut-out, a variant of the hurdle strategy leads to a successful entry. The left leg is first lifted into the vehicle from a standing position and the right leg is brought over the sill and seat using the two-handed support on the window frame and B-pillar. With a swinging movement, both legs are now brought into the foot space and the upper body slides behind the buttocks into the vehicle interior. Overall, however, Rigel (2005) points to the high frequency of the slipping strategy (Fig. 7.105).

46 Historical vehicles that require this strategy include the Mercedes-Benz 300 SL W198 (1952–57) and the Jaguar E-Type (1961–74). This strategy is also necessary when entering through the side window with firmly locked doors (racing cars, stock cars).

7.6.2.2 Exit Strategies

The movement strategies of the **exit** are similar to the boarding movements, but still differ due to the very different starting position, the need to support oneself inside the vehicle and to free oneself from the enclosing shell of the passenger compartment. In general, the driver's basic posture (seated, reclined posture, feet on the pedals, hands on the steering wheel, looking forward) is abandoned when the seat belt is opened, in which the person turns towards the side door and straightens slightly out of the seat back. Now unlock the door from the inside and slightly open it. A distinction can be made between whether the person then opens the door further with the right or left arm. In this situation, it is recommended that learner drivers turn around and observe the following traffic through the opening gap to see whether a vehicle or possibly a cyclist is approaching from behind. Since this situation is of particular importance only when parking at the roadside, this precautionary measure is omitted in row parking lots, which is why many people do not adopt this behaviour and get out of the vehicle carelessly even in critical situations. The attempt to induce this turning strategy by means of an opening bolt located further back on the inside of the door did not prove to be successful and is therefore hardly implemented today.

Exit: "Bridging" Strategy

With the bridging strategy ("exit-slipping"), the door is pushed open with the left hand or forearm, but depending on the available free space and traffic space, often only as much as is necessary to push the basin through. The left leg is placed on the road surface through the lower door gap, while the head pushes sideways upwards through the upper door gap into the open at almost the same time. In very tight conditions, the back and buttocks glide upwards on the seat backrest. While the left leg already has solid ground under the feet, the right leg remains inside the vehicle on the right side of the steering wheel. For a moment, the person getting out stands upright in the open door gap with his left leg stretched and spread. When the left foot is in a stable position, the right leg is pulled out with the thigh between the seat and the steering wheel and the right foot is placed next to the left. The person now moves sideways or backwards out of the door gap and the movement is completed when the door is closed. This bridging strategy can be observed in two variants: In the first variation, the right leg (as described) remains on the right side of the steering wheel until the leg is pulled out of the vehicle. Thus, this variant is similar to the fiddling strategy of the entry. In another variation, after opening the door, the right knee (as in the entry slipping strategy) is passed under the steering wheel, from the right to the left side of the steering wheel, so that both knees are close to each other on the left side of the steering wheel. The bridging strategy is used particularly in confined situations.

Exit: "Climbing" Strategy

This sequence of movements is similar to the one mentioned before, in so far as a one-legged stand with the left leg is sought ("exit-fiddler"). However, when "climbing"the upper body and thus also the head is tilted towards the door frame. This is usually done with a shock-like jerk out of the seat. The left hand supports the body in the upward movement and/or the right hand pulls the upper body upwards against the door. As soon as the buttocks are released from the seat, the right leg is pulled between the seat cushion and the steering wheel and the right foot is also placed on the ground. Variations result from the order in which the right knee is passed under the steering wheel. Here, the person getting out is standing on the side to the rear of the vehicle. If the right foot is now moved to the rear with a lunge step, the person can simultaneously turn around and step out of the door gap so that the door can be closed.

Exit: "Lifting" Strategy

With this strategy ("exit-sagger"), the door usually has to be opened wider than in the movement sequences mentioned above. As soon as the door has opened a sufficient gap, both feet are placed on the road almost simultaneously. This causes the person in the seat to turn a quarter of a turn, so that he or she sits at right angles to the direction of travel. By leaning forward in a seated position, the head is brought through the door opening and the entire body is lifted out of the seat with the help of the arms. Both feet stand at the same time and firmly on the ground. The body weight is carried by both legs. The hands are looking for the door or window frame, the B-pillar or the seat backrest. More rarely, both hands are supported on the thighs. If the person now stands upright with his/her back to the vehicle in the door gap, he/she must turn half a turn of his/her body. He/she moves sideways out of the opening and then closes the door. The lifting strategy can also differentiate whether both knees are on the right side of the steering wheel first, or whether the left foot is placed on the road first, followed by the right leg under the steering wheel and then the right foot on the ground. Here, too, older, more corpulent and weaker people prefer this stable and safe sequence of movements.

Different authors have made categorizations of the entry and exit process, which sometimes take somewhat different aspects as a basis for the naming, but come to the same result in the overall description. Sabbah (2010) has compiled and assigned a list to the categorizations made here (◘ Table 7.12).

Table 7.12 Overview of entry and exit strategies ("–" = without naming; Sabbah 2010)

	Rigel (2005)	Andreoni et al. (1997)	Chateauroux et al. (2007)	Reed et al. (2008)
Ingress	Hurdle jumping	One-foot forward motion	Right-leg-first	Head-first
	Fiddling	One-foot lateral motion		Hip-first
	Slipping	One-foot backward motion		
	Sagging	Two-foot trunk forward	Buttock-first	
		Two-foot trunk backward		
Egress	Exit hurdle jumping	–	–	–
	Exit fiddling	One-foot parallel to the vehicle	–	–
	Exit slipping	One-foot head forward	–	–
	Exit sagging	Two-feet lying strategy	–	–

7

7.6.3 Evaluation Methods

The different entry and exit strategies and their complexity make it difficult to evaluate these processes a priori in the concept phase. Nevertheless, this is necessary, because changes at a later stage of development are cost-intensive and often not feasible due to the associated intensive intervention in the exterior design. In order to interpret such a procedure correctly, the statement by Fuchs-Kittowski (1976) must be observed: "The principle of the same cause of the same effect no longer applies in physics, nor does it apply to highly complex living systems. It must be replaced by the principle same causes - same possible effect". Three methods are currently being discussed for this purpose in addition to the usual test person experiments on the finished mock-up representing the future vehicle, namely: the immediate evaluation of the geometric boundary conditions, the determination of a neutral movement causing little discomfort and the simulation of the entry and exit process with the aid of a human model, which is then used for the evaluation. The evaluation of the geometric boundary conditions ultimately defines a certain vehicle concept and does not allow an answer to be given as to whether a certain constellation is equally suitable for all persons. By defining a neutral movement, one practically describes an "entry hose"and examines whether it is affected by the given vehicle constellation. The simulation method, on the other hand, allows a human-related evaluation, i.e. which joints are particularly loaded and how this can be prevented by changing the geometric conditions. The aim of such a general modelling of motion behaviour is to be able to apply it ideally to any product.

7.6.3.1 Geometric Constraints

Figure 7.106 shows the internationally defined GCIE dimensions (Global Car manufacturers Information Exchange group) and the corresponding dimensions according to DIN 70020–1 and SAE J1100 which describe the door opening and are therefore relevant for the entry and exit process. It also includes some vehicle dimensions which are not directly defined in standards and can only be determined indirectly via dimension chains or also those which do not originate from public sources. All these measurements are referred to as "self-defined" and named with a small letter and a serial number (Table 7.13; Sabbah 2010).

The limiting parameters of the door flange opening are compared for a dimensional comparison of different vehicle concepts or competitor products. The main dimensions

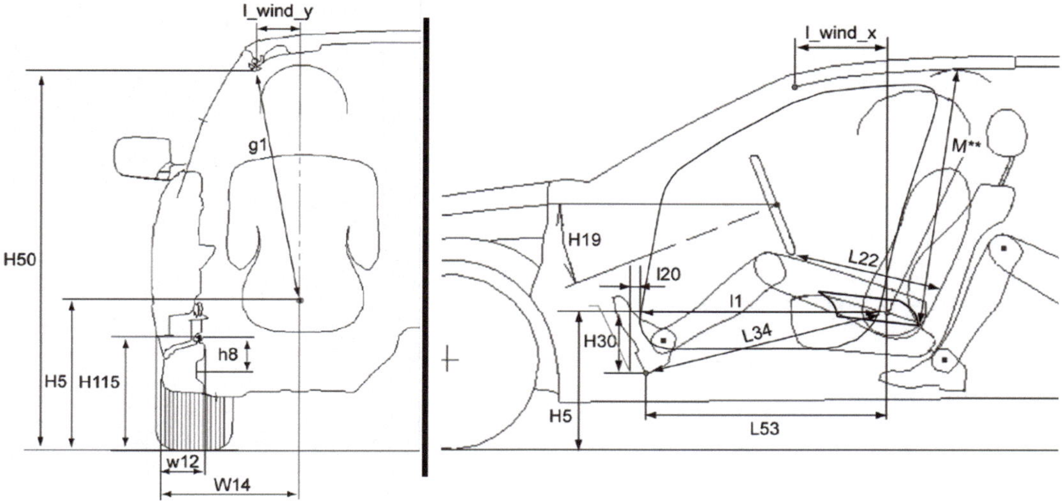

Fig. 7.106 Entry relevant main dimensions

Table 7.13 Vehicle parameters relevant for entry/exit (According to Sabbah 2010)

Short	Definition of the measure	norm
H5	Vertical dimension between SRP and stand plane	DIN 70020–1
H30	Vertical dimension between SRP u heel contact point	DIN 70020–1
H50	Vertical dimension between the lower edge of the roof frame and the standing plane (measured in the x-plane by the SRP).	DIN 70020–1
H74	Smallest distance between steering wheel rim and front edge of seat cushion (in SRP position)	DIN 70020–1
H115	Vertical dimension between the road level and the top of the sill (measured 330 mm before SRP)	GCIE 2004
h4	Vertical dimension between the A-pillar root and the stand plane	Self-defined
h6	Height angle of the A-pillar to the horizontal (specified as height dimension)	Self-defined
h8	Vertical dimension between the upper edge of the sill and the vehicle floor (measured in front of the A-pillar)	Self-defined
w12	Largest horizontal dimension in y-direction through the Centre (x-position) of the sill	Self-defined
W14	Largest horizontal dimension in y-direction between sill outer edge and SRP	GCIE 2004
w16	Horizontal dimension in y-direction between roof outer edge and seat Centre plane (measured by SRP)	Self-defined
I22	Smallest distance between steering wheel rim and backrest upholstery (in SRP position and 25° angle)	Self-defined
I1	Horizontal dimension in x-direction between A- and B-pillar	Self-defined

relevant for access are mainly determined by the sill cross-section and the position of the roof and columns. The dimensions of the sill represent a major obstacle. As soon as the buttocks are placed on the seat cushion, the upper roof frame and the window frame as an extension of the A-pillar limit the swivelling movement of the upper body and the possibility of comfortably swivelling the head into the vehicle interior without excessive rolling of the thoracic spine. The main dimensions relevant for getting out of the vehicle are additionally determined by the relative position of the A-pillar to the H-point, since this dimension determines how strongly the foot must be pulled towards the body before it can be put out of the vehicle onto the road. In addition, the inner step of the side sill and the side roof entry are decisive for the exit comfort.

Sabbah (2010) has carried out systematic discomfort tests on a mock-up with variably adjustable sill height, A-pillar inclination, roof edge height, roof edge offset, steering wheel and seat position in order to achieve optimum values that cause the least discomfort. He has varied the range of values observed in the market today in the range from sedans to SUVs. The experiments were carried out with subjects of a body height between the fifth and 95th percentile. The results are summarized in ◘ Table 7.14.

◘ **Table 7.14** Vehicle parameters that cause the least discomfort when entering and exiting the vehicle (Sabbah 2010)

Designation of the measure	Brief description	Optimum
Seat height above roadway to seat height above vehicle floor	H5: H30	520: 260
A-pillar inclination - vertical dimension	h4	950
A-pillar inclination - height angle to the horizontal	h6	40°
Roof edge to seat height	H50 - H5	789
Roof edge offset	W14 - w16	230
Steering wheel position: Lower steering wheel rim to vehicle floor	h7	500
Steering wheel position: Lower steering wheel rim to pedal level	l2	500

Irrespective of the distance between the road surface and the vehicle floor, the lowest sill height of 20 mm is expected to cause the least discomfort. Despite the large difference between the lowest and highest roof edge height H50 between 1202 and 1601 mm (difference 459 mm), the difference between the roof edge height and the height H5 of the seat reference point above the roadway level relevant for boarding and alighting shows only a range from 763 mm to 815 mm (difference: 52 mm), so that the mean value of 789 mm can be regarded as a good compromise. The tests on roof edge offset show that a roof edge further inside is perceived as more pleasant when entering, but more discomfort when exiting. However, since the discomfort values stated by the test persons differ only slightly, the mean setting of 230 mm can be recommended. The height of the lower edge of the steering wheel rim above the vehicle floor has a decisive influence on access, especially for larger men. For women, the longitudinal alignment of the lower steering wheel rim had the greatest influence. Overall, it could be observed in the experiments that, contrary to the general expectation, getting in caused a higher discomfort than getting out. The more precise analysis of the observed body angles leads to the conclusion that this effect is due to the fact that the entrance is from a "randomly chosen" starting point and that this may have to be compensated by a stronger torsion of the upper body when diving under the roof frame. The start position is defined when you get off the vehicle, allowing you to plan the process more precisely. Overall, larger volunteers indicate higher discomfort values, which can be attributed to the necessarily lower body angles. It is also interesting to note that for the ideal entry and exit setting, the seat is located at an average of 40 mm behind and 10 mm under the individual driving setting. In addition, sufficient clearance between the lower B-pillar and the hip area of the individual seat adjustment is required for the entry and exit process. To put it simply, the door cut-out of the vehicle should be large enough so that the seat and backrest surfaces adjusted for the driver - viewed from the side - lie within the door cut-out. This, however, creates a conflict of interest with access to the second row of seats in a (shorter) four-door vehicle.

7.6.3.2 Evaluation of the Entry Hose

The French research institute INRETS (now IFSTAR) developed the database-driven model RPx in collaboration with Renault. The basis for this is the recording of a multitude of entry procedures into existing vehicle concepts by means of a motion tracking system using test persons from fifth to 95th percentile height. For the corresponding processes there are statements of the test persons about the perceived discomfort. All motion recordings, the associated discomfort statements and the associated dimensions of the body concepts used are stored in a database. For the evaluation of a new body concept, the most similar concept is searched for in this database and the corresponding evaluation is output. Dofour and Wang (2005) have derived a concept for determining dynamic discomfort from this. This is based on the idea of a neutral body movement, which is obtained for each body joint by means of an averaging procedure from the aforementioned database. A deviation of the observed movement from the neutral course beyond a given level is evaluated as discomfort (◘ Fig. 7.107). The areas of discomfort are integrated over the entire initial movement path and output as a measure of the discomfort experienced.

Such a neutral movement or the recorded movement process in a body concept that has been evaluated as good can be represented by a human model and fitted into an existing body concept (◘ Fig. 7.108). This makes it obvious how much space each design manikin requires for entry and exit.

From Sabbah's investigations (2010), the effect of different positions of the influencing parameters on discomfort listed in ▶ Sect. 7.6.3.1 can be derived. A similar procedure is also described by Bothe (2010) based on the study paper by Neuendorf (1996). On this basis, an evaluation tool was developed, which is demonstrated using the example of the construction and position of the sill in ◘ Fig. 7.109. The connection between entrance height and entrance width is visualized by the color-coded evaluation area. In the case of a high entry height, the sill must therefore be at a shorter distance from the seat than in the case of low vehicles.

7.6.3.3 Evaluation of the Movement Process

The aim of the work of Cherednichenko (2007), as well as the previous work of Rigel (2005), was to simulate the entire movement process of the entrance using mathematical methods, so that the geometric influencing factors directly control the movement process, thus providing the basis for an individual assessment. For this purpose, BMW developed the Mock-Up VEMO (Variable Entry-level Model), with the help of which a variable entry-level geometry could be generated. At the same time, the mock-up had to be set up in such a way that the cameras could capture motion with as unobstructed a view of the scene as possible. ◘ Figure 7.110 shows the structure of VEMO and its basic adjustment possibilities.

Cherednichenko limits his modelling project to the hatching strategy because of the frequencies observed at Rigel (2005; see ◘ Fig. 7.105). Motion tracking is used to record the movement of the test person. The initial process is divided into five functional phases. For each of these phases a "leading" body part can be identified, which is primarily responsible for the fulfilment of the movement function. On the basis of observation by motion tracking, this leading part of the body is characterized by the fact that its movement

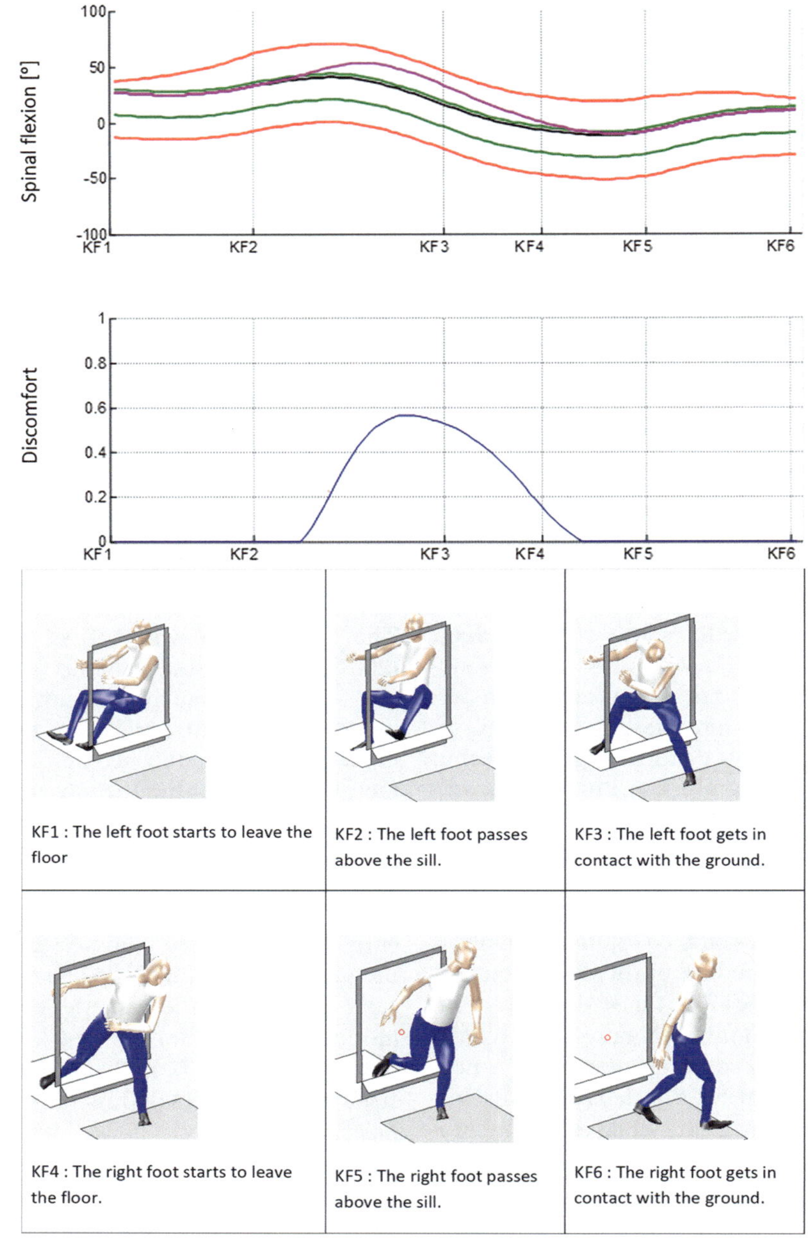

Fig. 7.107 Neutral movement (black line) and discomfort corridor. The green line characterizes the area of tolerable, low discomfort, the red line should not be crossed. (After Dufour & Wang, 2009)

in this situation takes place exactly in a plane plane (Fig. 7.111).

The neuronal coordination thus seems to be concerned with the direction and speed of this part of the body, the innervation of the remaining musculature takes place quasi autonomously on the basis of local muscle reflexes and under consideration of a stable posture with regard to the given equilibrium conditions. Using this method, the "leading body parts" shown in Table 7.15 and Fig. 7.112 can be defined for the individual function phases.

The plane in which the "leading body part" moves is determined by calculating a regression plane in the point cloud of a move-

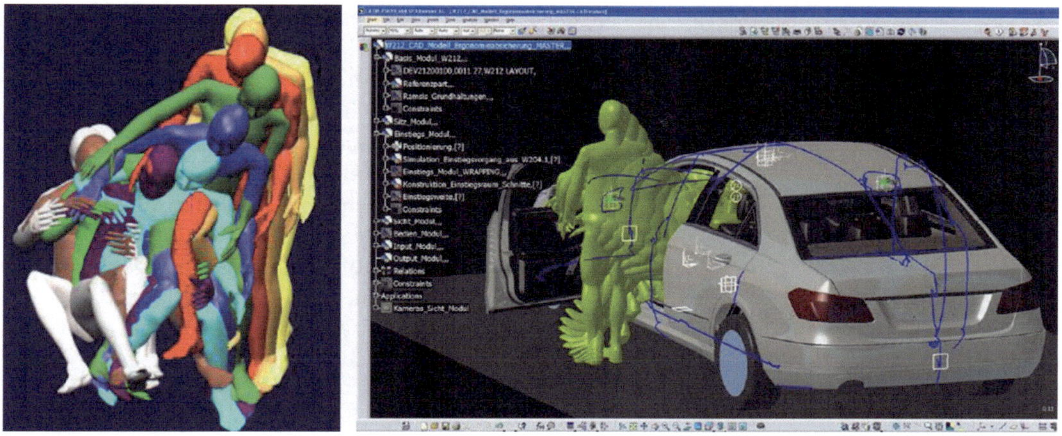

Fig. 7.108 Entry postures (movement hose) for a neutral movement. (From Sabbah 2010) and fitting into an existing vehicle. (From Bothe 2010)

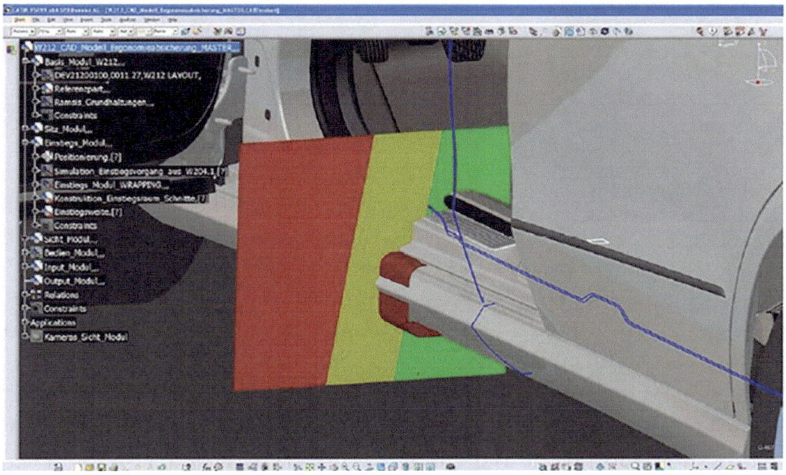

Fig. 7.109 Discomfort evaluation for entrance width at the sill. (Bothe 2010)

ment track. The deviation from this level does not exceed 10 mm. If a body part clearly moves out of this tolerance range, it can be assumed that it has finished its role as the leading body part. It turns out that approx. 80% of the total movement of such a body part takes place within this tolerance range. The remaining 20% at the beginning and end of the movement can be understood as a transition phase to the next leading part of the body.

In a test series, Cherednichenko adapted the VEMO to the individual anthropometric measurements of the test persons (simplified: a small car for small persons and a large car for large persons). It is noteworthy that under these conditions people of different anthropometry show the same behaviour.[47] This allows a general modeling of the initial movement. The first part of movement modeling is dedicated to the movement of the leading body parts. The modeling of the *Phase I* is

47 This is remarkable in that the ideal seat pressure distribution is also independent of the individual anthropometry. Obviously people show similar behaviour when the environment is adapted to their individual dimensions. The differently observed behaviour is essentially due to the fact that the same, given environment is different for each person.

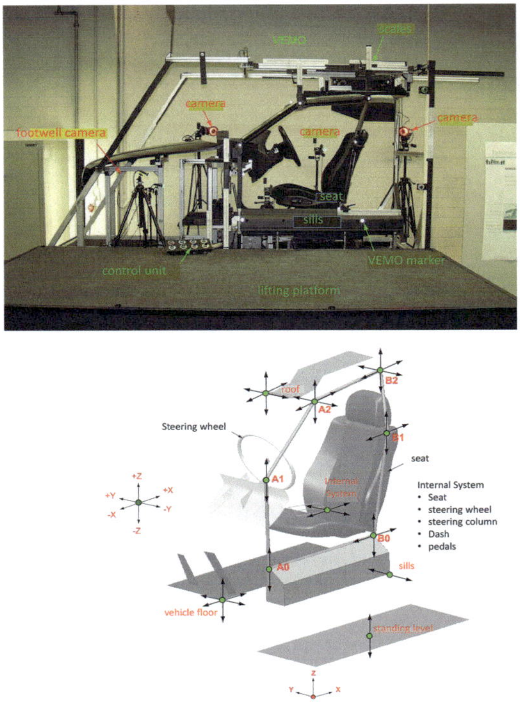

Fig. 7.110 Variable entry model (VEMO) for systematic research of entry behaviour. (Built at BMW, from Cherednichenko 2007)

reduced to the parameters that are relevant for the initial movement. These are the approach direction and the position of the left foot at the end of the approach phase.

In *Phase II*, "taking a stable supporting position", the plane of motion of the right foot depends both on the anthropometric measurements of the person and on the geometric entry configurations.The individual direction vectors of the motion plane are influenced differently by the parameters. It was possible to develop a mathematical procedure by which their alignment can be modelled via correlation analyses.

In preparation for a harmonious lowering of the pelvis to the seat in the subsequent Phase III, the position of the right footrest must be correct. This is done through visual support. From this it can be deduced (and is also observed in the experiment) that this process is severely impaired if the foot space cannot be accurately estimated due to visual impairment. The functional relationship between footrests and the sitting position can be regarded as an indication of the backward planning of the foot movement in the entry phase II, which is specific to the body dimension. The presented model allows the calculation of the plane of motion of the right foot and in this the plane of motion of the path in the form of a spline curve composed of two third degree polygons. This is assigned a velocity profile whose exact form depends on the geometric variables related to the anthropometric measures of the person (see Fig. 7.113).

The *Phase III* "taking up a sitting position" from the point of view of movement planning is the central goal of movement. The pelvis is the leading part of the body. The upper body follows this movement and, together with the upper extremities, has the task of maintaining balance and stabilising the sequence of movements through posture corrections. The unit vectors of the pelvic plane of motion show strong correlations with the body measurements of the test subjects, especially with the leading measurements body height, leg length and trunk length. This influence is taken into account in the motion model by the body dimension-centered vehicle parameters. The Body Mass Index (BMI) goes directly into the model, whereby obesity-related trajectory influences are taken into account. The trajectory of the basin in its plane can be completely described by a second order polygon. The modelling allows detailed statements on the influence of geometric variables on the necessary motion sequence. At this point, the advantage of functional modelling over database-driven modelling becomes obvious. Another striking feature is that the speed profile of the pelvis and the foot movements, which is characterized by a four-phase course, is almost identical. The organism thus seems to prefer a firm structure for leading body parts.

In the *Phase IV "transfer to frontal seating position."" the* functional goal of the movement is now the targeted shifting of the left foot to the footrest. The left foot is now the leading part of the body. However, the forces as well as the space conditions are usually not

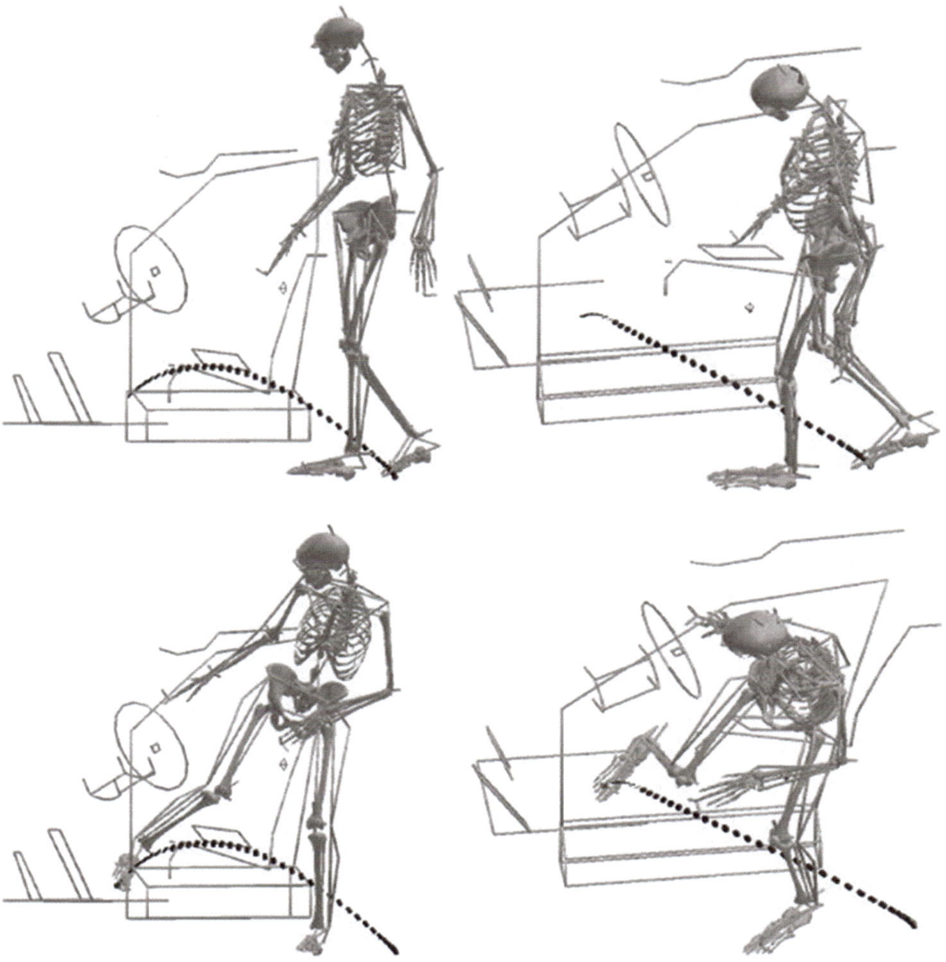

Fig. 7.111 Movement of the right foot in its phase as leading part of the body from different perspectives

Table 7.15 Phases of the boarding process and leading body parts (Cherednichenko 2007)

stage	Description of the	Leading body part
Phase I:	"Approach"	Left foot
Phase II	"Take a stable supporting position"	Right foot
Phase III:	"Take a sitting position"	Cymbals
Phase IV:	"Transfer to frontal seating position"	Left foot
Phase V:	"Take a driving posture"	Head

sufficient for an arbitrary movement of the leg. This is compensated by a correction movement in the seat, which serves to minimise the necessary force while making optimum use of the freedom of movement. The leadership function of the left foot must be divided into two sections, a necessity that generally always arises when a direct achievement of objectives is not possible. In detail, the movement path in the movement planes is simulated by spline functions. The modeling also includes the fact that a "comfort force" is obviously specified by the central nervous system for lifting the foot. If this force is exceeded (e.g. when the foot is raised very strongly), the movement is judged to be "uncomfortable".

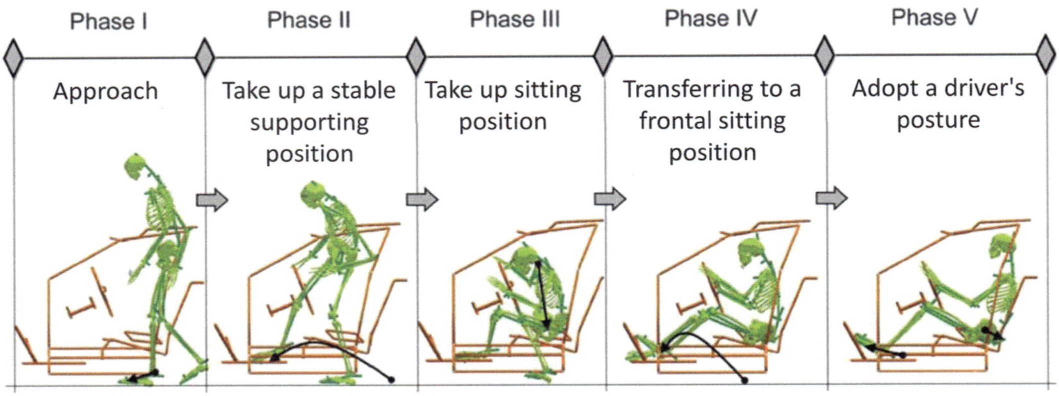

Fig. 7.112 Phases of an entry process. (Cherednichenko 2007)

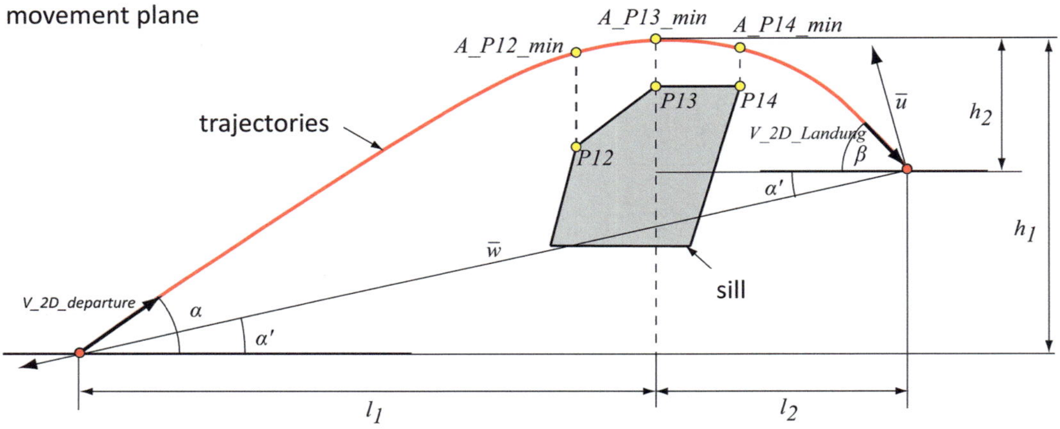

Fig. 7.113 Construction of the trajectory of the right foot above the sill. (For explanations see Cherednichenko 2007)

In the *Phase V* the boarding process is completed and the final driver posture is adopted. This refers to the driver-specific positioning of the feet, hands and upper body. Since good prognosis models exist for this driver posture and motion sequences in the driver posture have already been modelled with good success, this will not be discussed in more detail. Rather, the models mentioned can be used directly for modelling phase V.

The temporal phase synchronization shows that, on average, the entry process is completed within 4,6 s for all attempts. 15% of time takes up the approximation phase, the remaining phases each take up about the same amount of time, with the exception of the last phase which takes up the largest time share with 36.2%. It also clearly shows that the individual phases overlap at the seams, resulting in a smooth movement. According to the results of Sabbah (2010), a correct modelling of the time expenditure is also necessary because a close correlation between the time expenditure for the entry process and the perceived discomfort was observed.

While the parameters for the equations of motion of the leading body parts to achieve the goal are very narrowly defined, others can be varied within wide limits without significantly affecting the result. These parameters are called stability parameters. They are essentially these:

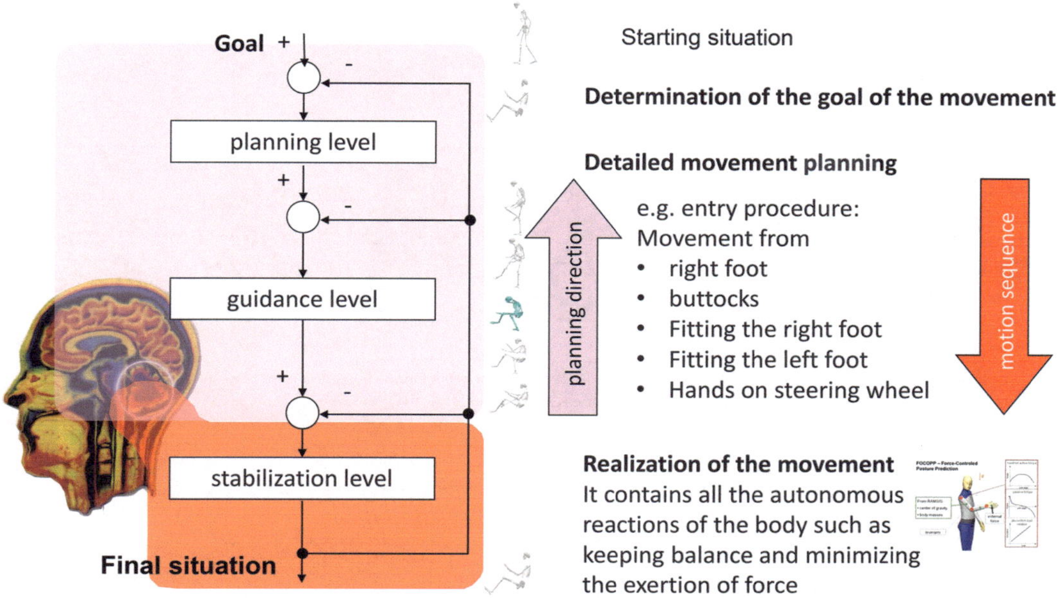

Fig. 7.114 Simulation of the entry process (slipper strategy) and mental analogy: Planning and guidance takes place in the cerebrum, stabilization in the cerebellum and basal ganglia. (see also ► Fig. 3.5)

- Support of movement coordination through head posture and visual movement control
- Balance stabilisation through hand movements and positioning
- Motion smoothing through torso and basin alignment

In all this it can be observed how complex motion sequences are additionally visually controlled.

The simulation of the total movement is built up backwards from the final state, taking into account the functional goals. Responsible for this are the leading body parts, which determine the movement of the entire body. Their movement is fully described by the statistical-mathematical model and estimated by means of the body dimension-centered vehicle dimensions. The whole body is regarded as a guided kinematic system. The structure of the motion simulation consists of the sub-steps:

1. Determination of the functional goal
2. Calculation of the support posture
3. Prediction of the movement of the leading body part
4. Prediction of the movement of the guided body parts
5. Stabilisation through stability variable.

The parallelism of this structure with the mental process is illustrated in Fig. 7.114.

The motion curve calculated in this way is visualized with the help of the RAMSIS human model. RAMSIS is connected to the trajectories of the leading body parts via so-called coupling points. These are the footrest points on the left and right foot and the hip centre, which is also required for calculating the sitting position in the seat and for the sitting posture (Fig. 7.115). The remaining posture is calculated using RAMSIS's own posture probability model or force posture model. The achievement of the movement goals is secured by support postures. They depict the states of movement at the boundaries of the phases of movement. The force-holding model mentioned above has so far been designed for static posture predictions. However, the initial movement is a dynamic process. By applying the d'Alembertian principle, the force-holding model can also be used for the simulation of dynamic processes.

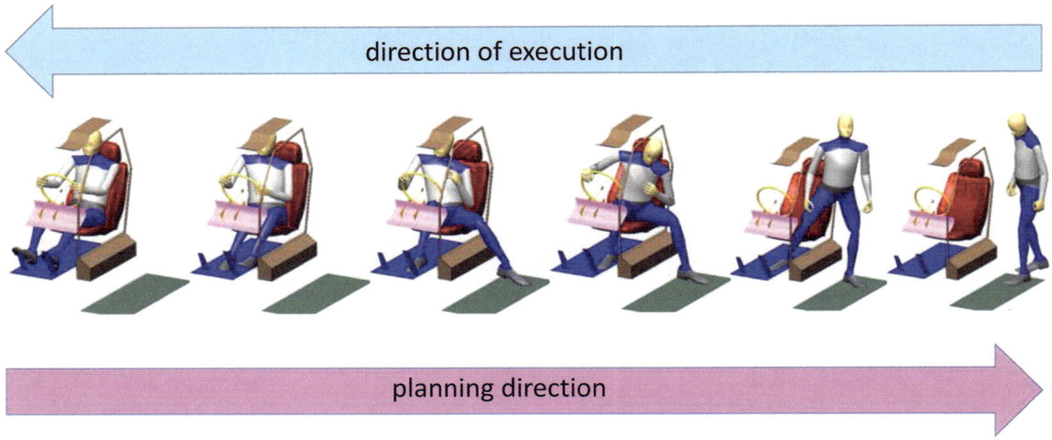

Fig. 7.115 Support postures for visualization of the initial process using RAMSIS

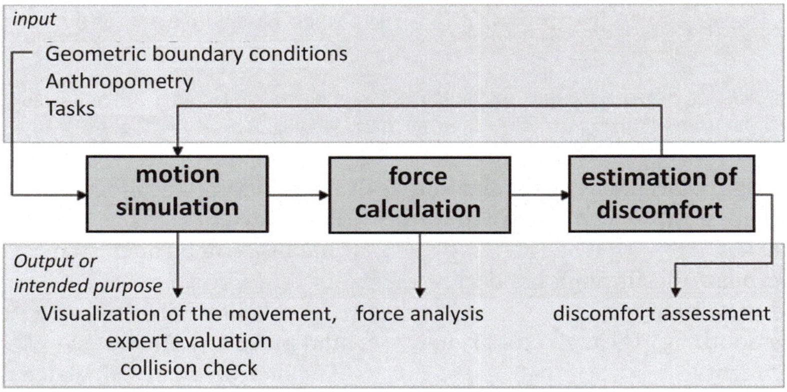

Fig. 7.116 Suggestion of discomfort prediction on the basis of movement simulation. (According to Cherednichenko 2007)

A measure of the discomfort of the calculated motion sequence can be found by calculating the forces and the degree of utilization of the respective maximum force (Sabbah 2010). According to the study by Zacher and Bubb (2004), the maximum discomfort found in any joint during such a movement process is decisive for the discomfort perceived overall.

On the basis of the procedure for movement simulation described here, Cherednichenko (2007) presents a general model for predicting discomfort during the entry procedure (Fig. 7.116). Then the input variables geometric boundary conditions, individual anthropometry and task (here entry according to the slipping strategy) determine the conditions for the motion simulation. This allows initial assessments by experts and collision checks due to the visualization in connection with the CAD model of the vehicle. With the aid of a multi-body system (either the RAMSIS force-holding model or e.g. ALASKA/Dynamicus, see also Sabbah 2010), the forces required for physical reasons to realize the simulated motion process can then be calculated using the moments of inertia of the human model used. The calculated forces in connection with the degree of utilization of the respective degrees of joint freedom are the basis for an estimation of the discomfort. Since the entire process depends on the geometric boundary conditions and the anthropometry of the design mannequin, the discomfort and the effect of changes on the discomfort can be estimated at an early stage of development.

7.6.4 Access to the Second and Third Rows of Seats

It can be assumed that the calculation and evaluation of access to the second row of seats for a four-door vehicle can, in principle, be carried out according to the same criteria as for access to the driver's seat. However, there are still no well-founded scientific studies available for this process. Essentially, the slipping strategy and the sagging strategy are suitable for this. For the former, it is important that there is sufficient space between the lower pivot point of the B-pillar and the rear seat, which can also be seen visually, so that the right foot is secure hold in the vehicle. For the application of the sagging strategy, which is more likely used by older people, the rear seat must not be positioned too far back between the wheel arches. However, the distance between the B-pillar and the rear bench must be large enough to allow the legs to swivel inwards unhindered. The upper roof frame may have a restrictive effect on access to the rear seat in particular, as a flattened roof at the rear appears attractive in today's design directions.

Access to the rear seat of a two-door vehicle has not yet been scientifically investigated. The size of the door opening and the distance between the front seat and the rear side of the body certainly play an important role here. Measures known as easy-entry increase this distance by allowing the front seat to be moved forwards either mechanically or by means of an electrical device.[48] In any case, the rear foot space must be large enough to allow a body turn. The roof must be high enough to provide the body with sufficient space for a short stay in a stooped posture. Here, too, the two entry strategies mentioned occur in a modified form. The following operations are described for entering the vehicle from the right. In the case of the fiddler strategy modification, the vehicle is first entered with the right foot. The entire body is pulled into the interior and then rotated to sit on the seat. In the case of a sagging strategy modification, the vehicle is first entered with the left foot. The foot is placed as far inside as possible on the vehicle floor. Now the person tries to pull the body inwards with the buttocks in front and push it directly on the seat.

Since today's vehicle concepts do not provide separate doors for the third row of seats, if any, access to the third row presents the same problems as access to the rear row of seats in a two-door vehicle. However, since no adjustable backrests are required for the second row of seats, it is possible to provide improved access by means of appropriate folding solutions.

While the placement of the inner door handles on a four-door vehicle is subject to similar criteria to those for the front seats, there is a specific problem for two-door vehicles and third row passengers. It must be ensured that the vehicle can be left without the assistance of a second person. To this end, it must be ensured that access to the door opener is enabled when the front seat slides forward. If this is not possible for the front seated person for design or ergonomic reasons, a special door opener must be provided for the rear row of seats.

7.7 Loading

Since the vehicle is used not only as a means of transport for people but also for goods, the useful volume of the rear area is of particular interest. Not only commercial vehicle users such as craftsmen or taxi and transport companies, but also the private user of the vehicle

48 Earlier solutions, in which the entire front seat could be folded forwards and thus opened up a larger access area, are no longer possible today due to the mandatory backrest adjustment. However, there are also parallelogram mechanisms in which the backrest and seat surface fold forward without requiring a shift in the seat rails!

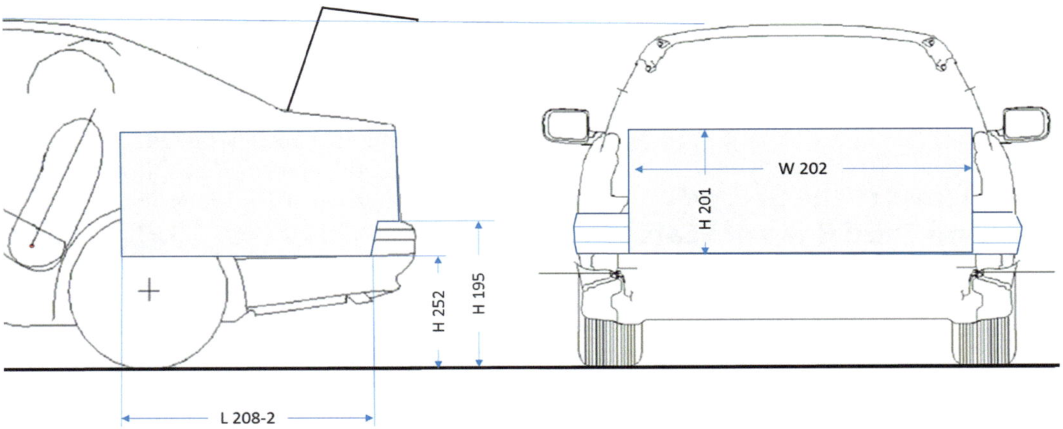

Fig. 7.117 The following variables have a decisive influence on the loading space

is dependent on a certain usable volume and the comfortable loadability, whether it is to store occasional purchases or larger objects such as beverage crates, moving boxes or for example a washing machine.

7.7.1 Geometry

According to Karwowski et al. (1993), in addition to the pure opening dimensions of the tailgate, the height, shape and position of the loading sill are of particular interest. Figure 7.117 shows the main parameters determining the loading space.

Depending on the overall vehicle concept, more or less attention is paid to the design of the rear area. Notchback saloons have a relatively low loading comfort because the boot lid covering the trunk is attached underneath the rear window and thus only permits a limited opening width. If the vehicle is equipped with a retractable soft top, the loading space is even further restricted by lowering the folding roof. Nevertheless, the loading edge should be as low as possible here as well. For reasons of body stiffness and to provide a mounting surface for the license plate, the area between the rear lights is often closed for design reasons. In hatchback limousines, where the tailgate is mounted above the rear window in the roof, the tailgate can be opened from the bumper over the entire available height, whereby a larger loading opening can be offered, both in height and in width. In station wagons, the tailgate is usually a little steeper and extends from the roof into the bumper, which means that the loading edge can be even lower.[49] Once the height and depth of the bumper has been overcome by the load, the tailgate seal and, if necessary, the projecting tailgate lock must be passed. The following step into the boot floor (H195 - H252) is higher or lower depending on the vehicle concept, but from an ergonomic point of view it should be as small as possible, ideally in one level with the edge of the bumper (H195 = H252). In order to resolve the contradiction between the largest possible loading volume and a level loading edge, the use of a movable loading floor is recommended[50] which can be adapted to suit the requirements of the load (Fig. 7.118). The loading sill (H195) from the stand to the highest point in the lower rear flange should not exceed a certain dimension. This measurement is given by a standardised loading situation in which a small person is loaded with a

49 The Fiat Tempra Station Wagon (1993–96) allowed the bumper to be folded down as a kind of "tailgate", which further reduced the loading sill. Citroën has the option of lowering the hydropneumatic chassis and thus also the loading edge by several centimetres in its estate cars (e.g. Citroën C5 Break (2004–08)).

50 For the Opel Insignia Sports Tourer (2013), an extendable loading floor is available as an option, which can also be used to easily bridge the depth of the bumper.

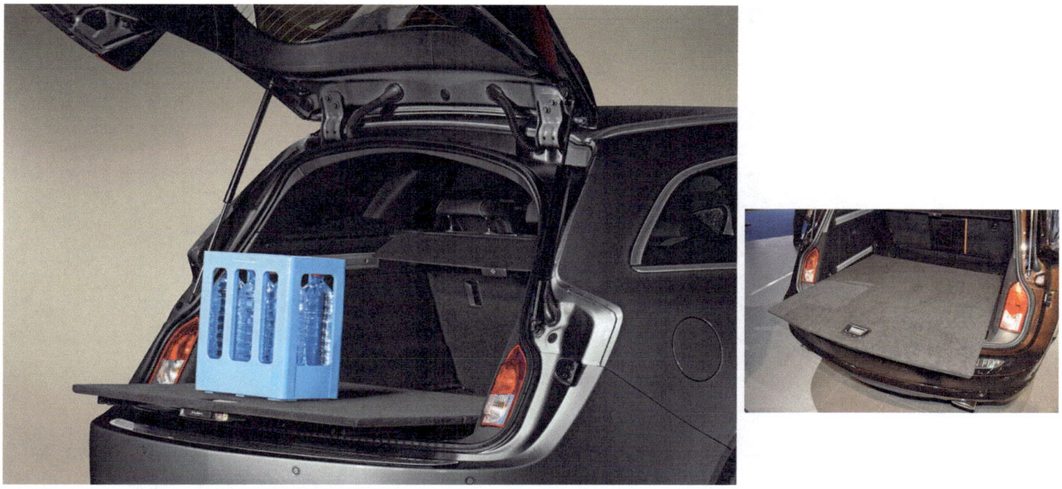

Fig. 7.118 Extendable loading floor of the Opel Insignia Sports Tourer (2013)

full beverages crate[51] approaches the tailgate opening and can place the crate - without lifting it substantially - in the rear compartment. A little woman[52] requires a loading sill of approx. 500 mm above the floor to place such a box, carried by hand, in the hold without lifting it. At 600 mm above the base level, 50% of men and 90% of women already have to lift the crate. At 700 mm all men and women must lift the crate so that the forearms are almost horizontal. With a crate weight from 10 to 20 kg, a considerable lifting capacity is therefore possible here[53] to perform. In principle, the height of the loading sill is chosen as low as possible based on the needs of the target group in the competitive vehicle environment.

Müller (2010) reports on systematic loading and unloading tests on a full rear-end vehicle in the mid-range segment. The motion sequences were recorded with a video camera in side and front view. The evaluation of these motion sequences shows no significant difference between loading and unloading. The dynamic space requirement is almost identical for both processes. Two different movement strategies were observed during the loading processes carried out with a 20 kg heavy hard-shell case. Each of these strategies can be divided into the phases "Luggage lifting" - "Orienting the body relative to the luggage" - "Parking the suitcase" and "Putting down and positioning the suitcase" (see Fig. 7.119).

- The so-called "one hand" strategy was observed in all male subjects. The suitcase is gripped with the right hand in a bent upper body position. It is then lifted from the ground by raising the upper body. The test person's body is oriented relative to the luggage volume and the left hand is placed on the side of the suitcase for guidance. After placing the suitcase on the loading area, it is tilted onto the side surface and finally pushed into the trunk with both hands.
- The "two-hand" strategy was observed in the female subject. The case is first gripped with the right hand and also lifted by raising the upper body. It is then gripped by the handle with her left hand. Here, too, the body is aligned relative to the luggage volume. It is placed on the loading sill with

51 A commercial beverage crate "Deutscher Brunnen" with twelve 0.7 litre glass bottles has the dimensions: length 353 mm, width 275 mm, height 345 mm, weighs empty approx. 6 kg and filled approx. 18 kg.

52 The little woman is assumed to be the fifth percentile of the body height of the German population.

53 According to a recommendation of the Federal Office for Occupational Safety and Health (baua), men should not lift more than 20 kg over a distance of 5 m (women 10 kg). In addition, the load evaluation doubles when the load is evaluated in terms of work technology.

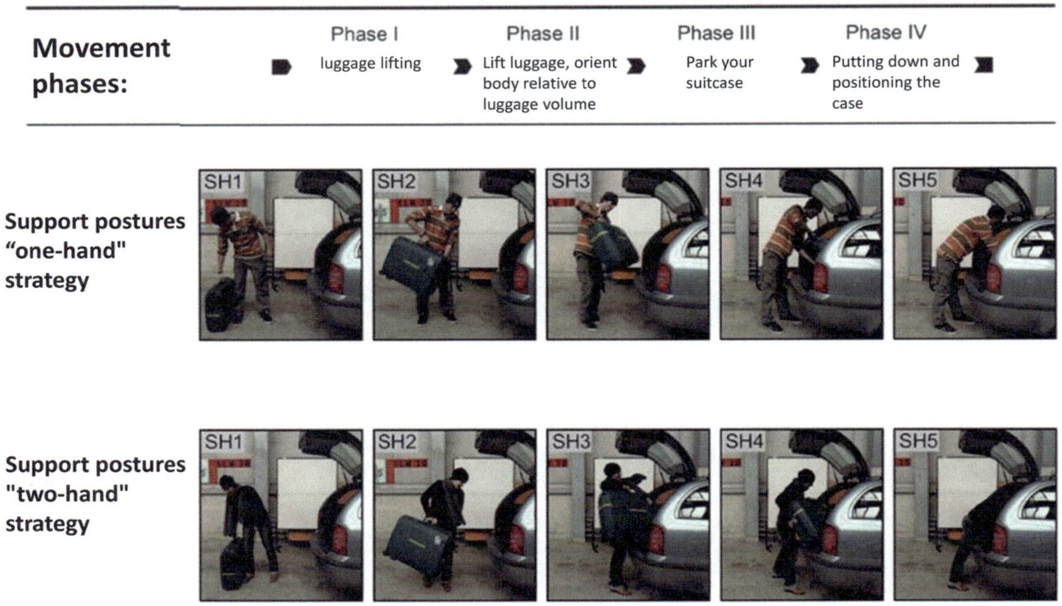

Fig. 7.119 Movement process when loading the boot volume with a suitcase using the "one-hand" strategy and the "two-hand" strategy. (From Müller 2010)

the support of the thigh and with the arms angled. It is then placed on the loading area with the right hand positioning the right edge of the suitcase.

With knowledge of these strategies described in detail and using the force posture model in RAMSIS, it is possible to model the movement sequences mentioned with the various design manikins (it is sufficient here to model the large man MLMM and the small woman FSMM). This allows the necessary movement volumes and, in particular, the minimum opening of the baggage compartment to be identified.

7.7.2 Operability

From an ergonomic point of view, the operability of the luggage compartment opening plays an important role. The height of the opening handle can be found as a compromise between the position of the free hanging hand of the small woman and the big man. It has to be considered that in many practical applications the baggage compartment has to be opened if heavy pieces of luggage are in the hand at the same time and cannot be stored for opening. A gesture control system has emerged as the solution to this problem, in which a corresponding movement of the left or right foot (observed, for example, by the rear view camera) results in automatic opening when the user carries the vehicle key with him. To handle the open tailgate, it must be folded upwards so that even the tall man does not bump into it when leaning forward. This condition can also be checked or interpreted with the aid of the corresponding design manikin. This open tailgate, however, is further restricted by the little woman: it must be ensured that she can reach the handle with her outstretched hand to close it. A compromise between these two extremes, to program different opening angles individually, is possible by using the relevant manikins. Another solution is to support opening and closing by electric motor. A problem then is the location of the switch that initiates closing. It must be installed in such a way that the user does not injure himself during the closing process. One possibility is to open and close the door remotely by actuating the key. However, even under these conditions, manual opening and closing must be possible. The simplest way to resolve this conflict is to attach a textile loop so that even small people can reach the tailgate to close it.

The requirements for access to the boot are not exclusively determined by the anthropometry of the driver. Many vehicles are used daily to transport children and young people, who often get off during a short stop and take their school bags or sports bags out of the vehicle themselves. If the driver has to leave the vehicle to close the tailgate, this can cause serious discomfort and, if necessary, dangerous situations for the user.

7.8 Consideration of Specific User Groups

7.8.1 Older Vehicle Users

The change in the so-called age pyramid and the associated increase in the number of older people and their need to fulfil their desire for mobility, which can only be satisfactorily fulfilled by the car, is a reason for the vehicle industry to take greater care of this group of people, especially as this group of people is also considered to be sufficiently solvent to buy new vehicles. The average age of new car buyers today is 50.6 years, with 29% older than 60 years (Dudenhöffer 2008). In order to concretize constructive specifications for this clientele, it is necessary to have more precise knowledge of the age-specific changes. However, a uniform description of the ageing process encounters the difficulty that the interindividual variance in the individual stages of life is greater than the age-related variance in performance (Schmidt and Lang 2007). Notwithstanding this, the World Health Organization distinguishes the following age groups:

- Adolescent or adolescent adult (15–30 years)
- Age of maturity (31–45 years)
- Conversion or middle age (46–60 years)
- Stage of life of the elderly person (61–75 years)
- Stage of life of the old person (76–90 years)
- Life stage of the very old person (more than 90 years)
- Durable (more than 100 years)

However, since a description of the ageing process based purely on the calendar age is unrealistic, Weineck (2004) differentiates between the two following ageing processes:

- Physiological aging that refers to the synchronous degenerative alteration of all organs and tissues, which is understood as "normal" aging.
- Pathological aging characterized by a strong insufficiency of an organ or system, accompanied by severe physical and mental loss.

The first category of ageing, which is of particular interest here, includes all age-specific changes, which are briefly described below. However, the definition of the "healthy elderly" also takes into account the most common pathological limitations, especially high blood pressure and osteoarthritis, which can lead to sudden and unexpected driver incapacity (heart attack, stroke) and local extreme movement restrictions.

Degenerative changes include, above all, the loss of sensory performance, in particular the reduction of the *accommodative capacity of the eye* (see ◘ Fig. 7.55) combined with an increase in reaction time. According to Hager (2009), of the over 75-year olds (multiple answers possible) have

- 70% cataract: progressive opacity of the eye lens, associated with loss of visual acuity, increased sensitivity to glare, poorer contrast perception and delayed adaptation to light and dark,
- 5% glaucoma: Too much intraocular pressure in the eye, causing the nerve fibres to be injured and partly die off. Result: narrowing of the visual field, in extreme cases blindness of the affected eye,
- 30% age-related macular degeneration (AMD): Cells on the macula (yellow spot, place of sharpest vision) die. Result: the sight in the middle of the field of vision decreases strongly.

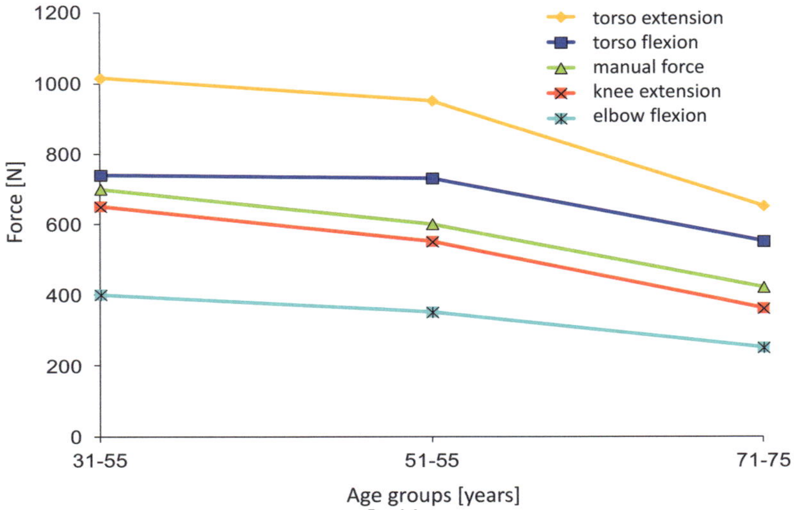

Fig. 7.120 Maximal isometric muscle strength as a function of age. (Viitasalo et al. 1985)

Also the *hearing* is subject to age-related changes. In particular, the detectable volume level range is reduced, especially the upper limit frequency decreases, the signal selection capability is reduced and the acoustic location is reduced. For the *haptic sensation* it can also be assumed that the responsible receptors are subject to age-related changes. For example, older people suffer from a loss of sensitivity and, for example, are more difficult to distinguish pressure differences (summarily quoted from Brenner 2013).

Furthermore *memory performance or intelligence* age-specific changes. A distinction is made between the so-called fluid intelligence, which shows itself in the speed of reaction and the ability to combine, and the crystalline intelligence, which represents quasi acquired knowledge. While the former already experiences a significant reduction from the age of 30, the latter remains virtually unchanged for a long time. In certain situations, "expert knowledge" acquired over many years can often even compensate the effect of reduced fluid intelligence performance. It is assumed that a noticeable loss of cognitive abilities only occurs on average from the age of 75. Many deficits in the area of changes in the sensory organs and cognitive abilities can be partially compensated by appropriate assistance systems. The use of assistance systems is therefore recommended by many experts, especially for older drivers (in this regard, see ► Chap. 9).

The reduction of physical strength and mobility is of particular interest for the age-appropriate design of vehicle packaging. Brenner (2013) compiles here some examples from the literature (Figs. 7.120, 7.121, 7.122 and 7.123). An overview of literature data on age-dependent changes in forces can be found in D'Souza (2014).

As the previous chapters have shown, human models play a decisive role in the ergonomic design of vehicle packaging. Even if these statements make reference to forces and ranges of motion, they do not yet include statements on age-dependent changes. Amereller (2014) systematically investigated age-dependent joint mobility in a population of over 300 subjects with the aim of establishing these results in the RAMSIS human model. When comparing the mobility of men and women, almost all joints - with the exception of the hip rotation - show a significantly higher mobility of women. The negative relationship between BMI and mobility claimed in the literature could also be partially demonstrated, although it should be noted that the test subjects with a BMI of 22.6 +/– 3.0 and therefore the expected effect could not be as large (see also ► Sect. 4.2.4.2).

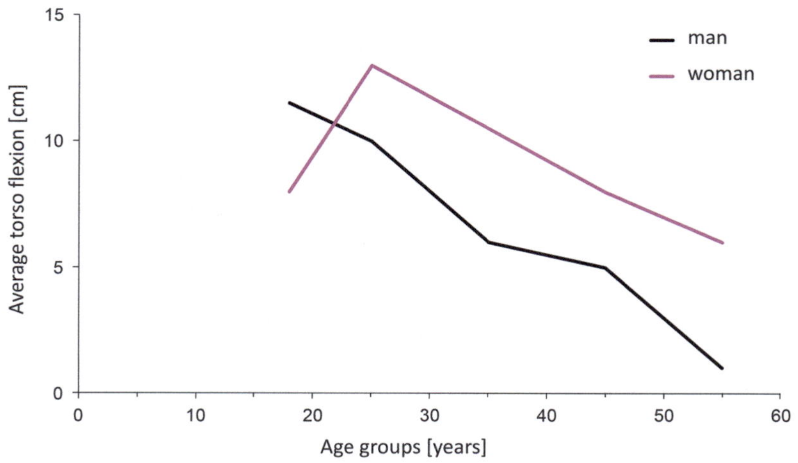

Fig. 7.121 Average torso flexion for determination of spinal column mobility. (Richter 1974)

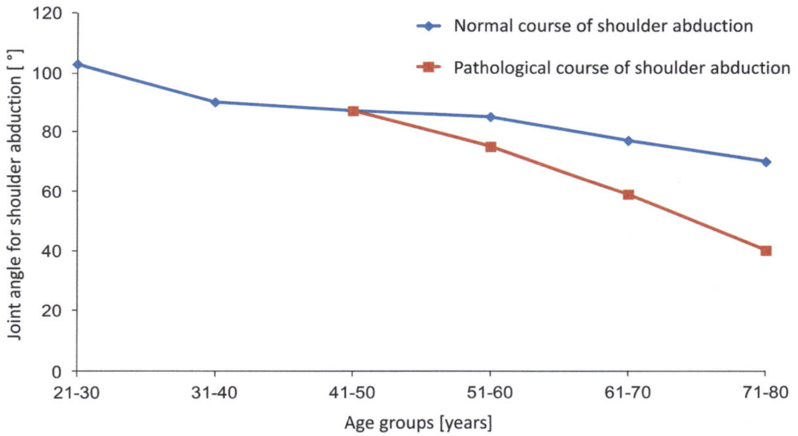

Fig. 7.122 Shoulder mobility in old age. (Clarke et al. 1975)

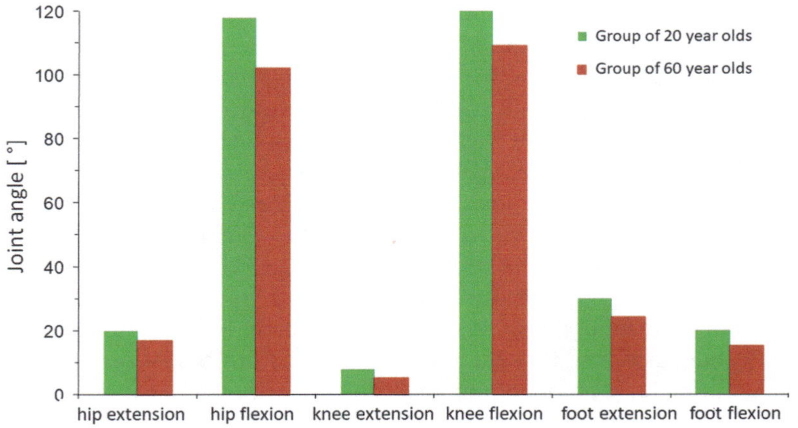

Fig. 7.123 Reduction of hip, knee and foot mobility in old age. (Lark et al. 2004)

Due to the complex relationships with respect to body forces, D'Souza (2014) does not refer directly to tabular force data, but deals more closely with the scaling approaches of force modelling in Anybody (see also ► Sect. 5.2.2.3, and ► Chap. 4), whereby the original scaling algorithms in Anybody lack the influence of age and gender as well as the addressing of different functional muscle groups. She examines the elbow joint with the age groups 50 to 59, 60 to 69 and 70 to 79 as representatives for the upper body parts and the knee joint for the lower body parts. In accordance with the results described in the literature, women show only about 50% of the corresponding forces of men. The maximum moments at the elbow joint as well as at the knee joint decrease with increasing age, whereby the decrease is stronger in men than in women. However, there are considerable individual variations with regard to these values. A factor analysis shows gender and body element length (50%) as the main influencing factors for both the forces in the elbow and knee joints, body mass (16%) in second place and age (12%) in third place. Equations were developed which allow a prediction of elbow moment and knee joint moment respectively and which can be implemented in Anybody.

7

In an IST analysis, Brenner (2013) has identified weak points that are particularly noticeable for older people but are also criticised by younger people. These are particularly high or wide door sills for getting in and out, the seat height, which should be optimally adapted to the individual anthropometry, the seat clamp height, the position of the A-pillar and the roof edge height as well as the cladding of the steering column. As far as accessibility is concerned, the belt in particular is criticised for being accessible only by turning the body and for its fixed lock not being freestanding. If the force required to operate a vehicle element is too great, this is displeasing to all user groups, regardless of age. This is especially true for opening, less for closing the doors, but also for other operating elements (e.g. ignition lock or operation of the handwheel for seat back adjustment). In terms of vision, particularly negative ratings are given for the lack of "outward vision", with older people additionally criticising necessary body movements. Other aspects relate to the size and position of switches and pictograms. In the "seats" area, it is noticeable that older people generally want to sit higher than younger people. Another point of criticism is the armrest, which is sometimes perceived as too far away from the body.

On the basis of these criticisms, Brenner has made modifications to seat boxes that were identical to the vehicle models used in the IST analysis to examine whether ergonomic improvements are perceived and whether improvements that satisfy older vehicle users may be rejected by younger users. As in the IST analysis, the three categories 30, 50 and 70 year olds were formed for the experiments. In the "entry and exit"cluster, all improvements in each age group lead to a more favourable assessment. It is noticeable that younger users are often already satisfied with a moderate improvement, whereby an increase that still leads to an improvement in the judgement for older users no longer has any effect. Brenner's work contains detailed information on the position of the radio climate control panel, the cup holder, the armrest (which, interestingly enough, is rather rejected by the older ones) and the release of control elements (e.g. seat adjustment, belt buckle). In general, it is noticeable that older users are usually more critical and make demands that are secondarily perceived as improvements by younger users.

From the results obtained, direct information on design improvements could also be obtained. For each question, it is necessary to **one** descriptive measure that can be assigned to the characteristic value of the valuation. So a descriptive measure for getting off is the distance w5 between the centre of the vehicle seat and the outer edge of the door sill (◘ Fig. 7.124). On the other hand, for the categories of sports cars and mid-range cars there are also clear age-specific dependencies with regard to the evaluation (◘ Fig. 7.125). The same applies to the evaluation of the seat height above the roadway H5, whereby an increase in the vehicle seat is initially evaluated positively, but this is reversed again with a further increase. In any case, the parallelism of the curves is interesting for the three age groups (◘ Fig. 7.126).

The example of the release "control element seat adjustment"shows that expectations regarding the respective vehicle segment must also be included in the evaluation of the dimensions. More space is also expected in a larger vehicle (good rating for sports cars already at 50 mm, for SUVs only at 65 mm). As the examples location/accessibility door handle inside (◘ Fig. 7.127) and location/accessibility boot loading height (◘ Fig. 7.128) show, there are also connections that are completely independent of the age of the test persons. In both cases, smaller distances independent of age and vehicle class lead to better evaluation.

From the investigations at Brenner (2013) it can be concluded: The average ratings of older users are either equal to or worse than those of younger users, but never better, and in many cases there is no difference in rating between younger and older users. From the investigation the following main fields of action can be defined, in which a special consideration of older users is necessary: entry and exit, effort, position and accessibility, visibility conditions and form of the control

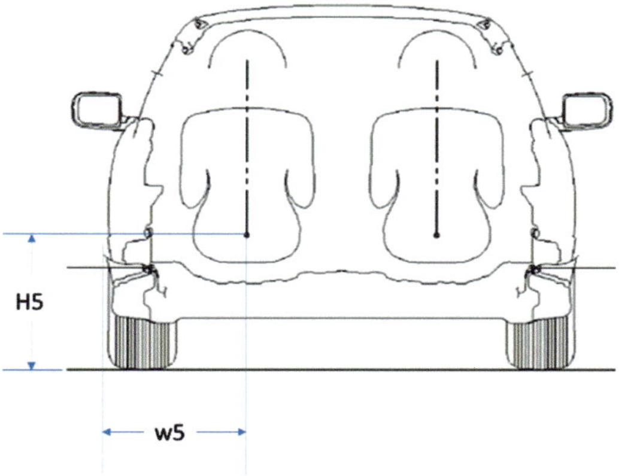

◘ **Fig. 7.124** Dimensions relevant for entry and exit (after Brenner 2013), H5 = height of SgRP above the carriageway, w5 = distance between centre of seat and outer edge of door sill

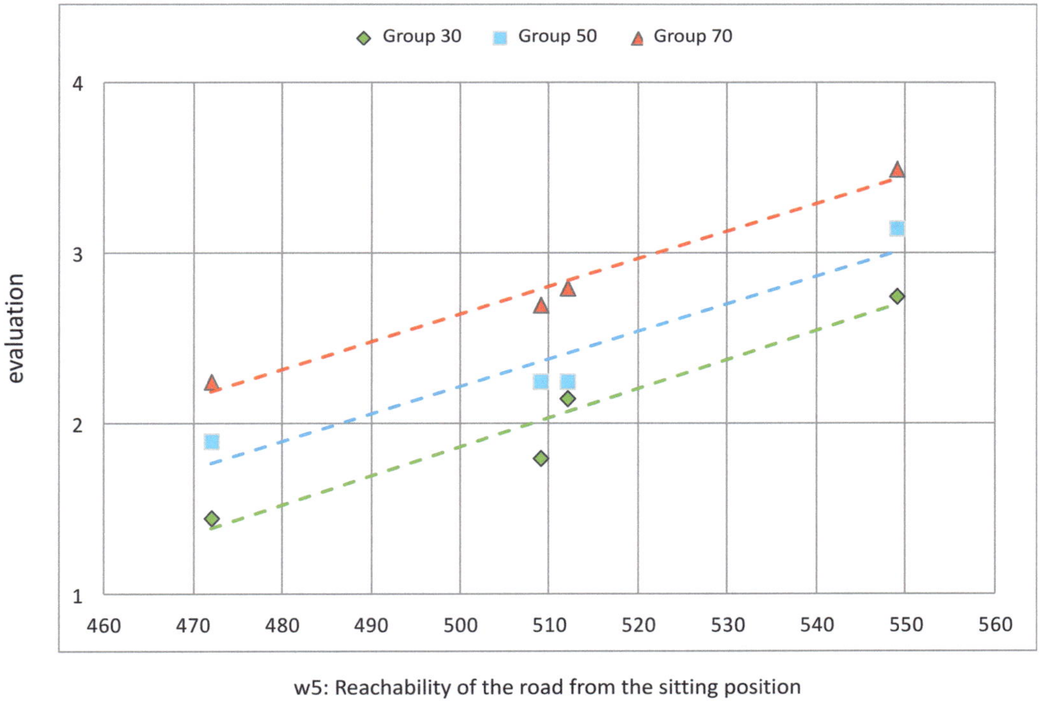

◘ **Fig. 7.125** Evaluation (1 = "very good"; 5 = "unsatisfactory") of the measure w5 "accessibility of the road from the sitting position" for sports cars and mid-range vehicles by three age groups. (From Brenner 2013)

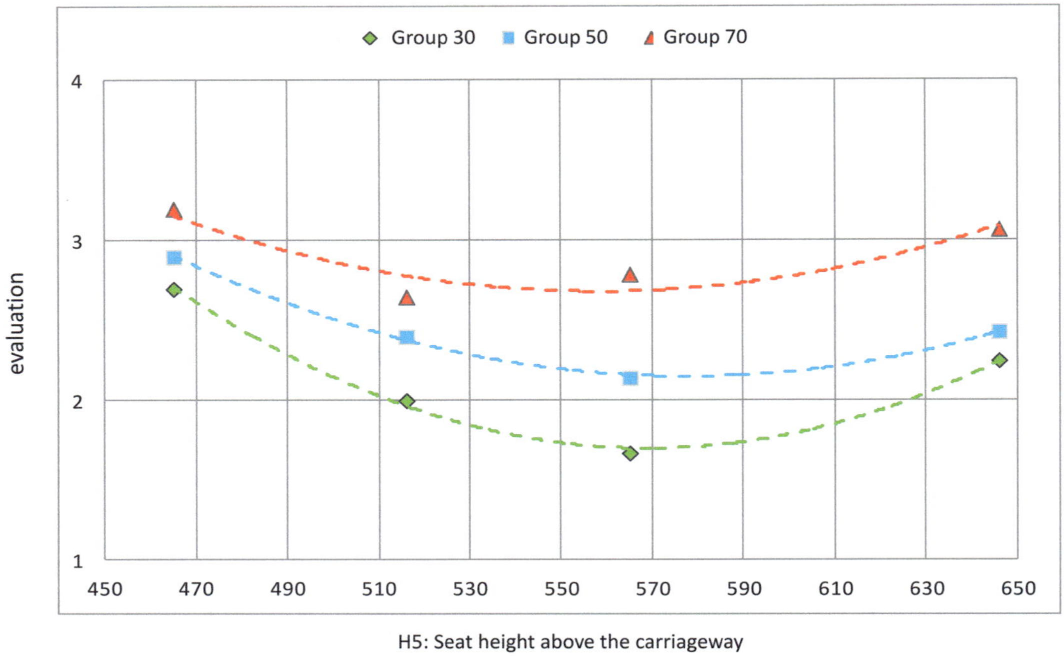

Fig. 7.126 Rating (1 = "very good"; 5 = "unsatisfactory") of the measure H5 "seat height above the road surface" for sports cars and mid-range vehicles by three age groups. (From Brenner 2013)

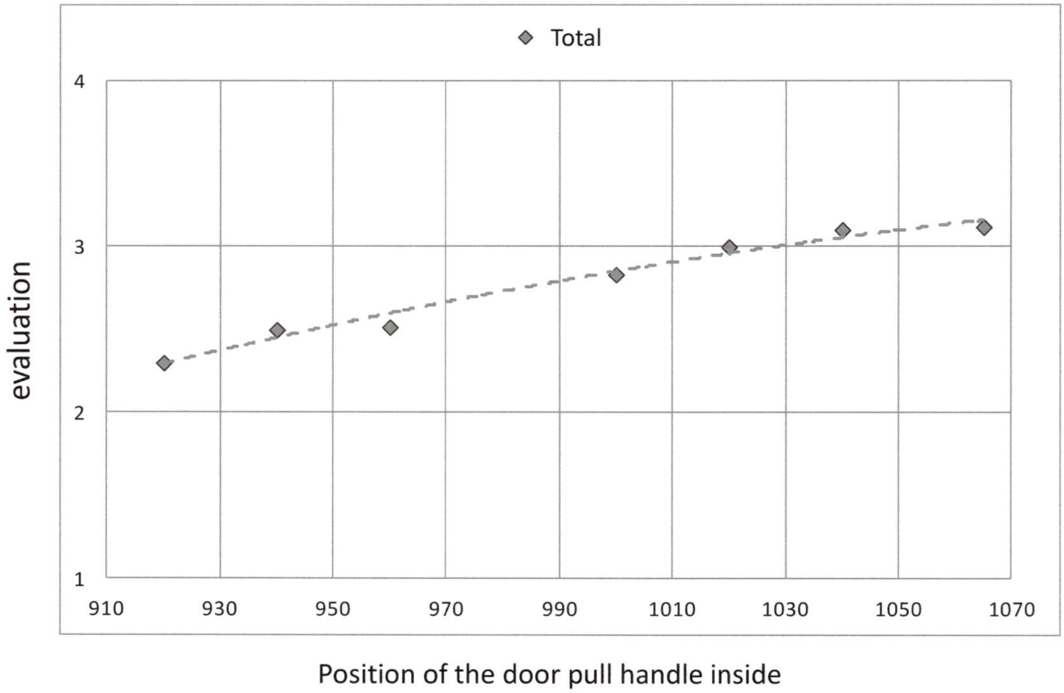

Fig. 7.127 Evaluation of the dimension "accessibility door handle" defined as radial distance of the door handle from the individually adjusted seat centre. (From Brenner 2013)

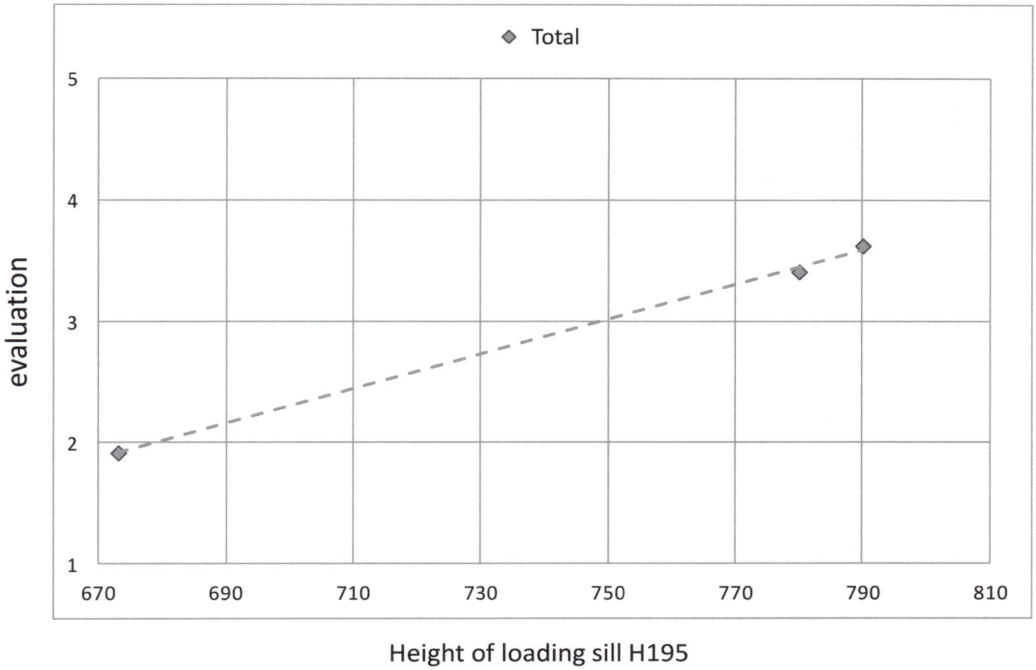

Fig. 7.128 Evaluation of the dimension H195 "loading sill". (From Brenner 2013)

elements. Overall, the study confirms the concept or the demand of "design for all", i.e. not to provide for differences for different age groups, but always to orient oneself to the highest demands.

Many manufacturers are making an effort to achieve senior-friendly design according to the motto "design for all". In addition, however, entry aids were also developed, which Mercedes-Benz offers in a special equipment program in the form of swivel and/or swivel seats in various variants (Fig. 7.129).

Fig. 7.129 Swing seat "Swing-Up". (Mercedes-Benz 2010)

The fact that there is officially no "senior vehicle"on the market is due to the fact that the image of such a vehicle is not sales-promoting and is rejected by those concerned as stigmatising. Even with existing restrictions, disabilities and ailments, vehicle users do not want this to be recognisable on the vehicle. The promising concept of a small car with small, manageable dimensions, a high, clearly arranged seating position and large, smooth-running sliding doors for access to the driver's seat only met with a limited market response. In 2005, the PSA Group had already launched the Peugeot 1007[54] (Fig. 7.130). launched such a vehicle on the market and consistently implemented the needs of families and senior citizens in the urban environment

54 Peugeot 1007 (2005): length 3731 mm, width 1686 mm, height 1620 mm, wheelbase 2315 mm.

Fig. 7.130 Peugeot 1007 (2005)

Fig. 7.131 Opel Mokka (2013)

(Fig. 7.130). Between 2005 and 2009, however, only a total of 120,000 units of the courageous concept were built, which was due more to the lack of acceptance by potential customers than to the technical implementation or quality. More successful in the segment of generational vehicles are small off-road vehicles which, as "small SUVs", offer equally manageable dimensions, good entry and exit and the appearance of a full-fledged automobile with fashionable accents. The Opel Mokka launched in 2013[55] (Fig. 7.131). has already in its first year achieved a similar production number as the Peugeot 1007 in five years and meets the almost identical target group families and "Empty Nesters".[56] These two examples illustrate very clearly that technically sensible concepts with ergonomic attributes require appropriate "packaging"in order to be perceived as added value and not as proof of purely rational necessities.

7.8.2 Children

An important usage scenario for the automobile is its use as a family vehicle. Here, solutions specially adapted to the special needs of the youngest occupants are a decisive market advantage. Additional child seats are available on the market in large numbers and thanks to the ISOFIX® fastening and Top Tether anchoring, they can also be used universally in a wide variety of vehicles. This satisfies the anthropometric requirements to the extent that the seat contour is adapted to the smaller body dimensions and the restraint system can also be effective for children and cause little injury. However, the ergonomic requirements for children go beyond the pure safety aspect. Already the entrance into the second row of seats confronts the independent child with body-related challenges. Even a schoolchild who is independent enough to open the door and get on the bench occasionally encounters technical solutions that adults can cope with but children have problems with. The design of the external door handle should be such that the child is able to unlock the door from the outside, both at the height of the arrangement and in the direction of operation, and in particular to apply the force to open it. Overcoming the sill requires a low entry height and a narrow wheel arch contour, which also requires the above mentioned "design for all". Many human models also provide child models that can be used to address the issues raised.

Children often find long car journeys boring. The interest in observing the landscape or reaching a certain destination is not particularly pronounced with them. Nevertheless, the unrestricted view through the side window to the outside is a requirement which is not fulfilled by many vehicles due to a modern parapet line rising to the rear for children.

55 Opel Mokka (2013): length 4278 mm, width 1774 mm, height 1646 mm, wheelbase 2555 mm.

56 Empty Nesters: As children grow up, their life situation changes, as do their parents' automotive needs. Now only two seats and little loading space are necessary, while vans and station wagons are practically necessary for young families.

Possibilities must be provided for compensating for the state of boredom. Therefore, electrical supply in the second and possibly third row of seats is important for the operation of mobile entertainment devices. A possibility to sleep must also be provided. In conjunction with passive safety, this is a problem that has not been solved satisfactorily in practice.

Frequently, small children in particular experience disturbances of well-being during the journey. This is caused by chassis vibrations transmitted to the child's body via the seat. If these oscillations correspond approximately to the natural frequency of the child's stomach (approx. 10 Hz), irritations of the peripheral nervous system (gastrointestinal tract) can occur in the form of nausea, sometimes even vomiting. However, vibrations are also transmitted from the chassis via the seat which respond to the natural frequency of the eyes (approx. 20 Hz) or the brain (approx. 18 Hz) and lead to irritation in the neuromuscular system in the absence of sufficient visibility of the vehicle environment, which can also cause nausea. For the adults in the front seats, the vibration excitations are far less harmless because of their different anthropometry and also because of the greater distance to the rear axle. They are also not always aware that sports suspensions and low-profile tyres additionally impair driving comfort for children on the rear seat. As in the case of sensitive adults, it is particularly important for the state of well-being that the perceived acceleration and the visually observed movement correspond. The accelerations, in particular the acceleration changes (jerk), which are unexpected and cannot be understood, can cause nausea. A visual control of the environment has a stabilizing effect as long as the eye can follow the moving objects. The necessary eye contact with the surroundings can be limited by the vehicle body (A, B, C and D pillars but also the parapet height), the seats with the headrests, but also blinds or tinted films on the side windows. A conscious distancing of the gaze from the outward facing view of the surroundings to the inner close-up area can therefore already lead to the listed disturbances of well-being. Reading in the passenger seats is a frequent cause of nausea, and playing with handheld consoles or working with portable media players, video devices or smartphones can also lead to these effects by looking far away from the surroundings. In this respect, it is also a question of individual sensitivity and educational intervention to what extent the above-mentioned means can actually be used to combat boredom.

7.9 Craftsmanship

The car is today the most expensive consumer good[57] in a private household. This statement applies both to the purchase of a new car and a used car. Due to the high investment costs, the buyer wants to see what he has spent his money on. He wants to have the impression that it is made by a good craftsman who understands his art. Bhise (2012) argues that craftmanship is a relatively new, increasingly important field of ergonomic design. The customer wants to get the impression that his vehicle is made by an excellent craftsman who has put all his skill into polishing the perceptible characteristics of the product in terms of appearance, feel, noise and smell. It thus addresses the so-called product quality. If one follows the definition of quality according to ISO 8402, which is described as "the totality of characteristics of a unit of consideration (here the product), the ability to satisfy defined and self-evident requirements", the problem lies in the definition of "self-evident" requirements, because they characterize also unspoken expectations of the customer. The quality impression mentioned today in all test reports relates to this aspect. The totality of this impression cannot be described by technically measurable quantities alone, such as uniform gap dimensions, the absence of paint defects and scratches, the deburring of plastic parts etc. Based on interviews with numerous vehi-

57 Due to the high loss in value to which a vehicle is exposed during its life cycle, it generally lacks the property of an investment good, unlike the purchase of an apartment or a house.

cle owners and discussions with engineers and designers from various vehicle manufacturers and suppliers, Bhise (2012) has compiled the following list of features and requirements that can describe the concept of craftsmanship in more detail:

Optical Quality

1. Perfect fit of different components in the visible area, characterized by
   - smooth lines, small gaps and avoidance of misalignments,
   - even gaps at all articulated joints, invisible axles
   - flush surface adjustments, parallel, smoothly rounded (deburred) edges,
   - low surface variability/unevenness (no distortion or distortion), etc.
2. Optical harmony (similar appearance and feel of adjacent components with similar materials in terms of colour and brightness, texture/granularity, shine/reflection properties, finish, etc.)
3. High-quality surfaces (e.g. no rust, fading paint, cracks, flaking, scratches).
4. No visible connections (e.g. no visible screws, clips, cables, etc.). Product surfaces should look clean and uninterrupted due to invisible or non-protruding joints. In this way the designer should convey the feeling for a "well made" product. On the other hand, even a few visible screws can give the impression of precise, good craftsmanship over the resulting "machine look".
5. Avoid annoying visual irritations (e.g.: garishly protruding light sources, reflection of brighter surfaces into the glazing or reflecting surfaces, waviness, distortions of mirrored objects)

Tactile Quality

1. Surfaces in the interior that are frequently touched by the user (e.g.: knobs, handles, seats, control panels, door panels, armrests, consoles, etc.) should give a pleasant feeling regarding the touch characteristics, which can be described and scaled by adjective pairs like soft/hard, smooth/rough, structured/non-structured, smooth/sticky, etc.)
2. A pleasant feeling of use of switches, e.g.: Feedback felt during switch movements, suppression of vibrations, no joint play, "crunchy" detent feeling (pronounced pressure point, objectifiable by force-displacement process), etc. (Please refer to the comments in ► Sect. 6.2.1.3 in this context).

Acoustic Quality

1. Pleasant sound caused by the technical functional units (determined and smooth, not hard or tinny, e.g.: healthy engine noise, saturated door slam, not beepy warning signals, crisp acoustic feedback of switch actuations etc.)
2. Absence of undesirable/annoying noises such as squeaking, rattling, roughness (see also ► Sect. 8.2.1.4 and 8.2.2).

Harmony

1. Harmony across all systems, subsystems, and components, e.g.: similarities in appearance and feel of radio and air conditioning; however, certain differences are necessary to distinguish between functions and thus reduce driver error during use (see ► Sect. 6.2). Nevertheless, the different handling characteristics should give the impression of design consistency, i.e. all systems should look and feel as if they have been designed by the same designer. At the same time, they should reflect the character of the respective car brand.
2. Harmony between the materials and their design within a vehicle to create a brand image.
3. Only a few different materials in the immediate vicinity, e.g.: many components made of different materials with many dividing lines placed within a small area should be avoided, because in this way the perception of confusion, mismatches, unevenness, etc. is created.

Odour Quality

1. Use odourless materials wherever possible, in particular avoid using materials with unpleasant and toxic odours.

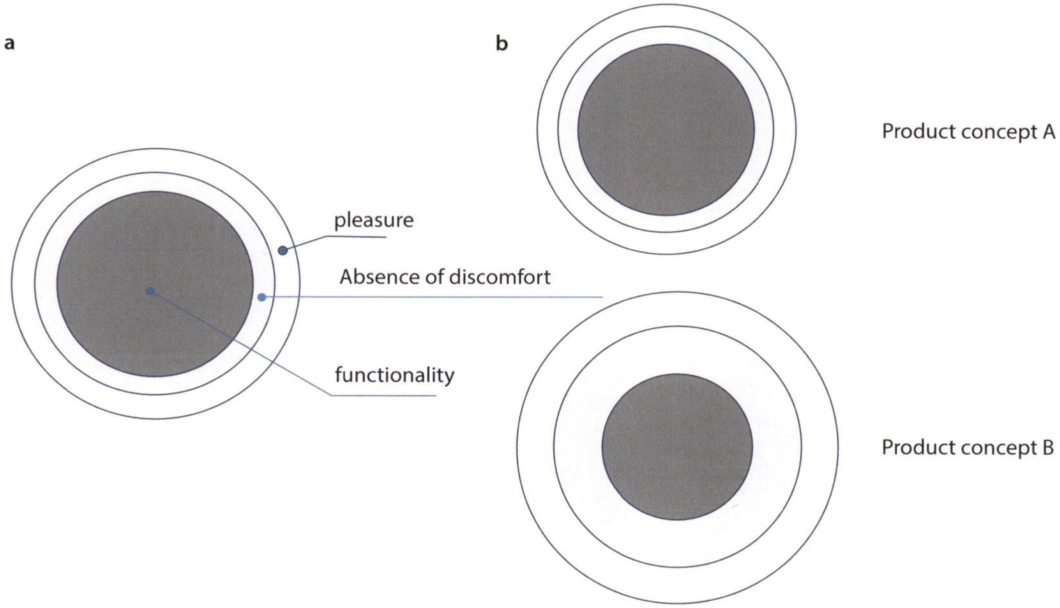

Fig. 7.132 Ring model of product conformity according to Peters (1987) and Levitt (1980), **a** Meaning of the rings, **b** two different product concepts

2. Use of material with pleasant smell, e.g.: smell of natural leather, flowery, fruity, spicy smell.

Many of the points mentioned above have nothing to do directly with the function or the ergonomic quality described in the chapters above. Much of this is in the sense of the comfort model described in ► Sect. 3.3.4 by Zhang et al. (1996) aspects of pleasure. In this context, Bhise (2012) refers to the ring model of the product compliance of Peters (1987) and Levitt (1980), which is briefly described here - slightly modified to the comfort model mentioned above. The core of each product is then described by its functionality. The size of the area representing the functionality qualitatively describes its extent (Fig. 7.132). It essentially characterises what is expected of a product (e.g.: a car is always available as a vehicle in its range of functions). In addition to this functionality, there is an overall ring of the absence of discomfort (e.g. no unfavourable posture is imposed, the view from the vehicle is not restricted). If discomfort were to occur, this would have a negative impact on the assessment. The outer ring characterizes the amount of pleasure (e.g. the interior pleases the eye through the harmonious coordination of materials and color; the vehicle allows the docking of a mobile phone). It therefore describes the extent to which the expectations of the user are exceeded. Figure 7.132 on the right shows the qualitative assessments of two different product concepts. Compared to product B, product A actually stands out due to its higher functionality. Nevertheless, product B achieves a significantly higher level of desirability because its appeal exceeds the expectations of the user.

The general problem of applying ergonomic rules in combination with product desirability is that practically only the discomfort can be reduced as much as possible. Aspects of pleasure are the classic areas of industrial design, but which must not be realized at the expense of avoiding discomfort. For this reason alone, close cooperation between designers and ergonomists is indispensable in the development of vehicle interiors. Finding new desires is a demand on the creativity of all those involved in development. As already mentioned elsewhere, customer surveys can only reveal existing

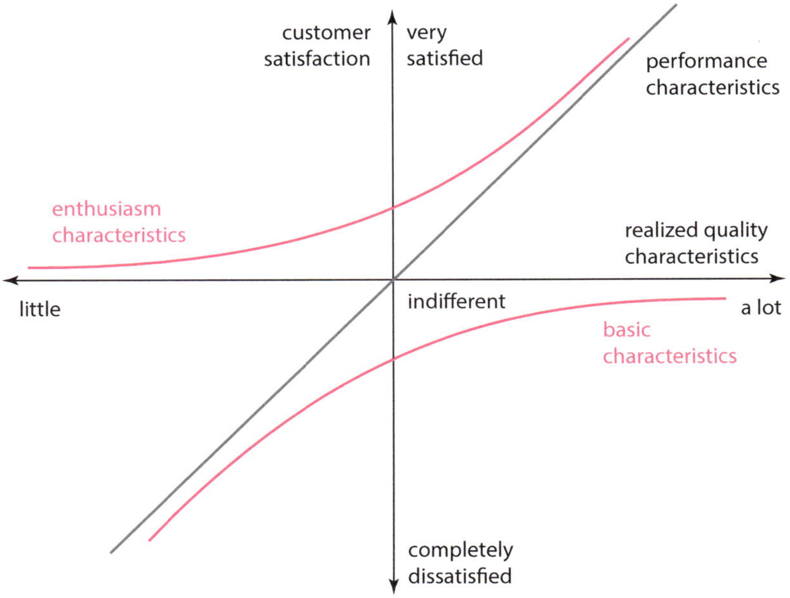

Fig. 7.133 Kano model. (Source: Wikipedia)

7

shortcomings. It cannot be expected that the customer "invents" something. However, it is possible to test the desirability of a new product idea by interviewing representatives of the target clientele.

Bhise (2012) cites the Kano method, which plays a relatively large role in marketing, as one way of quantifying this desirability. In principle, the method consists of determining the acceptance or rejection of a characteristic (e.g.: presence of a wooden strip) by means of a functional and a dysfunctional question on a five-stage rating scale (Klopp 2014).[58] Based on the combination of the respondent's answer to the functional and dysfunctional question, this answer is assigned to one of the following categories using an evaluation table developed by Kano:

- **A** (Attractive) = enthusiasm factor
- **M** (Must-Be) = base factor
- **O** (One-dimensional) = power factor
- **I** (indifferent) = irrelevant product features
- **R** (Reverse) = undesirable (leads to dissatisfaction)
- **Q** (Questionable) = misunderstood (e.g. answers in this category will not be considered)

The frequencies, which result from a survey with a large number of test persons, are converted into coefficients by means of different formulas, which enable a positioning in the diagram of Fig. 7.133. The area between the curves "Enthusiasm features" and "Basic features" represents the neutral area that characterizes the pure performance features (= functionality). Although high performance features are a prerequisite for high customer satisfaction, they do not generate enthusiasm. They characterize the "unexpressed wishes". A product characteristic that is located below the curve of the basic characteristics should be avoided. The further above the curve of the enthusiasm characteristics a product characteristic is localized due to the customer questioning, the more it is to be assumed that it concerns here a characteristic, with which the concerning product would distinguish itself before competition prod-

58 Functional question: if there was a wooden strip on the dashboard of your vehicle, what would you think? Answers: 1. would be very happy ...-... 5. would bother me very much.

Dysfunctional question: if there was no wooden trim on the dashboard of your vehicle, what would you think? Answers: 1. would be very happy ...-... 5. would bother me very much.

ucts. These are features that customers have not seen before and which, according to Bhise, trigger a "wow" effect in them (e.g. HUD in a small car). However, it must be noted that over time such "wow" characteristics become unspoken customer wishes, so that the manufacturer must constantly invent new "wow" characteristics.

References

Amereller, M.: Die Gelenkbeweglichkeit des Menschen im Altersgang als Fokus wissenschaftlicher Forschung im automobilen Kontext. Dissertation Technischen Universität München (2014)

Arlt, F., Bubb, H.: Simulation of target directed movements within the CAD man model RAMSIS. In: Proceedings of SEA Conference on Human Modeling The Hague, 18–20 May 1999. SAE International, (1999)

Arlt, F.: Untersuchungen zielgerichteter Bewegungen zur Simulation mit einem CAD-Menschmodell, Dissertation Technischen Universität München (1998)

Bäumler, M.: Ermittlung einer ergonomisch optimalen Pedalanordnung für Personenkraftwagen. Diplomarbeit an der Katholischen Universität Eichstätt (1992)

Becker, S.: Robson Forensic Inc. Parking Brake Use Study, SAE Paper, Bd. 2013-01-0199. (2013)

Bhise, V., Kridli, G., Mamoola, H., Devaraj, S., Pillai, A., Schulze, R.: Development of a Parametric Model for Advanced Vehicle Design SAE Technical Paper, Bd. 2004-01-0381. Society of Automotive Engineers, Inc., Warrendale (2004)

Bhise, V., Pillai, A.: A Parametric Model for Automotive Packaging and Ergonomic Design. In: Proceedings of the 2006 International Conference on Computer Graphics & Virtual Reality Las Vegas, Nevada, June 26–29, (2006)

Bhise, V.D.: Ergonomics in the Automotive Design Process. CRC Press Taylor & Francis Group, Boca Raton, London, New York (2012)

Bitter, T.: Objektivierung des dynamischen Sitzkomforts, Schriftenreihe des Instituts für Fahrzeugtechnik, TU Braunschweig, Dissertation, Shaker Verlag 2005 (2005)

Bothe, A.: Entwicklung eines parameterbasierten CAD-Modells zur Ergonomieabsicherung einer Mittelklasselimousine. Diplomarbeit am Institut für Konstruktionstechnik der Universität Stuttgart (2010)

Brackett, R.O., Pezoldt, V.J., Sherrod, M.G., Roush, L.K.: Human Factors Analysis of Automotive Foot Pedals. National Highway Traffic Safety Administration, Springfield (Virginia) (1989)

Brenner, M.: Nutzergruppenspezifische Ergonomie im Fahrzeug. Dissertation Technischen Universität München (2013)

Brückner, B.: Neue Ansätze für die Ergonomieabsicherung in der Fahrzeugentwicklung. Vortrag zum 3. Symposium Human Machine Interaction Design – Von der nutzergerechten bis zur energieeffizienten Gestaltung. Stuttgart (2011)

Bubb, H., Hudelmaier, J.: 1-Liter-Auto und Anthropometrische Gestaltung. In: Fahrzeugkonzepte für das 2. Jahrhundert Automobiltechnik 19.VDI/VW Gemeinschaftstagung. VDI-Berichte, Bd. 1653, VDI-Berichte, Düsseldorf (2001)

Cherednichenko, A.: Funktionales Modell der Einstiegsbewegung in einem PKW, Dissertation Technischen Universität München (2007)

Clarke, G.R., et al.: Preliminary studies in measuring range of motion in normal and painful stiff shoulders. Rheumatol. Rehabilit. **14**(1), 39–46 (1975)

Dassault Systems: ENOVIA V5 DMU: Using 3D Manikins to Optimize Vehicle Occupant Packaging for Vehicle Occupant Accomodation (2009)

Dirlich, T.: Experiments in Simulated Microgravity under Water: Preliminary Findings on Individual Neutral Postures. AHFE 2010, Miami (2010)

Dofour, F., Wang, X.: Discomfort Assessment of Car Ingress/Egress Motion using the Concept of Natural Movement SAE International Digital Human Modeling Conference (2005)

Dreßel, J., Kain, V.: Untersuchungsbericht über das reelle Sitzverhalten des Fahrers. Unveröffentlichte Studie. Institut für Ergonomie, München (1985)

D'Souza, S.F.: Physical efficiency as a function of age: Development of an age scaling model for biomechanical simulation. Dissertation Technischen Universität München (2014)

Dudenhöffer, F.: Demografische Entwicklung und schrumpfender Automarkt. Jahrbuch der Absatz und Verbrauchsforschung. **54**(1), 81–90 (2008) G. Nürnberg, Ed.

Flügel, B., Greil, H., Sommer, K.: Anthropologischer Atlas. Edition Wötzel, Frankfurt/Main (1986)

Franz, M., Kamp, I., Durt, A., Kilincsoy, Ü.: A light weight car seat shaped by human body war contour. Int. J. Human Factors Modell. Simulat. **2**(4), 314–326 (2011)

Fuchs-Kittowski, K.: Probleme des Determinismus und der Kybernetik in der molekularen Biologie. Tatsachen und Hypothesen über das Verhältnis des technischen Automaten zum lebenden Organismus. Gustav Fischer Verlag, Jena (1976). S. 185

Göhler, T.: Umblickverhalten im Pkw – Eine Validierung von RAMSIS-Augpunktlagen. Diplomarbeit – München, Untertürkheim: Lehrstuhls für Ergonomie der Technischen Universität München, DaimlerChrysler AG (1999)

Hager, K.: Sehen, Hören, Schmecken und Riechen. In: Kolb, G.F., Leischker, A.H. (eds.) Medizin des

alternden Menschen. Lehrbuch zum Gegenstandskatalog der neuen ÄApprO, pp. 85–93. Wissenschaftliche Verlagsgesellschaft Stuttgart, Stuttgart (2009)
Hammond, D.C., Roe, R.W.: SAE Controls Reach Study SAE Paper, Bd. 720199. Society of Automotive Engineers Inc., Warrendale, PA (1972)
Hammond, D.C., Mauer, D.E.: Razguanas: Control Reach: the Hand Reach of Drivers SAE Paper, Bd. 750357. Society of Automotive Engineers Inc., Warrendale, PA (1975)
Hartung, J.: Objektivierung des statischen Sitzkomforts auf Fahrzeugsitzen durch die Kontaktkräfte zwischen Mensch und Sitz. Dissertation Technischen Universität München (2005)
Hirao, A., Kitazaki, S., Yamazaki, N.: Development of a new driving posture focused on biomechanical loads 2006 Word Congress, Detroit, Michigan. SAE Technical Paper, Bd. 2006-01-1302 (2006)
Hirz, M., Hirschberg, W., Dietrich, W.: Integrated 3D-CAD Design in Early Automotive Development Processes. In: FISITA – World Automotive Congress 2008. The Future of Automobiles and Mobility, London (2008a)
Höllrich, M.: Untersuchung des Einstiegsvorgangs beim PKW mittels Bildverarbeitung und Softdummy. Unveröffentlichte Diplomarbeit. BMW, München (1992)
Hudelmaier, J.: Sichtanalyse im Pkw unter Berücksichtigung von Bewegungen und individuellen Körpercharakteristika. Dissertation Technischen Universität München (2003)
Human Solutions: RAMSIS: Beschreibung der Maße des RAMSIS Package Designer: Speyer, H.. Unveröffentlichtes Manuskript, Kaiserslautern (2000)
Human Solutions: Ramsis Package Designer: Speyer, Hartmut. Unveröffentlichtes Manuskript, Kaiserslautern (2002)
Human Solutions: RAMSIS User Guide, Version 3.8. Kaiserslautern (2008)
Kahlmeier, W., Marek, C.: Gern Groß. In: Der neue Ford Mondeo ATZ/MTZ- Sonderausgabe. (2000). Okt. 2000
Kamp, et al.: Chosen postures during specific sitting activities. Ergonomics. **54**(11), 1029–1042 (2011)
Karwowski, W., Yates, J.W., Pongpatana, N.: Ergonomic guidelines for design of a passenger car trunk. In: Automotive Ergonomics. Taylor, Francis, London (1993)
Kilincsoy, Ü., Wagner, A.-S., Bengler, K., Bubb, H., Vink, P.: In: Ahram, T., Karwowski, W., Marek, T. (Hrsg.) Comfortable Rear Seat Posture Preferred by Car Passengers Proceedings of the 5th International Conference on Applied Human Factors and Ergonomics AHFE 2014, Kraków, Poland, 19–23 July 2014. (2014)
Klopp, E.: Die Kano – Methode (2014). http://www.eric-klopp.de/texte/angewandte-psychologie/18-die-kano-methode
Kolich, M.: Occupant Preferred Back Angle Relative to Head Restraint Regulations. SAE 2010-01-0779, Troy, MI (2010)
Kolling, J.: Validierung und Weiterentwicklung eines CADMenschmodells für die Fahrzeuggestaltung. München: Technische Universität München, Lehrstuhl für Ergonomie (LfE), Dissertation (1998)
Krist, R.: Modellierung des Sitzkomforts, eine experimentelle Studie. Dissertation Katholische Universität Eichstätt (1993)
Lange, C.: Wirkung von Fahrerassistenz auf der Führungsebene in Abhängigkeit der Modalität und des Automatisierungsgrades. Technische Universität München, Dissertation (2008)
Lark, S.D., et al.: Knee and ankle range of motion during stepping down in elderly compared to young men. Eur. J. Appl. Physiol. **91**(2–3), 287–295 (2004)
Layer, J.: Untersuchung dynamischer Vorgänge am Menschen beim Einsteigen in einen PKW, unveröffentlichte Diplomarbeit. München, 1992, BMW (1992)
Levitt, T.: Marking Success through Differentiation of Anything. Harvard Business Review, January–February (1980)
Lorenz, S.: Assistenzsystem zur Optimierung des Sitzkomforts im Fahrzeug. Dissertation Technischen Universität München (2011)
Lorenz, D.: Äußere Einflussparameter auf Sitzpositionen im Fahrzeug. Dissertation an der Technischen Universität München, Audi Dissertationsreihe, Göttingen, Cuvillier (2013)
McGuire, N., Madley, S., Sevrence, S., Kostrzewa, J., Bilk, J., Gourd, D., Thyagarajan, R.: GENPAD – Ergonomic Packaging SAE Technical Paper, Bd. 2002-01-1241. Society of Automotive Engineers, Inc., Warrendale (2002)
Meldrum, J.F.: Automobil Driver Eye Position. Human Factors Research Dept, Ford Motor Company, Dearborn, MI (1965)
Mercedes-Benz (Daimler AG): Individuell ist bei uns Standard: Fahrhilfen und Sitzanpassungen – jetzt ab Werk (2010). http://www.mercedes-benz.de/content/media_library/germany/mpc_germany/de/mercedes-benz_deutschland/sonderfahrzeuge_ng/spezielle_branchen-/limited_mobility/broschuerendownload/fahrhilfen_broschuere0.object-Single-MEDIA.download.tmp/Web_10009_Broschuere_Fahrhilfen_Sitzmodi_neu.pdf, Zugegriffen: 09.10.2012
Mergl C.: Entwicklung eines Verfahrens zur Optimierung des Sitzkomforts auf Automobilsitzen. Dissertation Technischen Universität München (2005)
Mergl, C., Bubb, H., Vogt, C., Kress, H.: Ergonomic Layout Process for a Driver Working Place in Cars Society of Automotive Engineers, Inc., Warrendale. Bd. 2006-01-2313. SAE Technical Paper, (2006)
Michel, B.: Ergonomische Analyse der Fahrerumgebung im Fernverkehrs-Lkw. Dissertation Technischen Universität München (2014)
Mies: Experimentelle Bestimmungen der Kopfhüllkurven im Fahrzeug mit einer photogrammetischen

Methode. München: Technische Universität München, Institut für Ergonomie (1987)
Monnier, G.: Simulation de mouvements humains complexes et prédiction de l' inconfort associé: Application à l'évaluation ergonomique du bouclage de la ceinture de sécurité. L'institut national des sciences appliquées de Lyon, Lyon (2004)
Müller, A.: Systematische und nutzerzentrierte Generierung des Pkw-Maßkonzeptes als Grundlage des Interior- und Exteriordesigns. Dissertation an der Universität Stuttgart (2010)
National Aeronautics and Space Administration: Man-Systems Integration Standards, Rev. B, Vol. 1. NASA-STD-3000. Houston, Texas. National Aeronautics and Space Administration (1995)
Neuendorf, M.: Pilotstudie zur Ermittlung der komfortbestimmenden Parameter beim PKW-Ein- und Ausstieg. Studienarbeit – Berlin, Untertürkheim: Institut für Straßen- und Schienenverkehr der Technischen Universität Berlin, Mercedes Benz AG, 1996 (1996)
Nilsson, G.: Validity of Comfort Assessment in RAMSIS SAE Technical Paper, Bd. 1999-01-1900. Society of Automotive Engineers, Inc., Warrendale (1999)
Nordin: (2003)
Peters, T.: Triving on Chaos. Harper; Row, N. Y. (1987)
RAMSIS: RAMSIS 3.8 in CATIA V5 Handbuch - HUMAN SOLUTIONS GmbH (1990, 2011)
Reed, M.P., Manary, M.A., Flannagan, C.A., Schneider LW: A statistical method for predicting automobile driving posture. Hum. Factors. **4**(4), 4 (2002)
Rekittke, T., Brückner, B.: Mercedes-Benz Pkw RAMSIS Familie. Interne Unterlage, Daimler AG, Sindelfingen (2010)
Remlinger, W.: Analyse von Sichteinschränkungen im Fahrzeug. Dissertation, Technische Universität München (2013)
Richter, H.: Eine Testbatterie zur allgemeinen Beurteilung motorischer Leitparameter unter besonderer Berücksichtigung der Anforderungen des Freizeit- und Erholungssportes. Dissertation. Leipzig: DHfK (1974)
Rigel, S.: Entwicklung und Validierung einer Methode zur quantitativen Untersuchung der Ein- und Ausstiegsbewegung in einem Pkw, Dissertation Technischen Universität München (2005)
Sabbah, A. O.: Entwicklung eines dynamischen Diskomfortmodells am Beispiel des Ein- und Aussteigs. Dissertation Technischen Universität München (2010)
Schmidt, S., Seiberl, W., Schwirtz, A.: Influence of different shouder-elbow configurations on steering precision and steering velocity in automotive context. Applied Ergonomics (2014)
Schmidt, R.F., Lang, F.: Physiologie des Menschen: Mit Pathophysiologie, 30. Aufl. Springer-Lehrbuch. Springer Medizin Verlag, Berlin, Heidelberg (2007)
Schober, H.: Das Sehen Bd. 1. VEB Fachbuchverlag, Leipzig (1970)
Scholly, U.: Sicht im Pkw – Evaluationsmethoden in der Praxis, Tagungsunterlagen Herbstkonferenz der Gesellschaft für Arbeitswissenschaft (GfA), Sindelfingen, 28.-29.10.2006. (2006)
Seidl, A.: Das Menschmodell RAMSIS: Analyse, Synthese und Simulation dreidimensionaler Körperhaltungen des Menschen. Dissertation Technischen Universität München (1993)
Viitasalo, J.T., Era, P., Leskinen, A.-L., Heikkinen, E.: Muscular strength profiles and anthropometry in random samples of men aged 31-35, 51-55 and 71-75 years. TERG. **28**(11), 1563–1574 (1985)
Vogt, C.: Ergonomische Fahrzeuginnenraum-Auslegung und Erstellung eines Hochschulpraktikums. München. Technische Universität München, Lehrstuhl für Ergonomie (LfE), Diplomarbeit, 2003 (2003)
Vogt, C., Mergel, C., Bubb, H.: Interior layout design for passenger vehicles with RAMSIS. Human Factors Ergonom. Manufact. **15**(2), 197–212 (2005)
Wagner, P.-O.: Der Mensch als entscheidende Zielgröße im Produktentstehungsprozess. Vortrag auf der Tagung „Digital Humans in Application", Leipzig 12./13.11.2013 (2013)
Wallentowitz, H.: Aufbau von Kraftfahrzeugen. IKA, Aachen (1995)
Wang, X.: In: Duffy, V.G. (Hrsg.) Discomfort Evaluation and Motion Measurement, in Handbook of digital human modeling. Research for applied ergonomics and human factors engineering. CRC Press (Human factors and ergonomics), Boca Raton (2009)
Weineck, J.: Sportanatomie, 10. Aufl. Verlag PERIMED-spitta, Erlangen (2001)
Weineck, J.: Sportbiologie, 9. Aufl. Spitta-Verl., Balingen (2004)
Wilke, H.J.: Möglichkeiten zur Bestimmung der Wirbelsäulenbelastung und Konsequenzen für die Empfehlungen für das Sitzen. In: ergo mechanics, Interdisziplinärer Kongress Wirbelsäulenforschung, S. 78–103 (2004)
Wolf, H.: Ergonomische Untersuchung des Lenkgefühls an Personenkraftwagen. Dissertation Technischen Universität München (2009)
Xiaoxiang, S.: Evolution of Concept Vehicle Archetype Development – NX Automotive Product Update: Vortrag, 25.04.2007 (2007)
Zacher, I., und Bubb, H.: Strength Based Discomfort Model of Posture and Movement. In: Proceedings of the SAE DHMS 2004 (2004) Rochester, 04DHM-66
Zaindl, A.: Simulation von Kreuzungskonflikten durch Sichtverdeckung der A-Säule. Diplomarbeit am Lehrstuhl für Ergonomie der Technischen Universität München, 2009 (2009)
Zenk R.: Objektivierung des Sitzkomforts und seine automatische Anpassung. Dissertation Technischen Universität München (2008)
Zenk, R., Franz, M., Bubb, H.: Spine Load in the Context of Automotive Seating 2007. SAE, Bd. 2007-01-2485. SAE, Washington (2007)
Zhang, L., Helander, M.G., Drury, C.G.: Identifying factors of comfort and discomfort in sitting. In Human Factors. **38**(3), S. 377–S. 389 (1996)

Further Literature

Andreoni, G., Rabuffetti, M., Pedotti, A.: New approaches to car ergonomics evaluation oriented to virtual prototyping EURO-BME course on Methods & Technologies for the study of human activity & behaviour, Italy. **1997**, S. 19–S. 20 (1997)

Chatereaux, E., Wang, X., Tasbot, J.: A Database of Ingress/Egress Motions of elderly People SAE International conference and exposition of Digital Human Modelling for Design and Engineering, Seattle, 2007. Technical Paper, Bd. 2007-01-2493 (2007)

Franz M.: Comfort, experience, physiology and car seat innovation - Theory, Design and Evaluation. Dissertation thesis Technical University Delft (2010)

Hirz, M., Göber, T., Lang, M.: Integrated 3D-CAD Design Strategies in Vehicle and Engine Development Processes. In: 8. Internationales Stuttgarter Symposium. Automobil- und Motorentechnik. Dokumentation. Vieweg und Sohn, Wiesbaden (2008b)

Khan, S.M., Sundström, J.: Effects on vibration on sedentary activities in passenger trains. J. Low Frequen. Noise Vibrat. Active Cont. **26**(1), 43–55 (2007)

Meic-Sidic, T.: Entwicklung eines Bewertungsverfahrens zur Beurteilung der 360°-Rundumsicht aus dem Pkw. Diplomarbeit. Institut für Verbrennungsmotoren und Kraftfahrwesen der Universität Stuttgart, DaimlerChrysler AG, Stuttgart, Sindelfingen (2005)

Parkinson, M.B., Reed, M.P.: Optimizing Vehicle Occupant Packaging Technical Paper, Bd. 2006-01-0961. Society of Automotive Engineers, Inc., Warrendale (2006)

Reed, M.P., Huang, S.: Modeling Vehicle Ingress and Egress Using the Human Motion Simulation Framework Technical Paper, Bd. 2008-01-1896. SAE International conference and exposition of Digital Human Modelling for Design and Engineering, Pittsburgh (2008)

Reed, M.P., Manary, M.A., Flannagan, C.A., Schneider, L.W.: New Concepts in Vehicle Interior Design Using Aspect Technical Paper, Bd. 1999–01 0967. Society of Automotive Engineers, Inc., Warrendale (1999a)

Reed, M.P., Manary, M.A., Flannagan, C.A., Schneider, L.W.: Automobile Occupant Posture Prediction for Use With Human Models Technical P per, Bd. 1999-01-0966. Society of Automotive Engineers, Inc., Warrendale (1999b)

Rossbacher, P., Hirz, M., Harrich, A., Dietrich, W., Theiss, N.: The Potential of 3D-CAD based process – optimization in the automotive concept phase Technical Paper, Bd. 09IDM-0024. Society of Automotive Engineers, Inc., Warrendale (2009)

SAE J1100. 2005–09. Motor Vehicle Dimensions. Warrendale: Society of Automotive Engineers, Inc.. DIN 70020–1

Schmidtke, H., Bundesamt für Wehrtechnik und Beschaffung (Hrsg.): Handbuch der Ergonomie (HdE) mitergonomischen Konstruktionsrichtlinie und Methoden. Bd. 1–5. Bundesamt für Wehrtechnik und Beschaffung, Koblenz (o. J.) (n.d.)

Schmidtke, H. und Fraczek, I.: EKIDES, Ergonomische Datenbank auf Microsofts Access Basis. (n.d.)

Siemens PLM Software: Occupant Packaging Toolkit for Jack: Design and test vehicle interiors with digital humans (2010)

7

Design of Condition Safety

Heiner Bubb

Contents

8.1 Lighting – 470
8.1.1 Photometric Dimensions – 470
8.1.2 Outdoor Lighting – 471
8.1.3 Interior Lighting – 472

8.2 Sound – 475
8.2.1 Driving Noises – 475
8.2.2 Small Noises – 480
8.2.3 Useful Signals – 482

8.3 Vibrations – 483
8.3.1 Vibration Phenomena – 483
8.3.2 Perception of Vibrations – 483
8.3.3 Vibration Evaluation – 485
8.3.4 Comfort and Driving Safety – 490
8.3.5 Kinetosis – 492

8.4 Climate – 493
8.4.1 Climate, Performance and Comfort – 493
8.4.2 Climatic Comfort – 494
8.4.3 Environmental Conditions – 499
8.4.4 Technical Requirements – 501

8.5 Odour – 513

References – 515

H. Bubb et al. (eds.), *Automotive Ergonomics*, https://doi.org/10.1007/978-3-658-33941-8_8

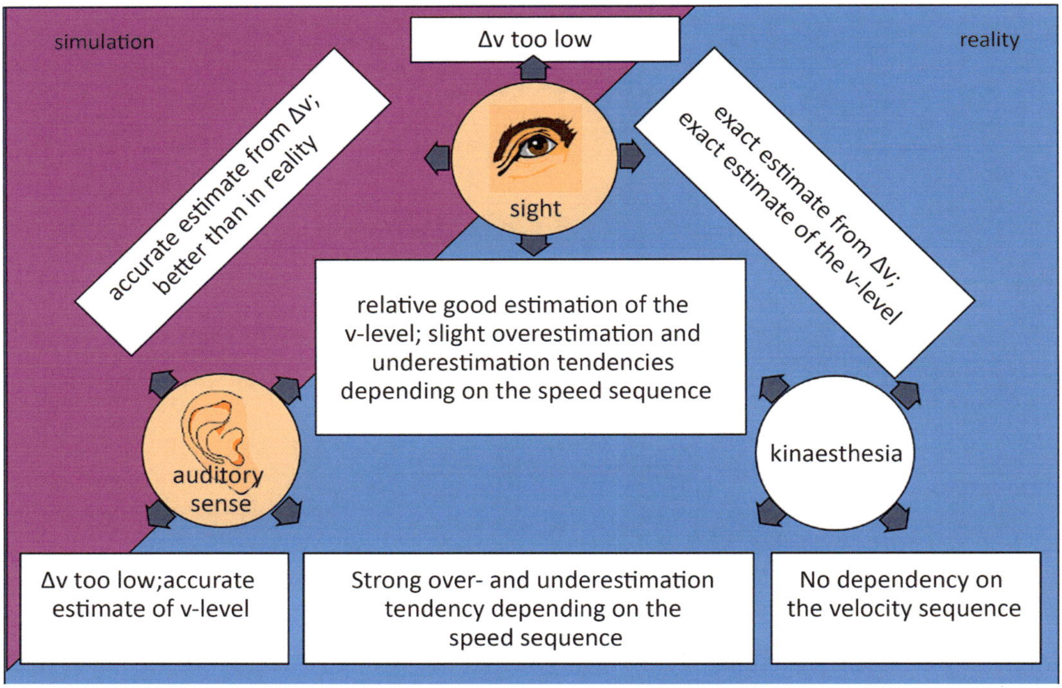

Fig. 8.1 Influences of the respective sensory impressions on the perception of speed. (Bubb 1977)

In all design measures, the dual character of these environmental factors must therefore be taken into account as feedback and impact factors.

8.1 Lighting

If 90% of the information is perceived via the optical sensory channel when driving a car, it is trivial to note that the lighting has a primary feedback character. On the other hand, the comfort pyramid shown in Fig. 3.72 shows that inadequate lighting, in addition to unpleasant odours, creates discomfort above all other environmental factors. The double character mentioned above therefore plays an important role in the design of the lighting.

8.1.1 Photometric Dimensions

Lighting dimensions play an important role in the assessment of lighting conditions. Fig. 8.2 shows a compilation. Although light is a form of energy and would therefore be the adequate unit of the power emitted by a light source Watt [W], the power emitted in the visible light range (wavelength between 390 and 770 nm) is evaluated by means of the so-called V(λ) curve, which takes into account that the eye is significantly less sensitive in the blue and red range than in the range of the centre of the sunlight at 555 nm (see also ► Sect. 3.2.1.1). The radiant power evaluated in this way is referred to as luminous flux and is quantified in the light unit lumen [lm]. The luminous flux emitted into the solid angle f/ r^2 is referred to as luminous intensity (unit of measurement candela [cd]). The luminous flux emanating from the light source falls on the illuminated surface of interest. The luminous flux incident per unit surface is referred to as illuminance (unit lux [lx]). Depending on the optical properties (location- and space-dependent absorption and reflection coefficients) of the illuminated surface, the light is now reflected in the various spatial directions. The beam of light falling into the eye characterizes the location point seen in each case. The amount of light emitted from this point is called luminance. It is calculated from the illu-

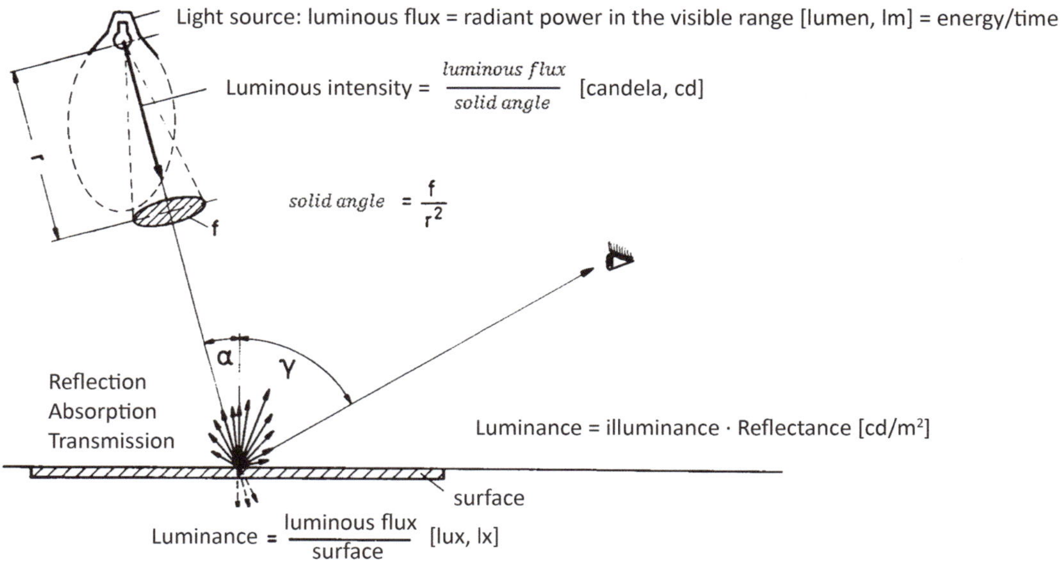

Fig. 8.2 Lighting dimensions

minance and the site-specific reflection coefficient. The unit of measurement the luminance is therefore cd/m².

The eye has an enormous ability to adapt to different lighting conditions (approx. 1:10[16]). As explained in ► Sect. 3.2.1.1, this adaptation is done through various mechanisms. A quick adjustment is done by narrowing and dilating the pupil, but this only allows an adjustment in the range of 1:4. The actual adaptation takes place on the one hand by changing the interconnection of the receptors to the retina of the eye (change in the receptive fields), which in principle takes time and is associated with a loss of local resolving power during dark adaptation. The adjustment effects described so far all occur in the so-called photopic region of the cone vision, which also allows a color resolution of the objects seen. In extreme darkness adaptation, light can only be perceived via the very light-sensitive rods. This area of vision is called scotopic. The transition area between rod vision and complete cone vision is called mesopic. Fig. 8.3 shows an assignment of these different visual ranges to the above-mentioned photometric units with some practical examples. In particular, it can be seen from the image that the brightness range of a car headlamp enables vision in the lower photoptical range.

8.1.2 Outdoor Lighting

It is not the framework here to deal with the various statutory approval regulations for vehicle lighting. ► Section 3.2.1.1 details the individual conditions that must be met in order for objects to be detected. The main problems of night driving are the luminance of the light reflected into the eye by the road and objects, possibly too high a luminance of the radiators themselves, which may also be sources of physiological and psychological glare. Since the eye has extreme adaptability to different lighting levels, the time required for adaptation plays a major role. Although in normal road traffic at night there is only mesopic adaptation (no complete dark adaptation, which still shows objectifiable improvements after half an hour, see above), it has to be considered that the adaptation process from light to dark is much slower (in the seconds to minutes range) than in the opposite direction (in the 100 ms to 1 s range). This plays a major role during the day, especially in the case of strongly changing lighting conditions (especially tunnel entrances and exits), but can practically not be absorbed by vehicle technology. Since, however, light adaptation is essentially controlled by the sensitivity of the rods in the fundus of the eye and these are

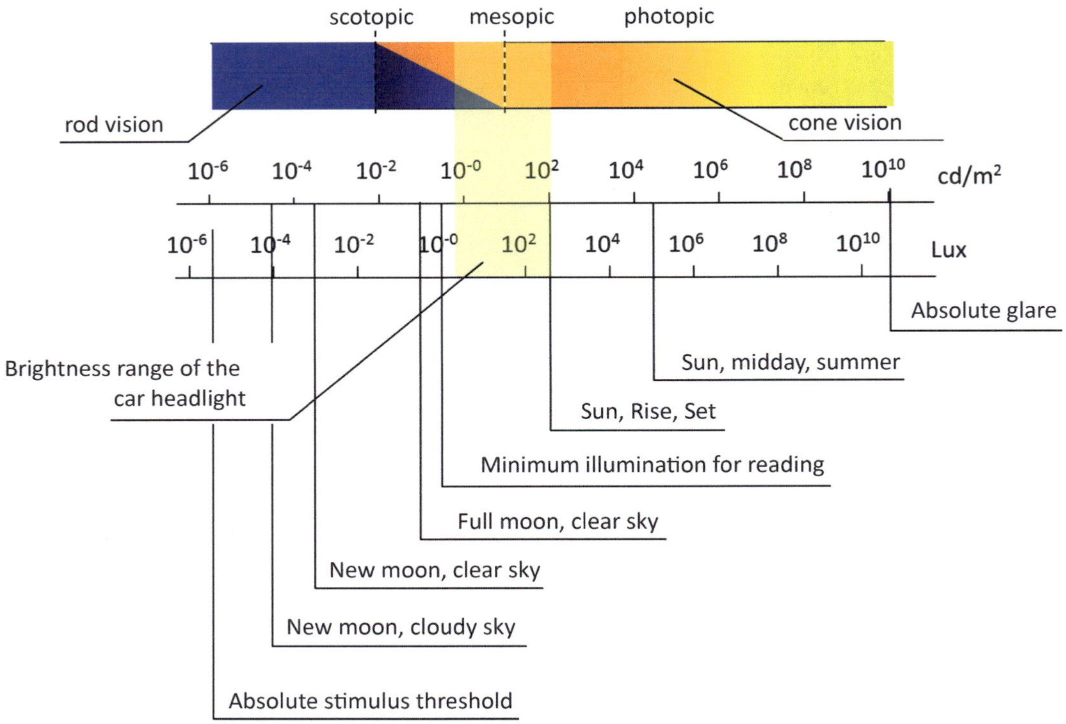

Fig. 8.3 Cone vision and rod vision as a function of luminance and illuminance

particularly sensitive to short-wave blue light, blue light sources, especially large-area light sources, should be avoided wherever possible in the vehicle interior in order to avoid undesired adaptation to apparently bright lighting conditions.[1] Otherwise, new developments in outdoor lighting, which can significantly improve night-time visual conditions, are dealt with in ▶ Sect. 9.2.1.3.

A special case of outdoor lighting is the automatic lighting of certain areas of the vehicle environment (e.g. lighting of door openers) when the driver approaches the vehicle in darkness. In order to prevent misuse by uninvolved pedestrians, these comfort functions can only be implemented in conjunction with the so-called keyless entry system, through which the vehicle recognizes the proximity of the authorized driver by means of a special transponder system.

1 In order to avoid undesired light adaptation, red light illumination was previously provided in the interior of submarines. Today it is known that so-called filtered white light (the extreme blue range is filtered out) has the best effect in terms of preventing light adaptation.

8.1.3 Interior Lighting

8.1.3.1 Instrument Lighting

Since many drivers feel disturbed by excessively bright instrument lighting when driving at night, it is absolutely necessary to make it dimmable. The control unit provided for this purpose (usually in the form of a knob or knurled wheel) must be easily accessible. It is usually installed near the light switch. With conventional instrument lighting (incident light), *dimming* is no problem and is part of the standard equipment. However, dimmability must also refer to the LED displays that are increasingly being used today. In particular, this must be provided for the central display (CID) and for the instrument cluster, if this is implemented in LCD form. Particularly with LCDs, the option between a day and a

	black	grey	white	magenta	blue	cyan	green	yellow	red
black		3	2	3	4	2	2	2	3
grey	1		2	4	3	4	4	3	4
white	1	2		2	1	4	4	5	1
magenta	3	5	2		3	3	3	3	4
light blue	1	3	3	5	2	5	1	3	3
medium blue	4	5	2	4		3	2	1	5
dark blue	1	4	1	2	5	2	2	1	4
cyan	1	2	5	3	2		5	5	3
light green	1	2	4	4	1	1	1	3	2
medium green	2	3	3	4	2	5		5	3
dark green	3	5	1	3	1	1	5	1	4
yellow	1	2	5	3	1	4	4		2
light red	1	2	4	3	2	4	4	5	2
medium red	3	5	2	5	4	3	4	3	
dark red	4	5	1	3	5	3	3	2	5

Note:
Chromatic Aberration:
The eye is short-sighted for blue, farsighted for red.

Legend:
1 = very good
2 = good
3 = satisfactory
4 = sufficient
5 = unsatisfactory
6 = inadequate

Fig. 8.4 Examples of good (2), medium (1) and bad (0) color contrast

night display version can be provided (day display: bright background – dark visual signs; night display: dark background – bright visual signs). However, automatic switching should be avoided, as irritating light stimuli may occur, especially in changing lighting conditions.

For good recognizability, visual signs must be marked with an *contrast* from 10: 1 (defined here as the ratio of the luminance of the visual sign to that of the background). In the case of external lamps, this is achieved by the graphic design – the contrast is maintained regardless of the intensity of the lighting (dark grey visual signs against a light grey background, for example, are unfavourable). However, the effective contrast of self emitting lamps depends on the ambient lighting. The light emission from LED displays, for example, would therefore have to be controlled depending on the outdoor illuminance.

In addition to brightness contrast, colour contrast is also very important for the recognition of visual signs. In principle, a high color contrast is achieved by a distance as large as possible between the color coordinates of the respective visual sign and the background in the color triangle of the Fig. 3.19. Fig. 8.4 gives examples. The correct choice of colour contrast is particularly important in conjunction with large LCD displays in the central instrument (CID) or, more recently, in the instrument cluster. It should also be noted in this context that due to the wavelength-dependent different refractive properties of the eye lens, the eye is short-sighted for blue light and farsighted for red light. This means that the direct contrast between red and blue visual signs should be avoided as far as possible (red visual signs against a blue illuminated background and vice versa are particularly unfavourable).

8.1.3.2 Interior Lighting

The interior lighting during a night-time drive is in principle even more disturbing than irritatingly bright instrument lighting. For various reasons, however, the desire for interior lighting occasionally arises (e.g. orientation on a map by the passenger, the desire of passengers to read while driving, etc.). Therefore, the design of the interior lighting requires spe-

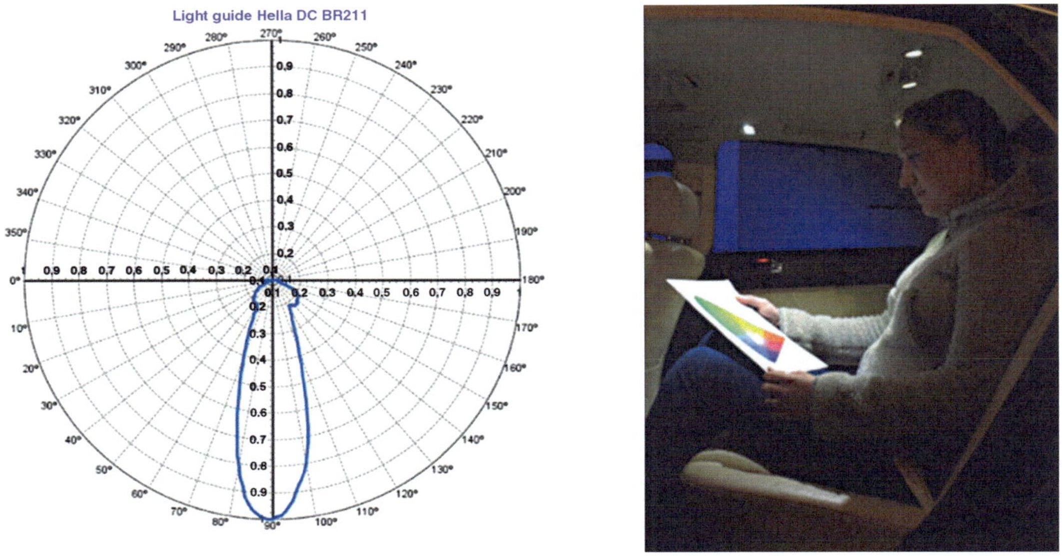

■ **Fig. 8.5** Spatial emission spectrum of an interior light with prism optics. (Hella: Nachtigall 2007)

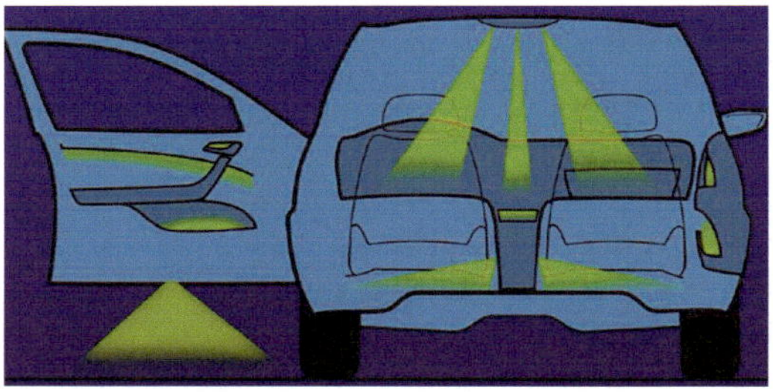

■ **Fig. 8.6** Areas of interior lighting. (Hella: Nachtigall 2007)

cial care. In particular, it shall be avoided that the light source of the interior lighting may be in the driver's field of vision. In addition, directed light (in the sense of a spotlight) must be used (■ Fig. 8.5). ■ Fig. 8.6 shows areas where interior lighting can be installed in the vehicle in a profitable way. Attention should also be paid to the light emitted by interior lighting and then reflected by objects. In this context, reference is made in particular to the comments in ► Sect. 7.3.4. The switch for activating the interior lighting should be easily accessible for the respective user. It should also be examined whether it makes sense for the driver to be able to switch off the interior lighting on his own initiative under certain circumstances.

In general, the night-time operation of controls in the vehicle cockpit may pose a problem that cannot always be solved satisfactorily by the backlighting of these controls that is common today. Hella has presented an innovative solution in the form of motion-guided light. The corresponding control element area for the driver is then illuminated glare-free (see above) when he approaches it with his hand (see ■ Fig. 8.7). It is then even possible to illuminate the following control panels adaptively in advance for better orientation (e.g.: Air conditioning – ventilation nozzles; Pietzonka 2004).

8.1.3.3 Ambient Lighting

For many people, S in an absolutely darkened room is unpleasant. In this context, the outstanding importance of light for the reduction of discomfort sensitivity is again pointed out. This unpleasant feeling can be compensated for by discreet ambient lighting. ■ Fig. 8.8 shows an example of such ambient lighting in the door leaf of the vehicle. In this context, particular care should be taken to ensure that this additional interior lighting neither impairs the view of the road nor creates any irritating reflections in the windscreen or side windows. Nachtigall (2007)

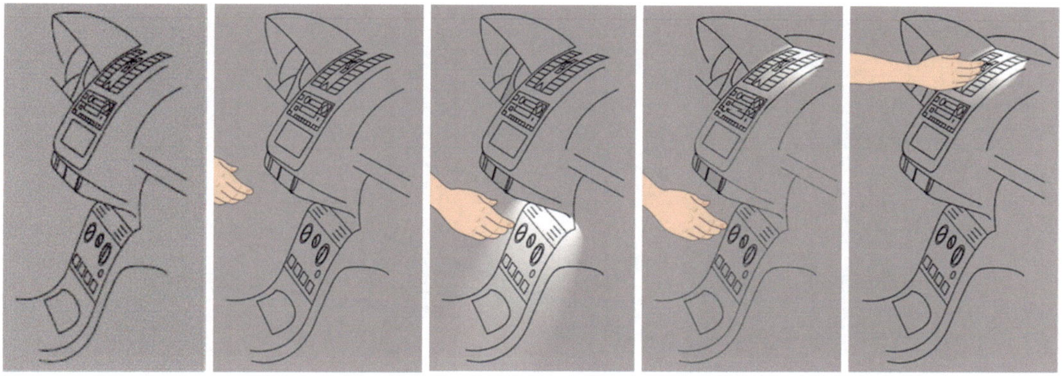

Fig. 8.7 "Motion-controlled light". (Hella: Pietzonka 2004)

Fig. 8.8 Ambient lighting in door leaf. For ambient lighting, blue light should be avoided, as light adaptation is mainly done by the blue light sensitive cones. (Source Hella)

reports that when asked what they saw as the main advantage of ambient lighting, about a quarter of respondents cited improved orientation and a pleasant atmosphere. 17% said they had a better sense of space and 12% even believed that this could help prevent fatigue.

8.2 Sound

8.2.1 Driving Noises

Irrespective of experience with technical systems, it is probably a fundamental object of human experience that increasing speed is always associated with increasing volume and increasing average frequency or individual frequency components of the associated noise (own movement, observation of animal movements and, of course, experience with technical noise development). In addition, sounds are perceived as pleasant (e.g. music) and annoying (the term is then "noise"). The sound generated by the vehicle thus cause a complex spectrum of sensations: they provide information about the driving condition (e.g. speed level, see above; feedback character), they can evoke joyful emotions ("great engine sound") and they can also be annoying (e.g. hissing joints, rattling noises). Vehicle acoustics is a complex and extensive field. Nor can it be dealt with in an almost exhaustive manner within the present framework. In this context, reference is made to e.g. the detailed work of Zeller (2011).

8.2.1.1 Noise Sources and Paths

Driving noise refers to all noises in the vehicle interior that are generated directly by the vehicle's movement. The noise sources are essentially the operating noise of the drive motor, wind noise at the body and rolling noises caused by the wheels and the chassis. Zeller (2011) states: "In the case of comfort-oriented mid-size sedans, rolling noise dominates the interior of the vehicle at low and medium driving speeds and low engine loads. This is perceived as airborne sound via the underbody and the side wall (panes) and as structure-borne sound via the chassis. As the load increases, the components of the engine noise become increasingly audible. In the low frequency range up to approx. 100 Hz, these are primarily airborne sound induced motor

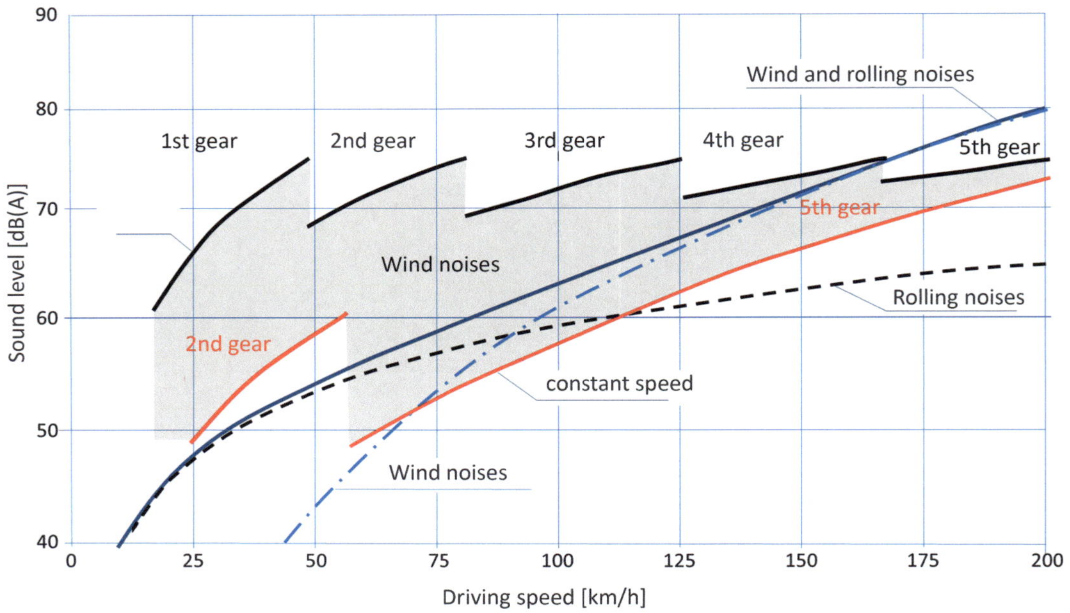

Fig. 8.9 Noise level in the vehicle interior as a function of driving speed and driving conditions. (Schematic, according to Zeller 2011)

orders.[2] The frequency range up to 400 Hz mainly includes structure-borne noise, which is excited by the gas and mass forces and introduced via the engine mount. At higher engine speeds, the mechanical engine noise becomes dominant in the area above 400 Hz, which is introduced as airborne noise via the front wall. Only at speeds above 80–100 km/h the rolling noise and later also the engine noise are increasingly "masked" by the wind noise. The noise generated by the motor depends considerably on its load condition. When the tyre rolls along the road, airborne and structure-borne noise is emitted due to the tread pattern of the tyre (rather singing, tonal noise) and due to unevenness of the road surface (rumbling noises). The wind noises are aerodynamically induced noises due to flow fluctuations on the outer skin of the vehicle. They cause a broadband noise that cannot be located directly. In order to produce as few discomforts as possible, this noise should contain as few tonal components as possible (e.g. high-frequency pipe tones). Fig. 8.9 shows a summary of the noise components described in the form of their contribution to the noise levels measured inside the vehicle.

8.2.1.2 Perception

The driver perceives not only the volume of the driving noise, but also its composition. This refers not only to the aspects of the audible individual noise sources, but also to the local assignment. All in all, the creation of the noise spectrum is an extremely complex matter. One reason for this is the narrow and complex structure of the vehicle interior. By reflecting the sound waves on the walls and on the various furnishings of this room, which can also be hard-acoustic and soft-acoustic, so-called standing waves are created if the corresponding distance happens to be a quarter of the wavelength of the respective sound component. Thus the vehicle interior is characterized by a confused mixture of such standing waves. One of the consequences of this is that very different spectra can be obtained depending on the position of the microphone of a sound analyzer in this room. Of course, this observation also applies to the

2 Engine orders: Frequencies which are caused by the speed-dependent so-called ignition peaks and engine vibrations.

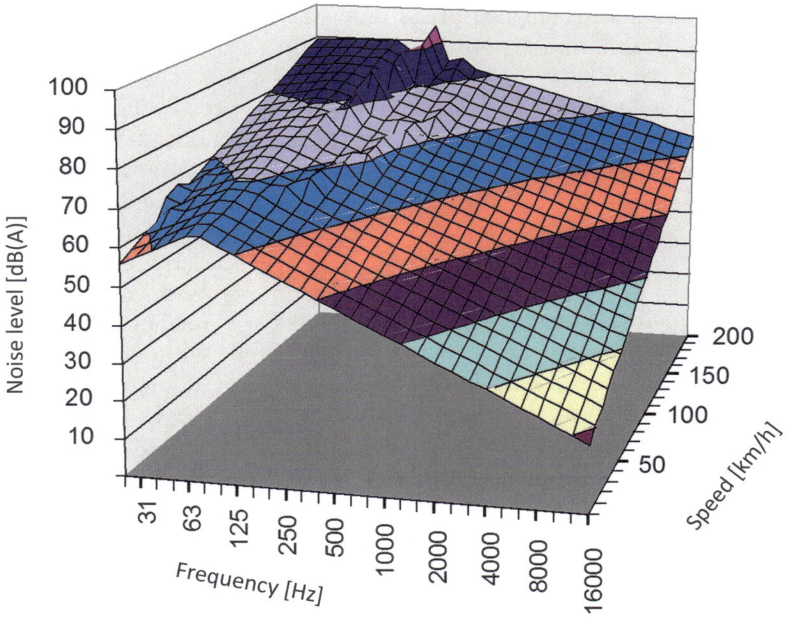

Fig. 8.10 Typical frequency composition of passenger car interior noise with 4-cylinder engine as a function of driving speed. (Bubb 1996)

human ear. If one now considers, however, that the hearing process consists in seeing the local temporal pattern of the travelling waves on the basilar membrane, as it were, whereby, among other things, similarities in noises are recognized (see ► Sect. 3.2.1.2), which are not easily explained by the frequency spectrum, it becomes understandable that for auditory perception – similar to other areas of perception – a constancy performance exists in the form that a sound can be recognized more or less independently of the properties of the sound chamber. At the same time, however, the type of sound chamber is also perceived. This means that a driver can easily recognize – even acoustically – whether he is in a larger or smaller vehicle. In particular, it is capable of detecting the engine noise of a convertible as such, regardless of whether the soft top is open or not. With some experience, he is still able to assign the individual noise components to different sources. For example, he detects the noise caused by the tyres independently of the aerodynamically induced noise due to flow fluctuations on the vehicle skin. These findings entitle us to interpret the frequency composition of a sound detected by a measuring microphone close to the ear. A prerequisite, however, is that the microphone is always positioned at the same point during repeated measurements. In detail, however, it is difficult to explain the performance of the human ear indicated above from the interpretation of this frequency spectrum alone.

Fig. 8.10 shows the frequency composition of a passenger car interior noise as a function of the driving speed (Bubb 1996).[3] The picture shows the typical dependency: with increasing speed the volume level also increases, i.e. the whole noise mountain range is raised (see also Fig. 8.9). Overall, the spectrum shows a triangular shape with a peak at about 70 Hz, whose position on the frequency axis is independent of speed and motor speed. This is probably due to the resonance characteristics of the vehicle cab. Below 70 Hz, the noise level is raised mainly as a function of the speed, mainly due to rumbling noises from the chassis. The range above the top of 70 Hz increases with engine speed (which can be separated if measurements are made with different gears). In the upper frequency range (>4000 Hz) hiss-

3 The above-mentioned experiments were carried out in the mid-eighties; today's interior noise levels are about 10 dB lower than these measurements. This means a halving of the perceived volume.

ing noises are added at higher speeds. Individual frequency peaks, whose position depends on the motor speed, are superimposed on this noise mountain. With increasing engine speed they migrate to higher frequencies. The lowest frequency n is calculated from the motor speed N [1/min] as follows.

$$n = N / 60[\mathrm{Hz}] \qquad (8.1)$$

In a four-cylinder engine, the 2n peak shows the highest volume level according to the ignition sequence. For a six-cylinder engine, this applies to the 3n peak. In relation to this, there are increasingly weaker peaks at integer multiples of the base peak. Depending on the design of the exhaust system, half-numbered orders can also be observed. In particular, the integer engine orders from the third order onwards, which are characteristic of six-cylinder in-line engines in particular, are perceived as "silky smooth", while the second order in four-cylinder engines is responsible for a rather "grumpy" sound impression. The 1.5-fold engine order for eight-cylinder engines is responsible for their characteristic "babble" (Zeller 2011). All in all, these peaks of higher order and especially their loudness behaviour as a function of speed and load make up the acoustic characteristics of a motor.

8

8.2.1.3 Interpretation of Driving Noise

In the study mentioned above (Bubb 1996), the assessment of the acceleration of vehicles with different engine characteristics was compared with their noise characteristics. It was found out specifically:

- A vehicle whose noise spectrum increases only slightly with speed, engine speed and torque is considered to be comfortable but not particularly powerful. A more pronounced peak level emphasis improves this assessment more than a general increase in the overall noise mountain.
- Vehicles which are characterized by a noise spectrum during the acceleration process which increases with speed, especially in the higher frequency range, cause a similar acceleration assessment as vehicles with increasing peak levels.

On the basis of this experience, a noise simulator was developed that could generate different noise compositions in a real vehicle that was specially insulated for the tests. The recommendations quoted below were therefore based on the objective performance of the vehicle, which was always the same and only the acoustic impression was modified. For a good feeling of acceleration the following resulted:

- With increasing motor speed, the frequency edge > 70 Hz should increase,
- with increasing load (= demanded torque of the motor) the frequency slope should decrease > 70 Hz, that means the peaks should be audible more clearly,
- With increasing speed and load, the overall volume level should increase only slightly,
- the frequency edge < 70 Hz should increase only slightly with speed and rpm.

A negative effect was observed when individual peaks stand out clearly from the noise spectrum. This is described as uncomfortable, but can give a sporty impression. It turned out to be particularly favourable if the peaks only protrude from the noise mountain during the acceleration process. A noise level that was too low (< 65 dB (A)) proved to be unfavourable, indicating the feedback character of the noise for the driver..

8.2.1.4 Sound Design

The importance of noise for the subjective enjoyment of a vehicle has recently gained enormous importance. Sound design represents the creative creation of sounds in order to convey a specific hearing impression to the customer. On the basis of factor-analytical methods applied to semantic profiles with which test persons assessed the acoustic characteristics of engine noise, four largely independent descriptive factors were found. It's these:

- Loudness (e. g.: "loud – quiet", "booming – muffled"),
- Dynamics (e. g.: "sporty – unsporty", "weak – powerful"),
- Hardness (e. g.: "rough – smooth", "uniform – pulsed"),
- Timbre (e. g.: "low – high", "blunt – sharp").

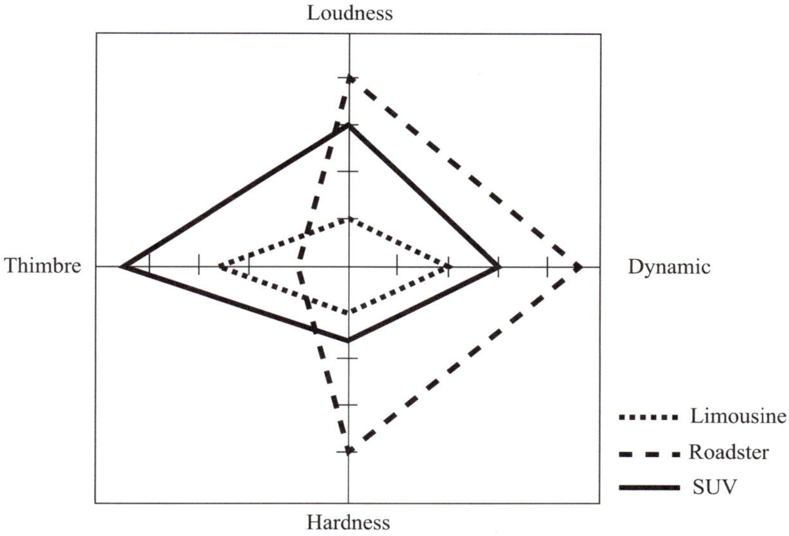

Fig. 8.11 Noise character profile for different vehicle types. (From Zeller 2011)

Zeller (2011) explains that these orthogonal perception dimensions obtained by means of statistical analyses are a further step towards psychoacoustic modelling, whereby further research is necessary in order to identify those parameters on the physical or psychoacoustic level which form the basis of the respective hearing phenomenon. On the basis of these perception dimensions, target profiles can at least be defined for different vehicle types, represented in the form of the spider diagram technique (see Fig. 8.11).

The acoustic comfort impression in the vehicle interior is mainly determined by the noise at constant speed and thus by the wind roll noise (see also Fig. 8.9). However, the vehicle dynamic impression during longitudinal acceleration is significantly influenced by the prominent engine noise. One measure of this is the level jump from the noise level at constant speed to that during acceleration under full load. The level of wind roll noise and engine noise must therefore be adjusted so that only the former can be heard at constant speed and, at the same time, the level jump of the engine noise can sufficiently emerge when accelerating. The maximum acceptable overall level must not be exceeded (Zeller 2011).

By means of active noise control (ANC), it is possible to at least partially compensate the interfering low frequencies by overlaying them with a corresponding counter spectrum by measuring the sound spectrum near the ear and applying various techniques (see Zeller 2011, page 209 ff.).[4] This method could be used, for example, to record the sound effects via the vehicle loudspeakers which have a positive influence on the perceived dynamics as described in ► Sect. 8.2.1.2. In this way, a comfortably quiet vehicle could be achieved at a constant speed, which is able to generate the emotionalising sound especially during acceleration processes. It would also be conceivable to meet different driver wishes with selectable sound programs. In particular, it is possible in this way, as is already happening in part today, to enhance the rather sober noise characteristics of a four-cylinder engine, which is preferable for consumption reasons, accordingly.

The future use of electric motors as a general drive source in vehicles places completely new demands on interior acoustics. As Fig. 8.12 shows, the engine noise of an electric vehicle is completely masked by wind and rolling noises from a speed of around 40 km/h and above. Many people who have had their first experience with electric vehicles have reported

4 In principle, this method only works for limited areas within the vehicle cabin. For reasons of energy conservation, the amplitude of the corresponding frequencies must be doubled at other points by the additional introduction of the compensation sound.

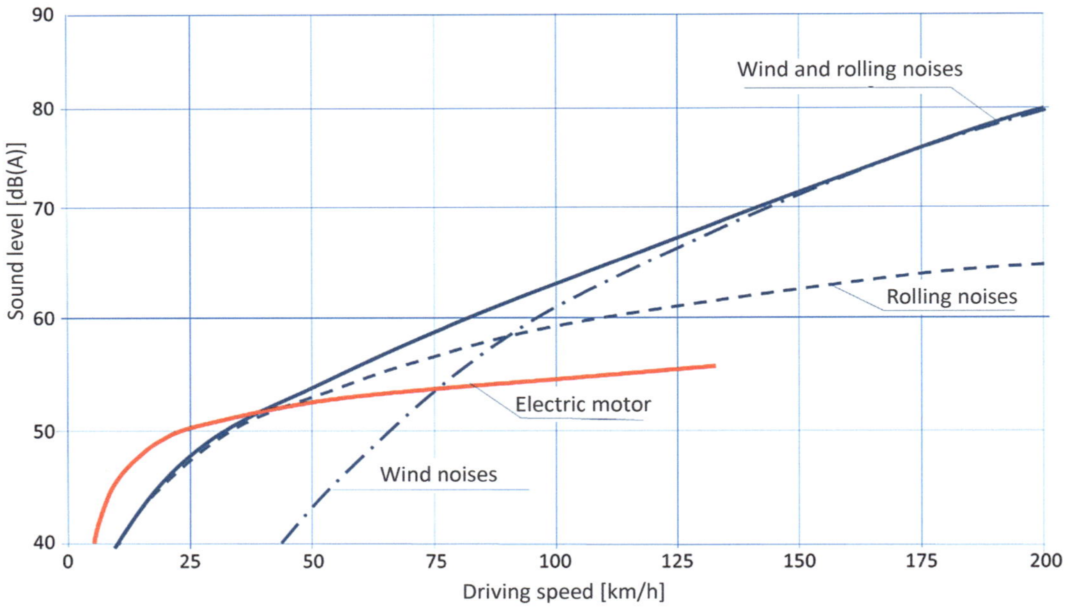

Fig. 8.12 Noise level in the vehicle interior as a function of driving speed and driving conditions with an electric drive. (According to Vogl 2010)

that they respond spontaneously to changes in the accelerator pedal position, but at the same time also complain about the "tram-like" driving noise. The even increase of volume level and frequency range with speed is at least compared to the usual sound of a vehicle with internal combustion engine needs getting used to. Obviously, the division of the entire speed spectrum into several upturns of the engine has a positive effect on the subjective feeling that "something is moving" (compare Fig. 8.9 with Fig. 8.12).[5] However, it is also questionable whether it can make sense to artificially introduce the noise of a combustion engine into the inherently quiet electric vehicle. All this shows that there is still a considerable need for research into the subjective experience of driving noise in electric vehicles.

With regard to the additional problems that arise in hybrid vehicles when the combustion engine is switched on or off during driving, reference is made to the treatise in Zeller (2011, p. 214).

8.2.2 Small Noises

With the reduction of noise levels in modern motor vehicles, the problem of annoying rattling, creaking or scraping noises has become more and more important, as these now stand out from the overall noise level and have thus become more and more dominant. In addition: if a disturbing noise is once consciously perceived, it remains in the foreground of attention and thus has an increasingly annoying effect.[6] Annoyance is a largely subjective phenomenon, which can vary greatly depending on the frequency of occurrence, intensity, localization, personal associations and noise character. In this context, it is also necessary to distinguish between disturbing noises which have nothing to do with the driving process or other functions of the vehicle and noises which are connected with certain functions and, if necessary, also give feedback on the use of the corresponding function as confirmation noise. Table 8.1 gives an exem-

5 The constant engine speed during the acceleration process in the initial designs of CVT transmissions was also judged negatively by most drivers.

6 A well-known example of this is the ticking of the clock, which disturbs sleep once attention has been drawn to it.

Table 8.1 Classification mechatronics of operating noises (according to Zeller 2011). Grey background: most likely perceived as annoying

	Noise	Confirmation sound
Short operating time	Steering booster Level control Secondary air pump Braking noises (rubbing, squeaking and ä.)	Window lifter/sliding roof Seat adjustment Mirror adjustment Defrosting plant
Long operating time	Motor fan Fuel pump Roll stabilization	Air-conditioning fan Windscreen wiper Seat fan

plary compilation. All noises that appear to the driver as if by chance and not directly influenced by him are perceived as annoying disturbing noises. This applies all the more, the longer the corresponding noises are perceptible. For many self-initiated actions, however, the corresponding sounds can be understood as feedback for the action performed. However, this is only valid if the operating time is correspondingly short (e.g. only extends to the corresponding duration of the operating element operation). If the operating time is long, the perception of the corresponding noise quickly turns into annoyance.

In contrast to these noises which can be assigned to a function, contact point noises in particular, which are caused by relative movements of components to each other, are perceived as extremely unpleasant. Technically, these noises can be divided into impact noises (rattling, buzzing), stick-slip noises (creaking, cracking, squeaking) and other noises (e.g.: loosening of adhesive joints, smacking). They are stimulated by vibrations in the drive train, unevenness of the road surface and by specific acoustic frequencies in the Hi-Fi system. Regarding the evaluation of these annoying noises, the simple requirement is: "Disturbing noises must not be evaluated, but turned off!". (Moosmayr 2011). However, it is necessary to prioritise the measures according to probability of occurrence and annoyance within the framework of a cost-benefit analysis. However, hearing impressions from different customer groups are only assessed to a limited extent in a uniform manner (Moosmayr 2008). While some test persons classify a certain noise as "very annoying", other test persons do not perceive the same noise at all. When made aware of this, they also classify it as "very annoying"(see also footnote 7). In addition, the assessment as background noise depends very much on the situation-related expectations. From the corresponding investigations by Moosmayr (2008) it can be summarised that the acceptance of background noise is significantly lower with low vehicle excitation (e.g. city driving at constant speed on asphalt) than with strong excitation (e.g. cobblestones or potholes). This again points to the fact that masking effects of driving noise can conceal useful and background noise (Grimm et al. 2007). The Zwicker method (Zwicker and Fastl 1990) for determining loudness makes it possible to objectify this effect. Moosmayr (2008) presents a new method for finding sources of background noise which, in combination with a hydropulse/shaker system, evaluates the background noises that occur, combines identical noises into clusters and automatically creates a uniform problem documentation.

The acoustic experience associated with opening and closing the vehicle door is a special feature with regard to noise evaluation. Consciously or unconsciously, it already conveys an impression of solidity, which is often used to draw conclusions about the quality of the entire vehicle. According to Liebing (2009), test persons prefer significantly higher sound levels for closing noises (preferred 58 to 62 dB (A)) than for opening noises (preferred 50 to 54 dB (A)). Levels above or below this level are increasingly rejected, regardless of the vehicle type. A rather dark sound image (emphasis on the bass range < 100 Hz) without audible clicks (noise shortly after the main sound with frequencies >3 kHz) with at most slight pops (similar to champagne corks) is desirable. Any kind of post-oscillation is rejected (appears to be "tinny") and reduces the quality impression of the door slamming noise (see Zeller 2011 for details).

8.2.3 Useful Signals

In contrast to the small noises described above, which should "hide" behind the noise mountain of the driving noise if possible, useful signals must protrude clearly from this noise mountain so that they can be perceived safely. Useful signals in this sense are all acoustic feedback signals (see also ▶ Sect. 6.2.1.1), signals from the infotainment systems installed on the vehicle (hi-fi system, voice instructions of the navigation system, telephone system, reading of e-mails and internet messages etc.), understanding of mobile infotainment systems (smartphones) and last but not least the possibility of talking to passengers of the vehicle.

In order to objectify the protrusion of useful signals against the noise mountain of driving noise, the methods developed for measuring speech intelligibility can be used. All common methods are based on the assumption that the entire auditory frequency range is divided into different frequency bands. Usually, the third-octave bands are used for this purpose, since they largely correspond to the 24 frequency groups found by Zwicker (bark), in which the loudness impression in the inner ear is formed (see Zwickler and Fastl 1990). Each of these frequency bands k carries a weighting factor g_k for comprehensibility or audibility. g_k is maximum if only the undisturbed useful signal is present in the respective frequency band and zero if it is completely covered by the driving noise. In each frequency band, the signal-to-noise ratio $(S/N)_k$ is calculated from the measured signal level S[dB(A)] and the noise level N[dB(A)]. Values > 15 dB or < 15 dB are limited to these numbers, as signal-to-noise ratios from 15 dB are already full speech intelligibility and – 15 dB represent absolute incomprehensibility. For simple signal tones, which are only presented in a narrow frequency band, the same limit values naturally apply. It is generally assumed that the signal tone must be at least +6 dB higher than the weighted volume level in the corresponding frequency band. The articulation index AI or the speech transmission index STI[7] is determined as a weighted sum over all frequency bands:

$$\text{AI or STI} = \sum_k g_k (S - N)k \tag{8.2}$$

The weighting factors g_k represent a frequency weighting dependent on the language material. For example, with whole sentences and multisyllable words, the low frequencies are relatively more important than with monosyllable speech material, which is characterized by less redundancy. The resulting index is a quantity normalized to the range from 0 to +1, which represents a measure of intelligibility and is strictly monotonously related to speech intelligibility for the speech material used.

From the dependence of speech intelligibility as well as the recognition of signal tones it is already clear that the volume level of the useful signal must be raised or lowered as a function of the volume level of the driving noise, taking into account the above premises. In principle, this also applies to hi-fi systems installed in the vehicle.

Especially in premium automobiles with high-quality hi-fi systems, the reproduction of audio signals represents a particularly outstanding point of complaint in customer surveys. As Zeller explains, this is usually not caused by disturbances in the electrical part of the audio system, but in many cases by mechanical interfering noises, which are reso-

7 AI and STI differ only in the determination of the S/N ratio (for details see Zeller, 2011, chapter 8.4)

natorily excited by the powerful audio system. The acoustic coupling of the loudspeakers to the body structure is the main cause of errors.[8] Otherwise, it is up to specialists to design the right choice of loudspeakers and their localisation in the vehicle interior. The individual taste with regard to the resulting spatial sound plays an outstanding role.

8.3 Vibrations

Both in the technical area and in the area of ergonomics, movement phenomena are treated under the generic term "vibrations". Just like the acoustic effects, the vehicle vibrations also have a dual character: on the one hand, they contribute considerably to discomfort when the driver or passengers feel "shaken" by the vehicle. On the other hand, the driver needs a sufficiently precise perception of the vehicle movement to ensure safe vehicle guidance (see ◘ Fig. 8.1). Acoustic and vibration discomfort are thus closely related disciplines in several respects, since in both cases the adequate stimuli are induced by structure-borne sound, but in different frequency ranges. While the vibration sensitivity of humans ranges from 0 Hz to a maximum of approx. 500 Hz (vibration sensitivity, see ► Sect. 3.2.1.3), an acoustic sensation only begins clearly via 20 Hz and reaches at least 10 kHz for older persons and 20 kHz for young persons.

8.3.1 Vibration Phenomena

The passengers in a vehicle are exposed to vehicle movements or vibrations in different ways. Due to road unevenness, the entire vehicle is set in motion, which is essentially represented as a lifting movement (translatory in z-direction), a pitching movement (rotatory about the y-axis) and a rolling movement (rotatory about the x-axis). These movements are damped by the suspension system, but may also be amplified in the natural frequency range. In addition, there are vibrations transmitted from the engine and power transmission to the vehicle body, as well as aerodynamic effects that can also cause the entire body to vibrate. These vibrations are mainly transmitted to the driver via the contact surfaces of the floor, steering wheel, seat and backrest. Other contact surfaces, such as armrests, are also transmission elements for vibrations. With the exception of the steering wheel, the last mentioned transmission paths also apply to the remaining passengers of the vehicle. ◘ Figure 8.13 shows the mentioned vibration sources and transmission paths.

Analogous to the acoustic categorization tone, sound and noise as well as impulse sound ("bang"), one can also distinguish between purely sinusoidal excitations, periodic excitations (superposition of several sinusoidal oscillations), stochastic and impulse-like excitations ("shock") in the area of oscillations. Sinusoidal and periodic excitations in the low frequency range should always be avoided in order to prevent resonance effects with human organs (see below). Stochastic excitations and pulse-like excitations differ fundamentally from their sensory quality.

As the frequency ranges of acoustic and vibration sensation overlap, there are certain phenomena that are both heard and felt. ◘ Fig. 8.14 gives an overview of various characteristic frequency ranges in the vehicle, which Knauer (2010) has compiled by evaluating different literature references.

8.3.2 Perception of Vibrations

The human organism perceives vibrations via various sensory organ systems. In particular, the vestibular organ, various nerve endings in the skin surface and the entire proprioception of the body position are involved (Griffin 1990; see also ► Sect. 3.2.1.3). Each of these systems has its maximum sensitivity at specific frequencies. In addition, the vibration judgment of test persons in a vehicle is significantly influenced by perception from other sensory organs, in particular from the optical

8 Mounting loudspeakers on flexible cladding parts can cause them to vibrate mechanically, thereby also reducing the acoustic efficiency of the loudspeaker.

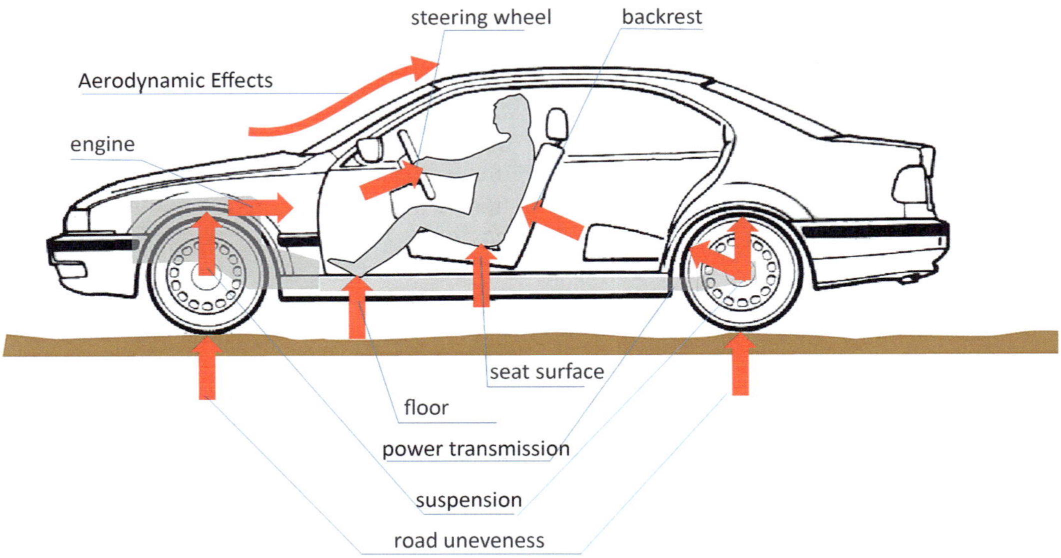

Fig. 8.13 Vibration sources and transmission paths on a vehicle moving on an uneven road. (According to Mansfield 2013)

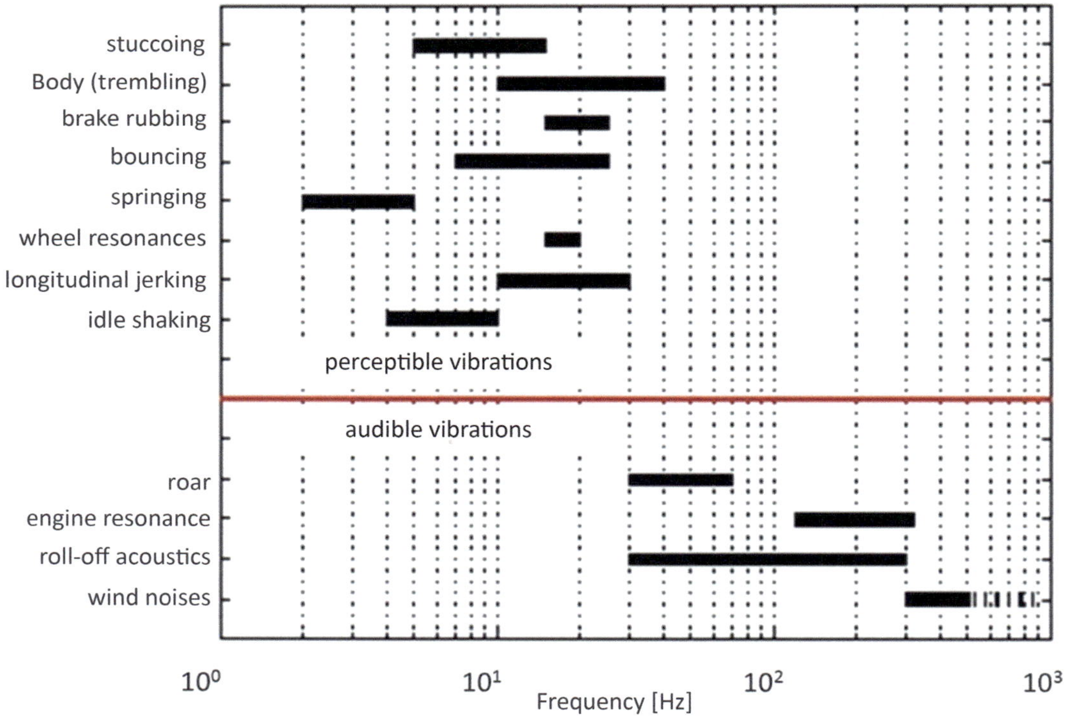

Fig. 8.14 Characteristic frequency ranges in the vehicle. (Knauer 2010)

and acoustic channel. It is therefore difficult to define a single determinant quantity that describes the influence of vibrations in the vehicle (Mansfield et al. 2007). The human response to vibrations has been extensively tested in the laboratory. According to this, the perception of vibrations is particularly sensitive to frequencies in which the human body

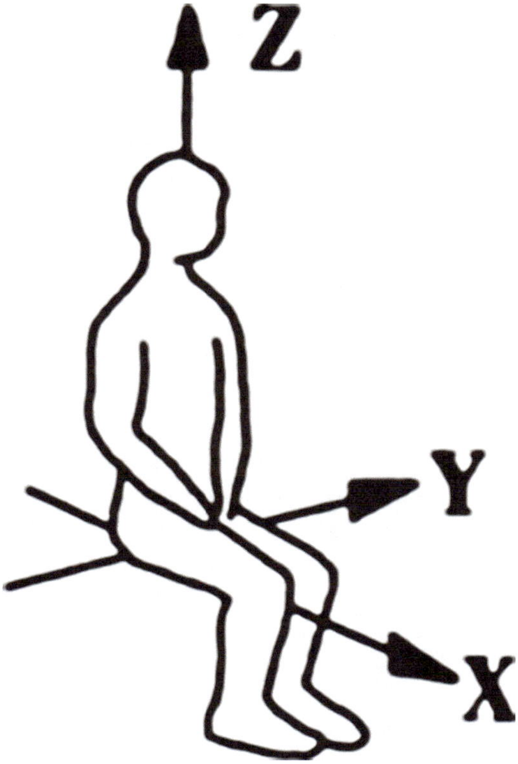

Fig. 8.15 Human coordinate system

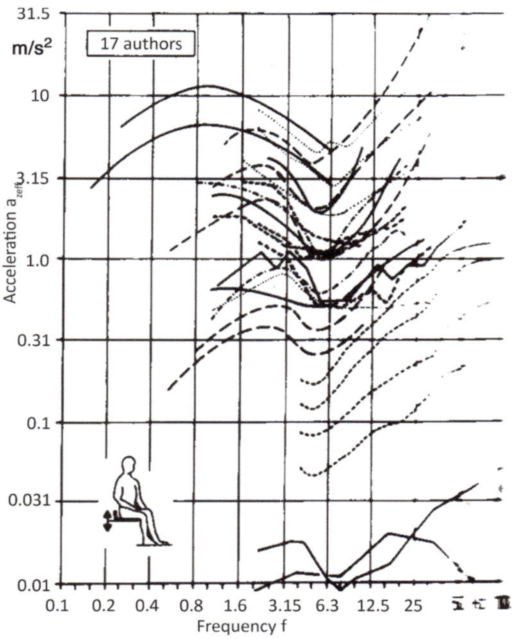

Fig. 8.16 Frequency-dependent curves of equally strong vibration perception – research results of 17 authors. (Dupuis 1993)

shows biomechanical resonances (Mansfield 2013). To objectify this, one uses the human-related coordinate system shown in Fig. 8.15. The effective value a_{eff} of the acceleration, which is defined as the square average of the measured acceleration a(t) over the exposure time T, is used as the key figure for the influence of vibration:

$$a_{eff} = \sqrt{\frac{1}{T}\int_0^T a(t)^2\,dt} \tag{8.3}$$

Dupuis and Zerlett (1984, 1986) compiled curves of equal strength of perception for the effects of vibrations on the seated person in the z-direction (Fig. 8.16).

From these results it can be seen that, independent of the vibration exposure a_{eff}, a sensitivity maximum lies between 4 Hz and 8 Hz, but is also sensitive to very low frequencies (< 0.4 Hz). Knauer (2010) compiles resonance ranges of the human organism from studies by Dupuis (1969), Dupuis et al. (1974), Hennecke (1994), Recknagel (1995) and Scheibe (1979) (Fig. 8.17). Dupuis (1993) shows in another compilation of literature that the transmission in the z-direction between fuselage and head has a resonance at about 4 Hz and that the eye has a resonance frequency of about 20 Hz.

From all this it can first be concluded that extremely low frequencies should be avoided (see also ► Sect. 8.3.5), that the frequency range around 4 Hz is particularly critical, that frequencies around 20 Hz should also be technically absorbed in order to avoid visual disturbances and that, as explained in more detail in ► Sect. 3.2.1.3, frequencies around 200 Hz should be dammed away as well as possible so that no unpleasant feeling of vibration arises. Within vehicle however, all these areas are subject to vibrations (see Fig. 8.14).

8.3.3 Vibration Evaluation

In VDI standard 2057-1 and ISO 2631-1 the currently valid evaluation procedures for mechanical vibrations are published. In the

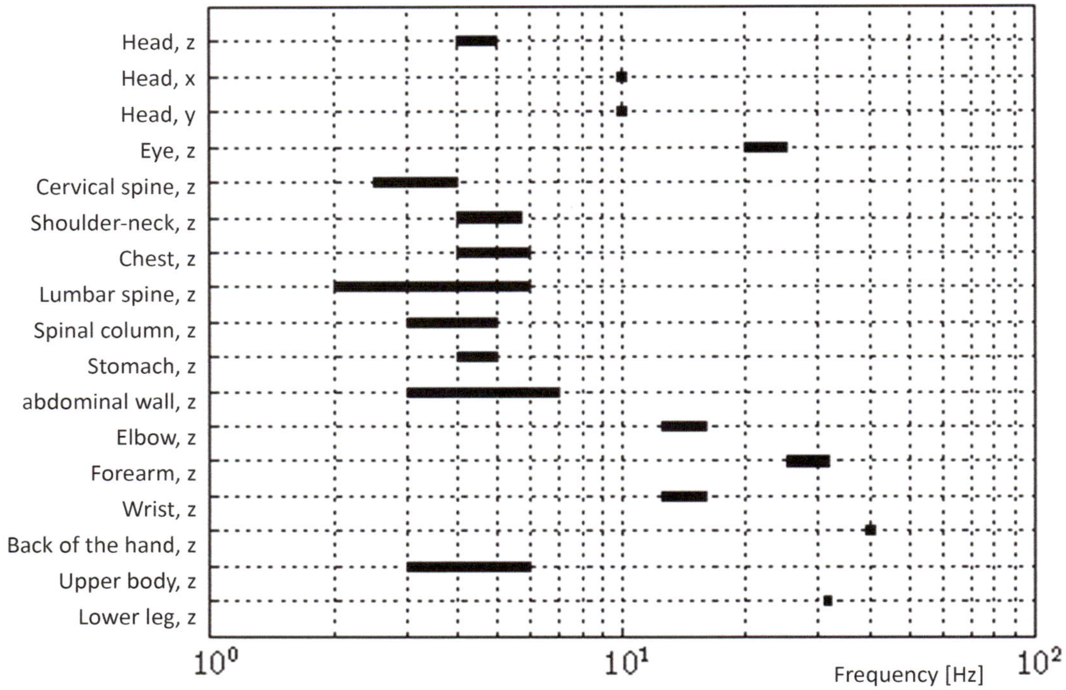

Fig. 8.17 Resonance ranges of various organs and body parts of the human organism compiled by Knauer (2010) from various literature sources

8

usual field of application in vehicle technology, the acceleration $a_w(t)$ transferred to the human body is measured by means of flat accelerometers mounted between the buttocks and the seat of the test person or at other points to be examined. Depending on the spatial direction of the vibration effect, the measured value is weighted frequency-dependently, whereby dependencies of the sensation, as shown in Fig. 8.16, are taken into account. From the values obtained in this way, the effective accelerations $\bar{a}_{wx}$, $\bar{a}_{wy}$ and $\bar{a}_{wz}$ are now calculated. The next step is the so-called multi-axis single-point excitation $\bar{a}_{wv}$ calculated. It is obtained by quadratic averaging of the weighted effective values in the individual spatial directions (vectorial addition). A further factor takes into account whether the excitation is via the seat or the backrest. After the excitations on the human body occur via several contact points, the last step the multiaxial multipoint excitation $\bar{a}_{wvges}$ is calculated. The procedure for this calculation and the individual formulas for it are compiled in Table. 8.2. In addition to the calculation formula for multi-axis multi-point excitation, there is also a tabular assignment of the values obtained to verbally defined discomfort values. A value $\bar{a}_{eff} > 0{,}5\,m/s^2$ can be used as an orientation which represents the trigger limit for damage to health caused by all-day exposure. However, this value is practically never reached in normal passenger car driving.

As Knauer (2010) criticizes, the evaluation of stochastic suggestions according to VDI 2057–1 cannot under all circumstances be reconciled with the judgements of test persons. In addition, the assessment of vibration phenomena containing shocks is not possible by such a purely spectral method. Neither the number of shocks nor their distribution in the time signal are taken into account. A large number of correlative approaches have been developed to improve the characteristic values obtained by the VDI standard using various frequency-dependent weighting factors. Among these, the Hennecke procedure (1994), which provides for a correction via a so-called instationarity factor, deserves special mention. This results in a significantly improved correlation

Table 8.2 Calculations of the evaluated vibration strength according to VDI 2057 (from Knauer 2010)

Single-axis single-point excitation

$$\bar{a}_w = \sqrt{\frac{1}{T}\int_0^T a_w^2(t)dt}$$

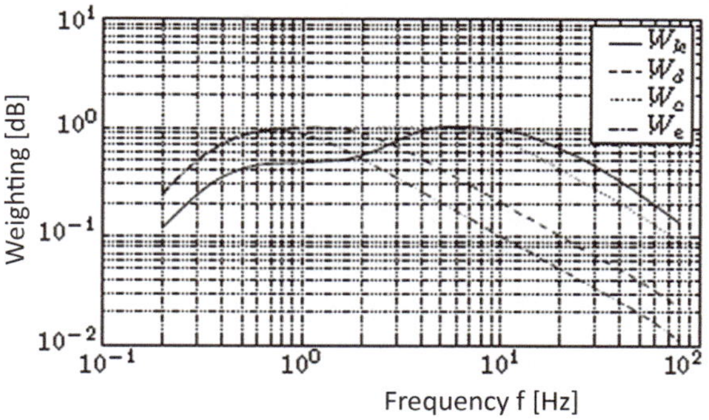

Measuring location	Direction	Factor
Seat surface	z	W_k
	x,y	W_d
Backrest	x	W_c
Foot platform	x,y,z	W_k
Rotation	r_x, r_y, r_z	W_e

Multi-axis single-point excitation

$$\bar{a}_{wv} = \sqrt{k_x^2\bar{a}_{wx}^2 + k_y^2\bar{a}_{wy}^2 + k_z^2\bar{a}_{wz}^2}$$

Measuring location	Direction	Factor
Seat surface	x,y,z	1.00
Backrest	x	0.80
	y	1.50
	z	1.40

Multi-axis multipoint excitation

$$\bar{a}_{wv_{ges}} = \sqrt{\bar{a}_{wv_1}^2 + \bar{a}_{wv_2}^2 + \ldots + \bar{a}_{wv_n}^2}$$

$\bar{a}_{wv_{ges}}$ $[m/s^2]$	discomfort level
< 0.315	not unconfortable
0.315 to 0.63	somewhat unconfortable
0.5 to 1	distinctly unconfortable
0.8 to 1.6	unconfortable
1.25 to 2.5	very uncomfortable
>2	extremely uncomfortable

to subjective assessments. The literature also describes numerous approaches that determine the comfort judgement with the aid of artificial neural networks. An examination of the known applications of neural networks on the objectivization of the evaluation of vibrations shows the highly experimental procedure in the selection of network parameters. Although very good results can be achieved, the transferability to new unknown data sets is difficult to verify (a compilation of all this work can be found in Knauer 2010).

In connection with vibration influences, the seat is traditionally considered to be of great importance, since it is assumed that good seat suspension combined with appropriate damping properties can absorb the transmission of hard impacts to humans. Since experiments with real persons encounter a variety of difficulties and, particularly because of the individual deviations mentioned above, it is difficult to objectify the behaviour of the seats, attempts are made to resort to technical solutions. In many cases,

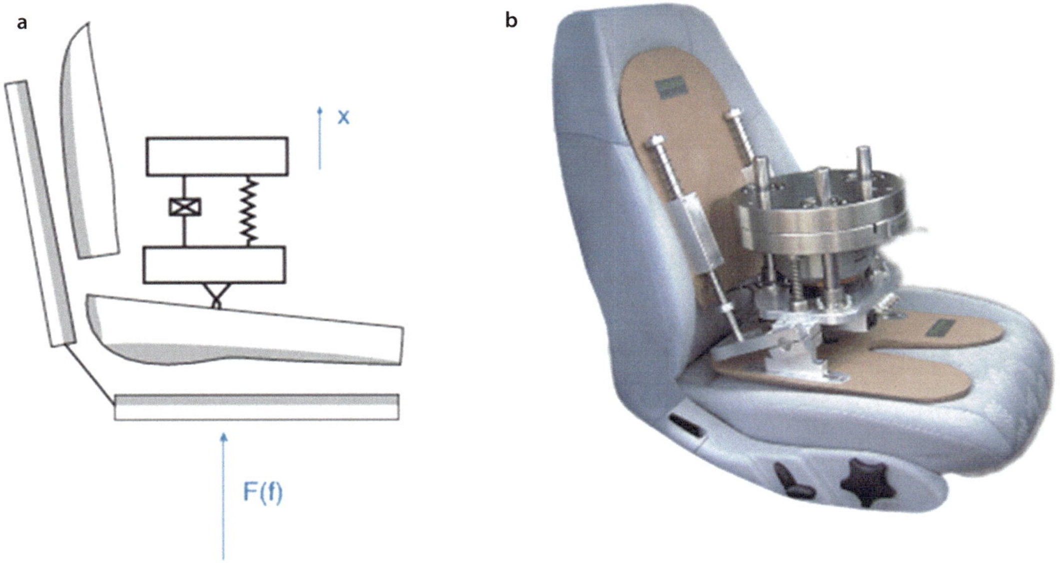

Fig. 8.18 Principle sketch **a** and **b** structure of the Memosik test specimen. (According to Wölfel)

so-called "water dummies" (plastic vessels roughly imitating the shape of a seated person, consisting of thigh and back, which are filled with water so that the seat load corresponds approximately to that of a test person) are used for this purpose and the transmission behavior between the vibration initiation measured on the seat rail and its contact surface to the seat is measured. Since this simple procedure cannot reproduce the complex oscillation behaviour of humans at all, attempts have been made to use appropriate mechanical reproductions of humans for this purpose. The most common and at the same time very practical method is the Memosik test specimen developed by Wölfel (Fig. 8.18). An active control system makes it possible to adapt the so-called mechanical impedance $I = F(\nu)/\dot{x}(\nu)$ to that of humans. This provides a means of objectifying the vibrational behaviour of the seat.

However, the investigations by Bitter et al. (2005) show a high correlation between the vibration influences, as measured on the seat rail, on the vehicle floor and on the steering wheel, and those on the seat surface. In addition, this study shows that there is obviously a kind of constancy performance with regard to static seat pressure and vibration behaviour, i.e. at least in the range relevant for passenger cars, occupants are clearly able to distinguish between the static seat pressure and the dynamic properties resulting from the movement of the vehicle. This observation contradicts the model postulated by Ebe and Griffin (2000), according to which a dynamic factor is based on the static factor, which causes an increase in the overall discomfort with increasing oscillation amplitude (and increasing time).

In all the methods described so far, it is necessary to resort to test persons whose influence on the result is not insignificant due to their individual biomechanical properties. In addition, all these procedures do not allow a prospective evaluation of the vibration influence, i.e. possible already in the planning phase. The greatest progress can therefore be expected in the application of biomechanical human models. Knauer (2010) gives an overview of the possible models (Table 8.3). Especially the model CASIMIR is able to predict the transfer function between seat and human being well, but an evaluation regarding the subjective discomfort is still pending.

However, some studies by Knauer (2010) can provide a basis for the prospective discomfort evaluation of vibrations. The experience with the unbalance of a wheel, which can be clearly perceived on a flat road, but whose perception "disappears in the noise"on

Table 8.3 Overview of biomechanical human models for the calculation of seat contact and vibration influences (from Knauer 2010)

	Circumference	Main points
Moes (2000)	Detailed FE model of a thigh and half of the pelvis	Model for the investigation of seating comfort Imaging of soft tissues and bone structure The measured pressure distribution could not be reproduced with different material laws Model not validated
Brosh and Arcan (2000)	2D-FE model of the seated person	Soft tissue behaviour determined by indentation tests on test persons and transferred to model Good comparison between calculation results and test data Model for investigating tissue properties, less suitable for sitting comfort
Schmale (2002)	MKS model of the human being and FE model of the seat	Man modeled as rigid body on soft seat Buttocks not deformable Pressure distribution on seats cannot therefore be determined
Verver (2004)	MKS total body model and FE model of the buttocks	Bone and soft tissues as FE model based on the analysis of a postmortem male body On real seat and on wooden plate validated model for the evaluation of seat comfort dyn. Seating comfort on the basis of the MKS model in the range 0–15 Hz investigated FE model validated under static conditions
Siefert (2013)	Casimir, FE-model of the seated person	Reproduction of the complete skeletal structure Reproduction of the soft tissues of the thighs, buttocks and the back Reproduction of the abdomen and back muscles Variable in percentile and posture Seat pressure distribution calculable Seat transfer functions determinable

a bad road, shows, for example, that there is also a masking effect in the area of vibration perception. At least with regard to harmonic oscillations, it can also be determined in the experiment that the level difference that can be perceived at the moment hardly depends on the frequency, but the frequency difference that can be perceived at the moment increases with increasing reference frequency. Also in the field of vibrations, the effect of attention is known from the evaluation of acoustic signals. It is repeatedly observed that only when the evaluator consciously concentrates on the corresponding stimulus is he able to detect an overthreshold stimulus. The perceptibility threshold therefore appears to be a useful criterion for discomfort relevance. This also applies to the assessment of transient events, such as those that occur when driving over a pothole in the form of an impulse. In trials, Knauer did not use the steering wheel to transmit the impulses, since the impulses perceived there would rather be referred to as steering shock. It is quite clear that an acceleration signal in the seat surface at least by a factor of 3.50 and in the seat back at least by a factor of 3.24 must be greater than the effective value of the background noise in order to be perceived as an impulse. The result seems quite plausible if you consider that with white noise, to which no outstanding individual impulses would be assigned, already 2% of the signals are 2.5 times higher than its RMS value. The decay process of an

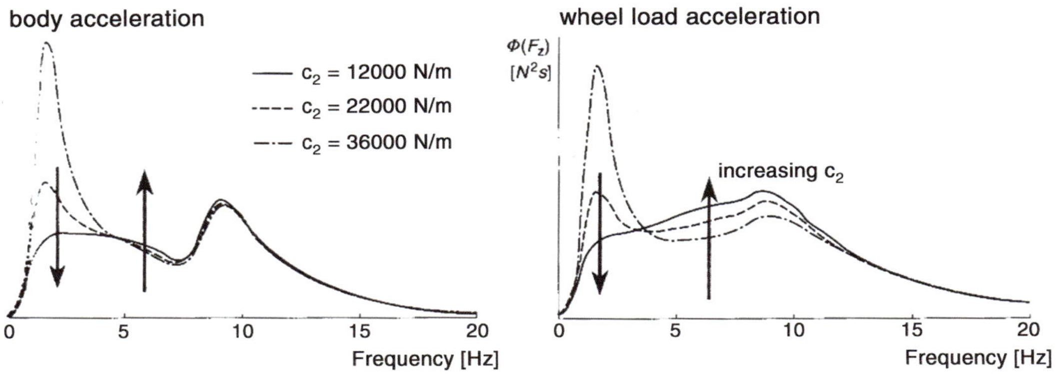

Fig. 8.19 Influence of spring stiffness c_2 on the power spectra of body acceleration and wheel load fluctuations (from Bootz et al. 2011)

Fig. 8.20 Transmission behaviour seat rail – seat surface for three different vehicle seats. (Bitter et al. 2005)

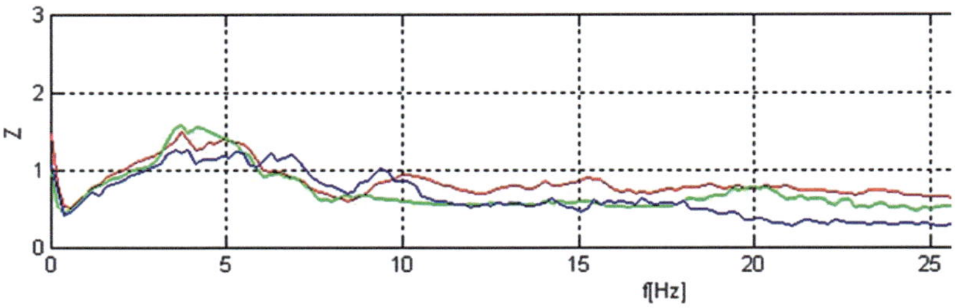

impulse also has an effect on the sense of comfort. This rebound process should be as short as possible.

8.3.4 Comfort and Driving Safety

The aim of a comfort-oriented chassis design is to keep the accelerations transmitted to humans as low as possible and at the same time to avoid natural vibrations as far as possible, i.e. the discomfort caused by acceleration forces should be minimized. Even a simple geometric consideration shows that large wheels and a long wheelbase can achieve a geometric average of the road unevenness transferred to the vehicle. Furthermore, elementary physical considerations indicate that a soft spring (low spring constant) in combination with a high mass (heavy vehicle) shifts the natural frequency of the system to desired low values (out of the range of human natural resonances, see Fig. 8.17). At the same time, correctly designed damping reduces overshoot when excited in the resonance range. In fact, however, the vehicle represents a complex vibration structure which, even in a simple description, must take into account the spring characteristics of the four tyres and the spring-damper connection of the four wheels to the vehicle body. Bootz et al. (2011) show on the basis of a simplified model that in the frequency range around approx. 1.5 Hz with increasingly softer suspension c_2 the build-up acceleration $\Phi(\hat{z}_2)$, whereby a second resonance peak at 10 Hz is not influenced by this. The wheel load fluctuations also decrease in the 1.5 Hz range, but increase slightly in the range up to 10 Hz (Fig. 8.19).

In all experiments reported by Bitter et al. (2005), an increase in resonance in the z-direction from seat rail to seat surface was observed at approx. 4 Hz (Fig. 8.20). Since there are many natural resonances of the human organism in this frequency range, excitation by the vehicle body in this range must be avoided. In this respect, the reduction of the build-up acceleration with softening suspension shown in Fig. 8.19 is a measure to be aimed at.

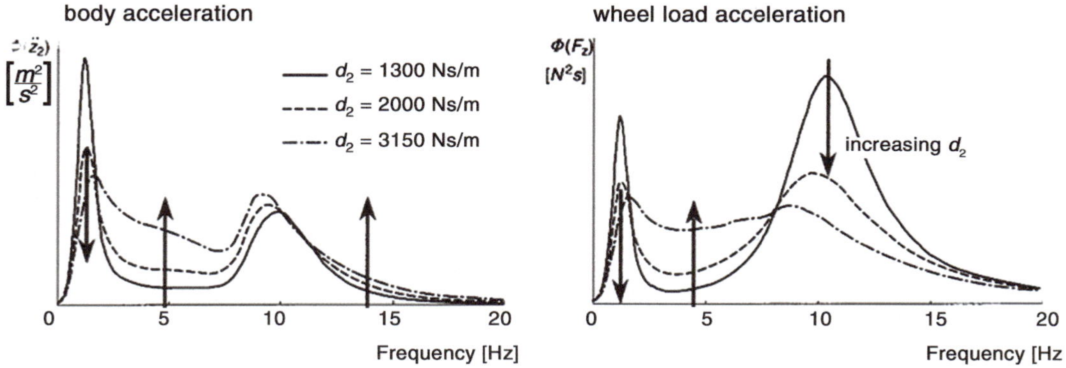

Fig. 8.21 Influence of dampers d_2 on the power spectra of body acceleration and wheel load fluctuations. (From Bootz et al. 2011)

However, the effects of pitch and roll oscillations must also be taken into account in this context. In addition, it plays a major role whether the vehicle is designed more comfortable for the driver or for passengers near the rear axle, so that finally a compromise must be found between the demand for low discomfort due to undesired body movement on the one hand and driving safety on the other, which is only achieved by good and even contact of the wheels with the road surface. For example, the optimum design for the chassis of sports vehicles can be based on small roll angles and lower load fluctuations at the expense of comfort, resulting in significantly harder springs, lower spring travel and correspondingly tighter vibration damping.

Fig. 8.21 shows on the left the power spectra of the vertical body acceleration of the vehicle for different damper characteristics when driving on a middle country road. It can be seen that in the frequency range of approx. 2 to 30 Hz the soft characteristic curve of the damper reduces the build-up acceleration. From 0.3 to approx. 1.5 Hz, however, the hard characteristic curve reduces the acceleration amplitudes and thus the vibration discomfort. With regard to driving stability, wheel load fluctuations in particular must be taken into account, which with increasing damping result in a reduction in the frequency range around 1.2 Hz and around 12 Hz, but at the same time in an increase in the range around 5 Hz (Fig. 8.21 right). In principle, these different requirements can be adjusted by means of adjustable damper systems. As part of the technical development, various adjustable damper systems were developed which are now almost universally used in vehicles of the higher performance classes (see further details in Bootz et al. 2011). An even better adaptation to different requirements is possible by the combination with active springs, by which – depending on the technical design – different loading conditions can be considered. Of all the possible designs for this, the air spring has proven to be particularly favourable because it can achieve an almost constant body natural frequency even under different loading conditions of the vehicle. Bootz et al. (2011) state that today, for cost reasons, most chassis of a conventional nature are still constructed from passive components, but that the next technical evolutionary step in the sector of controlled chassis up to the so-called "previewing" chassis, which contain a temporal and spatial forecast for lane guidance and road unevenness, has to be sought. The Magic Body Control recently introduced by Mercedes in the S-Class represents a first series development for the latter system. Up to a speed of 130 km/h, it is able to adjust the spring and damper rates to the road unevenness detected by a stereo camera, thus minimizing bodywork fluctuations without having to compromise driving safety. Since the system is based on optical detection by a camera, it automatically switches back to reactive active suspension in poor lighting conditions (night, rain, snow and fog).

The vibrations transmitted to the driver by the vehicle body not only affect discomfort, but possibly also driver performance. In their investigations, Baker and Mansfield (2010) found no significant differences between the control power without vibration and the effect of vibration similar to that of off-road driving. If, however, the subjective stress is examined using the NASA TL-X questionnaire, an increase in stress under the influence of vibrations is clearly evident. The same was observed for objective performance in speed maintenance. A particular problem is the handling of a touch screen under the influence of mechanical vibrations. As Moseley and Griffin (1986) state, visual perception is most affected when the display moves relative to the observer and least when the observer and display move in phase. In the vehicle, the target (touch screen) and driver are both in motion, but the relative motion depends on the respective frequency. At the resonance frequency of the driver's arm, for example, there may be significant relative movements between the driver and the controls. Especially with pointing tasks, there can be more than the double movement compared to that of the seat (Griffin 1990). An effective method to improve the driver's performance, to select a small target, is to allow the hand to contact an environmental object while operating. This is easily possible when operating controls on a traditional dashboard, but hardly feasible with a large touchscreen (an effect that, by the way, is barely noticeable when checking in a vehicle salesman's stand or showroom).

8.3.5 Kinetosis

Passengers in a vehicle often suffer from symptoms of the so-called motion sickness (kinetosis). The most common symptoms are: heat, dry mouth, headache, drowsiness, lethargy, unpleasant feelings in the stomach, seasickness and often – vomiting. The generally accepted theory for the occurrence of kinetosis is the "sensory conflict theory"or "sensory rearrangement theory". This theory is based on the idea of the combined processing from different sensory organs described in ► Sect. 3.1.3 (see also ◘ Fig. 3.11). The organism then assembles optical information (foveal and peripheral) and information from kinesthetic, haptic and proprioceptive perception into a single, inherently consistent motion perception. For the existence of everyday demands, perception of movement is also always connected with the expectation: "what information will I receive when I initiate a certain action? (see ► Sect. 3.2.2.4 and specifically ◘ Fig. 3.47). Mansfield (2005) has illustrated this theory by the images shown in ◘ Fig. 8.22 using the example of the driver and a passenger in the same vehicle who does not take care of the driving process (e.g. by reading a book). While the information perceived optically, kinesthetically (vestibular) and haptically (somatically) by the driver is completely consistent with his expectation of what will happen when he turns the steering wheel (a), for the accompanying reading passenger there are inconsistencies in his optical perception between the movement perceived consistently via the vestibular channel and the force via the haptic channel, which are also not compensated by conscious processing (b). This example is typical for many observations of kinetosis (Probst et al. 1982). It is also common experience that the same people who tend to have kinetosis as co-drivers have no complaints of this kind when they themselves act as drivers.

Mansfield (2013) suspects that the cause of these observations is a side effect of the body's defence against dangerous toxins. Many potentially dangerous, even lethal substances affect the sensory organ system before they cause permanent damage. It is then a sensible reaction of the organism to eliminate the cause of this effect from the food tract as quickly as possible. Excessive alcohol consumption is a well-known example of this reaction of the organism. However, this reaction mechanism, which developed over a long period of evolution, loses its meaning in connection with modern means of transport.

In order to prevent kinetosis, it therefore makes sense to provide passengers with as good a view of the road as possible (see ► Sect. 7.2.5.1 for the conflict arising for rear-seat passengers as a result of passive safety regulations). In addition to the aforemen-

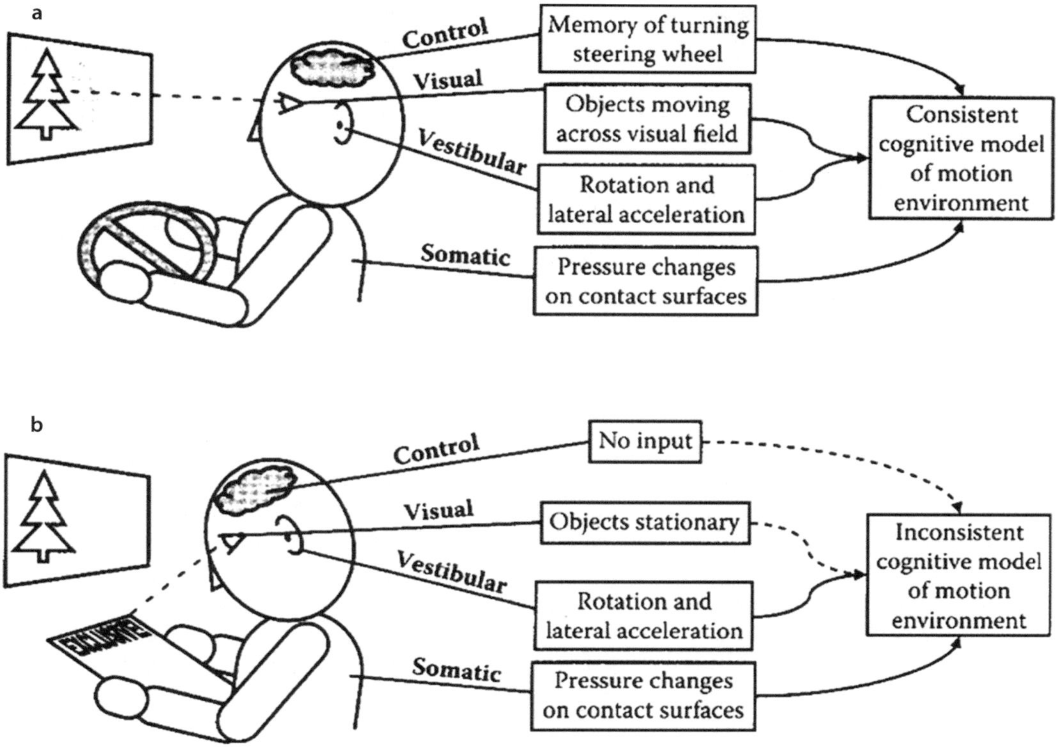

Fig. 8.22 Illustration of information paths for motion perception. (From Mansfield 2013 and 2005)

tioned perception conflict and possibly in connection with it, both laboratory and field studies have found that frequencies in the range of 0.2 Hz in particular cause kinetosis (Lawther and Griffin 1987). These are frequencies that are mainly caused by the formation of the road. So pitching, rolling and lifting movements within this frequency range should be avoided as far as possible. However, a particularly soft – apparently "comfortable" – suspension, which shifts the resonance frequency of the vehicle body to lower frequencies, could imply a rocking of the body movement in this region rather than a hard suspension. Many passengers who have a tendency to kinetosis report that they use soft-suspension vehicles[9] less than vehicles with hard suspension. Since one can assume that especially for these slow movements there is little experience with the coordination of vision and initiative, further scientific experiences have to be awaited, to what extent the previewing suspension (Macic-Body-Control by Mercedes) provides relief for sensitive persons or possibly causes the opposite.

9 A particularly critical vehicle in this context was the Citroën DS 19 with its soft hydropneumatic suspension.

8.4 Climate

For the whole complex issue of climate, see the excellent summaries in Temming (2003); Großmann (2013) and Hodder (2013).

8.4.1 Climate, Performance and Comfort

In contrast to the environmental factors lighting, sound and vibrations, the climate (collective term for the thermal environment of humans) has no feedback character about the driving process. However, as a homeostatic living being, humans have to maintain a body core temperature of 37 °C within relatively narrow limits under all external temperature conditions, the climatic conditions contribute considerably to their well-being and thus also to their performance. In the history of vehicle development, the climate in the vehicle cabin was given relatively late attention. In combination with higher speeds, closed vehicle cabins were initially introduced to protect passengers from disgusting weather conditions. To counter the fogging tendency of the windows in these cabins and to meet the needs of passengers at higher outside temperatures[10] various simple measures were introduced, in particular opening windows (crank windows, windscreen to be raised, opening windows, sliding windows for very simple vehicles).[11] It was not until the 1950s that it became common practice in Europe to use the waste heat from the combustion engine to heat the vehicle in winter. Since 1963 the legislator has prescribed the existence of a heating system. Due to the relatively moderate temperature conditions in Europe, it was customary until the turn of the millennium to deliver vehicles as standard only with this prescribed heating system. From the eighties, however, the desire for better air conditioning became ever greater, especially under summer conditions, possibly also with a view to the USA, where air conditioning systems have been offered as standard since 1953 (the company Packard Motor Car introduced a complete air conditioning system for the vehicle in the USA as early as 1939).

A large number of experiments in the world of work dealt with the relationship between climate and performance. There is a clear link between physical performance and climatic conditions, but this is irrelevant in view of the fact that the driver does not perform at all. Of interest, however, are the motor and cognitive abilities of the driver depending on the climate. For cold conditions, it can be assumed that at an air temperature of <10 °C, the movement speeds of the extremities decrease progressively. The speed of nerve conduction is also reduced at low temperatures, so that at least the output of mental performance is slowed down. Temming provides an excellent overview of heat conditions (2003; FAT Study No. 177). According to this study, cognitive abilities in particular are restricted at temperatures >30° C_{eff} (see 8.4.2), whereby younger persons are obviously less affected than older ones. This study was not able to prove a connection with traffic accidents and conditions that can only be attributed to the climate. Even the expectation that aggressiveness is related to climatic conditions cannot be clearly proven scientifically. However, on the basis of general findings from occupational science, it can be stated that under certain conditions any strain that leads to discomfort at least impairs cognitive performance simply because mental resources are tied to the discomfort and thus capacity for the actual driving task is lost.

10 Until the 1940s it was common practice to immobilise a vehicle in winter.

11 In the accessories trade there was also an additional pane of glass with electric wires to be glued to the windscreen, which was intended to create a clear view in this area.

8.4.2 Climatic Comfort

8.4.2.1 Thermoregulation

As already described in ► Sect. 3.2.1.4, the thermoregulation of the human organism takes place via the warm receptors located mainly in the hypothalamus and the cold receptors distributed on the skin surface. A climatically neutral feeling (comfort = no perceptible discomfort) is achieved when a temperature between 33 °C and 34 °C is reached on the skin surface. The fact that the temperature perceived as comfortable is significantly lower depends essentially on the effect of the clothing, which creates a microclimate in the area of the skin surface that provides this temperature. Based on all scientific findings it can be assumed that comfort is achieved when the effort of the organism to maintain the mentioned skin surface temperature is minimized. This effort depends to a large extent on the climatic conditions. These are first and foremost determined by the *air temperature*. Normally the body emits heat to the environment by convection (exchange of the air heated on the skin surface by colder air) and body contact (touching low-temperature objects; e.g. cold vehicle seat – this way, of course, heat can also be transferred to the body via the other way round e.g. seat heating). If the air temperature is too low, it can generate heat through body movement (due to the poor efficiency of the muscles, etc. e.g. cold shivering). If the air temperature is too high, the heat dissipation via the path of convection can be increased by increasing the *air velocity* (fan out of air). A further increase in heat dissipation is possible through the release of sweat. The heat necessary for the evaporation of this liquid is extracted from the environment, i.e. especially from the skin surface. The effect of evaporation can also be significantly improved by air movement. However, it depends considerably on how much moisture the air can absorb at all (at 100% humidity, no further moisture can be absorbed. In this case the sweat formation has no effect.) Thus, in addition to the already mentioned air movement, the *relative humidity* an essential factor defining the comfort climate. Like every body, the human body is also in an exchange of radiation with its surroundings (every body emits electromagnetic radiation according to its surface temperature, at the usual ambient temperatures essentially in the long-wave infrared range). If the surface temperature of the surrounding surfaces coincides with the surface temperature of the body (here the surface temperature on the clothing plays the decisive role), then radiation equilibrium prevails, that ver *rays* no heat exchange takes place. If the surface temperature of the surrounding surfaces is significantly higher than that of the body, heat is transferred to the body via this path, which may then have to be released again via the mechanisms mentioned above. Conversely, the body can also emit heat via the path of the radiation if it has a significantly lower temperature than the body surface (especially in winter this is done by cold body parts).

However, the physical process described depends to a large extent on how much heat is actually produced by the body and how the individual body is given the opportunity to retain or release the heat produced. Thus, in addition to the above conditions, the perceived comfort is considerably influenced by the individual energy metabolism and the clothing worn. In addition, the organism has the ability – as with other environmental influences – to acclimatise (adapt) to long-term conditions.

The ***energy turnover*** depends on age, sex and body dimensions (height and weight). Due to the large influence of the body dimensions, the energy conversion [W] is often related to the body surface [m^2] specified [W/m^2]. A medium adult male (70 kg, 1,75 m body height) with a surface area of about 1,8 m^2 has under normal driving conditions a heat emission of about 70 W/m^2. In poorer road conditions, this energy conversion can increase slightly due to the physical activity required (Hodder 2013). This information is valid for temperate climates. In hot environments, due to increased pulse rate and sweating, an increase of 5 to 10 W/m^2 (DIN EN 28996, 1993) can be observed.

The ***clothing*** can make a significant contribution to maintaining thermal comfort. The insulating effect of clothing means that, on the one hand, the effective heat emission of the body at a correspondingly low temperature maintains the conditions of the comfortable surface temperature of the skin. This effect can also be largely regulated by adapting the clothing to the temperature conditions. Under extreme conditions it can even be ensured that the heat transfer from the outside to the body (especially radiant heat) is kept within limits. The effect of clothing is defined by the internationally accepted unit of measurement "clothing unit"(abbreviation clo). The reference value of 1 clo applies by definition to the "typical office suit". Estimated values for typical clothing combinations can be found in ▫ Table 8.4.

Temming (2003) presents the following findings regarding the climatic ***acclimatisation*** different literature results together. After more than eight days it can be expected that a person exposed to the heat climate will experience a notable acclimatisation effect. However, all the literature references found refer to observations of heavy physical work in a heat climate. According to the unanimous opinion of most

Table 8.4 Insulation values of clothing combinations according to DIN ISO 7730 (1987)

Type of clothing combination	Insulation value	
	m²-°C/Watt	clo
Unclothed	0	0
Shorts (shorts)	0,015	0,1
Typical clothing in tropical areas: Underpants, shorts, short-sleeved shirt with open collar, light stockings and open shoes	0,045	0,3
Light summer clothes: Underpants, long, light trousers, short-sleeved shirt with open collar, light stockings and shoes.	0,08	0,5
Light work clothes: Underwear, cotton work shirt with long sleeves, work trousers, wool socks and shoes	0,11	0,7
Typical winter clothing for interiors: Underwear, shirt with long sleeves, trousers, jacket or pullover with long sleeves, thick stockings and shoes	0,18	1,0
Heavy, traditional, European office clothing: Cotton underwear with long legs and sleeves, shirt, suit consisting of trousers, vest and jacket, woollen stockings and heavy shoes	0,23	1,5

of the authors quoted, acclimatisation is due to the extent of sweating, which is itself controlled by the elevated body temperature. The duration of the heat exposure must be at least 1 hour daily in order to cause an increased body temperature and the associated increased sweating. Comparable acclimatisation gains can also be observed for older people. For persons sitting quietly and exposure times of at least 6 hours, however, only an increase of the tolerance limit by approx. 2 K is possible[12] to observe. It can therefore be assumed that the acclimatisation effects mentioned have no effect on the discomfort sensation in the vehicle.

8.4.2.2 Summary Climate Measures

Due to the complexity of the individual physical quantities influencing climate perception, there have been repeated attempts to combine them into a single quantity describing climate perception. The advantage of such a procedure is to recognize what effect the design change of a physical quantity has and then to derive appropriate measures from it. There are two different approaches to this. One method tries to find climate combinations that lead to the same temperature sensation (e.g. in the sense of curves with the same strength of perception in the acoustic range). The best known and oldest representative of this variant is the effective temperature according to Yaglou (1927). This is followed by a series of modified procedures, which are summarised under the term climate sum measures. The second method is to estimate, on the basis of a kind of energy balance, the proportion of people who feel comfortable with the climate combination in question. The best known system of this version was developed by Fanger (1967 et seq.). This is also followed by a series of modified procedures through which particular aspects of influence are to be taken into account. A comprehensive compilation and discussion of all these procedures can be found in Temming (2003).

12 the Celsius scale represents a measure at the interval scale level and therefore only has significance if one wishes to describe the temperature state through it. If one wants to indicate the difference of temperature values, it is therefore correct to use the temperature scale according to Kelvin (K = 273 + x °C). Of course, the same numerical values result.

Climate Sum Measures

In the twenties, Yaglou (1927) developed an integrating climate measure in the USA, which combines all combinations of air temperature, psychometric humid temperature (a measure for the relative humidity of the air, and the temperature of the air.[13]) and flow velocity of the air is indicated by a measurement number, which caused the test subjects to feel the same amount of heat during the experiments. Two nomograms were developed from the test results (one for clothed and one for unclothed persons), with which a single value describing the thermal sensation, the so-called "effective temperature t_{eff}". The diagrams are included in DIN 33403, Part 3 (2000). In the meantime, work has become known which presents the content of nomograms in the form of equations, so that the effective temperature for concrete climate combinations can also be calculated numerically (Müller and Gebhart 1999; Gebhardt et al. 1999). The effective temperature indicates the heat sensation that occurs at the corresponding dry temperature, no air movement and 0% relative humidity. For sitting, practically non-physical work, a range from t_{eff} = 19° C_{eff} to 23° C_{eff} is assumed for neutral temperature sensation. Using the nomograms mentioned above, Bubb & Schmidtke (1984) have compiled the air velocity required to achieve a neutral temperature sensation of 23°C_{eff} as a function of air temperature and relative humidity in a diagram (Figure 8.23).

The nomogram originally developed by Yaglou does not contain the influence of the radiation temperature. Various proposals have been made to take this handicap into account. The recommendation to measure the so-called Globe temperature instead of the dry temperature and to use it instead of the dry temperature in the Yaglou's nomogram has become widely accepted. The probe of the globethermometer is located in the center of a black painted copper ball with a diameter of 150 mm. The value determined in this way is called the "corrected effective temperature". Also concerning the size and definition of this globe thermometer different variants are described in the literature.

As already mentioned, a large number of other sum measures have been developed. In this regard, reference is made to the compilation in Hodder (2013) and especially Temming (2003).

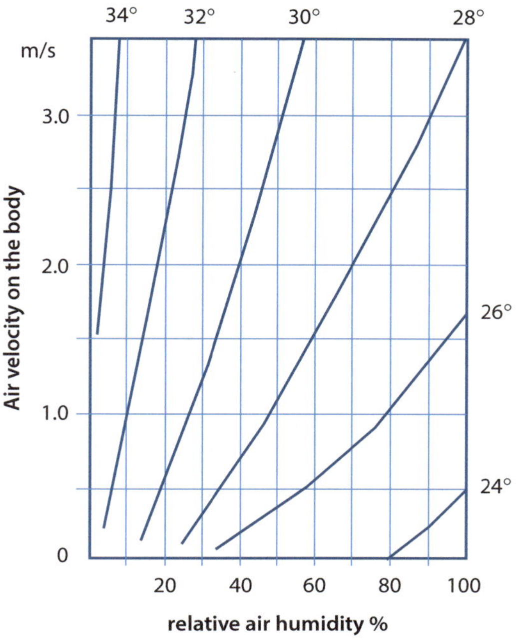

Fig. 8.23 The air velocity required to reach an effective temperature of t = 23°C_{eff} depending on the relative humidity and air temperature. (From: Bubb & Schmidtke 1984)

Comfort Equation According to Fanger

The comfort equation developed by Fanger (1970) is based initially on the body's heat balance: The heat generated by the body due to its basic metabolic rate q_{met} minus the heat flows emitted by the various mechanisms (heat conduction, convection, radiation, diffusion, evaporation of sweat and respiration)

13 In an apparatus known as a psychrometer, the temperature is measured with a thermometer, the sensor of which is surrounded by a moistened wad of cotton wool and fanned with prescribed air movement. Evaporative cooling removes heat from the thermometer so that it indicates a lower temperature than the so-called dry temperature measured in parallel. The lower the relative humidity, the greater the difference. The value obtained in this way is called the "wet bulb temperature".

Table 8.5 Perception and scale of PMV according to DIN EN ISO 7730

Sensation	*PMV*	Dissatisfied [%]
Hot	3	100
Warm	2	78
Slightly warm	1	26
Neutral	0	5
Slightly cool	−1	26
Cool	−2	78
Cold	−3	100

$\dot{q}_{ab,.n}$ represents the effort of the organism to maintain the body core temperature of 37 °C.

$$q_{met} - \sum_{n=1}^{n} \dot{q}_{ab,.n} = \Sigma q \tag{8.4}$$

On the basis of complex psychophysical experiments with test persons (see also ▶ Sect. 11.3.1.2), he investigated which effort can be assigned to which sensation. Using regression analytical methods, he has developed a "predicted average overall sensation" of the climatic conditions, the PMV (Predicted Mean Vote):

$$PMV = 0.303 \cdot e^{(-0.036 \cdot q_{met} + 0.028)} \cdot \Sigma q \tag{8.5}$$

The Fanger method is included in many standards (e.g. DIN EN ISO 7730) and thus represents the "state of the art" of climate assessment. For the calculation of the heat flows there are many detailed data there. Großmann (2013) also provides a clear presentation of this. ◘ Table 8.5 provides the verbal assignment to the *PMV*-values as well as the expected percentage of dissatisfied people in the individual categories after catchers.

According to Großmann's experience (2013), instead of eq. 8.5 in the automotive-relevant range from metabolic rates to 150 W/m² the following simplified version can also be used:

$$PMV = \frac{3{,}8}{\dot{q}_{met}} \cdot \Sigma q \tag{8.6}$$

In any case, it can be seen that, in the case of the balanced heat balance $\Sigma q = 0$ also the *PMV*-value becomes 0. However, according to the experience of Fanger, 5% of the test persons are still dissatisfied with the temperature conditions (see ◘ Table. 8.5). Mayer (1998) even found out in his own car-related experiment that the percentage of the dissatisfied even with $PMV = 0$ is still at 15%. In another publication, he notes that most people feel comfortable in the vehicle with a PMV = 0.5, whereby 18% of the people are still dissatisfied (Mayer et al. 2007).

The PMV value developed by Fanger actually refers to air conditioning in immobile work and recreation rooms. Due to the tight conditions in the vehicle, however, the model must be extended for this application. For this purpose, the human surface is divided into surface elements and the absorption of direct solar radiation is taken into account in order to assess the local individual heat flows. On the basis of Fanger's comfort studies, specific evaluation methods were developed which evaluate local comfort in individual parts of the body with the LMV (Local Mean Vote) instead of global comfort (Bureau et al. 2003; Frühauf 2002; Kühnel et al. 2003; example see ◘ Fig. 8.24). By the way, there are limits to a simple summation of the heat flows: e.g. a cold left and at the same time hot right foot could simulate a neutral feeling in the calculation. This is another reason why local assessments are important (more on this in Großmann 2013). A large number of publications deal with the modification of these comfort equations in order to better adapt them to the conditions of high thermal stress. In this context, reference is made to the detailed presentation in Temming (2003). Due to the uncertainty of the assessment by test persons, climate manikins are increasingly being used to assess the climate in cars in order to objectify the local assessment (◘ Fig. 8.25).

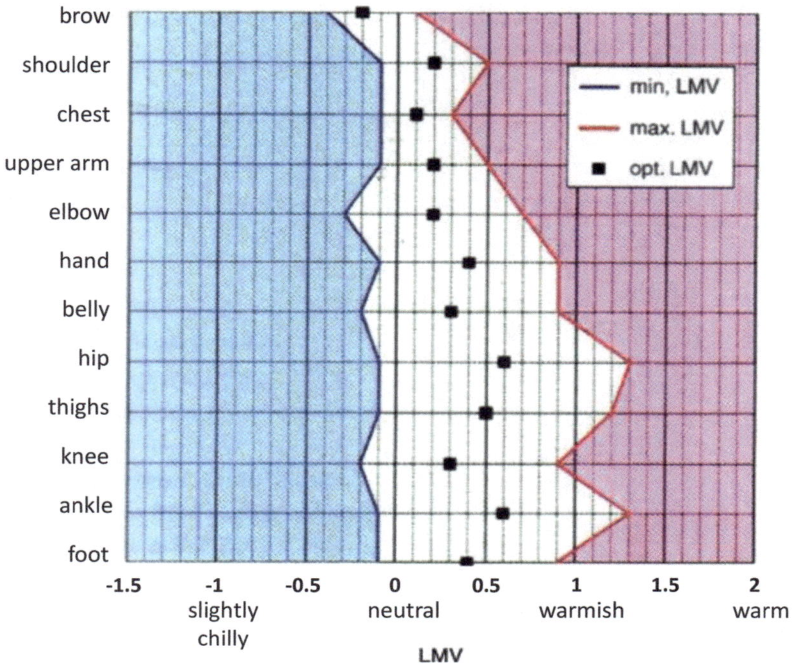

Fig. 8.24 Comfort evaluation on body parts with LMV. (From Wawzyniak 2011)

8.4.3 Environmental Conditions

Vehicles are used worldwide under extreme climatic conditions. Temming (2003) and his working group compile some interesting external climate data, with a focus on conditions in Germany. In this context, reference is also made to the Handbook of Fundamentals (ASHRAE 1997), which is updated every four years by the American Society of Heating, Refrigeriating and Air Conditioning Engineers. Although this basic work contains predominantly climate data for the USA, it also contains orientation data for practically all countries of interest in the world. According to DIN 1946–3 (2006–2007), outdoor air temperatures worldwide can vary between −45 °C and 55 °C. DIN 4710 (2003) compiles extensive meteorological data for Germany. The climate data guideline VDI 4710,1 (2008) describes non-European climate data for building services engineering. As an example, Großmann (2013) gives the values preseted in Table 8.6.

Vehicles are generally designed to operate in a temperature range between −20 °C and 40 °C. This also corresponds to the temperature conditions measured in Europe (e.g. Russia Kotlas: −18 °C, 30 frost days; Granada: 35 °C). However, there are also extreme deviations, for example in the winter of 2013/14 on the east coast of the USA −45 °C and in the same period in South America +40 °C. Extreme values are also possible in Germany: Großmann (2013) reports that an air temperature of 38.7 °C was measured in Manching (near Ingolstadt) on 27th July 1983 and even 40.2 °C on the same day in Gärmersdorf near Amberg. In Munich, the temperature measured at the airport normally varies over the year between −4 °C in January and 25 °C in July.

Apart from extreme winter conditions, at least for the temperate climatic conditions in Europe, winter operation is ensured by adequate heating up to an external temperature of −10 °C can still produce pleasant interior temperatures, possibly supported by an additional heating system. A particular challenge, however, are the summer climatic conditions. They therefore play a special role in the available publications. Temming (2003) compiles a large number of data sheets for summer outdoor temperatures. Among other things, he notes that local differences in particular arise not only on a large scale due to different alti-

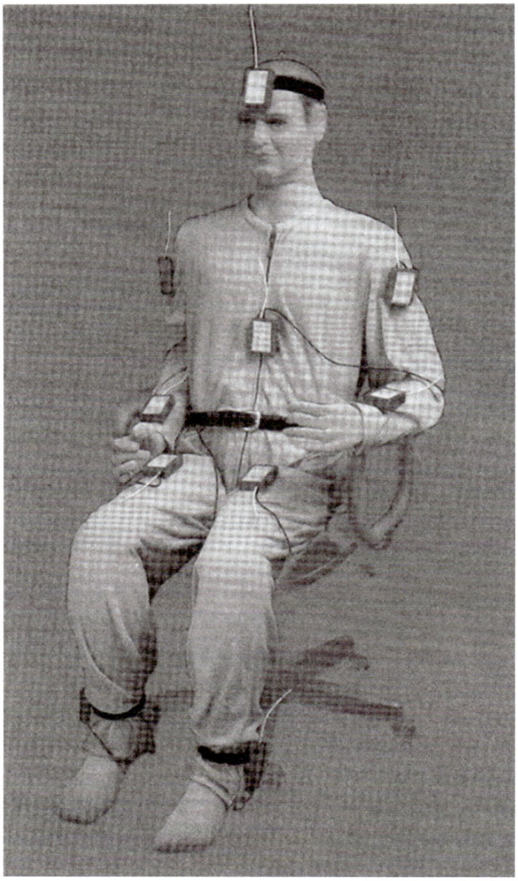

Fig. 8.25 Climate manikin DRESSMAN. (Mayer et al. 2007, source Fraunhofer Institute for Building Physics)

Table 8.6 Examples for air temperatures and relative humidity at different locations (from Großmann 2013).

Location	Air temperature °C	Relative air humidity %.
Phoenix (USA)	43	15
Munich	30	50
Tokyo	30	75

tudes or the effects of sea and land climates, but also as a result of small-scale peculiarities such as building density and vegetation. In this context, he quotes Völksch (1978), who, on the basis of extensive literature research, found that differences of up to 10 °C can be observed in comparison with the surrounding open land during low-exchange nocturnal weather conditions in large cities.

According to the physical connections that air with a higher temperature is able to absorb a larger amount of moisture, the highest moisture contents of the outside air (9 g/kg) are observed on cheerful days in July in contrast to the lowest in January (approx. 3.5 g/kg). According to Dufner et al. (1993), the moisture content of the air in rural areas is typically 8% higher in summer than in urban areas. The differences in both air temperature and moisture content between urban and rural areas are mainly attributed to the water vapour release of the vegetation and the differences in dew formation.

The heat input by the sun represents the greatest challenge, along with extreme cold for the air conditioning of the vehicle. The intensity of the sun's rays depends on the season, the time of day, the degree of latitude on earth and the cloud cover. Großmann (2013) provides detailed information and calculation examples. In general, it can be assumed that the maximum sun intensity in the summer months in Central Europe is between 800 and 1000 W/m^2. The long-wave heat rays of the sun pass through the windows of the vehicle into the interior of the vehicle and heat up the surfaces of the objects located there (dashboard, steering wheel, seats, rear shelf, etc.). The radiation emitted by the heated objects can only be partially emitted to the outside through the window panes, which partially absorb the heat radiation. For the most part, however, the air in the vehicle is heated by convection from the hot surfaces. In a stationary vehicle there is no air exchange with the environment, so that very high air temperatures occur in this way (so-called greenhouse effect). Temming (2003) reports temperatures between 42 °C to 51 °C in the footwell, 55 °C to 67 °C in the headspace and temperatures up to 100 °C on the sheet metal surfaces measured in a vehicle that stood for 2 hours in the midday sun at 26 °C in 52° north latitude (Kassel) and at 43 °C in 33° north latitude (Phoenix, Arizona, USA). Although these extreme temperatures

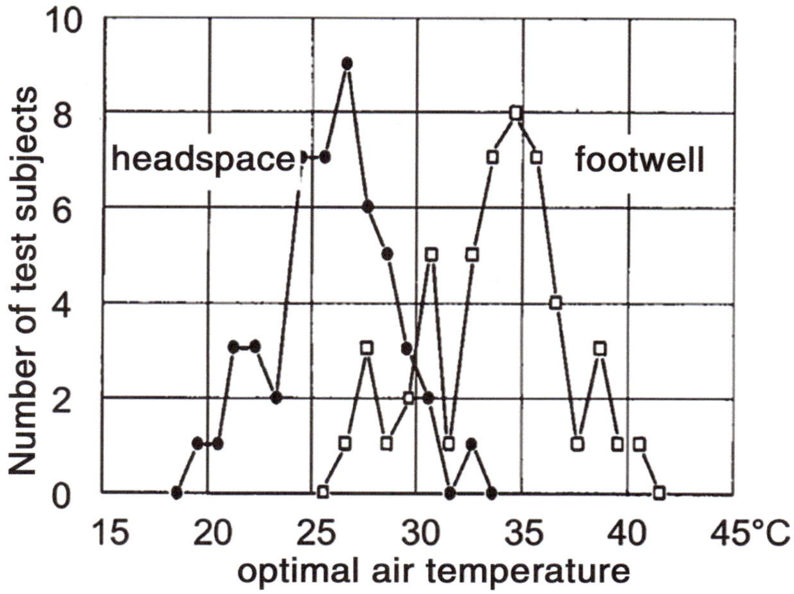

Fig. 8.26 Comfortable head and foot temperature in the car. (After Amano and Imai 1971)

can be reduced more or less quickly (for periods of several minutes) by the operation of the vehicle, since an exchange of the vehicle interior air can now take place via the ventilation system or open windows. However, due to the greenhouse effect described above, the temperature in the passenger cabin always rises by several degrees Celsius in summer (between 5 °C and 10 °C depending on the technical requirements), even during operation. With regard to the heat load in the interior, it is also important that the mentioned temperature rise does not contain the direct effect of the heat radiation, because the quoted values were mostly measured with simple thermometers (Temming 2003).

8.4.4 Technical Requirements

8.4.4.1 Driving Operation

According to Großmann (2013), it is usual to determine the mean interior air temperature in a vehicle as the arithmetic mean of the mean temperature at the level of the footwell and at head height. While according to DIN 1946–2 (1994–2001) an average interior temperature of 22 °C is recommended in residential areas of buildings, according to Großmann's experience the average temperature in the vehicle is higher both in winter and summer. There are several reasons for this.

Winter Operation

Especially in winter, passengers radiate heat to the cold windows. Since the distance between the side windows is very small in comparison to corresponding walls and windows in buildings, the air in the footwell must be considerably warmer than at head height for a comfortable climate in the vehicle. Großmann recommends an air temperature stratification of 5 K to 12 K. Temming (2003) gives temperature differences of 4 K to 8 K based on a literature compilation (see also Fig. 8.26). According to Großmann (2013), with an outside air temperature of −20 °C, an average indoor air temperature of 28 °C can be achieved in the stationary state.

Fig. 8.27 schematically shows the heat flow in a vehicle during winter operation. The air heated by the heating system flows into the interior. According to Großmann (2013), the air mass exchange under blower operation is between 7 to 9 kg/min. Weible and Kern (1984) give maximum values in the range of 8 to 10 kg/min for this purpose. It is also strongly dependent on the driving speed. The entire body mass is in heat exchange with this warm internal air, additional heating by the engine

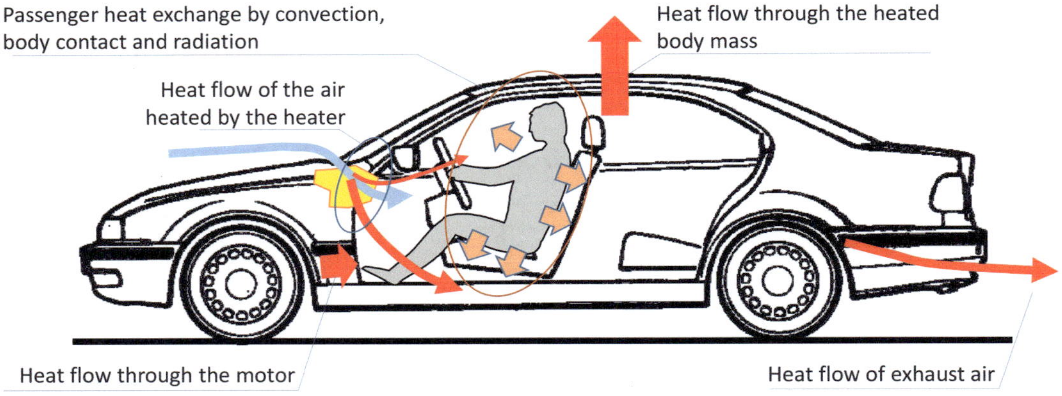

Fig. 8.27 Heat flow through the passenger compartment during winter operation

8

heat and dissipated heat to the cold external air.[14] Passengers emit heat through radiation to the body parts with low surface temperature and through contact, especially to the seat, but also to other objects that are touched (steering wheel, armrest). From this point of view, it is more comfortable if a vehicle is equipped with seat heating (this also applies to the rear seats; see also Brooks and Parson 1999). For the design it has to be considered that in winter many people use the vehicle with winter clothing, which has a high insulation value. In addition, there is also the insulation value of the seats, which according to Hodder (2013) can be set at 0.2 Clo. Since winter clothing may not be required for longer journeys, it is essential to regulate the output of the seat heating system.

For technical reasons, the heating power of a vehicle depends on many circumstances, such as driving condition (speed, uphill/downhill), surface temperature of the road, air temperature fluctuations, solar radiation. Therefore, regulation of the heating effect is necessary in any case. According to Petzold (1975), the perception threshold for temperature fluctuations is in the range of about 1 K. According to DIN 1946 (2006–2007), the time deviation from the setpoint of the room temperature should not exceed or fall below a value of 2 K. This makes a technical regulation of the outlet temperature of the heating system necessary.

As is relevant for the summer operation described below, it is also true for winter operation that the individual thermal comfort requirements vary greatly (see Table. 8.5). This means that individual regulation of the heating output for the individual seats is necessary. Today, a separate regulation for the driver's and passenger's seat is offered – usually in use with air conditioning – at least for the front seats. An air supply to the rear seat area and a regulation independent of the front area is additionally necessary in order to ensure individual thermal comfort for the rear seat passengers.

A problem of winter driving operation for the occupants is the low humidity caused by the low outside temperature (see also ► Sect. 8.4.3). As this air is heated by the heating system, the resulting relative humidity in the vehicle interior is correspondingly low. According to various recommendations, an air humidity between 30% and 50% is considered optimal (Sterling et al. 1985; quoted after Temming 2003, see also Deyhle and Bienert 2011). A technical humidification of the vehicle interior air would therefore make sense, especially since the disturbing electrostatic charge can also be largely reduced from an air humidity of approx. 40% (McIntyre 1980). In this context, however, reference should also be made to the requirements described in ► Sect. 8.4.4.3, since low humidity can ultimately prevent fogging of the windows.

14 For detailed calculations of these data, please refer to the detailed treatment of Großmann (2013).

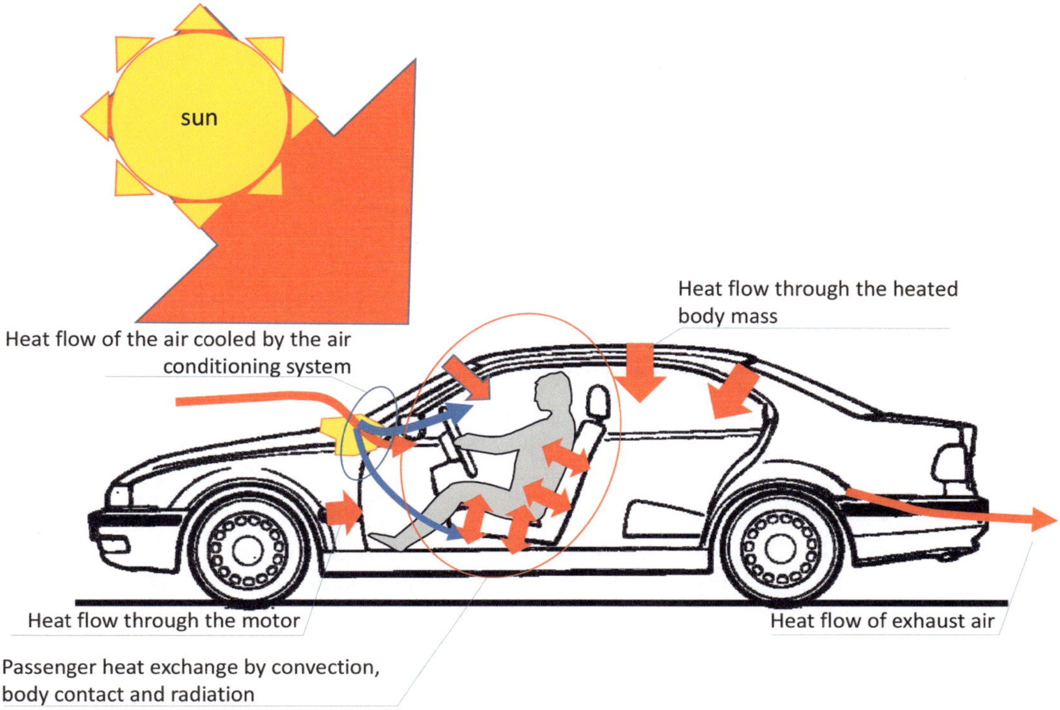

Fig. 8.28 Heat flow through the passenger compartment during summer operation

Summer Operation

In the temperate climate zones of Europe and many other parts of the world, general acclimatisation to summer climatic conditions cannot be assumed, and even if this occurs, the corresponding effect is relatively small (see also Sect. 8.4.1). Consequently, it is to be assumed that for summer driving conditions, similar requirements are to be placed on the comfort conditions in the vehicle interior as under winter driving conditions.

Fig. 8.28 schematically shows the heat flows through the passenger compartment during summer operation. Without air conditioning, the outside air enters the interior at an almost unchanged temperature. One of the effects of the heated sheet metal housing is that the air drawn in via the ventilation system is additionally heated, which is referred to as "summer air heating on the engine hood". Since the absorption of solar radiation also depends to a large extent on the paint colour, there are various effects here which have an effect especially in slow stop-and-go operation (Großmann 2013). Apart from the marginal cooling effect caused by the airstream, the body mass has approximately the same temperature as the outside air in stationary driving mode without sunlight. In addition, the sheet metal housing is heated by the motor in the front footwell and especially at the transmission tunnel. In general, an air conditioning system can achieve that the interior temperature is shifted into the comfort range despite a higher outside air temperature.[15]

The conditions change drastically when an additional heat flow is added by the thermal radiation of the sun. In any case, the exchange of radiation with the sun causes the body mass to heat up accordingly, which in turn contributes to the heating of the interior air. However, this is also associated with heat radiation (detailed information and considerations can be found in Temming 2003). Großmann (2013) reports that at an outside

15 Fig. 8.23 shows the sometimes extreme air velocities that are necessary without air conditioning in order to achieve an effective temperature of 23 Ceff.

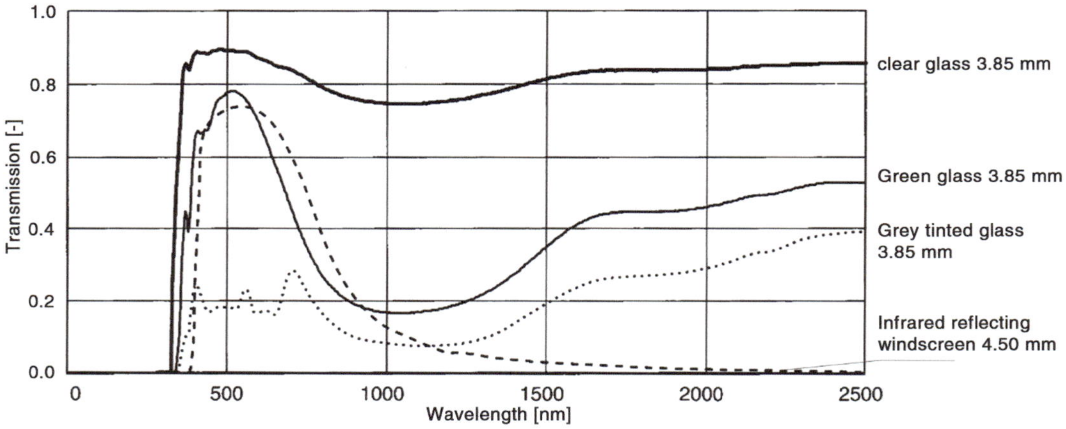

Fig. 8.29 Transmission of different glass types as a function of wavelength. (Source: NSG Group Pilkington Automotive, after Großmann 2013)

air temperature of 40 °C, solar irradiation of 1000 W/m^2 and a relative humidity of 30% was measured on the nozzles of the air conditioning system 8 °C. As a result of the solar radiation and other effects, the air heated up from these 8 °C to 27 °C as it passed through the passenger compartment, with the average air temperature at head height being 21 °C. The air temperature in the passenger compartment was then reduced to 21 °C.

The decisive influence, however, comes from the direct irradiation of the long-wave electromagnetic solar radiation through the windows. This not only causes the greenhouse effect described above, which is less significant due to the exchange of air in the moving vehicle than the direct irradiation of the body parts of the passengers. This influence naturally depends to a large extent on the weather conditions, the position of the sun and thus the time of day and in relation to it the direction of travel of the vehicle (for more details see Großmann 2013). Due to the large flat window areas, this influence has increased significantly compared to historical vehicles.

Heat-insulating glazing, which is intended to reduce the penetration of thermal radiation, has now become the standard. Absorbing and reflecting panes as well as mixed forms of them are used for this purpose. The absorbing discs contain particles (e.g. iron oxide in classic green glass), which absorb the heat radiation of the sun. This heat is released into the ambient air by the airstream. A thin layer (e.g. tin oxide) is evaporated onto the outside of the reflecting panes. This means that it is possible to prevent part of the solar radiation from penetrating the driver's cab even without wind (Großmann 2013). Fig. 8.29 shows as an example the effect of different glasses. The high absorption effect of the infrared reflecting glass for the long wavelengths is conspicuous. However, various authors also point out that the influence of radiant heat can be greatly reduced by choosing clothing (e.g. white clothing instead of black) (Schwab 1994). Roller and Goldman (1968) even found that the thickness of the clothing material can reduce the transmission of solar energy to the body. If the body is exposed to direct irradiation, the hair and clothing temperature can rise by 15 K to 18 K, while with correspondingly exposed bare skin only a temperature increase between 5 K and 6 K can be recorded (Clark 1981).

The most effective method to compensate for the thermal radiation effect is to fan the corresponding parts of the body with air cooled by the air conditioner. As Temming (2003) states, car occupants want air inlet temperatures in the passenger compartment to decrease as the outside temperature rises. Figure 8.30 shows the result of measurements over several months in an air-conditioned vehicle during longer test drives by 8 individual drivers. It clearly shows that

■ **Fig. 8.30** Individually adjusted air inlet temperatures in an air-conditioned passenger car. (From Temming 2003)

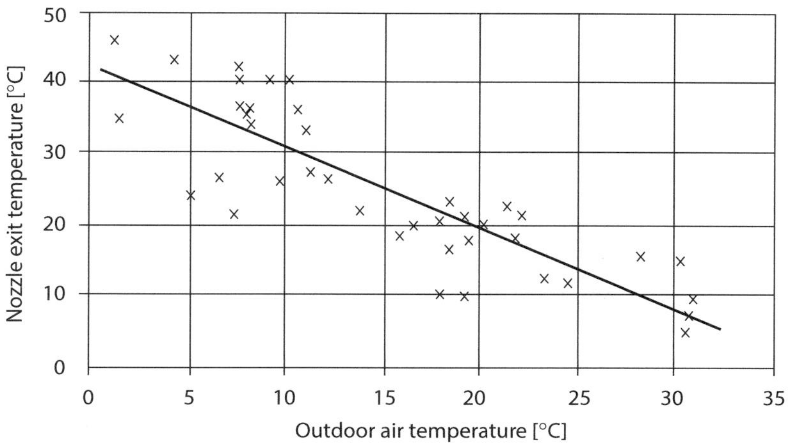

with low outside temperatures (winter conditions) high inlet temperatures of the heating system up to 40 °C are required, that already with outside temperatures below 20 °C the air conditioning system was in operation and that with increasing outside temperature the outlet temperature at the air nozzles was adjusted to values up to 10 °C and below. Deyhle and Bienert (2011) also recommend rather low air speeds combined with low exhaust temperatures for vehicle air conditioning. This requires relatively large outlet openings, since high air volumes have to be exchanged in the vehicle via relatively few air outlet cross-sections in order to meet the different thermal requirements. Hinz et al. (1983) require an air flow rate of up to 10 m^3/min for a large car in summer in order to achieve the required maximum high air velocity of 2 to 3 m/s in the interior (see also calculations by Großmann 2000). Asakai and Sakai (1974) indicate values between 1.5 to 2.1 m/s close to the body as necessary to ensure a cooling effect through vehicle ventilation in summer. On the other hand, speeds above about 3 m/s are rejected as uncomfortable. Zipp et al. (1977) have stated that the ventilation of the chest and lap area shows the best results both in terms of objectively measurable effects and subjective assessment. The air flow should be distributed as evenly as possible, which can be achieved by a larger number of nozzles or nozzles with a large outlet opening and by a greater distance of the nozzle from the passenger. In this context, however, it should also be borne in mind that there are quite different requirements with regard to preferences. For example, Europeans and Asians predominantly prefer rather indirect, diffuse air conditioning with low air speeds, while North Americans prefer directly directed large air mass flows combined with corresponding cooling capacity to air conditioning (Deyhle and Bienert 2010). These different requirements are met by special country variants of climate control and by so-called comfort diffusers, which allow adaptation to individual comfort requirements through positions between "diffuse" (draught-free), "conventional" and "spot" (direct air flow) (Fritsche and Feith 2007). Asakai and Sakai (1974) also hold the opinion that a pulsating current, preferably at about 1.2 Hz, can cause a pleasant feeling of coolness. More recent evaluation approaches (e.g. DIN 1946 Teil 2, 2006–2007) require a lower average air velocity at higher turbulence levels in order to avoid tensile sensations. According to Temming (2003), it remains to be seen for the time being whether the increased heat dissipation due to increased turbulence levels could also have positive effects at high ambient temperatures. An application of such considerations in the vehicle has not yet become known.

Contact with the seat in particular creates another problem zone for the occupants of the vehicle that influences their climatic sensation. The body enters into a contact heat exchange with the seat. Madsen (1994) found in a study with a climate manikin that a ventilated seat can improve the removal of heat.

The actual problem, however, is essentially that in a heated vehicle interior the passengers secrete sweat in order to maintain their body core temperature by evaporative cooling. This of course also happens in the back and buttocks area, where the body is in contact with the seat. If the seat is unable to absorb the moisture, an unpleasant feeling may arise, among other things, due to the fact that the sweat itself is not very heat-conductive and that heat accumulates between the skin and the contact surface. Fung and Parson (1993) have done extensive research with different seat materials. Their subjects were exposed to a heat environment (34 °C, humidity 35%) and were asked to evaluate the thermal sensation. The worst rated seats were those that were impermeable to moisture due to their surface material or the foam used. However, later research by Fung (1997) showed that the influence of clothing, the duration of the trial and intra-individual personal preferences made it impossible to find an ideal seat material. Cengiz & Babalik (2007) evaluated the thermal comfort of three different vehicle seats in a field test and found no significant differences between the seats under real conditions (quoted from Hodder 2013).

8

▪ Electric Vehicles

The air-conditioning (heating and cooling) of electric vehicles poses a particular challenge because, on the one hand, no excess process heat is available from the electric motor and, on the other hand, any energy required to create a comfortable interior climate consumes the energy resources carried along and thus reduces the range of the vehicles. Even a better thermal insulation of the interior, which could reduce heat dissipation to the outside in winter and heat input from the outside in summer, does not have a penetrating effect, because the demand for high air throughput in the small volume of the passenger cabin is still maintained. The technically possible use of combustion systems for air conditioning counteracts the environmentally friendly effect of electric propulsion. It remains a task of future ergonomic research to re-examine the climatic comfort needs of humans under this particular aspect in order to determine which measures can be used to achieve optimum effects with minimum energy input. One could make oneself thereby e. g. the fact to make use of that climatic comfort with a skin surface temperature between 33 and 34 °C adjusts itself. Heating and cooling elements mounted close to the body, which do not have to air-condition the entire room, seem to point the way to success. A very extreme idea in this context would be "climate clothing", in which Peltier elements are woven into the fabric as heating or cooling elements in conjunction with thermal sensors, in which a controller (microprocessor) carried along ensures the specified skin temperature. The energy required for this type of air conditioning is likely to be relatively low, so that a commercially available battery should suffice as a power supply for road operation. For driving, one would connect the clothes to the power supply of the vehicle.

8.4.4.2 Start of the Journey

From a climatic point of view, the entry (dis)-comfort is an essential criterion for the acceptance of the vehicle. In winter, the vehicle takes on the outside temperature after prolonged outdoor parking. When the weather is correspondingly cold, this creates a high discrepancy between the subjective expectation of a "cosy" interior and the actual conditions. In summer, the vehicle heats up considerably due to the greenhouse effect described above, especially in sunshine after prolonged parking. In the interior, the air temperature may then be far above any comfort requirements. In addition, the thermal radiation exchange with the sun causes the surfaces in the interior to reach partial temperatures, which in the worst case can even lead to burns.

▪ Winter Conditions

According to Großmann (2013), the combustion engine heats up after 10 to 20 minutes to such an extent that the occupants feel the transferred heat in the passenger compartment. ◘ Fig. 8.31 shows the course of the mean interior temperature during a controlled measurement in a climatic chamber with roller test bench (described in more detail in Großmann 2013). More practical measure-

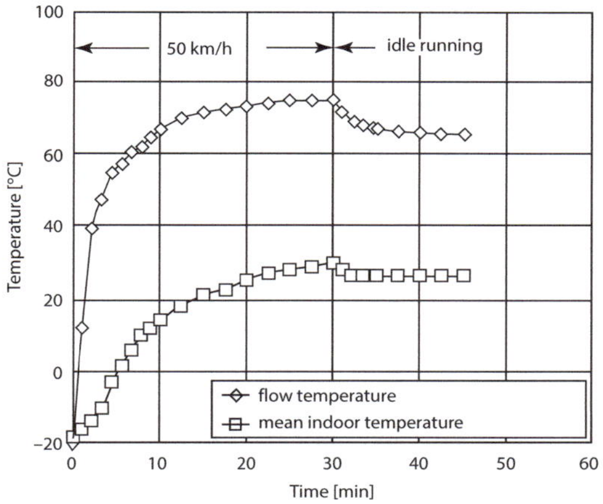

Fig. 8.31 Heating power measurements on a passenger car with a 1.8 l petrol engine with direct injection; outside air temperature −20 °C, 30 min Travel speed 50 km/h, then idle; air-side electric auxiliary heater with 800 W; the flow temperature is the cooling water temperature: measurement on a roller test bench. (From Großmann 2013)

ments were carried out by the magazine Auto-Motor-und-Sport in January 2013 (Übler 2013). The tested test vehicles stood one night at −15 °C outdoors. At the beginning of the test the outside temperature was −11 °C. The temperatures in the head area (interior mirror) and in the foot area (passenger foot area) were measured. The vehicles were moved in convoy in steady, restrained travel (max. 80 km/h). Fig. 8.32 shows the temperature curves for the vehicle with the fastest temperature rise (left) and the slowest temperature rise (right). In all vehicles, an almost linear increase in temperature can be observed, especially in the headspace. Even in the best vehicle, under these realistic conditions, which also provided more favourable temperatures than the laboratory measurement documented in Fig. 8.31, no temperature corresponding to the above comfort recommendations was reached after 20 minutes. Großmann (2013) therefore recommends various auxiliary heaters (e.g. electric auxiliary heating, heat storage, electric preheating of the coolant, use of residual engine heat), the technology of which is described in detail. According to his argumentation, the parking heater in particular represents the most convenient solution. Due to the sluggish rise in temperature under winter conditions, additional seat and steering wheel heating are recommended to prevent the discomfort caused by extremely low temperatures when starting up the cold vehicle.

Summer Conditions

A vehicle parked in the sun heats up considerably under summer conditions due to the greenhouse effect and can reach temperatures that can even lead to burns (see above and Table 8.7). Temming (2003) quotes a study by Shimizu et al. (1982), which shows heating curves when parking under summer heat radiation and cooling curves after the subsequent start of the journey using an air conditioning system (see Fig. 8.33). It is noteworthy that in the heating-up phase, the surface temperature of the outer skin (e.g. the roof surface) already reaches a constant level after about 20 minutes, while the interior temperatures continue to rise even after 40 minutes. Temming refers to studies according to which the increase in indoor temperatures continues even after 2 hours. In general, the air conditioning system reaches an indoor temperature of around 20 °C within 20 minutes of the start of the journey.

The heating effect depends considerably on the colour design of the vehicle. Großmann (2013) describes a test under controlled conditions in which two vehicles (a white one with a completely white interior and a black one with completely black interior equipment) are tested on the same vehicle.[16] The results are summarized in Table 8.7. Under the health aspect mentioned above, the temperatures at the steering wheel, front passenger seat and armrest in the front door are highlighted.

A particularly effective method of avoiding such extreme temperatures, which are associated with considerable discomfort, is, according to Großmann (2013), the use of stand ventilation with solar technology. With a so-called solar roof (sliding roof with inte-

16 Großmann points out in particular that the white dashboard surface and rear parcel shelf, which are favourable for reasons of radiation absorption, lead to extremely annoying reflections in the windscreen and rear window (rear-view mirror).

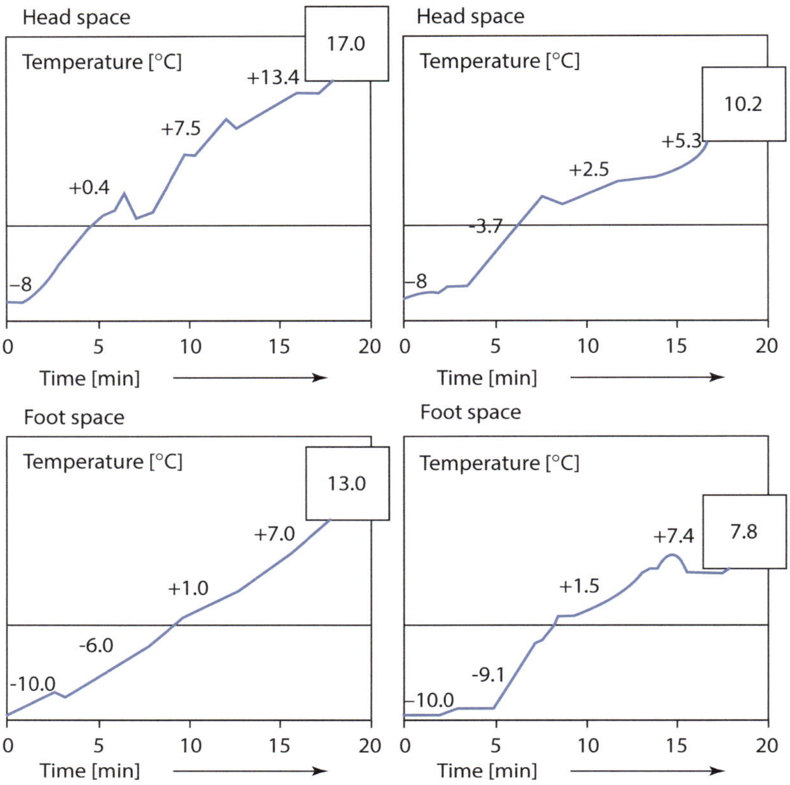

Fig. 8.32 Temperature curves in the head and foot area of two vehicles at an outside temperature of −11 °C and a real ride with a maximum of 80 km/h. (Übler 2013)

8

grated solar cells), electrical power is generated when the sun is shining, which is made available for blower operation and thus constantly exchanges the air in the parked vehicle. This prevents an essential part of the heating of the interior air by the high surface temperature of the components in the interior (Fig. 8.34).

8.4.4.3 Driving Safety Requirements

The previous discourse dealt with vehicle air conditioning with a special focus on the discomfort felt by drivers and passengers and on the climatic comfort. However, the air conditioning of the vehicle also has an essential safety aspect. The ability of air to absorb water vapour depends significantly on air pressure and temperature. At low temperatures it can absorb significantly less water vapor than at high temperatures. As a result, whenever warm air – possibly with high relative humidity – meets cold vehicle windows, the air cooled in the contact zone condenses the water contained. The pane is thus fogged with water and opaque. If the temperature of the panes is correspondingly low, ice will also form. This is, of course, especially important for the windscreen, but also for the front side windows and the rear window. Fogging is possible both on the inside and outside of the panes.

For fogging the windows of *on the outside* occurs when warmer outside air or air with a high moisture content hits the cold vehicle windows. This is generally the case when the vehicle is parked outdoors in winter weather conditions and is correspondingly cold. As is well known, it is then the driver's task to make the windows free of ice and fog before the vehicle starts. There may also be a sudden formation of external fog if the vehicle parked in a cold garage encounters the warmer and more humid outside air on its way out (under unfavourable conditions, this effect may also occur after a long tunnel journey). If there is no ice formation under these conditions, which is very rarely the case, a clear view can be restored very quickly by pressing the wiper.

Table 8.7 Component and air temperatures in a parked black car (including black interior) and a white car (including white interior) in sunlight, outside air temperature 30 °C, sun intensity 1000 W/m² (from Großmann 2013)

Designation of the measuring point	Temperatures in a white car [°C]	Temperatures in a black passenger car [°C]
Air at head height	67	75
Air in the footwell	44	46
Instrument panel, centre	72	99
Instrument panel, hat	78	102
Roof lining, centre	67	46
Hat shelf, centre	73	100
Trunk lid, interior	64	86
Front passenger seat	62	74
Armrest in the front door	61	73
Steering wheel	70	90
Gearshift	62	72
Rear seat backrest, centre	78	95

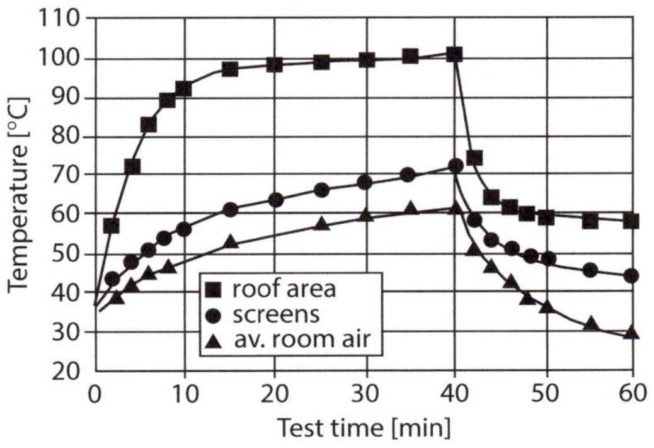

Fig. 8.33 Temperature curves in a stationary and air-conditioned passenger car with summer heat radiation. (After Shimizu et al. 1982; quoted in Temming 2003)

A fitting of the windows *inboard* occurs when there is warmer air inside the vehicle that carries more moisture and condenses on the cold windows. This is particularly the case at low outside temperatures, when the warm air breathing through the windows is subject to high relative humidity, but also when there are other moisture-carrying objects inside the vehicle (e.g. wet clothing, wet carpets, open bottles, etc.). As Großmann (2013) describes it, recirculation operation at low ambient temperatures is particularly critical. Occasionally, this can also occur during mountain passes, when the temperature of the outside air decreases with increasing altitude and the windows are cooled down accordingly. However, the water separated from the evaporator of the air conditioning system and stored there can also be the reason for the interior misting of the windows. Even at outside temperatures of 20 °C, the water stored in the evaporator may condense on the cool panes after a motorway drive and subsequent slow drive. Otherwise, however, an air conditioning system is favourable with regard to the tendency of the windows to fog up, as the cooling of the outside air is also associated with dehumidification, so that relatively dry air enters the vehicle interior. Internal fittings are particularly critical because they can occur suddenly. Due to the large distance of the windscreen from the driver due to the current design, it is also extremely difficult and even impossible to remove the fitting manually.

The traditional technical measure to make the panes fog-free and to keep them fog-free is to heat them. Usually, special ventilation nozzles are installed under the windscreen and in the lateral dashboard area, which direct the air heated by the heating system onto the windows. Various legal regulations exist for window de-icing and dehumidification (USA: FMVSS 103, EU 78/317/EEC; Australia ADR 15; quoted after Großmann 2013). Fig. 8.35 shows the measurement of the time curves up to freedom from fog in accordance with US regulation FMVSS 103. The driver's field of vision (see also ► Sect. 7.3.1) must then be free from ice and fog after 20 minutes at the latest (for a more detailed description, see Großmann 2013). In princi-

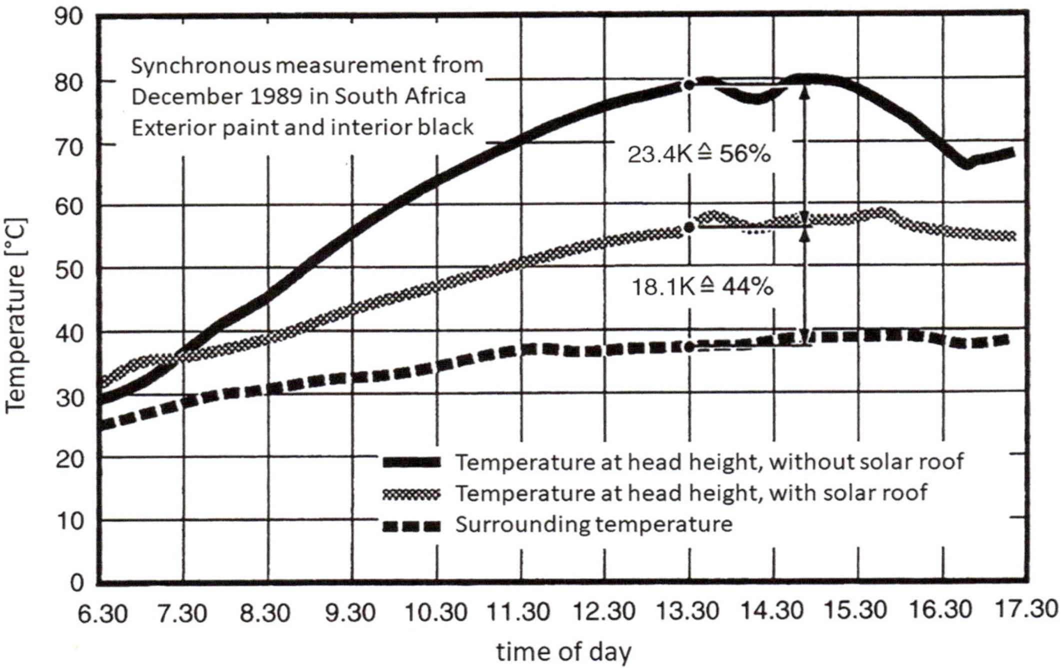

Fig. 8.34 Effect of a solar-powered stand ventilation system. Synchronous measurements of air temperatures at head height as a function of daily variation; Audi Coupé, 1989. (Ater Großmann 1992; quoted in Großmann 2013)

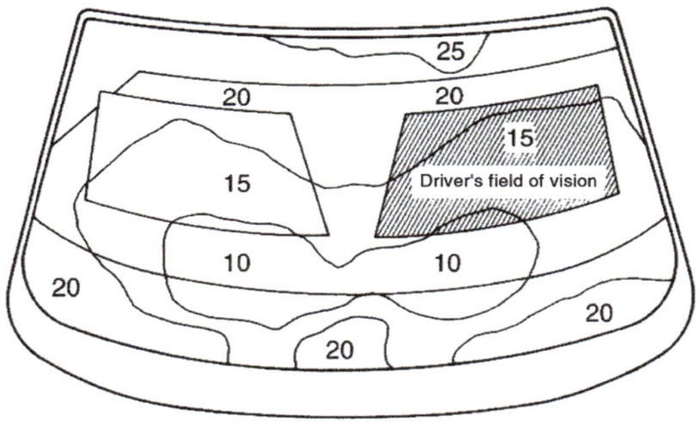

Fig. 8.35 Defrost lines as a function of time according to FMVSS 103. (Großmann 2005)

ple, the hotter the air flowing from the nozzles, the more effective de-icing and dehumidification is. Today, depending on the vehicle type, a completely defrosted windscreen can be reached in less than 7 minutes (Brinkkötter et al. 2007).

Another very effective method of heating the windows is electric heating. The best known is the electric heating of the rear window, which has been a standard feature since the seventies. In some vehicles, electric heating is also provided for the windscreen (e.g. Ford Focus). Instead of the few wires of the rear window, many very thin wires are used which are embedded in the foil of the laminated safety glass. Particularly in darkness, rain and oncoming traffic, however, optical diffraction effects on these wires can cause slight visual distortions, which is why this type of heating has not yet become generally accepted. According to Großmann (2013), this can be remedied by electrically conductive films, which are also integrated in laminated safety glass. It remains to be seen for future developments to what extent nanocoated panes can basically prevent fogging, because this avoids the crystallization points necessary for condensation of the water.

The annoying fogging and ice formation described also apply to the exterior mirrors. It is therefore a further safety benefit if electrical heating is also provided for this purpose. The water of the windscreen washer tends to travel at the outlet nozzles at correspondingly low temperatures. For this purpose, too, an externally temperature-controlled heating system should be provided wherever possible.

8.4.4.4 Vehicle Air Conditioning Design

After already noticing temperature differences from 1 K, a constant readjustment of the air-conditioning performance is necessary, depending on the driving conditions. A distinction is made between manual, semi-automatic and fully automatic systems. Manual systems, in which the individual technical elements such as heating/cooling capacity, air distribution and fan speed have to be set directly by the user, are nowadays usually only offered in small vehicles. In semi-automatic air conditioning systems, the interior temperature is measured by appropriate sensors and the necessary blow-out temperature is controlled via these sensors. In addition to the temperature, fully automatic air conditioning systems regulate the blower output and thus the air volume and, in the best versions, also the air distribution. Additional data such as outside temperature, driving speed, engine speed, coolant temperature and similar are included in the control process (Wawzyniak 2011).

As already explained in ► Sect. 8.4.2, the individual demands of passengers on the climatic conditions to achieve comfort vary greatly. The design of the vehicle air-conditioning system must take this into account. At the very least, different adjustment possibilities shall be provided for the driver and the passenger. In addition, it has been customary since the beginning of the introduction of heating systems in vehicles to provide separate regulation for the head and foot areas. However, heated or cooled air should also be led into the rear compartment, where it should be individually regulated, including special ventilation nozzles. Wawzyniak (2011) explains that for people with cold-sensitive feet, the so-called "variable layering" allows the footwell exit temperature to be raised without changing the basic setting.

There is no standard for the operation of air conditioning systems. The operation should therefore be integrated into the overall operating and design concept. However, actuators which are related to possibly rapidly changing conditions must be immediately available. For example, all controls that have something to do with safe vehicle guidance (removal of windscreen condensation and ice) must be immediately available and, if possible, can be initiated with a single operating step (e.g. "defrost button"). Also the adjustment of the temperature must not be hidden in any menu of the on-board computer. The term "blue" for cold and "red" for warm is internationally understood and should be placed on the reference surface of the actuator (not on the actuator itself, see ► Sect. 6.1.3; ◘ Fig. 6.12). With controlled air conditioning systems, it makes sense to display the set target temperature in degrees Celsius or Fahrenheit (can be changed if possible), as this enables the respective user to assign his individual preference to a specific number. Similarly, direct access to the air distribution (top – middle – bottom) and the intensity of the air velocity is recommended. In combination with the possibility of directly influencing the air flow at the outlet openings, the user can quickly achieve a satisfactory result even under extreme climatic conditions. More complex setting options for the air-conditioning system can then be accommodated in a submenu of the vehicle setting.

Modern air conditioning systems not only change the temperature of the incoming air, they also ensure air quality through appropriate filtering by largely keeping air impurities and unpleasant odours out of the cabin. In view of the increasing number of allergy sufferers, the use of pollen filters in particular makes sense (Herzog 2007). This shows that not only the fresh air, but also the circulating air must be filtered. Wawzyniak (2011) describes the different filter materials and their separation efficiencies required to absorb the different particle sizes. Since the separated particles remain suspended in the filter material, they must be replaced at regular intervals. The service life of the filters is generally approx. 30.000–50.000 km driving performance, but should not exceed two years. From an ergonomic point of view, it must be demanded in this context that this replacement can also be carried out independently.

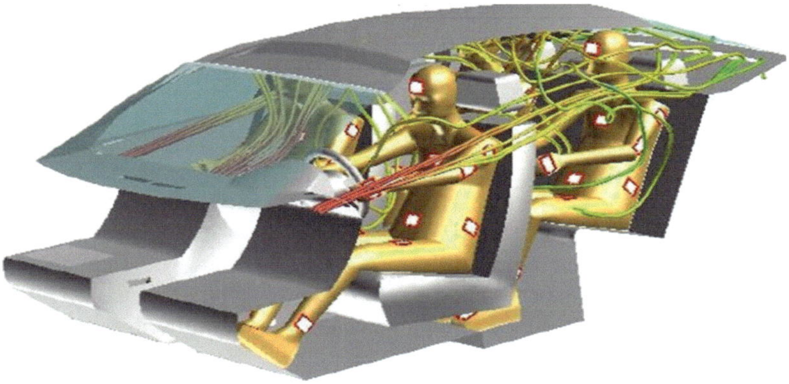

Fig. 8.36 CFD simulation of the flow conditions in the vehicle interior including virtual climate dummies. (Binder 2004)

This creates an additional acoustic entry, not least because of the fan required in every air-conditioning system. Due to considerable development effort, the noise nuisance caused by the air conditioning system is kept as low as possible (examples: Herzog 2007).

8.4.4.5 Air Conditioning Systems in the Product Development Process

The development of a climate control system for the vehicle is an extremely complex process. This is not only due to the requirements of the passengers discussed above, but also to the complex flow conditions in the interior of a vehicle. As in other areas of product development, simulation technology also plays an important role here. Computional Fluid Dynamic (CDF) analysis, which helps, among other things, to make a selection from a series of competing design drafts, is one method that can be used to optimize the flow of components at an early stage (Huco 2005). Fig. 8.36 shows a simulation of the flow conditions in the vehicle interior using this program system. Binder (2004) reports that the data on speed, temperature, radiation and body heat available in CFD analysis can be used to calculate all the data needed to determine thermal comfort using a specially developed virtual thermal dummy. It could be shown that the values calculated in the simulation show a high correlation to the values measured with the real climate dummy MARCO under comparable conditions (MARCO is very similar to the climate dummy DRESSMAN developed by the Fraunhofer Gesellschaft shown in Fig. 8.25).

Despite these very good simulation results, the degree of maturity of the vehicle air conditioning system must be examined at every stage during the entire development period from concept development to series production. Deyhle and Bienert (2011) report on a recording and evaluation system (PEA) specially developed by Porsche for objective evaluation, with the aid of which up to 200 air and fluid temperatures can be recorded in the test vehicle using thermocouples, enabling a comprehensive analysis of the functions of the air conditioning system under a wide variety of conditions, particularly with regard to automatic air conditioning control. A further development tool at all automotive companies is the climatic wind tunnel, in which reproducible and constant conditions can be generated and thus differences between different solution possibilities can be determined and optimizations with regard to the objectives can be carried out systematically.

Whether the development goal of "comfort in the interior" has really been achieved can only be determined by comparing the objective measurement results with the subjective assessments of test persons. For this purpose, field driver tests under different weather conditions in different climate zones are necessary. As Deyhle and Bienert (2011) argue, such surveys require on the one hand a statistically reliable basic population (e.g. 100 test persons) and on the other hand the broadest possible range of persons of dif-

Table 8.8 Evaluation scale according to ISO 1405-3 (2006)

ISO scale for temperature perception		Extended temperature perception scale	
		+5	**Extremely hot**
		+4	Very hot
+3	Hot	+3	Hot
+2	Warm	+2	Warm
+1	Slightly warm	+1	Slightly warm
±0	Neutral	±0	Neutral
−1	Slightly cool	−1	Slightly cool
−2	Cool	−2	Cool
−3	Cold	−3	Cold
		−4	Very cold
		−5	Extremely cold

Table 8.9 Discomfort verbalisation of the valuation figures in Table 8.8

"Uncomfortable scale.		Sultriness scale		Preference scale	
3	**Very uncomfortable**	**3**	**Very sultry**	**+3**	**Much warmer**
2	Uncomfortable	2	Sultry	+2	Warmer
1	Slightly uncomfortable	1	Slightly sultry	+1	Slightly warmer
0	Not uncomfortable	0	Not sultry	0	No change
				−1	Slightly cooler
				−2	Cooler
				−3	Much cooler

ferent ethnic origins. The subjects should be representative of the expected user population in terms of age, gender, driving experience and anthropometry. In order to arrive at statistically verified results, it is necessary to use questionnaires which make the subjective evaluation quantitatively describable. Hodder (2013) observes that over a long period of time the investigators often developed their own questionnaire techniques for this purpose. In recent years, however, standardized measurement and evaluation methods have been developed by ergonomists. The advantage of these standardised methods is, among other things, that valid protocols and measurement scales can be used for the evaluation, which also enable a comparison with the results found by another department. ISO 1405-3 (2006) deals directly with the subjective evaluation of vehicles. Table 8.8 shows the rating scales published in ISO. With regard to temperature perception, the scale shown on the left in the first table should normally be sufficient. For extreme environmental conditions, it may be useful to extend this scale from 7 to 11 points (right-hand part of the table). The "uncomfortable" and "sultriness" scales refer to the negative aspects associated with thermal discomfort and sweating (Table 8.9). The preference scale indicates how a person would like to feel. It may happen that a person e. g. reports in his temperature perception that it is warm, but that he/she nevertheless indicates "no change" in the preference scale.

With such attempts it is useful to add an overall survey in which the respondents express their satisfaction and acceptance by answering "yes" and "no". These answers can be used to quantify the degree of achievement across the board.

8.5 Odour

The comfort pyramid presented in ► Sect. 3.3.4 shows that odour is the decisive factor for discomfort perception and can mask all other influences if necessary. Due to the high adaptability of the sense of smell, the first impression plays an important role. If you enter a new vehicle, the perceived smell fundamentally influences the impression positively or negatively. It is therefore extremely important to ensure that odours to be rejected are

Fig. 8.37 Vehicle equipped with gas sensors to objectify the emissions of the components used. (Source: Daimler Global Media Site)

avoided under all circumstances. (See also the comments on "craftsmanship"in ► Sect. 7.9). Objectifying the odour with technical measuring methods encounters considerable difficulties because – due to the evolutionary character – there is a connection of high complexity between the triggering gaseous signal substance and the triggered odour perception (Boeker 2004). Today, there are a number of gas sensors that can be used to objectively determine the evaporation of substances. Since there is no possibility to verify the emotional effect of odours by objective measurements so far, the vehicle companies employ olfactory experts, who decide about the odour emissions of the different materials. For this purpose, the materials to be used in a new vehicle are heated separately in glass containers at 80 °C in order to simulate the maximum evaporation that can occur in a sunlit parked vehicle. After cooling down to 60 °C, the olfactory experts evaluate the odours in the glasses according to a school grading system (1 = odourless ... 6 = unbearable; Laukart 2011). Only materials with an average grade of 1 and 2 are normally considered for series production. The materials to be tested must be manufactured with the tools that are also used for series production. For a final test the complete vehicle will be equipped beside others with gas sensors (see ◘ Fig. 8.37). In detail, an inspection can be carried out according to various standards, for example VDA 276 or FAT AK 26. In addition, the materials used must be tested for possible inhalation allergies and potential skin contact allergies.

Due to the already mentioned high adaptability of the sense of smell, the odour nuisance no longer plays the serious role during the ride as it does when boarding. However, here too unpleasant odours can penetrate the consciousness again and again due to fanning effects. It is a triviality to demand that odours from the fuel system (petrol or diesel odours) must be eliminated in any case. However, the heating system and especially the air conditioning system can also be a cause of unpleasant odours. The condensate generated by the evaporator is separated during operation. However, a residual quantity remains in the evaporator even when the air conditioning system is switched off (Großmann 2013). In particular, if the air conditioning system is rarely operated, mould formation and the like can cause considerable odour and health nuisance. In general, materials which are deposited in the vehicle during use and which secrete perfumes (fruit, food and drink residues, ash residues, stale nicotine odour, dog odours, vomit, etc.) are the subject of odour nuisance, especially in older vehicles. However, unpleasant odours can also be caused by the filters themselves used for air cleaning, if they are clogged and possibly contaminated with moisture for some reason. In the internet different house prescriptions are called, with whose assistance one should be able to eliminate such smells again. A decisive prerequisite for all measures is that the vehicle is thoroughly cleaned beforehand. A treatment with ozone seems to be a particularly good effect. Ozone (O_3) is a radical that reacts among other things with microorganisms, i.e. also with odour molecules, thereby inactivating them. After such ozone treatment, the vehicle must be well ventilated so that the ozone which causes coughing is removed from the vehicle again.

But also an active treatment of the air while driving can have a positive effect on the reduction of odour nuisance. In particular, the positive effect of air ionization is reported here (Taxis-Reischl 1999). Artificial ionisation restores the natural concentration of negative air ions, which is lost when the weather changes or the hair dryer is used, for example. The effect of the electric field can cause suspended matter, especially very small particles such as dust, cigarette smoke, pollen and the like, to be excreted from the air. In addition, negative ionization is attributed to an increase in resilience, concentration, energy and vitality, although this has not been scientifically proven.

The artificial introduction of fragrances into the passenger compartment has also recently been added to the product range by various manufacturers. For example, Citroën offers a fragrance dispenser for the passenger compartment of various vehicles. The fragrance intensity can be regulated and also turned off. There are different fragrances to choose from, so that you can choose the fragrance you want. Mercedes also offers a fragrance dispenser for the new S-Class model, with four fragrances available. A special technology (clocked control) and the use of special fragrances ensure that the fragrance evaporates quickly after deactivation. With the emotional effect that every fragrance has in both the positive and negative sense, however, it must also be borne in mind that with additional scenting, unpleasant existing smells can only be masked, but not eliminated. When the masking fragrance loses its effect, the old unpleasant fragrances regain their effectiveness.

References

Literature Used

Amano, Y., Imai, H.: A car air conditioner. In: Gijutsu, J. (ed.) Die Kraftfahrzeugtechnik, vol. 25, pp. 1096–1101 (1971). Japanese

Asakai, M., Sakai, Y.: Colling effect of car ventilation. Bulletin of JSAE No. 6/1974, 75–82 (1974)

ASHRAE: Handbook of Fundamentals. American Society of Heating, Refrigerating and Air Conditioning Engineers, Inc., Atlanta, GA, USA (1997)

Baker, W., Mansfield, N.J.: Effects of horizontal whole-body vibration and standing posture on activity interference. ergonomics. **53**(3), 365–374 (2010)

Binder: New trends in air conditioning for passenger cars. Behr, Automotive Interior Symposium (2004)

Bitter, T., Fritzsche, F., Hartung, J.: Representation of the vibration behaviour of vehicle occupants – symbiosis of experiment and simulation FAT series, vol. 189 (2005)

Boeker, P.: Technical-sensory odour measurement 8th workshop "Odour and emissions at Kunststoffen", 29–30 March 2004. Kassel (2004). lecture manuscript

Bootz, A., Hohenöcker, O., Niklas, J., Seethaler, L., Sagan, E.: Landing gear design. In: Braess, H.-H., Seiffert, U. (eds.) Vieweg Handbook Automotive Engineering, 6th edn. ATZ/MTZ reference book. Vieweg + Teubner, Wiesbaden (2011)

Brinkkötter, C., Grünig, C., Kaufmann, M.: Optimization of the windscreen approach flow Lecture at the conference automotive interior-cockpit, Cologne, 9.10.2007 (2007)

Brooks, J.E., Parson, K.C.: An ergonomic investigation into human thermal comfort using an automobile seat heated with encapsulated carbonized fabric (ECF). ergonomics. **42**(5), 661–673 (1999)

Brosh, T., Arcan, M.: Modeling the body/chair interaction – an integrative experimental-numerical approach. Clin. Biomech. **15**, 217–219 (2000)

Bubb, H.: Analysis of speed perception in motor vehicles. Journal of Ergonomics 31. **1977/2**, 103–112. Dr. Otto Schmid KG, Cologne (1977)

Bubb, H.: Experiments to the contradiction "noise as strain or noise as feedback". In: Mital, A., Krueger, H., Kumar, S., Menozzi, M., Fernandez, J.E. (eds.) Advances in occupational ergonomices and safety, pp. 650–654. International Society for Occupational Ergonomics and Safety, Cincinnati, OH, USA (1996). II Proc. Of the XI.th Annu. Int. For Occupational Ergonomics and Safety 1996

Bubb, H., Schmidtke, H.: Ergonomic aspects in the design of motor vehicles. In: Wagner, H.-J. (ed.) Verkehrsmedizin – Including All Transport Sciences. Springer, Berlin, Heidelberg, New York, Tokyo (1984)

Bureau, C. et al: MARCO, Method to Assess Thermal Comfort. VTMS6, Paper C599/051/203 Brighton (2003)

Cengiz, T.G., Babalik, F.C.: An on-the-road experiment into the thermal comfort of car seats. Applied Ergonomics. **38**(3,5), 337–347 (2007)

Clark, R.P.: Human skin temperature and convective heat loss. In: Studies in Environment Science, Bioengineering, Thermal Physiology and Comfort, vol. 10. Elsevier Scientific Publishing Company, Amsterdam (1981)

Deyhle, H., Bienert, R.: Vehicle air conditioning from the customer's point of view: Subjective vs. objective evaluation Congress Progress in automotive interiors, Stuttgart, November 16–17, 2010 (2011)

DIN 1946: Air conditioning of passenger cars and trucks, Part 2 (2006–2007)
DIN 1946-2: Indoor air – Health requirements (VDI ventilation rules) (1994–2001)
DIN 33 403, Part 3: Assessment of the climate in the warm and hot areas on the basis of selected cumulative climate measures (draft) (2000)
DIN 4710: Statistics meteorological data for the calculation of the energy demand of heating and ventilation systems in Germany (2003)
DIN EN 28996: Ergonomics; Determination of heat generation in the human body. Beuth Verlag GmbH, Berlin (This standard contains the German translation of the International Standard ISO 8996:1990) (1993)
DIN ISO 7730: Moderate ambient climate. Determination of the PMV and PPD index and description of the conditions for thermal comfort. Beuth Verlag GmbH, Berlin (1995–2009)
Dufner, K.L., Bailey, D.T., Wolfe, D.E., Arya, S.P.: Determination of climate within metropolitan areas, Phase I summary. ASHRAE Transaction. **99**, 430–449 (1993)
Dupuis, H., Zerlett, G.: Stress on humans due to mechanical vibrations. Progress Report Whole Body Vibrations. Hauptverband der Gewerblichen Berufsgenossenschaften, St. Augustin, pp. 1–147 (1984)
Dupuis, H.: Zur physiologischen Belastungen des Menschen durch mechanische Schwingungen VDI Progress reports Series 11, Vol. 7. VDI Verlag GmbH, Number (1969)
Dupuis, H.: Mechanical vibrations and shocks. Chapter 4.4. In: Schmidtke, H. (ed.) Ergonomics. Carl Hanser, Munich, Vienna (1993)
Dupuis, H., Zerlett, G. (eds.): The effects of whole-body-vibration, pp. 1–161. Springer, Berlin Heidelberg, New York (1986)
Dupuis, H., Hartung, E., Hammer, W.: Investigations of vibration tolerance on the hand-arm system Ergonomic studies Reports, Vol. 2 Federal Office of Defense Technology and Procurement, (1974)
Ebe, K., Griffin, M.J.: Qualitative Models of Seat Discomfort Including Static and Dynamic Factors. ergonomics. **43**(6), 771–790 (2000)
Fanger, P.O.: Calculation of thermal comfort: Introduction of a basic comfort equation ASHRAE Transactions, Vol. 73, Part II (1967). Paper No. 2051, presented at the ASHRAE 74th meeting, Minneapolis, June 26–28, 1967
Fanger, P.O.: Thermal comfort. Analysis and application in environmental engineering. Danish Technical Press, Copenhagen (1970)
Fritsche, U., Feith, T.: Comfort nozzles for more climate comfort in the vehicle cabin. ATZ 2007(9), Koblenz (2007). Behr Special
Frühauf, F.: Thermal comfort in tomorrow's vehicle IIR conference "Innovative concepts for thermal management in motor vehicles", Stuttgart (2002)
Fung, N., Parson, K.C.: Some investigations into the relationship between car seat cover material and thermal comfort using human subjects. Journal of Coated Fabrics. **26** (1996)
Fung, W.: How to improve thermal comfort of the car seat. Paper presented at Proceedings of the 4th Bologna, Italy. International, Conference Comfort in the Automotive Industry, Recent Developments and Achievement (1997)
Gebhardt, H.J., Kampmann, B., Müller, B.H., Peters, H., Piekarski, C.: Systematic analysis of current climate sum measures for heat workplaces Publication series of the Federal Institute for Occupational Safety and Health – Research, Vol. Fb 829, Publication series of the BAuA, Dortmund/Berlin (1999)
Griffin, M.J.: Handbook of Human Vibration. Academic, London (1990)
Grimm, M., Kroschel, K., Schuller, B., Rigoll, G., Moosmayr, T.: Acoustic Emotion Recognition in Car Environment Using 3D Emotion Space Approach. In: Proc. DAGA (2007)
Großmann, H.: Grundlagen der Pkw-Klimatisierung: Extrema in Winter und im Sommer Klimatisierung von Kraftfahrzeugen, Technische Akademie Esslingen, 18–19 May 2000. (2000) Contribution to the training course
Großmann, H.: Heating, ventilation, air conditioning of passenger cars. In: Hucho, W.H. (ed.) Aerodynamics of the Automobile. Vieweg, Wiesbaden (2005)
Großmann, H.: Car Air Conditioning – Physical Basics and Technical Implementation. Springer Vieweg, Berlin, Heidelberg (2013)
Großmann, H.: Solar-powered stand ventilation for passenger cars. Development and market experience. In: Reichelt, J. (ed.) Car Air Conditioning. C. S. Müller, Karlsruhe (1992)
Hennecke, D.: For the evaluation of the vibration comfort of passenger cars with transient excitations. Dissertation at the Technical University Carolo-Wilhelmina zu Braunschweig (1994)
Herzog, A.: Cockpit – Innovations of the heating and air conditioning unit Conference automotive interior-cockpit, Cologne, 9.10.2007. (2007) Valeo: lecture
Hinz, L., Kern, J., Schweizer, G.: New generation of blowers for heating and air conditioning systems in motor vehicles. automotive magazine. **85**(9), 563–568 (1983)
Hodder, S.: Thermal environments and vehicles. Cap. 7. In: Gkikas, N. (ed.) Automotive Ergonomics – Driver-Vehicle Interaction. CRC Press Taylor & Francis Group, Boca Raton, London, New York (2013)
Huco, W.-H. (ed.): Aerodynamics of the Automobile, 5th edn. Vieweg Publishing House, Wiesbaden (2005)
ISO 1405-3: Evaluation of Thermal Environment in Vehicles, Part 3: Evaluation of Thermal Comfort Using Human Subjects (2006)
Knauer, P.: Objectification of vibration comfort with transient road excitation. Dissertation at the Technical University of Munich (2010)
Kühnel, W., Guilbaud, F., Proksch, Ch., Heckenberger, T., Heinle, D.: CFD Cabin Flow Analysis as Part

of the Product Development Process. ImechE paper C599/053/2003 presented at VTMS 6, Brighton, UK
Laukart, G.: Vehicle interior equipment. In: Braess, H.-H., Seiffert, U. (eds.) Vieweg Handbook Automotive Engineering, 6th edn. ATZ/MTZ reference book. Vieweg + Teubner, Wiesbaden (2011)
Lawther, A., Griffin, M.J.: Prediction of the incidence of motion sickness from the magnitude, frequence and duration of vertical oscillation. J. Acoust. Soc. Am. **82**(3), 957–966 (1987)
Liebing, R.: Acoustic evaluation methods for transient functional sounds. Dissertation at the University of Oldenburg (2009)
Madsen, T.L.: Thermal effects of ventilated car seats. International Journal of Industrial Ergonomics. **13**(3), 253–258 (1994)
Mansfield, N.J.: Human Response in Vibration. CRC Press, Boca Raton, FL (2005)
Mansfield, N.J.: Human response to vehicle vibration. In: Gkikas, N. (ed.) Automotive Ergonomics – Driver-Vehicle Interaction. CRC Press Taylor & Francis Group, Boca Raton, London, New York (2013)
Mansfield, N.J., Ashley, J., Rimell, A.N.: Changes in subjective ratings of impulsive steering wheel vibration due to changes in noise level: a cross-modal interaction. International Journal of Vehicle Noise and Vibration. **3**(2), 185–196 (2007)
Mayer, E.: Is the previous assignment of PMV and PPD still correct? Air and refrigeration technology. **34**(12), 575–577 (1998)
Mayer, E., Hellwig, R., Holm, A.: Thermal comfort in the vehicle. In: Hofhaus, J. (ed.) Car Air-Conditioning. V. Expert, Renningen (2007)
McIntyre, D.A.: Indoor Climate. Applied Science Publishers Ltd, London (1980)
Moes, C.: Geometric Model of the Human Body. In: Proceedings of the TMCE 2000 Delft, the Netherlands Third International Symposium on Tools an Methods of Competitive Engineering, 2000 Delft University Press, Delft (2000)
Moosmayr, T.: Objectification of transient interfering noises in the vehicle interior. Dissertation at the Technical University of Munich (2008)
Moosmayr, T.: Contact point noises. In: Zeller, P. (Ed.) Handbuch der Fahrzeugakustik – Grundlagen, Auslegung, Berechnung, Versuch, 2. Aufl. Vieweg+Teubner, Wiesbaden (2011). ATZ/MTZ reference book
Moseley, M.J., Griffin, M.J.: Effects of display vibration and whole-body vibration on visual performance. ergonomics. **29**(8), 977–983 (1986)
Müller, B.H., Gebhart, H.: Calculation models for the computer-aided determination of current climate totals for workplaces exposed to heat. Journal for Ergonomics. **53**(2), 107–113 (1999)
Nachtigall, K.: Automotive Interior Lighting Symposium automotive interior – cockpit, Cologne, 9/10/2007 (2007). lecture
Petzold, K.: Thermal load and room temperature – state and tendencies. Air and refrigeration technology. **11**(5), 232–236 (1975)
Pietzonka, S.: Interior lighting in the field of vision of new vehicle concepts Ludwigsburg, 21./22.09.2004 (2004). symposium
Probst, T., Krafczyk, S., Buchele, W., Brandt, T.: Visual prevention of motion sickness in cars. Archive for Psychiatry and Nervous Diseases. **231**(5), 409–421 (1982)
Recknagel, S.: Investigation of Mechanical Vibration Transmission at the Upper and Lower Extremities of Humans VDI Progress Reports Series 17, vol. 123. VDI Verlag GmbH, No (1995)
Roller, W.L., Goldman, R.F.: Prediction of solar heat load on man. J. Appl. Physiol. **24**, 7817 (1968)
Scheibe, W.: Assessment and Stress, Activity and Exposure of Humans to Continuous and Interrupted Exposure to Vertical Vehicle Speeds in Simulation and Field Experiments VDI Progress Reports Series 11, Vol. 31. VDI Verlag GmbH, No (1979)
Schmale, G., Stelze, W., Kreienfeld, T., Wolf, C.D., Härtel, T., Jödicke, R.: COSYMAN – A Simulation Tool for Optimization of Seating Comfort in Cars. In: Digital Human Modeling Digital Human Modeling Conference, Munich, June 18-20, 2002 VDI Berichte, Vol. 1675 (2002)
Schwab, R.: Influence of Solar Radiation on Thermal Comfort in Motor Vehicles FAT Publication Series, vol. 109, Forschungsvereinigung Automobiltechnik e.V., Frankfurt (1994)
Shimizu, S., Hara, H., Asakawa, F.: Analysis on air-conditioning heat load of passenger vehicle. JSAE Review. **82**, 80–90 (1982). Reprint in: International Journal of Vehicle Design, Vol.4, May 1983, No.3, pp. 292–311
Siefert, A.: Numerical Modelling and Experimental Validation of the Passive and Active Mechanical Properties of Human Tissue and its Implementation into a Whole-Body Model. Schenker Verlag, Reports from Biomechanics, Aachen (2013)
Sterling, E.M., Arundel, A., Sterling, T.D.: Criteria for human exposure to humidity in occupied buildings. ASHRAE Transactions. **91 Part 1**, 611–622 (1985)
Taxis-Reischl, B.: Heat load and driving behaviour. ATZ. **99**(9) (1999)
Temming, J.: Vehicle air conditioning and road safety. Impact of summer climates on the performance of drivers FAT Series, Vol. 177 (2003)
Badder, J.: Turn on the Heat! Auto-Motor-Und-Sport Magazine 2/2013, 54–59 (2013)
Verver, M.: Numerical Tools for Comfort Analysis of Automotive Seating, Eindhoven University of Technology, Diss., 2004 (2004)
Vogl, C.: New demands on interior acoustics through alternative drives. (Magna Steyer) Congress Progress in Automotive Interiors, Stuttgart, November 16–17, 2010 (2010)
Völksch, G.: On climate change in urban areas. Air and refrigeration technology. **14**(2), 88–90 (1978)
Wawzyniak, M.: Interior comfort/thermal comfort. In: Braess, H.-H., Seiffert, U. (Ed.) Vieweg Handbook Automotive Engineering, 6th ed. Vieweg + Teubner, Wiesbaden (2011). ATZ/MTZ reference book

Weible, R., Kern, J.: Control and regulation systems for heating and air-conditioning systems in motor vehicles VDI reports, vol. 515, p. 167–173 (1984)

Zeller, P.: Handbuch der Fahrzeugakustik – Grundlagen, Auslegung, Berechnung, Versuch, 2. Aufl. Vieweg+Teubner, Wiesbaden (2011). ATZ/MTZ reference book

Zipp, P., Rohmert, W., Klinkenhammer, K.: Ergonomic investigation of the position of the air outlet nozzles in the car to create optimum climatic conditions for the car occupants. Institute for Ergonomics, Darmstadt University of Technology, Darmstadt, pp. 1–64 (1977). research report

Zwicker, E., Fastl, H.: Psychacoustics. Facts and Models. Springer, Berlin, Heidelberg (1990)

Further Literature

78/317/EEG European Commission: Defrosting and demisting systems of glazed surfaces of motor vehicles (1977)

ADR 15 Australian Design Rule, Demisting of Windscreen

Bitter, T.: Objektivierung des dynamischen Sitzkomforts, Schriftenreihe des Instituts für Fahrzeugtechnik, TU Braunschweig, Dissertation. Shaker Verlag, Aachen (2005)

Fanger, P.O.: Local discomfort to the human body caused by non-uniform thermal environments. Annais of Occupational Hygiene. **20**, 285–291 (1977)

Fanger, P.O.: Maßstäbe für die thermische Behaglichkeit im PKW. In: Arbeitsplatz Auto, Bericht über das 5. Symposium Verkehrsmedizin des ADAC in Baden-Baden ADAC Schriftenreihe Straßenverkehr, Bd. 29, S. 151–156. (1984)

FMVSS 103 Federal Motor Vehicle Safety Standards, Windshield defrosting and defogging systems

Großmann, H.: Solarzellenbetriebene Standbelüftung für Pkw. Entwicklung und Markterfahrung. In: Reichelt, J. (ed.) Pkw-Klimatisierung, pp. 37–49. C.F. Müller Verlag, Karlsruhe (1982)

ISO2631-1: Evaluation of human exposure of whole-body vibration – Part 1: General requirements. International Organisation for Standardisation, Geneva (1997)

Moes, C., Horvath, I.: Estimation of the non-linear material properties for a finite elements model of the human body parts involved in sitting. In: Proceedings of the DETC 2002, 2002 ASME Design Engineering Technical Conference Montreal, 2002 (2002)

Pankoke, S.: Numerische Simulation des räumlichen Ganzkörperschwingungsverhaltens des sitzenden Menschen unter Berücksichtigung der individuellen Anthropometrie und Haltung Fortschritt-Berichte VDI. VDI Verlag GmbH, Düsseldorf (2003)

Siefert, A., Pankoke, S., Eckard, C.: Durchgängige virtuelle Prozesskette zur Optimierung von Fahrzeugsitzen unter dem Aspekt Humanschwingungen. In: Humanschwingungen VDI-Berichte, Bd. 2002, VDI Verlag GmbH, Düsseldorf (2007)

VDI2057-1: Beurteilung der Einwirkung mechanischer Schwingungen auf den Menschen, Ganzkörper-Schwingungen. Verein Deutscher Ingenieure, 2002

Driver Assistance

Heiner Bubb and Klaus Bengler

Contents

9.1 What Is Assistance? – 520

9.2 Driver Assistance and Driving Task – 522
9.2.1 Driver Assistance Systems Available Today for the Primary Driving Task – 523
9.2.2 Categorisation of Driver Assistance Systems for the Primary Driving Task – 548
9.2.3 Fatigue Warning – 554

9.3 Contribution of Driver Assistance Systems to Driving Safety – 558

9.4 Ergonomic Design – 563
9.4.1 Operation and Display – 563
9.4.2 Distinctness of the Modes of a Driver Assistance System – 567

References – 570

H. Bubb et al. (eds.), *Automotive Ergonomics*, https://doi.org/10.1007/978-3-658-33941-8_9

Since the mid-1980s a number of national and international projects have been launched (e.g.: PROMETHEUS, DRIVE, MOTIV, INVENT, RESPONSE, AKTIV) in which vehicle manufacturers, suppliers, universities as well as public and private research institutions have cooperated to develop systems which should support the driver in his task by means of the new possibilities gained by microelectronics (König 2012; Akamatsu et al. 2013). The projects were characterized by the cooperation of engineers, psychologists, physicians and ergonomists. On this basis, a large number of different technical developments resulted, some of which are also on the market today and some of which have become very important (driver assistance systems should be mentioned here and, especially in the field of driver information systems, navigation systems, which are available today as very inexpensive additional equipment and which now have a large market presence).

In this chapter, the expected experience of the effect and the interaction of the driver with driver assistance systems will be examined. It is not the approach of this chapter to describe the different technical designs and challenges of driver assistance systems. In this context, reference is made to the comprehensive work of Winner et al. (2012a).

9.1 What Is Assistance?

The first step is to define what is meant by driver assistance. If one starts from the closed loop paradigm, each machine ultimately represents a amplification of human will or human ability with regard to the task to be performed. The question then is: how many actions must be performed onto the machine as an operator – or from the point of view of system ergonomics – how much additional information beyond the task must be transferred to the machine in order to complete the desired task? To stick to a historical example: "Is it necessary to adjust the ignition timing to the engine speed in order to achieve different speeds of the vehicle with this?" Or – in development an important step further: "Is it necessary to take different accelerator pedal positions in order to maintain a desired speed depending on different driving resistances?" So is the automatic ignition timing adjustment (whatever technical measure is used) already an assistance system? Can cruise control be described as an assistance system? No user who criticizes the interventions in the driving process by ABS or ESP would be satisfied today with a non-synchronized transmission, would wish to have to permanently adjust the ignition timing or even operate small oil pumps at regular intervals while driving in order to constantly supply the moving parts of the vehicle with lubricant. So where is the dividing line to a so-called "assistance system"? It can certainly not be defined by the fact that certain adjustment processes are now electronically controlled; because the engine management or, for example, the brake system is now electronically controlled under the aspect of a traditional interaction between driver and vehicle. Probably at a later stage one will wonder about our current discussions about assistance systems, which – apparently – limit the driver's autonomy, because much of what seems innovative today has become self-evident.[1]

In order to define the idea of assistance, König et al. (2000) used the paradigm of the human assistant and asked which tasks a "perfect" assistant takes on in non-mobile professional life. ◘ Table 9.1 shows on the basis of this idea exemplarily a confrontation of activities in the occupation everyday life and with the driving a car.

Also from the field of professional motorist activities, such as the driver of a forwarding agency, the taxi or bus driver, the chauffeur of an important personality, but also the co-driver in a car rally, it is possible to analyse

1 It should be remembered here that even with the introduction of synchronised gearboxes there were fierce debates as to whether someone was still a capable driver who could no longer really put his foot down the throttle. A similar discussion is still virulent today in large parts of Europe – in contrast to the USA – regarding the automatic transmission!

Table 9.1 Comparison of the activities of a human assistant in everyday working life and the corresponding expectations of an assistance system when driving a car

Activities in everyday working life		**Driving a car**			
Main task	Trip planning Scheduling Hotel bookings Organizational activities	Driving task	Primary	Naviga-tion	Route planning Determining departure times Hotel search Search for special destinations
	Difficult, thought-intensive, but objectively limited subtasks moderate warning: "this operation could have the following negative consequences: …" Reference to information that the "boss" cannot know Tiresome routine activities "courageous" intervention if the "boss" makes a serious mistake (e.g. sending denials)			Guid-ance	Coping with complex driving tasks Crossing situations Parking and unparking Avoid imminent head-on collisions threatening sideways collisions deviation from the road Indication of hidden objects that influence the journey (e.g. by parked cars, walls, etc.) Boring motorway tripsStop and go traffic
				Stabili-sation Ung	Interception of violent Braking manoeuvres Steering manoeuvres
	Provision of work utensils necessary under certain conditions make routine calls		Second-ary		Wiper actuation Headlamp operation (switching on, fading in and out) Use turn signal
Secondary tasks	Maintenance and repair of working materials and equipment (tools, writing utensils, batteries, etc.) private errands, such as writing private letters, obtaining gifts, remembering important private holidays, etc.	Tasks	Tertiary		Indications of necessary Service work (e.g. cooling water, windscreen washer fluid, brake pads etc.) General inspection Entertainment Radio operation Choice of music (Reading) messages Interaction with the outside world Phone E-mail message exchange (etc.)
Vacation replacement		Fully automatic driving			

specific assistance activities which represent analogies to professional activities.

Another frequently asked question is: are assistance systems primarily only a contribution to the improvement of driving comfort/ reduction of discomfort or also a contribution to safety? In the sense of the comfort model presented in ▶ Sect. 3.3.4, any contribution that substitutes for an annoying activity is a measure to minimise discomfort. On the other hand, there is the aspect of comfort – i.e. pleasure – which can cause an activity that is perceived as discomfort under certain conditions to produce pleasant reactions under other conditions.[2] This already explains the dilemma that is always faced when designing an assistance system. With

2 ESP, for example, ensures that the course set by the steering wheel is maintained even under extreme conditions – as far as the physical conditions permit. If the conditions allow it (for example on a closed off area) it can be quite fun to induce lateral drifting, which is also characterized by a deviation of the steering wheel from the current direction of movement.

regard to safety, it can be said that any measure that avoids distraction or at least reduces it to an unavoidable minimum, that keeps mental stress in the optimal work area and that substitutes for a lack of human abilities, contributes to active safety. There is no longer any doubt that assistance systems that support drivers in difficult driving and traffic situations can make a significant contribution to accident prevention. Accordingly, the German Road Safety Council (DVR) in its current definition (published as part of the 12th DVR Forum "Safety and Mobility" in 2006) defines driver assistance systems (DAS) as systems that are suitable for supporting the driver in his driving task with regard to perception, scheduling and operation, which act in the three areas of navigation, vehicle guidance and stabilisation and make a significant contribution to accident prevention. The characteristic feature is that such systems monitor the vehicle, the driving behaviour of the vehicle, the driver and/or the environment by means of sensors and support the driver with additional information up to autonomous intervention in the driving of the vehicle. A similar definition was presented in the framework of the EU project "Advanced Driver Assistance Systems in Europe (ADASE)": *"Advanced Driver Assistance Systems (ADAS) are concepts to improve transport safety, efficiency and comfort without additional loads on resources (energy and land use) or on environment and quality of life".*

9.2 Driver Assistance and Driving Task

As can be seen in ◘ Table 9.1, assistance is possible at all levels of the driving task up to tertiary tasks and is also useful from the point of view of safety and partial comfort (see above). In ► Sect. 6.3, various assistance functions in the area of secondary driving tasks and especially for tertiary tasks have already been dealt with under the ergonomic aspect of interaction by means of displays and control elements. In principle, all assistance systems that provide support for the primary and secondary tasks are connected in parallel to the driver who has to perform these tasks himself in the unassisted system. However, the type of interaction varies with the assistance systems currently available. In the case of secondary tasks, the task is usually taken over by the automatic system when the assistance functions are switched on; the driver then has a quasi-monitive function with regard to this task and only has to intervene if the observed task fulfilment does not meet his expectations (e.g. automatic transmission, automatic lighting, automatic wiper). As already explained in ► Sect. 6.3, this interaction should be carried out in such a way that it is practically no different from the handset operation that would be necessary without assistance. With the assistance systems for the primary driving task, the type of interaction is different depending on the respective systems. This special feature will be discussed in more detail below.

The assistance systems described above are all based on the vehicle, i.e. from conditions on the vehicle or by measuring information from the driving environment, interactions with the vehicle are initiated or information is transmitted to the driver. A fundamentally different approach is to analyse the driver himself with regard to his ability to drive the vehicle safely and, if necessary, to give him hints or warnings to start or continue the journey. In this context, proposals have been repeatedly developed on how to prevent an alcoholic or otherwise drug-affected driver from using the vehicle. All the measures devised in this context, however, cause such awkward starting manoeuvres for the normal user capable of driving that this approach was never able to assert itself. However, the so-called fatigue warners, which are now offered as standard by some vehicle manufacturers, have gained a certain significance in the meantime.

9.2.1 Driver Assistance Systems Available Today for the Primary Driving Task

9.2.1.1 Historical Overview

Apart from early systems working on a mechanical basis (e.g. mechanical anti-lock braking system), which never went into series production, the development of driver assistance systems in today's sense (see above) began in the 70s of the last century, after correspondingly fast control systems working on an electronic basis were available. The first systems (ABS, ASR and ESP) related to the stabilization of the vehicle. According to the arguments of van Zanten and Kost (2012), it would be more appropriate here to rely on *Vehicle assistance systems* because they "help the vehicle to remain controllable, whereas driver assistance systems help the driver to correctly dose and coordinate the steering, propulsion and braking settings".[3]

The ABS anti-lock braking system went into series production as early as 1978. In 1991 it became mandatory for heavy commercial vehicles. From 1992 all Mercedes-Benz vehicles were equipped with ABS. Today this system is standard for practically every new car. The first electronic stability program ESP was produced in 1995. After the so-called "elk test" (double lane change) carried out by a Swedish journalist in 1997, a vehicle of the then new Mercedes-Benz A-Class overturned, as a consequence ESP was installed as standard in all vehicles of this company. The positive impact on accident statistics was so convincing that from November 2011 all newly registered passenger car and commercial vehicle models (except trucks) in the European Union will have to be equipped with ESP. From 2014, this will even apply to all new vehicles, regardless of when they were first registered. Newly registered trucks must now also be equipped with ESP.

In the later development the systems were added, which refer to the web guiding task (ACC, LCA, LKA, LDW etc. see below). They were introduced successively from the end of the 90s (first introduction in 1998), initially in luxury class vehicles as optional extras. With the increased use caused by series production and also caused by technical developments of low-cost components (e.g. systems based on laser technology instead of radar technology), these systems are gradually being offered more and more in the lower classes. However, there is no legal requirement for these assistance systems in the foreseeable future, which is why, from the customer's point of view, they remain expensive optional equipment, which of course hinders their distribution. From 2015, however, new heavy commercial vehicles will have to be equipped with anticipatory emergency braking and lane departure warning systems.

Apart from military applications, which date back to the Second World War, the development of navigation systems for vehicles also began in the 70s of the last century. The first systems available on the market did not yet use the data of the satellite-supported Global Positioning System (GPS).[4] A first system released by Honda in 1981 in cooperation with Alpine (*Electro Gyrocator*) was able to calculate a two-dimensional line representing the distance travelled using a mechanical rotation angle sensor and displacement sensors. The driver had to put a transparent map in front of a monitor showing this line and could see the distance covered. The system was not a navigation system in the true sense of the word and did not achieve any significant distribution. In 1984, VDO, in coopera-

3 One can see this argumentation as controversial: the improvements to engine management, transmission and the like mentioned at the beginning of the chapter would therefore also be vehicle assistance systems, although they take on tasks that the driver himself had to take on beforehand.

4 GPS has been developed by the US Department of Defense since the 1970s. It replaced the old satellite navigation system around 1985 *NNSS* of the US Navy, as well as the Vela satellites for the detection of nuclear weapons explosions. It has been fully operational since the mid-1990s. Since the artificial signal degradation was switched off on 2 May 2000, it has also ensured a positioning accuracy of around 10 metres for civilian purposes.

tion with the Falk road map publishing house, introduced the *city pilot* a system that was ultimately based on the compass and was able to calculate the direction of the target and the distance as the crow flies. This system also failed when it was launched on the market. 1983 presented Blaupunkt *EVA* (*electronic traffic controller for motorists*), the location of which was based solely on the acquisition of wheel sensor data. This relatively inaccurate system used voice output to convey driving instructions. But it was at least the first autarkic navigation system. A similar principle was used in the development of the Etak *navigator*. The latter, however, had to be updated over and over again via a cassette recorder due to the low storage capacity of the computer used at that time. In 1989 Blaupunkt published the *TravelPilot IDS* which, in addition to information from radar sensors, also used a digitally stored road map. In 1990, Pionier offered the first GPS-supported car navigation system. In 1994 BMW offered the first standard navigation system in a German car (Philips CARIN) for its 7 Series. Since that time, navigation computers have been marketed both as optional equipment, permanently installed in the vehicle and relatively expensive, but especially as an add-on from the aftermarket (practically 10 times cheaper than the fixed installations!) an unparalleled triumphal march in the vehicles. By now, almost all smartphones have a navigation system, which mostly uses the maps provided by Google.

A brief overview of the driver assistance systems available today can be found in Maier (2014), a detailed description of both the technical and the interaction possibilities by the driver in Winner et al. (2012b). The following description is partly based on these two literature sources.

9.2.1.2 Stabilisation Task

One of the oldest assistance systems at stabilization level is the ***automatic speed control (cruise control)***. A speed specified by the driver, which he derives from the completion of his guiding task, is kept within certain limits by the vehicle via a controller. In most systems, however, there is no brake intervention, with the result that the set speed can be exceeded when driving downhill – if the braking torque of the motor is not sufficient. As there is no other feedback than a glance at the speedometer, which, however, is omitted because of the trust in the reliability of the machine, this is an ergonomic inadequacy of these systems.[5] The cruise control is generally operated by means of an additional steering column lever or by means of buttons mounted on a steering wheel spoke.[6] Mostly by using that lever arm, one can take straight speed as nominal speed, by moving that lever arm up resp. down – at other manufacturers ahead resp. back – by increasing resp. decreasing speed. To date, no manufacturer-independent operating system has been established for cruise control. For safety reasons, the cruise control switches off automatically when the brakes are applied. From an ergonomic point of view, most systems fail to display the set speed, so that the driver may be surprised by the vehicle's speed when using the resume button.

Another system similar in its technical requirements to the Tempomats is the ***speed limiter***. Here the driver can define a speed limit which cannot be exceeded by the vehicle. This is useful, for example, if you want to avoid inadvertently exceeding statutory speed limits. Unfortunately, the speed limiter on most vehicles – as far as they have such equipment at all – is sometimes awkward to operate. This is sometimes done via the so-called on-board computer, which makes it practically impossible to operate the system while the vehicle is in motion and thus to adapt to changing speed regulations. Often, however,

5 A historical forerunner of cruise control is idle gas, which was often installed in more powerful vehicles of the 1930s ("motorway cars"). Since there is no regulation here, the speed achieved depends considerably on the instantaneous driving resistances. In European vehicles the idle gas could not assert itself in the long run.

6 A disadvantage of steering wheel operation is, by the way, that, for safety reasons, the general willingness of the system to accept tempo inputs must first be switched on, so that spontaneous operation as a result of the traffic situation may be made more difficult.

9

the operation is also integrated into that of the cruise control. To ensure that the driver can fully control the vehicle at all times (see below), exceeding the speed limiter speed limit is made possible by a kick-down function on the accelerator pedal (such as the kick-down function for an automatic transmission). Currently, the EU has plans to only allow vehicles with a built-in speed limiter in the future.

The ***Anti-lock braking system (ABS)*** prevents the wheels from locking during emergency braking by measuring and controlling the wheel speed and thus maintains the steerability of the vehicle even under these circumstances. The control process consists of reducing the brake pressure for each wheel independently at standstill until a wheel rotation is measured again (in less expensive systems, however, this independence only applies to the front wheels). This creates a pulsating pedal feeling for the user under these conditions, which he/she can interpret as feedback that the physical limit has now been reached, but which also leads to misinterpretation by the inexperienced driver and leads him to take his foot off the brake pedal in shock. In particular, the ABS system only slowly increases the braking pressure of the wheels on the road side with the higher static friction coefficient under different friction conditions on the left and right sides of the vehicle, so that even an inexperienced driver can prevent the vehicle from breaking away, which is induced by one-sided braking (Bosch 2004). Overall, the ABS system makes much better use of the maximum static friction than an experienced driver could achieve with the previously recommended "stutter brake". However, the ABS system can also have disadvantages, especially when braking on snow and gravel, because the "braking wedge" that builds up in front of the blocking tyre is avoided by the constant short release of the brake. This then leads to an extension of the braking distance (Engel 2009). Recent developments try to reduce or avoid this disadvantage.

The ***traction control (ASR)*** is practically a further development of the ABS system as it uses the same sensors (van Zanten and Kost 2012). In this case, acceleration shall measure whether the peripheral speed of one of the driving wheels is greater than the ground speed of the vehicle (normally measured on the non-driven wheels). If this is the case, the motor torque is reduced and the corresponding wheel is braked on the better systems. This effectively prevents the loss of lateral guidance that occurs when the wheel is spinning. The driver normally perceives the use of ASR when cornering by briefly slipping away, but this is intercepted immediately. When driving straight ahead, he can only determine the effect of ASR by a warning light (and the reduced drive force of the engine). Today, ASR is practically no longer available as an optional extra, as it is integrated into the ESP described below.

The ***Electronic Stability Control (ESC)*** or ***Electronic Stability Program (ESP)*** uses the functions of ABS and ASR, supplemented by the measurement of the yaw angle rate. It therefore also represents a further development of these two systems. The target yaw angle rate is calculated from the steering wheel position and the current speed. If this deviates from the measured yaw angle rate over a given limit, the drive torque is first reduced via the motor control. If this is not sufficient to stabilize the vehicle, a yaw moment is generated around the vertical axis by means of a wheel-specific brake intervention, which is intended to reduce the difference to the calculated yaw angle rate. This intervention can counteract unstable driving behaviour and prevent the vehicle from skidding within the given physical limits. The driver is thus helped to regain control of his vehicle during critical driving maneuvers (Bosch 2004). Further developed ESP systems even change the steering angle of the steered wheels in the given situation, whereby unintentional yawing during braking can be largely avoided even under different friction conditions on both sides of the vehicle (Raste 2012). The feedback for the driver about the use of ESP control is similar to that of traction control. Especially for very committed drivers who describe themselves as sporty, the rigorous reduction of engine power is perceived as disturbing in situations in which the driver consciously pushes the limits of driving dynamics. The ESP control can therefore be switched off, partly because this is technically necessary when the vehicle is fitted with snow chains, because these require a certain amount of slippage of the drive wheels in order to play off their effect at all. It is a matter of the so-called manufacturer's philosophy whether the switch-off is carried out completely or whether a reduced ESP function remains even after the switch-off.

Even the ***Brake Assist (BA)*** uses components of the systems described above.[7] It was developed because, beside others it was observed in driving simulator experiments at Daimler-Benz that many drivers in a dangerous situation only initiate emergency braking after an onset of comfort deceleration and that some drivers are even afraid to even carry out emergency braking (van Zanten and Kost 2012). The BA analyses the actuating speed of the brake pedal (possibly even the transfer speed from the accelerator pedal to the brake pedal). If this is below a certain threshold, the brake pressure is increased up to the control range of the ABS system and the maximum brake pressure is made available regardless of the actuating force of the brake pedal. However, as soon as the driver reduces the pressure on the brake pedal again, the brake pressure is reduced to the driver's specification (Maier 2014). Van Zanten and Kost (2012) describe that, in addition to this system, further measures to increase the braking effect in a hazardous situation are possible: For example, when it rains, the wetness can be removed from the brake discs by rhythmically applying the brake pads (approx. all 3 min) and sudden load changes can be used to apply the brake pads as a precaution and thus reduce the delay time until the braking effect begins as far as possible. There is no actual feedback for the driver about the effect of the BA. In an extremely dangerous situation he will be glad that the vehicle brakes so effectively. If there is no hazardous situation and the BA responds for the reasons mentioned above, the driver perceives a significantly faster braking reaction.

7 Also technically referred to as HBA for "Hydraulic Break Assist" (van Zanten and Kost 2012).

9

All three driver assistance systems described last have the characteristic that they respond within a very short time (<200 ms) and that the driver has no influence on the mode of operation of the system immediately after the response. This seems to be justified from a safety point of view, because within these extremely short times a reaction of the driver is not possible at all for physiological and psychological reasons. As Gelau et al. (2012) argue, this also does not contradict the Vienna Convention of 1968 (WÜ-StV), according to which "every vehicle and connected vehicles, when they are in motion, must have a driver" (Article 8 para. 1 WÜ-StV) and this vehicle driver "... must control his vehicle under all circumstances in order to be able to comply with the duty of care and to be able to carry out all vehicle movements incumbent upon him at all times ...". (Article 13 (1) WÜ-StV), because the systems described here would – as far as physically possible – ensure in time-critical situations that the regulation always corresponds to the driver's will. However, it is precisely for this reason that increased functional safety requirements come into play for these systems in order to ensure their availability and function.

9.2.1.3 Guidance Task

The guidance task results from the static and dynamic situation that arises for the driver in front of the vehicle. Assistance systems at the guidance level must therefore ensure that the driver is able to detect all objects that might influence his journey and if this is not possible for various reasons or if he is not prepared to replace this deficiency with technical measures. Measures that improve night-time visibility beyond what is possible with existing headlamp systems are therefore just as much a part of this as systems that directly control longitudinal and lateral dynamics. Because the technology for lateral and longitudinal dynamics is completely different and because these two dynamics are also experienced differently by the driver due to their strongly diverging time constants, the corresponding driver assistance systems were developed practically independently of each other. A characteristic feature of all systems for assistance in the guidance task is that the information necessary for its effect is not obtained solely from the reactions of the vehicle and input by the driver, but that objects at a greater distance in front of the vehicle are recorded in different ways and their position and position changes are included in the control.

Traffic Sign Detection

Traffic regulations set the external framework within which the guidance task is to be performed depending on the temporarily changing, current situations. Apart from the general regulations, indications and restrictions set by traffic signs play an important role. Traffic signs which impose a restriction for a particular section of road pose a problem in that they must not only be detected by the driver, but also the duration and conditions of their validity must be noted. The most important representatives of this type of traffic sign are local speed limits and overtaking bans. Local restrictions due to stopping and parking restrictions only play a role if the driver intends to park his vehicle. Another important class of traffic signs provides information on danger points and regulations regarding right of way regulations. Traffic sign observers, which are technically based on camera systems with downstream filter algorithms and display the result of this analysis to the driver in the instrument panel, can be an essential aid. Since many of the traffic signs mentioned are permanently bound to a local area – except for those which, for example, indicate a change in traffic guidance during construction work – the information for the respective traffic sign can also be obtained from the navigation system in addition to the camera detection. From an ergonomic point of view, the driver should only be shown the result of a traffic sign recognition if the conditions for the respective restriction or indication are met. In this respect, it makes sense to combine the processing of this information, for example, with the speed sensors, the time, the switching status of the light system/light sensor and the windscreen wiper/rain sensor. When displaying this result, the type of reaction of the driver resulting from the current driving situation must also be taken into account. For example, it was observed within the INVENT program that the display of a stop sign (here preferably in the HUD so-called STOP-Assistant, Mages et al. 2012) should not appear at a fixed spatial distance from the preferred road, but at a time interval resulting from the current speed and the assumption of a comfort delay (<3 m/s^2).

In principle, the displays in the graphic layout should correspond to the traffic signs visible in reality. ◘ Figure 9.1 shows on the left an example for the display of a speed limit in the analog instrument. On the right is a proposal from an ergonomic point of view, which can be implemented in this form with display panels in the instrument cluster, which work on the basis of a graphic display. In this case, the analogue display even makes sense because it functions as a limit value display: the driver can see at a glance how far he has approached the set limit.

Unfortunately, the traffic sign observers available today are still faulty. A study of the traffic sign detection Opel-Eye published by the magazine Auto-Motor-und-Sport showed

◘ **Fig. 9.1** Display of the speed limit by the traffic sign recogniser **a** Example Bosch, **b** Ergonomics proposal

a recognition rate of only 80%. The recognition rate drops in poor visibility conditions or at night even to a recognition rate of 80%. The driver is not relieved with such a recognition rate, but must pay even more critical attention to the signs in order to correct false indications (Gulde 2009). In fact, Maier (2014) critically states whether such systems will actually improve compliance with speed regulations and overtaking bans. An analysis carried out as part of the third investigation phase of the EU project SARTRE (Social Attituds to Road Traffic Risk in Europe) specifically on the subject of "Compliance with speed limits" showed that only 41% of the drivers surveyed consider compliance with speed limits to be important or comply with them. In Germany alone, around 25% of the drivers surveyed stated that they exceeded the speed limits in road traffic (Cauzard 2003). Probably only a combination with the ACC system described in the subsection "longitudinal dynamics"would bring about a significant improvement in this respect.

Vision Improvement Systems

Although the total number of night-time accidents involving personal injury decreased during the observation period between 1991 and 2002, probably due to the successive technical improvements of conventional headlamp systems, an accumulation of night-time accidents can still be observed in the dark winter months. Khanh and Huhn (2012) argue that with attentive driving and good environmental conditions at a speed of 100 km/h a stopping distance of 91 m (no emergency braking; composed of basic reaction time, decision, braking process with several stages) is required. The detection limit of high beam is approx. 150 m, so that obstacles under these conditions are detected sufficiently early at night. When driving with low beam, the maximum detectability limit with halogen headlights is 65 m, which would actually require a speed limit of 60 km/h to 75 km/h. In addition, the share of high beam in journeys with required lighting is estimated at approx. 5% (Maier 2014). Visibility-improving measures are therefore an important prerequisite for a further reduction in night-time accidents. According to Khanh and Huhn, two elementary requirements for lighting and vehicle systems can be derived from the properties of the optical information perceiving of humans (see also ► Sect. 3.2.1):

1. A homogeneous light distribution in front of the vehicle with good lateral illumination of the roadway and the roadway environment as well as plenty of light at as large a distance as possible could improve the contrast of relevant objects with the background and thus achieve a large distance from the vehicle. It is important that this light distribution adapts to the route as a function of the current driving task.
2. At the same time, the glare caused by oncoming traffic and the following traffic, the light of which is visible in the rear-view mirror, must be minimised or at best even eliminated. This requires a headlamp levelling system to be designed in such a way that the illuminance in the eye of oncoming and preceding traffic does not under any circumstances exceed the maximum permitted by official regulations.

Already the use of gas discharge lamps (so-called xenon light) raises the detection limit to approx. 85 m and thus the speed limit to approx. 90 km/h, whereby the feared psychological risk of glare could not be proven in experiments (Khanh and Huhn 2012). With the introduction of LED lighting systems, the performance limits of the gas discharge lamp could even be exceeded, although experts estimate that this technology will become more and more established due to higher energy efficiency despite higher costs (Götz et al. 2009).

Adaptive lighting systems, which are described as so-called AFS headlamps (Advanced Frontlighting System) in the ECE regulation 123 defined in 2007 (introduced by Hella in 2006 as the first manufacturer), do not yet belong to the area of assistance systems in the narrower sense. They contain the following functionalities:

- The ***city light*** is only used up to speeds of 50 km/h. It facilitates the early detection of pedestrians and objects through a widened and symmetrical light distribution.

- The ***highway light*** is based on today's dipped beam light and provides for more extensive illumination of one's own carriageway and improved illumination of the side area.
- The ***bend light*** is called *static* (from a certain steering angle an additional sideways directed headlight is switched on) and *dynamic* (the entire lighting unit is rotated about the vertical axis – usually up to ±18° – depending on the steering wheel position. With the so-called α/2 algorithm, the headlamp on the outer side of the curve is turned only half as much as the one on the inner side; Khanh and Huhn 2012). Extensive tests have also shown that the static bend lighting does not provide the driver with a visibility range greater than the required braking distance in bends (Grimm and Casenave 2007). Further developments in bend lighting include information from the navigation system and the front camera, which is necessary anyway for automatic switching (see below), in the control system. An even more effective further development in the future will be the inclusion of the glare-free high beam technology described below.
- The ***motorway light*** has a higher range in the center. However, glare is largely avoided due to the large distance between the road and oncoming traffic.
- The ***bad-weather light*** allows better orientation at the edge of the road, especially when the road is wet, thanks to a wide light distribution that is slightly swivelled outwards. At the same time, the glare of oncoming traffic from the light reflected from the road is reduced.

From an ergonomic point of view, it makes sense if the functions described are switched automatically depending on external conditions. An automatic switchover between low beam (driving) and high beam, which is controlled via information from the image processing of a camera image, is already available today as an optional extra from many manufacturers.

The lighting systems described above in accordance with the AFS description refer only to general traffic situations but not to a specific driving situation which may change at short notice. This is only achieved by the so-called ***assisted light distribution*** possible. The technical prerequisite for this is a camera system that can use image processing to detect the lighting conditions (headlights of oncoming traffic, rear lights of traffic ahead) and, if necessary, living objects (on the basis of an infrared camera) according to the situation, as well as a headlight system with which specific light sources can be specifically switched in intensity and direction. On this basis, the following assistance systems have been developed and are now also available in luxury vehicles (depicted after Khanh and Huhn 2012):

- ***Marker light:*** The information supplied by an infrared camera is used to switch on an additional spotlight and direct it at the object in question (◘ Fig. 9.2). In principle, the necessary detection technology is also used for the so-called night vision systems. From an ergonomic point of view, marking light is a far better solution than displaying or marking an object on a video screen which is not in the driver's field of vision and which, due to the reduced magnification of the image, does not allow correct angular assignment to reality (see also ► Sect. 6.2.1 and ► Fig. 6.16).
- ***Variable headlamp levelling:*** Depending on the distance of the vehicle to the surrounding traffic, the cut-off edge is varied vertically so that no glare can be caused. The principle can be illustrated by the example

◘ **Fig. 9.2** Marking light. (Source Hella KGa Hueck &Co)

of encountering oncoming traffic. If no oncoming traffic has been detected by the camera system, the system is in high beam mode. As soon as oncoming traffic is detected, the cut-off edge is continuously lowered (as opposed to the automatic switching between a fixed driving beam and a fixed dipped beam described above). If oncoming traffic comes close to one's own vehicle, the cut-off edge finally reaches the state of the dipped beam.

- ***Glare-free high beam:*** The headlight system is always in high beam mode. The camera system of the own vehicle constantly records the traffic area and calculates the angular positions as well as the distances of all vehicles located there. The light is then removed selectively and precisely at the points where oncoming vehicles and vehicles in front are located. The advantage over the principle of variable headlamp levelling is that light is retained in zones where glare cannot occur.

Longitudinal Dynamics

The ***Adaptive Cruise Control (ACC)*** often referred to as ***distance tempomat*** or ***distronic*** (Daimler) extends the functions of the cruise control by regulating the distance to a vehicle in front. In principle, the system must be activated by the driver, similar to cruise control. When the brake is applied, the system is automatically deactivated. On the open road, it maintains a speed preset by the driver similar to conventional cruise control. On the basis of radar distance measurement or – in newer (low budged; see below) developments on the basis of laser distance measurements or camera-based systems – it detects the distance between vehicles in front and initiates braking and acceleration processes based on the measured differential speed[8] with the aim of maintaining a specified safety distance from the vehicle in front. The braking decelerations automatically maintained by the system are below the value of -2 m/s^2 to -3 m/s^2. Also for the acceleration processes only moderate values are kept (see also ■ Table 9.2). Due to the limited range of the radar sensors (it is in the range of about 200 m; the alternative technical systems such as lasers or optical cameras are sometimes much worse), the maximum control speed is limited to 180 km/h for most manufacturers, and to 210 km/h for Audi (for the cheaper laser or camera-based systems introduced in the lower classes in the future, the maximum control speed is even limited to 130 km/h).

In addition to selecting the speed limit, the driver can also select the distance that the system automatically maintains. This distance is given in seconds, because it represents the distance s which at the given speed v, is traveled within this time distance t ($s = v \cdot t$).[9] Usually this distance can be set in three or four steps. The shortest distance for most manufacturers is at 1 s, the longest distance at 2 s.[10] If a preceding vehicle has been detected by the system, the driver usually receives visual feedback in the form of a pictorial display (■ Fig. 9.3). In order to avoid an uncomfortable behaviour of the adaptive cruise control, all manufacturers tolerate a short-term undercutting of the set safety distance, especially at previous high differential speeds. If the system reaches its system limits (e.g. if maintaining the safety distance can only be achieved by delays greater than -3 m/s^2) the driver is warned by an acoustic signal. In order to avoid inscrutable reactions of the system, all objects detected by the system that move towards the vehicle at the vehicle's own speed or higher (i.e. stationary and oncoming objects) are hidden for the calculation of the safety distance. One exception is a preceding vehicle, which was already detected by the system and now comes to a standstill – for example in front of a red traffic light. In this case, the ACC vehicle also brakes almost to a standstill (a moderate acoustic warning is given shortly before this).

8 Older systems did not brake, which can be quite confusing for the driver under given conditions.

9 Many users find this idea of a distance in time, which characterizes a distance, difficult.

10 In this context, it should be noted that in Germany a fine is imposed for falling below the safety distance of 0.8 s.

Table 9.2 Mean values of the manoeuvres carried out in the Strasser tests (2012)

	Characteristics of the ACC system	Collision avoidance	Overtaking manoeuvre	Reevers	Outrigger
Reaction time	Forward-looking		3.40 s	1.38 s	5.33 s
	Moderately		2.79 s	4.18 s	2.34 s
	Dynamic		1.48 s	4.87 s	2.49 s
Response distance	Forward-looking	138 m		48 m	
	Moderately	112 m		33 m	
	Dynamic	69 m		27 m	
Minimum distance	Forward-looking	38 m		39 m	
	Moderately	28 m		25 m	
	Vibrant	35 m		16 m	
Detection distance	Forward-looking	157 m			
	Moderately	124 m			
	Dynamic	80 m			
Maximum acceleration	Forward-looking		0.80 m/s^2		0.92 m/s^2
	Moderately		0.91 m/s^2		0.91 m/s^2
	Dynamic		1.59 m/s^2		1.59 m/s^2
Overtaking time	Forward-looking		13.49 s		14.76 s
	Moderately		12.07 s		11.56 s
	Dynamic		9.21 s		10.09 s
Maximum delay	Forward-looking	−1.37 m/s^2		−1.87 m/s^2	
	Moderately	−1.23 m/s^2		−1.26 m/s^2	
	Dynamic	−3.00 m/s^2		−1.52 m/s^2	
Minimum speed	Forward-looking	75.9 km/h			
	Moderately	75.9 km/h			
	Dynamic	72.2 km/h			

Situations can also occur in which the system does not react in a comprehensible manner from the driver's point of view, possibly resulting in critical driving situations. For example, if the ACC vehicle "loses" the vehicle in front, it may unexpectedly accelerate to the set speed for the driver (a typical situation for such "losing" is, for example, when the vehicle in front is no longer within the detection range of the radar sensors in a roundabout or in a tight bend). It may also happen that the system does not recognize certain road users, such as motorcyclists or cyclists, and therefore does not react to them. The ban on overtaking on motorways is also only being observed with the latest ACC systems. Another typical situation that can cause confusion for the driver is when the radar sensors of the overtaking ACC vehicle detect a slow-moving truck in a long left-hand bend on the motorway. Then, incomprehensibly for the driver, a braking manoeuvre is initiated and, at the same time, a warning is issued against an impending collision. These system-related

Fig. 9.3 Feedback on the operating status of the ACC (BMW E60, **a** or Audi **b** and Mercedes Distronic plus **d** using a pictorial display. In the optional HUD (BMW E60, **c**), the display of the set target speed (here 70 km/h) disappears after a few seconds

shortcomings lead to rejection by many users of the ACC system.

The ACC system is used in a very narrow range of tasks (longitudinal movement in rolling traffic >30–40 km/h) already represent a certain form of automatic driving (see also ► Sect. 9.4), since the driver, as long as the system is active, takes on a monitoring function with regard to the speed to be maintained and the distances to the vehicles in front. As Strasser (2012) states, ACC is the most widely used assistance system and the one best studied in extensive and international studies. It is not really surprising that when using the ACC system, a constant time interval is also objectively driven (Fancher et al. 1998; Rakha et al. 2001; Ma and Kaber 2005; Viti et al. 2007). Otherwise, however, the results of the various investigations are sometimes quite contradictory. In the studies by Hoedemaeker and Brookhuis (1998); Dragutinovic et al. (2005) and Freyer (2008) as well as in the simulator study by Lange et al. (2008), all of which were carried out on out-of-town roads, it was observed that the test persons drove faster with ACC than without. In contrast, studies by Stanton et al. (1997) and Filzek (2002) show no change in speed. Other studies (Tricot et al. 2004; Kovordányi 2005; Ojeda and Nathan 2006) even observe a reduction in speed. The reason for these different findings may lie in the fact that, due to the necessity to specify a target speed for the use of the ACC system, the driver may drive faster in relation to the situation than would be the case on the basis of the intuitive assessment, but that on the other hand predetermined speed limits are more likely to be adhered to by the cruise control function than without it. The study by Oei and Polak (2002) and Acarman et al. (2006) indirectly confirms that there was a significant reduction in speed in urban traffic. The distances driven are also sometimes smaller, but always in the safe range (Hoedemaeker and Brookhuis 1998; Dragutinovic et al. 2005; Eick and Debus 2005; Lee and Nam 2007; Popiv et al. 2008; Lange et al. 2008). As was shown in a long-term test, many drivers prefer to choose the shortest time gap (Sacher and Bubb 2006). After all, using ACC increases the mean time gap compared to unassisted driving (Kovordányi 2005; Seppelt et al. 2005). In addition, ACC apparently also results in less frequent lane changes on motorways, and thus more frequent driving on the left or middle lane (Tricot et al. 2004; Freyer 2008; Jenness et al. 2008). With ACC drivers initiate a lane change for an overtaking manoeuvre earlier than without, obviously to avoid disruptive system intervention (Alkim et al. 2007 and Freyer 2008).

The results with ACC vehicles described here are based almost without exception on investigations with drivers who came into contact with such systems for the first time. Weinberger (2001, see also Weinberger et al. 2001) has conducted studies with frequent drivers who were able to use such a vehicle for 4 weeks, in particular to investigate the effects of habits, since ACC driving differs considerably from conventional driving due to its automatic function and therefore requires an adaptation phase. The results were as follows: After 4 weeks of regular use of the new displays and controls, the verdict of all test subjects shifted without restriction to "very familiar with me", and after 2 weeks a stable state of judgement was reached. Also with regard to the transfer situations, such as "approaching a slower vehicle" or "necessary reaction to unexpected strong braking of the vehicle in front", the test persons were of the opinion that they were familiar with the system after approx. 1–2 weeks. This was also demonstrated objectively by the average minimum gap in the takeover situation, which did not change significantly after the second week and stood at approx. 0.9 s . In addition, the "time to collision"values in the takeover situation showed that ACC changes the behaviour of both drivers who classify themselves as "sporty" and those who describe themselves as "comfort-oriented" towards a similar driving style, whereby a stable state also emerged after approx. 2 weeks of experience (in the case of the test persons used here, this meant a driving experience of approx. 2600 km). The time portions, in order to move e.g. with overtaking procedures closer to the predecessors, increased during the habituation phase from approx. 11% to on the average 16.6%. Overall, the average time interval between ACC operating activities is 18 s, with one ACC period lasting on average 164 s (2 min 44 s). Based on observations of individuals, the average longest uninfluenced usage span was 52 s within a usage period of 263 s (4 min 24 s), the shortest was 6 s within a usage period of 196 s (3 min 16 s). Overall, there are large interindividual (between) but small intraindividual (within) differences; i.e. the usage behaviour is obviously determined more by personality variables than by traffic situations. Although it is hardly possible to speak of a fatigue effect due to understrain with the relatively short, uninfluenced usage ranges observed, the test persons assessed the gain in comfort through ACC positively from the outset. This judgement even improved with increasing experience with the system. This is explained by a shift of the task content by ACC: the driver is relieved of the control activities on the stabilization level.

Due to technical boundary conditions, situations repeatedly arise in which the driver has to take back control of the longitudinal dynamics of the vehicle from the ACC. A distinction must be made between two classes: on the one hand, those for which the transfer *all the time* (e.g.: approach to a stationary obstacle) and those where the driver only has to take over when certain limit values are exceeded (e.g.: strong braking of the vehicle in front). For the former, the decision to act is clearer and therefore easier for the driver to make. Subjectively, the subjects find that such decision-making processes take place unconsciously, that they rather wait longer to see whether the ACC system reacts and that they consider such situations to be generally unproblematic. With regard to these judgments, there is no effect with increasing experience. Delay values after driver intervention are generally weaker than $-3\ m/s^2$. The ACC vehicle is actually always decelerated more strongly than the vehicle in front.

On average, the decision to apply the brakes manually is most difficult to make when another vehicle pulls over and shears in. There is hardly any difference between the assessment of the decision in the case of "strong deceleration of the vehicle in front" and the "approach to another vehicle with clearly low speed". The easiest way to make a decision is to "approach a traffic light". This situation is also clearly classified as 'harmless', whereas the situation of 'strong deceleration of the vehicle in front' and 'reeving of another vehicle' is more likely to show a tendency towards the judgement 'dangerous'. At the time of the takeover, the "time-to-collision" value, which can be used as a measure for the driver's foresighted takeover, is actually always above 5 s, but in borderline cases can also be just below 2 s. The maximum deceleration after driver intervention is the same in all situations and is approx. $-2.8\ m/s^2$. The strongest average delay values are shown in the situations "Stop" and "Approach". The smallest gap in takeovers is shown by "reeving vehicle" with an average of 0.5 s; the 5-percentile value here is even 0.16 s. However, this is to view from the dynamic situation, i.e. this extremely short distance is quickly increased again.

With regard to the choice of the desired speed, the permissible maximum speed is the most important influencing factor. The system for city speeds (60 km/h) is only used to a limited extent (time share: 4.6%). The most frequently used speed is 140 km/h (time share 13%), whereby this value is mainly selected on motorway sections with speed limit at 120 km/h. The desired speed is changed every 1.5 min on average. The results found here actually contradict the findings of Buld and Krüger (2002), who observed vigilance problems with increasing automation. Although the subjective strain on the driver also decreased there, the fewer interventions were necessary, the less attentive and tired the drivers became (cf. also Lee and Nam 2007). Buld and Krüger (2003) even noted that the transition to monitoring only reacts to incentives that are directly related to the supported task. This can lead to a pulling effect in critical situations such as when the vehicle in front enters a bend too quickly, which can only be compensated for by a combination of lateral and longitudinal support. Alkim et al. (2007) even come to the conclusion that ACC promotes the performance of secondary activities and thus leads to a certain distraction of the driver from the primary driving task.

According to Weinberger (2001), the following behaviour groups can be clearly separated with regard to distance behaviour: one group (approx. 60%) of the test persons always chose the shortest time interval (1 s), the other the longest (2 s). Only very few test persons changed the distance value depending on the situation. However, the test subjects who keep larger distances "exceed" them more frequently, as they obviously often see the need to choose a shorter distance due to the situation. Most of the test persons stated that they kept larger distances with ACC than without it. The main criterion for changing the distance selection is the weather condition. The change of the target distance is altogether considerably less frequent than the selection of the desired speed. The distance at the beginning of the habituation phase is changed significantly more frequently (on average all 12.5 min) than at the end (all 22.7 min).

Customer acceptance of ACC systems depends to a large extent on the parameters selected to determine the behaviour of the system under changing driving conditions. Strasser (2012) has dealt with this problem and in particular wanted to find out whether specific different parameter settings are necessary depending on the vehicle type. In his practical experiments, which were carried out on a closed-off test area with a target vehicle in front, he specifically examined the following scenarios:

- *Drive up to column* (Ego vehicle approaches a slower moving column)
- *Overtaking manoeuvre* (Ego vehicle shears behind a slower moving target vehicle and overtakes it)
- *Reevers* (another target vehicle pushes itself into the distance maintained by the ACC system to the original target vehicle)
- *Outrigger* (target vehicle leaves the lane of the ego vehicle and the ego vehicle accelerates back to the set desired speed)

For each of the four scenarios, three different versions of the ACC system were defined, representing interpretations available on the market. These characteristics were identified by the terms "moderate", "forward-looking" and "dynamic". The characteristic values used by him to describe the behavior of the ACC vehicle are summarized in ▫ Table 9.2.

The tests showed that the subjective statements on the categories "sporty", "dynamic", "safe", "comfortable" and "favourite" can be clearly assigned to technical parameters in the individual manoeuvres considered. However, this assignment does not depend on the vehicle class. This can be explained by the fact that if the assistance system is used, the driver quasi slips into the role of a passenger and observes the correct behaviour of the assistance system from there. The following general statements can also be made from the test results:

- *Drive onto a column:* For all three criteria "sporty", "safety" and "comfort", the system should react early but with rather weak dynamics.
- *Overtaking:* The system should react promptly and overtake with moderate acceleration.
- *Reevers/S hearing in:* The aim of the vote should be for the system to react rather early with moderately strong momentum.
- *Outrigger:* The user wants the system to react quickly to the outrigger. How strong the deceleration should be depends on the type of driver.

As already mentioned, the *operation* of the ACC system is derived from that of the cruise control. As a result, one can distinguish between operation via a steering column lever (usually located on the left side) and operation via a steering wheel-integrated control. For reasons of spatial compatibility, the ACC control panel should also be located on the same side as the tachometer. In the case of steering wheel-integrated systems, the driver must switch on the basic readiness of the system to accept target speed and distance values via its own button. This may interfere with the spontaneous use of this system and sometimes lead to confusion if the entire system is accidentally shut down when the ACC function is cancelled. In contrast to conventional cruise control, the resumption of the set speed once set (so-called resume function) does not present a problem for the ACC systems, since the set speed is generally indicated by an LED display or an additional pointer in the analogue speedometer. In this context, it is particularly important that the resume operation differs significantly from other operations (for more details, see ▶ Sect. 6.2.2). When presetting the target speed, there is no clear recommendation for a compatible design, at least for the design of the steering column lever. If the driver can see the influence of the pointer on the analogue instrument, then a movement of the steering column lever upwards corresponds to the compatible increase in speed. If, however, he uses the influence of the vehicle as an internal model during operation, it is advisable to move the lever forward (= acceleration) to increase the speed. By the way, this problem would not arise if the speed were generally displayed digitally. When operating via buttons on the steering wheel, there are no compatibility problems if the increase in speed is achieved by a button at the top and the reduction by a button at the bottom. For the ACC operation, it has proved to be advantageous to provide for the change of the target speed normally in steps of 10 km/h (with some systems, a fine adjustment can be made in steps of 1 km/h or 5 km/h by briefly touching or weakly touching). In principle, similar compatibility problems arise when setting the time gaps, which is often realised by a vertically or horizontally mounted rocker arm: does a movement upwards mean a greater distance (internal compatibility) or a closer approach to the vehicle in front (= shortening of the distance; external compatibility)? A corresponding symbolism in the display can largely alleviate this problem (see also ◘ Fig. 9.3, top left and bottom right). For a vertically mounted rocker arm for the distance settings, it is advisable to support understanding by means of a symbol, as shown in ◘ Fig. 9.3, bottom right.

With the technical prerequisites of an ACC system, the ***Emergency Brake Assistant*** developed. The information, which uses radar sensors to measure the distance and differential speed to the vehicle ahead, is used to precondition the vehicle for imminent emergency braking by precharging the brake system and applying the brake discs. In contrast to the Brake Assist, which works at the stabilization level and performs this preconditioning only on the basis of the driver's reactions, the detection of the area in front of the vehicle is included here. This also makes it possible to warn the driver of the imminent danger. This warning also occurs when the ACC system is not being used. In addition, this will further shorten the brake response time (currently available on the market as "BAS plus"from Mercedes; "Forward Alert" from Ford). By generating just enough braking force within the physical limits to ensure timely and safe stopping (so-called target breaking), approx. 24% of subsequent accidents after rear-end collisions could be avoided or at least reduced in their effects or severity, according to estimates by Schittenhelm (2009). A further development of this emergency brake assistant is the ***Autonomous Braking System (AEBS)***. This is a multi-level system. Also on the basis of the detection of the surroundings by means of radar sensors or a camera system, a partial braking at reduced braking force with a constant but significantly reduced deceleration compared to an emergency braking is carried out when the distance to the vehicle in front falls below a set distance. If the distance falls below a speed-dependent defined target distance, the driver is first warned acoustically, visually or haptically (e.g. active accelerator pedal) (this function is actually similar to the ACC function, but takes place at significantly shorter distances). If there is no braking intervention after the warning, braking is triggered with a small delay. Only if a collision is no longer avoidable, an emergency braking with maximum deceleration is induced (realized as Pre-Safe of Mercedes, Daimler 2006, 2011; as "Active City Stop"of Ford or Volvo City Safety System; the systems of Ford and Volvo work only up to a speed of 30 km/h – recently with Ford even up to 50 km/h -, whereby up to 15 km/h rear-end collisions are completely avoided). LEXUS offers autonomous brake assistants which are connected to a driver monitoring system. A camera records the driver's line of vision. If the driver's gaze is turned away from the road when approaching an obstacle, an additional brake jerk is applied in addition to visual and audible warnings to draw the driver's attention to the critical situation (Lexus 2010). As Maier (2014) explains, the strategy for triggering emergency braking is a critical issue in the implementation of the autonomous emergency brake assistant. In order to avoid false actuations, the emergency brake is only triggered when the accident is unavoidable (see also the Ford and Volvo interpretations). This is justified on the one hand by questions relevant to liability law (e.g. a rear-end collision in subsequent traffic caused by a false tripping) and on the other hand by possible acceptance problems of the system by the customer in the event of false tripping (Gründl 2005). In terms of driving dynamics, on the other hand, if an obstacle suddenly occurs, it would be necessary to apply the brakes as early as possible.

As Winner et al. (2012b) point out, the expensive radar and, with some limitations, laser systems are more suitable for the higher speed range, while the more economical camera systems tend to cover the lower speed range and can therefore also find their way into the lower vehicle classes. They expect that with the advancing technical development these two system types will grow together and that in future the entire speed spectrum with regard to longitudinal dynamics can be supported by driver assistance systems in all vehicle classes.

9

Lateral Dynamics

As already explained in ► Sect. 2.5 (see ◘ Fig. 2.25), approximately a quarter of all accidents in Germany involving personal injury occur when changing lanes or leaving the road unintentionally (Weißmann 2009). Assistance systems that compensate for incorrect driver behaviour with regard to lateral dynamics can thus make a significant contribution to reducing accidents. ISO 17361 (2007) defines "lane departure warning systems as driver assistance systems designed to prevent a vehicle from accidentally leaving the lane". The technical prerequisite for lane departure warning systems is that the road in front of the vehicle is detected in a suitable manner. The simplest system was introduced by Citroën in 2005. Here, the light reflected from the edge strips of the road, which are illuminated by an infrared laser beamer, is detected by sensors on the front and rear bumpers. When driving over an edge strip, this is indicated to the driver by means of location-compatible vibration on the seat. However, the system only recognizes markings shortly before they are passed over and is not immune to ambiguities such as those that occur with construction site markings (Walter et al. 2012). Most systems on the market, on the other hand, use an electronic front camera. Using image processing, the lanes and the position of the vehicle between them are extracted from the captured image. As Walter et al. (2012) describe it, this is a multi-stage process. Due to the strongly fluctuating light conditions depending on time of day and weather as well as many other conditions, the detection rate is never 100% even with good road markings. None of the systems will work without clean road markings. The technical system requirements are also very high, depending on the different marking systems used in the different countries. The knowledge of the lanes and the own position between them makes it possible to calculate the so-called time-to-line crossing (TLC; see also ▶ Eq. 2.21 and ◘ Fig. 2.21) in an area up to 30 m in front of the vehicle and to derive measures to support the driver from this.

A distinction is made between the following variants:

- ***Lane Departure Warning (LDW):*** Two systems have to be distinguished here: With the so-called *DLC system* (Distance to Line Crossing) only the distance of the wheel to the lane is used for the warning without prediction into the future (the Citroën system described above belongs to this category). With the *TLC system* (Time to Line Crossing), the TLC value is used to calculate the time to cross the marker. The driver is warned if the time falls below a predefined time threshold. The TLC system enables the driver to react earlier, but increases the requirements for lane recognition (Walter et al. 2012). In both systems, the warning itself is optical, acoustic or haptic. The investigations by Gayko (2009) show that the haptic warning via the steering wheel (usually vibration, or vibrating seat) is preferable to optical or acoustic warnings, since the driver can assign the information directly to the hazardous situation and thus react intuitively. Another advantage of the haptic warning is that it can only be recognized by the driver and not by any of the other passengers.

  There are many situations where the LDW warning appears inappropriate and annoying to the driver (e.g. utting corners or overtaking without using the direction indicator). For this reason, various efforts are being made to create a so-called ***Advanced Lane Departure Warning System (ALDW)***. The aim is to suppress unnecessary warnings. For example, an attempt is made to detect such situations by evaluating the navigation system, vehicle acceleration, steering wheel angle and accelerator pedal position. Warnings on narrow roads can be suppressed or delayed by shifting the warning time. However, there are also situations in which a somewhat earlier warning makes sense. For example, it is advantageous to warn approaching the outer edge of the curve earlier than on the inside, because there is more space available, apart from oncoming traffic. The evaluation of the operation of distracting control elements (radio, mirror adjustment, navigation device, etc.) can be used to output a warning earlier (for more details, see Walter et al. 2012). According to a report by the Dutch Ministry of Transport in 2007, Walter et al. (2012) also note that although LDW systems are very frequently switched on by drivers who have such systems, they are considered less effective than, for example, the ACC system. On the other hand, it is expected that Lane Keeping support systems, which provide active steering intervention, will be attributed a higher benefit.
- ***Lane Keeping Support System (LKS):*** This is an extension of the LDW, as it uses the same technical measuring system, with the difference that if it is determined that the vehicle is leaving the lane, it not only gives an optical and acoustic warning, but also returns the vehicle to the intended course through targeted steering intervention. The steering intervention can be performed by a torque motor integrated into the steering system (generally only possible with an electric power steering system) or by using the superimposed steering system. The system is designed in such a way that this steering effect is only effective when the driver himself steers, so that the responsibility

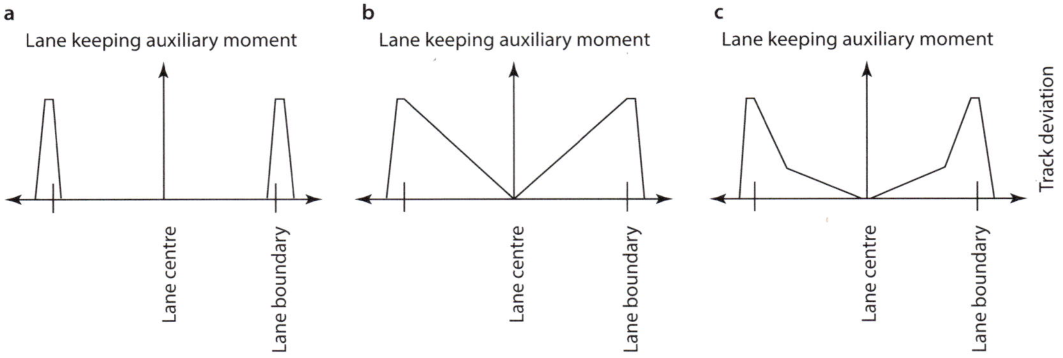

Fig. 9.4 Schematic representation of the possible courses of the auxiliary torque as a function of the lateral displacement in an LKS system (Gayko 2012)

9

remains with him[11] (Ishida et al. 2003, for example, suggest that 80% of the calculated steering torque should contribute to the driver's assistance, but the rest should be borne by the driver himself). Autonomous driving is not possible with this system. Figure 9.4 illustrates the characteristics of a lane departure warning system by means of the auxiliary torque as a function of the lateral position from the centre of the lane with assumed straight ahead travel. The process shown in Fig. 9.4a characterises a support only from the point of view of safety: only when the driver threatens to cross the lane boundary is he given a corresponding restoring torque on the steering wheel. Figure 9.4b illustrates a close guidance, because even small deviations of the vehicle from the middle of the lane lead to a perceptible corrective moment. Experience has shown that such an interpretation is rejected by many drivers (e.g. Penka 2001). The course shown in Fig. 9.4c characterizes a gentle support with small deviations from the center of the lane, which only lead to low discomfort sensations. Only when approaching the edge of the lane does a clearly perceptible intervention take place.

It is questionable, however, whether the LDW, which supports the driver before he leaves, can really be effective in the dreaded microsleep. As Gründl (2005) reports, this type of accident often only occurs *after* the driver wakes up on his own or due to the noise of the tyres when leaving the road and then tears the steering wheel out of fright. In such cases, a system that keeps the vehicle independently in the lane would make sense (Walter et al. 2012).

This effect is provided by a future extension of the LKS, the so-called ***Lane Departure Prevention (LDP).*** As with the LDW (or ALDW), the driver is first warned of the threat of leaving the lane. If no driver activity can then be detected, the vehicle is actively forced back onto the lane by one-sided braking intervention via the ESP system. This intervention, which is in the range of max. 200 ms, reduces the speed at the same time. In order to prevent false triggers to a large extent, the intervention only takes place when no driver activity is detected, similar to the LKS system. If this is recognizable, the intervention is stopped immediately. As Walter et al. (2012) explain, this LKS extension still places considerable demands on the development of adequate algorithms and on the fusion of object detections.

Both the LKS system available today and the LDP system just described are used for motorway-like roads with clearly visible lane

11 With the Audi Active Lane Assist, the lane-keeping system even works when the driver briefly takes his hands off the steering wheel. If he does this for a long time, he is warned with a gong and the system is deactivated shortly afterwards.

Fig. 9.5 Examples of assistance system control units (**a** Mercedes S-Class, **b** BMW 7 Series, **c** Volkswagen Golf)

Fig. 9.6 Feedback via the activated LDW or LK S system (**a** Volkswagen, **b** BMW, **c** Mercedes)

markings and straight sections or long bends. In any case, the LKS system must be activated by the driver. Today, this is done partly by means of an assistance activation unit installed in the dashboard (see Fig. 9.5), but partly also by means of a corresponding control element on the steering wheel or on a steering column lever.

The feedback on the operating status is given relatively uniformly by a pictorial display (Fig. 9.6). If the indicated road boundary appears alone, this means that the LKS system is switched on. The arrows on the lines symbolize that it is now active due to the outer constellation (see above). In vehicles equipped with only one LDW system, feedback shall be provided in a similar way.

Table 9.3 Value ranges of some parameters of available LKS systems (according to Gayko 2012)

Characteristic	**Typical values**
Lower speed limit	65–70 km/h
Upper speed limit	170–180 km/h
Maximum corrective steering torque	2–3 Nm
Supported track width/width	3–4 m
Minimum curve radius	230 m
Maximum lateral acceleration	2 m/s^2

Table 9.3 gives a summary of the value ranges in which the LKS systems available today can provide support. If these values are combined with the limitations mentioned at the beginning, which result from the high technical demands on camera and image processing, it becomes obvious that an LDW or LKS system can only be effective under very limited conditions and that at the same time it is to be expected very often that the system will not be available temporarily even during operation. From an ergonomic point of view, therefore, effective feedback that can be directly assigned to the driving task is indispensable. It is advisable to display the corresponding information in the HUD as manufacturers do when they have a HUD in their offer.

According to Bartels et al. (2012), more than 5% of accidents involving personal injury result from incorrect lane changes. The problem is complex: even with curved rear-view mirrors, so-called dead angles (areas in which a vehicle on the side cannot be detected by looking into the mirror) cannot be completely avoided. However, curved mirrors in

Table 9.4 Classification of the zone coverage of lane change assistants (according to ISO 17387)

Type	Blind spot monitoring left and right	Monitoring of the approach zone left and right	Information for the driver
I	x		Warning of blind spot vehicles
II		x	Warning of vehicles approaching from behind
III	x	x	Warning of dangerous lane changes

particular make it very difficult to estimate distances and in particular to perceive the approach speed of vehicles on the lane to be changed. In many cases, however, the driver is inattentive and changes lanes without being particularly protected against rear traffic. Through ***Lane Change Decision Aid Systems*** these dangers are to be mitigated. Technically, the lane change assistant is based either on a camera system (usually cameras housed in the left and right rearview mirrors) or on a radar system. The advantage of radar-based systems is the additional possibility of determining the differential speed to the approaching vehicle. The ISO 17387 standard therefore classifies warning strategies according to zone coverage (Table 9.4).

While Type I warning systems only inform about vehicles in the blind spot, Type II warning systems only react to approaching vehicles and therefore do not provide information about vehicles in the blind spot. Only type III warning systems cover the full spectrum. Due to the maximum range of the sensors, the Type II and III systems are also subdivided once again with regard to the maximum differential speed to the vehicle approaching from behind and the curve radius, which must not be undershot so that the approaching vehicle can be detected at all (see Table 9.5).

Due to the many restrictions mentioned above, different system states must be distinguished from each other for the use of a lane change assistant. These are illustrated in ISO 17387 in a system state diagram (Fig. 9.7). In most cases, the system must first be armed in a similar way to the assistance systems described above. If this is the case, the system is in the "inactive" state. Only when a certain

Table 9.5 Maximum differential speed of the vehicle approaching from the rear and minimum curve radius which must not be undercut for recognition (according to ISO 17387)

Type	Maximum differential speed with respect to the vehicle approaching from behind	Minimum curve radius
A	10 m/s or 36 km/h	125 m
B	15 m/s or 54 km/h	250 m
C	20 m/s or 72 km/h	500 m

criterion is fulfilled (e.g., when the ego vehicle has exceeded a certain speed limit – usually 30 km/h) the system is switched to the "active" state. From an ergonomic point of view, this condition should be indicated optically in the instrument panel (e.g. in the form of a pictorial display). Only when certain conditions are met (depending on the technology used, e.g. detection of a vehicle in a blind spot or a vehicle approaching from behind in the lane to the left or right of the ego vehicle) a discreet warning of level 1 is issued (usually in the form of a weaker light in or at the foot of the left or right wing mirror). Only when the system recognises that the driver wishes to initiate a lane change under these conditions (e.g. recognised by the operation of the indicator lever, evaluation of steering angle/torque, position of the own vehicle within the lane etc.) is a warning given at level 2 (e.g.: flashing light on the corresponding exterior mirror and possibly additional warning tone). If the driver nevertheless initiates a lane change under these conditions, it would be sensible

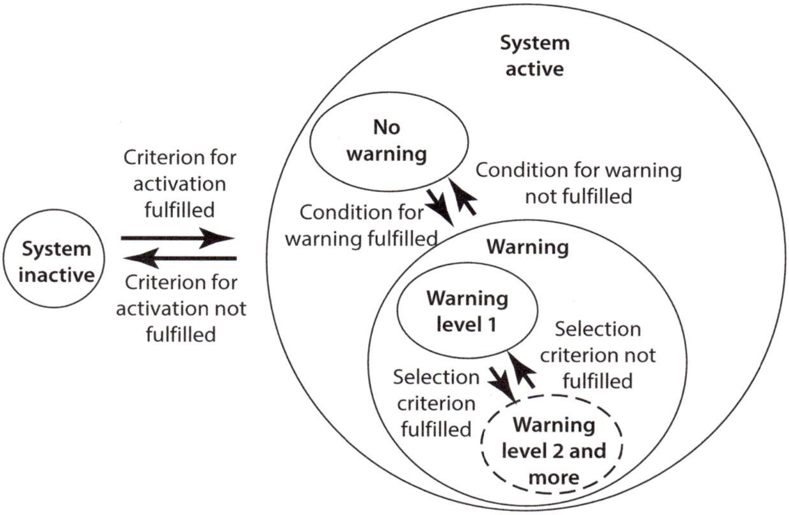

Fig. 9.7 System state diagram for a lane change assistant according to ISO 17387, quoted from Bartels et al. (2012)

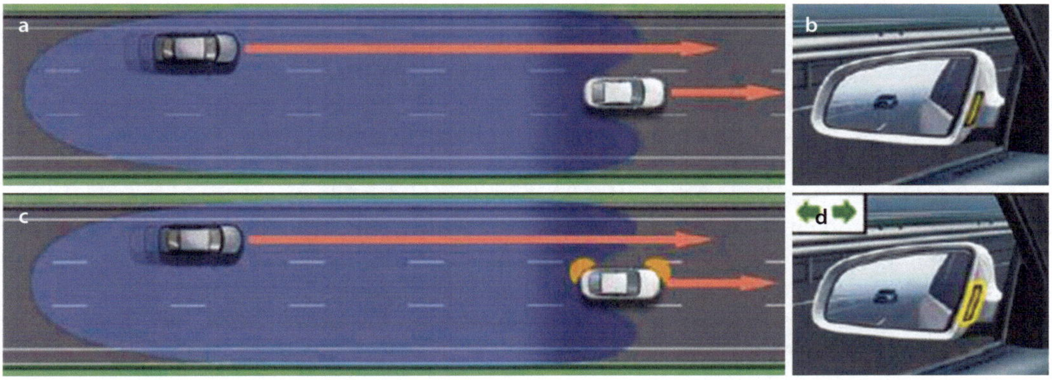

Fig. 9.8 Audi Side Assist: In the example shown here, an overtaking vehicle is in the detection area. **a, b** Warning level 1: A faint yellow light in the foot of the left exterior mirror indicates to the driver that the vehicle is in the fast lane, **c, d** Warning level 2: if the driver sets the indicator or indicates by a moment on the steering wheel that he intends to change lanes, he is made aware of the danger by a brief bright flashing of the lamp in the foot of the rear-view mirror

from an ergonomic and safety point of view to warn the driver of this step or prevent him from doing so by means of a moment on the steering wheel. However, due to the limited detection field, which is attached to the current lane change assistant for technical reasons, this step has not yet been taken.

Figure 9.8 uses the example of VW's "Audi Side Assist" and "Side Assist" to demonstrate the implementation of the warning system described in ISO 17387 (according to Bartels et al. 2012). Due to the many limitations limiting the effectiveness of the system, this point is specifically referred to in the operating instructions of vehicle manufacturers. In particular, it states that it is only a tool that may not recognise all vehicles and cannot replace the driver's attention. In addition, dirty sensors or adverse weather conditions (e.g. rain, snow or strong spray) can cause vehicles to be detected inadequately or not at all. Missing or false warnings may also occur if the ego vehicle is on the outer edge of a particularly wide lane or on the inner edge of the lane in the case of narrow lanes. Even systems capable of detecting the speed of approaching vehicles (Type II and III) can fail at extreme approach speeds (summarily quoted from Bartels et al. 2012).

Combined Systems

Systems that assist both the control of longitudinal and lateral dynamics can of course be implemented by the driver himself by keeping the ACC and lane keeping support systems in operation at the same time. However, due to the many and varied restrictions to which both systems are subject, their use is practically limited to motorway sections that are not too busy.

For the low speed range (<60 km/h), the systems have now been integrated to the semiautomatic (according to the BASt definition) ***Congestion Assistant*** or ***ACC Stop & Go***. In this case, braking, accelerating and steering take place completely independently. If the driver does not intervene (e.g. does not apply the brake), the vehicle automatically starts up again when the vehicle in front picks up speed again after a short time. If the traffic jam dissolves again (recognizable by speeds >60 km/h), the driver is prompted – generally – by an acoustic signal to take over the vehicle guidance again. Since the driver can turn away from the driving task during the phase of ACC Stop & Go activity, he is prompted after a certain standing time (usually about 3 s) to activate a so-called start-up trigger so that the vehicle can start again independently (Winner et al. 2012b). To do this, either briefly press the accelerator pedal, press the ACC control button on a BMW, or pull the cruise control lever on a Mercedes. It should be noted, however, that such a measure would partly counteract the ability of customers to use nerve-racking congestion journeys for other activities, as the driver would not be able to completely turn away from the driving task.

A further integration of the existing sensors – partly with an improved range – is possible with the ***Avoidance Maneuver Assistant***. In the event of an imminent rear-end collision (proximity to a stationary vehicle detected by the radar sensors of the ACC system at a speed that does not allow the vehicle to come to a standstill behind the vehicle even during maximum braking), the system automatically checks the available space in the left or right lane and automatically steers the vehicle into the free lane (Bosch study on driver assistance systems 2012). However, in future research, the controllability of such a system still has to be checked, because to implement this function relatively high steering torques are necessary, which can hardly be countered by the driver in the event of a false triggering and can therefore have serious consequences (for example oncoming traffic).

A special driving situation, which differs in many respects from those described so far, but which is necessarily associated in some way with every driving process, is the parking of the vehicle after the end of the journey. Everything related to this is represented by the expression ***standstill management***. Doisl (2008) defines parking as "... part of the primary driving task, in which the selection of the parking position, parking and safe parking, and conversely parking out, up to reintegration into the flowing traffic, is the main focus". The search for a parking space can also be described as the normal driving manoeuvre through the three levels of the primary driving task "navigation" (search for a suitable parking space), "guidance" (determination of the exact course into the parking space) and "stabilization" (realization of the determined course). A parking situation catalogue distinguishes between longitudinal, vertical and diagonal parking.

Both the system ergonomic analysis and the real tests show that communication with rear traffic is a major problem when looking for a parking space. In his experiments, Doisl (2008) used a sign that could be raised mechanically when looking for a parking space and visibly drew the attention of the following traffic to the imminent manoeuvre in the rear window (◘ Fig. 9.9). While without this warning, on average 46% of parking manoeuvres in the attempts were classified as dangerous, this percentage was reduced by half to 23% with the hint assistant, even though the symbol was absolutely new and had to be unknown to subsequent drivers. In addition, with this assistant 20% more often a comfortable distance for parking was maintained by the following traffic. In contrast to the unassisted search for a parking space, the assistant experienced the process in exactly the same way as the moderate stress caused by driving on the motorway and a normal city trip.

Through ***Parking Assistance Systems*** the driver should be supported in finding a suitable parking space and leading the vehicle safely and stringently into it (Kesler and Mangin 2007). Katzwinkel et al. (2012) describe the various parking assistance systems. This overview is referred to below. Afterwards, the following forms of assistance systems can be distinguished for parking:

- ***Informative parking assistance systems.*** These are the most widely used systems, generally based on ultrasonic sensors mounted in the front and rear of the vehicle. In the simpler systems, the measured distance is transmitted to the driver in an acoustically coded form, whereby the stereo characteristics of the car radio can be used to provide spatially compatible information. In more advanced systems, the information is presented in a pictorial display (see ► Sect. 6.2.1, ◘ Fig. 6.15a and 9.10, Park-Distance-Control, PDC). The described form of parking aid is normally switched on automatically when reverse gear is engaged. In addition, there must also be a button or switch by means of which the system can be switched on or off in response to the driver's initiative.

  With the aid of cameras integrated at the vehicle edges and image processing technology, it is possible to display an artificial top-down view of the vehicle on the central information display (CID). This means that the driver can at least see the immediate surroundings when looking at this display. Doisl (2008) recommends combining this display form with the parking distance display (◘ Fig. 9.11).

  From an ergonomic point of view, the systems discussed above are an important aid, especially from the point of view of today's rather confusing body shapes. Due to the geometry of the sensor cone, however, it can happen that detected objects disappear again, an effect that can lead to confusion and thus to the wrong reaction of the driver.
- ***Guided parking assistance.*** In this case, guide lines are displayed in the image of a rear view camera, which on the one hand show the target course of the parking process calculated on the basis of the above-mentioned measuring systems and on the other hand the future course based on the currently selected steering wheel position ("driving hose", ◘ Fig. 9.12). The image is usually displayed in the central information display (CID). With such systems, both parking in longitudinal and lateral parking spaces can be supported. Without image processing, which may automatically recognize what type of parking space it is, the driver must make the selection

◘ **Fig. 9.9** Reference to searching for a parking space for rear traffic (Doisl 2008)

◘ **Fig. 9.10** Optical display of distance information systems (**a**: a simple version, **b**: Park-Distance-Control, PDC)

Fig. 9.11 All-round camera image combined with parking distance display (PDC) (Doisl 2008)

Fig. 9.12 Actual and target driving hose display superimposed on the rear view camera for parking (example: BMW)

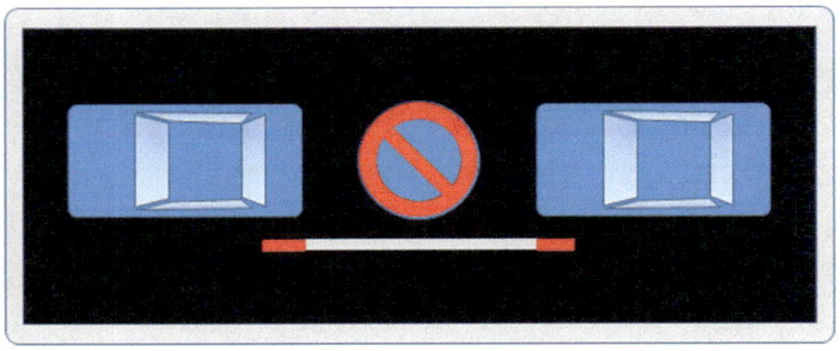

Fig. 9.13 Display about the possibility of parking (example Citroën Picasso)

manually. For longitudinal parking spaces, the correct auxiliary line is selected by setting the indicator. With the more advanced systems, the target steering angle or the steering angle difference is generally displayed in the CID, as well as the direction of travel to be taken, the respective stop points and the end of the parking process.

- From an ergonomic point of view, however, this very rich display in the CID is questionable, as there is a danger that the driver only drives after the display and no longer observes the real environment or mirrors.
- ***Semiautomatic and automatic systems.*** These systems can also include those in which longitudinal parking spaces are measured while driving past and the driver is shown whether parking is possible (e.g. by displaying "Parking space sufficiently large", "Parking space too small", see example in Fig. 9.13). Some systems often also indicate the degree of difficulty of the parking process (e.g. "easy", "normal", "difficult"). If necessary, such systems can even be used to provide instructions on how to "reset further" or "drive forwards", "parking test makes sense". The maximum pass-by speed is usually set to a value between 15 and 30 km/h. In principle, such systems must be switched on by a button of the driver searching for a parking space. If the maximum pass-by speed is exceeded, the system automatically switches itself off again. According to the investigations by Doisl (2008), however, parking searches take place at speeds of up to 45 km/h, so that the limit speed selected for the systems is set too low.

With fully developed semi-automatic systems, the driver is usually released from the lateral control. If the system has detected a sufficiently large parking space, it only has to influence the longitudinal

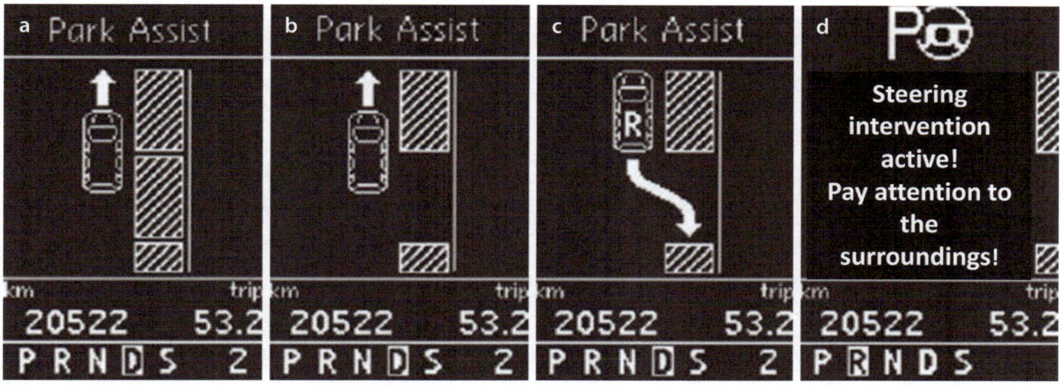

Fig. 9.14 Human machine interface for a semi-automatic parking system. From **a** no parking space, **b** parking space detected, **c** reverse gear engaged, **d** driver independent

dynamics of the vehicle by accelerating and braking, while the steering process is carried out automatically. Figure 9.14 shows an example of a human machine interface for such a semi-automatic parking system (VW system, quoted from Katzwinkel et al. 2012).

In the future, it is to be expected more and more that certain parking procedures will even be carried out fully automatically. For example, so-called garage parkers were shown, which can park fully automatically, only controlled by the driver standing outside the vehicle, with the help of reflection points previously installed in the garage. Such systems could possibly be used in car parks in the future to park vehicles very close to each other, with passengers having to get off before parking. The available parking space could be much better used in this way.

On the basis of system ergonomic rules and the findings of his extensive experiments, Doisl (2008) develops a concept recommendation for video-based parking assistance systems. The picture of the panoramic camera shows the respective recommendation for the driver's activity as well as any dangerous objects. This recommendation applies to both guided parking assistance and semi-automatic systems (Fig. 9.15).

9.2.1.4 Navigation Task

As already mentioned, assistance systems that support the navigation task are among the oldest developments in this field. The current state of development is correspondingly high, both in terms of operation and presentation of the navigation recommendation. In ► Sects. 6.2.1 and 6.3.1.3, detailed information is provided from a system ergonomic point of view. In particular, the observations made there point out that navigation systems are also operated intensively during the journey. The resulting requirements with regard to language operation are also described in the chapters mentioned. The significantly faster development cycles of the accessories industry – both the navigation systems to be retrofitted in the vehicle and, more recently, the mobile phones – make it possible, on the one hand, to use newer technologies such as voice recognition more quickly – but not adapted to the special acoustic conditions of the vehicle used – but, on the other hand, manual operation must be carried out via the relatively small touch-screen surface of these additional devices, which is questionable for safety reasons. It would be desirable to develop an interface via which the operation of such additional devices can be connected to the control elements installed in the vehicle (e.g. Touchpad or rotary pusher) (Various development approaches in this direction are currently

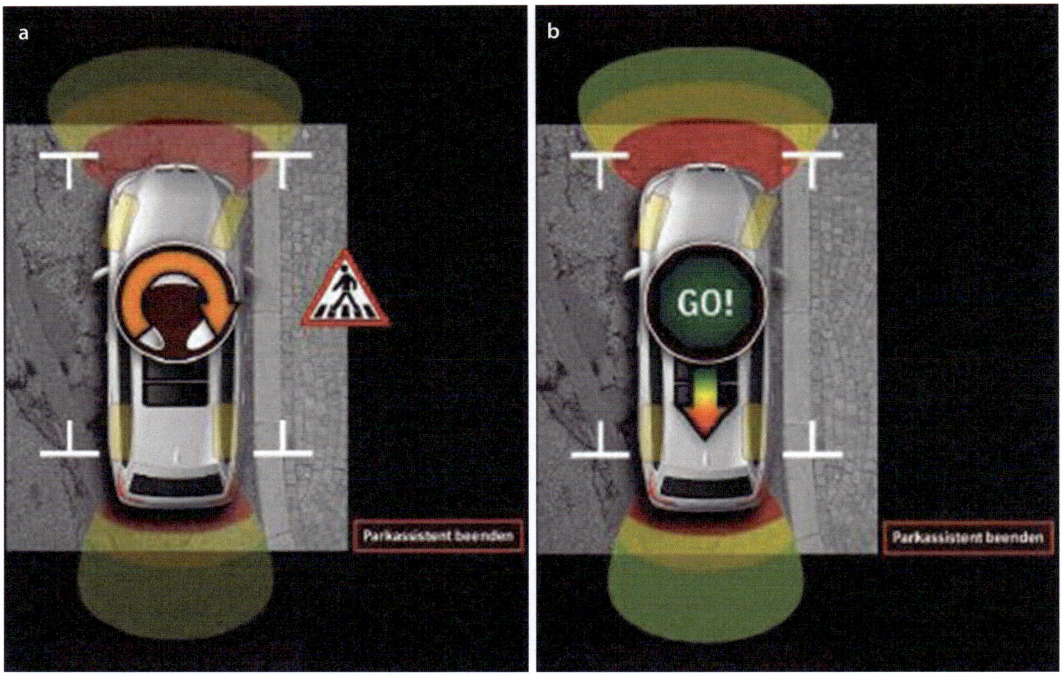

Fig. 9.15 Concept recommendation for a video-based parking assistance system (Doisl 2008) **a** "turn the steering wheel to the right to the stop – pedestrians to the right." **b** "go backwards!"

being presented by many car manufacturers or are now also in the regular product range.

In the following, some features are presented that complement the assistance function of navigation systems from an ergonomic point of view.

- In view of the increasing density of traffic and the difficulty of finding a suitable parking space for the vehicle in a foreign environment, navigation is reliant on support in ***parking search*** to attach special importance (e.g. finding a multi-storey car park with free spaces). In particular, it will be possible in future to extend navigation to areas outside the stored map system, i.e. to car parks or multi-storey car parks (Kleine-Besten et al. 2012). Then it would be possible to give navigation instructions about the direction in which the parking place is to be left.
- Kleine-Besten et al. (2012) describe the various options for "***updating***" the map material used in navigation systems. In particular, a distinction must be made here between so-called on-board cards, which are regularly replaced by the user himself with new variants, and systems which are made available to the vehicle by an external server. A special feature are the so-called ***learning cards.*** Among other things, systems are mentioned which can detect and correct errors in the database themselves. In particular for frequently used routes (e.g. the way to work), an economical, consumption-optimised route that takes the user's experience into account can be saved. Another application of the learning card is the personalization for certain driver groups, e.g. older drivers. For example, routes and intersections that have already been driven can be marked and individually evaluated in the digital map memory in order to mark "problem intersections", "stressful routes" or "routes that are easy to drive around" and to avoid them if necessary.
- For some professional groups, but also for private individuals, the automatic management of a ***logbook*** be of interest (some navigation devices available on the auxil-

iary market today already have a function that makes it possible to save the route travelled and later display it on the computer e.g. using Google Maps). Of course, in this case it must be guaranteed for data protection reasons that this function can be switched off at any time and that no recording of any kind is made "in the background".

As already mentioned several times, the narrower guidance task takes place in a section of track that is traversed within about 2 s. More than 4 s foresight is generally not possible due to the optical conditions. This means that all activities that result from events in another field must be assigned to the navigation. Kleine-Besten et al. (2012) refer in this context to navigation-supported assistance functions which can be generated and provided directly from the information stored in the navigation system. This includes, for example, the ***cornering warning device*** which indicates to the driver that the speed is too high in front of a forward bend, or ***hazard warning device*** which draws the driver's attention to special accident blackspots, kindergartens/schools etc.

Bidirectional communication with other vehicles represents an essential extension, which can be located in the transition area between navigation and guidance. This type of communication is becoming increasingly important from a technical point of view, particularly as a result of mobile radio technology. ***Car-to-Car (C2C)*** refers to the communication between vehicles. For example, it can be used to warn of a traffic jam (if the distance is long enough and the road network has options, an automatic navigation recommendation to its surroundings can also be made). It is also conceivable to transmit the danger detected by the vehicle sensors (e.g. smooth road detected by ABS or ESP) to following vehicles. Popiv (2011) has developed ergonomic display concepts for such information and tested their acceptance in simulator studies (◘ Fig. 9.16). Not only the aspect of safety was the subject of the investigation, but also the possibility of reducing the vehicle's energy consumption through early reaction. However, such information should not be expected to meet with acceptance by the driver under all circumstances. The situations in which the assistance system helped are characterised by the fact that

- the situation is not in the driver's field of vision,
- the development of the situation is not clear at the time when the driver first sees the situation.

Where the assistance systems had no effect, the situation was clear for the driver to see. In

◘ **Fig. 9.16** Bird-view indicator for approaching the end of a traffic jam (Popiv 2011)

urban traffic, advice from assistance systems that require a very early reaction in the form of accelerator throttling was not accepted. In rural traffic, the longest deceleration trips caused by engine braking take place at 13 s, in the case of multimodal switching (additional information via the active accelerator pedal) the assistance recommendation even 16 s, on the motorway 8 s. The test subjects are relatively tolerant of inaccuracies in the system, as long as situations that cannot be seen are pointed out.

Another form of bidirectional communication is the exchange of information between the vehicle and the infrastructure (*Car-to-Infrastructure, C2I*). For example, it is possible to transmit information from traffic signs equipped with transponders (e.g. speed limits, overtaking bans, in particular also information from traffic sign changing systems) and from traffic lights to the vehicle. Information about the green phase can also be conveyed in this way, enabling the vehicles to adapt to the speed of the so-called "green wave". In the research project Kolibri (Kooperative Lichtsignaloptimierung – Bayerische Pilotstudie, cooperative light signal optimization – Bavarian pilot study; Krause und Bengler 2013) a display system was investigated which displays the speed range on mobile phones which ensures that the vehicle remains in the green wave and, if necessary, indicates in good time that the foot should be removed from the accelerator pedal in order to largely use the engine brake for the next traffic light stop. Here, too, a bird-view-like display with consideration of system ergonomic requirements proved to be particularly effective (▫ Fig. 9.17).

It is to be expected that in the future it will be possible to use C2I technology to carry out external traffic planning in good time. The fundamental problem, however, is that in many cases only those road users can be integrated who have the appropriate technology. Kleine-Besten et al. (2012) cite the VICS (Vehicle Information and Communication System) as an example of the successful introduction of a modern traffic control system, which was introduced throughout Japan with public funding from 1996 to 2003. The data

▫ **Fig. 9.17** Display of Kolibri on mobile phone (Krause and Bengler 2013)

collected by the police and the road administration is passed on to the VICS centre. Drivers who have an appropriate navigation device in their vehicle are informed in real time about the current traffic situation in the form that the traffic flow on the individual roads is classified in colour (red, yellow, green) in overview and detailed maps, thus enabling them to assess the traffic situation (similar offers exist in Europe through commercial navigation device manufacturers such as TomTom and free from charge by Google Maps).

9.2.2 Categorisation of Driver Assistance Systems for the Primary Driving Task

As the description of the various assistance systems has already shown, they interact with the driver in very different ways – partly due to the technical implementation. Depending on the situation, some of these systems give the driver hints (e.g. the navigation systems), others warn him of a dangerous situation (e.g. Lane Departure Warners), others intervene autonomously under such conditions (e.g. Lane Departure Prevention) and again other systems enable the automation of partial tasks (e.g. ACC) or even the entire task (e.g. Traffic jam assistant). ▫ Table 9.6 gives an overview of the classification based on this depth of intervention.

Since both the warning of certain situations and the assumption of tasks also have

Table 9.6 Classification of driver assistance systems according to intervention depth

Assignment of driver assistance systems to the intervention depth					
	Informing	**Warning**	**Assisting**	**Partially autonomous**	**Fully autonomous**
Intervention depth	Navigation system…	Lane Departure warning system Collision warning …	Lane departure warning with steering intervention Collision warning with short brake intervention …	Tempomat ACC …	ABS ASR ESP …
Features	DAS gives the driver information	DAS warns the driver An action of the driver is required	DAS intervenes briefly in vehicle guidance Override by the driver possible at any time	DAS takes over parts of the driving task The driver monitors the system Override by the driver possible at anytime	DAS intervenes in driving dynamics Override by the driver not possible

From Maier (2014)

legal consequences, the Federal Highway Research Institute of Germany (BASt) made a proposal for the categorisation of assistance systems under this particular aspect (Gasser 2012). ◘ Table 9.7 shows this categorization, which knows the levels “Driver only” – “Assistant” – “partially automated” – “highly automated” – “fully automated”. Gasser notes that up to the “partial automation” stage there is compliance with § 1 of the StVO due to the functions that can be overridden at any time. In the higher levels of automation, however, the driver would at least violate his obligations under the road traffic regulations in the automatically controlled phases, which by definition allow the driver to dispense with permanent monitoring of the roadway and traffic environments. Only in the special case of an emergency stop system, which takes over the vehicle control for the short duration until the vehicle stops safely in the event of the driver’s unconsciousness, would there be no contradiction to the behavioural obligations despite the high degree of automation.

A categorization of driver assistance possibilities or driver assistance systems from the driver’s point of view must be oriented to the hierarchical order of driving tasks in order to define what is to be assisted (see also ► Sect. 9.2.1; the following treatise refers only to the primary driving task; information on secondary and tertiary tasks can be found in Sect. 6). For this purpose, the description of the driving task in ► Chap. 2 is reproduced in relation to the primary driving task. In any case, an assistance system – regardless of the expansion stage – is connected in parallel with the respective tasks to be performed by the driver. There are various alternatives for linking the information flow between the driver and the vehicle:

- The assistance system takes over *alternatively* the respective task to the driver. In the circuit diagram of ◘ Fig. 9.18 this is characterized by a switch. If the switch is placed on the assistance system, the assistance system acts as an automatic. The driver can deactivate the automatic function by pressing the switch. He thus retains the upper hand over the assistance system. There are various technical conditions by which the switch is actuated and the automatic sys-

Table 9.7 Name and classification of driver assistance functions according to a proposal developed for the BASt (as of 06.09.2010, not final, Gasser, 2012)

Nomenclature	Description degree of automation and expectations of the driver	Exemplary system characteristics
Driver only	Driver permanently (during the entire journey) extends the longitudinal guide (acceleration/deceleration) and the transverse guide (steering)	No (driver assistance) system intervening in longitudinal or lateral guidance active
Assisted	The driver permanently executes either the lateral or the longitudinal guidance. The other driving task is carried out by the system within certain limits. The driver must permanently monitor the system. The driver must be ready to take over complete control of the vehicle at all times.	Adaptive Cruise Control: longitudinal guidance with distance and speed control Parking assistant: Lateral guidance by Park Assistant (automatic steering in parking spaces. The driver controls the longitudinal guidance)
Semi-automated	The system provides longitudinal and lateral guidance (for a certain period of time and/or in specific situations) The driver must <u>permanently</u> monitor the system. The driver must be ready at all times to take full charge of the vehicle.	Highway assistant: automatic longitudinal and lateral guidance on motorways up to an upper speed limit Driver must permanently monitor and react immediately when asked to take over.
Highly automated	The system takes over lateral and longitudinal guidance (for a certain period of time in specific situations). The driver does <u>not</u> have to permanently monitor the system. If necessary, the driver is requested to take over the driving task with sufficient time reserve. System boundaries are all recognized by the system. The system is not able to create the minimum risk status from every initial situation.	Highway chauffeur: automatic longitudinal and lateral guidance on motorways up to an upper speed limit Driver does not have to monitor permanently and reacts with a certain time reserve after being asked to take over.
Fully automated	The system takes over lateral and longitudinal guidance completely in a defined application. The driver does <u>not</u> need to monitor the system. Before leaving the application, the system asks the driver with sufficient time reserve to take over the driving task. If this is not done, the system is returned to the minimum-risk state. System boundaries are all recognized by the system. In all situations, the system is able to return to the minimum-risk system status	Highway pilot: automatic longitudinal and lateral guidance on motorways up to an upper speed limit Driver does not have to monitor. If the driver does not respond to a request to take over, the vehicle brakes to a standstill.

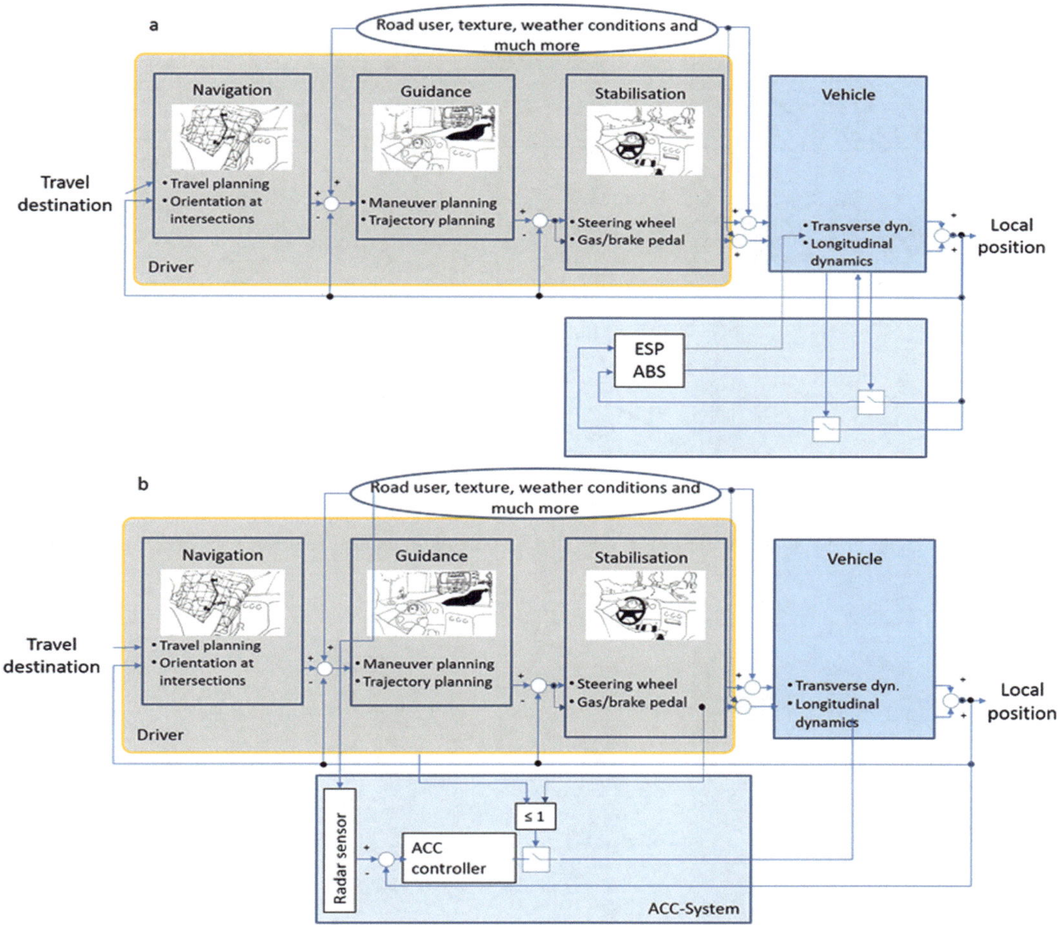

Fig. 9.18 Structure image of assistance systems which take over tasks as an alternative to the driver (**a** ABS/ESP, **b** ACC)

tem thus automatically overrides. At the guidance level, this is the functional principle of the ACC with regard to influencing the longitudinal dynamics. At the stabilisation level, this corresponds to ABS and ESP; however, here the principle of operation is modified in such a way that the assistance system is switched on on the basis of parameters measured on the vehicle; if the corresponding conditions no longer exist, the assistance system is also ineffective (see the hierarchical order of the driving task in 1.4 and 2.1.1).

- The assistance system takes over the respective task *simultaneously* and on equal authority with the driver. The information of the driver's intervention and the assistance system act on the vehicle via a sum point (Fig. 9.19). This switching principle can only be effective if at the same time the driver is given feedback on the influence the assistance system wishes to exert at the moment. If necessary, he can now counteract this by applying more force, thus overruling the assistance system at any time. At the vehicle guidance level, the lane-keeping system and even more so the lane-departure prevention are examples of this type of wiring. However, the system must be switched on at all for the driver here as well. The system is automatically turned off if the information recorded by the front camera is not sufficient for control.
- The assistance system does not interfere with the control process. Instead, depend-

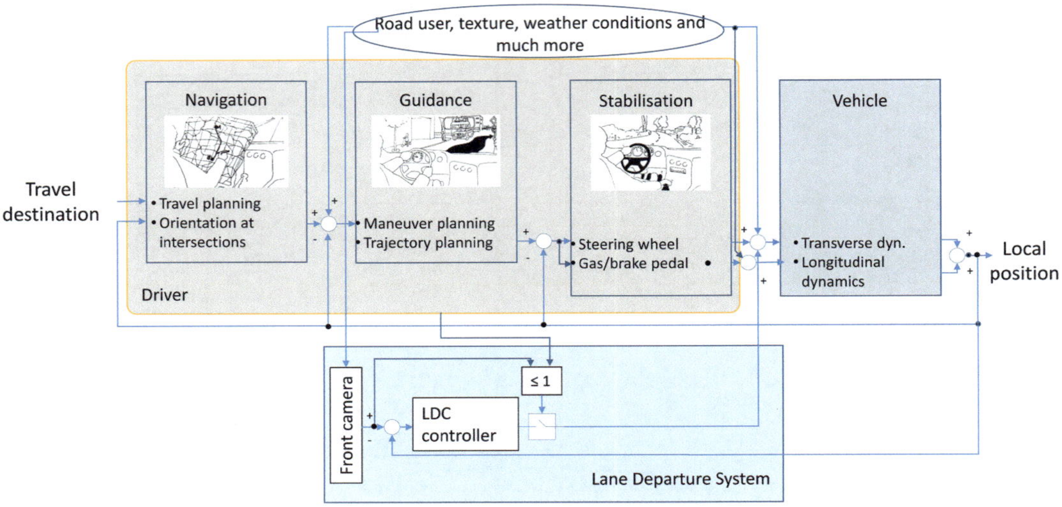

Fig. 9.19 Circuit principle of an assistant system that works simultaneously with the driver (example LDC)

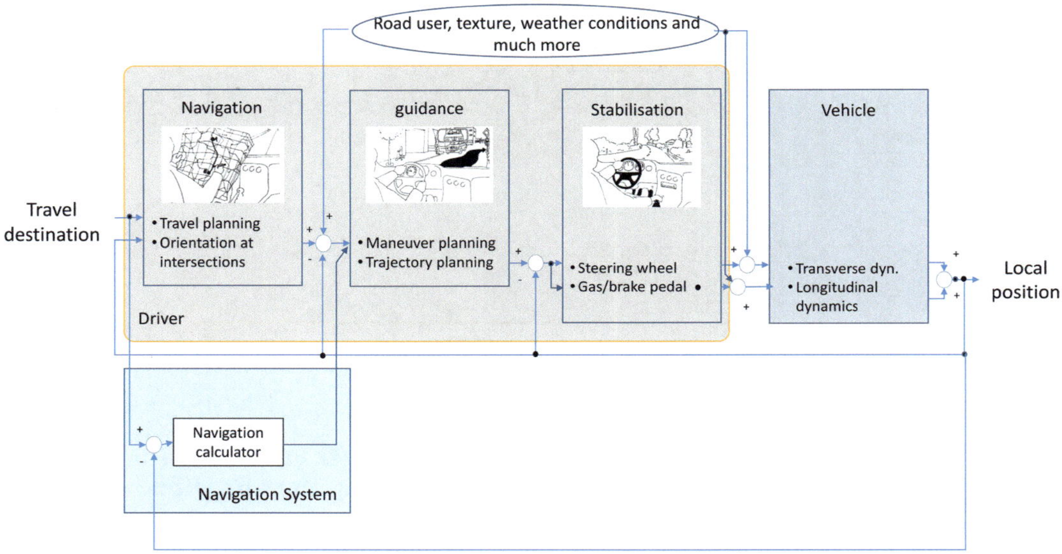

Fig. 9.20 Structural image of an assistance system with informative character

ing on the situation, there are notes or warnings. This means that the assistance system has only *informing or warning* character. In order to realize this information, corresponding displays are necessary. The modality of the provision of information depends on the urgency of the time: Information of low temporal urgency, but of interest for a longer period of time, should be presented optically, information with ad-hoc character acoustically or haptically. The classic example of an assistance system with an informative character is the navigation system (Fig. 9.20). At the guidance level, an example of an assistance system with a warning character is the emergency brake assistant. The reactions of the aforementioned assistance systems, both in terms of their informative and warning effect, are also communicated via advertisements.

Fig. 9.21 Example of a control panel for "manoeuvre assignment" (Franz et al. 2011)

This comparison reveals another difference in the driver's interaction with the assistance system:

- In the case of the alternative assistance system, the switch is operated in a specific traffic situation. Depending on their sequence, the driver gives the command to take over a specific sub-task. He orders a manoeuvre (in the case of the ACC the manoeuvre is called: "Maintain the specified speed or the distance to the vehicle in front"). The ordered manoeuvre is now performed independently by the system. In the following, the driver only has to monitor the correct execution (monitive system).
- In the case of the simultaneously working assistance system, it is constantly itself in action (active system) with regard to the respective assisted subtask. When the driver switches on this system, he asks it to support him in his work, which he still does himself.

This distinction was the basis for two DFG-funded research projects:

- In the "Conduct-by-Wire" project, the concept of "maneuver assignment" is being extended to other maneuvers. For certain sections of the route, it is conceivable to supplement the ACC system with automatic lane guidance (practically extending the congestion assistant to higher speed ranges) or, in particular, special manoeuvres resulting from the situation, such as "overtaking now", "changing lanes", "turning right at the next junction", "parking"and much more. In addition to the question of technical realisation, the ergonomic design of the interaction with the vehicle, i.e. the question of how the manoeuvre is to be commissioned, also plays an important role (Winner et al. 2006; Winner and Hakuli 2006). Figure 9.21 shows such a control panel in the form of a touch screen that would be mounted to the right of the steering wheel (Franz et al. 2011). The Conduct-by-Wire project thus addresses the driver's sequential, discrete interaction with the assistance system.
- The project "H-Mode"[12] pursues the simultaneous integration of driver and assistance system. The metaphor of horse and rider should characterize the fact that two intelligent systems are simultaneously acting, whereby a continuous transition from one system to the other is conceivable

12 "H" stands for "horse."

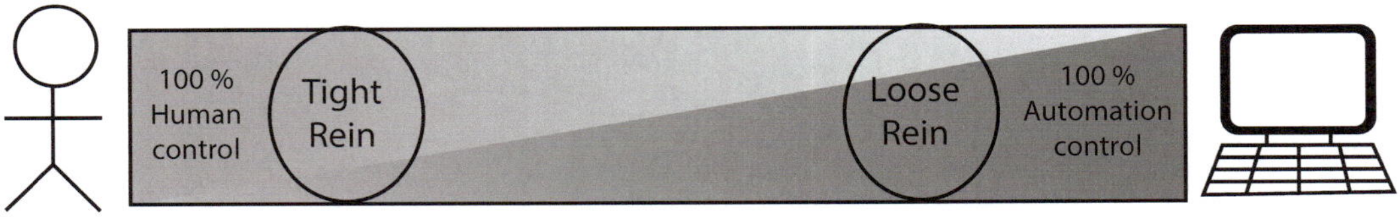

Fig. 9.22 Continuous transition between "Tight Rein" and "Loose Rein" in H-Mode

(so-called "transition"). "Loose rein" characterizes that the assistance system primarily takes over the task, while "tight rein" means that the driver "holds the reins in his hand". These two modes stand for the extremes "fully automated vehicle guidance" and "conventional manual vehicle guidance" (Flemisch et al. 2014; see Fig. 9.22). One version is designed to continuously capture the desire for the current level of automation through the force with which the driver acts on the control element. In order to implement H-mode, conventional vehicle control systems such as the Lane Keeping Assistant require an electrically controllable steering system and an active accelerator pedal. Alternatives include joystick steering or the so-called "joke", a steering wheel-like control element that can also be moved longitudinally to enable acceleration and deceleration inputs (Kienle 2015).

Damböck (2014) tries to bundle the different classification possibilities of driver assistance systems on the basis of a broad literature study (Table 9.8), whereby he essentially takes up the basic idea of the concept proposed by Gasser (2012) (Table 9.7). He adds the distinction between conventional and advanced driver assistance systems (Donner et al. 2007). It also takes into account which steps of the information processing process the individual systems comprise and to what extent the driver must remain involved in the driving task.

9.2.3 Fatigue Warning

Hörwick (2011) gives an overview of the possibilities for driver monitoring. There are two different approaches: On the one hand the driver's behaviour can be observed directly and on the other hand the driver can be instructed to perform operating actions, the correct processing of which is then analysed (Fig. 9.23).

9.2.3.1 Driver Monitoring by Forced Operation

The latter method, as a dead man's switch, is the main component of the so-called *safety control circuit* (SIFA) in rail traction vehicles. At regular intervals (in Germany every 30 s), the driver is asked to release a foot pedal and press it again. A similar measure is also being implemented in some European countries for public road passenger vehicles (buses, there in the form of a push button). Apart from the fact that this measure represents an additional tertiary task, which may distract from the actual driving task in difficult situations, studies from the railway sector show that the dead man's button is still operated correctly even in sleep stage C (Peter 1980). A suitable variation of this principle for the vehicle may be the so-called potential trigger (Wimmer et al. 2012). After that, it is assumed that the attention level decreases constantly from a given starting value during a monotonous ride. When a certain lower position is reached, the driver is prompted to react by an optical signal (if possible in the HUD). The optical presentation is intended to ensure that the inattentive driver is not "woken up" by the signal itself. The descent speed is determined by various parameters, which essentially consist of the duration of the view of the traffic situation, the use of infotainment systems and the criticality of the traffic situation (recorded by the ACC sensors or front camera). The technical measured variables recorded for this purpose are calculated by applying fuzzy technology to the mentioned sink rate of the attention level. The system is intended exclu-

Table 9.8 Nomenclature and description of driver assistance and automation systems based on the Gasser designation (2012)

Denomination	Automation/assistance skills	Role of the driver
Driver only[a]	No system active	The driver permanently drives the vehicle and has sole control over the vehicle.
Supported[b]	Support by conventional assistance systems (e.g. ESP)	The driver is permanently in control of the vehicle
Informed[b]	Information and, if necessary, warning of the driver by assistance systems with environment detection (e.g. Lane Departure Warning)	The driver is permanently in control of the vehicle
Assisted[a,b]	Automation systems with environment detection can actively intervene in vehicle guidance in certain situations to support the driver (e.g. LKAS)	The driver is permanently in control of the vehicle
Semiautomated[c]	The system permanently takes over lateral **or** longitudinal guidance within certain limits (e.g. ACC) System boundaries are not recognized	The driver only takes on one dimension of the driving task The driver must continuously monitor the functioning of the system and at all times be capable of performing the driving task in full.
Maneuver automated[b]	The system takes over the lateral **and** Longitudinal guidance for performing specified manoeuvres (e.g. CbW, Winner et al. 2006) System boundaries are not recognized	The driver commands the manoeuvres to be performed The driver must continuously monitor the functioning of the system and at all times be capable of performing the driving task in full.
Semi-automated[d]	The system takes over the lateral **and** Longitudinal guidance for a certain period of time and/or in specific situations (e.g. Highway patrol assistent, cf. Gasser 2012) System boundaries are not recognized	The driver must continuously monitor the functioning of the system and at all times be capable of performing the driving task in full.
Highly automated[d]	The system takes over the lateral **and** Longitudinal guidance for a certain period in specific situations (e.g. Autobahn chauffeur, cf. Gasser 2012) System boundaries are recognized and the driver is prompted to accept them The system is not in a position to determine the minimum risk condition in every situation. bring about If the driver has to take over the driving task, the system provides him with enough time.	The driver must be able to check the function of the system. **not** monitor permanently The driver must perform the driving task at system boundaries.

(continued)

Table 9.8 (continued)

Denomination	Automation/assistance skills	Role of the driver
Fully automated[d]	The system takes over the lateral **and** Longitudinal guidance in a defined application (e.g. Autobahnpilot, see Gasser 2012) System boundaries are recognized and the driver is prompted to accept them If the driver does not intervene, the system is able to bring about the minimum risk condition.	The driver must be able to check the function of the system. **not** keep a watch on The driver can take over the driving task at system boundaries.

From Damböck (2014)
[a]Meaning changed
[b]new automation level
[c]was renamed
[d]is taken over

9

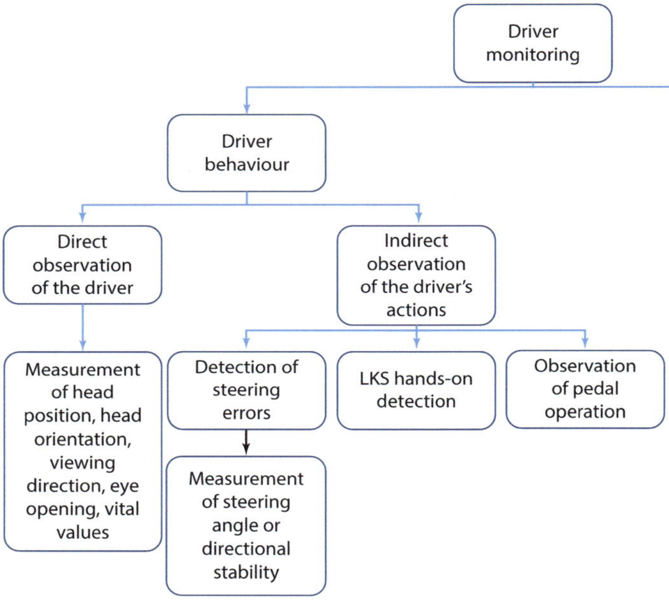

Fig. 9.23 Approaches to driver monitoring. (From Hörwick 2011)

sively for use in combination with highly automated driver assistance systems to ensure that the driver can still draw his attention to the traffic situation. According to this idea, the driver must also set the potential trigger when starting the DAS. If he does not do this, it is not possible to activate the DAS. The assistance system in action is switched off when the lower point of the attention level is reached. In this way, a certain "situation awareness" of the driver should be ensured even if he is distracted by the occupation with non-driving-relevant actions.

9.2.3.2 Observation of Driver Behaviour

In principle, the driver's behaviour itself can be monitored either directly by observing the driver or indirectly by recording the driver's operating actions. Within the framework of *direct driver observation* only video-based recording of the driver is possible for series production. The head position and its orientation, in particular the eye opening angle and the eyelid closure frequency as well as the viewing direction of the driver's eyes are recorded by image processing technology. Hörwick (2011) indicates several corporate and university developments that allow such monitoring. It is assumed that the driver's attention can be inferred from knowledge of the parameters mentioned and their temporal course. In particular, it must be assumed that an attentive driver must look through a defined area of the windscreen with a certain regularity in order to follow the traffic situation appropriately. Under the name "Driver Monitoring System", Lexus is the only automotive manufacturer in its LS series to offer a video-based driver monitoring system based on the recording of head position and alignment in series production. If the system detects an obstacle in front of the vehicle and at the same time the driver turns his head away from the road for too long, a visual and audible warning is issued. If he still does not react, a brake jerk is triggered to warn the driver. It should be noted that the video-based head position can be determined much better than that of the eyes (spectacle wearers and strongly fluctuating light conditions are limiting factors). Therefore, systems for the detection of the eyes are not yet in series use.

At the *indirect driver observation* one tries to infer the condition of the driver by monitoring the operating actions for – essentially – the primary driving task. The movement of the steering wheel and, if necessary, in conjunction with it, that of the driving and brake pedals is recorded. This is based on the assumption that a tired and inattentive driver makes small steering errors during manual driving, which he corrects or compensates in a characteristic manner (Daimler AG 2008). For the detection of the steering error, the steering wheel movement curve or the observation of the lateral position of the vehicle can be used (Altmüller 2007). Daimler's "Attention Assist", which is currently available on the market, is based on the measurement of the steering angle curve, whereby other parameters such as speed, travel time, longitudinal and lateral acceleration as well as indicator and pedal actuation and influences acting on the driver such as time of day, current traffic situation, crosswind and road condition are taken into account for the calculation of the warning time. Volvo's Driver Alert Control system, on the other hand, observes the lateral placement of the vehicle from the centre of the lane by means of a camera directed at the lane marking and inertial sensors. At speeds above 60 km/h, the travelled trajectory is reconstructed and it is assessed whether it runs controlled within the lane. In all cases, the driver is informed of the lack of attention by an acoustic display and simultaneously by an optical display (usually in the form of a coffee cup symbol, which is intended to encourage a break).

Simpler systems (currently implemented at VW and Honda) merely analyse whether the driver has his hands on the steering wheel. Sensors on the steering wheel record the steering torque currently applied by the driver. If no independent steering movement of the driver is detected, the Lane Keeping Support System (LKS) is switched off after a certain tolerance time (usually in the range of seconds).

The problem of all fatigue alarms – no matter on which principle they work – is the high rate of "false alarms" and "misses" (alarm, although there is no fatigue, missing alarm in case of fatigue, see also ► Sect. 11.2.6). In addition, fatigue is subjectively very well recognized before the actual microsleep. The pressure of time is often overwhelming, especially in business life, so that it is to be feared that the fatigue detector's advice is often disregarded and is therefore often even completely eliminated. In the case of truck drivers, disregarding the fatigue warning is also linked to the potentially unsuccessful search for suitable parking spaces (Eichinger 2011).

9.3 Contribution of Driver Assistance Systems to Driving Safety

Driver assistance systems should make a significant contribution to active safety (see also ► Sect. 9.1). Therefore, it makes sense to divide the course before a collision (pre-crash) into time zones. The area more than 10 s before the crash indicates normal driving. If a dangerous situation develops out of this, the driver has the possibility to avoid it until at least 1 s before the collision. In the period up to approx. 1 s before the actual crash, the possible manoeuvres may no longer be sufficient to avoid the accident. Everything that happens after the collision can no longer be calculated in advance. One can only try to keep the consequences as low as possible (passive safety area) by means of e.g. braking/deceleration. The effectiveness of driver assistance systems can be assigned to these time zones (◘ Fig. 9.24).

In order to quantify the safety benefits of driver assistance systems, Langwieder (2005), Knoll and Langwieder (2006) brought into play the terms "potential" and "maintenance", among others. The ***impact potential*** characterizes the frequency in relation to the

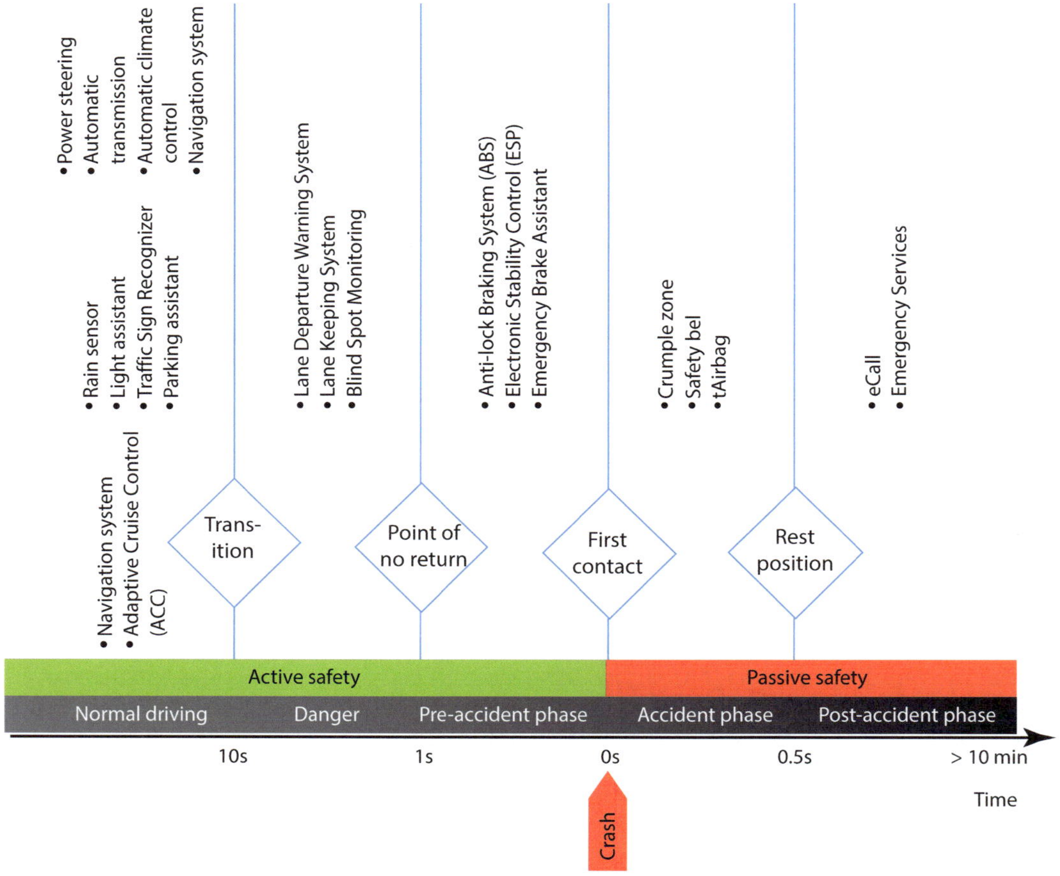

◘ **Fig. 9.24** Assignment of driver assistance systems to phases of more critical driving situations. (Modified after Maier 2014, based on Aparicio 2005; SEISS Final Report 2005; Eichberger et al. 2010)

total number of vehicle accidents considered of those accident situations for which a special driver assistance system could have a positive effect. It is defined by:

$$Impact\ potential = \frac{Number\ of\ systematically\ relevant\ acidents}{All\ vehicle\ accidents}$$

The ***active maintenance*** indicates the proportion of vehicle accidents with personal injury that could be avoided by the system with 100% market penetration. It is defined by:

$$Effective\ potential = \frac{Number\ of\ systematically\ relevant\ acidents}{All\ vehicle\ accidents}$$

Based on these formulas, ◘ Fig. 9.25 provides an estimate of the individual assistancesystems compiled by Maier (2014). In addition to the electronic stabilisation control (ESC), the effectiveness of which has already been statistically verified and which will therefore be mandatory for new vehicles both in the USA and in the EU in the future, the automatic emergency brake assistant is attributed the greatest potential in agreement with other authors. However, one can also see that the effect potential is consistently higher than the expected effect. This becomes particularly clear with the Lane Keeping Assist, whose effect potential also includes many other factors (e.g. alcoholisation, speed selection etc.) and which therefore, depending on the author, shows a wide dispersion range. Overall, however, it should be noted that the impact expectancy would have to be multiplied by the relative frequency of use in order to obtain a realistic assessment of the change

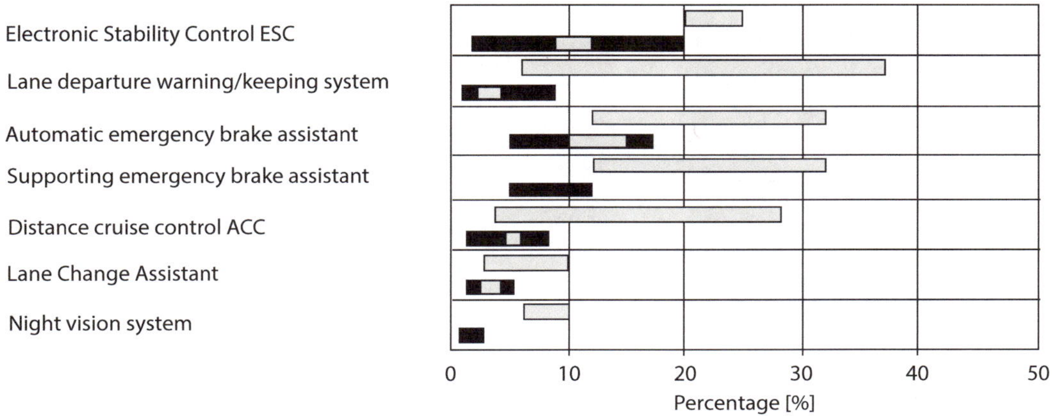

◘ **Fig. 9.25** Compilation of the areas of effect potential (grey) and effect maintenance of different assistance systems described by different authors (black; the bright area indicates the median value in each case); according to Maier (2014)

in accident incidence – even with complete market penetration.

In fact, the frequency of use is the weakness of all assistance systems. Estimates show that such systems are only switched on at a maximum of 25% of the journey time (Sacher 2009; Pereira et al. 2013). After many systems (e.g. ACC) automatically switch off after an intervention by the driver and then have to be consciously reactivated, they often remain switched off for a long time after such an incident. Another reason for the low ratings is the relatively high number of "false alarms", which can be explained by the technically sophisticated detection algorithms used to derive the correct reaction from the "noise of the signals".

In order to estimate at least the theoretical probability of the effectiveness of an assistance system, it is necessary to take into account how the information generated by the driver and the assistance system is brought together so that it can have an impact on the vehicle and thus on the driving result. Figure 9.26 shows the two possible alternatives: either the selection can be made by a switch, as is the case with most assistance systems today, or the information of the driver and the assistance system can be brought together equally via a sum point (see also Figs. 9.18 and 9.19). In Fig. 9.26,

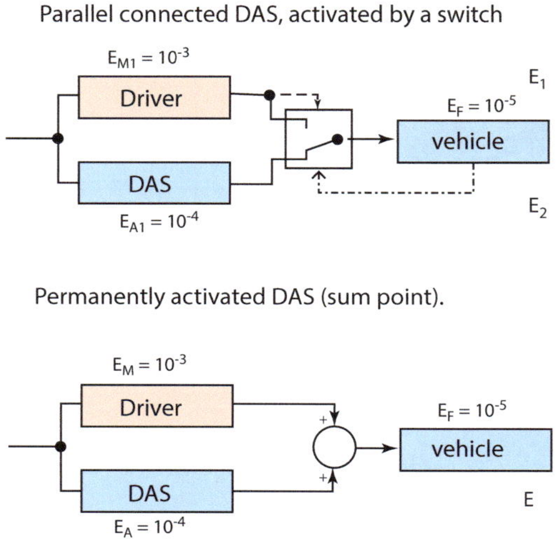

Fig. 9.26 Possible wiring of driver, assistance system and vehicle (schematic)

assumptions are made for the reliability of the individual systems: as usual, the error probability of the driver for the highly practiced activity of driving itself is calculated with $E_M = E_{M1} = 10^{-3}$ presumed.[13] For the vehicle, the error probability is set to $E_F = 10^{-5}$ as it can usually be assumed for conventional technical systems. For the assistance system, a relatively optimistic error probability of 10^{-4} is assumed. This assumption does not relate to the technical function of the system itself. Like that of the vehicle, this should be at 10^{57} or even better. Rather, it is intended to take into account that there are situation constellations for each assistance system that it cannot cope with. In the case of the system activated by the switch, two error branches must now be calculated: if the switch is on the driver, i.e. if the assistance system is switched off, the error probability E_1 of the first branch:

The error probability E is calculated for the activated assistance system E_2 with the assumptions made here:

$$\begin{aligned} E_2 &= 1-(1-E_A)(1-E_F) \\ &= 1-\left(1-10^{-4}\right)\left(1-10^{-5}\right) \\ &= 1\cdot 10^{-4} \end{aligned}$$

If the switch is actuated by the driver as shown in Fig. 9.26, because of its uncertain actuation probability of approx. 0.25, the possible improvement of the error probability by the assistance system is practically reduced to the level of the driver. For this reason, some authors doubt the benefits of assistance systems (e.g. Gründl 2005).

With assistance systems at the stabilisation level (ABS, ESP, emergency brake assistant), however, the switch is also operated on the basis of technically measured conditions. This makes sense especially under the tight time conditions here (in the 100 ms area), because under such circumstances an oversteering by the driver is not possible at all due to his reaction ability. Then for this short time the high

13 In an acute hazardous situation or if the driver is inattentive, this error probability can be significantlXy higher!

technical level is effective, which then can be even better than the 10^{-4}.

The calculation is different if the redundant information about a sum point is merged. In this case, the redundancy is fully effective (AND operation: $E_A \cdot E_M$). The calculation shows:

$$\begin{aligned} E &= 1-(1-E_A \cdot E_M)(1-E_F) \\ &= 1-\left(1-10^7\right)\left(1-10^{-5}\right) \\ &= 1 \cdot 10^{-5} \end{aligned}$$

The parallel connection of driver and assistance system allows the respective error possibilities to compensate each other and the error probability is even reduced to that of the technical system "vehicle".

The effect of the sum point can be realized via an active control element (see ◘ Fig. 9.27).

As with the conventional control element, the force is applied to it by the processing of the task. This force is used to influence the vehicle, which itself brings about the result again. However, the result is now reported back to the control element and a servomotor brings it into a position corresponding to the result. This means that the driver not only receives feedback on the result via the usual optical path, but also feels it on the control element (see also ► Sect. 6.4.3). The advantage of the active control element, however, is that information from the driver assistance system can be fed in between the control element and the vehicle in the sense of a sum point.. The fact that its effect on the vehicle can also be felt haptically via the servomotor means that the driver immediately receives information from the "will" of the driver assistance system, against which he can "fight" at any time via the sum point. The version of the permanently activated driver assistance system, as shown in ◘ Fig. 9.26, is thus implemented.

◘ Table 9.9 lists most of the assistance systems available today. It is examined at which level of the driving task these systems are effective, whether they only have a warning or indicating character, whether they switch on automatically via a switch or switch off automatically under certain conditions or whether they are switched on or off manually (Bubb 2013). Three systems stand out that have no direct influence whatever on the vehicle: the navigation system provides *advice* for the driver at the navigation level (see also ◘ Fig. 9.20) Its positive effect on safety is undoubted, except for the avoidance that occurs when navigation inputs are made during the journey. The Night Vision system can provide advices at the guidance level. However, the effectiveness of this system for safety is very questionable, since in a critical situation the information available there would have to be taken up by looking away from the road (see Bergmeier 2009). Systems in which the information from the night vision system is fed into an automated collision avoidance system must be evaluated differently. Although the fatigue warning is effective at all driving task levels via the driver, it is only effective in terms of safety if its warning is followed and the journey is interrupted for a rest period.

Assistance systems at the *stabilization level* are almost exclusively characterized by the fact that they are automatically switched on for a short time in an extreme situation and the driver has no possibility of intervening in

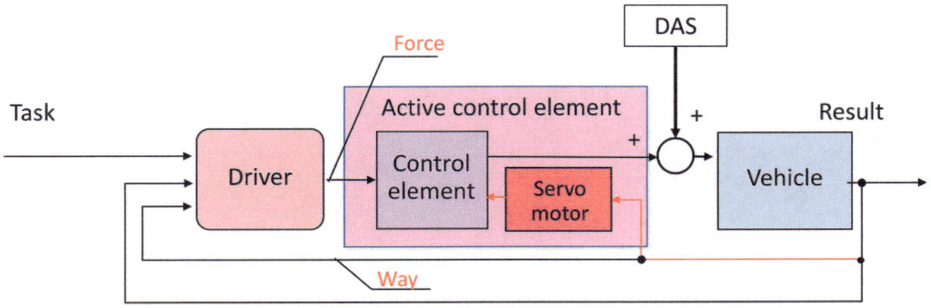

◘ **Fig. 9.27** Information flow with an active control element

Table 9.9 Analysis of the assistance systems available today from the point of view of their level of effectiveness and the way they work

Assistance systems		Effective at a level of			Mode of action				
		Navigation	Guidance	Stabilization	Warn./Note	Switch			Sum point.
						automat.		Hand On/ off	
						on	off		
C&L	Navigation system	x						x	
Cross.	Lane keeping system		x	x				x	x
	Lane keeping warning		x		x			x	
	Lane change assistant		x		x			x	
Longitudinal	Distance cruise control ACC		x	x			x	x	
	Automatic emergency brake sys.		x	x				x	
	Supporting emergency brake ass.			x		x			
C&L	Night vision system		x		x			x	
C	Electronic Stability Control ESC			x		x	x	(x)	
L	Anti-lock braking system ABS			x		x	x		
Fatigue warning		x	x	x	x			x	

the system during this phase. In the sense of ◘ Fig. 9.24 these systems become active in the phase before the collision. If the situation has been overcome by their intervention without a crash occurring, they automatically switch off again. The potential and expected effectiveness of these systems are high (◘ Fig. 9.25).

Assistance systems at the *guidance level* are expected to yield high safety gains. This is justified by the fact that unfavourable conditions created by the driver at the guidance level during the normal driving phase (e.g. too high speed, too small distance) can then no longer be compensated at the stabilization level (point of no return). However, not all assistance systems intervene at the stabilization level, which means that the driver has to intervene himself within an often very tight time window. This is also questionable for safety reasons.

The only system at the guidance level that realizes the sum point according to ◘ Table 9.9 is the lane keeping system. It provides feedback (vibration or even corrective reset torque) to the steering wheel when the driver approaches too close to the lane edge defined by the system. However, the driver can "override" this message at any time.

Otherwise, what is striking about the compilation of ◘ Table 9.9 is that all assistance systems have to be switched on manually, with the result that the low switch-on rate already mentioned becomes effective. Overall, the unsystematic distribution of the crosses in ◘ Table 9.9 points to the poorly coordinated operating and operating philosophy of the assistance systems from an ergonomic point of view, which can lead to confusion for the technically untrained driver. The question must therefore be asked how improved ergonomics in assistance systems can increase driver acceptance and thus improve safety.

A disadvantage of many of today's assistance systems is their mutual independence and often also the lack of ability to correctly consider the future course of the road. The SANTOS project (König et al. 2003) was a first attempt to overcome this shortcoming. Among other things, a speed recommendation based on the so-called V85[14] was made via the active accelerator pedal, but this met with rejection from many test drivers. In the sense of "anticipatory driving", which is gaining in importance again today because of the necessary economical energy consumption of combustion and electric motor vehicles, this reference to the information stored in the navigation system receives new impulses. However, the results of SANTOS and many other tests (Penka 2001; Lange 2008) also show that active access to driver behaviour must be chosen moderately and carefully in order not to jeopardise the acceptance of such systems. In particular, a certain self-adjustment by the driver shall be provided for In the case of lateral guidance, a wide unassisted corridor as opposed to a narrow corridor promotes acceptance. In the case of longitudinal guidance, the safety time gap must be selectable within limits by the driver.

9.4 Ergonomic Design

9.4.1 Operation and Display

According to ◘ Table 9.9, the ideal design of an assistance system from a safety point of view would look as follows:

- Effectiveness at the guidance level,
- reaching the stabilization level (automatic intervention),
- automatic switching on, if necessary also automatic switching off, which must be indicated however,
- cannot be switched off permanently by hand,
- realisation of the sum point in the form that the intention of the assistance system can be felt on the accelerator/brake pedal or steering wheel, but can be overridden at

14 85% of the observed drivers drive slower than this speed; for the V85 there are formulas from the field of road construction which allow to calculate them from the curve radius and the other track characteristics. It can thus be determined from the data of a navigation system.

any time (the assistance system is therefore not an automatic system in the actual sense).
- Technically almost error-free implementat ion/availability

The required realization of the sum point indicates a dilemma: while the steering wheel basically allows a haptically perceptible force/torque impact, which can be directed both in and against the driver's intention, this is only possible to a limited extent with accelerator and brake pedals. Due to the given technical development, the longitudinal dynamics of the vehicle are influenced not by one but by two control elements (apart from further control elements such as clutch and gear lever, which can be omitted), both of which are not +/− characteristics, such as the steering wheel. It is therefore not surprising that many scientific studies on an innovative vehicle control system have switched to a two-dimensional joystick-like control element, since assistance functions can be integrated here much more consistently (Bolte 1991; Eckstein 2001; Penka 2001). Recently, Kienle (2014) realized the already mentioned Joke, which is not so radically different in operation and appearance from the conventional operation and therefore meets with higher acceptance among test persons. If, however, a radical change is not made, the active accelerator pedal (AAP), which is capable of generating a restoring force that can be influenced by the system, has been available for some time for longitudinal dynamics. For example, the driver can be informed that he is driving too fast or that the distance to the vehicle in front is too small. If a correction is only possible by braking, in this case the AAP is moved completely to the zero position; however, a haptic feedback about the braking process that has now taken place is not possible (it would therefore make sense to indicate the activity of the brake by a light in the instrument panel that is coupled with the brake light). In the case of a braking process initiated by the driver, it is not necessary to switch off the system with this design, as there is no automatically controlled forward movement of the vehicle. In combination with a wide aisle Lane Keeping Assistant (LKS) (see Fig. 9.4a), this would create a driver-vehicle interaction that could be in continuous operation and would only inform the driver when objective limits are reached. Further driver support would be achieved if the speed limit set by the traffic sign recognition system or the navigation system were automatically adopted by the ACC system as the target speed. With the usual cruise control operation for setting the target speed, the driver could then make acute individual corrections. This would also render superfluous the speed limiter offered by some vehicle manufacturers, which otherwise requires its own control.[15] This automatic adoption of the maximum permitted speed has proved to be less stressful overall in tests and also provides the better objective values, but at the same time is subjectively judged to be "unsportsmanlike" (Lange 2008).

For technical reasons, there are many situations in which the functioning of the assistance system cannot be guaranteed (see the individual restrictions on the availability of assistance systems described in ► Sect. 9.2.1.3). If such limits are reached, this must be indicated to the driver by displays. Primarily the optical as well as the acoustic displays, which are already in use today in connection with the assistance systems, come into question. However, these displays are not situation-specific in every situation. As Lange (2008) was able to show, the "advice" of the assistance system is all the more accepted the better it is understood from the situation. He formulated the sentence for it: *"The driver should be haptically told what to do and visually why he should do it"*.

The most effective visual representation is achieved by the HUD (Head-Up-Display), both after long and other attempts, because it can convey the information without taking the view from the road. Whereas conventional HUD uses mirror technology to generate an

15 The proposal discussed here differs from the design implemented in the above-mentioned SANTOS project: while a concrete speed proposal was made there at any time via the AAP, only an indication of the maximum permitted speed is given here.

Fig. 9.28 Contact analog HUD (kHUD) with spacer bar (prototype Audi)

Fig. 9.29 Display of a necessary lateral guide correction in the cHUD (Lange 2008)

image approximately at the height of the bonnet on a vertical virtual plane at a distance of approx. 2–3 m, the contact-analog head-up display (cHUD) uses the virtual plane directly on the road so that objects can be displayed at the correct distance. The perceived position is then not dependent on the head position of the driver. (Bubb 1975; Schneid 2009). In the form suggested by Bubb and also implemented by Schneid, a crossbeam is displayed that moves in front of the vehicle, the distance of which, depending on the speed, corresponds to the distance travelled through in approx. 1.2–1.5 s. This distance should be selectable by the driver within certain limits and should correspond to the standard distance of the ACC system. Figure 9.28 shows this form of display using the example of a prototype developed by Audi (Schneid 2009). If the ACC control is used in the situation shown there, the active accelerator pedal would indicate to the driver through its back movement that the speed must be reduced and at the same time through a change in the bar colour why this is necessary. The cumulative point requirement is met by the driver deliberately reducing the distance to the vehicle in front against the "advice" of the ACC beyond the artificial pressure point of the accelerator pedal, or by reducing the pressure on the accelerator pedal he could also increase the distance or reduce the speed. Malfunctions of the ACC would also become apparent to the driver, for example by recognizing from the position of the beam that he is by no means on a collision course to a dangerous obstacle.

Also the insight into a necessary lateral guide correction is achieved by the cHUD. According to a proposal by Lange (2008), an arrow would appear on the distance bar simultaneously with the restoring torque generated by the lane keeping system, "pushing" it in the right direction (Fig. 9.29). This would allow the driver to intuitively understand why he is now feeling a corresponding additional moment on the steering wheel.

Even if the assistance system is not functioning, the cHUD display should be remain, without any regulatory intervention. The display of the distance bar, the lateral position of which is determined by the current position of the steering wheel, indicates where the vehicle will move to within the next 1.2–1.5 s. When the assistance system is switched on, the driver will thus experience the correct behaviour with regard to the outside world through the cHUD display, a behaviour which he should then habitually adopt himself when the assistance system is not in operation. Israel (2013) was able to show that the distance beam, which corresponds to the vehicle width, can also pass difficult bottlenecks on construction sites with a steady hand and reduced stress, even if the lane guidance assistant cannot be active. Figure 9.30 shows further suggestions of possible displays in the cHUD such as tracking, avoidance recommendation in case of collision warning and above all a

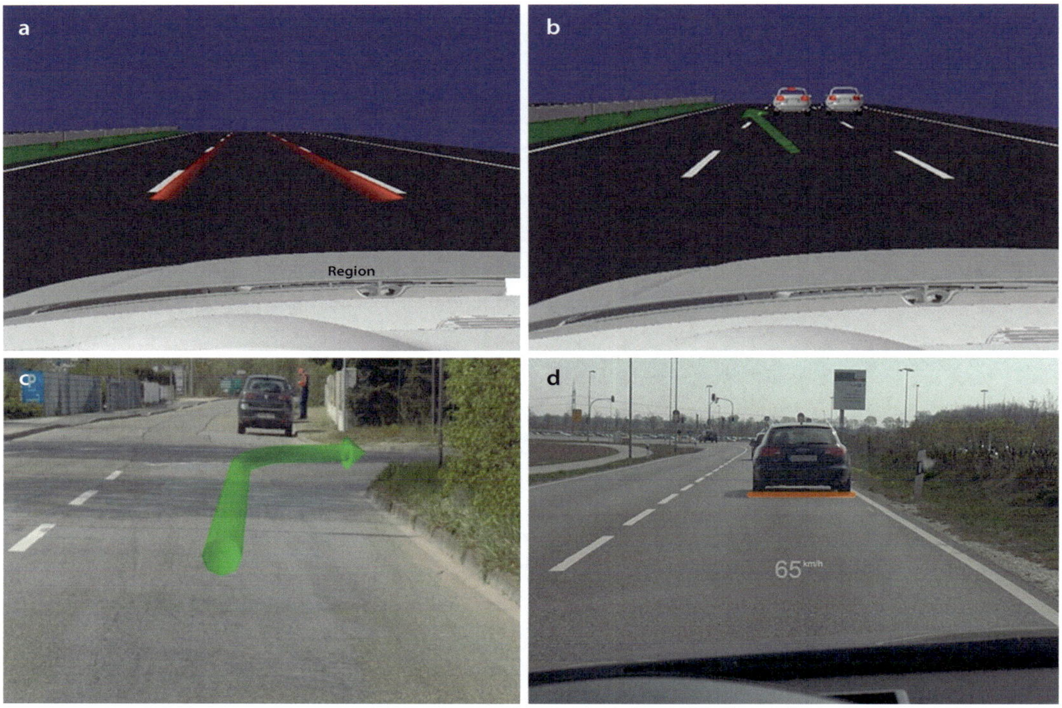

Fig. 9.30 Further possible display concepts for lane keeping **a**, avoidance recommendation in collision warning **b**, navigation **c** and safety distance **d** (Israel and Bubb 2010)

contact-analogue navigation display, which according to the investigations by Israel (2013) at difficult junctions causes a clearly reduced number of wrong decisions.

Damböck et al. (2012) and Weißgerber et al. (2012), in a version where the course proposed by the assistance system is indicated by a trajectory in the cHUD, demonstrated that in the event of a malfunction of this assistance system, drivers make a correction faster and more accurately than without such an indication.

With the interaction variants of assistance systems presented here, only a reduction of longitudinal and lateral guidance accidents can be expected. The problem of intersection accidents and in particular of dense and very heterogeneous traffic in city centres is therefore not yet addressed. The recently introduced systems (e.g. Ford Focus), which, if the safety distance in the speed range of city traffic (≤50 km/h) can, however, be integrated into the system presented here without contradiction. Intensive activities took place within the framework of the Urban project in cooperation with various institutes and industry. Also in this context the chance and the potential of the conventional HUD and of the the cHUD were explored. The sum point problem addressed remains virulent.

Often assistance systems are apostrophized as "fun brakes". As many examples show, such things are fun to act, where one penetrates to some kind of border and now tries to walk the tightrope, on the one hand, not to "fall down" when crossing the border and, on the other hand, not to fall back into the irrelevance of the unattractive (see also ► Sect. 3.2.2 and ◘ Fig. 3.51). For safety reasons, however, this subjective limit must not lie within the range of objective danger. The combination of haptic feedback and visual indication of the set limit suggested here could combine the effect of fun with that of safety. The fun should come from getting as close as possible to the boundary seen in the cHUD, which now no longer has to be felt somehow from experience (see ◘ Fig. 2.24), but is objectified by measurement. In detail, however, further research is necessary to achieve this goal.

9.4.2 Distinctness of the Modes of a Driver Assistance System

Already from the categorization approaches presented in ▶ Sect. 9.2.2 it emerges that the future development must start from different levels of the degree of automation provided by assistance systems. With regard to the acceptance, the integration of the operation and in particular the transition between the different assistance modes are of outstanding importance. Lindberg (2007, 2012) dealt with the question of how different driver assistance systems can be integrated in the lateral guidance area and how driver assistance systems can also be integrated to meet the user's expectations. This shows that the location of the support relative to the vehicle and the criticality of the event are possible category systems. Wimmer (2014) has dealt with the design of control elements and displays for different assistance modes, both anthropometric and system ergonomic. His focus is on the operation of comfort-oriented driver assistance systems that intervene continuously at the guidance level. The aim is beside others to create a clear and understandable demarcation between these modes and each other for the driver. On the basis of the detailed discussion given there and the considerations presented here, the following five modes can be distinguished from an ergonomic point of view:

- ***"Manual travel"MF*** The ESP, ABS Emergency Brake Assist and Lane Departure Warning systems, which operate on the stabilization level, are a matter of course here. The cHUD should always be available in the form described above.
- ***"Supported Driving UF*** The ACC systems, which intervene at the guidance level, in conjunction with the active accelerator pedal and lane guidance assistant, are an integral part of this system. Information about the activities of the two assistance systems is displayed in cHUD in the form described above.
- ***"Semi-automated driving"TA*** ACC provides the functions that are common today. Narrower lane guidance is activated at the lane guidance assistant. The display in the cHUD corresponds to the UF mode.
- ***"Fully automatic driving"VA*** In addition to the functions of TA, there is the possibility of a manoeuvre interface through which the driver can commission certain manoeuvres in the sense of conduct-by-wire (see ▶ Sect. 9.2.2). As far as technically possible, the cHUD visualizes the corresponding maneuvers.
- ***"Autonomous driving"AUTON*** The driver does not have to deal with driving and can carry out other independent activities. If the AUTON condition is no longer tenable, it must be requested in good time to accept it (according to the Damböck 2014 investigations, at least 6–8 s before the actual system failure, in order to ensure that the *Situation Awareness* can be secured. In particular, if one takes unforeseen system failures into account, this is a practically unfulfillable requirement).

In contrast to the clear separation between the individual modes, the H-Mode concept mentioned in ▶ Sect. 9.2.2 provides for a continuous transition between these modes.

Wimmer used four of these modes in his investigations. The UF mode is not intended for him. It does not provide for feedback in the cHUD in the above sense, but for feedback displays in the conventional HUD.. In the event of failure or when the DAS system limits are reached, the driver shall represent the fallback plane in all modes. The driver's attention must therefore always be monitored at all times. Wimmer suggests using the potential trigger for this (see ▶ Sect. 9.2.3).

As an example of the operation of switching between the different modes, Wimmer's study (2014) presents the variant rotary pusher, which received the greatest preference from the test persons in his experiments (see

Fig. 9.31).[16] The four modes are selected by means of a variable shape actuating element, which considerably extends the functions of a rotary pusher and which is accommodated in an anthropometrically favourable position in the centre console.

- An essential element of this control is the A+ key (① in Fig. 9.31a). By pressing this key, the next higher automatic level is requested. In MF mode, the control is recessed in the centre console, only the keypad is accessible to the operator.
- The TA mode is activated by pressing the A+ key. If this mode is available for sensor technology, the entire rotary pusher automatically rises from the centre console (Fig. 9.31b). To the right of the A+ key is the OFF key (② in Fig. 9.31), which can be used to deactivate the currently active automation mode across all modes. A short push operation causes the TA mode to return to standby, a long push operation returns the system to MF mode. If ACC has been deactivated for technical reasons – for example by actuating the brake – the automatic function can be resumed by pressing the RESUME button below (③ in Fig. 9.31). The driver adjusts the target speed by turning the ring. By pushing the rotary pusher forward with a monostable short push actuation, the distance taken by the ACC is reduced by one step or extended backward by one step with a corresponding actuation of the rotary pusher (movement A ↔ B).
- By pressing the A+ button again, the driver activates the VA mode when the system is available. Now the rotating ring lowers into the centre console. The remaining joystick, which tapers downwards, contains the described key unit on the upper side, the keys of which also have the meaning described above in this mode (Fig. 9.31c). Due to the smooth metallic surface and the changed actuation characteristics, the operating element in the VA mode differs both haptically and optically from the TA mode. The driver now uses this stick to "order" different manoeuvres. By pressing forward or backward the distance to the vehicle in front can be varied within the permitted limits (movement E/E+ ↔ F/F+) and by moving the joystick to the right or left the position within the lane can be varied (movement C/C+ ↔ D/D+). The joystick, which is designed as a self-centering, active control element, indicates the limits of such an assignment by increasing the haptically felt force. By overcoming a smaller "mountain of power" (e.g. C+), the order can be placed for a lane change and by moving the joystick forward (E+) an overtaking manoeuvre can be indicated. If the driver is looking for a parking space, corresponding space-compatible movements induce an automatic parking manoeuvre. The same applies, of course, to leaving the car park.
- Pressing the A+ key again switches the system to the AUTON mode (Fig. 9.31d). If the corresponding conditions are given from the sensor side, the embroidery element now lowers into the centre console. The rotary pusher is now completely recessed again as in the MF state with the difference that the OFF key is now available to the user to deactivate the automation.[17]

The UF function, which Wimmer did not investigate, could be easily integrated into the concept described here. It would not differ from the constellation for TA mode. Only the

16 Wimmer's work includes basic system ergonomic considerations on the different modes as well as anthropometric aspects that play a role in this context. Other operating concepts are also being discussed, but their presentation would go beyond the scope set here.

17 In the original concept, the active keys were supposed to be raised against the countersunk ring even in the AUTON state. However, this could not be realized in the prototypical realization.

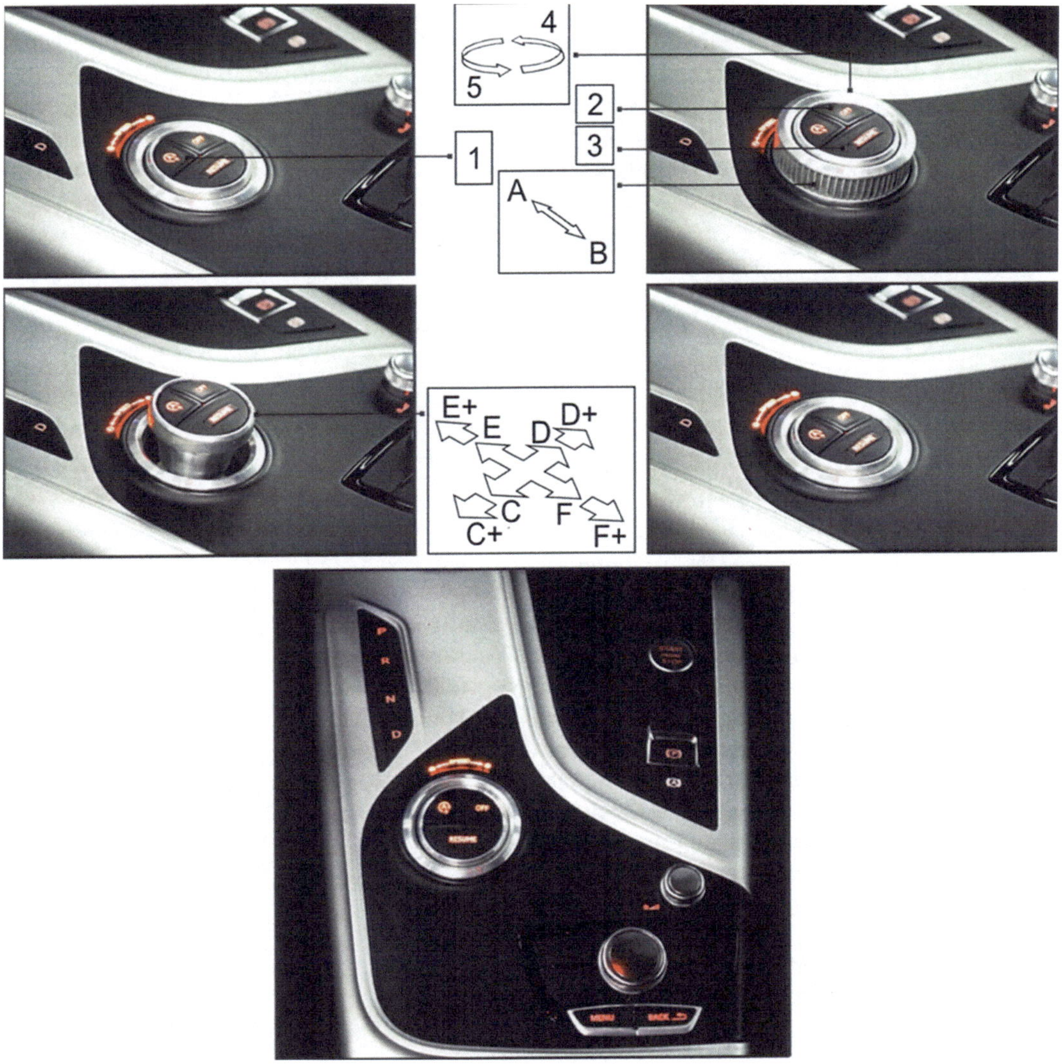

Fig. 9.31 Prototypical implementation of an integrative control element for switching between different driver assistance modes. Here in the form of a rotary pusher. (From Wimmer 2014). **a** MF mode. **b** TA mode. **c** Mode VA. **d** AUTON. **e** Integration of the control element in the center console

OFF and RESUME keys would be inactive under this condition. With the Wimmer concept, the vehicle is in MF mode after the start. For safety reasons, it would make sense to use the UF mode instead when starting the vehicle with the concept supplemented here. The driver would then have to actively switch back to MF mode if necessary, as is already the case today when the ESP is deactivated. The constant availability of the assistance function required under ► Sect. 9.3 would thus be much better guaranteed than if the driver had to switch up from a low to a higher assistance level.

In connection with the various modes, the considerations in ► Sect. 9.3 should be recalled, according to which a significant reduction in the overall error probability can only be achieved by simultaneous participation of the driver and the assistance system in the driving task. The extent to which this is still the case in AS and VA modes depends to a large extent on the frequency with which the driver is required to interact. Already for

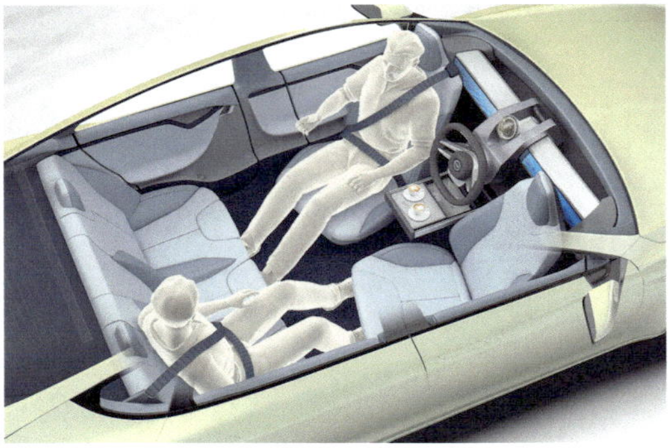

Fig. 9.32 Rinspeed concept study XchangeE (Rinderknecht 2014)

today's ACC, concerns are expressed in the literature that the driver will turn his attention away from the driving process and increasingly turn to tertiary tasks (see the argumentation in ► Sect. 9.2.1). It is also questionable whether the hoped-for safety gain can even be achieved with completely autonomous driving (AUTON mode). Nevertheless, the autonomous vehicle plays an important role in today's discussion. The Rinspeed concept study (Rinderknecht 2014; Fig. 9.32) is cited as an example of the design consequences that could result from this. Taking into account the already quoted work of Damböck (2014), according to which a minimum time of 6–8 s is required to return to the driving task when asked to take over the driving task from a tertiary activity, it seems impossible to take up a driving position again within such a short time if the vehicle interior is completely reoriented.

9

References

Literature Used

Abele, J., Kerlen, C., Krueger, S., Baum, H., Geißler, T., Grawenhoff, S., Schneider, J., Schulz, W.H., SEISS: In: VDI/VDE Innovation & Technik GmbH & Institute for Transport Economics at the University of Cologne (Hrsg.) Exploratory Study on the Potential Socio-Economic Impact of the Introduction of Intelligent Safety Systems in Road Vehicles. VDI Verlag GmbH, Teltow/Köln (2005)

Acarman, T., Liu, Y., Zgüner, Ö.Ü.: Intelligent cruise control stop and go with and without communication. In: Institute Of Electrical And Electronics Engineers, Inc (IEEE) (ed.) American Control Conference, pp. 4356–4361. IEEE, Minneapolis (2006)

Akamatsu, M., Green, P., Bengler, K.: Automotive technology and human factors research: Past, present, and future. Int. J. Veh. Technol., 526180, 27 pages. (2013). https://doi.org/10.1155/2013/526180

Alkim, P.T., Bootsma, G., Hoogendoorn, P.S.: Field operational test "The assisted driver". In: Institute of Electrical and Electronics Engineers, Inc (IEEE) (ed.) IEEE Intelligent Vehicles Symposium: Tagung Istanbul, Turkey, pp. 1198–1203. IEEE Computer Society Press, Istandbul (2007)

Altmüller, T.: Driver monitoring and drowsiness detection by steering signal analysis. Dissertation an der Universität der Bundeswehr München, Neubiberg (2007)

Aparicio, F.: EEVC WG19- Activities on primary and secondary safety interaction. In NHTSA (Hrsg.): Proceedings in the 19th International Technical Conference on the Enhanced Safety of Vehicles (ESV)(Paper Number 05-9462), Washington D.C. (2005)

Bartels, A., Steinmeyer, S., Brosig, S., Spichalsky, C.: Fahrstreifenwechselassistenz. In: Winner, Hakuli, H.S., Wolf, G. (eds.) Handbuch Fahrerassistenzsysteme – Grundlagen, Komponenten und Systeme für aktive Sicherheit und Komfort, 2. Aufl. ATZ/MTZ – Fachbuch, Wiesbaden (2012). Kap 36

Bergmeier, U.: Kontaktanalog markierendes Nachtsichtsystem – Entwicklung und experimentelle Absicherung. Dissertation an der technischen Universität München (2009)

Bolte, U.: Das aktive Stellteil – ein ergonomisches Bedienkonzept Fortschritts-Berichte VDI, Reihe 17 ″Biotechnik″. VDI-Verlag, Düsseldorf (1991)

Bosch: Sicherheits- und Komfortsysteme - Funktion, Regelung und Komponenten. Friedr. Vieweg & Sohn Verlag/GWV Fachverlage GmbH, Wiesbaden (2004)

Bosch-Studie zu Fahrerassistenzsystemen: Fahrerassistenzsysteme – wie viel Unterstützung wünschen deutsche Autofahrer? (2012). www.Bosch-kraftfahrzeugtechnik.de

Bubb, H.: Untersuchung über die Anzeige des Bremsweges im Kraftfahrzeug. Dissertation, Technische Universität München (1975)

Bubb, H.: Menschliche Zuverlässigkeit und sicheres Fahren – ein Widerspruch? In: Winner, H., Bruder, R. (Hrsg.) Maßstäbe des sicheren Fahrens Darmstädter Kolloquium, Ergonomie, Stuttgart, 6./7. März 2013. S. 107–132. (2013)

Buld, S., Krüger, H.-P.: Wirkung von Assistenz und Automation auf Fahrerzustand und Fahrsicherheit:

Projekt: EMPHASIS. Effort-Management und Performance-Handling in sicherheitsrelevanten Situationen (2002)
Buld, S., Krüger, H.-P.: Die Auswirkung von Teilautomation auf das Fahrverhalten. In: Grandt, M. (ed.) Entscheidungsunterstützung für die Fahrzeug- und Prozessführung DGLR-Bericht, Bd. 2003-04, pp. 241–253 (2003)
Cauzard, J.P.: SATRE 3 – Results: differences in attitude and opinions of European motorists with respect to traffic safety measures and regulations. In: Bundesanstalt für Straßenwesen (Hrsg.) Sicher fahren in Europa 5. Symposium, Bd. M161, Wirtschaftsverlag NW Verlag für neue Wissenschaft, Bremerhaven (2003)
Daimler: Zehn Jahre BremsassistentMercedes-Sicherheitssystem bewirkt deutlichen Rückgang schwerer Fußgänger- und Auffahrunfälle. Daimler AG, (2006). Internetquelle: http://media.daimler.com/dcmedia/0%2D%2D921-49-825551-1-0-0-0-0-1-12635-614226-0-1-0-0-0-0-0.html
Daimler AG: Bitte nicht einschlafen. Daimler HighTechReport: Faszination Technologie, 2/2008 (2008)
Daimler: Reagieren – Pre-SAFE®-Bremse und BAS+: Automatische Vollbremsung. Daimler AG, (2011). Internetquelle: http://www.daimler.com/dccom/0-5-1210220-49-1210348-1-0-0-1210338-0-0-13-0-0-0-0-0-0-0-0.html
Damböck, D.: Automationseffekte im Fahrzeug – von der Reaktion zur Übernahme. Dissertationsschrift an der Technischen Universität München (2014)
Damböck, D., Weißgerber, T., Kienle, M., Bengler, K.: Evaluation of a contact analog head-up display for highly automated driving 4th international conference on applied human factors and ergonomics (2012)
Doisl, C.: Systemergonomische Analyse von Anzeige- und Bedienkonzepten zur Unterstützung des Parkvorgangs. Dissertationsschrift an der Technischen Universität München (2008)
Donner, E., Winkle, T., Walz, R., Schwarz, J.: RESPONSE 3 Code of Practice für die Entwicklung, Validierung und Markteinführung von Fahrerassistenzsystemen. VDA, Technischer Kongress, Sindelfingen (2007)
Dragutinovic, N., Brookhuis, A.K., Hagenzieker, P.M., Marchau, A.V.: Behavioural effects of advanced cruise control use - a meta-analytic approach. Eur. J. Transp. Infrastruct. Res. **5**(4), 267–280 (2005)
Eckstein, L.: Entwicklung und Überprüfung eines Bedienkonzepts und von Algorithmen zum Fahren eines Kraftfahrzeugs mit aktiven Sidesticks Fortschr.-Ber. VDI, Reihe 12, Bd. 471. VDI-Verlag, Düsseldorf (2001)
Eichberger, A., Tomasch, E., Rohm, R., Steffan, H., Hirschberg, W.: Detailed analysis of the benefit of different traffic safety systems in fatal accidents. In: EVU (ed.) Proceedings of the Annual EVU Meeting, pp. 301–315. EVU-Jahrestagung, Prag (2010)
Eichinger, A.: Untersuchungskonzepte für die Evaluation von Systemen zur Erkennung des Fahrerzustandes. BASt-Bericht F 80 (2011)
Eick, E.-M., Debus, G.: Adaptation effects in an automated car following scenario. In: Underwood, G. (ed.) International Conference of Traffic and Transport Psychology: Tagung Amsterdam, Netherlands, pp. 243–255 (2005)
Engel, B.: ABS verlängert bei Schnee den Bremsweg. Welt online, 05.01.2009 (Internetquelle: abgerufen am 04.09.2010: http://www.welt.de/article2972775/ABS-verlaenger-bei-Schnee-den-Bremsweg (2009)
Fancher, P., Ervin, R., Sayer, J., Hagan, M., Bogard, S., Bareket, Z., Mefford, M., Haugen, J.: Intelligent Cruise Control Field Operational Test. University of Michigan, Transportation Research Institute, Ann Arbor (1998)
Filzek, B.: Abstandsverhalten auf Autobahnen – Fahrer und ACC im Vergleich. Darmstadt, Technische Universität Darmstadt, Fachbereich Maschinenbau, Dissertation, 2002 (2002)
Flemisch, F.O., Bengler, K., Bubb, H., Winner, H., Bruder, R.: Towards cooperative guidance and control of highly automated vehicles: H-Mode and Conduct-by-Wire. Ergonomics. **02**, 343–360 (2014)
Franz, B., Kauer, M., Schreiber, M., Blanke, A., Distler, S., Bruder, R., Geyer, S.: Maneuver-Based Driving Today and in the Future – Development of a New Human-Machine Interface for Conduct-by-Wire. Paper presented at the 6. VDI Tagung - Der Fahrer im 21. Jahrhundert, Braunschweig, Germany (November 8–9 2011)
Freyer, J.: Vernetzung von Fahrerassistenzsystemen zur Verbesserung des Spurwechselverhaltens von ACC. Neubiberg bei München, Universität der Bundeswehr, Fakultät für Luft- und Raumfahrttechnik, Institut für Arbeitswissenschaft, Dissertation, 2008 (2008)
Gasser, T.: Ergebnisse der Projektgruppe Automatisierung: Rechtsfolgen zunehmender Fahrzeugautomatisierung. Bundesanstalt für Straßenwesen, Bergisch Gladbach (2012). gasser@bast.de
Gayko, J.: Lane departure warning. In: Winner, H., Hakuli, S., Wolf, G. (Hrsg.) Handbuch Fahrerassistenzsysteme - Grundlagen, Komponenten und Systeme für aktive Sicherheit und Komfort, 2. Aufl., S. 543–553. ATZ/MTZ – Fachbuch, Wiebaden (2009)
Gayko, J.: Lane keeping support. In: Winner, H., Hakuli, S., Wolf, G. (Hrsg.) Handbuch Fahrerassistenzsysteme – Grundlagen, Komponenten und Systeme für aktive Sicherheit und Komfort, 2. Aufl. ATZ/MTZ – Fachbuch, Wiebaden (2012). Kap. 35
Gelau, C., Gasser, T., Seeck, A.: Fahrerassistenz und Verkehrssicherheit. In: Winner, H., Hakuli, S., Wolf, G. (Hrsg.) Handbuch Fahrerassistenzsysteme – Grundlagen, Komponenten und Systeme für aktive Sicherheit und Komfort, 2. Aufl. ATZ/MTZ – Fachbuch, Wiebaden (2012). Kap 3

Götz, M., Kleinkes, M., Pietzonka, S., Schiermeister, N.: Künftige Entwicklungen von LED-Scheinwerfern. ATZ. **09**(111), 620–627 (2009)

Grimm, M., Casenave, S.: DBL: a feature that adds safety to night time traffic. In: International Symposium on Automotive Lighting (ISAL), Technische Universität Darmstadt, pp. 355–363. Utz Verlag, München (2007)

Gründl, M.: Fehler und Fehlverhalten als Ursache von Verkehrsunfällen und Konsequenzen für das Unfallvermeidungspotenzial und die Gestaltung von Fahrerassistenzsystemen. Inauguraldissertation Universität Regensburg (2005)

Gulde, D.: Opel-Eye and Blaupunkt.Gerät im Vergleich. Auto Motor Sport online (2009). http://www.auto-motor-und-sport.de/testbericht/verkehrsschilder-erkennung-im-test-13242415.html, Zugegriffen: 04.06.2009

Hoedemaeker, M., Brookhuis, A.K.: Behavioural adaptation to driving with an adaptive cruise control (ACC). Transport. Res. F: Traffic Psychol. Behav. **1**, 95–106 (1998)

Hörwick, M.: Sicherheitskonzept für hochautomatisierte Fahrerassistenzsysteme. Dissertationsschrift an der Technischen Universität München (2011)

Ishida, S., Tanaka, J., Kondo, S., Shingyoji, M.: Development of a driver assistance system SAE 2003 World Congress & Exhibition, Detroit (2003)

ISO 17361: "Intelligent transportation systems. Lane departure warning systems. Performance requirements and test procedures". British Standards Institution ISO-Norm 17387, "Lane Change Decision Aid System" (2007)

ISO 17387: "Lane Change Decision Aid System", British Standards Institution ISO-Norm (2007)

Israel, B.: Potenziale eines kontaktanalogen Head-Up-Displays für den Serieneinsatz. Dissertation, Technische Universität München (2013)

Israel, B., Bubb, H.: Augmeted Reality im Fahrzeug – Möglichkeiten und Grenzen der Darstellung mit dem kontaktanalogen Head-up-Display. In: 5. Fachtagung USEWARE, Baden-Baden (2010)

Jenness, W.J., Lerner, D.N., Mazor, S., Osberg, S.J., Tefft, C.B.: Use of advanced in-vehicle technology by young and older early adopters. Survey results on adaptive cruise control systems (2008)

Katzwinkel, R., Auer, R., Brosig, S., Rohlfs, M., Schöning, V., Schroven, F., Schwitters, F., Wuttke, U.: Einparkassistenz. In: Winner, H., Hakuli, S., Wolf, G. (Hrsg.) Handbuch Fahrerassistenzsysteme - Grundlagen, Komponenten und Systeme für aktive Sicherheit und Komfort, 2. Aufl. ATZ/MTZ – Fachbuch, Wiesbaden (2012). Kap. 31

Kesler, M., Mangin, B.: Nutzerorientierte Auslegung von teilautomatisierten Einparkassistenzsystemen. In: Tagungsband zur 4. VDI-Tagung Fahrer im 21. Jahrhundert. VDI-Verlag GmbH, Braunschweig (2007)

Khanh, T.Q., Huhn, W.: Sichtverbesserungssysteme. In: Winner, H., Hakuli, S., Wolf, G. (Hrsg.) Handbuch Fahrerassistenzsysteme - Grundlagen, Komponenten und Systeme für aktive Sicherheit und Komfort, 2. Aufl. ATZ/MTZ – Fachbuch, Wiesbaden (2012). Kap. 30

Kienle, M.: Anzeige-Bedienkonzepte hochautomatisierter Fahrzeuge – vom manuellen zum hochautomatisierten Autofahren. Dissertation an der Technischen Universität München (2014)

Kienle, M.: Kooperative Interaktionskonzepte zum Fahren eines teilautomatisierten Fahrzeugs. Dissertation an der technischen Universität München (2015)

Kleine-Besten, T., Kersken, U., Pöchmüller, W., Schepers, H.: Navigation und Telematik. In: Winner, H., Hakuli, S., Wolf, G. (Hrsg.) Handbuch Fahrerassistenzsysteme - Grundlagen, Komponenten und Systeme für aktive Sicherheit und Komfort, 2. Aufl. ATZ/MTZ – Fachbuch, Wiesbaden (2012). Kap 39

Knoll, P., Langwieder, K.: Der Sicherheitseffekt von ESP in Realunfällen – Überlegungen zum volkswirtschaftlichen Nutzen von prädiktiven Fahrerassistenzsystemen. In: TUM/TÜV Süd (Hrsg.) Tagungsband der Tagung „Sicherheit durch Fahrerassistenz". TÜV-Süd, Garching (2006)

König, W.: Nutzergerechte Entwicklung der Mensch-Maschine-Interaktion von Fahrerassistenzsystemen. In: Winner, Hakuli, H.S., Wolf, G. (Hrsg.) Handbuch Fahrerassistenzsysteme - Grundlagen, Komponenten und Systeme für aktive Sicherheit und Komfort, 2. Aufl. ATZ/MTZ – Fachbuch, Wiesbaden (2012) Kap. 4

König, W., Weiss, K.-E., Gehrke, H., Haller, R.: S.A.N.T .O.S. Situationsangepasste und nutzertypzentrierte Optimierung von Systemen zur Fahrerunterstützung GfA Herbstkonferenz über Ergonomie und Verkehrssicherheit, Technische Universität München, 12.-13.10.2000. (2000)

König, W., Weiß, K.E., Mayser, Ch.: S.A.N.T.O.S – A Concept for Integrated Driver Assistance, Electronic Systems for Vehicles, Baden Baden (2003) sowie www.santos.web.de

Kovordányi, R.: When will advanced driver support systems be user-adaptive? The case of adaptive cruise control. In: AAAI Spring Symposium, 2005 (2005)

Krause, M., Bengler, K.: My phone, my car and I – and may be a traffic light assistant. In: Ege, R., Koszalka, L. (eds.) ICONS 2013, The Eighth International Conference on Systems, pp. 33–39 (2013)

Lange, C.: Wirkung von Fahrerassistenz auf der Führungsebene in Abhängigkeit der Modalität und des Automatisierungsgrades. Dissertation, Technische Universität München (2008)

Lange, C., Bubb, H., Tönnis, M., Klinker, G.: Sicherheitspotential und Verbrauchsreduzierung durch ein intelligent geregeltes aktives Gaspeda. In: TÜV SÜD AUTOMOTIVE GMBH, Lehrstuhl für Fahrzeugtechnik der Technischen Universität München (Hrsg.) 3. Tagung Aktive Sicherheit durch Fahrerassistenz: Tagung München. TÜV-Süd, München (2008)

Langwieder, K.: Wissenschaftlicher Erkenntnisstand zu ESP. Beitrag zu Konferenz 10 Jahre ESP, 23.02.2005. Kongressbericht, Berlin (2005)

Lee, W.-S., Nam, H.-K.: A study on driver acceptance of adaptive cruise control using a driving simulator. In: THE UNIVERSITY OF IOWA (Hrsg.) Driving Simulation Conference: North America 2007 Iowa City, 2007 (2007)

Lexus: LS - Driver Monitoring System. LEXUS (Internetquelle abgerufen am 18.12.2010 http://www.lexus.eu/range/ls/key-features/safety/safety-driver-monitoring-system.aspx (2010)

Lindberg, T.: Entwicklung einer ABK Metapher für gruppierte Fahrerassistenzsysteme. Dissertation an der technischen Universität Berlin (2012)

Lindberg, T., Schaller, T., Grandenegger, B.: Stauassistent – Unterstützung des Fahrers durch Übernahme der Quer- und Längsführung im Stau. In: der Fahrer des 21. Jahrhundert – Human Machine Interface, 4. VDI Fachtagung (2007)

Ma, R., Kaber, B.D.: Situation awareness and workload in driving while using adaptive cruise control and a cell phone. Int. J. Ind. Ergon. **35**(10), 939–953 (2005)

Mages, M., Hopstock, M., Klanner, F.: Kreuzungsassistenz. In: Winner, H., Hakuli, S., Wolf, G. (Hrsg.) Handbuch Fahrerassistenzsysteme - Grundlagen, Komponenten und Systeme für aktive Sicherheit und Komfort, 2. Aufl. ATZ/MTZ – Fachbuch, Wiesbaden (2012). Kap 37

Maier, F.: Fahrerassistenzsysteme in der Fahrausbildung. Dissertation an der Technischen Universität München (2014)

Oei, H.-L., Polak, H.P.: Intelligent speed adaptation (ISA) and road safety. IATSS Research. **26**, 45–51 (2002)

Ojeda, L., Nathan, F., Brouwer, F.T.R., Hoedemaeker, M.D. (Hrsg.): Driver Support and Information Systems: Experiments on Learning, Appropriation and Effects of Adaptiveness (2006)

Penka, A.: Vergleichende Untersuchung zu Fahrerassistenzsystemen mit unterschiedlichen aktiven Bedienelementen. Dissertation Technische Universität München (2001)

Pereira, M., Bruyas, M.-P., Kaufmann, C., Britschgi, V., Gil, J., Zaoral, A.: Reported use of speed control systems: cruise control and speed limiter IET intelligent transport systems (2013)

Peter, J. W.: Kann die Sicherheitsfahrschaltung SIFA die Dienstfähigkeit von Triebfahrzeugfahrern gewährleisten? Inaugural-Dissertation and der Philipps-Universität Marburg/Lahn, Marburg/Lahn (1980)

Popiv, D.: Enhancement of driver anticipation and its implications on efficiency and safety. Dissertationsschrift an der Technischen Universität München (2011)

Popiv, D., Lange, C., Bubb, H.: Driving with and without ACC: evaluated differences. In: 54. Kongress der Gesellschaft für Arbeitswissenschaft: Tagung München, 2008, S. 141–145 (2008)

Rakha, H., Hankey, J., Patterson, A., Van Aerde, M.: Field evaluation of safety impacts of adaptive cruise control. J. Intell. Transp. Syst. **6**(3), 225–259 (2001)

Raste, T.: Fahrdynamikregelung mit Brems- und Lenkeingriff. In: Winner, H., Hakuli, S., Wolf, G. (Hrsg.) Handbuch Fahrerassistenzsysteme – Grundlagen, Komponenten und Systeme für aktive Sicherheit und Komfort, 2. Aufl. ATZ/MTZ – Fachbuch, Wiesbaden (2012). Kap. 26

Rinderknecht, F.M.: Konzeptstudie XchangeE (2014). www.rinspeed.eu

Sacher, H.: Gesamtheitliche Analyse des Bedienverhaltens von Fahrzeugfunktionen in der täglichen Nutzung. Dissertation, Technische Universität München (2009)

Sacher, H., Bubb, H.: Wie nutzen Fahrzeugführer das adaptive cruise control (ACC)? In: TÜV SÜD AUTOMOTIVE GMBH, Lehrstuhl für Fahrzeugtechnik der Technischen Universität München (Hrsg.) 2. Tagung Aktive Sicherheit durch Fahrerassistenz: Tagung München 2006. TÜV-Süd, München (2006)

Schittenhelm, H.: The Vision of Accident free Driving – how efficient are we actually in avoiding or mitigating longitudinal real world accidents. In NHTSA (Hrsg.): Proceedings in the 21st International Technical Conference on the Enhanced Safety of Vehicles conference (ESV) (Paper Number 09-0510) Stuttgart (2009)

Schneid, M.: Entwicklung und Erprobung eines kontaktanalogen Head-Up-Displays im Fahrzeug. Dissertation, Technische Universität München (2009)

Seppelt, D.B., Lees, N.M., Lee, D.J.: Driver distraction and reliance: adaptive cruise control in the context of sensor reliability and algorithm limits. In: Third International Driving Symposium on Human Factors in Driver Assessment, Training and Vehicle Design: Tagung Rockport, USA 2005. S. 255–261 (2005)

Stanton, A.N., Young, M., McCaulder, B.: Drive-by-Wire: the case of driver workload and reclaiming control with adaptive cruise control. Saf. Sci. **27**(2/3), 149–159 (1997)

Strasser, P.: Vernetzung von Test- und Simulationsmethoden für die Entwicklung von Fahrerassistenzsystemen. Dissertation, Technische Universität München (2012)

Tricot, N., Rajaonah, B., Pacaux, M.-P., Popieul, J.-C.: Driver's behaviors and human-machine interactions characterization for the design of an advanced driving assistance system. In: Institute of Electrical and Electronics Engineers, INC (IEEE) (ed.) IEEE International Conference on Systems, Man and Cybernetics: Tagung The Hague, Netherlands, Bd. 4, pp. 3976–3981. IEEE Computer Society Press, Washington D.C. (2004)

van Zanten, A., Kost, F.: Bremsenbasierte Assistenzfunktionen. In: Winner, Hakuli, H.S., Wolf, G. (Hrsg.) Handbuch Fahrerassistenzsysteme - Grundlagen, Komponenten und Systeme für aktive

Sicherheit und Komfort, 2. Aufl. ATZ/MTZ – Fachbuch, Wiesbaden (2012)

Viti, F., Hoogendoorn, P.S., Alkim, P.T., Bootsma, G.: Driving behavior interaction with ACC: results from a field operational test in the Netherlands. In: Institute of Electrical and Electronics Engineers, INC (IEEE) (ed.) IEEE Intelligent Vehicles Symposium: Tagung Istanbul, Turkey 2007, pp. 745–750. IEEE Computer Society Press, Washington D.C. (2007)

Walter, M., Fechner, T., Hellmann, W., Thiel, R.: Lane departure warning. In: Winner, H., Hakuli, S., Wolf, G. (eds.) Handbuch Fahrerassistenzsysteme – Grundlagen, Komponenten und Systeme für aktive Sicherheit und Komfort, 2. Aufl. ATZ/MTZ – Fachbuch, Wiesbaden (2012). Kap. 34

Weinberger, M.: Einfluss von Adaptiven Cruise Control Systemen auf das Fahrerverhalten. Dissertationsschrift an der TU-München (2001)

Weinberger, M., Winner, H., Bubb, H.: Adaptive cruise control field operational test – the learning phase. JSAE Rev. **22**, 487 (2001). Elsevier

Weißgerber, T., Damböck, D., Kienle, M., Bengler, K.: Auswirkungen einer kontaktanalogen Anzeige auf die Querführung hochautomatisierter Fahrzeuge 28. VDI/VW-Gemeinschaftstagung Fahrerassistenz und integrierte Sicherheit, Wolfsburg (2012)

Weißmann, W: Die (un)heimlichen Helfer. Fahrschule, 6/209, 12–14 (2009)

Wimmer, M.: Entwicklung und Erprobung von Mensch-Maschine-Systemen zur automatischen Fahrzeugführung. Dissertationsschrift an der Universität der Bundeswehr München (2014)

Wimmer, M., Siedersberger, K.-H., Meuerle, J., Färber, B.: Evaluierung des Fahrerüberwachungs- und Interaktionskonzeptes Potentialtrigger für hochautomatisierte Fahrerassistenzsysteme. In: Kolrep, H., Jürgensohn, T. (Hrsg.) 28. VDI/VW-Gemeinschaftstagung Fahrerassistenz und integrierte Sicherheit Mensch-Maschine-Systeme der Reihe Forschungs-Berichte VDI, Bd. 22, S. 123–135 (2012)

Winner, H., Hakuli, S.: Conduct-by-wire – Following a new paradigm for driving into the future. In: Proceedings of FISITA World Automotive Congress (2006)

Winner, H., Hakuli, S., Bruder, R., Konigorski, U., Schiele, B.: Conduct-by-Wire – ein neues Paradigma für die Weiterentwicklung der Fahrerassistenz 4. Workshop Fahrerassistenzsysteme, Löwenstein (2006) (2006)

Winner, H., Hakuli, S., Wolf, G. (eds.): Handbuch Fahrerassistenzsysteme - Grundlagen, Komponenten und Systeme für aktive Sicherheit und Komfort, 2. Aufl. ATZ/MTZ – Fachbuch, Wiesbaden (2012a)

Winner, H., Danner, B., Steinle, J.: Adaptive cruise control. In: Winner, H., Hakuli, S., Wolf, G. (eds.) Handbuch Fahrerassistenzsysteme - Grundlagen, Komponenten und Systeme für aktive Sicherheit und Komfort, 2. Aufl. ATZ/MTZ – Fachbuch, Wiesbaden (2012b)

Further Literature

Deutscher Verkehrssicherheitsrat e. V.: Fahrerassistenzsysteme - Innovationen im Dienste der Sicherheit/13ten Schriftenreihe Verkehrssicherheit/12tes DVR Forum - Sicherheit und Mobilität – München 21.09.2006 - S. 36 (2006)

Dutch Ministry of Transport: "Roads to the future: the assisted driver". Report from the Dutch Ministry of Transport, April 2007 (2007)

König, W., Weiß, K.-E., Mayser, C.: S.A.N.T.O.S Situations – angepasste und Nutzer-Typ- zentrierte Optimierung von Systemen zur Fahrerunterstützung. Gemeinsamer Projektabschlussbericht Der Robert Bosch GmbH und der BMW group. BMFT, Bonn (2002)

Lindberg, T., Tönert, L., Rötting, M., Bengler, M.: Integration aktueller und zukünftiger Fahrerassistenzsysteme - wie lässt sich der Lösungsraum für die HMI-Entwicklung strukturieren? Der Mensch im Mittelpunkt technischer Systeme. 8. Berliner Werkstatt - Mensch-Maschine-Systeme. VDI Verlag, 187–188 und 457–463 [CD] (7.–9. Okt.2009)

9

Methods of Ergonomic Vehicle Development

Heiner Bubb

Contents

10.1 Ergonomics in the Vehicle Development Process – 576

10.2 Virtual Reality – 580

10.3 Simulation of Anthropometric Conditions – 581
10.3.1 Seat Box – 581
10.3.2 Variable Ergonomics Test Bench – 582
10.3.3 Application of Virtual Reality – 584

10.4 Simulation of Dynamic Driving and Traffic Aspects – 589
10.4.1 Motivation for Driving Simulators and their Technical Challenge – 589
10.4.2 Simulator Techniques – 591
10.4.3 Relevance of Driving Simulators of Different Levels – 596
10.4.4 Use of Augmented Reality in Real Vehicles – 599

10.5 Experiments in a Real Vehicle – 601
10.5.1 Trials on Closed off Terrain – 601
10.5.2 Experiments on Public Roads – 603
10.5.3 Customer Feedback – 604

References – 605

H. Bubb et al. (eds.), *Automotive Ergonomics*, https://doi.org/10.1007/978-3-658-33941-8_10

10.1 Ergonomics in the Vehicle Development Process

The product development process (PDP) of a new vehicle model is characterized by a high degree of complexity, which is primarily due to the interaction of numerous creative employees both within the automotive manufacturer's organization (OEM, Original Equipment Manufacturer) as well as in the supplier environment and with various service providers (Widmann 2011). In total, more than 1000 participants can be networked interdisciplinarily and locally in a simultaneous engineering process. It is not the task of this chapter to describe the different organisational forms of the PDP, but it is only briefly dealt with its temporal sequence and this with special regard to the use of ergonomics. ◘ Figure 10.1 shows the essential elements of this timeline. In the area before the actual product definition, research, advanced design, innovation management and scouting (trend research) are used to create, formulate, present and, to a large extent, experimentally research new ideas. These processes run continuously and permanently independent of vehicle model cycles. The attractiveness of a vehicle is ultimately determined not only by its targeted positioning in the targeted customer segment with a view to brand image, design, product features and price, but also by innovative technical content (Widmann 2011). Innovations shape the brand image (e.g.: Safety passenger compartment ↔ Mercedes; "quattro©" – permanent four-wheel drive ↔ Audi; double-clutch gearbox ↔ VW; straight six-cylinder engine in the lower mid-size class ↔ BMW). Increasingly, innovations in the field of ergonomics will also acquire such image-enhancing significance (e.g.: dashboard oriented towards the driver, HUD ↔ BMW). This is why many

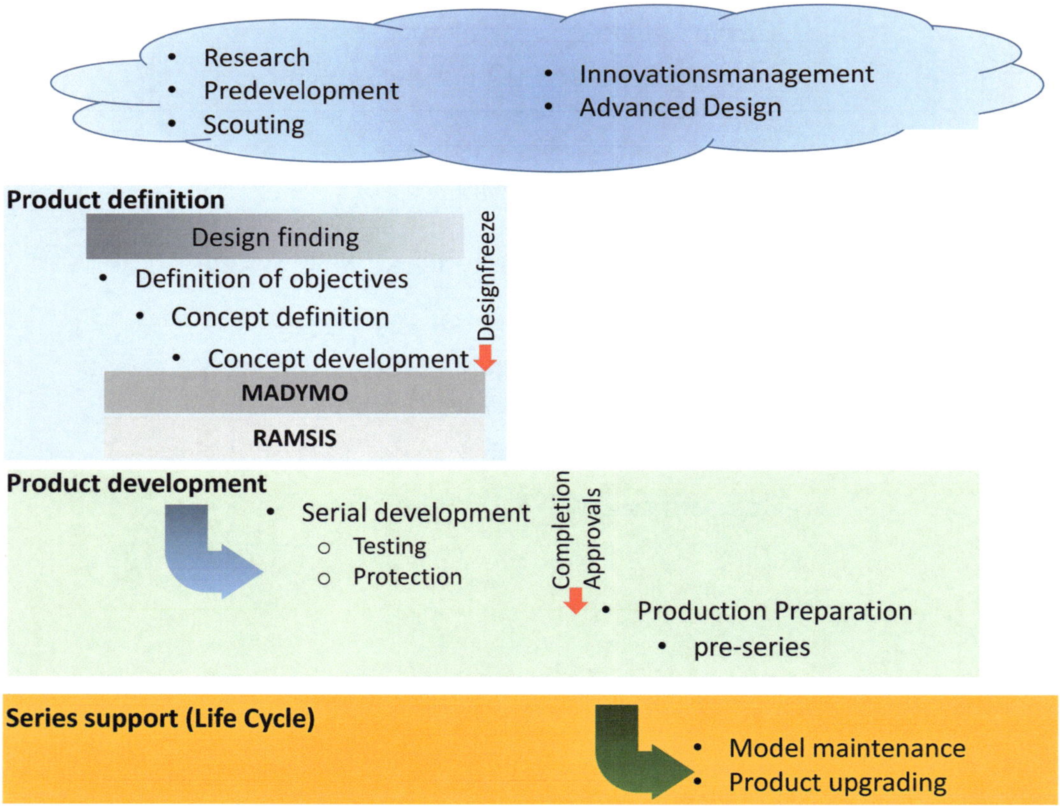

◘ **Fig. 10.1** Product development process: from research to series support (after Widmann 2011 and Wagner 2011)

innovations in the field of ergonomics are emerging today, especially in research (both company research and the involvement of research institutes and universities). In the area of pre-development, the first integration of such innovations into the vehicle environment takes place. The aim is to test and demonstrate the potential of vehicle innovations and their ability to develop series production.

The actual *product planning* begins with an analysis of future customer needs. These analyses are based both on the know-how gained within the organisation itself and on scenario research carried out by relevant institutes. In addition to many questions of future economic development in the global markets, the availability of resources, changes in tax policy etc., the question of changing the age pyramid on buyer behaviour plays a particularly important role today with regard to ergonomics. At the end of the product planning phase, the future vehicle is described with regard to its characteristics, cost blocks are allocated and the planning proof of profitability is provided. These objectives are to be assumed in the further course of the PDP. The available capacities must be used to ensure that they are achieved (see Fig. 10.1).

Within the framework of the *concept development* the vehicle is defined in detail (objectives and concept definition) on the basis of the various requirements defined beforehand. As far as the available technology is concerned, a distinction must be made in terms of ergonomics between the development of anthropometric packaging and the system-ergonomically oriented design of display and operating interaction (see Fig. 10.2).

The concept developer for the anthropometric area is faced with the task of developing a package that allows the realisation of the properties described above. He starts with a so-called technology model, which – independent of the design – represents a virtual complete vehicle. With a view to possible predecessors and competitor models, new concepts are devised or old ones improved. In this phase, the first and, above all, very basic ergonomic requirements can already be introduced and taken into account by means of virtual human models. By using CAD technology, it is possible to constantly parametrically adapt

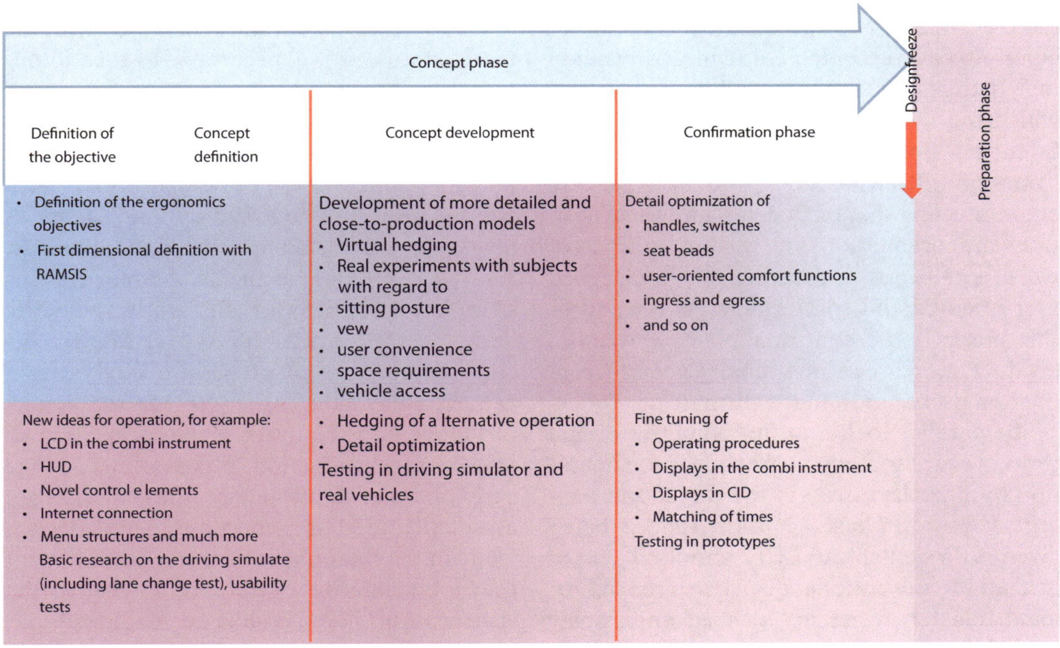

Fig. 10.2 Classification of ergonomics in the concept phase of vehicle development

the data of the still virtual vehicle model to the changing design models. The various methods of virtual reality (VR) also come into play here. As Widmann explains, this technical model serves the designers as a guideline for various model designs. In the next phase of concept development, more detailed and realistic models closer to series production are already available. In addition to other virtual valiation tests, these are also used to carry out real experiments with test persons. Models are made of hard foam for this purpose. They gradually contain all development statuses to be confirmed from the design, package and technical departments. In cooperation with the virtual hedging, accessibility and free spaces are checked, the belt design optimised and the view of the instruments ensured. For virtual validation, it makes sense to define fixed rules that ultimately ensure objectifiable and reproducible results. In contrast, investigations with test persons in the real seat boxes have to expect variance and also deviations from the virtual results. By means of categorised and detailed test plans, these variation can be kept within limits and, in particular, deviations can be made comprehensible (see ► Chap. 11). At the end of this phase, the coordination between these two procedures enables an ergonomically acceptable design and packaging. It thus defines the installation spaces in which the individual detail designs can then be further optimised. In the subsequent confirmation phase, handles and switches are ergonomically shaped in detail and their position and orientation are further optimised, while seat beads and armrests are designed user-oriented and comfortable. At the end of this phase is the confirmation of the entire product, which can now finally go into production preparation in the form of pre-series.

In parallel to the anthropometric design process described above, new ideas for operation and displays arise in the target and concept definition phase. Consideration is being given to whether an LCD should be used instead of conventional displays in the instrument cluster, whether a head-up display should be provided, alternative menu structures are being devised. New ideas are being developed for operating the automatic transmission and in particular for interaction with the driver assistance systems. All these new ideas must first be developed and argued theoretically from the point of view of system ergonomics. For the time being, they will then undergo an initial evaluation in rudimentary form (e.g. image sequences on Power Point level) in partly simple driving simulators (e.g. using the Lane Chance Test; see ► Sect. 10.4; Herrler 2006). In the next phase of concept development, more near-series realizations are now available in the area of displays and operating elements, which are to be verified in simulators in part, but in real vehicles with test persons in part. Analogous to the anthropometric development, fine-tuning of the operating procedures, the displays in the combi instrument and in the central display still needs to be carried out in the confirmation phase, and in particular the timing of the individual procedures needs to be brought to a final state – particularly with regard to the driver assistance systems.

During the entire development described, so-called synchro points or milestones are set, which act as a caesura at which all departments involved up to this point have to present a defined result. They offer the opportunity to compare the actual status achieved with the target status and, if necessary, to take initiatives to subsequently ensure the target status. As a final break the concept development is finished with the so-called designfreeze, which fixes a no longer changeable stage of development. This is now followed by series development, which is terminated by the final release. As Wagner (2011) explains, human models such as RAMSIS for the anthropometric design of the vehicle and MADYMO for the correct development of passive safety are in use throughout the entire phase from the start of the planning contract to the completion of release for production preparation. In the early phase of development, these models are essentially used for conception and design, while in the final phase of series development they are primarily used to check whether the specifications have actually been adhered to.

The consideration of the human being in the development of a highly complex technical system such as the motor vehicle represents a special challenge, because the "soft" system element "human being" is confronted with the "hard" technology. As a result, unfavourable technical conditions can be compensated by the flexibility of the user, which makes the fulfilment of ergonomic requirements seem less urgent.[1] Nevertheless, unfavourable anthropometric conditions, for example, lead to discomfort and in the long run even to pain. Particularly in connection with the handling of modern assistance systems, which can only be of use if they are also used, the question of the adaptation of the technology to the individual and day form dependent different inclinations and abilities plays an essential role. In order to take all these aspects into account as far as possible during development, typical development scenarios have emerged that are generally valid both for purely technical development and for human involvement. For the following categorization, a distinction is made on the basis of the paradigm of the control loop:

- *Software in the Loop (SIL):* The test object (e.g. the package of the vehicle, an assistance system) is simulated on a development computer. All other components of the control loop are also implemented in a simulated version, i.e. also the driver (in the case of the package selected as an example, anthropometric human models can be used, in the case of the assistance system a modelling of the human control behaviour is necessary; see ▶ Chap. 5). There is generally no requirement for real-time, so the test sequence can be interrupted at any time. SIL can be used at a very early stage of development, but requires considerable expert knowledge depending on the depth of the simulation.
- *Hardware in the Loop (HIL):* The test object with all its input and output interfaces is real, all other components of the control loop are simulated. This variant plays a subordinate role for the treatment of ergonomic questions. Various procedures have been developed to check the technical characteristics of the vehicle, by means of which the load caused by human influences is simulated in a reproducible manner (e.g. loading of the seat when getting in and out by a robot with a special movement program; "clutch robot" to check the stability of the clutch including clutch pedal with very frequent repetition of operation etc.). Still relatively rare is the version in which the driver is completely replaced by a control robot, which enables better reproducible test conditions for the test object.
- *Driver in the Loop (DIL):* The driver is real, the other components can be fully simulated or partially real and partially simulated. In the case of the examination of anthropometric questions, the use of so-called seat boxes and the techniques of virtual reality (VR; CAVE© or HMD; see ▶ Sect. 10.3) are used here. For questions of driver-vehicle interaction, the driving simulator is the means of choice. Especially the test object (e.g. a special variant of an assistance system) can be a real, a simulated or a rapid prototyping control unit. The main objectives of the DIL are ergonomic issues and the early exploration of customer acceptance.
- *Vehicle in the Loop (VIL):* Driver and vehicle are real, the environment, especially other road users, are simulated. A particularly elaborate variant of this test version is the use of Augmented Reality (AR), where the driving task, consisting of the driving route and other road users, is simulated and communicated to the driver using head-mounted technology, whose vehicle is actually moving on a closed-off test site (see ▶ Sect. 10.4.4). With this variant, the test object itself can be a real, simulated or rapid prototyping control unit, as with DIL.
- *Realtest (RT):* All components are real, whereby the test object itself can also be tested in the final installed version or in the form of a rapid prototyping control unit. Tests of this type may be carried out on a closed-off test site or, subject to special precautions, on public roads (see ▶ Sect. 10.5).

1 It should be noted in this context that the first vehicles did not even have an adjustable driver's seat. People of different anthropometric dimensions also adapt to the same chairs in the living area.

10.2 Virtual Reality

If a person is shown something in an experimental situation that is connected with the application given later in reality, but which is visibly different for him/her from the situations actually experienced in everyday life, the question must always be asked to what extent such a judgment is relevant to the application. When developing a new vehicle, you always had to test parts in the lab before you could do a test in the practical real environment. This statement also applies to the question of how the driver and thus the future customer perceives the vehicle. Hardware models are used to convey the exterior, which the design department first creates on a reduced scale and later in a 1: 1 representation of the outer skin. A proven means of exploring the spatial impression and possibilities of exploiting the space of a new vehicle is the so-called seat box, which, however, represents the interior in a unmoved state.

Due to the rapid progress of computer technology, the representation of virtual worlds has become possible with such precision that many decisions, which used to have to be made on the expensive models, can now be made at an early stage on the computer screen – of course with the aid of suitable presentation techniques (e. g. "powerwall"[2]).

The possibility of making virtual worlds generated in the computer available to the user in a vivid way has made the question of how such virtual worlds are experienced an object of diverse scientific research. Most of the results of this research can also be applied to the experience of hardware models of the future vehicle. For this reason, the treatment of the individual development techniques is preceded by the results of this scientific analysis of virtual worlds.

Riedl (2012) writes: "The term ***virtual reality*** is often used in the technical literature abbreviated ***VR***. This combination of two words with actually opposing meanings is not uncontroversial, but can generally be equated with a 'reality simulated by the computer' (Duden 2009)". Schrader (2003), Stäbler (2007) and other authors state that "a wide variety of terms such as Virtual Environment, Virtual Presence, Artifcial Reality or Cyberspace can be found in the literature". VR should be understood in the present context as a generic term for the entirety of virtual technologies. "These technologies usually allow the user to both view and manipulate virtual objects, but go beyond the possibilities of a normal screen workstation with three-dimensional graphic representation from the CAX range, for example through representation on stereoscopic displays or projection environments" (Riedl 2012). The main objective of virtual reality is to integrate the user into the computer-based three-dimensional world. However, this integration is not limited to visual information, as is often assumed, but should cover as many of the human sensory channels as possible. The more sensory channels are addressed by a VR system, the more comprehensively the user feels integrated into the virtual world, i.e. the greater the degree of so-called immersion (Hofmann 2002). The term ***immersion*** (lat. immergere = immerse, penetrate) describes the degree of immersion of a user in a virtual environment (Bartle 2003).

If the technology is suitably designed, the user may have the impression that he is part of the virtual world. This phenomenon is called ***presence***. It represents a cognitive state, the degree of which depends not only on the technical conditions but also on the individual characteristics of the user (i.e. his willingness to engage in the virtual or experimental situation). According to the model that Hofmann (2002) compiled by carefully weighing the models available in the literature to describe the perception of VR, presence comes about through the so-called immersion factors and the reality factors (◘ Fig. 10.3). Immersion factors include all influences that favour immersion in the virtual world, such as correct alteration of the depicted images depending on one's own position, additional acoustic, haptic and, if necessary, kinaesthetic and

2 The powerwall is a large-area projection device (generally with rear projection) that allows a virtual vehicle to be displayed in its original size. In most cases, a stereo projection is used to convey the spatial effect.

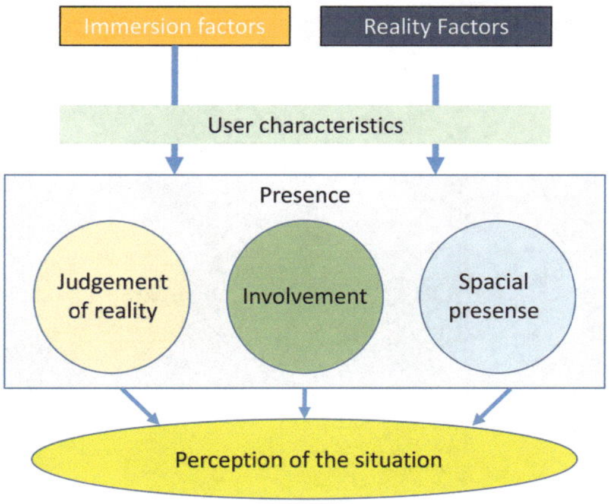

Fig. 10.3 Presence model modified according to Hofmann (2002)

olfactory stimuli. The reality factors describe the correctness of the stimuli, e.g. the detailed representation of surfaces (e.g. grained leather in the vehicle interior, textures on the road and the houses in the driving environment, leaves on the trees depicted, realistic engine noise etc.).

According to the proposals of Schubert et al. (1999) and Regenbrecht (1999), presence is then understood as a multidimensional cognitive state. This is composed of the "judgment of reality" (impression of the similarity of the virtual environment with a comparable real environment), the "involvement" (selective attention to the virtual environment) and the "spatial presence" (the user has the impression of being physically not in his real environment, but in the computer-generated virtual environment). Presence is not necessarily achieved by perfectly implementing all three components. For example, the willingness to get involved in the world of a video game (involvement) can lead to a pronounced sense of presence even if the judgement of reality is moderate and the spatial presence is practically non-existent. Applied to ergonomic vehicle development, this means, for example, that a test person can assess the spatial impression in a seat box (see above) even if the environment is a laboratory (distorted spatial presence) and the model is immobile (partial lack of reality judgement). However, a well-developed sense of presence is – however it comes about – the prerequisite for the situation to be assessed to be perceived correctly.

10.3 Simulation of Anthropometric Conditions

Although human models (so-called "soft dummies"; the RAMSIS human model plays an important role in the automotive sector today, see ▶ Sects. 5.2.2 and 7.2.1) are available in the computer, through which many questions concerning the anthropometric design of the driver's seat and the other passengers can already be answered in CAD during the conception phase – on behalf of possible test subjects, so to speak – it will not be possible to dispense with individual judgements, which are asked of selected test subjects in so-called seat boxes or mock-ups during vehicle development. Many of the hardware-based mock-up's today are connected with the various possibilities of VR technology and thus extend this type of verification to a considerable extent.

10.3.1 Seat Box

The creation of seat boxes for the evaluation of the interior conditions of a vehicle to be designed has a long tradition in the vehicle industry. Usually the basic structure of the seat boxes is made of wood (hence the name "..box"). By means of the special hard foam material widely used in design departments today, the aesthetic impression of the interior and exterior can be assessed even before a ready-to-run model is available for evaluation. In particular, aspects that cannot be assessed with purely geometrically oriented human models or can only be assessed with a great deal of intuition, such as the aesthetic impression of the spatial positioning of instruments and switches, the impression of the inclination of the windscreen in conjunction with the A-pillars, the door trim, the design of the armrests, the positioning and design of door openers, the seating feeling, which must be different for a sports car than, for example, for

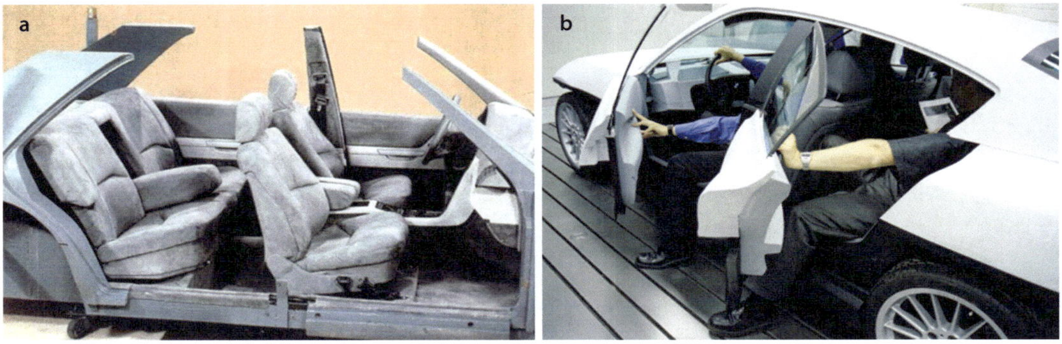

Fig. 10.4 Seat box (**a** Schrader 2003), **b** BMW

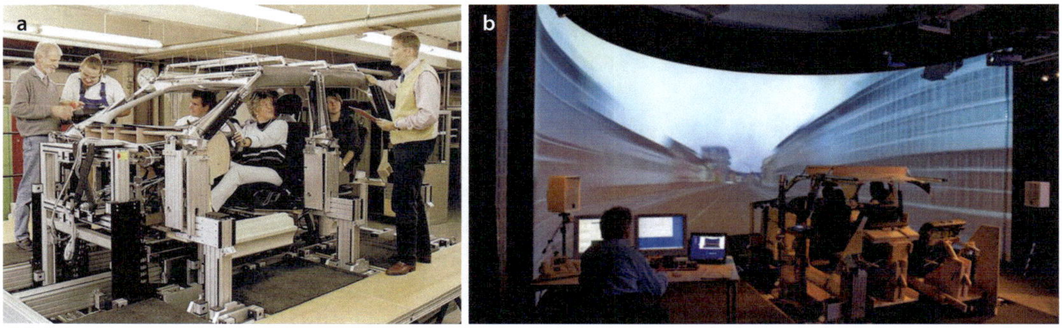

Fig. 10.5 **a** Electric motor variable mock-up (Daimler), **b** in conjunction with a driving simulator

a VAN, the overall impression of space and much more can be assessed with it. However, the construction of such seat boxes is costly. Once a seat box has been created, it can only ever be used to assess the dimensional concept for which it has been created. Changes, especially to the geometric dimensions, are only possible to a minor extent (Fig. 10.4).

Such seat boxes or even partial bodies of vehicles already in production are often used for many studies on driver posture. In particular, the position of the eye points is of interest in addition to other aspects such as the angular position of the extremities, pressure distribution in the seat etc. In order to achieve reasonably valid results under these conditions, which generate a feeling of presence only to a small extent, it is advisable to set up or project a picture – preferably of a straight street – in front of the seat box. It is important that the viewing angles are correct and that the horizon of the image is at eye level. This at least creates a certain spatial presence.

10.3.2 Variable Ergonomics Test Bench

Parallel to the use of CAD in design, variable mock-ups are also used, which can be adjusted by an electric motor and which are bidirectionally connected to the CAD system. The underlying idea is that the dimensions available in CAD can be transferred directly to the mock-up and that, conversely, changes to the mock-up that are requested by the test persons during an experiment can be registered in CAD and, if necessary, used for modified drawing templates. A variable driver's seat model built by Daimler AG serves as an example here: It is a movable base frame on which the stylized body of a motor vehicle with A, B and C pillars is located (see Fig. 10.5). The columns consist of variable square elements with internal telescopic drive. They carry a roof, the extension of which is also variable due to a roller blind construction. The position of the steering wheel, pedals, seats and floor can be adjusted by electric motors. A total of 80 such engines ensure that almost all the dimensions that determine the interior of a motor vehicle can be modified. The variability ranges from the flat seating position in a sports car to the configuration in a normal passenger car to very upright positions in VANs. The variable ergonomics test bench can be equipped with two front seats and two rear seats or a bench seat. Depending on the respective problem, the test bench can also be equipped with examination-specific attachments such as doors or column claddings. It can be connected to a driving simulator. Displacement sensors on the adjusting elements provide feedback on the respective setting status. Thus it is possible to establish the connection to the CAD system via a control computer in the form of a bidirectional interface. This means that any body concept developed in the CAD system can be transferred directly to the variable ergonomics test bench and vice versa. Thus all essential package dimensions, which are set on the ergonomics test bench, are directly available in the CAD system as key data. Since the ergonomics test bench is intended not only to present different vehicle concepts but also to enable the type-independent determination of the effects of a body change and comparison with competitor concepts, the following operating modes form the basis of the technical layout described on the operating side:

1. *Interactive mode:* Here, a specific body concept is developed by the test manager via a control panel in consultation with the test person. It is also possible to switch one of the dimensions to be varied to the horn button on the steering wheel and have it adjusted directly by the test person.
2. *Automatic mode:* either a CAD dataset or a dataset saved (e.g. from a previous test) is loaded here.

For the variable ergonomics test bench, Braun (1997) specifies the following possible applications:

- Creation of concept parameters (directly or by CAD data),
- Optimization of concept parameters with the help of test persons,
- Concept-specific studies (e.g. predecessor studies – follow-up studies, study of critical key data with test persons)
- Ergonomic basic investigations; for this purpose, the companies compile a pool of test subjects whose members are defined by the measurement of approx. 40 characteristic body measurements. This makes it possible to compile groups of test subjects according to anthropometric criteria that may be relevant to a particular question.

The technical design of such variable mock-ups, however, cannot in principle offer the aesthetic standard of the seat boxes discussed above. It is therefore necessary to answer the question of whether the variable mock-up, which is rather abstractly sober from a subjective point of view, reflects a similar sense of space as corresponding real vehicles. Braun (1997) therefore compared such a variable driver position model with corresponding real reference models (three conceptually different vehicle models: Mercedes S-Class, off-road vehicle, roadster), whereby the test persons answered certain questions randomly in the real and in the virtual model, without knowing that exactly the real vehicles were realized in the test bench. The result of this investigation is that in the virtual vehicle model approx. 88% of the questions were answered correctly, while in the real vehicle model this percentage was 97%. No differences were found in the locatability of actuators, while clear differences were found in the assessments, which also require haptic feedback (accessibility and operability). Overall, however, the test subjects in the virtual model needed significantly more time for the assessments. The vehicle models used had no influence on the judgement quality.

Variable mock-ups, such as one described here as an example, are used today in almost all development departments of the automotive industry to carry out basic ergonomics tests with the help of test persons.

10.3.3 Application of Virtual Reality

With the visualization possibilities offered by modern computer technology, it has become increasingly attractive to allow test persons to evaluate the future interiors of motor vehicles in a virtual world. The technology for optical presentation is the most advanced. In order to achieve the best possible immersion, acoustic, kinesthetic, haptic and olfactory output systems should also be used. With the exception of the simulation of acoustic stimuli, however, technological development is much less advanced here. The lack of haptic stimuli in particular is a major obstacle to the assessment of vehicle interiors. One helps oneself partly with an optical substitution, in which one marks the touching of surfaces optically, e.g. by red coloring of the surface. However, this measure can only be regarded as a largely insufficient substitution of the correct haptic stimulus (see below). There are also attempts to provide the test person with special gloves and to signal the touching of surfaces by an electrical stimulus. However, even in this way the complex haptic sensation created when a switch is operated cannot be achieved.

Two different technologies have essentially established themselves for the optical presentation: In the CAVE™ the computer-generated video images are projected onto five or even six surfaces surrounding the user by means of rear projection. With the Head Mounted Display (HMD) method, the test person sees the virtual world through eyepieces in front of their eyes, behind which there are two displays that are firmly connected to the helmet. Both technological variants have in common that the head position is measured (so-called tracking) and a correspondingly corrected image is shown to the user. Depending on the performance of the computer that generates the images, there may be noticeable delays between the new head position and the trailing visual impression, leading to unpleasant kinetosis in sensitive individuals (see also ▶ Sect. 3.1.3 and ◘ Fig. 3.11).

10.3.3.1 CAVE™

◘ Figure 10.6 shows the functional principle of the CAVE™. In order to achieve a spatial stereo impression for the viewer, two projectors are used per projection surface in one variant, which produce the images for the left and right eye in different polarization directions. In this case, the user must wear polarization glasses in order to see the correct image with each eye. The other variant uses only one projector per screen, but for each screen the perspective for the left and right eye is displayed alternately. In this case, the user must wear so-called shutter glasses, which are synchronized with the projection fields, so that the correct image is fed alternately to the left and right eye. With a 3D pointer he is able to mark certain points or even to move himself in virtual space, i.e. correctly said, to move the space around him as he would appear to him with his own movement. The latter application plays a rather subordinate role in connection with the assessment of vehicle interiors, since in the CAVE™ a real seat with a steering wheel is usually fitted, on which the test person takes place. An advantage of the CAVE™- principle in contrast to the HMD procedure described below, is that the test person can see his or her own body unaltered.

An essential question in the application of CAVE™ is, as with the variable ergonomics test bench, the extent to which the sense of space perceived in the virtual world corresponds to that of reality. Hofmann (2002) has particularly dealt with the question of the extent to which the distances perceived in this test rig are correct. He assumes that the human being in the real world uses their stimuli to derive actions in the space thus opened up on the basis of the inner models formed from them. This interaction can take place on a physiological and cognitive level. The sources of information for egocentric absolute distance perception are binocular disparity, convergence, and accommodation; for exocentric depth perception, motion parallax; and information such as relative brightness, masking, relative size, relative optical density,

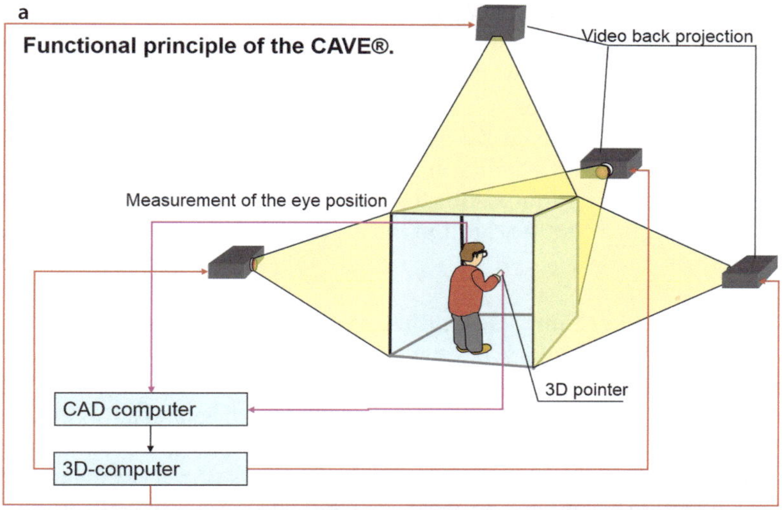

Fig. 10.6 Function principle of the CAVE© **a** and view from the point of view of an outside observer **b**

visual field height, linear perspective, atmospheric perspective, and kinetic depth. All these sources of information ultimately lead the observer to distance information that is free of contradictions. It can be assumed that this mechanism of action is also valid in the virtual environment. However, there the own rules of virtual space can lead to other perceptions of space. Due to the technology used, there are specific characteristics:

- In projection-based VR systems, virtual objects appear slightly translucent.
- In most VR systems, the perspective on the virtual environment is limited.
- The resolution of the representation is technically limited, which reduces the quality of stimuli for distance and depth perception.
- The representation of the virtual environment is often not adapted quickly enough to the user's head movements.
- Inaccuracies in the measurement of the user's head position can lead to subjectively distorted images.
- Accommodation and convergence information do not coincide, because to see sharply the user has to accommodate at the distance of the projection screens.

- When stereoscopy is performed using the shutter glasses system, afterimages may result in insufficient decoupling of the images for both eyes.
- With the polarization filter procedure, which is comfortable in terms of wear characteristics, double images can occur when the head rotates around the horizontal axis.

Hofmann (2002) found during his investigations: The image build-up rate was identified as an essential immersion factor. It causes a temporary distortion of the representation, which leads to the virtual objects being perceived enlarged up to 4%. This has a considerable effect on practical application when it comes to assessing interior concepts using VR for example. The following mechanism is assumed to explain this: The movement of the head causes temporary shifts of the virtual object at a low image build-up rate, which the user interprets as an apparent enlargement of the object. This also affects the variance of the measured values: It decreases with the systematic influence at a low image build-up rate, while it increases at a high build-up rate according to the natural variance of the movement.

For the effect of closeness to reality on spatial perception, the following mechanism is assumed on the basis of the experimental results: in a natural environment, the luminance of an object illuminated by extraneous light decreases with the fourth power of distance. The luminance is consistently higher in the case of the representation with low closeness to reality because of the missing textures. The surface must therefore appear closer to the user on the basis of his natural experience.

The following results can be summarised from Hofmann's investigations:

- All three presence components have a significant influence on spatial perception. With increasing involvement and spatial presence, the cockpits are perceived larger, with increasing judgement of reality smaller.
- The perception of the cockpit width (horizontal scaling) is not influenced by any of the presence components.

An essential effect is the conflict between the accommodation on the projection screens at a constant distance and the different distances mediated by the optical stimuli. When entering the virtual environment (CAVE™), the accommodative information is suppressed due to the unfamiliarity of the conflict situation. With increasing spatial presence and involvement, however, their weight increases and so does the size of the perceived distance. With the same objective image size on the retina, the corresponding object is therefore perceived larger. Although the perception mechanism assigns less weight to accommodation information in this conflict situation, since convergence, absolute motion parallax and vertical disparity in themselves provide consistent information in the VR environment, the perception mechanism does not assign as much weight to accommodation information in this conflict situation. The accommodation influence itself, however, is amplified by the accommodative convergence, the calibration of the horizontal disparity, by the importance of the depth of field, which is emphasized due to the low illuminance, and by accommodative stimuli of real objects in the CAVE™.

The conflict effects described play an important role especially for the first user of a VR. If used frequently, it can be assumed that the more the user accepts the virtual environment as his current, direct environment, the stronger the habituation effect. But this is precisely what makes up the spatial presence and involvement. A high judgement of reality has the opposite effect to the opposite, that under this influence the rules of perception, which are accustomed from everyday life, gain more weight. From this model conception, measures can be derived to prevent an undesirable misjudgement of the size in the VR environment. Since this is essentially due to the movement of the user, the following compensation options are available

- Restriction of the mobility of the user (mostly unsuitable)
- Maximization of the image build-up rate (may run into technical limitations if high detail accuracy is required)

- Control of the image build-up rate and prognosis of the resulting effect (calibration experiments are necessary, but inaccurate because the individual reaction is hardly predictable).

10.3.3.2 HMD Technology

With Head Mounted Display (HMD) technology, the test person wears a helmet that conveys the virtual world to both eyes separately via two eyepieces. This ensures an immediate stereoscopic view. ◘ Figure 10.7 shows the basic arrangement. A tracking system not only records the head position of the test person, which is necessary for the correct calculation of the perspective of the virtual world represented in the HMD, but also the position of the hands using data gloves. This is needed to control a human model (here RAMSIS) so that the subject can see the position of his own hands and arms in the HMD. The view through the eyepieces implies several limitations. As with the CAVE™ there is a perceptual conflict between the level of accommodation and the distance perceived via the other stimuli (it is therefore advisable to adjust the eyepieces so that they are accommodated at the mean distance of the objects in the vehicle interior, i.e. approx. 0.8–1.2 m).

◘ Figure 10.8 shows a variable ergonomics test bench which – similar to the one described under ► Sect. 10.3.2 – is bidirectionally connected to the CAD system. This makes it possible to make immediate changes in dialogue with the test persons. In the lower part of the picture the data helmet is shown. Until now, the heavy helmets have caused major problems for the application (Voss 2008; Riedl 2012). However, it can be assumed that with the increasing use of such helmets/data glasses in the consumer sector, they will become more comfortable, with a larger field of vision, better resolution and also available at an acceptable price. However, certain problems remain independent of this development.

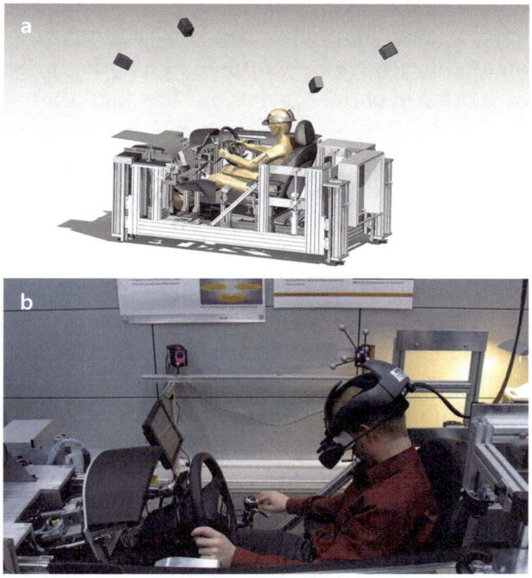

◘ **Fig. 10.8** Variable ergonomics test bench **a** and HMD in the trainer **b** from Riedl 2012)

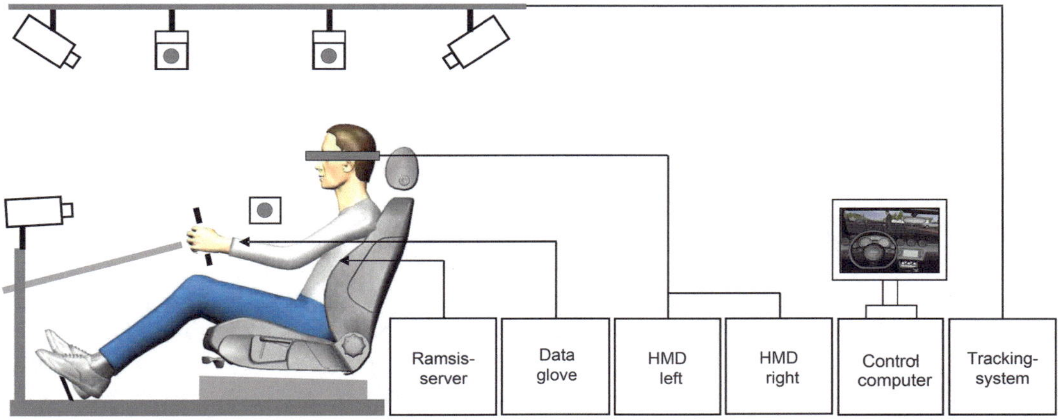

◘ **Fig. 10.7** Principle arrangement of an ergonomics test bench with HMD technology (Voss 2008)

As Riedl (2012) states in extensive comparisons, the correct alignment of the virtual horizon on the real horizon is indispensable, especially for the realistic positioning of the test person on the ergonomics test bench.

Voss (2008) has dealt with the necessary accuracy of the correspondence between virtual, optical and real, proprioceptive perception. According to the available results, a 5-percentile perception threshold of 2° for rotatory divergence and one of 6 mm for translatory divergence can be observed. Since this value is generally above the accuracy of the hand and scale measurement, a sufficient haptic sense of orientation in a VR environment can always be assumed. The advantage of HMD technology lies in the fact that a rudimentary haptic feedback can be simulated by, for example, preventing the penetration of the hand into the virtual dashboard shown at least roughly by a real board. However, as Voss also shows, this no longer applies to complex movements. A comparison was made using the example of a current vehicle model, which in the real case was a 1:1 mock-up made of wood or model construction foams ("seat box"). The experiment was based on visual and gripping examinations. In the virtual test situation, a collision detection system was integrated, which gave the test subject optical feedback when virtual components were touched. With regard to the visual examinations, there was no difference between the real mock-up and the virtual representation. This was also confirmed by Riedl (2012) in a later study independently of this. During the gripping tests, significant differences were found in the accessibility of the inner door actuation, the usability of the armrest and the accessibility of the glove box opener. Riedl also examined this again using the example of removing a cup from a cup holder. As a result, it can be stated that the virtual model presents the evaluation situation more critically than reality. People who have experience in the virtual model come to much more realistic assessments. Otherwise no difference between VR and reality could be detected at all experiments. The observed differences can therefore always be explained with the respective complex haptic feedback in reality vis-à-vis VR.

Riedl (2012) has developed a completely new test variant that can only be realized with HMD technology, the so-called *percentile simulation*. The user is put into the perspective of any body percentile in order to experience the vehicle interior from this perspective. The basic idea here is to take the individual body size of the assessor as the starting size and then to change the vehicle dimensions of the variable vehicle model in such a way that the assessor experiences the vehicle interior from the perspective of another body percentile (preferably 5th or 95th percentile) with regard to the visual conditions, the movement ranges and the feedback via the correspondingly positioned controls. The basis for this transfer process is the measurement of the current eye point and the position of the body extremities (hands), which are transferred in real time to a digital human model (RAMSIS), information which in turn is used to control the variable vehicle mock-up accordingly. In order to validate this approach, 25 test subjects should experience the situation from the perspective of the 5-percentile woman, describe the problems and then give a subjective assessment of the degree of support and immersion perception of the simulation. The evaluations of individual tasks were compared with the discomfort values that can be obtained with the RAMSIS digital human model. The assessments for most postures then agree well. There are major differences in the postures which, according to the test persons' experience, only have to be adopted for a short time (light switches, door openers) or which are only adopted when the vehicle is stationary (glove box openers). All in all, the test subjects attest the percentile simulation a high benefit for ergonomics studies.

In principle, both the CAVE™- as well as with the HMD method, the only statically simulated external view can be combined with a driving simulator (see also ◘ Fig. 10.5b). So far the limited computer technology made such applications practically impossible. With the progress of computer technology, however, a further field of application is opening up.

10.4 Simulation of Dynamic Driving and Traffic Aspects

10.4.1 Motivation for Driving Simulators and their Technical Challenge

The ever faster development cycles also make it appear profitable in chassis development to carry out more and more investigations at a stage when the future vehicle is only virtually available. In addition, the ever more complex traffic situation and the simultaneously increasing demands on safety make it necessary to conduct investigations with the person interacting with the vehicle. These investigations are often necessary at a time when the new vehicle is not even available as hardware. In particular, when it comes to the evaluation of innovative ideas for driver assistance systems, especially in the initial phase of development, which should increase road safety but also improve driving comfort, tests with real vehicles are almost impossible. For ethical reasons, such tests in particular cannot be carried out in a real vehicle in order to exclude any danger to the test subjects. These are all reasons to carry out experiments in simulators. With the increasingly improved possibilities of computer technology, such simulators have moved more and more into the realm of possibilities for many development areas. However, there is often the question of how much effort is actually needed to clarify certain questions.

According to ▶ Sect. 10.2, the immersion factors and the reality factors must be as high as possible in order to convey an intensive presence experience to the test person in the simulator, so that he or she shows behaviour that is reflected in reality. Essentially, the immersion factors are achieved by the correct temporal and spatial interlocking of the stimuli acting on the sensory organs and the reality factors by the best possible reproduction quality of the individual stimuli. As already shown in ▶ Chap. 3 (◘ Fig. 3.1), the four sensory channels of the visual, auditory, vestibular and skin senses are decisive for the reception of information when driving a vehicle. Today, computer-generated video technology is able to operate the visual senses with impressive scenes. In principle, very good acoustic simulation is also possible due to the progressive HiFi technology. By the use of original cabins including the control elements installed there also the skin sense (haptics) can be supplied with very realistic information. A fundamental problem is the operation of the sense of balance and, consequently, of the part of the skin that receives acceleration signals when touching the vehicle (seat, steering wheel, armrests, etc.). This acceleration is often achieved by tilting the simulator cabin (see ◘ Fig. 10.9). For example, in order to generate an acceleration of 0.2 g (comfort acceleration often occurring in road traffic), it is necessary to tilt by $\alpha = 13°$ (the reduction in gravity caused by this is practically not perceived), whereby the entire visual simulation must also be rotated at the same time. However, the tilting process itself is perceived as such via the vestibular organ. One tries to prevent this impression by a so-called washout strategy. To do this, the test subject is initially accelerated translatorily, but this acceleration is slowly reduced and replaced – unnoticed by the test subject – by the tilted position. Assuming the perception threshold for rotational acceleration of $b_{rot} < 0.41°/s^2$ this process takes approx. 8 s. The simultaneously reduced translation acceleration needs a way from 171 m for this time! This way should be provided by correspondingly dimensioned X-Y-sled (for translation acceleration in longitudinal and lateral direction). However, this dimensioning would only be sufficient for moderate comfort accelerations. After initial euphoria about the possibilities of simulation technology, the early optimization of running gears with the help of simulators, even with the involvement of test drivers, is therefore practically impossible from today's point of view.

Because of these restrictive considerations, the question of the quality of simulation results must therefore be asked at all. In the literature, a distinction is made in this context between the terms absolute and relative validity (Negele 2007; Abendroth et al. 2012). *Absolute validity* is characterised by the quali-

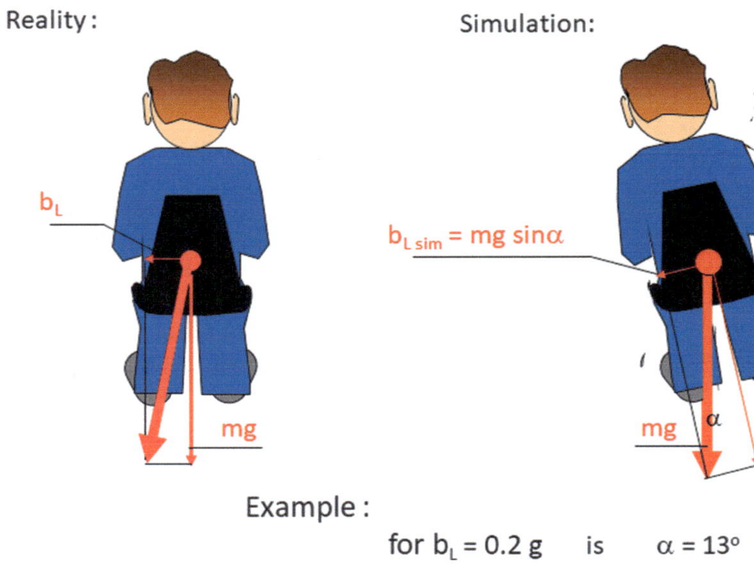

Fig. 10.9 Necessary tilting of the simulator cabin for creating the illusion of a lateral or longitudinal acceleration of 0.2 g

tative and quantitative correspondence of the parameters used to describe driver behaviour between simulator tests and corresponding real tests.[3] It requires at least the proof that a missing or insufficient stimulus (see above) has no influence on the result. Negele (2007) could not find any absolute validity in the sources he considered (Blana and Golias 1999; Buld and Krüger 2001) for fulfilling the primary driving task, i.e. speed selection and tracking. Only after a certain period of acclimatisation will the speed (in the initial phase, at least experienced drivers tend to drive too fast) be kept in a range that roughly corresponds to the real ride (see also the tests by Abendroth et al. 2012). Even according to the author's experience, test persons in so-called static driving simulators (without simulation of motion stimuli) tend to call up full power during acceleration and also to extreme braking manoeuvres. The tracking quality in such a simulator is usually much worse than in reality. Due to the lower optical image sharpness (see below) larger distances are kept in the simulator than in reality.

Negele (2007) states that relative validity[4] on the other hand does not necessarily require that the physical simulation parameters are in complete agreement with reality. Whether the relative validity is sufficient for the application depends on whether the test person recognizes and accepts the simulator run as the control task "driving a vehicle". Decisive for this is the development of a sufficient sense of presence, i.e. the experience of being in the simulated world and of moving in it. The following simple observations illustrate that this effect can be easily achieved: even if no rear-view mirror information is provided in a static driving simulator, it can be observed that the rear-view mirror is also looked into when changing lanes on the motorway and the shoulder gaze

3 In general, this proof cannot be provided simply because it is very time-consuming and only possible at all with suitable software to reproduce the same typological situation of reality in the simulator and it is practically impossible to reproduce the constantly changing traffic situation of reality faithfully in the simulator.

4 Relative validity exists if the characteristic values to be compared in the simulator show the same tendency as in the field test, but differ from each other in their absolute magnitude (Blana, 1996). According to the author's experience, it practically never happens that a tendency found in a simulator is reversed in reality.

is also performed. When the vehicle is stopped, it is often observed in a simulator with an original vehicle cab that the handbrake is pulled and that the test person is protected against the – non-existent – rear traffic when getting out of the simulator vehicle. All this shows that with sufficiently good immersion behaviour patterns are stimulated and recalled which have been acquired in real traffic. In this respect, questions which require the use of such internal models can be dealt with in the simulator and their results can be transferred with great caution to the behaviour in real driving situations. Driving simulators are therefore particularly suitable for the evaluation of secondary or tertiary tasks and for comparative studies.

10.4.2 Simulator Techniques

With regard to simulator technology, the hardware and software requirements for a driving simulator are briefly presented and described below (Negele 2007 contains a detailed discussion of the various techniques for driving simulators).

Since 80% of all information important for the driving process is transmitted via the optical channel, visual simulation is of the greatest importance. The distance of the image medium from the eye is decisive for the geometric design of the simulator. In order to achieve practical accommodation of the eye, a distance of at least 2.5–3 m is recommended. The change of view from the interior (e. g. combi instrument) to the driving scene is thus possible without effects on the immersion. For the horizontal and vertical visibility range, masking is a decisive influencing factor for the sense of presence. A 360° field of vision can often be dispensed with due to the usually concealed rear view (vehicle itself, view over rearview mirror) and in applications that do not require a high degree of presence perception and complete viewing angle. However, the horizontal width of the field of view plays a decisive role for the behavior in the simulator. Kapstein et al. (1996) found in a dynamic driving simulator, for example, that a braking process is more controlled from a visual field of 120° than from a significantly lower visual field. Also Jamson (2000) observed that up to this value of 120° the speed estimation becomes better, but there are hardly any improvements afterwards. The focus for the design is therefore clearly on the front view. For the design, the horizontal field of view is to be assumed, then the vertical field of view and the distance of the image medium from the eye are to be determined and thus the image size is to be defined (▪ Fig. 10.10).

However, the Head-Mounted Display (▶ Sect. 10.3.3.2), which is well usable for geometric ergonomics investigations, is currently not suitable for investigations in the field of driver-vehicle communication. The head tracking with the tracking of the simulation regarding the eye position of the test person would theoretically be a gain, but so far no development simulators are known for it, which is probably due to the sluggish recalculation of the current head-related position, which additionally leads to kinetose effects. For a multi-channel field of view, it is necessary to edit the overlapping areas (edge blending). Due to visual acuity (1 arc minute $= 1' \approx 0.01°$) of the human being and the distance to the image medium, the image resolution can be calculated (see also ▪ Fig. 10.10), which has a considerable effect on the costs. The resolution also depends on the requirement of readability of objects (traffic signs etc.). The upper limit for very good immersion is 2–3 arc minutes/pixel. A distinction is made between front and rear projection, whereby the latter normally only makes sense with the fixed screen concept for reasons of available space. For the image calculation, on the other hand, the delay- and jerk-free display is in the foreground, which is more important for a good immersion than the resolution. A distinction must be made between monitor and projection with screen for image display. While it is in principle possible with the projection principle to create a sufficiently large field of view for the driver in the simulator by using several projectors (usually three projectors are sufficient for the frontal area), this is also difficult with the large monitors available today. In particular, there is currently no method that makes the

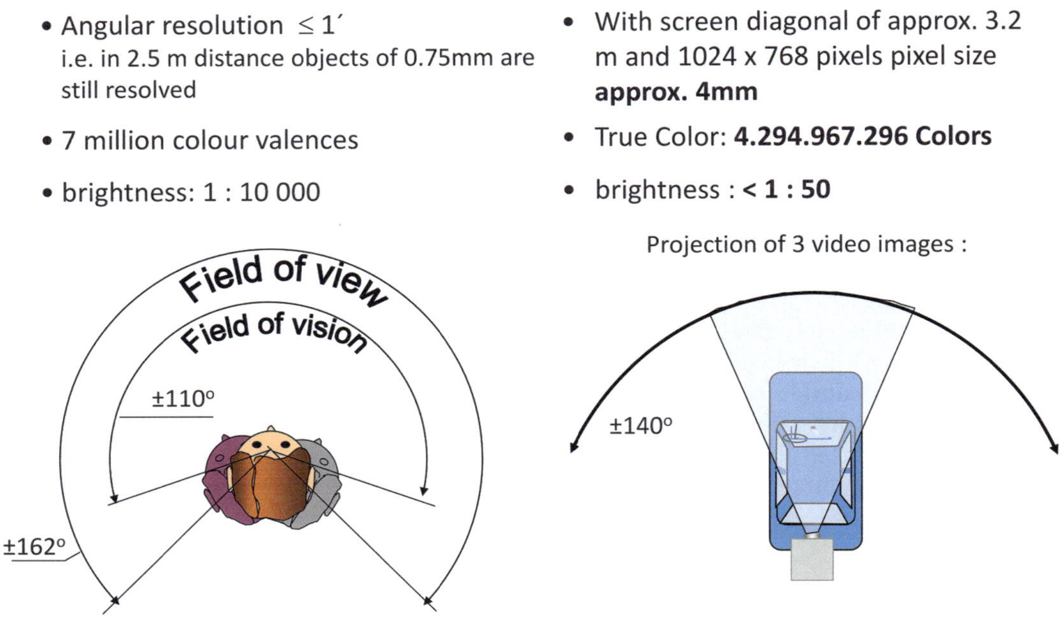

Fig. 10.10 Necessary field of view on the video images in a driving simulator

transition from one monitor to another virtually invisible to the viewer (edge blending). Irrespective of the imaging medium used, good immersion requires correct viewing angles. Figure 10.11 shows the corresponding geometric relationships. The minimum requirement is that the horizon on the image is exactly at eye level *h* of the test person.[5] The virtual eye level used in the software program must match this real eye level. The best way is to use a straight piece of road in the software program and read the position y_L of the virtual driver's eye to the left lane, and y_R to the right lane. At the distance d_S between the screen and the real driver's eye, a line across the road, which is at a virtual distance d from the virtual driver's eye, must be below the horizon line by the distance h_S It's on:

$$h_S = \frac{h \cdot d_S}{d} \tag{10.1}$$

The same applies to the horizontal spacings. y_{LS} and y_{RS} of the calibrating lateral line to the road edges:

$$y_{RS} = \frac{y_R \cdot d_S}{d} \quad \text{und} \quad y_{LS} = \frac{y_{RL} \cdot d_S}{d} \tag{10.2}$$

If only a small screen is available, a large Fresnel lens can be used to establish the correct viewing angle ratios (Fig. 10.12). For the necessary distance a of the screen from the Fresnel lens with the focal length *f* applies when the distance to be displayed is *d* on the screen h_B is below the horizon line:

$$a = \frac{h_B \cdot d}{h} \tag{10.3}$$

A further increase in immersion can be achieved by presenting appropriate information on the two exterior mirrors and in the interior rear-view mirror. In many cases, small screens will be used for this purpose. A much more realistic representation, however, can be achieved by using a projection screen for the

5 If the distance d_S is large enough of the screen of the driver's eye, it is sufficient to adjust the horizon height to the middle height of the driver's eye. It should be noted at this point that the human eye is extremely tolerant of falsifications of the viewing angle and that consequently errors in the setup are not detected by mere observation. For a good immersion the correct viewing angle is indispensable.

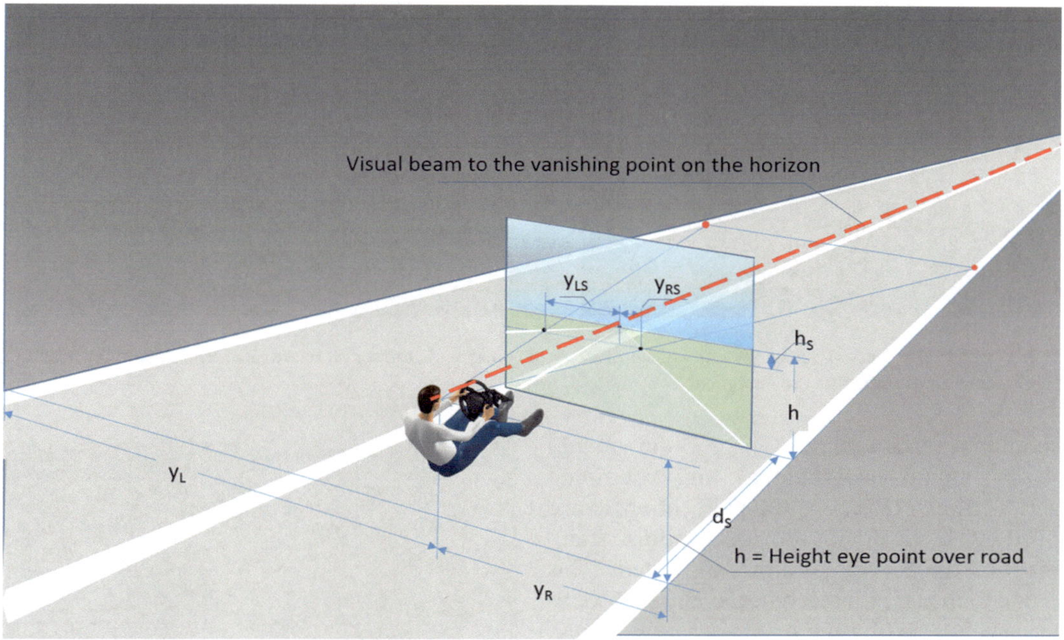

Fig. 10.11 Geometry ratios for setting the correct viewing angles

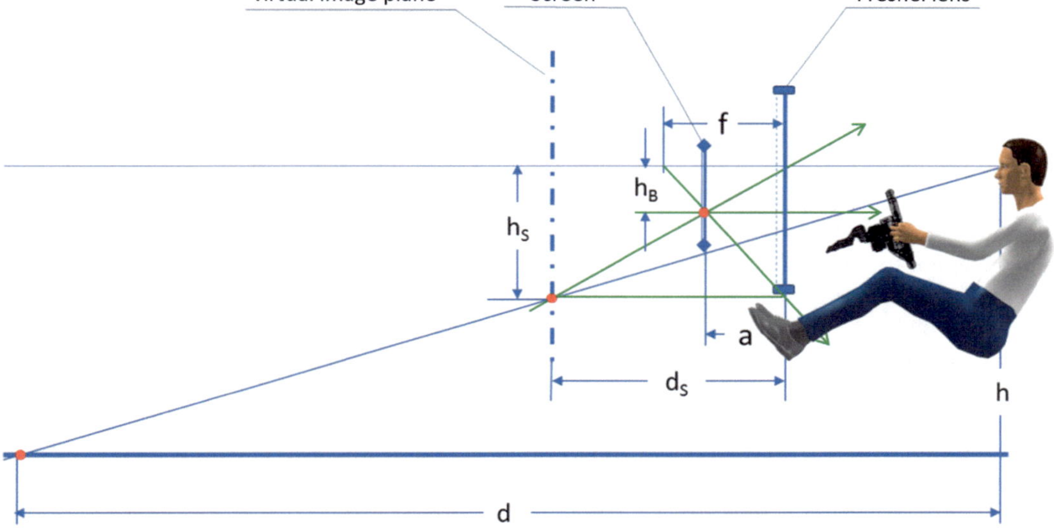

Fig. 10.12 Establish correct viewing angle ratios to improve immersion at small screen size (the green lines characterize the construction of the virtual image according to the rules of geometric optics)

rear-view mirrors as well, since the natural body and head movements for widening the viewing angle in the rear-view mirrors have a similar effect as in reality. Because of the different perspectives, however, you have to use three projectors in this case. Figure 10.13b gives an example of this.

As already mentioned, the *motion simulation* is an important immersion factor, while at the same time placing the highest demands on

Fig. 10.13 Projection device for rear-view mirrors **a** and rear view mirror **b** (Example: Driving simulator of the Institute of Ergonomics at the Technical University of Munich)

simulation technology.[6] The motion system shall simulate translational and rotational accelerations. However, the publications compiled by Negele do not provide a uniform statement on the benefits of a motion system. This is due to the fact that technical comparison values are falsified by the respective motion cueing, that there are usually no values directly available from the vehicle cabin, that the totality of the sensory impressions is different for different simulators and that the data obtained depend on the test sequence. The following statements apply to the design of a driving simulator: a well tuned motion system does not impair driver behaviour (the reference is driving in reality), has an effect on the stabilisation level of the primary driving task and has a positive effect on fulfilling a secondary task.

For the representation of the real acceleration there are three different methods.

1. A translational movement is converted into an inclination (inclination simulation, perception threshold 25°) and concluded by a return to the zero position (wash-out, perception threshold 2–6 °/s^2 see above). Figure 10.14 shows the NADS simulator as an example of a complex simulation technique. By using a hexapod, both rotational and translational movements are possible in a small area. For larger longitudinal accelerations, the entire simulation technology is mounted on an X-Y traversing carriage, which is used beside others to correctly map lane change processes.

Fig. 10.14 NADS simulator on hexapod and x-y traversing carriage for the simulation of kinesthetic stimuli using wash-out effects

2. With this method, the data from the vehicle dynamics are scaled with a factor in order to keep the necessary travel distances smaller. Often one tries to get along with the movement possibilities of a hexapod. However, the scaling factor cannot be kept constant over the entire frequency range.
3. Another method is the representation of translational acceleration by centrifugal forces. The mock-up is mounted on an XY slide with rotation platform. This allows a good representation of the translational acceleration. Currently, however, it is not possible to cover the entire acceleration range without a scaling factor.

If the movements acting on the driver are compared with the possibilities of the motion system, four significant execution stages can be derived. The build-up movement (pitching, roll-

6 Here, too, reference is made to the detailed work of Negele (2007).

ing and lifting movement), which even through small movement systems lead to very good results (see below). The body movement and path movement information at stabilization level improve tracking. The frequencies to be displayed are in the range of 1.5 Hz. The body movement and path movement information at high-frequency vehicle dynamics level can only be realized on larger movement systems (travel path ±1.5 m). The frequencies to be displayed are also in the range of 1.5 Hz, but must be kept longer (wash-out concealed by feedback of the build-up movement). The build-up movement and longer-lasting path movement information make the highest demands and require combined systems (see also ◘ Fig. 10.14). The translational traversing devices can only be used in a limited range compared to the unrestricted specifications of reality (see argumentation at the beginning of this subchapter). Our own experiences with simulator technology and experiments suggest that in motion simulation it is less a matter of the exact reproduction of the accelerations present in reality than of the exact temporal relations. For example, it can be perfectly sufficient to set the vehicle cab to hydraulically or electrically activatable pulsers and to convey a kinaesthetic feeling of reaction by briefly buckling the body when starting off or accelerating, but in correct relation to reality in terms of time (<200 ms when braking). The same applies to steering movements. Experience with simulated city passage suggests that the yaw movement must be reproduced correctly, which makes it necessary to mount the entire simulation setup on a turntable controlled by the simulation software.

The *noise simulation* in volume and direction represents a very important contribution to creating a good sense of presence within the simulation environment. In addition, the noise presentation has a positive influence on the control behaviour of humans. In addition to the driving condition-dependent noises, the various stationary and ambient noises that the driver associates with driving in the vehicle also play an important role. This also includes the "palpable" noises (<20 Hz). These can be generated by structure-borne sound transducers or hydropulse cylinders, whereby a representation via the motion system is perceived as unpleasant. Equalizers must be provided for distribution and possible overlapping. The positioning of all noise sources involved is very important. This means, for example, that the spectra simulating the engine noise should rather come from loudspeakers in the front area of the cabin, while noises that can be assigned to the chassis should come from rear loudspeakers that are mounted as low as possible.

The structure of a real *mock-ups* has a direct influence on the presence perception of the test persons. The task is decisive for the characteristics of the human-machine interface. Depending on the question, this can range from simple control elements to the real interior to freely programmable control and display elements.

Of great importance is the *steering simulation*. The model receives the input values of the restoring torque from the driving dynamics. Very high demands are placed on the moment (reaction time of the local holding reflex 0.02 s, force resolution of the hand 2% of the holding force, whereby the high vibrotactile sensitivity of the hand must be taken into account). In addition, it can be advantageous to apply the "Hardware in the loop" (HIL) concept (i.e. a steering system – generally electric – used in the series is used, which is controlled by the software of the virtual vehicle model). The accelerator pedal can be realized by an electronic pedal (standard). Various satisfactory spring solutions have been implemented for the brake pedal.

Although today in the gaming sector from the point of view of graphics and also the dynamic response at low prices astonishing *simulation programs* are available, which unfortunately are hardly adaptable for experimental purposes, the software of a driving simulator still represents a very expensive factor. The database for software that can be used in vehicle development is the virtual representation of the environment and contains the terrain description, road course, road characteristics, models, logical information and textures. High-quality models and textures improve the sense of presence and the control activity through a realistic impression of speed and distance. Three types of databases can be distinguished.

- The *static database* contains a single true-to-scale map of the entire landscape within which the entire simulation runs. Changes that may be desirable for experimental reasons are only possible to a very limited extent (usually limited to variable setting of traffic signs!).
- At the *modular static database* landscape and road course are composed of predefined, static and locally geometrically consistent modules.
- At the *dynamic database* a road course is not given, but a logic, which strings road sections together as part of surrounding modules during the simulation (Kaussner et al. 2001, 2004). This results in a road course that is dependent on driver behaviour and possibly also on the traffic situation and does not have to be geometrically consistent globally. With the two modular approaches, there are restrictions with regard to the design of the surrounding landscape due to the not necessarily given geometric consistency and the necessary "fit" between all modules. For these two types, a far-reaching view is often problematic due to the modular architecture of the landscape (Negele 2007).

10

The perspective calculation must be correct (see ◻ Fig. 10.11). The degree of reality can be divided into close range and far range if necessary, i.e. there are equipment which only has to be of high quality in the close range (stabilisation task) or only in the far range (indication stimuli for guidance task). The traffic simulation contains vehicle models which determine the course of the autonomous vehicles on the basis of a predefined set of rules. For certain scenarios, especially in connection with the possible effect of assistance systems, but also in connection with visual examinations (see the Remlinger 2013 investigations) on the driver, it must be possible to use vehicles depending on the event.

10.4.3 Relevance of Driving Simulators of Different Levels

Simulators, which enable a very good presence experience, are according to the explanations of the ▶ Sect. 10.4.2 very complex and thus also very expensive. Consequently, the question arises as to which reductions in technical effort can be accepted in connection with specific questions. According to Negele (2007), the following steps should be taken to answer this question:

1. Determination of the stimulus-response mechanisms underlying a specific driving task.
2. On the basis of step 1, a statement about the human being's ability to mentally compensate for missing perceptions in the simulator in order to achieve valid results.
3. Derivation of the simulator concept and the required technology.

Step 1: The necessary stimuli to be provided by a simulator can be determined on the basis of the different levels of the driving task. At the level of the *stabilization task,* which of the *primary driving task* in particular the impression of speed and the position of the vehicle on the road are of decisive importance. The speed impression is determined more by the image resolution in the central visual area than by peripheral movement stimuli, which in turn create a good impression of the direction of movement. Not only the quality of the image resolution is important, but especially the presence of reference points that move correctly according to their own speed. Furthermore, the important influence of the hearing is to be pointed out, whereby the speed level is estimated substantially. With regard to the correct perception of the position on the road, a vehicle cabin corresponding to the experience of the test person is of great

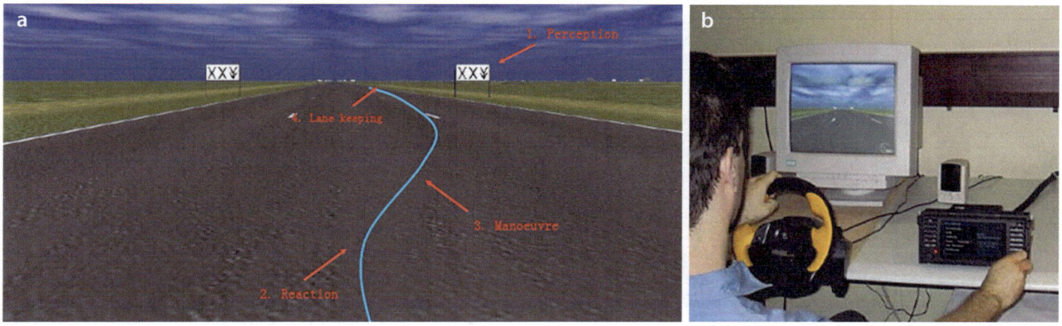

Fig. 10.15 Representation of the motion task in the Lane-Change-Test program **a** and frequently encountered experimental setup **b**

importance.[7] Since it is not possible to completely replicate the real acceleration stimuli with the methods available for simulation, simulators must always expect acceleration and braking behavior to deviate from reality. Of particular importance is the simulation of haptic stimuli by the control elements as true to the original as possible.

On the *guidance level* the formation of the target speed and the target course is based exclusively on the speed impression and on the optical estimation of the distance to the objects of the environment, from which mental values such as time-to-collision and time-to-line crossing are formed. This means that even higher demands can be placed on the resolution of the scenery depicted and on the various weather conditions. In particular, a realistic representation of the behaviour of other road users is of crucial importance. This also includes the depiction of the scenery in the rear-view mirror.

For investigations of *secondary driving tasks* the primary driving task should be reproduced so well with regard to faster learnability and the sense of presence that a longer lasting normal driving is possible without any problems. It is necessary to draw attention to the need for uniform, low-brightness illumination of the vehicle interior, which counteracts the disturbing perception of a day's journey when viewed from a dark vehicle interior.

The requirements relevant for secondary tasks also apply in principle to *tertiary tasks.* However, since tertiary tasks are not related to the driving task, simulators that largely deviate from the above requirements, e.g. in the form of tracking tasks, can also be used for their investigation. The so-called Lane-Change-Test has become a quasi-standard for this. This is a program which can be implemented on any PC and which is connected to the control elements of a game console (steering wheel, accelerator pedal and brake pedal). The driving task is to keep the vehicle on a lane of a three-lane motorway at a constant speed (e.g. 100 km/h) and to change lanes when prompted to do so by a traffic sign (see Fig. 10.15a). In general, the use of this program is not regulated by anthropometric conditions or viewing angle conditions, so the experimental set-up often consists of an office chair and a PC screen on an ordinary desk (see Fig. 10.15b). Under these conditions, it can hardly be assumed that there will be sufficient involvement or even a sufficient sense of presence. If at all – experiments with this arrangement can only be attributed relative validity.

Thus, there are good reasons for using more elaborated driving simulators even for questions on tertiary tasks, such as the dis-

7 In video games, the non-existent cabin is often compensated by displaying the ego vehicle in the lower part of the image that you are looking at in the sense of a bird view display. However, this form of representation is completely unsuitable for use in the scientific field or for early answers to questions in vehicle development, because it partially turns the task of driving, which is primarily to be understood as a compensation task, into a pursuit task.

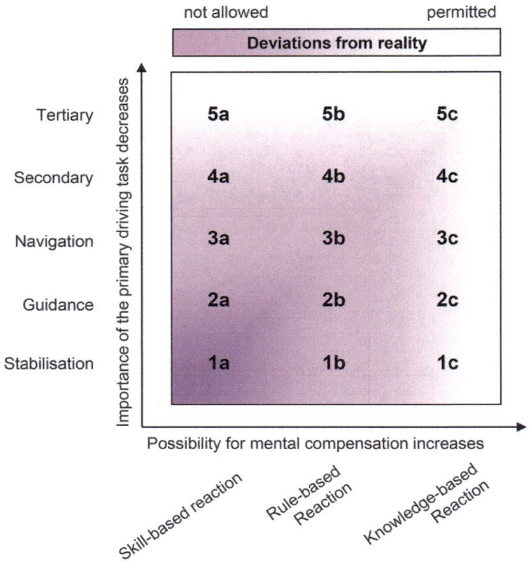

Fig. 10.16 Classification of automotive applications. (From Negele 2007)

pensability of learning a special tracking task or an atypical driving task and the facilitated mental transfer of the tested system into the real driving task.

Acting in road traffic is always also social acting. Many social conventions that apply to road traffic are not present in this form in the simulator. By coupling two simulators (the vehicle position in simulator 1 is shown as a foreign vehicle in simulator 2 and vice versa), as it is also used in combat aircraft simulators, one can try to introduce a little human individuality into the simulation, but since indirect information, such as the body posture of the "opposing" driver or his gestures cannot be simulated, one comes up against a further limit of simulation technology in addition to the problem of simulating kinesthetic stimuli. For the correct interpretation of simulator results, factors also play a role, such as the experience of the test subjects with simulators, the non-simulability of existential engagements and also the tendency of the test subjects to "cut a good figure".

Step 2: Negele (2007) has developed a system for the classification of vehicle technology applications that makes it possible to estimate the degree to which a deviation from reality is permissible. This system is based on comparing the different levels of the driving task with the different levels of driver behaviour according to Rasmussen (see Fig. 10.16). This results in 15 application classes. It can be generally inferred from the system that the greater the conscious cognitive effort required to complete a task, the more it is possible to deviate from a strict reference to reality. Based on the assumption that experienced normal drivers are usually used as test subjects in the field of vehicle development, some categories also remain unoccupied. For example, for the driver with normal experience, there are no actions on the stabilization level that are knowledge-based, as is the case with a novice driver (1c). This also means, for example, that a training simulator for beginners does not make any special demands on the movement simulation. Even skill-based navigation (3a), such as the daily route to work, can hardly be depicted in the simulator.

Despite this classification, it is essential for the developer to have a precise idea of the parameters to which the test result will react sensitively in order to take this into account in the details of the specification.

Step 3: In order to make it easier for vehicle developers to classify their applications, Negele describes the following 6 typical vehicle simulator designs based on the matrix:

1. The investigation *skill-based stabilization* places the highest demands on the primary driving task and thus on motion simulation. Typical for the simulator configuration in this case is the prioritization of the motion simulation before the visual simulation. If investigations on vertical dynamics are to be carried out, special requirements must again be met for the true-to-original simulation of lateral motion (for an example, see Fig. 10.14).
2. During the traffic analysis of the *rule-based guiding* the quality and flexibility of the database and the traffic simulation must be given priority. In particular, the software must enable traffic situations to be triggered by the driver's behaviour in order to confront each test person with the desired situation. In order to achieve the best possible involvement, the primary driving task must also be mapped in sufficient quality.

Movement information, on the other hand, is only relevant insofar as the primary driving task is almost as easy to master as in reality. A typical example of a simulator that meets these requirements is shown in ◘ Fig. 10.13.

3. For the simulation of tasks that require the *rule- and knowledge-based navigation* a driving simulator in connection with navigation can provide reproducible conditions for basic research on human orientation in an unfamiliar environment. This application must be distinguished from investigations of the human-machine interface, which is concerned with the operation of the navigation device. The requirements to the simulator are not as high as in the first mentioned cases regarding the visual reproduction. A movement simulation can be completely dispensed with. The gain in knowledge in such investigations is usually best achieved by questioning and observing the driver in conjunction with eye-tracking systems.
4. During the investigation *rule-based secondary tasks* (e.g. setting the direction indicator), the primary driving task serves primarily to create the necessary context for traffic situations. In such tasks, the sense of presence is particularly important, which is determined by the simulation of as many relevant sensory organs as possible and by the attitude of the test person himself. This places relatively high demands on visual, noise and movement simulation. If however secondary systems with no driving dynamics are investigated (e.g. switching of different wiper functions), there are only minor requirements for motion simulation.
5. With *knowledge-based secondary tasks* (e.g. user behaviour of driver assistance systems), the functions to be tested and their effects on the driving process are largely unknown. Within the framework of preliminary developments, such experiments can provide initial indications of the interaction between driver and driver assistance system and their acceptance. The deviations of the simulator generated from the real driving task can be greater. However, this does not apply if the test is intended to make statements about the practical suitability of a driver assistance system. The requirement profile for the simulator is therefore similar to that for rule-based secondary tasks.[8] However, as Zöller et al. (2013) argue, in order to investigate the integration of assistance systems into vehicle dynamics, the validity – if possible absolute – of driving simulator investigations has yet to be investigated.
6. During the investigation *rule- and knowledge-based tertiary tasks* (e.g. new infotainment systems), the decoupling from the actual primary driving task results in greater degrees of freedom in the performance of tasks and thus a higher personal dependency of behaviour. The driving task must therefore be designed in such a way that it brings as little scatter as possible into the results. These boundary conditions are very similar to those for rule- and knowledge-based secondary tasks. Investigations of this kind are often carried out with the Lane-Change test mentioned above (◘ Fig. 10.15). In order to achieve a minimum degree of involvement, however, at least the correct viewing angle ratios should be observed (e.g. by using a Fresnel lens, see also ◘ Fig. 10.12).

10.4.4 Use of Augmented Reality in Real Vehicles

Bock (2008) tried to compensate the main problem of driving simulators, i.e. to cause an insufficient feeling for driving and driving dynamics by the non-existent or only limited kinesthetic feedback, by using Augmented Reality (AR). He uses a real vehicle, which is moved on a closed off test area with given road course. The driver looks at the real road

8 It should be noted here that when used regularly in reality, there is a continuous transition from knowledge-based to rule-based to skill-based use. The integration of operating procedures for driver assistance systems depending on complex traffic situations can take two or more weeks (Weinberger 2001). Such a process running in reality can hardly be simulated with simulator technology.

through an HMD with semi-permeable mirrors (see ◘ Fig. 10.17).

These can be used to mirror virtual third-party vehicles whose position and movement is calculated by a driving simulator program in the test vehicle. A prerequisite for the functioning of this system is the 1:1 reconstruction of the real test track in the simulator software as well as the measurement of the current position of the vehicle on the track and the head position of the driver in the vehicle. The simulator program is now connected to the real or prototype assistance system, which reacts to the virtual vehicles depending on the position of the real vehicle (SIL). The interaction of the different components is shown in ◘ Fig. 10.18.

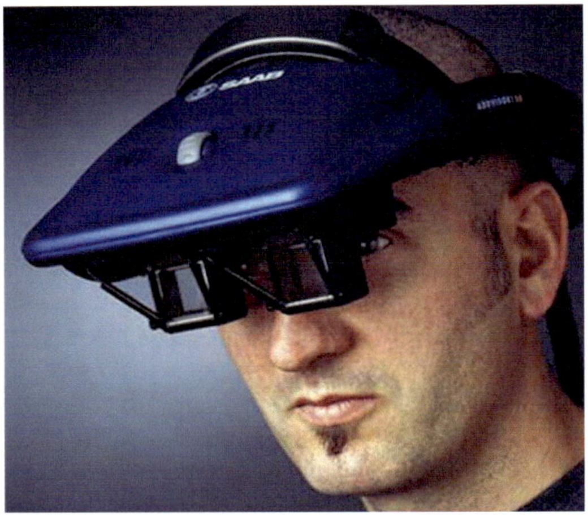

◘ **Fig. 10.17** Optical See Through HMD Saab AddVisor (SAABTECH 2006, from Bock 2008)

◘ Figure 10.19 reflects the view offered to the test subject by this arrangement. This enables him/her to experience the reaction of the assistance system to the virtual vehicles in real driving conditions. The proposal implemented by Bock represents a Vehicle-in-the-Loop (VIL) solution.

It is aimed primarily at test engineers, which is clear from the fact that the tests must be carried out on a closed-off site which must be precisely measured. By comparing situations that were encountered in a real vehicle and managed with this VIL system, Bock was able to show the extent to which valid state-

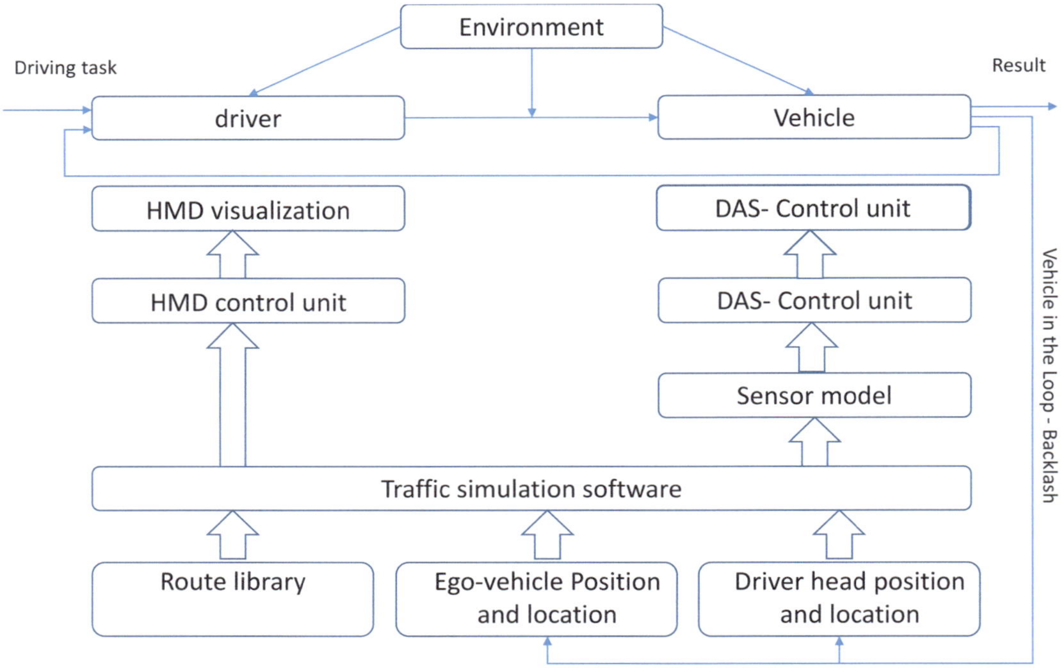

◘ **Fig. 10.18** Functional architecture of the Vehicle in the Loop test setup. (From Bock 2008)

Fig. 10.19 View of the test subject through the HMD on the real track with simulated virtual vehicles. (From Bock 2008)

Fig. 10.20 Using an air car to check the effectiveness of the emergency brake assistant

ments can be expected when using this system. He also carried out these experiments in situations where no assistance system was active. He was able to show that the distance to the virtual front vehicle was estimated to be comparable to the tests in the real vehicle. The test subjects react to the displayed vehicles with similar reaction times and accept them as road users. Mostly, they show comparable driver behaviour when driving with the virtual front vehicle as in the real test. In particular, there were no symptoms of the simulator disease as in the driving simulator. However, there were also difficulties in assessing the acceleration manoeuvres of the front vehicle in curves and in assessing the lane affiliation of distant vehicles. Only some of the test persons reacted to unexpectedly critical lane change maneuvers of the preceding virtual vehicle in a way comparable to the real situation. Bock argues that most of these differences are due to the still inadequate technical implementation of the HMD. In the course of the further development of HMDs, which has already been mentioned several times and is driven by the game software, these problems should be overcome in the future.

10.5 Experiments in a Real Vehicle

10.5.1 Trials on Closed off Terrain

Almost all automobile companies today have their own test site where vehicle prototypes can be tested and examined. In general, tests are carried out there by experienced test drivers who are specially trained for such tasks. Only in exceptional cases or on test sites belonging to research institutes such test drives are carried out with laypersons, which should, if possible, be representative for future customers. Since such experiments often deal with the reaction of test persons to unforeseen situations, they must be designed in such a way as to exclude any risk to persons. This often requires considerable technical effort. A frequently used procedure in this context is the use of so-called "air cars" (Fig. 10.20). These are inflatable vehicle models made of plastic, which have the dimensions and the outer appearance of a real vehicle, whereby a collision with this remains without consequences because of their lightness. Similarly, so-called "dummies", which also represent other road users (e.g. pedestrians, children), can be used. For some test situations it is necessary that such an air car or dummy moves. This movement can then be simulated with an often complex rope pull system, whereby it may depend on the reaction of the actual test vehicle. It is not the space here to describe the technical effort required for such a test constellation. In addition, this always depends considerably on the respective question. One example of how such interaction takes place is the EVITA system developed together with Honda R&D and the Institute of Automotive Engineering at the University of Darmstadt (Hoffmann 2008; see also Fig. 10.21). It consists of the

Fig. 10.21 Target vehicle with EVITA trailer, which prevents collision with the test vehicle even under extreme conditions. (Source: Fahrzeugtechnik, TU-Darmstadt)

combination of a towing vehicle, a trailer and the following test vehicle, in which the test person is located, whose reaction mode – e.g. in the case of distraction or in the case of a delayed reaction of an assistance system, which is to avoid a rear-end collision – is to be investigated. During the stationary subsequent drive, the trailer ("dummy target") brakes surprisingly for the test person driving the following vehicle. Regardless of whether the subject reacts to the manoeuvre in time or not, the trailer is actively pulled out of the collision area at the last moment by means of a cable system.

In many cases – as in this example – two vehicles are involved in such tests, a so-called target vehicle and the actual test vehicle in which the test person with the assistance system to be assessed is located. The situation to be investigated is induced by the target vehicle. For example, Strasser (2011) has defined the following test scenarios for the objectification of assistance system properties (here ACC) by subject judgements (see also ▶ Sect. 9.2.1.3):

- *Drive up to column:* For all three settings "sporty", "safety" and "comfort" the system should react early, but with rather weak dynamics.
- *Overtaking:* The system should react promptly and overtake with moderate acceleration.
- *Reevers:* The aim of the vote should be for the system to react rather early with moderately strong momentum.
- *Outrigger:* The user wants the system to react quickly to the outrigger. How strong the deceleration should be depends on the type of driver.

Wimmer (2014) has set up various interaction possibilities for future assistance systems in a real vehicle and investigated the different automation modes as well as the change between them with neutral test persons on a closed-off test site. Here, too, various test scenarios similar to those mentioned above were defined and the reaction of the test persons observed.

In addition to these 2-vehicle tests described above, however, tests with only one vehicle are also carried out on restricted terrain. Essentially, the reaction (time and error) during the operation of tertiary (rarely secondary) tasks is then observed. However, such experiments can also be carried out on public roads if the measurement setup allows the reaction to be observed. Other tests require the special conditions of the test site. For example, experiments on discomfort, caused by the influence of mechanical vibrations, are carried out on suitably prepared sections of the test site or the movements measured there are later simulated in the laboratory (Bitter 2005).

The studies mentioned here serve only as examples of what kind the experiments with test persons on closed test sites can be. In all this, it is important to point out the unusual situation in which the test persons find themselves: they usually know what it is all about. They are exceptionally concentrated in the unfamiliar surroundings and want to "make a good impression". The results must also be interpreted in the light of this aspect, because in the reality of daily car traffic, people fall into their usual patterns of behaviour and do not show the same attention as in the experiments, so that the reactions there may differ greatly from those of the experimental conditions.

10.5.2 Experiments on Public Roads

Because of the desired closeness to reality and also because of the availability, many tests are carried out on public roads. In this case, the first priority is to ensure that the safety of the test persons and other road users is not impaired more than would be the case if they were to participate in normal road traffic. The experiments in public road traffic are particularly necessary when it comes to measuring effects that only occur after a longer period of time (e.g. long-term discomfort) or where the established habits play a special role (e.g. willingness to be distracted by tertiary tasks). Trial runs of four or more hours in the simulator cannot be expected of any test person and are practically impossible for external test persons to carry out on test sites.

Many tests on public roads therefore focus on the aspect of comfort, especially sitting comfort. Usually the driver's seat is then covered with a seat pressure mat (e. g. Zenk 2008). Driver posture is also frequently observed with various camera systems (e. g. Kolling 1997; Estermann 1999).

The aim of the experiments on public roads is beside others to record driver behaviour in everyday driving life. In general, a test vehicle is prepared for this, which has appropriate recording facilities (often data from the CAN bus such as speed, lateral acceleration, also distance to the vehicle in front if an ACC is on board). The experiments are accompanied by a test leader who, depending on the research question, may also ask the subjects to take action, such as reading certain information from a display or performing acoustically set tasks. In many cases such journeys are accompanied by trained driving instructors (e.g. Schweigert 2003) who register the behaviour of the test persons in relation to the traffic situation. Physiological data of the test persons are also frequently recorded (the heart rate is particularly popular [9]). The attention paid by the driver can be particularly well objectified by eye-tracking systems (vehicle-mounted or head-based). Particular attention must be paid to the safety of these additional measurements carried out on test persons. A characteristic feature of all these experiments is that, in addition to the indicated objective data, subjective data are also collected by means of questionnaires, whereby attempts are made to correlate these subjective impressions with the objective observations.

The previously mentioned experimental constellations all have the disadvantage that an observer is present who impairs the behaviour of the test persons by his presence alone. For this reason, experiments have been developed that do without such an investigator. Sacher (2006) for example equipped two test cars with different measuring sensors, which were made available to the test persons for several weeks. Each time these test vehicles were started, a PC was also started which recorded all measured values. This gives a relatively objective picture of user behaviour without observation. However, this advantage is bought at the price of the fact that the allocation of the conduct to external tasks – be they those resulting from the individual use of the vehicle or those resulting from the traffic situation – is not possible or only possible indirectly. In the above-mentioned experiments, for example, a more precise recording

9 In this context, it should be noted that heart rate, like many other physiological parameters, is at best an indicator of changes in mental stress, but does not represent it correctly in absolute terms.

(e.g. of the allocation to certain road sections or the observation of traffic events by recording camera images) was prohibited for data protection reasons.

These last mentioned disadvantages appear more and more, the less one influences the behaviour of the driver by observation. A recently discussed possibility is the use of data recorded by smartphones and exchanged with the radio stations of the mobile radio network. So far, the best that can be obtained in this way is information about traffic flows or speed distributions in certain sections of the route. Information about the use of certain vehicle-specific systems or even the behaviour of the driver in the vehicle is therefore not possible.

The extreme of this transition from ever greater closeness to reality at the expense of ever less detailed knowledge of individual behaviour is accident statistics. Here, too, the aim is to use so-called in-depth accident analyses.[10] However, the ability to subsequently analyse the behaviour leading to the accident depends on the willingness and precision of those involved in the accident to remember.

10.5.3 Customer Feedback

Feedback from buyers naturally plays a dominant role in the acceptance of new developments. All automotive companies therefore have departments that systematically evaluate customer complaints using statistical methods. The experience gained there not only serves to improve the current series, but also provides suggestions for new developments. In addition to this use of internal information, the published survey results, which are carried out as part of the TÜV/Dekra surveys or by automobile clubs, also play an important role. However, the quite general questions used there often only provide very rough clues. Studies of particular interest in the automotive industry are conducted by J.D. Power[11] a global market information provider that annually delivers data and solutions for performance improvement, social media and customer satisfaction. The studies carried out in Germany on the satisfaction of vehicle owners with 1–3 year old vehicles are called the "Vehicle Ownership Satisfaction Study^{SM}" (VOSS for short). The questions are divided into the following categories: "Vehicle attractiveness" (27%), "Maintenance costs" (25%), "Vehicle quality and reliability" (24%) and "Service satisfaction" (23%) (the figures in brackets indicate the significance for user satisfaction according to the results of J.D Power 2013 VOSS in Germany). In particular, the categories vehicle quality/reliability and vehicle attractiveness contain questions that relate directly to aspects of vehicle ergonomics.

Another important method of finding customer acceptance is the so-called car clinics, which are carried out by specialised market research companies on behalf of vehicle manufacturers. These tests are often performed long before a new model is introduced to the market. The test subjects, who are selected according to representative criteria, are bound to secrecy. During the tests, new prototypes of a planned vehicle, usually worked out down to the last detail, are presented and the image and acceptance with regard to design, materials and quality impression as well as the purchase motives are examined in detailed interviews. In special cases, test drives to evaluate the driving characteristics are even carried out as part of these investigations. Virtual car clinics are also being tested recently. Using

10 In this form of analysis, the police, who were called to an accident, informed a team of scientists, consisting of technicians, psychologists and doctors, who make their investigations independent of the police. The accident victims who have agreed to be interviewed are assured in advance that the statements are confidential and will not be passed on to the public accident recording authorities or the subsequent negotiating partners in court. In this way, statements can be recorded and observations made that are closer to the real events than the official police versions.

11 J.D. Power is headquartered in Westlake Village, California and has offices in North America, Europe and Asia Pacific. The company is part of McGraw-Hill Financial, a financial data company that provides credit ratings, benchmarks, and analytics services for the global capital and commodity markets.

different visualization techniques (3D films, powerwall and similar), models and details are presented to the study participants for evaluation. These studies can even be carried out online and thus guarantee a better regional distribution.

With all the techniques mentioned, however, the following must be taken into account: “The customer does not invent anything”. At best, it is to be expected that these survey and interview techniques will result in an optimal mixture of already known solutions.

References

Literature Used

Abendroth, B., Schreiber, M., Bruder, R., Maul, S., Maul, D.: Neue Ansätze zur Beurteilung der Fahrsimulatorvalidität. Zeitschrift für Arbeitswissenschaft. **1**(1), 1–11 (2012)

Bartle, R.A.: Designing Virtual Worlds. New Riders Publishing, Indianapolis (2003)

Bitter, T.: Objektivierung des dynamischen Sitzkomforts, Schriftenreihe des Instituts für Fahrzeugtechnik, TU Braunschweig, Dissertation, Shaker Verlag 2005 (2005)

Blana, E.: Driving simulator validation: A literature review. Insitute for Transport Studie Working Paper, Bd. 480. University of Leeds, Leeds (1996)

Blana, E., Golias, J.: Behavioural validation of a fixed-base driving simulator. In: Proceedings of DSC 1999 (1999)

Bock, T.: Vehicle in the Loop – Test- und Simulationsumgebung für Fahrerassistenzsysteme Audi Dissertationsreihe, Bd. 10. (2008)

Braun, M.: Entwicklung eines variablen Ergonomie-Prüfstandes mit CAD-Anbindung und Validierung einer Arbeitsmethodik. Dissertation an der Technischen Universität München. (1997)

Buld, S., Krüger, H.P.: Validierung Quer- und Längsführung, EMPHASIS Effort-Management und Performance-Handling in sicherheitsrelevanten Situationen Meilensteinbericht. Interdisziplinäres Zentrum für Verkehrswissenschaften, Universität Würzburg, Würzburg (2001)

Duden: Die deutsche Rechtschreibung – Das umfassende Standardwerk auf der Grundlage der aktuellen amtlichen Regeln. Bibliografisches Institut und F.A. Brockhaus, 25. Auflage, vollständig neu bearbeitete und erweiterte Auflage (2009)

Estermann, S.: Komfortmodellierung für den 3D-Softdummy RAMSIS unter realen Fahrbedingungen. Dissertation an der Technischen Universität München (1999)

Herrler, M.: Systemergonomie und Usability-Test bei bildschirmbasierten Bedienoberflächen im Automobil. Dissertationsschrift an der Technischen Universität München (2006)

Hoffmann, J.: Das Darmstädter Verfahren zum Testen und Bewerten von Frontalkollisionsgegenmaßnahmen. Dissertation an der TU-Darmstadt (2008)

Hofmann, J.: Raumwahrnehmung in virtuellen Umgebungen, 1. Aufl. Deutscher Universitäts-Verlag, Wiesbaden (2002)

Jamson, H.: Driving simulator validity: Issues of field of view and resolution. In: Proceeding Driving Simulation Conference Paris, 2000. S. 57–64 (2000)

Kapstein, N., van der Horst, A., Hoekstra, W.: The effect of field of view and scene content on the validity of driving simulator for behavioural research. TNO-TM., Soesterberg (1996)

Kaussner, A., Grein, M., Krüger, H.-P.: An achitecture for driving simulator databases with generic and dynamically changing road networks. In: Proceedings of DSC 2001 (2001)

Kaussner, A., Mark, C., Grein, M.: Fahrsimulator-Datenbasen mit dynamisch veränderbaren Straßennetzwerken. IZVW, Würzburg (2004)

Kolling, J.: Validierung und Weiterentwicklung eines CAD-Menschmodells für die Fahrzeuggestaltung. Dissertation an der Technischen Universität München (1997)

Negele H. J.: Anwendungsgerechte Konzipierung von Fahrsimulatoren für die Fahrzeugentwicklung. Dissertation an der Technischen Universität München (2007)

Regenbrecht, H.: Faktoren für Präsenz in Virtueller Architektur. Dissertation an der Bauhaus-Universität Weimar (1999)

Remlinger, W. M.: Analyse von Sichteinschränkungen im Fahrzeug. Dissertation an der Technischen Universität München (2013)

Riedl, M.: Potential eines virtuellen Fahrerplatzmodells in der Fahrzeugentwicklung. Dissertation an der Technischen Universität München (2012)

SAAB-tech: AddVisor150 – High Performance Head-mounted Display. http://saabtech.se (Juni 2006)

Sacher, H.: Wie nutzen Fahrzeugführer das adaptive cruise control? In: Bericht zur 2. Tagung "Aktive Sicherheit durch Fahrerassistenz" 4.-5. April, 2006. TÜV Süd, München (2006)

Schrader, K.: Entwurf und Realisierung eines Ergonomie-Mock-Ups unter Verwendung von Methoden der Virtuellen Realität, Dissertation an der Technischen Universität Braunschweig (2003)

Schubert, T., Friedmann, F., Regenbrecht, H.: Embodied presence in virtual environments. In: Paton, R., Neilson, I. (eds.) Visual Representations and Interpretations, pp. 269–278. Springer-Verlag, London (1999)

Schweigert, M.: Fahrerblickverhalten und Nebenaufgaben. Dissertation an der Technischen Universität München (2003)

Stäbler, M.: Systemvergleich zweier VR-Systeme vom Datenfluss bis zum Echtzeit-Raytracing, Fachhochschule Ansbach, Studiengang Multimedia und Kommunikation, Diplomarbeit, 2007 (2007)

Strasser, P. S. B.: Vernetzung von Test- und Simulationsmethoden für die Entwicklung von Fahrerassistenzsystemen. Dissertation an der Technischen Universität München (2011)

Voss, T.: Untersuchungen zur Beurteilungs- und Entscheidungssicherheit in virtuellen Umgebungen. Dissertation an der Technischen Universität München (2008)

J.D. VOSS.: http://www.jdpower.com/press-releases/2013-germany-vehicle-ownershipsatisfaction-study-voss (2013)

Wagner, P.-O.: Ergonomie und Komfort. In: Braess, H.-H., Seiffert, U. (Hrsg.) Vieweg Handbuch Kraftfahrzeugtechnik, 6. Aufl. AZT/MTZ- Fachbuch. Vieweg + Teubner Verlag/Springer Fachmedien, Wiesbaden (2011)

Weinberger, M.: Einfluss von ACC- Systemen auf das Fahrverhalten. Dissertation an der Technischen Universität München (2001)

Widmann, U.: Produktentstehungsprozess. In: Braess, H.-H., Seiffert, U. (Hrsg.) Vieweg Handbuch Kraftfahrzeugtechnik, 6. Aufl. AZT/MTZ- Fachbuch. Vieweg + Teubner Verlag/Springer Fachmedien, Wiesbaden (2011)

Wimmer, M.: Entwicklung und Erprobung von Mensch-Maschine-Systemen zur automatisierten Fahrzeugführung. Dissertation an der Universität der Bundeswehr, München (2014)

Zenk R.: Objektivierung des Sitzkomforts und seine automatische Anpassung. Dissertation an der Technischen Universität München (2008)

Zöller, I., Diedrich, C., Abendroth, B., Bruder, R.: Fahrsimulatorvalididtät – Systematisierung und quantitative Analyse bisheriger Forschungen. Zeitschrift für Arbeitswissenschaft. **4**(4), 197–206 (2013)

Measurement Techniques

Klaus Bengler, Heiner Bubb, Christian Lange, Carmen Aringer-Walch, Nicole Trübswetter, Antonia Conti-Kufner, and Markus Zimmermann

Contents

11.1 Necessity of Experiments – 608
11.1.1 Scientific Requirements for Experiments – 608
11.1.2 Assessment of Usability – 609
11.1.3 Design of Experiments – 610

11.2 Objective Measurements – 611
11.2.1 Recording of Individual Anthropometry – 611
11.2.2 Recording of Posture and Body Movement – 614
11.2.3 Capturing Contact Forces – 617
11.2.4 Eye Movement – 619
11.2.5 Performance Measurements – 627
11.2.6 Physiological Parameters – 632

11.3 Subjective Measurements – 634
11.3.1 Psychophysics – 634
11.3.2 Interview – 639
11.3.3 Standardised Questionnaires – 647

References – 648

N. Trübswetter in cooperation with A. Conti and M. Zimmermann

H. Bubb et al. (eds.), *Automotive Ergonomics*, https://doi.org/10.1007/978-3-658-33941-8_11

11.1 Necessity of Experiments

Klaus Bengler

Every new development of an object, including a vehicle, is only justified if it is characterised by an innovation ahead of the existing one, e.g. a predecessor model or previous methods. The innovation is based on an idea which, on the basis of previous experience and existing knowledge, assumes some kind of improvement. Since innovations generally require high technical and financial expenditure for their realisation, it is necessary to ensure that the expected effect (e.g. improved safety, higher usability, better attractiveness etc.) is actually achieved before the costly realisation of a production and offering on the market.

Many demands on the ergonomic quality of the automotive product are attempted on the basis of theoretical knowledge and increasingly with the help of computer use. Despite the increasing use of these techniques, which guarantee an early optimization in the product development process, one is however dependent on an experimental examination of the acceptance of a new product or of partial aspects of this product by the using humans. On the one hand one does not want to decide without an examination and judgement of people in many areas (especially in the field of packaging) and on the other hand the corresponding theoretical knowledge is often not yet available (e.g. with new functions) or not realized in the form of virtual models (e.g. many cognitive characteristics of humans with the exception of the visual perception of information). Particularly when it comes to safety problems, appropriate experimental safeguards are indispensable.

This protection is provided by methods of evaluation and testing. The evaluation of the usability of new information systems plays an important role precisely because of the progressive development of information technology systems. In addition, the well-known methods of classical experimental design are of eminent importance for all questions of anthropometric conditions and the effectiveness of new displays and controls as well as driver-vehicle interaction.

11.1.1 Scientific Requirements for Experiments

The objective of any experimental design must be to produce results that meet the criteria of the *objectivity* (the result is independent of the test conditions and of the person carrying out the test), the *reliability* (a repetition of the experiment comes to the same result) and the *validity* (it is really measured what is subject of the investigation).

11.1.1.1 Objectivity

The objectivity of a measurement means that a measurement is independent of the test administrator and, as far as possible, of the test conditions. Particularly in the field of vehicle testing, many different tests are carried out by different groups of people in the course of development and internal tests are repeated at a later point in time by external testing institutes. Here it is very important that the influence of the test manager is minimised as far as possible by standardised procedures, instructions and measurement set-ups. A consideration of the reliability and objectivity e. g. of the Lane-change-Test can be found in Bengler et al. 2010.

11.1.1.2 Reliability

Reliability is the accuracy with which a measurement is taken. This includes the variance of the measured values for repeated measurements, but also the comparability of two parallel measurements.

In the field of vehicle ergonomics, a repeated measurement with a test person must lead to comparable results. For example, disciplined anthropometric measurements (see ► Sect. 11.2.1) generally show a high degree of reliability. This is also the case for dual-task procedures (see ► Sect. 11.2.5), provided that the test persons were trained accordingly before the start of the measurement. Otherwise, in a large number of tests dealing with the operability of functions or vehicle guidance, a limited reliability (stability) can be observed with repeated measurements. This is often due to the fact that the test persons learn to handle the system or vehicle in the course of the experiments and that the measured initial values thus differ significantly from later measured values of the same test person.

This can be remedied by training runs carried out accordingly. It must also be ensured that the object to be evaluated (e.g. the seat to be measured during a seat evaluation) or the measuring equipment (e.g. stretch marks or restoring torques) does not change in the course of the tests.

The selection of the test participants can also influence the reliability. The aim is to draw a sample that is as representative as possible by carefully pre-selecting the test persons and to document this sample in a comprehensible manner. The homogeneity of a sample can be checked with the so-called split-half calculation. For this purpose, the sample is randomly divided into two halves, which are then examined for their correlation. A verification of the reliability in connection with the validity can be done with the help of statistical methods. In principle, it should be noted that a lack of reliability always has a negative effect on the achievable validity. (see ► Chap. 12).

11.1.1.3 Validity

The validity describes the agreement of a measured value with a criterion to be measured. The level of validity is determined by the correlation of a measurement with an external criterion to be measured. For example, it could be demonstrated that the required opening intervals in an occlusion test (see ► Sect. 11.2.5.3) represent a valid estimate for the total glance durations measured by eye tracking (see ► Sect. 11.2.4) in a driving test. Especially for the evaluation of traffic safety, the validity of a measurement represents a decisive characteristic, since in laboratory or driving simulation tests valid proof is to be provided that vehicle systems can be used safely in real road traffic. Another example is the validation of stress estimation questionnaires (NASA TLX, DALI) based on physiological measurements or driving behavior data. Deviating experiments may well call this proof of validity into question.

11.1.2 Assessment of Usability

Especially in connection with interaction concepts and functionalities for tertiary tasks, the usability (*usability*) is a central property. But the interaction with primary and secondary functions of the vehicle is also based on this. Following the common definitions (ISO 9241), usability is described by means of three partial dimensions:

- *Effectiveness:* The degree to which the objectives have been achieved,
- *Efficiency:* The relationship between resources used and objectives achieved,
- *Satisfaction:* The degree to which the interaction contributes to the satisfaction of the user.

High usability has an overall positive effect on the use of the vehicle in the various situations, since it means that the tasks set can be solved with high probability, little effort and high acceptance.

Not under all circumstances, high usability also means freedom from distraction, since it is precisely in the area of tertiary functions that tasks can be solved that may conflict with the primary driving task. For this reason, ISO 17287 defines *suitability* as a further property for the interaction concepts for tertiary functions that are used in addition to the primary driving task. According to this, before information systems can be integrated into a vehicle, the intended purpose and the context of use must first be defined. Subfunctions not intended for use while driving shall be identified and measures taken to prevent the use of such functions while driving shall be described. Failures of the function are also to be displayed clearly for the driver. All evaluation results arising in the course of the development of an information system shall be recorded and documented. It is precisely this approach that is intended to guarantee suitability in the case of apps and software products that find their way into the vehicle, for example on mobile end devices. This consideration of suitability beyond usability thus represents a necessary quality assurance for the in-vehicle interaction in terms of road safety.

The evaluation of usability can be done with different methods. These include expert evaluation, heuristic evaluation and various survey methods.

11.1.2.1 Expert Evaluation

Performed by a small number of experts, possibly using checklists (e.g. TRL checklist), this form of assessment represents a very time- and cost-saving form of evaluation. However, it must be precisely documented what the assessor's expertise consists of and in what form it can be proven.

11.1.2.2 Heuristic Evaluation

This form of evaluation is based on a small sample of representative users and is based on the assumption that the majority of usability problems are highly likely to be discovered by just a few users and can be solved in an iterative procedure. Especially in these iterations paired with repeated small sample tests of future users the potential of this method lies in very early development phases. (Nielsen 1994).

11.1.2.3 Standardised Questionnaires

These tools are particularly suitable for use in comparative tests of various concepts and functional implementations. They can be used especially in the context of heuristic evaluations and enable quantitative documentation of the increasing conceptual maturity and improvements during development.

As standardised questionnaires for determining usability, they are used specifically in the vehicle environment:

- **System Usability Scale (SUS)**

Suitable for evaluating the usability of interactive systems, the SUS consists of 10 items that are evaluated by the test person using a 5-point scale (so-called Likert scale). The scale ranges from "voice full to" to "voice not closed at all". The SUS score is given on a scale of 0–100 points (see also Brooke 1996).

- **Post-Study System Usability Questionnaire (PSSUQ)**

Also designed to evaluate interactive systems, the PSSUQ consists of 19 items that are answered on a 7-point Likert-like scale. The totality of the PSSUQ subscales is used to assess usability. The subscales deal with the properties System Usefulness, Information Quality, Interface Quality, and Satisfaction (Lewis) in a differentiated way.

11.1.3 Design of Experiments

For the test, the product to be tested is manufactured according to the current state of development (e.g. seat box) or the component to be tested is reconstructed in such a way that the relevant interesting property can be investigated (e.g. a novel driver-vehicle interaction).

Prerequisite for each experiment is an assumption, which one would like to confirm or reject by the experiment (e. g.: "Large persons, in contrast to small persons, have problems entering a given new vehicle model"). In order to arrive at such a judgment verifiably, the presumption must take the form of a *hypothesis* or a combination of hypotheses in such a way that numerical values (e.g.: necessary entry times for large and small persons, EMG values making a statement on muscle load, survey results) are available which allow the decision to be made for or against the assumption of the respective hypothesis according to statistical rules (see ► Sect. 12.3.2). These numerical values are *dependent variables* because they are dependent on the given experimental conditions, the so-called *independent variables* (in the present example, the given access opening and the anthropometric and other characteristics of the available subjects). It is an essential part of the experimental design to keep these independent variables as constant as possible and at least to record them, since they can have a significant influence on the experimental result. Thus, every trial planning also includes the organizational procedure, i.e. the time schedule, the local conditions, the supply of the trial team, trial subjects and trial object, emergency plans, etc. Of particular interest in the context discussed here is the recording of the test persons' properties of the. Objective and subjective measurement methods are available for this as well as for the recording of the dependent variables, i.e. of the reaction of the test persons in the experiment. These procedures, which are of particular interest in connection with ergonomic issues, are described below.

11.2 Objective Measurements

Heiner Bubb

In contrast to subjective measurements, objective measurements are those which are not based on interviews with test persons and which are assumed to be essentially task-related (in the broadest sense). For the objective recording of the driver-vehicle interaction, in addition to the recording of the control element actuation, which can generally be read from the CAN bus data, the posture and movement of the body, the body contact with the vehicle and data that say something about the mental stress are of interest. This is shown quite directly in the reaction of the eyes. But other physiological data are also of interest, such as heart rate, skin resistance, adrenaline release and electroencephalogram (EEG), electromyogram (EMG) and electrooculogram (EOG).

11.2.1 Recording of Individual Anthropometry

11.2.1.1 Use of Measuring Tape and Callipers

The traditional method of recording body measurements uses measuring instruments or similar tools named after Martin (1914/1987). These essentially consist of various callipers and a slide compass (special calliper), as well as a calibrated measuring tape (see ▸ Sect. 4.2). Almost only distance measures from bone to bone are measured. Measuring instructions are described beside others in detail in DIN 33402, Part 1. Anthropometric measures are generally taken on people standing or sitting on a table with their legs hanging freely. The body posture is stretched to the maximum and the head is held in the so-called "Frankfurter Horizontale", i.e. in such a posture that the line connecting the lower edge of the bony eye socket with the upper edge of the auditory canal runs parallel to the standing or sitting surface. The measurements are taken on the largely unclothed test person (swimwear).

However, the mere recording of bone-to-bone distances is not sufficient to capture the diversity of human dimensions. Therefore, at least body circumferences and body weight should also be recorded. However, the transfer of such conventionally collected data to digital human models is problematic because joint to joint spacings are the determining measures that cannot be reliably determined from the bone-to-bone spacings.

11.2.1.2 Non-contact Processes (Computer Anthropometry)

In the non-contact measurement of human dimensions, optoelectronic measuring instruments are connected to a computer, which both controls the measurement procedure and calculates the data obtained. These methods are also known as computer anthropometry. The simplest method, the so-called *shadow crack procedure* the subject to be measured stands in front of a wall illuminated in white by rear projection and is recorded by an electronic camera. If the illumination and camera sensitivity are set correctly, you will get a shadow image of the body on the CCD of the camera. In a connected computer, relatively simple image processing technology is now sufficient to extract the coordinates of the border line of this silhouette. For the correct recording of body measurements, defined body positions must be observed (see ◘ Fig. 11.1). In particular, separate mounts for front view and side view are required. The body height is obtained directly as the distance of the highest point of the body from the floor, whereby the perspective image, which is determined by

Fig. 11.1 Anthropometric measurement using shadow cracking techniques. (Source: HumanSolution)

the distance of the camera from the wall and by the optical system, is taken into account. By determining the maximum distances of the limit line in relation to the body height at defined heights, further body measurements (e.g. shoulder width, hip width, crotch height etc.) can be automatically recorded if the accuracy requirement is not too high.

11.2.1.3 Superimpose Method (PCMAN)

As part of the development of the RAMSIS human model, which could originally only be installed on workstations, a three-dimensional human model running on the PC was also developed, which is identical to RAMSIS in its geometric design, but has no other functions apart from the option of freely setting the dimensions and angles. This model was used to transfer the individual anthropometric measures of test persons to the RAMSIS model and to further realize the evaluation of driver postures necessary for the RAMSIS development. The corresponding software program was then further developed under the name PCMAN at the Chair of Ergonomics at

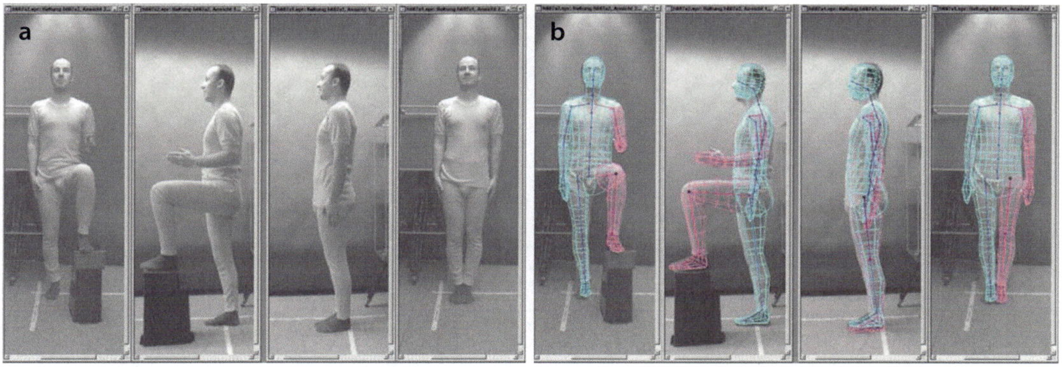

Fig. 11.2 Photogrammetric recording of the test subject **a** and overlay with the computer dummy **b** using the PCMAN procedure

the TU Munich. The special feature is that an interface to RAMSIS exists for this model, which allows both anthropometric measurements and angle values to be transferred to RAMSIS. For anthropometric measurement, the procedure consists of recording the subject by two cameras whose optical axes are perpendicular to each other. A calibration of the images captured by the cameras makes it possible to project the dummy present in the computer onto virtual planes using software in such a way that the corresponding dummy images can be superimposed with the photos of the real test subject. Now it is possible to adapt the dummy to the proportions of the test subjects (see Fig. 11.2). The further anthropometric measuring program provides for various fixed postures, through which the position of the joints and, to a certain extent, their variation depending on the posture are recorded. By modifying the outer model of the dummy so that its contour is as close as possible to the contour of the person being photographed, a good image of the dummy is obtained on the computer.[1]

1 In principle, the PCMAN method can also be combined with the shadow crack method. Due to the lack of information about the individual body parts, however, this adjustment process is very uncomfortable for the user and also leads to worse results. The idea to enable an automatic adjustment with the combination of shadow crack method and PCMAN has not proven to be successful so far.

11.2.1.4 Bodyscanning

The most common method today to record anthropometric measurements quickly and without contact is the method of laser scanning, also known as (body) scanning. A light plane is generated via a laser with widening optics. The test person to be measured stands in this plane of light and thus forms a visible intersection line with it. A video camera, located at a fixed distance from the laser defining the light plane, captures this intersection line. Using computer-based image processing, the position of the pixels on the camera image is recorded and converted into the spatial coordinates in the computer using simple trigonometry. By moving the entire apparatus, consisting of laser and camera, electromotively along a linear guide, the plane of light is drawn through the entire body to be measured. This process requires approx. 120 s. Thus one receives the coordinates of the spatial position of the body surface. Since only one sector of the body can be measured from the angle of a camera, four devices as described above are usually used to measure the whole body. These are arranged in a square layout around the body and moved synchronously (see Fig. 11.3). Recently, there are also hand-held scanners on the market that make it possible to capture individual body segments. In any case, separate algorithms are necessary, with the help of which anthropometric data can be obtained from these scans. Within the framework of the SizeGERMANY project, algorithms for data acquisition were

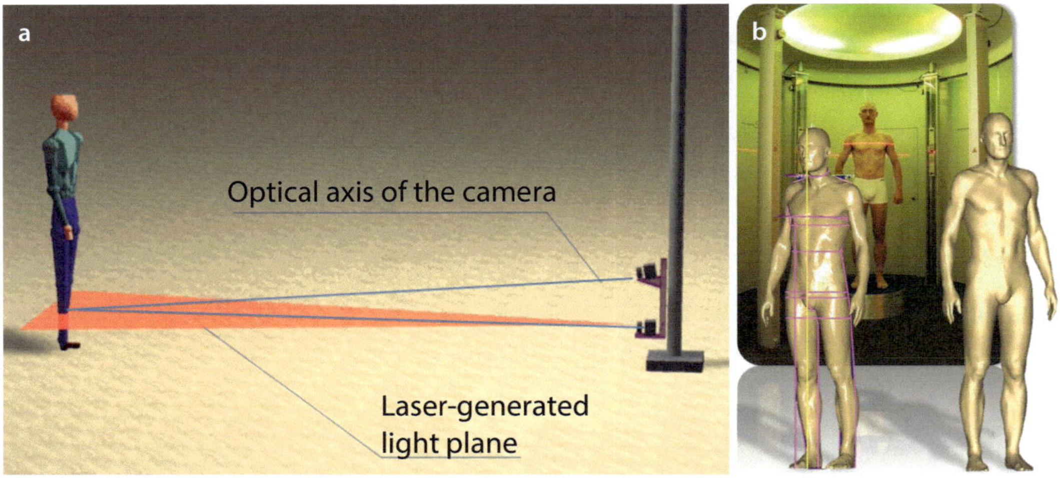

Fig. 11.3 Principle of Laserscan Technology **a** and practical design of a laser scanner **b**

developed which made it possible to carry out statistical analyses of over 12,000 test persons. Such body scanners are used today in many automotive companies to record the anthropometric data of the test persons employed there.

11.2.2 Recording of Posture and Body Movement

The objectification of body posture and movement as a temporal sequence of postures is done with known individual anthropometry in principle by recording the joint angles.

11.2.2.1 Mechanoelectric Methods

In the mechanoelectric processes, measuring sensors, so-called goniometers, are adapted to the joints from the outside, which themselves contain a joint that is adjusted via the respective body position. The relative position of the joint is usually measured using potentiometers, the data is then transferred to the computer and used there to track an often quite simple human model ("stick figure"). It goes without saying that only two-dimensional movements can be recorded with this method. In addition, there is the disadvantage that the test person is connected to the computer via a cable harness. Despite these disadvantages, this method is often preferred because of its robustness, especially since the wiring harness in the vehicle does not represent a particular obstacle when correctly laid. A modification of this method is the application of fiber optic elements at the respective joints, which change their light transmission depending on the curvature. In particular the movement of fingers, but also that of the spinal column can be recorded quite well online.

11.2.2.2 Marker-Based Methods

The most widely used methods use markers to capture posture and motion. Sometimes ultrasound transmitters are used as markers (three receiving detectors distributed in space are necessary for spatial localization of the transmitter), electromagnetic methods, but mostly active or passive light sources are used (such methods are distributed by many manufacturers, e.g. Vicon, Polhemus, Peak, Qualisis, Simimotion etc.) The passive light source method is most commonly used in the form of well reflecting balls attached to the extremities and the trunk (these are better than flat markers because they are more likely to be detected regardless of the position of the subject in space). In some cases, special systems consisting of four spatially configured markers are also used, whereby this arrangement helps to precisely capture the orientation of the body part to which this arrangement is attached (Fig. 11.4) and to generate an image on – at

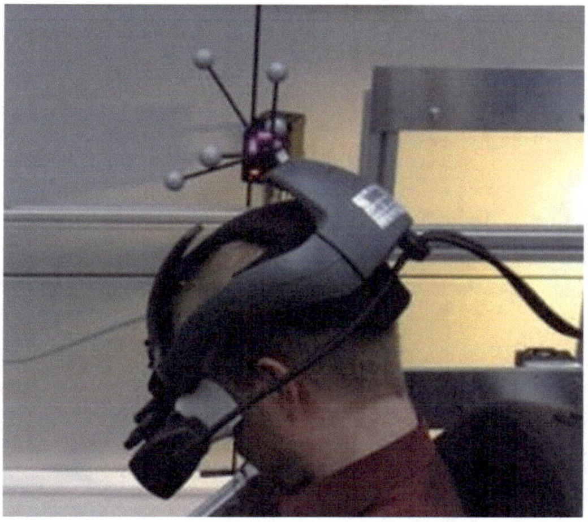

Fig. 11.4 Marker arrangement for detecting the spatial orientation of a body element

least – two cameras installed in the room. In addition to the normal lighting, the scenery is illuminated with infrared light. Cameras that are only sensitive in the infrared light spectrum are used for recording. This makes it possible to detect the position of the markers on the CCD of the camera very reliably. Using a simple geometric calculation, the computer calculates the spatial position of the markers with knowledge of the camera focal length. Since the markers are not coded and the identification of the respective marker on all images is necessary for the calculation of the spatial coordinates, incorrect calculations can occur if several markers on one image exist, which is absolutely necessary for the motion capture. Various algorithms have been developed to reduce the probability of such misalignments (e.g. by calculating the probable motion of a marker image based on the previous motion). Nevertheless, they pose a major problem in practical measurements, especially when markers are temporarily obscured by objects in the working environment.

In order to move from the spatial marker positions to human-related data, two methods are practically possible: in the simplest case, the markers are attached as precisely as possible to the position of the respective joints. In the computer, the corresponding spatial points are then connected with each other in such a way that the representation of the test person is given by a stick figure. Now it is quite easy to calculate angles between distances, which represent posture resp. motion depending on time. Since the markers cannot be placed in the joints, but only next to them and only imprecisely, this method results in considerable inaccuracies.

With the second method, a human model exists in the computer which must be as well adapted as possible to the individual anthropometric measures of the test person. With this method, the markers can be attached virtually anywhere on the respective body elements (this method is commercially offered by Vicon, Fig. 11.5). In a special calibration procedure, however, the positions of the corresponding virtual markers must be determined before the actual measurement. In the following movement recording, the human model is tracked by means of the measured marker positions via these virtual markers. The angle values of this human model show the posture and movement much more accurately than with the first method (we would like to refer here to the work of Rigel 2005 and Cherednischenko 2007). However, both methods have in common that the dimensions of the respective body elements can vary during the movement. In order to compensate the ultimately inaccurate correspondence of the joint points assumed by the markers or by the human model with the real joint points, the error caused by this can be reduced by starting from a human model with rigid dimensions adapted to the individual anthropometry and connecting the virtual markers elastically to this human model by means of a program (Rigel 2005).

11.2.2.3 Markerless Procedures

Markerless measurements are always based on optical detection. The subject is observed by at least two cameras. This is the main application of the PCMAN. The test subject is recorded from different perspectives in the experimental position of interest (e.g. sitting in a motor vehicle mock-up) with the aid of several cameras (two to three cameras proved to be useful) (see Fig. 11.6, left). If the optical parameters of the cameras are known, it is possible to calculate a central-perspective

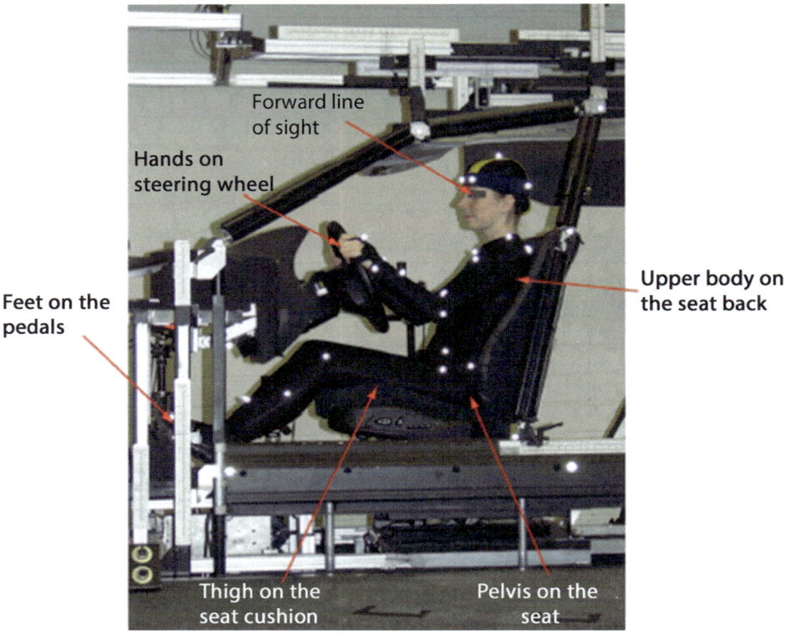

Fig. 11.5 Recording scenery for motion detection using marker and camera. (Cherednischenko 2007)

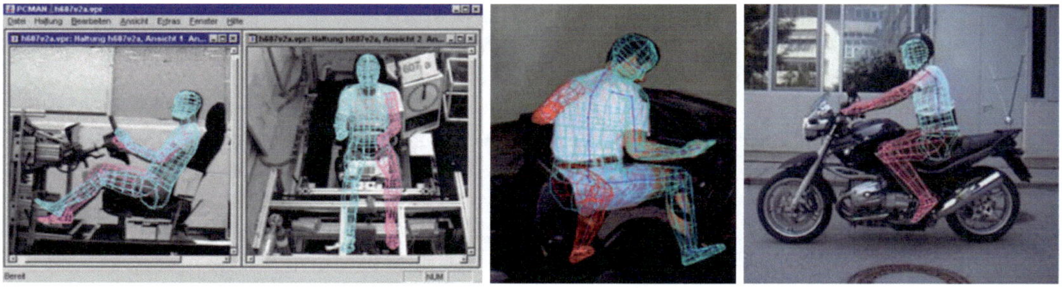

Fig. 11.6 Configuration for posture and motion analysis using PCMAN

image of the dummy, which has already been adapted to the test subject, on the image planes given by the cameras in the computer with the same parameters, so that by superimposing the camera images with the corresponding calculated images and by animating the dummy, the model can be adapted to the test subject's posture. As shown in Fig. 11.6, this method can be used to record any posture and describe it in the form of mathematical angle values. This method is used in various automotive companies to record body posture in seat boxes, but also in moving vehicles, when getting in or on a motorcycle. Thanks to the very small cameras available today, this can also be carried out in confined vehicle spaces.

A particular challenge, however, is the measurement of movement. In principle, it can be understood as a sequence of individual postures. However, the effort required to capture individual images using the overlay technique is extremely high. For this reason, the use of the marker method described above has become generally accepted in anthropometric measurement technology for the recording of body movements (e.g. for investigations of the entry and exit process). Although these methods quickly provide precise data, they have the disadvantage that the test subjects feel impaired by the markers when they are glued on during the movement experiments. Within the framework of the DFG project MeMoMan a markerless method was developed, which is based on the superposition method shown in Fig. 11.6, but which is currently still in research. In the meantime, Simimotion has presented a similar process on this basis, which is commercially available.

11.2.3 Capturing Contact Forces

If the user of a vehicle sits in the vehicle, performs movements there, but also if he gets into or out of the vehicle, he exerts forces on the vehicle at the contact points. For many investigations, these forces are of interest because they may allow statements to be made about the effort that the user has to expend in order to perform the respective tasks. Of particular interest in this context, of course, are the forces in the seat and in the backrest, because it is hoped that these will provide information on seating (dis-)comfort.

11.2.3.1 Measurement of Seat Pressure Distribution

In seat research, the seat pressure measuring mat is an indispensable aid today. These are matrix-shaped mats at whose intersection points the slight change in distance exerted by pressure is converted into an electrical value via different physical principles (change in resistance, change in capacitance or change in light transmission). This, supported by the computer program connected to the mat, produces a pressure distribution image, which normally shows the distribution of the pressures in the seat and backrest surface in color-coded form (Figs. 11.7 and 11.8).

A general technical problem here is that the mat – similar to a sheet of paper – can only be bent about one axis, so that when the measuring mat fits into a real seat, wrinkles always occur that can influence the measurement result. In principle, the pressure values cannot be used directly to make statements about the perceived discomfort. There are physiologically induced limit values that must not be exceeded at all. Diebschlag et al. (1992) state that pressures between 0.08 and 0.16 N/cm^2 lead to an impression of the venous blood flow and pressures >0.42 N/cm^2 cause a critical impression of the blood supply at all, an effect which is subjectively experienced as "falling asleep of the legs". As the investigations by Hartung (2006) and Mergl (2006) show, individually related values of load distribution, gradient and maximum pressure, especially in the area of the thigh, are decisive for minimizing discomfort. This creates another problem: The values obtained with the measuring mat are seat-related and also depend on the positioning of the mat on the seat. In order to arrive at an evaluation of the discomfort, however, it is necessary to interpret these values in relation to the test person. Hartung (2006) has therefore developed a special body-grid method with the help of which the values obtained can be related to the individual test person and his current position in the seat (see Fig. 11.9).

Fig. 11.7 Seat pressure mat in a vehicle seat

11.2.3.2 Measurement of Other Contact Forces

In addition to the important distribution of pressure in the seat and backrest that influences long-term discomfort, short-term or long-term contact forces also play a role, which must be taken into account in experiments with different objectives. Apart from dynamic movements during boarding and alighting, handling in the luggage compartment or in the engine compartment, these are mainly contact forces in the following areas during conventional driving (compilation of the results of Fröhmel 2010 tests):

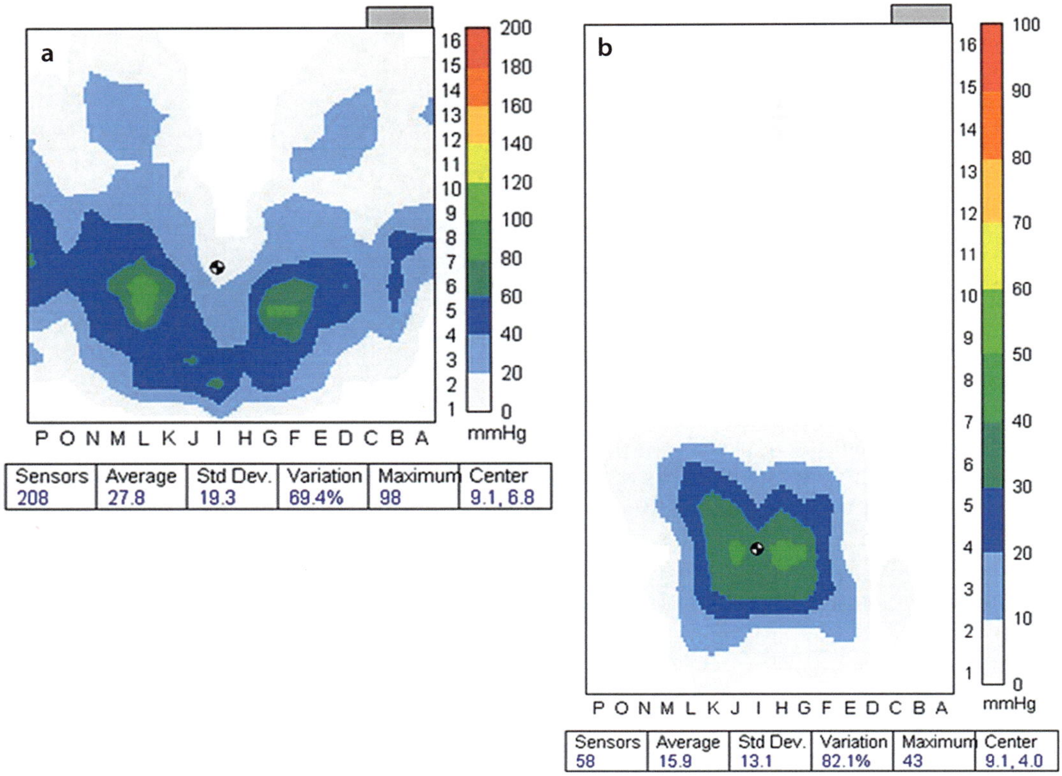

Fig. 11.8 Example of a pressure distribution image. **a** Seat, **b** back

Footwell:
- Left foot on footrest and clutch pedal
- right foot on brake and accelerator pedal
- left and right foot on the floor

Legroom:
- left knee at the driver's door in front
- right knee at the centre console in front

Arm area:
- Left elbow or forearm on the armrest door
- left elbow or forearm resting on the parapet of the driver's door
- right elbow or forearm on the armrest centre console
- right hand on the gear lever
- left and right hand on steering wheel

Seating area:
- seat and back on the seat
- Head on the headrest

For the measurement of these contact forces, two fundamentally different methods can be considered: either measurements directly on humans or measurements using force transducers attached to the vehicle. When force transducers are attached to the test persons, it should be noted that they must not influence their habits if possible. The weight and dimensions of the sensors must therefore be as small as possible and lines must be avoided as far as possible or laid in such a way that the test person is not affected. In principle, a similar consideration applies to the attachment of measurement sensors to the vehicle: they must not change the geometry and anthropometric conditions of the vehicle, because the resulting change in the geometric conditions of the test persons would also necessitate other applications of force. Fröhmel (2010) describes different types of force transducers as well as their advantages and disadvantages and compiles their suitability for measure-

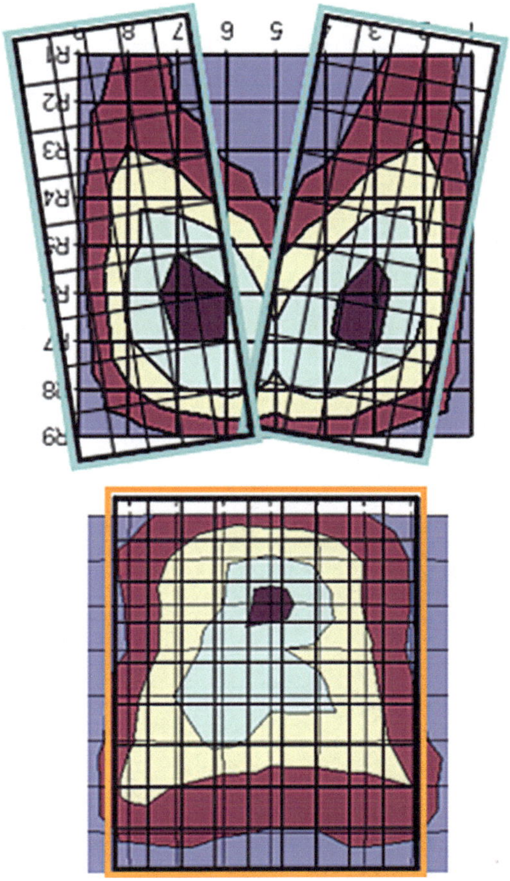

Fig. 11.9 Adaptation of a body-grid adapted to the individual anthropometry of the test person to the measured pressure distribution. (After Hartung 2006)

ments of driver posture in a table (Table 11.1).

It results from this that measurements on the vehicle are in many respects cheaper than those on humans. Fröhmel compiles the following sensor measuring ranges from data sheets of the Handbook for Ergonomics (Schmidtke 1989) and internal measurements:

Measuring range of pedals and footrest:

- Footrest: 0 ... 800 N
- Clutch pedal: 0 ... 800 N
- Brake pedal: 0 ... 1500 N
- Accelerator pedal: 0 ... 150 N

Measuring range floor (it is assumed that each floor plate is mounted on three sensors)

- the ground sensor: 0 ... 200 N

Seat measuring range (it is assumed that the seat is mounted on four sensors)

- per seat sensor: 0 ... 500 N

Measuring range steering wheel (it is assumed that the steering wheel is mounted on a three-dimensional force transducer)

- per space coordinate: 0 ... 500 N

The stated values refer to static load. In the case of the test arrangement described by Fröhmel, these force transducers were installed in a static mock-up. During the boarding process, all sensors were secured so that no dynamic forces could be transmitted that would have destroyed the force transducers. For this reason, a corresponding measurement in a moving vehicle is practically impossible. Estermann (1999) asked the test subjects to take a seat in a mock-up immediately after the test drive, with a similar question, and the setting in the real vehicle was transferred exactly to this. The test subject should adopt the same posture as in the test vehicle, which could be verified using the PCMAN method.

11.2.4 Eye Movement

Christian Lange

As mentioned at different points in the previous chapters, eye tracking is a quasi-objective method of measuring attention and visual effort. One can essentially distinguish between measurement with head-based measuring instruments and non-contact measurements. In addition, the electrooculogram (EOG) method is available, which registers the potential difference between the cornea and retina by means of surface electrodes attached near the eye. In the angle range until 40° this method provides good results. However, the EOG measurement is very susceptible to artifacts due to day-dependent voltage fluctuations, as well as potential changes due to blink of the eye and facial musculature (see also Electromyogram, EMG).

11

Table 11.1 Evaluation matrix of different force measurement methods for driver posture (from Fröhmel 2010); coding: +++ "very well suited" ... – - – "very poorly suited"

Basis Human	Influencing the VP		Sensor characteristics				Measurement	
	Weight	Dimen-sions/movement	Expenses	Sturdiness	Calibration/measurement accuracy	Avail-ability	Of surface load	Of secondary forces
Gaiters with foil cushions Filling medium gas	+	- – –	- – –	- – –	–	- – – –	+	–
Gaiters with foil cushions Filling medium liquid	- – – –	- – –	- – –	- – –	–	- – – –	+	–
Gaiters with light sensors	+	- – –	- – –	++	- – – –	- – – –	+	–
Foil measurement technology (measuring glove/foot pressure sensor mat)	++	–	+	- – –	- – –	++	++	- – – –
Suit with miniature force sensors	- – –	- – –	++	–	+++	++	- – – –	+++
Base vehicle								
Foil measuring technology (foil sensor mats)	+++	++	+++	–	- – – –	+++	+++	+++
Force sensors	+++	+++	++	++	+++	++	- – – –	++

11.2.4.1 Head-Based Methods

A visual field camera is attached to a helmet (also glasses etc.) which is fastened to the head of the test person as stable as possible. The image of this camera is superimposed on the direction of view of the eye in the connected computer. The following procedures are used to record eye movement:

- Limbus, pupil or eyelid registration: The eye is illuminated extensively with infrared light. A second eye camera attached to the helmet uses a mirror that only reflects infrared light and is therefore completely transparent to visible light, or directly captures the image of the eye (◘ Fig. 11.10). By a corresponding calibration procedure the eye image is enlarged (zoomed) so far that it can be superimposed directly on the visual field image. In another version of this technique, the position of the dark pupil in the eye image is recorded using image processing technologies and the position thus obtained is marked in the visual field image (◘ Fig. 11.11).
- Cornea reflex method: An infrared light source connected to the helmet directs a focused beam onto the eye. The facet-like surface of the eye reflects this beam, which is captured by a camera also attached to the helmet (so-called Purkinje image technique). The position of his image detected there can be superimposed on the field of view image after a calibration procedure and thus indicates the current viewing direction (the result is similar to that of ◘ Fig. 11.11b).
- Point of Regard Measurement: An infrared light beam is directed onto the eye and creates a reflex there. A second camera attached to the helmet records this reflex and the current position of the dark pupil. Image processing is used to capture the positions of the images of both objects. From their relative position to each other, the position of the eye and thus the viewing direction can be calculated in the computer and superimposed on the visual field image (the result is similar to that of ◘ Fig. 11.11b).

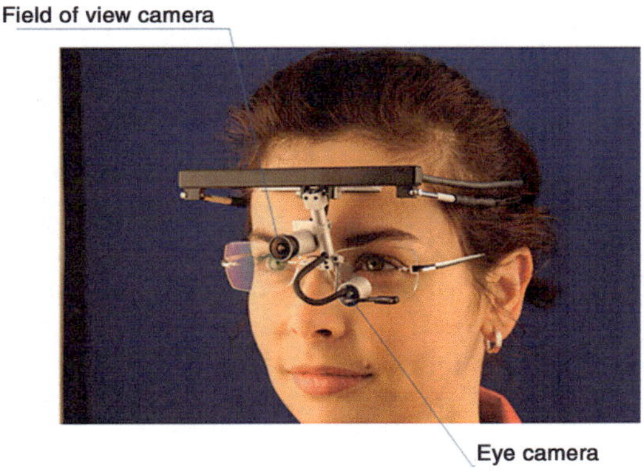

◘ **Fig. 11.10** Blickerfassungssystem Dikablis

In contrast to procedures based on image processing technologies, the first simple superimposition method is very robust against disturbances. In any case, however, it requires human interpretation in the evaluation of the individual images. In contrast to this method,

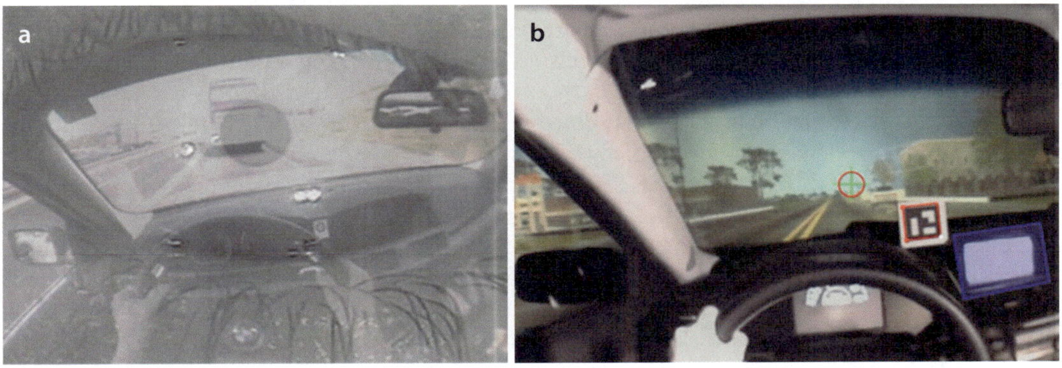

◘ **Fig. 11.11** Marking of the object viewed only by the pupil image **a**, by means of image processing **b**

procedures with image processing technologies make it possible to capture both eyes and thus also convergences of the eye axes. However, image processing is very sensitive to light fluctuations and is therefore only suitable for use in the open field to a limited extent.

11.2.4.2 Non-contact Methods

With the non-contact methods, one or two cameras permanently mounted in the room (usually on the dashboard surface in the vehicle) observe the test person. A pulsed infrared radiator is directed at the head of the test person and produces a corneal reflex on both eyes. The images from the two observation cameras are analyzed in a computer using image processing. The head position and head posture are also recorded on an optical basis or, if necessary, by means of an electromagnetic head tracking system, since the eye position (in contrast to the head-based methods) is not static in relation to the camera coordinate system and is therefore only known after appropriate image processing. In a similar way to the helmet-based procedure described above in third place, the respective position of the reflex, pupil and iris in each camera image is determined. Depending on the system, this happens only on one eye (e.g.: ETS system) or on both eyes (stereo system, e.g. Facelab). From the calculated head position and recognized orientation points (e.g. corner of the mouth, eyebrows, nose, ...) within the face, with the help of the calculated relative position of the eyes, the viewing direction of the eye (or both eyes) in space is determined. General eye-based metrics such as saccades, fixations, or pupil diameter can be derived from all systems. Depending on the system and the expansion stage, the calculated view vector can still be used in several ways. Firstly, the direction of view can be used to draw conclusions about the viewing direction of familiar areas, such as a mirror view, the display area in the dashboard or a view of the scenery behind the windscreen (see Fig. 11.12). If the environment is fixed and the positions of the possible static observation points (AOIs = areas of interest) are known by a CAD model, a fully automatic evaluation is possible. Secondly, the view vector for dynamic environments (e.g. the course of the road) and the dynamic AOIs contained therein can be continuously marked on the one hand in the image of a camera (or the video recording of a simulation) permanently connected to the scenery (e.g. facing forwards in the vehicle) (see Fig. 11.13).

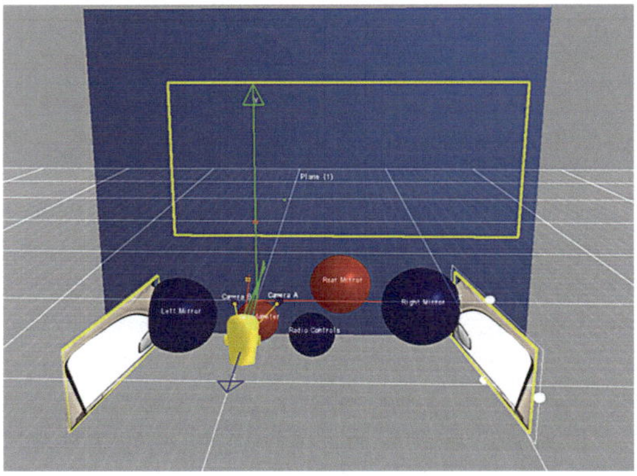

Fig. 11.12 CAD model of the environment (mirror, canvas of a driving simulation, ...), the recognized head (yellow), as well as the calculated view vector (and the green intersection with the scenery on the canvas) of a non-contact system

Because of the many interlocking image processing processes and coordinate transformations (from camera coordinates to head coordinates, on to eye coordinates, and finally to the gaze vector), non-contact methods are sensitive to light fluctuations, vibrations or larger head rotations and gaze deflections. Although they would be particularly helpful for use in vehicles, their application is therefore rather difficult there.

11.2.4.3 Technology Comparison

In the following, a technology comparison of the eye-tracking methods is presented (Table 11.2). It builds on the systematisation of Taylor et al. (2013) and was supplemented by exemplary systems. The abbreviations in the second column are explained in the following text.

Typically, all systems generate angular errors between 0.5° and 1° after calibration in the ideal case. The systems are usually real-time capable, i.e. the gaze data can be retrieved with low latency via network channels (TCP/UDP).

Fig. 11.13 Combined recording of the scenario (virtual reality on a screen), several AOIs, and the view (green dot)

All eye trackers allow eye based measurements (**A**) such as the occurrence of fixation, saccades, and usually also the detection of pupil diameter.

Most non-contact eye trackers (B-ET) as well as pure head trackers (B-KT) enable visual head-based measurements (**K**) and to make statements about the head position and the head direction (e.g. on scenery or a shoulder look). For head-based systems, this functionality can usually be realized via contactless head tracking (K-ET-S-HT). However, in contrast to contactless eye trackers, this does not allow statements to be made about the availability of the user (**V** e.g. eyes open), as contactless systems usually only recognize the pupil, but not the facial geometry.

The strength of all eye trackers are of course eye based measurements (**B**) of gaze interactions, e.g. gaze directions (or e.g. fixations) to certain coordinates. All eye trackers support these measurements relative to the head and usually provide more accurate angles or coordinates (again relative to the head position), which are then used to overlay the relative view coordinates of head-bound scenery cameras (K-ET-S). Such records, however, make only rough statements about which objects were viewed, or they require a complex evaluation or conversion, such as in world coordinates. If contact-free eyetrackers are coupled with head tracking systems (K-ET-S-HT) or if contact-free eyetrackers (B-ET) are used, these are capable of visual interaction in world coordinates (**BW**) (e.g. within the simulated scenery or a real room). This enables a fast computer-based evaluation if the positions of the AOIs within the world are known. This is usually the case for a shelf (in market research), a monitor (for software evaluation) or a projection screen (in a driving simulation). The last two examples show why a desktop recording or a virtual reality recording (B-ET-VR) can be useful here: If software or a simulated environment is evaluated, it is usually also known which graphic (static and dynamic) elements were output on the displays at which position; this simplifies automated evaluation.

The strength of non-contact eye trackers is the fast detection of gaze interaction with defined static (**BS**) AOIs (e.g. fixation on mirrors or non-moving areas). These are predefined, which allows a fast computer-based evaluation. The strength of head-based eye trackers, on the other hand, is their simple evaluation of gaze interaction with defined dynamic (**BD**) AOIs (e.g. of a traffic light) in the scenery, since the gaze data relative to the head position always match the fixed head-based scenery camera. In order to automate the evaluation, markers are either positioned in the environment before the experiment, or image processing algorithms are applied after-

Table 11.2 Technology comparison

Category	Functions	Systems	Tracking	Special features
Head-based eye tracker (K-ET)	A B MO €€	SR Research EyeLink II	P C 500 Hz (1000 Hz)	Two-eye system, heavy system (K).
Head-based eye tracker with head-mounted scenery camera (K-ET-S)	A B BD MO €€€	Ergoneers Dikablis	P 50 Hz Interlaced	Single-eye system, scenery camera (fieldcam) integrated ex works. Marking of AOIs in the scenery by means of markers is possible (simplifies the evaluation).
		TobiiGlasses	P 30 Hz Scenery camera: 56°.	Single-eye system, scenery camera and microphone integrated ex works.
		SR Research EyeLink II + Scene Camera + Eyeworks	Scene camera: 250 Hz	See above
		SMI Eye Tracking Glasses 2.0	P C 30–60 Hz Scenery camera: 60°.	Two-eye system, HD scenery camera and microphone. AOIs can be marked in the scenery using object recognition (simplifies evaluation).
Head-based eye tracker with head-mounted scenery camera and head tracking (K-ET-S-HT)	A K B BD BS BW MO €€€€	SMI Eye Tracking Glasses 2.0 + SMI 3D Eye Tracking Package		6D head tracking via standardized VPRN interface (A.R.T., Vicon, Kinect, WorldViz, ...). Additional shutter glasses functionality for VR environments possible.
		Ergoneers Dikablis + Vicon Nexus Motion Capturing	Dikablis: See above Vicon: 120–1000 Hz	Fusion of the 3D human position with the 3D view vector in real time. Coupling with human movements.
Non-contact headtracker (B-HT)	A *B BS BD* BW K V KO €	Seeing machine face API	±90° head rotation	Works with any webcam.

Table 11.2 (continued)

Category	Functions	Systems	Tracking	Special features
Non-contact eye tracker (B-ET)	A K V B BS BW KO €€	Seeing machine face LAB	P I C 60 Hz ±45° cover	Two small cameras that can be positioned relatively freely. Fusion of the 3D head position with the 3D view vector in real-time. Multiple systems can be connected to increase the tracking range.
		Smart Eye Pro	P I C 60–120 Hz ±45° cover	Two small cameras that can be positioned relatively freely. Up to 8 cameras per system to increase the tracking range.
		SR Research EyeLink 1000 Plus	P 500 Hz (1000 Hz, 2000 Hz)	Various camera versions (desktop, LCD, tower, primate research, long range). Only video overlay of the desktop possible (BW), no knowledge of the head position (K), not suitable for the vehicle, large camera.
		SMI Red	P C 60 Hz (500 Hz) ±20° cover	Cameras are pre-assembled in a strip and are large (good for desktop applications, not suitable for in-vehicle use), small horizontal angle cover.
		Tobii X/X2	P 60–120 Hz	Cameras are pre-assembled in a strip and partly large (good for desktop applications, unsuitable for use in vehicles), small horizontal angle cover.
		Eyetech VT2	P 80+ Hz	Cameras are pre-assembled in a strip and partly large (good for desktop applications, unsuitable for use in vehicles), small horizontal angle cover.
		LC Technologies EyeFollower	P 120 Hz ±45° cover	4 cameras, pulsed infrared light, very large housing (34x18x23 cm).
		Mirametrix S2	60 Hz	Cameras are pre-mounted in a strip and large (good for desktop applications, unsuitable for in-vehicle use), small horizontal angle cover, relatively inexpensive (€)
		ITU Gauze Tracker	P C	Free eye tracking with any webcam. Head-based mode also possible.
		Open Eyes	P	Free Eyetracking with assembly instructions for hardware.

(continued)

Table 11.2 (continued)

Category	Functions	Systems	Tracking	Special features
Touch-free eyetracker with static scenery camera (B-ET-S)	A K V B BS BD BW KO €€€	Seeing machine face LAB + Scene Camera	Scenery camera: 30 Hz, 95°, 95°, black	
		Smart Eye Pro + Scene Camera		Normal scenery camera or panorama camera possible.
Non-contact eye tracker with screen-based recording or virtual reality (B-ET-VR)	A K V B BS BD BW KO €€€	Eyetracker from Arrington Research, EyeTech, LC Technologies, Mirametrix, Seeingmachines, SMI, SR Research, Tobii + Eyetracking Inc. Eyeworks	20 Hz	Screen recording or recording of the virtual reality with 20 Hz and superimposition of the view coordinates. Mark and move dynamic AOIs in a timeline.

wards that can dynamically track objects. A third possibility is the semi-automatic marking of objects using geometric elements (circle, square, ...) and the gradual movement of these elements. However, head-based eye trackers must treat static AOIs like dynamic ones, which requires additional effort for evaluation. Non-contact eye trackers cannot usually capture the entire field of view of the test person, as either the recording is limited to generated information (desktop or virtual reality, B-ET-VR), or a vehicle-bound scenery camera (B-ET-S) only provides a limited image section.

Ultimately, head-mounted and non-contact systems are diametrically opposed in two respects: For the former, the user is mobile (**M**) and little restricted in its freedom of movement, with the latter the use is comfortable (**KO**), as the test person may not have to carry heavy equipment on his body.

Different tracking technologies are used, whereby pupil tracking (P) and iris tracking (I) are less accurate than corneal reflex extracting (C). However, corneal reflex extracting is only advantageous if an infrared source is used whose position is clearly defined (both in head-based and non-contact systems). A sampling rate of 50 or 60 Hz is usually sufficient, but faster sampling rates allow better/stable eye-based measurements, such as saccade detection or high-frequency changes in pupil diameter.

Depending on the expansion stage, the device categories are cheap (€) to very expensive (€€€€), whereby considerable manufacturer-specific differences can be found within the categories.

11.2.4.4 Interpretation of Gaze Data

For the analysis of gaze data, so-called areas of interest (AOI's, see above) must often be defined in the environment. These can be objects on the road (traffic signs, other road users), the environment (advertising signs, buildings, objects of nature, irrelevant passengers or vehicles) or objects in the vehicle (instruments, mirrors, controls). According to ISO/TS 15007–2 (2001), the deflection effect of the driving task but also the attention to special objects can be described by the following parameters:

- **Total glance duration of an AOI** (Summation of all single glance durations on an AOI during the observation period).
- **Maximum glance duration on an AOI**
- **Number of glances on an AOI**
- **Average glance duration on an AOI** (Total glance duration on an AOI related to the number of glances on that AOI).

- **Percentage glances on an AOI** (Total duration of the AOI in relation to the duration of the observation period).
- **Glance sequence of an AOI** (Number of views of one AOI per time unit [second]).

These characteristic values can be automatically evaluated for AOIs that are locally fixed, such as a display or a control element. For mobile AOIs, such as pedestrians or other road users, however, the evaluation must be done manually or image processing recognition signs must be used to detect specific objects.

There are also visual parameters which represent the driver's stress. It's these:

- **Average fixation period:** According to Menn et al. (2005), the average fixation duration increases with mental or physical fatigue.
- **Amplitude of the average saccade angle:** according to Menn et al. (2005), the amplitude of the average saccade angle decreases from 5.5 ° (free distance) to 4.6° e.g. during tunnel travel (= increased stress).
- **Visual search activity** = average saccade angle per time unit (Saito 1992): According to Menn et al. (2005), the visual search activity decreases from 17.8°/s in tunnels to 2.5°/s in free stretches (B). The value is calculated by taking the sum of all individual journeys for the corresponding section of the route and adding it to the duration of the section.
- **Percentage of environmental views:** Schweigert (2003) and Thompson (2005) show that these characteristic values decrease when the driver is exposed to a secondary activity (see also ► Fig. 3.55). Environmental glances are defined as all glances that are not part of the primary driving task, such as glances into the sky or onto trees. This does not include glances at traffic signs or other road users.

With the exception of the latter parameter "percentage of environmental views", all driver stress parameters can be calculated automatically because they are not related to the environment, but represent pupil activity measures (Lange and Bubb 2008).

In this context, some comments will be made on the quality of eye tracking data. In general, anomalies in the data that could influence the results should be reported when they occur in the context of eye tracking experiments. These include measurement interruptions due to changing lighting conditions or dark eyes of the test person. If, for various reasons, data is subsequently corrected or excluded from evaluation, this must be documented.

Before an automated evaluation of eye tracking data is started, it is recommended to check the quality of the data, because depending on certain test persons or test situations beside others there can be significant dropouts in the measurement. In accordance with ISO Standard 15,007–2 we speak of *excellent* data, if the availability of the measured frames is above 95% and of a *good* measurement with availability up to 85%. Below 70% is considered an unacceptable measurement in this context.

11.2.5 Performance Measurements

Klaus Bengler

In the sense of the man-machine system, the driver uses the vehicle to master the driving task. This task must be carried out with a certain level of quality in order to comply with road safety and traffic rules. According to the general ideas of ergonomics, this quality is influenced on the one hand by the individual characteristics and abilities of the driver and on the other hand by the difficulty of the driving task, the available technical material (vehicle with all its characteristics) and by distractions from the activity. For this reason, test measurements of driving performance provide important information about the above-mentioned influences. On the basis of a normative model (traffic rules, instructions, safety considerations), different assessments are made on the basis of driving metrics. In this context, special reference is made to ► Sect. 2.1.3, which deals with the application of the definition of quality and performance to the field of driving tasks. Many of the metrics presented in the following do not completely fulfill the relationships presented there. Nevertheless, they are widely described and used in literature. In order to identify these deviations, the following simple statements are referred to:

- Quality, defined as the degree to which the task is fulfilled, is a dimensionless quantity. It is often given as a percentage value.
- Performance as the quality produced in time therefore has the dimension 1/time (generally 1/s, more rarely 1/min).
- All dimensional variables (e.g. m, km/h etc.) are to be categorised as indicators of quality or performance.

11.2.5.1 Metrics for Measurement and Evaluation of Vehicle Longitudinal Control

Vehicle longitudinal control metrics are often used to evaluate the driver's interaction with driver information systems or driver assistance systems. The main categories are as follows:

- speed
- vehicle following behaviour
- pedal usage

▪ Speed

There is an original connection between speed and measurable road safety, which is justified by the high degree to which the speed driven is included in the damage balance in the event of a possible accident due to driving physics (Nilsson 1984). The measurement of speed can be carried out very well via appropriate vehicle interfaces (CAN bus), via measurement systems installed in the vehicle (COREVIT) and meanwhile also very precisely and reliably via mobile measurement platforms e.g. on the basis of a smartphone. This data is often used for evaluation purposes:

- the average speed driven
- the standard deviation of the speed (variability)
- the maximum or minimum speed.

All these measures can be related to a time duration or, depending on the question, to a distance covered (e.g. during the solution of a tertiary task). It is important for the interpretation, however, that the speed itself is not a sufficient quality measure (in the sense of the lower the better), but that it can always only be correctly interpreted in relation to a given task (quality = result/task; see also the detailed argumentation in ► Sect. 2.1.5). Comparisons with baseline data or normatively specified speed values are frequently used for the assessment. The corresponding excess indicates a lack of situational awareness or also overuse phenomena. Even acceleration values far outside the range accepted as convenient ($>3\,m/s^2|$) can in this sense be evaluated as deviations from the standard (= target value = task).

■ **Vehicle Following Behaviour**

The distance between two vehicles is defined as the measured distance from the front bumper to the rear bumper, often measured by a radar or lidar sensor. In this case, the following can also be evaluated

- the mean distance
- the variability of the distance
- the minimum distance

If the distance is measured in metres in relation to the speed of the vehicles, then it is possible to calculate further metrics:

- the time headway: distance[m]/v_{ego} = time interval
- the time to collision (TTC): distance [m]/ ($v_{ego} - v_{ahead}$)

The TTC is an important measure primarily for safety reasons. Many traffic rules mention the time interval between two vehicles as a criterion, whereby values below 1.5 s are often drawn as a limit.

■ **Pedal Usage**

Frequent pedal use – both when changing between accelerator and brake pedals and frequently changing gears (using the clutch pedal) – can be seen as an indicator of a hectic and unbalanced driving style. However, since this behaviour is essentially influenced by the route and the traffic situation, a meaningful interpretation of this data is only possible if data of a baseline journey carried out under the same and controlled conditions are available.

11.2.5.2 Metrics of Lateral Control – Steering Behaviour

Measurements based on steering behaviour also provide differentiated information about the strain on the driver, possible assistance requirements and possible distractions caused by tertiary tasks.

On the basis of control engineering approaches and the basic models of the driving task, data on steering behaviour and lateral stabilisation can be evaluated in a similar way to data on longitudinal control (see ► Sect. 2.1.4). In addition, there are approaches in which the frequency spectrum of the steering inputs is also taken into account. Accordingly, high cognitive stress often results in a shift of the steering frequency spectrum compared to corresponding baseline measurements. While higher stress in the primary driving task is reflected in an increase in small, high-frequency steering inputs, higher cognitive stress is expressed precisely in the absence of these correction movements, which are small in their amplitude (McDonald and Hoffman 1980).

The most common measures to describe the steering behaviour and its quality are:

- the standard deviation of the lateral deviation
- the steering wheel reversal rate, steering angle reversal points [n]
- the number of lane exceedances [n]
- the steering entropy
- the steering angle speed [°/s]
- the absolute track position [m].
- Time to Line Crossing (TLC) [s]
- Standard deviation of the steering angle [°]
- Share of high-frequency steering movements [%]

With regard to the standard deviation of the lateral deviation, the absolute track position and the time to line crossing, reference is made to the illustrations in ► Sects. 2.1.4 and 2.4.1 (► Fig. 2.21).

Knappe et al. (2007) gives details on the sensitivity of these measured variables.

11.2.5.3 Measurement of Visual Stress and Interruptibility – Occlusion

In connection with driver information systems, the question to their distracting effect frequently arises and, above all, as to the compatibility of interaction with the requirements of the primary driving task. To be able to answer these questions already in the early phases of the development process, the occlusion method is a good choice. The method is aimed at ensuring that interactions and, above all, eye contact with displays in the interior are of short duration and can be easily interrupted. The method thus checks whether driver information systems can be used well even with brief glances.

During an occlusion test, the respondent is asked to perform a defined tertiary interaction while stationary and to solve an instructed task (e.g. entering a navigation target). The task is processed in an experimental condition with full eye contact and in a further experimental condition with interrupted eye contact. This is produced by shutter glasses, which – following the ISO standard 16,673 – open for 1.5 sec. and "opaque"[2] switches. The Total Task Time (TTT) in both conditions and the Total Shutter Open Time (TSOT) in the occlusion condition are measured. On the basis of these measurements, the TSOT can be used very well to determine the necessary eye-turn, which can also be determined by eye movement measurement. If the TSOT of the occlusion condition is divided by the TTT of the unoccluded test condition, the parameter R (resumability) is obtained, which allows a statement to be made about the interruptibility of the interaction. High R values (R > 0,7 for this procedure) speak for operating concepts that can hardly be interrupted and which should be avoided as far as possible.

The occlusion method has been validated in driving simulation tests and has proven to be extremely reliable (Baumann et al. 2004; Foley 2010). For practical use, it also impresses with its very simple, time-saving data collection and evaluation.

11.2.5.4 Detection Tasks to Assess Driver Attention

Detection tasks are often used to assess the driver's available attention. If only a limited capacity of human information processing is assumed, in particular for decision processes (see also ► Sect. 3.2.2.5), it is believed that the stress caused by the driving task can be measured by an additional task to be performed while driving. One then regards the reduction in power in this second task as an indicator of the stress exerted by the primary task. Secondary tasks in this sense can be acoustic tasks (e.g. reading out monosyllabic or multisyllabic words; the respondent's task is to indicate when one of these words identifies a living being) or optical tasks (e.g. presenting simple figures on a display; the respondent's task is to react when a certain type of figure appears). A particularly frequently used variant for measuring the available attention is to place a small lamp array at the edge of the windscreen in the near peripheral area and to let a small lamp light up in a random sequence. It is then the respondent's task to indicate in a suitable way (verbally or by pressing a button on the steering wheel) that he has perceived the light. In this way, for example, the increasing "tunnel vision" with increased stress caused by the driving task can be objectified and quantified.

The fundamental problem with all dual-task measurements is still that one cannot be sure whether the test person actually regards

2 The LCD-based lens of the shutter glasses becomes milky and therefore opaque by applying a voltage.

the driving task as the primary task under all circumstances. Particularly in the case of very light driving tasks (e.g. on a remote country road with few bends and no traffic), it is by no means certain that the driver will not concentrate entirely on the secondary task. At least for such dual-task settings it is essential to also measure the quality/performance in the primary task. Conti et al. (2013a) describes an experiment that investigates the corresponding influences of instruction and task difficulty.

The so-called Detection Response Task (DRT) is currently being developed for the evaluation of secondary tasks that contain very little visual-manual components but very many cognitive components. Above all speech operation, hearing or mental operations such as mental arithmetic are the subject of the evaluation. While the test person carries out a secondary activity, stimuli (tactile or visual) are presented, which are to be reacted to as quickly as possible at the touch of a button. The measured reaction time and the missed reactions represent the degree of cognitive stress for the estimator. The tactile stimuli are presented, for example, by vibration in the area of the neck or wrist. The presentation of the visual stimuli in the form of an LED array or an LED fixed centrally to the head. The experiments can be carried out in the form of second task experiments (DRT + secondary task) or third task experiments (driving task – field or simulator – + DRT + secondary task) (Young et al. 2013; Merat and Jamson 2008; Jahn et al. 2005; Conti et al. 2012).

The advantage of the DRT lies in its ability to evaluate even very short secondary tasks with high sensitivity. It is a valid measuring instrument especially in the area of cognitive tasks.

11.2.5.5 Measurement of the Distraction Effect of Tertiary Tasks – Lane Change Test

While in the case of the occlusion method, the tertiary task is examined in a single-task condition, the lane-change test represents a standardized dual-task situation. The Lane Change Test (LCT) is produced by a simple computer simulation, but can also be implemented in the driving simulator. Although it looks like a simple driving simulation, the LCT is basically a simple reaction, decision and tracking task. In these tasks it simulates the requirements of the driving task. The test person must change lanes as quickly as possible on an unused lane with three lanes. He/she will receive the instruction from a corresponding signposting (see ► Sect. 10.4.3 and ► Fig. 10.15). The evaluation determines the deviation of the driven trajectory from a normative model (optimal trajectory) related to the driven route (see ◘ Fig. 11.14). The area integral of the deviation gives an estimated value (MDEV) for the deflection by a secondary task performed during the LCT and is subject to assessment. The smaller the MDEV value, the less distracting the secondary task will be. The LCT is also suitable for use during development – it is a very economical test method. The exact implementation regulations can be found in the ISO-26022 standard; the basic structure is described in Bengler et al. (2010).

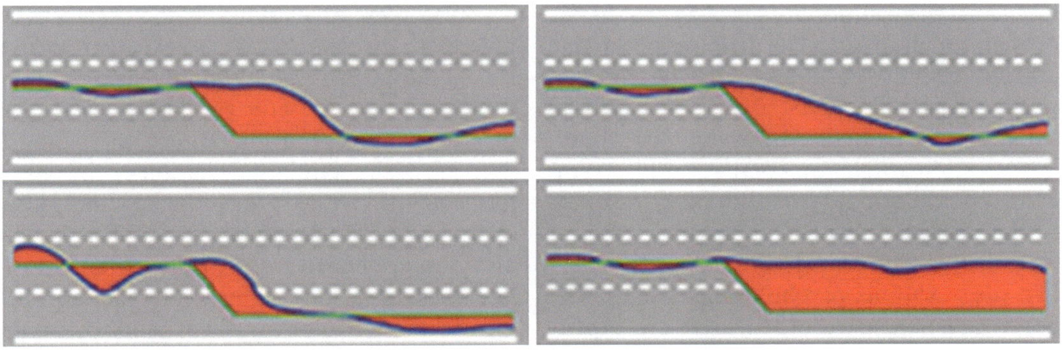

◘ **Fig. 11.14** Error definition during Lane Change Test (from ISO 26022)

11.2.6 Physiological Parameters

At first glance, physiological measurements appear to offer the prospect of direct measurement of human exposure and, due to technical developments in the field of sensor technology and measurement data acquisition, they have come within a gratifying range. However, it should be noted that significant artefacts can occur in the environment of the driving task, which can lead to a falsification of the values. The motor activity while driving the vehicle and the possible change in the interior temperature can certainly lead to effects in the heart rate or skin galvanic resistance range. Also the interindividual differences are to be considered, which can be physiologically justified but also occasionally explained by differences in driving experience.

In the field of gaze measurement, it seems promising to consider pupil change as an indicator of increased cognitive activity. The corresponding measurements must be carried out at high frequency and disturbing influences of the ambient brightness are still being reported, which can produce artefacts. The interested reader will find further information at Marshall (2002) and Dlugosch et al. (2013).

The application of physiological measurements, analysis and, above all, interpretation of physiological measurement series therefore requires the greatest care and experience.

11.2.6.1 Heart Rate

In dynamic muscle work, the increased energy turnover of the muscles is satisfied by an increase in the heart rate, as more oxygen and nutrients are transported to the muscles per time unit. In evolutionary development, it has proved beneficial to also increase heart rate during mental stress, since in the original endangering wilderness a changing mental situation usually had to be overcome by physical exertion (flight or combat). This "remnant" of evolutionary development can also be used in situations that represent a psychological stress, such as driving a car, to objectify the resulting psychological strain. The heart rate is also recorded in many experiments, beside others also because it can be recorded relatively easily. Today, electrodes glued to the skin are mainly used for this purpose, since the nervous pulse generator for heart innervation associated with electrical activity also spreads to the skin. An even simpler method is the earlobe measurement. Here a sensor is clamped to the earlobe, which contains a weak light source on one side and a photodiode on the other side. In step with the heart rate, more or less blood is also pumped through the earlobe, so that the slight change in light transmission can be used to measure the heart rate. Both measuring methods are in principle strongly artefact-dependent (muscle activity can transport additional electrical impulses over the skin surface; strong movement of the head results in changes of the measured light intensity at the photodiode), but modern error compensation algorithms can compensate these errors very well, especially under the condition of relatively quiet sitting at the driver's seat.

For the interpretation of the heart rate, however, it must be taken into account that the increase in pulse rate under changed mental stress is extremely dependent on individual circumstances and sensitivities, which means that, in contrast to the evaluation of physical effort, the observed change in pulse rate is only an indicator of the changed mental stress. The level of the pulse frequency rise thus says little about the level of strain. Nevertheless, it can be of interest for many research questions to even observe that a psychological reaction is present.

The statement made here also applies to measured variables derived from the heart rate. One of the most important of these is the so-called pulse arrhythmia. If the organism is unloaded, the pulse rate is not exactly regular. Its regularity increases with increasing physical exertion. Even with psychological stress, one can notice that the pulse arrhythmia decreases. The standard deviation of the pulse frequency serves as a measure for the pulse arrhythmia. However, even for the pulse arrhythmia defined in this way, its numerical size correlates only slightly with the level of mental strain and obviously depends on other uncontrollable influences.

11.2.6.2 Skin Conductive Resistance

Another physical parameter that reacts to changes in mental stress is skin resistance. It is measured by sticking two electrodes to the surface of the skin (usually on the forearm) and allowing a slight measuring current to flow between these electrodes. For similar reasons, which apply to increasing the heart rate, sweat is precautionarily released from the organism during mental stress in order to compensate for the increased heat release of the muscles during the expected physical exertion. The saline sweat shows a lower electrical resistance than the dry skin, which can be determined by the measuring current mentioned. Since the contact resistance at the electrodes is different with every new adhesive application, even with careful application, data from the skin resistance measurement can only be compared with each other within one measuring cycle anyway. Here, too, the above statement applies that the amount of perspiration depends on many other factors, so that it is also only an indicator of changed mental strain.

The skin resistance can also be used to objectify whether the test person has his hands on the steering wheel or not. If a slight electrical potential difference is created between the steering wheel and the seat, a slight current flows through the driver. If the driver does not have his hands on the steering wheel, this circuit is interrupted. It can be used to technically monitor at least the driver's physical readiness to react in certain assistance systems that require the driver to remain responsive.

11.2.6.3 Electroencephalogram (EEG)

Electroencephalography (EEG) allows researchers to better understand and determine the mental state or activity of subjects. In this method, the extracellular cortical electrical activity that is produced in response to a specific task or tasks is amplified and then measured (for more detailed descriptions, see Niedermeyer and Lopez da Silva 2005; Speckmann and Elger 2005). In order to make such investigations possible, electrodes are typically used for[3] which are attached to an elastic hood and placed on the scalp with a conductive gel.

The recorded activities correspond to the potential difference between the measuring electrode (e.g. active cortical areas) and the reference electrode (e.g. less active areas). The EEG measurements show very precise measurement results, which partly compensate the disadvantage of the low spatial resolution. Although technical improvements in source localization have been achieved by "Low Resolution Electromagnetic Tomography" (LORETA; Pascual-Marqui et al. 1994), currently the maximum resolution is not limited by specific brain structures but by intracranial sources of electrical activity of the brain.

Event related potentials (EKP) and power spectral density (PSD) are often used to quantify the EEG signal. EKPs, which are the brain's stimulus response to a specific stimulus, are mainly used to investigate specific questions regarding the speed of perceptual, cognitive and motor components. On the other hand, PSDs are better suited for e.g. Block-Design brain states, which are determined with this method. Brain frequencies can be categorized as follows: delta (1–4 Hz); theta (4–7 Hz); alpha (8–13 Hz); beta and gamma (20–60 Hz) (Steriade 2005). Although the EEG was originally used primarily in the field of clinical environments (Niedermeyer 2005), this method is increasingly gaining acceptance in practical areas such as ergonomics (see Parasuraman and Rizzo 2008; for further information on the EEG see Gevins and Smith 2008). Modern EEG systems are now available in various designs and sizes, in mobile versions and even without the use of conductive gels.[4]

---

3 recently active electrodes..

4 e.g.: ► www.emotiv.com. Last update 30/April 2014. ► http://neurosky.com/products-markets/eeg-biosensors/; Brainproducts (► www.brainproducts.com). Last update 30/April 2014.

EEG systems can be used to measure a variety of different user conditions such as mental stress, vigilance or driver fatigue. The increase of the central executive function and the mental stress derived from it, is connected with the increase of the theta amplitude in the frontal head areas (Sauseng et al. 2007).

Schmidt et al. (2009) and Simon et al. (2011) have found that the alpha spindle rate increases and the P3 amplitude decreases when test subjects drive for 4 hours, which means a reduction in vigilance. In addition, increased delta and theta activities serve as indicators of fatigue in other studies. In particular, Lal and Craig (2002) found that fatigue was mainly accompanied by an increase in delta and theta activity when the subjects performed a monotonous driving task for 2 hours or until fatigue. Some minor changes in alpha and beta activity have also been reported (Lal and Craig 2002). EEG systems have even been studied as Brain Computer Interfaces (BCIs) outside the vehicle to establish a direct link between the cortical signals and the respective machine. Wolpaw et al. (1991) used Mue rhythm amplitudes to realize one-dimensional cursor movements. Currently there are projects like "BrainDriver" (2011) and "Think and Drive"[5] who are concerned with controlling cars with neuro-signals.

11

11.2.6.4 Electromyogram (EMG)

Muscle activity is generated by α or γ innervation (see also ► Sect. 3.2.3). Since each innervation is also associated with electrical activity, the resulting potential differences can be observed by electrodes attached to the skin surface near the muscle of interest. If more muscle effort is required, a higher innervation rate is necessary, so that the measured potential is a measure of the muscle effort. The recording of these potentials is called electromyography (EMG). Of course, the contact resistance of the electrodes varies from adhesive to adhesive. This means that EMG data recorded on different days cannot be compared in terms of amplitude. In any case, only relative comparisons within a test cycle are possible. In the automotive sector, it would be of particular interest to measure the activity of the back muscles, especially with a view to assessing discomfort. Unfortunately, no statements can be made with EMG in this respect, since the contact resistance between electrode and skin is extremely changed by the contact pressure. This means that measurements are not possible in the area where the back is in contact with the backrest.

11.2.6.5 Adrenaline Release

The release of adrenaline in critical situations is a very reliable indicator for stress situations. However, this method is not very suitable for use in vehicles, as the adrenaline content can only be determined initially by blood tests. Since adrenalin also shows itself in saliva, it can be detected without bloodshed, albeit with a considerable time delay to the onset of the stress situation. This method is therefore more suitable for an integrative objectification of stress situations than for a direct correlation between the situation and the reaction.

11.3 Subjective Measurements

11.3.1 Psychophysics

11.3.1.1 Psychophysical Laws

The model of the stress and strain concept applied in ergonomics states that a certain physically or sociologically describable situation leads to an individual strain depending on the characteristics, abilities and sensitivities of the person on whom this situation affects. "Stress" can be both positive (e.g. pleasure, well-being, enthusiastic reaction, acceptance) and negative (discomfort, rejection). Many attempts made in the development process of an automobile relate to finding quasi-legal relationships between the physical stimuli exerted by a particular vehicle concept and the sensations these stimuli

5 AutoNOMOUS Lab (2011). BrainDriver. Last update 30/April 2014 from: ► http://www.autonomos.inf.fuberlin.de/subprojects/braindriver; Last update 30/April 2014. Think and Drive – Brain Driven Hybrid Vehicle from: ► https://www.facebook.com/Brain.driven.hybrid.vehicle. Last update 30/April 2014.

evoke, since such relationships enable a user-oriented concept to be developed at an early stage of the development process. To establish such connections is the subject of classical psychophysics.

The general aim of psychophysics is to establish a lawful connection between the form of

$$E = f(R) \tag{11.1}$$

to formulate. The following is stated E for the subjective sensation and R for the objective stimulus. Weber (1834) made basic experiments on the basis of estimating weights. For many stimuli, he found the connection that the increase in stimulus ΔR which is necessary to feel just a change, from the stimulus height R depends. He formulated this connection in the form:

$$\frac{\Delta R}{R} = k \tag{11.2}$$

k is called *Weber's constant*. It is stimulus-specific and depends on the test conditions (e.g. the chronological order of presentation of the stimuli).

Fechner (1860) developed a law of psychophysics from Weber's observations, which should describe the desired relationship (11.1) between sensation E and stimulus R. He made the – strictly speaking contradicting Weber's observations – border crossing for this purpose $\Delta R \rightarrow dR$ and could thus integrate Weber's equision 11.2. The result is the well-known Weber-Fechner law of psychophysics:

$$E = c \cdot \log \frac{R}{R_0} \tag{11.3}$$

R_0 is the stimulus threshold found in the experiment, which just leads to a sensation, c a constant of your choice.[6] Later investigations by Stevens, especially in the acoustic field, showed, however, that according to Weber-Fechner's law, a doubling of the sensation by no means corresponds to a doubling of the perceived volume. Stevens found much better results following a power approach of form:

$$E = c\left(\frac{R}{R_b}\right)^n \tag{11.4}$$

The following is included R_b any reference stimulus that causes the expression R/R_b is made dimensionless (normally the unit of measurement of the stimulus R is used for this). n is an exponential factor that depends on stimulus and experimental conditions.[7]

11.3.1.2 Determination of Thresholds

An essential part of psychophysics provides experimental methods with the help of which absolute stimulus thresholds (the stimulus that just leads to a sensation)[8] and difference in stimulus thresholds (the increase or decrease in stimulus needed to feel a change)[9] to find out. The classical methods for threshold determination date back to Fechner (1860). To be able to work on the so-called *constant method* to find out the absolute threshold experimentally, the test person is presented with previously determined stimuli R in random order, to which he must indicate whether he perceives the stimulus or not in the sense of a yes/no decision. If the aim is to find out the difference threshold, a reference stimulus R

---

6 In acoustics, the sound intensity I = R represents the stimulus. At a frequency of 1000 Hz the currently perceivable intensity I_0 with $10^{-12}\,W/m^2$ is specified. c = 10 is set. The volume level E = L, which is referred to as the sensation and is dimensionless in itself, is then given in decibels [dB]. Between the sound pressure level Δp measurable at a microphone and the intensity I there is proportionality $I \approx (\Delta p)^2$. The currently perceivable sound pressure is specified with $\Delta p_0 = 2 \cdot 10^{-12}\,N/m^2$. This makes the connection: $L = 10 \cdot \log \frac{I}{I_0} = 20 \cdot \log \frac{\Delta p}{\Delta p_0}$

7 In the acoustic field, Stevens found that (at a tone from 1000 Hz) from a volume level of L = 40 dB, an increase of 10 dB corresponds to a doubling of the perceived volume E. This allows the constants c and n to be calculated for Steven's power law. It reads: $E = \frac{1}{16} \cdot \left(\frac{\Delta p}{\Delta p_0}\right)^{0.6}$

8 e. g.: Which acceleration, depending on the speed level, leads to the sensation of a speed change?

9 e. g.: Which deceleration of acceleration is just felt during an acceleration process?

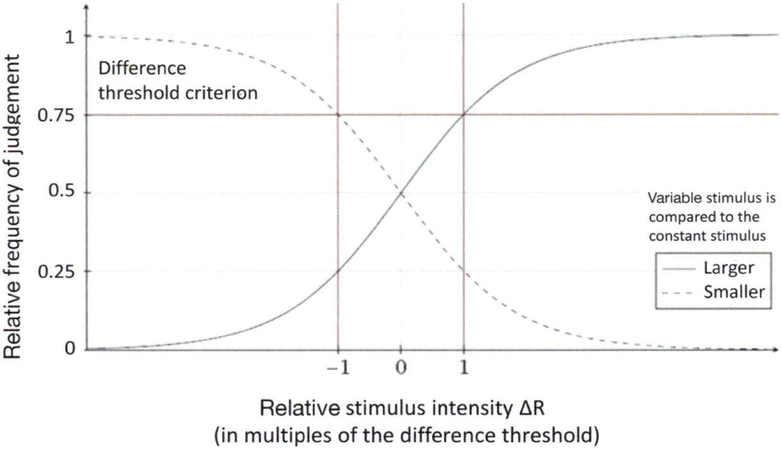

Fig. 11.15 Idealized course of an experiment according to the constant method for difference threshold determination with the response categories "larger/smaller". (From Kühner 2014)

and a comparative stimulus $R + \Delta R$ are also presented to the test person in random sequence. Depending on the experimental approach, he/she then has the response options "greater/smaller", "greater/equal" or "greater/equal/smaller". Figure 11.15 shows the idealized course of such an experiment with the answer "larger/smaller". As the differential stimulus increases, the probability for the answer "greater" increases, while the probability for the answer "smaller" decreases. By setting a 75% limit, one then determines for ΔR an area with 50% uncertain answers (so-called 50% criterion).

In contrast to the constancy method, the *border procedures* does not present the entire spectrum of possible stimuli to the subject. Rather, stimuli ascending and descending in several passes are presented to it to determine the absolute threshold or to determine the difference threshold until a change in the yes/no response occurs. The respective threshold is then determined by averaging these abort stimulus strengths.

A modern variant of the boundary method is the so-called *simple staircase* or *up/down procedure* which was first presented by Békésy (1947) and Dixon and Mood (1948) (Fig. 11.16). Starting from a starting value,

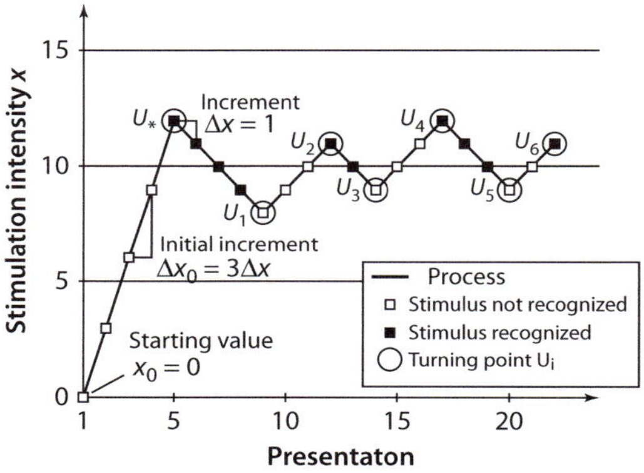

Fig. 11.16 Exemplary course of a simple staircase procedure. (From Kühner 2014)

the stimulus strength is first increased in larger steps until the test subject first receives the stimulus. R notices or perceives that the stimulus $R + \Delta R$ is greater than the comparative stimulus R_0. From this turning point U_* the stimuli are now reduced in smaller steps until the test person determines that the stimulus is no longer perceptible or smaller than the comparative stimulus. After this turning point U_1 the whole is repeated in the opposite direction until the next reversal point U_2 etc. The threshold value results from the arithmetic mean of the stimulus intensities at the normally 6

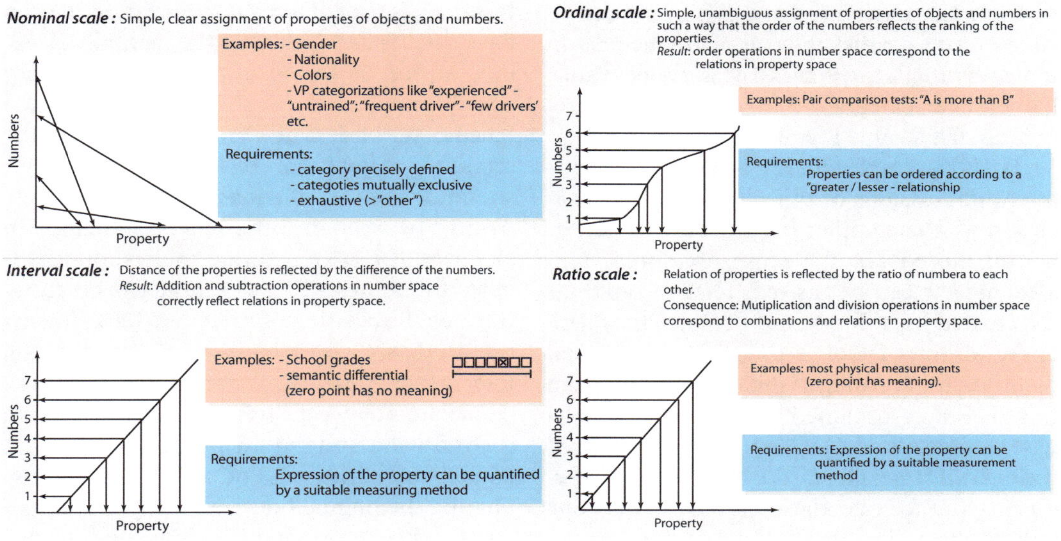

Fig. 11.17 Scale level depending on the question that describes the quality of the assignment of properties (here generally "sensation") and number

reversal points, the first reversal point being U_* is often excluded from the threshold value determination. Of course, the procedure can also be carried out in the opposite direction, i.e. coming from larger values. If one uses both methods, one will usually get slightly different limit values.

In the further course of the development, many different adaptive methods were developed, which differ mainly in the rules for adjusting the stimulus intensity. This compilation is an extract from Kühner's treatise on psychophysics (2014), which contains many further references.

11.3.1.3 Automotive Applications

It is true that stimulus thresholds and stimulus difference thresholds are of interest for many questions. In the automotive application, however, psychophysical functions of the type of eq. 11.1 are often of interest. In the simplest case, it is a matter of approving or rejecting certain forms of execution; in other cases it is a matter of comparing whether a certain variant is perceived as better/worse than another variant. Frequently, however, a direct statement about the extent of an evaluation of an object is required. As also explained in ► Chap. 12 and also in ► Sect. 11.3.2.1, the type of data collection or the question is the basis for the so-called *scale level* of the respective statement. Figure 11.17 shows a comparison of these different scale levels.

A lot of data is stored on *nominal scale level*. The test subjects are either clearly assigned to categories based on their characteristics (e.g. gender, age, driving experience etc.) or the test subjects themselves are asked to assign their feelings to a previously defined category. It is then possible to determine the absolute frequency in each of these categories and, if the number of test persons is large enough, also the relative frequency.[10] The result is often presented in the form of histograms, so-called pie charts or spider diagrams. Depending on the question, it may be necessary to use statistical methods to check whether the differences between the individual categories are significant or whether equality can be assumed (Kap. 12: χ^2-test,

10 The relative frequency or the percentage suggests that this statement refers to a larger stratum of the population. As shown in ► Chap. 12, such a generalization requires a representative sample or, in the case of a random sample, at least a sufficiently large number of subjects. In most practical cases, however, this is not actually the case (even with test persons in the range between 30 and 50!).

► Table 12.5). In some cases it is of interest to characterise the distribution of frequencies in the individual categories by a single value, namely the category that occurs most frequently. This value is called *modal value* .

If the attempt consists in comparing different impressions or objects or forms of execution with each other (in the sense of better/worse; more/less), the corresponding evaluated objects can be ranked. If such rankings are created by different test persons, they generally differ. There are various statistical methods for forming a middle ranking from this. The data obtained on the basis of pairwise comparisons correspond to the level of the *ordinal scale*. Here, too, it is often of interest to characterize the result with just one specification. This is done by specifying the so-called 50th percentile. This is the characteristic that 50% of the test persons show a lower/worse and therefore 50% of the test persons show a higher/better sensation. This value is also called *median*. However, the pair comparison (e.g. larger/smaller comparison) is often the basis for the threshold value determination described in ► Sect. 11.3.1.2. Assuming the validity of Weber-Fechner's law, a psycho-physical connection between physical stimulus and sensation can be determined according to eq. 11.3. Due to the questionability of Fechner's derivation and Stevens' experience, however, it should be considered whether this is opportune.

In many cases, a quantitative assessment of the quality of a sensation is required from the test persons. This is done either by specifying descriptive adjectives with boxes (e.g.: "very good" – "good" – "moderate" – "bad" – "very bad") or by drawing a line at the ends of which there are two opposing statements (see ► Sect. 11.3.2.3). The respondent is now asked to give his or her opinion by ticking one of these boxes, or by characterizing his or her sensation by a cross on the line. By applying a scale, the position of the cross can be converted into a numerical value. This scale assumes that the distances between sensations in the lower range are entered in the same way as in the upper range. Since the equality of numbers reflects the equality of sensations, this scale is called *interval scale*. Since one can have doubts as to whether the mentally imagined distance, for example, between "very good" and "good" is the same as between "mediocre" and "bad", one often modifies the semantic differential by inserting a corresponding number of squares without concrete naming between the opposing statements.[11] In any case, it is assumed that the respondent is able to represent his sensation correctly through the position of this cross. Experiments on the basis of such data of sensation on interval scale level allow in principle to show a connection between stimulus and sensation in a coordinate system[12]. One can now gain a connection in the sense of eq. 11.1 by approximating the point cloud created by the experiments with an optimal function. Also in this case one must assume a mathematical basic form, whose parameters are then determined by regression analytical procedures. In general, one will assume that the basic forms are as simple as possible (linear, quadratic, cubic context; in many cases it is quite sufficient to assume the validity of Stevens' power function). It often makes sense to try out different basic mathematical forms and to assume the descriptive psychophysical law that has the greatest regression coefficient. Because of the assumed equidistance, it is permitted to make additions and subtractions for data at the interval scale level that correspond to corresponding distances at the sensory level. Therefore, the *arithmetical averages* is a measure that describes the central tendency of interval-scale distributed data.

11 Many test subjects obviously find it difficult to understand the application of semantic differentials correctly. It is the experience of many experimenters that this is made much easier by offering clear positions (due to the phenomenon of the "tendency towards the centre", it is also recommended to offer an even number of boxes if possible; see ► Sect. 11.3.2.3).

12 In principle, this relationship can also be multidimensional. If one has to assume that several influencing variables lead to a sensation (e.g. this is the case with temperature sensation, where the influences air temperature, humidity, air movement and radiation influence lead to a "felt temperature"), the experimental effort increases accordingly.

In contrast to the interval scale, the *ratio scale* the number "0" is an equivalent on the property or sensation side. Since one can generally assume that at least in complex contexts the case of "no sensation" does not occur and no further profit would be associated with it, the ratio scale hardly plays a role in psychophysical experiments.

In many practical cases, one cannot be sure whether the data collected are actually at the level assumed by the nature of the question. In this case, statistical analysis is often carried out using methods that are actually intended for a lower scale level, and conservative statements are thus made.

In all psychophysical experiments, the test persons must always give an answer in some form (e.g. "difference present: yes/no"). Due to the large scatter of such experiments, the application of statistical methods is absolutely necessary in order to protect oneself from premature conclusions (see ► Chap. 12). For many experimenters who come from a technical background and who are used to an experiment leading to unambiguous results, the objective physiological data described in ► Sect. 11.2.6 are particularly trusted. Apart from the fact that physiological data also show a considerable dispersion in experiments, it should be pointed out that sensations must never be equated with unseen physiological data. Sensations can ultimately only be found by asking – in whatever form – the test person.[13]

11.3.2 Interview

Carmen Aringer-Walch

In order to formulate questions whose answers are assigned numbers and which can thus be used for statistical analysis, it is necessary to know what one actually wants to ask. In many cases this is not easy to determine (e.g.: what is "sporty"?). Particularly in connection with new areas of application or previously unknown technical systems, it is very difficult to anticipate the possible reactions of the users in advance and, as a result, to catch them by corresponding questions. In these cases, so-called qualitative survey methods are a good method to "stake out the terrain" at all. In the field of automotive ergonomics, it is often a question of collecting knowledge, attitudes and user experience in dealing with technical systems, as well as evaluating the subjectively perceived benefits, and also of gaining knowledge about existing usage barriers. (Trübswetter and Bengler 2013).

In contrast to quantitative approaches, which build on existing theories, qualitative methods are suitable for the investigation of still little described subject areas. They offer the advantage of gaining deep insights into new research areas and generating new questions and hypotheses (Flick et al. 2009). Various methods are available for open or partially standardised qualitative surveys. The procedures most frequently used in practice in the field of ergonomics include guideline and expert interviews as well as group discussions. Each method offers its advantages and disadvantages, depending on the object of investigation as well as the time and financial framework.

The so-called content analysis is suitable for the evaluation of the qualitative data material. Again, a distinction is made between quantitative and qualitative content analysis. While quantitative analysis methods primarily count the frequency of occurring text elements, qualitative methods analyse the entire text on the basis of a category system.

One method that has proven itself scientifically and is frequently used is content analysis according to Mayring (2008). ◘ Figure 11.18 shows the process model of the content analysis. First, the source material of the analysis, i.e. the number and extent of the transcripts, must be determined and the development situation must be described. The next step is to define the concrete questions of the analysis, which determine the direction of

13 For example, it cannot be assumed at all that there is a direct correlation between the local pressure values measured at the contact surface between the test person and the seat and the perceived discomfort (Hartung 2006; Mergl 2006).

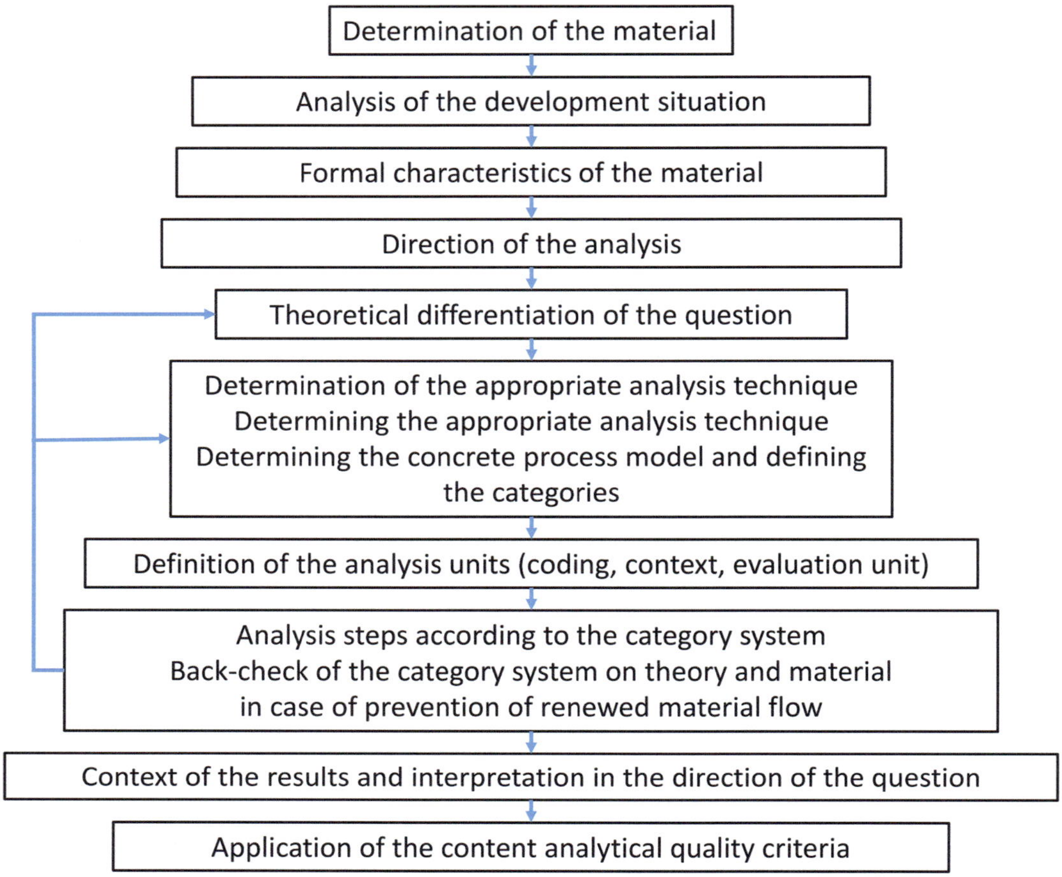

Fig. 11.18 Process model of qualitative content analysis. (Based on Mayring 2008, p. 60)

the analysis (Mayring 2008, p. 50). If possible, these should build on existing theories. Both the analysis technique and the individual analysis units are determined within the framework of a process model. The next step is to create a category system. This is done on the basis of deductive, i.e. theory-based, or inductive, i.e. category formation derived from the transcripts, or a combination of both. Mayring (2008) distinguishes three approaches: The summary, the explication and the structuring. A detailed description of these analytical techniques can be found in Mayring (2008). Once the category system has been established, the actual coding of the text material can be carried out. In order to analyse the results in the direction of the problem, it is advisable to derive main categories on the basis of frequency analyses. This shows what the central findings of the survey are.

Most frequently, however, automotive studies are carried out or supplemented by questionnaire procedures. For example, NASA-TLX is used to interview test persons about the subjectively perceived strain after an experiment, to survey the discomfort sensation in car seats or to interview drivers about their attitude or acceptance of driver assistance systems. Even though in many cases standardised questionnaires/tests such as NASA-TLX can be used, it is often necessary to construct your own questionnaire for a specific topic, for example if you want to record the attitudes of test persons with regard to a concrete fact or if you supplement an expert study such as the group discussion or the interview with a questionnaire.

In principle, questionnaires can cover different aspects. In psychology, they are frequently used as test instruments and, for example, record the development of personality traits (e.g. anxiety etc.; cf. Bortz and Döring 2006, p. 253) or the performance of a person summarily via a score. In the following, we will focus on the creation of questionnaires that record the behaviour or attitudes of individuals. In principle, the construction of such questionnaires is subject to the same quality criteria (objectivity, reliability, validity) as a (personality) test; however, more stringent requirements are placed on tests, which is why reference is made to further literature for the test construction (Bühner 2011).

In the following, the focus is on an overview-like presentation of the construction *more standardized* questionnaires as part of a survey, i.e. the sequence of the questions and the possible answers are predominantly given. These questionnaires can be used orally (by the test management) or in writing (subjects fill in the form themselves). It must be ensured that the questionnaire for written interviews is self-explanatory and can be processed without further instructions, and that the interviewers' instructions and formulations are identical in the context of an oral interview. Scholl (2009, p. 78) mentions as concrete examples that interviewers must not vary text and questions, interviewers must adhere to given rules when asking questions and the interview style should be neutral (neither too authoritarian nor too personal). The behaviour patterns should be discussed or trained with the interviewers in advance, and information for the interviewers can also be added to the questionnaire (see Porst 2008, p. 145 ff.). Whether an oral (with interviewer) or written data aquisition (subjects fill out the questionnaire independently) is used depends on the concrete question. If one assumes that test persons need time to form an opinion about the facts or to answer answers more honestly without an interviewer, a written questioning can be more meaningful. A survey with an interviewer is suitable for complex issues that may involve inquiries (for an overview of the advantages and disadvantages, see Scholl 2009, p. 60).

11.3.2.1 Methodological Approach

▪ From the Theoretical Construct to the Questions

The construction of a questionnaire is based on a theoretical question and hypotheses, which are answered with the help of the questionnaire. If the questionnaire e. g. is to record how the users assess the usability of a new instrument in comparison to the predecessor model and if the hypothesis exists that the new model performs better, the concept of usability must be presented more precisely in order to ask the appropriate questions for testing the hypothesis. This process of translating theoretical questions into a questionnaire or an empirical instrument is called operationalisation (cf. Scholl 2009, p. 144). A first step can be the research of already existing questionnaire concepts on a question, whereby it must be ensured that in particular the results of the quality characteristics (reliability, objectivity and validity) are not transferred unchecked to one's own investigation and that they are adapted to the target group – for example with regard to the formulations (cf. Bortz and Döring 2006, p. 253). The research of literature and definitions in this phase can provide information as to whether the concept of usability possibly has several dimensions, such as the intuitive operation of the system or the appropriate feedback to the user, which must be captured. Further ideas for approaching a topic can be provided by qualitative methods such as interviews with experts or test persons (cf. Raab-Steiner and Benesch 2012, p. 48). For the concrete collection of questions, brainstorming in a team is a good option (cf. Bortz and Döring 2006, p. 253 f.). After the questions have been limited to thematic aspects, the task and answer format is defined.

▪ Question Content and Answer Formats

At this point, a distinction is made between the questions of a questionnaire in terms of content and form (cf. Porst 2008, p. 51). With regard to the content, only a selection is presented here, since the form of content is generally freely selectable (ibid.). Nevertheless, with regard to the interpretation of the data, it makes sense to consider which format the content of a question is based on; a selection is presented below (cf. Scholl 2009, p. 147 ff.):

- ***Factual questions*** (relate to fixed, constant circumstances)

Example: "Do you own a car?" or all demographic questions (age, occupation, etc.)

- ***Knowledge questions*** (the knowledge of the interviewees is determined)

"What assistance systems are available on the market?"

- ***Valuation issues*** (a subjective assessment is made about an issue)

"How do you like the new system?"

- ***Behavioural questions*** (refer to present or past behaviour, for example)

"How often have you used the Lane Change Assistant in the last two weeks?"

- ***Questions of intent*** (relate to future behavior)

"Could you imagine using this system more frequently in the future?"

In addition to the content format, questions can also be differentiated with regard to the answer format. In principle, a distinction is made between open and closed questions; some authors also mention half-open questions or mixed forms (for example, Porst 2008, p. 55 or Raab-Steiner and Benesch 2012, p. 51).

11.3.2.2 Open Questions

No defined selection options are given for open questions. Respondents have the opportunity to freely answer a question. In a written survey this is signalled by lines, in an oral survey it must be ensured that the interviewer writes down the answer as comprehensively and correctly as possible (cf. Porst 2008, p. 54 f.). In contrast to the closed question, the answer must be categorised retrospectively, using methods such as content analysis (cf. ◘ Figure 11.18). This means a higher effort for the evaluation. Bortz and Döring (2006, p. 254) also point out that respondents often write only short and incomplete sentences in the fields provided due to fear of spelling mistakes. Open answers are particularly suitable if there is no manageable number of possible answers or if the subject area is still relatively unknown and if a target group is interviewed that can and wants to express himself on the subject (e.g. experts) (cf. Scholl 2009, p. 162). In addition, open questions in longer questionnaires can serve to maintain motivation and prevent respondents from being steered in a given direction by answer categories (cf. Porst 2008, p. 64). An example of an open question might be:

Please formulate here your concrete ideas for an optimization of the seat!

11.3.2.3 Closed Questions

Closed questions are frequently used in questionnaire design due to the relatively simple evaluation options. The respondents are given a limited number of possibilities within which the answer can move. A distinction is made between answers with one or more possible choices (multiple answers). The possibilities to answer closed questions, especially content categories, must be exhaustive, i.e. the respondent must be able to find himself in the answers. If, for example, the frequency of car use were surveyed with the following categories: "daily", "2–3 times a week" and "less than 2 times a week", those test persons who do not drive daily, but at least more frequently than 3 times a week, would have difficulties assigning themselves. Closed questions are in principle more appropriate when there is a limited and accessible range of possibilities and the topic is more or less known (cf. Porst 2008, p. 63 f.). ◘ Figure 11.19 shows an example of a closed question with several possible answers (the assistance systems listed serve only as an example).

In addition to the two response formats open and closed, there are also *half-open questions* or *mixed forms.* Like the example in ◘ Fig. 11.19, these have response categories, but also contain an open category. In such a case, the above example could be supplemented by the answer "Further, namely:_____". In this category, respondents who could not assign themselves to a given category or want to add

The following selection lists assistance systems currently available on the market. Please tick all assistance systems you know (even if only by hearsay)	
o	Anti-lock Braking System (ABS)
o	Parking aid
o	Lane Change Assistant
o	Cruise control

Fig. 11.19 Example of a closed question with multiple possible answers

further aspects can answer. Porst (2008, p. 57) recommends this answer option on the one hand if the possible answers to a question can be roughly estimated but not completely identified. On the other hand, this form of questioning allows the respondents to position themselves through their own presentation in the case of unclear assignments, which has a positive effect on motivation. These answers must in turn be extra coded and evaluated.

If a closed answer format is used, the next step is to ask how the choices are presented. Within the framework of a statistical analysis of closed questions, a question in the questionnaire becomes a variable (e.g. usage behaviour) which can take on defined characteristics (1 = very frequent, 2 = frequent or 1 = male; 2 = female) as a result of the possible answers (cf. Scholl 2009, p. 164). "If a respondent's response is measured as the expression of a variable, the response specifications are called a scale." (ibid.)

Scale Levels

Variables can be distinguished in terms of their scale level, as already explained in ► Sect. 11.3.1.3 and therefore briefly presented here. The scale or data level must be taken into account in the statistical evaluation, since not all calculations are possible with every level. A distinction is made between the following types: **Nominal scaled variables** are used to form categories with regard to equality and inequality, usually by collecting gender (male, female) in this way. The example with the assistance systems also provides nominal scaled data. The use of **ordinal scaled variables** sequences can be formed, e.g. the highest school leaving certificate acquired (1 = Hauptschule, 2 = ...). **Interval scaled variables** allow further calculations (e.g. averaging, standard deviation, factor analysis), since the distances between the values are the same. Also allow **proportional variables** the formation of ratios as they occur in length and weight.

Number of Possible Responses

If there are only two possible answers to a question (yes/no, right/wrong ...) one speaks of a dichotomous answer format. These questions can be quickly evaluated due to the clear answer pattern, but their usefulness for further calculations is limited by the low variance of the answers (cf. Raab-Steiner and Benesch 2012, p. 55); they must therefore be used in a targeted manner. Dichotomous answers are suitable, for example, as filter questions ("Do you own a car?"). – If you answered "Yes", please continue with question 6) or several dichotomous questions can be combined into one index (see Scholl 2009, p. 168). Most questionnaires use scales with more than two possible answers (e.g. rating scales). "This multi-level nature of the scales corresponds on the one hand to the need of interviewees for a differentiated presentation of their position, but on the other hand also to the evaluators' need for the broadest possible variation of possible evaluation procedures." (Porst 2008, p. 75).

Rating Scales

If opinions, attitudes, preferences, behaviour etc. are recorded in the course of a survey, rating scales are frequently used (cf. Scholl 2009, p. 167). "Rating scales indicate (by numbers, verbal descriptions, examples, etc.) marked sections of a continuum of features that the judges are to rate as equal [...]." (Bortz and Döring 2006, p. 177). If no concrete scale steps are offered and the answer can be freely given at any point on a continuum, one speaks of an analogue scale (cf. Raab-Steiner and Benesch 2012, p. 59).

■ Unipolar and Bipolar Rating Scales

The structure of such scales can be *unipolar* or *bipolar*. Unipolar scales run from a zero point in one direction (e.g. from "never" to "frequently", ◘ Fig. 11.20), while bipolar scales are characterized by respective opposites at the ends of the scale (e.g. "I completely reject" to "I completely agree"; ◘ Fig. 11.21) and the zero point is either fictitious (with an even number of scale gradations) or actually (with an odd number of scale gradations) in the middle of the scale (cf. Raab-Steiner and Benesch 2012, p. 56 f.; Porst 2008, p. 90). "If it is difficult to find a suitable counter term to a term, one uses instead of *more bipolar* scales *unipolar* Rating scales. This applies in particular to features with a natural zero point, such as the extent of annoyance from noise." (Bortz and Döring 2006, p. 177).

■ Designation of the Scale Points

If rating scales are used, there is a concrete reference to the scale level, because it is assumed that rating scales and the respondents' judgements are to be understood as interval-scaled (ibid. and p. 181 for the discussion of problems of measurement theory). The designation of the different levels is possible in different ways. One *numerical designation* divides the scale into equal distances, each marked with a digit. The handling must be explained in the introduction. The following shows a possible design for a unipolar and a bipolar scale with verbal endpoints. Bortz and Döring (2006, p. 177) and Rohrmann (1978, p. 223) mention that numerical scales are abstract, which is why it should be ensured that the target group can handle this format when using them.

For a *verbal labelling* each step of the scale is given a printout, as in the example in ◘ Fig. 11.22. "This has the advantage that the interpretation of the scale points is intersubjectively more uniform [...]" (Moosbrugger and Kelava 2007, p. 52). Care must be taken to ensure that the verbal distances are also regarded by the respondents as equal and are not arbitrarily assigned (the distance between "rarely" and "occasionally" should be perceived as just as large as the distance between "often" and "very often").

In his research, Rohrmann (1978) provides evidence for an equidistant formulation of five-stage scales with different possible applications:

- **Frequency** ("How often do you drive?"):

"never – rarely – occasionally – often – always", whereby depending on the question "never" can be replaced by "very rarely" and "always" by "very often" (which is suitable for this example, as respondent drives "always", see also ◘ Figs. 11.22 and 11.23) no.

- **Intensity** ("Are you satisfied with the new system?"):

"not at all – little – mediocre – predominantly – totally"

- **Likelihood** ("Will you buy yourself a new car in the next two years?"): "not at all – probably not – maybe – quite probably – quite sure".

Never	1	2	3	4	5	6	Very often

◘ **Fig. 11.20** Example of a numeric, unipolar scale with verbalized endpoints

I absolutely refuse	-3	-2	-1	0	1	2	3	I fully agree

◘ **Fig. 11.21** Example of a numeric, bipolar scale with verbalized endpoints

never	rarely	occasionally	often	always
0	1	2	3	4

◘ **Fig. 11.23** Example of a verbal-numeric scale

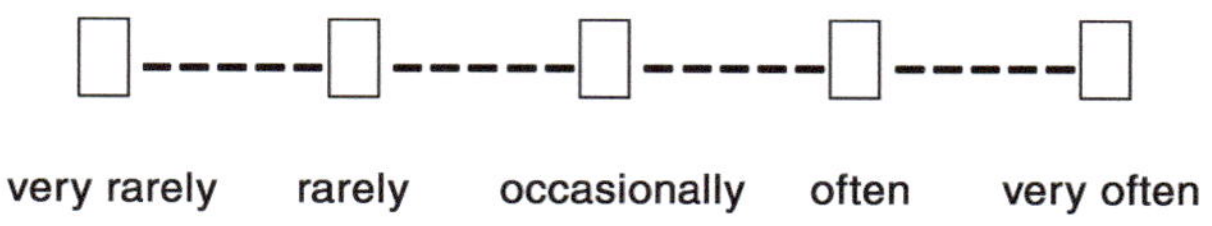

◘ **Fig. 11.22** Example for verbal scale points

- **Evaluation of statements** ("I like to drive fast!"): "doesn't apply at all – applies little – applies partly, partly – applies quite well – applies completely")

In addition to numerical and verbal marking, rating scales can also be represented symbolically (for example by plus/minus or smileys) (cf. Bortz and Döring 2006, p. 177). Another possibility of the scale designation is the scale anchoring by examples at certain selected positions of the scale (ibid. p. 180). Scale designations can also be combined with each other, so that a numerical designation is also displayed in addition to a verbal designation. Bühner (2011, p. 116) mentions this variant in order to make it easier for respondents to process the data. In such a case, the designations should correspond logically and make it easier for respondents to interpret a scale. For example, a five-step scale with the two end points "never" and "always" should not contain digits of −2 (never) to +2 (always) but better from 0 (never) to 4 (always) (see for this and for optical scale designations): Moosbrugger and Kelava 2007, p. 53).

Number of Stages

The number of levels of rating scales can be even or odd. As a result, either a mean category exists or the test persons have to give an answer in one direction of the scale. They may be "forced" to a tendency on a scale with an even number of steps, which they may not represent at all. On the other hand, the argument against a middle category is that this category is not only marked with a cross in the case of a balanced opinion on a fact, but also if the interviewees are uncertain in their answer and for this reason switch to the middle of the scale (cf. Scholl 2009, p. 168). However, the aspects mentioned are likely to be less important on a scale with more levels (for example, six or seven) (see ibid.). Bortz and Döring (2006, p. 180) recommend not offering a neutral middle category if one expects a "tendency towards the middle"of the interviewees (see for further errors of judgement p. 183ff.). Moosbrugger and Kelava (2007, p. 54) state that the arguments as a whole tend to argue against the use of a category of funds. One subsequent consideration concerns the additional category 'I do not know'. This category should be included in particular if some of the respondents are unlikely to be able to give an assessment of the facts (ibid.).

Besides the decision for an even or odd number of steps, the decision about the number of steps itself has to be made. "Here is [...] a trade-off to be struck between accuracy and reliability. The more levels are given, the more accurate the scale, but the choice of a particular level becomes less reliable and random because the respondent has too many choices". (Scholl 2009, p. 167). Porst (2008, p. 92) recommends the use of five to seven-step response formats, as at a certain point respondents can only differentiate their assessment to a limited extent. Scales with very many points (e.g. 100) show that the interviewees predominantly select decimal numbers or numbers divisible by five (cf. Bortz and Döring 2006, p. 180). For the CP 50 scale for measuring discomfort (50 scale points), the assessment is therefore carried out in two steps: In a first step the test persons are asked to roughly assess the perceived discomfort (based on 5 levels), in a second step they assign themselves to a concrete number from 1 to 10 within this level.

Analogue Scales

If analogue scales are used, the judgement of the test persons can be freely assessed on a continuum and very precisely evaluated (Fig. 11.24). Moosbrugger and Kelava (2007, p. 51) state that the use of analogue scales is increasing due to computer-based evaluation options, but that, overall, they are still rarely used because the "differentiation of the measurement does not usually correspond to the differentiation of the judgement" (ibid.). The subsequent evaluation of the judgments obtained in this way is carried out by means of categorisation. For example, the

Fig. 11.24 Example of an analog scale. (For example, percentage specifications could also be used)

number of categories to be created can be calculated using the Sturges rule.

In addition to the rating scales presented here, there are a number of other possibilities and forms, for which, however, reference is made to the relevant literature (e.g. Bortz and Döring 2006).

11.3.2.4 Formulation of Questions

When formulating a question, the first consideration is whether there is a question ("Do you think that the purchase of electric cars should be financially supported by the legislator?") or a statement ("The purchase of electric cars should be financially supported by the legislator!"). In order to capture attitudes and positions, statements are more suitable to capture concrete facts, the question form (cf. Bortz and Döring 2006, p. 254).

Questionnaires should include terms with *unclear meaning* (cf. for the following aspects Bühner 2011, p. 136 ff.). The statement "I like to drive sporty" could be interpreted positively by respondents in the sense of "good, controlled, dynamic" as well as "too fast, careless" and is therefore unclear. *Terms used* (e.g. Head-Up-Display) must be known by the interviewees and must be aligned with the target group (customers, engineers, employees); this can be ascertained by a pre-test in advance. In each question only ***one*** content aspect must be illustrated. The statement "I was able to assess and control the system well" refers to two aspects, assessment and control, and should therefore be replaced by two questions. *Double negations* are to be avoided, as they reduce comprehensibility, for example "I never felt I could not control the system". Is applied to a *time span* should be as concrete as possible. The question: "Have you used your brake lane assistant lately" is not clear about "lately". It would be better, for example, to ask for the last 2 weeks in concrete terms. Also *generalizations* as "always", "all", "never" should be avoided because of their generality.

Porst (2000) recommends the following aspects in his "10 Commandments of Question Formulation", which are listed here at the end:

1. *You should use simple, unambiguous terms that are understood by all respondents in the same way!*
2. *You should avoid long and complex questions!*
3. *You should avoid hypothetical questions!*
4. *You should avoid double stimuli and negations!*
5. *You should avoid insinuations and suggestive questions!*
6. *You should avoid questions that are aimed at information that many respondents are suspected not to have!*
7. *You should use questions with a clear temporal reference!*
8. *You should use answer categories that are exhaustive and disjunctive (non-overlapping)!*
9. *You should make sure that the context of a question does not affect its answer!*
10. *You should define unclear terms!*

11.3.2.5 Design of the Questionnaire

The design of the questionnaire must be geared to the target group (age, level of education, previous knowledge, etc.), particularly with regard to complexity and formulation. For written surveys, a short introductory description of the research project and a reference to the anonymity of the survey are usual at the beginning. Furthermore, notes on the handling of scales can be listed. If interviewers are used in the survey, the questionnaire will contain instructions applicable to all interviewers, and in the case of a written survey, the instructions for respondents must be clear (e.g. whether one or more crosses are possible). The visual design of the questionnaire should be clear (clearly separate questions, no page breaks in a question, shade every second line of longer questionnaires to avoid slipping in the line, appropriate font size, sufficient space for open questions, etc.; cf. Scholl 2009, p. 176 ff).

When determining the order of the questions, care should be taken to ensure that at the beginning of the questionnaire there are easily answered questions. This also applies to the end of longer questionnaires, which is why the demographic questions (if they are needed for the course of the survey, then they are also at the beginning of the questionnaire) are often asked here (ibid. p. 175). In the case of longer questionnaires, thematic blocks should be formed so that the interviewees can concentrate on one subject area and do not have to jump in their thoughts (e.g. user behaviour, likes, etc.) (cf. also Porst 2008, p. 142 f.). In addition, the order of the questions should be chosen as far as possible so that they do not influence each other (cf. Scholl 2009, p. 175). If, for example, a sample of assistance systems is given to respondents in question 2 and they are asked in question 10 to indicate which assistance systems they know, then question 10 should be influenced by question 2.

"Answer patterns should be varied as often as practicable to avoid fatigue effects, but not too often to avoid confusing the respondent." (ibid. p. 177). For example, longer "questionnaires" of rating scales should be alternated with other formats in doses (open question, nominally scaled selection question) in order to avoid symptoms of fatigue. In addition, items in questionnaires can be reversed in isolated cases and formulated positively instead of negatively in order to attract attention. Before using a questionnaire, it is strongly recommended to test it on a small sample as part of a "pre-test", for example using the "loud thinking" method (cf. Raab-Steiner and Benesch 2012, p. 61) and then to revise it again.

11.3.3 Standardised Questionnaires

Nicole Trübswetter

11.3.3.1 Workload (NASA-TLX)

To determine the subjectively felt strain of a task, the questionnaire NASA TLX (Hart and Staveland 1988) or the Driver Attentional Load Index (DALI) derived from it (Pauzié and Pachiaudi 1997) can be used. Both the primary driving task and secondary/tertiary tasks or a combination of all can be considered as tasks.

In the sense of a multidimensional measurement, the strain (workload) caused by the processing of a task is to be reproduced by the respondent on six subscales on the basis of assessments. The subscales include the factors mental, physical, temporal stress, performance, frustration and effort to solve.

The individual judgments are then calculated into an overall-work-load-index (OWI) with a range from 0 to 100. The subjective individual assessments x_i with their respective weighting factors w_i multiplied. The weighting factors are again determined in a pair comparison before the subjective judgements are asked ("Which of the two dimensions has the greatest influence on your workload?"). The weighting factor w_i is the sum of the preferences of the subscale i in this pair comparison.

$$OWI = \frac{1}{15}\sum_{i=1}^{6} w_i \cdot x_i \qquad (11.5)$$

11.3.3.2 Rating Scale Mental Effort (RSME)

The Rating Scale of Mental Effort (RSME) by Zijlstra and Van Doorn (1985) is a one-dimensional, graphical scale that measures the effort subjectively experienced by test persons in performing a task (■ Fig. 11.25). The RSME value is measured on a 150 mm long scale with nine verbal anchor points, which is based on a subjective rating of *hardly strenuous* by the time *extremely strenuous*. It can be used both directly after and during a journey. Compared to other evaluation methods, the RSME is extremely sensitive (Verwey and Veltman 1996). In addition, the RSME is very reliable and very well reflects the stress caused by driving tasks or secondary tasks in the vehicle ergonomics environment. This allows short-term load peaks and permanent additional loads to be identified during the journey. In comparison to the TLX, it represents a short, direct questioning of the test person about the strain experienced, which can be answered very quickly and easily.

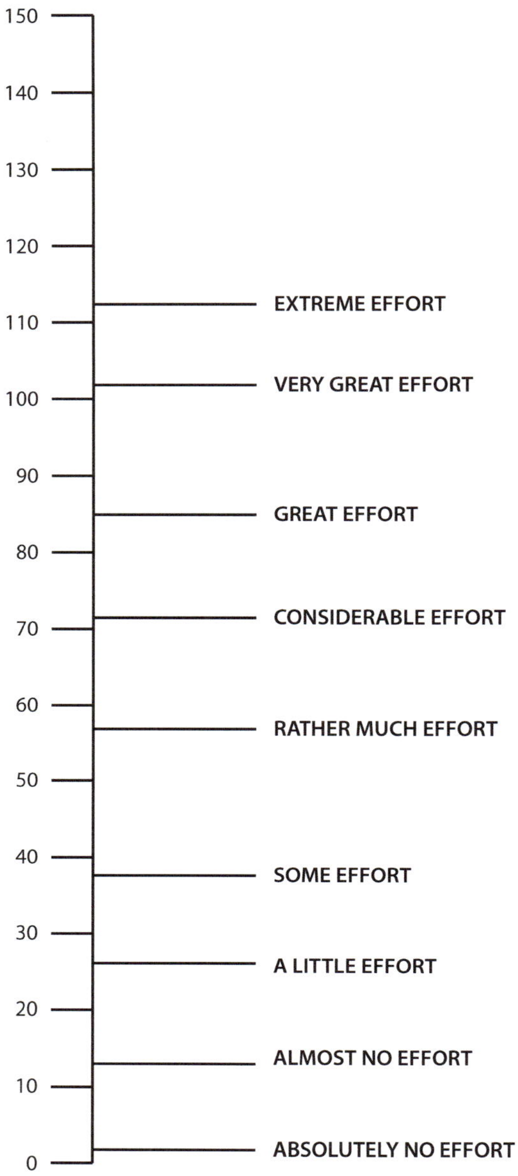

Fig. 11.25 RSME scale. (Zijlstra and Van Doorn 1985)

References

Literature Used

Baumann, M., Keinath, A., Krems, J.F., Bengler, K.: Evaluation of in-vehicle HMI using occlusion techniques: experimental results and practical implications. Appl. Ergon. **35**, 197–205 (2004)

Békésy, G.: A new audiometer. Acta otolaryng. **35**(411–422) (1947)

Bengler, K., Mattes, S., Hamm, O., Hensel, M.: Lane change test: preliminary results of a multi-laboratory calibration study. In: Rupp, G.L. (Hrsg.) Performance Metrics for Assessing Driver Distraction: The Quest for Improved Road Safety. Chpt. 14, S. 243–253. SAE International, Warrendale, PA (2010)

Bortz, J., Döring, N.: Forschungsmethoden und Evaluation, 4. Aufl. Springer, Heidelberg (2006)

Brooke, J.: SUS: a ″quick and dirty″ usability scale. In: Jordan, P.W., Thomas, B., Weerdmeester, B.A., McClelland, A.L. (Hrsg.) Usability Evaluation in Industry. Taylor and Francis, London (1996)

Bühner, M.: Einführung in die Test- und Fragebogenkonstruktion, 3. Aufl. Pearson Studium, München (2011). aktualisierte Auflage

Cherednischenko, A.: Funktionales Modell der Einstiegsbewegung in einen PKW. Dissertation an der Technischen Universität München (2007)

Conti, A.S., Dlugosch, C., Schwarz, F., Bengler, K.: Driving and Speaking: Evaluations by the Head-Mounted Detection Response Task. In: Proceedings of the Seventh International Driving Symposium on Human Factors in Driver Assessment, Training, and Vehicle Design (2013a)

Conti, A.S., Dlugosch, C., Vilimek, R., Keinath, A., Bengler: An Assessment of Cognitive Workload using Detection Response Tasks. In: Proceedings 4th Conference on Applied Human Factors and Ergonomics (AHFE) International Conference San Francisco, U.S., July 2012 (2012)

Diebschlag, W., Heidinger, F., Dupuis, H., Hartung, E., Meiller, H.: Ergonomie des Sitzens: Arbeitsmedizinische und technische Aspekte Sitzgestaltung Die Bibliothek der Technik, Bd. 68. Verlag Moderne Industrie, Reichenschwand (1992)

Dixon, W.J., Mood, A.M.: A method for obtaining and analyzing sensitivity data. Journal of the American Statistical Assoziation. **43**(241), 109–126 (1948)

Dlugosch, C., Conti, A.S., Bengler, K.: Driver Distraction through Conversation Measured with Pupillometry. In: Proceedings of the Seventh International Driving Symposium on Human Factors in Driver Assessment, Training, and Vehicle Design (2013)

Estermann, S.: Komfortmodellierung für den 3D-Softdummy RAMSIS unter realen Fahrbedingungen. Dissertation an der Technischen Universität München (1999)

Fechner, G.T.: Elemente der Psychophysik. Breitkopf & Härtel, Leipzig (1860). Bände 2

Flick, U., Kardorff, E., von Steinke, I. (Hrsg.): Qualitative Forschung: Ein Handbuch, 6. Aufl. Rowohlt-Taschenbuch-Verl., Reinbek bei Hamburg (2009)

Foley, J.P.: Now you can see it. Now you don't. Visual occlusion as a surrogate distraction measurement technique. In: Rupp, G.L. (Hrsg.) Performance Metrics for Assessing Driver Distraction: The Quest for Improved Road Safety, S. 123–134. SAE International, Warrendale, PA (2010). Chapter 9

Fröhmel, C: Validierung des RAMSIS-Krafthaltungsmodells. Dissertation an der TU-München (2010)
Gevins, A., Smith, M.E.: Electroencephalography in neuroergonomics. In: Parasuraman, R., Rizzo, M. (Hrsg.) Neuroergonomics – The Brain at Work, 1. Aufl., S. 15–31. Oxford University Press, USA (2008)
Hart, S.G., Staveland, L.: Development of a multi-dimensional workload rating scale: results of empirical and theoretical research. In: Hancock, P.A., Meshkati, N. (Hrsg.) Human Mental Workload, S. 139–183. Elsevier, Amsterdam, The Netherlands (1988)
Hartung, J.: Objektivierung des statischen Sitzkomforts auf Fahrzeugsitzen durch die Kontaktkraft zwischen Mensch und Sitz, Dissertation an der Technischen Universität München (2006). http://www.gesis.org/fileadmin/upload/forschung/publikationen/gesis_reihen/howto/how-to2rp.pdf, Zugegriffen: 22.04.2014
ISO/TS 15007–2: Road vehicles – Measurement of driver visual behavior with respect to transport information and control systems – Part 2: Equipment and procedures (2001)
Jahn, G., Oehme, A., Krems, J., Gelau, C.: Peripheral detection as a workload measure in driving: effects of traffic complexity and route guidance system use in a driving study. Transp. Res. F. **8**, 255–275 (2005)
Nielsen, J.: Usability Engineering. Morgan Kaufmann, San Francisco, S. 155–163 (1994)
Knappe, G., Keinath, A., Bengler, K.: Driving simulators as an evaluation tool – assessment of the influence of the field of view and secondary tasks on lane keeping and steering performance ESV conference, Lyon (2007)
Kühner, M.: Haptische Unterscheidbarkeit mechanischer Parameter bei rotatorischen Bedienelementen. Dissertation an der Technischen Universität München (2014)
Lal, S.K.L., Craig, A.: Driver fatigue: electroencephalography and psychological assessment. Psychophysiology. **39**(3), 313–321 (2002). https://doi.org/10.1017.S0048577201393095
Lange, C., Bubb, H.: Trajectory preview in vehicles by means of contact analogue display: an evaluation using eye movement tracking. Zeitschrift für Arbeitswissenschaft. **2008**(03), 134–143 (2008)
Marshall, S.P.: The Index of Cognitive Activity: measuring cognitive workload. In: Proceedings of the IEEE 7th Conference on Human Factors and Power Plants, S. 7-5–7-9 (2002)
Martin: Lehrbuch der Anthropologie, 4. Aufl. R. Knußmann, Stuttgart (1987). 1. Aufl. (1914)
Mayring, P.: Qualitative Inhaltsanalyse – Grundlagen und Techniken, 10. Aufl. BelzVerlag, Weinheim und Basel (2008)
McDonald, W.A., Hoffman, E.R.: Review of relationships between steering wheel reversal rate and driving task demand. Hum. Factors. **22**(6), 733–739 (1980)
Menn, M., Studer, N., Cohen, A.: Eye movement behaviour when driving through the Gotthard tunnel: A pilot study ECEM13, European Conference on Eye Movements. Mental Workload, Bern, 2005. Elsevier, Amsterdam, The Netherlands (2005)
Merat, N., Jamson, A.H.: The effect of stimulus modality on signal detection: implications for assessing the safety of in-vehicle technology. Hum. Factors. **50**, 145–158 (2008)
Mergl, C.: Entwicklung eines Verfahrens zur Optimierung des Sitzkomforts auf Automobilsitzen. Dissertation am Lehrstuhl für Ergonomie der Technischen Universität München. (2006)
Moosbrugger, H., Kelava, A.: Testtheorie und Fragebogenkonstruktion. Springer, Heidelberg (2007)
Niedermeyer, E.: Historical aspects. In: Niedermeyer, E., Lopes da Silva, F. (Hrsg.) Electroencephalography: Basic Principles, Clinical Applications, and Related Fields, 5. Aufl., S. 1–15. Lippincott Williams & Wilkins, Philadelphia, PA (2005)
Niedermeyer, E., Lopez da Silva, F.: Electroencephalography: Basic principles, Clinical Applications, and Related Fields, 5. Aufl. Lippincott Williams & Wilkins, Philadelphia, PA (2005)
Nilsson, G.: Speeds, accident rates and personal injury consequences for different road types. VTI, Linköping (1984). Rept. VTI-277
Parasuraman, R., Rizzo, M.: Introduction to neuroergonomics. In: Parasuraman, R., Rizzo, M. (Hrsg.) Neuroergonomics – the Brain at Work, 1. Aufl., S. 3–12. Oxford University Press, USA (2008)
Pascual-Marqui, R.D., Michel, C.M., Lehmann, D.: Low resolution electromagnetic tomography: a new method for localizing electrical activity in the brain. International Journal of Psychophysiology : Official Journal of the International Organization of Psychophysiology. **18**(1), 49–65 (1994) http://www.ncbi.nlm.nih.gov/pubmed/7876038
Pauzié, A., Pachiaudi, G.: Subjective evaluation of the mental workload in the driving context. In: Rothengatter, T., Carbonell Vaya, E. (Hrsg.) Traffic and Transport Psychology: Theory and Application, S. 173–182. Pergamon, Oxford (1997)
Porst, R.: Question Wording –Zur Formulierung von Fragebogen-Fragen ZUMA How-to-Reihe, Bd. 2. (2000)
Porst, R.: Fragebogen. Ein Arbeitsbuch. VS Verlag für Sozialwissenschaften, Wiesbaden (2008)
Raab-Steiner, E., Benesch, M.: Der Fragebogen. Von der Forschungsidee zur SPSS-Auswertung, 3. Aufl. facultas.wuv, Wien (2012)
Rigel, S.: Entwicklung und Validierung einer Methode zur quantitativen Untersuchung der Ein- und Ausstiegsbewegung in einen Pkw. Dissertation an der Technischen Universität München. (2005)
Rohrmann, B.: Empirische Studien zur Entwicklung von Antwortskalen für die sozialwissenschaftliche Forschung. Z. Sozialpsychol. **9**, 222–245 (1978)
Saito, S.: Does fatigue exist in a quantitative measurement of eye movements? Ergonomics. **35**, 607–615 (1992)
Sauseng, P., Hoppe, J., Klimesch, W., Gerloff, C., Hummel, F.C.: Dissociation of sustained attention

from central executive functions: local activity and interregional connectivity in the theta range. Eur. J. Neurosci. **25**(2), 587–593 (2007). https://doi.org/10.1111/j.1460-9568.2006.05286.x

Schmidt, E.A., Schrauf, M., Simon, M., Fritzsche, M., Buchner, A., Kincses, W.E.: Drivers' misjudgement of vigilance state during prolonged monotonous daytime driving. Accid. Anal. Prev. **41**(5), 1087–1093 (2009). https://doi.org/10.1016/j.aap.2009.06.007

Schmidtke, H.: Handbuch der Ergonomie mit ergonomischen Konstruktionsrichtlinien. Zweite, überarbeitete und erweiterte Auflage. Bundesamt für Ausrüstung, Informationstechnik und Nutzung der Bundeswehr, Koblenz (1989)

Scholl, A.: Die Befragung, 2. Aufl. UVK, Konstanz (2009). überarbeitete Auflage

Schweigert, M.: Fahrerblickverhalten und Nebenaufgaben. Dissertation an der Technischen Universität München (2003)

Simon, M., Schmidt, E.A., Kincses, W.E., Fritzsche, M., Bruns, A., Aufmuth, C., Bogdan, M., Rosenstiel, W., Schrauf, M.M.: EEG alpha spindle measures as indicators of driver fatigue under real traffic conditions. Clin. Neurophysiol. **122**, 1168–1178 (2011)

Speckmann, E.-J., Elger, C.E.: Introduction to the neurophysiological basis of the EEG and DC potentials. In: Niedermeyer, E., Lopes da Silva, F. (Hrsg.) Electroencephalography: Basic Principles, Clinical Applications, and Related Fields, 5. Aufl., S. 17–29. Lippincott Williams & Wilkins, Philadelphia, PA (2005)

Steriade, M.: Cellular substrates of brain rhythms. In: Niedermeyer, E., Lopes da Silva, F. (Hrsg.) Electroencephalography: Basic Principles, Clinical Applications, and Related Fields, 5. Aufl., S. 31–83. Lippincott Williams & Wilkins, Philadelphia, PA (2005)

Taylor, T., Pradhan, A.K., Divekar, G., Romoser, M., Muttart, J., Gomez, R., Fisher, D.L.: The view from the road: the contribution of on-road glance-monitoring technologies to understanding driver behavior. Accid. Anal. Prev. **58**(0), 175–186 (2013). https://doi.org/10.1016/j.aap.2013.02.008

Thompson, L.: Development and Evaluation of a Driver Assistance Concept for Speed and Lane Assistance, Master Thesis, Lehrstuhl für Ergonomie der TU München (2005)

Trübswetter, N., Bengler, K.: Why Should I Use ADAS? Advanced Driver Assistance Systems and the Elderly: Knowledge, Experience and Usage Barriers. In: Proceedings of the 7th International Driving Symposium on Human Factors in Driver Assessment, Training and Vehicle Design, Bolton Landing New York, June 17–20. (2013)

Verwey, W.B., Veltman, H.A.: Detecting short periods of elevated workload: a comparison of nine workload assessment techniques. J. Exp. Psychol. Appl. **2**(3), 270–285 (1996)

Weber, E.H.: De Pulsu, Resorptione, Auditu Et Tactu: Annotationes Anatomicae et Physiologicae. Koehler, Leipzig (1834)

Wolpaw, J.R., McFarland, D.J., Neat, G.W., Forneris, C.A.: An EEG-based brain-computer interface for cursor control. Electroencephalogr. Clin. Neurophysiol. **78**, 252–259 (1991) http://www.ncbi.nlm.nih.gov/pubmed/1707798

Young, R.A., Hsieh, L., Seaman, S.: The Tactile Detection Response Task: Preliminary validation for measuring the attentional effects of cognitive load. In: Proceedings of the Seventh International Driving Symposium on Human Factors in Driver Assessment, Training, and Vehicle Design Iowa City, USA: University of Iowa. S. 71–77 (2013)

Zijlstra, F.R.H., Van Doorn, L.: The construction of a scale to measure perceived effort. University of Technology, Delft (1985). Technical Report

Further Literature

Akamatsu, M., Green, P., Bengler, K.: Automotive Technology and Human Factors Research: Past, Present, and Future. In International Journal of Vehicular Technology, 2013 (2013). Article ID 526180, 27 pages, https://doi.org/10.1155/2013/526180

Bengler, K., Bubb, H., Totzke, I., Schumann, J., Flemisch, F.: Automotive. In: Sandl, P., Stein, M. (Hrsg.) Information Ergonomics – a Theoretical Approach and Practical Experience in Transportation. Springer, Berlin, Heidelberg (2012)

Bengler, K., Praxenthaler, M., Theofanou, D., Eckstein, L.: Investigation of Visual Demand in Different Driving Simulators within the ADAM Project Driving Simulator Conference 2004 Europe, Paris., S. 91–104 (2004). Proceedings

Breuer, J., Bengler, K., Heinrich, C., Reichelt, W.: Development of advanced driver attention metrics (ADAM). In: Strasser, H., Kluth, K., Rausch, H., Bubb, H. (Hrsg.) Proceedings of the Annual Spring Conference of the GfA on the occasion of the 50th Anniversary of the Foundation of the Gesellschaft für Arbeitswissenschaft and the XVII Annual ISOES Conference. Quality of work and products in enterprises of the future, S. 37–40. ergonomia, Stuttgart (2003)

Conti, A.S., Dlugosch, C., Bengler, K.: The Effect of Task Instruction on Detection Response Task Performance. In: de Waard, D., Brookhuis, K., Wiczorek, R., di Nocera, F., Brouwer, R., Barham, P., Weikert, C., Kluge, A., Gerbino, W., Toffetti, A. (Hrsg.) Proceedings of the Human Factors and Ergonomics Society Europe Chapter 2013 Annual Conference (2013b). http://hfes-europe.org

DIN 33402: Körpermaße des Menschen. Part 1–3. Berlin: Beuth (1978)

DIN EN ISO 9241, Teil 10: Ergonomische Anforderungen für Bürotätigkeiten mit BildschirmRgeräten – Teil 10: Grundsätze der

11

Dialoggestaltung (ISO 9241-10:1996); Deutsche Fassung EN ISO 9241-10:1996. Berlin: Beuth (1996)
DIN EN ISO 9241, Teil 11: Ergonomische Anforderungen für Bürotätigkeiten mit Bildschirmgeräten – Teil 11: Anforderungen an die Gebrauchstauglichkeit – Leitsätze. (ISO 9241-11:1999); Deutsche Fassung EN ISO 9241-12:1998. Berlin: Beuth (1999)
DIN EN ISO 9241, Teil 110: Ergonomie der Mensch-System-Interaktion –Teil 110: Grundsätze der Dialoggestaltung. Deutsche Fassung EN ISO 9241-110:2006, Normenausschuss Informationstechnik und Anwendungen, Berlin, 2008 (2006)
DIN EN ISO 9241, Teil 12: Ergonomische Anforderungen für Bürotätigkeiten mit Bildschirmgeräten – Teil 12: Informationsdarstellung (ISO 9241-12:1998); Deutsche Fassung EN ISO 9241-12:1998. Berlin: Beuth (2000)
Engström, J., Åberg, N., Johansson, E., Hammarbäck, J.: Comparison between Visual and Tactile Signal Detection Tasks Applied to the Safety Assessment of in-Vehicle Information Systems (IVIS). In: Proceedings of the Third International Driving Symposium on Human Factors in Driver Assessment, Training and Vehicle Design Rockport, Maine (2005)
Flügel, B., Greil, H., Sommer, K.: Anthopologischer Atlas, Grundlagen und Daten. Edition Wörtzel, Frankfurt a. M. (1986)
Gradenegger, B., Rauch, N., Krueger, H.-P., Bengler, K.: Situation awareness in driving with in-vehicle information systems. In: Rupp, G.L. (Hrsg.) Performance Metrics for Assessing Driver Distraction: The Quest for Improved Road Safety, S. 217–231. SAE International, Warrendale, PA (2010)
ISO 26022 (2010): Road vehicles – ergonomic aspects of transport information and control systems – simulated lane change test to assess in-vehicle secondary task demand (Norm Nr. ISO 26022). Geneva, CH: ISO
Keinath, A., Baumann, M., Gelau, C., Bengler, K., Krems, J.: Occlusion as a technique for evaluating in-car displays. In: Harris, D. (Hrsg.) Engineering Psychology and Cognitive Ergonomics, 5. Aufl., S. 391–397. Ashgate, Brookfield (2001)
Kincses, W.E., Hahn, S., Schrauf, M., Schmidt, E.A.: Mobiles EEG zur Messung der mentalen Fahrerbeanspruchung. ATZ. **110**(3), 204–209 (2008)
Krems, J., Keinath, A., Baumann, M., Gelau, C., Bengler, K.: Evaluating visual display designs in vehicles: advantages and disadvantages of the occlusion technique. In: Camarinha-Matos, L., Afsarmanesh, H., Erbe, H.-H. (Hrsg.) Advances in Network Enterprises, S. 361–368. Kluwer, Dordrecht (2000a)
Krems, J.F., Keinath, A., Baumann, M., Gelau, C., Bengler, K.: Evaluating visual display designs in vehicles: advantages and disadvantages of the occlusion technique. In: Camarinha-Matos, L.M., Afsarmanesh, H., Erbe, H.-H. (Hrsg.) Advances in Network Enterprises, S. 361–368. Kluwer, Dordrecht (2000b)
Kühner, M., Wild, J., Bubb, H., Bengler, K., Schneider, J.: Haptic perception of viscous friction of rotary switches. In: Jones, L.A., Harders, M., Yokokohji, Y. (Hrsg.) IEEE World Haptics Conference WHC 2011, Istanbul, Turkey, 21–24 June 2011. S. 587–591. IEEE Computer Society, Istanbul (2011)
Lange, C.: Wirkung von Fahrerassistenz auf der Führungsebene in Abhängigkeit der Modalität und des Automatisierungsgrades. Dissertation an der Technischen Universität München (2008)
Lewis, J.R.: IBM computer usability satisfaction questionnaires: psychometric evaluation and instructions for use. International Journal of Human-Computer Interaction. **7**(1), 57–78 (1995)
Zijlstra, F.R.H.: Efficiency in Work Behavior: a Design Approach for Modern Tools. Doctoral thesis. Delft University of Technology, Delft, The Netherlands (1993). http://repository.tudelft.nl/view/ir/uuid%3Ad97a028b-c3dc-4930-b2ab-a7877993a17f/

Statistical Methods

Mark Vollrath

Contents

12.1 Basic Questions: Distribution Vs. Examination of Differences – 654

12.2 Expression of Characteristics: Confidence Intervals – 655
12.2.1 Methodology: Sampling – 655
12.2.2 Statistics: Determination of Characteristic Value – 656

12.3 Differences Between Conditions: Significance Tests – 659
12.3.1 Methodology: Experimental Designs – 659
12.3.2 Statistics: Significance Tests – 663
12.3.3 Statistics: Presentation of Results – 669

12.4 External and Internal Validity – 672

References – 673

H. Bubb et al. (eds.), *Automotive Ergonomics*, https://doi.org/10.1007/978-3-658-33941-8_12

12.1 Basic Questions: Distribution Vs. Examination of Differences

In the field of vehicle ergonomics, two main questions can be distinguished, which can be solved with the help of statistical methods:

- Determination of the expression of certain characteristics in the relevant population
- Examination of the difference between different conditions

Both questions have in common that on the one hand a certain methodical procedure is necessary for the answer (sampling procedure, design of experiments). On the other hand the appropriate statistical procedures have to be chosen (confidence intervals, significance tests, see Fig. 12.1). This will be explained first.

The first question is always important when certain customer characteristics have to be taken into account when designing control elements or displays. Controls in the vehicle shall be so placed that they can be reached by the driver's hand without the driver having to change his sitting position. Here the essential characteristic is the arm length. A display should be placed at a height that can be easily seen by any driver. In this case, the eye height above the seat is a relevant property.

12

The methodological approach in this case focuses on the selection of a representative sample. Usually you want to make a statement for a certain driver population, e.g. for the German driver. Since it is not possible to examine all persons in this population, a sample must be taken. Which concrete persons have to be examined in order for the results to be representative for the German driver population?

On the one hand, the statistical methods for this question are concerned with how best to describe the distribution of the property. The mean value can be searched for in order to find the best case for most people. However, minimum or maximum values may also be meaningful to demonstrate that the solution chosen is also appropriate for people with the corresponding extreme characteristics. On the other hand, it is a question of the accuracy of the estimation of these parameters. This depends essentially on the size of the sample: The more persons examined, the more accurately the conditions in the population can be estimated.

The second question about differences is relevant when comparing variants or design alternatives and when examining to what extent certain conditions (e.g. a warning system) lead to changes compared to control conditions (e.g. a drive without a warning system) (e.g. a faster braking reaction of the driver). In each case, it is used to compare groups of people with each other, which can also be the same people at two different times (repeated measurement). In terms of the methodological approach, the focus here is on experimental design. How are the different groups "treated" so that a possibly found difference can actually be attributed to the interesting variation of influencing variables?

Another very important aspect here is the sample size. The smaller the difference, the more volunteers are needed to actually detect it in the test. Depending on the experimental

	Question	
	Expression of characteristics	**Differences between conditions**
Methodology	**Sampling procedure**	**Design of experiments**
Statistics	**Confidence intervals**	**Significance tests**

Fig. 12.1 Basic questions with the associated methodological and statistical aspects. For further explanation, see text

design and the quality of the data, different methods are used to prove the effect. The selection of the appropriate and sensitive procedure is the essential point of the statistics for these differential questions.

12.2 Expression of Characteristics: Confidence Intervals

12.2.1 Methodology: Sampling

The aim of sampling for this question is to obtain a sample that is as representative as possible, i.e. a sample that reflects the conditions in the population as well as possible. On the one hand, this concerns the procedure for drawing the sample and, on the other hand, the necessary number of persons.

The best method of sampling is *random selection*. With this method, each person in the population has the same chance of being included in the investigation. Thus, all influencing variables on the characteristic to be measured are distributed in the sample in the same way as in the population. This is all the more the case the more people are drawn. For example, many characteristics depend on the sex of the person. If 1000 persons are randomly drawn from the German population, the gender ratio in such a sample will most probably correspond to that in the population as a whole. If, on the other hand, only a small sample of two persons is drawn, the probability is about 50% that only men or only women will be examined and the sought-after characteristic, e.g. the mean body size, incorrectly estimated.

However, the random selection of a sufficiently large sample from the population is rarely possible for practical reasons in the field of vehicle ergonomics. The examinations must be carried out e. g. with a certain vehicle at a certain place. It is possible that the results should be treated confidentially, so that only employees of the company can be considered as participants. And the cost of the examination is so high that only about 30 people can be examined. Against this background, the question arises as to how to arrive at the best samples under these circumstances.

It is important, especially for small samples, that as many characteristics as possible that may influence the relevant characteristic are taken into account in the sample selection. This is called a *stratified sample*. Central characteristics are certainly age and sex. One would include approximately the same number of men as women and persons of different age groups in the study in order to specifically consider the influence of these characteristics. Other relevant features in the field of vehicle ergonomics are body height and weight, as well as driving performance for many questions. In order to map the range well with respect to these characteristics, one would like to try to map the possible expressions of the characteristics and their combinations with at least 3-10 persons. However, this leads to very large samples for only a few characteristics. If one takes both sexes into account, three age groups, three classes of body size and three groups with different driving experience are formed, one would have $2 \times 3 \times 3 \times 3 = 54$ combinations. If you want to examine 10 persons per combination, you would need 540 persons. From a practical point of view, this access with the help of stratified samples, which take into account the combination of characteristic values, is usually only possible with the inclusion of 2–3 characteristics.

In addition, the inclusion of different characteristics makes the sample more heterogeneous. The relevant properties thus scatter more, which makes reliable estimation more difficult. From this point on, it may make sense to first work with a *homogeneous sample* in order to obtain a good estimate for at least this type of test person even with a relatively small number of test persons, and then to extend this to other, again homogeneous samples in further steps.

In summary, the significance of the survey with regard to the population as a whole depends to a large extent on the sampling. If relatively homogeneous, locally limited samples are examined, the extent to which these results can be transferred to other groups of the population must be considered when interpreting the results. If statistical methods are applied to these results, they may be able to estimate relatively accurately the "true" val-

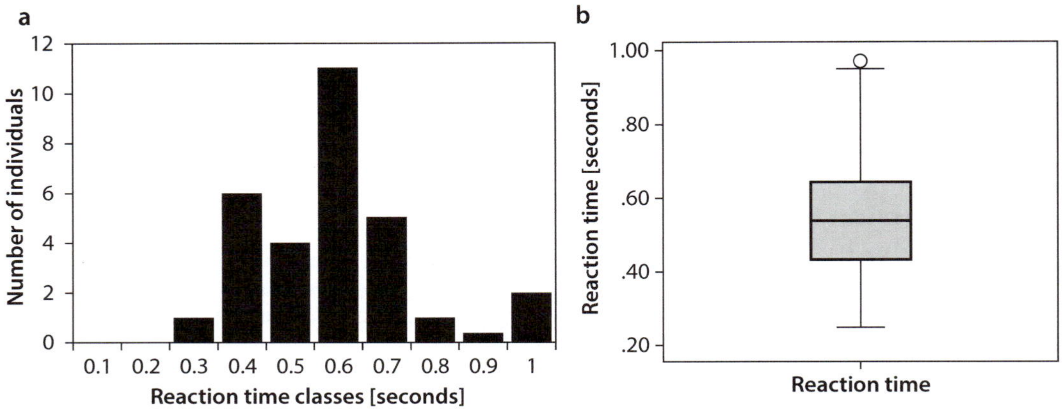

Fig. 12.2 Distribution of reaction times for a lane change task. Shown in **a** as histogram the number of persons in the classes of reaction times displayed on the x-axis. In **b**, the same data is displayed as a box plot

ues in the population. However, this estimate only applies to the part of the population corresponding to the sample. Therefore, in order to assess the relevant results, it is important not only to have information on the sample size and the resulting estimated distributions, but also on the type of sampling and the main characteristics of the sample. This is the only way to assess the extent to which the results are not only accurate but also representative.

12.2.2 Statistics: Determination of Characteristic Value

After the data collection, it is advisable to first present the collected data descriptively. This can be done for example as frequency distribution or histogram (see Fig. 12.2a). Here, meaningful categories of the characteristic are created and the number of values per category is displayed. One recognizes thereby very well the kind of the distribution (e.g. symmetrical or oblique, single or multiple peaks) and receives a first impression of the measured orders of magnitude. Also "outliers" can be recognized quite well (see, for example, the two large reaction times in Fig. 12.2, left).

A more compressed way of displaying the data is the boxplot (see Fig. 12.2b). The grey box contains the mean 50% of the measured values. The black horizontal line represents the median (see below). With the vertical strokes the 1.5-fold of the range of the box is applied upwards and downwards, but only up to the last available value. All values outside this range are drawn as individual points and are therefore easy to identify as outliers. This type of display is also very suitable for comparing several conditions with the help of box plots arranged next to each other.

The next step is to try to describe the relevant properties of these distributions using characteristic values. As Table 12.1 shows, this describes the typical values on the one hand, and the width or dispersion of the distribution on the other. As the column on the right shows, the interpretation of the individual measurements is slightly different. In addition to this different information content, these measures are also best suited for different types of data.

Four scale levels are distinguished according to the information content of the numbers used (see also ▶ Sect. 11.3.1.3). For calculations, the gender is often classified into the values "1: male" and "2: female". For these two numbers, it is only useful to interpret the equality of the measured values. The fact that the 2 is twice as large as the 1 is correct for the numbers, but not for the categories for which the numbers stand. Such a classification is called a *nominal scale*. A useful parameter is the mode (or modal value). A mode of 2 means in the example that women are more frequently included in the sample than men. The range can also be useful

Table 12.1 Overview of essential characteristic values of distributions

Characteristic value	Computation	Interpretation
Mean	$M\frac{\overset{n}{\bullet}\underset{1}{}\,x}{n}$ x: measured values n: number that measured values	Typical value of the sample. Sum of deviations from this value is minimal
Median	The value above which 50% of the measured values lie (interpolation for categorical values)	Typical value of the sample.50% of the values are below / above
Mode	Most frequently measured value	Typical value of the sample. The value that occurs most frequently
Standard deviation	$S=\frac{\sqrt{\sum_{1}^{n}(x-M)^2}}{n-1}$ X: measured values M: mean N: number of values	Deviation of the values Mean deviation from the mean value
Interquartile distance	Range between the value below which 75% of the values lie and that below which 25% of the values are	Deviation of the values. Width of the range in which the middle 50% of the values lie
Range	Maximum – minimum	Distribution of the values. Width of the range in which the values lie

here to describe the number of categories used. Finally, it is possible to indicate the percentage of occurrence of the different categories ("The sample contains 45% men").

The second scale level is the *ordinal scale*. Here the order of the numbers can be interpreted. If a font size is judged as "1: small" and "2: large" by test persons for example, the statement that judgement 2 is larger than judgement 1 is significant – a font size judged as "large" is larger than one judged as "small". Again, not significant is the statement that this large font size is twice as large as the small one, since 2 is twice as large as 1. For data at this level, the median is a meaningful description of the typical value and interquartile distance for variance.

The *interval scale* is the third scale level. The distances between the measured values can also be compared here. If you judge the volume of a warning tone with "1: very quiet", "2: quiet", "3: medium", "4: loud" and "5: very loud", you can first interpret the difference as well as the order of the numbers. In addition, it makes sense to say that the difference between 3 (medium) and 1 (very quiet) is greater than the difference between 2 (quiet) and 1 (very quiet). Whether on the other hand 2 (quiet) is twice as loud as 1 (very quiet) is doubtful. Here, too, this relation of the numbers must not be interpreted. Meaningful characteristic values for the interval level are the mean value and the standard deviation.

As the name suggests, you can use the *ratio scale* and interpret the numbers as ratios. This is often the case with physical data. A measurement of reaction times is an example of data at the ratio level. A reaction time of 500 ms is twice as long as one of 250 ms. Also for this scale level, mean and standard deviation are a good description of the typical value and dispersion.

Overall, at a certain scale level only certain interpretations of the numbers and thus only certain characteristic values are meaningful. At the higher scale level, the characteristic values of the lower levels can also be used and interpreted and sometimes provide interesting additional information.

In addition to describing the data of the sample by specific characteristic values, it is often a matter of estimating the conditions in the population with the help of the sample. Especially for the mean value, but also for the percentage share of certain categories, the question arises as to how precisely the calculation of the sample characteristics reflects the conditions in the population. This is answered with the help of confidence intervals. A confidence interval indicates the range in which 95% (sometimes also 99%) of the population characteristics that may have generated the sample characteristic lie. Or: With a probability of 95% (or 99%), the true value in the population lies in this range. From a methodological point of view, this of course only applies if the sample is a representative selection of the population. Confidence intervals only indicate how accurate the estimate based on the sample is, but not how well the sample reflects the population.

12

The formula of the confidence interval for a mean is as follows:

$$\Delta_{crit} = M \pm z_{\alpha/2} \cdot \widehat{\sigma_M} \quad (12.1)$$

The following is included

$$\widehat{\sigma_M} = \sqrt{\frac{\sum_1^n (x-M)^2}{n \cdot (n-1)}} \quad (12.2)$$

M in the formulas stands for the mean value. $\widehat{\sigma_M}$ is the standard error of the mean value. The z value stands for the corresponding value of a standard normal distribution that includes the mean 95% (99%) of this distribution. Formally, this corresponds to the z-value with $\alpha = 5\%/2$ or $\alpha = 1\%/2$. The following numerical values can be found in the corresponding tables:

$$95\%: \ z_{\alpha/2} = 1.96$$

$$99\%: \ z_{\alpha/2} = 2.58$$

For the data from ◘ Fig. 12.2, the mean value is M = 0.53 seconds with n = 30 persons and a standard error of the mean value of 0.03. This results in $\Delta_{crit} = 0.53 \pm (1.96 \cdot 0.03) = 0.53 \pm 0$ .06. The 95% confidence interval therefore ranges from 0.47 to 0.59 seconds. Thus, a satisfactorily accurate estimation of the mean population reaction time is already achieved with 30 test persons.

Eq. 12.2 also makes it immediately clear which role the sample size n plays. The larger the sample, the smaller the standard error, which in turn directly determines the width of the confidence interval. This can be used to calculate the required sample size for a given accuracy if the mean value and standard deviation are known e.g. from a pilot study. The formula is as follows:

$$N_{necessary} = \frac{1.96^2 \cdot sd^2}{Accuracy^2_{requested}} \quad (12.3)$$

In the above example, the standard deviation was sd = 0.17. If the mean value of the population is to be estimated with an accuracy of ±0.1 second, then according to the formula the result is a $N_{necessary} = 11$.

You can also calculate confidence intervals for percentage values. The basic formula is comparable:

$$\Delta_{crit(\%)} = P \pm z_{\alpha/2} \cdot \widehat{\sigma_\%} \quad (12.4)$$

Where P is the empirically calculated percentage. The standard error results as:

$$\widehat{\sigma_\%} = \sqrt{\frac{P \cdot (100-P)}{n}} \quad (12.5)$$

In the above example, the reaction time of 11 of the 30 = 37% of the subjects was in the category between 0.5 and 0.6 seconds. What is the confidence interval of this percentage? The standard error is:

$$\widehat{\sigma_{\%}} = \sqrt{\frac{37 \cdot 63}{30}} = \sqrt{77.7} = 8.8$$

This calculates

$$\Delta_{\text{crit}(\%)} = 37\% \pm (1.96 \cdot 8.8) = 37\% \pm 17$$

The 95% confidence interval thus ranges from 20% to 54%. Here, too, it is possible to specify by conversion which sample would be necessary to achieve a certain accuracy.

$$N_{necessary} = \frac{1.96^2 \cdot (P \cdot (100 - P))}{Accuracy_{requested}^2}$$

So if you want to estmate the percentage with an accuracy of ±5% the formula above calculates $N_{\text{requested}} = 358$.

12.3 Differences Between Conditions: Significance Tests

12.3.1 Methodology: Experimental Designs

The second type of question compares at least two conditions. The general question is to what extent certain influencing factors systematically change the measured values. The scientific concern here is the search for causal laws, i.e. for cause-effect relationships. To make this clear, a distinction is made between independent variables (IV, causes) and dependent variables (DV, measured values, see also ▶ Sect. 11.1.3). The relationship is represented schematically as follows:

$$\text{IV} \rightarrow \text{DV} \quad \text{or} \quad \text{DV} = \text{f}(\text{IV})$$

The IV systematically causes certain changes in the DV, the measured values. The measured values are therefore a function of the independent variables. In ergonomics, it can be demonstrated for example that a certain display variant in the head-up display with a warning tone leads to faster reaction times than the conventional display with a warning tone in the combined display. Here IV is the type of display in two steps (HUD vs. convetional display), DV is the reaction time. It is assumed that the mean value of the reaction times of a group of drivers with the HUD is smaller than in a group with a conventional display.

Even more complex questions can be represented in this scheme. It can be assumed that, in addition to the location of the display, the presence of a warning tone is also essential for the effect of the warning. To check this, the second IV would be to introduce the warning tone in the steps "without" and "with". In order to investigate the effect of both IVs alone and in combination, both IVs would now have to be combined, resulting in four experimental groups. Independent variables are also referred to as "factors" in order to distinguish experimental designs according to the number of IVs studied. This leads to the description as "single factorial", "two factorial", etc. experimental design (see ◘ Table 12.2).

The number of levels of the factors must be distinguished from the number of factors. An influencing factor is often examined in two levels (e.g. without vs. with). But also the comparison of several levels (warning in the conventional display, in the HUD, in the centre console) is not unusual. For each design, therefore, the number of levels shall be given in addition to the number of factors. This is often done in the form "a × b × c × ... factorial experimental design", where a, b and c are the levels of the respective IV. The above example with the two factors "location of the display" and "warning sound" could be described as 2 × 2 factorial experimental design.

The experimental designs also differ on the DV side. Here it is very important how many DVs are examined. If reaction time is the only measured DV, it is a univariate plan. Mostly, however, several DVs are recorded, e.g. also subjective evaluations. If only one global judgement ("How good was the display?") is recorded as the only additional DV, it is a bivariate plan. In general, multiple DVs are called multivariate plans.

The scale level of the measurements is also important for further evaluations (see

Table 12.2 Overview of essential aspects of experimental designs. For further explanation, see text

Criterion	Significance	Description of the
Number of IV	How many influencing factors are investigated?	Single factorial Two-factor … Multifactorial
Number of levels of IV	Which aspects of the influencing factors are investigated?	2 levels 3 levels …
Number of DV	How many parameter are measured?	Univariate Bivariate … Multivariate
Scale level of DV	What is the meaning of the measured numbers?	Nominal scale Ordinal scale Interval scale Ratio scale
Measurement repetition	Does each test person receive only one or more levels of IV?	Independent Mixed Dependent

► Sect. 12.2.2). As described above, the question arises as to which of the information contained in the figures can also be interpreted. This leads to the different types of characteristic values described above, but also has further evaluations in the statistical comparisons (see ► Sect. 12.3.2).

The last criterion is the distinction between between-subjects and within-subjects designs. In the case of between-subjects designs, each subject receives only exactly one level or combination of the independent variables. In the case of within-subject designs (also called repeated measurements), each subject provides measured values in all conditions. With multi-factor plans, it is also possible to examine individual IVs with repeated measurements, while others can be examined with independent groups. The decision to do so shall be taken on the basis of substantive considerations. If can be assumed that experiencing one condition influences the reactions in another condition, independent plans should be chosen. So if the different warnings are to be investigated with the help of an unexpectedly occurring critical situation in the driving simulator, a dependent plan makes little sense, since the subject already knows this critical situation after the first condition and this would no longer be unexpected for him in the second condition. Whenever learning effects are assumed, repeated measurements should be avoided. The same applies if the tests are very tiring for the test person, so that a decrease in motivation and performance is to be feared. This could also distort the results, making independent plans more useful.

Then why repeat measurements at all? The advantage of within-subject designs is that each subject can be compared with him/herself. Especially when people react quite differently, such a change can always be expressed in terms of the individual typical value. Small effects can thus be discovered with just a few test persons, since the differences between persons that are not of interest are eliminated in this way. Plans with repeated measurements are therefore advantageous both in terms of the number of test persons required and their sensitivity to discover effects.

However, in these plans possible time effects have to be controlled. Since both learning and fatigue can in principle never be completely ruled out, it is important to ensure that this works evenly in the various conditions. This control of time effects is achieved by varying the order of conditions for each subject. There are essentially two possibilities: The complete permutation of all possible sequences and the technique of the Latin square. Figure 12.3 provides an overview of this.

Under two conditions, i.e. only one IV, two groups of subjects are formed, randomly assigned to the two groups to ensure that the groups are comparable. The first group receives treatment A at first, then B. In the second group, the order is reversed.

Under three conditions, a multiple of 6 test persons is required. Each of these 6 subjects receives a different sequence of conditions. The smallest meaningful number for

2 Conditions
Cross-over

	T1	T2
G1	A	B
G2	B	A

3 Conditions
Complete
permutation

	T1	T2	T3
P1	A	B	C
P2	A	C	B
P3	B	A	C
P4	B	C	A
P5	C	A	B
P6	C	C	C

	T1	T2	T3	T4
P1	A	B	C	D
P2	B	C	D	A
P3	C	D	A	B
P4	D	A	B	C

Random numbers Columns 2 / 4 / 1 / 3

	T1	T2	T3	T4
P1	A	D	A	C
P2	B	C	B	D
P3	C	B	C	A
P4	D	C	D	B

Random numbers Series 4 / 1/ 3 / 2

4 Conditions
Latin Square

	T1	T2	T3	T4
P1	A	C	D	B
P2	B	D	A	C
P3	D	B	C	A
P4	C	A	B	D

Fig. 12.3 Overview of techniques of control of time. The respective test conditions (A..D) for subjects (P) or groups (G) are shown. The different times are designated as T1..T4

this case are 12 test persons with whom one can already prove effects by the dependent plan.

Combining several IVs with each other quickly creates more conditions. For a 2 × 2 plan with complete repeated measurement, four conditions are present. If a third factor with also 2 levels is introduced, the number increases to 8. A 3 × 3 plan contains 9 conditions and so on. Here it is usually no longer possible to present all conceivable sequences in a complete permutation. An alternative here is the technique of the Latin square, in which each subject receives all conditions and the sequences are chosen in such a way that each condition is equally frequent at all times across the various subjects. Not all sequence effects can be controlled, but the simple time effects. This technique also requires multiples of the number of conditions. With four conditions (as shown in Fig. 12.3), at least 4 subjects are required. The base square for this case is shown in the middle of the figure. In order to introduce a certain randomness here, a random sequence of the four points in time is generated, in the example 2/4/1/3. The columns in the second square are then sorted in this order. The original column 2 becomes the new column 1 and so on. The same is done with the lines where the order 4/1/3/2 was drawn here. So the new line 1 is the old line 4 and so on. This procedure is repeated for all groups of four subjects to be examined. Again, 12 subjects are the lower limit of what appears to make sense in this dependent plan. For 5 or more conditions, additional squares can be created accordingly.

The advantages of between-subjects designs and designs with repeated measurements are shown in Table 12.3. Independent plans are insensitive to learning and fatigue effects. Due to the greater differences between subjects, effects cannot be discovered as easily as in within-subjects designs. On the other hand, these effects are more robust, can be better replicated and are more significant. With repeated measurements, considerably fewer test persons are required and even small effects can be discovered. However, time effects may influence the effects of UV and the effects may be limited to the specific sample and thus poorly transferable to other individuals. Depending on the issue at hand and the practical framework conditions, the plan to be chosen must be weighed up accordingly.

In order to make the procedure and the experimental design easy to understand for the readers of the corresponding reports, a schematic representation is recommended. ◘ Figure 12.4 shows an example of a three-factor plan. As the first IV1, the driver age is examined in two levels, distinguishing young and old drivers. Further there is as IV2 the warning tone with the conditions "without" and "with". Finally, as the third IV3, the location of the warning is investigated in three levels, with each subject experiencing all three locations (repeated measurement). In addition to the IV, the number of test persons can be seen in the cells. The numbering makes it clear at which point a repeated measurement was introduced and where independent groups are examined. Since the IV3 is examined with repeated measurements, a complete permutation with 12 test persons was used here.

The presentation of the experimental design is also so important because the questions of the experiment can be derived directly from it. As shown above, the aim is to examine the extent to which the IVs lead to a systematic change in the DVs. If one compares a IV with two levels (without and with warning tone), the question arises whether the characteristic values of the corresponding two groups differ (see ◘ Table. 12.4). At three levels of IV, one can examine whether IV leads to differences between the three groups at all. Further one is interested in which of the groups differ from each other.

It gets more complex with two or more IV. With two IVs, one is interested on the one hand in the effect of each individual factor, and on the other hand in the interaction of the factors. Does the warning sound have a different effect when combined with the HUD than with the conventional display? Such an effect is called interaction. The same applies to three-factor plans, where the interaction between all three factors is added to the individual effects and two-fold interactions. This increasing complexity leads to the fact that already the results of four factor plans are dif-

◘ **Table 12.3** Advantages of independent plans and plans with repeat measurements

Between subjects design	**Within-subjects design (repeated measurements)**
Robust to learning and fatigue effects	Only few test persons are needed
When an effect is discovered, it is more replicable and more significant	Even small effects can be detected

		IV1 "Driver's age"			
		Young drivers		Old drivers	
		IV2 "warning sound"			
IV3 "Location of warning"	HUD	VP 1..12	VP 13..24	VP 25..36	VP 37..48
	Conventional display	VP 1..12	VP 13..24	VP 25..36	VP 37..48
	Centre console	VP 1..12	VP 13..24	VP 25..36	VP 37..48

◘ **Fig. 12.4** Example of a three-factor experimental design with repeated measurements on the "Warning location" factor (VP stands for subject)

■ Table 12.4 Experimental designs and related questions. For further explanation, see text

Experimental design	Questions
1 IV, 2 levels	Do the two groups differ?
1 IV, 3 and more levels	Does the IV work? Which groups are different?
2 IV	Does the IV 1 work? Does the IV 2 work? Does IV 1 have a different effect depending on the characteristics of IV 2 and vice versa? (interaction)
3 IV	Are IV 1, IV 2 and IV 3 working? Is there an interaction between IV 1 and IV 2, IV 1 and IV 3 or IV 2 and IV 3? Is there an interaction between the three IVs?

ficult to interpret in practice and from there one can only recommend to concentrate on the most relevant three IV per investigation and in case of doubt to carry out several investigations. These problems of interpretation are described in more detail in ▶ Sect. 12.3.2. Before doing so, however, it is important to present the statistical validation of the results under the keyword "significance tests".

12.3.2 Statistics: Significance Tests

Why is an examination with several subjects necessary for the investigation of differences between different conditions? The reason lies in the diversity of persons described above. Not everyone reacts with the same speed, so that a group of people always get a distribution of the measured values, although they are examined under the same conditions. However, it also follows from this that differences will always occur when comparing two groups, even if the groups are treated equally. If the groups are treated differently, the question arises as to whether the differences found can be explained by the random error or whether they arise systematically as an effect of the influencing factors investigated. This question examines the statistics with the help of significance tests. Essentially, these questions answer the following question:

- Question of statistics: Are the differences found between the various conditions so great that it can be assumed that the influencing factors investigated have an effect?

To answer this question, it is formulated somewhat differently:

- How probable are the differences found under the assumption that this is only due to the (random) differences between persons, but not the factors examined?

The advantage of this formulation is that the assumption it describes can be converted into a statistical model. If you assume that only random differences are the cause, you can create a distribution of possible differences ("What results would you find if you repeated the investigation 100 times?"). If, for example 10 test persons were examined in two groups and their reaction time measured, these 20 measured values can be randomly assigned to two groups of 10 measured values each and the difference between the mean values of the two groups calculated. If this is repeated frequently, a distribution of possible differences is obtained on the assumption that only random differences (in this case distributing the 20 values randomly into two groups) had an effect. Such a distribution is shown in ■ Fig. 12.5. It can be seen that under random conditions the same results are relatively frequent in both groups (reaction time difference = 0), while large deviations in positive and negative direction are less frequent.

With the help of this random distribution, the above question can now be answered by indicating how likely it is that the difference actually found in the experiment will occur under random conditions. One takes the probability of this and more extreme differences, which corresponds to the area under the curve. This is shown in the ■ Fig. 12.5 filled out accordingly. In order to decide whether this result indicates that the influencing vari-

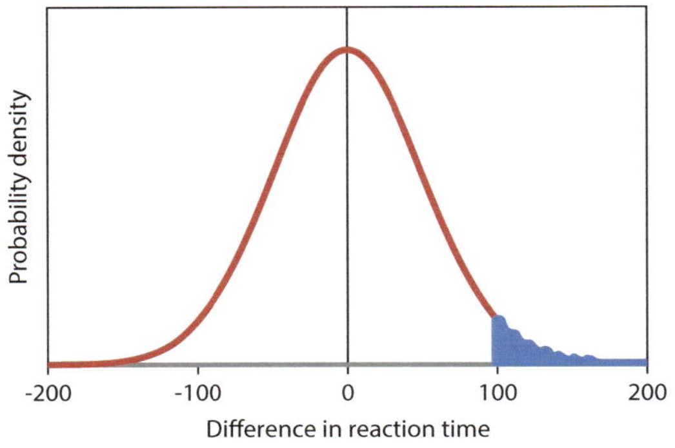

Fig. 12.5 Example of a random distribution of reaction time differences. The area in which the actually found reaction time difference and more extreme differences lie is shown in full

able actually acted, a decision rule is introduced:

- If the difference found is extremely unlikely under random conditions, then it is assumed that the influencing factor (IV) has worked, i.e. the difference is systematic and not random.

To decide whether something is improbable, a so-called significance level α is defined:

- "Unlikely" or "significant" usually corresponds to a significance level of $\alpha = 5\%$.

You can also find the convention to use a $\alpha = 1\%$. This is often referred to as "highly significant" results. The various significance tests (see below) now indicate the probability of the found result under random conditions, usually as a relative frequency of e.g. $p = 0.023$. Since this p is smaller than the significance level α (which is usually given in percent), it is decided that the result cannot be explained well by random differences, i.e. the influencing factor has worked or "the result is significant". In summary, a significant result means that this is very difficult to explain by random differences. A proof of the effect in the very strict sense is of course not possible, because always a certain amount of uncertainty remains – even under random conditions this result could have occurred, albeit very rarely.

Statistically, the assumption that only random differences had an effect is considered to be a null hypothesis ("H_0"). The alternative hypothesis ("H_1") assumes that there is an effect. A distinction is made between a specific and an unspecific alternative hypothesis. The specific alternative hypothesis indicates the direction of the effect, e.g. that the reaction times become shorter with a new warning system. With the unspecific alternative hypothesis, on the other hand, one suspects a difference, which can, however, be in both directions. Here for instance is formulated: "The reaction times in the experimental group are different than those of the control group". The significance test can also be understood as a decision about these hypotheses. If the result is significant, the null hypothesis can be rejected. This indicates the presence of an effect. If no significant result is found, the null hypothesis must be maintained. So you couldn't see any effect. It is important to note that this does not automatically mean that there is no effect. This is related to two types of errors shown in Fig. 12.6 and explained below.

The first type of error is the alpha error. If one finds a significant result in an experiment, although the null hypothesis actually applies, i.e. in reality (i.e. in the population) this difference is not present, one makes an erroneous decision: On the basis of the result of the significance test one concludes that there is an effect, which is not true. This is shown graphically in Fig. 12.7. If the null hypothesis is correct, the possible results of studies are distributed according to the red curve. If one finds in a study a reaction time difference of e.g. +150 ms that is significant according to the statistical test, one decides to reject the null hypothesis, although in reality it is true in the population. One therefore concludes that an effect exists even though none exists.

To protect yourself from this alpha error, you can choose a lower significance level, e.g. 1% instead of 5%. This reduces the probability of wrongly choosing an effect. Another possibility is to replicate the effect under conditions as similar as possible. If a significant result is also found in the repetition, the probability of a wrong decision is significantly lower overall. If, for example, two studies are carried out and a significant result is found at

	Reality/total population	
	No difference Null hypothesis is correct	Effect existing Alternative hypothesis is correct
Significance test		
Significant	Alpha error	Correct decision
Non-significant	Correct decision	Beta error

Fig. 12.6 Correct and incorrect decisions in the significance test. For further explanation, see text

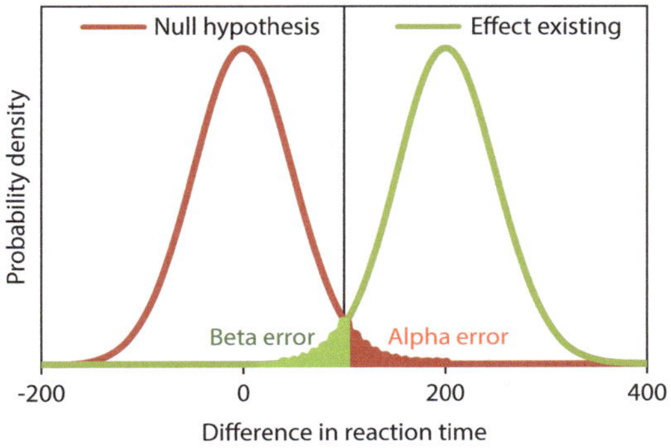

Fig. 12.7 Alpha and beta errors for the example of reaction times. The red curve shows the distribution of the possible results (difference between the mean values of the two investigated groups, "reaction time difference") under the assumption of the null hypothesis, the green curve the distribution of possible results in the "reality", i.e. the population in which an effect is present

α = 5%, the probability is that both studies will become significant, even though in reality there is no difference is p = 0.05*0.05 = 0.0025, i.e. only 0.25%. In this way one can minimize the probability of an erroneous interpretation in the sense of claiming that there is a difference, although this is not the case in the population. If, for example, you want to introduce a new warning system in the vehicle, which is, however, associated with considerable costs, you can be sure in this way that there is actually a benefit for the driver.

However, this minimization of the alpha error should be seen in the light of the fact that it is associated with an increase in a second type of error, the so-called beta error. This is illustrated in Fig. 12.7 with the help of the green curve. This shows the distribution of test results in the event that an effect is actually present in the population, which can be described in the example as an extension of the reaction times around 200 ms. Since only a part of the population is examined in each study, it is possible that a random sample is examined in which this effect does not show up, so that the reaction time difference e.g. is only +50 ms. According to the decision criterion shown in the figure, the null hypothesis cannot then be rejected, since this result is not quite probable even if there is no effect. This erroneous decision is referred to as a beta error. Here it is concluded on the basis of the test result that there is no effect, although in reality there would actually be a difference in the population. The more you try to minimize the alpha error, the bigger the beta error will be, the less you will discover an effect even though it is present.

Therefore, if you want to be sure not to overlook any effect, it makes sense to select the significance level alpha relatively large. This is particularly the case if one wants to prove that two variants are equivalent. If a new warning concept for example has been developed, which is associated with significantly lower costs than an older, relatively expensive solution, the equivalence of the variants is to be demonstrated in the experiment. In this case, the interest is to confirm the null hypothesis. Here it is important not to overlook it if there is a difference between the variants. Usually, the significance level is then set at 25%. In addition, there are also special types of significance tests, the so-called equivalence tests, which can be used to prove equality. A detailed description can be found at Wellek (Wellek 2010).

Statistical testing for difference thus assumes that distributions of possible differences are present under random conditions. One could now create the corresponding random distributions for each experimental design with the parameters measured there in order to achieve this statistical estimation. As this would be relatively time-consuming, either the measured values are transformed into ranks or categories and corresponding tables are drawn up for the various experimental designs. In each case, it must be taken into account how many independent variables were examined in which steps and how many test persons were involved. Or one transforms the measured values in such a way that they correspond to certain statistical distributions, which are then used to read the probabilities in a comparable way as in the example in ◘ Fig. 12.5. In the latter case, one speaks of distribution-based methods, while the first are called distribution-free or non-parametric methods. "Distribution-free" means that no theoretical distribution is referred to. This designation is therefore not completely correct, since an empirically provided distribution (e. g. of ranks) is used. From this point of view, the term non-parametric procedure is preferable, since "parameters" refers to the essential parameters of the theoretical distribution (e.g. mean value and standard deviation for a normal distribution). Such parameters are not required for non-parametric procedures.

Data quality shall be taken into account when deciding on the procedure to be used. Distribution-based methods are only really meaningful from the interval level, since differences are calculated in the calculation of the characteristic values that are only significant at this scale level. Furthermore, a certain sample size is required (e.g. more than 30), since the values are distributed sufficiently similar to these theoretical distributions only for larger samples. Finally, the question always arises as to whether the measured values are actually distributed sufficiently similarly to the theoretical distribution. It can be partly checked whether the empirical values correspond to certain prerequisites. However, these tests are often relatively sensitive and indicate deviations from assumptions which, however, practically do not lead to any substantial change in the statistical evaluation. The non-parametric methods are also applicable at the lower scale levels, but not quite as sensitive to detect significant effects. Non-parametric evaluations are, however, difficult, especially in the case of multifactorial plans, since the assessment of the interaction of several factors is often made by adding effect estimates, which in turn does not appear to be useful at the ordinal level. Therefore, parametric methods are also frequently used in practice, although the prerequisites are doubtful.

In the presentation of the results of the statistical tests, the corresponding test quantities into which the empirical characteristic values have been converted are given on the one hand in order to be able to reproduce the calculation. The value of the test variable itself also includes an indication of the general conditions, which essentially corresponds to the number of test persons examined or the measured values used. This is hidden (in a slightly transformed form) in the so-called degrees of freedom (df). On the other hand, the result of the statistical test is reported as a p-value (as shown above).

Special programs such as SPSS, R (as open source variant) or toolboxes in Matlab are used for the calculation. A detailed description of the individual tests would go beyond the scope of this chapter. A detailed description can be found in corresponding textbooks of statistics, e.g. at Bortz und Schuster (2010) or Sedlmeier und Renkewitz (2008). An overview of the most important procedures can be found in ◘ Table 12.5. The parameters listed in the right-hand column can be found in the corresponding editions of the statistics programmes. When selecting the individual tests, it is important to consider whether a test plan with repeated measurements is available or whether independent groups have been compared. In the first case, the measured values must also be arranged in such a way that measured values for the various conditions (columns of the data matrix) are available for each test person (row of the data matrix). In the case of independent groups, group membership is coded

Table 12.5 Overview of important statistical methods and the presentation of test results

Experimental design	Test	Scale level	Repeat measurement?	Result formulation
1 IV, 2 levels	Independent sample t-test	Interval level	Without	t(18) = 2.3, p = 0.017
	Dependent sample t-test	Interval level	With	t(9) = 3.5, p = 0.003
	U-test	Ordinal level Distribution-free	Without	U(9) = 14, p = 0.009
	Wilcoxon test	Interval level Distribution-free	With	T(10) = 10, p = 0.042
	Sign test	Ordinal level Distribution-free	With	x(10) = 1, p = 0.011
	Chi-square test	Nominal level	Without	X^2 = 5.05, p = 0.025
	McNemar test (2 Categories)	Nominal level	With	X^2 = 4.5, p = 0.033
1 IV, 3 and more levels	Single factorial analysis of variance	Interval level	Without	F(2,27) = 4.7, p = 0.03
	H test	Ordinal level	Without	H(1) = 7.06, p = 0.007
	Friedman Rank Variance Analysis	Ordinal level	With	$X^2r^{(2)}$ = 6.1, p = 0.047
	Chi-square test	Nominal level	Without	X^2(3) = 12.5, p = 0.005
	Q test by Cochran (2 Categories)	Nominal level	With	Q(3) = 8.3, p = 0.003
2 IV	Analysis of variance (ANOVA)	Interval level	Without	HW IV 1: F(1,30) = 4.7, p = 0.038 HW IV 2: F(2,30) = 5.2. p = 0.011 WW F(2,30) = 4.4, p = 0.021

using a separate variable. The scale level of the measurements must also be observed. For smaller samples (per group n ≤ 10), non-parametric testing should be used if possible, since violations of the requirements of distribution-based testing are very important for small samples. In order to assess the test variables, it is necessary to specify the degrees of freedom, since the significance (expression of the p-values) depends on them. For the degrees of freedom, either the number of subjects is relevant (marked "n" in the table) and/or the number of levels of the independent variable (marked "k" and "l" in the table). The degrees of freedom can also be found in the corresponding editions of the statistical programs.

The interpretation of statistically significant effects is simple when comparing two groups – these two groups differ. If several levels of a IV are investigated, the test parameter indicates whether at least two of the investigated groups differ. You can then either decide graphically where the differences lie, or make corresponding comparisons in pairs to decide this statistically. These pairwise comparisons are usually performed automatically by the statistics programs.

Another interesting aspect is the interpretation of the results of the two- and multi-factor variance analyses. For the main effects, i.e. the effects of the individual factors examined, it is indicated in each case whether the corresponding factor leads to significant differences irrespective of the characteristics of the other factors. Furthermore, the interactions between two factors each and, depending on the experimental design, between three and more factors are examined. If there are interactions, the main effects can sometimes no longer be interpreted depending on the direction of the effects. This will be shown at following chapter.

A central property of the statistical tests can finally be described at the distribution in ◘ Fig. 12.5. The less the difference values scatter, the narrower this distribution is and the more likely it will be to decide that a particular difference is very unlikely under random conditions. The consequence of this is that smaller effects can already be discovered with larger samples. The larger the sample, the more reliably and accurately the true value of the group is estimated (see above). This means that the difference between the values of the two groups is also less prone to error, and thus spreads less in the distribution. In parametric tests, this is also taken into account by the fact that when calculating a corresponding test variable, a measure of the difference between the groups (e.g. the difference between the group averages) is usually placed in relation to the measurement errors (e.g. the pooled variance within the groups). This becomes particularly clear with the test variable of the variance analysis, where the F-value (test variable) is a fraction of primary variance and error variance. Primary variance describes the difference in the measured values resulting from the different independent variables, error variance the random differences between the test persons.

In addition to enlarging the sample in order to capture the differences with a lower measurement error, this provides a second possibility for experiment planning in order to better recognize significant effects. For this purpose, either homogeneous groups of test persons are used in order to minimize the error variance, or each test person is compared to him/herself under different conditions (dependent test plans or test plans with repeated measurements). From a statistical point of view, this also explains why experimental designs with repeated measurements can very well detect even small effects, even if only a relatively small number of test persons are examined.

The consideration of the appropriate sample size is also related to the statistical validation of effects. The term "power" is used to describe how well a statistical test is suitable for statistically proving an existing effect. The power of a test is be calculated as power = 1 – ß where β is the beta error described above. The greater the probability of incorrectly rejecting an effect, the smaller the power, i.e. the ability of a test to detect an actually existing effect. As an investigator, one is correspondingly interested in using a test that is as powerful as possible. For example, parametric tests are usually more powerful than distribution-free methods. However, the main determinant of power is the sample size. The more people examined, the more likely it is that smaller effects can be statistically proven. If you know how the measured values with which you decide on the effect are distributed and how large the examined effect is in reality, then you can estimate how many test persons are needed before the examination to be able to statistically prove this effect. As a rule, however, the effect size is not known before the start of a study (otherwise the study would not be needed either). One can then fall back on conventions and, with the help of corresponding programs, calculate estimated values for suitable sample sizes for small, medium and large effects, e.g. for the test corresponding to the experimental design. A program very frequently used in this context is the freely available G*Power (Faul et al. 2009). An estimation of a meaningful sample size before the start of the investigation is very useful to ensure that a relevant difference can be demonstrated at all with the selected number of subjects.

Table 12.6 Example of a tabular display of the results of a two-factor analysis of variance

	UV 1 Modality: verbal vs. manual		UV 2 System: navigation vs. telephone number		Interaction modality x system	
	F (1,29)	p	F (1,29)	p	F (1,29)	p
SDLP	*9.1*	*0.005*	*4.6*	*0.040*	*5.2*	*0.029*
Reaction time	*56.9*	*0.000*	0.9	0.345	*11.3*	*0.002*

For the F-value, the degrees of freedom are given in brackets. SDLP is the Standard Deviation of Lane Position

12.3.3 Statistics: Presentation of Results

The significance test can only indicate whether the independent variable has had an effect. This is the necessary prerequisite for differences to be interpreted at all. If the test is not significant, the differences found may be random and should therefore not be described as effects. From this point on, the first step in interpreting the effects is to indicate where effects occurred at all. For this purpose, the results of the statistical tests are listed according to Table 12.5. In the case of a large number of dependent variables or complex experimental designs, a tabular display can be useful here. Table 12.6 shows an example of a two-factor variance analysis. In the comparison of navigation system and the selection of a telephone number (IV 2: system) the difference between verbal and manual operation (IV 1: modality) was examined. The standard deviation of lane position (SDLP) and a reaction time to road signs requiring steering were measured. You can find the F- and p-values of the analysis of variance. Since the degrees of freedom for each test were the same, they are given in brackets after the F-value. The significant results at $\alpha = 5\%$ are shown in bold. It can be seen that in SDLP both main effects and the interaction are significant, in reaction time the main effect of the modality and the interaction.

Starting from such a result, the significant effects are then displayed and described graphically or in a table. The best type of representation, especially for two-factor experimental designs, is the line graph, since this makes the different interpretation possibilities most clearly visible. Two types of representation are possible for two-factor experimental designs, as shown in Fig. 12.8. On the left side the blue line represents the navigation system, the red line the telephone. You can see that both systems have the lines running down, i.e. the SDLP is smaller for linguistic operation. With speech the lane keeping performance is better. This description corresponds to the main effect of the IV 1 "modality". It can also be seen that there is no main effect for the IV 2 "system", as the navigation system leads to a larger SDLP when operated manually, while there are no differences when operated verbally. Thus, the main effect of this IV cannot be interpreted, although it is significant. This different effect of the system depending on the modality corresponds to the significant interaction: The effect of one IV can only be interpreted as depending on the other IV. Each interaction can be interpreted in two directions: As an effect of IV 1 depending on the levels of IV 2 and vice versa. For example: The improvement of the SDLP through verbal operation is significantly stronger for the navigation system than for the telephone (Interpretation 1). When used manually, the SDLP is significantly worse with the navigation system than on the phone. This difference cannot be found in verbal operation (Interpretation 2). Both interpretations can be clearly seen in the left side of Fig. 12.8. Looking at the right side of the illustration, the second interpretation is more important here. The red line is parallel to the x-axis, while the blue line falls. Depending on which effects are

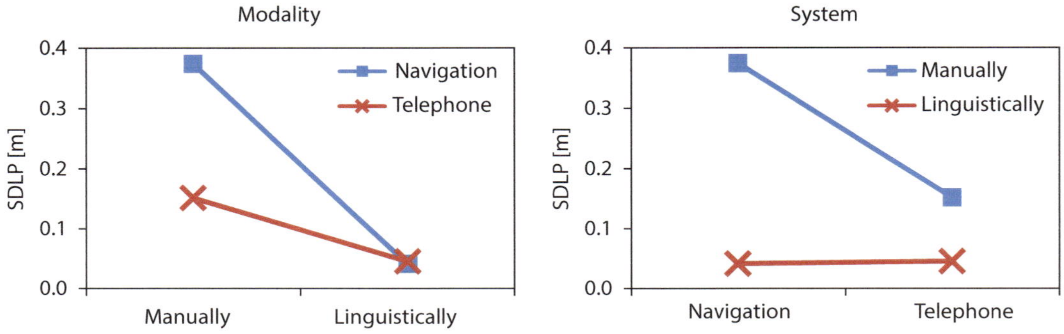

▪ **Fig. 12.8** Presentation of the effects for the example in ▪ Table 12.6. The mean values are shown. For explanation, see text

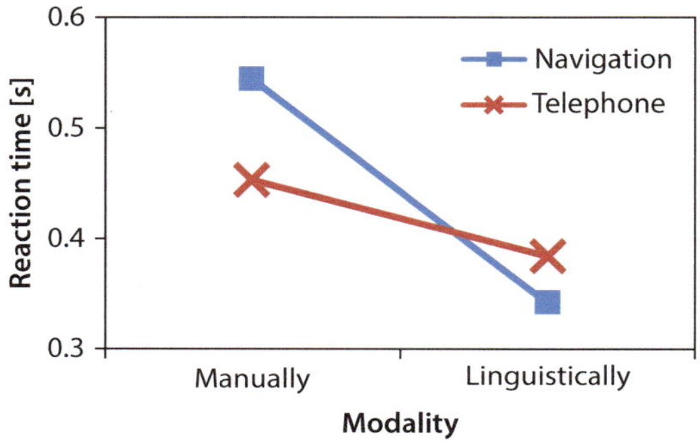

▪ **Fig. 12.9** Results of the reaction time for the example from ▪ Table 12.6. The mean values are shown. For explanation, see text

significant and significant for the reader, the right or left illustration may be more meaningful.

▪ Figure 12.9 shows the results of the reaction time. Here the main effect of the modality and the interaction were significant. It can be seen that the reaction time is shorter with verbal operation than with manual operation. The interaction shows that the interaction with the navigation system benefits more from the verbal operation than the operation of the telephone. Or: With manual operation, the response time suffers more from the navigation system. The telephone is a bit worse when using the verbal operation.

This somewhat detailed example illustrates the different roles of the statistical test results and the description of the results. Not every statistically significant result can be interpreted, as the SDLP example shows. For the interpretation a graphical description of the data is necessary. However, only those effects that were actually significant may be interpreted in the graphics. From this point of view it is very important to choose the right type of presentation.

In principle, different types of effect patterns can occur in two-factor experimental designs, which are often misinterpreted. ▪ Figure 12.10 provides an overview of the most important types of effects. The example is a fictitious experiment to read fiction (entertainment) vs. a textbook either on the screen or on paper. The speed of reading was evaluated as a performance quality. The following four cases are important:

- Main effects only: At the top left you can see that the performance on paper is better for both types of text. Further one recognizes that the performance for the entertainment reading is better. Thus, here one finds the two main effects, but no interaction.
- Disordinal interaction: At the top right it becomes clear that no main effect can be interpreted, even if it were significant. The effect of the type of reading depends on the medium. With the digital medium, entertainment is better, on paper, textbook reading. The effect of the medium thus depends on the type of reading. Digital is better for entertainment reading, paper better for text books. Therefore, only the interaction is to be interpreted here.
- Ordinal interaction: At the bottom left you can see that both main effects and the interaction may be interpreted. On paper, reading is better than with the digital ver-

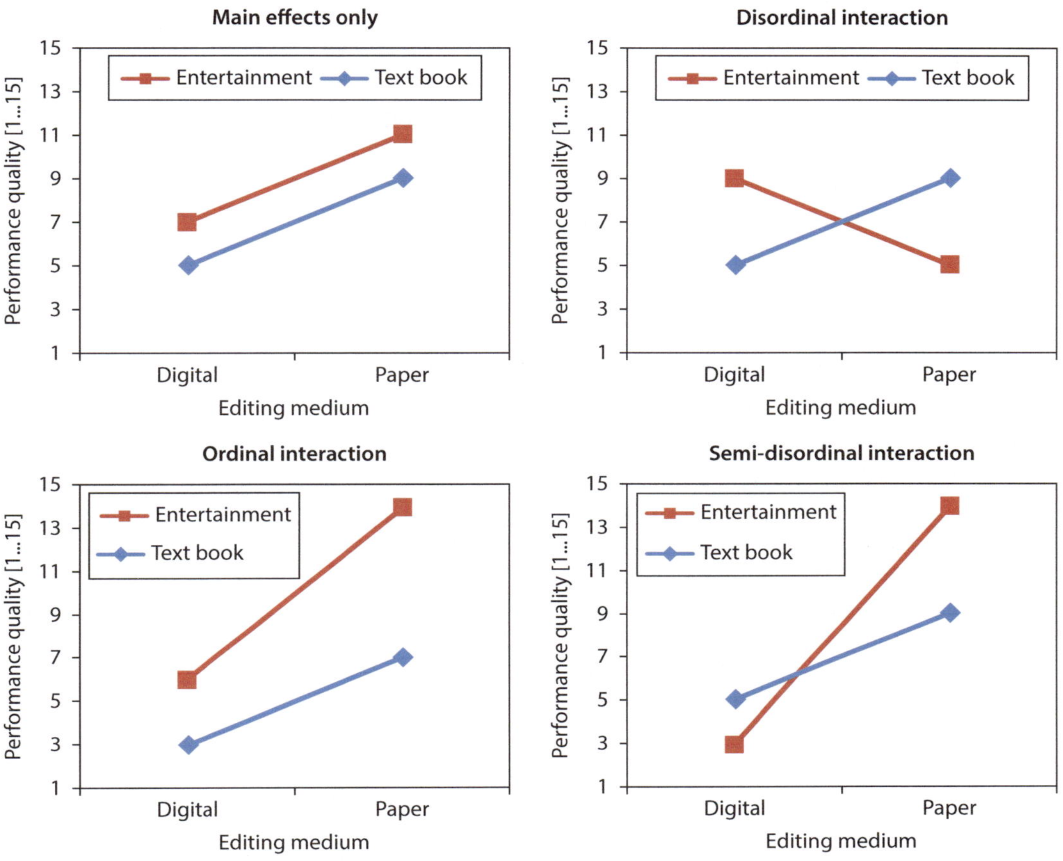

Fig. 12.10 Overview of types of interactions. For interpretation, see text

sion. Rading fiction (entertainment) is faster than reading a textbook (main effects). Furthermore, the advantage of entertainment reading on paper is stronger than with a digital medium. Or: the effect of paper is stronger in entertainment reading than in a text book (interaction).

- Semi-disordinate interaction: In the case shown at the bottom right, the main effect of the medium may be interpreted. On paper, performance is better than digital. The type of reading does not have a uniform effect and cannot therefore be interpreted as the main effect. The interaction is in turn interpretable if it is significant. The effect of the medium is stronger for entertainment reading than for text books. On a digital medium the textbook reads better, while on paper the entertainment reading is better to read.

These examples make it clear that the scientific content is not in the significance test, but in the graphical or tabular presentation of the measured values. The significance test is necessary for deciding what may be interpreted. With the corresponding presentation, it becomes clear to the reader what the effects mean. A list of the results of the significance tests is worthless without descriptive statistics and graphs.

Another useful way of displaying the data is via bar graphs. Often mean values and standard deviations are shown here (see Fig. 12.11). Adjacent bars are very well suited for direct comparisons. The standard deviations are helpful in relativizing the size of the differences. As shown above, statistical testing compares the effect with a measure of random error to investigate significance. In a way, this is analogous to the representation of the mean values and the standard deviation.

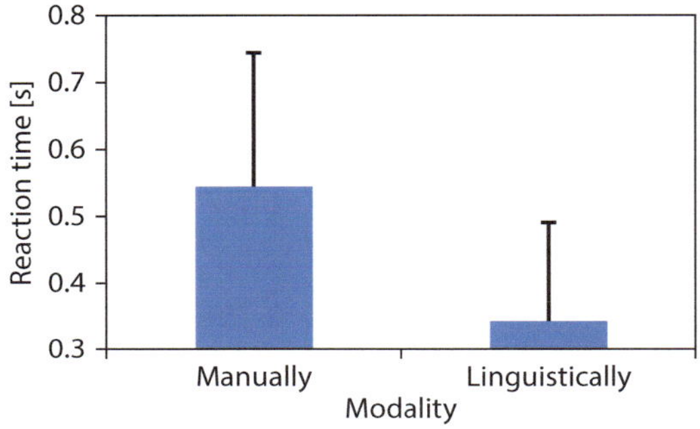

Fig. 12.11 Mean value of the reaction time and standard deviation depending on the modality

Overall, it should be noted that graphics contain the relevant information in a clear and easy to understand manner. Axes must always be meaningful and labelled with units. It is not advisable to use colored backgrounds, 3-D representations, etc., as the data points will be pushed into the background compared to the graphic design elements and will then often be difficult to recognize. Also against this background line graphics are a very effective way of displaying.

12

12.4 External and Internal Validity

The description of the two central statistical approaches revealed a common ground. In both cases, an essential question is the representativeness of the results. This is also referred to as "external validity", i.e. transferability. This depends above all on the drawing of an appropriate sample. Whenever not only statements about quantities to be measured directly are to be made, but these are to be interpreted, a second aspect of validity arises here: Are the characteristics valid, do they really measure what they are to measure? What question do you have to ask a driver to predict that he will buy a particular vehicle? This aspect of external validity is described in more detail in ▸ Sect. 11.1.1. In addition, there is a third aspect of external validity, which results from the examination situation. Are the data obtained e.g. in a driving simulator, representative for driving in one's own vehicle in normal traffic? On the one hand, this is about the closeness to reality of the study, on the other hand it is about the attitude of the participants. The better they succeed in conveying an understanding of the meaning and purpose of the investigation, the more they will be able to behave "normally". Instruction, the clarification of the aims of the experiment, plays a central role here.

Under certain circumstances, however, it may also be necessary not to properly inform the test participants beforehand. If for example the effect of a collision warning system should be investigated, it is important that the drivers are surprised by a critical event similar to the one in real traffic. Therefore, it may be necessary to distract the attention of the drivers via a cover story in order to achieve a surprise effect in the driving simulator as well. For ethical reasons, the participants are to be informed in detail after the attempt. They must also be able to exclude their data from the experiment. In principle, any deception is ethically questionable and its use thoroughly weighed up.

In summary, three aspects of external validity can be distinguished:

- Representative sample
- Valid measurement methods
- Realistic situations with "normal" behaviour

In the second type of statistical question, the search for the effect of influencing factors, "internal validity" is added. It is a question of whether an effect found is actually undoubtedly attributable to the influence of the independent variable. This is only possible from the logic of the experiment, if the investigated groups differ only in the independent variables, otherwise they are treated completely identically. As described above, this is always a problem with repeated measurement designs when there may be fatigue or practice effects. If for example the trip with the visual warning system would always be the second trip, a practice effect could also explain the better reaction time. In order to ensure that the effect is actually due to the IV, the sequence of treatments will therefore always be permuted (see Fig. 12.3). The internal validity thus depends on how well the influence of interference variables can be eliminated or kept constant in the various conditions. This is the reason to standardize the test procedure including the instructions as much as possible in order to ensure a comparable treatment of all participants. Ultimately, however, internal validity cannot be guaranteed by a flowchart. As an investigator, one should always ask oneself whether a certain result could not also be explained by other factors than the influence of the independent variables. This critical way of thinking is an essential prerequisite for good research.

References

Bortz, J., Schuster, C.: Statistik für Human- und Sozialwissenschaftler (Lehrbuch mit Online-Materialien). Springer, Berlin (2010)

Faul, F., Erdfelder, E., Buchner, A., Lang, A.-G.: Statistical power analyses using G*Power 3.1: tests for correlation and regression analyses. Behav. Res. Methods. **41**, 1149–1160 (2009)

Sedlmeier, P., Renkewitz, F.: Forschungsmethoden und Statistik in der Psychologie. Pearson, München (2008)

Wellek, S.: Testing Statistical Hypotheses of Equivalence and Noninferiority, 2. Aufl. Taylor & Francis, Boca Raton (2010)

Outlook

Klaus Bengler and Heiner Bubb

Contents

13.1 Electric Mobility – 676

13.2 Automation – 677

13.3 Mobility Behaviour – 678

References – 678

H. Bubb et al. (eds.), *Automotive Ergonomics*, https://doi.org/10.1007/978-3-658-33941-8_13

13.1 Electric Mobility

Klaus Bengler

The transition from the internal combustion engine to the hybrid vehicle or electric vehicle thus makes a new consideration of driver and driving behaviour appear reasonable (Helmbrecht et al. 2014). The information about the range and the influence of the driving style on the range requires additional displays and above all a differentiated consideration of pedal concepts which, for example, support the recuperation of energy without using the brake pedal via one-pedal-drive. Eberl (2014) investigated the effects of different manifestations of a recuperation function on driving experience (see too ► Sect. 6.4.2). In general will the introduction of electric drivetrains lead to an increase of automatic gear shift in combination with different one-pedal drive and recuperation concepts.

Compared to battery electric vehicles the majority of users prefers to use hybrid vehicles. In this case it is important that the HMI provides adequate and understandable information how to use these vehicles efficiently. The driving style plays a dominant role whether the electric potential of a hybrid vehicle can be used to reduce its CO-2 footprint. It is important that this learning process is supported by standardized symbology and interaction logic in cars provided by different brands.

It is already evident that the electrification of the drive also changes the experience of longitudinal dynamics and gives ergonomics new scope for design (Müller et al. 2014; Helmbrecht et al. 2014). There is therefore a great need for a better understanding of the experience of longitudinal dynamics (acceleration, deceleration, sailing). The value ranges of human perceptual performance with respect to longitudinal vehicle dynamics characteristics vary relatively strongly depending on the test condition and are mainly based on experimental laboratory investigations (Meiry 1965; Benson et al. 1986; Kingma 2005) and only in rare cases on real vehicle investigations (Rockwell and Snider 1965). Müller has taken up the topic again using adaptive psychophysical procedures and a test vehicle with current measurement and calibration software (Müller et al. 2013). Here the difference thresholds for vehicle longitudinal accelerations could be specified more precisely and the effects of different influencing variables quantified (Müller et al. 2014). Müller's experiments provide practical confirmation of the statements made by Mallery et al. (2010) and Bigler and Cole (2011) that the vestibular apparatus responsible for the perception of acceleration stimuli, in contrast to other sensory channels, has no quantified Weber constant and can therefore maintain low perception thresholds even with increasing stimulus sizes and across different experimental conditions. The development of appropriate test facilities is also necessary in order to be able to compare different designs. One goal is to increase the proportion of recuperation phases without active braking and the proportion of sailing phases and to enhance the positive experience, especially in the field of electric driving with its enormous longitudinal dynamic potential. The design of the braking system and the interaction with the brake pedal, as well as the design of the recuperation function and the interaction with the accelerator pedal or better the control unit for acceleration and deceleration, are at the centre of attention here. The driver must be given immediate and unambiguous feedback in these new vehicle concepts, so that an anticipative driving style is generally supported, which will play an even greater role than before (Rommerskirchen et al. 2014). The driver needs additional information about the vehicle status (range, recuperation status and battery status), but also the connectivity of the vehicle and its configuration before the start of the journey are becoming more important. This will also increase the importance of mobile devices that allow remote access to the vehicle. Especially connectivity is of more importance for electric cars compared to their combustion equivalent, as users want to monitor and manage the charging process remotely.

The increasing energetic consideration of the vehicle and its operation naturally has significant effects on air conditioning and vehicle weight. In the case of air conditioning, reliable findings are available (Fanger 1972; Fiala et al. 2010; Zhang 2003; see also ► Sect. 8.4). However, an increase in efficiency and optimisation of the air-conditioning concepts is necessary, since the waste heat from the combustion engine can no longer be used in the event that heating is required. In principle, the switch to more decentralized and local heating components appears to be promising, but there must be no noticeable loss of comfort. Heated seats, steering wheel and further surfaces are of much higher importance in EVs. This has to go along with more sophisticated HMI concepts to adjust temperature and climate adjustments per seat. Initial findings show that it is quite possible to close the gap in terms of ergonomics by using alternative air conditioning concepts for heating and cooling. Furthermore, it is necessary to extend the existing digital human models in such a way that a model-based validation of air conditioning is already possible in the early phase (Janta et al. 2014; Stuke and Bengler 2014; see too ► Sect. 8.4.4.5). Here, too, the technical development in this area must be accompanied by classical ergonomic design principles, but also necessitates a reappraisal. Increasingly, air conditioning concepts are being discussed which are no longer based on pure air exchange and air flow, but which also consider radiative and conductive heating and cooling elements. If this would replace the space-intensive design of the HVAC (Heating, Ventilation and Ait Conditioning) and its air duct, the space gained could be reused. In addition, this could mean a distinction in the design of heating and cooling concepts, since different parts of the body have different degrees of influence on the global comfort of people. However, their design and positioning is not covered by the current modelling and evaluation tools. At best, the use of digital human models makes it possible to position the vehicle relative to the drivers' postures in the interior (Janta et al. 2014).

The extension of existing human models, in particular by evaluating local comfort and its influence on global perception in dynamic, inhomogeneous environments, requires comprehensive investigations in the vehicle and in the climate chamber. Within the framework of the TUM projects MUTE (Both et al. 2010; Kremser et al. 2011; Lorenz et al. 2011) and VisioM some concepts for the heating case have already been investigated and tested. In the cooperation program TUM CREATE similar experiments were considered for the application of the e-vehicle EVA (Stuke and Bengler 2014) and for the cooling case in tropical cities.

13.2 Automation

Increasing automation of individual transport and the possible transition to cooperative vehicle guidance will become increasingly important in the future. Here, too, technological development represents a paradigm shift for the interaction between driver and vehicle, which must not impair safety and comfort under any circumstances (Flemisch et al. 2014; Bengler et al. 2014).

While driver assistance systems such as ACC and Lane Departure Warning have already improved comfort and safety, further integration of these systems and an expansion of their functional range can be expected. If, however, the degree of automation also increases, the role of the driver will change to that of system supervisor. In this case, specific ergonomic measures must be taken to ensure that the interaction concepts are such that the driver is able to resume driving in a timely and competent manner (Gold 2016; Vogelpohl et al. 2016). In any case, the monitoring of such a vehicle places different and sometimes even higher demands on the driver than the manual driving of the vehicle. The displays and controls designed for manual vehicle steering must also be reconsidered with regard to their suitability for semi-automated and highly automated vehicle guidance (see too ► Sect. 9.4).

13.3 Mobility Behaviour

Heiner Bubb

Many future scenarios assume that future mobility behaviour, at least in conurbations, will no longer be characterised to a large extent by the individual ownership of vehicles. Such scenarios suggest, for example, that a smartphone – similar to the way a taxi is waved at the side of the road today – can be used to call an autonomously driving vehicle, which is then available for the personal driving request and automatically withdraws again after the driving order has been completed (Gnatzig 2015). For other mobility contracts, for example, a suitable vehicle could be ordered for a weekend excursion. Various combinations of public and private transport, including the use of bicycles and/or e-scooter, are also conceivable. In this way, it is hoped that the proportion of stationary traffic can be significantly reduced and that the available space in cities can essentially be reserved for rolling traffic. In all these scenarios, one of which is only highlighted here, the smartphone practically plays a central role as the key to the mobile world. This results in completely new requirements for ergonomics regarding the software-ergonomic design of such user-smartphone interactions. Irrespective of the technical problem of creating interfaces for different smartphone and vehicle technologies so that the above version actually functions smoothly, there are also ergonomic visions, such as the automatic adaptation of the ordered vehicle to the individual wishes of the driver – starting with the individualisation of the cockpit to the automatic adaptation of the driver's seat, steering wheel and mirror settings to the respective driver; data that may be stored in the smartphone and transferred to the called vehicle. This aspect already plays an important role in many scenarios, which are by no means as ambitious as the one described above and which can also be implemented with today's readily available technology. In all this, however, a statement by Weiser (1991) must be taken into account in terms of ergonomics, who said: "The most profound technologies are those that disappear", i.e. that user friendly technology fulfills the users wishes without him having to worry about details of operation.

The indicated technological developments illustrate that the automobile is still subject to significant innovative changes and that the development outlined in Akamatsu et al. (2013) will continue. On the one hand, ergonomic requirements can be translated into new concepts, on the other hand, new methodological and conceptual questions are posed to ergonomics. A representative survey of users in Europe commissioned by Bosch shows that around 60% of those questioned cannot imagine doing without a car. A further 22% would be prepared to make only a partial renunciation (Ehrenfeuchter 2020). For all the rational reasons that speak in favor of largely foregoing a vehicle in personal ownership, this emotional reason must not be forgotten: many see the vehicle as a private retreat that also promises individual and spontaneous mobility without much planning. Only technical solutions that can satisfy this need will prevail in the future.

References

Akamatsu, M., Green, P., Bengler, K.: Automotive technology and human factors research: past, present, and future. Int. J. Veh. Technol., 526180, 27 pages, (2013). https://doi.org/10.1155/2013/526180

Bengler, K., Dietmayer, K., Färber, B., Maurer, M., Stiller, C., Winner, H.: Three decades of driver assistance systems: review and future perspectives. IEEE Intell. Transport. Syst. Mag. **6**(4), 6–22 (2014). https://doi.org/10.1109/MITS.2014.2336271

Benson, A.J., Spencer, M.B., Stott, J.R.R.: Thresholds for the detection of the direction whole-body, linear movement in the horizontal plane. Aviat. Space Environ. Med., 1088–1096 (1986)

Bigler, R., Cole, D.: A review of mathematical models of human sensory dynamics relevant to the steering task. In: Proceedings of the 22nd IAVSD Symposium on the Dynamics of Vehicles on Roads and Tracks (2011)

Both, J., Kremser, F., Rüger, O.: Neue Grenzen, neue Freiheiten – das MUTE Elektrofahrzeug der TU München. Automobil Industrie INSIGHT. **55**(B01046), 72–75 (2010)

Eberl, T.: Charakterisierung und Gestaltung des Fahr-Erlebens der Längsführung von Elektrofahrzeugen.

Dissertation Technische Universität München (2014)

Ehrenfeuchter, M.: Antriebstechnik: so fahren wir morgen. Auto-Motor-Sport, Heft. **21**, 40–43 (2020)

Fanger, P.O.: Thermal Comfort: Analysis and Applications in Environmental Engineering. McGraw-Hill, New York (1972)

Fiala, D., Psikuta, A., Jendritzky, G., Paulke, S., Nelson, D.A., van Marken Lichtenbelt, W.D., Frijns, A.J.H.: Physiological modeling for technical, clinical and research applications. Front. Biosci. **52**, 939–968 (2010)

Flemisch, F.O., Bengler, K., Bubb, H., Winner, H., Bruder, R.: Towards a cooperative guidance and control of highly automated vehicles: H-mode and conduct-by-wire. Ergonomics. **57**(3), 343–360 (2014)

Gnatzig, S.: Trajektorienbasierte Teleoperation von Straßenfahrzeugen auf der Basis eines Shared-Control-Ansatzes. Dissertation, Technische Universität München (2015)

Gold, C. G.: Modeling of Take-Over Performance in Highly Automated Vehicle Guide. Dissertation, Technische Universität München (2016)

Helmbrecht, M., Olaverri-Monreal, C., Bengler, K., Vilimek, R., Keinath, A.: How Electric Vehicles Affect Driving Behavioral Patterns (2014). ID: T-ITS-14-01-0033.R1

Janta, M., Bengler, K., Senner, V.: Komfort durch lokale Klimatisierung in Elektroautos GfA-Frühjahrskongress, Gestaltung der Arbeitswelt der Zukunft., S. 398–400 (2014)

Kingma, H.: Thresholds for perception of direction of linear acceleration as a possible evaluation of the otolith function. BMC Ear Nose Throat Disord. **5**(1), 5 (2005). https://doi.org/10.1186/1472-6815-5-5

Kremser, F., Pietsch, R., Wilden, W., Lienkamp, M., Bengler, K.: Anthropometrische Innenraumauslegung eines Elektrofahrzeugs der Subcompact-Klasse. In: Gesellschaft für Arbeitswissenschaft e.V. (GfA) (Hrsg.) Mensch, Technik, Organisation – Vernetzung im Produktentstehungs- und -herstellungsprozess, Bericht zum 57. Arbeitswissenschaftlichen Frühjahrskongress vom 23.-25. März an der Technischen Universität Chemnitz, S. 239–242. GfA-Press, Dortmund (2011)

Lorenz, D., Remlinger, W., Kremser, F., Matz, S., Bubb, H., Bengler, K.: Sichtauslegung eines kompakten Elektrofahrzeugs mit ‚RAMSIS kognitiv' 2. Automobiltechnisches Kolloquium, Antriebstechnik, Fahrzeugtechnik, Speichertechnik. VDI-Berichte. VDI Verlag GmbH, Düsseldorf (2011)

Mallery, R.M., Olomu, O.U., Uchanski, R.M., Militchin, V.A., Hullar, T.E.: Human discrimination of rotational velocities. Exp. Brain Res. **204**(1), 11–20 (2010). https://doi.org/10.1007/s00221-010-2288-1

Meiry, J.L.: The Vestibular System and Human Dynamic Space Orientation (Doctoral Thesis). Massachusetts Institute of Technology (1965)

Müller, T., Hajek, H., Radić-Weißenfeld, L., Bengler, K.: Can You Feel The Difference? The Just Noticeable Difference of Longitudinal Acceleration September 2013. Proceedings of the Human Factors and Ergonomics Society Annual Meeting., S. 1219–1223 (2013)

Müller, T., Hajek, H., Frank, T., Bengler, K.: Das menschliche Auflösungsvermögen von Dynamikveränderungen im Fahrzeug Conference on Future Automotive Technology, Garching b. München, 17–18. März 2014 (2014)

Rockwell, T., Snider, J.: An Investigation of Variability in Driving Performance on the Highway: Final Report. Ohio State University, Columbus, OH (1965)

Rommerskirchen, C., Helmbrecht, M., Bengler, K.: The impact of an anticipatory eco-driver assistant system in different complex driving situations on the driver behavior. Intell. Trans. Sys. Mag. IEEE. **6**(2), S. 45–S. 56 (2014). https://doi.org/10.1109/MITS.2014.2307078

Stuke, P., Bengler, K.: New Approach on Vehicle Interior Cooling to Increase Cmfort and Reduce Energy Consumption CoFAT 2014, München. (2014)

Vogelpohl, T.; Vollrath, M.; Kühn, M., Hummel, T.; Tina Gehlert, T.: Übergabe von hochautomatisiertem Fahren zu manueller Steuerung. Unfallforschung der Versicherer GDV 1 (2016)

Weiser, M.: The Computer for the 21st Century. Scientific American Special Issue on Communications, Computers, and Networks, September, 1991 (1991)

Wenzelis, A. M.: Optimiertes Wankverhalten durch aktive Fahrwerkssysteme und empfindungsoptimierte Objektivierung. Dissertation, Technische Universität München (2018)

Zhang, H.: Human Thermal Sensation and Comfort in Transient and Non-Uniform Thermal Environments. PhD thesis, University of California, Berkley (2003)

Supplementary Information

Index – 683

H. Bubb et al. (eds.), *Automotive Ergonomics*, https://doi.org/10.1007/978-3-658-33941-8

Index

A

Ability 54, 66, 69, 75, 80, 97, 102, 113, 120, 146, 147, 172, 184, 220, 395, 410, 454, 461, 471, 495, 508, 520, 522, 542, 560, 563, 596, 604, 631, 668
- cognitive 17, 101, 118, 213, 225, 227, 250, 281, 309, 325, 356, 400, 409, 413, 454, 494, 580, 581, 608, 631
- motor 3, 9–11, 17, 18, 37, 40, 66, 68, 94, 113, 114, 116, 120, 121, 162, 163, 172, 185–187, 196, 203, 211, 262, 295, 302, 316, 328, 331, 387, 416, 417, 428, 452, 475–478, 480, 481, 494, 503, 506, 524, 525, 537, 563, 582, 583, 616

ABS system 4, 8, 520, 523, 525, 526, 547, 551, 560, 567
Absolute threshold 69, 70
Absorption and reflection coefficients 470
ACC Stop & Go 542
Acceleration
- control 44, 46
- restrained 507
- signal 489, 589
- worth 355, 375, 404, 413

Accelerator 8, 9, 20, 26, 114, 221, 284, 290, 291, 301, 326, 328, 329, 350, 364, 367, 375, 377–379, 387, 389, 433, 434, 480, 520, 525, 526, 536, 537, 542, 548, 554, 563–565, 618, 676
- active 3–6, 8, 16, 21, 33, 54, 58, 73, 112, 115, 120, 137, 163, 170, 172, 205, 206, 273, 299, 308, 328, 330, 334, 403, 479, 488, 491, 515, 522, 532, 536, 537, 539, 540, 548, 554, 555, 558, 559, 563–565, 567, 568, 615, 633
- electronic 8, 42, 120, 229, 279, 328, 330, 348, 425, 523, 536, 559

Acceptance limit
- empirical 150
- physical 149

Access to 2nd and 3rd row of seats 392, 393, 395
Accessibility
- border 53–54
- surface 246

Acclimatisation 495, 496, 503, 590
Accommodation 79, 80, 345, 389, 402, 411, 419, 421, 584, 585, 587
Actin 171
Action sequences 114, 120
Active noise control (ANC) 479
Active steering 9
Active steering intervention 537
Actuator 34, 35, 263, 266, 311, 511
Adaptability of the sense of smell 513, 514
Adaptation 17, 26, 35, 51, 58, 59, 77, 81, 84, 86, 95, 96, 138, 179, 245, 251, 353, 361, 366, 370, 372, 374, 387, 389, 390, 412, 413, 453, 471, 472, 491, 505, 533, 579
Adaptation mechanism 72
Adaptive cruise control (ACC) 5, 10, 152, 530
Adaptive lighting systems 528
Additional tasks 128–130, 133
Adenosine diphosphate (ADP) 171
Adenosine triphosphate (ATP) 171
Adequate stimulus 69
Adjustment range 358, 361, 362, 364–369, 379, 381, 382, 396, 400
Advanced Driver Assistance Systems in Europe (ADASE) 522
Advanced lane departure warning system (ALDW) 537
Advancing 268, 536
Aesthetic impression of 581
Affordance 271, 295
AFS headlights 528
Age 83, 115, 120, 136, 173–176, 178, 179, 182–184, 191, 199, 203, 205–211, 213, 345, 402, 410, 411, 419, 432, 453–459, 495, 513, 655
- pyramid 206, 453

Agent for tank 310
Age-specific change
- hearing 471, 478, 481, 483, 498, 511
- memory performance 454
- visual performance 81

Aggressiveness 494
Agonist 121, 172
Air
- conditioning
  - fully automatic 511
  - manual 511
  - semi-automatic 511
- damp 14
- sound 481
- temperature 99, 101, 494, 495, 497, 499–504, 506, 507, 509, 510

Airbag 3, 54, 381
α/2 algorithm 529
Alpha error 664, 665
α innervation 68, 121, 122
α motoneurons 120–122
Alpine 523
Alternative hypothesis 664
- unspecific 664

Ambient lighting 6, 473–475
Amplitude characteristic 39
Analogue scales 645
Analysis of variance 669
Anatomy 162, 163, 176, 249
Angle detectors 72, 103, 111
Animation 241
Anonymity of the survey 646
Antagonist 121, 172
Antagonistic principle 121
Antagonists 172, 175, 206
Anthropometric 16
- layout 17
- workplace design 16

Anthropometry 176, 179, 197, 235, 308, 358, 364, 371, 427, 443, 448, 449, 453, 456, 461, 513, 611, 614, 615
Anticipation agent 232

Anticipatory control 47
Anti-lock braking system 523
AnyBody 200, 204, 205, 239, 242, 250, 456
Application class 350, 355, 357, 361, 382, 395, 400, 406, 411, 417
Appraisal
– of the entry hose 438
– the fitness for use 427
– geometric boundary conditions 427, 438, 448
Approach zone 313, 540
Archways 92, 93
Area of interest (AOI) 124, 139
Arithmetic mean 178, 501, 636
Armrest 186, 190, 195, 347, 384–386, 389, 392, 456, 462, 483, 502, 507, 509, 578, 581
Armrest centre console 618
Arrow axis (sagittal axis) 162
Articulation index (AI) 482
Aspect sciences 15
Assignments
– content 272
– dynamic 96
– design 260
– monitive 34, 522, 553
Assistance activation unit 539
Assistance system 599
– informative 543
– simultaneous 551–553, 557, 565
– warning 19
Assistant mode
– autonomous driving (AUTON) 567
– fully automatic driving (VA) 567
– manual driving (MF) 567
– semi-automated driving (TA) 567
– supported driving (UF) 567
Assumption 55, 57, 204, 205, 241, 379, 431, 482, 527, 548, 557, 560, 598, 663, 664
Astigmatism 79, 81
Atlas vertebra 164, 165, 167
Atrophied back muscles 370
Atrophy 175
Attempt
– on public roads 309
– vehicle 3
Attention assist 557
Attenuation 90
Attitude
– body 349, 350, 354, 356–361, 363–365, 368–370, 375, 381, 382, 384, 385, 389, 390, 392, 394, 396, 397, 406, 412, 418, 420, 422, 427, 428, 431, 432, 434, 440, 445
AttraktDiff2 146
Auditory area 90
Augmented reality in the real vehicle 599
Automated evaluation 623, 627
Automatic ignition timing adjustment 520
Automatic speed control (cruise control) 524
Automatic toggle 292
Automation 266, 534, 548, 549, 554–556, 567, 568, 677
Automotive Seat and Package Evaluation and Comparision Tool (ASPECT program)
Autonomous braking system (AEBS) 536
Autonomy 520
Auxiliary heating 507
Availability 54, 176, 181, 184, 564, 569
Available space 187, 678
Averting 124, 129, 142, 295, 347, 401
Avoidance actions 56–58
Avoidance maneuver assistant 542
Awareness 266
Axons 69, 101, 121

B

Back contour 370, 374, 406
Backrests
– setting 449
– surface 370, 441, 617
Bad-weather light 529
Bang-Bang model 225, 226
Barley corn 192
BAS plus 535
Basal ganglia 68, 447
Baseline ride 629
Basic cockpit 273
Basilar membrane 88, 89, 477
Beginner's risk 115
Behaviour patterns 180, 432, 591, 641
Belt run 382, 395
Bicycle model 229
Binocular vision 396
Biomechanical human models 488, 489
Bird-view display 29, 306
Blaupunkt EVA (electronic traffic controller for drivers) 524
Blind spot 10, 80, 407, 408, 540
Blind spot effect 280
Body
– builder 246, 359, 366
– cancer 245
– construction 138, 234
– dimensions 176, 178, 179, 187, 197, 444, 495
– height (stature) 176, 180, 185, 187, 245
– movement 289, 595, 616
– point 361, 362, 423
– scanner 176, 177, 187, 191, 244, 614
Body mass index (BMI) 194, 444
Body-grid 617
Bodystyles 345
BOEMAN 238
Bonnet 363, 565
Boolean algebra 53, 55
Border area 143, 314
Box plot 656
Brainstorming in a team 641
Brake assist (BA) 526, 535, 567
Brake booster 7, 387, 389
Brake-by-wire 330
Brake lining 367
Brake pedal 8, 9, 26, 155, 186, 261, 284, 289, 291, 300, 311, 326, 328, 329, 333, 375, 377, 379, 525, 526, 557, 563, 564, 597, 629, 676
Brake squealing 282

Brakes 7, 8, 19, 66, 142, 524, 526, 530, 533, 536, 566
Branching point 33
Breathing air of the passengers 509
Buckling procedure 382
Bunion point, ball of foot (BOF) 350
Business trip 258, 259
Butterfly door 430
Buttock knee depth 188

C

Cabriolet 345
CAD technology 577
CALSPAN-3D-CVS (Crash-Victim-Simulator) 239
Capturing contact forces 617
Car radio 13, 543
Car-sharing 259
Car-to-car (C2C) 547
Car-to-car communication (Car2Car) 58–59
Car-to-infrastructure (C2I) 548
Cascade prediction model 355–357
CASIMIR 249, 488
Cataract 453
Categorisation of driver assistance systems for the primary driving task
– impulse sound 483
– noise 483
– sound 483
CATIA automotive-extensions vehicle architecture (CAVA) 357, 383
– safety 357
– vision 357
– wiper 357
Cause-effect-principle 34
Cause-effect relationships 659
CAVE 579, 584
Cell
– complex 72
– hypercomplex 72, 73, 88, 103, 111, 112
Center-point-feeling, situational 319
Central processing 69
Centralis 79, 102, 111
Centre console 185, 186, 195, 296, 334, 345, 347, 354, 392, 418, 436, 568, 618, 659
Centrifugal force 45, 422
Centrum collum diaphysis angle (CCD) 184
Cerebellum 68, 121, 447
Cervical vertebrae 167
Characteristics 14, 16, 17, 29, 32–35, 41, 46, 54, 65, 69, 71, 75, 77, 79, 81, 86, 89, 95, 100, 101, 103, 111, 121, 123, 124, 164, 224, 303, 354, 359, 366, 461, 462, 464, 465, 477–479, 490, 522, 526, 534, 538, 539, 543, 557, 563, 564, 568, 609, 655, 660, 662
Charging process 2, 676
Charging station 303
Chassis 6
Chauffeur vehicle 354, 424
Checklist, TRL 610
Children 150, 177, 245, 345, 395, 425, 428, 453, 460, 461
– seat 195
Chunk 115, 136, 229
Ciliary muscle 79
City light 528
City pilot 524
City safety system 536
Classes 195
Classification
– directionally compatible 332
– functional 332
Clients
– acceptance of ACC systems 534
– complaint 482
– feedback 71
Climate
– clothing 506
– mannequin 449
– wind tunnel 512
Climate dummy MARCO 512
Climb through 437
Clothing unit (clo) 495
Clutch pedal 20, 186, 290, 377, 379, 389, 391
CO_2-reduction 2
Coccygeal vertebrae 167
Cockpit design 17, 59
Coding
– analog 68
– digital 68, 69
– spatial 282
Cognitive architecture 230
Cold receptors 97, 98, 494
Colour 80, 85, 86, 102, 111, 180, 231, 272, 425, 426, 462, 503, 507, 548, 565
– contrast 2, 17, 37, 48, 58, 66, 76, 80–82, 84, 86, 95, 101, 110, 111, 123, 125, 144, 170, 184, 206, 248, 262, 281, 316, 333, 347, 392, 412, 434, 453, 473, 481, 482, 493, 500, 520, 528, 532, 535, 543, 567, 578, 610, 611, 621, 639
– giving 400, 410
COMAND-system 13
Comfort
– diffuser 505
– driving safety 169
– impression, more audible 478
Comfort equation according to Fanger 497, 498
Comfort pyramid 101, 145, 470, 513
Communication with the rear traffic 540, 542, 543
Compatibility
– external 267
– primary 267
– secondary 268
– spatial 267
Compensation display 264, 265
Compensation task 29, 44–46, 264, 265
Competence 139, 259, 329
Complete permutation 660, 662
Compromise
– of design alternatives 654
– of variants 389
Computer anthropometry 611

Computer-aided anthropological mathematical system for passenger simulation designation 244
Computional fluid dynamic (CDF) analysis 512
Concept development 512, 578
Condition 52, 53, 55, 260, 318, 377, 379, 410, 425, 432, 452, 475, 502, 506, 534, 535, 540, 555–557, 567, 569, 660, 661, 676
Confidence interval 654, 655, 658
Confirmation phase 578
Conflict 56, 58, 492, 493
Congestion assistant 542, 553
Connectivity 302, 676
Constancy performance 75, 77, 80, 97, 110, 288, 477, 488
Constant method 635
Constitutional biology 179
Construction 350, 369, 405, 429, 441, 446, 527, 536, 563, 565, 588
Construction of standardized questionnaires 641
Constructivity 77
Contact surface for door and tunnel cladding 424
Contact us
– surface 285, 311, 424, 488, 506
– zone 313, 508
Contact-analog head-up display 279
Contactless process 623
Context 2, 4, 5, 7, 16, 26, 34, 41, 42, 46, 55, 60, 65, 69, 73, 97, 105, 107, 116, 120, 125, 141, 143, 151, 154, 155, 190, 192, 196, 201, 347, 354, 356, 357, 359, 373, 395, 396, 419, 462, 463, 473–475, 480, 491, 493, 498–500, 502, 505, 506, 511, 520, 522, 530, 535, 547, 566, 568, 617, 668
Contraction type
– concentric 172
– eccentric 172
Control
– active 42, 46, 222, 288, 325, 332, 333, 335, 336, 488, 561
– analog 289
– hand-operated 292
– integrated 291
– rotatory 285
– translational 288
Control engineering 26, 28, 34, 42, 43, 49–53, 113, 220
Control loop 16, 26–29, 35, 36, 41, 42, 47, 68, 120–122, 223, 258, 432, 579
– muscle 120, 122
– picture 26
Controller hand 35
Convection 98, 494, 495, 497, 500
Cooperative vehicle guidance 677
Coordinate systems 348, 349, 638
Cornea reflex method 621
Cornering 6, 36, 97, 133, 323, 347, 392, 422, 423, 525
Cornering warning device 547
Corpus callosum 65
Cortex, more prefrontal 68
Corti organ 88
Coulomb friction 318
Cover story 672
CP-50 scale 370
Craftsmanship 461–465, 514
Crash behaviour 381
Critical driving 525, 531, 558
Crosswind 557
Crosswind influence 26, 557
Cruise control system (cruise control) 7, 10, 333, 520, 524, 525, 530, 532, 535, 542, 564
Crumple zone 3, 54
Current speed 133, 304, 525, 527
Curve [dB(A)] 90
Cutting frequency 221, 222
Cyberspace 580

D

Daimler-Scholly process 248, 403, 404
d'Alembert principle 248, 447
Damper characteristic curve 490, 491
Dark adaptation 83, 85, 86, 471
Dark adjustment 82
Data
– descriptive 656
– objective 136, 603
– subjective 136
Data quality 666
Database 243
– dynamic 423, 441
– modular static 596
– static 596
Daylight reflection 412, 413
D-behavior 37, 96
Dead time 222
Decibel 39, 635
Decision mechanism 101, 114–120
Decision rule 664
Defrost button 511
Degree of utilization (force) 448
Degrees of freedom (DOF) 164–166, 173, 198, 206, 290, 666, 667
De-icing and dehumidification 509, 510
Dependability 54
Depth perception 86, 97, 102, 584, 585
Derivatives 191–194
Design for all 211, 459, 460
Detectors for specific lengths 72, 103
Detents 285
Development 579
Diaphysis 163
Diarthroses 168
Dichotomous answer format 643
Difference in proportions 181
Different priority 418
Differential equations 33, 36, 40, 41, 43
Differential threshold 322, 636
Difficulty of assessment 136
Diffraction (flexion) 510
Digits 276
Dilemma 2, 4, 142, 521, 564
Dimension concept 348, 355–357

Dimension conception 348, 350, 353, 357
Dimensionality 262, 263, 290
Dimensions 178
Dimensions of the Hand 190
Dimmability 472
DIN 348, 396, 438, 439
Direct glare 412, 414
Direct irradiation of the body parts of the passengers 504
Direct view 396, 398, 400, 402–404, 406
Direction 91
Direction of travel 109, 149, 312, 315, 345, 348, 354, 392, 395, 406, 426, 428, 437, 504, 544
Direction orientation 43
Directional constancy 110
Directional stability 323
Disability glare 413
Discomfort, more dynamic 441
Discomfort values, verbally defined 486
Disembark 431
Disordinal interaction 670
Display
– accuracy 273
– acoustic 273, 282, 530, 537, 542, 545, 557, 564
– analog 49, 274–276, 300
– contact-analogue 273
– mode 264
– photorealistic 273
– pictorial situation analogues 273
– separate 264
– visual 273, 305, 306
Distance between pedals 375–377
Distance perception 584
Distance to line crossing 537
Distractions 21, 47, 53, 58, 59, 124, 131, 132, 154, 260, 522, 534, 629
Distronic 5, 10, 530
Disturbances 26, 27, 29, 34, 35, 42, 461, 482, 485
DLC system 537
Door
– fairing 347
– front hinged 430
– rear hinged 428, 429
– sill 354, 456, 457
Double negations 646
Drag torque, combustio 329
Drainage channels 313
Drawing templates 235–237, 582
Drive
– hose 138, 304, 305, 438, 443, 543, 544
– mission 259
– mistake 55, 56, 155, 521
– noise 34, 72, 78, 90, 91, 145, 149, 231, 266, 298, 461, 475–483, 488, 489, 538, 560
– simulator 110, 132, 133, 135, 136, 138–141, 150, 285, 287–289, 304, 306, 332, 335, 375, 403, 478, 526, 532, 547, 590, 598–600, 603, 660, 672
– style 152, 301, 303, 328, 404, 533, 676
Drive slip control (ASR, ASC) 8
Driver 230, 231
– alert control 557
– assistance system 136, 139, 148, 152, 186, 305, 335, 520, 521, 532, 534, 535, 539, 546, 547, 549, 551–554, 556, 559–561, 563–566, 569, 579, 599
– information system (FIS) 13
– monitoring 536, 540, 554, 556, 557
– observation 10, 75, 86, 123, 129, 140, 150, 173, 264, 276, 299, 301, 360, 365, 415, 441, 475, 476, 488, 528, 557, 604, 622, 626
– workplace 347, 349, 356, 359, 424
Driver attentional load index (DALI) 647
Driver in the Loop (DIL) 579
Driving for fun 259
Driving operation 380, 502
Driving safety 558–561, 563
Driving style 152, 301, 303, 328, 404, 533, 676
Driving task 6, 18–22, 26, 29, 30, 34, 43, 49, 51, 77, 114, 120, 124, 128–130, 134, 149, 186, 229, 246, 258, 263, 280, 294, 298, 299, 310, 322, 324, 393, 424, 494, 521, 522, 528, 534, 539, 542, 549, 554–557, 561, 569, 570, 579, 596, 609, 626, 630, 631, 634
– primary 2, 3, 16–20, 49, 88, 112, 113, 120, 122, 124, 141, 163, 181, 211, 267, 273, 284, 292, 308, 311, 354, 470, 522, 534, 542, 549, 557, 594, 598, 599, 609, 627, 630, 668
– secondary 3, 18, 19, 22, 113, 124, 134, 142, 186, 267, 269, 273, 276, 293, 390, 481, 522, 534, 549, 609
– tertiary 21, 26, 113, 124, 129, 131, 132, 294, 310, 522, 549, 554, 570, 591, 609, 629, 631
Dual-view displays 412
Dummy 244, 512, 601
Duration of task processing (Total Task Time TTT) 630
DYNAMICUS 241

E

Eardrum 87, 98–100
Ease of use 145, 354
Easy-entry 449
Edge blending 591, 592
Edge detectors 103
Effective accelerations 486
Effective temperature 496, 497, 503
Effectiveness 58, 145, 515, 541, 558–563, 608
Efficiency 2, 15, 98, 483, 495, 522, 528, 677
Elders
– persons 453
– vehicle user 432
Electric
– drive 300, 328, 330, 480
– heating 510
– mobility 676
– vehicle 2, 3, 302, 303, 306, 328, 354, 479, 480, 506, 676
Electro Gyrocator 523
Electroencephalography (EEG) 633
Electromyogram (EMG) 611, 619
Electronic 187
Electronic stability control (ESC) 525
Electronic stability program (ESP) 8, 523, 525

Electrooculogram (EOG) 611, 619
Electrostatic charge 502
Element
– with memory 33
– without memory 33
Emergency seat 395
Emergency brake assistant 535, 536, 552, 559, 560
Eye movements and movement of attention (EMMA) 229
Emotional perception 68
End stop 287, 289
Endolymphs 87, 92
Endomorph 179
Energy
– balance 496
– recovery 326, 329
– turnover 495, 632
Engine
– control 8, 228, 229, 301
– order 478
– power 9, 291, 525
Engineering 15, 224, 345, 396, 499
Engineering psychology 15
Entry 37, 187, 196, 213, 347, 354, 375, 427, 428, 430–432, 434, 436–441, 443, 444, 446–449, 456, 457, 459, 460, 512, 610, 616
Entry and exit procedures 347, 354
Entry relevant main dimensions 439
Entry strategy
– hurdler (hurdle jumping) 434–436
– plumps (Sagging) 434, 435
– slip (Slipping) 432–434
Entry-(Dis)-comfort 506
Environmental influences 16, 27, 32, 34, 414, 495
Epiphyses 163
Equilibrium 92, 198, 442
ERGOMAN 239
Ergonomics
– anthropometric 17, 353, 354, 356–358
– cognitive 17
– communication 17
– functional 17
– Information 17, 265
– macro 14–15
– micro 14–15
– software 17
– system 520
Error rate 55
Estimated time 48
Etak 524
Evaluation of acceleration 537, 540
Event related potentials (EKP) 633
Evolutionary development 632
Exhaustion 148, 149
Exit 20, 187, 196, 213, 347, 354, 427, 428, 430–432, 434, 436–441, 443, 444, 446–449, 456, 457, 460, 511
– process 616
– strategy
  – climbing (climbing) 437
  – lifting (lifting) 437–438
  – spreading (bridging) 437
Expectation conformity 268
Experimental design
– more dependent 46
– more independent 610
– more two-factor 659
Experts valuation 641, 642
Exponential distribution 55
Exterior lighting 13, 540, 541
Eye-based metrics 622
Eye contact to the surroundings 461
Eye movement 16, 80, 110, 111, 124, 135, 136, 211, 213, 234, 289, 402, 619, 621, 630
– deliberate 565
Eye movements and movement of attention (EMMA) 229
Eyes
– dominance 186
– ellipse 235, 350, 396, 398, 421
– height 177, 188, 654
– muscle 211
– point location 470
– sitting height 187

F

Factors
– levels of 46, 124, 196, 522, 542, 549, 567, 659, 667
– number of 626, 645, 660
Failure 53–55, 58–60, 567
Failure rate 55
Fanning effects 514
Fashion and zeitgeist 144
Fatigue
– physical 627
– psychic 77
Fatigue warners 522
Favours 104, 117, 152, 180, 332, 348
Feedback 16, 26, 27, 29, 41–45, 47, 70–72, 94, 95, 107, 114, 116, 122, 123, 221, 261, 266, 283, 289, 291, 295, 299, 308, 311, 319, 332, 336, 353, 373, 462, 470, 475, 478, 480–482, 493, 524–526, 530, 539, 551, 561, 563, 564, 566, 567, 599, 641, 676
– acoustic 462, 482
– variable haptic 296
Feel comfortable 496, 498
Fiber optic elements 614, 615
Fibrous ring (anulus fibrosus) 168, 169
Field of application 248, 486
Field of vision 11, 13, 102, 109–111, 124, 142, 211, 226, 304, 357, 397, 404, 407, 453, 474, 509, 529, 547, 591
– aside 74
– forward 396
50th percentile man 94, 178, 181, 182, 359, 371
Filler neck
– for windscreen washer water 354
Finite element methods (FEM) 241
Fixation 81, 111, 125, 127–129, 133, 135, 211, 623
Fixation field 211

Floor 345, 350, 354, 356, 357, 367, 395, 397, 398, 428, 439, 440, 449–451
Focus of expansion (FoE) 107, 108, 110
Fog orientation 43, 223
Fogging
– inboard 509
– on the outside 508
Foot
– attitude 22, 121, 149, 199, 360, 368, 369, 379, 640
– and shoe 191
Footprint 676
Force
– atlases 201
– measuring machine 196–198, 200, 201, 204, 349, 350, 364, 366
Force servo mechanism 122
Foreign radiator 11
Foresight 43, 47, 48
Fovea 79, 80, 102, 111, 124
Fragrances 101, 515
Frame position 365, 366
Frankfurt horizontal 176, 187
Frequency areas 249
Frequency distribution 84, 393, 656
Fresnel lens 592
Friction hysteresis 285
Front view 451, 612
Frontal plane 162
Fuel cell technology 3
Functional visual field (FVF) 230
Funtion monitive 522

G

Gain reserve 43
γ innervation 68, 121, 634
γ motoneurons 120
Garage parkers 545
Garments 194
Gaze aversion 143, 309
Gaze averting time 130
Gaze behavior type 125, 126, 128, 138, 150
Gaze tracking system 134
Generalizations 646
Generic parametric design (GENPAD) 357
German Road Safety Council (DVR) 522
Gesture
– arbitrary 94, 95, 110, 113, 121, 236, 445
– involuntary 231, 299
Gesture control 299, 452
G-factor 350, 416
Glare
– disability glare 413
– discomfort glare 413
– physiological 81, 83, 413
– psychological 413, 471
Glare-free high beam 529, 530
Glass 79, 80, 397, 403, 404, 412, 415, 426, 427, 451, 494, 504, 510, 514
– dome car 402, 403
– house 353, 461, 514, 524
– roof 101, 179, 183, 192, 347, 353, 354, 364–367, 415, 422, 426–428, 430, 431, 434, 439, 440, 449, 450, 456, 481, 507
Glaucoma 453
Global Car Manufacturers Information Exchange Group (GCIE) exchange lists 348, 349, 438, 439
Global customers 345
Global positioning system (GPS) 19, 523, 524
Globe temperature 497
Glove 194, 354, 418, 588
Gong 538
Goniometer 379, 614
G*Power 668
Grasp 97, 102, 124, 295, 382
Grasping 347, 432
Greenhouse 353, 354, 403, 412, 500, 501, 504, 506, 507
Grip 380, 417, 418
– enclosing 426, 436
Gripping spaces 239, 273, 350, 357, 416–419, 421, 426, 428
Guided interview 310
Gyrus 66

H

H30 dimension 350, 351, 384, 392
Handedness 185, 392
Hands on the steering wheel 360, 380, 419, 436, 557, 633
Haptics 283, 425
Hardware in the Loop (HIL) 579, 595
Harmony 76, 462
Hazard warning device 547
Head
– position 92, 280, 350, 422, 557, 565, 600, 623
– support 59
– and torso movements 406
Head-based methods 621, 622
Head mounted display (HMD) 587
Heading control (HC) 10
Headlamp orientation 43, 46, 222
Headlights 11, 103, 308, 528, 529
Head-mounted display 591
Head-up display (HUD) 13, 49, 59, 77, 79, 143, 281, 356, 410–412, 465, 527, 532, 539, 554, 564–567
– contact analogue 49, 304, 306, 566
Hearing 88–91, 454, 477–479, 481
Heart rate 68, 632
Hedonistic aspect 325
Height (stature) 178
Heiner, B. 205, 239, 465, 515, 570, 611
Heterogeneous 655
Heuristic evaluation 610
Highway light 529
Hinge 164, 165, 428
– capsule 164
– cartilage 163, 164, 169, 184
– conical 403, 404

– flat 91, 163, 165, 168, 187, 189, 246, 280, 369, 374, 377, 400, 413, 416, 425, 486, 488, 504
– hinged 390, 392, 428–430
– receptor 69, 71, 88, 98, 111
– saddle 165, 166
– spherical 79
Hinged door 429
Hip point (H point) 349, 369
Hippocampus 68, 113
Histograms 637, 656
Historical overview 7–14, 523, 524
HMD technology 588
Hohenstein Institutes 244
Homeostasis 119
Honda 46, 523, 557
Hood release lever 391, 392
H-point adjustment field 350
H-point measuring machine 349, 350, 364
Human builder 241, 242, 357
Human engineering 14
Human error probability (HEP) 56
Human Factors & Ergonomics Society (HFES) 15
Human Factors Engineering Committee 348, 349
Human failure 150
Human solutions GmbH 177
Human templates 179
Human-machine interaction (MMI) 15, 16, 27, 42, 296
Human-machine-systems (MMS) 16, 33
Hybrid concepts 3
Hyperflexion 169, 369
Hypothalamus 101, 494
Hypothesis 138, 610
Hysteresis 223, 314, 319, 330
– restrained 507

I

Ia fibres 121
Ib fibre 122
I-behavior 37
I-board 14
Ice formation 508, 510
Idle gas 524
Idle stroke 286
I-Drive 11–13, 131
Ignition timing adjustment 7
Illuminance 470–473, 528
Illumination 80, 472, 528, 529, 597
Image build-up rate 586
Image processing technology 543, 557, 611
Immersion factors 580, 589
Improvement 57
Improvement measures 57–60
Impulse 65, 69–72, 98, 103, 121, 124, 483, 489, 490, 563
Indirect view 353, 407, 409
Industrial design 146, 463
Inertia 8, 37, 40, 42, 68, 74, 93, 94, 121, 286, 448
Inertia of the hand-arm system 220
Inertial 94, 557
Information 14, 95, 353, 354, 369, 373, 377, 392, 395, 404, 407, 409, 411–413, 456, 470, 475, 492, 493, 495, 500, 503, 520–522, 524, 527–529, 535, 537, 539, 540, 543, 545, 547–549, 551, 552, 554, 555, 560, 561, 563, 564, 567
– agents 310
– processing agent 233
– recording 73, 103, 119, 220
Inner models 113–117, 122–124, 132, 137, 141–144, 332, 333
Inner skeleton model 241
Innervation rate 634
Innovations 7, 577, 608
Input 32, 34, 36–41, 57, 64, 68, 70, 71, 234, 298, 347, 448, 500, 506, 524, 527, 554, 561, 579
Instructions 5, 306, 482, 524, 541, 544, 546, 672, 673
Instruments
– lighting 472–473
Integration 16, 213, 356, 542, 553, 567, 677
Interaction, multimodal 289, 300
Interface 11, 265, 367, 387, 409, 545, 567, 610
Interior
– bright 426
– dark 425, 426
Internal compatibility 268
International Ergonomic Association (IEA) 14, 15
Interpretation of gaze data 626
Interquartile distance 657
Interrelations ordinal 670
Interval scale 496, 657
Intervertebral discs 167–170, 184, 369, 371
Inverse kinematics 203, 242, 355
Investigation
– rule- and knowledge-based navigation 599
– secondary driving tasks 21
Involvement 52, 173, 581
Ischial tuberosities 370
Isochrones 213, 401

J

JACK 241, 243
Jill 243
Joke 554, 564
Joystick 143, 309, 331, 335, 554, 564, 568

K

Kalman filter 224
Kamms' circle 314
Kano method 464
Keyless entry system 472
Kick-down function 525
Kiel doll 236
Kinetosis 74, 395, 492, 493
Knee room 354, 362, 379, 381, 392, 395, 424, 432–434
Kneecap 163, 166, 188
Kolibri 548
Kyphosis 167

L

Lagrange function 248
λ/4 effect 415, 416
Landoltring 410
Lane change
– assistant 10, 325, 540, 541
– decision aid systems 540
– support 10
Lane departure prevention (LDP) 538, 548, 551
Lane departure warning (LDW) 10, 304, 308, 325, 523, 536–542, 555, 567
Lane guidance assistant 5, 6, 19, 59, 565, 567
Lane-change-test 597, 599, 608
Lateral forces 311, 315
Lateral support 374
Latin square 660, 661
Layout
– humane 17, 149, 150, 349
– of the questionnaire 136, 646
Leading body part 445
Learning 152, 432, 660
Learning cards 546
Learning process 113, 114, 232, 283, 291, 676
LED lighting systems 528
Left hand drive (LHD) 185, 186, 375, 384, 385, 387, 388, 390–392, 432, 472, 473, 578
Leg and foot space 436
Length of extremities 190
Length servo mechanism 120–122
Levels 662
Lidar 10
Lidar sensor 629
Lifting capacity 451
Lifting plate 428, 429
Light
– adaptation 84, 86, 471, 472, 475
– current 84
– distribution 528, 529
– fluctuation 622
– motion-guided 474
Lightweight construction, seat 2
Likert scale 610
Limbic system 68, 101, 113
Line graphics 672
Liquid crystal displays (LCDs) 11, 281, 412
Load 7, 16, 121, 136, 147, 168–170, 172, 196, 345, 348, 371, 372, 450, 451, 475, 478, 479, 488, 491, 501, 526, 619
– condition 476
Loading 2, 169, 187, 354, 449–453, 457, 459, 460, 491
– condition 259
– floor 450, 451
Load-strain concept 16
Local mean vote 498
Local speed limits 527
Lockability 290
Locomotion 17, 105–107, 163
Longitudinal axis 44, 162, 165, 172, 323, 387
Longitudinal dynamics 19, 26, 30–31, 34, 43, 47, 48, 221, 222, 226, 261, 325, 332, 333, 526, 528, 530–536, 544–545, 551, 564, 676
Lordosis 167
Lordosis support 167
Loudness 478, 481, 482
Lumbar vertebrae 167, 189
Luminance 81–84, 412, 413, 470–473, 586

M

Macular degeneration 453
Macular organs (statolithic organs) 92, 97
MADYMO 248, 578
Magic body control 491
Magnetic resonance imaging 66
Main effects 668–670
Manageability 145, 381
Maneuver planning 19
Maneuverability
– active 600
– age-dependent 454
– gender-dependent 206
Maneuver planning 19
Manual brake intervention 9
Marker-based methods 614
Marker light 529
Markerless measurements 615
Markov analysis 132
Masking effect 90, 91, 145, 286, 403, 481, 489
Maximum control speed 530
Maximum forces 172, 173, 175, 196–198, 201–203, 448
Maximum pressure 370, 372, 617
Mean 4, 29, 31, 48, 56, 57, 94, 127, 143, 149, 178, 179, 190, 203, 316, 374, 395, 440, 498, 501, 506, 531, 532, 535, 655, 663, 670, 671, 677
Mean saccade 135
Mean time to an failure (MTTF) 55
Measure 2–6, 8, 10, 29, 31, 42, 46, 47, 50, 53, 57–60, 90
– inner 179
Measurement 98, 203
– dynamic 198
– static isometric 198
– unacceptable 627
Measuring instruments 176, 177, 611, 619
Mechanoelectric method 614
Mechanoreceptors 64, 68, 94, 97
Mechatronics 481
Median 31, 94, 178, 559, 656
Median plane (sagittal plane) 162, 167, 204, 394
Medulla oblongata 68
Meissen corpuscle 96
Memory 112, 113, 117, 141
Memosik 488
Mental stress 134, 522, 634
Menu
– structure 13, 578
– tree 309, 310
Merkel cells 94
Mesomorph 179
Metabolism 98, 164, 495
Microsleep 47, 538, 557
Microvans 345
Middle ear cavity 87

Migration 143
Milestones 578
Minimum
– opening width 431
– size 81
Mirror
– aspherically curved 408
– interior rear 592
– outdoor 507
Mistakes 152, 432
MMI system 13
Mobile entertainment devices 461
Mock-up variable 583
Modal value 31, 127–129
Mode
– automatic 583
– more interactive 11
Model
– adaptive 223
– anthropometric 245
– biomechanical 238, 248–250
– cross-over 222, 224
– dual mode 223
– regulatory 222
Modelling
– the expenditure of time 446
– psychoacoustic 479
Moisture content 500, 508
Monitive 33
Monopoint system 192
Monotony 148, 149, 265, 307
Moose test 9
Motion
– apparatus 205
– coding 272, 293
– cueing (motion perception) 594
– detectors 72, 103, 111
– free space 205
– lifting 169
– measurement 93
– parallax 86, 105–107
– pitch 483, 493
– range more joint specific 205
– rolling 483
– room more task specific 198, 205
– rotational 93, 94
– simulation 447, 448
– strategy 427
– translational 93
– type (motion coding) 272, 292, 293
– wobbly 38
Motivation 19, 113, 117, 173, 642
Motor function
– coarse 121
– fine 121
Motor unit 121
Motorway light 529
Movable loading floor 450
Move-eye-and-head function 407
Multi-agent system 233
Multi purpose vehicle (MPV) 345
Multi-storey car parks 546
Muscle
– cells 170, 171
– cross section (PCSA) 172
– exertion 68, 79, 632, 633
– fiber 121, 171, 172
– pain 370, 384
– spindles 68, 94, 97, 121, 122
Musculature, striated 121
MUTE 677
Myofibrils 171
Myosin 171

N

NASA TLX 136, 492, 609, 640, 647
Natural frequency 461, 483, 491
– of the child's stomach 461
– of the system 490
Navigation 11, 13, 14, 21, 26, 28, 29, 49, 69, 154, 213, 230, 277, 281, 290, 294, 298, 306, 374, 392, 412, 520–522, 524, 537, 545–548, 561, 566, 599
– system 19, 21, 151, 154, 259, 265, 273, 283, 303, 305, 306, 374, 482, 520, 523, 524, 527, 529, 537, 545–548, 552, 561, 563, 564, 669
– task 18, 19, 22, 26, 49, 300, 305, 310, 542, 545–548
Navigator 524
Nervous system 67, 72, 162, 170, 171, 198, 445, 461
Neural network 205, 226, 487
Neuromuscular delay time 221
Neuron cell 65, 69, 70, 72, 102, 103, 112
Neutral body movement 441
Neutral driving behaviour 119
Night reflection 413
Night vision 529, 561
Night-vision device 11, 279, 304
Noise 6, 16
– annoying 7, 306, 413, 462, 475, 480, 481, 507, 510, 521, 537
– focal point 396
– rattling 10, 149, 462, 475, 480, 481
– scraping 480
– small 480–482
– spectrum 16, 474–480, 536, 540, 615, 629
Nominal scale 643, 656, 660
Nominal scale level 637
Non-contact methods 622
Normal-sighted 401
Normal weight 195
Normatively specified speed values 628
Notchback sedan 345, 450
Nozzles 504, 505, 510
Nucleus 168
Nucleus pulposus 167
Null hypothesis 664, 665
Number of nozzles 505
Number of possible responses 643
NX general packaging 357

Nystagmus 81

O

Obesity 180, 183, 194, 195, 359, 364, 427, 428, 432, 444
Objective measurements 512, 514, 611
Objectivity 608
Observation 41, 55
– camera 622
Occlusion method 630, 631
Odor, objectification 53, 614, 634
Oil rod, oil level 355
Olfactory bulb 101
Olfactory experts 514
Olfactory mucosa 101
One-pedal drive 328, 676
One-pedal interaction 328
Operability of the luggage compartment opening 452
Operating and display components 353, 354, 409–412, 416–420
Operating element, variable in shape 190
Operating noise of the drive 475
Operating safety 260
Operation
– finger 294
– foot 155
– one-handed 273
– sequential 262
– simultaneous 262, 293
– stereotypical 268
Optical flow 105–111, 123, 125, 323
Optical presentation 554, 584
Optimization method 356, 357
Order
– spatial 261, 262
– temporal 262
Ordinal scale 657, 660
Orientation 91
Original equipment 576
Original equipment manufacturer (OEM) 576
Ossa longa 163
Ossa plana 163
Ossicle chain 87
Outcome
– exact 26
– representative 119
Outlet 502, 505, 510, 511
Output 17, 32, 36–41, 48, 64, 68, 70, 71, 77, 98, 234, 356, 441, 494, 502, 511, 524, 537, 584
– interfaces 579
Outrigger 531, 534, 535
Overall vehicle architecture (OVA) 357
Overall-work-load-index (OWI) 647
Oversteer 9, 46
Overstrain due to understrain 147

P

Pacini corpuscles 94, 95
Package designer 356
Packaging 243, 345, 347–350, 353–358, 460
Pain receptors 98, 169
Pairwise comparisons 638, 667
Paper pilot 220
Parallax 584
Parametric model for automotive packaging and ergonomic design 357
Parapet height 365, 461
Parking 2, 7, 22, 46, 129, 134, 149, 186, 265, 268, 277, 280, 299, 305, 347, 354, 375, 390, 419, 425, 430, 431, 436, 506, 507, 521, 527, 542–546, 553, 557, 568
– longitudinal 544
– oblique 172
– perpendicular 93
Parking assistance
– automatic 520–522, 524, 525, 529, 530, 532, 533, 542, 544–547, 549, 553, 559, 568
– guided 543
– semi-automatic 544
Parking brake 186, 292, 390, 391, 418, 419
Parking search 546
Passenger 3, 5, 8, 45, 66, 124–141, 154, 187, 190, 192, 348, 354, 365, 381, 390, 392–395, 397, 412, 422, 428–430, 434, 436, 461, 473, 477, 486, 488, 492, 501–507, 509, 511, 515, 523, 534, 554
PATH DRIVer cognitive 232
P-behavior 96
PCMAN 246, 366, 394
PD controller 221
Pedal 7–9, 20
– reference point (PRP) 350
– use 629
PELOPS 232
Percentage load distribution 370
Percentage of environmental views 627
Percentile
– 50-percentile man 350
– simulation 588
Perceptibility threshold 489
Perception
– agent 233
– conflict theory 587
– deep 286
– haptic 97
– kinesthetic 43, 45, 97
– threshold 93, 95, 101, 322, 502, 676
Perilymph's 87, 94
Peripheral visual field (PVF) 230
Phase reserve 222
Phases
– angle 38
– edge 42
Philips CARIN 524
Photopic area 471
Physical measures 6
Physiological parameters 632
PID controller 37, 229
Pilot study 548, 658
Pioneers 8
Plateau method 198
Plot 199

Point cloud 442
Point of no return 563
Point of regard measurement 621
Polarizing sunglasses 414
Pollen filters 511
Popometer 97, 323
Portal doors 429, 430
Position
– hands on the steering wheel 380
– upright 162
Position coding 272, 292
Position controls 45, 385
Position of the joints 244, 613
Positioning point (PHPT) 367, 369
Post for safety 482
Postural comfort prognosis 246
Potential trigger 554, 556, 567
Power 2, 3, 7, 8, 17, 64, 74, 79, 80, 89, 103, 120, 174, 185, 197, 295, 470, 471, 483, 490–492, 502, 506–508, 668
Power spectral density (PSD) 633
Power steering 7, 9, 537
Powerwall 580
Praxeology 15
Precision model 221
Pre-crash 558
Prediction of discomfort in the entry process 446
Predictors 176
Pre-safe 536
Presence 33, 53, 56, 58, 127, 131, 136, 464, 596, 597
Present 3, 27, 65, 72, 84, 103, 106, 121, 123, 124, 127, 131, 132, 138, 142, 154, 203, 205, 224, 299, 419, 475, 482, 535, 583, 613, 639, 656, 661, 664, 665
– of the future 3, 124, 132, 271, 347, 403
– of the past 123, 124, 132, 389
Pressure
– button 202
– distribution diagram 7
– point 328, 462, 565
– switch 123, 288
Pre-test 646
Previewing chassis 491
Pro engineer 357
Probabilistics 34
Probability pots 246
Proceedings
– distribution-based 666
– joint-specific 203
– non-parametric 666
– para 666
Product
– ergonomics 17
– planning 577
– quality 461
Production ergonomics 17
Programming
– the route 305
– while driving 306
Prolapse 167, 169
Pronation 208, 211
Proportions 5, 129, 179, 181, 183, 184, 350, 358–360, 362, 365, 381, 412, 414, 419, 427, 496
Propulsion 523
Protection 53, 149, 177, 357, 608
Protective distances 432
Pseudo-three-dimensionality 278
Psychological units 115
Psychophysics 14, 145, 634, 635, 637
Pulposus 168
Pulsating pedal feeling 525
Pupil 79, 188, 471
Pupil diameter 622, 623
Push to Talk 298

Q

Qualitative survey methods 639
Quality
– acoustic 462
– optical 462
– sensation 638
– tactile 462
Quantitative 590
Quasi-linear human model 220
Question
– behavioural 119, 150, 152, 549
– closed 642
– deliberate 124, 151, 299
– factual 262
– knowledge 646
– open 642, 647
– valuation 456, 513

R

Radar 5, 10, 17, 523, 524, 530, 531, 535, 536, 540, 542
Radiation
– equilibrium 495
– exchange 506
Ramp method 198
Ranking 154
Rating 457, 458
Rating scale
– bipolar 643
– unpolar 644
Ratio scale 660
Reachability
– the pedals 345, 347, 353, 354, 358, 360, 362, 365, 374, 375, 377, 387, 389, 400
Reading glasses 410
Reafference principle 110, 111
Realistic anthropological mathematical system for interior-comfort simulation (RAMSIS)
– cognitively 410
– family 359, 369
– next generation 245
– right-hand-drive car
  – market 392
  – problem 384, 385, 387, 389, 390, 392

– vehicle 375
– sits, seat belt design 356
REALMAN 250
Rear and side windows 353
Rear wall door 354
Receptors 64, 69, 83, 91, 92, 97, 100–103, 114, 116, 169, 454, 471
Recognition
– gestural 271, 299
Recognition threshold 101
Recuperation 2, 3, 21, 326, 328, 676
Redundancy 56, 73, 75, 77, 198, 482, 561
Reference 19, 22
Reference variable 28, 35, 43, 48, 49, 120–122, 221
Reflection
– diffuse 412
– diffuse-directed 412
– directed 416
– image 414
Reflective glare (reflected glare) 412
Regression analytical methods 498
Regulations 22, 23, 34, 43, 51, 350, 353, 357, 381, 385, 388, 390, 395–400, 402, 404, 407, 408, 431, 502, 509, 511, 524, 526, 528
Reliability 16, 17, 53–60, 149, 150, 266, 524, 560, 608, 609, 645
Remaining range 259, 306
Remembrance
– long-term 23, 68, 112, 113, 119, 141, 149, 197, 370, 495, 532, 617
– secondary 3, 19, 20, 22, 113, 124, 134, 142, 186, 269, 273, 276, 293, 308, 390, 481, 521, 522, 534, 591, 609, 627, 647
– sensory 16, 42, 45, 64, 65, 67–69, 72–75, 77, 92, 93, 96, 97, 100, 101, 103, 104, 110, 112, 141, 143, 145, 162, 226, 271, 273, 285, 321, 335, 453, 454, 470, 483, 484, 492, 580
– short-term 23, 41, 68, 112, 115, 117–119, 124, 530, 617
– tertiary 21, 26, 113, 124, 129, 131, 132, 294, 308, 310, 521, 522, 554, 570, 609, 629, 631
– tertiary driving task 609
Representative sample 178, 654, 672
Research chair 370
Resonance 39, 41, 42, 477, 483, 485, 490, 492, 493
– biomechanical 485
Response behaviour 37
Restrictions 4, 14, 51, 132, 203, 284, 353, 359, 360, 362–367, 369, 375, 377, 379–382, 387, 389, 392, 412, 453, 459, 527, 533, 540, 542, 564, 586
Retina 74, 79–81, 83, 86, 102, 103, 105, 110, 111, 413, 471
Retinal flow 111
Reverse 365, 387, 406, 419, 464, 543, 545
Right-hand drive (RHD) 185, 186, 375, 384, 386, 387, 390, 392
Road contact 4, 6, 311, 313, 316
Roll acceleration 94
Rolling noises 475, 476, 479
Roof
– grips 428
– interior headlining 354
Rotary encoder 284
Rotary pusher 131, 545, 567–569
Rotary switch 284, 291
Rotational acceleration forces 93
Routine violations 151
R-point 350
Ruffini corpuscles 94, 96

S

Saccades 80, 81, 86, 134–136, 154
Saccules 92
Sacral vertebrae 167
Safari door 390
Safety and security
– active 16
– belt 353
– drive circuit (SIFA) 554
– passive 353, 357, 364, 395, 430, 461
– primary 54
– system 530, 566
Safety belts 430
Safework 241, 242
– Human Builder 242
Sailing 329, 676
SANTOS 319, 563, 564
Sarcomeres 171, 172
Satisfaction 14, 143, 146, 464, 513, 609, 610
Scales
– anchoring 645
– level 638, 656, 666
Scaling of muscle forces 205
Scatter 639, 655, 668
Scenario research 577
Scenarios 579
Schematic representation 70, 347, 356, 538, 662
Scheme, univariate 659
Scissor door 430
Seasickness 74, 395, 492
Seat
– adjustment 293, 353, 356, 358, 365, 367–369, 372, 373, 375, 381, 396, 400, 418, 425, 426, 441, 456, 457
  – automatic 372
– for children 461
– contour 183
– design 16
– dwarf 181–183
– heating 495, 502
– more ventilated 505
– place
  – more compromised 345
– and steering wheel 507
Seat pressure
– distribution 370–372, 443
– measuring mat 617
Seat tilt 372
– adjustable 372
Seat width (Seated seat width) 189

Seating giant 181, 183
Seating position 350, 353, 365, 366, 372, 400, 425, 426, 431, 444, 445, 459
– symmetric 375, 400
Seating positions 400
Seating reference point (SgRP) 349, 350, 353, 356, 358, 364, 366, 367, 384, 392, 393, 396, 398, 416, 440, 457
Secondary tasks 26, 77, 130
Secular acceleration 184
Seedometer 300
Self-explanatory 69, 291
Self-reflex 223
Self-reflex arc 68
Semi-disordinate
– interrelations 671
Senior vehicle 459
Sense organ system 483
Sense of smell 101, 514
Sense of space 426, 475, 583, 584
Sensitivity 80, 91, 96, 101, 111, 145, 221, 287, 453, 454, 461, 471, 474, 483, 485, 660
Sensor 4, 5, 8–11, 20, 371, 497, 514, 522–525, 527, 530, 531, 535, 536, 540–543, 547, 554, 557, 568, 618, 619, 632
– sensory conflict theory 492
Sensory maximum 483
Service 12, 22, 194, 354
Servo brake 329
Seven-segment display 273
Sex 173, 176, 183, 205, 210, 211, 366, 495, 655
Shading system 426
Shadow cracking 612
Shoe
– size 191–193, 347
– type 192
Shoulder look 128, 139, 623
Shutter 630
Shutter glasses 584
Sickness 345, 492
Side assist 541
Siemens PLM 203, 357
Sight
– direct 353
– indirect 347, 353
– onto the road 397, 440
– target 353
Signal-to-noise ratio $(S/N)_k$ 482
Significance 69, 76, 99, 132, 134, 162, 418, 496, 522, 604, 654, 655, 663, 664, 671
– test 659, 662, 663, 665
SIMPACK 248
Simulation
– noise 595
– programs 595
– steering 595
Simulator 47, 246
– technology 591, 595
Simultaneous engineering 576
Simultaneous space threshold 95
Single-point excitation
– multi-axial 486
Sitting 6, 163, 167, 178, 180, 182, 183, 189, 350, 353, 358–361, 364–375, 377–382, 384, 385, 387–395, 424, 429, 431, 444, 445, 457, 474, 489, 496, 497
– posture 164, 169, 190, 350, 365, 368, 369, 371, 384, 424, 447
Situation awareness 556, 567
SizeAMERICA 245
SizeGERMANY 177, 181, 182, 191, 194, 245, 613
Skill-based stabilization 598
Skin sense (haptics) 589
Sliding door 429, 430
Slip angle 46
Slippage 526
Smartphone 628, 678
Social Attituds to Road Traffic Risk in Europe (SARTRE) 528
Social conventions 150
Society of Automotive Engineers (SAE) 236, 348–353, 356–358, 364, 366, 367, 392, 396–399, 416, 417, 419, 421, 422, 438
Soft dummies 581
Soft tissues 176
Software-ergonomic design 678
Software in the Loop (SIL) 579
Soil diagram 220
Solar roof 507
Solutions GmbH 244
Somatotypes 179, 180
Sound
– air conditioning nuisance 512
– design 478–480
– programs 479
Space requirement
– for dynamic cornering 422
– more dynamic 420–424, 451
– more resulting 420, 423
– more static 420–424
Spatial presence 581, 582
Specific angle or length receptor 274
Speech transmission index 482
Speed control 10, 45, 263
Speed limiter (speed limiter) 151, 418, 524, 525, 564
Speedometer 128–133, 213, 524, 535
Spider diagram 479
Spinous process (Processus spinosus) 168
Split-half calculation 609
Spotlight headlights 11
Spreading (abduction) 165
SPSS, R (as open source variant) 666
SSDRIVE 232
Stabilisation task 19, 26, 230, 300, 524–526, 596
Stand ventilation with solar technology 507
Standard
– deviation 29, 178, 199, 208, 394, 395, 628, 629, 657, 671
– normal distribution 658
– posture 393
Standardized method 36
Standardized questionnaires 360
Standards & regulations 356
Standing bearing 138, 142, 404

Standing waves 476
Standstill management 542
Start-up assistant 9
Statements 45, 53, 57, 178, 360, 377, 403, 426, 438, 441, 444, 454, 461, 534, 535, 580, 617, 628, 657, 676
States 116
States of the world 116
Static friction 8, 525
Static holding work 370
Static seat pressure 488
Station wagon 345, 348, 450, 460
Stationary circular trip 315
Statistical rules 137
Steer-by-Wire (SbW) 42, 311, 331
Steering
– column fairing 347, 432, 433
– direct 46, 324
– feel
  – strictly speaking 635
– horn 201
– indirect 46, 324
– intervention 537
– power
  – course 10
– ratio
  – nonlinear 324
  – speed-dependent 46, 324, 325
– torque
  – parking process 315
– wheel
  – banana 410
  – diameter 316, 381
– wheel rim (thickness) 381
– wheel torque
  – course (articulation) 10
Step 145, 361, 364, 389, 418, 428, 431, 436, 437, 440, 450, 479, 486, 491, 511, 530, 535, 554, 568, 644, 645, 656
Stereotypes 268
Stick figure 614, 615
Stimulation 95, 96, 112, 141, 329
Stopping distance 48, 49, 303, 528
Storage compartments 345, 418
Strain
– mental 261, 632, 633
– psychic 77
– subjectively perceived 640
– under the influence of vibrations 492
– visual 147–149
Strategy
– hybrid 3, 28
– one-hand 452
– two-hand 451, 452
Stretching (extension) 121
Structural engram 112, 141
Structure-borne 31
Structure-borne sound 87, 475, 483, 595
Stutter brake 8, 525
Subjective benefit 116
Subjective data 136
Subjective estimation 116, 119
Subscales 647
Suffering 143, 145, 146, 259, 322
Sum point 26–28, 33, 56, 551, 560, 561, 563, 564, 566
Summer air heating 503
Summer operation 502, 503
Summer temperature rise 504
Sun visor 347, 397, 414, 415, 418, 419, 427
Superimposed steering system 9, 324, 336, 537
Superimpose-method (PCMAN) 244, 612, 615
Supination 165, 200
Surface haptics, operating or actuating haptics 284
Sustainability (suitability) 17, 276, 618, 677
Switches 116, 185, 290, 390, 456, 462, 524, 544, 553, 568, 581, 630
Synapses 69, 121
Synchro points 578
Synovial 162, 164
System
– active 34, 553
– border 611
– deviation 27, 35
– dynamic 22, 38
– ergonomics 16, 17
– monitive 265
– post-study system usability questionnaire (PSSUQ) 610
– reliability 34, 53
– structure 32, 34, 53
– system usability scale 610
– theory 31

T

Tactile sense 75, 284
Tailbone (Os coccygeum) 167
Tank
– lid 354, 450
– opening flap 390
Target vehicle 534
Target velocity 48
Taste buds 100
Tea trolley effect 312
Technical psychology 15
Technical requirements 364, 501–513, 524
Technology 588
Technology Comparison 622
Tecmath 244
Temperature threshold 98–100
Temporal aspect 34, 262, 321
Temporal relations 595
TEMPUS 243
Tendency towards the middle 645
Tendon, elements (gold elements) 122
Test
– driver 314, 315
– specimens 349
– terrain 68
Thermoreceptors 97, 98
Thoracic vertebrae 167
Threading into a street 404

Three-color theory 80
Three-finger grip 418
Threshold 69, 70, 77, 89–91, 93–96, 98, 99, 101, 205, 526, 537, 676
Timbre 478
Time
– span 142
– to collision 49, 50, 533, 629
– to intersection (TTI) 231
– to line crossing (TLC) 537
– windows 407, 422
– zone 321, 558
Toolboxes in Matlab 666
TOP event 57–60
Torque characteristic 285
Total shutter open time (TSOT) 630
Touchpad 306, 545
Touchscreens 293, 295, 416, 419, 492
Tracking
– experiment 37, 113, 220, 627
– pursuit 45
Tracking variable 28, 35, 222
Traction control (ASR) 8, 525
Traffic
– conflict 56
– psychology 15
– sign observer 527
– situation 18, 21, 116, 123, 125, 128, 129, 136, 141, 301, 329, 375, 402, 408, 522, 524, 529, 533, 548, 553, 554, 556, 557, 589, 598, 603
Training measures 149
Trajectory 350, 444, 446, 557, 566
Transition function 35–37, 41
Transition probability 58
Translational acceleration force 92
Transmitter 68, 614
Transverse axis 162
Transverse disparation 86
Transverse dynamics 19
Trip progression 258
Trunk length 179, 182, 188, 432, 444
Tulip 412
TUM CREATE 677
Tunnel vision 148
Turbulence degree 505
Tustin 41
Tutorial 373
Two-door vehicle 354, 449
Type
– athletic 179
– of grip 190
– of gripping 190
– leptosomeric 179

U

Understeer 9, 46
Underwater test 360
Unlocking device 428
Urban project 566
Usability (usability) 609
Utriculus 92

V

V(λ) curve 470
Validity 29, 150, 211, 424, 527, 672, 673
Value 39
Variable entry geometry 441
Variable ergonomics test bench 583
Variable headlamp levelling 529
Variant
– eyepoint fixed 364, 368, 381, 396–400, 403, 407, 413–415
– fixed heel point 362
Varifocals 356, 401, 410
Vehicle door 481
Vehicle in the Loop (VIL) 579
Vehicle information and communication system (VICS) 548
Vehicle ownership satisfaction study 604
Vehicleforeign vehicle 299
Ventilation nozzles 474, 509, 511
Ventilation of the chest and lap area 505
Version
– enhanced 357
– standard 357
Vertebral arch 168
Vertical or longitudinal axis 43
Vestibular
– apparatus 676
– organ 64, 68, 92–94, 97, 121
Video technology 589
Vienna Convention of 1968 526
Viewing time 128–130, 139, 140
Vigilance 117, 147, 634
Village window 604
Virtual environment 585
Virtual human models 577
Virtual reality 248, 579, 584, 623, 626
Virtual thermal dummy 512
Viscose damping 318
Visibility
– binocular 400
– monocular 400
Vision
– indirect 111, 211, 239, 357, 396, 407, 409
– peripheral 111, 125, 129, 133
Visual acuity 84, 86, 108, 111, 410, 411, 453
– dynamic 108, 109
Visual field, peripheral 109, 111, 231
Visual impression of the driver 400
Volume level 89, 90, 282, 454, 477, 478, 480, 482

W

Warm receptors 98, 494
Warning 149, 353, 391, 418, 462, 521, 525, 530, 531, 535–537, 540–542, 548, 552, 555, 557, 561, 565, 566, 662, 672

Warning limits 149
Warnings 10, 59
Wash-out strategy 589
Waste heat from the combustion engine 494, 677
Water tank 389
Weak points 456
Web guiding task 523
Weber-Fechner's Law of Psychophysics 635
Weightless condition 360
WERNER 239
Wheel 7, 8, 345, 347, 353, 354, 356, 358, 363–365, 368, 374–376, 379–382, 384–387, 391, 392, 396, 398, 400, 409, 410, 416, 422, 428, 431, 432, 434, 436, 437, 439, 440, 449, 460, 472, 483, 488, 489, 492, 500, 502, 507, 509, 521, 524, 525, 529, 535, 537–539, 541, 543, 546, 553, 554, 557, 563–565
Wheel load fluctuations 490, 491
Wheelbase 490
Wheelbase 2315 459
Wheelbase 2555 460
Wheels 2, 4, 8, 9, 475, 490, 491
Whole-body vibrations 370
Wind noise 475, 476
– on the body 475
Window regulator switch 185, 390
Windscreen washer system 292, 510
Windshield 411, 414
Windshield wiper 7
Winter operation 499, 501, 502
Wiper system 389, 390
Wobbling 38
Work
– memory 66, 115, 141, 147
– quality 29
– science 14
Wrinkles 617

X

X-by-Wire 330
Xenon light 528

Y

Yuhara 47

GPSR Compliance

The European Union's (EU) General Product Safety Regulation (GPSR) is a set of rules that requires consumer products to be safe and our obligations to ensure this.

If you have any concerns about our products, you can contact us on ProductSafety@springernature.com

In case Publisher is established outside the EU, the EU authorized representative is:

Springer Nature Customer Service Center GmbH
Europaplatz 3
69115 Heidelberg, Germany

Batch number: 10777416

Printed by Printforce, the Netherlands